Weici Tang and
Gerhard Eisenbrand

Handbook of Chinese Medicinal Plants

Related Titles

Barceloux, D. G.

Medical Toxicology of Natural Substances

Foods, Fungi, Medicinal Herbs, Plants, and Venomous Animals

2009

ISBN: 978-0-471-72761-3

Khan, I. A., Abourashed, E. A.

Encyclopedia of Common Natural Ingredients Used in Food, Drugs, and Cosmetics

2009

ISBN: 978-0-471-46743-4

Kayser, O., Quax, W. J. (eds.)

Medicinal Plant Biotechnology

From Basic Research to Industrial Applications

2007

ISBN: 978-3-527-31443-0

Ahmad, I., Aqil, F., Owais, M. (eds.)

Modern Phytomedicine

Turning Medicinal Plants into Drugs

2006

ISBN: 978-3-527-31530-7

Liang, X.-t., Fang, W.-s. (eds.)

Medicinal Chemistry of Bioactive Natural Products

2005

ISBN: 978-0-471-66007-1

Zhou, J., Xie, G., Yan, X., Milne, G. W. A. (eds.)

Traditional Chinese Medicines

Molecular Structures, Natural Sources and Applications

2004

ISBN: 978-0-566-08427-0

Weici Tang and Gerhard Eisenbrand

Handbook of Chinese Medicinal Plants

Chemistry, Pharmacology, Toxicology

Volume 2

WILEY-
VCH

WILEY-VCH Verlag GmbH & Co. KGaA

The Authors

Prof. Dr. Weici Tang

Prof. Dr. Gerhard Eisenbrand

University of Kaiserslautern
Department of Chemistry
Division of Food Chemistry and Toxicology
Erwin-Schrödinger-Str. 52
67663 Kaiserslautern
Germany

with collaboration from
Dr. Karl-Heinz Merz and
Dipl. Chem. Ingrid Hemm

1st Reprint 2011

Library of Congress Card No.: applied for

British Library Cataloguing-in-Publication Data
A catalogue record for this book is available from the British Library.

Bibliographic information published by the Deutsche Nationalbibliothek
The Deutsche Nationalbibliothek lists this publication in the Deutsche Nationalbibliografie; detailed bibliographic data are available on the Internet at http://dnb.d-nb.de.

Cover Adam Design, Weinheim
Typesetting Thomson Digital, Noida, India
Printing and Binding Strauss GmbH, Mörlenbach

Printed in the Federal Republic of Germany
Printed on acid-free paper

ISBN: 978-3-527-32226-8

Contents

Volume 1

Preface *XI*

Abrus cantoniensis *1*
Acanthopanax gracilistylus *5*
Acanthopanax senticosus *9*
Achyranthes bidentata *16*
Aconitum carmichaeli* and *Aconitum kusnezoffii *23*
Acorus calamus* and *Acorus tatarinowii *31*
Agrimonia pilosa *36*
Ailanthus altissima *45*
Akebia quinata*, *Akebia trifoliata* and *Akebia trifoliata* var. *australis *50*
Albizia julibrissin *53*
Alisma orientalis *57*
Allium sativum *63*
Aloe barbadensis* and *Aloe ferox *80*
Alpinia galanga *88*
Alpinia katsumadai *93*
Alpinia officinarum *96*
Alpinia oxyphylla *102*
Amomum compactum*, *Amomum kravanh* and *Amomum tsao-ko *106*
Andrographis paniculata *110*
Anemarrhena asphodeloides *121*
Angelica dahurica* and *Angelica dahurica* var. *formosana *127*
Angelica pubescens* f. *biserrata *139*
Angelica sinensis *143*
Apocynum venetum *153*
Arctium lappa *158*
Ardisia japonica* and *Ardisia crenata *163*
Areca catechu *169*
Aristolochia contorta* and *Aristolochia debilis *176*

Handbook of Chinese Medicinal Plants: Chemistry, Pharmacology, Toxicology
Gerhard Eisenbrand and Wei-Ci Tang

ISBN: 978-3-527-32226-8

Arnebia euchroma and *Arnebia guttata* 186
Artemisia annua 192
Artemisia argyi 202
Artemisia capillaris and *Artemisia scoparia* 207
Asarum heterotropoides var. *mandshuricum*, *Asarum sieboldii* and *Asarium sieboldii* var. *seoulense* 214
Asparagus cochinchinensis 222
Aster tataricus 225
Astragalus membranaceus and *Astragalus membranceus* var. *mongholicus* 229
Atractylodes chinensis, *Atractylodes lancea* and *Atractylodes macrocephala* 240
Belamcanda chinensis 246
Bolbostemma paniculatum 250
Brassica juncea and *Sinapis alba* 253
Brucea javanica 258
Buddleja officinalis 267
Bupleurum chinense and *Bupleurum scorzonerifolium* 271
Caesalpinia sappan 278
Carthamus tinctorius 283
Cassia acutifolia and *Cassia angustifolia* 287
Cassia obtusifolia and *Cassia tora* 292
Celosia argentea and *Celosia cristata* 296
Centella asiatica 299
Centipeda minima 305
Chrysanthemum indicum 311
Chrysanthemum morifolium 314
Cimicifuga dahurica, *Cimicifuga foetida* and *Cimicifuga heracleifolia* 318
Cinnamomum cassia 323
Cissampelos pareira var. *hirsuta* 330
Cistanche deserticola and *Cistanche tubulosa* 334
Citrus aurantium and *Citrus sinensis* 342
Citrus reticulata 361
Clematis chinensis, *Clematis hexapetala* and *Clematis manshurica* 368
Cnidium monnieri 371
Codonopsis pilosula, *Codonopsis pilosula* var. *modesta* and *Codonopsis tangshen* 376
Coix lacryma-jobi var. *ma-yuen* 380
Coptis chinensis, *Coptis deltoidea* and *Coptis teeta* 384
Coriolus versicolor 397
Cornus officinalis 404
Corydalis bungeana and *Corydalis decumbens* 411
Corydalis yanhusuo 416
Crocus sativus 421
Croton tiglium 428
Curculigo orchioides 432
Curcuma kwangsiensis, *Curcuma longa*, *Curcuma phaeocaulis* and *Curcuma wenyujin* 435

Cuscuta chinensis 448
Cyperus rotundus 451
Dalbergia odorifera 454
Daphne genkwa 460
Datura metel 464
Dendrobium candidum, *Dendrobium fimbriatum* var. *oculatum* and *Dendrobium nobile* 468
Desmodium styracifolium 471
Dianthus chinensis and *Dianthus superbus* 473
Dictamnus dasycarpus 475
Dioscorea futschauensis, *Dioscorea hypoglauca*, *Dioscorea nipponica*, *Dioscorea opposita* and *Dioscorea septembloba* 479
Dipsacus asperoides 485
Eclipta prostrata 488
Ephedra equisetina, *Ephedra intermedia* and *Ephedra sinica* 491
Epimedium brevicornum, *Epimedium koreanum*, *Epimedium pubescens*, *Epimedium sagittatum* and *Epimedium wushanense* 496
Erigeron breviscapus 504
Eriobotrya japonica 508
Eucommia ulmoides 511
Eugenia caryophyllata 519
Eupatorium fortunei 526
Euphorbia kansui and *Euphorbia pekinensis* 528
Euphorbia lathyris 531
Evodia rutaecarpa, *Evodia rutaecarpa* var. *bodinieri* and *Evodia rutaecarpa* var. *officinalis* 534
Foeniculum vulgare 545
Forsythia suspensa 550
Fraxinus chinensis, *Fraxinus rhynchophylla*, *Fraxinus stylosa*, and *Fraxinus szaboana* 555
Fritillaria cirrhosa, *Fritillaria delavayi*, *Fritillaria przewalskii* and *Fritillaria unibracteata* 561
Ganoderma lucidum and *Ganoderma sinense* 566
Gardenia jasminoides 577
Gastrodia elata 582
Gentiana crassicaulis, *Gentiana dahurica*, *Gentiana macrophylla*, *Gentiana manshurica*, *Gentiana rigescens*, *Gentiana scabra*, *Gentiana straminea*, and *Gentiana triflora* 587
Ginkgo biloba 591
Glehnia littoralis 599
Glycyrrhiza glabra, *Glycyrrhiza inflata* and *Glycyrrhiza uralensis* 601
Gynostemma pentaphyllum 614
Hedysarum polybotrys 621
Hippophae rhamnoides 624
Houttuynia cordata 629

Huperzia serrata *633*
Ilex chinensis **and** ***Ilex cornuta*** *639*
Illicium verum *645*
Inula britannica, Inula helenium, Inula japonica **and** ***Inula linariifolia*** *651*
Isatis indigotica *657*
Juncus effusus *666*
Kaempferia galanga *668*
Kochia scoparia *671*

Volume 2

Laminaria japonica **and** ***Ecklonia kurome*** *675*
Lamiophlomis rotata *680*
Lentinus edodes *682*
Leonurus japonicus *688*
Ligusticum chuanxiong *692*
Ligustrum lucidum *701*
Linum usitatissimum *705*
Liriope muscari **and** ***Liriope spicata*** **var.** ***prolifera*** *713*
Lonicera confusa, Lonicera hypoglauca, Lonicera japonica **and** ***Lonicera macranthoides*** *715*
Luffa cylindrica *723*
Lycium barbarum **and** ***Lycium chinense*** *727*
Lycopus lucidus **var.** ***hirtus*** *735*
Magnolia biondii, Magnolia denudata **and** ***Magnolia sprengeri*** *738*
Magnolia officinalis **and** ***Magnolia officinalis*** **var.** ***biloba*** *741*
Melia azedarach **and** ***Melia toosendan*** *752*
Menispermum dauricum *757*
Mentha haplocalyx *763*
Momordica cochinchinensis *768*
Momordica grosvenori *771*
Morinda officinalis *774*
Morus alba *777*
Myristica fragrans *782*
Nardostachys chinensis **and** ***Nardostachys jatamansi*** *787*
Nelumbo nucifera *791*
Notopterygium forbesii **and** ***Notopterygium incisum*** *798*
Oldenlandia diffusa *801*
Ophiopogon japonicus *804*
Oryza sativa *808*
Paeonia lactiflora **and** ***Paeonia veitchii*** *811*
Paeonia suffruticosa *820*
Panax ginseng *827*
Paris polyphylla **var.** ***chinensis*** **and** ***Paris polyphylla*** **var.** ***yunnanensis*** *847*

Perilla frutescens *850*
Periploca sepium *858*
Peucedanum praeruptorum *862*
Phellodendron amurense **and** ***Phellodendron chinense*** *868*
Phyllanthus emblica *872*
Physalis alkekengi **var.** ***franchetii*** *875*
Phytolacca acinosa **and** ***Phytolacca americana*** *878*
Picrasma quassioides *884*
Picria fel-terrae *888*
Picrorhiza scrophulariiflora *890*
Pinellia ternata *893*
Piper kadsura *896*
Piper longum **and** ***Piper nigrum*** *898*
Plantago asiatica **and** ***Plantago depressa*** *905*
Platycladus orientalis *908*
Platycodon grandiflorum *911*
Pogostemon cablin *917*
Polygala sibirica **and** ***Polygala tenuifolia*** *919*
Polygonatum cyrtonema**,** ***Polygonatum kingianum*****,** ***Polygonatum odoratum*** **and** ***Polygonatum sibiricum*** *923*
Polygonum cuspidatum *927*
Polygonum multiflorum *938*
Polyporus umbellatus *946*
Poria cocos *949*
Portulaca oleracea *954*
Prunella vulgaris *957*
Prunus **spp.** *962*
Pseudolarix kaempferi *967*
Pseudostellaria heterophylla *971*
Psoralea corylifolia *974*
Pterocarpus santalinus *979*
Pueraria lobata **and** ***Pueraria thomsonii*** *981*
Pulsatilla chinenesis *989*
Punica granatum *992*
Pyrola calliantha **and** ***Pyrola decorata*** *995*
Pyrrosia lingua**,** ***Pyrrosia petiolosa*** **and** ***Pyrrosia sheareri*** *999*
Rabdosia rubescens *1001*
Raphanus sativus *1006*
Rehmannia glutinosa *1008*
Rhaponticum uniflorum *1015*
Rheum officinale**,** ***Rheum palmatum*** **and** ***Rheum tanguticum*** *1017*
Rhododendron dauricum **and** ***Rhododendron molle*** *1024*
Ricinus communis *1027*
Rosa chinensis**,** ***Rosa laevigata*** **and** ***Rosa rugosa*** *1031*
Rubia cordifolia *1033*

Salvia miltiorrhiza 1040
Sanguisorba officinalis and ***Sanguisorba officinalis*** var. ***longifolia*** 1056
Santalum album 1059
Saposhnikovia divaricata 1062
Sarcandra glabra 1065
Saururus chinensis 1067
Schisandra chinensis and ***Schisandra sphenanthera*** 1071
Schizonepeta tenuifolia 1081
Scrophularia ningpoensis 1084
Scutellaria baicalensis 1086
Scutellaria barbata 1100
Sedum sarmentosum 1106
Selaginella pulvinata and ***Selaginella tamariscina*** 1108
Senecio scandens 1113
Sesamum indicum 1120
Siegesbeckia glabrescens, Siegesbeckia orientalis and ***Siegesbeckia pubescens*** 1129
Sinomenium acutum and ***Sinomenium acutum*** var. ***cinereum*** 1133
Smilax glabra 1140
Sophora flavescens 1144
Sophora japonica 1154
Spatholobus suberectus 1164
Stemona japonica, Stemona sessilifolia and ***Stemona tuberosa*** 1167
Stephania cepharantha 1170
Stephania tetrandra 1176
Tribulus terrestris 1185
Trichosanthes kirilowii and ***Trichosanthes rosthornii*** 1192
Trigonella foenum-graecum 1202
Tussilago farfara 1206
Typha angustifolia and ***Typha orientalis*** 1210
Uncaria hirsuta, Uncaria macrophylla, Uncaria rhynchophylla, Uncaria sessilifructus and ***Uncaria sinensis*** 1213
Vaccaria segetalis 1222
Verbena officinalis 1224
Viscum coloratum 1229
Vitex negundo var. ***cannabifolia, Vitex trifolia***, and ***Vitex trifolia*** var. ***simplicifolia*** 1234
Zanthoxylum bungeanum and ***Zanthoxylum schinifolium*** 1237
Zanthoxylum nitidum 1242
Zingiber officinale 1249
Ziziphus jujuba and ***Ziziphus jujuba*** var. ***spinosa*** 1259

Index 1265

Laminaria japonica and *Ecklonia kurome*

Thallus Laminariae or Thallus Eckloniae (Kunbu) is the dried thallus of the brown alga *Laminaria japonica* Aresch. (Laminariaceae) or *Ecklonia kurome* Okam. (Alariaceae). It is used in the treatment of scrofula, goiter, hyperthyroidism, and edema. *Laminaria japonica*, kelp, is a well-known edible seaweed.

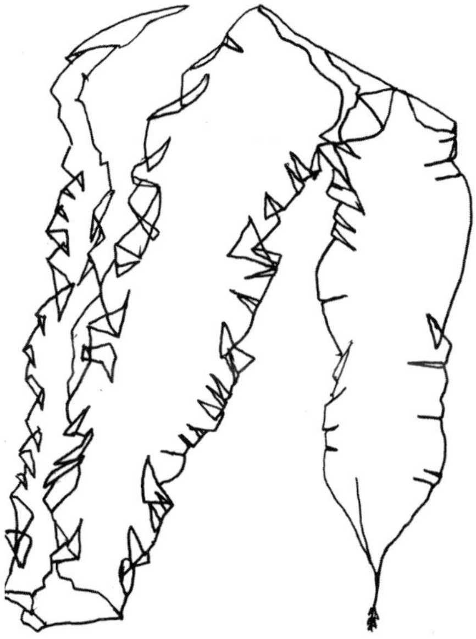

Laminaria japonica Aresch.

Chemistry

The thallus of *L. japonica* and of *E. kurome* is known to be rich in iodine. The Chinese Pharmacopoeia requires a quantitative determination of iodine in Thallus Eckloniae and Thallus Laminariae by iodometric titration. The content of iodine in Thallus Eckloniae should be not less than 0.2%, and in Thallus Laminariae not less than 0.35%.

Some phlorotannins with a dibenzo-*p*-dioxin skeleton have been isolated from *E. kurome*. These consist of phloroglucinol moieties linked by aryl–aryl, aryl–ether bonds, or bonds of a mixed type. The phlorotannins were identified as eckol, phloroeckol, dieckol [1], 6,6′-bieckol, 8,8′-bieckol, and phlorofucofuroeckol A [2].

Fucose-containing sulfated polysaccharides B-I, B-II, C-I, and C-II have been isolated from *E. kurome*. Polysaccharides B-I and B-II are composed of fucose, galactose, mannose, xylose, glucuronic acid, and ester sulfate, while polysaccharides C-I and C-II are composed of fucose, galactose, glucuronic acid, and ester sulfate in different molar ratios [3]. A structural examination showed polysaccharide C-I to consist mainly of 3-linked and 3,4-disubstituted fucopyranosyl residues, in addition to nonreducing terminal fucofuranosyl and

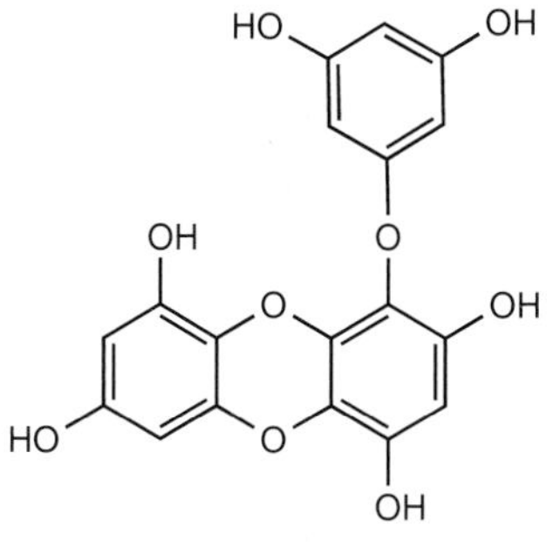

Eckol

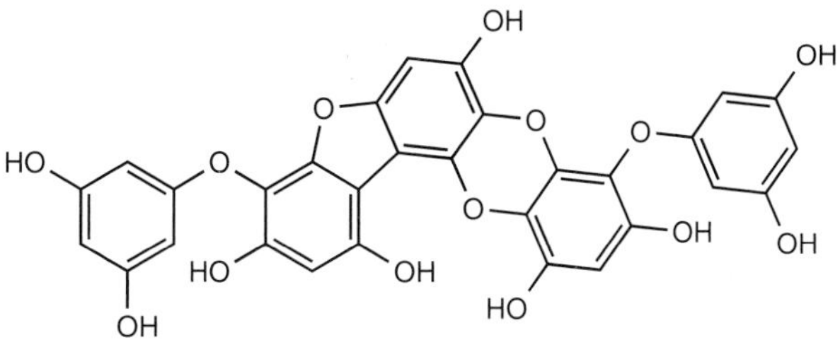

Phlorofucofuroeckol A

Handbook of Chinese Medicinal Plants: Chemistry, Pharmacology, Toxicology
Weici Tang and Gerhard Eisenbrand

ISBN: 978-3-527-32226-8

fucopyranosyl residues, 2,3-di- and 2,3,4-tri-substituted fucopyranosyl residues, and galactopyranosyl residues with various glycosidic linkages [4]. Polysaccharide C-II showed the presence of 3-*O*- and 3,4-*O*-disubstituted fucopyranosyl residues, in addition to small proportions of nonreducing terminal fucofuranosyl and fucopyranosyl groups, as well as 2,3-di-*O*- and 2,3,4-tri-*O*-substituted fucopyranosyl and galactopyranosyl residues with various glycosidic linkages [5].

A ribosome-inactivating protein, lamijapin, was isolated from *L. japonica*. Lamijapin, the first ribosome-inactivating protein to be obtained from a marine alga, was found to have a molecular mass of approximately 36 kDa, and thus was slightly larger than the other known single-chain ribosome-inactivating proteins from higher plants [6].

Pharmacology and Toxicology

The methanolic extract of *E. kurome* has been found to inhibit the action of α_2-macroglobulin and α_2-plasmin inhibitors, counteracting the fibrinolytic enzyme system. Phlorotannins were found to be responsible for the inhibition of α_2-macroglobulin and α_2-plasmin inhibitors [1]. For example, the IC_{50} of phlorofucofuroeckol A for inhibiting α_2-macroglobulin inhibitor was 1.0 $\mu g\ ml^{-1}$, and for inhibiting α_2-plasmin inhibitor was 0.3 $\mu g\ ml^{-1}$ [2].

The sulfated polysaccharides B-I, B-II, C-I, and C-II were reported to show anticoagulant activities [3]. The fucan sulfate C-II significantly inhibited the clotting of fibrinogen by thrombin, even in the absence of protease inhibitors, and the amidolytic activity of the protein only in the presence of heparin cofactor II. The fucan sulfate C-II was not adsorbed on an antithrombin III-agarose column, and its anticoagulant activity in antithrombin III-depleted plasma was equal to that in normal plasma. The mechanism of antithrombin action of the fucan sulfate C-II was suggested to be mediated by heparin cofactor II, and not by antithrombin III. It also was suggested that the polysaccharide linked to fibrinogen, thus blocking the thrombin action. The direct thrombin inhibition was reported to be very weak [7]. The heparin cofactor II-mediated antithrombin activity of the fucan sulfate C-II was dependent on its sulfate content and molecular weight. The inhibitory effect of the polysaccharide on fibrinogen clotting by thrombin might be attributable to steric hindrance by its binding to fibrinogen [8].

The fucan sulfate C-II was further found to significantly inhibit the generation of thrombin in both the intrinsic and extrinsic pathways, although the intrinsic inhibitory effect by fucan sulfate C-II was more marked than the extrinsic effect. In contrast, fucan sulfate C-II was a good inhibitor of factor Xa generation in the intrinsic pathway, but a poor inhibitor in the extrinsic pathway. In purified systems, fucan sulfate C-II also inhibited formation of the prothrombin-activating complex (i.e., prothrombinase), but not its activity. This indicates that, in addition to its antithrombin activity, fucan sulfate C-II also has an inhibitory effect on thrombin generation by blocking the formation of prothrombinase, and also prevents the generation of intrinsic factor Xa [9].

When the toxicity of fucoidan was studied in rats, no significant toxicological signs were observed following oral administration at a daily dose of 300 $mg\ kg^{-1}$. Neither were any toxicological signs identified at daily fucoidan doses of 900 and 2500 $mg\ kg^{-1}$, except for a prolongation of blood coagulation time [10].

The fucoidan C-I (C-I-H) was reported to enhance the activation of glutamine- and lysine-plasminogen by high-molecular-weight urokinase-type plasminogen activator (u-PA). C-I-H also moderately potentiated activation by single- and two-chain tissue-type plasminogen activators (t-PA). These effects were more pronounced than those of heparin. The fucoidan C-I-H promoted the generation of plasmin in the plasminogen activation by u-PA and t-PA, but not the activity of generated plasmin. Kinetic analyses suggested that C-I-H enhanced u-PA-mediated plasminogen activation by increasing the affinity of the activator for glutamine- and lysine-plasminogen, and by increasing the molecular activity of the activator [11].

Oral administration of fucoidan to rats with chronic renal failure induced by partial nephrectomy, or by cryoinjury, caused a significant decrease in the elevated levels of serum creatinine and urea nitrogen, in dose-dependent manner. Histopathological changes of the renal tubules and interstitium were markedly alleviated by fucoidan, and the mesangial areas were also greatly reduced [12]. Fucoidan, at daily doses of 50, 100, and 200 mg kg^{-1} for four weeks, also inhibited the formation of proteinuria and renal failure in rats with active Heymann's nephritis [13].

The pretreatment of streptozotocin-induced diabetic rats with an aqueous extract of *L. japonica* at an oral daily dose of 100 mg kg^{-1} for five days led to a significant reduction in blood glucose levels and hepatic lipid peroxidation, increased the glutathione content, and decreased the activity of glutathione reductase and glutathione peroxidase. The extract also caused a significant suppression of the increased activity of xanthine oxidase in diabetic rat liver; this suggested that an extract of *L. japonica* might prevent hyperglycemia in diabetic rats through its antioxidant activity [14]. In hyperlipemic quails, after administration of the polysaccharide from *L. japonica* for two weeks, the serum total cholesterol, triglyceride, and low-density lipoprotein were each significantly decreased, and the content of high-density lipoprotein significantly increased. The extent of aortic atherosclerosis and atherosclerotic lesions was also found to be diminished [15]. A hypotensive effect of *L. japonica* was also demonstrated in normotensive rats [16].

Laminaria tents have been used to dilate the cervix for the interruption of pregnancy and for other intrauterine procedures [17]. The induction of cervical dilation by a small *L. japonica* tent was demonstrated clinically in pregnant primigravida women who requested a first-trimester abortion [18]. A clinical trial in women requesting second-trimester abortion using prostaglandin E_2 (PGE_2) suppositories, employing intracervical *L. japonica*, has also been reported. The placement of *L. japonica* at 24 h before the PGE_2 vaginal suppository-induced abortion led not only to a significantly shorter induction-to-delivery time but also to a reduction in the number of suppositories required to complete the abortion [19].

A significant antimutagenic activity was demonstrated using a hot water extract of *L. japonica*. Suppressive effects on umu gene expression of the SOS response against DNA damage in *Salmonella typhimurium* TA1535/pSK1002 were described. The extract showed a strong antimutagenic activity against 2-acetylaminofluorene (AAF) and 3-amino-1,4-dimethyl-5*H*-pyrido[4,3-b] indole (Trp-P-1) in the presence of a metabolic activation system. In contrast, weak – but significant – inhibitory effects on the mutagenicity of *N*-methyl-*N'*-nitro-*N*-nitrosoguanidine (MNNG) and furylfuramide (AF-2) were observed in the absence of the metabolizing enzymes. The antimutagenic activities were found to be associated with the nonpolysaccharide fraction of low molecular weight [20]. Here, female rats fed a diet containing 2% *L. japonica* var. *ochotensis* for 152 days, followed by a basic diet for 59 or 60 successive days, received 7,12- dimethylbenz[*a*]anthracene (DMBA) intragastrically (20 mg kg^{-1}) at 27 days after the start of feeding. A reduction in the incidence of mammary tumor (9/18; 50%) compared to controls (20/29; 69%) was observed [21]. A dietary supplementation with of *L. japonica* var. *ochotensis* was also shown to be effective in inhibiting the growth of sarcoma-180 cells implanted subcutaneously into mice [22].

Lamjapin, the ribosome-inactivating protein from *L. japonica*, inhibited protein synthesis in a rabbit reticulocyte lysate, with an IC_{50} of approximately 0.7 n*M*. RNA was depurinated at multiple sites in the rat ribosomes. Lamjapin exhibited the same base and position requirements as did the ribosome-inactivating proteins from higher plants [6]. The phlorotannins from *E. kurome* were found to inhibit some strains of food-borne pathogenic bacteria, several strains of methicillin-resistant *Staphylococcus aureus*, and also *Streptococcus pyogenes*. The minimum bactericidal concentrations of dieckol and 8,8′-bieckol against *Campylobacter jejuni* were each 30 n*M* [23].

References

1 Fukuyama, Y., Kodama, M., Miura, I., Kinzyo, Z., Kido, M., Mori, M., Nakayama, Y., and Zakahashi, M. (1989) Anti-plasmin inhibitor, part III. Structure of an anti-plasmin inhibitor, eckol, isolated from the brown alga *Ecklonia kurome* Okamura and inhibitory activities of its derivatives on plasma plasmin inhibitors. *Chem. Pharm. Bull. (Tokyo)*, **37**, 349–353.

2 Fukuyama, Y., Kodama, M., Miura, I., Kinzyo, Z., Mori, H., Nakayama, Y., and Takahashi, M. (1990) Anti-plasmin inhibitor. VI. Structure of phlorofucofuroeckol A, a novel phlorotannin with both dibenzo-1,4-dioxin and dibenzofuran elements, from *Ecklonia kurome* Okamura. *Chem. Pharm. Bull. (Tokyo)*, **38**, 133–135.

3 Nishino, T., Yokoyama, G., Dobashi, K., Fujihara, M., and Nagumo, T. (1989) Isolation, purification, and characterization of fucose-containing sulfated polysaccharides from the brown seaweed *Ecklonia kurome* and their blood-anticoagulant activities. *Carbohydr. Res.*, **186**, 119–129.

4 Nishino, T., Kiyohara, H., Yamada, H., and Nagumo, T. (1991) An anticoagulant fucoidan from the brown seaweed *Ecklonia kurome*. *Phytochemistry*, **30**, 535–539.

5 Nishino, T., Nagumo, T., Kiyohara, H., and Yamada, H. (1991) Structural characterization of a new anticoagulant fucan sulfate from the brown seaweed *Ecklonia kurome*. *Carbohydr. Res.*, **211**, 77–90.

6 Liu, R.S., Yang, J.H., and Liu, W.Y. (2002) Isolation and enzymatic characterization of lamjapin, the first ribosome-inactivating protein from cryptogamic algal plant (*Laminaria japonica* A). *Eur. J. Pharmacol.*, **269**, 4746–4752.

7 Nishino, T., Aizu, Y., and Nagumo, T. (1991) Antithrombin activity of a fucan sulfate from the brown seaweed *Ecklonia kurome*. *Thromb. Res.*, **62**, 765–773.

8 Nishino, T. and Nagumo, T. (1992) Anticoagulant and antithrombin activities of oversulfated fucans. *Carbohydr. Res.*, **229**, 355–362.

9 Nishino, T., Fukuda, A., Nagumo, T., Fujihara, M., and Kaji, E. (1999) Inhibition of the generation of thrombin and factor Xa by a fucoidan from the brown seaweed *Ecklonia kurome*. *Thromb. Res.*, **96**, 37–49.

10 Li, N., Zhang, Q., and Song, J. (2005) Toxicological evaluation of fucoidan extracted from *Laminaria japonica* in Wistar rats. *Food Chem. Toxicol.*, **43**, 421–426.

11 Nishino, T., Yamauchi, T., Horie, M., Nagumo, T., and Suzuki, H. (2000) Effects of a fucoidan on the activation of plasminogen by u-PA and t-PA. *Thromb. Res.*, **99**, 623–634.

12 Zhang, Q., Li, Z., Xu, Z., Niu, X., and Zhang, H. (2003) Effects of fucoidan on chronic renal failure in rats. *Planta Med.*, **69**, 537–541.

13 Zhang, Q., Li, N., Zhao, T., Qi, H., Xu, Z., and Li, Z. (2005) Fucoidan inhibits the development of proteinuria in active Heymann nephritis. *Phytother. Res.*, **19**, 50–53.

14 Jin, D.Q., Li, G., Kim, J.S., Yong, C.S., Kim, J. A., and Huh, K. (2004) Preventive effects of *Laminaria japonica* aqueous extract on the oxidative stress and xanthine oxidase activity in streptozotocin-induced diabetic rat liver. *Biol. Pharm. Bull.*, **27**, 1037–1040.

15 Li, C., Gao, Y., Li, M., Shi, W., and Liu, Z. (2005) Effect of *Laminaria japonica* polysaccharides on lowing serum lipid and anti-atherosclerosis

in hyperlipemic quails. *Zhong Yao Cai*, **28**, 676–679.

16 Chiu, K.W. and Fung, A.Y. (1997) The cardiovascular effects of green beans (*Phaseolus aureus*), common rue (*Ruta graveolens*), and kelp (*Laminaria japonica*) in rats. *Gen. Pharmacol.*, **29**, 859–862.

17 Peeples, W.J., Given, F.T. Jr, and Bakri, Y.N. (1983) The use of *Laminaria japonica* in intracavitary radiation therapy when anesthesia is contraindicated. *Int. J. Radiat. Oncol. Biol. Phys.*, **9**, 1405–1406.

18 Darney, P.D. and Dorward, K. (1987) Cervical dilation before first-trimester elective abortion: a controlled comparison of meteneprost, laminaria, and hypan. *Obstet. Gynecol.*, **70** (1), 397–400.

19 Atlas, R.O., Lemus, J., Reed, J. 3rd, Atkins, D., and Alger, L.S. (1998) Second trimester abortion using prostaglandin E_2 suppositories with or without intracervical *Laminaria japonica*: a randomized study. *Obstet. Gynecol.*, **92**, 398–402.

20 Okai, Y., Higashi-Okai, K., and Nakamura, S. (1993) Identification of heterogenous antimutagenic activities in the extract of edible brown seaweeds, *Laminaria japonica* (Makonbu) and *Undaria pinnatifida* (Wakame) by the umu gene expression system in *Salmonella typhimurium* (TA1535/pSK1002). *Mutat. Res.*, **303**, 63–70.

21 Yamamoto, I., Maruyama, H., and Moriguchi, M. (1987) The effect of dietary seaweeds on 7,12-dimethyl-benz[*a*]anthracene-induced mammary tumorigenesis in rats. *Cancer Lett.*, **35**, 109–118.

22 Yamamoto, I., Maruyama, H., Takahashi, M., and Komiyama, K. (1986) The effect of dietary or intraperitoneally injected seaweed preparations on the growth of sarcoma-180 cells subcutaneously implanted into mice. *Cancer Lett.*, **30**, 125–131.

23 Nagayama, K., Iwamura, Y., Shibata, T., Hirayama, I., and Nakamura, T. (2002) Bactericidal activity of phlorotannins from the brown alga *Ecklonia kurome*. *J. Antimicrob. Chemother.*, **50**, 889–893.

Lamiophlomis rotata

Herba Lamiophlomis (Duyiwei) is the dried whole herb of *Lamiophlomis rotata* (Benth.) Kudo (Lamiaceae). It is used as an analgesic against rheumatic pain and as a hemostatic agent for the treatment of traumatic bleeding.

Lamiophlomis rotata (Benth.) Kudo

A galenic preparation from Herba Lamiophlomis, "Duyiwei Jiaonang," is listed in the Chinese Pharmacopoeia. The preparation is supplied as capsules containing an aqueous extract of Herba Lamiophlomis; each capsule contains 0.3 g of extract. It is used as a hemostatic agent and analgesic for the treatment of traumatic bleeding, menorrhalgia, gumboil, and rheumatic pain. The Chinese Pharmacopoeia notes that "Duyiwei Jiaonang" capsules should be used in pregnancy only with caution.

Chemistry

Herba Lamiophlomis is known to contain flavones, with luteolin as the major component, and iridoid glucosides. Iridoid glucosides isolated from the whole plant of *L. rotata* were identified as lamiophlomiol A and B [1], lamiophlomiol C [2], 8-*O*-acetylshanzhiside methyl ester, 6-*O*-acetylshanzhiside methyl ester, penstemoside, and 7, 8-dehydropenstemoside [3]. Lamiophlomiol A and B are epimers.

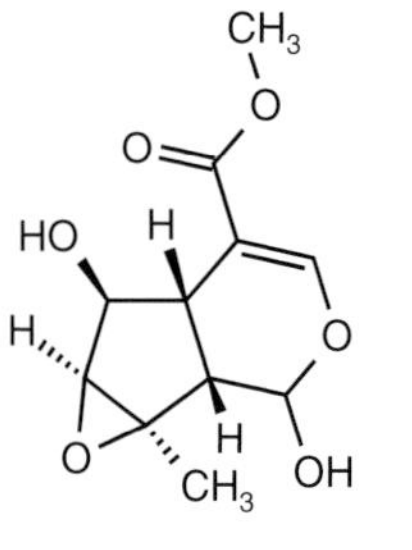

Lamiophlomiol A, B

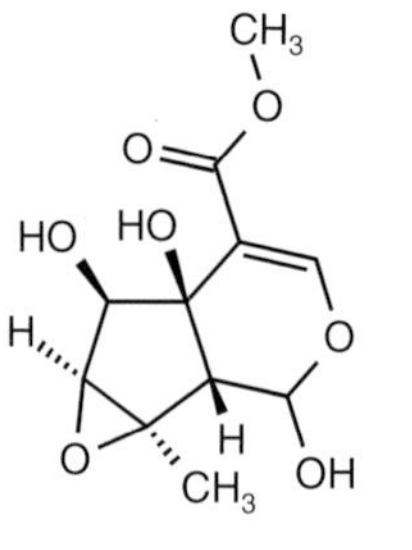

Lamiophlomiol C

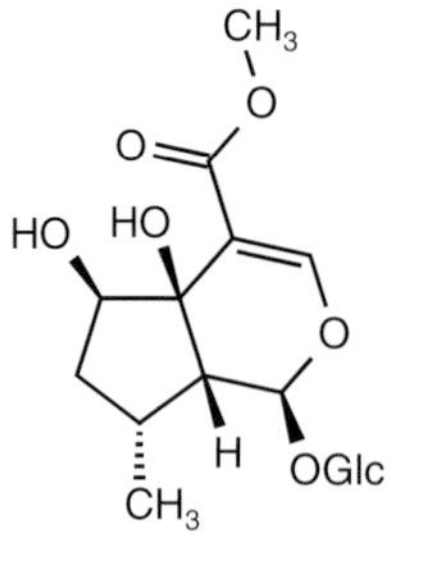

Penstemoside

Handbook of Chinese Medicinal Plants: Chemistry, Pharmacology, Toxicology
Weici Tang and Gerhard Eisenbrand

ISBN: 978-3-527-32226-8

The Chinese Pharmacopoeia requires a quantitative determination of luteolin in Herba Lamiophlomis by HPLC. The content of luteolin in Herba Lamiophlomis should be not less than 0.15%. The Chinese Pharmacopoeia further requires a quantitative determination of the *n*-butanol-soluble part of the "Duyiwei Jiaonang" capsule by gravimetric analysis, a quantitative determination of the total flavones by spectrophotometric analysis, and a quantitative determination of luteolin by HPLC. The *n*-butanol-soluble part should be not less than 8.0%; the content of total flavones in a capsule not less than 26 mg, calculated as rutin; and the content of luteolin not less than 0.8 mg.

Pharmacology and Toxicology

Herba Lamiophlomis was used mainly as an analgesic and hemostatic drug [4]. Recent results have been reported from an experimental study on blood clotting parameters after the oral administration of Herba Lamiophlomis extract to rats at doses of 3, 1.5, and 0.75 g kg^{-1}. After seven days, dose levels of 3 and 1.5 g kg^{-1} had each significantly reduced the thrombin time, and caused significant hyperfibrinogenemia.

The results indicated that an aqueous extract of Herba Lamiophlomis exerted an hemostatic effect in rats after oral application, which was both dose- and time-dependent [5].

References

1 Yi, J.H., Zhong, C.C., Luo, Z.Y., and Xiao, Z.Y. (1991) Studies on the chemical constituents from the roots of *Lamiophlomis rotata* (Benth.) Kudo, a medical plant in Xi-Zang (Tibet). *Yao Xue Xue Bao*, **26**, 37–41.

2 Yi, J.H., Zhong, C.C., Luo, Z.Y., and Xiao, Z.Y. (1992) Structure of lamiophlomiol C. *Yao Xue Xue Bao*, **27**, 204–206.

3 Yi, J.H., Huang, X.P., Chen, Y., Luo, Z.Y., and Zhong, C.C. (1997) Studies on the iridoid glucosides of the root of *Lamiophlomis rotata* (benth.) kudo, a medicinal plant in Xi zang (Tibet). *Yao Xue Xue Bao*, **32**, 357–360.

4 Li, M., Jia, Z., and Zhang, R. (2004) A review on the study of *Lamiophlomis rotata* (Benth.) Kudo: an analgesic and hemostatic drug. *Zhong Yao Cai*, **27**, 222–224.

5 Li, M.X., Jia, Z.P., Shen, T., Zhang, R.X., Zhang, H.X., and Li, Z.Y. (2006) Effect of Herba *Lamiophlomis rotata* extract on rats blood conglomeration parameters by oral administration. *Zhong Yao Cai*, **29**, 160–163.

Lentinus edodes

Xianggu (Lentinus) is the dried fruiting body of *Lentinus edodes* (Berk.) Sing. (Tricholomataceae). Lentinus is not officially listed in the Chinese Pharmacopoeia. It is a common edible mushroom in China, Japan, and other Asian countries. Lentinus might be used as an immunostimulatory agent.

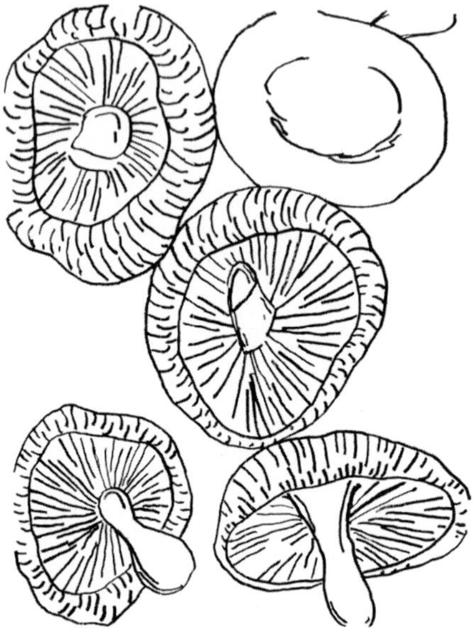

Lentinus edodes (Berk.) Sing.

Chemistry

Lentinan, a polysaccharide from *L. edodes*, is one of the most extensively studied fungal polysaccharides [1, 2]. The isolation of lentinan was reported as early as 1970 [3]. Lentinan is a (1 → 3)-β-glucan that is highly branched, with (1 → 3)-β- and (1 → 6)-β-linked glucose residues existing mainly as linear triple-helical structures in aqueous solution, in physiological saline, and as a single flexible chain in dimethyl sulfoxide [4, 5]. The polysaccharide L-II from the fruiting body of *L. edodes* was found to consist of D-glucopyranose and to have a molecular weight of 203 kDa [6]. Besides polysaccharides, a protein lentin was isolated from the fruiting bodies of Lentinus. The molecular mass of lentin was found to be 27.5 kDa, and the N-terminal sequence similar to that of endoglucanase [7].

A butyric acid derivative, eritadenine, from *Lentinus* was structurally elucidated as 2(*R*),3(*R*)-dihydroxy-4-(9-adenyl)-butyric acid [8]. A sulfur compound characteristic for the *Lentinus* flavor, lenthionine, was identified as 1,2,3,5,6-pentathiepane [9].

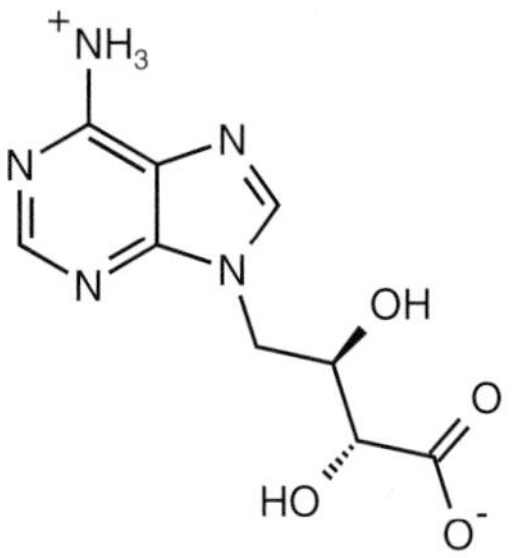

Eritadenine

Pharmacology and Toxicology

Lentinan is relatively nontoxic, with LD_{50} values in mice (ICR) and rats (CD) being essentially the same estimated as 250–500 mg kg^{-1} after intravenous (i.v.) administration, and in excess of 2500 mg kg^{-1} by intraperitoneal (i.p.), subcutaneous (s.c.) and oral administration. No remarkable toxic signs specific to lentinan were observed after i.p., s.c., and oral treatments; however, cyanosis, convulsion and death were observed in animals receiving i.v. lentinan, at higher dosages [10]. An extract of Lentinus at concentrations of 0.5, 1.0, and 1.5 mg ml^{-1} did not show mutagenicity in the micronucleus test in HEp-2 cells *in vitro*.

Handbook of Chinese Medicinal Plants: Chemistry, Pharmacology, Toxicology
Weici Tang and Gerhard Eisenbrand

ISBN: 978-3-527-32226-8

Moreover, the extract exerted an antimutagenic activity against the genotoxicity of the alkylating agent methyl methanesulfonate [11].

Lentinan was reported to show immunostimulatory and antitumor activities in clinical studies. A multicenter prospective study with lentinan, used in combination with cytostatic agents, in patients with advanced unresectable and recurrent gastric cancer revealed survival prolongation and an improvement in the quality of life. The median survival of lentinan-treated patients was significantly longer than that of untreated patients (297 versus 199 days) [12]. Lentinan at a daily i.v. dose of 2 mg per patient, when combined with 5′-deoxy-5-fluorouridine [13], cisplatin, 5-FU [14], and tegafur [15], was effective in the postoperative therapy of gastric cancer. A >50% increase in interleukin (IL)-1β production in the peripheral blood of gastric cancer patients associated with lentinan treatment was observed [16, 17], as well as an enhanced induction of lymphokine-activated natural killer (NK) cell activity [18]. The addition of lentinan to maintenance therapy with 5-FU did not provide any prognostic benefit in hepatocellular carcinoma patients [19]. The treatment of breast cancer patients with 5-FU, epirubicin and cyclophosphamide as adjuvant chemotherapy, markedly decreased the NK cell activity and leukocyte count. However, such decreases caused by the chemotherapy were not apparent in patients who received oral *L. edodes* mycelia simultaneously [20].

Lentinan exhibited significant immunostimulatory effects and showed antitumor activities in experimental animals, augmenting helper T-cell-mediated cytotoxic T-cell activity, NK cell activity, and humoral immune responses. It also activated the nonspecific cytotoxicity of macrophages *in vitro* [21] and *in vivo* [22]. Beneficial effects of lentinan on cellular immune function were also observed in cyclophosphamide-treated mice [5]. Among the cytokine genes, mRNA expression levels of IL-1α, IL-1β, tumor necrosis factor-α (TNF-α), interferon (IFN)-γ and macrophage colony-stimulating factor in mouse peritoneal cells and splenocytes were markedly induced [23]. The application of 5′-deoxy-5-fluorouridine, in combination with lentinan to rats inoculated with AH66 ascites hepatoma resulted in a significantly better inhibition of tumor growth, as compared to 5′-deoxy-5-fluorouridine alone. Lentinan induced pyrimidine nucleoside phosphorylase activity in the tumor tissue, and increased the susceptibility of tumor cells to 5′-deoxy-5-fluorouridine [24]. Lentinan decreased the glutathione-*S*-transferase (GST)-II and GST-III contents in colon 26 adenocarcinoma tissues transplanted into mice, and also enhanced the tumor susceptibility to cisplatin [25]. β-(1 → 3)-Glucans were reported to be more effective on T cells than on B cells [26]. It was suggested that lentinan affected the tumor vascular system, which in turn resulted in an induction of hemorrhagic necrosis, depending on the T cells [27, 28]. The binding of lentinan to human monocytes might also bear some influence here [29].

Synthetic β-(1 → 6)-branched β-(1 → 3) glucohexaose and its analogues containing an α-(1 → 3)-linked bond, exhibited antitumor activity and immune stimulatory effects similar to those of lentinan [30]. In mice, the spleen weight was increased after an i.p. injection of these synthetic oligosaccharides. Administration of the oligosaccharides or lentinan enhanced mouse spleen mRNA production of TNF-α, but not that of IL-2. It enhanced the proliferation of mouse splenocytes induced by Concanavalin A (Con A). In contrast, the *in vitro* administration of the oligosaccharides did not show such effects. It was suggested that the antitumor effects of the synthetic oligosaccharides might occur through an induction of the splenocyte-mediated immune response [31]. It was further reported that the O-sulfonated derivatives of the water-insoluble (1 → 3)-α-D-glucans from different types of fruiting bodies of *L. edudes* exhibited a higher cytotoxic activity *in vitro* and a higher antitumor activity *in vivo* against S180 solid tumor than did those of the native glucans [32].

Antitumor activity of the polysaccharide L-II was also reported. The treatment of mice bearing S180 tumors with L-II at doses of 1, 5, and 10 mg kg^{-1} for 10 days resulted in a significant increase in spleen and thymus weights, a delayed-type hypersensitivity and phagocytosis of macrophages, and a significant inhibition of tumor growth. The serum concentrations of TNF-α and IFN-γ, but not of IL-2, were significantly increased in treated mice. Moreover, L-II increased nitric oxide (NO) production and catalase activity in macrophages. The antitumor activity of L-II in mice transplanted with S180 tumor cells thus appeared to be mediated by an immunomodulation of the T-cell- and macrophage-dependent immune system responses [6].

In vitro, an ethyl acetate fraction of the extract of Lentinus was shown to be growth inhibitory against MDA-MB-453 and MCF-7 human breast carcinoma cells, MCF-10F human nonmalignant breast epithelial cells, and RPMI-8226 and IM-9 myeloma cells. At a concentration of approximately 50 μg ml^{-1}, apoptosis was found in 50% of the population of four human tumor cell lines, supposedly mediated through an up-regulation of the pro-apoptotic Bax protein. Cell cycle analysis revealed that the ethyl acetate fraction induced cell cycle arrest in S phase, associated with the induction of cdk inhibitors p21 and the suppression of cdk4 and cyclin D1 activity. Compared to malignant tumor cells, nonmalignant cells were less sensitive [33].

The i.p. treatment of mice with lentinan at a dose of 10 mg kg^{-1} affected the number, plastic-adherence, and peroxidase activity of peritoneal cells. Lentinan-stimulated peritoneal macrophages showed cytotoxicity against several murine and human metastatic tumors, including Lewis lung carcinoma and two human melanomas [34]. Lentinan treatment of peritoneal carcinomatoses induced in BDIX rats by an i.p. injection of syngeneic cells from a colon carcinoma cell line (five i.p. injections, two days apart, at a daily dose of 2 mg kg^{-1}, starting on day 14 after tumor cell injection) significantly inhibited the growth of carcinomatoses and increased the life span of the animals [35]. The oral administration of *L. edodes* mycelia to mice suppressed the postoperative liver metastasis of primary colorectal cancer. It increased the survival period, elevated the activities of NK cells and macrophages in the liver and increased IL-1β levels in the liver and spleen [36]. A combined use of lentinan and IL-2 was found to result in synergistic antitumor and antimetastatic effects in mice carrying spontaneously metastatic 3-methylcholanthrene-induced fibrosarcoma [37].

The extract from Lentinus mycelia was reported to show hepatoprotective effects in mice with liver injury induced by *N*-nitrosdimethylamine. It decreased the serum aspartate aminotransferase and alanine aminotransferase levels, partially inhibited the overaccumulation of collagen fibrils, and suppressed the overexpression of genes for α-smooth muscle actin and/or heat-shock protein 47. The extract also inhibited morphologic changes and the proliferation of isolated rat hepatic stellate cells in a concentration-dependent manner, without cytotoxicity. A direct interaction between the extracts and hepatic stellate cells seemed to be important for the hepatoprotective activity. Polyphenols present in the extract were considered to be potentially responsible for the hepatoprotective effects [38]. The extract was also found to exhibit hepatoprotective activity against liver injury in rats induced by D-galactosamine, and to protect cultured primary hepatocytes against D-galactosamine [39].

Eritadenine from Lentinus was reported to exert hypocholesterolemic activity and to have a wide range of effects on lipid metabolism. For example, eritadenine increased the concentration of liver microsomal phosphatidylethanolamine, decreased liver microsomal Δ6-desaturase activity, and modulated the fatty acid and molecular species profile of liver and plasma lipids. Dietary supplementation of eritadenine to rats at a dose of 50 mg kg^{-1} caused a significant inhibition of liver microsomal Δ6-desaturase activity and a decreased mRNA expression for the enzyme. In addition, a significant correlation between the Δ6-desaturase activity and phosphatidylethanolamine concentration (but not phosphatidylcholine concentration) was observed. Thus, eritadenine might suppress the activity of liver microsomal Δ6-desaturase by altering the microsomal phospholipid profile, as represented by an increase in phosphatidylethanolamine concentration. The effects of eritadenine appeared to be mediated by the regulation of gene expression [40]. The treatment of rats with guanidinoacetic acid resulted in a dose-dependent decrease in hepatic *S*-adenosylmethionine concentrations and the activity of cystathionine β-synthase, as well as increases in hepatic *S*-adenosylhomocysteine and homocysteine concentrations. The dietary supplementation of eritadenine in rats was found to suppress hyperhomocysteinemia induced by guanidinoacetic acid. Eritadenine also restored the decreased hepatic concentration of *S*-adenosylhomocysteine and cystathionine β-synthase activity. Thus, eritadenine might elicit its effect by both slowing homocysteine production and increasing cystathionine formation [41].

The essential oil from Lentinus was found to inhibit platelet aggregation induced by arachidonic acid and U-46619, the analogue of thromboxane A_2. The effect of lenthionine, the sulfur-containing flavor in the essential oil, was almost equivalent to that of the essential oil. Platelet aggregation inhibitory activity of the essential oil from lentinan might therefore be mainly attributed to lenthionine [9].

The protein lentin from Lentinus was reported to inhibit the growth of a variety of fungal species, including *Physalospora piricola*, *Botrytis cinerea* and *Mycosphaerella arachidicola*. Moreover, lentin was also found to inhibit human immunodeficiency virus-1 (HIV-1) reverse transcriptase and the proliferation of leukemia cells [7].

References

1 Tang, W.C., Hemm, I., and Bertram, B. (2003) Recent development of antitumor agents from Chinese Herbal Medicines. Part II. High molecular compounds. *Planta Med.*, **69**, 193–201.

2 Ooi, V.E. and Liu, F. (2000) Immunomodulation and anti-cancer activity of polysaccharide-protein complexes. *Curr. Med. Chem.*, **7**, 715–729.

3 Chihara, G., Hamuro, J., Maeda, Y., Arai, Y., and Fukuoka, F. (1970) Fractionation and purification of the polysaccharides with marked antitumor activity, especially lentinan, from *Lentinus edodes* (Berk.) Sing. (an edible mushroom). *Cancer Res.*, **30**, 2776–2781.

4 Zhang, L., Li, X., Xu, X., and Zeng, F. (2005) Correlation between antitumor activity, molecular weight, and conformation of lentinan. *Carbohydr. Res.*, **340**, 1515–1521.

5 Wang, G.L., and Lin, Z.B. (1996) The immunomodulating effect of lentinan. *Acta Pharm. Sin.*, **31**, 86–90.

6 Zheng, R., Jie, S., Hanchuan, D., and Moucheng, W. (2005) Characterization and immunomodulating activities of polysaccharide from *Lentinus edodes*. *Int. Immunopharmacol.*, **5**, 811–820.

7 Ngai, P.H., and Ng, T.B. (2003) Lentin, a novel and potent antifungal protein from shitake mushroom with inhibitory effects on activity of human immunodeficiency virus-1 reverse transcriptase and proliferation of leukemia cells. *Life Sci.*, **73**, 3363–3374.

8 Enman, J., Rova, U., and Berglund, K.A. (2007) Quantification of the bioactive compound eritadenine in selected strains of shiitake mushroom (*Lentinus edodes*). *J. Agric. Food Chem.*, **55**, 1177–1180.

9 Shimada, S., Komamura, K., Kumagai, H., and Sakurai, H. (2004) Inhibitory activity of shiitake flavor against platelet aggregation. *Biofactors.*, **22**, 177–179.

10 Moriyuki, H., and Ichimura, M. (1980) Acute toxicity of lentinan in mice and rats. *J. Toxicol.*, **5** (Suppl.), 1–9.

11 Miyaji, C.K., Poersch, A., Ribeiro, L.R., Eira, A. F., and Colus, I.M. (2006) Shiitake (*Lentinus edodes* (Berkeley) Pegler) extracts as a modulator of micronuclei induced in HEp-2 cells. *Toxicol. In Vitro*, **20**, 1555–1559.

12 Nakano, H., Namatame, K., Nemoto, H., Motohashi, H., Nishiyama, K., and Kumada, K. (1999) A multi-institutional prospective study of lentinan in advanced gastric cancer patients with unresectable and recurrent diseases: effect on prolongation of survival and improvement of quality of life. *Hepatogastroenterology*, **46**, 2662–2668.

13 Takita, M., Onda, M., Tokunaga, A., Shirakawa, T., Ikeda, K., and Hiramoto, Y. (1998) Successful treatment of hepatic metastasis of gastric cancer with 5′-DFUR and lentinan. *Gan To Kagaku Ryoho*, **28**, 129–133.

14 Mio, H. and Terabe, K. (1997) Clinical effects of postoperative immunochemotherapy with a combination of 5-FU, CDDP and lentinan for stage IVb gastric carcinoma and long-term pharmacokinetic studies on CDDP and 5-FU. *Gan To Kagaku Ryoho*, **24**, 337–342.

15 Taguchi, T. (1987) Clinical efficacy of lentinan on patients with stomach cancer: end point results of a four-year-follow-up survey. *Cancer Detect. Prev.*, **1** (Suppl.), 333–349.

16 Takeshita, K., Hayashi, S., Tani, M., Kando, F., Saito, N., and Endo, M. (1996) Monocyte function associated with intermittent lentinan therapy after resection of gastric cancer. *Surg. Oncol.*, **5**, 23–28.

17 Arinaga, S., Karimine, N., Takamuku, K., Nanbara, S., Nagamatsu, M., and Ueo, H. (1992) Enhanced production of interleukin 1 and tumor necrosis factor by peripheral monocytes after lentinan administration in patients with gastric carcinoma. *Int. J. Immunopharmacol.*, **14**, 43–47.

18 Arinaga, S., Karimine, N., Takamuku, K., Nanbara, S., Inoue, H., and Nagamatsu, M. (1992) Enhanced induction of lymphokine-activated killer activity after lentinan administration in patients with gastric carcinoma. *Int. J. Immunopharmacol.*, **14**, 535–539.

19 Suto, T., Fukuda, S., Moriya, N., Watanabe, Y., Sasaki, D., and Yoshida, Y. (1994) Clinical study of biological response modifiers as maintenance therapy for hepatocellular carcinoma. *Cancer Chemother. Pharmacol.*, **33** (Suppl.), 145–148.

20 Nagashima, Y., Sanpei, N., Yamamoto, S., Yoshino, S., Tangoku, A., and Oka, M. (2005) Evaluation of host immunity and side effects in breast cancer patients treated with adjuvant chemotherapy (FEC therapy). *Gan To Kagaku Ryoho*, **32**, 1550–1552.

21 Chihara, G. (1983) Preclinical evaluation of lentinan in animal models. *Adv. Exp. Med. Biol.*, **166**, 189–197.

22 Li, J.F., Guo, J.W., and Huang, X.F. (1996) Study on the enhancing effect of Polyporus polysaccharide, mycobacterium polysaccharide and lentinan on lymphokine-activated killer cell activity *in vitro*. *Chin. J. Integr. Trad. West Med.*, **16**, 224–226.

23 Liu, F., Ooi, V.E., and Fung, M.C. (1999) Analysis of immunomodulating cytokine mRNAs in the mouse induced by mushroom polysaccharides. *Life Sci.*, **64**, 1005–1011.

24 Ogawa, T., Ohwada, S., Sato, Y., Izumi, M., Nakamura, S., and Takeyoshi, I. (1999) Effects of 5′-DFUR and lentinan on cytokines and PyNPase against AH66 ascites hepatoma in rats. *Anticancer Res.*, **19**, 375–379.

25 Murata, T., Hatayama, I., Kakizaki, I., Satoh, K., Sato, K., and Tsuchida, S. (1996) Lentinan enhances sensitivity of mouse colon 26 tumor to cis-diamminedichloroplatinum (II) and decreases glutathione transferase expression. *Jpn. J. Cancer Res.*, **87**, 1171–1178.

26 Haba, S., Hamaoka, T., Takatsu, K., and Kitagawa, M. (1976) Selective suppression of T cell activity in tumor-bearing mice and its improvement by lentinan, a potent anti-tumor polysaccharide. *Int. J. Cancer*, **18**, 93–104.

27 Suzuki, M., Takatsuki, F., Maeda, Y.Y., Hamuro, J., and Chihara, G. (1994) Antitumor and immunological activity of lentinan in comparison with LPS. *Int. J. Immunopharmacol.*, **16**, 463–468.

28 Mitamura, T., Sakamoto, S., Suzuki, S., Yoshimura, S., Maemura, M., and Kudo, H. (2000) Effects of lentinan on colorectal carcinogenesis in mice with ulcerative colitis. *Oncol. Rep.*, **7**, 599–601.

29 Oka, M., Hazama, S., Suzuki, M., Wang, F., Wadamori, K., and Iizuka, N. (1996) *In vitro* and *in vivo* analysis of human leukocyte binding by the antitumor polysaccharide, lentinan. *Int. J. Immunopharmacol.*, **18**, 211–216.

30 Ning, J., Zhang, W., Yi, Y., Yang, G., Wu, Z., Yi, J., and Kong, F. (2003) Synthesis of β-(1 → 6)-branched β-(1 → 3) glucohexaose and its analogues containing an α-(1 → 3)-linked bond with antitumor activity. *Bioorg. Med. Chem.*, **11**, 2193–2203.

31 Yan, J., Zong, H., Shen, A., Chen, S., Yin, X., Shen, X., Liu, W., Gu, X.
and Gu, J. (2003) The β-(1 → 6)-branched β-(1 → 3) glucohexaose and its analogues containing an α-(1 → 3)-linked bond have similar stimulatory effects on the mouse spleen as lentinan. *Int. Immunopharmacol.*, **3**, 1861–1871.

32 Unursaikhan, S., Xu, X., Zeng, F., and Zhang, L. (2006) Antitumor activities of *O*-sulfonated derivatives of (1 → 3)-α-D-glucan from different *Lentinus edodes*. *Biosci. Biotechnol. Biochem.*, **70**, 38–46.

33 Fang, N., Li, Q., Yu, S., Zhang, J., He, L., Ronis, M.J., and Badger, T.M. (2006) Inhibition of growth and induction of apoptosis in human cancer cell lines by an ethyl acetate fraction from shiitake mushrooms. *J. Altern. Complement Med.*, **12**, 125–132.

34 Ladanyi, A., Timar, J., and Lapis, K. (1993) Effect of lentinan on macrophage cytotoxicity against metastatic tumor cells. *Cancer Immunol. Immunother.*, **36**, 123–126.

35 Jeannin, J.F., Lagadec, P., Pelletier, H., Reisser, D., Olssson, N.O., and Chihara, G. (1988) Regression induced by lentinan, of peritoneal carcinomatoses in a model of colon cancer in rat. *Int. J. Immunopharmacol.*, **10**, 855–861.

36 Morinaga, H., Tazawa, K., Tagoh, H., Muraguchi, A., and Fujimaki, M. (1994) An *in vivo* study of hepatic and splenic interleukin-1β mRNA expression following oral PSK or LEM administration. *Jpn. J. Cancer Res.*, **85**, 1298–1303.

37 Suzuki, M., Kikuchi, T., Takatsuki, F., and Hamuro, J. (1994) Curative effect of combination therapy with lentinan and interleukin-2 against established murine tumors, and the role of CD8-positive T cells. *Cancer Immunol. Immunother.*, **38**, 1–8.

38 Akamatsu, S., Watanabe, A., Tamesada, M., Nakamura, R., Hayashi, S., Kodama, D., Kawase, M., and Yagi, K. (2004) Hepatoprotective effect of extracts from *Lentinus edodes* mycelia on dimethylnitrosamine-induced liver injury. *Biol. Pharm. Bull.*, **27**, 1957–1960.

39 Watanabe, A., Kobayashi, M., Hayashi, S., Kodama, D., Isoda, K., Kondoh, M., Kawase, M., Tamesada, M., and Yagi, K. (2006) Protection against D-galactosamine-induced acute liver injury by oral administration of extracts from
Lentinus edodes mycelia. *Biol. Pharm. Bull.*, **29**, 1651–1654.

40 Shimada, Y., Yamakawa, A., Morita, T., and Sugiyama, K. (2003) Effects of dietary eritadenine on the liver microsomal Δ6-desaturase activity and its mRNA in rats. *Biosci. Biotechnol. Biochem.*, **67**, 1258–1266.

41 Fukada, S., Setoue, M., Morita, T., and Sugiyama, K. (2006) Dietary eritadenine suppresses guanidinoacetic acid-induced hyperhomocysteinemia in rats. *J. Nutr.*, **136**, 2797–2802.

Leonurus japonicus

Herba Leonuri (Yimucao) is the dried, above-ground part of *Leonurus japonicus* Houtt. (Lamiaceae). It is used in the treatment of gynecological disorders such as menorrhalgia, menostasia, and other irregular menstruation. It is also used as a diuretic in the treatment of edema and acute nephritis.

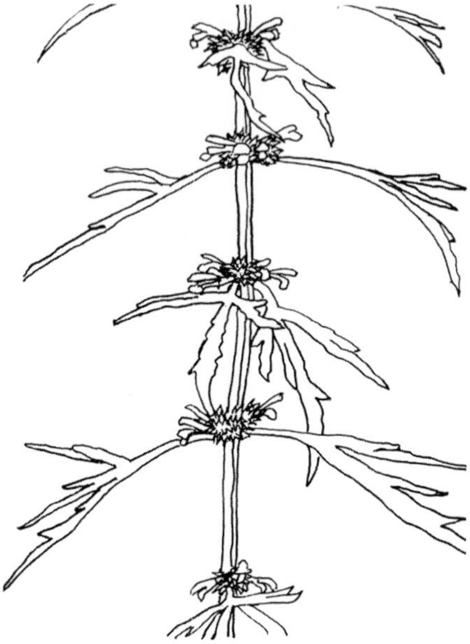

Leonurus japonicus Houtt.

Fructus Leonuri (Chongweizi) is the dried ripe fruit of *Leonurus japonicus*. It is mainly used in the treatment of menstrual disorders, eye inflammation, dizziness, and headache. The Chinese Pharmacopoeia notes Fructus Leonuri to be used with care in patients with pupil dilation.

Three galenic preparations of Herba Leonuri are listed in the Chinese Pharmacopoeia: Extractum Leonuri Liquidum (Yimucao Liujingao); "Yimucao Koufuye," the oral solution prepared from Herba Leonuri (10 ml each); and "Yimucao Gao," the extract of Herba Leonuri. The Chinese Pharmacopoeia notes that Herba Leonuri and its three galenic preparations are contraindicated in pregnancy.

Chemistry

Herba Leonuri was reported to contain alkaloids, with stachydrine and leonurine as the major components [1]. Stachydrine is a cyclic imino acid inner salt, while leonurine is a guanidinobutanol ester of syringic acid. The total alkaloid content in the above-ground part of *L. japonicus* herb ranges from 0.1% to 2%. Younger and more succulent plants showed higher alkaloid contents [2]. The highest alkaloid content was determined in overwintering young seedlings, but this decreased with advancing maturation [3]. The content of stachydrine in *L. japonicus* herb at different growth periods was 0.6%-1.7%, and that of leonurine was 0.02%-0.1% [1]. A diterpene compound from *L. japonicus* was identified as prehispanolone [4]. The above-ground part of *Leonurus heterophyllus* Sweet and *Leonurus sibiricus* L. contained stachydrine and leonurine, also used as Herba Leonuri and Fructus Leonuri [5]. In the Chinese Pharmacopoeia Ed. 1995, *Leonurus heterophyllus* Sweet was the official plant species for Herba Leonuri and Fructus Leonuri.

The Chinese Pharmacopoeia requires a qualitative determination of stachydrine hydrochloride in Herba Leonuri and all galenic preparations by thin-layer chromatographic comparison with reference substance; a quantitative determination of stachydrine hydrochloride in Herba Leonuri, in Extractum Leonuri Liquidum, and in the extract "Yimucao Gao" by thin-layer chromatography–densitometry; and a quan-

Handbook of Chinese Medicinal Plants: Chemistry, Pharmacology, Toxicology
Weici Tang and Gerhard Eisenbrand

ISBN: 978-3-527-32226-8

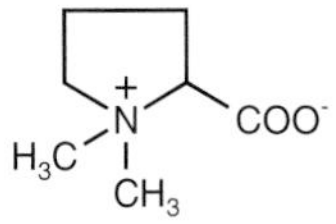

Stachydrine

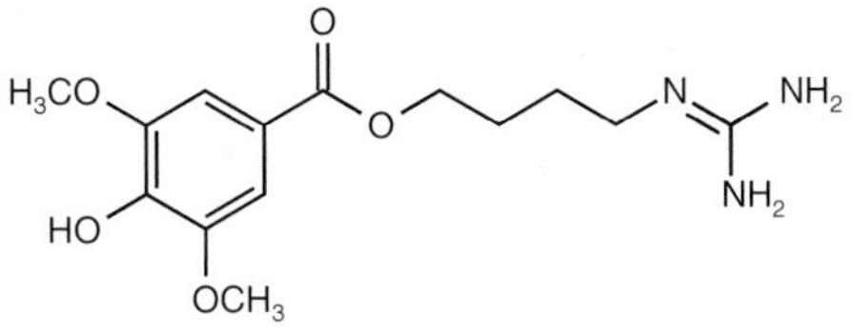

Leonurine

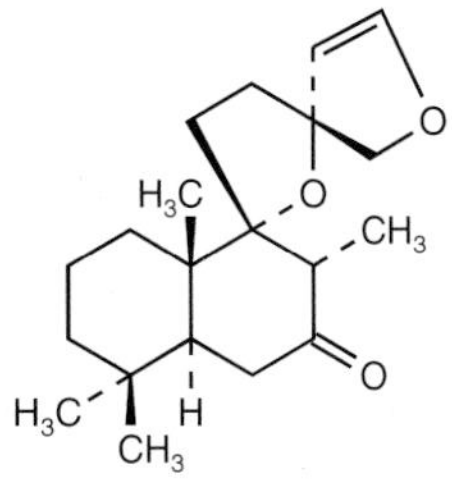

Prehispanolone

titative determination of stachydrine hydrochloride in the oral solution "Yimucao Koufuye" by spectrophotometry. The content of stachydrine hydrochloride in Herba Leonuri should be not less than 0.5%; in Extractum Leonuri Liquidum not less than 0.2%; in the oral solution "Yimucao Koufuye" not less than 1.0 mg ml^{-1}; and in the extract "Yimucao Gao" not less than 3.6 mg g^{-1}.

Pharmacology and Toxicology

Herba Leonuri was used mainly in traditional Chinese medicine to treat gynecological disorders [6]. Pharmacological studies, however, reported improvements in the coronary circulation, improvements in cardiac functions, as well as effects on the microcirculation, cardiovascular hemodynamics, blood viscosity, and peripheral perfusion [7]. Platelet aggregation was also inhibited [8]. Clinical studies indicated that Herba Leonuri improved hemorheological parameters. The intravenous (i.v.) treatment of 105 patients with *Leonurus* extract for 15 days resulted in a therapeutic effect in 95%, including a decreased blood viscosity and fibrinogen volume and a reduction in platelet aggregation [9]. The patients (n = 105) with hyperviscosity syndrome included 60 cases of hypertension, 12 cases of cerebral thrombosis, 21 cases of coronary disease, nine cases of cerebral atherosclerosis, and three cases of diabetes mellitus. The clinical efficacies of Herba Leonuri for the treatment of cerebral ischemia [10] and thrombocytopenic purpura [11] were also reported.

The i.v. administration of *Leonurus* extract to dogs for five days resulted in a decrease in blood urea nitrogen and an increase in renal blood flow [12]. The aqueous extract of Herba Leonuri was found to enhance the phenylephrine-induced contraction of the aorta with an intact endothelium, but not after the endothelium had been removed. As with *N*-nitro-L-arginine methyl ester (an inhibitor of nitric oxide synthase), the extract also significantly inhibited the relaxation induced by acetylcholine in the intact aorta. An i.v. injection of the aqueous extract to rats caused a transient increase in blood pressure (for 5 min), indicating the presence of a vasoconstrictive component in the extract [13]. A methanolic extract of the aerial parts of *L. sibiricus*, when injected intraperitoneally in mice at doses of 250 and 500 mg kg^{-1}, showed a significant analgesic effect against acetic acid-induced writhing. The extract, when given orally at doses of 200 and 400 mg kg^{-1}, also showed a significant anti-inflammatory activity against paw edema in rats induced by carrageenan [14].

Herba Leonuri was reported to show a uterine contractive efficacy in isolated rat uterus which was comparable to that of oxytocin [15, 16]. Stachydrine was reported to serve as the oxytocin-like constituent in Herba Leonuri [17]. A decoction of Herba Leonuri showed a stimulatory action on the mouse uterus *in vitro*; this effect was suggested to be related to stimulating H_1-receptor and α-adrenergic receptors [18]. An oral dose of Herba Leonuri given to 121 normal fertile women resulted in an increase in the intrauterine pressure in 40% of cases [19]. Leonurine was found to stimulate uterine contractions [20], and to show an initial inhibitory effect on rabbit small intestine, followed by a prolonged stimulatory action. However, in the isolated rabbit heart only an inhibitory effect was observed [4].

Prehispanolone was reported to be a platelet activating factor (PAF) receptor antagonist. Prehispanolone and several of its derivatives inhibited the binding of PAF to rabbit platelets, with potencies closely related to the inhibition of PAF-induced aggregation. Structure–activity studies revealed that the tetrahydrofuran ring in prehispanolone was critical for its interaction with the PAF receptor. By hydrogenating the dihydrofuran ring and replacing the keto group of prehispanolone with a hydroxyl group, the compound obtained was more stable and more active than prehispanolone in respect of PAF receptor antagonism [21].

References

1 Luo, S.R. and Mai, L. (1986) Analysis of alkaloids in *Leonurus heterophyllus*. *Chin. J. Pharm. Anal.*, **6**, 47–48.

2 Xu, L.Q. and Dong, Y.W. (1985) Quantitative determination of total alkaloids in Herba leonuri by colorimetric method with ammonium chromium thiocyanate. *Chin. Trad. Herbal Drugs*, **16**, 487–488.

3 Hui, Y.M., Zhang, C.Y., Sun, D.Q., and Wu, Y. P. (1982) Qualitative and quantitative analysis of Yi Mu Cao (*Leonurus heterophyllus*) in different growth stages. *Bull. Chin. Mater. Med.*, **7**, 8–9.

4 Hon, P.M., Lee, C.M., Shang, H.S., Cui, Y.X., Wong, H.N.C., and Chang, H.M. (1991) Prehispanolone, a labdane diterpene from *Leonurus heterophyllus*. *Phytochemistry*, **30**, 354–356.

5 Luo, S.R. (1985) Separation and determination of alkaloids of *Leonurus*. *Bull. Chin. Mater. Med.*, **10**, 32–35.

6 Zhang, C.F., Jia, Y.S., Wei, H.C., Zhu, X.M., Hui, Y.M., Zhang, C.Y., Mo, Q.Z., and Gong, B. (1982) Studies on actions of extract of motherwort. *J. Trad. Chin. Med.*, **2**, 267–270.

7 Wang, D.R. (1987) Experimental studies and clinical applications of Yimucao. *Zhejiang J. Trad. Chin. Med.*, **22**, 340–344.

8 Chang, C.F. and Li, C.Z. (1986) Experimental studies on the mechanism of anti-platelet aggregation action of motherwort. *Chin. J. Integr. Trad. West. Med.*, **6**, 39–40.

9 Zou, Q.Z., Bi, R.G., Li, J.M., Feng, J.B., Yu, A. M., Chan, H.P., and Zhen, M.X. (1989) Effect of motherwort on blood hyperviscosity. *Am. J. Chin. Med.*, **17**, 65–70.

10 Yu, G.Z., Wang, S.M., Jia, Z.X., Wang, X.R., and Yan, H.L. (1990) Studies of hemorheology and microcirculation in the treatment of ischemic cerebrovascular disease with *Leonurus heterophyllus*. *Chin. J. Pathophysiol.*, **6**, 257.

11 Han, D.W., and Xu, R.L. (1988) Progress in the research on blood activation and hemostasis removal. *Abstr. Chin. Med. (Hong Kong)*, **2**, 466–483.

12 Gu, T.C., Du, L.Z., Long, C.Y., and Chen, M.F. (1988) Treatment of acute renal failure in dogs with *Leonurus heterophyllus*. *Acta Univ. Med. Sec. Shanghai*, **8**, 219–223.

13 Pang, S., Tsuchiya, S., Horie, S., Uchida, M., Murayama, T., and Watanabe, K. (2001) Enhancement of phenylephrine-induced contraction in the isolated rat aorta with endothelium by H_2O-extract from an Oriental medicinal plant Leonuri herba. *Jpn. J. Pharmacol.*, **86**, 215–222.

14 Islam, M.A., Ahmed, F., Das, A.K., and Bachar, S.C. (2005) Analgesic and anti-inflammatory activity of *Leonurus sibiricus*. *Fitoterapia*, **76**, 359–362.

15 Yang, M., Yang, S., Jin, Z., and Zhu, S. (2002) Study on the biological assay of Herba Leonuri: analysis the dosage response curve of Herba Leonuri and oxytocin and establishment of adequate potency pattern. *Zhong Yao Cai*, **25**, 409–411.

16 Yang, M., Yang, S., Jin, Z., and Guo, Y. (2002) Study on the biological assay of Herba Leonuri. I. Establishment of standard uterus models

and optimizing uterus environment conditions. *Zhong Yao Cai*, **25**, 333–336.

17 Tang, Y., Wang, X.Q., and Lu, Y.R. (1989) Determination of stachydrine in four Chinese proprietary drugs containing *Leonurus heterophyllus*. *Chin. J. Pharm. Anal.*, **9**, 22–24.

18 Shi, M., Chang, L., and He, G. (1995) Stimulating action of *Carthamus tinctorius* L., *Angelica sinensis* (Oliv.) Diels and *Leonurus sibiricus* L. on the uterus. *China J. Chin. Mater. Med.*, **20**, 173–175.

19 Chan, W.C., Wong, Y.C., Kong, Y.C., Chun, Y. T., Chang, H.T., and Chan, W.F. (1983) Clinical observation on the uterotonic effect of I-mu Ts'ao (*Leonurus artemisia*). *Am. J. Chin. Med.*, **11**, 77–83.

20 Yu, C.F. (1981) Studies on the chemical constituents of the Chinese traditional medicine I-Mu-Tsao. II. Structure of "alkaloid A". *Acta Pharm. Sin.*, **39**, 94–96.

21 Lee, C.M., Jiang, L.M., Shang, H.S., Hon, P.M., He, Y., and Wong, H.N. (1991) Prehispanolone, a novel platelet activating factor receptor antagonist from *Leonurus heterophyllus*. *Br. J. Pharmacol.*, **103**, 1719–1724.

Ligusticum chuanxiong

Rhizoma Chuanxiong (Chuanxiong) is the dried rhizome of *Ligusticum chuanxiong* Hort. (Apiaceae). It is used as an analgesic agent for the treatment of menstrual disorders, amenorrhea, dysmenorrhea, abdominal pain, swelling, pain by traumatic injuries such as headache, and rheumatic arthralgia.

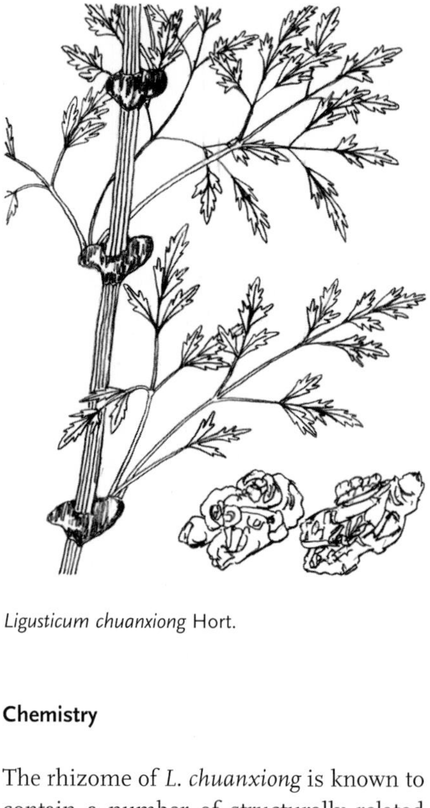

Ligusticum chuanxiong Hort.

Chemistry

The rhizome of *L. chuanxiong* is known to contain a number of structurally related phthalides, with ligustilide as the major component. Minor phthalides were identified as chuanxiongol, ligusticum lactone, cnidilide, neocnidilide, ligustilidiol, senkyunolide, senkyunolides B–N and Q–S, and 3-butyl-isobenzofuranone [1, 2]. Ligustilide was also found to be the major component in the essential oil of *L. chuanxiong*, with a content of about 60%. The contents of 3-butyl-isobenzofuranone and sabinene in the essential oil were reported as 5% and 6%, respectively [3]. Phthalides are also contained in Radix Angelicae Sinensis, the root of *Angelica sinensis* (Oliv.) Diels.

Phthalide dimers isolated from the rhizome of *L. chuanxiong* were identified as senkyunolides O and P, diligustilide (levistolide A), tokinolide B, and riligustilide [4]. Among the alkaloids, ligustrazine (chuanxiongzine, tetramethylpyrazine) seemed to be one of the compounds in the rhizome of *L chuanxiong* with marked biological activities. Further alkaloids were identified as perlolyrine, L-isoleucyl-L-valine anhydride, L-valyl-L-valine anhydride, uridine, trimethylamine hydrochloride, and choline chloride [1].

Ligustrazine phosphate (Linsuan Chuanxiongqin) and its galenic preparations Ligustrazine Phosphate Tablets (Linsuan Chuanxiongqin Pian, 50 mg each tablet), Ligustrazine Phosphate Injection (Linsuan Chuanxiongqin Zhusheye, 50 mg per 2 ml), and Ligustrazine Phosphate Capsules (Linsuan Chuanxiongqin Jiaonang, 50 mg each capsule) are listed in the Chinese Pharmacopoeia, Volume II. These are used for vasodilation.

Another item of *Ligusticum* species officially listed in the Chinese Pharmacopoeia is Rhizoma et Radix Ligustici. Rhizoma et Radix Ligustici (Gaoben) is the dried root and rhizome of *Ligusticum sinensis* Oliv. or *Ligusticum jeholense* Nakai et Kitag. It is used as an analgesic in the treatment of cold, headache, and rheumatic pain. The major active principles in the root and rhizome of *L. sinensis* or *L. jeholense* were reported to be ferulic acid and phthalides, including ligustilide, cnidilide, neocnidilide, ligusticum

Handbook of Chinese Medicinal Plants: Chemistry, Pharmacology, Toxicology
Weici Tang and Gerhard Eisenbrand

ISBN: 978-3-527-32226-8

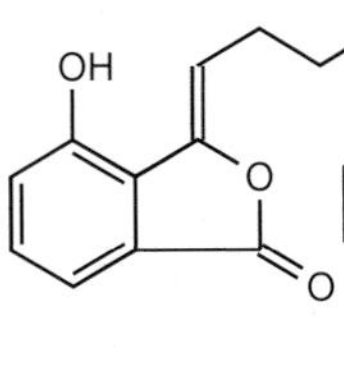
Chuanxiongol

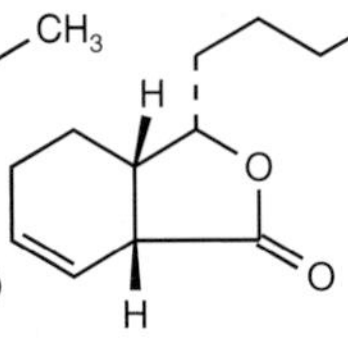
Cnidilide

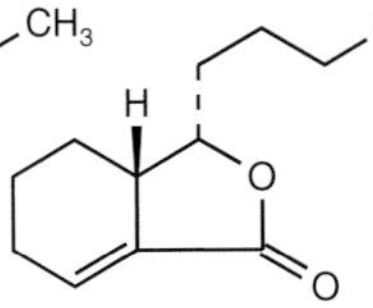
Neocnidilide

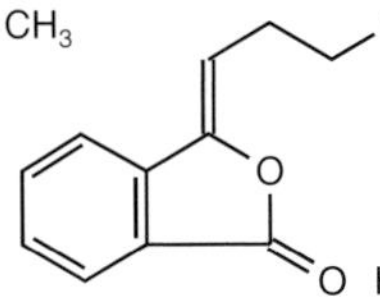
Ligustilide

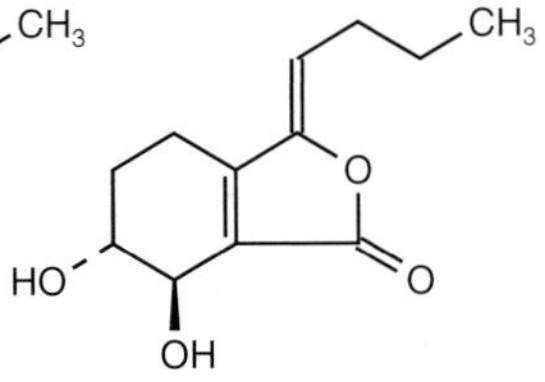
Ligustilidiol

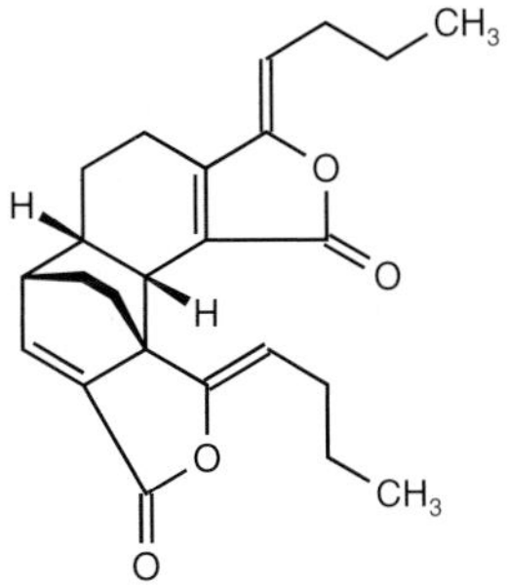
Diligustilide

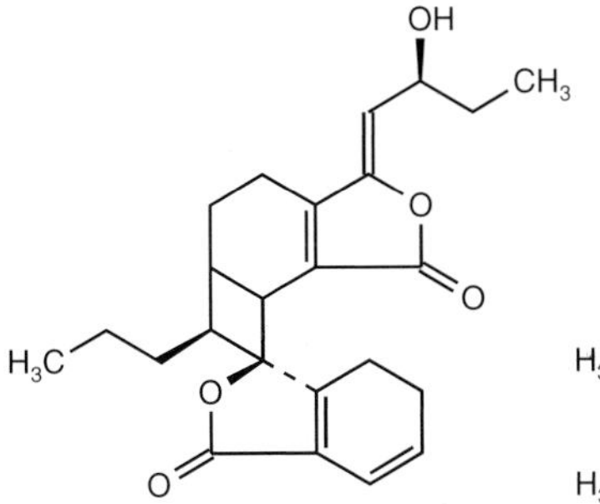
Riligustilide

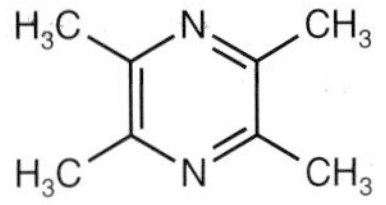
Ligustrazine

lactone, senkyunolides [5], and two stereoisomers of diligustilide [6]. The major components in the essential oil from *L. jeholense* are terpenes β-phellandrene (33%) and 4-terpineol acetate (20%), whereas only 6% accounts for ligustilide; the major components in the essential oil of *L. sinensis* are neocnidilide (26%), cnidilide (11%), and limonene (14%) [7]. The Chinese Pharmacopoeia requires a quantitative determination of ferulic acid in Rhizoma Ligustici by HPLC. The content of ferulic acid in Rhizoma et Radix Ligustici should be not less than 0.05%.

Pharmacology and Toxicology

The biological activities of the rhizome of *L. chuanxiong* are mainly caused by phthalides and ligustrazine. The rhizome is used clinically for the treatment of asthma and cardiovascular and cerebrovascular diseases. The extract of *Ligusticum* rhizome inhibited bronchospasm in asthma patients, and significantly decreased the plasma level of thromboxane B_2 (TXB_2) [8]. Patients with transient ischemic attack appeared to respond significantly better to *Ligusticum* rhizome than to aspirin [9] or to low-molecular-weight dextran [10]. The treatment of patients with cerebrovascular or cardiovascular diseases with *Ligusticum* rhizome led to an effective inhibition of platelet activation, a significant decrease in the plasma levels of β-thromboglobulin, platelet factor 4 and TXB_2, increased plasma levels of 6-keto-prostaglandin $F_{1\alpha}$ (6-keto-$PGF_{1\alpha}$), and normalized TXA_2–PGI_2 imbalance [11, 12].

Experimental studies provided results similar to those observed in clinical trials. The extract of *Ligusticum* rhizome inhibited bronchospasm in guinea pigs induced by histamine and acetylcholine and caused a remarkable decrease in the plasma level of TXB_2 [8]. Treatment of rabbits under acute experimental cerebral ischemia with *Ligusticum* rhizome also resulted in an inhibition of platelet activation, a decrease in the plasma levels of β-thromboglobulin and platelet factor 4 and TXB_2, an increase in plasma 6-keto-$PGF_{1\alpha}$, and a correction of any TXA_2–PGI_2 imbalance [13]. Thus, in atherosclerotic rats treated with *Ligusticum* rhizome, the hemodynamic parameters were not significantly different from those of healthy controls [14]. The treatment of rabbits with acute, glycerol-induced renal failure with *Ligusticum* rhizome led to a reduction in plasma TXB_2 levels, slightly raised the plasma level of 6-keto-$PGF_{1\alpha}$, and maintained the 6-keto-$PGF_{1\alpha}$/TXB_2 ratio at normal levels [15].

The phthalides present in the rhizome of *L. chuanxiong* were reported to show anti-inflammatory activity, a smooth muscle-relaxing activity, and an inhibition of platelet activation. Both, ligustilide and senkyunolide were shown to inhibit the production of tumor necrosis factor-α (TNF-α) and to suppress TNF-α mRNA transcription. In addition, ligustilide and senkyunolide exhibited significant inhibitory effects on TNF-α-mediated nuclear factor-κB (NF-κB) activation [16]. Phthalides, such as ligustilide and butylidenephthalide, were identified as the active principles of the *L. chuanxiong* rhizome that would induce vasorelaxation [17]. Ligustilide and senkyunolide each exhibited relaxing effects in isolated rat aorta that had been precontracted with phenylephrine, serotonin, and KCl. The vasorelaxation was unaffected by endothelium removal, by inhibitors of adenylate cyclase and by guanylate cyclase, or by nonselective K^+ channel blockers [18]. Mechanistic studies on the vasodilatation effected by ligustilide in rat mesenteric artery showed that the vasorelaxation was induced by inhibiting the voltage-dependent and receptor-operated calcium channels, as well as receptor-mediated Ca^{2+} influx and release [19]. Ligustilide, when administered orally, was also shown to reduce the contraction of isolated rat aorta induced by phenylephrine, but had no effect on the systolic blood pressure of spontaneously hypertensive rats [20].

In rats with focal cerebral ischemia induced by occlusion of the middle cerebral artery for 24 h, the oral administration of ligustilide at doses of 20 or 80 mg kg^{-1} at 2 h after ischemia, reduced the cerebral infarction volumes by 48% and 85%, respectively. Treatment with ligustilide also reduced brain swelling, and significantly improved behavioral deficits; this indicated that ligustilide could exert neuroprotective effects in transient cerebral ischemia in rats [21]. A neuroprotective efficacy of ligustilide was also demonstrated in mice with transient forebrain cerebral ischemia induced by a 30-min bilateral common carotid artery occlusion. The i.p. injection of ligustilide to mice at the start of reperfusion caused a significant reduction in the infarction volume determined 24 h after ischemia. Treatment with ligustilide also caused a significant decrease in the level of malondialdehyde, and increased the activities of glutathione peroxidase and superoxide dismutase in the ischemic brain tissues. In addition, ligustilide induced a marked increase in Bcl-2 expression as well as a significant decrease in Bax and caspase-3 in the ischemic cortex. This indicated that the antioxidant and anti-apoptotic properties of ligustilide might contribute to the neuroprotective potential in cerebral ischemic damage [22].

Ligustilide was reported to potently inhibit the proliferation of vascular smooth muscle cells and cell cycle progression. Furthermore, ligustilide suppressed reactive oxygen species (ROS) production and extracellular signal-related kinases (ERK), c-Jun N-terminal protein kinase (JNK), and p38 mitogen activated protein kinase (MAPK). The results showed that the antiproliferative effect of ligustilide was associated with a decrease in ROS, which in turn resulted in a suppression of the MAPK pathway [23]. Both, ligustilide and butylidenephthalide were also found to suppress the proliferation of rat smooth muscle cells stimulated by basic fibroblast growth factor (bFGF). However, neither ligustilide nor butylidenephthalide had any effect on the normal growth of smooth muscle cells, which indicated that the phthalides might inhibit the abnormal proliferation of smooth muscle cells [24].

The phthalide dimers from the rhizome of *L. chuanxiong* were reported to prevent liver fibrosis, as observed in tests with rat and human hepatic stellate cell lines. *Z,Z'*-6,8',7,3'-Diligustilide and levistolide A each significantly suppressed proliferation in both rat and human hepatic stellate cell lines, as stimulated by platelet-derived growth factor. Such inhibition of proliferation was found to be associated with a reduction of α-smooth muscle actin and collagen expression. The cell cycle-promoting proteins – cyclins D1, D2, E, A, and B1 – were each down-regulated, while the inhibitory proteins p21 and p27, and JNK phosphorylation were all up-regulated. In T6 hepatic stellate cells, the two phthalide dimers induced apoptosis through the activation of caspase-9 and caspase-3, an increase in cytosolic cytochrome c release, and a down-regulation of Bcl-2 and Akt phosphorylation. The two phthalide dimers were suggested to inhibit the platelet-derived growth factor-activated proliferation of hepatic stellate cells through mechanisms of cell cycle inhibition and apoptosis [25].

The pharmacokinetics of ligustilide following oral administration to rats has been reported. The maximum serum concentration of ligustilide ($1.5\,\mu g\,ml^{-1}$) occurred at 40 min after oral administration; the subsequent area under the concentration–time curve (AUC) was $34\,h\cdot\mu g\,ml^{-1}$. The data obtained from the pharmacokinetic study indicated that ligustilide was easily absorbed, but slowly eliminated between 3 and 12 h after oral administration. It is possible that ligustilide might be capable of crossing the blood–brain barrier, since concentrations in the rat cerebellum, cerebrum, spleen and kidney were higher than those in other organs [26].

Ligustrazine was found to be effective for the treatment of ischemic vascular diseases, especially as a pulmonary vasodilator. It has been widely used to treat cardiovascular diseases in China [27]. Experimental studies have shown that ligustrazine may have a beneficial effect on pulmonary vascular changes induced by chronic hypoxia in rats [28]. Acute pulmonary alveolar hypoxia in dogs caused an increase in plasma endothelin-1 levels which was significantly reduced by an injection of ligustrazine. This indicated that ligustrazine can inhibit the release of endothelin-1, which might correlate with the blockade of Ca^{2+} channels [29]. Similar results were observed in rabbits [30]. An *in vitro* study showed that ligustrazine inhibited endothelin 1 gene expression in vascular endothelial cells stimulated by angiotensin II. Ligustrazine also inhibited the angiotensin II-induced increase of intracellular ROS and ERK phosphorylation [31]. The intracoronary administration of endothelin-1 caused significant myocardial ischemia through coronary vasoconstriction, which could be inhibited by ligustrazine [32]. Both, the mean pulmonary artery pressure and pulmonary vascular resistance in dogs with acute hypoxic pulmonary hypertension were significantly reduced by injection of ligustrazine [33]. Portal hypertension in rats induced by partial portal vein ligation can be treated with ligustrazine, at doses between 3 and $30\,mg\,kg^{-1}$. After treatment, the total peripheral resistance was significantly reduced and the cardiac index slightly increased [34]. In cirrhotic rats, ligustrazine limited (in dose-dependent fashion) the decrease in portal venous pressure, and also reduced the mean arterial pressure and total peripheral resistance [35].

The clinical treatment of chronic cor pulmonale patients with ligustrazine significantly decreased the pulmonary arterial pressure, pulmonary vascular resistance, and heart rate. It also increased the cardiac output and improved the right-heart cardiac function. These effects of ligustrazine appeared rapidly but were not longlasting [36]. Ligustrazine was also used to treat coronary disease, as it very effectively protected the myocardium from ischemia and reperfusion injury [37]. In experimental studies, the i.v administration of ligustrazine to guinea pigs increased the coronary flow and decreased the myocardial contractile force. In dogs, ligustrazine lowered the arterial blood pressure and increased the left ventricular pressure, heart rate, coronary blood flow, and myocardial oxygen consumption [38]. It was also reported that ligustrazine induced cardioprotective effects in rats during ischemia–reperfusion; these were related to a reduction in TNF-α content by an inhibition of free radical production in the heart [39]. In the isolated rat heart, pretreatment with ligustrazine enhanced PGI_2 outflow and attenuated the release of TXA_2 during normoxia, hypoxia, and reoxygenation [40]. The i.v. administration of ligustrazine to rats led to a significant reduction in the levels of plasma lipid peroxidation, and also increased the superoxide dismutase activity [41].

Ligustrazine is a true Ca^{2+} antagonist, in that it not only blocks the entry of extracellular Ca^{2+} but also inhibits the release of intracellularly stored Ca^{2+} in vascular smooth muscle cells [42]. Ligustrazine is known to be an inhibitor of cAMP-phosphodiesterase, and the intracellular accumulation of cAMP is considered a further important factor for the action of ligustrazine on vascular smooth muscle. In addition, ligustrazine might increase cGMP levels in isolated rat aortic strips, as ligustrazine-induced vasodilatation was abrogated by the blockade of cGMP-dependent protein kinases; this suggested that the relaxation of rat aortic strip by ligustrazine was induced in a cGMP-dependent manner [43]. Also in dogs, the cardiovascular effects of ligustrazine were considered to be related to an inhibition of phosphodiesterase activity [44].

Ligustrazine has been shown to have neuroprotective effects. The i.p. administration of ligustrazine to adult rats at a daily dose of 80 mg kg^{-1} at 2 h after focal cerebral ischemia, as induced by left middle cerebral artery occlusion, resulted in an increased number of proliferating cells in the cortex and striatum. These results suggested that ligustrazine might promote self-repair after ischemia [45]. The effects of ligustrazine in rats with focal cerebral ischemia seemed partly mediated by a free radical-scavenging activity and by an inhibition of neutrophil activation, which resulted in a reduced infarction volume in ischemia–reperfusion brain injury [46]. Ligustrazine also exhibited, in concentration-dependent fashion, a significant neuroprotective effect against ischemic deficits caused by a reduction of behavioral disturbance. Both, neuronal loss and brain infarction in the ischemic side of rats were markedly lowered by treatment with ligustrazine. Internucleosomal DNA fragmentation, caspase-8, caspase-9, and caspase-3 activation, and cytochrome c release induced by cerebral ischemic and reperfusion were each reduced by ligustrazine treatment. The suppression of any inflammatory reaction, the reduction of neuronal apoptosis, and the prevention of neuronal loss might each contribute to the neuroprotective effect of ligustrazine [47]. Neuroprotective effects of ligustrazine were also observed in rabbits following spinal cord ischemia. Here, ligustrazine exerted preventive effects against spinal cord ischemia–reperfusion injury by reducing apoptosis through regulating Bcl-2 and Bax expression [48]. *In vitro*, ligustrazine caused a marked block in H_2O_2-induced cell apoptosis by regulating Bcl-2 family members, suppressing cytochrome c release, and activating the caspase cascade in PC12 cells [49].

Recently, ligustrazine was reported to prevent NRK-52E rat renal tubular cell apoptosis induced by gentamicin. Ligustrazine was found to inactivate the gentamicin-stimulated activities of caspase-3, caspase-8, and caspase-9, to inhibit the gentamicin-induced release of cytochrome c, and to raise the expression of Bcl-xL. Ligustrazine also inhibited gentamicin-induced TNF-α expression, inactivated the transcription factor NF-κB, and reduced apoptotic injury in rat renal tubular cells; this suggested that gentamicin-induced oxidative stress and apoptotic injury in rat renal tubular cells might be counteracted by ligustrazine [50]. Ligustrazine was given intraperitoneally to rats at a daily dose of 100 mg kg^{-1} for 15 days after a single i.v. injection of anti-glomerular basement membrane antibody. A significant decrease in malondialdehyde content was observed, together with significant increases in the levels of glutathione, superoxide dismutase, glutathione peroxidase, and catalase in the kidney tissues. The protective activity of ligustrazine against nephritis, as induced by an accelerated anti-glomerular basement membrane antibody production, might be due to its antioxidant properties and an inhibition of ROS production [51].

An antiplatelet aggregation activity of ligustrazine was also demonstrated in shear-induced platelet aggregation in rats and rabbits [52], and in dogs during cardiopulmonary bypass and arterial thrombosis [53]. The anti-platelet aggregation activity of ligustrazine on human platelets was caused by an inhibition of intracellular Ca^{2+} mobilization (mainly from internal stores), and by an enhancement of intracellular cAMP following phosphodiesterase inhibition [54]. The inhibition of platelet aggregation and promotion of disaggregation contributed to the antithrombotic activity of ligustrazine. The antiplatelet activity of ligustrazine was also considered to involve an inhibition of phosphoinositide breakdown at lower concentrations, and an inhibition of TXA_2 formation at higher concentrations, leading to the inhibition of platelet aggregation via binding to the glycoprotein IIb/IIIa complex [55]. Ligustrazine was further reported to prevent the death of mice with endotoxemia by decreasing the biosynthesis of PAF [56]. The inhibition of platelet aggregation by ligustrazine was also shown clinically, in patients with acute ischemic stroke [57]. Ligustrazine also demonstrated an antimetastatic effect in patients with lung carcinoma of different types; the mechanism of this effect was thought to be mediated by an inhibition not only of blood platelet adhesion but also of the activity of coagulation factors [58].

Pharmacokinetic studies of orally administered ligustrazine in healthy volunteers revealed that ligustrazine was absorbed rapidly, distributed widely in the body, and eliminated at a fairly rapid rate with $t_{1/2\alpha}$ of 0.5 h and $t_{1/2\beta}$ of 3 h [59]. The major metabolite of ligustrazine in human urine was identified as 3,5,6-trimethylpyrazine-2-carboxylic acid [60]. A similar metabolic pathway was observed in rabbits [61]. Concentrations of ligustrazine in various regions of the rat brain, such as the cerebral cortex, brainstem, striatum, hippocampus, cerebellum and midbrain, did not differ significantly at 15 min after a single i.v. injection (10 mg kg^{-1}). In rats, the mean plasma concentration of ligustrazine was approximately fivefold greater than in the brain tissues [62]. When comparing the pharmacokinetic characteristics of ligustrazine in healthy dogs and in those with pathological changes to their hemodynamics, serum ligustrazine concentrations, peak drug concentration time, AUC, and $t_{1/2\alpha}$ were each increased significantly in the diseased animals [63].

References

1 Cao, F.Y., Liu, W.X., Wen, Y.S., He, Z.R., and Qin, W.J. (1983) Studies on chemical constituents of *Ligusticum chuanxiong. Chin. Trad. Herbal Drugs*, **14**, 241–242.

2 Naito, T., Ikeya, Y., Okada, M., Mitsuhashi, H., and Maruno, M. (1992) Two phthalides from *Ligusticum chuangxiong. Phytochemistry*, **31**, 639–642.

3 Huang, Y.Z. and Pu, F.D. (1988) Studies on the chemical components of essential oil from the rhizome of *Ligusticum sinense* Oliv. cv. *chuanxiong* Hort. *Acta Pharm. Sin.*, **23**, 426–429.

4 Naito, T., Katsuhara, T., Niitsu, K., Ikeya, Y., Okada, M., and Mitsuhashi, H. (1991) Phthalide dimers from *Ligusticum chuangxiong* Hort. *Heterocycles*, **32**, 2433–2442.

5 Luo, Y.M., Zhang, J.H., Pan, J.G., Yao, S.L., Huang, H.L., Zhu, Y., and Li, Q.S. (1994) Five constituents isolated from the rhizome of *Ligusticum sinense. Chin. Pharm. J.*, **29**, 714–716.

6 Luo, Y., Zhang, J., Pan, J., and Li, Q. (1995) Two phthalide dimers from rhizome of *Ligusticum sinense* Oliv. cv. *chuanxiong. China J. Chin. Mater. Med.*, **20**, 39–41.

7 Dai, B. (1988) Comparison of chemical constituents of essential oil from four species of gaoben (*Ligusticum*) by GC-MS analysis. *Acta Pharm. Sin.*, **23**, 361–369.

8 Shao, C.R., Chen, F.M., and Tang, Y.X. (1994) Clinical and experimental study on *Ligusticum wallichii* mixture in preventing and treating bronchial asthma. *Chin. J. Integr. Trad. West. Med.*, **14**, 465–468.

9 Chen, D.R. (1992) Clinical and experimental study of *Ligusticum wallichii* and aspirin in the treatment of transient ischemic attack. *Chin. J. Integr. Trad. West. Med.*, **12**, 672–674.

10 Chen, K.J. and Chen, K. (1992) Ischemic stroke treated with Ligusticum chuanxiong. *Chin. Med. J. (Engl.)*, **105**, 870–873.

11 Liu, Z. (1991) Effects of *Ligusticum wallichii* on the plasma levels of β-thromboglobulin, platelet factor 4, thromboxane B_2 and 6-keto-PGF_1 in patients with acute cerebral infarction. *Chin. J. Integr. Trad. West. Med.*, **11**, 711–713.

12 Yu, Z., Chen, K.J., Qian, Z.H., Weng, W.L., Yu, Y.Q., Tu, X.H., Ma, H.M., Wu, Y.S., and Zhang, H. (1987) Effect of chuanxiong granule on platelet function and prostaglandin metabolism in coronary disease patients. *Chin. J. Integr. Trad. West. Med.*, **7**, 8–11.

13 Liu, Z. (1990) Effects of *Ligusticum wallichii* on the plasma levels of -thromboglobulin, platelet factor 4, thromboxane B_2 and 6-keto-PGF_1 in rabbits under acute experimental cerebral ischemia. *Chin. J. Integr. Trad. West. Med.*, **10**, 543–544.

14 Wang, J., Shi, Y.M., and Zheng, H.M. (1993) Experimental study of *Ligusticum wallichii* on cerebrovascular hemodynamic parameters. *Chin. J. Integr. Trad. West. Med.*, **13**, 417–419.

15 Hu, W.L. and Ma, Y.J. (1993) Effect of glycerol-induced acute renal failure in rabbit with *Ligusticum wallichii* on thromboxane B_2, 6-keto-prostaglandin $F_{1?}$/thromboxane B_2. *Chin. J. Integr. Trad. West. Med.*, **13**, 549–550.

16 Liu, L., Ning, Z.Q., Shan, S., Zhang, K., Deng, T., Lu, X.P., and Cheng, Y.Y. (2005) Phthalide lactones from *Ligusticum chuanxiong* inhibit lipopolysaccharide-induced TNF-α production and TNF-α-mediated NF-κB activation. *Planta Med.*, **71**, 808–813.

17 Liang, M.J., He, L.C., and Yang, G.D. (2005) Screening, analysis and *in vitro* vasodilatation of effective components from *Ligusticum chuanxiong. Life Sci.*, **78**, 128–133.

18 Chan, S.S., Cheng, T.Y., and Lin, G. (2007) Relaxation effects of ligustilide and senkyunolide A, two main constituents of *Ligusticum chuanxiong*, in rat isolated aorta. *J. Ethnopharmacol.*, **111**, 677–680.

19 Cao, Y.X., Zhang, W., He, J.Y., He, L.C., and Xu, C.B. (2006) Ligustilide induces vasodilatation via inhibiting voltage dependent calcium channel and receptor-mediated Ca^{2+} influx and release. *Vasc. Pharmacol.*, **45**, 171–176.

20 Du, J.R., Yu, Y., Yao, Y., Bai, B., Zong, X., Lei, Y., Wang, C.Y., and Qian, Z.M. (2007) Ligustilide reduces phenylephrine induced-aortic tension *in vitro* but has no effect on systolic pressure in spontaneously hypertensive rats. *Am. J. Chin. Med.*, **35**, 487–496.

21 Peng, H.Y., Du, J.R., Zhang, G.Y., Kuang, X., Liu, Y.X., Qian, Z.M., and Wang, C.Y. (2007) Neuroprotective effect of Z-ligustilide against permanent focal ischemic damage in rats. *Biol. Pharm. Bull.*, **30**, 309–312.

22 Kuang, X., Yao, Y., Du, J.R., Liu, Y.X., Wang, C. Y., and Qian, Z.M. (2006) Neuroprotective role of Z-ligustilide against forebrain ischemic injury in ICR mice. *Brain Res.*, **1102**, 145–153.

23 Lu, Q., Qiu, T.Q., and Yang, H. (2006) Ligustilide inhibits vascular smooth muscle

cells proliferation. *Eur. J. Pharmacol.*, **542**, 136–140.

24 Liang, M.J. and He, L.C. (2006) Inhibitory effects of ligustilide and butylidenephthalide on bFGF-stimulated proliferation of rat smooth muscle cells. *Yao Xue Xue Bao*, **41**, 161–165.

25 Lee, T.F., Lin, Y.L., and Huang, Y.T. (2007) Studies on antiproliferative effects of phthalides from *Ligusticum chuanxiong* in hepatic stellate cells. *Planta Med.*, **73**, 527–534.

26 Shi, Y., He, L., and Wang, S. (2006) Determination of ligustilide in rat blood and tissues by capillary gas chromatography mass spectrometry. *Biomed. Chromatogr.*, **20**, 993–998.

27 Zhang, K.J. and Zhang, Y.X. (1995) Recent advances in experimental and clinical study on ligustrazine in treating respiratory disease. *Chin. J. Integr. Trad. West. Med.*, **15**, 638–640.

28 Yuan, X.J. (1988) Inhibitory effect of ligustrazine on acute and chronic pulmonary hypertension in rats. *Acta Acad. Med. Sin.*, **10**, 31–35.

29 Cao, W., Zeng, Z., Zhu, Y.J., Luo, W., Demura, H., Naruse, M., and Shi, Y. (1998) Effects of tetramethylpyrazine, a Chinese medicine, on plasma endothelin-1 levels during acute pulmonary hypoxia in anesthetized dogs. *J. Cardiovasc. Pharmacol.*, **31** (Suppl. 1), S456–S459

30 Zeng, Z., Zhu, W., Zhou, X., Jin, Z., Liu, H., Chen, X., Pan, J., Demura, H., Naruse, M., and Shi, Y. (1998) Tetramethylpyrazine, a Chinese drug, blocks coronary vasoconstriction by endothelin-1 and decreases plasma endothelin-1 levels in experimental animals. *J. Cardiovasc. Pharmacol.*, **31** (Suppl. 1), S313–S316

31 Lee, W.S., Yang, H.Y., Kao, P.F., Liu, J.C., Chen, C.H., Cheng, T.H., and Chan, P. (2005) Tetramethylpyrazine downregulates angiotensin II-induced endothelin-1 gene expression in vascular endothelial cells. *Clin. Exp. Pharmacol. Physiol.*, **32**, 845–850.

32 Pu, Z., Zhu, W., Jing, Z., Zeng, Z., and Zuo, W. (1996) Effect of tetramethyl pyrazine on coronary vasoconstriction induced by endothelin-1 in dogs. *Acta Acad. Med. Sci.*, **18**, 133–137.

33 Li, L., Li, Z.C., Yan, H.X., Hu, H.Y., Chen, L., Yang, W.X., and Zheng, Y. (2005) Protective effect of tetramethylpyrazine on the respiratory inhibition induced by hypoxia in rats. *Sichuan Da Xue Xue Bao Yi Xue Ban*, **36**, 533–536.

34 Chang, F.C., Huang, Y.T., Hong, C.Y., Lin, J.G., and Chen, K.J. (1999) Haemodynamic effects of chronic tetramethylpyrazine administration on portal hypertensive rats. *Eur. J. Gastroenterol. Hepatol.*, **11**, 1027–1031.

35 Huang, Y.T., Chang, F.C., Chen, K.J., and Hong, C.Y. (1999) Acute hemodynamic effects of tetramethylpyrazine and tetrandrine on cirrhotic rats. *Planta Med.*, **65**, 130–134.

36 Wei, D.G., Li, G.S., and Sun, B.T. (1996) Effects of ligustrazine on hemodynamics in chronic pulmonary heart disease patients. *Chin. J. Integr. Trad. West. Med.*, **16**, 730–732.

37 Lin, L.N., Wang, W.T., and Xu, Z.J. (1997) Clinical study on ligustrazine in treating myocardial ischemia and reperfusion injury. *Chin. J. Integr. Trad. West. Med.*, **17**, 261–263.

38 Song, Z.J., Chen, Q., and Li, B.L. (2004) Protective effect of tetramethylpyrazine and L-arginine on rats with acute myocardial infarction. *Zhongguo Zhong Xi Yi Jie He Za Zhi*, **24**, 912–914.

39 Zhou, Y., Hu, C.P., Deng, P.Y., Deng, H.W., and Li, Y.J. (2004) The protective effects of ligustrazine on ischemia-reperfusion and DPPH free radical-induced myocardial injury in isolated rat hearts. *Planta Med.*, **70**, 818–822.

40 Feng, J., Liu, R., Wu, G., and Tang, S. (1997) Effects of tetramethylpyrazine on the release of PGI_2 and TXA_2 in the hypoxic isolated rat heart. *Mol. Cell. Biochem.*, **167**, 153–158.

41 Feng, J., Wu, G., and Tang, S. (1999) The effects of tetramethylpyrazine on the incidence of arrhythmias and the release of PGI_2 and TXA_2 in the ischemic rat heart. *Planta Med.*, **65**, 268–270.

42 Pang, P.K., Shan, J.J., and Chiu, K.W. (1996) Tetramethylpyrazine, a calcium antagonist. *Planta Med.*, **62**, 431–435.

43 Tsai, C.C., Lai, T.Y., Huang, W.C., Liu, I.M., and Liou, S.S. (2005) Role of cGMP signals in tetramethylpyrazine induced relaxation of the isolated rat aortic strip. *Life Sci.*, **77**, 1416–1424.

44 Lin, C.I., Wu, S.L., Tao, P.L., Chen, H.M., and Wei, J. (1993) The role of cyclic AMP and phosphodiesterase activity in the mechanism of action of tetramethylpyrazine on human and dog cardiac and dog coronary arterial tissues. *J. Pharm. Pharmacol.*, **45**, 963–966.

45 Qiu, F., Liu, Y., Zhang, P.B., Tian, Y.F., Zhao, J. J., Kang, Q.Y., Qi, C.F., and Chen, X.L. (2006) The effect of ligustrazine on cells proliferation in cortex and striatum after focal cerebral ischemia in adult rats. *Zhong Yao Cai*, **29**, 1196–1200.

46 Hsiao, G., Chen, Y.C., Lin, J.H., Lin, K.H., Chou, D.S., Lin, C.H., and Sheu, J.R. (2006) Inhibitory mechanisms of tetramethylpyrazine in middle cerebral artery occlusion (MCAO)-induced focal cerebral ischemia in rats. *Planta Med.*, **72**, 411–417.

47 Kao, T.K., Ou, Y.C., Kuo, J.S., Chen, W.Y., Liao, S.L., Wu, C.W., Chen, C.J., Ling, N.N., Zhang, Y.H., and Peng, W.H. (2006) Neuroprotection by tetramethylpyrazine against ischemic brain injury in rats. *Neurochem. Int.*, **48**, 166–176.

48 Fan, L.H., Wang, K.Z., Cheng, B., Wang, C.S., and Dang, X.Q. (2006) Anti-apoptotic and neuroprotective effects of tetramethylpyrazine following spinal cord ischemia in rabbits. *BMC Neurosci.*, **7**, 48.

49 Cheng, X.R., Zhang, L., Hu, J.J., Sun, L., and Du, G.H. (2007) Neuroprotective effects of tetramethylpyrazine on hydrogen peroxide-induced apoptosis in PC12 cells. *Cell Biol. Int.*, **31**, 438–443.

50 Juan, S.H., Chen, C.H., Hsu, Y.H., Hou, C.C., Chen, T.H., Lin, H., Chu, Y.L., and Sue, Y.M. (2007) Tetramethylpyrazine protects rat renal tubular cell apoptosis induced by gentamicin. *Nephrol. Dial. Transplant.*, **22**, 732–739.

51 Fu, H., Li, J., Li, Q.X., Xia, L., and Shao, L. (2007) Protective effect of ligustrazine on accelerated anti-glomerular basement membrane antibody nephritis in rats is based on its antioxidant properties. *Eur. J. Pharmacol.*, **563**, 197–202.

52 Liao, F.L. and Li, B. (1997) Inhibition of shear-induced platelet aggregation by Chinese herbal medicines. *Clin. Hemorheol. Microcirc.*, **17**, 315–318.

53 Wu, G.X., Wu, J.C., Ma, H.T., and Ruan, C.G. (1992) Inhibitory effects of tetramethylpyrazine on platelets during cardiopulmonary bypass and arterial thrombus formation in dogs. *Acta Pharm. Sin.*, **13**, 330–333.

54 Liu, S.Y. and Sylvester, D.M. (1994) Antiplatelet activity of tetramethylpyrazine. *Thromb. Res.*, **75**, 51–62.

55 Sheu, J.R., Kan, Y.C., Hung, W.C., Ko, W.C., and Yen, M.H. (1997) Mechanisms involved in the antiplatelet activity of tetramethylpyrazine in human platelets. *Thromb. Res.*, **88**, 259–270.

56 Fu, Y. and Hu, Y.M. (1999) Prophylactic effects of tetramethyl pyrazine on mice with endotoxemia and its relationship with platelet-activating factor. *Acta Pharm. Sin.*, **20**, 529–532.

57 Gao, B.T. (1989) The effects of ligustrazine, aspirin and beta-histine on platelet aggregation in patients with acute ischemic stroke. *Chin. J. Nerv. Mental Dis.*, **22**, 148–151.

58 Chen, S.X., Wang, L.X., and Xing, L.L. (1997) Effects of tetramethylpyrazine on platelet functions of advanced cases of lung carcinoma. *Chin. J. Integr. Trad. West. Med.*, **17**, 531–533.

59 Cai, W., Dong, S.N., and Lou, Y.Q. (1989) HPLC determination of tetramethylpyrazine in human serum and its pharmacokinetic parameters. *Acta Pharm. Sin.*, **24**, 881–886.

60 Ye, Y., Wang, S., and Jiang, J. (1996) Studies on the metabolites of tetramethylpyrazine in human urine. *Acta Acad. Med. Sin.*, **18**, 288–291.

61 Chen, X., and Dong, S.N. (1996) Studies on the metabolites of tetramethylpyrazine in rabbits. *Acta Pharm. Sin.*, **31**, 617–621.

62 Liang, C.C., Hong, C.Y., Chen, C.F., and Tsai, T.H. (1999) Measurement and pharmacokinetic study of tetramethylpyrazine in rat blood and its regional brain tissue by high-performance liquid chromatography. *J. Chromatogr. B. Biomed. Sci. Appl.*, **724**, 303–309.

63 Huang, X., Jiang, Y.P., and Zang, Y.M. (1996) Study on pharmacokinetic characteristics of tetramethylpyrazine and hemodynamics of heart blood stasis in dogs. *Chin. J. Integr. Trad. West. Med.*, **16**, 352–354.

Ligustrum lucidum

Fructus Ligustri Lucidi (Nüzhenzi) is the dried ripe fruit of *Ligustrum lucidum* Ait. (Oleaceae). It is used for the treatment of vertigo and tinnitus.

Ligustrum lucidum Ait.

Chemistry

The fruit of *L. lucidum* contains oleanolic acid and ursolic acid as the major components [1]. The Chinese Pharmacopoeia requires a qualitative determination of oleanolic acid in Fructus Ligustri Lucidi by thin-layer chromatographic comparison with reference substance, and a quantitative determination by thin-layer chromatography–densitometry. The content of oleanolic acid in Fructus Ligustri Lucidi should be not less than 0.6%. Minor triterpenes were identified as ursolic acid methyl ester and tormentic acid (2α,19-dihydroxyursolic acid) [2]. The content of oleanolic acid in the external, medial, and inner skin, and in the seed, were determined as 1.6%, 1.8%, 1.0%, and 0.7%, respectively [3]. The unripe fruit, collected in August, contained more oleanolic acid than the ripe fruit, collected in December [4].

p-Hydroxyphenethyl β-D-glucopyranosides from the fruits were identified as salidroside, oleoside dimethyl ester, ligustroside, oleuropein, nuezhenide, isonuezhenide, neonuezhenide, lucidumosides A, B, C, and D [5, 6], nuezhenidic acid, and ligustrosidic acid [7]. The aglycone *p*-hydroxybenzene ethanol (tyrosol) and its α-D-glucopyranoside and polysaccharide were also isolated from the fruits of *L. lucidum* [8].

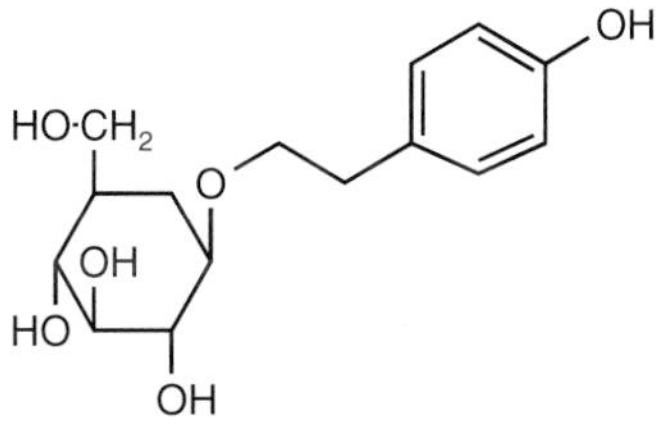

Salidroside

About 0.01% essential oil has been obtained from the fruit of *L. lucidum*. The major components of the essential oil were identified as α-pinene (5.3%), β-pinene (3.8%), limonene (6.7%), 4-terpineol (3.1%), 2-phenyl-1-ethanol (4.2%), and eugenol (6.1%) [9].

Pharmacology and Toxicology

The fruits of *L. lucidum* are mainly used as a tonic. It was reported that the fruits exhibit preventive effects against experimental atherosclerosis in rabbits [10]. More important was the immunostimulatory activity of the fruits. The fruit was found to increase the clearance rate of intravenously administered charcoal particles in mice, and to

Handbook of Chinese Medicinal Plants: Chemistry, Pharmacology, Toxicology
Weici Tang and Gerhard Eisenbrand

ISBN: 978-3-527-32226-8

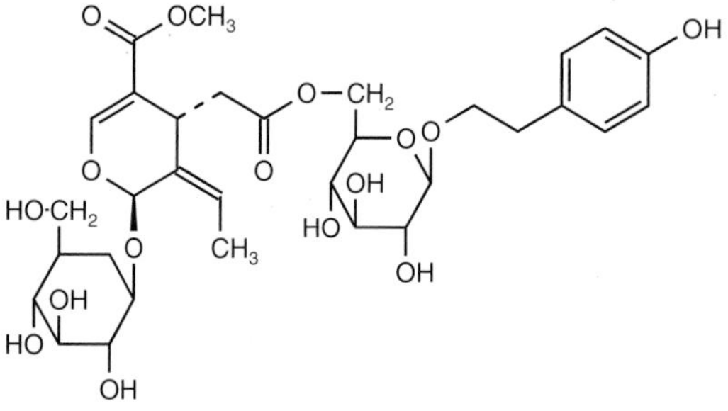

Nuezhenide

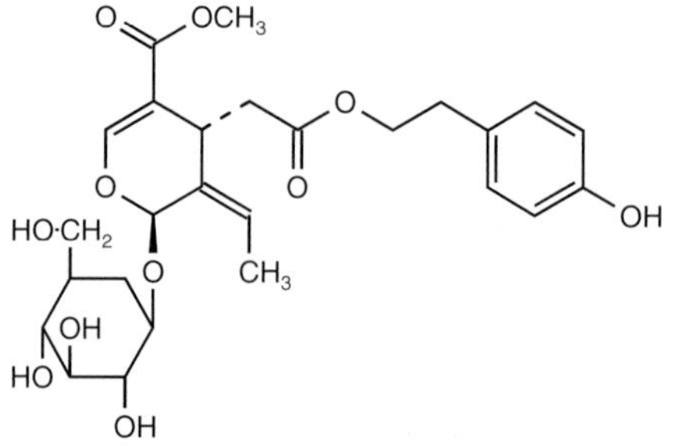

Ligustroside

antagonize the immunosuppressive action of prednisolone. It also protected mice against carbon tetrachloride (CCl_4) intoxication. The *in vivo* antioxidant activities of the fruits were found to be associated with oleanolic acid, as evidenced in a test against the hepatotoxicity of CCl_4 in mice. The hepatoprotective action of oleanolic acid by pretreatment might be mainly mediated by the enhancement of hepatic glutathione regeneration capacity, particularly under conditions of CCl_4-induced oxidative stress. [11]. Clinical studies showed that an extract from the fruits of *L. lucidum* effected an immune restoration in nine of 13 cancer patients, with an increase in local graft-versus-host reaction from 32 mm^3 to 118 mm^3. This indicates that the fruits might contain potent immune stimulants, and this may provide the rational basis for their therapeutic use as biological response modifiers [12]. However, it was also reported that oral administration of 240 mg of the extract from *L. lucidum* from day 1 to day 12 did not prevent myelosuppression in rats caused by an intravenous (i.v.) dose of 75 mg kg^{-1} cyclophosphamide on day 1. No difference between the treated and control animals in terms of the absolute neutrophil count and the platelet count was observed [13].

An extract of the fruits of *L. lucidum* was also reported to show free hydroxyl radical scavenging activity [14]. In broilers, dietary supplementation with *L. lucidum* significantly decreased the serum concentration of malondialdehyde and elevated the activity of serum glutathione reductase. Dietary supplementation with *L. lucidum* also significantly improved the antibody titer against Newcastle disease virus and lymphocyte proliferation; this indicated an antioxidant and immune stimulating effect of *L. lucidum* [15]. Dietary supplementation with 1% of the fruit of *L. lucidum* was also found to enhance immune function and antioxidant status of laying hens during heat stress [16].

An aqueous extract of the fruits of *L. lucidum* was reported to enhance spontaneous [^{3}H]thymidine incorporation into the mononuclear cells in 14 healthy volunteers, with a stimulation index of 3.1. The extract also significantly augmented the proliferation of lymphocytes from healthy subjects *in vitro*, as induced by suboptimal concentrations of phytohemagglutinin, concanavalin A (Con A), and pokeweed mitogen. The extract also significantly augmented the phytohemagglutinin and Con A responses of mononuclear cells from 14 cancer patients, in a dose-dependent manner. The optimal concentration for stimulating the mononuclear cells of cancer patients was 100 μg ml^{-1}, compared to 10 μg ml^{-1} for the mononuclear cells of healthy donors [17].

A significant dose-related depression of the oxidative burst in a murine macrophage cell line J774 occurred on incubation with murine renal cell carcinoma (Renca) and murine bladder tumor cells or the cell-free tumor extracts. Depression was found to be partially or completely reversed by the presence of 50–100 μg ml^{-1} of the extract of *L. lucidum*; this suggested that *L. lucidum* might exert its antitumor activity via an abolition of tumor-associated macrophage suppression [18] and via the augmentation of phagocyte and lymphokine-activated killer cell activities [19].

The fruit of *L. lucidum* also exerted cancer chemopreventive properties. It reduced the mutagenic activity of aflatoxin B_1 (AFB_1) in *Salmonella typhimurium* TA100 [20], and of benzo[*a*]pyrene (BaP) in *S. typhimurium* TA98 [21] in the presence of an activating system. It significantly inhibited AFB_1 binding to DNA, reduced AFB_1 DNA adduct formation, and also significantly decreased the formation of organosoluble metabolites of AFB_1 [20]. Oleanolic acid, ursolic acid and nuezhenide were found to be the antimutagenic components against BaP mutagenesis [21]. The fruit of *L. lucidum* and oleanolic acid were also found to protect against the formation of micronuclei in the bone marrow of mice induced by cyclophosphamide and urethane [22]. Moreover, the extract of *L. lucidum* was found to exert growth inhibition against a number of human and murine cancer cell lines of different tissues, including breast, lung, pancreas and prostate, but not of normal human mammary epithelial cells [23].

Salidroside was reported to have an anti-inflammatory activity. Interaction of salidoside with the cyclooxygenase and 5-lipooxygenase pathways of the arachidonate metabolism have been studied in calcium-stimulated mouse peritoneal macrophages and human platelets. Salidoside exerted a preferential effect on the cyclooxygenase pathway, inhibiting release of the cyclooxygenase metabolites prostaglandin E_2 (PGE_2) with an IC_{50} value of 45 μ*M* and, to a lesser extent, reducing thromboxane B_2 (TXB_2) levels [24]. The effect of salidroside on anoxia and reoxygenation damages has been studied in cultured myocytes from neonatal rat hearts. Following anoxia for 3 h and reoxygenation for 1 h, the beating of the myocardial cells was slowed and lactate dehydrogenase (LDH) liberation by the myocardial cells increased. An electron microscopic examination of the myocardial cells revealed localized defects of the cell membrane, dilatation of the endoplasmic reticulum, and swelling of the mitochondria. At 1 h before anoxia, the addition of salidroside at concentrations of 10 and 30 μg ml^{-1} increased the beat rate of myocardial cells and depressed the LDH release from myocytes; the myocardial ultrastructure was reported to be normal during both anoxia and reoxygenation [25].

Oleoside dimethyl ester, oleuropein, neonuezhenide, and lucidumosides B and C were reported to exhibit strong antioxidant effects against the hemolysis of red blood cells, as induced by free radicals [26].

References

1 Hu, Y.A., Wang, Y.M., Luo, G.A., and Wei, W. (2000) High-speed separation of oleanolic acid and ursolic acid from the fruit of *Ligustrum lucidum* Ait. and *Crataegus* by cyclodextrin-modified micellar electrokinetic chromatography. *Anal. Lett.*, **33**, 357–371.

2 Cheng, X.F., He, M.F., Zhang, Y., and Meng, Z.M. (2000) Chemical constituents of fructus ligustri lucidi. *J. China Pharm. Univ.*, **31**, 169–170.

3 Li, M.L., Liu, M.L., and Hong, W.H. (1995) Determination of oleanolic acid content in different portion of Fructus Ligustri Lucidi. *China J. Chin. Mater. Med.*, **20**, 216–217.

4 Mi, H.M., Cao, Y.B., Song, H.T., and Zheng, D. J. (1995) Quantitative analysis of oleanolic acid from the fruits of glossy privet (*Ligustrum lucidum*) in different growing periods. *J. Trad. Herbal Drugs*, **26**, 258–260.

5 He, Z.D., Dong, H., Xu, H.X., Ye, W.C., Sun, H.D., and But, P.P. (2001) Secoiridoid

constituents from the fruits of *Ligustrum lucidum*. *Phytochemistry*, **56**, 327–330.
6 Shi, L., Ma, Y., and Cai, Z. (1998) Quantitative determination of salidroside and specnuezhenide in the fruits of *Ligustrum lucidum* Ait by high performance liquid chromatography. *Biomed. Chromatogr.*, **12**, 27–30.
7 Wu, L.J., Xiang, T., Hou, B.L., Liang, W., Yin, S., and Zhou, X.C. (1998) Chemical constituents from fruits of *Ligustrum lucidum*. *Acta Bot. Sin.*, **40**, 83–87.
8 Xu, J.H., Xie, Y.G., Ji, H.L., and Yang, X.Y. (1999) Influences of different preparation methods on content of tyrosol and polysaccharide in fructus ligustri lucidi. *Zhejiang Yike Daxue Xuebao*, **28**, 206–208.
9 Hayashi, S., Narita, M., Kameoka, H., and Arai, T. (1995) Components of essential oil from the dried fruit of *Ligustrum lucidum* Aiton. *Yukagaku*, **44**, 380–386.
10 Peng, Y. (1983) Prevention of experimental atherosclerosis in rabbits with *Ligustrum lucidum* fruit. *Bull. Chin. Mater. Med.*, **8**, 32–34.
11 Yim, T.K., Wu, W.K., Pak, W.F., and Ko, K.M. (2001) Hepatoprotective action of an oleanolic acid-enriched extract of *Ligustrum lucidum* fruits is mediated through an enhancement on hepatic glutathione regeneration capacity in mice. *Phytother. Res.*, **15**, 589–592.
12 Sun, Y., Hersh, E.M., Talpaz, M., Lee, S.L., Wong, W., Loo, T.L., and Mavligit, G.M. (1983) Immune restoration and/or augmentation of local graft versus host reaction by traditional Chinese medicinal herbs. *Cancer*, **52**, 70–73.
13 Khoo, K.S. and Ang, P.T. (1995) Extract of Astragalus membranaceus and Ligustrum lucidum does not prevent cyclophosphamide-induced myelosuppression. *Singapore Med. J.*, **36**, 387–390.
14 Li, H. and Wang, Q. (2004) Evaluation of free hydroxyl radical scavenging activities of some Chinese herbs by capillary zone electrophoresis with amperometric detection. *Anal. Bioanal. Chem.*, **378**, 1801–1805.
15 Ma, D., Li, Q., Du, J., Liu, Y., Liu, S., and Shan, A. (2006) Influence of mannan oligosaccharide, *Ligustrum lucidum* and *Schisandra chinensis* on parameters of antioxidative and immunological status of broilers. *Arch. Anim. Nutr.*, **60**, 467–476.
16 Ma, D., Shan, A., Chen, Z., Du, J., Song, K., Li, J., and Xu, Q. (2006) Effect of *Ligustrum lucidum* and *Schisandra chinensis* on the egg production, antioxidant status and immunity of laying hens during heat stress. *Arch. Anim. Nutr.*, **59**, 439–447.
17 Sun, Y., Hersh, E.M., Lee, S.L., McLaughlin, M., Loo, T.L., and Mavligit, G.M. (1983) Preliminary observations on the effects of the Chinese medicinal herbs *Astragalus membranaceus* and *Ligustrum lucidum* on lymphocyte blastogenic responses. *J. Biol. Response Mod.*, **2**, 227–237.
18 Rittenhouse, J.R., Lui, P.D., and Lau, B.H. (1991) Chinese medicinal herbs reverse macrophage suppression induced by urological tumors. *J. Urol.*, **146**, 486–490.
19 Lau, B.H., Ruckle, H.C., Botolazzo, T., and Lui, P.D. (1994) Chinese medicinal herbs inhibit growth of murine renal cell carcinoma. *Cancer Biother.*, **9**, 153–161.
20 Wong, B.Y., Lau, B.H., Tadi, P.P., and Teel, R. W. (1992) Chinese medicinal herbs modulate mutagenesis, DNA binding and metabolism of aflatoxin B_1. *Mutat. Res.*, **279**, 209–216.
21 Niikawa, M., Hayashi, H., Sato, T., Nagase, H., and Kito, H. (1993) Isolation of substances from glossy privet (*Ligustrum lucidum* Ait.) inhibiting the mutagenicity of benzo[*a*]pyrene in bacteria. *Mutat. Res.*, **319**, 1–9.
22 Hang, B.Q., Dai, Y., Wu, G.Z., Dili, N., Zhao, L., and Tan, L.W. (1987) Protective effects of Fructus Ligustri Lucidi and oleanolic acid against chromosomal damage induced by cyclophosphamide and urethane. *J. China Pharm. Univ.*, **18**, 222–224.
23 Shoemaker, M., Hamilton, B., Dairkee, S.H., Cohen, I., and Campbell, M.J. (2005) *In vitro* anticancer activity of twelve Chinese medicinal herbs. *Phytother. Res.*, **19**, 649–651.
24 Diaz Lanza, A.M., Abad Martinez, M.J., Fernandez Matellano, L., Recuero Carretero, C., Villaescusa Castillo, L., Silvan Sen, A.M., and Bermejo Benito, P. (2001) Lignan and phenylpropanoid glycosides from *Phillyrea latifolia* and their *in vitro* anti-inflammatory activity. *Planta Med.*, **67**, 219–223.
25 Ye, Y.C., Chen, Q.M., Jin, K.P., Zhou, S.X., Chai, F.L., and Hai, P. (1993) Effect of salidroside on cultured myocardial cells anoxia/reoxygenation injuries. *Acta Pharmacol. Sin.*, **14**, 424–426.
26 He, Z.D., But, P.P.H., Chan, T.W., Dong, H., Xu, H.X., Lau, C.P., and Sun, H.D. (2001) Antioxidative glucosides from the fruits of *Ligustrum lucidum*. *Chem. Pharm. Bull. (Tokyo)*, **49**, 780–784.

Linum usitatissimum

Semen Lini (Yamazi) is the dried ripe seed of *Linum usitatissimum* L. (Linaceae). It is used as a laxative. The seed is known as flaxseed or linseed, and is also used as a food supplement. The oil is used for nutritional purposes.

Linum usitatissimum L.

Chemistry

The seed of *L. usitatissimum* is a widely used laxative agent, not only in China but also worldwide. It is rich in fatty oil and plant mucilage. In addition, lignan derivatives, cyanogen glycosides and phenylpropane glucosides were isolated from flaxseed. Flaxseeds were found to contain 30–45% fatty oil composed of 51–55% linolenic acid (α-linolenic acid), 17% linolic acid (*Z,Z*-9,12-octadecadienic acid), 22% oleic acid, and low concentrations of saturated fatty acids [1].

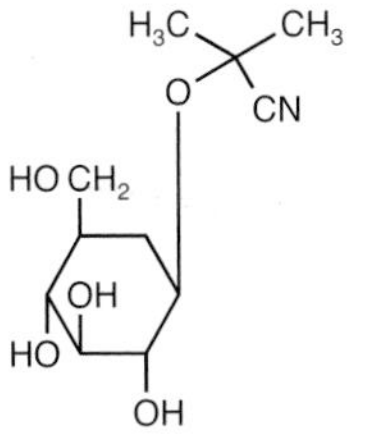

Linamarin

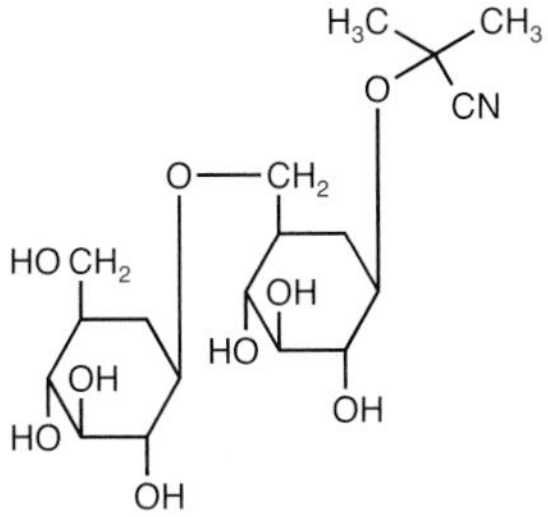

Linustatin

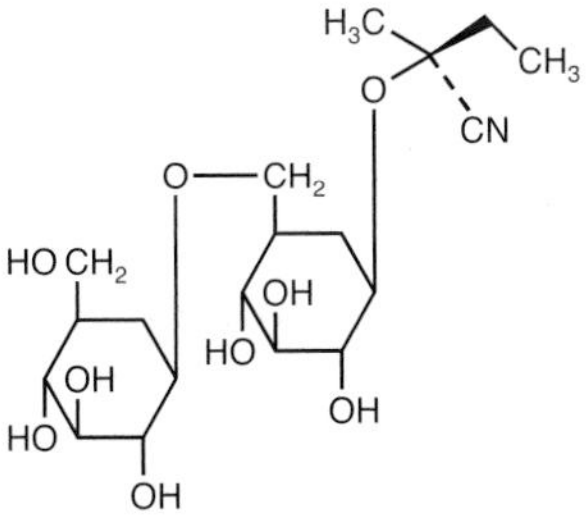

Neolinustatin

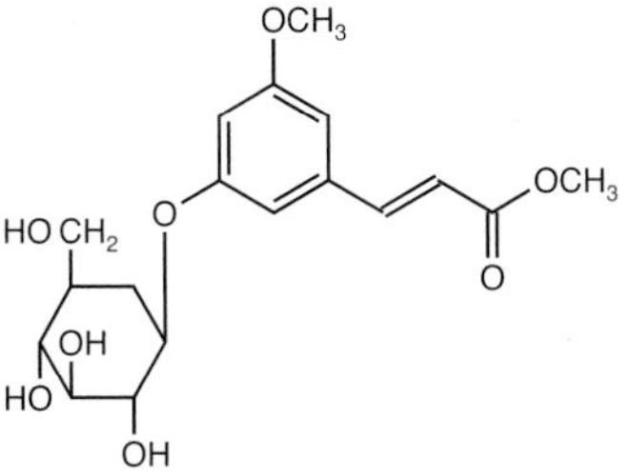

Linusitamarin

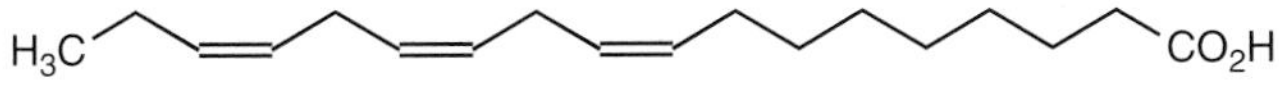

Linolenic acid

Handbook of Chinese Medicinal Plants: Chemistry, Pharmacology, Toxicology
Weici Tang and Gerhard Eisenbrand

ISBN: 978-3-527-32226-8

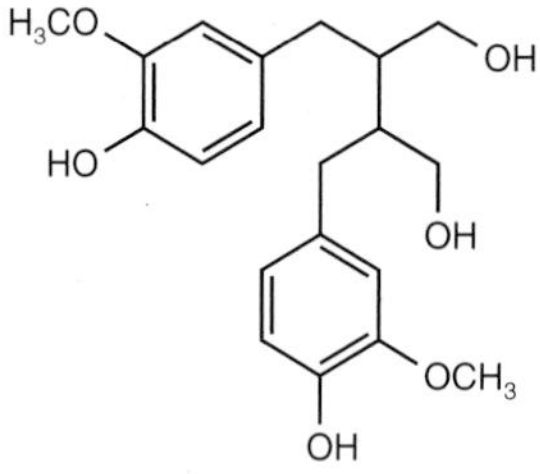

Secoisolariciresinol

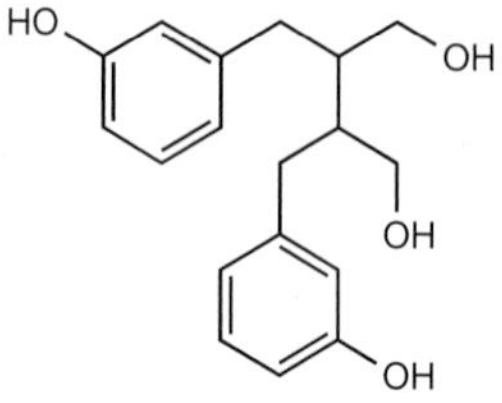

Enterodiol

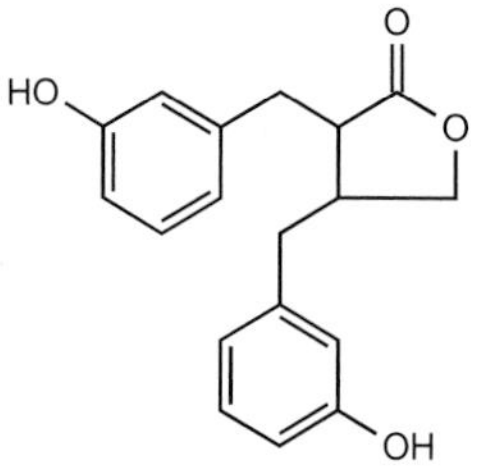

Enterolactone

Flaxseed is one of the richest dietary sources of linolenic acid, and is also a good source of soluble fiber mucilage. The plant mucilage, localized in the peel and epidermis of the seeds, is composed of two different rhamnogalacturonans and an arabinoxylan. Rhamnose, galactose, arabinose, xylose, galacturonic acid, and glucuronic acid were identified as the sugar units [1]. Cyanogenic glycosides in flaxseeds were identified as linamarin, linustatin, and neolinustatin. These cyanogenic glycosides can release toxic hydrogen cyanide in the presence of water by auto- or enzymatic hydrolyses [2]. A phenylpropanoid glucoside from the defatted meal of flaxseed was identified as linusitamarin [3].

The major lignan in flaxseed has been identified as secoisolariciresinol diglycoside [4]. This is transformed into enterodiol and enterolactone *in vivo* via biotransformation by bacterial enzymes in the mammalian colon [5, 6]. Enterodiol and enterolactone could be produced by fermenting flaxseed or secoisolariciresinol diglycoside with a human fecal inoculum [6]. In rats fed a high-fat diet with flaxseed at concentrations up to 5%, or with secoisolariciresinol diglycoside at concentrations up to 22 μ*M*, a linear increase in enterodiol and enterolactone plasma levels was followed by a plateau. The urinary enterodiol and enterolactone levels of secoisolariciresinol diglycoside-fed rats were only 20% of the urinary levels of flaxseed-fed rats. This indicates the presence of other precursors that might cause an incomplete conversion of secoisolariciresinol diglycoside to enterodiol and enterolactone [6].

The fecal excretion of secoisolariciresinol in premenopausal women consuming flaxseed powder was increased significantly. Matairesinol excretion in the feces of premenopausal women consuming flaxseed powder was also markedly increased, indicating the presence of matairesinol in flaxseeds [7]. Matairesinol is a lignan related to enterolactone, and is the major lignan component present in Fructus Forsythiae, the dried fruit of *Forthysia suspensa* (Thunb.) Vahl (Oleaceae). An increased urinary excretion of enterodiol and enterolactone, formed from secoisolariciresinol diglycoside, was observed in premenopausal women consuming flaxseeds [8]. Elevated enterodiol and enterolactone levels were also observed in plasma after the consumption of flaxseed [9].

Pharmacology and Toxicology

Flaxseed is often used as a dietary supplementation. A modest flaxseed intake was found to effectively reduce hypercholesterolemic atherosclerosis, without markedly lowering serum cholesterol levels; this effect was presumably due to a suppression of the enhanced production of oxygen free radicals by polymorphonuclear leukocytes. Dietary flaxseed supplementation also prevented hypercholesterolemia-related heart attack and strokes [10]. Flaxseed with a very low linolenic acid content was also found to significantly exhibit antiatherogenic activity in rabbits [11].

The administration of a diet containing linolenic acid as a precursor of eicosapentaenoic acid to healthy volunteers caused significant increases in linolenic acid concentrations in serum and in peripheral blood mononuclear cell lipids. It also caused a significant increase in the eicosapentaenoic and docosapentaenoic acid contents of peripheral blood mononuclear cell lipids [12]. The supplementation of linolenic acid to hamsters was reported to significantly decrease hepatic cholesterol, but no effect was observed on heart and kidney cholesterol levels [13]. Increased concentrations of cellular eicosapentaenoic acid are thought to be beneficial in coronary heart disease, hypertension, and inflammatory disorders [14].

In healthy female volunteers who consumed 50 g of ground, raw flaxseed each day for four weeks, elevated contents of linolenic acid and long-chain n-3 fatty acids, in both plasma and erythrocyte lipids, as well as an increased urinary excretion of thiocyanate, were observed. Flaxseed also lowered the serum levels of total cholesterol and low-density lipoprotein (LDL) [15]. In lupus nephritis patients, the administration of flaxseed resulted in a significant reduction in total and LDL cholesterol, of blood viscosity, of inhibition of platelet aggregation, of a significant decline in serum creatinine and in a concomitant increase in creatinine clearance [16].

Experimental studies revealed that the dietary application of flaxseed exerted potent antiatherogenic effects in rabbits. Such dietary application was also studied in LDL receptor-deficient mouse as an animal model that more closely represented the human atherosclerotic condition. Supplementation of the cholesterol-enriched diet with 10% ground flaxseed to LDL receptor-deficient mice lowered plasma cholesterol and saturated fatty acids, and inhibited plaque formation in the aorta and aortic sinus compared to control mice. The increased expression of proliferating cell nuclear antigen (PCNA) and the inflammatory markers interleukin (IL)-6, mac-3, and vascular cell adhesion molecule-1 (VCAM-1) in the aortic tissue of mice fed cholesterol or coconut oil was significantly reduced or normalized by flaxseed. These results showed that dietary flaxseed might counteract atherosclerosis in LDL receptor-deficient mice through a reduction of circulating cholesterol levels, and via antiproliferative and anti-inflammatory actions [17]. Dietary flaxseed and flax oil attenuated the decline in renal function and reduced glomerular injury, with favorable effects on blood pressure, plasma lipids, and urinary prostaglandins in rats with 5/6 renal ablation [18].

The consumption of linolenic acid-rich oils might offer protective effects against cardiovascular disease via their ability to inhibit platelet aggregation [19]. The dietary consumption of flax oil caused a marked rise in arterial compliance. A rapid functional improvement in the systemic arterial circulation was observed, despite an increased oxidizability of LDL [20]. Dietary flaxseed supplementation to rats, produced significant changes in the fatty acid composition of the serum and peritoneal exudate cells, and also inhibited oxygen generation by peritoneal exudate cells [21]. Linolenic acid was further reported to be an effective modulator of thromboxane and prostacyclin biosynthesis [22]. In healthy volunteers, the intake of flax oil in domestic food preparations for four weeks also inhibited the synthesis of tumor necrosis factor-α (TNF-α) and IL-1β [23].

The lignan complex from flaxseed was given at a daily oral dose of $40\,mg\,kg^{-1}$ to hypercholesterolemic rabbits, as induced by high-cholesterol diet for two months. The development of atherosclerosis, associated with a decrease in serum total cholesterol, LDL, the total cholesterol:HDL ratio, serum malondialdehyde, and aortic malondialdehyde were each found to be reduced. Serum HDL levels were elevated in both hypercholesterolemic and normocholesterolemic rabbits; however, the lignan complex did not affect levels of total cholesterol, LDL, and serum malondialdehyde in normocholesterolemic rabbits. These results suggested that the lignan complex from flaxseed reduced the extent of hypercholesterolemic atherosclerosis in rabbits, associated with marked decreases in oxidative stress, serum total cholesterol, LDL, and with an elevation of serum HDL [24]. The lignan complex at the given dose did not show any adverse effect on the erythrocyte count, hematocrit, hemoglobin, white blood cell counts, granulocytes, lymphocytes, monocytes and platelets in both normocholesterolemic and hypercholesterolemic rabbits [25]. The oral treatment of rats with secoisolariciresinol diglucoside at a daily dose of $20\,mg\,kg^{-1}$ for two weeks increased the capillary density and myocardial function. Isolated preparations from treated rats showed an increased level of aortic flow and functional recovery after 2 h of reperfusion following a 30 min period of ischemia. Treatment with secoisolariciresinol diglucoside also reduced the infarct size and decreased cardiomyocyte apoptosis. In human coronary arteriolar endothelial cells, treatment with secoisolariciresinol diglucoside at concentrations of 50 and $100\,\mu M$, caused a significant increase in tubular morphogenesis and also increased the expression of vascular endothelial growth factor (VEGF), angiopoietin-1, and phosphorylated endothelial nitric oxide synthase [26]. In contrast, a lignan complex from flaxseed rich in secoisolariciresinol diglucoside given to healthy postmenopausal women at a daily dose of 500 mg for six weeks had no effect on plasma concentrations of total cholesterol, LDL, HDL, and triglyceride. However, enterolactone concentrations in the serum and in urine were significantly higher following the lignan complex intervention period [27].

Flaxseed, enterolactone, and enterodiol have each been suggested to have a protective role against both breast and colon cancer [28, 29]. They were also reported to exhibit growth-inhibitory activity against human breast cancer cells *in vitro* [30]. An extract of flaxseed was tested to determine its effect on the proliferation and hormone production of Jeg3 cells, an estrogen receptor (ER)-positive trophoblast tumor cell line. The results showed a concentration-dependent inhibition of the proliferation of Jeg3 cells by flaxseed extract. Both, matairesinol and biochanin A appeared to be active principles in the extract [31]. Weak estrogenic or anti-estrogenic [32] and anti-aromatase activities [33] were ascribed to flaxseed and to the lignans. The supplementation of a high-fat diet with flaxseed or secoisolariciresinol diglycoside to rats was reported to produce a dose-related cessation or lengthening of the estrous cycle [34]. Enterolactone and enterodiol were further reported to exert growth-inhibitory activities against human colon tumor cell lines, including LS174T, Caco-2, HCT-15 and T-84, via mechanisms other than anti-estrogenic activity, because these tumor cell lines were not estrogen-sensitive and the growth was not affected by the presence of 17β-estradiol [35]. Enterolactone was also reported to inhibit the growth of prostate cancer cells. A study using LNCaP human prostate carcinoma cells showed that enterolactone selectively suppressed the growth of LNCaP cells by triggering apoptosis. Mechanistic studies revealed that such induced apoptosis was characterized by a concentration-dependent loss of mitochondrial membrane potential, a release of cytochrome c, and the cleavage of procaspase-3 and poly(ADP-ribose)polymerase [36]. Flaxseed also inhibited metastasis and decreased extracellular VEGF in human breast cancer xenografts in nude mice [37].

Flaxseed is one of the richest sources of mammalian lignan precursors, such as secoisolariciresinol diglycoside. The mammalian lignans – especially enterolactone and enterodiol – have attracted much attention due to their tumor-preventive activities. Flaxseed was reported to significantly reduce chromosomal aberrations in mammary epithelial cells in rats initiated with 7,12-dimethylbenz[*a*]anthracene (DMBA) and mammary tumor size at the promotion stage [38], as well as at the later stages of carcinogenesis [39]. Secoisolariciresinol diglycoside was found to show effects similar to flaxseed, which suggested that the protective effect of flaxseed against mammary carcinoma might be caused by the lignans [29].

As enterolactone and enterodiol are known to be formed in the colon, they would be expected to exert their primary effects in that organ. Flaxseed meal and defatted flaxseed flour given to rats for four weeks significantly reduced the total number of aberrant crypts and aberrant crypt foci in the colon induced by azoxymethane, which suggested that flaxseed feeding might reduce the risk for colon carcinogenesis [28]. There were no significant differences between the flaxseed meal and defatted flaxseed flour. This indicated that flaxseed may have a protective effect against colon cancer that was due, in part, to secoisolariciresinol diglycoside, and that such protective effect was associated with an increased β-glucuronidase activity [40]. The results of epidemiologic studies further supported the hypothesis that mammalian lignans have cancer-protective effects [41]. Dietary flaxseed supplementation was also reported to cause a significant suppression of the lung metastases of B16BL6 murine melanoma cells in mice [42].

Secoisolariciresinol was also reported to show hepatoprotective activity. The intraperitoneal (i.p.) pretreatment of mice with secoisolariciresinol at doses of 10 and 50 mg kg^{-1} at 12 h and 1 h before intoxication with D-galactosamine and lipopolysaccharide (LPS), led to a significant reduction in DNA fragmentation and also prevented chromatin condensation, apoptotic body formation, and hepatitis. Pretreatment with secoisolariciresinol significantly inhibited the elevation of serum levels of TNF-α and interferon-γ (IFN-γ) as important mediators of hepatocyte apoptosis. Secoisolariciresinol also exhibited a significant concentration-dependent protective effect on cell death in primary cultured mouse hepatocytes induced by D-galactosamine and TNF-α, and protected murine L929 fibrosarcoma cells from TNF-α-mediated cell death. These results indicated that secoisolariciresinol prevented D-galactosamine and LPS-induced hepatic injury by inhibiting hepatocyte apoptosis through the blocking of TNF-α and IFN-γ production by activated macrophages and a direct inhibition of TNF-α induced apoptosis [43].

Enterolactone was found to undergo extensive glucuronidation with rhesus monkey liver microsomes to form O-glucuronides at both phenolic hydroxy groups. In addition to glucuronidation, enterolactone was also a good substrate for oxidative metabolism; the major products were identified as monohydroxylated products of enterolactone. The incubation of enterolactone with liver microsomes containing UDPGA, NADPH, and *N*-acetylcysteine resulted in the formation of glucuronides. The incubation of enterolactone with human and rhesus monkey hepatocytes resulted in glucuronides as the major metabolites, and sulfates and monohydroxylated products as minor metabolites [44].

Following the oral administration of secoisolariciresinol diglucoside to healthy volunteers at a dose of 1.3 μM kg^{-1}, enterolignans appeared in the plasma at 8–10 h after ingestion. Enterodiol reached a maximum plasma concentration at about 15 h, whereas maximum enterolactone levels were reached at about 20 h. The mean elimination half-life of enterodiol (4.4 h) was shorter than that of enterolactone (12.6 h), and the mean area under the curve of enterolactone was twice that of enterodiol. Within 3 days of dosing, up to 40% of the secoisolariciresinol diglucoside was excreted as enterolignans via the urine, with the majority (58%) as enterolactone [45].

References

1 Cunnane, S.C., Ganguli, S., Menard, C., Liede, A.C., Hamadeh, M.J., Chen, Z.Y., Wolever, T. M., and Jenkins, D.J. (1993) High α-linolenic acid flaxseed (*Linum usitatissimum*): some nutritional properties in humans. *Br. J. Nutr.*, **69**, 443–453.

2 Chadha, R.K., Lawrence, J.F., and Ratnayake, W.M. (1995) Ion chromatographic determination of cyanide released from flaxseed under autohydrolysis conditions. *Food Addit. Contam.*, **12**, 527–533.

3 Luyengi, L., Pezzuto, J.M., Waller, D.P., Beecher, C.W., Fong, H.H., Che, C.T. and Bowen, P.E. (1993) Linusitamarin, a new phenylpropanoid glucoside from *Linum usitatissimum*. *J. Nat. Prod.*, **56**, 2012–2015.

4 Obermeyer, W.R., Musser, S.M., Betz, J.M., Casey, R.E., Pohland, A.E., and Page, S.W. (1995) Chemical studies of phytoestrogens and related compounds in dietary supplements: flax and chaparral. *Proc. Soc. Exp. Biol. Med.*, **208**, 6–12.

5 Xie, L.H., Ahn, E.M., Akao, T., Abdel-Hafez, A. A., Nakamura, N., and Hattori, M. (2003) Transformation of arctiin to estrogenic and antiestrogenic substances by human intestinal bacteria. *Chem. Pharm. Bull. (Tokyo)*, **51**, 378–384.

6 Rickard, S.E., Orcheson, L.J., Seidl, M.M., Luyengi, L., Fong, H.H., and Thompson, L.U. (1996) Dose-dependent production of mammalian lignans in rats and *in vitro* from the purified precursor secoisolariciresinol diglycoside in flaxseed. *J. Nutr.*, **126**, 2012–2019.

7 Kurzer, M.S., Lampe, J.W., Martini, M.C., and Adlercreutz, H. (1995) Fecal lignan and isoflavonoid excretion in premenopausal women consuming flaxseed powder. *Cancer Epidemiol. Biomarkers Prev.*, **4**, 353–358.

8 Lampe, J.W., Martini, M.C., Kurzer, M.S., Adlercreutz, H., and Slavin, J.L. (1994) Urinary lignan and isoflavonoid excretion in premenopausal women consuming flaxseed powder. *Am. J. Clin. Nutr.*, **60**, 122–128.

9 Atkinson, D.A., Hill, H.H., and Shultz, T.D. (1993) Quantification of mammalian lignans in biological fluids using gas chromatography with ion mobility detection. *J. Chromatogr.*, **617**, 173–179.

10 Prasad, K. (1997) Dietary flax seed in prevention of hypercholesterolemic atherosclerosis. *Atherosclerosis*, **132**, 69–76.

11 Prasad, K., Mantha, S.V., Muir, A.D., and Westcott, N.D. (1998) Reduction of hypercholesterolemic atherosclerosis by CDC-flaxseed with very low α-linolenic acid. *Atherosclerosis*, **136**, 367–375.

12 Kelley, D.S., Nelson, G.J., Love, J.E., Branch, L. B., Taylor, P.C., Schmidt, P.C., Mackey, B.E., and Iacono, J.M. (1993) Dietary α-linolenic acid alters tissue fatty acid composition, but not blood lipids, lipoproteins or coagulation status in humans. *Lipids*, **28**, 533–537.

13 Yang, L., Leung, K.Y., Cao, Y., Huang, Y., Ratnayake, W.M., and Chen, Z.Y. (2005) α-Linolenic acid but not conjugated linolenic acid is hypocholesterolaemic in hamsters. *Br. J. Nutr.*, **93**, 433–438.

14 Mantzioris, E., James, M.J., Gibson, R.A., and Cleland, L.G. (1995) Differences exist in the relationships between dietary linoleic and α-linolenic acids and their respective long-chain metabolites. *Am. J. Clin. Nutr.*, **61**, 320–324.

15 Cunnane, S.C., Hamadeh, M.J., Liede, A.C., Thompson, L.U., Wolever, T.M., and Jenkins, D.J. (1995) Nutritional attributes of traditional flaxseed in healthy young adults. *Am. J. Clin. Nutr.*, **61**, 62–68.

16 Clark, W.F., Parbtani, A., Huff, M.W., Spanner, E., de Salis, H., Chin-Yee, I., Philbrick, D.J., and Holub, B.J. (1995) Flaxseed: a potential treatment for lupus nephritis. *Kidney Int.*, **48**, 475–480.

17 Dupasquier, C.M., Dibrov, E., Kneesh, A.L., Cheung, P.K., Lee, K.G., Allexander, H.K.,

Yeganeh, B.K., Moghadasian, M.H., and Pierce, G.N. (2007) Dietary flaxseed inhibits atherosclerosis in the LDL receptor-deficient mouse in part through antiproliferative and anti-inflammatory actions. *Am. J. Physiol. Heart Circ. Physiol.*, **293**, H2394–2402.

18 Ingram, A.J., Parbtani, A., Clark, W.F., Spanner, E., Huff, M.W., Philbrick, D.J., and Holub, B.J. (1995) Effects of flaxseed and flax oil diets in a rat-5/6 renal ablation model. *Am. J. Kidney Dis.*, **25**, 320–329.

19 Allman, M.A., Pena, M.M., and Pang, D. (1995) Supplementation with flaxseed oil versus sunflowerseed oil in healthy young men consuming a low fat diet: effects on platelet composition and function. *Eur. J. Clin. Nutr.*, **49**, 169–178.

20 Nestel, P.J., Pomeroy, S.E., Sasahara, T., Yamashita, T., Liang, Y.L., Dart, A.M., Jennings, G.L., Abbey, M., and Cameron, J.D. (1997) Arterial compliance in obese subjects is improved with dietary plant n-3 fatty acid from flaxseed oil despite increased LDL oxidizability. *Arterioscler. Thromb. Vasc. Biol.*, **17**, 1163–1170.

21 Babu, U.S., Bunning, V.K., Wiesenfeld, P., Raybourne, R.B., and O'Donnell, M. (1997) Effect of dietary flaxseed on fatty acid composition, superoxide, nitric oxide generation and antilisterial activity of peritoneal macrophages from female Sprague-Dawley rats. *Life Sci.*, **60**, 545–554.

22 Ferretti, A. and Flanagan, V.P. (1996) Antithromboxane activity of dietary α-linolenic acid: a pilot study. *Prostaglandins Leukot. Essent. Fatty Acids*, **54**, 451–455.

23 Caughey, G.E., Mantzioris, E., Gibson, R.A., Cleland, L.G., and James, M.J. (1996) The effect on human tumor necrosis factor alpha and interleukin 1β production of diets enriched in n-3 fatty acids from vegetable oil or fish oil. *Am. J. Clin. Nutr.*, **63**, 116–122.

24 Prasad, K. (2005) Hypocholesterolemic and antiatherosclerotic effect of flax lignan complex isolated from flaxseed. *Atherosclerosis*, **179**, 269–275.

25 Prasad, K. (2005) Effect of chronic administration of lignan complex isolated from flaxseed on the hemopoietic system. *Mol. Cell. Biochem.*, **270**, 139–145.

26 Penumathsa, S.V., Koneru, S., Thirunavukkarasu, M., Zhan, L., Prasad, K., and Maulik, N. (2007) Secoisolariciresinol diglucoside: relevance to angiogenesis and cardioprotection against ischemia-reperfusion injury. *J. Pharmacol. Exp. Ther.*, **320**, 951–959.

27 Hallund, J., Ravn-Haren, G., Buegel, S., Tholstrup, T., and Tetens, I. (2006) A lignan complex isolated from flaxseed does not affect plasma lipid concentrations or antioxidant capacity in healthy postmenopausal women. *J. Nutr.*, **136**, 112–116.

28 Serraino, M. and Thompson, L.U. (1992) Flaxseed supplementation and early markers of colon carcinogenesis. *Cancer Lett.*, **63**, 159–165.

29 Serraino, M. and Thompson, L.U. (1992) The effect of flaxseed supplementation on the initiation and promotional stages of mammary tumorigenesis. *Nutr. Cancer*, **17**, 153–159.

30 *Hirano, T., Fukuoka, K., Naito, T., Hosaka, K., Mitsuhashi, H., and Matsumoto, Y. (1990) Antiproliferative activity of mammalian lignan derivatives against human breast cancer cell line, ZR-75-1. *Cancer Invest.*, **8**, 595–602.

31 Waldschlaeger, J., Bergemann, C., Ruth, W., Effmert, U., Jeschke, U., Richter, D.U., Kragl, U., Piechulla, B., and Briese, V. (2005) Flaxseed extracts with phytoestrogenic effects on a hormone receptor-positive tumor cell line. *Anticancer Res.*, **25** (A), 1817–1822.

32 Kurzer, M.S. and Xu, X. (1997) Dietary phytoestrogens. *Annu. Rev. Nutr.*, **17**, 353–381.

33 Wang, C., Makela, T., Hase, T., Adlercreutz, H., and Kurzer, M.S. (1994) Lignans and flavonoids inhibit aromatase enzyme in human preadipocytes. *J. Steroid Biochem. Mol. Biol.*, **50**, 205–212.

34 Orcheson, L.J., Rickard, S.E., Seidl, M.M., and Thompson, L.U. (1998) Flaxseed and its mammalian lignan precursor cause a lengthening or cessation of estrous cycling in rats. *Cancer Lett.*, **125**, 69–76.

35 Sung, M.K., Lautens, M., and Thompson, L.U. (1998) Mammalian lignans inhibit the growth of estrogen-independent human colon tumor cells. *Anticancer Res.*, **18**, 1405–1408.

36 Chen, L.H., Fang, J., Li, H., Demark-Wahnefried, W., and Lin, X. (2007) Enterolactone induces apoptosis in human prostate carcinoma LNCaP cells via a mitochondrial-mediated, caspase-dependent pathway. *Mol. Cancer Ther.*, **6**, 2581–2590.

37 Dabrosin, C., Chen, J., Wang, L., and Thompson, L.U. (2002) Flaxseed inhibits metastasis and decreases extracellular vascular endothelial growth factor in human breast cancer xenografts. *Cancer Lett.*, **185**, 31–37.

38 Serraino, M. and Thompson, L.U. (1991) The effect of flaxseed supplementation on early risk markers for mammary carcinogenesis. *Cancer Lett.*, **60**, 135–142.

39 Thompson, L.U., Rickard, S.E., Orcheson, L.J., and Seidl, M.M. (1996) Flaxseed and its lignan and oil components reduce mammary tumor growth at a late stage of carcinogenesis. *Carcinogenesis*, **17**, 1373–1376.

40 Jenab, M. and Thompson, L.U. (1996) The influence of flaxseed and lignans on colon carcinogenesis and β-glucuronidase activity. *Carcinogenesis*, **17**, 1343–1348.

41 Nesbitt, P.D. and Thompson, L.U. (1997) Lignans in home made and commercial products containing flaxseed. *Nutr. Cancer*, **29**, 222–227.

42 Yan, L., Yee, J.A., Li, D., McGuire, M.H., and Thompson, L.U. (1998) Dietary flaxseed supplementation and experimental metastasis of melanoma cells in mice. *Cancer Lett.*, **124**, 181–186.

43 Banskota, A.H., Nguyen, N.T., Tezuka, Y., Le Tran, Q., Nobukawa, T., Kurashige, Y., Sasahara, M., and Kadota, S. (2004) Secoisolariciresinol and isotaxiresinol inhibit tumor necrosis factor-α-dependent hepatic apoptosis in mice. *Life Sci.*, **74**, 2781–2792.

44 Dean, B., Chang, S., Doss, G.A., King, C., and Thomas, P.E. (2004) Glucuronidation, oxidative metabolism, and bioactivation of enterolactone in rhesus monkeys. *Arch. Biochem. Biophys.*, **429**, 244–251.

45 Kuijsten, A., Arts, I.C., Vree, T.B., and Hollman, P.C. (2005) Pharmacokinetics of enterolignans in healthy men and women consuming a single dose of secoisolari-ciresinol diglucoside. *J. Nutr.*, **135**, 795–801.

Liriope muscari and *Liriope spicata* var. *prolifera*

Radix Liriopes (Shanmaidong) is the dried tuber of *Liriope muscari* (Decne) Baily or *Liriope spicata* (Thunb.) Lour. var. *prolifera* Y.T. Ma (Liliaceae). It is used as an expectorant, sedative, and laxative.

Liriope muscari (Decne.) Baily.

Chemistry

Radix Liriopes was often confused with Radix Ophiopogonis (Maidong in Chinese), the dry tuber of *Ophiopogon japonicus* (Thunb.) Ker-Gawl. (Liliaceae). Commercial drug resources of "Maidong" from different provinces were identified involving 23 species and three varieties from *Ophiopogon* and *Liriope*. The most widespread and mostly used have been found to be *O. japonicus* and *L. muscari*. Species used locally in several provinces including *Ophiopogon bodinieri*, *O. bodinieri* var. *pygmaeus*, *O. intermedius*, *Ophiopogon mairei*, *Ophiopogon szechuanensis*, *Ophiopogon stenophyllum*, *Liriope platyphylla*, and *Liriope spicata* [1]. However, the ether fraction of the ethanol extract and the volatile oil of *O. japonicus* were found to be distinct from that of *L. muscari* and *L. spicata* var. *prolifera* [2].

The tuber of *L. muscari* and *L. spicata* var. *prolifera* was reported to contain steroid glycosides with ruscogenin as the aglycone. Ruscogenin glycosides from liriope roots were identified as lirioproliosides A, B, C, and D, ophiopogonin A, and related saponins [3–7]. Yamogenin 1-*O*-α-L-rhamnopyranosyl-(1 → 2)-[β-D-xylopyranosyl-(1 → 3)]-β-D-glucopyranoside [4] and β-sitosterol 3-*O*-β-D-glucopyranoside, stigmasterol 3-*O*-β-D-glucopyranoside were also isolated from liriope root [6].

Pharmacology and Toxicology

Ruscogenin 1-*O*-β-D-glucopyranosyl-(1 → 2)-[β-D-xylopyranosyl-(1 → 3)]-β-D-fucopyranoside (Formula p805) was reported to significantly prolong the survival time of mice under hypoxic conditions [3]. An aqueous extract of the tubers of *L. spicata* var. *prolifera* was found to exert an immunostimulatory activity in mice. In mice, the extract increased the spleen weight, enhanced the clearance rate of intravenously administered charcoal particles, and antagonized the leukopenia caused by cyclophosphamide [8]. The intravenous (i.v.) administration of an aqueous root extract of *L. spicata* at a dose level of 1.75 $g\,kg^{-1}$ to anesthetized cats was found to show hemodynamic effects by increasing the ventricular contractile force; it might also contribute to increasing the cardiac pump function [9].

Ruscogenin glycoside (Lm-3) isolated from *L. muscari* as ruscogenin 1-*O*-[β-D-glucopyranosyl(1 → 2)][[β-D-xylopyranosyl

Handbook of Chinese Medicinal Plants: Chemistry, Pharmacology, Toxicology
Weici Tang and Gerhard Eisenbrand

ISBN: 978-3-527-32226-8

(1 → 3)]-β-D-fucopyranoside, and its aglycone, ruscogenin, were reported to improve liver injury in mice when administered during the effector phase of a delayed-type hypersensitivity reaction. The pretreatment of nonparenchymal cells, but not hepatocytes, with Lm-3 or ruscogenin *in vitro* caused a concentration- and time-dependent protection against the damage. Lm-3 showed a stronger inhibition than ruscogenin, with IC_{50} values of 0.6 n*M* and 0.4 μ*M*, respectively. However, neither Lm-3 nor ruscogenin blocked the hepatotoxic potential of carbon tetrachloride (CCl_4), when preincubated with hepatocytes. In addition, Lm-3 and ruscogenin inhibited concanavalin A (Con A)-induced lymphocyte proliferation at high concentrations [10]. The saponin Lm-3 was also reported to exert an anti-inflammatory activity; here, it inhibited the adhesion of Jurkat cells activated by anti-CD3 to type I collagen, and also inhibited cell attachment to fibronectin and laminin. Lm-3 also inhibited Jurkat cell adhesion activated by phorbol 12,13-dibutyrate as a protein kinase C (PKC) activator. A similar inhibition by Lm-3 of the phorbol 12,13-dibutyrate-induced adhesion to collagen was also observed in lymphocytes freshly isolated from mice with contact dermatitis [11]. Allergic contact dermatitis caused by ruscogenin and its glycosides was reported [12].

References

1 Yu, B.Y., Xu, G.J., Jin, R.L., and Xu, L.S. (1991) Drug resources and identification of commercial drugs on "Maidong". *J. China Pharm. Univ.*, **22**, 150–153.

2 Liu, X.M., Sun, H.X., and Zeng, X.W. (1993) Application of cluster analysis in the chemical taxonomy of "Maidong". *China J. Chin. Mater. Med.*, **18**, 585–587.

3 Yu, B.Y., Yin, X., Rong, Z.Y., Yang, T.M., Zhang, C.H., and Xu, G.J. (1994) Biological activity of ruscogenin 1-*O*-β-D-glucopyranosyl-(1 → 2)-[β-D-xylopyranosyl-(1 → 3)]-β-D-fucopyranoside from tuberous roots of *Liriope muscarli*. *J. China Pharm. Univ.*, **25**, 286–288.

4 Yu, B.Y., Hirai, Y., Shoji, J., and Xu, G.J. (1990) Comparative studies on the constituents of Ophiopogonis tuber and its congeners. VI. Studies on the constituents of the subterranean part of, *Liriope spicata* var. *prolifera* and *L. muscarli* (1). *Chem. Pharm. Bull. (Tokyo)*, **38**, 1931–1935.

5 Ling, D.K., Liu, L.Z., and Zhu, Y.X. (1991) Chemical constituents of creeping liriope (*Liriope spicata*). *Chin. Trad. Herbal Drugs*, **22**, 489–490.

6 Liu, W., Wang, Z.L., and Liang, H.Q. (1989) Studies on the chemical constituents of *Liriope spicata* Lour (Thunb), var. *prolifera* Y.T. Ma. *Acta Pharm. Sin.*, **24**, 749–754.

7 Yu, B.Y., Qiu, S.X., Zaw, K., Xu, G.J., Hirai, Y., Shoji, J., Fong, H.H., and Kinghorn, A.D. (1996) Steroidal glycosides from the subterranean parts of *Liriope spicata* var. *prolifera*. *Phytochemistry*, **43**, 201–206.

8 Yu, B., Yin, X., Xu, G., and Xu, L. (1991) Quality of tuberous root of *Liriope spicata* (Thunb.) Lour. var. *prolifera* Y. T. Ma and *Ophiopogon japonicus*, (L.F.) Ker-Gawl.: comparison of immune function. *China J. Chin. Mater. Med.*, **16**, 584–585.

9 Gao, G.Y., Li, C.X., Duan, P., Han, X.Y., and Yang, Y.Q. (1989) Hemodynamic effects of a water soluble extract of *Liriope spicata* Lour on anesthetized cats. *China J. Chin. Mater. Med.*, **14**, 552–554.

10 Wu, F., Cao, J., Jiang, J., Yu, B., and Xu, Q. (2001) Ruscogenin glycoside (Lm-3) isolated from *Liriope muscari* improves liver injury by dysfunctioning liver-infiltrating lymphocytes. *J. Pharm. Pharmacol.*, **53**, 681–688.

11 Liu, J., Chen, T., Yu, B., and Xu, Q. (2002) Ruscogenin glycoside (Lm-3) isolated from *Liriope muscari* inhibits lymphocyte adhesion to extracellular matrix. *J. Pharm. Pharmacol.*, **54**, 959–965.

12 Ramirez-Hernandez, M., Garcia-Selles, J., Merida-Fernandez, C., and Martinez-Escribano, J.A. (2006) Allergic contact dermatitis to ruscogenins. *Contact Dermatitis*, **54**, 60.

Lonicera confusa, *Lonicera hypoglauca*, *Lonicera japonica* and *Lonicera macranthoides*

Flos Lonicerae Japonicae (Jinyinhua) is the dried flower buds of *Lonicera japonica* Thunb. (Caprifoliaceae). It is used as an antibacterial and antiphlogistic agent for the treatment of abscess, laryngitis, pharyngitis, infection of upper respiratory tract, dysentery, common cold, and fever.

Lonicera japonica Thunb.

Flos Lonicerae (Shanyinhua) is the dried flower buds of *Lonicera hypoglauca* Miq., *Lonicera confusa* DC., or *Lonicera macranthoides* Hand.-Mazz. It is used as an antibacterial and antiphlogistic agent in the treatment of abscess, laryngeal catarrh, infection of the upper respiratory tract, dysentery, common cold, and fever.

Caulis Lonicerae Japonicae (Rendongteng) is the dried cane of *L. japonica*. It is used as an antiphlogistic and bacteriostatic in the treatment of fever, dysentery, abscess, acute arthritis, and rheumatic swelling.

Chemistry

The flower and stem of *L. japonica* are known to be rich in chlorogenic acid, with a content of about 6% in the flowers. The contents of chlorogenic acid in the canes and leaves were 1–1.5% and 0.3–3%, respectively [1]. Isochlorogenic acid was also isolated from the flowers and canes of *L. japonica* [2]. The content of isochlorogenic acid in the leaf was given as 1.0–5% [1]. Chlorogenic acid and isochlorogenic acid are both caffeic acid esters of quinic acid; chlorogenic acid represents the 3-caffeic acid ester of quinic acid, while isochlorogenic acid is a mixture composed of three dicaffeic acid esters of quinic acid. The leaves of some varieties of *L. japonica* were found to contain more chlorogenic acid and isochlorogenic acid than the corresponding

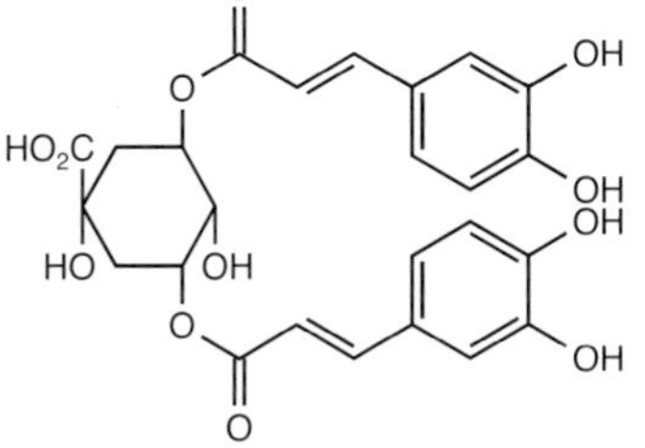

Isochlorogenic acid a

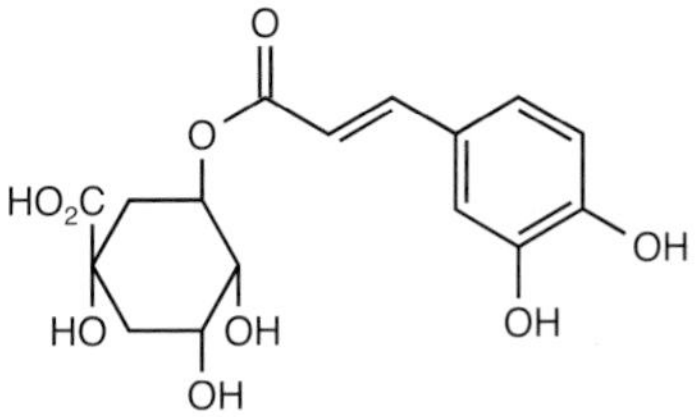

Chlorogenic acid

Handbook of Chinese Medicinal Plants: Chemistry, Pharmacology, Toxicology
Weici Tang and Gerhard Eisenbrand

ISBN: 978-3-527-32226-8

flowers. The acid levels in the leaves peaked in August and September [3].

The Chinese Pharmacopoeia requires a qualitative determination of chlorogenic acid in Flos Lonicerae Japonicae and in Flos Lonicerae by thin-layer chromatographic comparison with reference substance; a quantitative determination of chlorogenic acid in Flos Lonicerae, in Flos Lonicerae Japonicae and in Caulis Lonicerae Japonicae by HPLC; as well as a quantitative determination of luteolin 7-*O*-glucoside in Flos Lonicerae Japonicae by HPLC. The content of chlorogenic acid in Flos Lonicerae Japonicae and in Flos Japonicae should be not less than 1.5%, and in Caulis Lonicerae Japonicae not less than 0.1%. The content of luteolin 7-*O*-glucoside in Flos Lonicerae Japonicae should be not less than 0.1%.

The flavones and flavone glycosides of the flowers or aerial parts of *L. japonica* were identified as lonicerin, loniceraflavone, corymbosin, hydnocarpin, quercetin, astragalin, isoquercitrin, rhoifolin, diosmetin, luteolin, chrysoeriol, chrysoeriol 7-*O*-β-D-glucopyranoside, isorhamnetin 3-*O*-β-D-glucopyranoside, kaempferol 3-*O*-β-D-glucopyranoside, quercetin 3-*O*-β-D-glucopyranoside, and hyperoside [4, 5], and the biflavones as ochnaflavone and ochnaflavone $4'$-*O*-methylether [6]. Protocatechuic acid and methyl caffeate were also isolated from the flowers of *L. japonica*.

Iridoid glucosides from *L. japonica* were identified as loganin, secoxyloganin, secologanin dimethylacetal, vogeloside, and epivogeloside [7].

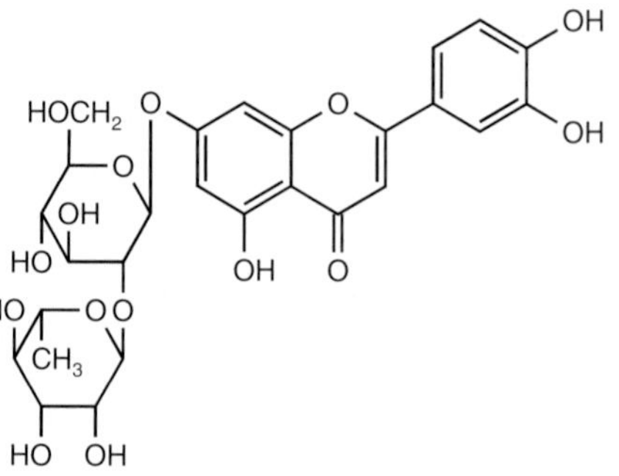

Lonicerin

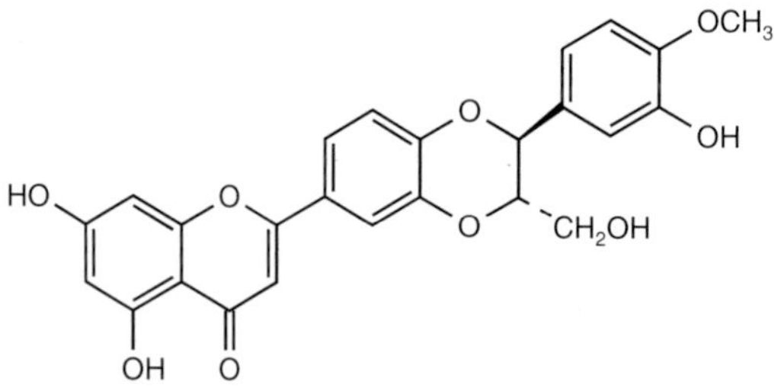

Hydnocarpin

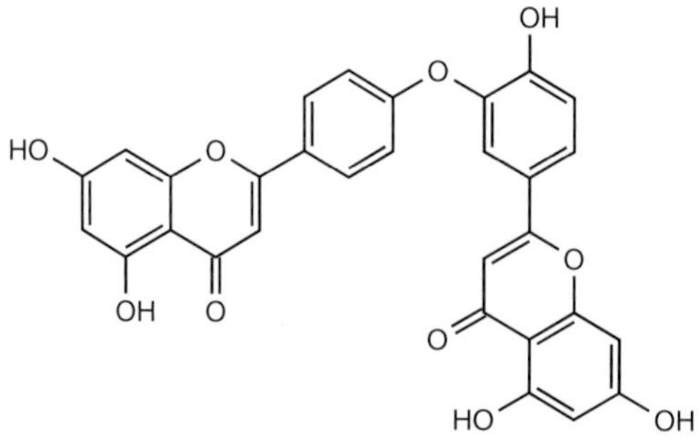

Ochnaflavone

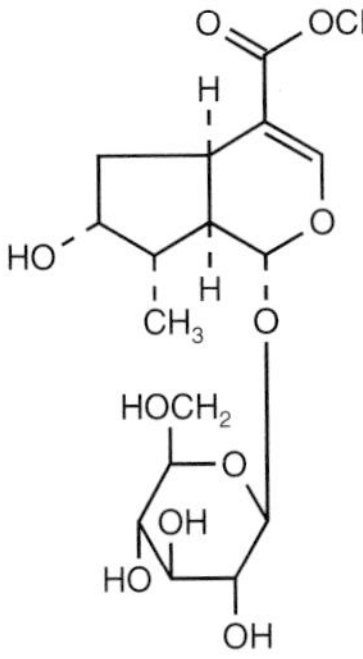
Loganin

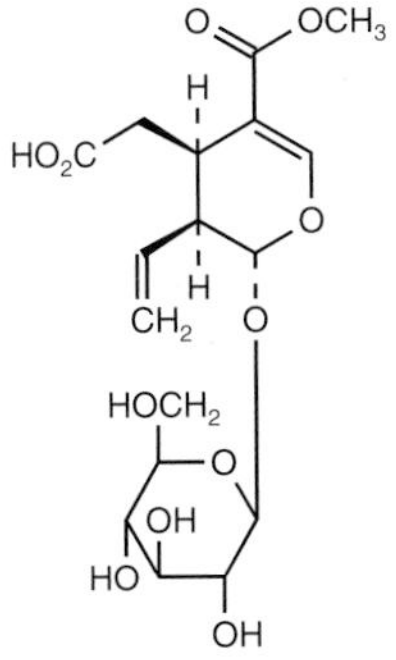
Secoxyloganin

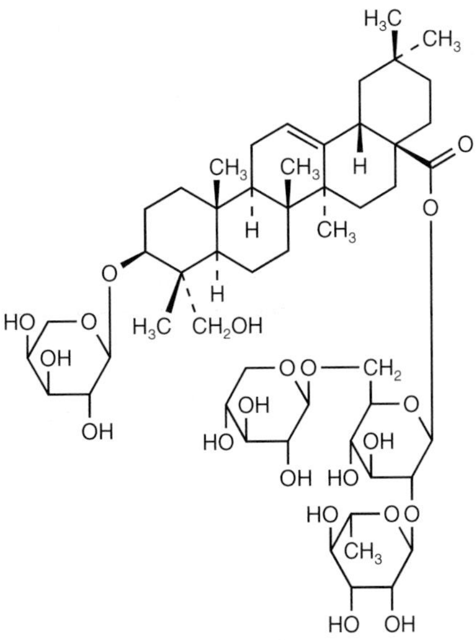
Loniceroside A

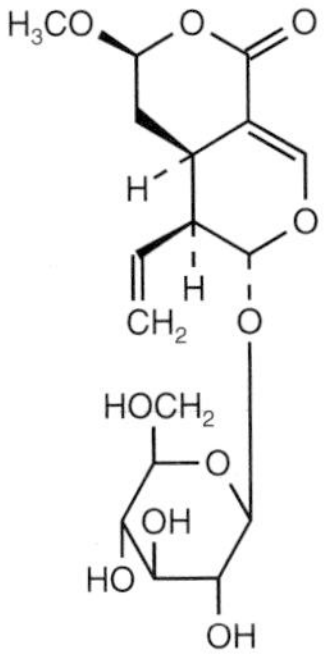
Vogeloside

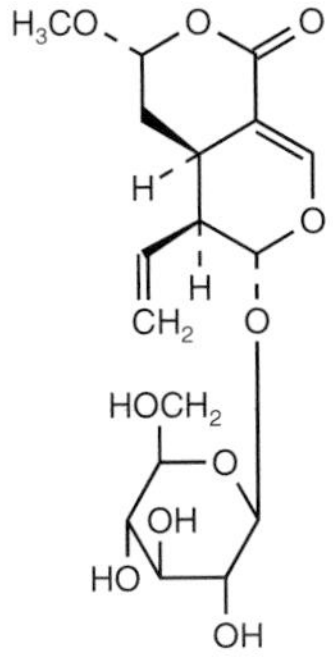
Epivogeloside

A number of oleanolic acid and hederagenin saponins, including lonicerosides A and B, were isolated from the aerial part of *L. japonica* [8]. Oleanolic acid 28-*O*-α-L-rhamnopyranosyl-(1 → 2)-[β-D-xylopyranosyl-(1 → 6)]-β-D-glucopyranosyl ester and hederagenin 3-*O*-α-L-arabinopyranoside were isolated from Flos Lonicerae [5].

The major components of the essential oil from the flowers of *L. japonica* were found to be linalool, geraniol, aromadendrene, eugenol, α-farnesene and germacrene D [9]. Minor volatile components were identified as pinene, α-terpineol, β-phenylethyl alcohol, carvacrol, and dihydrocarveol [10]. Linalool, jasmone, jasmin lactone, methyl jasmonate, and methyl epi-jasmonate were the important components that characterize the volatiles of the flower [11].

Pharmacology and Toxicology

The toxicity of *L. japonica* to rodents was found to be very low. A single oral dose of $5\,g\,kg^{-1}$ of the ethanol extract of the leaf of *L. japonica* at given to rats did not cause mortality or significant changes in the general behavior, nor did it affect the gross appearance of the internal organs. In a subacute toxicity study, the oral treatment of rats with an ethanol extract at a daily dose of $1\,g\,kg^{-1}$ for 14 days did not cause any changes in body or organ weights; neither were any adverse effects seen in the hematological analysis and pathological examination [12].

Chlorogenic acid was absorbed from the rat ileum plus jejunum during 45 min to a level of 8%. A minor fraction (1.2%) was recovered in the gut effluent as caffeic acid, which showed the presence of trace esterase activity in the gut mucosa. No chlorogenic acid was detected in either the plasma or bile [13]. The bioavailability of chlorogenic acid was largely dependent on its metabolism by the gut microflora. In rats fed a diet supplemented with chlorogenic acid at a daily dose of 250 μ*M* for eight days, only 0.8% of the administered chlorogenic acid was excreted in the urine. The total urinary excretion of caffeic acid from the hydrolysis of chlorogenic acid, and the methylated metabolites, ferulic and isoferulic acids, accounted for less than 0.5% of the ingested dose. In contrast, the metabolites of microbial origin (which included *m*-coumaric acid and derivatives of phenylpropionic acid, benzoic acid and hippuric acid) represented the major compounds in both the urine and plasma. Hippuric acid largely originated from transformation of the quinic acid moiety, all other metabolites from the caffeic acid moiety. Together, these microbial metabolites accounted for approximately 57% of the chlorogenic acid intake [14].

After oral administration of chlorogenic acid at single doses of 200, 400, or 600 mg kg^{-1}, the absorption half-lives were 10, 19, and 28 min, respectively, the distribution half-lives 12, 31, and 39 min, respectively, and the elimination half-lives 231, 337, and 421 min, respectively. The area under the concentration–time curve (AUC) was not proportional to the administered dose. Within the range of doses examined, the absorption pharmacokinetics of chlorogenic acid in rats was based on nonlinear kinetics [15]. Chlorogenic acid appeared to be poorly absorbed from the digestive tract after oral administration to rats [16]. Following oral administration of an extract from Flos Lonicerae to rabbits, the plasma concentration of chlorogenic acid showed two peaks; the first (0.84 $\mu g\ ml^{-1}$) at 35 min, and the second (0.37 $\mu g\ ml^{-1}$) at 273 min [17].

The flower and aerial parts of *L. japonica* were known to be widely used as anti-inflammatory herbal medicine. A study demonstrated that an aqueous extract of the flower exhibited different activities on cyclooxygenase (COX), when extracted with boiled or nonboiled water. The boiled water extract directly inhibited both COX-1 and COX-2 activities, while the nonboiled water extract stimulated COX-1. These results indicated that different active principles, including flavones and flavone glycosides, saponins and chlorogenic acid and isochlorogenic acids, may be responsible for the pharmacological properties of the flowers [18]. An aqueous extract from the *L. japonica* flower inhibited the increase in nuclear factor NF-κBp65 and degradation of inhibitory IκBα in the liver of rats challenged with lipopolysaccharide (LPS). An immunohistochemical analysis of rat hepatocytes showed that LPS-induced inflammatory responses, which involved the degradation of IκBα and induction of NF-κBp65, tumor necrosis factor-α (TNF-α) and inducible nitric oxide synthase (iNOS), were partially inhibited by pretreatment with the extract. This indicated that the flower extract exerted its anti-inflammatory effect through the regulation of NF-κB activation [19].

An anti-inflammatory effect of the extract of *L. japonica* was also observed in mice against proteinase-activated receptor 2-mediated paw edema, as induced by the injection of trypsin. The extract also significantly inhibited myeloperoxidase activity and TNF-α expression in the paw tissue [20], as well as TNF-α expression in mast cells induced by trypsin. It also inhibited the phosphorylation of extracellular signal-regulated kinase (ERK) [21].

Luteolin, as one of the major flavones from *L. japonica*, and at a concentration of 50 μ*M*, inhibited the proliferation of human lung carcinoma CH27 cells, induced cell cycle arrest at S-phase and apoptosis, and also induced significant DNA condensation with apoptotic body formation. The expression of apoptotic markers, such as caspase-3 and apoptosis-inducing factor protein, was changed [22], concomitant with an activation of antioxidant enzymes such as superoxide dismutase and catalase, but not through the production of reactive oxygen species (ROS) or disruption of the mitochondrial membrane potential [23].

The biflavone ochnaflavone from *L. japonica* was shown to inhibit angiotensin II-induced hypertrophy and serum-induced smooth muscle cell proliferation. In human aortic smooth muscle cells, ochnaflavone potently inhibited DNA synthesis in the presence of TNF-α. Ochnaflavone also inhibited TNF-α-induced matrix metalloproteinase-9 (MMP-9) secretion in human aortic smooth muscle cells, transcriptionally regulated at the NF-κB and activation protein-1 (AP-1) sites in the MMP-9 promoter [24]. Ochnaflavone also inhibited the COX-2-dependent phases of prostaglandin D_2 (PGD_2) generation in mouse bone marrow-derived mast cells, with an IC_{50} value of 0.6 μM. The decrease in the quantity of PGD_2 product was accompanied by a decrease in the COX-2 protein level. In addition, ochnaflavone consistently inhibited the production of leukotriene C_4 (LTC_4) in a concentration-dependent manner, indicating COX-2 and 5-lipooxygenase inhibitory activities [25]. In RAW264.7 murine macrophage cells, ochnaflavone inhibited the production of nitric oxide (NO) in a concentration-dependent manner. Expression of iNOS stimulated by LPS, mediated by ERK1/2 via transcription factor NF-κB regulation was also blocked [26].

Chlorogenic acid is widely recognized as an antioxidant. It was reported to inhibit the initiation of lipid chain peroxidation by free radicals [27], and to effectively inhibit the iron-induced lipid peroxidation of microsomes from bovine liver [28] and rat liver [29] by forming a chelate with iron. Chlorogenic acid also suppressed the hydroxyl radical formation by 3-hydroxyanthranilic acid, ferric chloride, and hydrogen peroxide. This inhibitory effect on hydroxyl radical formation was postulated as due to chelation with iron ions. Chlorogenic acid inhibited the absorption of iron from the intestine of rats *in vitro* [30]. Given to spontaneously hypertensive rats in the diet at a concentration of 0.5% for eight weeks (ca. 300 mg kg^{-1} per day) it inhibited the development of hypertension. It was suggested that dietary chlorogenic acid reduced oxidative stress and improved NO bioavailability by inhibiting production of ROS in the vasculature, as well as endothelial dysfunction, vascular hypertrophy, and hypertension [31].

Chlorogenic acid was found to be an inhibitor of the rat hepatic glucose 6-phosphatase system that plays a major role in the homeostatic regulation of blood glucose and is responsible for the formation of endogenous glucose. In addition, chlorogenic acid was found to be a specific inhibitor of glucose-6-phosphate translocase [32]. Furthermore, chlorogenic acid inhibited *in vitro* the Na^+-dependent D-glucose transport in the brush border membrane vesicles isolated from the rat small intestine [33].

Chlorogenic acid markedly reduced the *N*-nitrosation of 2,3-diaminonaphthalene [34] and morpholine [35] by nitrite under simulated gastric juice conditions, supposedly by scavenging the nitrosating agent. Chlorogenic acid was antimutagenic against aflatoxin B_1 (AFB_1) in *Salmonella typhimurium* TA100 [36], and reduced the mutagenicity of cigarette smoke condensate in *S. typhimurium* in the presence of an activating system [37]. It also reduced the AFB_1-activating and DNA adduct-generating activities of liver microsomes of rats after a three-week period of feeding [38]. The oral administration of chlorogenic acid at a dose of 150 mg kg^{-1} was shown to inhibit nitrosourea-induced DNA damage in the bone marrow and colon epithelial cells of mice that had simultaneously received methylurea and sodium nitrite [39]. Chlorogenic acid also significantly reduced the frequencies of micronuclei in mouse bone marrow induced by whole body γ-irradiation [40].

Chlorogenic acid, when administered in the diet, significantly reduced aberrant crypt foci in the colon of rats induced by the subcutaneous injection of azoxymethane, and also caused a regression in the aberrant crypt foci as precursor lesions of colorectal cancer [41]. Chlorogenic acid also significantly reduced the number of hyperplastic liver cell foci and the incidence of colon tumors in hamsters given a single i.v. injection of methylazoxymethanol acetate [42]. An inhibition of tongue carcinogenesis (as induced by 4-nitroquinoline-1-oxide) in rats by concurrent administration of chlorogenic acid was also reported [43]. After topical application to mice it was weakly active as an inhibitor of epidermal lipooxygenase activity and 12-*O*-tetradecanoyl-phorbol 13-acetate (TPA)-induced ear inflammation [44].

Isochlorogenic acid and related compounds, including 4,5-dicaffeoylquinic acid and 1,5-dicaffeoylquinic acid, were found to be inhibitors of human immunodeficiency virus (HIV)-1 replication in T-cell lines at concentrations ranging from 1 to 6 μ*M*. These compounds inhibited HIV integrase (an essential enzyme that mediates integration of the HIV genome into the host chromosome) *in vitro*, at submicromolar concentrations. Molecular modeling of these ligands with the core catalytic domain of the integrase indicated an energetically favorable reaction, with the most potent inhibitors filling a groove within the predicted catalytic site of the integrase. The calculated change in internal free energy of the ligand–integrase complex correlated with an ability of the compounds to inhibit HIV-1 integrase *in vitro*; this suggested that the dicaffeoylquinic acids act as a class of potent and selective inhibitors of HIV-1 integrase [45].

Chlorogenic acid was also reported to be cytotoxic against human oral squamous cell carcinoma (HSC-2) and salivary gland tumor (HSG) cell lines. At millimolar concentrations, it produced radicals under alkaline conditions, acting as a pro-oxidant, whereas at lower concentrations it scavenged superoxide and hydroxyl radicals. Chlorogenic acid produced large DNA fragments and nuclear condensation in tumor cells, and also activated caspase [46].

References

1 Zhang, Y.Q., Cheng, B.S., Li, H.B., and Hu, Z.W. (1991) Chlorogenic acid in the root, stem and leaves of *Lonicera japonica*. *Chin. Pharm. J.*, **26**, 145–147.

2 Li, B.T.D (1986) Comparative analysis of the chlorogenic acid and isochlorogenic acid in flower and cane of Japanese honeysuckle (*Lonicera japonica* Thunb). *Chin. Trad. Herbal Drugs*, **17**, 10–11.

3 Gao, W.L. and Liu, Q.L. (1989) Quantitative determination of chlorogenic acid and isochlorogenic acid by TLC-densitometry. *J. Shenyang Coll. Pharm.*, **6**, 99–103.

4 Huang, L.Y., Lu, Z.Z., Li, J.B., and Zhou, B.N. (1996) Studies on the chemical constituents of Japanese Honeysuckle (*Lonicera japonica*). *Chin. Trad. Herbal Drugs*, **27**, 645–647.

5 Choi, C.W., Jung, H.A., Kang, S.S., and Choi, J.S. (2007) Antioxidant constituents and a new triterpenoid glycoside from Flos Lonicerae. *Arch. Pharm. Res.*, **30**, 1–7.

6 Chang, H.W., Baek, S.H., Chung, K.W., Son, K.H., Kim, H.P., and Kang, S.S. (1994) Inactivation of phospholipase A2 by naturally occurring biflavonoid, ochnaflavone. *Biochem. Biophys. Res. Commun.*, **205**, 843–849.

7 Kawai, H., Kuroyanagi, M., and Ueno, A. (1988) Iridoid glucosides from *Lonicera japonica* Thunb. *Chem. Pharm. Bull. (Tokyo)*, **36**, 3664–3666.

8 Son, K.H., Jung, K.Y., Chang, H.W., Kim, H.P., and Kang, S.S. (1994) Triterpenoid saponins from the aerial parts of *Lonicera japonica*. *Phytochemistry*, **35**, 1005–1008.

9 Schlotzhauer, W.S., Pair, S.D., and Horvat, R.J. (1996) Volatile constituents from the flowers of Japanese honeysuckle (*Lonicera japonica*). *J. Agric. Food Chem.*, **44**, 206–209.

10 Zhang, L., Peng, G.F., Lin, H.B., and Zhong, F.X. (1995) Chemical constituents of essential oil from flowers of *Lonicera japonica*. *Chin. Pharm. J.*, **30**, 651–653.

11 Ikeda, N., Ishihara, M., Tsuneya, T., Kawakita, M., Yoshihara, M., Suzuki, Y., Komaki, R., and Inui, M. (1994) Volatile components of honeysuckle (*Lonicera japonica* Thunb.) flowers. *Flavour Fragrance J.*, **9**, 325–331.

12 Thanabhom, S., Jaijoy, K., Thamaree, S., Ingkaninan, K., and Panthong, A. (2006) Acute and subacute toxicity study of the ethanol extract from *Lonicera japonica* Thunb. *J. Ethnopharmacol.*, **107**, 370–373.

13 Lafay, S., Morand, C., Manach, C., Besson, C., and Scalbert, A. (2006) Absorption and metabolism of caffeic acid and chlorogenic acid in the small intestine of rats. *Br. J. Nutr.*, **96**, 39–46.

14 Gonthier, M.P., Verny, M.A., Besson, C., Remesy, C., and Scalbert, A. (2003) Chlorogenic acid bioavailability largely depends on its metabolism by the gut microflora in rats. *J. Nutr.*, **133**, 1853–1859.

15 Ren, J., Jiang, X., and Li, C. (2007) Investigation on the absorption kinetics of chlorogenic acid in rats by HPLC. *Arch. Pharm. Res.*, **30**, 911–916.

16 Azuma, K., Ippoushi, K., Nakayama, M., Ito, H., Higashio, H., and Terao, J. (2000) Absorption of chlorogenic acid and caffeic acid in rats after oral administration. *J. Agric. Food Chem.*, **48**, 5496–5500.

17 Yang, H., Yuan, B., Li, L., Chen, H., and Li, F. (2004) HPLC determination and pharmacokinetics of chlorogenic acid in rabbit plasma after an oral dose of Flos Lonicerae extract. *J. Chromatogr. Sci.*, **42**, 173–176.

18 Xu, Y., Oliverson, B.G., and Simmons, D.L. (2007) Trifunctional inhibition of COX-2 by extracts of *Lonicera japonica*: direct inhibition, transcriptional and post-transcriptional down regulation. *J. Ethnopharmacol.*, **111**, 667–670.

19 Lee, J.H., Ko, W.S., Kim, Y.H., Kang, H.S., Kim, H.D., and Choi, B.T. (2001) Anti-inflammatory effect of the aqueous extract from *Lonicera japonica* flower is related to inhibition of NF-κB activation through reducing I-κBα degradation in rat liver. *Int. J. Mol. Med.*, **7**, 79–83.

20 Tae, J., Han, S.W., Yoo, J.Y., Kim, J.A., Kang, O. H., Baek, O.S., Lim, J.P., Kim, D.K., Kim, Y.H., Bae, K.H., and Lee, Y.M. (2003) Anti-inflammatory effect of *Lonicera japonica* in proteinase-activated receptor 2-mediated paw edema. *Clin. Chim. Acta*, **330**, 165–171.

21 Kang, O.H., Choi, Y.A., Park, H.J., Lee, J.Y., Kim, D.K., Choi, S.C., Kim, T.H., Nah, Y.H., Yun, K.J., Choi, S.J., Kim, Y.H., Bae, K.H., and Lee, Y.M. (2004) Inhibition of trypsin-induced mast cell activation by water fraction of *Lonicera japonica*. *Arch. Pharm. Res.*, **27**, 1141–1146.

22 Leung, H.W., Wu, C.H., Lin, C.H., and Lee, H. Z. (2005) Luteolin induced DNA damage leading to human lung squamous carcinoma CH27 cell apoptosis. *Eur. J. Pharmacol.*, **508**, 77–83.

23 Leung, H.W., Kuo, C.L., Yang, W.H., Lin, C.H., and Lee, H.Z. (2006) Antioxidant enzymes activity involvement in luteolin-induced human lung squamous carcinoma CH27 cell apoptosis. *Eur. J. Pharmacol.*, **534**, 12–18.

24 Suh, S.J., Jin, U.H., Kim, S.H., Chang, H.W., Son, J.K., Lee, S.H., Son, K.H., and Kim, C.H. (2006) Ochnaflavone inhibits TNF-α-induced human VSMC proliferation via regulation of cell cycle, ERK1/2, and MMP-9. *J. Cell. Biochem.*, **99**, 1298–1307.

25 Son, M.J., Moon, T.C., Lee, E.K., Son, K.H., Kim, H.P., Kang, S.S., Son, J.K., Lee, S.H., and Chang, H.W. (2006) Naturally occurring biflavonoid, ochnaflavone, inhibits cyclooxygenases-2 and 5-lipoxygenase in mouse bone marrow-derived mast cells. *Arch. Pharm. Res.*, **29**, 282–286.

26 Suh, S.J., Chung, T.W., Son, M.J., Kim, S.H., Moon, T.C., Son, K.H., Kim, H.P., Chang, H.W., and Kim, C.H. (2006) The naturally occurring biflavonoid, ochnaflavone, inhibits LPS-induced iNOS expression, which is mediated by ERK1/2 via NF-κB regulation, in RAW264.7 cells. *Arch. Biochem. Biophys.*, **447**, 136–146.

27 Kono, Y., Kobayashi, K., Tagawa, S., Adachi, K., Ueda, A., Sawa, Y., and Shibata, H. (1997) Antioxidant activity of polyphenolics in diets. Rate constants of reactions of chlorogenic acid and caffeic acid with reactive species of oxygen and nitrogen. *Biochim. Biophys. Acta*, **1335**, 335–342.

28 Kono, Y., Kashine, S., Yoneyama, T., Sakamoto, Y., Matsui, Y., and Shibata, H. (1998) Iron chelation by chlorogenic acid as a natural antioxidant. *Biosci. Biotechnol. Biochem.*, **62**, 22–27.

29 Yoshino, M. and Murakami, K. (1998) Interaction of iron with polyphenolic compounds: application to antioxidant characterization. *Anal. Biochem.*, **257**, 40–44.

30 Gutnisky, A., Rizzo, N., Castro, M.E., and Garbossa, G. (1992) The inhibitory action of chlorogenic acid on the intestinal iron absorption in rats. *Acta Physiol. Pharmacol. Ther. Latinoam.*, **42**, 139–146.

31 Suzuki, A., Yamamoto, N., Jokura, H., Yamamoto, M., Fujii, A., Tokimitsu, I., and Saito, I. (2006) Chlorogenic acid attenuates hypertension and improves endothelial function in spontaneously hypertensive rats. *J. Hypertens.*, **24**, 1065–1073.

32 Hemmerle, H., Burger, H.J., Below, P., Schubert, G., Rippel, R., Schindler, P.W., Paulus, E., and Herling, A.W. (1997) Chlorogenic acid and synthetic chlorogenic acid derivatives: novel inhibitors of hepatic glucose-6-phosphate translocase. *J. Med. Chem.*, **40**, 137–145.

33 Welsch, C.A., Lachance, P.A., and Wasserman, B.P. (1989) Dietary phenolic compounds: inhibition of Na^+-dependent D-glucose uptake in rat intestinal brush border membrane vesicles. *J. Nutr.*, **119**, 1698–1704.

34 Kono, Y., Shibata, H., Kodama, Y., and Sawa, Y. (1995) The suppression of the N-nitrosating reaction by chlorogenic acid. *Biochem. J.*, **312** (Pt 3), 947–953.

35 Li, P., Wang, H.Z., Wang, X.Q., and Wu, Y.N. (1994) The blocking effect of phenolic acid on *N*-nitrosomorpholine formation *in vitro*. *Biomed. Environ. Sci.*, **7**, 68–78.

36 Francis, A.R., Shetty, T.K., and Bhattacharya, R. K. (1989) Modification of the mutagenicity of aflatoxin B_1 and *N*-methyl-*N*′-nitro-*N*-nitrosoguanidine by certain phenolic compounds. *Cancer Lett.*, **45**, 177–182.

37 Romert, L., Jansson, T., Curvall, M., and Jenssen, D. (1994) Screening for agents inhibiting the mutagenicity of extracts and constituents of tobacco products. *Mutat. Res.*, **322**, 97–110.

38 Aboobaker, V.S., Balgi, A.D., and Bhattacharya, R.K. (1994) *in vivo* effect of dietary factors on the molecular action of aflatoxin B_1: role of non-nutrient phenolic compounds on the catalytic activity of liver fractions. *In Vivo*, **8**, 1095–1098.

39 Aeschbacher, H.U. and Jaccaud, E. (1990) Inhibition by coffee of nitrosourea-mediated DNA damage in mice. *Food Chem. Toxicol.*, **28**, 633–637.

40 Abraham, S.K., Sarma, L., and Kesavan, P.C. (1993) Protective effects of chlorogenic acid, curcumin and β-carotene against γ-radiation-induced *in vivo* chromosomal damage. *Mutat. Res.*, **303**, 109–112.

41 Morishita, Y., Yoshimi, N., Kawabata, K., Matsunaga, K., Sugie, S., Tanaka, T., and Mori, H. (1997) Regressive effects of various chemopreventive agents on azoxymethane-induced aberrant crypt foci in the rat colon. *Jpn. J. Cancer Res.*, **88**, 815–820.

42 Tanaka, T., Nishikawa, A., Shima, H., Sugie, S., Shinoda, T., Yoshimi, N., Iwata, H., and Mori, H. (1990) Inhibitory effects of chlorogenic acid, reserpine, polyprenoic acid (E-5166), or coffee on hepatocarcinogenesis in rats and hamsters. *Basic Life Sci.*, **52**, 429–440.

43 Tanaka, T., Kojima, T., Kawamori, T., Wang, A., Suzui, M., Okamoto, K., and Mori, H. (1993) Inhibition of 4-nitroquinoline-1-oxide-induced rat tongue carcinogenesis by the naturally occurring plant phenolics caffeic, ellagic, chlorogenic and ferulic acids. *Carcinogenesis*, **14**, 1321–1325.

44 Conney, A.H., Lysz, T., Ferraro, T., Abidi, T.F., Manchand, P.S., Laskin, J.D., and Huang, M.T. (1991) Inhibitory effect of curcumin and some related dietary compounds on tumor promotion and arachidonic acid metabolism in mouse skin. *Adv. Enzyme. Regul.*, **31**, 385–396.

45 McDougall, B., King, P.J., Wu, B.W., Hostomsky, Z., Reinecke, M.G., and Robinson, W.E. JJr (1998) Dicaffeoylquinic and dicaffeoyltartaric acids are selective inhibitors of human immunodeficiency virus type 1 integrase. *Antimicrob. Agents Chemother.*, **42**, 140–146.

46 Jiang, Y., Kusama, K., Satoh, K., Takayama, E., Watanabe, S., and Sakagami, H. (2000) Induction of cytotoxicity by chlorogenic acid in human oral tumor cell lines. *Phytomedicine*, **7**, 483–491.

Luffa cylindrica

Retinervus Luffae Fructus (Sigualuo) is the fibrovascular bundle of the dried ripe fruits of *Luffa cylindrica* (L.) Roem. (Cucurbitaceae). It is used for the treatment of paralytic diseases and as an antitussive in the treatment of chronic bronchitis. The unripe fruit is a common vegetable.

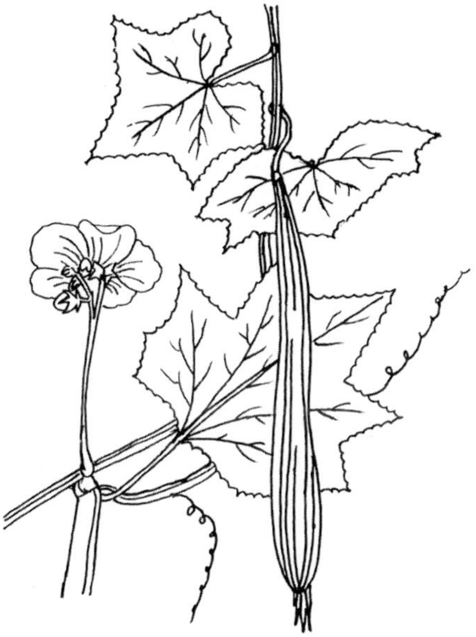

Luffa cylindrica (L.) Roem.

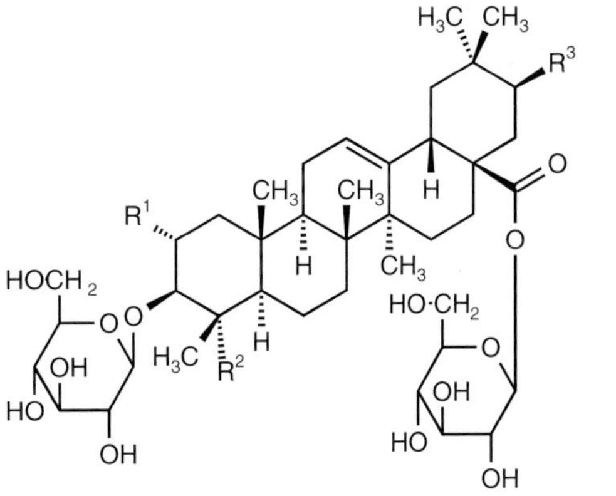

Lucyoside A:	$R^1 = H$	$R^2 = CH_2OH$	$R^3 = OH$
Lucyoside E:	$R^1 = H$	$R^2 = CH_2OH$	$R^3 = H$
Lucyoside B:	$R^1 = OH$	$R^2 = CH_2OH$	$R^3 = H$
Lucyoside D:	$R^1 = OH$	$R^2 = CHO$	$R^3 = H$
Lucyoside F:	$R^1 = H$	$R^2 = CHO$	$R^3 = H$

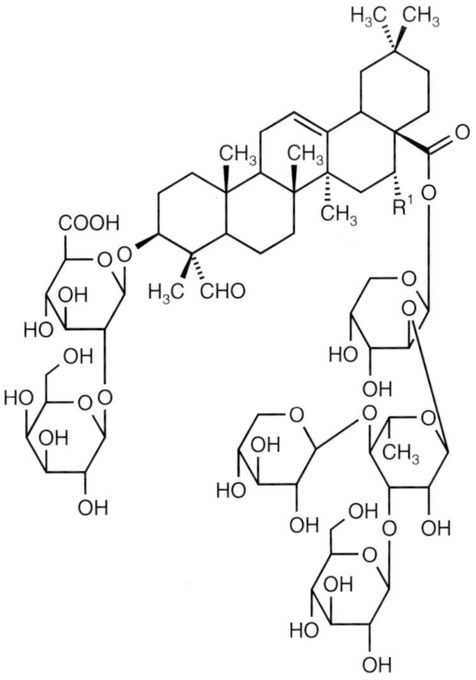

Lucyoside N:	$R^1 = OH$
Lucyoside P:	$R^1 = H$

Chemistry

The fruits, leaves and stem of *L. cylindrica* contain triterpenes and triterpene saponins as the major components. Saponins were identified as lucyosides A–P [1, 2], oleanolic acid, oleanolic acid 3-*O*-β-D-glucopyranoside, hederagenin 3-*O*-β-D-glucopyranoside [3], and ginsenosides Re and Rg_1 [1]. The aglycones of lucyosides were determined to be oleanolic acid, hederagenin, 21β-hydroxyhederagenin, arjunolic acid, machaerinic acid, maslinic acid, gypsogenin, 2α-hydroxygypsogenin, and 21β-hydroxygypsogenin.

Handbook of Chinese Medicinal Plants: Chemistry, Pharmacology, Toxicology
Weici Tang and Gerhard Eisenbrand

ISBN: 978-3-527-32226-8

Luffins, such as luffin-α [4], luffin-β [5], luffin-S [6], luffin P1 [7, 8], and luffacylin [9] are ribosome-inactivating proteins (RIPs) from the fruit and seed of *L. cylindrica*. Luffins are basic glycoproteins and possess a molecular weight of approximately 30 kDa. Luffin-α consists of 248 amino acid residues with a relative molecular weight of 27 021 Da, excluding the attached sugar chains present at each Asn residue of positions 28, 33, 77, 84, 206, and 227. The molecular mass of luffin P1 was determined as 5226.1 Da [8].

The RIPs, which are widespread throughout the plant kingdom, are a group of proteins capable of inactivating eukaryotic protein synthesis by attacking the 28S ribosomal RNA [10]. One of the first RIPs to be isolated, purified and sequenced was trichosanthin, from the root of *Trichosanthes kirilowii* [11]. α-Kirilowin, β-kirilowin, trichokirin are other RIPs that have been isolated from the seed of *T. kirilowii*. α-Kirilowin, which possesses a molecular weight of 27.5 kDa, did not show any crossreactivity with trichosanthin. A sequence comparison of the first 10 residues of β-kirilowin with trichosanthin and trichokirin indicated 60–70% identity [12]. α-Kirilowin was reported to have a molecular weight of 28.8 kDa. The amino acid composition of α-kirilowin grossly resembled that of β-kirilowin and other RIPs. The N-terminal sequence of α-kirilowin was identical to that of β-kirilowin, at least for the first 10 residues [13].

Additional recognized RIPs from Chinese herbal medicines include: cochinchinin and momorcochin S from the seed of *Momordica cochinchinensis* (Cucurbitaceae) [14]; cinnamomin and camphorin from the seeds of *Cinnamomum camphora* (Lauraceae) [15]; pokeweed antiviral proteins from the leaves and seeds of *Phytolacca americana* (Phytolaccaceae) [16]; and ricin from the seeds of *Ricinus communis* (Euphorbiaceae) [17].

Pharmacology and Toxicology

An aqueous extract of the seeds of *L. cylindrica* was reported to show strong activity against platelet aggregation and blood coagulation, and to exhibit fibrinolytic activity. Many saponins from the seeds, such as lucyoside N and lucyoside P, also exhibited fibrinolytic effects [18].

Luffins were reported to exhibit abortifacient, antitumor, ribosome-inactivating and immunomodulatory activities [19]. Luffin-α has been observed to inhibit the proliferation and to induce the apoptosis of human melanoma cells *in vitro* [20]. The protein luffin exhibited tenfold the inhibitory activity against protein synthesis in a rabbit reticulocyte lysate test (IC_{50} 0.42 ng ml^{-1}) compared to that of ricin A chain. In contrast, its cytotoxicity against murine leukemia L 1210 cells was very weak, corresponding to 1×10^{-6} to 1×10^{-5} that of ricin [21].

Luffin P1 was found to show a strong inhibitory activity on protein synthesis in a cell-free rabbit reticulocyte lysate, with an IC_{50} value of approximately 0.9 n*M*. The mechanism of luffin P1 was equivalent to that of the ribosome-inactivating protein, trichosanthin [8]. Interestingly, luffin P1 also showed a trypsin inhibitory activity, with an IC_{50} value of 22 μ*M* [7]. Luffacylin inhibited translation in a rabbit reticulocyte lysate system with an IC_{50} of 140 p*M*, and reacted positively in the N-glycosidase assay for ribosome-inactivating proteins [9].

Luffin elicited a concentration-dependent suppression of the mitogenic response of murine splenocytes to concanavalin A (Con A), with an IC_{50} of about 830 p*M*. Luffin was also found capable of inhibiting the replication of phage M13 in *Escherichia coli* at a concentration of 3.5 μ*M* [22]. The ribosome-inactivating proteins were also found to possess strong anti-human immunodeficiency virus (HIV) activity. Here, luffin was able to inhibit HIV-1 integrase. In contrast, it is only a very weakly suppresed HIV-1 reverse transcriptase and HIV-1 protease. Luffin evoked a strong concentration-dependent inhibition of the 3′-end processing and strand-transfer activities of HIV-1 integrase, which indicated that its anti-HIV activity might be due to an inhibition of HIV-1 integrase [23].

Interferon synthesis was induced in rabbits by the intravenous injection of an extract from the sprout of *L. cylindrica*. The serum interferon level peaked at 2 h or 4 h after the administration of 6 mg kg^{-1} or 1.2 mg kg^{-1} of the extract, respectively. Based on its stability at pH 2, and its thermostability, this interferon was identified as type I. Under optimal conditions, mice challenged with a lethal dose of Japanese encephalitis virus were protected by the sprout extract, the active component of which was shown to be the RNA fraction; the polysaccharide fraction proved to be inactive [24].

References

1 Takemoto, T., Arihara, S., Yoshikawa, K., Kusumoto, K., Yano, I., and Hayashi, T. (1984) Studies on the constituents of Cucurbitaceae plants. VI. On the saponin constituents of *Luffa cylindrica* Roem (1). *Yakugaku Zasshi*, **104**, 246–255.

2 Xiong, S.L., Fang, Z.P., and Zeng, X.Y. (1994) Chemical constituents of *Luffa cylindrica* (L.) Roem. *Chung Kuo Chung Yao Tsa Chih*, **19**, 233–234.

3 Liang, L., Lu, L.E., and Cai, Y.C. (1993) Three chemical components from the leaf of *Luffa cylindrica*. *Acta Pharm. Sin.*, **28**, 836–839.

4 Chen, R.S., Leung, H.W., Dong, Y.C., and Wong, R.N. (1996) Modeling of the three-dimensional structure of luffin-α and its simulated reaction with the substrate oligoribonucleotide GAGA. *J. Protein Chem.*, **15**, 649–657.

5 Kataoka, J., Habuka, N., Miyano, M., Masuta, C., and Koiwai, A. (1992) Nucleotide sequence of cDNA encoding β-luffin, another ribosome-inactivating protein from *Luffa cylindrica*. *Plant Mol. Biol.*, **19**, 887–889.

6 Gao, W., Ling, J., Zhong, X., Liu, W., Zhang, R., Yang, H., Cao, H., and Zhang, Z. (1994) Luffin-S: a small novel ribosome-inactivating protein from *Luffa cylindrica*. Characterization and mechanism studies. *FEBS Lett.*, **347**, 257–260.

7 Li, F., Xia, H.C., Yang, X.X., Hu, W.G., Li, Z., and Zhang, Z.C. (2003) Purification and partial characterization of luffin P1, a peptide with translational inhibitory activity and trypsin inhibitory activity, from seeds of *Luffa cylindrical*. *Sheng Wu Hua Xue Yu Sheng Wu Wu Li Xue Bao (Shanghai)*, **35**, 847–852.

8 Li, F., Yang, X.X., Xia, H.C., Zeng, R., Hu, W. G., Li, Z., and Zhang, Z.C. (2003) Purification and characterization of Luffin P1, a ribosome-inactivating peptide from the seeds of *Luffa cylindrica*. *Peptides*, **24**, 799–805.

9 Parkash, A., Ng, T.B., and Tso, W.W. (2002) Isolation and characterization of luffacylin, a ribosome inactivating peptide with anti-fungal activity from sponge gourd (*Luffa cylindrica*) seeds. *Peptides*, **23**, 1019–1024.

10 Wang, R.H., Zheng, S., Chen, X., and Shen, B. F. (1992) Inhibition of protein synthesis in cell-free system by single chain ribosome-inactivating proteins. *Chin. Biochem. J.*, **8**, 395–399.

11 Wang, Y., Ling, J.F., and Zhu, L.X. (1976) Preliminary studies on an abortifacient plant protein, trichosanthin. *Acta Zool. Sin.*, **22**, 137–143.

12 Dong, T.X., Ng, T.B., Yeung, H.W., and Wong, R.N.S. (1994) Isolation and characterization of a novel ribosome-inactivating protein, -kirilowin, from the seeds of *Trichosanthes kirilowii*. *Biochem. Biophys. Res. Commun.*, **199**, 387–393.

13 Wong, R.N.S., Dong, T.X., Ng, T.B., Choi, W. T., and Yeung, H.W. (1996) β-Kirilowin, a novel ribosome-inactivating protein from seeds of *Trichosanthes kirilowii* (family Cucurbitaceae): a comparison with β-kirilowin and other related proteins. *Int. J. Pept. Protein Res.*, **47**, 103–109.

14 Huang, B., Ng, T.B., Fong, W.P., Wan, C.C., and Yeung, H.W. (1999) Isolation of a trypsin inhibitor with deletion of *N*-terminal pentapeptide from the seeds of *Momordica cochinchinensis*, the Chinese drug mubiezhi. *Int. J. Biochem. Cell Biol.*, **31**, 707–715.

15 Zhang, A.H., Tang, S., and Liu, W. (2001) Substrate-structure dependence of ribotoxins on cleaving RNA in *C. camphora* ribosome. *J. Nat. Toxins*, **10**, 119–125.

16 Kurinov, I.V., Mao, C., Irvin, J.D., and Uckun, F.M. (2000) X-ray crystallographic analysis of pokeweed antiviral protein-II after reductive methylation of lysine residues. *Biochem. Biophys. Res. Commun.*, **275**, 549–552.

17 Despeyroux, D., Walker, N., Pearce, M., Fisher, M., McDonnell, M., Bailey, S.C., Griffiths, G.D., and Watts, P. (2000) Characterization of ricin heterogeneity by electrospray mass spectrometry, capillary electrophoresis, and resonant mirror. *Anal. Biochem.*, **279**, 23–36.

18 Wang, J.D., Narui, T., Yoshikawa, K., Arihara, S., and Okuyama, T. (1991) Hematological studies on naturally occurring substances. V. Studies on the anti-platelet aggregative, anti-blood coagulative and fibrinolytic promotive saponins of *Luffa cylindrica* Roem. *Shoyakugaku Zasshi*, **45**, 215–219.

19 Ng, T.B., Chan, W.Y., and Yeung, H.W. (1992) Proteins with abortifacient, ribosome inactivating, immunomodulatory, antitumor and anti-AIDS activities from Cucurbitaceae plants. *Gen. Pharmacol.*, **23**, 579–590.

20 Poma, A., Marcozzi, G., Cesare, P., Carmignani, M., and Spano, L. (1999) Antiproliferative effect and apoptotic response *in vitro* of human melanoma cells to liposomes containing the ribosome-inactivating protein luffin. *Biochim. Biophys. Acta*, **1472** (1–2), 197–205.

21 Kishida, K., Masuho, Y., and Hara, T. (1983) Protein synthesis inhibitory protein from seeds of *Luffa cylindrica* Roem. *FEBS. Lett.*, **153**, 209–212.

22 Wang, H.X. and Ng, T.B. (2001) Studies on the anti-mitogenic, anti-phage and hypotensive effects of several ribosome inactivating proteins. *Comp. Biochem. Physiol. C. Toxicol. Pharmacol.*, **128**, 359–366.

23 Au, T.K., Collins, R.A., Lam, T.L., Ng, T.B., Fong, W.P., and Wan, D.C. (2000) The plant ribosome inactivating proteins luffin and saporin are potent inhibitors of HIV-1 integrase. *FEBS. Lett.*, **471**, 169–172.

24 Xu, Z.X., Zhou, Z.Q., Qu, F.Z., Tong, L.L., and Li, L.Q. (1985) Extract of *Luffa cylindrica* sprout (L042) as an interferon inducer. *Chin. J. Microbiol. Immunol.*, **5**, 130–132.

Lycium barbarum and *Lycium chinense*

Fructus Lycii (Gouqizi) is the dried ripe fruit of *Lycium barbarum* L. (Solanaceae). It is used as a general tonic for the treatment of debility, dizziness and tinnitus, and diabetes.

Lycium barbarum L.

Cortex Lycii (Digupi) is the dried root bark of *L. barbarum* or *Lycium chinense* Mill. It is used as a tonic and hemostatic agent for the treatment of night sweating, cough, hemostypsis, and diabetes.

Chemistry

The fruits of *L. barbarum* are known to contain carotenoids [1], betaine, and polysaccharides [2]. The major carotenoids were identified as zeaxanthine and its dipalmitate. In addition, the fruits contained about 1–2.7% free amino acid, with proline as the major component. Taurine and γ-aminobutyric acid were also isolated [3]. Interestingly, a novel stable precursor of ascorbic acid, 2-*O*-(β-D-glucopyranosyl)ascorbic acid, was isolated from both the ripe fresh fruit and dried fruit of *L. barbarum*. The content of this ascorbic acid precursor was found to be about 0.5% in the dried fruit [4].

The isolation of a number of polysaccharides from the fruit of *L. barbarum* was reported. The homogeneous acidic glycoproteins LBP-I [5], LBP-II, LBP-III, and LBP-IV [6] were found to be composed of protein, galacturonic acid and monosaccharides, including rhamnose, galactose, glucose, arabinose, mannose, and xylose, in different molecular ratios. A glycoprotein LbGP with a molecular mass of 88 kDa was composed of arabinose, galactose, and glucuronic acid. Glycans with a molecular mass of 40 kDa were connected to core proteins via O-linkages [7]. The homogeneous peptidoglycans $LBPA_1$, $LBPA_3$, $LBPC_2$, and $LBPC_4$ possessed molecular masses of 18, 66, 12, and 10 kDa, respectively [8]. $LBPC_2$ is a $\beta(1 \rightarrow 4)(1 \rightarrow 6)$ peptidoglycan, composed of xylose, rhamnose and mannose, whereas $LBPC_4$ is

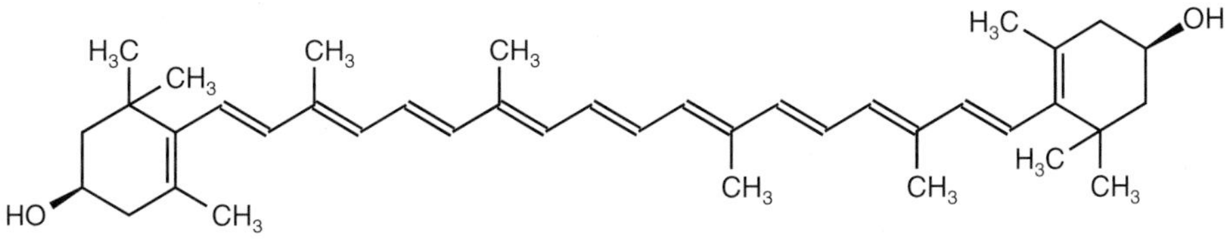

Zeaxanthin

Handbook of Chinese Medicinal Plants: Chemistry, Pharmacology, Toxicology
Weici Tang and Gerhard Eisenbrand

ISBN: 978-3-527-32226-8

an α(1 → 4)(1 → 6) peptidoglycan [9]. The polysaccharide content in different samples of *L. barbarum* was reported to range from 5% to 8% [10].

The Chinese Pharmacopoeia requires a quantitative determination of polysaccharide in Fructus Lycii by spectrophotometry, and a quantitative determination of betaine in Fructus Lycii by thin-layer chromatography–densitometry. The content of polysaccharide in Fructus Lycii should be not less than 1.8% (calculated as glucose), and the content of betaine not less than 0.3%.

A number of calystegine alkaloids have been isolated from the root bark of *L. chinense*. Calystegines are characterized as nortropan alkaloids, with a high degree of hydroxylation and an unusual aminoketal function at the bridgehead position [11]. Calystegines from the root bark of *L. chinensis* were identified as calystegines A_3, A_5, A_6, A_7, B_1, B_2, *N*-methyl-calystegine B_2, calystegines B_3, B_4, B_5, C_1, *N*-methyl-calystegine C_1, calystegine C_2, and calystegine N_1 [12]. Chemically, the calystegines were mainly classified into three groups – A, B, and C – on the basis of the number of hydroxy substituents in the nortropane ring, with three, four, or five hydroxy substituents, respectively. Calystegine N_1 is a calystegine with an amino group in the place of a bridgehead hydroxy group [13].

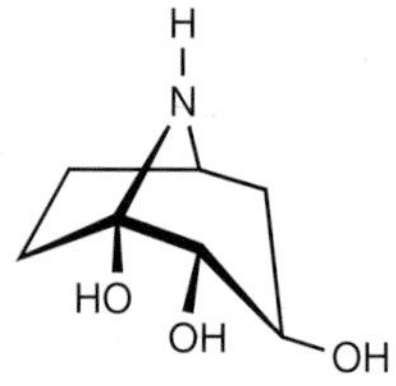

Calystegine A_3

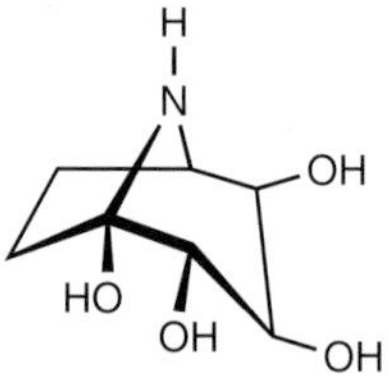

Calystegine B_2

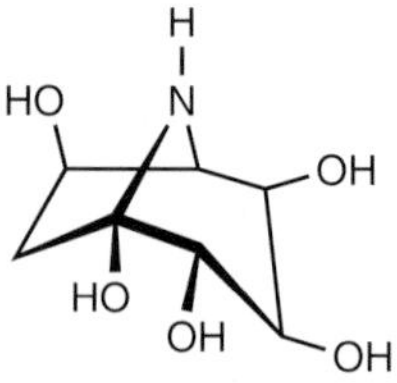

Calystegine C_1

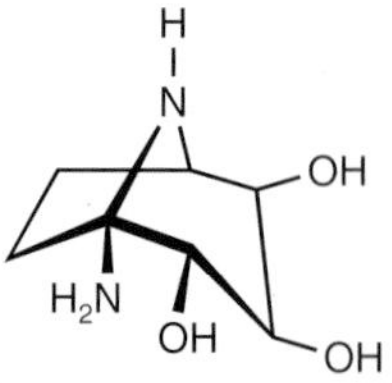

Calystegine N_1

In addition, β-sitosterol, betaine hydrochloride, scopoletin and ferulic acid ester were identified in the root of *L. chinense* [14]. A number of phenolic amides were isolated from the root bark of *L. chinense*. These were structurally elucidated as dihydro-*N*-caffeoyltyramine, *trans*-*N*-feruloyloctopamine, *trans*-*N*-caffeoyltyramine, and *cis*-*N*-caffeoyltyramine [15]. A lignan glycoside, (+)-lyoniresinol 3α-*O*-β-D-glucopyranoside, was also isolated from the root bark of *L. chinense* [16].

Pharmacology and Toxicology

The fruit of *Lycium* was mainly used as a general tonic. The fruit and its polysaccharides were reported to exhibit immunostimulatory effects. The fruit extract markedly increased the expression of interleukin-2 (IL-2) receptor (α- and β-chains) on the membrane of tonsilar mononuclear cells [17]. The supernatant containing IL-2 of splenocytes from adult mice (aged 2 months), when incubated with polysaccharides, increased proliferation of the lymphocytes. The effect of IL-2 from aged animals (16 months) in promoting the proliferation of lymphocytes was less than that of adult animals. However, *L. barbarum* polysaccharides raised the effect of IL-2 of aged mice to the same level as that of adult mice [18]. The fruit extract did not induce nonactivated lymphocyte proliferation, but stimulated lymphocyte proliferation activated by mitogens [19].

The polysaccharide from *L. barbarum* fruits, at a concentration of $100\,\mu g\,ml^{-1}$, significantly enhanced the cell membrane fluidity of rabbit red blood cells, and also accelerated the action of concanavalin A (Con A) on cell fluidity. At the same concentration, the polysaccharide also significantly increased the membrane protein kinase C (PKC) activity, but had no effect on cytoplasmic PKC activation in lymphocytes activated by Con A [20]. At concentrations of 50 to 400 $\mu g\,ml^{-1}$, the polysaccharide increased cellular cAMP and cGMP levels in mouse lymphocytes, in a concentration-dependent manner [21].

The polysaccharide, when given at a daily intraperitoneal (i.p.) dose of 5–10 $mg\,kg^{-1}$ to mice for seven days, increased the spleen lymphocyte proliferation induced by Con A, but not by lipopolysaccharide, as detected by [^{3}H]thymidine incorporation. The number of plaque-forming cells in the spleen in sheep red cell-immunized mice was increased by the polysaccharide [22]. A single i.p. injection of the fruit polysaccharides, at a dose of 10 $mg\,kg^{-1}$, induced the proliferation of splenocytes in mice aged 12 months. The lymphokine-activated killer (LAK) cell activities of the splenocytes of aged mice after i.p. treatment with the polysaccharide were significantly higher than those of control animals. A combined i.p. application of polysaccharides and IL-2 resulted in a synergistic induction of LAK cell activities from aged mice *in vitro* [23]. Furthermore, the polysaccharide was reported to partially or completely reverse the immunosuppressive effect of cyclophosphamide in mice [24]. The polysaccharide from *L. barbarum* was found to exert a therapeutic effect in mice against myelosuppression induced by mitomycin C. It was also effective in the recovery of red blood cells and platelet counts, but did not show any effect on neutropenia [25]. The polysaccharide from *L. barbarum* was also reported to stimulate the recovery from myelosuppression in mice caused either by carboplatin, or by radiation. The recovery of red blood cells and platelet counts may be a result of the stimulation of peripheral blood mononuclear cells to produce granulocyte colony-stimulating factor (G-CSF) [26]. The glycopeptide from *L. barbarum* was also found to stimulate the proliferation of murine splenic lymphocytes *in vitro* [27].

The i.p. injection of a root extract of *L. chinensis* for seven consecutive days also exerted a radioprotective effect on the bone marrow, with enhanced numbers of hematopoietic stem cells being observed. The recovery of leukocyte, erythrocyte and thrombocyte counts was significantly stimulated by the root extract. It was considered that such a radioprotective action might be induced by an enhanced regeneration of the hematopoietic stem cells, due either to an enhanced postirradiation repair or to an increased proliferation of the hematopoietic stem cells [28].

Following the oral administration of a polysaccharide–protein complex from *L. barbarum* at doses of 5, 10, and 20 $mg\,kg^{-1}$ for 10 days, a significant inhibition of tumor growth, and increases in macrophage phagocytosis, spleen lymphocyte proliferation, T lymphocyte activity, IL-2 mRNA expression level was observed, together with a reduction in lipid peroxidation in mice bearing S180 sarcoma cells. The results indicated that the polysaccharide–protein complex not only exerted a highly significant effect on tumor growth but also improved the immune function [29].

A direct cytotoxic activity of *L. barbarum* was also reported. A hot-water extract of the fruits of *L. barbarum* was found to inhibit the proliferation of H-4-II-E rat hepatocellular carcinoma and HA22T/VGH human hepatocellular carcinoma cells. Following a 24 h incubation with the extract, numbers of apoptotic cells and the expression of p53 protein were significantly increased in H-4-II-E cells hepatocellular carcinoma cells [30].

A group of 79 advanced cancer patients were treated in a clinical trial with LAK/IL-2, in combination with *L. barbarum* fruit polysaccharides. The initial results from 75 evaluable patients indicated objective regression of cancer in patients with malignant melanoma, renal cell carcinoma, colorectal carcinoma, lung cancer, and nasopharyngeal carcinoma. The response rate of patients treated with LAK/IL-2 plus polysaccharide was 41%, whereas that of patients treated with LAK/IL-2 alone was 16%. The mean period of remission in patients treated with LAK/IL-2 plus polysaccharides was also significantly longer. The use of LAK/IL-2 plus polysaccharides treatment led to more pronounced increases in natural killer (NK) cells and LAK cell activity than when LAK/IL-2 was used without polysaccharides. These findings indicate that polysaccharides from *L. barbarum* fruit could be used as an adjuvant in cancer treatment [31]. In 171 cancer patients who had received radiotherapy, the polysaccharides also caused a significant increase in the T-lymphocyte blastogenetic rate, and in the phagocytotic rate of macrophages, [32].

The results of experimental studies have shown that the polysaccharides can enhance the antineoplastic activity of irradiation and carmustine in G422 tumor-bearing mice [33]. Furthermore, the polysaccharides exhibited radiosensitizing effects in mice transplanted with Lewis lung cancer [34]. In mice bearing S180 tumors, the tumor weight was significantly decreased, in a dose-dependent manner. The number of splenocytes, activated T cells, NK cell activity and tumor necrosis factor-α (TNF-α) levels in tumor-bearing mice were also increased [35].

In *Salmonella typhimurium* TA100, in the presence of an activating system, the polysaccharide was found to be antimutagenic against the genotoxicity of 2-aminofluorene, and to reduce the frequency of sister chromatid exchanges (SCEs) in human lymphocyte, as induced by mitomycin C [36]. The fragmentation of DNA in murine thymocytes induced by dexamethasone was inhibited by the polysaccharide [37], whereas the weights of the spleen and thymus, as well as reticuloendothelial phagocytosis in normal mice, were increased [38]. A combined use of the polysaccharide fraction and *Corynebacterium parvum* caused a synergistic enhancement of the phagocytotic activity of peritoneal macrophages in mice [39].

The polysaccharide fraction from the fruit was reported to reduce the blood pressure in a two-kidney/one-clip rat model of renovascular hypertension. The results showed that the hypotensive effects of the polysaccharide might be mediated by the increased production of endothelium-derived relaxation factor (EDRF) [40]. The fruit of *L. barbarum* and its polysaccharide were also found to reduce the content of total cholesterol and triglycerides in hypercholesterolemic rabbits [41]. In experiments with cultured neonatal rat cardiomyocytes, the glycopeptide from *L. barbarum* was found capable of increasing the survival ratio and also inhibiting any elevation of the intracellular free calcium concentration in cardiomyocytes, as induced by hypoxia and high potassium concentrations. It was postulated that the glycopeptide might act on the L-type calcium channels [42]. In alloxan-induced diabetes in rabbits, the polysaccharide was also found to reduce levels of blood glucose [43].

The extract of the fruit, its polysaccharide and betaine all exhibited protective effects against the lipid peroxidation of erythrocyte membranes, as induced by hydrogen peroxide [44]. The extract, when given to mice, caused a significant increase in erythrocyte superoxide dismutase activity and a decrease in the malondialdehyde content [45]. The antiperoxidation activity of *L. barbarum* polysaccharide was also demonstrated in *Xenopus laevis* oocyte cell membranes, using electrophysiological technique, following the incubation of cells in a solution containing a free radical-producing system [46].

The "comet assay" with rat testicular cells showed that a pretreatment of these cells with *L. barbarum* polysaccharides at different concentrations (50–400 $\mu g\, ml^{-1}$) led to a significant decrease in oxidation damage in the form of DNA strand breakage induced by H_2O_2. This suggested that the polysaccharide might exert a free radical-scavenging activity and thus inhibit the DNA damage in cells caused by oxidative stress [47]. In a study with primary rat cortical neurons, an extract of *L. barbarum* was also reported to exert a neuroprotective effect; here, the extract effectively inhibited β-amyloid peptide neurotoxicity [48]. The polysaccharide from *L. barbarum* was further reported to protect against rat testis damage induced by heat exposure; the result was a significant increase in testis and epididymis weights, an improved superoxide dismutase (SOD) activity, and raised sex hormone levels in the damaged rat testes. In a study conducted with mouse testicular cells, the polysaccharide effectively protected against DNA damage induced by H_2O_2. The results of an *in vivo* study revealed that the polysaccharide might improve copulatory performance and the reproductive function of hemicastrated male rats, as well as improving sperm quantity and quality. These data would provide the scientific basis for the use of *L. barbarum* fruits as a traditional remedy for male infertility [49]. Betaine was further reported to increase the release of [^{3}H]mucin in primary hamster tracheal surface epithelial cells metabolically radiolabeled with [^{3}H] glucosamine. The effect of betain on mucin release may be associated with a direct action on the mucin-secreting cells of the respiratory tract [50].

The fruit of *L. chinense* was also reported to exert hepatoprotective activity against carbon tetrachloride (CCl_4) induced hepatotoxicity in rats. Pretreatment with the fruits led to a significant lowering of serum aspartate and alanine aminotransferase and alkaline phosphatase of intoxicated rats. The hepatoprotective effects of the fruits were confirmed by histological examination. Pretreatment of the rats with the fruits also inhibited the elevation of malondialdehyde levels, and prevented the depletion of reduced glutathione content and catalase activity in rat liver. Moreover, expression of the cytochrome P450 isoenzyme CYP2E1 (mRNA and protein) was significantly reduced in the liver of treated rats compared to that of controls. The hydroxide radical-scavenging activity of the fruit was further determined using electron spin resonance (ESR) spin-trapping. It was suggested that the hepatoprotective effect of *L. chinense* fruits might be related to the antioxidative activity and to the regulation of CYP2E1 expression [51]. Zeaxanthin and its dipalmitate were each reported to show protective activity against carbontetrachloride CCl_4-induced hepatotoxicity in mice [52].

The phenolic amides from the root bark of *L. chinense* were found to exert antifungal activity. Dihydro-*N*-caffeoyltyramine, *trans*-*N*-feruloyloctopamine, and *trans*-*N*-caffeoyltyramine each showed inhibitory activity against *Candida albicans* at concentrations of 5–10 $\mu g\, ml^{-1}$, whereas *cis*-*N*-caffeoyltyramine was effective at a concentration of 40 $\mu g\, ml^{-1}$. All of the phenolic amides were found to block the dimorphic transition of the fungus [15]. (+)-Lyoniresinol 3α-*O*-β-D-glucopyranoside from the root bark of *L. chinense* also demonstrated a potent antimicrobial activity against clinically isolated methicillin-resistant *Staphylococcus aureus* and human pathogenic fungi, without any hemolytic effect [16].

References

1 Li, G.Z., Peng, G.H., Chen, L., and Zhang, M. H. (1997) Determination of β-carotene in Fructus Lycii by nonaqueous reversed-phase high performance liquid chromatography. *Sepu*, **15**, 537–538.

2 Li, Z., Luo, Q., and Zhang, S.H. (1996) Study on Fructus Lycii and *Lycium barbarum* polysaccharide. *Shipin Kexue (Beijing)*, **17**, 9–12.

3 Xie, H. and Zhang, S.H. (1997) Determination of taurine in *Lycium barbarum* L. by high-performance liquid chromatography with OPA-urea pre-column derivatization. *Sepu*, **15**, 54–56.

4 Toyoda-Ono, Y., Maeda, M., Nakao, M., Yoshimura, M., Sugiura-Tomimori, N., and Fukami, H. (2004) 2-*O*-(β-D-Glucopyranosyl) ascorbic acid, a novel ascorbic acid analogue isolated from *Lycium* fruit. *J. Agric. Food Chem.*, **52**, 2092–2096.

5 He, J., Liang, Y.X., and Zhang, S.H. (1996) Isolation, purification and identification of polysaccharide I from *Lycium barbarum*. *Zhiwu Ziyuan Yu Huanjing*, **5**, 61–62.

6 Sun, Z.D. and Zhang, S.H. (1996) Extraction, isolation of *Lycium barbarum* polysaccharides and their physical and chemical characteristics. *Huazhong Nongye Daxue Xuebao*, **15**, 603–607.

7 Tian, G.Y. and Wang, C. (1995) Structure elucidation of a high MW glycan of a glycoprotein isolated from the fruit of *Lycium barbarum* L. *Acta Biochim. Biophys. Sin.*, **27**, 493–498.

8 Zhao, C.J., Li, R.Z., He, Y.Q., and Cui, G.H. (1996) Studies on the chemistry of Gouqi polysaccharides. *J. Beijing Med. Univ.*, **29**, 231–232, 240.

9 Zhao, C.J., He, Y.Q., Li, R.Z., and Cui, G.H. (1996) Chemistry and pharmacological activity of peptidoglycan from *Lycium barbarum*. *Chin. Chem. Lett.*, **7**, 1009–1010.

10 Wang, Q., Chen, S.Q., Zhang, Z.H., Yu, X.S., Gong, S.L., and Wu, J.K. (1991) Quantitative analysis of polysaccharides in Gouqizi. *Chin. Trad. Herbal Drugs*, **22**, 67–68.

11 Ducrot, P.H. and Lallemand, J.Y. (1990) Structure of the calystegines: new alkaloids of the nortropane family. *Tetrahedron Lett.*, **31**, 3879–3882.

12 Asano, N., Kato, A., Miyauchi, M., Kizu, H., Tomimori, T., Matsui, K., Nash, R.J., and Molyneux, R.J. (1997) Specific α-galactosidase inhibitors, *N*-methylcalystegines: structure/activity relationships of calystegines from *Lycium chinense*. *Eur. J. Biochem.*, **248**, 296–303.

13 Asano, N., Kato, A., Yokoyama, Y., Miyauchi, M., Yamamoto, M., Kizu, H., and Matsui, K. (1996) Calystegine N_1, a novel nortropane alkaloid with a bridgehead amino group from *Hyoscyamus niger*: structure determination and glycosidase inhibitory activities. *Carbohydr. Res.*, **284**, 169–178.

14 Zhou, X., Xu, G., and Wang, Q. (1996) Chemical constituents in the roots of *Lycium chinense* Mill. *China J. Chin. Mater. Med.*, **21**, 675–676, 704.

15 Lee, D.G., Park, Y., Kim, M.R., Jung, H.J., Seu, Y.B., Hahm, K.S., and Woo, E.R. (2004) Antifungal effects of phenolic amides isolated from the root bark of *Lycium chinense*. *Biotechnol. Lett.*, **26**, 1125–1130.

16 Lee, D.G., Jung, H.J., and Woo, E.R. (2005) Antimicrobial property of (+)-lyoniresinol-3α-*O*-β-D-glucopyranoside isolated from the root bark of *Lycium chinense* Miller against human pathogenic microorganisms. *Arch. Pharm. Res.*, **28**, 1031–1036.

17 Du, S.Y. and Qian, Y.K. (1995) Effect of extract of *Lycium barbarum* on the IL-2R expression of human lymphocytes. *Chin. J. Microbiol. Immunol.*, **15**, 176–178.

18 Deng, H.B., Cui, D.P., Jiang, J.M., Feng, Y.C., Cai, N.S., and Li, D.D. (2003) Inhibiting effects of Achyranthes bidentata polysaccharide and *Lycium barbarum* polysaccharide on nonenzyme glycation in D-galactose induced mouse aging model. *Biomed. Environ. Sci.*, **16**, 267–275.

19 Hu, G.J., Bai, H.Q., Du, S.Y., Zhang, H., Shi, Q., and Qian, Y.K. (1995) The regulation effect of Chinese medicine *Lycium barbarum* on lympho-proliferation and lymphocyte subpopulation. *Chin. J. Immunol.*, **11**, 163–166.

20 Zhang, X., Li, J., Liang, H.B., Wang, L., and Qian, Y.K. (1997) Effects of *Lycium barbarum* polysaccharide on the cell membrane fluidity and protein kinase C *in vitro*. *J. Beijing Med. Univ.*, **29**, 118–120.

21 Zhang, X., Xiang, S.L., Cui, X.Y., and Qian, Y.K. (1997) Effects of *Lycium barbarum* polysaccharide (LBP) on lymphocyte signal transduction system in mice. *Chin. J. Immunol.*, **13**, 289–292.

22 Geng, C.S., Wang, G.Y., Lin, Y.D., Xin, S.T., and Zhou, J.H. (1988) The effect of barbary wolfberry (*Lycium barbarum*) polysaccharide on [^{3}H]thymidine incorporation into splenic lymphocytes and on suppressor T-lymphocytes in mice. *Chin. Trad. Herbal Drugs*, **19**, 313–315.

23 Cao, G.W. and Du, P. (1992) Influence of *Lycium barbarum* polysaccharides and interleukin-2 *in vivo* on the induction of two kinds of LAK cells from aged mice *in vitro*. *Chin. J. Microbiol. Immunol.*, **12**, 390–392.

24 Wang, B.K., Xing, S.T., and Zhou, J.H. (1990) Effect of *Lycium barbarum* polysaccharides on the immune responses of T, CTL and NK cells in normal and cyclophosphamide-treated mice. *Chin. J. Pharmacol. Toxicol.*, **4**, 39–43.

25 Gong, H., Shen, P., Jin, L., Xing, C., and Tang, F. (2004) Therapeutic effects of *Lycium barbarum* polysaccharide (LBP) on mitomycin C (MMC)-induced myelosuppressive mice. *J. Exp. Ther. Oncol.*, **4**, 181–187.

26 Gong, H., Shen, P., Jin, L., Xing, C., and Tang, F. (2005) Therapeutic effects of *Lycium barbarum* polysaccharide (LBP) on irradiation or chemotherapy-induced myelosuppressive mice. *Cancer Biother. Radiopharm.*, **20**, 155–162.

27 Du, G., Liu, L., and Fang, J. (2004) Experimental study on the enhancement of murine splenic lymphocyte proliferation by *Lycium barbarum* glycopeptide. *J. Huazhong Univ. Sci. Technol. Med. Sci.*, **24**, 518–520, 527.

28 Hsu, H.Y., Yang, J.J., Ho, Y.H., and Lin, C.C. (1999) Difference in the effects of radioprotection between aerial and root parts of *Lycium chinense*. *J. Ethnopharmacol.*, **64**, 101–108.

29 Gan, L., Hua, Z.S., Liang, Y.X., and Bi, X.H. (2004) Immunomodulation and antitumor activity by a polysaccharide-protein complex from *Lycium barbarum*. *Int. Immunopharmacol.*, **4**, 563–569.

30 Chao, J.C., Chiang, S.W., Wang, C.C., Tsai, Y. H., and Wu, M.S. (2006) Hot water-extracted *Lycium barbarum* and *Rehmannia glutinosa* inhibit proliferation and induce apoptosis of hepatocellular carcinoma cells. *World J. Gastroenterol.*, **12**, 4478–4484.

31 Cao, G.W., Yang, W.G., and Du, P. (1994) Observation of the effects of LAK/IL-2 therapy combining with *Lycium barbarum* polysaccharides in the treatment of 75 cancer patients. *Chin. J. Oncol.*, **16**, 428–431.

32 Liu, J.N., Cheng, B.Q., Zhang, J.R., Tan, X.R., and Ji, Y.Z. (1996) Effect of *Lycium* polysaccharide on immune responses of cancer patients following radiotherapy. *Zhonghua Fangshe Yixue Yu Fanghu Zazhi*, **16**, 18–20.

33 Sun, W.J., Xu, W.L., Zhang, Y.X., Huang, R.H., and Duan, G.S. (1994) Therapeutic effects of *Lycium barbarum* polysaccharides in combination with irradiation and carmustine in G422 tumor-bearing mice. *Chin. J. Clin. Oncol.*, **21**, 930–932.

34 Lu, C.X. and Cheng, B.Q. (1991) Radiosensitizing effects of *Lycium barbarum* polysaccharide for Lewis lung cancer. *Chin. J. Integr. Trad. West. Med.*, **11**, 611–612.

35 Liu, J.L., Zhang, L.H., and Qian, Y.K. (1996) Tumor inhibition of *Lycium barbarum* polysaccharide on S180-bearing mice. *Chin. J. Immunol.*, **12**, 115–117.

36 Tao, M.X. and Zhao, Z.L. (1992) Effect of *Lycium barbarum* polysaccharide against genetic damage *in vitro*. *Chin. Trad. Herbal Drugs*, **23**, 474–476.

37 Liu, Y.T., Zhou, H., Bai, X.W., and Yin, J.Z. (1996) The regulation of *Lycium barbarum* polysaccharide on apoptosis of mouse thymocytes *in vitro*. *J. Beijing Med. Univ.*, **28**, 111–112, 133.

38 Sun, W.J., Sui, D.Y., Yu, X.F., Lu, Z.Z., and Hou, C.Z. (1996) Pharmacological studies of polysaccharide-proteins from *Lycium barbarum*. *J. Norman Bethune Univ. Med. Sci.*, **22**, 486–487.

39 Zhang, Y.X., Xing, X.T., and Zhou, J.H. (1989) Effects of *Lycium barbarum* polysaccharides and their combination with *Corynebacterium parvum* on the tumoristatic activity of peritoneal macrophages in mice. *Chin. J. Pharmacol. Toxicol.*, **3**, 169–173.

40 Jia, Y.X., Dong, J.W., Wu, X.X., Ma, T.M., and Shi, A.Y. (1998) Effect of *Lycium barbarum* polysaccharide on vascular tension in two-kidney, one clip model of hypertension. *Acta Physiol. Sin.*, **50**, 309–314.

41 Luo, Q., Yan, J., Li, J.W., and Zhang, S.H. (1997) Effect of *Lycium barbarum* and its polysaccharides on decreasing serum lipids in rabbits. *Yingyang Xuebao*, **19**, 415–419.

42 Xu, S.L., Huang, J., and Tian, G.Y. (2005) Effects of LbGp on the intracellular free calcium concentration of cardiomyocytes induced by hypoxia and KCl. *Zhongguo Zhong Yao Za Zhi*, **30**, 534–538.

43 Luo, Q., Li, J.W., and Zhang, S.H. (1997) Effect of *Lycium barbarum* polysaccharides-X on reducing blood glucose in diabetic rabbits. *Yingyang Xuebao*, **19**, 173–177.

44 Ren, B., Ma, Y., Shen, Y., and Gao, B. (1995) Protective action of *Lycium barbarum* L. (LbL) and betaine on lipid peroxidation of erythrocyte membrane induced by hydrogen peroxide. *China J. Chin. Mater. Med.*, **20**, 303–304.

45 Liu, X.F., Gao, W.H., Zhao, Z.L., Dai, F.Q., Li, X.D., and Gao, C.L. (1994) Extracts of the fruits of *Ziziphus jujuba* and *Lycium barbarum*: actions against lipid peroxidation in mice. *Chin. J. Prev. Med.*, **28**, 254.

46 Zhang, B., Zhang, X., and Li, W. (1997) The injury of *Xenopus laevis* oocytes membrane and its acetylcholine receptor by free radical and the protection of *Lycium barbarum* polysaccharide. *Chung Kuo Ying Yung Sheng Li Hsueh Tsa Chih*, **13**, 322–325.

47 Huang, X., Yang, M., Wu, X., and Yan, J. (2003) Study on protective action of *Lycium barbarum* polysaccharides on DNA impairments of testicle cells in mice. *Wei Sheng Yan Jiu*, **32**, 599–601.

48 Yu, M.S., Leung, S.K., Lai, S.W., Che, C.M., Zee, S.Y., So, K.F., Yuen, W.H., and Chang, R. C. (2005) Neuroprotective effects of anti-aging oriental medicine *Lycium barbarum* against β-amyloid peptide neurotoxicity. *Exp. Gerontol.*, **40**, 716–727.

49 Luo, Q., Li, Z., Huang, X., Yan, J., Zhang, S., and Cai, Y.Z. (2006) *Lycium barbarum* polysaccharides: Protective effects against heat-induced damage of rat testes and H_2O_2-induced DNA damage in mouse testicular cells and beneficial effect on sexual behavior and reproductive function of hemicastrated rats. *Life Sci.*, **79**, 613–621.

50 Lee, C.J., Lee, J.H., Seok, J.H., Hur, G.M., Park, J.J., Bae, S., Lim, J.H., and Park, Y.C. (2004) Effects of betaine, coumarin and flavonoids on mucin release from cultured hamster tracheal surface epithelial cells. *Phytother. Res.*, **18**, 301–305.

51 Ha, K.T., Yoon, S.J., Choi, D.Y., Kim, D.W., Kim, J.K., and Kim, C.H. (2005) Protective effect of *Lycium chinense* fruit on carbon tetrachloride-induced hepatotoxicity. *J. Ethnopharmacol.*, **96**, 529–535.

52 Kim, H.P., Kim, S.Y., Lee, E.J., Kim, Y.C., and Kim, Y.C. (1997) Zeaxanthin dipalmitate from *Lycium chinense* has hepatoprotective activity. *Res. Commun. Mol. Pathol. Pharmacol.*, **97**, 301–314.

Lycopus lucidus var. *hirtus*

Herba Lycopi (Zelan) is the dried, above-ground part of *Lycopus lucidus* Turcz. var. *hirtus* Regel (Lamiaceae). It is used for the treatment of amenorrhea, dysmenorrhea, abdominal pain, and also as a diuretic agent to treat edema.

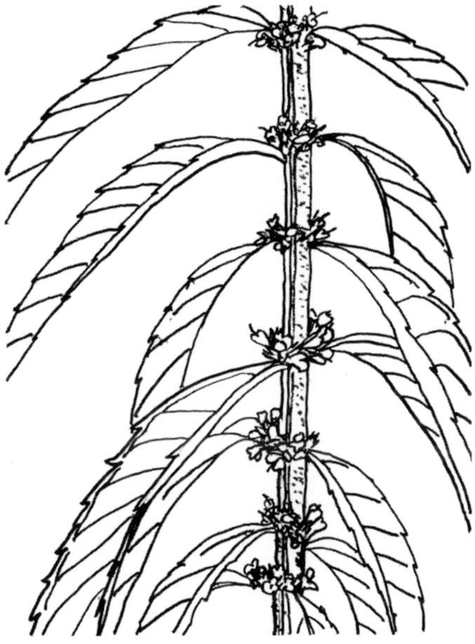

Lycopus lucidus Turcz. var. *hirtus* Regel

Chemistry

Herba Lycopi is known to contain ursolic acid as one of the major constituents. The Chinese Pharmacopoeia requires a qualitative determination of ursolic acid in Herba Lycopi by thin-layer chromatographic comparison with reference substance. Ursolic acid was known to occur in a number of medicinal plants. Further items containing ursolic acid as main ingredient officially listed in the Chinese Pharmacopoeia are Herba Cynomorii and Herba Glechomae. Herba Cynomorii (Suoyang) is the dried fleshy stem of *Cynomorium songaricum* Rupr. (Cynomoriaceae), and is used as a tonic for the treatment of impotence, and constipation. Herba Glechomae (Lianqiancao) is the dried, above-ground part of *Glechoma longituba* (Nakai) Kupr. (Lamiaceae), and is used as a diuretic and antiedemic agent for the treatment of jaundice, traumatic injuries, and pain caused by carbuncle. The Chinese Pharmacopoeia requires a qualitative determination of ursolic acid in both Herba Cynomorii and Herba Glechomae by thin-layer chromatographic comparison with reference substance. Further triterpenes from *L. lucidus* were identified as betulinic acid [1] and oleanolic acid [2].

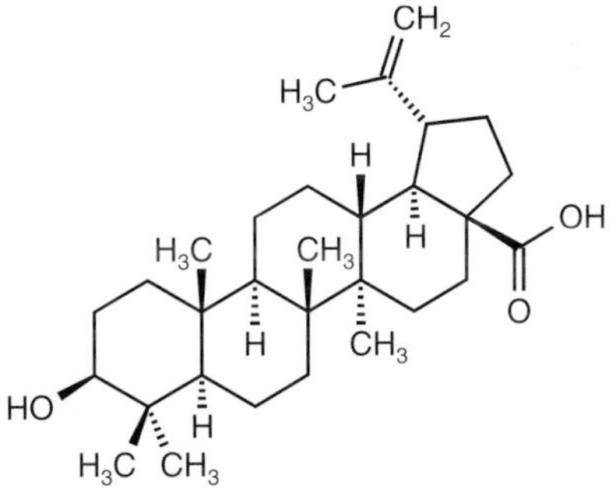

Betulinic acid

The phenolic compounds, rosmarinic acid and its methyl and ethyl esters, and the flavone luteolin and its 7-*O*-β-D-glucuronide methyl ester, were also isolated from the aerial part of *L. lucidus* [3].

Pharmacology and Toxicology

An aqueous extract of *L. lucidus* was reported to decrease mast cell-mediated, immediate-type allergic reactions, which are involved in many allergic conditions such as asthma and allergic rhinitis. The extract, when administered to mice, inhibited the anaphylactic reaction induced by compound 48/80. The extract decreased the passive cutaneous anaphylaxis activated by

Handbook of Chinese Medicinal Plants: Chemistry, Pharmacology, Toxicology
Weici Tang and Gerhard Eisenbrand

ISBN: 978-3-527-32226-8

anti-dinitrophenyl (DNP) IgE, and dose-dependently reduced histamine release from rat peritoneal mast cells activated by compound 48/80 or anti-DNP IgE. Furthermore, the extract decreased the secretion of tumor necrosis factor-α (TNF-α) and interleukin-6 (IL-6) in human mast cells stimulated by 12-*O*-tetradecanoylphorbol 13-acetate (TPA) plus calcium ionophore A23187. The inhibitory effect of the extract on the proinflammatory cytokine was dependent on p38 mitogen-activated protein kinase (MAPK) and nuclear factor-κB (NF-κB). The extract inhibited the degradation of IκBα and nuclear translocation of NF-κB induced by TPA plus A23187, and specifically blocked the activation of p38 MAPK, but not that of c-Jun N-terminal kinase and extracellular signal-regulated kinase (ESRK). This indicated that the extract of *L. lucidus* inhibited mast cell-derived immediate-type allergic reactions and the involvement of proinflammatory cytokines, p38 MAPK, and NF-κB [4].

An active fraction from the extract of *L. lucidus*, designated F04, was reported as capable of promoting the blood circulation and overcoming blood stasis. In rats, administration of the active fraction F04 at doses of $0.6\,g\,kg^{-1}$ and $0.3\,g\,kg^{-1}$ resulted in significant improvements in erythrocyte deformability and the inhibition of erythrocyte aggregation. A trend towards an improvement in liquidity of the cell membrane was also observed [5]. It was concluded that F04 might significantly improve rheological properties of erythrocytes [6]. The active fraction F04 was also found to inhibit platelet aggregation *in vitro* in rats, and *in vivo* in mice [7]. The increase of the maximum platelet aggregation rate induced by ADP in rats was inhibited by F04 at doses of 0.4 and $0.2\,g\,kg^{-1}$. As compared with controls, the thrombus weight in F04 treated rats was reduced. In addition, treatment with F04 also inhibited the thrombosis in artery–vein bypass [8].

Ursolic acid was reported to show anticarcinogenic activities, including an inhibition of skin tumorigenesis [9], and also to induce tumor cell differentiation [10] and inhibit tumor promotion [11, 12]. Ursolic acid was also found to exert an antiangiogenic effect in the chick chorioallantoic membrane [13], to inhibit tumor cell growth both *in vitro* and *in vivo* [14], and to exhibit an anti-invasive activity [15].

Betulinic acid has been reported to be a selective inducer of apoptosis in a variety of human cancer cell lines, and to exert both anti-inflammatory and immunomodulatory properties. The treatment of monocytes and tissue macrophages with betulinic acid induced the production of TNF-α and IL-1β, in a concentration-dependent manner. The treatment of macrophages with betulinic acid enhanced the expression of surface CD40 molecules, indicating that betulinic acid might induce the activation of macrophage and the production of proinflammatory cytokines [1]. Betulinic acid was further found to exhibit potent inhibitory activities against human Acyl-CoA: cholesterol acyltransferase ACAT-1 and ACAT-2, with IC_{50} values of about 16 and 29 μ*M*, respectively [2].

Rosmarinic acid and its esters, and luteolin and its 7-*O*-β-D-glucuronide methyl ester, were also found to have potent antioxidative activities [3].

References

1 Yun, Y., Han, S., Park, E., Yim, D., Lee, S., Lee, C.K., Cho, K., and Kim, K. (2003) Immunomodulatory activity of betulinic acid by producing pro-inflammatory cytokines and activation of macrophages. *Arch. Pharm. Res.*, **26**, 1087–1095.

2 Lee, W.S., Im, K.R., Park, Y.D., Sung, N.D., and Jeong, T.S. (2006) Human ACAT-1 and ACAT-2

inhibitory activities of pentacyclic triterpenes from the leaves of *Lycopus lucidus* Turcz. *Biol. Pharm. Bull.*, **29**, 382–384.

3 Woo, E.R. and Piao, M.S. (2004) Antioxidative constituents from *Lycopus lucidus*. *Arch. Pharm. Res.*, **27**, 173–176.

4 Shin, T.Y., Kim, S.H., Suk, K., Ha, J.H., Kim, I., Lee, M.G., Jun, C.D., Kim, S.Y., Lim, J.P., Eun, J.S., Shin, H.Y., and Kim, H.M. (2005) Anti-allergic effects of *Lycopus lucidus* on mast cell-mediated allergy model. *Toxicol. Appl. Pharmacol.*, **209**, 255–262.

5 Shi, H.Z., Gao, N.N., Li, Y.Z., Yu, J.G., Fan, Q. C., and Bai, G.E. (2002) Effects of active fractions from *Lycopus lucidus* L. F04 on erythrocyte rheology. *Space. Med. Med. Eng. (Beijing)*, **15**, 331–334.

6 Shi, H.Z., Gao, N.N., Li, Y.Z., Yu, J.G., Fan, Q. C., Bai, G.E., and Xin, B.M. (2005) Effects of L. F04, the active fraction of *Lycopus lucidus*, on erythrocytes rheological property. *Chin. J. Integr. Med.*, **11**, 132–135.

7 Tian, Z., Gao, N., Li, L., Yu, J., and Luo, X. (2001) Effect of two extract fractions from *Lycopus lucidus* on coagulation function. *Zhong Yao Cai*, **24**, 507–508.

8 Shi, H.Z., Gao, N.N., Li, Y.Z., Yu, J.G., Fan, Q.C., and Bai, G.E. (2004) Effects of active fractions from *Lycopus lucidus* L. F04 on platelet aggregation and thrombus formation. *Space. Med. Med. Eng. (Beijing)*, **17**, 313–317.

9 Huang, M.T., Ho, C.T., Wang, Z.Y., Ferraro, T., Lou, Y.R., Stauber, K., Ma, W., Georgiadis, C., Laskin, J., and Conney, A.H. (1994) Inhibition of skin tumorigenesis by rosemary and its constituents carnosol and ursolic acid. *Cancer Res.*, **54**, 701–708.

10 Lee, H.Y., Chung, H.Y., Kim, K.H., Lee, F J.J., and Kim, K.W. (1994) Induction of differentiation in the cultured F9 teratocarcinoma stem cells by triterpene acids. *J. Cancer Res. Clin. Oncol.*, **120**, 513–518.

11 Ohigashi, H., Takamura, H., Koshimizu, H., Tokuda, H., and Ito, Y. (1986) Search for possible antitumor promoters by inhibition of 12-*O*-tetradecanoylphorbol 13-acetate-induced Epstein-Barr virus activation: ursolic acid and oleanolic acid from an antiinflammatory Chinese medicinal plant *Glechoma hederaceae* L. *Cancer Lett.*, **30**, 143–151.

12 Tokuda, H., Ohigashi, H., Koshimizu, H., and Ito, Y. (1986) Inhibitory effects of ursolic acid and oleanolic acid on skin tumor promotion by 12-*O*-tetradecanoylphorbol 13-acetate. *Cancer Lett.*, **33**, 279–285.

13 Sohn, K.H., Lee, H.Y., Chung, H.Y., Young, H. S., Yi, S.Y., and Kim, K.W. (1995) Antiangiogenic activity of triterpene acids. *Cancer Lett.*, **94**, 213–218.

14 Young, H.S., Lee, C.K., Park, S.W., Park, F K.Y., Kim, K.W., Chung, H.Y., Yokozawa, F T., and Oura, H. (1995) Antitumor effects of ursolic acid isolated from the leaves of *Eriobotrya japonica*. *Nat. Med.*, **49**, 190–192.

15 Cha, H.J., Bae, S.K., Lee, H.Y., Lee, O.H., Sato, H., Seiki, M., Park, B.C., and Kim, K.W. (1996) Anti-invasive activity of ursolic acid correlates with the reduced expression of matrix metalloproteinase-9 (MMP-9) in HT1080 human fibrosarcoma cells. *Cancer Res.*, **56**, 2281–2284.

Magnolia biondii, Magnolia denudata and *Magnolia sprengeri*

Flos Magnoliae (Xinyi) is the dried flower buds of *Magnolia biondii* Pamp., *Magnolia denudata* Desr., or *Magnolia sprengeri* Pamp. (Magnoliaceae). It is used for the treatment of common cold, headache, and nasal catarrh.

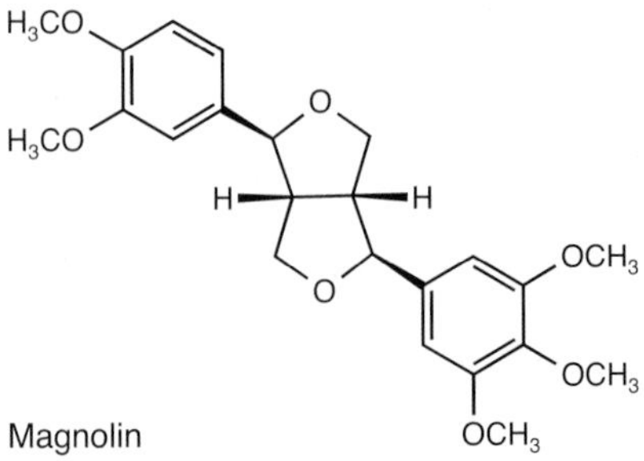

Magnolin

Magnolia denudata Desr.

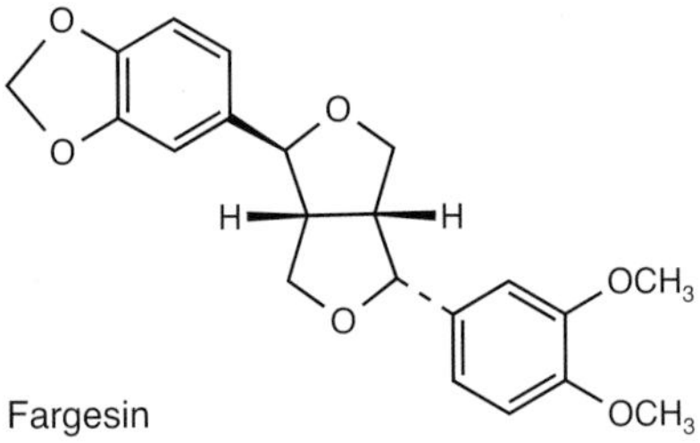

Fargesin

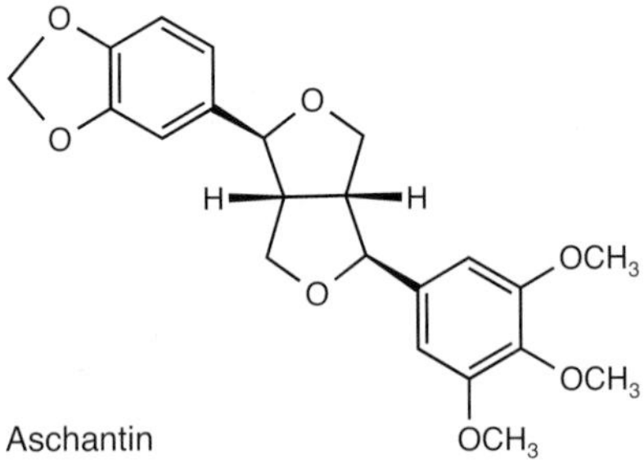

Aschantin

Chemistry

The flower buds of *M. biondii*, *M. denudata* and *M. sprengeri* contain essential oils. The essential oil of the flower buds of *M. biondii*, collected from different areas, was found to contain 1,8-cineole, sabinene, *d*-limonene, and terpinen-4-ol as the major components [1]. The major components of the volatile oil from the flower buds and twigs of *M. sprengeri* have been identified

Handbook of Chinese Medicinal Plants: Chemistry, Pharmacology, Toxicology
Weici Tang and Gerhard Eisenbrand

ISBN: 978-3-527-32226-8

as sabinene, β-pinene, *p*-cymene, bornyl acetate, *trans*-caryophyllene, caryophyllene oxide, and β-eudesmol [2]. The Chinese Pharmacopoeia requires a quantitative determination of the essential oil in Flos Magnoliae. The essential oil content in Flos Magnoliae should be not less than 1.0% ($ml\,g^{-1}$).

Lignan derivatives isolated from the flowers of *M. biondii* were identified as lirioresinol B dimethylether, pinoresinol dimethylether, magnolin, fargesin, aschantin, and demethoxyaschantin [3]. A number of bicyclo[3.2.1]octane and 6-oxabicyclo[3.2.2]nonane-type neolignans were also isolated from *M. denudata* [4]. The Chinese Pharmacopoeia requires a qualitative determination of magnolin in Flos Magnoliae by thin-layer chromatographic comparison with reference substance, and a quantitative determination by HPLC. The content of magnolin in Flos Magnoliae should be not less than 0.4%.

The bark of *M. biondii, M. denudata*, and *M. sprengeri* did not contain magnolol and honokiol, which are the major constituents in Cortex Magnoliae Officinalis. The bark of *Magnolia* species used for Flos Magnoliae can not, therefore, be used as substitute for Cortex Magnoliae Officinalis [5].

Pharmacology and Toxicology

Lignan and neolignan derivatives from *M. biondii* were reported to show antiplatelet aggregation activity induced by a platelet-activating factor [4]. Magnolin was inhibitory towards the production of tumor necrosis factor-α (TNF-α) in the murine macrophage cell line, RAW264.7, when stimulated by lipopolysaccharide [6].

The extract of the flower buds of *M. denudata* has been used for the treatment of allergic diseases. A study using RBL-2H3 mast cells showed that the extract induced cell apoptosis. Changes in cell morphology, the generation of DNA fragmentation, cell cycle arrest, and activation of caspase-3 were observed in cells incubated with the extract. The extract also caused a reduction in the mitochondrial membrane potential, and induced the release of cytochrome c to the cytosol. In addition, the extract up-regulated the expression of Bax and increased Bax protein content. The Bax protein was translocated from the cytosol to the mitochondria at early time points after treatment, which indicated that the flower buds of *M. denudata* could induce both mitochondria-dependent and caspase-dependent mast cell apoptosis [7].

References

1 Zhang, J., Mao, S.Z., Wu, C.Y., and Zeng, S.R. (1986) FFAP (carbowax 20m-2- nitroterephthalic acid) glass capillary column for the separation and identification of the components in the essential oils of Chinese herb "Xinyi". *Anal. Chem.*, **14**, 325–329.

2 Fang, H.J., Song, W.Z., and Yan, Y.P. (1987) Analysis and comparison of the constituents of the volatile oil from the flower buds and twigs of *Magnolia sprengeri* Pamp. *Acta Pharm. Sin.*, **22**, 908–912.

3 Ma, Y. and Han, G. (1995) Biologically active lignins from *Magnolia biondii* Pamp. *China J. Chin. Mater. Med.*, **20**, 102–104, 127.

4 Kuroyanagi, M., Yoshida, K., Yamamoto, A., and Miwa, M. (2000) Bicyclo[3.2.1]octane and 6-oxabicyclo[3.2.2]nonane type neolignans from *Magnolia denudata*. *Chem. Pharm. Bull. (Tokyo)*, **48**, 832–883

5 Zhou, G.Z. and Zhu, Z.F. (1985) Comparison tests of Shanxi's Jiang Po (*Magnolia*) and Hou Po (*Magnolia officinalis* and *M. officinalis* var. *biloba*) by TLC and GC. *Chin. Trad. Herbal Drugs*, **16**, 104–106.

6 Chae, S.H., Kim, P.S., Cho, J.Y., Park, J.S., Lee, J.H., Yoo, E.S., Baik, K.U., Lee, J.S., and Park,

M.H. (1998) Isolation and identification of inhibitory compounds on TNF-α production from *Magnolia fargesii*. *Arch. Pharm. Res.*, **21**, 67–69.

7 Kim, G.C., Lee, S.G., Park, B.S., Kim, J.Y., Song, Y.S., Kim, J.M., Yoo, K.S., Huh, G.Y., Jeong, M.H., Lim, Y.J., Kim, H.M., and Yoo, Y.H. (2003) Magnoliae flos induces apoptosis of RBL-2H3 cells via mitochondria and caspase. *Int. Arch. Allergy Immunol.*, **131**, 101–110.

Magnolia officinalis and *Magnolia officinalis* var. *biloba*

Flos Magnoliae Officinalis (Houpohua) is the dried flower buds of *Magnolia officinalis* Rehd. et Wils. or *Magnolia officinalis* Rehd. et Wils. var. *biloba* Rehd. et Wils. (Magnoliaceae). It is mainly used as a stomachic.

Magnolia officinalis Rehd. et Wils.

Cortex Magnoliae Officinalis (Houpo) is the dried bark of the stem, branch, and root of *M. officinalis* and *M. officinalis* var. *biloba*. It is used as a stomachic for the treatment of epigastric disorders, vomiting, diarrhea, abdominal distension and constipation; it is also used as an antiasthmatic agent.

Chemistry

The major constituents in the bark of *M. officinalis* are known to be neolignans, with magnolol and honokiol as the major components [1]. The Chinese Pharmacopoeia requires a qualitative determination of magnolol and honokiol in Cortex Magnoliae Officinalis and in Flos Magnoliae Officinalis by thin-layer chromatographic comparison with reference substances, and a quantitative determination by HPLC. The total content of magnolol and honokiol in Cortex Magnoliae Officinalis should be not less than 2.0%, and that in Flos Magnoliae Officinalis not less than 0.2%. Bark samples

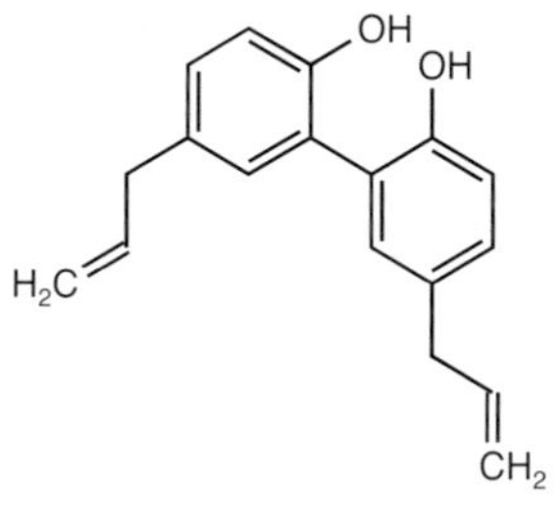

Magnolol

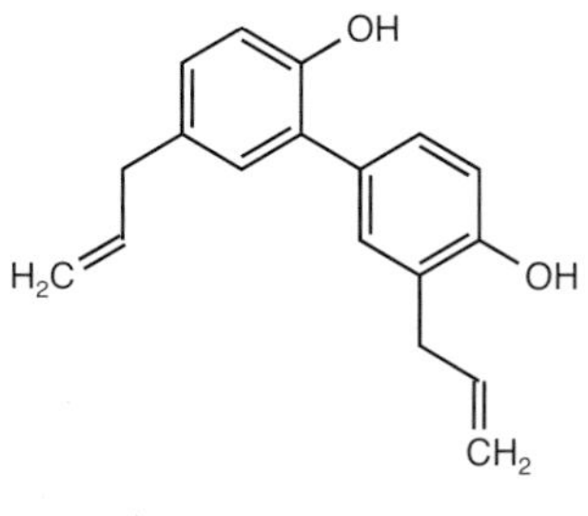

Honokiol

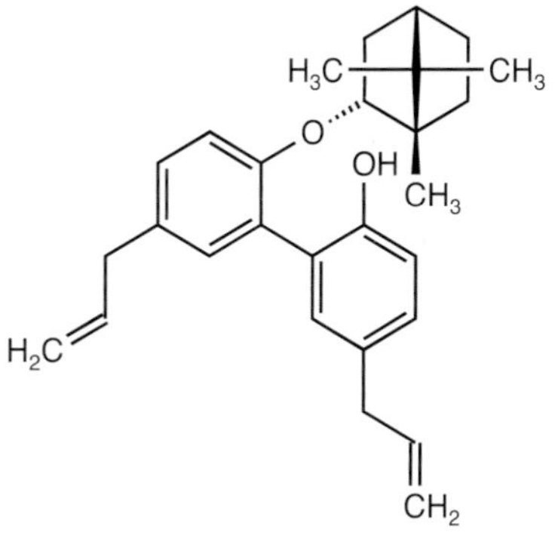

Bornylmagnolol

Handbook of Chinese Medicinal Plants: Chemistry, Pharmacology, Toxicology
Weici Tang and Gerhard Eisenbrand

ISBN: 978-3-527-32226-8

of *M. officinalis* from different areas were found to contain between 2% and 11% magnolol, and between 0.3% and 4.6% honokiol [1]. The root bark contained more magnolol and honokiol than the stem bark [2]. Minor hydroxybiphenyls from the bark of *M. officinalis* were identified as 5,5′-di-(2-propenyl)- and 4,4′-di-(2-propenyl)-2-hydroxy-3,2′,6′-trimethoxy-1,1′-biphenyl [3], piperitylmagnolol [4], and bornylmagnolol [5].

Mognolol, honokiol, α-eudesmol, and β-eudesmol were isolated from the bark of *M. officinalis* var. *biloba* [6]. The major components of the essential oil from the flowers of *M. officinalis* were identified as 1,8-cineol, camphor, and terpinen-4-ol [7].

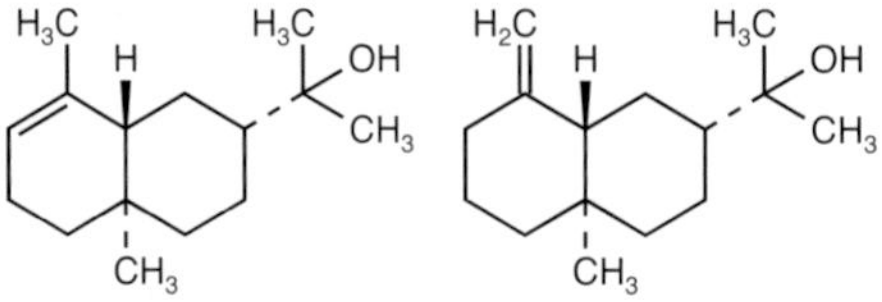

α-Eudesmol β-Eudesmol

Another item with *Magnolia* species officially listed in the Chinese Pharmacopoeia is Flos Magnoliae. Flos Magnoliae (Xinyi) is the dried flower buds of *Magnolia biondii* Pamp., *Magnolia denudata* Desr., or *Magnolia sprengeri* Pamp.

Pharmacology and Toxicology

Magnolol and honokiol are known to be the major active components of Cortex Magnoliae Officinalis, which exerts a variety of biological and pharmacological effects, including antimicrobial, anti-inflammatory, antiallergic, antiasthmatic and antioxidant activities, as well as cardiovascular and cerebrovascular protective activities. Both, magnolol and honokiol have been reported to exhibit significant antimicrobial activities against Gram-positive bacteria and fungi.

Biotransformation and pharmacokinetics of magnolol and honokiol have been reported. Following a single oral dose of [^{14}C]magnolol to rats, peak blood levels of radioactivity occurred at 15 min and again at 8 h, indicative of an enterohepatic circulation of magnolol and its metabolites. Although the radiolabel was mainly located in the gastrointestinal tract and liver, some was present also in the kidneys, pancreas, and lungs. Pharmacokinetic studies with magnolol in rabbits following i.v. administration at 5 mg kg^{-1} showed that plasma concentration–time profile complied with an open, two-compartment model. The elimination half-life of magnolol was approximately 15 min, and the total body clearance 76 ml min^{-1} kg^{-1}. The mean concentration of magnolol in the brain at 10 min after i.v. injection in rats (at 5 mg kg^{-1}) was approximately fourfold higher than that of plasma; the concentrations of magnolol were similar among the various regions of the brain [8]. Data obtained following rectal administration of the bark extract of *M. officinalis* in rabbits showed magnalol to be better absorbed rectally than orally [9].

On repeated oral administration of magnolol, the pattern of the fecal metabolites was changed significantly. Tetrahydromagnolol, 5-(1-propenyl)-5′-propyl-2,2′-dihydroxybiphenyl, 5-(2-propenyl)-5′-propyl-2,2′-dihydroxybiphenyl, isomagnolol (the 1-propenyl analogue of magnolol), and 5-(2-propenyl)-5′-(1-propenyl)-2,2′-dihydroxybiphenyl were identified as metabolites [10]. Both, 8,9-dihydroxydihydromagnolol, as a metabolite of magnolol [11], and unchanged magnolol [12], were identified in the urine of humans treated with the bark of *M. officinalis*. Pharmacokinetic studies of honokiol in rats after i.v. administration at doses of 5 or 10 mg kg^{-1} revealed a biphasic process consisting of a rapid distribution phase followed by a slower elimination phase, and a two-compartment model [13].

The extract of magnolia bark with magnolol and honokiol as the active principles showed potent antibacterial action against a cariogenic bacterium, *Streptococcus mutans*, with a minimum inhibitory concentration (MIC) of 6 $\mu g\,ml^{-1}$. The antibacterial action of both compounds was stronger than that of berberine [14]. A number of human pathogenic fungi, including *Trichophyton mentagrophytes*, *Microsporium gypseum*, *Epidermophyton floccosum*, *Aspergillus niger*, *Cryptococcus neoformans*, and *Candida albicans* were found to be sensitive to magnolol and honokiol, with MICs in the range of 25 to 100 $\mu g\,ml^{-1}$ [15]. *Helicobacter pylori*, isolated from the gastric mucosa of patients with chronic atrophic gastritis, was found to be sensitive to the extract of *M. officinalis in vitro* [16]. Piperitylmagnolol, magnolol, and honokiol were all found to be active against vancomycin-resistant *Enterococci* and methicillin-resistant *Staphylococcus aureus*, with MICs in the range of 6 to 25 $\mu g\,ml^{-1}$. Among the three lignans tested, piperitylmagnolol was the most potent compound. Piperitylmagnolol and honokiol each exerted bactericidal effects [4].

Magnolol was also found to exert anti-inflammatory effects. It inhibited a passive cutaneous anaphylactic reaction, neurogenic inflammation, dorsal skin and ear edema in mice [17]. Anti-inflammatory and analgesic effects of magnolol were also demonstrated in mouse hind-paw edema induced by carrageenan, and in a writhing response induced by acetic acid. Magnolol was further reported to reduce the lethality of an endotoxin challenge. Mechanistic studies revealed that the anti-inflammatory effect of magnolol is neither mediated by glucocorticoid nor through releasing steroid hormones from adrenal gland. It was proposed to be dependent on reducing the level of eicosanoid mediators [18]. Magnolol, at an intraperitoneal (i.p.) dose of 10 $mg\,kg^{-1}$, reduced polymorphonuclear leukocyte infiltration into the pleural cavity of mice, as induced by the calcium ionophore A23187. In an isolated rat peripheral neutrophil suspension, magnolol suppressed the A23187-induced formation of thromboxane B_2 and leukotriene B_4 [19]. Honokiol was also reported to inhibit the production of leukotriene C_4 and leukotriene B_4 in rat basophilic leukemia cells stimulated by A23187, but did not inhibit either phospholipase A_2 activity or leukotriene C_4 synthase and leukotriene A_4 hydrolase activities. These results indicated that honokiol could block leukotriene synthesis by inhibiting 5-lipoxygenase activity. In addition, honokiol was found to inhibit the immunoglobulin IgE-mediated production of these leukotrienes in rat basophilic leukemia cells [20].

The i.p. administration of magnolol at doses of 25 to 100 $mg\,kg^{-1}$ to normothermic rats and to febrile rats, as induced by interleukin-1β (IL-1β), produced a decrease in colon temperature and an increase in the foot skin temperature. The decrease in the body temperature of rats induced by magnolol was shown to be caused by a reduction of serotonin release from the hypothalamus [21].

Both, honokiol and magnolol were found to protect the myocardium of rats against ischemic damage caused by coronary ligation, and also to suppress ventricular arrhythmia during ischemia and reperfusion [22]. It was suggested that an increased nitric oxide (NO) synthesis was involved in the protective effect of magnolol and honokiol against arrhythmia during myocardial ischemia [23]. Magnolol and honokiol were also reported to exert free radical-scavenging activities, and to inhibit the damage by hydroxyl radicals in aqueous media [24]. These compounds also inhibited UV-induced mutation in *Salmonella typhimurium* TA102, based on the scavenging of hydroxyl radicals generated by UV irradiation [25].

Magnolol also showed an antioxidant activity and suppressed lipid peroxidation in rat liver mitochondria [26]. Lipid peroxidation was inhibited by magnolol and honokiol in rat heart mitochondria induced by ADP and $FeSO_4$, with an IC_{50} value of about 0.1 μ*M*. The inhibition of lipid peroxidation by magnolol in the plasma, liver, and lung of septic rats was also reported [27]. Magnolol and honokiol each protected mitochondrial respiratory chain enzyme activity against NADPH-induced peroxidative stress, and also protected the red cells against oxidative hemolysis [28]. Magnolol also significantly inhibited the generation of malondialdehyde as the end product of lipid peroxidation in sperm, and protected sperm motility by inhibiting lipid peroxidation induced by $FeSO_4$ [29]. Honokiol was reported to inhibit the oxygen consumption and malondialdehyde formation during iron-induced lipid peroxidation in liver mitochondria, and to protect rat hepatocytes from ischemia–reperfusion damage [30]. The bark of *M. officinalis* showed a stronger antioxidative effect against the oxidation of linoleic acid [31].

Magnolol was also found to inhibit the adhesion of human neutrophils to a fibrinogen-coated surface activated by 12-*O*-tetradecanoylphorbol 13-acetate (TPA), in a dose-dependent manner. This inhibitory effect of magnolol on neutrophil adhesion to the extracellular matrix (ECM) was believed to be mediated, at least in part, by an inhibition of the accumulation of ROS, which in turn suppressed the up-regulation of Mac-1 that was essential for neutrophil adhesion [32]. Magnolol also inhibited TPA-activated rat neutrophil aggregation, with an IC_{50} of 24 μ*M*. Magnolol suppressed the activity of neutrophil cytosolic and rat brain protein kinase C (PKC) over the same range of concentrations, which suggested that the inhibition of TPA-induced rat neutrophil aggregation by magnolol was most likely attributable to the direct suppression of PKC [33].

Honokiol was able to inhibit arterial thrombosis through a stimulation of prostacyclin generation and endothelial cell protection [34]. Magnolol and honokiol were each also reported to inhibit rabbit the platelet aggregation induced by collagen; likewise, this inhibition was observed in an *ex vivo* study [35]. Honokiol also effectively protected rat aortic endothelial cells in culture against oxidized low-density lipoprotein (LDL) injury, and significantly increased levels of 6-keto-prostaglandin F1α in a serum-free medium. The intravenous (i.v.) application of honokiol to rats also significantly prolonged the thrombus occlusion time as stimulated by an electric current; this indicated that honokiol might represent a potent inhibitor of arterial thrombosis [36]. The i.p. administration of magnolol to mice caused a marked and dose-dependent prolongation of the tail bleeding time [37]. Honokiol and related synthetic biphenyls also inhibited the platelet aggregation induced by different aggregating agents [38]. Magnolol and honokiol each reversibly inhibited phosphocholine acetyltransferase as a key enzyme in the biosynthesis of platelet-activating factor (PAF) in rat spleen microsomes and in the membrane fractions of human polymorphonuclear leukocytes (PMNL). The PAF production in human PMNL, as stimulated by the calcium ionophore A23187, was also suppressed by magnolol and honokiol [39].

One major indication of the bark of *M. officinalis* is the treatment of asthma. In experimental studies, magnolol and honokiol each markedly inhibited histamine release from rat peritoneal mast cells, as induced by compound 48/80, with approximate IC_{50}-values of 1 $\mu g\,ml^{-1}$ for magnolol and 2.8 $\mu g\,ml^{-1}$ for honokiol [40]. Both, magnolol and honokiol were found to possess a marked and longlasting muscle relaxant activity. The i.p. administration of magnolol to mice at a dose of 100 $mg\,kg^{-1}$ produced a strong muscle relaxation for 2 h. It was reported that magnolol caused two phases of relaxation – a fast phase and a slow phase – which counteracted the contraction of rat thoracic aorta induced by norepinephrine (noradrenaline). The fast relaxation was completely antagonized by hemoglobin, and disappeared in the de-endothelialized aorta, whereas the slow relaxation remained unaffected. Both, magnolol and honokiol also inhibited the high K^+-induced, Ca^{2+}-dependent contraction of rat aorta in a concentration-dependent manner. It was postulated that magnolol and honokiol could relax the vascular smooth muscle by releasing endothelium-derived relaxing factor, and also by inhibiting Ca^{2+} influx through voltage-gated Ca^{2+} channels [41]. In cultured smooth muscle cells of the human trachea, magnolol stimulated the large-conductance, Ca^{2+}-activated K^+ channels, which may contribute to its anti-asthmatic activity [42].

Magnolol and honokiol were both reported to exhibit pharmacological effects on the cardiovascular system. The bark of *M. officinalis* also exerted hypocholesterolemic activity. In rabbits fed a high-fat and cholesterol diet, the bark of *M. officinalis* ($300\,mg\,kg^{-1}$) and lovastatin ($6\,mg\,kg^{-1}$, serving as a positive control) both significantly decreased plasma lipids, oxidative stress, and arterial lesions. They also markedly lowered the expression of Fas ligand, caspase 8, and caspase 9 in the aortic arches [43]. Magnolol was able to act as an anti-inflammatory agent during endothelial injuries. The pretreatment of endothelial cells with magnolol inhibited the IL-6-induced Tyr705 and Ser727 phosphorylation of signal transducer and activator of transcription protein 3 (STAT3), a transcription factor which is involved in inflammation and the cell cycle, without affecting the phosphorylation of JAK1, JAK2, and extracellular-regulated kinases (ERKs). Magnolol also suppressed the IL-6-induced promoter activity of intracellular cell adhesion molecule (ICAM)-1. As a result, a reduced monocyte adhesion to IL-6-activated endothelial cells was observed [44]. Magnolol significantly induced apoptosis in rat vascular smooth muscle cells via the mitochondrial death pathway. This effect of magnolol was found to be mediated through a down-regulation of Bcl-2 protein levels, both *in vivo* and *in vitro*, which in turn indicated that magnolol might be used for the treatment of atherosclerosis and restenosis [45]. Magnolol was also reported to inhibit the intimal hyperplasia and monocyte chemotactic protein-1 expression in the balloon-injured aorta of cholesterol-fed rabbits [46].

Studies on the action mechanisms of honokiol for the growth inhibition of vascular smooth muscle cells revealed that honokiol inhibited cell viability and DNA synthesis in cultured cells. This inhibition was associated with G_1 cell cycle arrest. Honokiol was further found to down-regulate the expression of cyclins and cyclin-dependent kinase (CDK), and to up-regulate the expression of p21WAF1, a CDK inhibitor. Among the honokiol-mediated signaling pathways involved in the growth inhibition of vascular smooth muscle cells, honokiol induced a marked activation of p38 MAP kinase, which participates in p21WAF1 induction, and this subsequently led to a decrease in the levels of cyclin D1/CDK4 and cyclin E/CDK2 complexes. As a result, growth of the vascular smooth muscle cells was inhibited [47].

Honokiol was reported to exert a protective effect against focal cerebral ischemia–reperfusion damage in rats that paralleled a reduction in ROS production by neutrophils. Honokiol also showed antioxidative and anti-inflammatory effects in rat neutrophils activated by TPA or *N*-formyl-methionyl-leucyl-phenylalanine (fMLP) by inhibition of NADPH oxidase, myeloperoxidase and cyclooxygenase activities, and by an increase in glutathione peroxidase activity; this suggested that honokiol functioned as a potent inhibitor and scavenger of ROS [48]. The improvement effected by honokiol in focal cerebral ischemia–reperfusion damage in rats could be attributed to its antioxidant, anti-inflammatory, and antiplatelet aggregation actions, through inhibiting lipid peroxidation and reducing both neutrophil activation and ROS production [49].

Honokiol was seen to show neurotrophic activity on rat cortical neurons in culture at concentrations of 0.1 to 10 μM [50]. Honokiol and magnolol each effectively protected cultured rat cerebellar granule cells against neuron toxicity as induced by glucose deprivation, excitatory amino acids, and hydrogen peroxide. These results indicated that the neuroprotective effects of honokiol and magnolol might be related to their antioxidative actions [51]. A protective effect of magnolol against cerebral ischemic injury due to heatstroke was also reported in rats [52]. Magnolol had no effect on K^+-stimulated serotonin release in hippocampal slices, but did elicit a concentration-related inhibition of serotonin release from cortical slices. The inhibitory effect of magnolol on K^+-stimulated serotonin release from the cortex was not affected by antagonists of various serotonin receptor subtypes, nor by the voltage-dependent sodium channel blocker, tetrodotoxin. These results suggested that the suppression of brain serotonin release by magnolol was site-specific, and that the inhibition of cortical serotonin release by magnolol was not effected via the serotonin autoreceptors at the serotoninergic terminals [53].

The aqueous extract of *M. officinalis* was further reported to inhibit the proliferation, DNA synthesis, and expression of platelet-derived growth factor (PDGF)-BB, *CDK1* and *CDK2* genes, and transforming growth factor-β1 (TGF-β1) protein in human mesangial cells. Mesangial cell proliferation, as mediated by PDGF-BB, TGF-β1, and CDKs, was the common feature of glomerulosclerosis [54]. Honokiol was identified as the active component of the bark used to inhibit mesangial cell proliferation [55]. The oral administration of magnolol to type 2 diabetic rats was shown to cause significant decreases in fasting blood glucose levels, plasma insulin, urinary protein and creatinine clearance. It also prevented glomerular enlargement, thus indicating that magnolol might prevent or retard the development of diabetic nephropathy as a complication of diabetes [56].

Magnolol, honokiol, and bornylmagnolol were further found to be inhibitors of Epstein–Barr virus early antigen activation, as induced by TPA; such results indicated that these neolignans might serve as potent antitumor promoters [57]. Honokiol was also shown to exhibit a powerful growth-inhibitory activity against the human promyelocytic leukemic cell line HL-60, with an IC_{50} <100 ng ml^{-1}. Likewise, honokiol strongly suppressed the incorporation of [^{3}H]thymidine, [^{3}H]uridine, and [^{3}H]leucine into HL-60 cells. These findings suggested that honokiol might inhibit HL-60 cell growth by blocking the cellular synthesis of DNA, RNA, and protein [58].

Piperitylmagnolol, magnolol, honokiol, magnolol monoacetate and magnolol diacetate were each found to be cytotoxic against OVCAR-3 human ovarian adenocarcinoma cells, HepG2 hepatocellular carcinoma cells, and HeLa cervical epithelioid carcinoma cells, with IC_{50} values ranging from 3.3 to 13.3 μg ml^{-1}. The acetylated magnolol derivatives were much less potent [4]. Magnolol was found to inhibit the proliferation of human HL-60 cells and Jurkat T leukemia cells by inducing apoptosis, but did not cause apoptosis in the neutrophils and peripheral blood mononuclear cells of healthy donors. The activation of caspase-9, -3, and -2, and the cleavage of poly(ADP-ribose)polymerase, were detected during apoptosis induced by magnolol. Magnolol induced a reduction of the mitochondrial transmembrane potential and the release of cytochrome c into the cytoplasm [59]. Magnolol also inhibited tumor growth and tumor metastasis *in vivo*, this being due to an inhibition of tumor cell invasion [60]. The barks of *M. officinalis* and *M. officinalis* var. *biloba* were also found to inhibit matrix metalloproteinase-9 (MMP-9), which degrades type IV collagen. The latter constitutes the major structural component of the basement membrane and extracellular membrane, and has been shown to be overexpressed in tumor tissues [61].

The honokiol-induced apoptosis in CH27 human squamous lung cancer cells was accompanied by an up-regulation of Bad and down-regulation of Bcl-XL. However, honokiol had no effect on the levels of Bcl-2, Bcl-XS, Bag-1, Bax, and Bak proteins. In addition, honokiol induced the release of mitochondrial cytochrome c to the cytosol, followed by a sequential activation of the caspases. These data indicated that regulation of the Bcl-2 family, the accumulation of cytosolic cytochrome c, and the activation of caspase-9 and caspase-3 might represent the effector mechanisms of magnolol-induced apoptosis in human squamous lung cancer CH27 cells [62]. Similar cytotoxic and mechanistic effects of honokiol were observed in B-cell chronic lymphocytic leukemia (B-CLL) cells [63], and in rat hepatic stellate cells [64].

Magnolol, at concentrations of 3–10 μM, inhibited cell proliferation and DNA synthesis in COLO-205 and Hep-G2 human cancer cells, in a dose-dependent manner; however, no such effect occurred in human untransformed cells such as keratinocytes, fibroblasts, and human umbilical vein endothelial cells. At these concentrations, magnolol was not cytotoxic, but rather exerted an inhibitory effect on cell proliferation. Studies of [^{3}H]thymidine incorporation and flow cytometric analyses revealed that magnolol decreased the DNA synthesis and arrested the cells at the G_0/G_1 phase. At a concentration of 100 μM, apoptosis was observed in COLO-205 human colon cancer cells and in Hep-G2 human hepatoma cells, but not in cultured human fibroblasts and umbilical vein endothelial cells. The i.p. injection of magnolol caused a significant regression of solid tumors formed in nude mice after implantation with COLO-205 cells [65]. Mechanistic studies revealed that apoptosis in Hep G2 cells induced by magnolol was associated with a series of intracellular events, including an increase of cytosolic free Ca^{2+}, an increase in the translocation of cytochrome c from the mitochondria to the cytosol, the activation of caspases 3, 8, and 9, and the down-regulation of Bcl-2 protein [66].

The inhibitory effect of magnolol and honokiol on human fibrosarcoma HT-1080 cells invasion has been studied in a reconstituted basement membrane model. The results showed that neither magnolol nor honokiol affected the adhesion of HT-1080 cells to the basement membrane; however, both compounds inhibited tumor cell migration at high concentrations. Magnolol and honokiol also inhibited the activity of matrix metalloproteinase (MMP)-9, which is secreted by HT-1080 cells and degrades the ECM as a part of the invasive process. This indicates that an inhibition of MMP-9 might be partly responsible for the inhibitory activity of magnolol and honokiol on tumor cell invasiveness [67]. The inhibitory effects of magnolol on tumor metastasis *in vivo* were demonstrated in L5178Y-ML25 lymphoma liver and spleen metastases, and also in B16-BL6 melanoma spontaneous lung metastases. The i.p. administration of magnolol at doses of 2 or 10 mg kg^{-1} led to a significant suppression of liver, spleen, and lung metastases. This antimetastatic activity of magnolol was considered to be associated with the compound's ability to inhibit tumor cell invasion [68].

Honokiol was also found to inhibit the proliferation of transformed endothelial cells *in vitro*. *In vivo*, honokiol was highly effective against angiosarcoma in nude mice [69]. Further investigations into the mechanisms of the antiproliferative, anti-angiogenic, and anti-invasive activities of honokiol revealed that the tumor necrosis factor-α (TNF-α)-induced activation of the transcription factor nuclear factor-κB (NF-κ B) in several cancer cell lines was blocked by honokiol. Honokiol did not directly affect the NF-κB-DNA binding, but rather inhibited the TNF-α-stimulated phosphorylation and degradation of the cytosolic NF-κB inhibitor IκBα by suppressing the intrinsic and TNF-α-stimulated upstream IκB kinases [70]. Honokiol was further reported to down-regulate the expression of P-glycoprotein at the mRNA and protein levels in MCF-7/ADR, a human breast cancer cell line overexpressing multidrug resistance (MDR). The down-regulation of P-glycoprotein was accompanied by a partial recovery of the intracellular drug accumulation, and of the sensitivity towards adriamycin [71].

References

1 Li, A.J., Guo, X.F., Wang, X.M., Chen, C.B., Shi, Y.H., Sui, N.H., and Du, J.Q. (1983) Determination of phenolic substances in Hou Po by HPLC. *Chin. J. Pharm. Anal.*, **3**, 1–3.

2 Li, A.J., Guo, X.F., Feng, H.L., Chen, C.B., and Fang, Z.X. (1985) Contents of magnolol and honokiol in different parts of *Magnolia officinalis* and the effect of processing on their contents. *Bull. Chin. Mater. Med.*, **10**, 154–157.

3 Baek, N.I., Kim, H., Lee, Y.H., Park, J.D., Kang, K.S., and Kim, S.I. (1992) A new dehydrodieugenol from *Magnolia officinalis*. *Planta Med.*, **58**, 566–568.

4 Syu, W.J., Shen, C.C., Lu, J.J., Lee, G.H., and Sun, C.M. (2004) Antimicrobial and cytotoxic activities of neolignans from *Magnolia officinalis*. *Chem. Biodivers.*, **1**, 530–537.

5 Konoshima, T., Kozuka, M., Tokuda, H., Nishino, H., Iwashima, A., Haruna, M., Ito, K., and Tanabe, M. (1991) Studies on inhibitors of skin tumor promotion, IX. Neolignans from *Magnolia officinalis*. *J. Nat. Prod.*, **54**, 816–822.

6 Song, W.Z., Liu, Y.L., and Ji, Q.Y. (1984) Studies on medicinal plants of Magnoliaceae. IV. Study on active constituents of rootbark of "Ao Ye" magnolia (*Magnolia biloba*). *Chin. Trad. Herbal Drugs*, **15**, 450–451.

7 Xu, Z.L., Pan, J.G., and Zhao, Z.Z. (1989) Studies on the essential oils of flos magnoliae. *China J. Chin. Mater. Med.*, **14**, 294–296.

8 Tsai, T.H., Chou, C.J., and Chen, C.F. (1996) Pharmacokinetics and brain distribution of magnolol in the rat after intravenous bolus injection. *J. Pharm. Pharmacol.*, **48**, 57–59.

9 Tan, Y., Lu, W., Zhao, S., Hu, Y., and Ma, Z. (1995) Research on rectal administration of bark of official Magnolia. *China J. Chin. Mater. Med.*, **20**, 30–32.

10 Hattori, M., Endo, Y., Takebe, S., Kobashi, K., Fukasaku, N., and Namba, T. (1986) Metabolism of magnolol from Magnoliae Cortex. II. Absorption, metabolism and excretion of (ring-^{14}C) magnalol in rats. *Chem. Pharm. Bull. (Tokyo)*, **34**, 158–167.

11 Homma, M., Oka, K., Yamada, T., Niitsuma, T., Ihto, H., and Takahashi, N. (1992) A strategy for discovering biologically active compounds with high probability in traditional Chinese herb remedies: an application of saiboku-to in bronchial asthma. *Anal. Biochem.*, **202**, 179–187.

12 Homma, M., Oka, K., Taniguchi, C., Niitsuma, T., and Hayashi, T. (1997) Systematic analysis of post-administrative saiboku-to urine by liquid chromatography to determine pharmacokinetics of traditional Chinese medicine. *Biomed. Chromatogr.*, **11**, 125–131.

13 Tsai, T.H., Chou, C.J., Cheng, F.C., and Chen, C.F. (1994) Pharmacokinetics of honokiol after intravenous administration in rats assessed using high-performance liquid chromatography. *J. Chromatogr. B Biomed. Appl.*, **655**, 41–45.

14 Namba, T., Tsunezuka, M., and Hattori, M. (1982) Dental caries prevention by traditional Chinese medicines. Part II. Potent antibacterial action of Magnoliae Cortex extracts against *Streptococcus mutans*. *Planta Med.*, **44**, 100–106.

15 Bang, K.H., Kim, Y.K., Min, B.S., Na, M.K., Rhee, Y.H., Lee, J.P., and Bae, K.H. (2000) Antifungal activity of magnolol and honokiol. *Arch. Pharm. Res.*, **23**, 46–49.

16 Bae, E.A., Han, M.J., Kim, N.J., and Kim, D.H. (1998) Anti-*Helicobacter pylori* activity of herbal medicines. *Biol. Pharm. Bull.*, **21**, 990–992.

17 Wang, J.P., Raung, S.L., Chen, C.C., Kuo, J.S., and Teng, C.M. (1993) The inhibitory effect of magnolol on cutaneous permeability in mice is probably mediated by a nonselective vascular hyporeactivity to mediators. *Naunyn Schmiedeberg's Arch. Pharmacol.*, **348**, 663–669.

18 Wang, J.P., Hsu, M.F., Raung, S.L., Chen, C. C., Kuo, J.S., and Teng, C.M. (1992) Anti-inflammatory and analgesic effects of magnolol. *Naunyn Schmiedeberg's Arch. Pharmacol.*, **346**, 707–712.

19 Wang, J.P., Ho, T.F., Chang, L.C., and Chen, C. C. (1995) Anti-inflammatory effect of magnolol, isolated from *Magnolia officinalis*, on A23187-induced pleurisy in mice. *J. Pharm. Pharmacol.*, **47**, 857–860.

20 Hamasaki, Y., Kobayashi, I., Zaitu, M., Tsuji, K., Kita, M., Hayasaki, R., Muro, E., Yamamoto, S., Matsuo, M., Ichimaru, T., and Miyazaki, S. (1999) Magnolol inhibits leukotriene synthesis in rat basophilic leukemia-2H3 cells. *Planta Med.*, **65**, 222–226.

21 Hsieh, M.T., Chueh, F.Y., and Lin, M.T. (1998) Magnolol decreases body temperature by reducing 5-hydroxytryptamine release in the rat hypothalamus. *Clin. Exp. Pharmacol. Physiol.*, **25**, 813–817.

22 Hong, C.Y., Huang, S.S., and Tsai, S.K. (1996) Magnolol reduces infarct size and suppresses ventricular arrhythmia in rats subjected to coronary ligation. *Clin. Exp. Pharmacol. Physiol.*, **23**, 660–664.

23 Tsai, S.K., Huang, C.H., Huang, S.S., Hung, L. M., and Hong, C.Y. (1999) Antiarrhythmic effect of magnolol and honokiol during acute phase of coronary occlusion in anesthetized rats: influence of L-NAME and aspirin. *Pharmacology*, **59**, 227–233.

24 Lo, Y.C., Teng, C.M., Chen, C.F., Chen, C.C., and Hong, C.Y. (1994) Magnolol and honokiol isolated from *Magnolia officinalis* protect rat heart mitochondria against lipid peroxidation. *Biochem. Pharmacol.*, **47**, 549–553.

25 Fujita, S. and Taira, J. (1994) Biphenyl compounds are hydroxyl radical scavengers: their effective inhibition for UV-induced mutation in *Salmonella typhimurium* TA102. *Free Radic. Biol. Med.*, **17**, 273–277.

26 Chiu, J.H., Wang, J.C., Lui, W.Y., Wu, C.W., and Hong, C.Y. (1999) Effect of magnolol on *in vitro* mitochondrial lipid peroxidation and isolated cold-preserved warm-reperfused rat livers. *J. Surg. Res.*, **82**, 11–16.

27 Kong, C.W., Tsai, K., Chin, J.H., Chan, W.L., and Hong, C.Y. (2000) Magnolol attenuates peroxidative damage and improves survival of rats with sepsis. *Shock*, **13**, 24–28.

28 Haraguchi, H., Ishikawa, H., Shirataki, N., and Fukuda, A. (1997) Antiperoxidative activity of neolignans from *Magnolia obovata*. *J. Pharm. Pharmacol.*, **49**, 209–212.

29 Lin, M.H., Chao, H.T., and Hong, C.Y. (1995) Magnolol protects human sperm motility against lipid peroxidation: a sperm head fixation method. *Arch. Androl.*, **34**, 151–156.

30 Chiu, J.H., Ho, C.T., Wei, Y.H., Lui, W.Y., and Hong, C.Y. (1997) In vitro and in vivo protective effect of honokiol on rat liver from peroxidative injury. *Life Sci.*, **61**, 1961–1971.

31 Zhou, Y., and Xu, R. (1992) Antioxidative effect of Chinese drugs. *China J. Chin. Mater. Med.*, **17**, 368–369.

32 Shen, Y.C., Sung, Y.J., and Chen, C.F. (1998) Magnolol inhibits Mac-1 (CD11b/CD18)-dependent neutrophil adhesion: relationship with its antioxidant effect. *Eur. J. Pharmacol.*, **343**, 79–86.

33 Wang, J.P., Hsu, M.F., Raung, S.L., Chang, L. C., Tsao, L.T., Lin, P.L., and Chen, C.C. (1999) Inhibition by magnolol of formylmethionyl-leucyl-phenyl alanine-induced respiratory burst in rat neutrophils. *J. Pharm. Pharmacol.*, **51**, 285–294.

34 Zhang, X., Chen, S., and Wang, Y. (2007) Honokiol up-regulates prostacyclin synthase protein expression and inhibits endothelial cell apoptosis. *Eur. J. Pharmacol.*, **554**, 1–7.

35 Hu, H., Zhang, X.X., Wang, Y.Y., and Chen, S. Z. (2005) Honokiol inhibits arterial thrombosis through endothelial cell protection and stimulation of prostacyclin. *Acta Pharmacol. Sin.*, **26**, 1063–1068.

36 Ou, H.C., Chou, F.P., Lin, T.M., Yang, C.H., and Sheu, W.H. (2006) Protective effects of honokiol against oxidized LDL-induced cytotoxicity and adhesion molecule expression in endothelial cells. *Chem. Biol. Interact.*, **161**, 1–13.

37 Teng, C.M., Ko, F.N., Wang, J.P., Lin, C.N., Wu, T.S., Chen, C.C., and Huang, T.F. (1991) Antihaemostatic and antithrombotic effect of some antiplatelet agents isolated from Chinese herbs. *J. Pharm. Pharmacol.*, **43**, 667–669.

38 Takeya, T., Takeuchi, N., Kasama, T., Mayuzumi, K., Fukaya, H., and Tobinaga, S. (1990) Biphenyls, a new class of compound that inhibits platelet-activating factors. *Chem. Pharm. Bull. (Tokyo)*, **38**, 559–561.

39 Homma, M., Minami, M., Taniguchi, C., Oka, K., Morita, S., Niitsuma, T., and Hayashi, T. (2000) Inhibitory effects of lignans and flavonoids in saiboku-to, a herbal medicine for bronchial asthma, on the release of leukotrienes from human polymorphonuclear leukocytes. *Planta Med.*, **66**, 88–91.

40 Ikarashi, Y., Yuzurihara, M., Sakakibara, I., Nakai, Y., Hattori, N., and Maruyama, Y. (2001) Effects of the extract of the bark of *Magnolia obovata* and its biphenolic constituents magnolol and honokiol on histamine release from peritoneal mast cells in rats. *Planta Med.*, **67**, 709–713.

41 Ko, C.H., Chen, H.H., Lin, Y.R., and Chan, M. H. (2003) Inhibition of smooth muscle contraction by magnolol and honokiol in porcine trachea. *Planta Med.*, **69**, 532–536.

42 Wu, S.N., Chen, C.C., Li, H.F., Lo, Y.K., Chen, S.A., and Chiang, H.T. (2002) Stimulation of the BK(Ca) channel in cultured smooth muscle cells of human trachea by magnolol. *Thorax*, **57**, 67–74.

43 Chang, W.C., Yu, Y.M., Hsu, Y.M., Wu, C.H., Yin, P.L., Chiang, S.Y., and Hung, J.S. (2006) Inhibitory effect of *Magnolia officinalis* and lovastatin on aortic oxidative stress and apoptosis in hyperlipidemic rabbits. *J. Cardiovasc. Pharmacol.*, **47**, 463–468.

44 Chen, S.C., Chang, Y.L., Wang, D.L., and Cheng, J.J. (2006) Herbal remedy magnolol suppresses IL-6-induced STAT3 activation and gene expression in endothelial cells. *Br. J. Pharmacol.*, **148**, 226–232.

45 Chen, J.H., Wu, C.C., Hsiao, G., and Yen, M.H. (2003) Magnolol induces apoptosis in vascular smooth muscle. *Naunyn Schmiedeberg's Arch. Pharmacol.*, **368**, 127–133.

46 Chen, Y.H., Lin, S.J., Chen, J.W., Ku, H.H., and Chen, Y.L. (2002) Magnolol attenuates VCAM-1 expression *in vitro* in TNF-α-treated human aortic endothelial cells and *in vivo* in the aorta of cholesterol-fed rabbits. *Br. J. Pharmacol.*, **135**, 37–47.

47 Lee, B., Kim, C.H., and Moon, S.K. (2006) Honokiol causes the p21WAF1-mediated G_1-phase arrest of the cell cycle through inducing p38 mitogen activated protein kinase in vascular smooth muscle cells. *FEBS Lett.*, **580**, 5177–5184.

48 Liou, K.T., Shen, Y.C., Chen, C.F., Tsao, C.M., and Tsai, S.K. (2003) The anti-inflammatory effect of honokiol on neutrophils: mechanisms in the inhibition of reactive oxygen species production. *Eur. J. Pharmacol.*, **475**, 19–27.

49 Liou, K.T., Shen, Y.C., Chen, C.F., Tsao, C.M., and Tsai, S.K. (2003) Honokiol protects rat brain from focal cerebral ischemia-reperfusion injury by inhibiting neutrophil infiltration and reactive oxygen species production. *Brain Res.*, **992**, 159–166.

50 Fukuyama, Y., Nakade, K., Minoshima, Y., Yokoyama, R., Zhai, H., and Mitsumoto, Y. (2002) Neurotrophic activity of honokiol on the cultures of fetal rat cortical neurons. *Bioorg. Med. Chem. Lett.*, **12**, 1163–1166.

51 Lin, Y.R., Chen, H.H., Ko, C.H., and Chan, M. H. (2006) Neuroprotective activity of honokiol and magnolol in cerebellar granule cell damage. *Eur. J. Pharmcol.*, **537**, 64–69.

52 Chang, C.P., Hsu, Y.C., and Lin, M.T. (2003) Magnolol protects against cerebral ischemic injury of rat heatstroke. *Clin. Exp. Pharmacol. Physiol.*, **30**, 387–392.

53 Tsai, T.H., Lee, T.F., Chen, C.F., and Wang, L.C. (1995) Modulatory effects of magnolol on potassium-stimulated 5-hydroxytryptamine release from rat cortical and hippocampal slices. *Neurosci. Lett.*, **186**, 49–52.

54 Lee, B.C., Doo, H.K., Lee, H.J., Jin, S.Y., Jung, J. H., Hong, S.J., Lee, S.H., Kim, S.D., Park, J.K., Leem, K.H., and Ahn, S. Y. (2004) The inhibitory effects of aqueous extract of *Magnolia officinalis* on human mesangial cell proliferation by regulation of platelet-derived growth factor-BB and transforming growth factor-β1 expression. *J. Pharmacol. Sci.*, **94**, 81–85.

55 Chiang, C.K., Sheu, M.L., Hung, K.Y., Wu, K. D., and Liu, S.H. (2006) Honokiol, a small molecular weight natural product, alleviates experimental mesangial proliferative glomerulonephritis. *Kidney Int.*, **70**, 682–689.

56 Sohn, E.J., Kim, C.S., Kim, Y.S., Jung, D.H., Jang, D.S., Lee, Y.M., and Kim, J.S. (2007) Effects of magnolol (5,5′-diallyl-2,2′-dihydroxybiphenyl) on diabetic nephropathy in type 2 diabetic Goto-Kakizaki rats. *Life Sci.*, **80**, 468–475.

57 Konoshima, T., Kozuka, M., Tokuda, H., Nishino, H., Iwashima, A., Haruna, M., Ito, K., and Tanabe, M. (1991) Studies on inhibitors of skin tumor promotion, IX. Neolignans from *Magnolia officinalis*. *J. Nat. Prod.*, **54**, 816–822.

58 Hirano, T., Gotoh, M., and Oka, K. (1994) Natural flavonoids and lignans are potent cytostatic agents against human leukemic HL-60 cells. *Life Sci.*, **55**, 1061–1069.

59 Zhong, W.B., Wang, C.Y., Ho, K.J., Lu, F.J., Chang, T.C., and Lee, W.S. (2003) Magnolol

induces apoptosis in human leukemia cells via cytochrome c release and caspase activation. *Anticancer Drugs*, **14**, 211–217.

60 Ikeda, K., and Nagase, H. (2002) Magnolol has the ability to induce apoptosis in tumor cells. *Biol. Pharm. Bull.*, **25**, 1546–1549.

61 Seo, U.K., Lee, Y.J., Kim, J.K., Cha, B.Y., Kim, D.W., Nam, K.S., and Kim, C.H. (2005) Large-scale and effective screening of Korean medicinal plants for inhibitory activity on matrix metalloproteinase-9. *J. Ethnopharmacol.*, **97**, 101–106.

62 Yang, S.E., Hsieh, M.T., Tsai, T.H., and Hsu, S.L. (2003) Effector mechanism of magnolol-induced apoptosis in human lung squamous carcinoma CH27 cells. *Br. J. Pharmacol.*, **138**, 193–201.

63 Battle, T.E., Arbiser, J., and Frank, D.A. (2005) The natural product honokiol induces caspase-dependent apoptosis in B-cell chronic lymphocytic leukemia (B-CLL) cells. *Blood*, **106**, 690–697.

64 Park, E.J., Zhao, Y.Z., Kim, Y.H., Lee, B.H., and Sohn, D.H. (2005) Honokiol induces apoptosis via cytochrome c release and caspase activation in activated rat hepatic stellate cells *in vitro*. *Planta Med.*, **71**, 82–84.

65 Lin, S.Y., Liu, J.D., Chang, H.C., Yeh, S.D., Lin, C.H., and Lee, W.S. (2002) Magnolol suppresses proliferation of cultured human colon and liver cancer cells by inhibiting DNA synthesis and activating apoptosis. *J. Cell. Biochem.*, **84**, 532–544.

66 Lin, S.Y., Chang, Y.T., Liu, J.D., Yu, C.H., Ho, Y.S., Lee, Y.H., and Lee, W.S. (2001) Molecular mechanisms of apoptosis induced by magnolol in colon and liver cancer cells. *Mol. Carcinog.*, **32**, 73–83.

67 Nagase, H., Ikeda, K., and Sakai, Y. (2001) Inhibitory effect of magnolol and honokiol from *Magnolia obovata* on human fibrosarcoma HT-1080. Invasiveness *in vitro*. *Planta Med.*, **67**, 705–708.

68 Ikeda, K., Sakai, Y., and Nagase, H. (2003) Inhibitory effect of magnolol on tumour metastasis in mice. *Phytother. Res.*, **17**, 933–937.

69 Bai, X., Cerimele, F., Ushio-Fukai, M., Wagas, M., Campbell, P.M., Govindarajan, B., Der, C.J., Battle, T., Frank, D.A., Ye, K., Murad, E., Dubiel, W., Soff, G., and Arbiser, J.L. (2003) Honokiol, a small molecular weight natural product, inhibits angiogenesis *in vitro* and tumor growth *in vivo*. *J. Biol. Chem.*, **278**, 35501–35507.

70 Tse, A.K., Wan, C.K., Shen, X.L., Yang, M., and Fong, W.F. (2005) Honokiol inhibits TNF-α-stimulated NF-κB activation and NF-κB-regulated gene expression through suppression of IKK activation. *Biochem. Pharmacol.*, **70**, 1443–1457.

71 Xu, D., Lu, Q., and Hu, X. (2006) Down-regulation of P-glycoprotein expression in MDR breast cancer cell MCF-7/ADR by honokiol. *Cancer Lett.*, **243**, 274–280.

Melia azedarach and *Melia toosendan*

Cortex Meliae (Kulianpi) is the dried stem bark and root bark of *Melia toosendan* Sieb. et Zucc. or *Melia azedarach* L. (Meliaceae). It is used as an anthelmintic agent, and externally for the treatment of scabies. The Chinese Pharmacopoeia notes Cortex Meliae to be used in patients with hepatitis or nephritis only with caution.

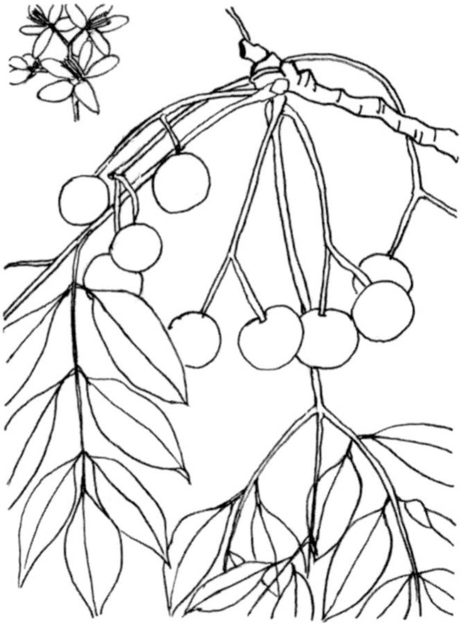

Melia toosendan Sieb. et Zucc.

Fructus Toosendan (Chuanlianzi) is the dried ripe fruit of *Melia toosendan*. It is used as an anthelmintic agent.

Chemistry

The stem bark or root bark of *M. toosendan* [1] and *M. azedarach* [2] were reported to contain limonoid derivatives with toosendanin (chuanliansu) and isotoosendanin (isochuanliansu) as the major components. It is a close relative to Azadirachtanin from Azadriachta indica (Meliaceae), the Indian Neem tree. The content of toosendanin in the bark of *M. toosendan* was reported to range from 0.1% to 0.2%, and was dependent on the season and age of the trees, with the highest contents between December and March, and the lowest between July and September. Both, toosendanin and isotoosendanin possess both a hemiacetal structure.

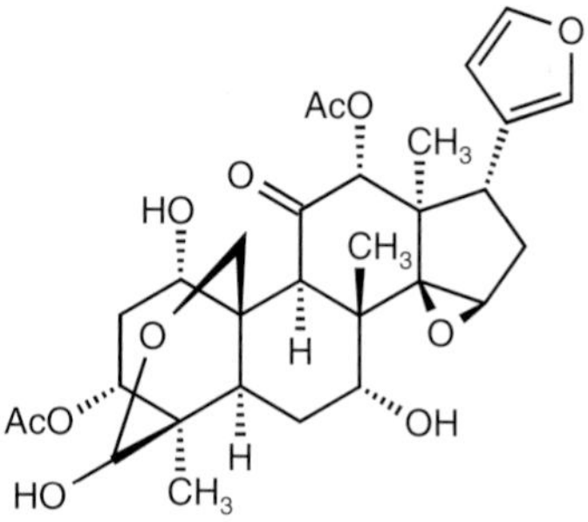

Toosendanin

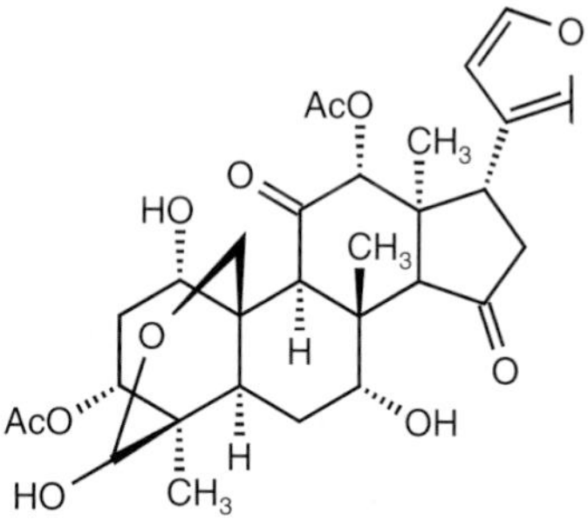

Isotoosendanin

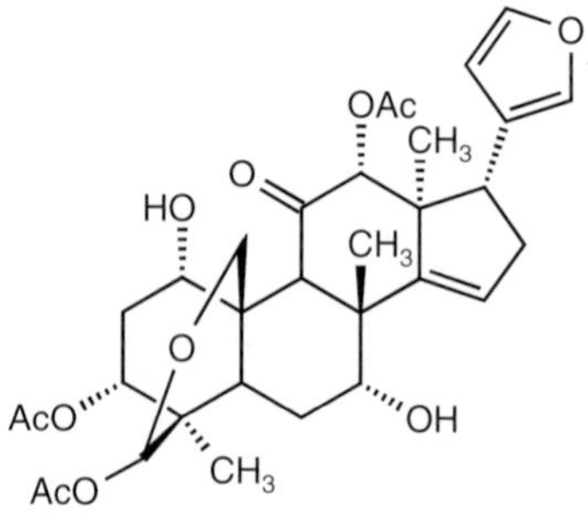

Azadirachtanin

Handbook of Chinese Medicinal Plants: Chemistry, Pharmacology, Toxicology
Weici Tang and Gerhard Eisenbrand

ISBN: 978-3-527-32226-8

Minor limonoids from the root bark of *M. azedarach* were identified as tigloyl, cinnamoyl, and acetyl methoxymeliacarpinins [3, 4]; trichilin-type limonoids trichilin A 1-acetate, trichilin B 1,12-diacetate, trichilin D, trichilin H, 3-deacetyltrichilin H, 1-acetyl-3-deacetyltrichilin H, meliatoxin B_1 [5], and a limonoid glycoside 3-deoxo-3β-glucosyloxygedunin [6]. Trichilin A is the 12-epimer of trichilin B; trichilin H is trichilin B 12-acetate.

The bark of *M. azedarach* was also found to contain triterpenes, with kulinone as the major component. The minor triterpenes were identified as kulactone, kulolactone, and methyl kulonate. Structurally related triterpene compounds, 21-*O*-acetyl-toosendantriol [7] and 21-*O*-methyltoosendanpentol [8] were isolated from the fruits of *M. toosendan*. Anthraquinone glycosides 1,3,8-trihydroxy-2-methylanthraquinone 3-*O*-β-D-galactopyranoside and 1,3,5-trihydroxy-8-methoxy-2-methylanthraquinone 3-*O*-α-L-rhamnopyranoside were also isolated from the bark of *M. azedarach* [9].

Pharmacology and Toxicology

[^{3}H]Toosendanin, when administered to monkeys intravenously, intramuscularly or orally at a dose of 0.25 mg kg^{-1}, revealed two-compartment kinetics. The respective

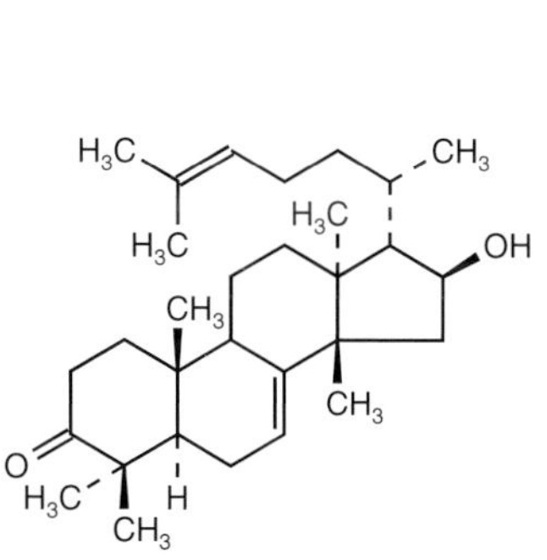

Kulinone

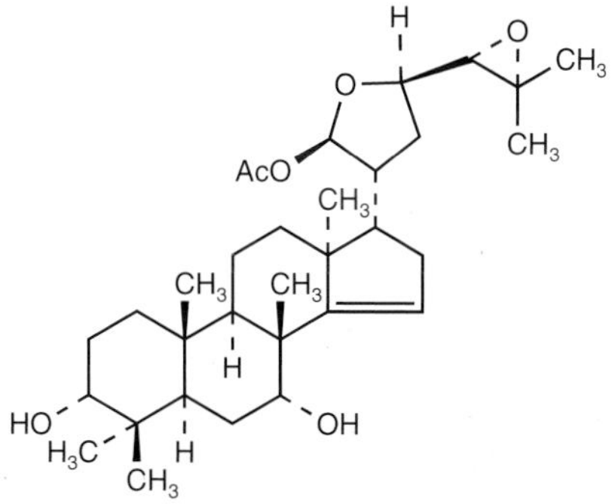

21-*O*-Acetyl-toosendantriol

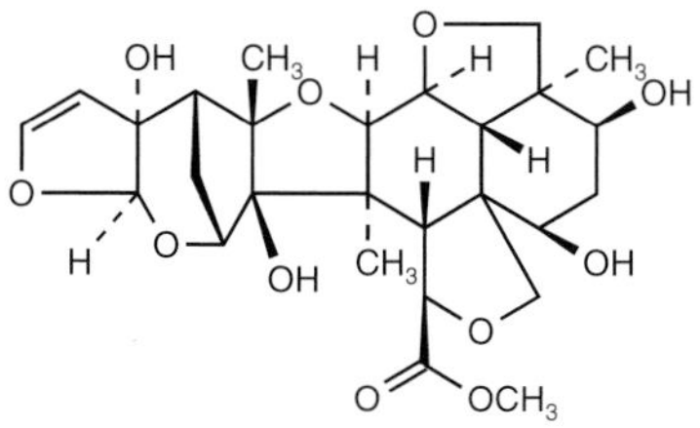

Meliacarpinin

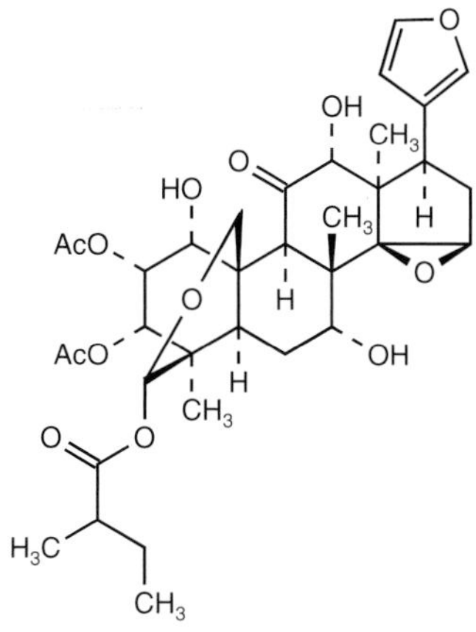

Trichilin B

half-lives following intravenous (i.v.), intramuscular (i.m.) and oral administration were $t_{1/2\alpha}$ (distribution) 0.2, 0.6, and 1.3 h, and $t_{1/2\beta}$ (elimination) 6.6, 18.0, and 25.4 h. Other kinetic parameters showed that toosendanin was rapidly absorbed and distributed among various organs, but eliminated slowly. The combined urinary and fecal excretion at 24 h after administration was approximately 51% after i.v., 27% after i.m., and 47% after oral administration. The extent of excretion at 11 days after each dose route ranged from 75% to 80%. The tissue distribution of toosendanin showed the highest concentrations in the gallbladder and liver, followed by the spleen, stomach, kidneys, and intestine; the lowest concentrations were in the bone [10].

The LD_{50} value of toosendanin in mice was reported as about 14 mg kg^{-1} after intraperitoneal (i.p.), i.v., and subcutaneous (s.c.) injection, and 240 mg kg^{-1} after oral administration. In rats, the LD_{50} was 98 mg kg^{-1} after i.p. dosing, and in rabbits was 4 mg kg^{-1} after i.v. injection. An increase in serum glutamic-pyruvic transaminase (GPT) activity was the most predominant indication of subacute toxicity. The toxic effects of toosendanin were reversible [11].

Toosendanin has been used as an anthelmintic against ascaris (roundworms), and has been found to induce neurotransmitter release as well as being active against the botulism neurotoxin [12, 13]. An extract of *M. toosendan* was found to provide an effective inhibition of schistosomiasis in mice [14]. The survival rate of mice poisoned with lethal doses of botulismotoxin was increased by more than 80% after the i.v, s.c. or oral administration of toosendanin, within 6 h after intoxication. Greater than 50% of monkeys poisoned with different types of botulismotoxin survived when toosendanin was administered 24 h after intoxication. A combined use of toosendanin and antiserum to botulismotoxin A resulted in a marked decrease in the amount of antiserum required for detoxication [15]. A mechanistic study revealed that incubation with toosendanin did not alter the amount of 25 kDa synaptosomal-associated protein, but rendered the synaptosomes completely resistant to its cleavage, as mediated by botulismotoxin A. Following the binding of botulismotoxin A to the synaptosomes, toosendanin partially antagonized the toxin-mediated cleavage of synaptosomal-associated protein of 25 kDa; this suggested that the antibotulism effect of toosendanin was due to the compound blocking the approach of the toxin to its enzymatic substrate [16]. A related mechanism of toosendanin against botulismotoxin was also observed in studies using PC12 cells. Here, the results showed that channel formation was delayed, and the channel size reduced in the PC12 cell membrane treated with toosendanin. This suggested that the protective effect of toosendanin against botulismotoxin may be mediated via an interference with toxin translocation [17].

Studies on the actions of toosendanin on the membrane current of mouse motor nerve terminals revealed that it partially blocked the voltage-dependent fast K^+ current and irreversibly increased the voltage-dependent slow Ca^{2+} current. The increase in the Ca^{2+} current of the nerve terminal accounted for the facilitation of transmitter release and the antibotulinus effect of toosendanin [18]. The inhibitory activity of toosendanin on large-conductance Ca^{2+}-activated K^+ channels was also reported in neurons freshly isolated from the hippocampal CA1 region of the rat. This effect might be mediated by reducing the open probability and unitary current amplitude of the channel [19]. However, the presynaptic blocker toosendanin, as a selective activator of the voltage-dependent calcium channels, did not interfere with the secretory machinery itself [20].

Toosendanin is known to be a presynaptic blocker [21], acting selectively and irreversibly on the release of acetylcholine from the nerve terminals. It was suggested that toosendanin binding sites should exist in presynaptic nerve terminals [22]. Studies on the action potentials and contractile force in guinea pig papillary muscles showed that toosendanin selectively inhibits the inward rectifier K^+current with a positive inotropic effect, resulting from a delay in Ca^{2+}channel inactivation secondary to a delay in ventricular repolarization [23]. In the absence of extracellular Ca^{2+}, toosendanin enhanced noradrenaline (norepinephrine) release through the liberation of intracellular Ca^{2+} stores from rat hippocampal slices [24]. Toosendanin also inhibited the delayed rectifier K^+ current in NG108-15 differentiated neuroblastoma pluglioma hybrid cells [25]. It also increased intracellular free-Ca^{2+}concentrations in NG108-15 cells; such enhancement was due to the influx of extracellular $Ca^{2+}$and related to L-type Ca^{2+} channels [26]. Toosendanin also acted as an agonist of L-type Ca^{2+}channels in neonatal rat ventricular cells [27].

In PC12 rat pheochromocytoma cells, toosendanin was found to potently promote cell differentiation and the outgrowth of neuronal processes at concentrations of 0.1–1 μM in culture medium after 24–48 h. This effect was mediated by the activation of protein kinase A and extracellular signal-regulated kinases [28]. Toosendanin also induced cell apoptosis; the number of apoptotic cells was increased significantly when the incubation time in the toosendanin-containing medium was prolonged [29]. The apoptosis-inducing effect of toosendanin in PC12 cells was associated with shrinkage of the cytosol, the condensation and fragmentation of nuclei, and the formation of a DNA ladder. Toosendanin was also shown to decrease PC12 cell viability in a time- and concentration-dependent manner, and to cause the release of cytochrome c from the mitochondria into the cytosol, which then led to the activation of caspase [30].

In human cancer cell lines, toosendanin significantly suppressed cell proliferation. Typically, the IC_{50} values were less than 0.2 μM, but U937 tumor cells were the most sensitive with an IC_{50} of 5.4 nM. Flow cytometric analysis revealed that the treatment of U937 cells with toosendanin resulted in a concentration- and time-dependent accumulation of cells in S phase. Cell apoptosis in U937 cells was evidenced by the typical condensed and fragmented nuclei, DNA fragmentation, and exposure of phosphatidylserine on the outer leaflet of the plasma membrane [31].

References

1 Guo, F., Hou, Y.G., Zhu, N.J., and Fu, H. (1984) Crystal structure of chuanliansu $C_{30}H_{38}O_{11}\cong 3H_2O$. *Jiegou Huaxue*, **3**, 91–94.

2 Xie, J.X. and Yuan, A.X. (1985) Molecular structure of isochuanliansu isolated from traditional Chinese medicine – the bark of *Melia toosendan* and *Melia azedarach*. *Acta Pharm. Sin.*, **20**, 188–192.

3 Takeya, K., Qiao, Z.S., Hirobe, C., and Itokawa, H. (1996) Cytotoxic azadirachtin-type limonoids from *Melia azedarach*. *Phytochemistry*, **42**, 709–712.

4 Nakatani, M., Huang, R.C., Okamura, H., Iwagawa, T., Tadera, K., and Naoki, H. (1995) Three new antifeeding meliacarpinins from Chinese *Melia azedarach* Linn. *Tetrahedron*, **51**, 11731–11736.

5 Takeya, K., Quio, Z.S., Hirobe, C., and Itokawa, H. (1996) Cytotoxic trichilin-type limonoids from *Melia azedarach*. *Bioorg. Med. Chem.*, **4**, 1355–1359.

6 Saxena, M. and Srivastava, S.K. (1986) A new limonoid glycoside from the stem bark of *Melia azedarach* Linn. *Indian J. Chem.*, **25**, 1087–1088.

7 Nakanishi, T., Inada, A., Nishi, M., Miki, T., Hino, R., and Fujiwara, T. (1986) The structure of a new natural apotirucallane-type triterpene

and the stereochemistry of the related terpene. X- ray and carbon-13 NMR spectral analyses. *Chem. Lett.*, 69–72.

8 Inada, A., Konishi, M., and Nakanishi, T. (1989) Phytochemical studies on meliaceous plants. V. Structure of a new apotirucallane-type triterpenen, 21-*O*-methyltoosendan-pentol from fruits of *Melia toosendan* Sieb. et Zucc. *Heterocycles*, **28**, 383–387.

9 Srivastava, S.K. and Mishra, M. (1985) New anthraquinone pigments from the stem bark of *Melia azedarach* Linn. *Indian J. Chem.*, **24**, 793–794.

10 Zou, J., Jia, G.R., and He, X.Y. (1982) Pharmacokinetic study of toosendanin. *Chin. Trad. Herbal Drugs*, **13**, 408–410.

11 Li, P.Z., Shi, X.C., Xu, Z.H., Li, J.F., Sun, G.Z., and Qin, B.Y. (1982) Pharmacological and toxicological studies on toosendanin. *Chin. Trad. Herbal Drugs*, **13**, 29–32.

12 Shi, Y.L. and Wang, W.P. (2006) Biological effects of toosendanin, an active ingredient of herbal vermifuge in Chinese traditional medicine. *Sheng Li Xue Bao*, **58**, 397–406.

13 Shi, Y.L. and Li, M.F. (2007) Biological effects of toosendanin, a triterpenoid extracted from Chinese traditional medicine. *Prog. Neurobiol.*, **82**, 1–10.

14 Zhao, C.X. (1984) Effect of *Melia azedarach* extract on schistosomiasis in mice. *Chin. Med. J. (Engl.)*, **97**, 910–912.

15 Shi, Y.L. and Wang, Z.F. (2004) Cure of experimental botulism and antibotulismic effect of toosendanin. *Acta Pharmacol. Sin.*, **25**, 839–848.

16 Zhou, J.Y., Wang, Z.F., Ren, X.M., Tang, M.Z., and Shi, Y.L. (2003) Antagonism of botulism toxin type A-induced cleavage of SNAP-25 in rat cerebral synaptosome by toosendanin. *FEBS Lett.*, **555**, 375–379.

17 Li, M.F. and Shi, Y.L. (2006) Toosendanin interferes with pore formation of botulinum toxin type A in PC12 cell membrane. *Acta Pharmacol. Sin.*, **27**, 66–70.

18 Xu, Y. and Shi, Y. (1993) Action of toosendanin on the membrane current of mouse motor nerve terminals. *Brain Res.*, **631**, 46–50.

19 Wang, Z.F. and Shi, Y.L. (2001) Inhibition of large-conductance Ca^{2+}-activated K^+ channels in hippocampal neurons by toosendanin. *Neuroscience*, **104**, 41–47.

20 Cui, Z.J. and He, X.H. (2002) The pre-synaptic blocker toosendanin does not inhibit secretion in exocrine cells. *World J. Gastroenterol.*, **8**, 918–922.

21 Wang, Z.F. and Shi, Y.L. (2001) Modulation of inward rectifier potassium channel by toosendanin, a presynaptic blocker. *Neurosci. Res.*, **40**, 211–215.

22 Shen, G.G., Zhuo, X.L., and Shi, Y.L. (1994) The binding sites of [^{3}H]toosendanin in rat cerebral cortex homogenate. *Acta Physiol. Sin.*, **46**, 546–552.

23 Gao, X.D., Tang, S.B., Lu, J., and Shi, Y.L. (1994) Effects of toosendanin on electric and mechanical properties of guinea pig papillary muscles. *Acta Pharmacol. Sin.*, **15**, 147–151.

24 Hu, H.Y., Zhou, C.W., and Shi, Y.L. (1996) Toosendanin facilitates [^{3}H]noradrenaline release from rat hippocampal slices. *Nat. Toxins*, **4**, 92–95.

25 Hu, Q., Huang, F., and Shi, Y. (1997) Inhibition of toosendanin on the delayed rectifier potassium current in neuroblastoma x glioma NG108-15 cells. *Brain Res.*, **751**, 47–53.

26 Xu, T.H., Ding, J., and Shi, Y.L. (2004) Toosendanin increases free-Ca^{2+} concentration in NG108-15 cells via L-type Ca^{2+} channels. *Acta Pharmacol. Sin.*, **25**, 597–601.

27 Li, M.F. and Shi, Y.L. (2004) Toosendanin, a triterpenoid derivative, acts as a novel agonist of L-type Ca^{2+} channels in neonatal rat ventricular cells. *Eur. J. Pharmacol.*, **501**, 71–78.

28 Yu, J.C., Min, Z.D., and Ip, N.Y. (2004) Melia toosendan regulates PC12 cell differentiation via the activation of protein kinase A and extracellular signal-regulated kinases. *Neurosignals*, **13**, 248–257.

29 Tang, M.Z., Wang, Z.F., and Shi, Y.L. (2003) Toosendanin induces outgrowth of neuronal processes and apoptosis in PC12 cells. *Neurosci. Res.*, **45**, 225–231.

30 Tang, M.Z., Wang, Z.F., and Shi, Y.L. (2004) Involvement of cytochrome c release and caspase activation in toosendanin-induced PC12 cell apoptosis. *Toxicology*, **201**, 31–38.

31 Zhang, B., Wang, Z.F., Tang, M.Z., and Shi, Y. L. (2005) Growth inhibition and apoptosis-induce effect on human cancer cells of toosendanin, a triterpenoid derivative from Chinese traditional medicine. *Invest. New Drugs*, **23**, 547–553.

Menispermum dauricum

Rhizoma Menispermi (Beidougen) is the dried rhizome of *Menispermum dauricum* DC. (Menispermaceae). It is used as an analgetic and antiphlogistic agent for the treatment of laryngopharyngeal diseases, diarrhea, and rheumatic pain.

Menispermum dauricum DC.

A galenic preparation, "Beidougen Pian," the tablets of the total alkaloids extracted from Rhizoma Menispermi, is officially listed in the Chinese Pharmacopoeia. It is used as an antipyretic, expectorant, and mucolytic agent for the treatment of laryngopharyngeal diseases, tonsillitis, and chronic bronchitis. The tablets "Beidougen Pian" contain 15 mg or 30 mg of the total alkaloid, calculated as dauricine.

Chemistry

The rhizome of *M. dauricum* is known to contain alkaloids as the active principles, with the bisbenzyltetrahydroisoquinoline-type alkaloid dauricine as the major component. The two benzyltetrahydroisoquinoline moieties in dauricine are connected via an ether bridge between the benzyl moieties. Related alkaloids derived by partial O-demethylation of dauricine in the rhizome were identified as dauricinoline, dauricoline, daurinoline, daurisoline, *N*-demethyldauricine, and dauriciline [1, 2]. The Chinese Pharmacopoeia requires a quantitative determination of total alkaloids in "Beidougen Pian" tablets by acidimetric titration. The content of total alkaloids in each tablet should be 90–110% of the given dose.

Minor alkaloids from the rhizome of *M. dauricum*, other than the bisbenzyltetrahydroisoquinoline-type, were identified as dimethylcorytuberine, isocorydine, menisperine, methylmenisperine, magnoflorine, sinomenine, cheilanthifoline, stepharine, stepholidine, bianfugecine, bianfugedine,

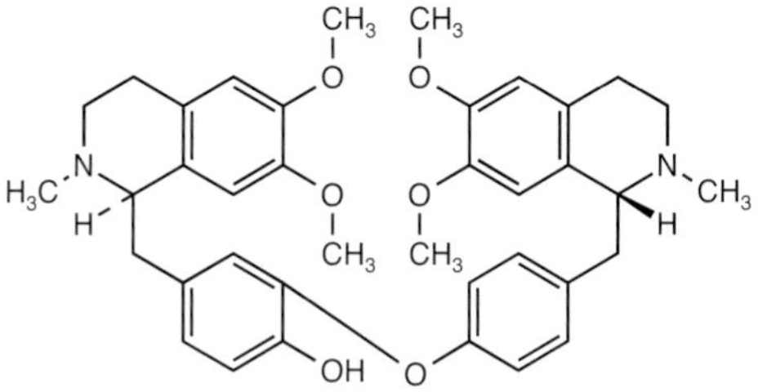

Dauricine

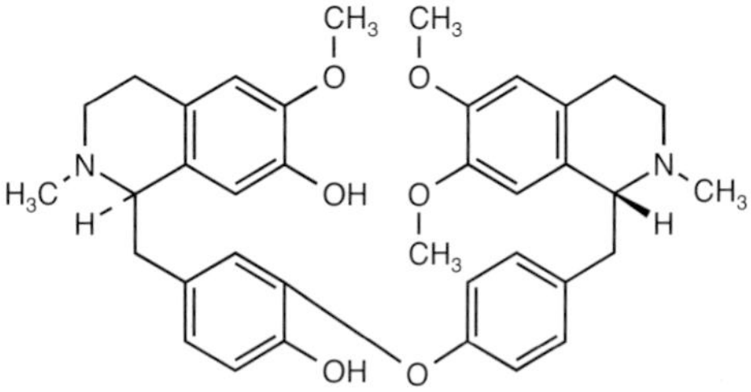

Daurisoline

Handbook of Chinese Medicinal Plants: Chemistry, Pharmacology, Toxicology
Weici Tang and Gerhard Eisenbrand

ISBN: 978-3-527-32226-8

bianfugenine, menisporphine, and dauriporphinoline [3–5].

line alkaloids tetrandrine, fangchinolin and berbamine as the active principles; these

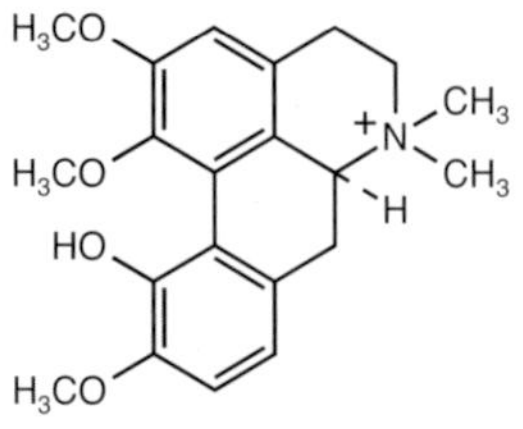

Menisperine

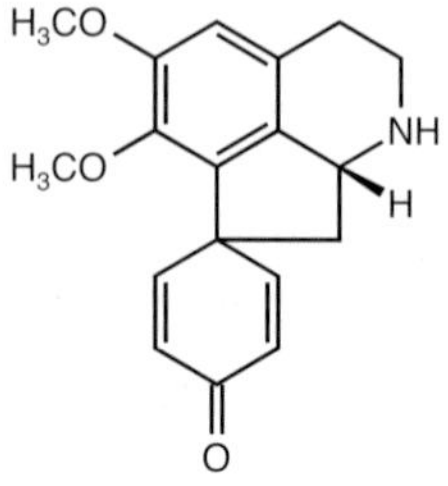

Stepharine

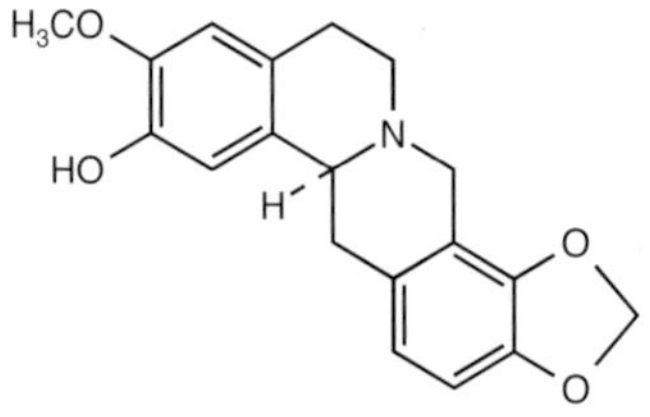

Cheilanthifoline

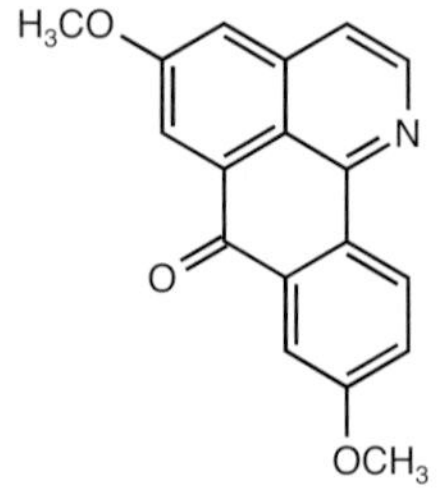

Bianfugecine

Five important items with menispermaceous plants are listed in the Chinese Pharmacopoeia: Herba Cissampelotis; Rhizoma Menispermi; Radix Stephaniae Tetrandrae; Caulis Sinomenii; and Radix Tinosporae. Herba Cissampelotis (Yahunu) is the dried whole plant of *Cissampelos pareira* L. var. *hirsuta* (Buch. ex DC.) Forman. Radix Stephaniae Tetrandrae (Fangji) is the dried root of *Stephania tetrandra* S. Moore. Caulis Sinomenii (Qingfengteng) is the dried stem of *Sinomenium acutum* (Thunb.) Rehd. et Wils. or *Sinomenium acutum* (Thunb.) Rehd. et Wils. var. *cinereum* Rehd. et Wils. Radix Tinosporae (Jinguolan) is the dried tuber of *Tinospora capillipes* Gagnep. or *Tinospora sagittata* (Oliv.) Gagnep. The most important alkaloids related to the biological activities from the rhizome of *M. dauricum* are the bisbenzylisoquinoline alkaloids dauricine and daurisoline with one ether bridge. In contrast, the root of *Stephania tetrandra* contains the bisbenzylisoquinoline alkaloids tetrandrine, fangchinolin and berbamine as the active principles; these have two ether bridges between the benzylisoquinoline moieties. The whole plant of *Cisammpelos pareira* var. *hirsuta* contains bisbenzylisoquinoline alkaloids having a tubocurarane skeleton (such as hayatine and cycleanine) as the active principles. The most important alkaloid in the stem of *Sinomenium acutum* is sinomenine, an alkaloid with morphinan skeleton.

Pharmacology and Toxicology

The oral administration of dauricine to rats at a daily dosage up to 300 mg kg^{-1} for two to three months, or 600 mg kg^{-1} for 18 days, caused no significant changes in cardiac function; however, dose-dependent damage to the liver was observed at a dose level of 150 mg kg^{-1} for two to three months, and mild renal damage at 300 mg kg^{-1} [6]. Teratogenic effects of dauricine were reported on the development and behavior of the offspring following prenatal treatment of rats with an oral dose >80 mg kg^{-1} [7].

At 1 h after a single oral dose of dauricine to rats at 150 mg kg^{-1}, approximately 50% of the dose had been absorbed from the gastrointestinal tract and distributed rapidly into the organs, including liver, lungs, kidneys, spleen, and brain, indicating that dauricine crossed the blood– brain barrier. The dauricine contents of the liver, lungs and brain were maximal at 1 h after administration, whereas peak levels occurred in the spleen only after 12 h. Dauricine concentrations were relatively high in the liver, spleen and kidneys, but low in the brain [8]. The pharmacokinetics of intravenously administered [^{3}H]dauricine in rats fitted an open, two-compartment model with $t_{1/2\alpha}$ of about 0.3 h and $t_{1/2\beta}$ of about 3 h. After 72 h, the urinary and fecal radioactivity each accounted for only about 10% of the total [^{3}H]dauricine [9]. The pharmacokinetic parameters of intravenously administered dauricine in rabbits with acute renal failure induced by $HgCl_2$ did not differ from those of normal rabbits; however, the elimination of dauricine in rabbits with acute liver injury induced by carbon tetrachloride (CCl_4) was prolonged compared to that of normal rabbits [10]. The liver was found to be the major organ for dauricine metabolism; the major metabolite of dauricine was *N*-desmethyldauricine [11].

Following intravenous (i.v.) administration, the LD_{50} of daurisoline methylbromide in mice was reported as 1.3 mg kg^{-1}. In rabbits, the mean dose required to induce respiratory paralysis was 3.7 mg kg^{-1}. Both, neostigmine and calcium glucuronate, were antagonistic to the effects of daurisoline methylbromide, while tubocurarine was synergistic [12]. The concentration profiles of daurisoline in rabbit plasma after an i.v. injection can be described by an open, two-compartment model [13].

Dauricine was reported to be an ion channel blocker. Investigations of the effects of dauricine on the action potentials and slow inward currents of guinea pig ventricular papillary muscles showed dauricine to exert an inhibitory effect on Na^+, K^+, and Ca^{2+} channels [14, 15]. The binding site of dauricine in the sodium channel was the gate-related receptor [16]. In guinea pig cardiomyocytes, dauricine inhibited the L-type calcium current [17]. A study on the binding activity of bisbenzylisoquinoline alkaloids at specific dopaminergic binding sites to rat striatic membranes showed the most active compounds to have only one ether bridge. An analysis of three-dimensional representations also indicated that the bisbenzylisoquinoline alkaloids with one ether bridge might be related to strong activity due to their stereochemistry [18].

Dauricine was also reported to exhibit antiarrhythmic and hypotensive actions. It markedly protected myocardial infarction and antagonized acute ischemic arrhythmia in dogs [19] and cats [20]. Dauricine exerted an antagonistic effect on CsCl-induced early after-depolarization, and triggered arrhythmia in rabbit heart *in vivo* [21]. It was postulated that the antiarrhythmic mechanisms of dauricine might include an influence on arachidonic acid metabolism in platelets [22] and the reduction of conduction in the ischemic zone [23]. Clinical studies also confirmed the antiarrhythmic efficacy of dauricine [24]. Intravenously injected dauricine induced hypotension in anesthetized cats in a dose-dependent manner; such an effect might be caused partly by arterial dilation [25].

Dauricin was also found to protect the rat brain from ischemic damage following focal cerebral ischemia/reperfusion. The neuroprotective effect of dauricine might be partly due to an inhibition of the acute inflammation process induced by ischemia/reperfusion [26]. The pretreatment of male rats with dauricine by intragastric gavage at doses of 42 and 84 mg kg^{-1} twice daily for three days resulted in neuroprotection against transient focal cerebral ischaemia induced by a 90 min temporary occlusion of the middle cerebral artery. Dauricine significantly diminished DNA fragmentation, reduced cytochrome c release, and inhibited the activation of caspase-9 and caspase-3 in the penumbra; this suggested that the infarct-reducing effects of dauricine might be due partly to an inhibition of cell apoptosis via a mitochondrial pathway [27]. In primary cultures of rat cortical neurons, dauricine was found to inhibit not only the increase in intracellular Ca^{2+} concentration but also the decrease in mitochondrial membrane potential induced by hypoxia and hypoglycemia [28].

Dauricine was found to inhibit rat platelet aggregation induced by arachidonic acid and ADP, and to inhibit human platelet aggregation induced by arachidonic acid, ADP and adrenaline *in vitro* [22], as well as by collagen, arachidonic acid and ADP *in vitro* and *in vivo* [29]. Dauricine was also reported to inhibit platelet activating factor (PAF) release from mouse peritoneal macrophages induced by calcimycin [30]. It prevented experimental thrombosis formation, most likely due to an inhibition of platelet adhesion and platelet aggregation [31], and due to increased prostaglandin-I_2 (PGI_2) and nitric oxide (NO) levels [32]. Dauricine was further reported to inhibit the redistribution of platelet membrane glycoprotein IV, and to inhibit the release of intracellular α-granule thrombospondin induced by thrombin [33].

Dauricine inhibited the cyclooxygenase (COX) pathway metabolites of arachidonic acid, but did not affect the plasma concentration of 6-keto-prostaglandin (6-keto-$PGF_{1\alpha}$) in patients; neither did it influence cAMP levels in the platelets and plasma of rabbits [34]. An *in vitro* treatment of bovine anterior cerebral arterial smooth muscle cells with dauricine resulted in a significant decrease in levels of thromboxane B_2 (TXB_2) and 6-keto-$PGF_{1\alpha}$. Stimulating effects on the production of TXB_2 and 6-keto-$PGF_{1\alpha}$ by leukotriene C_4, leukotriene D_4, or PAF in smooth muscle cells were also markedly inhibited by pretreatment with dauricine [35]. Dauricine also inhibited the production of PAF in bovine cerebral microvascular endothelial cells [36]. Furthermore, the stimulatory effects of leukotrienes and PAF on DNA synthesis, and the proliferation of bovine cerebral microvascular smooth muscle cells, were each inhibited by dauricine [37].

Daurisoline is the didemethyl derivative of dauricine, and may be used for the treatment of epilepsy, hypertension, and asthma. It was found to be a calcium channel blocker, and especially blocked the P-type calcium channels [38]. Daurisoline was reported to inhibit the increase of cytosolic free Ca^{2+} in fetal rat cerebral cells induced by Ca^{2+} agonists, especially by glutamate [39]. Studies on the effects of daurisoline on Ca^{2+} uptake and γ-aminobutyric acid (GABA) release from synaptosomes of rat cerebral cortex, and on the contractile activity of rat aorta, revealed that daurisoline produced a concentration-related inhibition of Ca^{2+} uptake and GABA release stimulated by high K^+ concentrations. Daurisoline was believed to be a potent blocker of Ca^{2+} channels in neurons [40]. An excessive accumulation of glutamate in the extracellular space is known to be toxic to neurons. The significantly increased output of glutamate under *in vitro* ischemic conditions might be effectively lowered by daurisoline [41]. Daurisoline was also found to potently inhibit both glutamate release and Ca^{2+} influx in the hippocampus of rats [42].

Daurisoline and its related alkaloids were reported to act as potent calmodulin antagonists, and to inhibit calmodulin-dependent cyclic nucleotide phosphodiesterases. Lipophilic substituents increased the binding affinities to calmodulin and also increased the activity [43]. Daurisoline inhibited platelet aggregation induced by ADP [44]. An antiarrhythmic activity of daurisoline in guinea pigs was also reported [45]. Protection by daurisoline of cultured hippocampal neurons against glutamate toxicity was reported to be due to a reduction in NO production [46].

References

1 Pan, X.P. (1992) A new alkaloid from *Menispermum dauricum* DC: *N*-desmethyldauricine. *Acta Pharm. Sin.*, **27**, 788–791.

2 Pang, Y.P., Chen, Y.W., Li, X.J., and Long, J.G. (1991) A new alkaloid of *Menispermum dauricum* DC; dauriciline. *Acta Pharm. Sin.*, **26**, 387–390.

3 Hu, S.M., Xu, S.X., Yao, X.S., Cui, C.B., Tezuka, Y., and Kikuchi, T. (1993) Dauricoside, a new glycosidal alkaloid having an inhibitory activity against blood-platelet aggregation. *Chem. Pharm. Bull. (Tokyo)*, **41**, 1866–1868.

4 Zhao, S.X., Ye, W.C., Tan, N.H., Zhao, H.R., and Xia, Z.C. (1989) A novel oxoisoaporphine alkaloid from the rhizome of *Menispermum dauricum*. *J. China Pharm. Univ.*, **20**, 312.

5 Takani, M., Takasu, Y., and Takahashi, K. (1983) Studies on constituents of medicinal plants. XXIII. Constituents of the vines of *Menispermum dauricum* DC. (2). *Chem. Pharm. Bull. (Tokyo)*, **31**, 3091–3093.

6 Chen, Y.X., Zhang, Y., Dong, D.C., Cai, Q.Y., Xiong, X.K., Hu, C.J., Leng, D.M., Guo, L.Y., and Zhou, Z.H. (1986) Subacute toxicity of dauricine in rats. *Acta Univ. Med. Tongji*, **15**, 133–136.

7 Wang, G.Z., Jiang, Y., and Hu, C.J. (1992) Behavioral teratogenicity of phenolic alkaloid of *Menispermum dauricum* in rats. *Chin. J. Pharmacol. Toxicol.*, **6**, 157–158.

8 Dai, Z.S., Hou, S.X., Guo, L.Y., and Hu, C.J. (1983) Absorption, distribution and excretion of dauricine in rats. *Chin. Pharm. Bull.*, **18**, 278–279.

9 Dai, Z.S., Yi, M.G., Li, J., and Hu, C.J. (1983) Study on the pharmacokinetics of ^{3}H-dauricine. *Acta Acad. Med. Wuhan*, **12**, 290–291.

10 Gui, Y.T., Du, Z.H., Zeng, F.D., and Hu, C.J. (1993) First-pass effect of dauricine. *Acta Pharmacol. Sin.*, **14**, 173–175.

11 Chen, S., Liu, L., Yang, Y., Dai, Z., and Zeng, F. (2000) Metabolism of dauricine and identification of its main metabolites. *J. Tongji Med. Univ.*, **20**, 253–256.

12 Gong, T., and Wu, Z.Y. (1979) Pharmacological effects of daurisoline methyl bromide, a preliminary report. *Acta Pharm. Sin.*, **14**, 439–442.

13 Gu, S.F., Shi, S.J., and Chen, H. (2003) Determination of daurisoline in rabbit plasma and study on its pharmacokinetics. *Yao Xue Xue Bao*, **38**, 908–910.

14 Li, S.N. and Zhang, K.Y. (1992) Effects of dauricine on action potentials and slow inward currents of guinea pig ventricular papillary muscles. *Acta Pharmacol. Sin.*, **13**, 535–537.

15 Qian, J.Q. (2002) Cardiovascular pharmacological effects of bisbenzylisoquinoline alkaloid derivatives. *Acta Pharmacol. Sin.*, **23**, 1086–1092.

16 Wu, Y.J. and Fang, D.C. (1992) Mechanisms of action of sodium channel blocker: gate-related receptor hypothesis. *Sci. China*, **35**, 1222–1231.

17 Guo, D.L., Zhou, Z.N., Zeng, F.D., and Hu, C. J. (1997) Dauricine inhibited L-type calcium current in single cardiomyocyte of guinea pig. *Acta Pharmacol. Sin.*, **18**, 419–421.

18 Cortes, D., Figadere, B., Saez, J., and Protais, P. (1992) Displacement activity of bisbenzylisoquinoline alkaloids at striatal ^{3}H-SCH 23390 and ^{3}H-raclopride binding sites. *J. Nat. Prod.*, **55**, 1281–1286.

19 Zhu, J.Q., Zeng, F.D., and Hu, C.J. (1992) Protective and anti-arrhythmic effects of dauricine and verapamil on acute myocardial infarction in anesthetized dogs. *Acta Pharmacol. Sin.*, **13**, 249–251.

20 Zhu, J.Q., Zeng, F.D., and Hu, C.J. (1991) Effects of dauricine on the electrophysiological characteristics and arrhythmia induced by acute coronary artery ligation and reperfusion in anesthetized cats. *Chin. Pharm. Bull. (Tokyo)*, **7**, 23–26.

21 Xia, J.S., Tu, H., Li, Z., and Zeng, F.D. (1999) Dauricine suppressed CsCl-induced early afterdepolarizations and triggered arrhythmias in rabbit heart *in vivo*. *Acta Pharmacol. Sin.*, **20**, 513–516.

22 Tong, L. and Yue, T.L. (1989) Effect of dauricine on rat and human platelet aggregation and metabolism of arachidonic acid in washed rat platelets. *Acta Pharm. Sin.*, **24**, 85–88.

23 Zhu, J.Q., Zeng, F.D., and Hu, C.J. (1990) Effects of dauricine on transmembrane potential of ischemic and non-ischemic

Purkinje fibers and ventricular muscles from infarcted canine hearts. *Acta Pharmacol. Sin.*, **11**, 506–509.

24 Zeng, W.C., Zeng, F.D., Leng, D.M., Hu, C.J., Liu, Y.W., and Mao, H.Y. (1990) Cardiac electrophysiologic effects of dauricine. *Chin. J. Clin. Pharmacol.*, **6**, 178–182.

25 Zeng, F.D., Zeng, W.C., Leng, D.M., and Hu, C.J. (1984) Effects of dauricine and verapamil on systolic time intervals and blood pressure in anesthetized cats. *Acta Acad. Med. Wuhan*, **13**, 205–208.

26 Yang, X.Y., Jiang, S.Q., Zhang, L., Liu, Q.N., and Gong, P.L. (2007) Inhibitory effect of dauricine on inflammatory process following focal cerebral ischemia/reperfusion in rats. *Am. J. Chin. Med.*, **35**, 477–486.

27 Li, Y.H. and Gong, P.L. (2007) Neuroprotective effects of dauricine against apoptosis induced by transient focal cerebral ischaemia in rats via a mitochondrial pathway. *Clin. Exp. Pharmacol. Physiol.*, **34**, 177–184.

28 Li, Y.H. and Gong, P.L. (2007) Neuroprotective effect of dauricine in cortical neuron culture exposed to hypoxia and hypoglycemia: involvement of correcting perturbed calcium homeostasis. *Can. J. Physiol. Pharmacol.*, **85**, 621–627.

29 Liu, J.T. and Qiu, P.L. (1986) Effect of dauricine on platelet aggregation in rabbits. *J. Xian Med. Univ.*, **7**, 31–34.

30 Zeng, G.Q. and Rui, Y.C. (1990) Inhibitory effect of dauricine on platelet activating factor released from calcimycin-induced mouse peritoneal macrophages. *Acta Pharmacol. Sin.*, **11**, 346–350.

31 Liu, J.T., Qiu, P.L., and Gao, X.L. (1991) Effect of dauricine on experimental thrombosis formation. *J. Xian Med. Univ.*, **12**, 33–35.

32 Kong, X.Y. and Gong, P.L. (2005) Effect of phenolic alkaloids of Menispermum dauricum on thrombosis and platelet aggregation. *Yao Xue Xue Bao*, **40**, 916–919.

33 Guo, T., Zhang, Y.Z., Liu, D.X., Zou, P., and Shen, D. (1999) Dauricine inhibits redistribution of platelet membrane glycoprotein IV and release of intracellular α-granule thrombospondin induced by thrombin. *Acta Pharmacol. Sin.*, **20**, 533–536.

34 Ding, Y.X., Zhou, J., Ye, X.M., Gu, S.F., and Zeng, F.D. (1991) Study of the effect of dauricine on platelet aggregation and its mechanism. *Chin. Pharmacol. Bull.*, **7**, 271–274.

35 Zeng, G.Q. and Rui, Y.C. (1990) Inhibitory effects of dauricine and anisodamine on production of prostaglandins on bovine cerebral arterial smooth muscle cells. *Acta Pharmacol. Sin.*, **11**, 530–533.

36 Sun, D.X., Huang, T.G., Zeng, G.Q., and Rui, Y.C. (1993) Platelet activating factor production in bovine cerebral microvascular endothelial cells and its drug inhibition. *Acta Pharmacol. Sin.*, **14**, 26–30.

37 Zeng, G.Q., Ju, D.W., Sun, D.X., and Rui, Y.C. (1993) Dauricine and anisodamine inhibited leukotrienes- and platelet activating factor-induced DNA synthesis and proliferation of bovine cerebral microvascular smooth muscle cells in culture. *Acta Pharmacol. Sin.*, **14**, 329–331.

38 Waldmeier, P.C., Wicki, P., Frostl, W., Bittiger, H., Feldtrauer, J.J., and Baumann, P.A. (1995) Effects of the putative P-type calcium channel blocker, *R,R*-(−)-daurisoline on neurotransmitter release. *Naunyn Schmiedeberg's Arch. Pharmacol*, **352**, 670–678.

39 Che, J., Zhang, J., Qu, Z., and Peng, X. (1995) Effects of daurisoline on cytosolic free calcium in fetal rat cerebral cells. *Chin. Med. J. (Engl.)*, **108**, 265–268.

40 Lu, Y.M. and Liu, G.Q. (1990) The effects of (−)-daurisoline on Ca^{2+} influx in presynaptic nerve terminals. *Br. J. Pharmacol.*, **101**, 45–48.

41 Lu, Y.M., Lu, B.F., Zhao, F.Q., Yan, Y.L., and Ho, X.P. (1993) Accumulation of glutamate is regulated by calcium and protein kinase C in rat hippocampal slices exposed to ischemic states. *Hippocampus*, **3**, 221–227.

42 Lu, Y.M. and Liu, G.Q. (1991) Effects of l-daurisoline on quinolinic acid-induced Ca^{2+} influx in hippocampus neurons in freely moving rats. *Acta Pharmacol. Sin.*, **12**, 301–304.

43 Sun, Y., Hu, Z.Y., and Xu, L.M. (1990) Studies on the interaction of daurisoline alkaloid derivatives and calmodulin by fluorescence spectroscopy. *Second Messengers Phosphoproteins*, **13**, 51–57.

44 Hu, S.M., Xu, S.X., Yao, X.S., Cui, C.B., Tezuka, Y., and Kikuchi, T. (1993) Dauricoside, a new glycosidal alkaloid having an inhibitory activity against blood-platelet aggregation. *Chem. Pharm. Bull. (Tokyo)*, **41**, 1866–1868.

45 Wang, Z.X., Zhu, J.Q., Zeng, F.D., and Hu, C.J. (1994) Antagonism of daurisoline on arrhythmogenic delayed after depolarization. *Acta Pharm. Sin.*, **29**, 647–651.

46 Liu, J.G., Li, R., and Liu, G.Q. (1999) l-S.R-daurisoline protects cultured hippocampal neurons against glutamate neurotoxicity by reducing nitric oxide production. *Acta Pharmacol. Sin.*, **20**, 21–26.

Mentha haplocalyx

Herba Menthae (Bohe) is the dried, above-ground part of *Mentha haplocalyx* Briq. (Lamiaceae). It is used for the treatment of headache, influenza, upper respiratory tract infection and other epidemic diseases, as well as to treat ulcers in the mouth, rubella, measles, and conjunctivitis.

Mentha haplocalyx Briq.

Oleum Menthae Dementholatum (Bohe-su You) is the partly dementholated essential oil of *Mentha haplocalyx*, obtained from the fresh stem and leaf by steam distillation. It is used as an aromatic and flavoring agent, carminative (to prevent gastrointestinal flatus). It is applied to the skin or mucous membrane to provide a cooling feeling and to relieve pain or other uncomfortable symptoms.

Mentholum (Bohenao) is the pure substance menthol. It has the same uses as Oleum Menthae Dementholatum.

Chemistry

The above-ground part of *M. haplocalyx* is known to contain an essential oil, with menthol, a cyclic monoterpene alcohol, and (−)-menthone, a cyclic monoterpene ketone, as the major components. The Chinese Pharmacopoeia requires a qualitative determination of menthol in Herba Menthae by thin-layer chromatographic comparison with reference substance, and a quantitative determination of essential oil in Herba Menthae. The content of essential oil in Herba Menthae should be not less than 0.8% (ml g^{-1}). The Chinese Pharmacopoeia further requires a quantitative determination of menthol and (−)-menthone in Oleum Menthae Dementholatum by gas chromatography. The content of (−)-menthone in Oleum Menthae Dementholatum should be 18.0–26.0%, and of menthol 28.0–40.0%. The Chinese Pharmacopoeia gives the following physico-chemical parameters for Oleum Menthae Dementholatum: relative density 0.888–0.908; refractive index 1.456–1.466; rotation −17 to −24°.

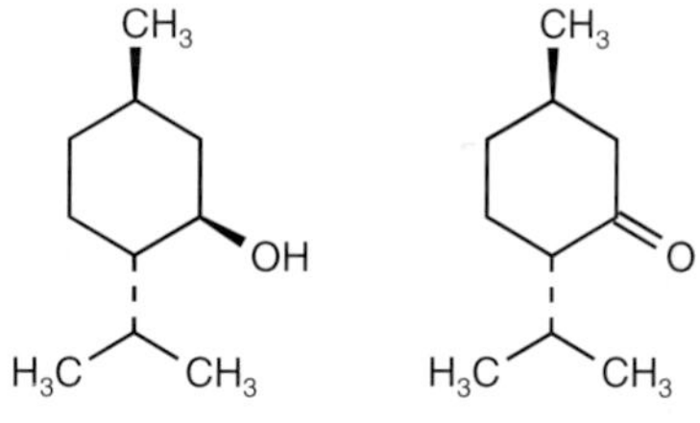

(-)-Menthol (-)-Menthone

Pharmacology and Toxicology

Toxicity studies conducted *in vitro* revealed that menthol, at a concentration of 0.5 m*M*, markedly inhibited the receptor-mediated respiratory stimulation of isolated brown adipocytes, while intracellular mitochondrial functions were unaffected. However, in isolated rat liver mitochondria, menthol (0.5 m*M*) was found to cause an increase in respiratory rate and osmotic swelling,

Handbook of Chinese Medicinal Plants: Chemistry, Pharmacology, Toxicology
Weici Tang and Gerhard Eisenbrand

ISBN: 978-3-527-32226-8

indicating a leakage of the mitochondrial membrane. It was suggested that one effect of menthol was a deterioration of biological membranes [1]. The oral administration of menthol at daily doses of 200, 400, and 800 mg kg^{-1} to rats caused a significant increase in the absolute and relative liver weights, and vacuolization of the hepatocytes. No sign of encephalopathy was observed in rats given menthol. In rats, the no-effect daily level for menthol was assessed as less than 200 mg kg^{-1} [2].

(−)-Menthol did not show mutagenicity in *Salmonella typhimurium* TA97a, TA98, TA100, and TA102 in the presence of an activating system [3]. However, (−)-menthol was found to cause chromosome aberrations in Chinese hamster ovary (CHO) cells over a narrow dose range, with a steeply increasing cytotoxicity [4]. In contrast, menthol was reported to exhibit a chemopreventive effect against the induction of mammary carcinoma in rats caused by 7,12-dimethylbenz[*a*]anthracene (DMBA) in the diet [5]. Menthol also inhibited the growth of viral H-*Ras*-transformed rat liver epithelial cells (WB-ras cells) at concentrations of approximately 1 m*M*. These cells, however, were not necessarily more sensitive to menthol compared to nontransformed and viral *raf*-transformed rat liver epithelial cells [6]. Menthol, when administered simultaneously with an oral dose of 2-aminofluorene to Sprague–Dawley rats, caused an increased excretion of 2-aminofluorene metabolites in the urine and feces; the major metabolite excreted in both the urine and feces was 9-hydroxy-acetylaminofluorene. The major residual metabolite of 2-aminofluorene in the liver was 9-hydroxy-acetylaminofluorene. When 2-aminofluorene was given for 24 h with menthol to Sprague–Dawley rats, the rate of carcinogen acetylation was decreased in the bladder, blood, colon, kidneys and liver [7]. Furthermore, menthol was reported to inhibit the proliferation of human gastric SNU-5 cancer cells, to inhibit topoisomerase I, IIα, and IIβ and to promote the levels of nuclear factor-κB (NF-κB) gene expression [8].

When [3-^{3}H]-(−)-menthol was given orally to male rats at a dose of 500 mg kg^{-1}, the major proportion of the dose (71%) was recovered in 48 h, with approximately equal parts in the urine and feces. In bile duct-cannulated male rats, 74% of the dose was recovered (67% in the bile, 7% in the urine). The major biliary metabolite was menthol glucuronide, which undergoes enterohepatic circulation. The urinary metabolites resulted from hydroxylation at the C-7 methyl group at C-8 and C-9 in the isopropyl moiety; this resulted in a series of mono- and dihydroxy-menthols and carboxylic acids, some of which were excreted in part as glucuronic acid conjugates [9]. Metabolites isolated and characterized from the urine of rats after oral administration at a dose of 800 mg kg^{-1} were identified as *p*-menthane-3,8-diol, *p*-menthane-3,9-diol, 3,8-oxy-*p*-menthane-7-carboxylic acid, and 3,8-dihyroxy-*p*-menthane-7-carboxylic acid with *p*-menthane-3,8-diol and 3,8-dihyroxy-*p*-menthane-7-carboxylic acid as the major urinary metabolites. The repeated oral administration of l-menthol to rats for three days resulted in an increased liver microsomal cytochrome P450 content (by almost 80%). Rat liver microsomes readily converted menthol to *p*-menthane-3,8-diol in the presence of NADPH and O_2. This activity was significantly higher in microsomes obtained from phenobarbital-induced rats [10].

Menthol, when given orally to 12 subjects as a single 100 mg dose in a capsule, was found to be rapidly metabolized, with menthol glucuronide being identified as the only metabolite in the plasma or urine. The average urinaryrecovery of menthol as the glucuronide was 45% [11, 12]. Menthol β-D-glucuronide was found to be a potential prodrug for the colonic delivery of menthol in the large intestine under *in vivo* conditions. Menthol β-D-glucuronide was stable at various pH conditions, from 4.5 to 7.4 over 4 to 24 h at 37°C. At pH 1.5, menthol β-D-glucuronide was slowly hydrolyzed (20%) over a 4 h period. It was stable in the luminal contents of the rat stomach, proximal small intestine, and distal small intestine. In rats, the rate of hydrolysis of menthol β-D-glucuronide was about 6 nmol min^{-1} mg^{-1} in the luminal contents of the cecum, and about 2 nmol min^{-1} mg^{-1} in the luminal contents of the colon. The hydrolysis rate was lower in human feces samples (0.5 nmol min^{-1} mg^{-1}). Menthol β-D-glucuronide had a log octanol/buffer partition coefficient of –1.6, which suggested poor absorption from the lumen of the gastrointestinal tract [13].

The pharmacological effects of the aboveground part and of the essential oil of *M. haplocalyx* are mainly caused by menthol. Menthol is widely used in a variety of commercial products and foods; an example is its use as an anti-inflammatory agent for the treatment of chronic inflammatory disorders such as bronchial asthma, colitis, and allergic rhinitis. It significantly suppressed the production of inflammation mediators such as leukotriene B, prostaglandin E and interleukin 1β (IL-1β) by monocytes from healthy volunteers stimulated by lipopolysaccharides *in vitro* at concentrations of 7–8 μg ml^{-1} [14]. Menthol was reported to be an effective antitussive agent; inhalation caused a reduction of evoked cough in healthy subjects [15]. The antitussive effects of menthol have also been observed in conscious guinea pigs [16].

The use of menthol in dermatology is ubiquitous, where it is frequently part of topical antipruritic, antiseptic, analgesic, and cooling formulations [17]. Transient receptor potential melastatin-8 (TRPM8) has been identified in the cell bodies of sensory neurons as a low-temperature and menthol-activated cation channel, mediating cooling-induced autonomic and behavioral heat-gain responses in experimental animals [18, 19]. The application of menthol to the skin or mucous membrane provided a cooling feeling and relieved pain or other uncomfortable symptoms. Menthol had a selective potentiating action on cutaneous cold receptors, and shifted the temperature response curve towards higher temperatures in animal experiments. The nasal inhalation of menthol reduced respiratory discomfort associated with loaded breathing in normal subjects; this indicated that a stimulation of cold receptors in the upper airway might alleviate the sensation of respiratory discomfort [20]. Similar effects were also observed in an experimental study in guinea pigs [21] and dogs [22]. TRPM8 was found to be dependent on extracellular Ca^{2+} ions, indicating that an elevation of the intracellular Ca^{2+} concentration induced by Ca^{2+} influx caused the desensitization. The cooling-activated TRPM8 induced the Ca^{2+}-dependent protein kinase C (PKC) isoenzymes to desensitize the receptor itself, indicating that menthol-induced desensitization of TRPM8 was mediated by Ca^{2+}-dependent PKC activation [23].

The effects of menthol on rat liver microsomal enzymes involved in the biotransformation of xenobiotic substances have been tested with ethoxyresorufin *O*-deethylase as a marker for the cytochrome P450 isoenzyme CYP1A1, methoxyresorufin *O*-demethylase as a marker for CYP1A2, and pentoxyresorufin *O*-depentilase as a selective marker for CYP2B1. Menthol proved to be an *in vitro* inhibitor of pentoxyresorufin *O*-depentilase, with an IC_{50} of 10.6 μ*M*. The inhibitory effects of menthol on ethoxyresorufin *O*-deethylase and methoxyresorufin *O*-demethylase were observed only at high concentrations [24]. The results from *in vitro* studies also showed that menthol inhibited the microsomal oxidation of nicotine to cotinine, and the CYP2A6-mediated 7-hydroxylation of coumarin [25].

Menthol was found to be a potent inhibitor of K^+ depolarization-induced and electrically stimulated responses in guinea pig ileum smooth muscle (IC_{50} of 8–28 μg ml^{-1}) and electrically stimulated atrial and papillary muscles of rat and guinea pig (IC_{50} of 10–68 μg ml^{-1}). Similar potencies were demonstrated against K^+ depolarization-induced $^{45}Ca^{2+}$ uptake in synaptosomes, and against K^+ depolarization-induced uptake in chick retinal neurons. Menthol also inhibited specific [^{3}H]nitrendipine binding to smooth and cardiac muscle and neuronal preparations. The binding of menthol at a concentration of 78 μg ml^{-1} was competitive against [^{3}H]nitrendipine in both smooth muscle and synaptosome preparations. This indicated that menthol exerted Ca^{2+} channel-blocking properties, and these might underlie its use in the treatment of irritable bowel syndrome [26]. Most of the Ca^{2+} channel blockers exerted analgesic properties. Menthol, when administered orally at doses of 3–10 mg kg^{-1}, produced a dose-dependent increase in the pain threshold in the mouse hot-plate test and abdominal constriction induced by acetic acid. However, (+)-menthol did not show any antinociceptive effect, even at doses of 50 mg kg^{-1} [27].

A pregnancy-terminating effect of *Mentha* oil in rabbits was also reported. *Mentha* oil was applied to one horn of the uterus in rabbits at gestation days 6 and 9. An examination of the uterus at gestation day 12 showed that *Mentha* oil was able to inhibit implantation and interrupt early pregnancy. Contraction of uterus *in vivo* was not stimulated by *Mentha* oil given intrauterinely. However, the effect of oxytocin on the uterus might be synergistic with that of *Mentha* oil. No changes were found in plasma estrogen or progesterone levels before and after administration of *Mentha* oil, but the level of human chorionic gonadotropin (HCG) was decreased significantly. On histological examination, a degeneration and necrosis of the chorion was observed following the administration of *Mentha* oil. It was postulated that chorion damage might represent one of the mechanisms of *Mentha* oil for terminating early pregnancy [28].

References

1 Bernson, V.S. and Pettersson, B. (1983) The toxicity of menthol in short-term bioassays. *Chem. Biol. Interact.*, **46**, 233–246.

2 Thorup, I., Wurtzen, G., Carstensen, F J., and Olsen, P. (1983) Short term toxicity study in rats dosed with puregone and menthol. *Toxicol. Lett.*, **19**, 207–210.

3 Gomes-Carneiro, M.R., Felzenszwalb, I., and Paumgartten, F.J. (1998) Mutagenicity testing (+/−)-camphor, 1,8-cineole, citral, citronellal, (−)-menthol and terpineol with the *Salmonella*/microsome assay. *Mutat. Res.*, **416**, 129–136.

4 Hilliard, C.A., Armstrong, M.J., Bradt, C.I., Hill, R.B., Greenwood, S.K., and Galloway, S.M. (1998) Chromosome aberrations in vitro related to cytotoxicity of nonmutagenic chemicals and metabolic poisons. *Environ. Mol. Mutagen.*, **31**, 316–326.

5 Russin, W.A., Hoesly, J.D., Elson, C.E., Tanner, M.A., and Gould, M.N. (1989) Inhibition of rat mammary carcinogenesis by monoterpenoids. *Carcinogenesis*, **10**, 2161–2164.

6 Ruch, R.J. and Sigler, K. (1994) Growth inhibition of rat liver epithelial tumor cells by monoterpenes does not involve Ras plasma membrane association. *Carcinogenesis*, **15**, 787–789.

7 Lin, J.P., Cheng, K.C., Chen, G.W., Yang, M.D., Chiu, T.H., and Chung, J.G. (2003) Effects of (−)-menthol on the distribution and metabolism of 2-aminofluorene in various

tissues of Sprague-Dawley rats. *In Vivo*, **17**, 441–455.

8 Lin, J.P., Lu, H.F., Lee, J.H., Lin, J.G., Hsia, T. C., Wu, L.T., and Chung, J.G. (2005) (−)-Menthol inhibits DNA topoisomerases I, II α and β and promotes NF-κB expression in human gastric cancer SNU-5 cells. *Anticancer Res.*, **25** (B), 2069–2074.

9 Yamaguchi, T., Caldwell, J., and Farmer, P.B. (1994) Metabolic fate of [^{3}H]-l-menthol in the rat. *Drug Metab. Dispos.*, **22**, 616–624.

10 Madyastha, K.M., and Srivatsan, V. (1988) Studies on the metabolism of l-menthol in rats. *Drug Metab. Dispos.*, **16**, 765–772.

11 Gelal, A., Jacob, P. 3rd, Yu, L., and Benowitz, N. L. (1999) Disposition kinetics and effects of menthol. *Clin. Pharmacol. Ther.*, **66**, 128–135.

12 Kaffenberger, R.M., and Doyle, M.J. (1990) Determination of menthol and menthol glucuronide in human urine by gas chromatography using an enzyme-sensitive internal standard and flame ionization detection. *J. Chromatogr.*, **527**, 59–66.

13 Nolen, H.W. 3rd, and Friend, D.R. (1994) Menthol β-D-glucuronide: a potential prodrug for treatment of the irritable bowel syndrome. *Pharm. Res.*, **11**, 1707–1711.

14 Juergens, U.R., Stober, M., and Vetter, H. (1998) The anti-inflammatory activity of L-menthol compared to mint oil in human monocytes *in vitro*: a novel perspective for its therapeutic use in inflammatory diseases. *Eur. J. Med. Res.*, **3**, 539–545.

15 Morice, A.H., Marshall, A.E., Higgins, K.S., and Grattan, T.J. (1994) Effect of inhaled menthol on citric acid induced cough in normal subjects. *Thorax*, **49**, 1024–1026.

16 Laude, E.A., Morice, A.H., and Grattan, T.J. (1994) The antitussive effects of menthol, camphor and cineole in conscious guinea pigs. *Pulm. Pharmacol.*, **7**, 179–184.

17 Patel, T., Ishiuji, Y., and Yosipovitch, G. (2007) Menthol: a refreshing look at this ancient compound. *J. Am. Acad. Dermatol.*, **57**, 873–878.

18 Tajino, K., Matsumura, K., Kosada, K., Shibakusa, T., Inoue, K., Fushiki, T., Hosokawa, H., and Kobayashi, S. (2007) Application of menthol to the skin of whole trunk in mice induces autonomic and behavioral heat-gain responses. *Am. J. Physiol. Regul. Integr. Comp. Physiol.*, **293**, R2128–R2135

19 Voets, T., Droogmans, G., Wissenbach, F U., Janssens, A., Flockerzi, V., and Nilius, B. (2004) The principle of temperature-dependent gating in cold- and heat-sensitive TRP channels. *Nature*, **430**, 748–754.

20 Nishino, T., Tagaito, Y., and Sakurai, Y. (1997) Nasal inhalation of l-menthol reduces respiratory discomfort associated with loaded breathing. *Am. J. Respir. Crit. Care Med.*, **156**, 309–313.

21 Sekizawa, S., Tsubone, H., Kuwahara, M., and Sugano, S. (1996) Nasal receptors responding to cold and l-menthol airflow in the guinea pig. *Respir. Physiol.*, **103**, 211–219.

22 Sant'Ambrogio, F.B., Anderson, J.W., and Sant'Ambrogio, G. (1992) Menthol in the upper airway depresses ventilation in newborn dogs. *Respir. Physiol.*, **89**, 299–307.

23 Abe, J., Hosokawa, H., Sawada, Y., Matsumura, K., and Kobayashi, S. (2006) Ca^{2+}-dependent PKC activation mediates menthol-induced desensitization of transient receptor potential M8. *Neurosci. Lett.*, **397**, 140–144.

24 De-Oliveira, A.C., Fidalgo-Neto, A.A., and Paumgartten, F.J. (1999) *In vitro* inhibition of liver monooxygenases by beta-ionone, 1,8-cineole, (–)-menthol and terpineol. *Toxicology*, **135**, 33–41.

25 MacDougall, J.M., Fandrick, K., Zhang, X., Serafin, S.V., and Cashman, J.R. (2003) Inhibition of human liver microsomal (S)-nicotine oxidation by (–)-menthol and analogues. *Chem. Res. Toxicol.*, **16**, 988–993.

26 Hawthorn, M., Ferrante, J., Luchowski, E., Rutledge, A., Wei, X.Y., and Triggle, D.J. (1988) The actions of peppermint oil and menthol on calcium channel dependent processes in intestinal, neuronal and cardiac preparations. *Aliment. Pharmacol. Ther.*, **2**, 101–118.

27 Galeotti, N., Di Cesare Mannelli, L., Mazzanti, G., Bartolini, A., and Ghelardini, C. (2002) Menthol: a natural analgesic compound. *Neurosci. Lett.*, **322**, 145–148.

28 Yang, S.J., Lu, Y.F., Wang, Q.J., and Liu, H.Y. (1991) Effect of *Mentha* oil on terminating pregnancy in rabbits and its mechanism. *Chin. Trad. Herbal Drugs*, **22**, 454–457, 478.

Momordica cochinchinensis

Semen Momordicae (Mubiezi) is the dried ripe seed of *Momordica cochinchinensis* (Lour.) Spreng. (Cucurbitaceae). It is used for the treatment of ulcer, mastitis, carbuncle, anal fistula, hemorrhoids, eczema, and neurodermatitis. The Chinese Pharmacopoeia notes that Semen Momordicae should be used with caution in pregnancy.

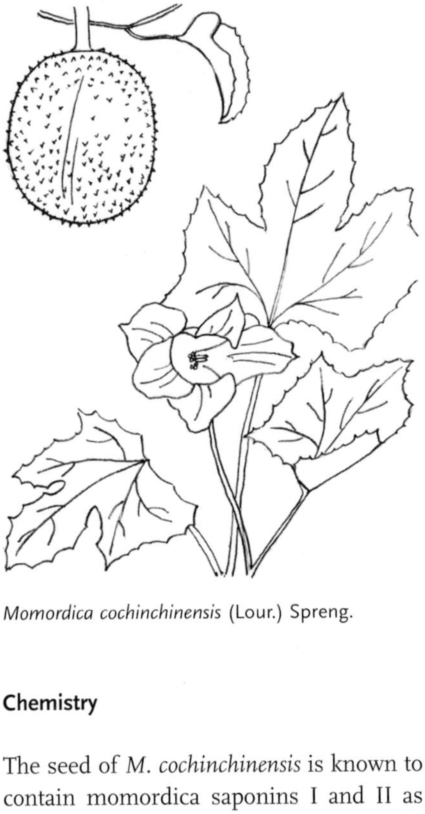

Momordica cochinchinensis (Lour.) Spreng.

Chemistry

The seed of *M. cochinchinensis* is known to contain momordica saponins I and II as the major saponin components [1]. The aglycone of momordica saponin I is gypsogenin, whereas that of momordica saponin II is quillaic acid; both triterpenes are derived from oleanane and are characterized by a carbonyl group at C-23. Momordica saponin II differs from momordica saponin I by an 16α-hydroxy substituent. In addition to oleanolic acid, momordic acid derived from oleanane with an 1-oxo group, was isolated from the seeds of *M. cochinchinensis* [2].

The seeds of *M. conchinchinensis* were found to be rich in fat oil; the content of fat oil in the seeds was 40%. Besides palmitic acid, stearic acid, oleic acid, and linoleic acid, a rare fatty acid, α-eleostearic acid [(*E*,*Z*,*E*)-9,11,13-octadecatrienoic acid], was also identified in the seed oil [3, 4].

As a member of the Cucurbitaceae, *M. cochinchinensis* was also found to contain ribosome-inactivating proteins, as do other members of this plant family, such as *Trichosanthes kirilowii* and *Luffa cylindrica* [5]. Ribosome-inactivating proteins from the seeds of *M. cochinchinensis* were designed as cochinchinin and momorcochin S. Cochinchinin is a single-chain glycoprotein with a molecular weight of 29 kDa and a sugar content of 3.7% [6]. Momorcochin S is a glycoprotein with a molecular weight of about 30 kDa [7]. The isolation of trypsin inhibitors containing 34 amino acid residues with three disulfide bridges and a cyclic polypeptide backbone from the seeds of *M. cochinchinensis* was reported [8, 9].

Pharmacology and Toxicology

Cochinchinin was reported to inhibit protein synthesis in a rabbit reticulocyte lysate, with an ID_{50} of 30 ng ml^{-1}. It was also a very potent inhibitor of protein synthesis in Thy 1.1-positive cells (SL-2) of mice (ID_{50} 3 ng ml^{-1}). The LD_{50} was determined as 16 mg kg^{-1} in mice, by intraperitoneal administration [6]. Momorcochin S also inhibited protein synthesis by a rabbit reticulocyte lysate and phenylalanine polymerization by isolated ribosomes, and altered rRNA in a similar manner as ricin A chain and related proteins. Momorcochin S could be considered as an isoform of momorcochin

Handbook of Chinese Medicinal Plants: Chemistry, Pharmacology, Toxicology
Weici Tang and Gerhard Eisenbrand

ISBN: 978-3-527-32226-8

from the fresh roots of *M. cochinchinensis* [10, 11]. It was linked to a monoclonal antibody against human plasma cells, and the resulting immunotoxin was selectively toxic to target cells [7]. Both, the ribosome-inactivating protein and the immunotoxins, induced apoptosis in the CD30-positive L540 [12], HT29 and A431 [13] cell lines.

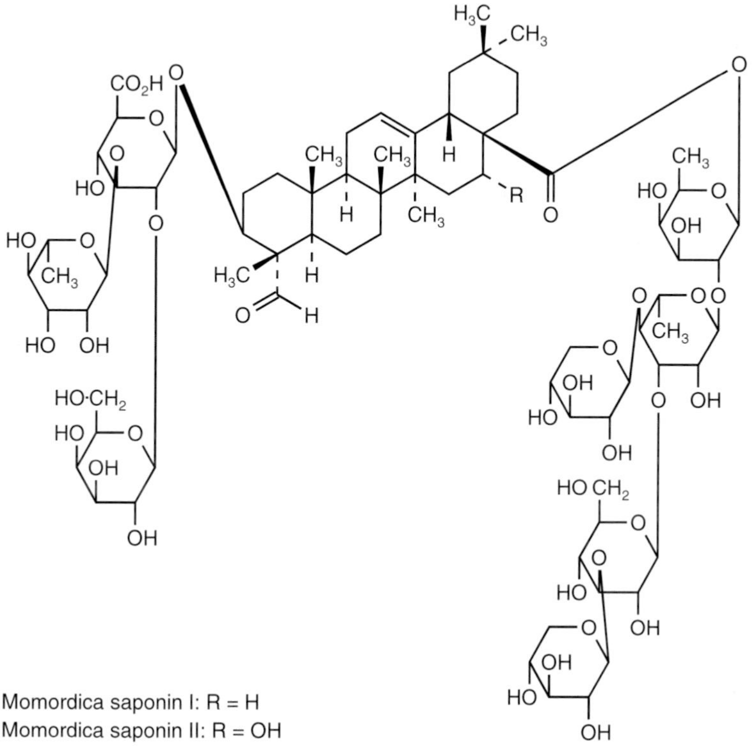

Momordica saponin I: R = H
Momordica saponin II: R = OH

References

1 Iwamoto, M., Okabe, H., Yamauchi, T., Tanaka, M., Rokutani, Y., Hara, S., Mihashi, K., and Higuchi, R. (1985) Studies on the constituents of *Momordica cochinchinensis* Sreng. I. Isolation and characterization of the seed saponins, momordica saponins I and II. *Chem. Pharm. Bull. (Tokyo)*, **33**, 464–478.

2 Murakami, T., Nagasawa, M., Itokawa, F H., Tachi, Y., and Tanaka, K. (1966) The structure of a new triterpene, momordic acid, obtained from *Momordica cochinchinensis* Sprenger. *Tetrahedron Lett.*, **7**, 5137–5140.

3 Huang, M.Q. (1986) Fatty acids in the kernel oil of *Momordica cochinchinensis*. *Guihaia*, **6**, 297–299.

4 Hopkins, C.Y., Chisholm, M.J., and Ogrodnik, J.A. (1969) Identity and configuration of conjugated fatty acids in certain seed oils. *Lipids*, **4**, 89–92.

5 Wang, R.H., Zheng, S., and Shen, B.F. (1993) Screening of single chain ribosome-inactivating proteins from plants. *Junshi Yixue Kexueyuan Yuankan*, **17**, 246–249.

6 Zheng, S., Li, G., and Yan, S.M. (1992) Purification and characterization of cochinchinin. *Acta Biochim. Biophys. Sin.*, **24**, 311–316.

7 Bolognesi, A., Barbieri, L., Carnicelli, D., Abbondanza, A., Cenini, P., Falasca, A.I., Dinota, A., and Stirpe, F. (1989) Purification and properties of a new ribosome-inactivating protein with RNA N-glycosidase activity suitable for immunotoxin preparation from

the seeds of *Momordica cochinchinensis*. *Biochim. Biophys. Acta*, **993**, 287–292.

8 Huang, B., Ng, T.B., Fong, W.P., Wan, C.C., and Yeung, H.W. (1999) Isolation of a trypsin inhibitor with deletion of N-terminal pentapeptide from the seeds of *Momordica cochinchinensis*, the Chinese drug mubiezhi. *Int. J. Biochem. Cell Biol.*, **31**, 707–715.

9 Hernandez, J.F., Gagnon, J., Chiche, L., Nguyen, T.M., Andrieu, J.P., Heitz, A., Trinh Hong, T., Pham, T.T., and Le \Nguyen, F D. (2000) Squash trypsin inhibitors from *Momordica cochinchinensis* exhibit an atypical macrocyclic structure. *Biochemistry*, **39**, 5722–5730.

10 Yeung, H.W., Ng, T.B., Wong, N.S., and Li, W. W. (1987) Isolation and characterization of an abortifacient protein, momorcochin, from root tubers of *Momordica cochinchinensis* (family Cucurbitaceae). *Int. J. Pept. Protein Res.*, **30**, 135–140.

11 Ng, T.B., Chan, W.Y., and Yeung, H.W. (1992) Proteins with abortifacient, ribosome inactivating, immunomodulatory, antitumor and anti-AIDS activities from Cucurbitaceae plants. *Gen. Pharmacol.*, **23**, 579–590.

12 Bolognesi, A., Tazzari, P.L., Olivieri, F., Polito, L., Falini, B., and Stirpe, F. (1996) Induction of apoptosis by ribosome-inactivating proteins and related immunotoxins. *Int. J. Cancer*, **68**, 349–355.

13 Dosio, F., Brusa, P., Crosasso, P., Fruttero, C., Cattel, L., and Bolognesi, A. (1996) Synthesis of different immunotoxins composed by ribosome inactivating proteins non-covalently bound to monoclonal antibody. *Farmaco*, **51**, 477–482.

Momordica grosvenori

Fructus Momordicae (Luohanguo) is the dried fruits of *Momordica grosvenori* Swingle (Cucurbitaceae). It is used for the treatment of cough and pharyngitis, and also as a laxative and a sweetening agent.

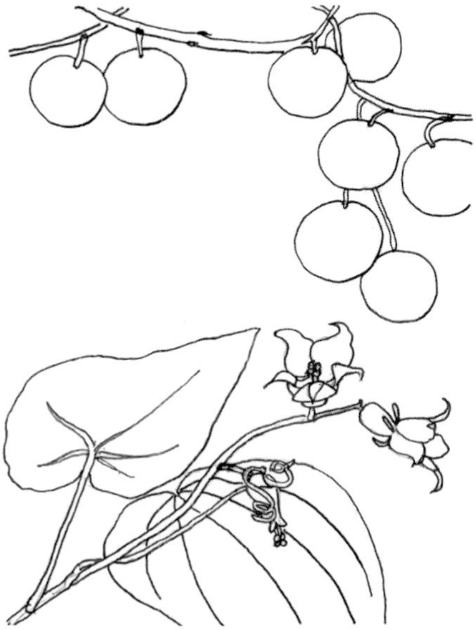

Momordica grosvenori Swingle

Chemistry

The fruit of *M. grosvenori* is well known to have an intense sweet taste. The major component of the sweet fraction was identified as mogroside V, a glycoside containing the aglycone mogrol [1–3]. The concentrations of mogroside V in the fruits were given as 0.8–1.3%. The highest mogroside V levels were found in the endocarp of the fruit rather than in the peel or in the seed [4, 5]. Minor glycosides related to mogroside V were identified as siamenoside I (sweet), 11-oxo-mogroside V (sweet), mogroside II-E, mogroside III-E (both tasteless), mogroside III, and mogroside IV (both sweet). The sweetness relative to sucrose of siamenoside I was estimated to be 560-fold, making this the sweetest of the cucurbitane glycosides isolated so far. The relative sweetness of mogroside V was about 300-fold [6, 7]. All of the minor components differ from mogroside V only in terms of the sugar composition [8].

Flavone glycosides kaempferol 3,7-*O*-di-α-L-rhamnopyranoside and grosvenorin (kaempferol 3-*O*-α-L-rhamnopyranosyl-7-*O*-β-D-glucopyranosyl-(1 → 2)-α-L-rhamnopyranoside) were isolated from the fresh fruits of *M. grosvenori* [9].

Pharmacology and Toxicology

The extract of the dried fruit has been used for the treatment of pharyngitis, cough, and gastrointestinal disorders. It has also been used as a sweetener [8, 10]. The fruit extract was found to be nonmutagenic and nontoxic for mice by oral administration [11]. Antioxidative effects of *M. grosvenori* fruit extract *in vitro* and *in vivo* have been reported [12].

The oral administration of Fructus Momordicae at daily doses of $150\,mg\,kg^{-1}$ or $300\,mg\,kg^{-1}$ for 30 days effectively regulated the immune imbalance in diabetic mice induced by alloxan, principally by up-regulating the $CD4^{+}$ T-lymphocyte subsets, and remodeling the intracellular cytokine profiles (reducing the expression of proinflammatory Th1 cytokines towards a beneficial Th2 pattern). This was ascribed to its induction and up-regulation of heme oxygenase-1, indicating that Fructus Momordicae might modulate an imbalance of the cellular immune system and prevent the progression of diabetes mellitus via induction of heme oxygenase-1 protein expression [13].

Handbook of Chinese Medicinal Plants: Chemistry, Pharmacology, Toxicology
Weici Tang and Gerhard Eisenbrand

ISBN: 978-3-527-32226-8

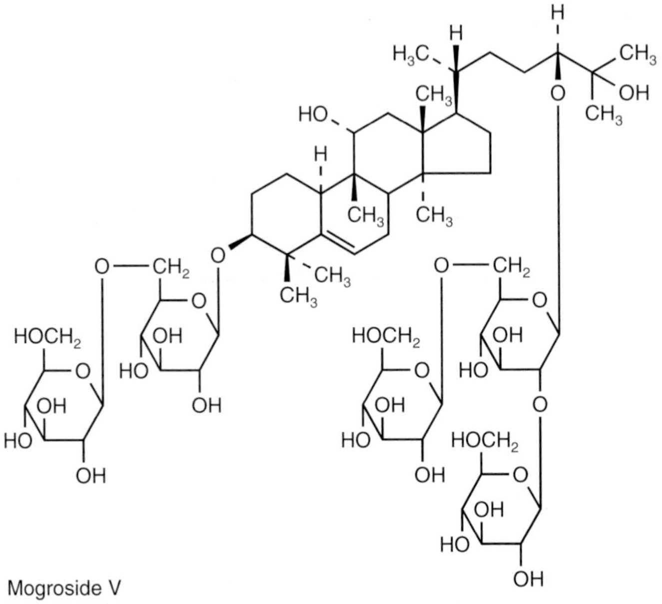

Mogroside V

The sweeteners, mogroside V and 11-oxo-mogroside V, were reported to exhibit significant inhibitory effects on the two-stage carcinogenesis of mouse skin tumors induced by peroxynitrite or 7,12-dimethylbenz[*a*]anthracene (DMBA) as initiator and 12-*O*-tetradecanoylphorbol 13-acetate (TPA) as promoter. They also inhibited the induction of Epstein–Barr virus (EBV) early antigen expression in Raji cells by TPA [14].

References

1 Takemoto, T., Arihara, S., Nakajima, T., and Okuhira, M. (1983) Studies on the constituents of Fructus Momordicae. II. Structure of mogrosides. *Yakugaku Zasshi*, **103**, 1167–1173.

2 Chen, D.H., Si, J.Y., Chang, Q., Li, C.J., and Shen, L.G. (1992) Studies and uses of natural nonsugar sweeteners from Luohanguo (fruit of *Siraitia grosvenorii*). *Tianran Chanwu Yanjiu Yu Kaifa*, **4**, 72–77.

3 Chang, Q., Chen, D.H., Si, J.Y., and Shen, L.G. (1994) NMR spectra of mogroside V. *Bopuxue Zazhi*, **11**, 43–48.

4 Makapugay, H.C., Nanayakkara, N.P.D., Soejarto, D.D., and Kinghorn, A.D. (1985) High-performance liquid chromatographic analysis of the major sweet principle to Lo Han Kuo fruits. *J. Agric. Food Chem.*, **33**, 348–350.

5 Chang, Q., Chen, D.H., Si, J.Y., and Shen, L.G. (1995) Determination of total saponin in *Siraitia grosvenorii*. *China J. Chin. Mater. Med.*, **20**, 554–555.

6 Matsumoto, K., Kasai, R., Ohtani, K., and Tanaka, O. (1990) Minor cucurbitane glycosides from fruits of *Siraitia grosvenori* (Cucurbitaceae). *Chem. Pharm. Bull. (Tokyo)*, **38**, 2030–2032.

7 Li, Y.Q., Wang, W.S., and Wang, C. (1993) Separation and determination of triterpene glucoside from Fructus momordicae. *Shipin Kexue*, **161**, 66–70.

8 Wang, Yaping and Chen, Jianyu (1992) Chemical constituents of grosvenor momordica (*Momordica grosvenori*). *Chin. Trad. Herbal Drugs*, **23**, 61–62.

9 Si, J.Y., Chen, D.H., Chang, Q., and Shen, F L. G. (1994) Isolation and structure determination of flavonol glycosides from the fresh fruits of *Siraitia grosvenori*. *Acta Pharm. Sin.*, **29**, 158–160.

10 Nabors, L.O., and Inglett, G.E. (1986) A review of various other alternative sweeteners. *Food Sci. Technol.*, **17**, 309–323.

11 Kinghorn, A.D., Soejarto, D.D., Katz, N.L., and Kamath, S.K. (1983) Studies to identify, isolate, develop and test naturally occurring noncariogenic sweeteners that may be used as

dietary sucrose substitutes. Report NIDR/CR-85/01; Order No. PB85-158079/GAR, 36 pp. (CA 103: 86674).

12 Wang, X.Y., Liu, J.K., Zhao, Y., Sanada, F M., Okada, S., and Mori, A. (1996) The antioxidant and antistress activities of the extract of Fructus momordica, in *Proceedings International Symposium on Natural Antioxidants: Molecular Mechanisms and Health Effects* (eds L. S Packer, M.G. S Traber and W.J., S Xin), AOCS Press, Champaign, Ill.

13 Song, F., Chen, W., Jia, W., Yao, P., Nussler, A. K., Sun, X., and Liu, L. (2006) A natural sweetener, *Momordica grosvenori*, attenuates the imbalance of cellular immune functions in alloxan-induced diabetic mice. *Phytother Res.*, **20**, 552–560.

14 Takasaki, M., Konoshima, T., Murata, Y., Sugiura, M., Nishino, H., Tokuda, H., Matsumoto, K., Kasai, R., and Yamasaki, K. (2003) Anticarcinogenic activity of natural sweeteners, cucurbitane glycosides, from *Momordica grosvenori*. *Cancer Lett.*, **198**, 37–42.

Morinda officinalis

Radix Morindae Officinalis (Bajitian) is the dried root of *Morinda officinalis* How (Rubiaceae). It is used for treatment of rheumatism, impotence, infertility, and menstrual disorders.

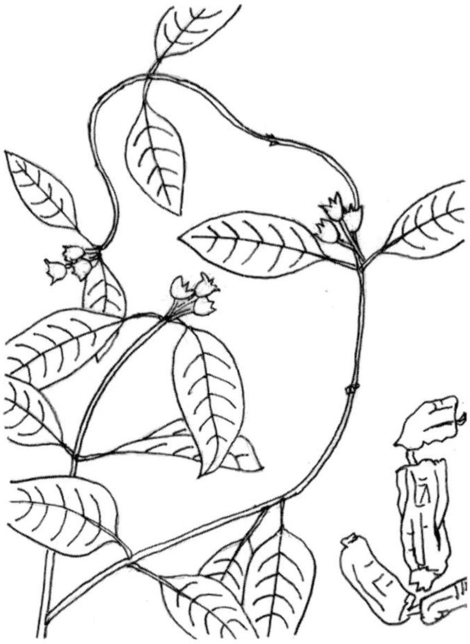

Morinda officinalis How

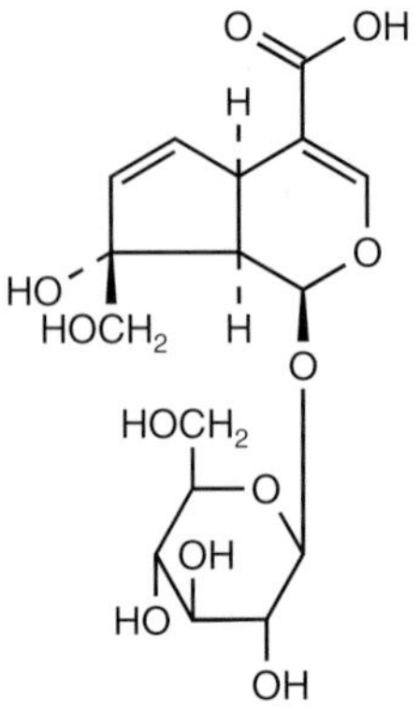

Monotropein

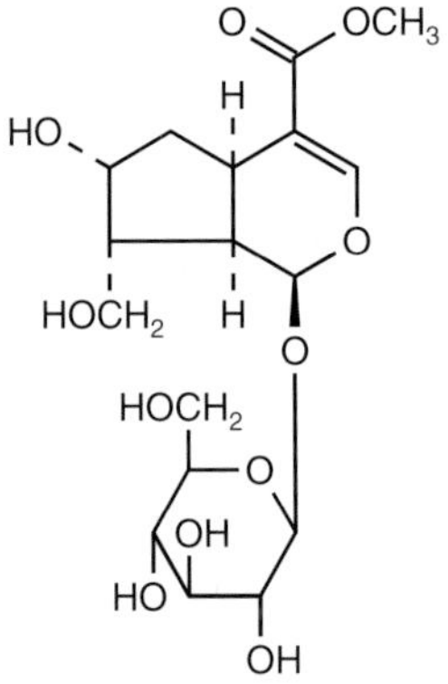

Morofficinaloside

Chemistry

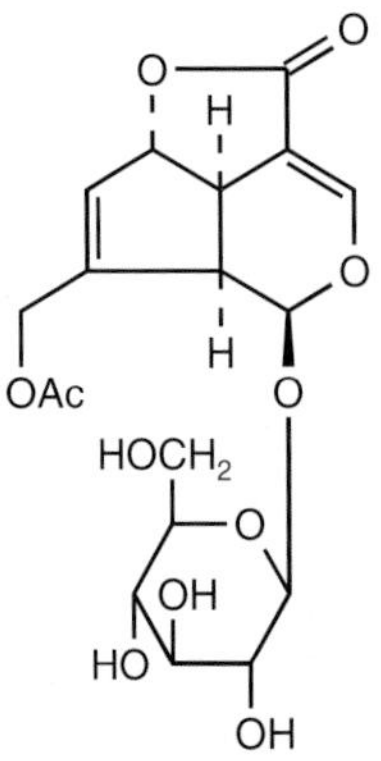

Asperuloside

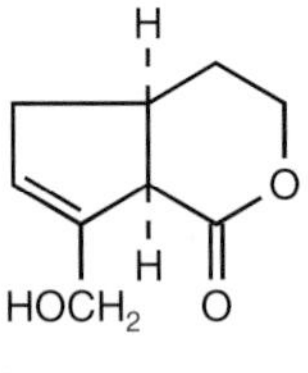

Morindolide

The root of *M. officinalis* was reported to contain anthraquinones, iridoids, triterpenes and phytosterols. Anthraquinone derivatives from the root were identified as rubiadin, rubiadin-1-methyl ether, 1-hydroxyanthraquinone, 1-hydroxy-2-methylanthraquinone, 1,6-dihydroxy-2,4-dimethox-

Handbook of Chinese Medicinal Plants: Chemistry, Pharmacology, Toxicology
Weici Tang and Gerhard Eisenbrand

ISBN: 978-3-527-32226-8

yanthraquinone, 1,6-dihydroxy-2-methoxyanthraquinone, 1-hydroxy-2-methoxyanthraquinone, and physcion [1]. Iridoids from the root were identified as monotropein, asperuloside tetraacetate [2], morindolide and morofficinaloside [3]. Other compounds from the root were identified as β-sitosterol [2], succinic acid, nystose, fructofuranosylnystose, inulin-type hexasaccharide, and heptasaccharide [4]. Several anthraquinone derivatives, including 2-methyl-anthraquinone, rubiadin 1-methylether were isolated from the bark of *M. officinalis* together with β-sitosterol and 24-ethylcholesterol [5].

Pharmacology and Toxicology

The methanol extract of the root of *M. officinalis* was found to exert anti-inflammatory and analgesic activities. The extract potently inhibited the production of nitric oxide (NO), prostaglandin E_2 (PGE_2) and tumor necrosis factor-α (TNF-α) in RAW 264.7 murine macrophage cells stimulated by lipopolysaccharide (LPS). It suppressed the expression of inducible nitric oxide synthase (iNOS), cyclooxygenase-2 (COX-2) and TNF-α at the mRNA level in a concentration-dependent manner. Furthermore, the extract inhibited the nuclear factor-κB (NF-κB) activation induced by LPS. This was associated with the prevention of degradation of the inhibitor κB (IκB) protein, and attenuation of p65 protein into the nucleus. The *in vivo* anti-inflammatory effects of the extract in rats were demonstrated by an inhibition of paw edema induced by carrageenan. The antinociceptive effect of the extract was assessed in mice against an abdominal constriction induced by acetic acid. Anti-inflammatory effects were observed in rats (100 and 200 mg kg^{-1} per day), and antinociceptive effects in mice (100 and 200 mg kg^{-1} per day) [6]. One of the active compounds of the root of *M. officinalis* was identified as monotropein. Pretreatment of rats with monotropein at daily doses of 20 and 30 mg kg^{-1} significantly reduced the acute paw edema induced by carrageenan [7].

An extract of the root of *M. officinalis* was given orally (150 mg kg^{-1} twice daily for 10 days) to rats in which diabetes had been induced with streptozotocin. The fasting serum glucose levels of hepatic and renal thiobarbituric acid-reactive substances were reduced, while hepatic superoxide dismutase and catalase activities, as well as glutathione levels, were significantly increased [8].

The oral administration of an aqueous extract of the root of *M. officinalis* to mice exposed to γ-irradiation was found to significantly increase the reduced leukocyte counts. It was reported that the root of *M. officinalis* had an anti-fatigue efficacy, improving the immunological status of young mice and reducing the excitability of the parasympathetic nervous system of mice with hypothyroidism. The extract did not demonstrate any acute toxicity, mutagenicity, or sex-hormone-like actions [9]. Rubiadin, however, is known to be both mutagenic and carcinogenic. This is an anthraquinone derivative that is widely distributed among plants of the family Rubiaceae, and is one of the major anthraquinone derivatives present in the root of *Rubia cordifolia* L., a further Chinese herbal medicine officially listed in the Chinese Pharmacopoeia [10].

Asperuloside was reported to exhibit anticlastogenic activity in Chinese hamster ovary (CHO) cells against chromosome aberrations induced by mitomycin C. The α-unsaturated carbonyl group was considered to play an important role in this anticlastogenicity [11]. In addition, asperuloside inhibited the activity of phosphatidylinositol 3-kinase as a growth factor signaling enzyme, with an IC_{50} value of 2 μ*M* [12]. Succinic acid, nystose, fructofuranosylnystose, inulin-type hexasaccharide, and heptasaccharide were all found to have antidepressant activities [4].

References

1 Yang, Y.J., Shu, H.Y., and Min, Z.D. (1992) Anthraquinones isolated from *Morinda officinalis* and *Damnacanthus indicus*. *Acta Pharm. Sin.*, **27**, 358–364.

2 Chen, Y.W. and Xue, Z. (1987) Chemical constituents of *Morinda officinalis*. *Bull. Chin. Mater. Med.*, **12**, 613–614.

3 Yoshikawa, M., Yamaguchi, S., Nishisaka, H., Yamahara, J., and Murakami, N. (1995) Chemical constituents of Chinese natural medicine, Morindae radix, the dried roots of *Morinda officinalis* How.: structures of morindolide and morofficinaloside. *Chem. Pharm. Bull. (Tokyo)*, **43**, 1462–1465.

4 Cui, C., Yang, M., Yao, Z., Cao, B., Luo, F Z., Xu, Y., and Chen, Y. (1995) Antidepressant active constituents in the roots of *Morinda officinalis* How. *China J. Chin. Mater. Med.*, **20**, 36–39.

5 Li, S., Ouyang, Q., Tan, X.Z., Shi, S.S., Yao, Z. Q., Xiao, H.B., Zhang, Z.Y., Wang, F B.T., and Zhou, Z.X. (1991) Chemical constituents of *Morinda officinalis* How. *China J. Chin. Mater. Med.*, **16**, 675–676.

6 Kim, I.T., Park, H.J., Nam, J.H., Park, Y.M., Won, J.H., Choi, J., Choi, B.K., and Lee, K.T. (2005) *In-vitro* and *in vivo* anti-inflammatory and antinociceptive effects of the methanol extract of the roots of *Morinda officinalis*. *J. Pharm. Pharmacol.*, **57**, 607–615.

7 Choi, J., Lee, K.T., Choi, M.Y., Nam, J.H., Jung, H.J., Park, S.K., and Park, H.J. (2005) Antinociceptive anti-inflammatory effect of monotropein isolated from the root of *Morinda officinalis*. *Biol. Pharm. Bull.*, **28**, 1915–1918.

8 Soon, Y.Y. and Tan, B.K. (2002) Evaluation of the hypoglycemic and anti-oxidant activities of *Morinda officinalis* in streptozotocin-induced diabetic rats. *Singapore Med. J.*, **43**, 77–85.

9 Qiao, Z.S., Wu, H., Su, Z.W., Li, C.H., Wang, L. H., Yi, N.Y., Xia, Z.Q., and Bian, F Y.J. (1991) Comparison with the pharmacological actions of *Morinda officinalis*. *Chin. J. Integr. Trad. West. Med.*, **11**, 415–417.

10 Bertram, B., Hemm, I., and Tang, W. (2001) Mutagenic and carcinogenic constituents of medicinal herbs used in Europe or in the USA. *Pharmazie*, **56**, 99–120.

11 Nakamura, T., Nakazawa, Y., Onizuka, S., Satoh, S., Chiba, A., Sekihashi, K., Miura, F A., Yasugahira, N., and Sasaki, Y.F. (1997) Antimutagenicity of Tochu tea (an aqueous extract of *Eucommia ulmoides* leaves): 1. The clastogen-suppressing effects of Tochu tea in CHO cells and mice. *Mutat. Res.*, **388**, 7–20.

12 Frew, T., Powis, G., Berggren, M., Abraham, R. T., Ashendel, C.L., Zalkow, L.H., Hudson, C., Qazia, S., Gruszecka-Kowalik, E., and Merriman, R. (1994) A multiwell assay for inhibitors of phosphatidylinositol-3-kinase and the identification of natural product inhibitors. *Anticancer Res.*, **14**, 2425–2428.

Morus alba

Cortex Mori (Sangbaipi) is the dried root bark of *Morus alba* L. (Moraceae). It is used as an antiasthmatic and antiedemic agent for the treatment of cough, asthma, and oliguria.

Folium Mori (Sangye) is the dried leaf of *M. alba*. It is used for the treatment of common cold, cough, vertigo, eye inflammation, dizziness, and headache.

Morus alba L.

Fructus Mori (Sangshen) is the ripe aggregate fruit of *M. alba*. It is used as a general tonic and sedative for the treatment of vertigo, tinnitus, palpitation, insomnia, diabetes, and constipation.

Ramulus Mori (Sangzhi) is the dried young branches of *M. alba*. It is used for the treatment of arthritis and rheumatic diseases.

Chemistry

The root bark of *M. Alba*, the mulberry tree, was reported to contain prenylflavones, with mulberrin (kuwanon C), cyclomulberrin, mulberrochromene (morusin), and cyclomulberrochromone (cyclomorusin) as the major components [1]. A great number of minor prenylflavones, prenylflavanones, and Diels–Alder adducts of two chalcones, or of chalcone with flavone or flavanone such as mulberranol, morusinol, kuwanons, mulberrofurans, albanol B, and albafurans, were isolated from mulberry root bark [2].

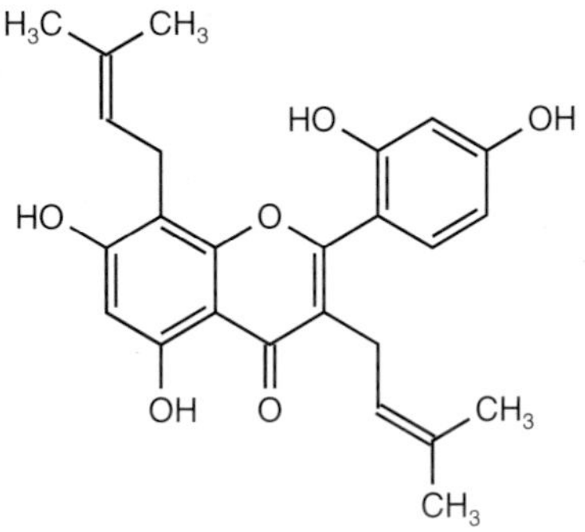

Mulberrin

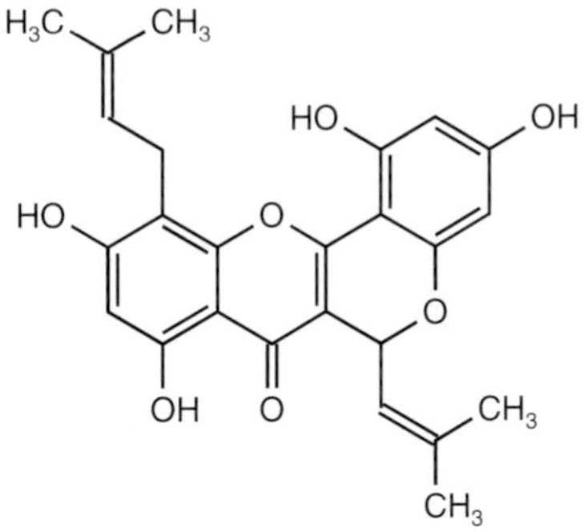

Cyclomulberrin

Flavones rutin, quercetin, quercetin glycosides and kaempferol glycosides [3], insect molting hormones β-ecdysone, and inokosterone [4] were isolated from the mulberry leaves. The Chinese Pharmacopoeia requires a quantitative determination of rutin in Folium Mori by HPLC. The content of rutin in Folium Mori should be not less than 0.1%. A series of cyclic polyols

Handbook of Chinese Medicinal Plants: Chemistry, Pharmacology, Toxicology
Weici Tang and Gerhard Eisenbrand

ISBN: 978-3-527-32226-8

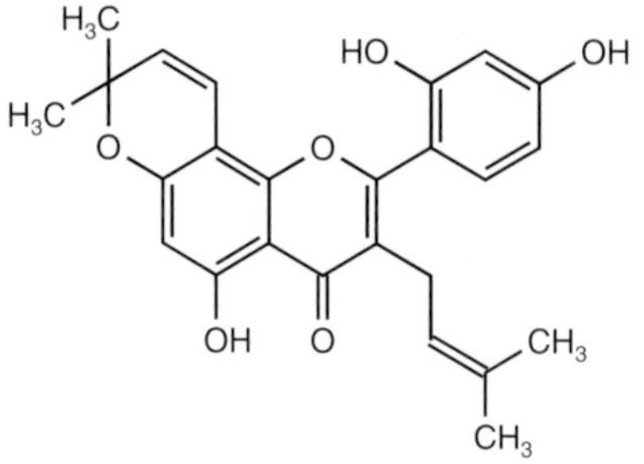

Mulberrochromene

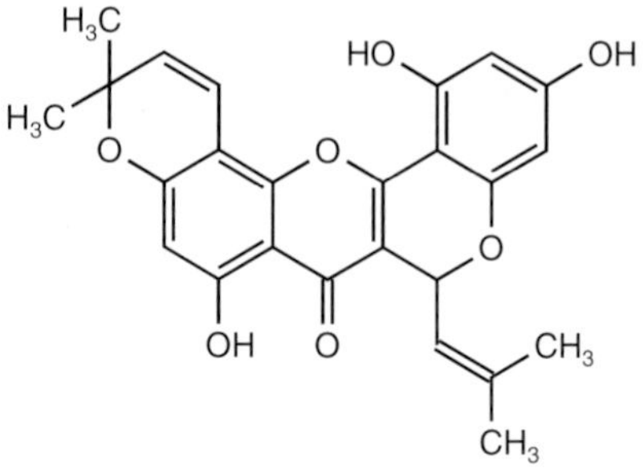

Cyclomulberrochromene

containing nitrogen from mulberry root and leaf were identified as deoxynojirimycin, *N*-methyl-deoxynojirimycin, fagomine, *epi*-fagomine, calystegin B_2, calystegin C_1, and a number of glycosides derived from deoxynojirimycin [5]. The mulberry fruits were reported to contain a series of anthocyanins, such as cyanidin 3-rutinoside and cyanidin 3-glucoside [6].

Pharmacology and Toxicology

The anthocyanins, cyanidin 3-rutinoside and cyanidin 3-glucoside from mulberry fruits were reported to exert a concentration-dependent inhibitory effect on the migration and invasion of highly metastatic A549 human lung carcinoma cells, without cytotoxicity. Both, cyanidin 3-glucoside and cyanidin 3-rutinoside were found to inhibit the expression of matrix metalloproteinase-2 (MMP-2) and urokinase plasminogen activator (u-PA), and to enhance the expression of tissue inhibitor of matrix metalloproteinase-2 (TIMP-2) and plasminogen activator inhibitor (PAI) in A549 cells. In addition, treatment of the tumor cells with cyanidin 3-rutinoside and cyanidin 3-glucoside resulted in an inhibition of the activation of Jun and nuclear factor-κB (NF-κB), demonstrating that anthocyanins might decrease the *in vitro* invasiveness of cancer cells [6].

The mulberry leaf is known to possess a number of biological activities, such as anti-inflammatory, antioxidant, antiatherosclerotic, hypolipidemic, and hypoglycemic effects. A methanolic extract of the leaf of *M. alba* was found to inhibit nitric oxide (NO) production in RAW264.7 macrophages activated by lipopolysaccharide (LPS), without any appreciable cytotoxic effect, at concentrations ranging from 4 to $100\,\mu g\,ml^{-1}$. LPS-induced prostaglandin E_2 (PGE_2) production was significantly reduced only by a butanol extract. In addition, a mulberry leaf extract significantly decreased the production of tumor necrosis factor-α (TNF-α), thus suppressing inflammatory mediators [7]. Quercetin 3-(6-malonyl)-glucoside, as a major flavone glycoside from the leaf, was found to attenuate atherosclerotic lesions in low-density lipoprotein (LDL) receptor-deficient mice [8].

The extracts of mulberry root bark were reported to show analgesic, diuretic, antitussive, antiedemic, sedative, anticonvulsive, and hypotensive activities in rodents and dogs [9]. A hot-water extract of the root bark of *M. alba* exhibited activity against anaphylactic reactions, inhibited the activation of rat peritoneal mast cells mediated by anti-chicken γ-globulin IgE, and also inhibited the systemic anaphylactic shock induced by compound 48/80 in mice. The extract also significantly inhibited the passive cutaneous anaphylaxis induced by anti-chicken γ-globulin IgE, but was not cytotoxic to rat peritoneal mast cells. Moreover, the extract concentration-dependently inhibited mast cell degranulation, histamine release and calcium uptake induced by the compound 48/80 or anti-chicken γ-globulin IgE, and elevated the level of intracellular cAMP in mast cells [10].

An aqueous extract of the mulberry leaf was found to inhibit the expression of lectin-like oxidized LDL receptor-1 at both protein and mRNA levels of cultured bovine aortic endothelial cells stimulated by TNF-α and LPS. It did not affect the expression of lectin-like oxidized LDL receptor-1 induced by transforming growth factor-β (TGF-β). Furthermore, a mulberry leaf extract inhibited the TNF-α-induced activation of nuclear factor-κB (NF-κB) and phosphorylation of inhibitory factor of NF-κBα (IκBα), indicating that the extract suppressed TNF-α- and LPS-induced lectin-like oxidized LDL receptor-1 gene expression by inhibiting NF-κB activation [11].

The flavones from mulberry leaves, especially quercetin and quercetin 3-*O*-β-D-glucopyranosyl-(1 → 6)-β-D-glucopyranoside, were found to exhibit significant radical scavenging effects [3]. Quercetin 3-*O*-β-D-glucopyranoside and quercetin 3,7-di-*O*-β-D-glucopyranoside from the leaves exerted significant growth-inhibitory effects against human promyelocytic leukemia HL-60 cells at a concentration of 0.2 m*M*. Quercetin 3,7-di-*O*-β-D-glucopyranoside also induced differentiation of the HL-60 cells in terms of expressing CD66b and CD14 antigens [12].

An aqueous extract of the leaves or root barks exhibited hypoglycemic activity in streptozotocin-induced diabetic mice [13]. A mulberry leaf extract containing 0.24% 1-deoxynojirimycin inhibited the sucrase activity of four human intestinal samples by 96%, and that of maltase and isomaltase by 95% and 99%, respectively. The activities of trehalase and lactase were inhibited by 44% and 38%, respectively [14]. An active component from the root bark was found to be a glycoprotein named moran A; this elicited marked hypoglycemic effects in both normal and alloxan-induced hyperglycemic mice when given intraperitoneally [16]. The neutral sugar part of moran A was composed of rhamnose, arabinose, mannose, galactose, and glucose, while the amino acid components were glycine, glutamic acid, aspartic acid, alanine, proline, threonine, serine, hydroxyproline, cystine, and other minor amino acids [13]. From the root bark, another glycoprotein – moran 20 K – with a molecular mass of approximately 20 kDa was isolated. This exhibited a similar hypoglycemic activity in streptozotocin-induced diabetic mice as did moran A [17]. A fraction of a 70% alcohol extract of the mulberry root bark containing morusin, cyclomorusin, neocyclomorusin, kuwanon E, moracin M, betulinic acid and methyl ursolate, was given to streptozotocin-induced diabetic rats at $600\,mg\,kg^{-1}$ for 10 days. This led to a significant reduction in the serum level of glucose, an increase in the insulin level, and a decreased serum content of thiobarbituric acid reactive substance. These findings suggested that the extract fraction might protect the pancreatic β-cells from degeneration and also diminish lipid peroxidation [18].

The mulberry leaf was reported to contain γ-aminobutyric acid, and so might exert a neuroprotective action against cerebral ischemia. Anaerobic treatment of the leaves might enhance the accumulation of γ-aminobutyric acid in the leaf tissues. Mulberry leaves treated in this way protected PC12 cells against oxidative damage induced by hydrogen peroxide. The neuroprotective effect of the treated mulberry leaves was also demonstrated *in vivo* against brain injury in rats induced by middle cerebral artery occlusion, as demonstrated by a significant reduction in the infarct volume of the brain [19].

An aqueous extract of Cortex Mori exhibited cytotoxic activity on K-562, B380 human leukemia cells and B16 mouse melanoma cells at concentrations in excess of 1 mg ml^{-1}, inducing cell apoptosis. Further studies showed that the extract bound to the tubulins; this resulted in a marked inhibition of assembly – but not disassembly – of microtubules. The suggestion was that the aqueous extract of Cortex Mori had induced the apoptosis of tumor cells by inhibiting microtubule assembly [20].

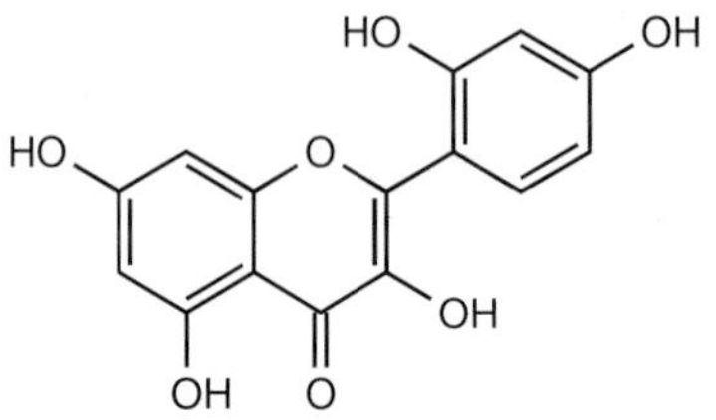

Morin

Morin from mulberry twigs was found to be a hypouricemic agent. Hyperuricemia is known to be associated with a number of pathological conditions such as gout. A lowering of the elevated serum levels of uric acid might be achieved by using xanthine oxidase inhibitors and by inhibiting renal urate reabsorption. Morin was shown to exert a potent inhibitory action on urate uptake in the rat renal brush-border membrane vesicles, with an IC_{50} value of 2 μM; this in turn indicated that morin had acted on the kidney to inhibit urate reabsorption. Morin was also shown to be an inhibitor of xanthine oxidase. In rats made hyperuricemic by oxonate treatment, morin exhibited hypouricemic effects [21]. A kinetic analysis of the uptake inhibition by morin indicated that morin appeared to act as a competitive inhibitor of urate uptake on the human urate transporter [22].

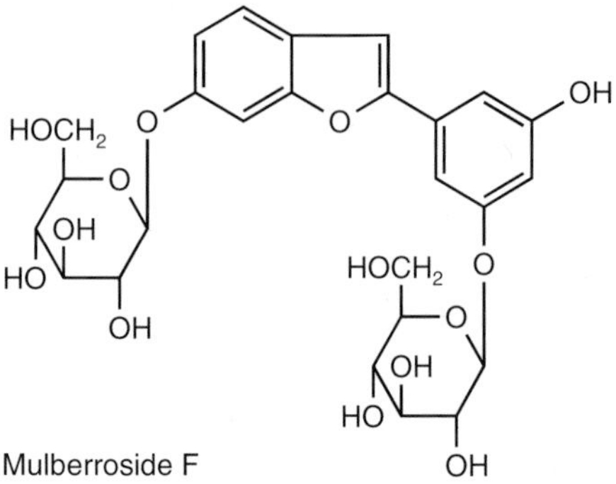

Mulberroside F

Mulberroside F from the leaves of *M. alba* was reported to inhibit melanin biosynthesis. The extract inhibited the tyrosinase activity that converts dopa to dopachrome in the biosynthetic process of melanin. Mulberroside F also exhibited superoxide scavenging activity, indicating a protection against auto-oxidation [23]. 2-Oxyresveratrol from Ramulus Mori was also found to inhibit tyrosinase in a competitive manner, but was not effective on tyrosinase synthesis and gene expression [24].

References

1 Nomura, T. and Fukai, T. (1979) On the structure of mulberrin, mulberrochromene, cyclomulberrin and cyclomulberrochromene. *Heterocycles*, **12**, 1289–1295.

2 Hano, Y. and Nomura, T. (1986) Constituents of the cultivated mulberry tree. Part 36. Structure of mulberrofuran P, a novel 2-arylbenzofuran derivative from the cultivated mulberry tree (*Morus alba* L.). *Heterocycles*, **24**, 1381–1386.

3 Kim, S.Y., Gao, J.J., Lee, W.C., Ryu, K.S., Lee, K. R., and Kim, Y.C. (1999) Antioxidative flavonoids from the leaves of *Morus alba*. *Arch. Pharm. Res.*, **22**, 81–85.

4 Chen, F., Nakashima, N., Kimura, I., Kimura, M., Asano, N., and Koya, S. (1995) Potentiating effects on pilocarpine-induced saliva secretion, by extracts and *N*-containing sugars derived from mulberry leaves, in streptozotocin-diabetic mice. *Biol. Pharm. Bull.*, **18**, 1676–1680.

5 Asano, N., Oseki, K., Tomioka, E., Kizu, F H., and Matsui, K. (1994) *N*-containing sugars from *Morus alba* and their glycosidase inhibitory activities. *Carbohydr. Res.*, **259**, 243–255.

6 Chen, P.N., Chu, S.C., Chiou, H.L., Kuo, W.H., Chiang, C.L., and Hsieh, Y.S. (2006) Mulberry anthocyanins, cyanidin 3-rutinoside and cyanidin 3-glucoside, exhibited an inhibitory effect on the migration and invasion of a human lung cancer cell line. *Cancer Lett.*, **235**, 248–259.

7 Choi, E.M. and Hwang, J.K. (2005) Effects of *Morus alba* leaf extract on the production of nitric oxide, prostaglandin E_2 and cytokines in RAW264.7 macrophages. *Fitoterapia*, **76**, 608–613.

8 Enkhmaa, B., Shiwaku, K., Katsube, T., Kitajima, K., Anuurad, E., Yamasaki, M., and Yamane, Y. (2005) Mulberry (*Morus alba* L.) leaves and their major flavonol quercetin 3-(6-malonylglucoside) attenuate atherosclerotic lesion development in LDL receptor-deficient mice. *J. Nutr.*, **135**, 729–734.

9 Yamatake, Y., Shibata, M., and Nagai, M. (1976) Pharmacological studies on root bark of mulberry tree (*Morus alba* L.). *Jpn. J. Pharmacol.*, **26**, 461–469.

10 Chai, O.H., Lee, M.S., Han, E.H., Kim, F H.T., and Song, C.H. (2005) Inhibitory effects of *Morus alba* on compound 48/80-induced anaphylactic reactions and anti-chicken γ-globulin IgE-mediated mast cell activation. *Biol. Pharm. Bull.*, **28**, 1852–1858.

11 Shibata, Y., Kume, N., Arai, H., Hayashida, K., Inui-Hayashida, A., Minami, M., Mukai, E., Toyohara, M., Harauma, A., Murayama, T., Kita, T., Hara, S., Kamei, K., and Yokoda, M. (2007) Mulberry leaf aqueous fractions inhibit TNF-α-induced nuclear factor κB (NF-κB) activation and lectin-like oxidized LDL receptor-1 (LOX-1) expression in vascular endothelial cells. *Atherosclerosis*, **193**, 20–27.

12 Kim, S.Y., Gao, J.J., and Kang, H.K. (2000) Two flavonoids from the leaves of *Morus alba* induce differentiation of the human promyelocytic leukemia (HL-60) cell line. *Biol. Pharm. Bull.*, **23**, 451–455.

13 Chen, F., Nakashima, N., Kimura, I., and Kimura, M. (1995) Hypoglycemic activity and mechanisms of extracts from mulberry leaves (folium mori) and cortex mori radicis in streptozotocin-induced diabetic mice. *Yakugaku Zasshi*, **115**, 476–482.

14 Oku, T., Yamada, M., Nakamura, M., Sadamori, N., and Nakamura, S. (2006) Inhibitory effects of extractives from leaves of *Morus alba* on human and rat small intestinal disaccharidase activity. *Br. J. Nutr.*, **95**, 933–938.

15 Kimura, T., Nakagawa, K., Kubota, H., Kojima, Y., Goto, Y., Yamagishi, K., Oita, S., Oikawa, S., and Miyazawa, T. (2007) Food-grade mulberry powder enriched with 1-deoxynojirimycin suppresses the elevation of postprandial blood glucose in humans. *J. Agric. Food Chem.*, **55**, 5869–5874.

16 Hikino, H., Mizuno, T., Oshima, Y., and Konno, C. (1985) Validity of the oriental medicines. 80. Antidiabetes drugs. 4. Isolation and hypoglycemic activity of moran A, a glycoprotein of *Morus alba* root barks. *Planta Med.*, **51**, 159–160.

17 Kim, E.S., Park, S.J., Lee, E.J., Kim, B.K., Huh, H., and Lee, B.J. (1999) Purification and characterization of Moran 20K from *Morus alba*. *Arch. Pharm. Res.*, **22**, 9–12.

18 Singab, A.N., El-Beshbishy, H.A., Yonekawa, M., Nomura, T., and Fukai, T. (2005) Hypoglycemic effect of Egyptian *Morus alba* root bark extract: effect on diabetes and lipid peroxidation of streptozotocin-induced diabetic rats. *J. Ethnopharmacol.*, **100**, 333–338.

19 Kang, T.H., Oh, H.R., Jung, S.M., Ryu, J.H., Park, M.W., Park, Y.K., and Kim, S.Y. (2006) Enhancement of neuroprotection of mulberry leaves (*Morus alba* L.) prepared by the anaerobic treatment against ischemic damage. *Biol. Pharm. Bull.*, **29**, 270–274.

20 Nam, S.Y., Yi, H.K., Lee, J.C., Kim, J.C., Song, C.H., Park, J.W., Lee, D.Y., Kim, J.S., and Hwang, P.H. (2002) Cortex Mori extract induces cancer cell apoptosis through inhibition of microtubule assembly. *Arch. Pharm. Res.*, **25**, 191–196.

21 Yu, Z., Fong, W.P., and Cheng, C.H. (2006) The dual actions of morin (3,5,7,2′,4′-pentahydroxyflavone) as a hypouricemic agent: uricosuric effect and xanthine oxidase inhibitory activity. *J. Pharmacol. Exp. Ther.*, **316**, 169–175.

22 Yu, Z., Fong, W.P., and Cheng, C.H. (2007) Morin (3,5,7,2′, 4′-pentahydroxyflavone) exhibits, potent inhibitory actions on urate transport by the human urate anion transporter (hURAT1) expressed in human embryonic kidney cells. *Drug Metab. Dispos.*, **35**, 981–986.

23 Lee, S.H., Choi, S.Y., Kim, H., Hwang, F J.S., Lee, B.G., Gao, J.J., and Kim, S.Y. (2002) Mulberroside F isolated from the leaves of *Morus alba* inhibits melanin biosynthesis. *Biol. Pharm. Bull.*, **25**, 1045–1048.

24 Lee, K.T., Lee, K.S., Jeong, J.H., Jo, B.K., Heo, M.Y., and Kim, H.P. (2003) Inhibitory effects of Ramulus mori extracts on melanogenesis. *J. Cosmet. Sci.*, **54**, 133–142.

Myristica fragrans

Semen Myristicae (Roudoukou) is the dried seed of *Myristica fragrans* Houtt. (Myristicaceae). It is used for treatment of gastrointestinal disorders, anorexia, vomiting, and diarrhea. It is also used as a common spice.

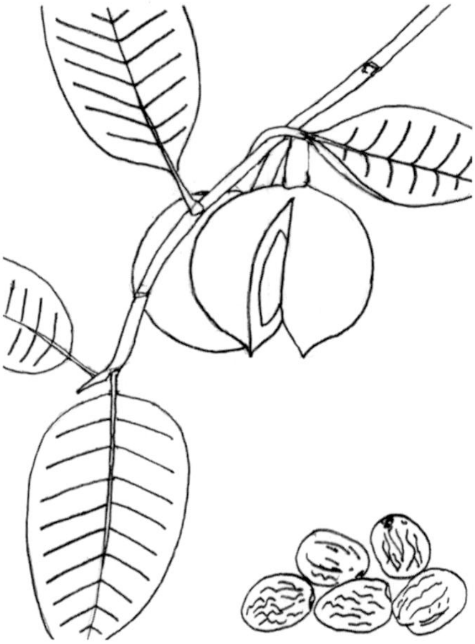

Myristica fragrans Houtt.

Chemistry

The seed of *M. frangrans,* nutmeg, is known to contain an essential oil with myristicin as the principal component [1]. Minor propenyl-benzene derivatives were identified as safrole, [2], eugenol, and isoeugenol [3]. The Chinese Pharmacopoeia requires a quantitative determination of essential oil in Semen Myristicae. The content of essential oil in Semen Myristicae should be not less than 6.0% (ml g^{-1}).

Lignans and neolignans from nutmeg were identified as fragransols A, B, and C, dehydrodiisoeugenol [4], and licarin B [5]. Related neolignan and lignan derivatives were also isolated from mace, the aril of *M. fragrans,* and identified as fragransols A–D, fragransins D_1, D_2, D_3, and E_1, and myristicanols A and B [6]. The diarylalkanes from nutmeg and mace were identified as malabaricones A, B, and C; all were derived from 1,9-diphenyl-1-nonanone [7].

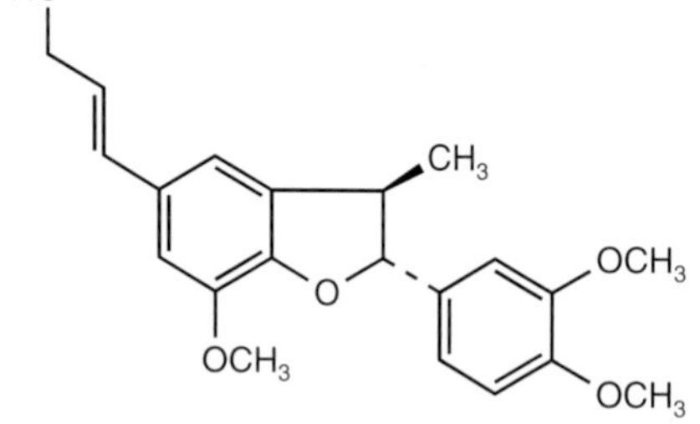

Fragransol C

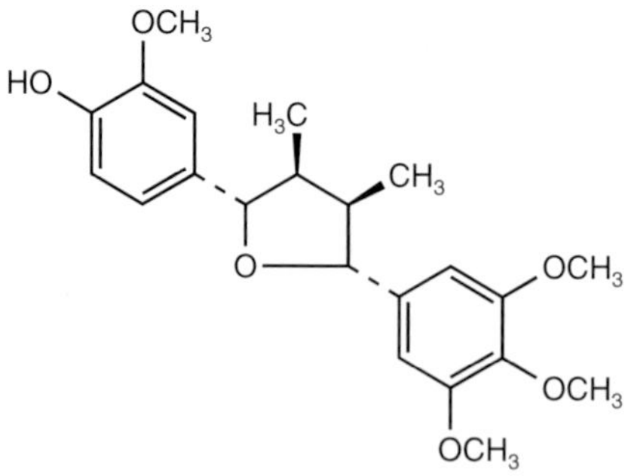

Fragransin D_1

Handbook of Chinese Medicinal Plants: Chemistry, Pharmacology, Toxicology
Weici Tang and Gerhard Eisenbrand

ISBN: 978-3-527-32226-8

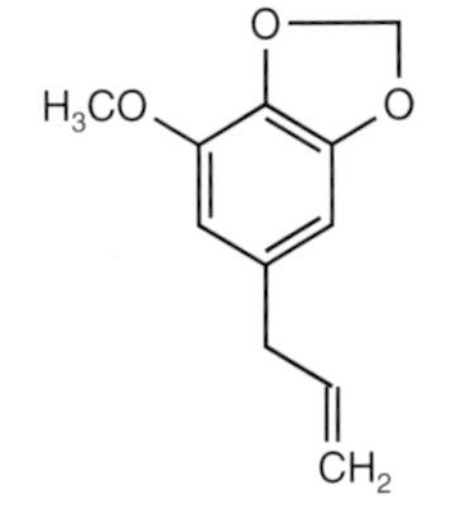

Myristicin

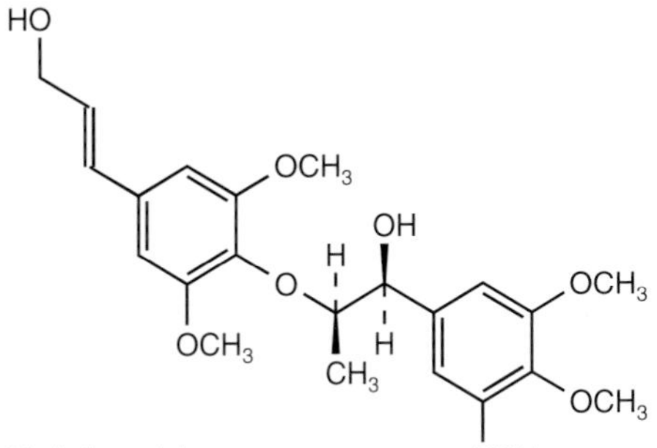

Myristicanol A

Pharmacology and Toxicology

The essential oil of nutmeg was found to exhibit considerable inhibitory effects against different genera of bacteria, including animal and plant pathogens, food poisoning and spoilage bacteria [8, 9]. Nutmeg also showed anticariogenic activity against *Streptococcus mutans* [10]. The chloroform extract of nutmeg showed anti-inflammatory, analgesic, and antithrombotic activities in rodents. The extract inhibited the carrageenan-induced rat paw edema, reduced writhing reactions in mice induced by acetic acid, and also protected against the formation of thrombosis in mice induced by ADP and adrenaline [11]. The anti-inflammatory effect of mace was ascribed to myristicin [12]. The essential oil of nutmeg inhibited rabbit platelet aggregation *in vitro*, with eugenol and isoeugenol as the most active components [3].

The ethanol extract of nutmeg was reported to exert hypolipidemic activity in rabbits with experimentally induced hyperlipidemia. After oral treatment of hyperlipidemic rabbits with nutmeg extract at a daily dose of $500\,mg\,kg^{-1}$ for 60 days, total cholesterol, low-density lipoprotein (LDL) cholesterol, and triglyceride levels were each significantly lowered, whereas high-density lipoprotein (HDL) cholesterol levels were not significantly affected [13]. The feeding of nutmeg extract to rabbits also prevented an accumulation of cholesterol, phospholipids and triglycerides in the liver, heart, and aorta. The fecal excretion of cholesterol and phospholipids was significantly increased in rabbits fed with the seed extract [14].

Myristicin was found to induce rat hepatic cytochrome P450 enzymes. The intraperitoneal (i.p.) treatment of rats with myristicin at a dose of $500\,\mu mol\,kg^{-1}$ caused two- to 20-fold increases in the activity of liver P450 isoenzymes CYP1A1/2, CYP2B1/2, and CYP2E1, which were consistent with the increases in protein levels. The levels of mRNA of CYP1A1/2, CYP2B1/2 – but not of CYP2E1 – were also shown to be increased [15]. Mace also significantly increased acid-soluble thiol groups in the liver of young adult male and female mice, when given 2% mace in the diet for 10 days [16]. A transmammary modulation of xenobiotic-metabolizing enzymes in the liver of mouse pups by mace was investigated. An aqueous suspension of mace, at oral daily doses of 25 or 100 mg per animal, was given to the dams from day 1 of lactation and continued for two or three weeks. The result was a significant elevation of hepatic thiol content, glutathione *S*-transferase (GST) and glutathione reductase activities, and cytochrome P450 content [17].

Nutmeg oil was found to be an antifertility agent and to induce chromosomal translocations in treated mice and their F_1 males at high doses of 60, 80, 100, and $400\,mg\,kg^{-1}$ for eight successive weeks.

The greatest reduction in fertility occurred in animals treated at 400 mg kg^{-1}. Chromosomal translocation was observed in F_1 males, derived from males treated with oil of nutmeg at dose levels of 60, 80, and 100 mg kg^{-1} [18].

Myristicin, like safrole, is a 2-propenylbenzene. Since safrole was identified as a mutagen and carcinogen, the mutagenic and carcinogenic properties of myristicin have been intensively investigated. The metabolism of myristicin resembled that of safrole. The acute toxicity of myristicin appeared to be low, with no toxic effects observed in rats given myristicin orally at a dose of 10 mg kg^{-1}. In contrast, a dose of 6–7 mg kg^{-1} is reported sufficient to cause psychopharmacological effects in humans. A covalent DNA-binding capacity has been demonstrated. Based on the available data, it seems unlikely that an intake of myristicin from essential oils and spices in food, estimated as a daily dose of a few milligrams per person, might cause adverse effects in humans. It is, however, at present not possible to make an adequate risk assessment. Satisfactory studies regarding genotoxicity and chronic toxicity, including reproductive toxicity and carcinogenicity, are lacking [19].

In an early study, one of several widely consumed cola drinks was given to mice in place of their drinking water for up to eight weeks. Significant levels of liver DNA lesions developed in a time-dependent manner, as measured by ^{32}P-post-labeling. No such lesions were detected in mice given tap water, or one of three non-cola beverages. Lesions induced by cola drinks were reported to be chromatographically identical (by ^{32}P radiolabel patterning) to those found in mice treated with extracts of nutmeg or mace, or with myristicin. In addition, small amounts of lesions possibly derived from safrole were observed. Liver DNA lesions were also detected in fetal liver when pregnant mice were given myristicin [20].

In contrast, mace was reported to show a chemopreventive action against carcinogenesis in the uterine cervix. Carcinogenesis was induced by the implantation inside the canal of the uterine cervix of young adult virgin mice of 600 μg 3-methylcholanthrene in beeswax. Mace, when administered orally at 10 mg per day per animal for seven days before, and for 90 days after, the carcinogen treatment, caused a significant reduction in the incidence of cervical carcinoma [21]. Mace also significantly inhibited the formation of papilloma in mice skin induced by a topical application of 150 μg 7,12- dimethylbenzo[*a*]anthracene (DMBA), and by the repeated application of 1% croton oil two weeks later. When animals received a diet containing 1% mace, the incidence of skin papilloma was reduced to 50%, while the average number of tumors per tumor-bearing mouse fell from 5.7 to 1.7 [22].

Myristicin exerted a hepatoprotective activity in rats with liver damage induced by the simultaneous administration of lipopolysaccharide (LPS) and D-galactosamine. The plasma aminotransferase activities were markedly reduced, and myristicin also markedly suppressed the enhancement of serum TNF-α concentrations and hepatic DNA fragmentation in mice induced by LPS and D-galactosamine. This suggested that the hepatoprotective activity of myristicin might be due to the inhibition of TNF-α release from macrophages [23].

Based on *in vitro* metabolic studies with myristicin, two metabolites were isolated and identified as 5-(2-propenyl)-1-methoxy-2,3-dihydroxybenzene and 1′-hydroxymyristicin. These two metabolites were also found in the urine following oral administration of myristicin to rats. Enzymatic hydrolysis of the urine suggested that the metabolites had been excreted in conjugated form [24]. The oxidation of myristicin to 5-(2-propenyl)-1-methoxy-

2,3-dihydroxybenzene as the major metabolite of myristicin was catalyzed by purified bacterial recombinant CYPs 3A4 and 1A2. Thus, CYP3A4 (and possibly other CYP3A enzymes) and CYP1A2 appear to play important roles in the formation of 5-(2-propenyl)-1-methoxy-2,3-dihydroxybenzene [25].

References

1 Li, T., Zhou, J., Jiang, W., and Li, C. (1990) Effects of processing on volatile oil constituents in nutmeg and on the contents of myristicin. *China J. Chin. Mater. Med.*, **15**, 471–473.

2 Archer, A.W. (1988) Determination of safrole and myristicin in nutmeg and mace by high-performance liquid chromatography. *J. Chromatogr.*, **438**, 117–121.

3 Janssens, J., Laekeman, G.M., Pieters, L.A., Totte, J., Herman, A.G., and Vlietinck, F A.J. (1990) Nutmeg oil: identification and quantitation of its most active constituents as inhibitors of platelet aggregation. *J. Ethnopharmacol.*, **29**, 179–188.

4 Juhasz, L., Kurti, L., and Antus, S. (2000) Simple synthesis of benzofuranoid neolignans from *Myristica fragrans. J. Nat. Prod.*, **63**, 866–870.

5 Kim, Y.B., Park, I.Y., and Shin, K.H. (1991) The crystal structure of licarin-B, ($C_{20}H_{20}O_4$), a component of the seeds of *Myristica fragrans. Arch. Pharm. Res.*, **14**, 1–6.

6 Hada, S., Hattori, M., Tezuka, Y., Kikuchi, T., and Namba, T. (1988) Constituents of mace. III. New neolignans and lignans from the aril of *Myristica fragrans. Phytochemistry*, **27**, 563–568.

7 Orabi, K.Y., Mossa, J.S., and el-Feraly, F.S. (1991) Isolation and characterization of two antimicrobial agents from mace (*Myristica fragrans*). *J. Nat. Prod.*, **54**, 856–859.

8 Dorman, H.J. and Deans, S.G. (2000) Antimicrobial agents from plants: antibacterial activity of plant volatile oils. *J. Appl. Microbiol.*, **88**, 308–316.

9 Nakasimhan, B. and Dhake, A.S. (2006) Antibacterial principles from *Myristica fragrans* seeds. *J. Med. Food*, **9**, 395–399.

10 Chung, J.Y., Choo, J.H., Lee, M.H., and Hwang, J.K. (2006) Anticariogenic activity of macelignan isolated from *Myristica fragrans* (nutmeg) against *Streptococcus mutans. Phytochemistry*, **13**, 261–266.

11 Olajide, O.A., Ajayi, F.F., Ekhelar, A.I., Awe, S. O., Makinde, J.M., and Alada, A.R. (1999) Biological effects of *Myristica fragrans* (nutmeg) extract. *Phytother. Res.*, **13**, 344–345.

12 Ozaki, Y., Soedigdo, S., Wattimena, Y.R., and Suganda, A.G. (1989) Antiinflammatory effect of mace, aril of *Myristica fragrans* Houtt., and its active principles. *Jpn. J. Pharmacol.*, **49**, 155–163.

13 Ram, A., Lauria, P., Gupta, R., and Sharma, V. N. (1996) Hypolipidaemic effect of *Myristica fragrans* fruit extract in rabbits. *J. Ethnopharmacol.*, **55**, 49–53.

14 Sharma, A., Mathur, R., and Dixit, V.P. (1995) Prevention of hypercholesterolemia and atherosclerosis in rabbits after supplementation of *Myristica fragrans* seed extract. *Indian J. Physiol. Pharmacol.*, **39**, 407–410.

15 Jeong, H.G. and Yun, C.H. (1995) Induction of rat hepatic cytochrome P450 enzymes by myristicin. *Biochem. Biophys. Res. Commun.*, **217**, 966–971.

16 Kumari, M.V. and Rao, A.R. (1989) Effects of mace (*Myristica fragrans*, Houtt.) on cytosolic glutathione *S*-transferase activity and acid soluble sulfhydryl level in mouse liver. *Cancer Lett.*, **46**, 87–91.

17 Chhabra, S.K. and Rao, A.R. (1994) Transmammary modulation of xenobiotic metabolizing enzymes in liver of mouse pups by mace (*Myristica fragrans* Houtt.). *J. Ethnopharmacol.*, **42**, 169–177.

18 Pecevski, J., Savkovic, D., Radivojevic, D., and Vuksanovic, L. (1981) Effect of oil of nutmeg on the fertility and induction of meiotic chromosome rearrangements in mice and their first generation. *Toxicol. Lett.*, **7**, 239–243.

19 Hallstrom, H. and Thuvander, A. (1997) Toxicological evaluation of myristicin. *Nat. Toxins*, **5**, 186–192.

20 Randerath, K., Putman, K.L., and Randerath, E. (1993) Flavor constituents in cola drinks induce hepatic DNA adducts in adult and fetal mice. *Biochem. Biophys. Res. Commun.*, **192**, 61–68.

21 Hussain, S.P. and Rao, A.R. (1991) Chemopreventive action of mace

(*Myristica fragrans,* Houtt) on methylcholanthrene-induced carcinogenesis in the uterine cervix in mice. *Cancer Lett.*, **56**, 231–234.

22 Jannu, L.N., Hussain, S.P., and Rao, A.R. (1991) Chemopreventive action of mace (*Myristica fragrans,* Houtt) on DMBA-induced papillomagenesis in the skin of mice. *Cancer Lett.*, **56**, 59–63.

23 Morita, T., Jinno, K., Kawagishi, H., Arimoto, Y., Suganuma, H., Inakuma, T., and Sugiyama, K. (2003) Hepatoprotective effect of myristicin from nutmeg (*Myristica fragrans*) on lipopolysaccharide/d-galactosamine-induced liver injury. *J. Agric. Food Chem.*, **51**, 1560–1565.

24 Lee, H.S., Jeong, T.C., and Kim, J.H. (1998) *In vitro* and *in vivo* metabolism of myristicin in the rat. *J. Chromatogr. B. Biomed. Sci. Appl.*, **705**, 367–372.

25 Yun, C.H., Lee, H.S., Lee, H.Y., Yim, S.K., Kim, K.H., Kim, E., Yea, S.S., and Guengerich, F.P. (2003) Roles of human liver cytochrome P450 3A4 and 1A2 enzymes in the oxidation of myristicin. *Toxicol. Lett.*, **137**, 143–150.

Nardostachys chinensis and *Nardostachys jatamansi*

Radix et Rhizoma Nardostachyos (Gansong) is the dried root and rhizome of *Nardostachys chinensis* Batal. or *Nardostachys jatamansi* DC. (Valerianaceae). It is used internally for the treatment of epigastric and abdominal distension with anorexia and vomiting, and externally to treat toothache and foot swellings.

Nardostachys chinensis Batal.

Chemistry

The root and rhizome of *N. chinensis* and *N. jatamansi* are known to contain an essential oil. The Chinese Pharmacopoeia requires a quantitative determination of essential oil in Radix et Rhizoma Nardostachyos; the content should be not less than 2.0% (ml g^{-1}). A series of sesquiterpenes were isolated from the root and rhizome of *N. chinensis*, with nardosinone as the major component [1]. Further sesquiterpenes were identified as isonardosinone, nardonoxide, kanshones A, B, C, D and E, gansongone, nardosinonediol, desoxonarchinol A, nardoperoxide, isonardoperoxide,

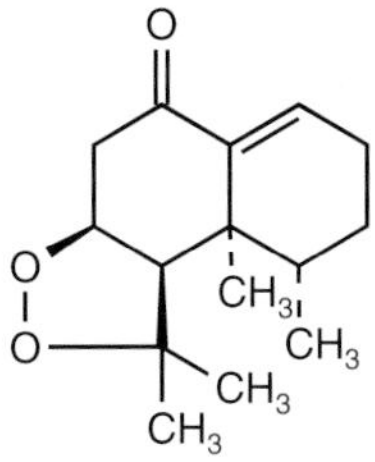

Nardosinone

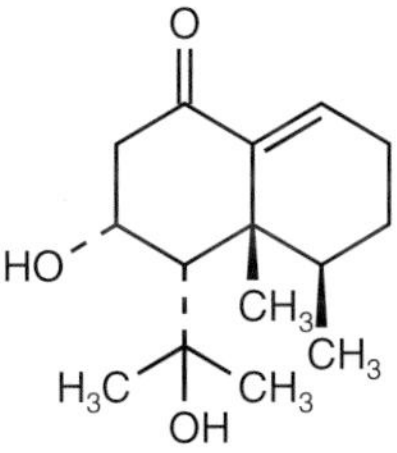

Nardosinonediol

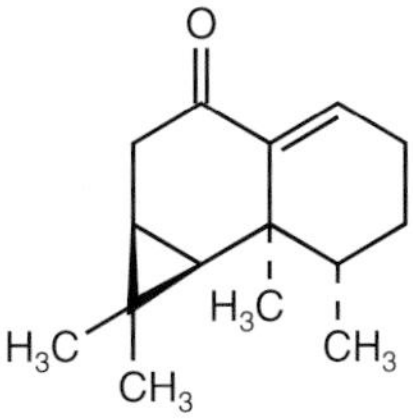

Gansongone

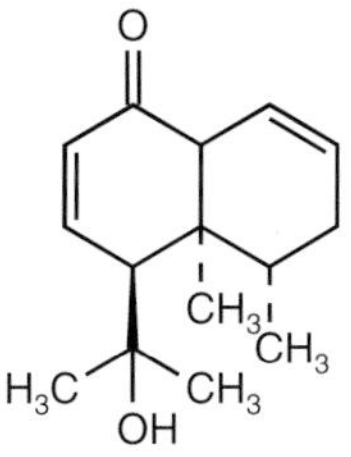

Kanshone A

Handbook of Chinese Medicinal Plants: Chemistry, Pharmacology, Toxicology
Weici Tang and Gerhard Eisenbrand

ISBN: 978-3-527-32226-8

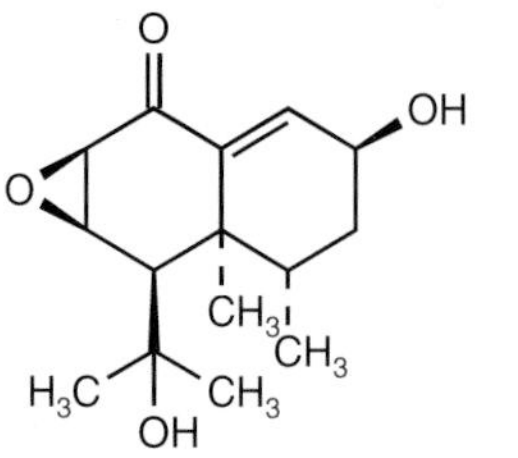

Kanshone D

nardoxide, debilon, eudesmendiol, nardostachnol, aristolene, tetradehydroaristolene, β-maaliene, narchinol A, desoxonarchinol A, patchouli alcohol and β-patchoulene, nardostachone, and aristolenol [2–7]. Sesquiterpenes from *N. jatamansi* were identified as jatamols A and B, spirojatomol, jatamansone, nardostachnol, dehydroaristolan-2-one, valeranone, calarenol, nardostachone, valeranal, nardol, norseychelanone, patchouli alcohol, seychellene, α-patchoulene and β-patchoulene, seychellane, and jatamansic acid [8, 9].

Neolignan and lignan derivatives from *N. jatamansi* were identified as *erythro*-1-(3,4-dimethoxyphenyl)-2-(2-methoxy-4(*E*)-propenylphenoxy)-propan-1-ol, virolin, *erythro*-1-(4-hydroxy-3-methoxyphenyl)-2-(2-methoxy-4(*E*)-propenylphenoxy)-propan-1-ol, pinoresinol, and 1-hydroxypinoresinol [10]. An iridoid derivative from *N. chinensis* was identified as nardostachin [11]. The presence of the alkaloid actinidine in the rhizome of *N. jatamansi* was reported [12].

Pharmacology and Toxicology

The pharmacological actions of the root and rhizome of *N. chinensis* and *N. jatamansi* were reviewed to be sedative, smooth-muscle relaxative, hypotensive, antiarrhythmic, antihypoxic, anti-ischemic, and antibacterial. The clinical indications involved epigastric pain and abdominal cramping, depression, neurasthenia, epilepsy, palpitation, angina pectoris, and hypertension [13]. The essential oil of *N. jatamansi* was found to exert high activity against *Salmonella typhi, Salmonella paratyphi, Xanthomonas campestris, Bacillus pumilus, Bacillus anthracis* and *Colletotrichum* sp., but to be comparatively inactive against *Pseudomonas mangiferae* [14]. The essential oil also showed inhibitory effects against fungi *Aspergillus flavus, Aspergillus niger,* and *Fusarium oxysporum* at a concentration of 0.1 μl ml^{-1} [15]. The endoperoxides nardoperoxide and isonardoperoxide were found to show strong antimalarial activity against *Plasmodium falciparum*, with EC_{50} values of 1.5 and 0.6 μ*M*, respectively [5].

The acute and subchronic oral administration of an alcohol extract of the roots of *N. jatamansi* to rats did not cause any changes in norepinephrine (noradrenaline) and dopamine levels in the brain, but caused significant increases in the levels of serotonin, 5-hydroxyindoleacetic acid, γ-aminobutyric acid, and taurine. A 15-day treatment of rats with the alcohol extract resulted in a significant increase in the levels of central mono-

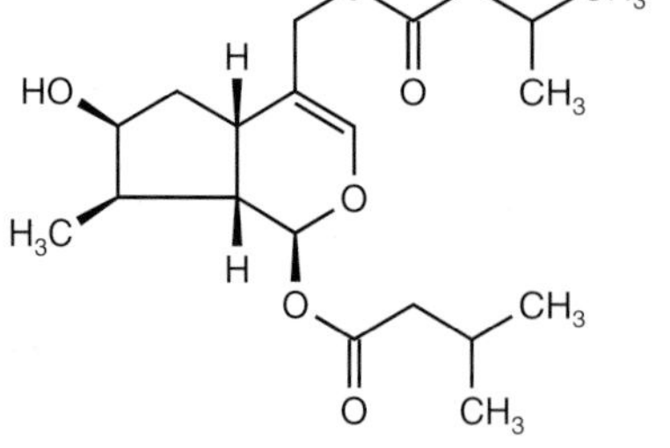

Nardostachin

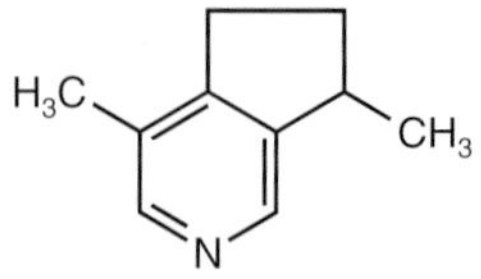

Actinidine

amines [16]. An ethanol extract of the root of *N. jatamansi* showed anticonvulsant activity and neurotoxicity in rats. It was shown that the root extract significantly increased the seizure threshold in rats against maximal electroshock, as indicated by a decrease in the extension : flexion ratio. The root extract also showed minimal neurotoxicity at doses that increased the seizure threshold [17]. The ethanol extract, given at an oral daily dose of 200 mg kg^{-1} for eight successive days to both young and aged mice, significantly improved learning and memory in the young mice and also reversed the amnesia induced by diazepam and scopolamine. Furthermore, the extract reversed aging-induced amnesia due to the natural aging of mice. As scopolamine-induced amnesia was reversed, it was possible that the memory improvement might be caused by a facilitation of cholinergic transmission in the brain [18]. The methanol extract of the rhizome of *N. jatamansi* was also found to inhibit acetylcholinesterase *in vitro*, with an IC_{50} value of approximately 47 μg ml^{-1} [19]. Nardosinone was found to be a neuritogenic substance. It did not exhibit a neurotrophic activity but rather caused a marked enhancement of the nerve growth factor-mediated neurite outgrowth from PC12D cells, most likely by amplifying an up-stream step of mitogen-activated protein kinase in the nerve growth factor receptor-mediated intracellular signaling pathway [20].

The root of *N. jatamansi* was further reported to show protective activity against cardiac damage in rats, as induced by doxorubicin. The intraperitoneal administration of doxorubicin to rats at a dose of 15 mg kg^{-1} resulted in myocardial damage that was manifested by elevations of serum lactate dehydrogenase, creatine phosphokinase, aspartate aminotransaminase, and alanine aminotransaminase. The animals also showed significant changes in antioxidant enzymes, including superoxide dismutase, catalase, glutathione peroxidase and glutathione-*S*-transferase (GST), and lipid peroxidation levels. Pretreatment of these rats with *N. jatamansi* extract significantly prevented the changes caused by doxorubicin, and restored the enzyme activity and lipid peroxides to near-normal levels [21].

References

1 Ruecker, G., Luo, S.D., and Olbrich, A. (1990) Peroxides as plant constituents. VII. Formation of the sesquiterpene peroxide nardosinone from an aristolane precursor. *Arch. Pharm. (Weinheim)*, **323**, 171–175.

2 Luo, S.D., Mayer, R., and Ruecker, G. (1987) Nardonoxide, a new nardosinane-type sesquiterpene ether from *Nardostachys chinensis*. *Planta Med.*, **53**, 332–334.

3 Bagchi, A., Oshima, Y., and Hikino, H. (1988) The validity of the Oriental medicines. CXXX. Sesquiterpenoids. 63. Kanshones D and E, sesquiterpenoids of *Nardostachys chinensis* roots. *Phytochemistry*, **27**, 3667–3669.

4 Luo, S.D., Olbrich, A., Mayer, R., and Ruecker, G. (1987) Gansongone, a new aristolane ketone from *Nardostachys chinensis* and structure revision of an aristolenol. *Planta Med.*, **53**, 556–558.

5 Takaya, Y., Kurumada, K., Takeuji, Y., Kim, H. S., Shibata, Y., Ikemoto, N., Wataya, Y., and Oshima, Y. (1998) Novel antimalarial guaiane-type sesquiterpenoids from *Nardostachys chinensis* roots. *Tetrahedron Lett.*, **39**, 1361–1364.

6 Itokawa, H., Masuyama, K., Morita, H., and Takeya, K. (1993) Cytotoxic sesquiterpenes from *Nardostachys chinensis*. *Chem. Pharm. Bull. (Tokyo)*, **41**, 1183–1184.

7 Masuyama, K., Morita, H., Takeya, K., and Itokawa, H. (1993) Eudesm-11-en-2,4α-diol from *Nardostachys chinensis*. *Phytochemistry*, **34**, 567–568.

8 Bagchi, A., Oshima, Y., and Hikino, H. (1991) Jatamols A and B: sesquiterpenoids of *Nardostachys jatamansi* roots. *Planta Med.*, **57**, 282–283.

9 Bagchi, A., Oshima, Y., and Hikino, H. (1990) Sesquiterpenoids. LXIV. Spirojatomol, a new

skeletal sesquiterpenoid of *Nardostachys jatamansi* roots. *Tetrahedron*, **46**, 1523–1530.

10 Bagchi, A., Oshima, Y., and Hikino, H. (1991) Neolignans and lignans of *Nardostachys jatamansi* roots. *Planta Med.*, **57**, 96–97.

11 Bagchi, A., Oshima, Y., and Hikino, H. (1988) Validity of the Oriental medicines. CXXIII. Nardostachin, an iridoid of *Nardostachys chinensis*. *Planta Med.*, **54**, 87–88.

12 Hirose, Y., Yonemitsu, K., and Sonoda, T. (1978) Chemical studies on the components of *Nardostachys jatamansi* de Candolle. I. *Shoyakugaku Zasshi*, **32**, 121–122, (CA 90: 51471).

13 Zhang, W.G. (1985) Pharmacological actions and clinical applications of gansong. *Jiangsu J. Trad. Chin. Med.*, **6**, 425–427.

14 Rao, F JT. (1986) Antimicrobial properties of the essential oil of *Nardostachys jatamansi*. *PAFAI J.*, **8**, 27–28.

15 Mishra, D., Chaturvedi, R.V., and Tripathi, F SC. (1995) The fungitoxic effect of the essential oil of the herb *Nardostachys jatamansi* DC. *Trop. Agric. (Trinidad)*, **72**, 48–52.

16 Prabhu, V., Karanth, K.S., and Rao, A. (1994) Effects of *Nardostachys jatamansi* on biogenic amines and inhibitory amino acids in the rat brain. *Planta Med.*, **60**, 114–117.

17 Rao, V.S., Rao, A., and Laranth, F KS. (2005) Anticonvulsant and neurotoxicity profile of *Nardostachys jatamansi* in rats. *J. Ethnopharmacol.*, **102**, 351–356.

18 Joshi, H. and Parle, M. (2006) *Nardostachys jatamansi* improves learning and memory in mice. *J. Med. Food*, **9**, 113–118.

19 Vinutha, B., Prashanth, D., Salma, K., Sreeja, S.L., Pratiti, D., Padjama, R., Rsdhika, S., Amit, A., Venkateshwarlu, K., and Deepak, M. (2007) Screening of selected Indian medicinal plants for acetylcholinesterase inhibitory activity. *J. Ethnopharmacol.*, **109**, 359–363.

20 Li, P., Matsunaga, K., Yamamoto, K., Yoshikawa, R., Kawashima, K., and Ohizumi, Y. (1999) Nardosinone was reported to be a novel enhancer of nerve growth factor in neurite outgrowth from PC12D cells. *Neurosci. Lett.*, **273**, 53–56.

21 Subashini, R., Yogeeta, S., Gnanapragasam, A., and Devaki, T. (2006) Protective effect of *Nardostachys jatamansi* on oxidative injury and cellular abnormalities during doxorubicin-induced cardiac damage in rats. *J. Pharm. Pharmcol.*, **58**, 257–262.

Nelumbo nucifera

Folium Nelumbinis (Heye) is the dried leaf of *Nelumbo nucifera* Gaertn. (Numphaeaceae). It is used as a hemostatic agent.

Nelumbo nucifera Gaertn.

Nodus Nelumbinis Rhizomatis (Oujie) is the dried nodes of the rootstock of *N. nucifera*. It is used as a hemostatic agent for the treatment of hematemesis, hemaptysis, hematuria, and abnormal uterine bleeding.

Plumula Nelumbinis (Lianzixin) is the dried young cotyledon and radicle from the ripe seed of *N. nucifera*. It is used as a sedative and hemostatic.

Receptaculum Nelumbinis (Lianfang) is the dried receptacle of *N. nucifera*. It is used as a hemostatic for the treatment of abnormal uterine bleeding, hematuria, and hemorrhoidal bleeding.

Semen Nelumbinis (Lianzi) is the dried mature seed of *N. nucifera*. It is used as a sedative and general tonic for the treatment of palpitation and insomnia.

Stamen Nelumbinis (Lianxu) is the dried stamen of *N. nucifera*. It is used as an astringent.

Folium Nelumbinis, Nodus Nelumbinis Rhizomatis and Receptaculum Nelumbinis are also used as hemostatic agent after carbonization.

Chemistry

The seed and rhizome of *N. nucifera*, the lotus seed and rhizome, are known to contain mainly starch and polysaccharides. A polysaccharide from the seeds was found to be highly branched with 3,4-linked D-galactopyranosyl units at the branch point [1]. Betulinic acid was also identified in the rhizome [2].

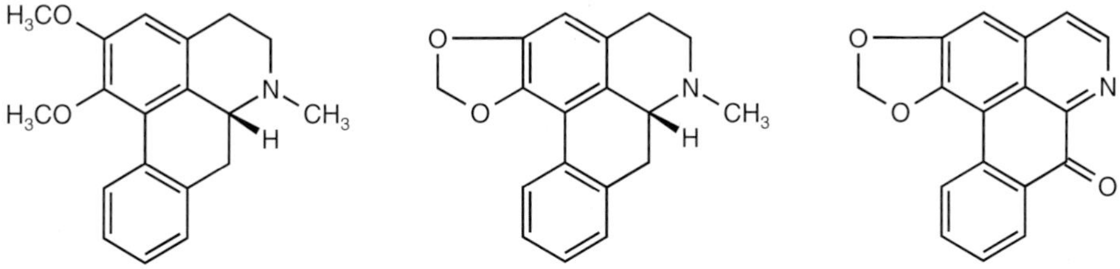

Nuciferine   Remerine   Liriodenine

The lotus leaves are known to contain a number of aporphine-type alkaloids, with nuciferine as the major component. Minor alka-

Handbook of Chinese Medicinal Plants: Chemistry, Pharmacology, Toxicology
Weici Tang and Gerhard Eisenbrand

ISBN: 978-3-527-32226-8

loids were identified as remerine, nornuciferine, *N*-nornuciferine, anonaine (*N*-demethylremerine), liriodenine, dehydroremerine, dehydronuciferine, dehydroanonaine, *N*-methylcoclaurine, asimilobine, and lirinidine [3, 4]. In addition, the benzyltetrahydroisoquinoline-type alkaloid methylcoclaurine, armepavine and the proaporphine-type alkaloid pronuciferine were isolated from lotus leaves [5]. The Chinese Pharmacopoeia requires a quantitative determination of nuciferine in Folium Nelumbinis by HPLC. The content of nuciferine in Folium Nelumbinis should be not less than 0.1%.

The flavor components of the leaf were analyzed and the major components identified as *cis*-hex-3-enol (40%), diphenylamine (8%), longifolene (6%), hexanol (4%), and also benzene (3%) [6].

The lotus cotyledon was reported to contain alkaloids as the major active constituents, including leonurine, nuciferine, pronuciferine, and anonaine, as well as bisbenzyltetrahydroisoquinoline alkaloids liensinine, isoliensinine, and neferine [7]. Leonurine is the 4-hydroxy-3,5-dimethoxybenzoic acid ester of 4-guanidino-*n*-butanol, and is a constituent of Herba Leonuri, the dried above-ground part of *Leonurus japonicus* Houtt. (Lamiaceae). The Chinese Pharmacopoeia requires a qualitative determination of leonurine in Plumula Nelumbinis by thin-layer chromatographic comparison with leonurine perchlorate, and a quantitative determination by HPLC. The content of leonurine in Plumula Nelumbinis should be not less than 0.2%. The content of liensinine as major component in the cotyledon was reported as approximately 1% [8].

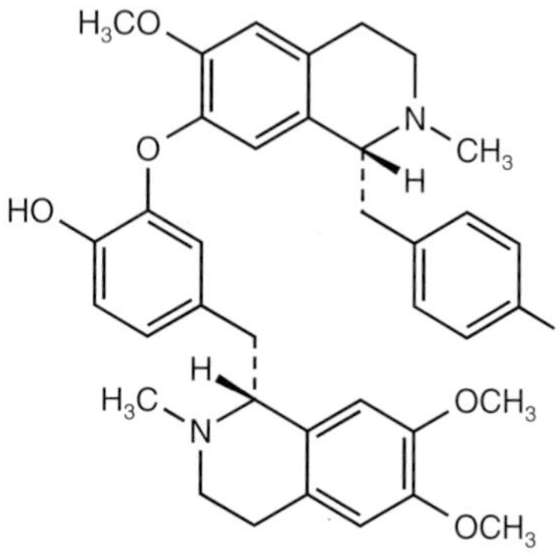

Liensinine: R = OH
Neferine: R = OCH_3

The fatty acid esters of β-sitosterol and β-sitosterol 3-*O*-β-D-glucopyranoside were also isolated from the lotus cotyledon [9]. In addition, the flavone glycosides rutin and hyperin were isolated from Plumula Nelumbinis [10].

A series of flavones were isolated from Stamen Nelumbinis and identified as kaempferol and kaempferol glycosides, myricetin 3′,5′-dimethylether 3-*O*-β-D-glucopyranoside, quercetin 3-*O*-β-D-glucopyranoside, and isorhamnetin glycosides. Besides flavones and flavone glycosides, other compounds were also isolated from Stamen Nelumbinis and identified as adenine, *myo*-inositol, arbutin, and β-sitosterol glucopyranoside [11].

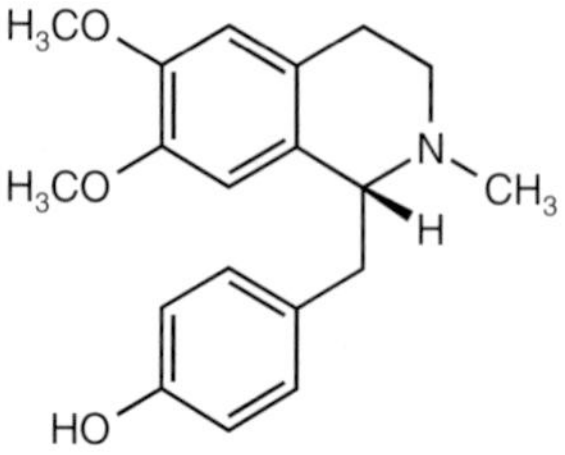

Armepavine

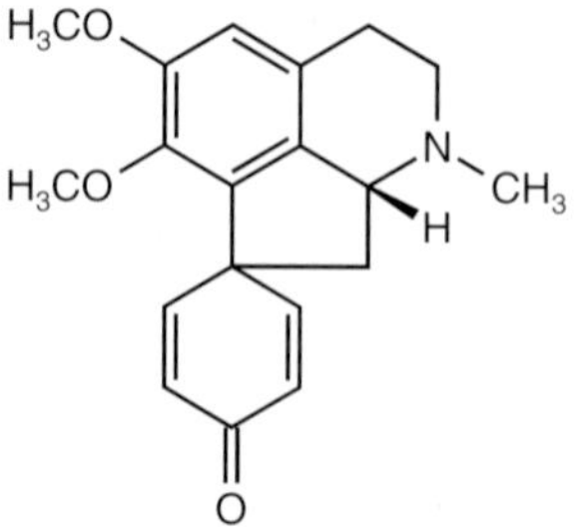

Pronuciferine

Pharmacology and Toxicology

A methanol extract from the leaves of *N. nucifera* exhibited potential antioxidant activity, and protected Caco-2 cells against H_2O_2-mediated cytotoxicity. A concentration-dependent protective effect against reactive oxygen species (ROS)-induced cytotoxicity was observed when Caco-2 cells were incubated with 10 m*M* H_2O_2 in combination with the methanol extract at concentrations of 0.1–0.3 mg ml^{-1}. The results showed that the extract exhibited scavenging activities on free radicals and hydroxyl radicals, which might be one of the mechanisms of the protective effect against oxidative damage. In addition, the extract was found to exhibit concentration-dependent antioxidant activities against hemoglobin-induced linoleic acid peroxidation and Fenton reaction-mediated plasmid DNA oxidation [12]. The methanol extract of Stamen Nelumbinis also showed strong antioxidant activity, scavenging peroxynitrite and 1,1-diphenyl-2-picrylhydrazyl free radicals, and inhibiting total ROS generation in kidney homogenate. Kaempferol glycosides were found to be the active principles [13].

An extract of the leaves of *N. nucifera* was found to possess an antiobesity effect. The treatment of mice with obesity induced by a high-fat diet for five weeks resulted in a prevention of increase in body weight, adipose tissue weight and liver triacylglycerol levels [14]. Quercetin 3-*O*-α-arabinopyranosyl-(1 → 2)-β-galactopyranoside, (+)-catechin, hyperoside, isoquercitrin, and astragalin from the leaves were found to stimulate lipolysis in the white adipose tissue of mice. This indicated the effect of the leaves in preventing diet-induced obesity to be due to various flavones. An activation of the β-adrenergic receptor pathway was suggested to be involved [15].

Early findings report nuciferine to possess neuroleptic property in rats, as well as weak anti-inflammatory, analgetic and antitussive activities in rodents. Nuciferine was believed to exert its pharmacological activities in association with dopamine-receptor blockade [16, 17]. The oral administration of neferine to spontaneously hypertensive rats caused a longlasting effect of lowering the systolic blood pressure [18]. Neferine markedly inhibited the phase I contraction of rat aortic rings treated with phenylephrine by release of intracellular Ca^{2+}, and inhibited the phase II contraction by entry of extracellular Ca^{2+} [19]. Neferine was also reported to significantly inhibit the spontaneous contractile force of portal vein and papillary muscle of the rat left ventricle [20]. Neferine clearly decreased the intracellular Ca^{2+}-dependent contraction of isolated aortic ring of rabbit induced by noradrenaline (norepinephrine), but not the extracellular Ca^{2+}-dependent component [18]. The effects of neferine on transmembrane potential in rabbit sinoatrial nodes and cultured myocardial cells of rats and guinea pigs indicated an inhibitory effect on the slow transmembrane Na^+, K^+, and Ca^{2+} current of myocardium [21]. Neferine was also reported to have an antiarrhythmic action in cats, and to decrease the arterial blood pressure [22].

Liensinine was found to be more potent than neferine on the cardiovascular system in mice after intravenous (i.v.) administration [23], and also exhibited antiarrhythmic activity in rats. The antiarrhythmic mechanism may be related to the blockade of Ca^{2+} and Na^+ influx [24]. The calcium antagonistic effect of liensinine was also demonstrated on the slow action potential in myocardium and slow inward current in the isolated sinoatrial node of rabbits [25]. An inhibition of platelet aggregation, an inhibition of lipid peroxidation, and a scavenging of oxygen free radicals by the alkaloids from lotus embryo, were reported [26].

Isoliensinine was reported to inhibit the proliferation of porcine coronary arterial smooth muscle cells stimulated by angiotensin II. Isoliensinine, at a concentration of 0.1 μ*M*, decreased the angiotensin II-stimulated overexpression of platelet-derived growth factor-β (TGF-β) and basic fibroblast growth factor (bFGF), and caused a decline in Fos, Myc, and Hsp70 overexpression. This indicates that an antiproliferative effect on the coronary arterial smooth muscle cells was related to a decrease in the overexpression of growth factors TGF-β and bFGF, and the protooncogenes Fos, Myc, and Hsp70 [27]. Isoliensinine was further reported to inhibit the pulmonary fibrosis in mice induced by bleomycin. An intratracheal injection of bleomycin to mice at a single dose of 0.1 mg per animal resulted in a significant increase in the hydroxyproline content and in histologically proven lung injury. Treatment of these mice with isoliensinine at oral doses of 10, 20, and 40 mg kg^{-1} caused a remarkable suppression in the increase in hydroxyproline content, and also supressed the lung histological injury. Supposedly, isoliensinine might enhance superoxide dismutase (SOD) activity and decrease malondialdehyde (MDA) levels in lung tissues and serum. Isoliensinine also significantly inhibited the overexpression of tumor necrosis factor-α (TNF-α) and TGF-β induced by bleomycin [28].

The alkaloid armepavine, from the leaf of *N. nucifera*, was found to suppress T-cell proliferation. The effects of armepavine were studied in MRL/MpJ-lpr/lpr mice, which have disease features similar to human systemic lupus erythematosus (SLE) [29]. T-cell immune responses are known to play important roles in the pathogenesis of SLE. After oral treatment of the MRL/MpJ-lpr/lpr mice for six weeks, armepavine was shown to have prevented lymphadenopathy and prolonged the life span of the mice. These effects seemed to be mediated by an inhibition of splenocyte proliferation, a suppression of interleukin-2 (IL-2), IL-4, IL-10, and interferon-γ (IFN-γ) gene expression, a reduction in glomerular hypercellularity and immune complexes deposition, and a decrease in urinary protein and anti-double-stranded DNA autoantibody production. Taken together, these findings suggested that armepavine might serve as an immunomodulator for the management of autoimmune diseases [29]. Armepavine was also found to inhibit the proliferation of human peripheral blood mononuclear cells induced by phytohemagglutinin, and to suppress the expression of IL-2 and IFN-γ genes, without direct cytotoxicity [30].

The seeds of *N. nucifera* were found to significantly increase the release of serotonin in rat hippocampus under normal conditions. Oral administration of the seeds to rats with a decreased hippocampal release of serotonin, as induced by chronic mild stress for eight weeks, led to a normalization of serotonin release in the hippocampus [31]. The treatment of rats with *N. nucifera* seeds also caused a significant reversal of the reduced sucrose intake caused by chronic mild stress. A mechanistic study showed the seeds to reverse the decrease in 5-HT_{1A} receptor binding in the CA2 and CA3 regions of the hippocampus, in the I to II regions of the frontal cortex, and in the hypothalamus of rat brain induced by chronic mild stress. These data indicated that *N. nucifera* seeds might exert their antidepression effect by enhancing 5-HT_{1A} receptor binding [32].

An alcoholic extract of *N. nucifera* seeds was found to exert an antioxidant activity in experimental studies, both *in vitro* and *in vivo*. The seed extract exhibited a powerful free radical-scavenging activity towards 1,1-diphenyl-2-picryl hydrazyl radical and nitric oxide (NO), with IC_{50} values of 6 μg ml^{-1} and 85 μg ml^{-1}, respectively. A study of the acute toxicity of the extract in mice failed to show any signs of toxicity with oral doses of up to 1 g kg^{-1}. Oral administration of the extract to rats at daily doses of 100 and 200 mg kg^{-1} for four days prior to carbon tetrachloride (CCl_4) intoxication caused a significant, dose-dependent increase in the level of SOD and catalase, and a significant decrease in the level of thiobarbituric acid reactive substances (TBARS) in both the liver and kidneys [33]. The alcohol extract also protected hepatocytes against cytotoxicity caused by CCl_4 and aflatoxin B_1 (AFB_1) [34]. The procyanidins from the seedpod of *N. nucifera* consisted of monomers, dimers and tetramers, with catechin and epicatechin as the base units. These are found to exhibit a strong antioxidant activity, to inhibit lipoxygenase activity, and to scavenge hydroxy and peroxy radicals [35].

The oral administration of an alcoholic extract of lotus rhizome [36] and of lotus stalk [37] significantly lowered the normal body temperature and yeast-provoked elevation of body temperature in rats, in a dose-dependent manner. The lotus rhizome extract was also found to markedly reduce the blood sugar levels of normal, glucose-fed hyperglycemic and streptozotocin-induced diabetic rats [38]. The extract exhibited a sedative effect, causing a reduction in spontaneous activity and a decrease in the exploratory behavioral pattern. The extract also enhanced the sleeping time of mice, as induced by pentobarbitone [39]. A methanol extract of the stamens of *N. nucifera* was shown to exert an inhibitory effect on rat lens aldose reductase. This is a key enzyme of the polyol pathway, also playing an important role in the complications associated with diabetes. The active compounds were identified as flavone glycosides bearing a 3-*O*-α-L-rhamnopyranosyl-(1 → 6)-β-D-glucopyranoside group such as kaempferol 3-*O*-α-L-rhamnopyranosyl-(1 → 6)-β-D-glucopyranoside and isorhamnetin 3-*O*-α-L-rhamnopyranosyl-(1 → 6)-β-D-glucopyranoside. Kaempferol [3-*O*-α-L-rhamnopyranosyl-(1 → 6)-β-D-glucopyranoside] and isorhamnetin [3-*O*-α-L-rhamnopyranosyl-(1 → 6)-β-D-glucopyranoside] exhibited high activity in inhibiting rat lens aldose reductase *in vitro*, with IC_{50} values of 5.6 and 9.0 μ*M*, respectively [11].

References

1 Das, S., Ray, B., and Ghosal, P.K. (1992) Structural studies of a polysaccharide from the seeds of *Nelumbo nucifera*. *Carbohydr. Res.*, **224**, 331–335.

2 Mukherjee, P.K., Saha, K., Das, J., Pal, M., and Saha, B.P. (1997) Studies on the anti-inflammatory activity of rhizomes of *Nelumbo nucifera*. *Planta Med.*, **63**, 367–369.

3 Kunimoto, J., Yoshikawa, Y., Tanaka, S., Imori, Y., Isoi, K., Masada, Y., Hashimoto, K., and Inoue, T. (1973) Alkaloids of *Nelumbo nucifera*. XVI. *Phytochemistry*, **12**, 699–701.

4 Shoji, N., Umeyama, A., Saito, N., Iuchi, A., Takemoto, T., Kajiwara, A., and Ohizumi, Y. (1987) Asimilobine and lirinidine, serotonergic receptor antagonists, from *Nelumbo nucifera*. *J. Nat. Prod.*, **50**, 773–774.

5 Kunitomo, J., Nagai, Y., Okamoto, Y., and Furukawa, H. (1970) Alkaloids of *Nelumbo nucifera*. XIV. Tertiary base. *Yakugaku Zasshi*, **90**, 1165–1169.

6 Fu, S.Y., Huang, A.J., Liu, H.W., Sun, Y.L., Wu, Q., and Xia, Y.Y. (1992) Studies of flavor components of lotus leaf. I. Analysis of natural flavor components of lotus leaf. *Acta Sci. Nat. Univ. Peking*, **28**, 699–705.

7 Nishibe, S., Tsukamoto, H., Kinoshita, H., Kitagawa, S., and Sakushima, A. (1986) Alkaloids from embryo of the seed of *Nelumbo nucifera*. *J. Nat. Prod.*, **49**, 548.

8 Hu, X.M., Zhou, B.H., Luo, S.D., Cai, H.S., and Yin, W.H. (1993) Quantitative determination of liensinine in the embryo Nelumbinis (*Nelumbo nucifera* Gaertn.) by TLC-scanning. *China J. Chin. Mater. Med.*, **18**, 167–168.

9 Lou, H.X., Yuan, H.Q., Ji, M., and Xu, D.Y. (1995) Sitosterol esters from embryo of the seed of *Nelumbo nucifera*. *J. Shandong Med. Univ.*, **33**, 346–348.

10 Wang, J.L., Hu, X.M., Yin, W.H., and Cai, H. S. (1991) Alkaloids of Plumula nelumbinis. *China J. Chin. Mater. Med.*, **16**, 673–675.

11 Lim, S.S., Jung, Y.J., Hyun, S.K., Lee, Y.S., and Choi, J.S. (2006) Rat lens aldose reductase inhibitory constituents of *Nelumbo nucifera* stamens. *Phytother. Res.*, **20**, 825–830.

12 Wu, M.J., Wang, L., Wang, C.Y., and Yen, J.H. (2003) Antioxidant activity of methanol extract of the lotus leaf (*Nelumbo nucifera* Gertn.). *Am. J. Chin. Med.*, **31**, 687–698.

13 Jung, H.A., Kim, J.E., Chung, H.Y., and Choi, J.S. (2003) Antioxidant principles of *Nelumbo nucifera* stamens. *Arch. Pharm. Res.*, **26**, 279–285.

14 Ono, Y., Hattori, E., Fukaya, Y., Imai, S., and Ohizumi, Y. (2006) Anti-obesity effect of *Nelumbo nucifera* leaves extract in mice and rats. *J. Ethnopharmacol.*, **106**, 238–244.

15 Ohkoshi, E., Miyazaki, H., Shindo, K., Watanabe, H., Yoshida, A., and Yajima, H. (2007) Constituents from the leaves of *Nelumbo nucifera* stimulate lipolysis in the white adipose tissue of mice. *Planta Med.*, **73**, 1255–1259.

16 Bhattacharya, S.K., Bose, R., Ghosh, P., Tripathi, V.J., Ray, A.B., and Dasgupta, B. (1978) Psychopharmacological studies on (–)-nuciferine and its Hofmann degradation product atherosperminine. Psychopharmacology, **59**, 29–33.

17 Curtis, D.R., Lodge, D., and Bornstein, J.C. (1979) Nuciferine and central glutamate receptors. *J. Pharm. Pharmacol.*, **31**, 795–797.

18 Hu, W.S., Guo, L.J., Feng, X.L., Ye, S.J., and Jiang, M.X. (1991) An analysis of anti-hypertensive and vasodilating effects of neferine. *Chin J. Pharmacol. Toxicol.*, **5**, 111–113.

19 Yao, W., Zhu, Q.Y., and Su, D.F. (1991) Effects of tetrandrine and neferine on the phenylephrine-induced contraction in rat aortic rings. *Chin. Pharmacol. Bull.*, **7**, 357–359.

20 Zhu, Q.Y., Jin, G., Yan, Z.L., Miao, C.Y., and Su, D.F. (1992) Inhibitory effects of neferine and tetrandrine on portal vein and papillary muscle in rats. *Acta Pharmacol. Sin.*, **13**, 359–361.

21 Li, G.R., Li, X.G., and Lu, F.H. (1989) Effects of neferine on transmembrane potential in rabbit sinoatrial nodes and clusters of cultured myocardial cells from neonatal rats. *Acta Pharmacol. Sin.*, **10**, 328–331.

22 Li, G.R., Qian, J.Q., and Lu, F.H. (1990) Effects of neferine on heart electromechanical activity in anaesthetized cats. *Acta Pharmacol. Sin.*, **11**, 158–161.

23 Li, Q.J., Zeng, G.Y., Pan, J.X., Li, C.L., Guo, X. D., and Wu, Y.J. (1992) The total alkaloids from the green embryo in the seed of *Nelumbo nucifera*: pharmacology on cardiovascular system and active component. *J. Beijing Med. Univ.*, **24**, 61–62.

24 Wang, J.L., Nong, Y., and Jing, M.X. (1992) Effects of liensinine on hemodynamics in rats and the physiologic properties of isolated rabbit atria. *Acta Pharm. Sin.*, **27**, 881–885.

25 Wang, J.L., Nong, Y., Xia, G.J., Yao, W.X., and Jiang, M.X. (1993) Effects of liensinine on slow action potential in myocardium and slow inward current in canine cardiac Purkinje fibers. *Acta Pharm. Sin.*, **28**, 812–816.

26 Lu, W.Q., Zhen, H.H., and Ge, X. (1996) Study of lianzixin, the embryo and young green leaf of seed of *Nelumbo nucifera* Gaetn. *Chin. Trad. Herbal Drugs*, **27**, 438–440.

27 Xiao, J.H., Zhang, Y.L., Feng, X.L., Wang, J.L., and Qian, J.Q. (2006) Effects of isoliensinine on angiotensin II-induced proliferation of porcine coronary arterial smooth muscle cells. *J. Asian Nat. Prod. Res.*, **8**, 209–216.

28 Xiao, J.H., Zhang, Y.L., Chen, H.L., Feng, X. L., and Wang, J.L. (2005) Inhibitory effects of isoliensinine on bleomycin-induced pulmonary fibrosis in mice. *Planta Med.*, **71**, 225–230.

29 Liu, C.P., Tsai, W.J., Shen, C.C., Lin, Y.L., Liao, J.F., Chen, C.F., and Kuo, Y.C. (2006) Inhibition of (S)-armepavine from *Nelumbo nucifera* on autoimmune disease of MRL/MpJ-lpr/lpr mice. *Eur. J. Pharmacol.*, **531**, 270–279.

30 Liu, C.P., Kuo, Y.C., Shen, C.C., Wu, M.H., Liao, J.F., Lin, Y.L., Chen, C.F., and Tsai, W.J. (2007) (S)-Armepavine inhibits human peripheral blood mononuclear cell activation by regulating Itk and PLCγ activation in a PI-3K-dependent manner. *J. Leukoc. Biol.*, **81**, 1276–1286.

31 Kang, M., Pyun, K.H., Jang, C.G., Kim, H., Bae, H., and Shim, I. (2005) Nelumbinis Semen reverses a decrease in hippocampal 5-HT release induced by chronic mild stress in rats. *J. Pharm. Pharmacol.*, **57**, 651–656.

32 Jang, C.G., Kang, M., Cho, J.H., Lee, S.B., Kim, H., Park, S., Lee, J., Park, S.K., Hong, M., Shin, M.K., Shim, I.S., and Bae, H. (2004) Nelumbinis Semen reverses a decrease in 5-HT_{1A} receptor binding induced by chronic mild stress, a depression-like symptom. *Arch. Pharm. Res.*, **27**, 1065–1072.

33 Rai, S., Wahlie, A., Mukherjee, K., Saha, B.P., and Mukherjee, P.K. (2006) Antioxidant activity of *Nelumbo nucifera* (sacred lotus) seeds. *J. Ethnopharmacol.*, **104**, 322–327.

34 Sohn, D.H., Kim, Y.C., Oh, S.H., Park, E.J., Li, X., and Lee, B.H. (2003) Hepatoprotective and free radical scavenging effects of *Nelumbo nucifera*. *Phytomedicine*, **10**, 165–169.

35 Ling, Z.Q., Xie, B.J., and Yang, E.L. (2005) Isolation, characterization, and determination of antioxidative activity of oligomeric procyanidins from the seed pod of *Nelumbo nucifera* Gaertn. *J. Agric. Food Chem.*, **53**, 2441–2445.

36 Mukherjee, P.K., Das, J., Saha, K., Giri, S.N., Pal, M., and Saha, B.P. (1996) Antipyretic activity of *Nelumbo nucifera* rhizome extract. *Indian J. Exp. Biol.*, **34**, 275–276.

37 Sinha, S., Mukherjee, P.K., Mukherjee, K., Pal, M., Mandal, S.C., and Saha, B.P. (2000) Evaluation of antipyretic potential of *Nelumbo nucifera* stalk extract. *Phytother. Res.*, **14**, 272–274.

38 Mukherjee, P.K., Saha, K., Pal, M., and Saha, B.P. (1997) Effect of *Nelumbo nucifera* rhizome extract on blood sugar level in rats. *J. Ethnopharmacol.*, **58**, 207–213.

39 Mukherjee, P.K., Saha, K., Balasubramanian, R., Pal, M., and Saha, B.P. (1996) Studies on psychopharmacological effects of *Nelumbo nucifera* Gaertn. rhizome extract. *J. Ethnopharmacol.*, **54**, 63–67.

Notopterygium forbesii and *Notopterygium incisum*

Rhizoma et Radix Notopterygii (Qianghuo) is the dried rhizome and root of *Notopterygium forbesii* Boiss. or *Notopterygium incisum* Ting ex H.T. Chang (Apiaceae). It is used for the treatment of enteritis and cough, headache, common cold, and rheumatic arthralgia.

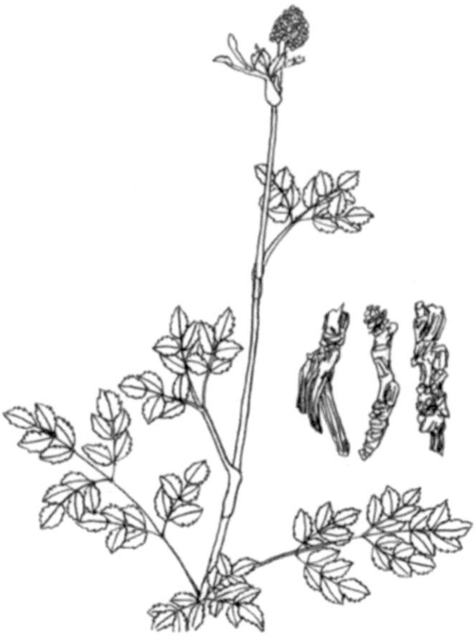

Notopterygium forbesii Boiss.

Chemistry

The root and rhizome of *N. forbesii* is a further herbal medicine containing furocoumarins as active components, in addition to Radix Angelicae Dahuricae (*Angelica dahurica*), Radix Peucedani (*Peucedanum praeruptorum* and *Peucedanum decursivum*), and Fructus Psoraleae (*Psoralea corylifolia*). Ferulic acid, cnidilin, bornyl ferulate, phenethyl ferulate, ostruthin, farcarindiol, *p*-hydroxphenethyl anisate, 4-hydroxy-3′,5′-dimethoxystilbene, β-sitosterol *O*-β-D-glucopyranoside, and the furocoumarins bergamottin, bergapten, bergaptol, bergaptol *O*-β-D-glucopyranoside, demethylfuropinnarin, isoimperatorin, nodakenin, and 6′-*O*-feruloylnodakenin, nodakenetin, notopterol, and phellopterin were isolated from the underground part of *N. forbesii* [1–3]. The underground part of *N. forbesii* contained large amounts of nodakenin (2%) as the major component, in addition to *p*-hydroxphenethyl anisate (0.7%), bergaptol *O*-β-D-glucopyranoside (0.2%), 6′-*O*-feruloylnodakenin (0.7%), and notopterol (0.08%) [4]. Related chemical composition was also found in the rhizome and root of *N. incisum* [5–9]. Notopterol (1.2%) [4] and nodakenin [10] seemed to be the major furocoumarins in the rhizome and root of *N. incisum*. The Chinese Pharmacopoeia requires a quantitative determination of essential oil in Rhizoma et Radix Notopterygii. The content of essential oil in Rhizoma et Radix Notopterygii should be not less than 2.8% (ml g^{-1}).

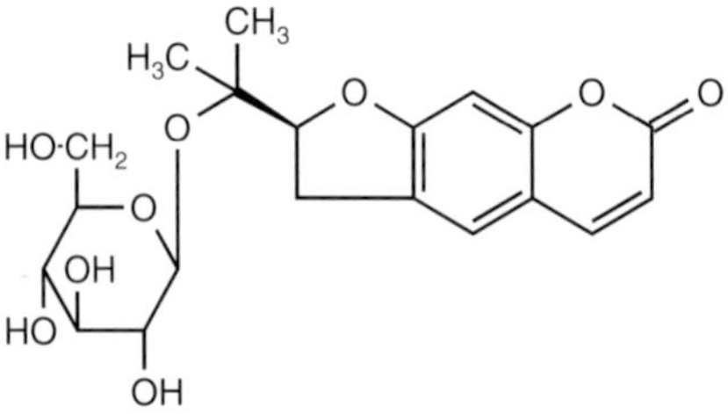

Nodakenin

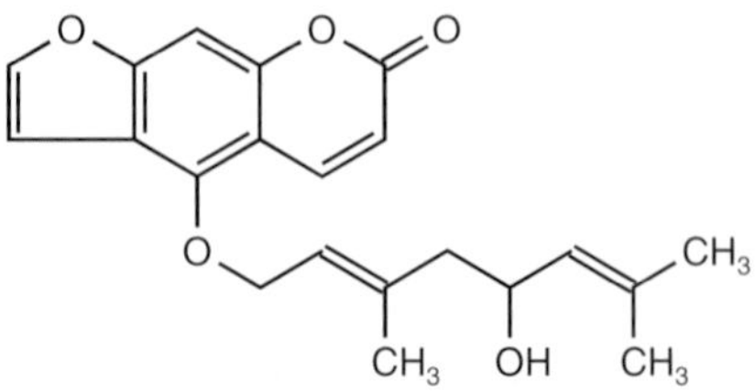

Notopterol

Handbook of Chinese Medicinal Plants: Chemistry, Pharmacology, Toxicology
Weici Tang and Gerhard Eisenbrand

ISBN: 978-3-527-32226-8

Pharmacology and Toxicology

Notopterol was identified as the analgesic component of *N. incisum* active against acetic acid-induced writhing in mice. Notopterol also showed anti-inflammatory activity, and inhibited vascular permeability in mice. An observed intensive prolongation of pentobarbital-induced sleep time appeared to be caused by the inhibitory effect of notopterol on drug metabolism in the liver [11]. The root and rhizome of *N. incisum* and *N. forbesii* have been found to antagonize arrhythmia induced by aconitine [12, 13].

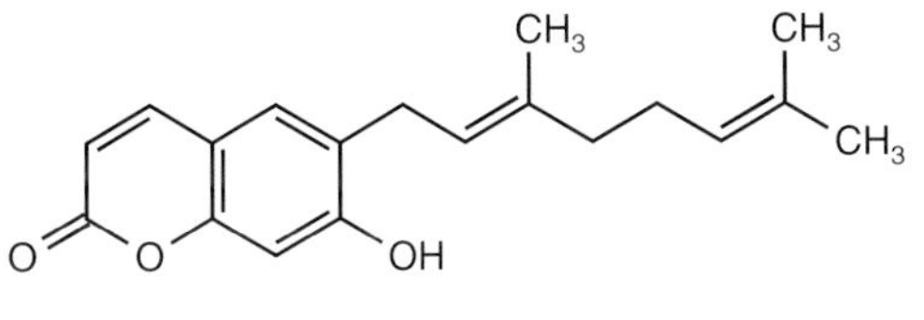

Ostruthin

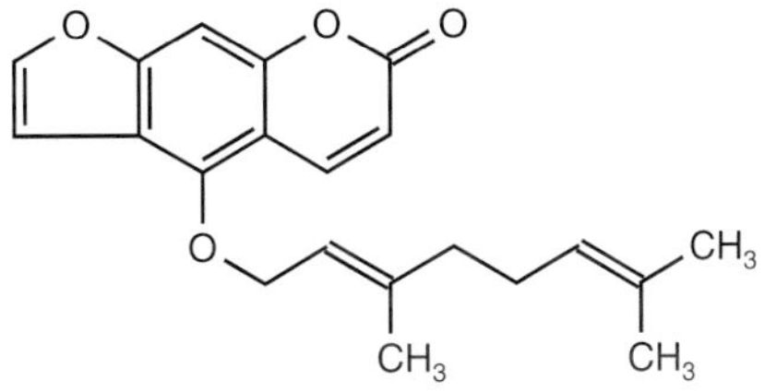

Bergamottin

The intraperitoneal (i.p.) administration of carbon tetrachloride (CCl_4) to mice led to significant increases in thiobarbituric acid-reactive substances, free malondialdehyde, lipid-conjugated dienes and fluorescent lipid peroxidation products in the liver. Subchronic pretreatment with oral doses of the methanol extract of either the underground part of *N. incisum*, or that of *N. forbesii*, caused an appreciable suppression of the formation of CCl_4-induced lipid peroxidation products. The suppressing potency by *N. incisum* was reported to be stronger [14].

The *n*-hexane extract of the underground parts of *N. incisum* inhibited, *in vitro*, the activity of 5-lipoxygenase and cyclooxygenase. Two major constituents, phenethyl ferulate and falcarindiol, and one minor compound, bornyl ferulate, were identified as the active principles. It was concluded that phenethyl ferulate was mainly responsible for inhibiting the cyclooxygenase activity, and falcarindiol for the 5-lipoxygenase activity [15]. Ostruthin and bornyl ferulate were reported to induce comparable, neurite-like structures in 20% of rat PC12 cells at 2 $\mu g\,ml^{-1}$, but showed cytotoxicity at concentrations higher than 3 $\mu g\,ml^{-1}$ [16]. A boiling water extract of *N. incisium* was found to induce chromosomal aberration and micronuclei in mice [17].

Bergamottin is one of the most abundant furocoumarins present in grapefruit juice, and has been proposed as a major intestinal inhibitor of the cytochrome P450 isoenzyme CYP3A4 [18]. Substantial inhibitory effects of bergamottin on CYP3A4 are observed below a concentration of 10 μM [19]. Bergamottin was also found to be a mechanism-based inactivator of CYP2B6 and CYP3A5 in a reconstituted system. The inactivation of both CYP2B6 and CYP3A5 was NADPH-dependent, and irreversible. The incubations of CYP2B6 and CYP3A5 with 20 μM bergamottin at 37 °C for 20 min resulted in an approximate 60% loss in catalytic activity, accompanied by a significant loss in intact heme. A study using [^{14}C]bergamottin showed that bergamottin was bound irreversibly to the apoprotein in the bergamottin-inactivated samples. Two major CYP2B6 metabolites of bergamottin were identified as 6′,7′-dihydroxybergamottin and bergaptol, whereas CYP3A5 generated three additional metabolites [20].

References

1 Wang, S. and Wang, T. (1996) Chemical constituents of *Notopterygium forbesii* Boiss. *China J. Chin. Mater. Med.*, **21**, 295–296.

2 Xin, L.J. and Ling, L.Q. (1988) Chemical constituents of *Notopterygium forbesii*. *Acta Bot. Sin.*, **30**, 562–564.

3 Yang, X.W., Yang, Z.K., Gu, Z.M., Zhou, G.C., Hattori, M., and Namba, T. (1994) Chemical constituents of underground parts of *Notopterygium forbesii*. *Chin. Pharm. J.*, **29**, 141–143.

4 Gu, Z.M., Zhang, D.X., Yang, X.W., Hattori, M., and Namba, T. (1990) Isolation of two new coumarin glycosides from *Notopterygium forbesii* and evaluation of a Chinese crude drug, qiang-huo, the underground parts of *N. incisum* and *N. forbesii*, by high-performance liquid chromatography. *Chem. Pharm. Bull. (Tokyo)*, **38**, 2498–2502.

5 Sun, Y.F. (1985) Isolation and identification of chemical constituents from an alcoholic extract of *Notopterygium incisium*. *Bull. Chin. Mater. Med.*, **10**, 31–33.

6 Xiao, Y.Q., Sun, Y.F., and Liu, X.H. (1994) Chemical constituents of *Notopterygium incisum* Ting. *China J. Chin. Mater. Med.*, **19**, 421–422. 447.

7 Sun, Y.F., Xiao, Y.Q., and Liu, X.H. (1994) Chemical constituents of *Notopterygium incisum* Ting. *China J. Chin. Mater. Med.*, **19**, 357–358. 383–384.

8 Sun, Y.F., Xiao, Y.Q., and Liu, X.H. (1994) Chemical constituents of *Notopterygium incisum* Ting. III. Chemical constituents isolated and identified from petroleum ether extracts of *N. incisium* Ting. *China J. Chin. Mater. Med.*, **19**, 99–100. 127.

9 Xiao, Y.Q., Baba, K., Taniguchi, M., Liu, X.H., Sun, Y.F., and Kozawa, M. (1995) Three new coumarins from *Notopterygium incisum*. *Acta Pharm. Sin.*, **30**, 274–279.

10 Qin, Y., Zhang, H., and Sun, Y. (1996) Determination of nodakenin in the rhizome or root of *Notopterygium incisum* Ting by TLC-scanning. *China J. Chin. Mater. Med.*, **21**, 486–487. 511.

11 Okuyama, E., Nishimura, S., Ohmori, S., Ozaki, Y., Satake, M., and Yamazaki, M. (1993) Analgesic component of *Notopterygium incisum* Ting. *Chem. Pharm. Bull. (Tokyo)*, **41**, 926–929.

12 Zhu, X. and Chu, R. (1990) A comparison of anti-arrhythmic effects of four kinds of rhizoma seu radix Notopterygii. *China J. Chin. Mater. Med.*, **15**, 366–368, 385.

13 Qin, C.L. and Jiao, Y. (1987) Effects of aqueous extracts of rhizoma seu radix Notopterygii (*Notopterygium incisum* Ting ex H.T. Chang) on experimental arrhythmias. *Bull. Chin. Mater. Med.*, **12**, 45–47, 60.

14 Yang, X.W., Gu, Z.M., Wang, B.X., Hattori, M., and Namba, T. (1991) Comparison of anti-lipid peroxidative effects of the underground parts of *Notopterygium incisum* and *N. forbesii* in mice. *Planta Med.*, **57**, 399–402.

15 Zschocke, S., Lehner, M., and Bauer, R. (1997) 5-Lipoxygenase and cyclooxygenase inhibitory active constituents from Qianghuo (*Notopterygium incisum*). *Planta Med.*, **63**, 203–206.

16 Qi, J., Ojika, M., and Sakagami, Y. (1999) Differentiation in a rat PC12 cell line induced by ostruthin and (−)-bornyl ferulate, constituents of a Chinese herbal medicine. *Biosci. Biotechnol. Biochem.*, **63**, 1501–1502.

17 Yin, X.J., Liu, D.X., Wang, H.C., and Zhou, Y. (1991) A study on the mutagenicity of 102 raw pharmaceuticals used in Chinese traditional medicine. *Mutat. Res.*, **260**, 73–82.

18 Paine, M.F., Criss, A.B., and Watkins, P.B. (2004) Two major grapefruit juice components differ in intestinal CYP3A4 inhibition kinetic and binding properties. *Drug Metab. Dispos.*, **32**, 1146–1153.

19 Girennavar, B., Poulose, S.M., Jayaprakasha, G.K., Bhat, N.G., and Patil, B.S. (2006) Furocoumarins from grapefruit juice and their effect on human CYP3A4 and CYP1B1 isoenzymes. *Bioorg. Med. Chem.*, **14**, 2606–2612.

20 Lin, H.L., Kent, U.M., and Hollenberg, P.F. (2005) The grapefruit juice effect is not limited to cytochrome P450 (P450) 3A4: evidence for bergamottin-dependent inactivation, heme destruction, and covalent binding to protein in P450s 2B6 and 3A5. *J. Pharmacol. Exp. Ther.*, **313**, 154–164.

Oldenlandia diffusa

The dried whole plant of *Oldenlandia diffusa* (Willd.) Roxb. (Rubiaceae) is listed in the appendix of the Chinese Pharmacopoeia.

Oldenlandia diffusa (Willd.) Roxb.

Chemistry

The whole plant of *O. diffusa* is known to contain β-sitosterol. Iridoid glycosides from the whole plant of *O. diffusa* were identified as geniposidic acid, scandoside, feretoside, oldenlandoside III, asperulosidic acid, deacetylasperulosidic acid [1], (*E*)6-*O*-*p*-coumaroyl scandoside methyl ester and its 10-methyl ether [2].

Pharmacology and Toxicology

The effects of *O. diffusa* on lymphocytes were studied *in vitro* using murine spleen cells. It was found to markedly stimulate the proliferation of murine spleen cells. B cells seemed to be the responder cells, because the response was depleted by the treatment of spleen cells with anti-immunoglobulin antibody and complement and after nylon wool column purification. *Oldenlandia diffusa* enhanced the production of immunoglobulins and enhanced the induction of alloantigen-specific cytotoxic T lymphocytes. However, it had no effect on natural killer (NK) cells. Furthermore, *O. diffusa* stimulated macrophages to produce interleukin-6 (IL-6) and tumor necrosis factor (TNF). Electroelution of the proteins from a SDS–PAGE gel showed that the active components of *O. diffusa* had an apparent molecular weight of between 90 and 200 kDa, and were sensitive to pronase E and $NaIO_4$ treatment, suggesting that they were glycoprotein in nature [3].

Oldenlandia diffusa has been used in traditional Chinese medicine for treating liver, lung, and rectal tumors. A dose-dependent augmentation of oxidative burst as an indicator of phagocytic function in a murine macrophage cell line J774 was observed with the whole plant. The plant also inhibited murine renal cell carcinoma (Renca) growth in Balb/c mice after subcutaneous transplantation at a daily oral dose of 4 mg per animal. This indicated that *O. diffusa* was capable of enhancing macrophage function *in vitro*, and inhibiting tumor growth *in vivo* [4]. The extract of the whole plant of *O. diffusa* caused a weak stimulation of the proliferation of human lymphocytes and also enhanced cytotoxic T-lymphocyte activity, but failed to enhance NK cell activity. The extract stimulated the production of immunoglobulin by B cells, and the production of IL-1 by monocytes *in vitro* [5]. The extract of *O. diffusa* markedly promoted the proliferation of spleen cells in mice. It also stimulated the cytotoxicity of human and

Handbook of Chinese Medicinal Plants: Chemistry, Pharmacology, Toxicology
Weici Tang and Gerhard Eisenbrand

ISBN: 978-3-527-32226-8

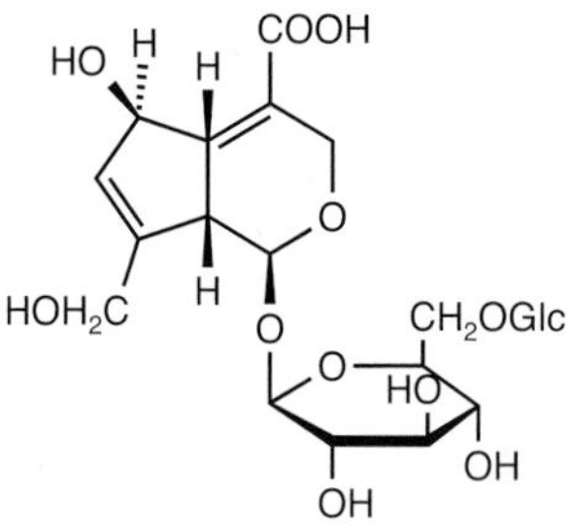

Scandoside

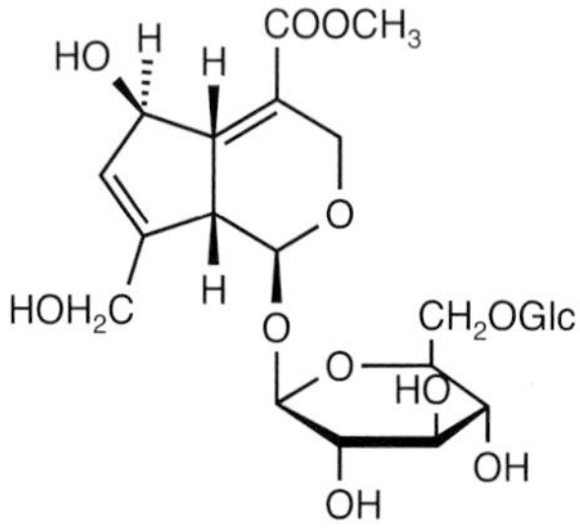

Feretoside

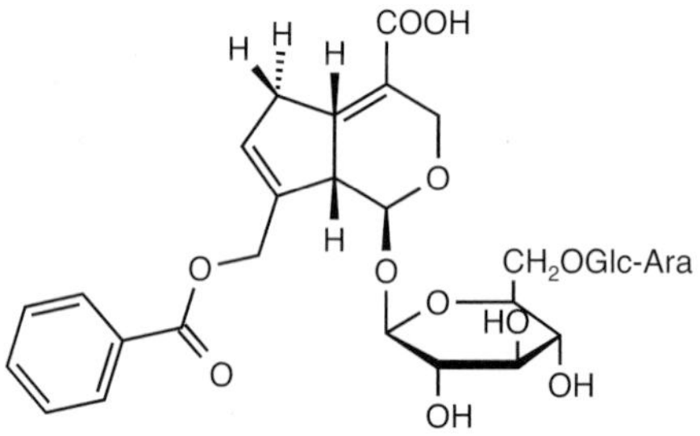

Oldenlandoside III

mice NK cells to tumor cells, antibody production by B cells, cytokine production by monocytes, and phagocytosis towards tumor cells [6].

An aqueous extract of *O. diffusa* was reported to significantly inhibit the mutagenesis of benzo[*a*]pyrene 7,8-dihydrodiol (BaP 7,8-DHD) and benzo[*a*]pyrene 7,8-dihydrodiol-9,10-epoxide (BPDE) in *Salmonella typhimurium* TA100, in the presence of a rat liver S-9 mix as the metabolic activation system, and in a concentration-dependent manner. The extract also inhibited the mutagenesis of BPDE in a concentration-dependent manner in the absence of a S-9 mix. However, *O. diffusa* significantly enhanced the DNA binding of both BaP 7,8-DHD and BPDE, inhibited the formation of organosoluble metabolites of BaP 7,8-DHD, and decreased the formation of water-soluble conjugates of BaP 7,8-DHD and BPDE. The total radioactivity in the water-soluble sulfate and glutathione conjugates was increased after addition of the extract of *O. diffusa*, whereas the glucuronide fraction was decreased. These findings led to the suggestion that the extract acted as a blocking agent through a scavenging mechanism [7].

An aqueous extract of *O. diffusa* was found to exhibit strong antiproliferative activity against a series of cancer cell lines (IC_{50} values ranged from 7 to 25 mg raw herb material per ml after 48 h), but only very limited cytotoxicity (10% inhibition) on normal pancreatic cells, even at a concentration of 50 mg ml^{-1}. Apoptosis was observed in B16-F10 cells after treatment with the extract. Oral administration of the herbal extract to C57Bl/j mice at a dose corresponding to 5 g raw material per kg body weight on days 3 to 12 after the transplantation of cancer cells effectively inhibited the growth of B16-F10 cells in the lungs, with a 70% reduction in lung metastases [8]. The incubation of HL60 human promyelocytic leukemia cells with an ethanol or aqueous extract of *O. diffusa* resulted in a concentration-dependent inhibition of cancer cell growth. Flow cytometry analyses revealed that the extracts induced a cell cycle arrest at sub-G_1, indicating the induction of apoptosis. A significant elevation in the activities of caspases 2 and 3 was detected [9].

The aqueous extract of *O. diffusa* also inhibited the mutagenesis of aflatoxin B_1 (AFB_1) in *S. typhimurium* TA100 in the presence of a rat liver S-9 mix in a concentration-dependent manner. The extract significantly inhibited AFB_1 binding to DNA, reduced AFB_1-DNA adduct formation, and also significantly decreased the formation of organosoluble metabolites of AFB_1 [10]. *Oldenlandia diffusa* inhibited the mutagenicity of BaP in *S. typhimurium* TA100 in the presence of either a noninduced or a β-naphthoflavone-induced S-9 mix. This suggested that *O. diffusa* may exert antimutagenic activity by inhibiting the CYPIA1-mediated metabolism of BaP [11].

Oldenlandis diffusa also consistently inhibited the mutagenicity of AFB_1 bioactivated by either a noninduced or a dexamethasone-induced S-9 mix. These effects correlated with the inhibition of cytochrome P450-linked aminopyrine *N*-demethylase activity in dexamethasone-induced microsomes, and with an inhibition of dexamethasone-induced S-9-mediated metabolism of $[^3H]AFB_1$. Since dexamethasone treatment has been associated with an induction of the CYP3 enzyme family, it was further suggested that *O. diffusa* might exert an antimutagenic activity towards AFB_1 through an inhibition of the CYP3-mediated metabolism of AFB_1 [12].

The iridoid glycosides geniposidic acid, scandoside and deacetylasperulosidic acid were reported to inhibit the oxidation of low-density lipoprotein (LDL) at a concentration of 20 μg ml^{-1} [1].

References

1 Kim, D.H., Lee, H.J., Oh, Y.J., Kim, M.J., Kim, S.H., Jeong, T.S., and Baek, N.I. (2005) Iridoid glycosides isolated from *Oldenlandia diffusa* inhibit LDL-oxidation. *Arch. Pharm. Res.*, **28**, 1156–1160.

2 Liang, Z.T., Jiang, Z.H., Leung, K.S., and Zhao, Z.Z. (2006) Determination of iridoid glucosides for quality assessment of Herba Oldenlandiae by high-performance liquid chromatography. *Chem. Pharm. Bull.*, **54**, 1131–1137.

3 Yoshida, Y., Wang, M.Q., Liu, J.N., Shan, B.E., and Yamashita, U. (1997) Immunomodulating activity of Chinese medicinal herbs and *Oldenlandia diffusa* in particular. *Int. J. Immunopharmacol.*, **19**, 359–370.

4 Wong, B.Y., Lau, B.H., Jia, T.Y., and Wan, C.P. (1996) *Oldenlandia diffusa* and *Scutellaria barbata* augment macrophage oxidative burst and inhibit tumor growth. *Cancer Biother. Radiopharm.*, **11**, 51–56.

5 an, B.E., Yoshida, Y., Sugiura, T., and Yamashita, U. (1999) Stimulating activity of Chinese medicinal herbs on human lymphocytes *in vitro*. *Int. J. Immunopharmacol.*, **21**, 149–159.

6 Shan, B.E., Zhang, J.Y., and Du, X.N. (2001) Immunomodulatory activity and anti-tumor activity of *Oldenlandia diffusa in vitro*. *Zhongguo Zhong Xi Yi Jie He Za Zhi*, **21**, 370–374.

7 Wong, B.Y., Lau, B.H., and Teel, R.W. (1992) Chinese medicinal herbs modulate mutagenesis, DNA binding and metabolism of benzo[*a*]pyrene 7,8-dihydrodiol and benzo[*a*] pyrene 7,8-dihydrodiol-9 10-epoxide. *Cancer. Lett.*, **62**, 123–131.

8 Gupta, S., Zhang, D., Yi, J., and Shao, J. (2004) Anticancer activities of *Oldenlandia diffusa*. *J. Herb. Pharmacother.*, **4**, 21–33.

9 Yadav, S.K. and Lee, S.C. (2006) Evidence for *Oldenlandia diffusa*-evoked cancer cell apoptosis through superoxide burst and caspase activation. *Zhong Xi Yi Jie He Xue Bao*, **4**, 485–489.

10 Wong, B.Y., Lau, B.H., Tadi, P.P., and Teel, R. W. (1992) Chinese medicinal herbs modulate mutagenesis DNA binding and metabolism of aflatoxin B_1. *Mutat. Res.*, **279**, 209–216.

11 Wong, B.Y., Lau, B.H., Yamasaki, T., and Teel, R.W. (1993) Modulation of cytochrome P-450IA1-mediated mutagenicity DNA binding and metabolism of benzo[*a*]pyrene by Chinese medicinal herbs. *Cancer. Lett.*, **68**, 75–82.

12 Wong, B.Y., Lau, B.H., Yamasaki, T., and Teel, R.W. (1993) Inhibition of dexamethasone-induced cytochrome P450-mediated mutagenicity and metabolism of aflatoxin B_1 by Chinese medicinal herbs. *Eur. J. Cancer Prev.*, **2**, 351–356.

Ophiopogon japonicus

Radix Ophiopogonis (Maidong) is the dried tuber of *Ophiopogon japonicus* (Thunb.) Ker-Gawl. (Liliaceae). It is used for the treatment of cough, fidgetiness and insomnia, diabetes, constipation, and diphtheria.

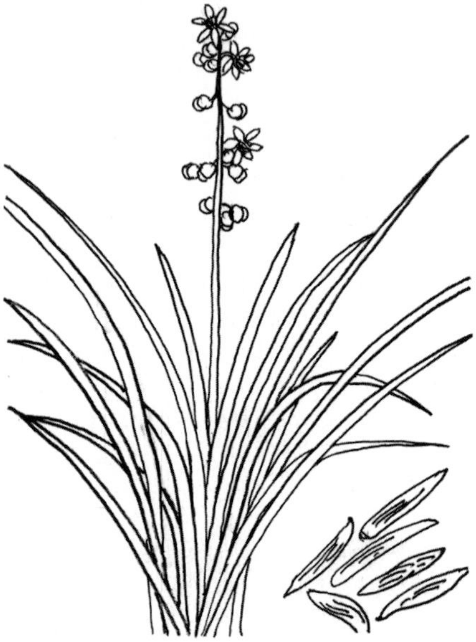

Ophiopogon japonicus (Thunb.) Ker-Gawl.

Chemistry

The root of *O. japonicus* was reported to contain a series of homoisoflavanones, with methylophiopogonanones A and B as the major components [1]. Minor homoisoflavanones were identified as 6-aldehydoisoophiopogonones A and B [2, 3].

Saponins from the root of *O. japonicus* were found mainly derived from the sapogenin ophiogenin, and identified as ophiopogonins A and B [4]. Saponins derived from ruscogenin [5], 25*S*-ruscogenin, and 23,24-dihydroruscogenin were also isolated, for example, ophiopogonin D′ (ruscogenin 1-*O*-α-L-rhamnopyranosyl-(1 → 2)-[β-D-xylopyranosyl(1 → 3)]β-D-fucopyranoside). Diosgenin glycosides such as ophiopogonin D (diosgenin 3-*O*-α-L-rhamnopyranosyl-(1 → 2)-[β-D-xylopyranosyl-(1 → 3)]-β-D-glucopyranoside) were also isolated from the root [6]. Ophiopogonin B and ophiopogonin D seemed to be the major saponins [7].

The essential oil from the root of *O. japonicus* was isolated in a yield of about 0.1%. The major components in the essential oil were identified as longifolene (18.5%), β-patchoulene (9.6%), guajal (5.2%), cyperene (2.6%), α-patchoulene (1.8%) and α-humulene (1.8%) [8]. In addition, borneol *O*-β-D-glucopyranoside, borneol *O*-β-D-apiofuranosyl-(1 → 6)-β-D-glucopyranoside [9], and calcium bisborneolhemisulfate [10] were isolated from the root. In addition, polysaccharides were also isolated from the root of *O. japonicus*. It should be mentioned that only the root of *O. japonicus* should be used as Maidong; however, other

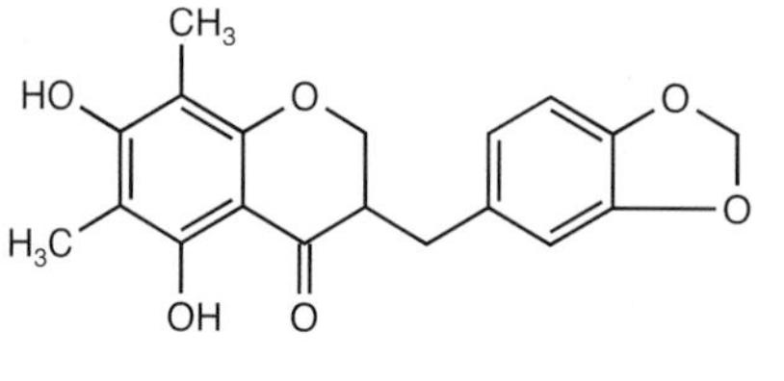

Methylophiopogonanone A

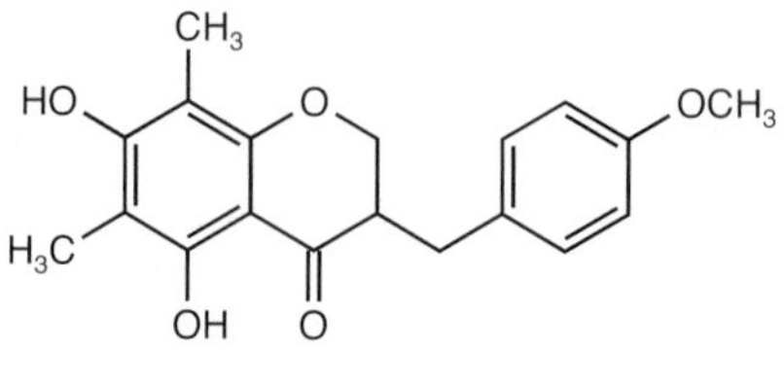

Methylophiopogonanone B

Handbook of Chinese Medicinal Plants: Chemistry, Pharmacology, Toxicology
Weici Tang and Gerhard Eisenbrand

ISBN: 978-3-527-32226-8

Ophiopogon species, including *Ophiopogon bodinieri* var. *pygmaeus, Ophiopogon intermedius, Ophiopogon mairei, Ophiopogon szechuanensis,* and *Ophiopogon stenophyllum,* as well as *Liriope* species such as *Liriope muscarli* and *Liriope spicata* var. *prolifera* and *Liriope platyphylla* were found in commercial Maidong [11].

Pharmacology and Toxicology

The total saponin from the root of *O. japonicus* was reported to be one of the active

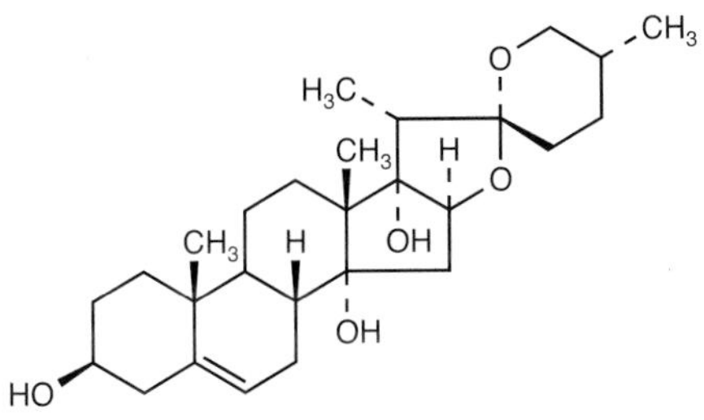

Ophiogenin

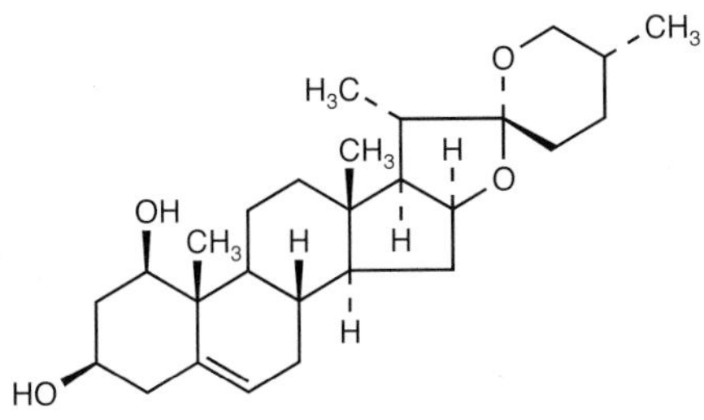

Ruscogenin

components. It increased the myocardiac contraction and cardiac output in the isolated toad heart. Both, myocardial contraction and coronary circulation were also increased by the saponins in isolated guinea pig heart [12]. An injection prepared from the root of *O. japonicus* was found to show a beneficial effect on anesthetized rabbits when given intravenously at 15 and 30 min after experimental myocardial infarction, and also to cause a significant elevation in the cAMP/cGMP ratio [13]. The total saponin exhibited antiarrhythmic effect in dogs induced by chloroform-epinephrine (adrenaline), $BaCl_2$, and aconitine. In dogs, the incidence of ventricular arrhythmia produced by ligation of the left anterior descending coronary artery was effectively decreased by treatment with the total saponin, without any changes in the hemodynamic indices. The antiarrhythmic properties of the total saponin might be related to the blocking of sodium and calcium channels [14]. A saponin monomer from the root of *O. japonicus,* DT-13, was found to show potent cardioprotective effects, with DT-13 exerting decreasing effects on the L-type calcium currents of single adult rat cardiac myocytes under hypoxia. It was suggested that there might be a relationship between the cardioprotective effects of DT-13 and L-type calcium channels under the condition of hypoxia [15].

An ethanol extract of the root of *O. japonicus* was found to significantly inhibit venous thrombosis induced by a tight ligation of the inferior vena cava for 6 h in mice and for 24 h in rats, by single oral administration at doses of 12.5 and 25 mg kg^{-1}, without any obvious effects on other coagulation parameters. A histological analysis of the inferior vena cava indicated that the extract protected endothelial cells from anoxic injury and also alleviated inflammatory changes in the vein wall. The extract significantly enhanced the viability of ECV304 cells injured by sodium dithionite, when administered before and after the anoxic induction. The extract also caused a remarkable inhibition of the adhesion of HL-60 cells to ECV304 cells injured by recombinant human TNF-α. This indicated that an ethanol extract of the *O. japonicus* root inhibited the venous thrombosis associated with endothelial cell-protective and antiadhesive activities [16].

An aqueous extract from the root of *O. japonicus*, given orally to mice at doses of 25 and 50 mg kg^{-1}, was reported to significantly inhibit ear swelling induced by xylene, and paw edema induced by carrageenan. Moreover, the extract also remarkably suppressed carrageenan-induced pleural leukocyte migration in rats, and zymosan A-evoked peritoneal total leukocyte and neutrophil migration in mice, without having any effect on pleural prostaglandin E_2 (PGE_2) levels. The two active compounds were identified as ruscogenin and ophiopogonin D. In the extract, ruscogenin and ophiopogonin D reduced, in concentration-dependent manner, the 12-*O*-tetradecanoylphorbol 13-acetate (TPA)-induced adhesion of HL-60 cells to ECV304 cells. However, these compounds showed no inhibitory effect on TPA-induced cyclooxygenase-2 (COX-2) mRNA expression in ECV304 cells. Notably, ruscogenin and ophiopogonin D also decreased zymosan A-induced peritoneal leukocyte migration [17].

A polysaccharide isolated from the root of *O. japonicus* was reported to significantly prolong the survival time of mice under normobaric hypoxia conditions. Notably, the spleen weight was increased, and the clearance rate of intravenously administered carbon particles enhanced. The polysaccharide also promoted the production of serum-specific antibody hemolysin in mice, and antagonized leukopenia in mice irradiated by γ-rays [18]. A fructan, Opaw-2, which comprised fructose and glucose in a molar ratio of 30 : 1 and was isolated from the root of *O. japonicus*, was shown to cause a significant stimulation of lymphocyte proliferation, in concentration-dependent manner [19].

References

1 Zhu, Y.X., Yan, K.D., and Tu, G.S. (1988) Separation and determination of homoisoflavonoids of *Ophiopogon* by reversed-phase high-performance liquid chromatography. *J. Chromatogr.*, **437**, 265–267.

2 Zhu, Y.X., Yan, K.D., and Tu, G.S. (1987) Isolation and identification of homoisoflavanones from maidong (*Ophiopogon japonicus* (Thunb) Ker-Gawl.). *Acta Pharm. Sin.*, **22**, 679–684.

3 Liu, C.J., Zeng, Q., Liu, D., Zhang, J., and Yang, Y.D. (1991) Studies on homoisoflavonoid content of *Ophiopogon japonicus*. *Chin. Trad. Herbal Drugs*, **22**, 60.

4 Adinolfi, M., Parrilli, M., and Zhu, Y.X. (1990) Trepenoid glycosides from *Ophiopogon japonicus* roots. *Phytochemistry*, **29**, 1696–1699.

5 Yu, B.Y., Xu, G.J., Jin, R.L., and Xu, L.S. (1987) Determination of ruscogenin in Maidong. *J. China Pharm. Univ.*, **18**, 117–119.

6 Yang, Z., Xiao, R., and Xiao, Z.Y. (1986) Chemical constituents of *Ophiopogon japonicus* grown in Sichuan. *West. China J. Pharm. Sci.*, **1**, 177.

7 Yu, B.Y., Xu, G.J., and Yasuaki, H. (1991) Simultaneous determination of ophiopogonin B and ophiopogonin D in *Ophiopogon japonicus* by HPLC. *Chin. Trad. Herbal Drugs*, **22**, 345–346.

8 Zhu, Y.X., Liu, L.Z., Wang, W., Ling, F D.K., and Sun, Z.P. (1991) Studies on chemical constituents of essential oil of *Ophiopogon japonicus*. *Chin. J. Pharm. Anal.*, **11**, 21–23.

9 Zhu, Y.X., Liu, L.Z., Ling, D.K., and Wang, F W. (1989) Constituents of *Ophiopogon japonicus*. *China J. Chin. Mater. Med.*, **14**, 359–360.

10 Nakanishi, H. and Kaneda, N. (1987) Studies on the components of *Ophiopogon* tuber (China) II. *Yakugaku Zasshi*, **107**, 780–784.

11 Yu, B.Y., Xu, G.J., Jin, R.L., and Xu, L.S. (1991) Drug resources and identification of commercial drugs on "Maidong". *J. China Pharm. Univ.*, **22**, 150–153.

12 Mo, Z.J., Jiang, G.C., Ran, L., Huang, F K., Yang, Z.W., Xiao, R., and Xiao, F Z.Y. (1991) Pharmacological studies on the active constituents of *Ophiopogon japonicus*. *West. China J. Pharm. Sci.*, **6**, 13–15.

13 Li, W.P. and Fang, J. (1989) Effect of Maidong injection on cAMP and cGMP level during myocardial infraction in rabbits. *Chin. J. Integr. Trad. West. Med.*, **9**, 100.

14 Chen, M., Yang, Z.W., Zhu, J.T., Xiao, Z.Y., and Xiao, R. (1990) Anti-arrhythmic effects and electrophysiological properties of *Ophiopogon* total saponins. *Acta Pharmacol. Sin.*, **11**, 161–165.

15 Tao, J., Wang, H., Zhou, H., and Li, S. (2005) The saponin monomer of dwarf lily turf tuber, DT-13, reduces L-type calcium currents during hypoxia in adult rat ventricular myocytes. *Life. Sci.*, **77**, 3021–3030.

16 Kou, J., Yu, B., and Xu, Q. (2005) Inhibitory effects of ethanol extract from Radix Ophiopogonis japonicus on venous thrombosis linked with its endothelium-protective and anti-adhesive activities. *Vascul. Pharmacol.*, **43**, 157–163.

17 Kou, J., Sun, Y., Lin, Y., Cheng, Z., Zheng, W., Yu, B., and Xu, Q. (2005) Anti-inflammatory activities of aqueous extract from Radix Ophiopogon japonicus and its two constituents. *Biol. Pharm. Bull.*, **28**, 1234–1238.

18 Yu, B.Y., Yin, X., Zhang, C.H., and Xu, F G.J. (1991) Immunity study on polysaccharide from tuberous roots of *Ophiopogon japonicus*. *J. China Pharm. Univ.*, **22**, 286–288.

19 Wu, X., Dai, H., Huang, L., Gao, X., Tsim, K.W., and Tu, P. (2006) A fructan, from Radix Ophiopogonis, stimulates the proliferation of cultured lymphocytes: structural and functional analyses. *J. Nat. Prod.*, **69**, 1257–1260.

Oryza sativa

Fructus Oryzae Germinatus (Daoya) is the dried germinated ripe fruit of *Oryza sativa* L. (Poaceae). It is used as a stomachic.

Oryza sativa L.

Chemistry

The rice bran oil contains, as a biologically active compound, oryzanol (γ-oryzanol); this is a mixture of ferulic acid esters of sterol and triterpene alcohols present at a level of 1–2% in the oil [1].

Pharmacology and Toxicology

In clinical studies, oryzanol was reported to be effective against the syndromes of autonomic nervous imbalance and climacteric disorders. In experimental studies, oryzanol given subcutaneously to rats in daily doses of 1 to 100 $mg\,kg^{-1}$ for five days, caused a significant reduction in the ulcer index induced by water-immersion stress. The effect was also observed in adrenalectomized and also sham-operated rats. It was suggested that the antiulcerogenic action of oryzanol was due to a participation of the autonomic nervous system. Oryzanol was shown to have a suppressant effect on the central nervous system, although its properties appeared different from those of tranquilizers [2]. An eight-day treatment with oryzanol in rats caused a significant inhibition of fasting ulcers, while a five-day pretreatment had a slight alleviating effect on ulcers induced by pyloric ligation. A 10-day treatment with oryzanol in acetic acid-induced ulcers in rats caused a lowering of the serum level of gastrin, and enhanced 11-hydroxycorticosterone levels. It has been suggested that the monoaminergic neuron system was involved in the antiulcer action of oryzanol. The norepinephrine (noradrenaline) content of the brain was slightly, but significantly, increased following the administration of oryzanol, but norepinephrine in the gastric area was unaffected. The turnover rate of norepinephrine in the brain tended to decrease with the administration of oryzanol. Based on these results, it seemed likely that successive doses of γ-oryzanol would increase brain norepinephrine levels by inhibiting the degradation or stimulating the release of norepinephrine [3].

Oryzanol also served as natural antioxidant [4] and was reported to lower cholesterol levels in the blood, thus reducing the risk of coronary heart disease [5]. All of the components of oryzanol exhibited significant antioxidant activities that were higher than those of any vitamin E components. The highest antioxidant activity was found for 24-methylenecycloartanyl ferulate [6]. Oryzanol, when given to hypercholesterol-

Handbook of Chinese Medicinal Plants: Chemistry, Pharmacology, Toxicology
Weici Tang and Gerhard Eisenbrand

ISBN: 978-3-527-32226-8

emic hamsters at 1% in the diet, caused a significant decrease in cholesterol absorption and aortic fatty streaks [7]. In experimental studies conducted in rats, the effects of oryzanol on hypolipidemia [8], hypocholesterolemia [9] and coronary atherosclerosis [10] were each reported.

The potential mutagenicity of oryzanol has been investigated in various tests. Oryzanol showed a negative response in the bacterial DNA repair test (Rec-assay), the bacterial reverse mutation tests (Ames test), and in the rat bone marrow chromosome aberration test. Oryzanol also showed a negative response in the metabolic cooperation inhibition test using wild-type (6-thioguanine-sensitive) Chinese hamster V79 cells in co-culture with 6-thioguanine resistant cells to test for tumor-promoting activity, as observed for TPA [11]. Oryzanol did not show any carcinogenic potential in $B6C_3F_1$ mice, nor in F344 rats of both genders, following oral administration at daily doses of 200, 600, or 2000 mg kg^{-1} for 78 weeks [12, 13].

In contrast, oryzanol was shown to exert chemopreventive activities. In female Sprague–Dawley rats pretreated with 7,12-dimethylbenz[*a*]anthracene (DMBA) at 50 mg kg^{-1} for 35 weeks, dietary oryzanol (1%) reduced the incidence of DMBA-induced mammary cancers [14]. Oryzanol also suppressed 12-*O*-tetradecanoylphorbol 13-acetate (TPA)-induced ear inflammation in mice, and markedly inhibited the tumor-promoting effect of TPA in mice initiated by DMBA [15]. However, oryzanol did not modify the 3,2′-dimethyl-4-aminobiphenyl (DMAB)-induced prostate carcinogenesis in male F344 rats, when given in the diet at a concentration of 2% for 20 weeks [16]. In contrast, oryzanol was reported to enhance the lung tumor incidence in rats initiated with two intraperitoneal injections of *N*-nitroso-*N*,*N*-di-(2,2′-dihydroxy)-*n*-propylamine (1000 mg kg^{-1}), followed by two intragastric administrations of *N*-nitroso-*N*-ethyl-*N*-hydroxyethylamine (1500 mg kg^{-1}), and then three subcutaneous injections of DMAB (75 mg kg^{-1}). At one week after the last injection, the rats had received a diet containing 1% oryzanol for 32 weeks [17].

Oryzanol was also reported to elicit anabolic effects, such as to increase testosterone production, and to stimulate human growth hormone release. The results of animal studies, however, indicated that oryzanol, whether given subcutaneously or intravenously, induced anti-anabolic or anti-catabolic activities. Following the oral administration of oryzanol, less than 5% is absorbed from the intestinal tract, with the majority being excreted in the feces. The i.v. or s.c. injection of oryzanol in rats has been shown to suppress luteinizing hormone release, to reduce growth hormone synthesis and release, and to increase the release of dopamine and norepinephrine in the brain [18].

References

1 Scavariello, E.M. and Arellano, D.B. (1998) γ-Oryzanol: an important component in rice bran oil. *Arch. Latinoam. Nutr.*, **48**, 7–12.

2 Hiraga, F Y., Nakata, F N., Jin, F H., Ito, F S., Sato, F R., Yoshida, F A., Mori, F T., Ozeki, F M. and Ikeda, F Y. (1993) Effect of the rice bran-derived phytosterol cycloartenol ferulic acid ester on the central nervous system. *Arzneim.-Forsch.*, **43**, 715–721.

3 Kaneta, F H., Kujira, F K., Shigenaga, F T. and Itaya, F K. (1979) Effects of γ-oryzanol on norepi-nephrine contents in the brain and stomach of rats. *Nippon Yakurigaku Zasshi*, **75**, 399–403.

4 Kim, S.J., Han, F D., Moon, K.D. and Rhee, J.S. (1995) Measurement of superoxide dismutase-like activity of natural antioxidants. *Biosci. Biotechnol. Biochem.*, **59**, 822–826.

5 Sugano, F M. and Tsuji, F E. (1997) Rice bran oil and cholesterol metabolism. *J. Nutr.*, **127**, 521S–524S.

6 Xu, F Z., Hua, F N. and Godber, J.S. (2001) Antioxidant activity of tocopherols, tocotrienols, and γ-oryzanol components from rice bran against cholesterol oxidation accelerated by 2,2′-azobis(2-methylpropionamidine) dihydrochloride. *J. Agric. Food Chem.*, **49**, 2077–2081.

7 Rong, F N., Ausman, L.M. and Nicolosi, R.J. (1997) Oryzanol decreases cholesterol absorption and aortic fatty streaks in hamsters. *Lipids*, **32**, 303–309.

8 Sakamoto, F K., Tabata, F T., Shirasaki, F K., Inagaki, F T. and Nakayama, F S. (1987) Effects of γ-oryzanol and cycloartenol ferulic acid ester on cholesterol diet induced hyperlipidemia in rats. *Jpn. J. Pharmacol.*, **45**, 559–565.

9 Seetharamaiah, G.S. and Chandrasekhara, F N. (1989) Studies on hypocholesterolemic activity of rice bran oil. *Atherosclerosis*, **78**, 219–223.

10 Zhang, Q.H. (1986) Effects of γ-oryzanol on experimental coronary atherosclerosis in rats. *Chin. J. Cardiol.*, **14**, 190–287, 319.

11 Tsushimoto, F G., Shibahara, F T., Awogi, F T., Kaneko, F E., Sutou, F S., Yamamoto, F K. and Shirakawa, F H. (1991) DNA-damaging, mutagenic, clastogenic and cell-cell communication inhibitory properties of γ-oryzanol. *J. Toxicol. Sci.*, **16**, 191–202.

12 Tamagawa, F M., Otaki, F Y., Takahashi, F T., Otaka, F T., Kimura, F S. and Miwa, F T. (1992) Carcinogenicity study of γ-oryzanol in B6C3F1 mice. *Food Chem. Toxicol.*, **30**, 49–56.

13 Tamagawa, F M., Shimizu, F Y., Takahashi, F T., Otaka, F T., Kimura, F S., Kadowaki, F H., Uda, F F. and Miwa, F T. (1992) Carcinogenicity study of γ-oryzanol in F344 rats. *Food Chem. Toxicol.*, **30**, 41–48.

14 Hirose, F M., Hoshiya, F T., Akagi, F K., Futakuchi, F M. and Ito, F N. (1994) Inhibition of mammary gland carcinogenesis by green tea catechins and other naturally occurring antioxidants in female Sprague-Dawley rats pretreated with 7, 12-dimethylbenz[*a*]anthracene. *Cancer. Lett.*, **83**, 149–156.

15 Yasukawa, F K., Akihisa, F T., Kimura, F Y., Tamura, F T. and Takido, F M. (1998) Inhibitory effect of cycloartenol ferulate, a component of rice bran, on tumor promotion in two-stage carcinogenesis in mouse skin. *Biol. Pharm. Bull.*, **21**, 1072–1076.

16 Nakamura, F A., Shirai, F T., Takahashi, F S., Ogawa, F K., Hirose, F M. and Ito, F N. (1991) Lack of modification by naturally occurring antioxidants of 3, 2′-dimethyl-4-aminobiphenyl-initiated rat prostate carcinogenesis. *Cancer. Lett.*, **58**, 241–246.

17 Hirose, F M., Ozaki, F K., Takaba, F K., Fukushima, F S., Shirai, F T. and Ito, F N. (1991) Modifying effects of the naturally occurring antioxidants γ-oryzanol, phytic acid, tannic acid and n-tritriacontane-16, 18-dione in a rat wide-spectrum organ carcinogenesis model. *Carcinogenesis*, **12**, 1917–1921.

18 Wheeler, K.B. and Garleb, K.A. (1991) γ-Oryzanol-plant sterol supplementation: metabolic, endocrine, and physiologic effects. *Int. J. Sport Nutr.*, **1**, 170–177.

Paeonia lactiflora and *Paeonia veitchii*

Radix Paeoniae Alba (Baishao) is the dried root of *Paeonia lactiflora* Pall. (Ranunculaceae). It is used as an analgesic, hemostyptic, and bacteriostatic agent for the treatment of headache, dizziness, abdominal pain, anemia, menstrual disorders, and night sweating.

Paeonia lactiflora Pall.

Radix Paeoniae Rubra (Chishao) is the dried root of *Paeonia lactiflora* or *Paeonia veitchii* Lynch. It is used as analgesic, hemostyptic, and bacteriostatic agent for the treatment of eye inflammation, dysmenorrhea, mass formation in the abdomen, traumatic injuries, boils, and sores.

The Chinese Pharmacopoeia notes that both Radix Paeoniae Alba and Radix Paeoniae Rubra are incompatible with Rhizoma et Radix Veratri.

Chemistry

Radix Paeoniae Alba and Radix Paeoniae Rubra, the peony root, are known to contain paeoniflorin, a glucoside of a pinane-type monoterpene as the major active component [1]. The structure of paeoniflorin, which involves a cage-like pinane skeleton, is unique among natural products. *Paeonia lactiflora* and *P. veitchii* are two species with the highest paeoniflorin contents [2]. The Chinese Pharmacopoeia requires a qualitative determination of paeoniflorin in Radix Paeoniae Alba and in Radix Paeoniae Rubra by thin-layer chromatographic comparison with reference substance, and a quantitative determination by HPLC. The content of paeoniflorin in Radix Paeoniae Alba should be not less than 1.6%, in Radix Paeoniae Rubra not less than 1.8%, and in processed Radix Paeoniae Rubra not less than 1.5%.

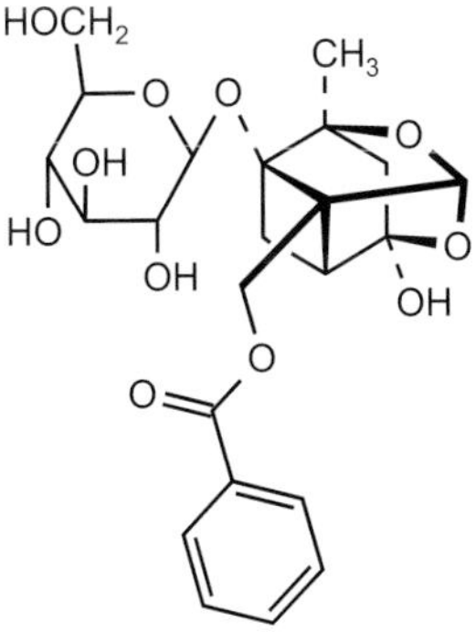

Paeoniflorin

Minor constituents from the peony root were identified as albiflorin, oxypaeoniflorin, benzoylpaeoniflorin, benzoyloxypaeoniflorin, galloylpaeoniflorin, debenzoylpaeoniflorin [3], and lactiflorin [4]. A number of monoterpene glycoside esters related to paeoniflorin were also isolated from peony root [5].

Monoterpenes and monoterpene esters from the peony root were identified as paeoniflorigenone, paeonilactones A, B, and C [6, 7].

Handbook of Chinese Medicinal Plants: Chemistry, Pharmacology, Toxicology
Weici Tang and Gerhard Eisenbrand

ISBN: 978-3-527-32226-8

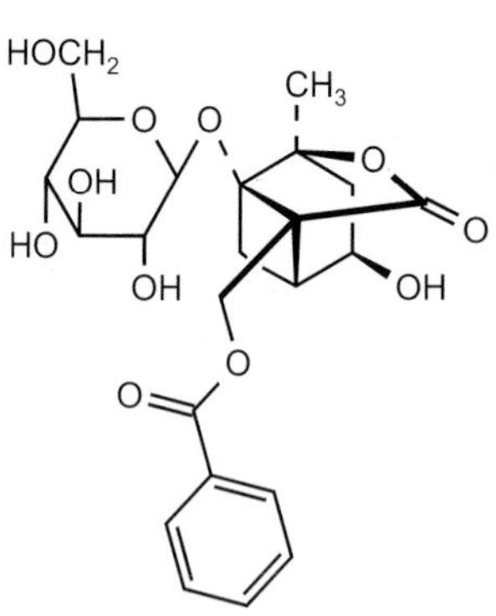

Albiflorin

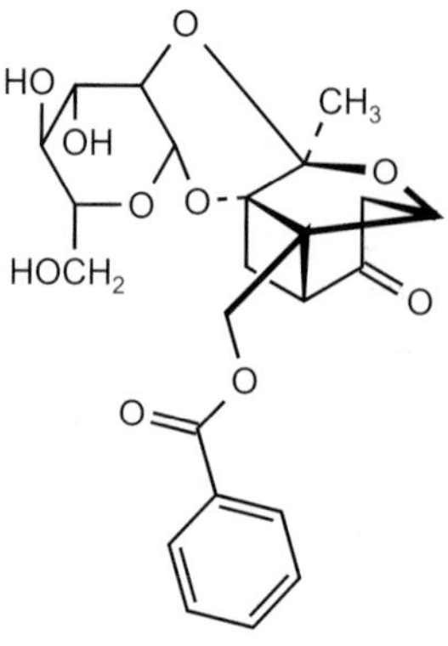

Lactiflorin

Gallotannins, such as gallic acid, methyl gallate, 1,2,3,4,6-penta-*O*-galloyl-β-D-glucose were isolated from the peony root [8].

A neutral polysaccharide peonan SA and an acidic polysaccharide peonan SB were isolated from the peony root. Peonan SA with a molecular mass of 112 kDa is composed of L-arabinose, D-galactose and D-glucose, while peonan SB with a molecular mass of 250 kDa is composed of L-arabinose, D-galactose, and D-galacturonic acid, in addition to small amounts of peptide moieties. About 40% of the hexuronic acid residues in peonan SB exist as the methyl ester [9]. Another acidic polysaccharide, peonan PA, from the peony root is composed of L-arabinose, D-galactose, and D-galacturonic acid, in addition to small amounts of *O*-acetyl groups and peptide moieties with a molecular mass of 60 kDa [10].

Pharmacology and Toxicology

After oral administration of paeoniflorin at a dose of 20 mg kg^{-1} to rats, its plasma concentration reached a maximum at 30 min, and paeonimetaboline I at 140 min. Estimation of the area under the curve (AUC) for paeoniflorin and paeonimetaboline I indicated that paeonimetaboline I was

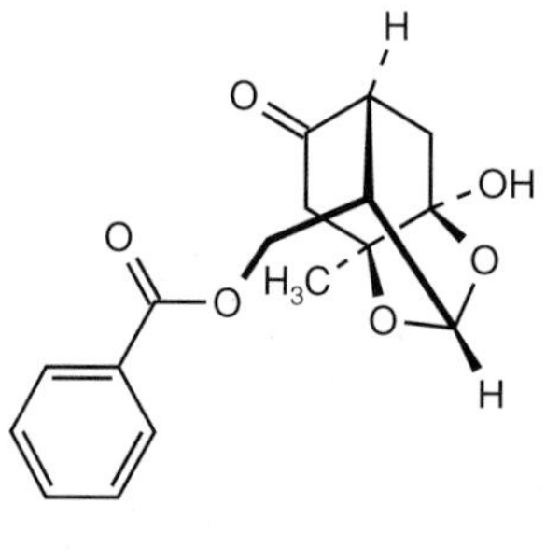

Paeoniflorigenone

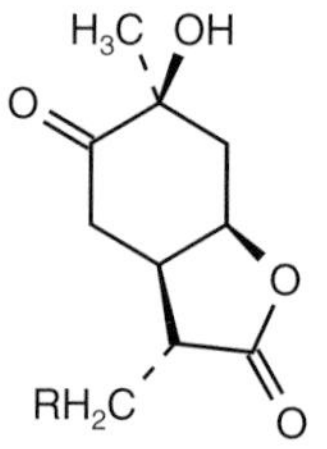

Paeonilactone A: R = H
Paeonilactone C: R = OCO-Ph

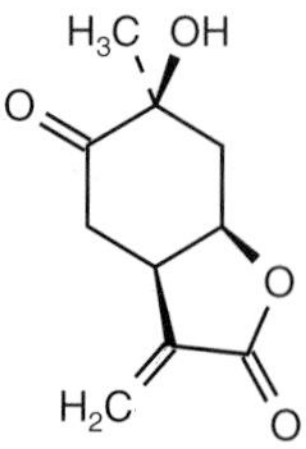

Paeonilactone B

the major compound present in the plasma. Paeoniflorin seemed to be poorly absorbed from the gastrointestinal tract, resulting in an extremely low bioavailability. Paeoniflorin was transformed to paeonimetaboline I by intestinal bacteria, to be subsequently absorbed from the gastrointestinal tract [11]. The pharmacokinetics of paeoniflorin in mice after oral administration of peony root extract was also reported. The plasma concentration–time curve of paeoniflorin resulted in mean terminal half-lives of 94 min [12]. After intravenous (i.v.) injection of paeoniflorin to rabbits at a dose of 25 mg kg^{-1}, the plasma concentration–time curve was found to fit a two-compartment model. The half-lives $t_{1/2\alpha}$ and $t_{1/2\beta}$ were reported as 4 min and 66 min, respectively. After oral administration of paeoniflorin to rats at a dose of 550 mg kg^{-1}, about 10.6% of the dose was recovered in the feces within 24 h, and about 1% in the urine. After i.v. injection of paeoniflorin to rats at a dose of 55 mg kg^{-1}, about 9% of the dose was recovered from the bile within 7 h [13].

In contrast, after oral administration to rats of a decoction of Radix Paeoniae Alba, paeoniflorgenin as a deglucosylated metabolite of paeoniflorin was identified in the serum. The paeoniflorin levels in serum were below the detection limit throughout the study. The C_{max}, t_{max}, and AUC of paeoniflorgenin were 8.0 μg ml^{-1}, 10 min, and 487.0 μg min ml^{-1}, respectively. Paeoniflorin was hydrolyzed into paeoniflorgenin through incubation with feces of rabbit, rat, pig, or human. It was suggested that paeoniflorin was not absorbed after oral administration. Hydrolysis in the intestinal tract by bacterial enzymes yielded its aglycone paeoniflorgenin which was absorbed, and circulated in the blood [14].

The pharmacokinetics of paeoniflorin have been studied after an i.v. administration of an extract of the root of *P. lactiflora* to rats with cerebral ischemia induced by occlusion of the bilateral carotid arteries for 2 h, followed by reperfusion. The extract was administered immediately after reperfusion, at a dose corresponding to 60 mg kg^{-1} paeoniflorin; the same dose was injected into normal rats. The half-lives $t_{1/2\alpha}$ and $t_{1/2\beta}$, the AUC, mean retention time, and clearance (CL) in normal rats (estimated by an open, two-compartmental model) were 0.7 min and 18.8 min, 5338.7 μg min ml^{-1}, 18.1 min, and 0.016 mg kg^{-1}·min, respectively. In ischemia–reperfusion rats, the corresponding parameters were 2.0 min and 24.5 min, 9626.0 μg min ml^{-1}, 29.8 min, and 0.007 mg kg^{-1}·min, respectively. These results showed that ischemia–reperfusion significantly increased the AUC values, decreased the CL values, and prolonged the terminal half-life of paeoniflorin. This suggested that ischemia–reperfusion injury might play an important role in the pharmacokinetics of paeoniflorin [15]. The concentrations of paeoniflorin in the cerebral cortex after ischemia–reperfusion were lower after dosing, and declined more slowly than under normal conditions [16].

Peony root and paeoniflorin were reported to exert anti-inflammatory, antispasmodic, and sedative activities. Studies on contractile and noncontractile Ca^{2+} mobilization of the nerve-stimulated mouse skeletal muscle in the presence of neostigmine revealed that paeoniflorin might have complementary effects on intracellular Ca^{2+} mobilization, blocking neuromuscular transmission [17]. Paeoniflorin inhibited the contraction of isolated rat atria induced by veratrine due to the blockade of Ca^{2+} channels [18]. It also inhibited the contraction of isolated mouse vas deferens induced by veratrine, in both the epididymal and the prostatic portions [19]. Therefore, it is reasonable to consider that peony root is incompatible with Rhizoma et Radix Veratri. The methanol extract of peony root was found to be anticholinergic in rats, with paeoniflorin being one of the active constituents [20]. Paeoniflorin reversed the inhibition of long-term potentiation in rat hippocampal slices, mediated by the muscarinic M_1 receptor. This effect of paeoniflorin might be involved in its protective effect on spatial cognitive impairment caused by cholinergic dysfunction [21]. Paeoniflorin also inhibited Na^+/K^+-ATPase of rabbit erythrocytes in a concentration-dependent manner, and enhanced the activity of adenyl cyclase [22]. It was also reported that peony root extract elevated the basal and calmodulin-stimulated Ca^{2+}/Mg^{2+}-ATPase activities of erythrocyte membrane in hyperlipidemic rabbits induced by cholesterol [23].

Clinical trials using the extract of the peony root for the treatment of patients ($n = 42$) with decompensated chronic cor pulmonale by i.v. administration resulted in a significant improvement of the clinical features of blood stasis, of hemorheologic parameters, and myocardial oxygen consumption [24]. Paeoniflorin was found to exhibit hypotensive effects in guinea pigs, possibly based on peripheral vasodilation. In contrast, paeoniflorin was also found to reverse guanethidine-induced hypotension in rats by i.v. administration, in a dose-dependent manner. The hypertensive mechanism of paeoniflorin was believed to be associated with activation of adenosine A_1 receptors in rat brain [25]. Using isolated guinea pig ileal synaptosomes, paeoniflorin was found to stimulate the release of noradrenaline (norepinephrine) in a concentration-dependent manner. It was suggested that paeoniflorin might stimulate the depolarization of membranes, resulting in a calcium-dependent and cAMP-related release of noradrenaline from noradrenergic nerve terminals [26]. Paeoniflorin further activated adenosine A_1 receptors to increase the translocations of protein kinase C (PKC) and glucose transporter, two major signals for glucose uptake, from the cytosol through the membrane of white adipocytes in rats [27].

Oral administration of the peony root extract reduced the area of atherosclerosis in the aorta of cholesterol-fed rabbits. The levels of plasma lipid peroxides and thromboxane B_2 (TXB_2) and the contents of cholesterol, phospholipid, and calcium in the media of the aorta were significantly lowered after treatment with peony root, but the level of plasma 6-keto-prostaglandin F_1 (6-keto-PGF_1) was significantly higher. The inhibition of lipid peroxides and regulation of thromboxane and prostaglandin balance might be an important mechanism of the antiatherogenic effects of peony root [28]. The administration of peony root extract to rats fed a high-fat diet was found to increase endothelium-dependent relaxation and to increase superoxide dismutase activity in erythrocytes. Peony root protected against the increase in endothelial superoxide anion and endothelial dysfunction caused by hypercholesterolemia [29].

The extract of the peony root has been shown to inhibit thrombosis and platelet aggregation and to increase fibrinolytic activity, promoting thrombolysis. It was found to prolong the prothrombin time and activated partial thromboplastin time; to significantly inhibit thrombin; to activate plasminogen; and to reduce urokinase activity *in vitro*. The inhibition of thrombin and activation of plasminogen might be important factors in promoting blood circulation and removing blood stasis [30]. The antithrombotic effect of paeoniflorin has also been evaluated *in vivo* using a photochemical reaction thrombosis model. The results showed that paeoniflorin significantly prolonged thrombosis time. The antithrombotic effect of paeoniflorin seemed to be related to an inhibition of arachidonic acid metabolism, increased tissue-type plasminogen activator (tPA) activity, and protective effects against free radicals [31].

A sedative effect of paeoniflorin has been demonstrated in rats. The i.v. administration of paeoniflorin to rats caused a loss of the righting reflex, and also prolonged sleeping duration induced by hexobarbital. Paeoniflorin inhibited the writhing symptoms in mice induced by the intraperitoneal (i.p.) administration of acetic acid, and showed weak hypothermic activities [32]. The total glycosides from peony root in mice or rats inhibited the convulsions induced by electric shock or by intoxication with strychnine [33].

Paeoniflorin significantly attenuated the learning impairment of aged rats, but not of young rats, and might be useful as a cognitive enhancer [34]. It also improved the radial maze performance of rats impaired by scopolamine [35]. It was shown that the α_1-adrenergic and β_1-adrenergic systems, but not the α_2-adrenergic system, were involved in the antagonistic effect of paeoniflorin on scopolamine-induced deficit in radial maze performance in rats [36]. Paeoniflorin derivatives were further found to protect memory impairment induced by scopolamine in mice. Structure–activity relationship studies revealed that the cage-like pinane skeleton, the benzoyl and the glucosyl moieties, are important structural elements against scopolamine-induced amnesia [37].

Albiflorin and pentagalloylglucose appeared to exert anticonvulsant activity due to an inhibition of the seizure-related decrease of extracellular calcium and consequent intracellular calcium increase, whereas paeoniflorin was relatively less potent [38]. Neuron damage in the CA1 area of the hippocampus, and frequent spike discharges induced by application of metallic cobalt to rat cerebral cortex, were completely prevented by the oral administration of peony root extract for 30 days prior to cobalt application. The gallotannin fraction showed a marked but incomplete protective action, whereas the combination of gallotannin fraction and paeoniflorin showed complete protective action [39]. The neuroprotective effect of paeoniflorin was also studied in cerebral ischemic rat. In Sprague-Dawley rats with transient ischemia induced by a 1.5-h occlusion of the middle cerebral artery, the subcutaneous (s.c.) administration of paeoniflorin at doses of 2.5 and 5 mg kg^{-1} produced a dose-dependent decrease in both neurological impairment and histologically measured infarction volume. A study on the mechanism indicated that activation of the adenosine A1 receptor might be involved in the neuroprotection by paeoniflorin in cerebral ischemia in rats [40].

The total glycoside of peony root was found to enhance tumor necrosis factor (TNF) production in rat peritoneal macrophages [41]. In rats, the total glycoside from peony root increased plasma corticosterone levels after small oral doses of 10–50 mg kg^{-1}, and decreased plasma corticosterone levels at high doses of 100–200 mg kg^{-1} [42]. Peony root was further reported to exert protective effects against the hepatotoxicity of D-galactosamine in rats [43]. It affected the conversion between androst-4-enedione and testosterone, inhibited testosterone synthesis, and stimulated aromatase-mediated estradiol synthesis by direct action on the rat proestrous ovary [44].

Both, paeoniflorin and debenzoylpaeoniflorin were found to exhibit significant hypoglycemic efficacy in normoglycemic and diabetic rats induced by streptozotocin. Neither compound affected the plasma insulin levels in normoglycemic rats, indicating an insulin-independent hypoglycemic mechanism. On the basis that paeoniflorin and debenzoylpaeoniflorin each reduced the elevation of blood sugar in glucose-challenged rats, it was considered that these compounds increased glucose utilization [3].

An ethanol extract of the peony root and its active components, gallic acid and methyl gallate, exhibited a significant free radical scavenging effect against 1,1-diphenyl-2-picryl hydrazine radicals, and also inhibited lipid peroxidation. These compounds strongly inhibited DNA damage in NIH-3T3 fibroblasts induced by H_2O_2, as assessed by single-cell gel electrophoresis. *In vivo* experiments showed that oral administration of the extract, gallic acid and methyl gallate potently inhibited the formation of micronucleated reticulocytes in mouse peripheral blood induced by a $KBrO_3$ [45]. 1,2,3,4,6-Penta-*O*-galloyl-β-D-glucose inhibited the activities of inducible nitric oxide synthase (iNOS) and cyclooxygenase-2 (COX-2) in Raw 264.7 cells stimulated by lipopolysaccharide (LPS) [46].

1,2,3,4,6-Penta-*O*-galloyl-β-D-glucose and related galloylglucoses, were each found to relax PGF_2-contracted rat aorta with intact endothelium, but failed to relax aortic rings without endothelium [47]. 1,2,3,4,6-Penta-*O*-galloyl-β-D-glucose strongly inhibited Na^+,K^+-ATPase activity, with an IC_{50} of 2.5 μ*M*, whereas galloylpaeoniflorin exerted weakly inhibitory activity, and albiflorin, oxypaeoniflorin and paeoniflorin were ineffective. Peony root suppressed the cleavage of pUC18 DNA induced by phenylhydroquinone, and scavenged the generated superoxide and hydroxy radical. The most potent constituent was found to be pentagalloylglucose [48]. The polysaccharides peonan SA, peonan SB and peonan PA were reported to show remarkable potentiating activity on the reticuloendothelial system, and to exert considerable anticomplement activity [9, 10]. 1,2,3,4,6-Penta-*O*-galloyl-β-D-glucose was also found to suppress hepatitis B virus (HBV) by measurement of HBV DNA and hepatitis B surface antigen (HBsAg) levels in the extracellular medium of HepG2.2.15 cells after treatment for eight days. It decreased the level of extracellular HBV in a concentration-dependent manner, with an IC_{50} of 1.0 μg ml^{-1} [49].

The extract of the root of *P. lactiflora* was also found to induce death of HL-60 cells. A reduced mitochondrial transmembrane potential of HL-60 cells, induction of cell apoptosis, and an increase in caspase-3 activity were observed within 12 h after incubation with the extract [50]. The extract also inhibited the proliferation of HepG2- and p53-deficient Hep3B human hepatoma cells, induced cell cycle arrest at the sub-G_1 phase, and induced cell apoptosis. Thus, the induction of apoptosis was not p53 pathway-dependent [51]. Furthermore, paeoniflorin was reported to inhibit the proliferation of human T-cell leukemia Jurkat cells, and to induce apoptosis, mediated through the reduction of mitochondrial membrane potential, activation of caspase, and fragmentation of DNA [52].

Paeoniflorin was further found to show radioprotective activity in thymocytes subjected to ^{60}Co γ-radiation. The pretreatment of thymocytes with paeoniflorin at concentrations of 50–200 μg ml^{-1} reversed the DNA damage and cell death induced by γ-radiation, and attenuated irradiation-induced reactive oxygen species (ROS) generation. Paeoniflorin showed several anti-apoptotic characteristics, including the ability to diminish cytosolic Ca^{2+} concentrations, inhibit caspase-3 activation, and to upregulate Bcl-2 and downregulate Bax in γ-irradiated thymocytes. Extracellular regulated kinase (ERK), c-Jun NH_2-terminal kinase (JNK), and p38 kinase activated by γ-irradiation were each partly blocked by the pretreatment of cells with paeoniflorin. The radioprotective effect of paeoniflorin in thymocytes appeared to be mediated by scavenging ROS, and attenuated the activation of mitogen-activated protein kinases [53].

Paeoniflorin, oxypaeoniflorin and benzoylpaeoniflorin were each converted to three metabolites by the intestinal bacteria, with paeonimetaboline I as the major metabolite, and paeonimetaboline II as the minor metabolite [54]. A number of fecal bacteria such as *Bacteroides, Bifidobacterium, Clostridium, Lactobacillus*, and *Streptococcus* sp. *Bacteroides fragilis thetaotus* formed preferably the 7S-paeonimetaboline I, while *Lactobacillus xylosus* and *Lactobacillus acidophilus* formed preferably the 7R-paeonimetaboline I. The other bacteria formed both epimers in almost equal amounts. Similarly, the fecal flora from different subjects showed potent metabolic ability, predominantly producing the 7S epimer [55].

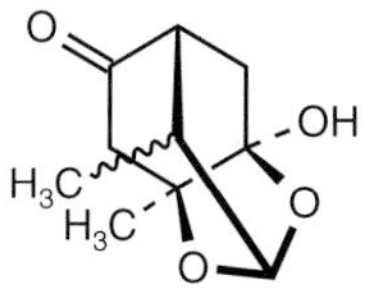

Paeonimetaboline I

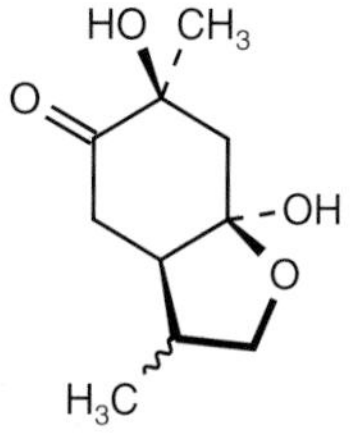

Paeonimetaboline II

References

1 Ho, L.Y., Feng, R.Z., and Xiao, P.G. (1980) Studies of the medicinal plants of the family Ranunculaceae in China. Part IV. The occurrence of paeoniflorine in the genus Paeonia. *Acta Pharm. Sin.*, **15**, 429–433.
2 Zhou, M., Cai, H., Huang, Z., and Sun, Y. (1998) HPLC method for the determination of paeoniflorin in *Paeonia lactiflora* Pall and its preparations. *Biomed. Chromatogr.*, **12**, 43–44.
3 Hsu, F.L., Lai, C.W., and Cheng, J.T. (1997) Antihyperglycemic effects of paeoniflorin and 8-debenzoylpaeoniflorin, glucosides from the root of *Paeonia lactiflora. Planta Med.*, **63**, 323–325.
4 Yu, J., Elix, J.A., and Iskander, M.N. (1990) Lactiflorin, a monoterpene glycoside from paeony root. *Phytochemistry*, **29**, 3859–3863.
5 Tanaka, T., Kataoka, M., Tsuboi, N., and Kouno, I. (2000) New monoterpene glycoside esters and phenolic constituents of Paeoniae radix, and increase of water solubility of proanthocyanidins in the presence of paeoniflorin. *Chem. Pharm. Bull. (Tokyo)*, **48**, 201–207.
6 Shimizu, M., Hayashi, T., Morita, N., Kimura, I., Kimura, M., Kiuchi, F., Noguchi, H., Iitaka, Y., and Sankawa, U. (1981) Paeoniflorigenone, a new monoterpene from paeony roots. *Tetrahedron Lett.*, **22**, 3069–3070.
7 Hayashi, T., Shinbo, T., Shimizu, M., Arisawa, M., Morita, N., Kimura, M., Matsuda, S., and Kikuchi, T. (1985) Paeonilactone-A, -B, and -C, new Monoterpenoids from Paeony root. *Tetrahedron Lett.*, **26**, 3699–3702.
8 Satoh, K., Nagai, F., Ushiyama, K., Yasuda, I., Seto, T., and Kano, I. (1997) Inhibition of Na^+, K^+-ATPase by 1,2,3,4,6-penta-*O*-galloyl-D-glucose, a major constituent of both Moutan Cortex and Paeoniae Radix. *Biochem. Pharmacol.*, **53**, 611–614.
9 Tomoda, M., Matsumoto, K., Shimizu, N., Gonda, R., and Ohara, N. (1993) Characterization of a neutral and an acidic polysaccharide having immunological activities from the root of *Paeonia lactiflora. Biol. Pharm. Bull.*, **16**, 1207–1210.
10 Tomoda, M., Matsumoto, K., Shimizu, N., Gonda, R., Ohara, N., and Hirabayashi, K. (1994) An acidic polysaccharide with immunological activities from the root of *Paeonia lactiflora. Biol. Pharm. Bull.*, **17**, 1161–1164.
11 Takeda, S., Isono, T., Wakui, Y., Mizuhara, Y., Amagaya, S., Maruno, M., and Hattori, M. (1997) In-vivo assessment of extrahepatic metabolism of paeoniflorin in rats: relevance to intestinal floral metabolism. *J. Pharm. Pharmacol.*, **49**, 35–39.
12 Chen, L.C., Lee, M.H., Chou, M.H., Lin, M. F., and Yang, L.L. (1999) Pharmacokinetic study of paeoniflorin in mice after oral administration of Paeoniae radix extract. *J. Chromatogr. B Biomed. Sci. Appl.*, **735**, 33–40.
13 Chen, G.L., Chen, C.H., and Xu, S.Y. (1992) Studies on pharmacokinetics of paeoniflorin in rabbits and rats. *Chin. Pharmacol. Bull.*, **8**, 278–280, 284.
14 Hsiu, S.L., Lin, Y.T., Wen, K.C., Hou, Y.C., and Chao, P.D. (2003) A deglucosylated metabolite of paeoniflorin of the root of *Paeonia lactiflora* and its pharmacokinetics in rats. *Planta Med.*, **69**, 1113–1118.
15 He, X., Xing, D., Ding, Y., Li, Y., Xu, L., and Du, L. (2004) Effects of cerebral ischemia-reperfusion on pharmacokinetic fate of

paeoniflorin after intravenous administration of Paeoniae Radix extract in rats. *J. Ethnopharmacol.*, **94**, 339–344.

16 Cao, C., He, X., Wang, W., Zhang, L., Lin, H., and Du, L. (2006) Kinetic distribution of paeoniflorin in cortex of normal and cerebral ischemia-reperfusion rats after intravenous administration of Paeoniae Radix extract. *Biomed. Chromatogr.*, **20**, 1283–1288.

17 Dezaki, K., Kimura, I., Miyahara, K., and Kimura, M. (1995) Complementary effects of paeoniflorin and glycyrrhizin on intracellular Ca^{2+} mobilization in the nerve-stimulated skeletal muscle of mice. *Jpn. J. Pharmacol.*, **69**, 281–284.

18 Tsai, H.Y., Lin, Y.T., Chen, Y.F., and Chen, C.F. (1997) The interactions of paeoniflorin and veratrine on isolated rat atria. *J. Ethnopharmacol.*, **57**, 169–176.

19 Chen, Y.F., Lin, Y.T., Tan, T.W., and Tsai, H.Y. (2002) Effects of veratrine and paeoniflorin on isolated mouse vas deferens. *Phytomedicine*, **9**, 296–301.

20 Kobayashi, M., Ueda, C., Aoki, S., Tajima, K., Tanaka, N., and Yamahara, J. (1990) Anticholinergic action of paeony root and its active constituents. *Yakugaku Zasshi*, **110**, 964–968.

21 Tabata, K., Matsumoto, K., and Watanabe, H. (2000) Paeoniflorin, a major constituent of peony root, reverses muscarinic M_1-receptor antagonist-induced suppression of long-term potentiation in the rat hippocampal slice. *Jpn. J. Pharmacol.*, **83**, 25–30.

22 Liu, X.T., Tang, H.F., and Xu, Y.F. (1993) Effects of paeoniflorin, matrine and oxymatrine on membrane enzymes *in vitro*. *Chin. Pharm. J.*, **28**, 658–660.

23 Zheng, L.L., Yan, X.F., and Zhang, Y.Z. (1996) Effect of *Paeonia lactiflora* on platelet cytosolic free calcium and erythrocyte membrane Ca^{2+}-Mg^{2+}-ATPase activity in hyperlipid rabbits. *Chin. J. Integr. Trad. West. Med.*, **16**, 295–296.

24 Jia, Y.B. and Tang, T.Q. (1991) Clinical study on *Paeonia Lactiflora* injection in treating chronic cor pulmonale with pulmonary hypertension. *Chin. J. Integr. Trad. West. Med.*, **11**, 199–202.

25 Cheng, J.T., Wang, C.J., and Hsu, F.L. (1999) Paeoniflorin reverses guanethidine-induced hypotension *via* activation of central adenosine A1 receptors in Wistar rats. *Clin. Exp. Pharmacol. Physiol.*, **26**, 815–816.

26 Liu, T.P., Liu, M., Tsai, C.C., Lai, T.Y., Hsu, F.L., and Cheng, J.T. (2002) Stimulatory effect of paeoniflorin on the release of noradrenaline from ileal synaptosomes of guinea pig *in vitro*. *J. Pharm. Pharmacol.*, **54**, 681–688.

27 Lai, C.W., Hsu, F.L., and Cheng, J.T. (1998) Stimulatory effect of paeoniflorin on adenosine A-1 receptors to increase the translocation of protein kinase C (PKC) and glucose transporter (GLUT 4) in isolated rat white adipocytes. *Life Sci.*, **62**, 1591–1595.

28 Zhang, Y.Z. and Yan, X.F. (1990) Effects of nifedipine and Paeonia lactiflora on plasma TXB_2 and 6-keto-PGF_1 in cholesterol-fed rabbits. *Chin. J. Integr. Trad. West. Med.*, **10**, 669–671.

29 Goto, H., Shimada, Y., Tanaka, N., Tanigawa, K., Itoh, T., and Terasawa, K. (1999) Effect of extract prepared from the roots of *Paeonia lactiflora* on endothelium-dependent relaxation and antioxidant enzyme activity in rats administered high-fat diet. *Phytother. Res.*, **13**, 526–528.

30 Wang, Y. and Ma, R. (1990) Effect of an extract of *Paeonia lactiflora* on the blood coagulative and fibrinolytic enzymes. *Chin. J. Integr. Trad. West. Med.*, **10**, 101–102.

31 Ye, J., Duan, H., Yang, X., Yan, W., and Zheng, X. (2001) Anti-thrombosis effect of paeoniflorin: evaluated in a photochemical reaction thrombosis model in vivo. *Planta Med.*, **67**, 766–767.

32 Zhang, Y., Ming, L., Wang, Y., Ma, H.X., Ma, C. G., and Xu, S.Y. (1994) Anticonvulsant action of total glycosides of *Paeonia lactiflora* root. *Chin. Pharmacol. Bull.*, **10**, 372–374.

33 Ohta, H., Matsumoto, K., Shimizu, M., and Watanabe, H. (1994) Paeoniflorin attenuates learning impairment of aged rats in operant brightness discrimination task. *Pharmacol. Biochem. Behav.*, **49**, 213–217.

34 Watanabe, H. (1997) Candidates for cognitive enhancer extracted from medicinal plants: paeoniflorin and tetramethylpyrazine. *Behav. Brain Res.*, **83**, 135–141.

35 Ohta, H., Ni, J.W., Matsumoto, K., Watanabe, H., and Shimizu, M. (1993) Peony and its major constituent, paeoniflorin, improve radial maze performance impaired by scopolamine in rats. *Pharmacol. Biochem. Behav.*, **45**, 719–723.

36 Ohta, H., Matsumoto, K., Watanabe, H., and Shimizu, M. (1993) Involvement of β-adrenergic systems in the antagonizing effect of paeoniflorin on the scopolamine-induced deficit in radial maze performance in rats. *Jpn. J. Pharmacol.*, **62**, 345–349.

37 Abdel-Hafez, A.A., Meselhy, M.R., Nakamura, N., Hattori, M., Watanabe, H., Murakami, Y., El-Gendy, M.A., Mahfouz, N.M., and Mohamed, T.A. (1998) Effects of paeoniflorin derivatives on scopolamine-induced amnesia using a passive avoidance task in mice; structure-activity relationship. *Biol. Pharm. Bull.*, **21**, 1174–1179.

38 Sugaya, A., Suzuki, T., Sugaya, E., Yuyama, N., Yasuda, K., and Tsuda, T. (1991) Inhibitory effect of peony root extract on pentylenetetrazol-induced EEG power spectrum changes and extracellular calcium concentration changes in rat cerebral cortex. *J. Ethnopharmacol.*, **33**, 159–167.

39 Tsuda, T., Sugaya, A., Ohguchi, H., Kishida, N., and Sugaya, E. (1997) Protective effects of peony root extract and its components on neuron damage in the hippocampus induced by the cobalt focus epilepsy model. *Exp. Neurol.*, **146**, 518–525.

40 Liu, D.Z., Xie, K.Q., Ji, X.Q., Ye, Y., Jiang, C.L., and Zhu, X.Z. (2005) Neuroprotective effect of paeoniflorin on cerebral ischemic rat by activating adenosine A1 receptor in a manner different from its classical agonists. *Br. J. Pharmacol.*, **146**, 604–611.

41 Wang, B., Chen, M.Z., and Xu, S.Y. (1995) Effect of total glycosides of peony (TGP) on tumor necrosis factor produced by peritoneal macrophages in rats. *Chin. Pharmacol. Bull.*, **11**, 36–38.

42 Zhou, H., Zhang, A.P., Chen, M.Z., Xu, S.Y., and Wang, G.L. (1994) The modulatory effects of total glycosides from *Paeonia lactiflora* on hypothalmo-pituitary-adrenal axis and immune function in rats. *Chin. Pharmacol. Bull.*, **10**, 429–432.

43 Qi, X.G. (1991) Protective mechanism of *Salvia miltiorrhiza* and *Paeonia lactiflora* for experimental liver damage. *Chin. J. Integr. Trad. West. Med.*, **11**, 102–104.

44 Takeuchi, T., Nishii, O., Okamura, T., and Yaginuma, T. (1991) Effect of paeoniflorin, glycyrrhizin and glycyrrhetic acid on ovarian androgen production. *Am. J. Chin. Med.*, **19**, 73–78.

45 Lee, S.C., Kwon, Y.S., Son, K.H., Kim, H.P., and Heo, M.Y. (2005) Antioxidative constituents from *Paeonia lactiflora*. *Arch. Pharm. Res.*, **28**, 775–783.

46 Lee, S.J., Lee, I.S., and Mar, W. (2003) Inhibition of inducible nitric oxide synthase and cyclooxygenase-2 activity by 1,2,3,4,6-penta-*O*-galloyl-β-D-glucose in murine macrophage cells. *Arch. Pharm. Res.*, **26**, 832–839.

47 Takagi, K. and Harada, M. (1969) Pharmacological studies on herb paeony root. I. Central effects of paeoniflorin and combined effects with licorice component F_M 100. *Yakugaku Zasshi*, **89**, 879–886.

48 Okubo, T., Nagai, F., Seto, T., Satoh, K., Ushiyama, K., and Kano, I. (2000) The inhibition of phenylhydroquinone-induced oxidative DNA cleavage by constituents of Moutan Cortex and Paeoniae Radix. *Biol. Pharm. Bull.*, **23**, 199–203.

49 Lee, S.J., Lee, H.K., Jung, M.K., and Mar, W. (2006) In vitro antiviral activity of 1,2,3,4,6-penta-*O*-galloyl-β-D-glucose against hepatitis B virus. *Biol. Pharm. Bull.*, **29**, 2131–2134.

50 Nishida, S., Kikuichi, S., Yoshioka, S., Tsubaki, M., Fujii, Y., Matsuda, H., Kubo, M., and Irimajiri, K. (2003) Induction of apoptosis in HL-60 cells treated with medicinal herbs. *Am. J. Chin. Med.*, **31**, 551–562.

51 Lee, S.M., Li, M.L., Tse, Y.C., Leung, S.C., Lee, M.M., Tsui, S.K., Fung, K.P., Lee, C.Y., and Waye, M.M. (2002) Paeoniae Radix, a Chinese herbal extract, inhibit hepatoma cells growth by inducing apoptosis in a p53 independent pathway. *Life Sci.*, **71**, 2267–2277.

52 Tsuboi, H., Hossain, K., Akhand, A.A., Takeda, K., Du, J., Rifa'I, M., Dai, Y., Hayakawa, A., Suzuki, H., and Nakashima, I. (2004) Paeoniflorin induces apoptosis of lymphocytes through a redox-linked mechanism. *J. Cell. Biochem.*, **93**, 162–172.

53 Li, C.R., Zhou, Z., Zhu, D., Sun, Y.N., Dai, J. M., and Wang, S.Q. (2007) Protective effect of paeoniflorin on irradiation-induced cell damage involved in modulation of reactive oxygen species and the mitogen-activated protein kinases. *Int. J. Biochem. Cell. Biol.*, **39**, 426–438.

54 Shu, Y.Z., Hattori, M., Akao, T., Kobashi, K., Kagei, K., Fukuyama, K., Tsukihara, T., and Namba, T. (1987) Metabolism of paeoniflorin and related compounds by human intestinal bacteria. II. Structures of 7S- and 7R-paeonimetabolines I and II formed by *Bacteroides fragilis* and *Lactobacillus brevis*. *Chem. Pharm. Bull. (Tokyo)*, **35**, 3726–3733.

55 He, J.X., Akao, T., and Tani, T. (2002) Development of a simple HPLC method for determination of paeoniflorin-metabolizing activity of intestinal bacteria in rat feces. *Chem. Pharm. Bull. (Tokyo)*, **50**, 1233–1237.

Paeonia suffruticosa

Cortex Moutan (Mudanpi) is the dried root bark of *Paeonia suffruticosa* Andr. (Ranunculaceae). It is used as an analgesic, hemostyptic, and bacteriostatic agent. It is also used to treat fever, night sweating, hemoptysis, hematemesis, dysmenorrhea, carbuncles, and traumatic injuries.

Paeonia suffruticosa Andr.

Chemistry

Like the peony root, Cortex Moutan, the moutan root bark also contains paeoniflorin and related compounds oxypaeoniflorin, benzoylpaeoniflorin, benzoyloxypaeoniflorin, galloylpaeoniflorin, and galloyloxypaeoniflorin [1, 2]. Further monoterpene glycosides related to paeoniflorin from the moutan root bark were identified as mudanpiosides A–F [3].

The most important constituent of the moutan root bark was known to be paeonol [4]. A number of glycosides of paeonol were also isolated from the root bark and identified as paeonoside, paeonolide, apiopaeonoside [5] and suffruticosides A–E [2]. Paeoniflorin was found to occur ubiquitously as a characteristic constituent in *Paeonia* species, whereas paeonol and its glycosides were only found in the root of *P. suffruticosa* and its substitutes. The Chinese Pharmacopoeia requires a qualitative determination of paeonol in Cortex Moutan by thin-layer chromatographic comparison with reference substance, and a quantitative determination by HPLC. The content of paeonol in Cortex Moutan should be not less than 1.2%.

Another item listed in the Chinese Pharmacopoeia containing paeonol as major component is Radix et Rhizoma Cynanchi Paniculati. Radix et Rhizoma Cynanchi Paniculati (Xuchangqing) is the dried root and rhizome of *Cynanchum paniculatum* (Bge.) Kitag. (Asclepiadaceae). It is used as an antiphlogistic, antirheumatic and analgetic agent for the treatment of toothache and traumatic disease. The Chinese Pharmacopoeia requires a qualitative determination of paeonol in Radix Cynanchi Paniculati by thin-layer chromatographic comparison with reference substance, and a quantitative determination by HPLC. The content of paeonol in Radix Cynanchi Paniculati should be not less than 1.3%.

Further constituents from the moutan root bark were identified as 1,2,3,4,6-penta-*O*-galloyl-β-D-glucose [6], mudanpinoic acid A, a hexacyclic triterpene, mudanoside B, a gallic acid glycoside [7], the monoterpenes paeonisuffrone, paeonisuffral [8], paeonisothujone, deoxypaeonisuffrone, and isopaeonisuffral [9], benzoic acid, resacetophenone, paeoniflorigenone, β-sitosterol, betulinic acid, oleanolic acid, quercetin, β-sitosterol β-D-glucopyranoside, and caffeic acid stearyl ester [7].

Handbook of Chinese Medicinal Plants: Chemistry, Pharmacology, Toxicology
Weici Tang and Gerhard Eisenbrand

ISBN: 978-3-527-32226-8

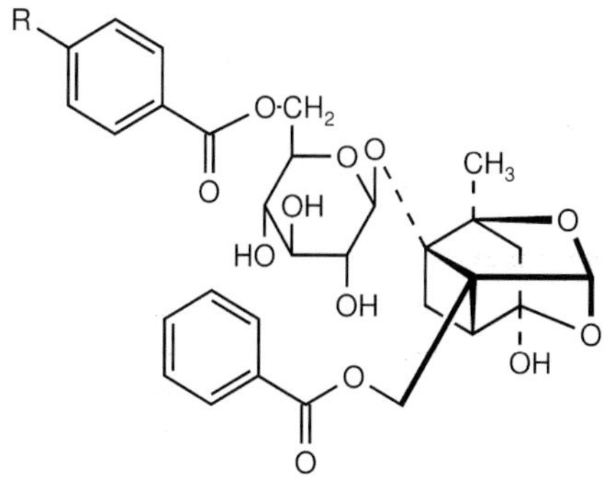

Mudanpioside A: R = OCH_3
Mudanpioside C: R = OH

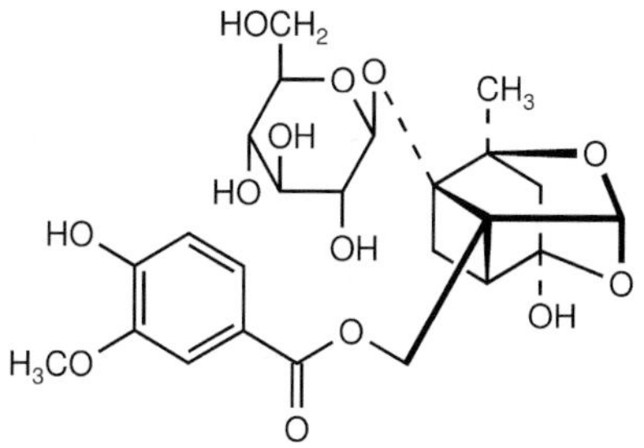

Mudanpioside E

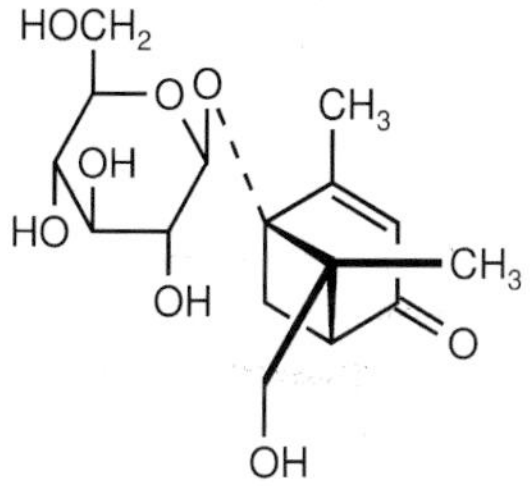

Mudanpioside F

Pharmacology and Toxicology

The pharmacokinetics of paeonol in rats after intravenous (i.v.) administration at various doses followed a two-compartment model. The area under the curve (AUC) increased proportionally with the dose. There was no dose-related difference in the elimination half-life or volume of distribution [10]. Paeonol was found to show a greater permeation from artificial gastric juice into artificial plasma when it was applied as a decoction or as a freeze-dried extract of moutan root bark, than when applied as purified paeonol [11]. The urinary metabolites of paeonol after oral administration to rats were identified as 2,4-dihydroxyacetophenone 5-*O*-sulfate, resacetophenone 2-*O*-sulfate, 2-hydroxy-4-methoxyacetophenone 5-*O*-sulfate, paeonol 2-*O*-sulfate, resacetophenone, and as unchanged paeonol. Among these metabolites, the sulfates of resacetophenone, 2-hydroxy-4-methoxyacetophenone and paeonol were detected in the plasma. The bile of rats treated orally with paeonol was found to contain 2-hydroxy-4-methoxyacetophenone 5-*O*-sulfate, which suggested the presence of an enterohepatic circulation of paeonol [12]. The plasma concentrations of paeonol in rats after an oral dose of a moutan cortex decoction also fitted a two-compartment model, with first-order absorption. The mean terminal half-life of paeonol was reported as about 81 min [13].

Paeonol was reported to exert immunostimulatory activity and to enhance specific cellular immunity by raising peripheral blood lymphocytes, releasing leukocyte-migratory inhibition factors, and increasing the phagocytosis of polymorphonuclear neutrophils [14]. Paeonol was also found to exhibit analgesic effects and inhibitory

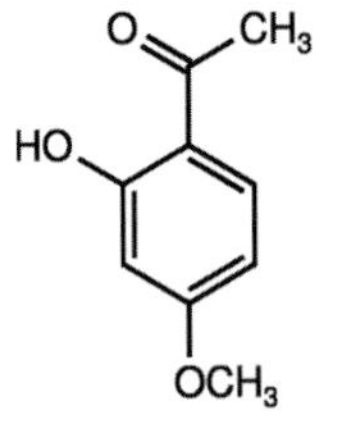

Paeonol

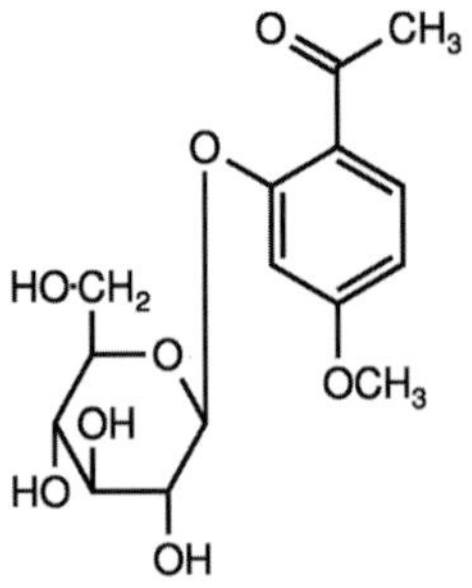

Paeonoside

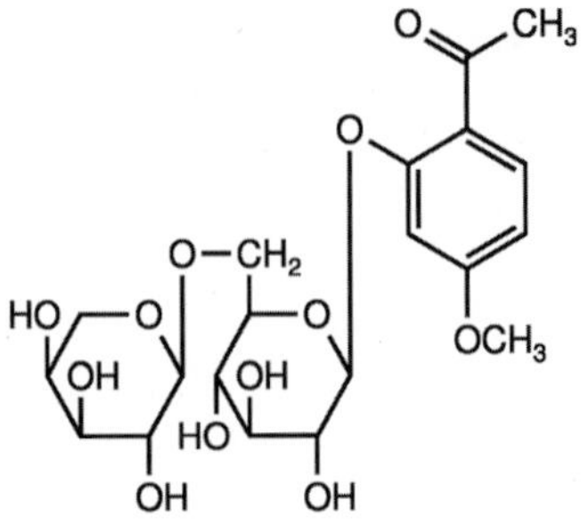

Paeonolide

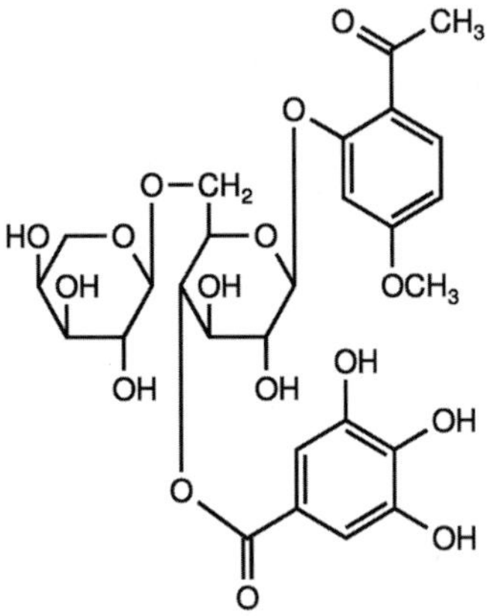

Suffruticoside D

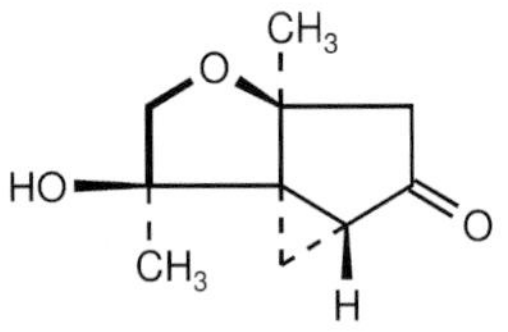

Paeonisothujone

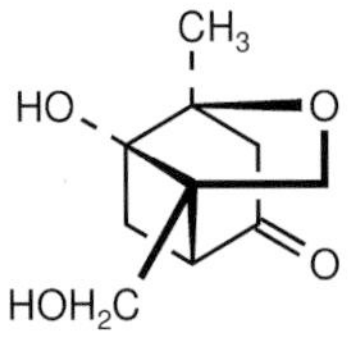

Paeonisuffrone

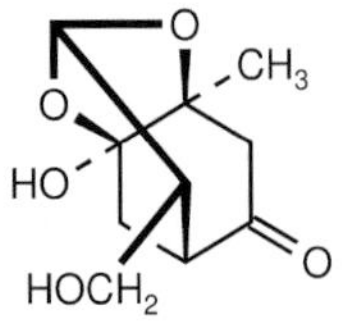

Isopaeonisuffral

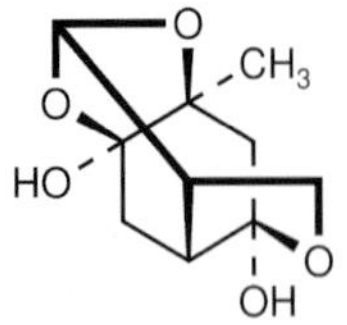

Paeonisuffral

action on gastrointestinal motility [15]. In addition, paeonol attenuated colitis in mice induced by trinitrobenzene sulfonic acid. Myeloperoxidase activity and inducible nitric oxide synthase (iNOS) production in the colon were reduced by paeonol. In colon cancer-derived CW-2 cells, paeonol inhibited iNOS protein and mRNA expression stimulated by tumor necrosis factor-α (TNF-α) and interferon-γ (IFN-γ). Furthermore, paeonol reduced TNF-α-induced nuclear factor-κB (NF-κB) activation in CW-2 cells [16].

The methanol extract of moutan root bark prevented disseminated intravascular coagulation induced by enterotoxin in rats. The methanol extract, the glycosidic fraction, and paeonol were each found to inhibit rabbit platelet aggregation induced by ADP or collagen, in concentration-dependent fashion [17]. Paeonol was found to selectively inhibit the aggregation of rabbit platelets induced by arachidonic acid, and also to inhibit the formation of thromboxane A_2 (TXA_2) and prostaglandin D_2 (PGD_2) from arachidonic acid [18]. Moutan root bark and its glycosidic fraction also showed anti-inflammatory effects. The intraperitoneal (i.p.) injection of paeonol inhibited hind paw edema induced by carrageenan, egg white, formaldehyde, histamine, serotonin, or bradykinin in rats, and also inhibited hind paw edema induced by carrageenan in adrenalectomized rats. Treatment with paeonol also inhibited ear swelling induced by xylene, and the increase of peritoneal capillary permeability induced by endotoxin in mice [19].

Treatment with paeonol in rats dose-dependently inhibited TNF-α and interleukin-1β (IL-1β) formation, but enhanced IL-10 production in the paw exudates, both at the early and late phases after carrageenan injection. Paeonol also dose-dependently decreased the formation of PGE_2 in rat paw exudates, with a greater inhibition at the late phase. However, an inhibition of NO generation was observed only during the late phase, accompanied by an inhibition of iNOS and cyclooxygenase-2 (COX-2) protein expression in paw tissue. Elevated myeloperoxidase activity in carrageenan-injected paws was also dose-dependently reduced by paeonol; this suggested that the mechanisms by which paeonol exerted its anti-inflammatory and analgesic effects might be associated with decreased production of proinflammatory cytokines, NO and PGE_2, and an increased production of IL-10 as an anti-inflammatory cytokine in carrageenan-injected rat paws. In addition, the attenuation of elevated iNOS and COX-2 protein expression, and of neutrophil infiltration in carrageenan-injected paws, might also be involved in the effects of paeonol [20].

Paeonol was found to significantly and reversibly inhibit voltage-gated K^+, Na^+, and Ca^{2+} currents in neuroblastoma and glioma hybrid cells in a concentration-dependent manner [21]. Inhibitory effects of the moutan root bark on the action potential of cultured rat myocardial cells were also reported [22]. The i.p. injection of paeonol to rats for 15 days significantly prevented the myocardial ischemia–reperfusion damage produced by occlusion of the left coronary artery. Ultrastructural damage of the myocardium induced by oxygen free radicals was also prevented by paeonol [23]. The exposure of rat ventricular myocytes to paeonol resulted in a concentration-dependent inhibition of the peak L-type Ca^{2+} channel current, with a half-maximum inhibition of 561 μM. The protection by paeonol against myocardial injury might be due to its blocking effect on L-type Ca^{2+} channel currents [24].

The effects of paeonol on cerebral infarction have been studied in rats with cerebral infarction established by occluding both common carotid arteries and the right middle cerebral artery for 90 min, followed by a 24 h period of reperfusion. Paeonol at doses of 15 and 20 mg kg^{-1} given prior to, and at a dose of 20 mg given after artery occlusion, markedly reduced the area of cerebral infarction. Paeonol pretreatment reduced the neurodeficit score, and reduced the lucigenin-chemiluminescence counts after a 2 h period of reperfusion. Numbers of ED1 (mouse anti rat CD68) and IL-1β immunoreactive cells were also found to be reduced in the cerebral infarction region; this suggested that paeonol had suppressed the formation of superoxide anion, scavenged radicals, inhibited microglia activation, and suppressed IL-1β in ischemia-reperfusion-injured rats [25].

Aqueous extracts of the moutan root bark, when given in the drinking water to male and female obese mice, inhibited their growth, especially in male animals. Treatment with the moutan root bark resulted in a significant decline in food intake and an increase in glucose tolerance. Only a minimal difference was apparent between animals with or without treatment with regards to the weights of the heart, liver, kidney, lung, spleen and major endocrine organs in both genders, and in the pattern of estrous cycles in females [26]. The extract of root bark of *P. suffruticosa* was also found to exert hypoglycemic activity, and significantly reduced thiobarbituric acid-reactive substances levels in normal rats compared to diabetic rats induced by streptozotocin [27]. Paeonol at oral doses of 200 and 400 mg kg^{-1} was also found to improve oral glucose tolerance in neonatal diabetic rats induced by streptozotocin [28].

Paeonol was also reported to inhibit the proliferation of Bel-7404 hepatocellular carcinoma cells, and to induce cell apoptosis, presumably regulated by phosphatidylinositol-3-kinase [29]. Paeonol also suppressed the growth of the human hepatoma cell line HepG2, arrested the cell cycle at S-phase, and induced cell apoptosis. Furthermore, the antiproliferative activity of some conventional antineoplastic agents (including cisplatin, doxorubicin and 5-fluorouracil) against HepG2 cells was significantly enhanced when used in combination with paeonol. The cell cycle arrest at S-phase of HepG2 cells by paeonol might represent one of the mechanisms for its ability to enhance the cytotoxicity of some chemotherapeutics [30].

The extract of moutan root bark and paeonol was also found to exhibit antimutagenic activity. For example, it decreased the mutation frequency in *Escherichia coli* WP2s induced by 4-nitroquinoline-1-oxide [31]. In contrast, an aqueous extract of moutan root bark was found to be mutagenic in a chromosomal aberration test and in a micronucleus assay in mice, but not mutagenic in a test using *Salmonella typhimurium* TA98 or TA100 with S-9 mix [32]. Paeonol promoted *N*-acetyltransferase (NAT) activity in human colon tumor cells, and increased the formation of the *N*-acetyl-2-aminofluorene–DNA adduct [33].

Both, antioxidative and free radical-scavenging activities of the aqueous extract of moutan root bark have been described. The extract inhibited rat erythrocyte hemolysis, and inhibited lipid peroxidation in rat kidney and brain homogenates [34, 35]. The extract and paeonol suppressed the phenylhydroquinone-induced cleavage of pUC18 DNA, scavenging superoxide and hydroxy radicals generated by phenylhydroquinone [36]. Galloyloxypaeoniflorin, galloylpaeoniflorin and suffruticosides A–D each showed potent radical-scavenging and antioxidative effects [2].

References

1 Yu, J., Lang, H.Y., and Xiao, P.G. (1985) The occurrence of paeoniflorins and paeonols in Paeoniaceae. *Acta Pharm. Sin.*, **20**, 229–234.

2 Yoshikawa, M., Uchida, E., Kawaguchi, A., Kitagawa, I., and Yamahara, J. (1992) Galloyl-oxypaeoniflorin, suffruticosides A, B, C, and D, five new antioxidative glycosides, and suffruticoside E, A paeonol glycoside, from Chinese moutan cortex. *Chem. Pharm. Bull. (Tokyo)*, **40**, 2248–2250.

3 Lin, H.C., Ding, H.Y., Shung, T., and Wu, P.L. (1996) Monoterpene glycosides from *Paeonia suffruticosa*. *Phytochemistry*, **41**, 237–242.

4 Ding, A., Guo, R., and Rong, J. (1996) Determination of paeonol in carbonized bark of *Paeonia suffruticosa* Andr. by HPLC. *China J. Chin. Mater. Med.*, **21**, 23–24.

5 Yu, J., Lang, H.Y., and Xiao, P.G. (1986) A new compound, apiopaeonoside, isolated from the root of *Paeonia suffruticosa*. *Acta Pharm. Sin.*, **21**, 191–197.

6 Takechi, M. and Tanaka, Y. (1982) Antiviral substances from the root of *Paeonia* species. *Planta Med.*, **45**, 252–253.

7 Lin, H.C., Ding, H.Y., and Wu, Y.C. (1998) Two Novel Compounds from *Paeonia suffruticosa*. *J. Nat. Prod.*, **61**, 343–346.

8 Yoshikawa, M., Harada, E., Kawaguchi, A., Yamahara, J., Murakami, N., and Kitagawa, I. (1993) Absolute stereostructures of paeonisuffrone and paeonisuffral, two new labile monoterpenes, from Chinese Moutan Cortex. *Chem. Pharm. Bull. (Tokyo)*, **41**, 630–632.

9 Yoshikawa, M., Harada, E., Minematsu, T., Muraoka, O., Yamahara, J., Murakami, N., and Kitagawa, I. (1994) Absolute stereostructures of paeonisothujone, a novel skeletal monoterpene ketone, and deoxypaeonisuffrone, and isopaeonisuffral, two new monoterpenes, from Moutan Cortex. *Chem. Pharm. Bull.*, **42**, 736–738.

10 Tsai, T.H., Chou, C.J., and Chen, C.F. (1994) Pharmacokinetics of paeonol after intravenous administration in rats. *J. Pharm. Sci.*, **83**, 1307–1309.

11 Tani, T., Inoue, K., Arichi, S., and Ohno, T. (1987) Biopharmaceutical studies on crude drug preparations. I. Permeation of paeonol in a decoction and dry extract of *Paeonia suffruticosa* root cortex using an absorption simulator. *J. Ethnopharmacol.*, **21**, 37–44.

12 Yasuda, T., Kon, R., Nakazawa, T., and Ohsawa, K. (1999) Metabolism of paeonol in rats. *J. Nat. Prod.*, **62**, 1142–1144.

13 Wu, X., Chen, H., Chen, X., and Hu, Z. (2003) Determination of paeonol in rat plasma by high-performance liquid chromatography and its application to pharmacokinetic studies following oral administration of Moutan cortex decoction. *Biomed. Chromatogr.*, **17**, 504–508.

14 Li, F.C., Zhou, X.L., and Mao, H.L. (1994) A study of paeonol injection on immune functions in rats. *Chin. J. Integr. Trad. West. Med.*, **14**, 37–38.

15 Sun, F.Z., Cai, M., and Lou, F.C. (1993) Analgesic effect and gastro-intestinal motility inhibitory action of 3-hydroxy-4-methoxy-acetophenone from *Cynanchum paniculatum* (Bunge) Kitagawa. *China J. Chin. Mater. Med.*, **18**, 362–363.

16 Ishiguro, K., Ando, T., Maeda, O., Hasegawa, M., Kadomatsu, K., Ohmiya, N., Niwa, Y., Xavier, R., and Goto, H. (2006) Paeonol attenuates TNBS-induced colitis by inhibiting NF-κB and STAT1 transactivation. *Toxicol. Appl. Pharmacol.*, **217**, 35–42.

17 Shi, L., Fan, P.S., Fang, J.X., and Han, Z.X. (1988) Inhibitory effect of paeonol on experimental atherosclerosis and platelet aggregation in rabbit. *Acta Pharmacol. Sin.*, **9**, 555–558.

18 Lin, H.C., Ding, H.Y., Ko, F.N., Teng, C.M., and Wu, Y.C. (1999) Aggregation inhibitory activity of minor acetophenones from *Paeonia* species. *Planta Med.*, **65**, 595–599.

19 Wu, G.Z., Hang, B.Q., Hang, J.X., and Ling, G. X. (1989) Anti-inflammatory effect of paeonol and its action mechanism. *J. China Pharm. Univ.*, **20**, 147–150.

20 Chou, T.C. (2003) Anti-inflammatory and analgesic effects of paeonol in carrageenan-evoked thermal hyperalgesia. *Br. J. Pharmacol.*, **139**, 1146–1152.

21 Hu, Q. and Shi, Y.L. (1994) Inhibition of voltage-gated K^+, Na^+ and Ca^{2+} currents in neuroblastoma × glioma hybrid cells by paeonol. *Acta Physiol. Sin.*, **46**, 575–580.

22 Ma, Y.L. and Li, L.D. (1986) Effects of quinidine and *Paeonia suffruticosa* on the action potential of cultured myocardial cells. *Chin. J. Integr. Trad. West. Med.*, **6**, 292–293.

23 Zhang, W.G. and Zhang, Z.S. (1994) Effects of paeonol on ischemia reperfusion damage and lipid peroxidation in rat heart. *Acta Pharm. Sin.*, **29**, 145–148.

24 Zhang, G.Q., Hao, X.M., Zhou, P.A., and Wu, C.H. (2003) Effect of paeonol on L-type calcium channel in rat ventricular myocytes. *Methods Fund. Exp. Clin. Pharmacol.*, **25**, 281–285.

25 Hsieh, C.L., Cheng, C.Y., Tsai, T.H., Lin, I.H., Liu, C.H., Chiang, S.Y., Lin, J.G., Lao, C.J., and Tang, N.Y. (2006) Paeonol reduced cerebral infarction involving the superoxide anion and microglia activation in ischemia-reperfusion injured rats. *J. Ethnopharmacol.*, **106**, 208–215.

26 Nagasawa, H., Iwabuchi, T., and Inatomi, H. (1992) Protection by tree-peony (*Paeonia suffruticosa* Andr) of obesity in (SLN × C3H/He) F1 obese mice. *In Vivo*, **5**, 115–118.

27 Jung, C.H., Zhou, S., Ding, G.X., Kim, J.H., Hong, M.H., Shin, Y.C., Kim, G.J., and Ko, S. G. (2006) Antihyperglycemic activity of herb extracts on streptozotocin-induced diabetic rats. *Biosci. Biotechnol. Biochem.*, **70**, 2556–2559.

28 Lau, C.H., Chan, C.M., Chan, Y.W., Lau, K.M., Lau, T.W., Lam, F.C., Law, W.T., Che, C.T., Leung, P.C., Fung, K.P., Ho, Y.Y., and Lau, C.B. (2007) Pharmacological investigations of the anti-diabetic effect of Cortex Moutan and its active component paeonol. *Phytomedicine*, **14**, 778–784.

29 Zhang, C.H., Hu, S.Y., Li, Y.H., and Cao, M.Q. (2006) Anti-tumor effect and mechanism of Paeonol on the hepatocellular carcinoma cell

line Bel-7404. *Zhong Nan Da Xue Xue Bao Yi Xue Bao,* **31**, 682–686. 695.

30 Xu, S.P., Sun, G.P., Shen, Y.X., Wei, W., Peng, W.R., and Wang, H. (2007) Antiproliferation and apoptosis induction of paeonol in HepG2 cells. *World J. Gastroenterol.*, **13**, 250–256.

31 Fukuhara, Y. and Yoshida, D. (1987) Paeonol: a bio-anti-mutagen isolated from a crude drug, Moutan cortex. *Agric. Biol. Chem.*, **51**, 1441–1442.

32 Yin, X.J., Liu, D.X., Wang, H.C., and Zhou, Y. (1991) A study on the mutagenicity of 102 raw pharmaceuticals used in Chinese traditional medicine. *Mutat. Res.*, **260**, 73–82.

33 Chung, J.G. (1999) Paeonol promotion of DNA adduct formation and arylamines *N*-acetyltransferase activity in human colon tumour cells. *Food Chem. Toxicol.*, **37**, 327–334.

34 Liu, F. and Ng, T.B. (2000) Antioxidative and free radical scavenging activities of selected medicinal herbs. *Life Sc.*, **66**, 725–735.

35 Zhang, H.Y., Ge, N., and Zhang, Z.Y. (1999) Theoretical elucidation of activity differences of five phenolic antioxidants. *Acta Pharmacol. Sin.*, **20**, 363–366.

36 Okubo, T., Nagai, F., Seto, T., Satoh, K., Ushiyama, K., and Kano, I. (2000) The inhibition of phenylhydroquinone-induced oxidative DNA cleavage by constituents of Moutan Cortex and Paeoniae Radix. *Biol. Pharm. Bull.*, **23**, 199–203.

Panax ginseng

Radix et Rhizoma Ginseng (Renshen) is the dried root and rhizome of *Panax ginseng* C.A. Mey. (Araliaceae). It is used as a tonic for the treatment of general weakness, insomnia in chronic diseases, heart failure, and cardiogenic shock. Wild ginseng, the mountain ginseng, is called "Shanshen" in Chinese. Cultivated ginseng, the garden ginseng, is called "Yuanshen" in Chinese.

Panax ginseng C.A. Mey.

Radix et Rhizoma Ginseng Rubra (Hongshen) is the steamed and dried root and rhizome of *Panax ginseng*. It is also used as a general tonic and for the treatment of heart failure and cardiogenic shock.

Folium Ginseng (Renshenye) is the dried leaf of *Panax ginseng*. It is used as a general tonic.

The Chinese Pharmacopoeia notes that Radix Ginseng, Radix Ginseng Rubra and Folium Ginseng are incompatible with Rizoma et Radix Veratri.

Chemistry

Ginseng, one of the best-known traditional Chinese medicines, contains ginsenosides as major active components. Ginsenosides detected are mainly derived from the aglycones 20(*S*)-protopanaxadiol and 20 (*S*)-protopanaxatriol. Only ginsenoside Ro possesses oleanolic acid as aglycone [1].

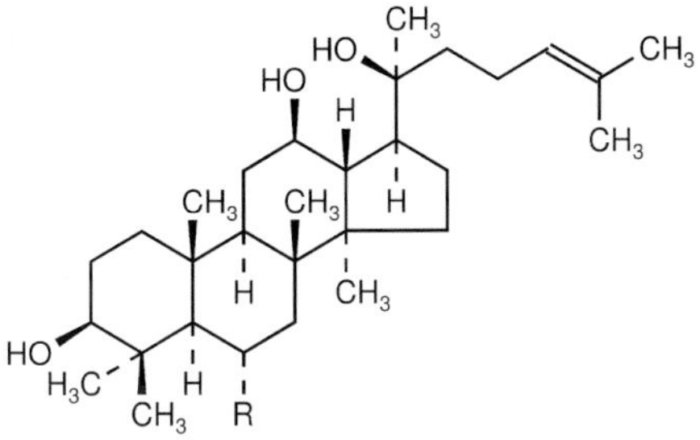

20(S)-Protopanaxadiol: R = H
20(S)-Protopanaxatriol: R = OH

Saponins from ginseng root with protopanaxadiol as aglycone were reported to be ginsenosides Ra_1, Ra_2, Ra_3, Rb_1, Rb_2, Rb_3, Rc, Rd, Rg_3, Rh_2, Rs_1, Rs_2, Rs_3, notoginsenoside R_4, quinquenoside R_1, malonylginsenosides Rb_1, Rb_2, Rc, Rd and koryoginsenoside R_2, with ginsenoside Rb_1 as major component [2]. Saponins from ginseng root with protopanaxatriol as aglycone were reported to be ginsenosides Re, Rf, Rg_1, Rg_2, Rh_1, notoginsenoside R_1 and koryoginsenoside R_1, with ginsenosides Re and Rg_1 as the major components [3]. The content of total ginsenoside in wild ginseng was reported as 3.3–5%, and in cultivated ginseng as 2.4–2.6% [4]. The Chinese Pharmacopoeia requires a qualitative determination of ginsenosides Rb_1, Re, Rf, and Rg_1 in Radix et Rhizoma Ginseng by thin-layer chromatographic (TLC) comparison with reference substances, and a quantitative determination of ginsenosides Rb_1, Re, and Rg_1 in Radix et Rhizoma Ginseng by HPLC. The total content of

Handbook of Chinese Medicinal Plants: Chemistry, Pharmacology, Toxicology
Weici Tang and Gerhard Eisenbrand

ISBN: 978-3-527-32226-8

ginsenosides Re and Rg_1 in Radix et Rhizoma Ginseng should be not less than 0.3%, and the content of ginsenoside Rb_1 not less than 0.2%.

Almost all ginsenosides isolated from the ginseng root are also found in red ginseng [5]. However, 20(*R*)-ginsenoside Rg_2, ginsenoside Rg_3, 20(*R*)-ginsenoside Rh_1, and ginsenoside Rh_2 are characteristic saponins found in red ginseng. These are considered to be degradation products by heating and hydrolysis during the processing of red ginseng. Ginsenosides Rf_2, Rh_3 and Rh_4 [6, 7] were isolated from red ginseng. The Chinese Pharmacopoeia requires a qualitative determination of ginsenosides Rb_1, Re, Rf, and Rg_1 in Radix et Rhizoma Ginseng Rubra by TLC comparison with reference substances, and a quantitative determination of ginsenosides Rb_1, Re, and Rg_1 in Radix et Rhizoma Ginseng Rubra by HPLC. The total content of ginsenosides Re and Rg_1 in Radix et Rhizoma Ginseng Rubra should be not less than 0.25%, and the content of ginsenoside Rb_1 not less than 0.2%.

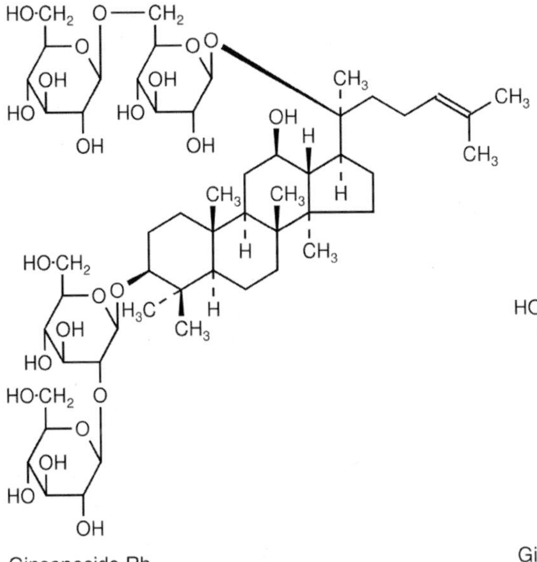

Ginsenoside Rb_1

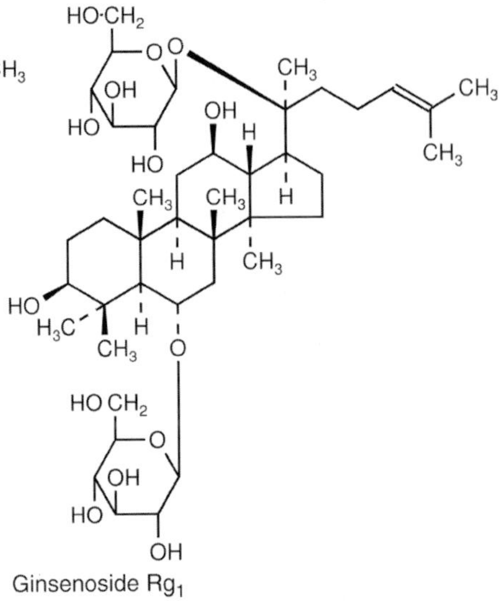

Ginsenoside Rg_1

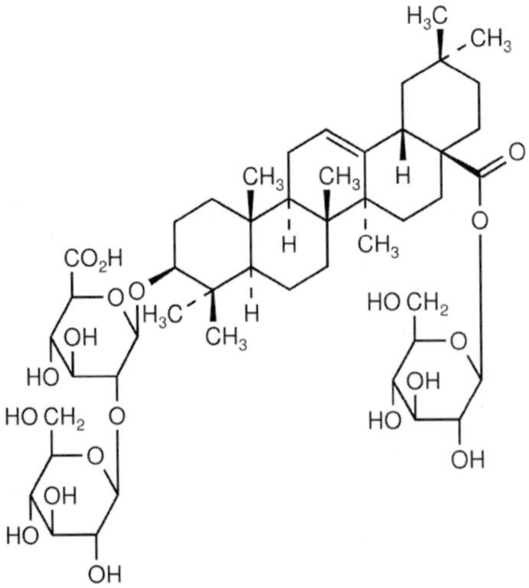

Ginsenoside Ro

Ginsenosides Rb_1, Rb_2, Rc, Rd, Re, Rf_2, Rg_1, Rg_2, Rg_3, 20-glucoginsenosides Rf, 20 (*R*)-ginsenoside Rh_2, and ginsenosides F_1, F_2, F_3, F_4 and La were isolated from ginseng leaves [8, 9]. Related ginsenosides were also isolated from ginseng flowers [10] and ginseng fruits [11]. The Chinese Pharmacopoeia requires a qualitative determination of ginsenosides Re and Rg_1 in Folium Ginseng by TLC comparison with reference substances, and a quantitative determination of ginsenosides Re and Rg_1 in Folium Ginseng by HPLC. The total content of ginsenosides Re and Rg_1 in Folium Ginseng should be not less than 2.25%. The ginseng leaves contain much more ginsenosides as compared to the ginseng root.

The volatile component of ginseng root contained a series of sesquiterpenes identified as eremophilene, β-gurjunene, *trans*-caryophyllene, *cis*-caryophyllene, muurolene, γ-patchoulene, β-endesmol, β-farnesene, β-bisabolene, aromadendrene, alloaromadendrene, β-guaiene, γ-elemene, mayurone, senecrassidiol, ginsenol, and panasinsanol A and its epimer panasinsanol B [12].

Acetylenic compounds from ginseng root were identified as panaxynol, panaxydol, panaxytriol, and ginsenoynes A–K [13]. The contents of panaxynol, panaxydol and panaxytriol in red ginseng were reported to be about 250 to $320\,\mu g\,g^{-1}$ [14]. Similar to panaxynol, panaxydol and panaxytriol, panaxyne epoxide, and ginsenoynes are all derived from 4,6-heptadiyne.

Ginseng root was reported to contain a number of hypoglycemic peptide glycans named panaxans [15]. An acidic polysaccharide, ginsan, with a molecular mass of 150 kDa [16], as well as ginsenans S-IA and S-IIA [17] and ginsenans PA and PB [18], were isolated from ginseng root. Polysaccharides were also isolated from stem [19] and leaves [20]. A neuroexcitotoxic non-protein amino acid, β-*N*-oxalo-L-α,β-diaminopropionic acid and its isomer α-*N*-oxalo-L-α,β-diaminopropionic acid were also isolated from ginseng root. β-*N*-Oxalo-L-α, β-diaminopropionic acid was verified as a natural substance, whilst its isomer α-*N*-oxalo-L-α,β-diaminopropionic acid was very likely an artifact generated during the separation process [21]. The isolation of

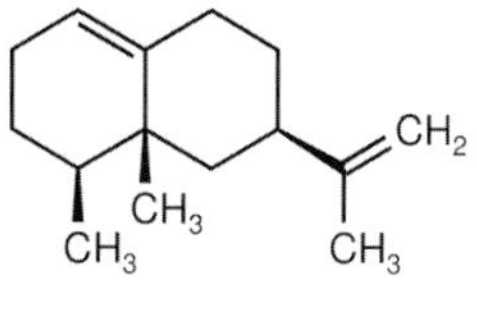

Eremophilene

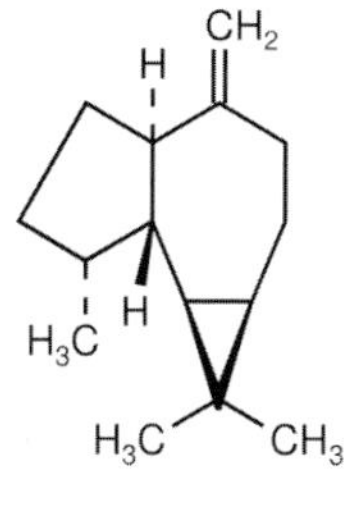

Aromadendrene

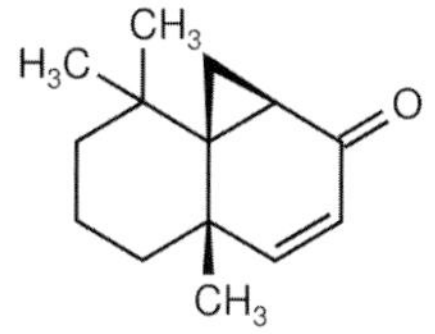

Mayurone

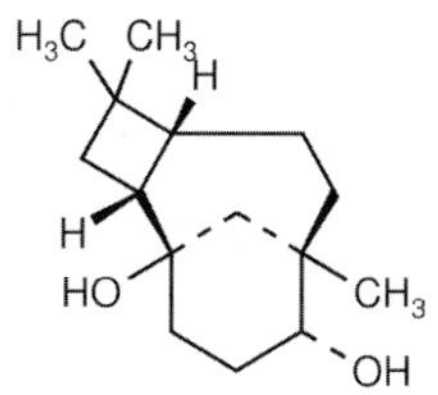

Senecrassidiol

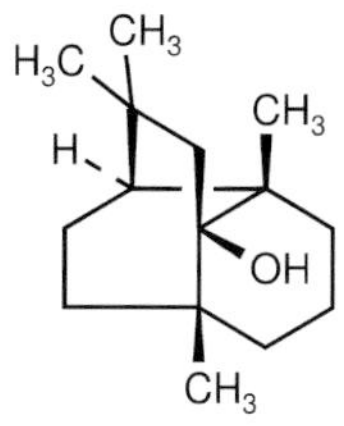

Ginsenol

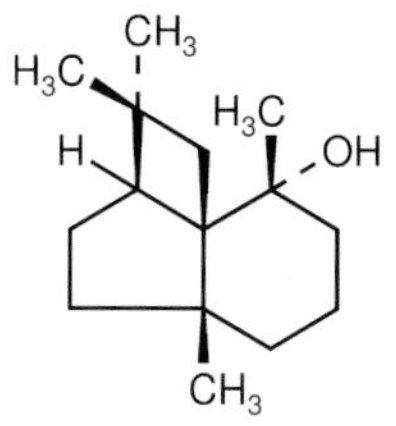

Panasinsanol A

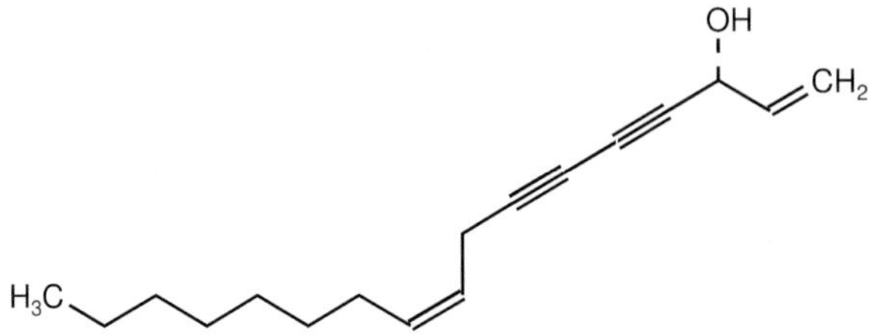

Panaxynol

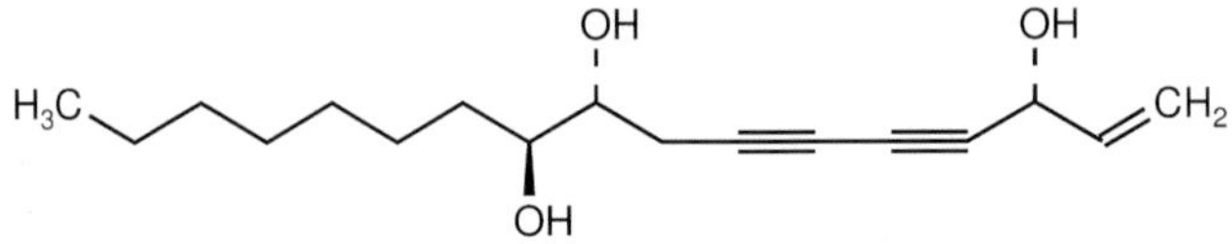

Panaxytriol

polypeptides from ginseng root was reported [22].

Four further items containing *Panax* species are officially listed in the Chinese Pharmacopoeia: Radix et Rhizoma Notoginseng; Radix Panacis Quinquefolii; Rhizoma Panacis Japonici; and Rhizoma Panacis Majoris. All of these *Panax* species contain ginsenosides and related saponins as the active principles. Radix et Rhizoma Notoginseng (Sanqi) is the dried root and rhizome of *Panax notoginseng* (Burk.) F. H. Chen. It is used mainly as a hemostatic agent for the treatment of different types of bleeding, and also for the treatment of traumatic diseases. The Chinese Phar-

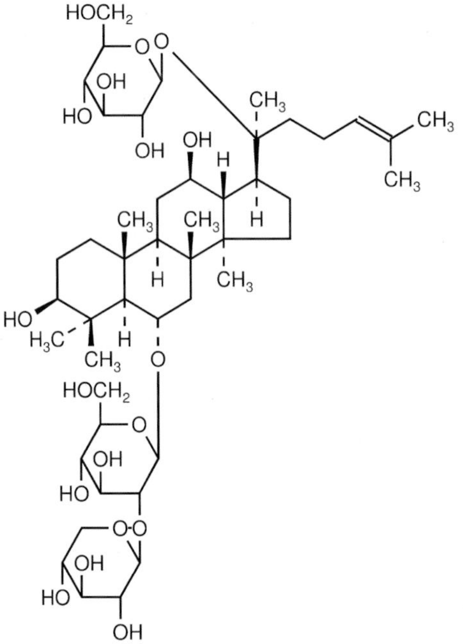

Notoginsenoside R_1

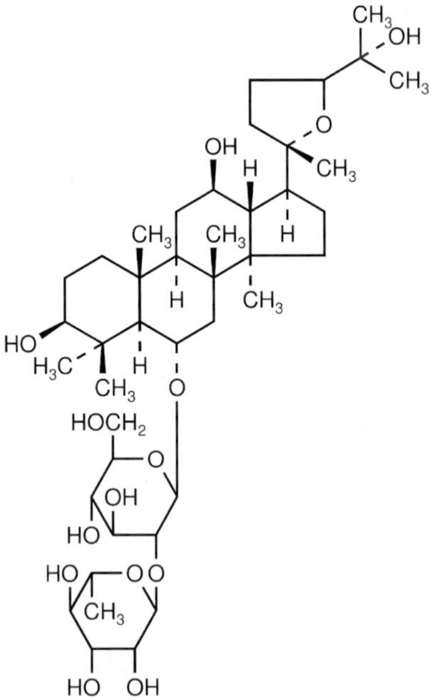

Pseudoginsenoside F_{11}

macopoeia notes that Radix et Rhizoma Notoginseng should be used in pregnancy only with caution. The Chinese Pharmacopoeia requires a qualitative determination of ginsenosides Rb_1, Re, Rg_1 and notoginsenoside R_1 in Radix et Rhizoma Notoginseng by TLC comparison with reference substances, and a quantitative determination of ginsenosides Rb_1, Rg_1, and notoginsenoside R_1 in Radix et Rhizoma Notoginseng by HPLC. The total content of ginsenosides Rb_1, Rg_1, and notoginsenoside R_1 in Radix et Rhizoma Notoginseng should be not less than 5.0%. The root and rhizome of *P. notoginseng* contained much more ginsenosides Rb_1 and Rg_1 than ginseng root. Notoginsenoside R_1 is used as a marker saponin for Radix et Rhizoma Notoginseng [23].

A galenic preparation of *P. notoginseng* officially listed in the Chinese Pharmacopoeia is "Sanqi Pian"; the tablets are prepared from Radix et Rhizoma Notoginseng. This is used mainly as a hemostatic and analgesic agent for the treatment of different types of bleeding, and of traumatic diseases. The Chinese Pharmacopoeia notes that "Sanqi Pian" is contraindicated in pregnancy. The Chinese Pharmacopoeia requires a qualitative determination of ginsenosides Rb_1, Rg_1 and notoginsenoside R_1 in "Sanqi Pian" by TLC comparison with reference substances, and a quantitative determination of ginsenosides Rb_1, Rg_1, and notoginsenoside R_1 in the "Sanqi Pian" by HPLC. The total content of ginsenosides Rb_1, Rg_1, and notoginsenoside R_1 in 0.25 g and in 0.5 g "Sanqi Pian" should be not less than 10.0 mg and 20.0 mg, respectively.

Another galenic preparation of *P. notoginseng* with the name "Qiye Shen'an Pian" that is officially listed in the Chinese Pharmacopoeia is the dragees containing total saponins extracted from the leaf of *P. notoginseng*. This exerts sedative activity, promotes blood circulation and relieves palpitation. Dragees of "Qiye Shen'an Pian" are each required to contain 50 mg or 100 mg extract. The Chinese Pharmacopoeia requires a qualitative determination of ginsenosides Rb_1 and Rb_3 in the "Qiye Shen'an Pian" by TLC comparison with reference substances, and a quantitative determination of the total saponins in "Qiye Shen'an Pian" by HPLC. The content of the total saponins in "Qiye Shen'an Pian" should be not less than 5.0 mg and 10.0 mg, respectively, calculated as ginsenoside Rb_3. The Chinese Pharmacopoeia also requires a quantitative determination of the total saponins in the extract by HPLC. The content of the total saponins in the extract should be not less than 10.0%, calculated as ginsenoside Rb_3.

Radix Panacis Quinquefolii (Xiyangshen) is the dried root of *Panax quinquefolium* L. It is used as a general tonic. *Panax quinquefolium*, known as American ginseng, is mainly produced by cultivation. Radix Panacis Quinquefolii was listed in the Chinese Pharmacopoeia Edition 2000 for the first time. The Chinese Pharmacopoeia requires a qualitative determination of pseudoginsenoside F_{11} and ginsenosides Rb_1, Re, and Rg_1 in Radix Panacis Quinquefolii by TLC comparison with reference substances, and a quantitative determination of ginsenosides Rb_1, Re, and Rg_1 in Radix Panacis Quinquefolii by HPLC. The content of ginsenosides Rb_1, Re, and Rg_1 together in Radix Panacis Quinquefolii should be not less than 2.0%. Pseudoginsenoside F_{11} is used as a marker saponin for Radix Panacis Quinquefolii [24]. The Chinese Pharmacopoeia further requires a differential analysis to exclude Radix et Rhizoma Ginseng in Radix Panacis Quinquefolii by TLC comparison. The total saponin content in American ginseng was reported to be in the range of 3–4% [25].

Rhizoma Panacis Japonici (Zhujieshen) is the dried rhizome of *Panax japonicus* C.A. Mey. It is used as a tonic, analgesic, mucolytic, and hemostatic agent for the treatment of general weakness after diseases, of consumptive cough with hemostypsis, and of traumatic injuries. Sapogenins in saponins from the rhizome of *P. japonicus* were identified as protopanaxadiol, protopanaxatriol, and oleanolic acid [26]. The Chinese Pharmacopoeia requires a qualitative determination of oleanolic acid, panaxadiol, and panaxatriol in Rhizoma Panacis Japonici by TLC comparison with reference substances after extraction and hydrolysis with sulfuric acid. Panaxadiol and panaxatriol are artifacts formed from protopanaxadiol saponins and protopanaxatriol saponins during acid hydrolysis.

Rhizoma Panacis Majoris (Zhuzishen) is the dried rhizome of *Panax japonicus* C.A. Mey. var. *major* (Burk.) C.Y. Wu et K.M. Feng and *Panax japonicus* C.A. Mey. var. *bipinnatifidus* (Seem.) C.Y. Wu et K.M. Feng. It is used as a tonic and hemostatic agent for the treatment of cough in consumptive diseases, of traumatic injuries and bleeding, and of hemoptysis and hematemesis. The major constituents in the rhizome of *P. japonicus* var. *major* and *P. japonicus* var. *bipinnatifidus* are known to be saponins, with protopanaxadiol, protopanaxatriol, and oleanolic acid as the aglycone [27, 28]. The Chinese Pharmacopoeia requires a qualitative determination of oleanolic acid and panaxadiol in Rhizoma Panacis Majoris by TLC comparison with reference substances after extraction and hydrolysis with sulfuric acid.

Pharmacology and Toxicology

Ginseng has been used in traditional Chinese medicine since ancient times as a general tonic. Systematic pharmacological investigations have revealed that ginseng has a wide spectrum of biological activities, including effects on the cardiovascular, immune, and nervous systems, as well as activity as an antiaging, anticarcinogenic, antitumor agent; it is also used as an antidiabetic agent.

The pharmacokinetics of ginsenoside Rb_1 as a representative of protopanaxadiol saponins has been studied. Only a small amount of ginsenoside Rb_1 was absorbed from the digestive tract after oral administration to rats. After intravenous (i.v.) administration to rats, the serum level of ginsenoside Rb_1 decreased biexponentially with a half-life $t_{1/2\beta}$ of 14.5 h. The high persistence of ginsenoside Rb_1 in serum and tissues was a result of high plasma protein binding. Ginsenoside Rb_1 was gradually excreted into the urine, but not into bile [29]. The appearance of a major metabolite of ginsenoside Rb_1 in rat plasma after oral administration was found to be associated with the activity of intestinal bacteria. When ginsenoside Rb_1 was administered orally to pathogen-free rats, neither this nor any other metabolite was detected in the plasma, intestinal tract, or feces. Most of the ginsenoside Rb_1 administered was recovered from the intestinal tract, especially from the cecum and feces [30].

The pharmacokinetics of ginsenoside Rg_1 as a representative of protopanaxatriol saponins has also been studied in rats. Ginsenoside Rg_1 was absorbed rapidly from the upper parts of the digestive tract; serum levels reached peak values after 30 min, while maximal levels in tissues were reached within 1.5 h. Ginsenoside Rg_1 was not found in the brain. Excretion into the urine and bile occurred at a ratio of 2:5. Ginsenoside Rg_1 was not metabolized to any significant degree in the liver, but decomposition and metabolism in the rat stomach and large intestine were confirmed [31]. Ginsenoside Rg_1 showed a half-life of 27 min after i.v. administration into minipigs; here, the pharmacokinetics were best described by a one-compartment model [32].

The pharmacokinetics of ginsenosides Rb_2, Re, and Rg_1 were also studied in rabbits. Ginsenoside Rb_2 showed a significantly longer half-life, higher plasma protein binding, and lower metabolic and renal clearance than ginsenoside Re and ginsenoside Rg_1. All three ginsenosides were slowly absorbed after intraperitoneal (i.p.) administration, but were not identified in rabbit plasma or urine after oral administration. These observed differences in the pharmacokinetics of these ginsenosides might be due to their different extents of protein binding. Ginsenoside Rb_2 was shown to be more toxic than ginsenoside Rg_1 after i.p. administration to mice. After oral administration, none of these ginsenosides was shown to be toxic. Both, panaxadiol and panaxatriol, proved to be more toxic and to have larger volumes of distribution than the ginsenosides [33].

A number of metabolites were obtained following the incubation of ginsenoside Rg_2 with gastric juice; three of the metabolites were identified as 20(*R*)-ginsenoside Rg_2, 25-hydroxy-ginsenoside Rg_2, and 25-hydroxy-20(*R*)-ginsenoside Rg_2 [34].

The LD_{50} value of the total ginsenosides from stems and leaves in mice by i.p. administration was reported as 670 mg kg^{-1} [35]. The total ginseng saponin did not show hemolytic activity. Generally, ginsenosides of the protopanaxadiol type were not hemolytic but rather exhibited an antihemolytic effect, whereas the ginsenosides of protopanaxatriol and oleanolic acid types were strongly hemolytic [36].

Ginseng was reported to have antiaging and adaptogenic properties. Total ginsenoside, when given intraperitoneally or orally to young mice or rats, resulted in a marked increase in body weight. A significant increase in the protein and RNA contents of the muscles and liver in treated rats was noted. The promotion of animal growth by ginsenoides seemed to be mediated by an influence on RNA and protein syntheses [37]. The i.p. administration of ginsenosides Rb_2, Rc, Re, or Rg_1 to rats at a dose of 5–10 mg kg^{-1} caused increases in DNA, RNA, protein, and lipid formation in the bone marrow cells. Ginsenosides Rb_2, Rc, and Rg_1 caused decreases in cAMP but increases in cGMP in bone marrow at 20 min after i.p. administration. [38]. The clinical observation of 358 persons of middle and old age revealed that the oral administration of ginseng saponin exerted marked antiaging effects, adjusted organic metabolism, improved physiological functions, promoted memory, elevated leukocyte counts, and improved immune function [39]. The oral administration of saponins to mice caused a significant inhibition of lipid peroxide formation in brain and liver, and significantly increased the superoxide dismutase content of blood [40].

Experimental studies also showed that a prolonged treatment of rats with ginseng extract increased the capillary density and oxidative capacity of muscles with greater aerobic potential, in a manner similar to the performance of physical exercise [41]. The treatment of 27-month-old rats with ginseng extract for 13 days at a daily dose of 30 mg kg^{-1} significantly improved their avoidance behavior [42]. Ginsenoside Re was able to protect rats against cerebral ischemia and reperfusion damage. The oral treatment of rats with ginsenoside Re for seven days before middle cerebral artery occlusion significantly ameliorated lipid peroxidation by raising the activities of superoxide dismutase and glutathione peroxidase, and also reduced the malondiadehyde content of the brain [43]. Ginsenoside Rb_1 also exerted a neuroprotective effect, reducing infarct and neuronal deficit in rats following transient cerebral ischemia. Further studies revealed that the prevention of ischemic neuronal death induced by transient cerebral ischemia by ginsenoside Rb_1 was related to an increased expression of anti-apoptotic genes and a modulation of the expression of glial-derived neurotrophic factor [44].

Ginseng is known to have immunostimulatory activity. The oral treatment of mice with ginseng extract for five to six days enhanced antibody formation to sheep red cells [45]. The phagocytic activity of the guinea pig reticuloendothelial system (RES) was increased after oral administration of total ginsenoside from the plant leaves at a daily dose of 400 mg kg^{-1} for three days. Serum levels of specific antibodies, and of immunoglobulin G (IgG), IgA, and IgM, were also increased in mice following an i.p. injection of total ginsenoside [46]. Protopanaxatriol-type ginsenoside-F_3 was reported to exert an immunoenhancing activity in murine spleen cells. Ginsenoside-F_3, at a concentration of 10 μ*M*, not only promoted murine spleen cell proliferation but also increased the production of interleukin-2 (IL-2) and interferon-γ (IFN-γ), while decreasing the production of IL-4 and IL-10. Ginsenoside-F_3 also enhanced the nuclear factor-κB (NF-κB) DNA-binding activity in murine spleen cells induced by concanavalin A (Con A) [47].

Ginsenoside Rg_1, when administered orally to mice at a dose of 10 mg kg^{-1} for three consecutive days before immunization, increased the number of spleen plaque-forming cell, the titers of serum hemagglutinins, and the number of antigen-reactive T cells. It also increased the splenocyte natural killer cell activity and the production of IL-1 by macrophages. Ginsenoside Rg_1 also partly restored the impaired immune reactivity of mice caused by cyclophosphamide treatment [48]. Protopanaxatriol saponins promoted the gene expression of human interleukins in macrophages *in vitro* [49]. Total ginsenoside stimulated the RNA anabolism of bone marrow stromal cells, and enhanced the activity of IL-3 and IL-6 in the supernatant of the cells [50]. Ginsenoside Rg_1 significantly stimulated the phenotype of lymphocytes and increased the fluidity of the lymphocyte membrane of elderly persons [51]. The administration of ginseng polysaccharides to normal mice resulted in a dose-dependent increase in IL-2 and IFN production and splenic natural killer cell activity [52]. The acidic polysaccharides, especially ginsenan S-IIA, were found to be potent inducers of IL-8 production by human monocytes, the induction being accompanied by increase in IL-8 mRNA expression [53]. Ginsenan PA from ginseng root was found to be a phagocytosis-activating polysaccharide [54]. Mouse spleen cells became cytotoxic to a wide range of tumor cells after a 5-day incubation with polysaccharide ginsan. The latter also induced the expression of mRNA for IL-2, IFN-γ and IL-1α in murine macrophages [55].

Ginseng saponin, when administered intraperitoneally to rats, caused an increase in the plasma levels of adrenocorticotropic hormone (ACTH) and corticosterone [56]. Total ginsenosides obtained from stems and leaves given intraperitoneally to rats caused an increase in adrenal intracellular cAMP concentrations; this effect was potentiated by ACTH [57]. Ginsenoside Rg_1 has been shown to bind to the glucocorticoid receptor, and a concentration-dependent induction of a luciferase reporter gene was observed in rat hepatoma FTO2B cells in response to ginsenoside Rg_1 [58]. The latter compound competed for dexamethasone binding to the glucocorticoid receptors, with a specific affinity of 1–10 μ*M*, and activated a glucocorticoid-responsive element. The concentration-dependence patterns of ginsenoside Rg_1 and dexamethasone for these effects were almost identical; however, a two to three orders of magnitude higher concentration of ginsenoside Rg_1 was required to achieve the response of dexamethasone. At the cellular level, the growth of FTO2B cells was suppressed by ginsenoside Rg_1 and also by dexamethasone [59]. An increase in plasma corticosterone levels was observed in cold-stressed mice and rats after oral treatment with ginseng saponins [60].

Ginseng was reported to show cardiovascular effects. The i.v. injection of total ginsenside into dogs caused a rapid decrease in the peak left ventricular and arterial systolic pressures. The heart rate and renal arterial blood flow were also decreased, whereas renal vasoresistance was markedly increased. The vasoconstrictive effect of ginsenoside was not inhibited by α-adrenoreceptor or serotonin receptor blockade [61]. Protective effects of ginsenosides on ischemic–reperfused myocardium were observed in dogs during open-heart surgery with hypothermic cardiopulmonary bypass. Ginsenosides might prevent and reduce the generation of oxygen free radicals that play a significant role in the genesis of myocardial ischemia and reperfusion during the surgical process [62]. Total ginsenoside increased the myocardial tolerance of animals to hypoxia, including dogs, guinea pigs, rats, and mice. The treatment of an ischemic mouse myocardium with ginseng extract normalized the anoxia-induced alterations in lactic dehydrogenase and succinic dehydrogenase activities [63]. Ginsenosides, when given intravenously to rabbits at a daily dose of 30–50 mg kg^{-1} for 14 days, exerted protective effects against experimentally induced sinus node dysfunction [64].

Ginsenoside Rg_1 was reported to inhibit the proliferation of vascular smooth muscle cells stimulated by TNF-α. Further studies showed that protein kinase C-ζ (PKC-ζ) and p21 pathways might be involved in the inhibitory effects of ginsenoside Rg_1 on the proliferation of vascular smooth muscle cells [65]. Ginsenoside Rg_1 was also reported to inhibit the proliferation of human arterial smooth muscle cells. Treatment of human arterial smooth muscle cells with ginsenoside Rg_1 resulted in an inhibition of cell proliferation, an induction of cell cycle in the G_1 phase, a downregulation of cyclin D1 and an upregulation of p53, and also in the expression of p21(WAF/CIP1) and p27 (KIP1). Ginsenoside Rg_1 markedly inactivated the extracellular signal-regulated kinases (ERK1/2) and protein kinase B (PKB), indicating that the inhibition of human arterial smooth muscle cell proliferation by ginsenoside Rg_1 was associated with ERK and PI3K/PKB pathways [66].

Ginsenosides Rg_1 and Re, with protopanaxatriol as aglycone, were found to cause endothelium-dependent relaxation in the rat aorta. This vasoactive effect was associated with the formation of cGMP. In contrast, protopanaxadiol ginsenoside Rb_1 did not affect vascular tone, nor the production of cGMP in the rat aorta [67]. Both, the protoparaxadiol and protoparaxatriol saponins were found to decrease the action potential of isolated myocardial cells in a concentration-dependent manner, similar to Mn^{2+} as a known calcium channel blocker, which suggested that the saponins might be potential calcium antagonists [68]. The action potentials of free radical-damaged rat myocardiocytes *in vitro* could be restored to normal by adding ginsenosides Rb_1, Rb_2, and Rb_3 at a concentration of 30 µg ml^{-1}. In normal myocardial cells, ginsenosides Rb_1, Rb_2, and Rb_3 at a concentration of 20 µg ml^{-1} inhibited the action potentials and spontaneous contractility [69].

Ginsenosides have been shown to promote nitric oxide (NO) release in endothelial cells, to induce relaxation of the rabbit corpus cavernosum, and to enhance NO release from endothelial cells elicited by other vasoactive substances [70]. Ginsenosides given to rabbits were shown to protect against myocardial ischemia and reperfusion damage, concomitant with increased 6-keto-prostaglandin $F_{1\alpha}$ (6-keto-$PGF_{1\alpha}$) and decreased lipid peroxidation. In perfused rabbit lung and in isolated rabbit aortic rings, ginsenosides protected the pulmonary and aortic endothelium against injury induced by free radicals. Ginsenoside Rb_1, and especially ginsenoside Rg_1, caused the relaxation of pulmonary vessels, which could be eliminated by *N*-nitro-L-arginine as an inhibitor of NO synthase (NOS). Thus, cardiovascular protection provided by ginsenosides might be partly mediated by the release of NO [71]. Ginsenosides increased renal blood flow, stimulated the endogenous release of NO in the kidney, and affected NOS in the kidney tissues of rats [72]. It was suggested that the antioxidant and organ-protective actions of ginseng were linked to enhanced NO synthesis in the endothelium of lung, heart, and kidney, and also in the corpus cavernosum [73]. The oral administration of ginsenoside Rg_3 to rats for 15 consecutive days caused a significant decrease in the elevated serum levels of nitrite/nitrate, glutamic-oxaloacetic transaminase, glutamic-pyruvic transaminase, and creatinine induced by lipopolysaccharide (LPS). It dose-dependently reduced thiobarbituric acid-reactive substance levels in the serum, liver, and kidney. The treatment of rats with ginsenoside Rg_3 also significantly decreased the elevated expression of NF-κB, cyclooxygenase-2 (COX-2), inducible nitric oxide synthase (iNOS) stimulated by LPS; this suggested that ginsenoside Rg_3 prevented LPS-induced acute oxidative damage in the liver and kidney [74].

In structure–activity studies using the hemolysis of human erythrocytes induced by 2,2′-azo-bis(2-amidinopropane) hydrochloride, both protopanaxadiol and protopanaxatriol played prooxidative roles. Ginsenosides without sugar moieties at position 20 (such as ginsenosides Rg_3, Rh_2, and Rg_2) acted as prooxidants. Ginsenoside Rh_1, which bears a glucose at position 6 instead of position 20, was found to be antioxidant [75]. Interestingly, among the ginsenosides, only Rd induced COX-2 expression and increased prostaglandin E_2 production in RAW264.7 murine macrophages. A further study revealed that Rd increased both the DNA binding and nuclear levels of CCAAT/enhancer binding protein (C/EBP)α/β and cAMP response element binding protein (CREB) in RAW264.7 cells. Thus, the ginsenoside Rd-mediated transcriptional activation of the COX-2 gene was regulated by C/EBP and CREB, which indicated that the activation of C/EBP and CREB by ginsenoside Rd was essential for the induction of COX-2 in RAW264.7 cells [76].

The oral administration of ginsenoside Rg_1 at a dose of 20 mg kg^{-1} inhibited thrombosis formation in rats [77]. Dietary supplementation with 25 mg (25 ppm of the total diet) of a lipophilic, nonsaponin fraction from ginseng to rats also showed an antithrombotic effect. The lipophilic fraction increased the cGMP content directly and the cAMP content indirectly, and inhibited platelet aggregation induced by either thrombin or by collagen [78]. The nonsaponin fraction also inhibited the aggregation of human platelets induced by thrombin *in vitro*, in a concentration-dependent manner. The nonsaponin fraction elevated the cGMP level in human platelets, inhibited Ca^{2+} influx into platelets, and inhibited thromboxane A_2 (TXA_2) production, whereas cAMP formation was only marginally affected. The lipophilic fraction appeared to regulate the levels of cGMP and TXA_2, thus inhibiting platelet aggregation induced by thrombin [79].

Protective effects of total ginsenosides on ischemia–reperfusion injury were also observed in the rat brain. Total ginsenosides increased cerebral blood flow, reduced calcium accumulation, potassium loss and cerebral edema in the cortex and hippocampus during reperfusion, thus reducing the mortality of rats. A close relationship between the effects of total ginsenoside on calcium accumulation and ultrastructural changes was observed [80]. Among the ginsenosides, Rb_1 protected the brain from ischemic–reperfusion injuries, but Rg_1 was ineffective [81].

The oral administration of total ginseng saponins decreased serum cholesterol and triglyceride levels in both normal rats and rats fed on a diet high in cholesterol [82]. Ginsenoside Rb_2, when given intraperitoneally to rats fed a high-cholesterol diet, induced a significant decrease in serum levels of total cholesterol, free cholesterol, low-density lipoprotein (LDL) -cholesterol, and triglycerides. In contrast, a significant increase in high-density lipoprotein (HDL) - cholesterol levels was observed after treatment [83]. The oral administration of ginseng extract also significantly reduced serum total cholesterol and triglycerides and significantly reduced fatty liver formation in hepatectomized rats [82]. Ginsenosides from the stem and leaf, when injected into rabbits, decreased the serum total cholesterol and triglyceride, increased HDL levels, decreased the content of lipid and malondialdehyde in the aorta walls, and decreased plasma levels of 6-keto-$PGF_{1\alpha}$, thromboxane B_2 (TXB_2), arterial wall TXB_2/6-keto-$PGF_{1\alpha}$, and 6-keto-$PGF_{1\alpha}$ [84]. The polyacetylene analogues, panaxynol, panaxydol, panaxydiol, and panaxytriol, were each found to be inhibitors of acyl-CoA:cholesterol acyltransferase from rat liver [85] and human cholesteryl ester transfer protein [86].

Oral administration of ginsenoside Rb_1 and its major intestinal metabolite compound K (20(S)-protopanaxadiol 20-*O*-β-D-glucopyranoside) both exhibited significant protection against the hepatotoxicity of *tert*-butyl hydroperoxide (t-BHP), and decreased serum levels of alanine aminotransferase (ALT) and aspartate aminotransferase (AST). Compound K, but not ginsenoside Rb_1, also exhibited a protective effect in HepG2 cells against the cytotoxicity of t-BHP. The i.p. administration of ginsenoside Rb_1 also did not inhibit the elevation of plasma ALT and AST induced by t-BHP in mice; this indicated that compound K, produced from ginsenosides in the intestine, was responsible for the hepatoprotective effect [87]. Most likely, ginsenoside Rg_3 and its metabolite ginsenoside Rh_2 both exerted hepatoprotective effects in mice against the hepatotoxicity of t-BHP. Ginsenoside Rh_2, but not ginsenoside Rg_3, showed a cytoprotective effect in HepG2 cells against the cytotoxicity of t-BHP. Thus, ginsenoside Rg_3, as a main saponin in ginseng, might be considered as a prodrug [88].

Total ginseng saponins exhibited hypoglycemic activity in alloxan-induced diabetic mice when administered before and after alloxan injection [89]. The pure ginsenoside Rb_2 decreased the blood glucose levels in streptozotocin-induced diabetic rats. In addition, animals treated with ginsenoside Rb_2 showed a significant rise in hepatic glucokinase activity and a significant decrease in glucose 6-phosphatase activity [90]. The antidiabetic effects of ginsenoside Re have been demonstrated in adult male C57BL/6J ob/ob mice. Diabetic ob/ob mice with fasting blood glucose levels of approximately 230 mg dl^{-1} received i.p. injections of ginsenoside Re at daily doses of 7, 20, and 60 mg kg^{-1} for 12 consecutive days. A dose-related effect of ginsenoside Re on fasting blood glucose levels was subsequently observed. The hypoglycemic effect of the ginsenoside persisted even at three days after treatment cessation, and there were no significant changes in either body weight or body temperature [91]. Ginsenoside Re was further reported to exert hypoglycemic and hypolipidemic effects in diabetic rats induced by streptozotocin [92].

In rats, ginseng was also able to stimulate insulin release and to increase insulin synthesis [93]. The ginseng extract showed a stimulatory effect on glucose uptake in sheep erythrocytes, in concentration-dependent manner [94]. Panaxans from the ginseng root exerted significant hypoglycemic actions in normal and alloxan-induced hyperglycemic mice [15]. However, panaxan A did not affect plasma insulin levels and insulin sensitivity in mice [95]. Ginseng polypeptide was shown to decrease blood sugar levels and liver glycogen content when injected intravenously to rats at doses of 50–200 mg kg^{-1}, without affecting total blood lipids. In mice treated subcutaneously with ginseng polypeptide at daily doses of 50 and 100 mg kg^{-1} for seven successive days, the blood glucose and liver glycogen levels were markedly reduced. In addition, the ginseng polypeptide was found to decrease experimental hyperglycemia induced by the injection of adrenaline, glucose, and alloxan [96].

Ginseng was reported to exert preventive effects against the genotoxicity, mutagenicity and carcinogenicity of genotoxic substances and cancerogens, as well as against ionizing radiation. A ginseng extract was able to inhibit DNA synthesis in V79 cells induced by UV radiation or methyl methanesulfonate. The extract also was shown to decrease mutation frequency in V79 cells at the hypoxanthine-guanine phosphoribosyl transferase (HGPRT) locus following exposure to methyl methanesulfonate. The ginseng extract was found to exert an inhibitory effect on the transformation of NIH 3T3 cells initiated by 3-methylcholanthrene, methyl methanesulfonate, and 1-methyl-3-nitro-1-nitroso-guanidine (MNNG) [97]. Ginseng was also reported to exhibit radioprotective effects on mammalian cells both *in vitro* and *in vivo*. The experimental results indicated that a water-soluble extract of whole ginseng provided a better protection against radiation-induced DNA damage than did the ginsenosides. The radioprotective mechanism of ginseng might be linked to its antioxidative capability by scavenging the free radicals responsible for DNA damage [98].

In anti-carcinogenicity studies *in vivo*, ginseng was reported to significantly decrease the incidence of lung adenomas in newborn mice induced by the s.c. injection of benzo[*a*]pyrene (BaP) [99], and also of liver cancer in rats induced by *N*-nitroso-*N*,*N*-diethylamine (NDEA) [100]. A red ginseng extract was found to significantly inhibit papilloma formation in two-step carcinogenesis in mouse skin initiated by 7,12-dimethylbenz[*a*]anthracene (DMBA) and promoted by croton oil [101]. An ethanol-insoluble fraction of ginseng was found to be effective in inhibiting lung tumor incidence in mice induced by a single s.c. injection of BaP. This fraction stimulated splenocyte proliferation and generated activated killer cells *in vitro*, which suggested that its anti-carcinogenic effects were the result of an immunomodulation [102]. An epidemiological study suggested that ginseng consumers had a decreased risk of cancer compared to nonconsumers. The decreased risk was observed in line with a rise in the frequency of ginseng intake. The preventive effect of ginseng was found not to be organ-specific [103].

The i.p. administration of an aqueous extract of ginseng to guinea pigs at a daily dose of 100 or 200 mg kg^{-1} for 14 or 28 days showed both protective and therapeutic effects against testicular damage caused by 2,3,7,8-tetrachlorodibenzo-*p*-dioxin (TCDD) given intraperitoneally to animals before or after treatment [104]. The aqueous extract prevented apoptosis in hair follicles, and accelerated the recovery of hair medullary cells in irradiated mice; this indicated that an aqueous extract of ginseng might exert a potent protective effect on the hair follicles [105]. Ginseng and the water-soluble, nonsaponin fraction of ginseng exerted radioprotective effects in mice [106, 107]. Panaxatriol, when given intraperitoneally to rats, alleviated any dysfunction of the reproductive endocrine axis induced by X-ray irradiation [108].

Ginsenoside Rh_2 was found to inhibit growth in a series of cultured tumor cells, including B16 melanoma cells [109], human lung adenocarcinoma cells (SPC-A-1) [110], and MCF-7 human breast carcinoma cells [111], and also to induce the differentiation of B16 melanoma cells [112]. Ginsenosides from the leaves and stems inhibited the growth and proliferation of the gastric carcinoma cell line BGC-823 [113]. The tumor cells, including B16, Meth-A, HeLa-S3, and K562, in different phases of the cell cycle showed different responses to ginsenoside Rh_2 [114]. Ginsenoside Rh_2 was reported to inhibit the growth of human ovarian cancer cells *in vitro* by coincubation, and in nude mice by oral administration. Oral treatment with ginsenoside Rh_2 resulted in an induction of apoptosis of tumor cells, in addition to an augmentation of natural killer activity in spleen cells from tumor-bearing nude mice [115].

The growth-inhibiting effect of ginsenoside Rh_2 against MCF-7 cells was found to be reversible, inducing a G_1-phase arrest in the cell cycle. Ginsenoside Rh_2 treatment downregulated the protein level of cyclin D3, but upregulated the expression of p21WAF1/CIP1, which is an inhibitor of cyclin-dependent kinases (Cdk). The increased levels of p21 protein were associated with an increased binding of p21. Ginsenoside Rh_2 markedly reduced phosphorylated retinoblastoma protein, and enhanced the association of unphosphorylated retinoblastoma protein with the transcription factor E2F-1 [111].

Both, ginsenoside Rh_1 and Rh_2 were found to significantly induce differentiation in B16 cells at a concentration of $10\,\mu g\ ml^{-1}$ *in vitro*. Melanin synthesis by B16 cells treated with ginsenoside Rh_2 was increased, whilst flow cytometric investigations showed that B16 cells treated with ginsenoside Rh_2 were blocked at G_1 phase [116]. Ginsenosides from the stem and leaves of P. ginseng induced the differentiation of acute, nonlymphocytic leukemia cells from 58 patients in primary culture [117]. Ginsenoside Rh_2 inhibited the proliferation of human lung adenocarcinoma A549 cells, induced cell-cycle arrest at G_1 phase and cell apoptosis, accompanied by a downregulation of the protein levels and kinase activities of cyclin-D1, cyclin-E and Cdk6, and by the upregulation of pRb2/p130; caspase-2, caspase-3, and caspase-8 were each markedly activated [118].

Ginsenosides Rh_2 and Rh_3 each induced the differentiation of HL-60 cells into granulocytes, and arrested the cell cycle at G_1/S phase. During differentiation by ginsenoside Rh_2, Ca^{2+}/phospholipid-dependent PKC activity was increased in both the cytosol and total cell extract. No significant change was observed in the cytosolic PKC_α isoform, but the PKC_β isoform was gradually increased with prolonged treatment. Although the PKC_γ isoform was not detected in the cytosol of untreated cells, a small amount was detected five days after treatment [117]. The induction of differentiation *in vitro* by ginsenosides Rh_1 and Rh_2 in F9 teratocarcinoma cells was also reported [119]. Ginsenoside Rh_2 was also found to induce cell differentiation in SMMC-7721 hepatocarcinoma cells. At concentrations of 10 or $20\,\mu g\ ml^{-1}$, the production and secretory extent of α-fetoprotein were each significantly reduced. The activities of γ-glutamyltranspeptidase and heat-resistant alkaline phosphatase were both remarkably declined, whereas the secretion of both albumin and alkaline phosphatase activity was enhanced; this indicated that ginsenoside Rh_2 might induce the differentiation of hepatocarcinoma cells [120].

Ginsenosides Rh_1 and Rh_2 were further reported to inhibit the cellular proliferation of NIH 3T3 fibroblasts. Both compounds caused effective reductions in phospholipase C activity, resulting in a decrease in intracellular levels of diacylglycerol. The incubation of cells with ginsenosides Rh_1 and Rh_2 was found to reduce intracellular PKC activity [121]. In SK-HEP-1 cells, ginsenoside Rs_3 was found to efficiently arrest the cell cycle at the G_1/S phase at lower concentrations, but to induce apoptosis at higher concentrations. These cell growth-inhibiting and apoptosis-inducing effects were confirmed by MTT assays, together with flow cytometric analyses, morphological changes, and DNA fragmentation. Immunoblotting showed that ginsenoside Rs_3 had significantly elevated protein levels of p53 and p21WAF1 prior to inducing apoptosis [122].

Ginsenoside Rg_3, 20(*R*)-ginsenoside Rg_3 and ginsenoside Rh_2 promoted the uptake of mitomycin C into the tumor cells. Mitomycin C treatment, when combined with red ginseng extract, showed stronger antitumor effects against the ascites form of Ehrlich ascites carcinoma and rat ascites hepatoma AH 130 *in vivo* [123]. Ginsenoside Ra_1 was also found to increase the antitumor activity of recombinant human tumor necrosis factor, both *in vitro* and *in vivo* [124].

The oral administration of ginseng extract (1 mg per mouse) and ginsenosides Rb_1, Rb_2, and Rc (0.5 mg per mouse), significantly inhibited metastasis to the lung caused by the i.v. injection of B16-BL6 melanoma cells in syngeneic mice. However, these compounds barely inhibited the invasion and migration potential of B16-BL6 melanoma and HT1080 fibrosarcoma cells *in vitro*. An intestinal bacterial metabolite inhibited metastasis to the lung of melanoma cells, as well as *in vitro* tumor cell invasion and migration, at either nontoxic or marginally toxic concentrations. Pharmacokinetics studies of ginsenoside Rb_1 and the M1 metabolite, after oral administration, revealed that intact Rb_1 was not detected in serum for 24 h after redosing, whereas serum levels of M1 reached a maximum (ca. $8\,\mu g\,ml^{-1}$) at 8 h after Rb_1 administration, and at 2 h after M1 administration. These results indicated that the *in vivo* antimetastatic effect of orally administered ginsenosides might be mediated via their metabolites [125]. Inhibitory effects on tumor metastasis in mice by ginsenoside Rb_2, 20(*R*)-ginsenoside Rg_3 and ginsenoside Rg_3 were also reported [126]. An inhibition of B16-BL6 melanoma angiogenesis by ginsenoside Rb_2 cells in syngeneic mice was observed as well [127].

Ginseng diyne compounds were found to inhibit the growth of various types of cultured cell lines in a dose-dependent manner. The cell growth-inhibitory activity of these compounds was much stronger against malignant cells than against normal cells [128]. Both, panaxydol and panaxynol inhibited the proliferation of all three human renal cell carcinoma cell lines A498, Caki-1, and CURC II, in a concentration-dependent manner *in vitro* [129]. Panaxytriol localized to the mitochondria in human breast carcinoma cells (Breast M25-SF) rapidly inhibiting cellular respiration and disrupting the cellular energy balance in tumor cells. ATP depletion resulting from a direct inhibition of mitochondrial respiration might represent a critical early event in the cytotoxicity of panaxatriol [130]. Panaxytriol was also found to potentiate the cytotoxicity of mitomycin C against a human gastric carcinoma cell line, MK-1 [131]. Some analogues of ginseng diyne derivatives were also synthesized and tested for their antiproliferative activity against L1210 cells. An analysis of the structure–activity relationship revealed that the presence of heptadec-1-ene-4,6-diyn-3-ol was essential for an antiproliferative activity, and that the epoxy and alkyl groups in the structure contributed to an enhancement of the antiproliferative activity [132]. The ginseng volatile oil was also found to be effective in inhibiting growth of the human gastric carcinoma in cell line SGC-823 *in vitro* [133].

References

1 Sanada, S., Kondo, N., Shoji, J., Tanaka, O., and Shibata, S. (1974) Studies on the saponins of Ginseng. I. Structures of ginsenoside-Ro, -Rb_1, -Rb_2, -Rc and -Rd. *Chem. Pharm. Bull. (Tokyo)*, **22**, 421–428.

2 Kim, D.S., Chang, Y.J., Zedk, U., Zhao, P., Liu, Y.Q., and Yang, C.R. (1995) Dammarane saponins from *Panax ginseng*. *Phytochemistry*, **40**, 1493–1497.

3 Sanada, S., Kondo, N., Shoji, J., Tanaka, O., and Shibata, S. (1974) Studies on the saponins of Ginseng. II. Structures of ginsenoside-Re, -Rf, and -Rg_2. *Chem. Pharm. Bull. (Tokyo)*, **22**, 2407–2412.

4 Wu, G.X., Wei, Y.D., Song, C.C., Wang, C.R., Ma, X.Y., Xu, J.D., Zhang, D.X., and Jiang, X. K. (1988) Comparative analysis of the contents of ginseng saponins in wild ginseng and cultivated ginseng. *Chin. Pharm. Bull.*, **23**, 397–398.

5 Xu, S.X., Wang, N.L., and Li, Y.H. (1986) Chemical constituents of red ginseng (II). *Acta Pharm. Sin.*, **21**, 356–360.

6 Park, J.D., Lee, Y.H., and Kim, S.I. (1998) Ginsenoside Rf_2, a new dammarane glycoside from Korean red ginseng (*Panax ginseng*). *Arch. Pharm. Res.*, **21**, 615–617.

7 Baek, N.I., Kim, D.S., Lee, Y.H., Park, J.D., Lee, C.B., and Kim, S.I. (1996) Ginsenoside Rh_4, a genuine dammarane glycoside from Korean red ginseng. *Planta Med.*, **62**, 86–87.

8 Zhao, Y.Q., Yuan, C.L., Fu, Y.Q., Wei, X.J., Zhu, H.J., Chen, Y.J., Wu, L.J., and Li, X. (1990) Chemical studies of minor triterpene compounds from the stems and leaves of *Panax ginseng*. *Acta Pharm. Sin.*, **25**, 297–301.

9 Zhang, S.L., Yao, X.S., Chen, Y.J., Cui, C.B., Tezuka, Y., and Kikuchi, T. (1989) Ginsenoside La, a novel saponin from the leaves of *Panax ginseng*. *Chem. Pharm. Bull. (Tokyo)*, **37**, 1966–1968.

10 Shao, C.J., Xu, J.D., Jiang, X.K., and Cheng, G. R. (1984) Studies on chemical constituents of flower buds of *Panax ginseng* - isolation and identification of ginsenoside-Rb_3 and -Rc. *Bull. Chin. Mater. Med.*, **9**, 172–173.

11 Zhao, Y.Q., Yuan, C.L., and Lu, H.R. (1991) Isolation and identification of 20(R)-ginsenoside Rh_2 (an anti-cancer constituent) from the fruits of *Panax ginseng* C.A. Meyer. *China J. Chin. Mater. Med.*, **16**, 678–679.

12 Iwabuchi, H., Kato, N., and Yoshikura, M. (1990) Studies on the sesquiterpenoids of *Panax ginseng* C.A. Meyer. IV. *Chem. Pharm. Bull. (Tokyo)*, **38**, 1405–1407.

13 Hirakura, K., Morita, M., Nakajima, K., Ikeya, Y., and Mitsuhashi, H. (1992) The constituents of *Panax ginseng*. Part 3. Three acetylene compounds from roots of *Panax ginseng*. *Phytochemistry*, **31**, 899–903.

14 Matsunaga, H., Katano, M., Yamamoto, H., Fujito, H., Mori, M., and Takata, K. (1990) Cytotoxic activity of polyacetylene compounds in *Panax ginseng* C.A. Meyer. *Chem. Pharm. Bull. (Tokyo)*, **38**, 3480–3482.

15 Konno, C., Sugiyama, K., Kano, M., Takahashi, M., and Hikino, H. (1984) Validity of the oriental medicines. LXX. Antidiabetes drugs. 1. Isolation and hypoglycemic activity of panaxans A, B, C, D, and E, glycans of *Panax ginseng* roots. *Planta Med.*, **50**, 434–436.

16 Lee, Y.S., Chung, I.S., Lee, I.R., Kim, K.H., Hong, W.S., and Yun, Y.S. (1997) Activation of multiple effector pathways of immune system by the antineoplastic immunostimulator acidic polysaccharide ginsan isolated from *Panax ginseng*. *Anticancer Res.*, **17**, 323–331.

17 Tomoda, M., Hirabayashi, K., Shimizu, N., Gonda, R., Ohara, N., and Takada, K. (1993) Characterization of two novel polysaccharides having immunological activities from the root of *Panax ginseng*. *Biol. Pharm. Bull.*, **16**, 1087–1090.

18 Tomoda, M., Takeda, K., Shimizu, N., Gonda, R., Ohara, N., Takada, K., and Hirabayashi, K. (1993) Characterization of two acidic polysaccharides having immunological activities from the root of *Panax ginseng*. *Biol. Pharm. Bull.*, **16**, 22–25.

19 Zhou, Y.F., Tai, G.H., and Zhang, Y.S. (1992) Water soluble polysaccharides from the stem of *Panax ginseng* C.A. Meyer. II. Purification and a structural analysis of S-2A. *Acta Biochim. Biophys. Sin.*, **24**, 22–26.

20 Gao, Q.P., Kiyohara, H., Cyong, J.C., and Yamada, H. (1991) Chemical properties and anti-complementary activities of heteroglycans from the leaves of *Panax ginseng*. *Planta Med.*, **57**, 132–136.

21 Long, Y.C., Ye, Y.H., and Xing, Q.Y. (1996) Studies on the neuroexcitotoxin β-N-oxalo-L-α,β-diaminopropionic acid and its isomer α-*N*-oxalo-L-α,β-diaminopropionic acid from the root of *Panax* species. *Int. J. Pept. Protein Res.*, **47**, 42–46.

22 Wu, Q.F., Wei, J.J., and Xu, J.D. (1991) Purification and identification of red ginseng polypeptides. *Acta Pharm. Sin.*, **26**, 499–504.

23 Li, Y.H., Li, X.L., Hong, L., Liu, J.Y., and Zhang, M.Y. (1991) Determination of panaxadiol and panaxatriol on the root of *Panax notoginseng* and "Yunnan Baiyao" by capillary supercritical fluid chromatography. *Acta Pharm. Sin.*, **26**, 764–767.

24 Chan, T.W., But, P.P., Cheng, S.W., Kwok, I. M., Lau, F.W., and Xu, H.X. (2000)

Differentiation and authentication of *Panax ginseng, Panax quinquefolius,* and ginseng products by using HPLC/MS. *Anal. Chem.*, **72**, 1281–1287.

25 Ma, X., Lu, R., Song, J., and Chen, Z. (1990) Effects of concentration and ratio of N,P and K on seedlings of *Panax quinquefolium* L. China. *J. Chin. Mater. Med.*, **15**, 78–81. 125.

26 Cai, P., Xiao, Z.Y., and Wei, J.X. (1982) Studies on the chemical constituents of Zhu Jie Shen (*Panax japonicus*). *J. Chin. Trad. Herbal Drugs*, **13**, 1–2.

27 Wang, D.Q., Fan, J., Wang, X.B., Feng, B.S., Yang, C.R., Zhou, J., Ning, Y.C., and Tao, J.X. (1988) Saponins of the rhizome of *Panax japonicus* C.A. Meyer var. *major* (Burk.) Wu et Feng, collected in Qinling Mountain (Shaanxi). *Acta Bot. Sin.*, **30**, 403–408.

28 Wang, D.Q., Fan, J., Wang, X.B., Feng, B.S., Yang, C.R., Zhou, J., Ning, Y.C., Feng, Y.P., and Tao, J.X. (1988) Saponins of the rhizome of *Panax japonicus* var. *bipinnatifidus* and their significance for chemotaxonomy. *Acta Bot. Yunnan.*, **10**, 101–104.

29 Odani, T., Tanizawa, H., and Takino, Y. (1983) Studies on the absorption, distribution, excretion and metabolism of ginseng saponin. III. The absorption, distribution and excretion of ginsenoside-Rb_1 in the rat. *Chem. Pharm. Bull. (Tokyo)*, **31**, 1059–1066.

30 Akao, T., Kida, H., Kanaoka, M., Hattori, M., and Kobashi, K. (1998) Intestinal bacterial hydrolysis is required for the appearance of compound K in rat plasma after oral administration of ginsenoside Rb1 from *Panax ginseng*. *J. Pharm. Pharmacol.*, **50**, 1155–1160.

31 Odani, T., Zanizawa, H., and Takino, Y. (1983) Studies on the absorption, distribution, excretion and metabolism of ginseng saponins. II. The absorption, distribution and excretion of ginsenoside Rg_1 in the rat. *Chem. Pharm. Bull. (Tokyo)*, **31**, 292–298.

32 Jenny, E., and Soldati, F. (1985) Pharmacokinetics of ginsenosides in the mini pig, in *Advances in Chinese Medicinal Material Research* (eds H.M. Chang, H.W. Yeung, W.W. Tso and A. Koo), World Science, Singapore, pp. 499–507.

33 Chen, S.E., Sawchuk, R.J., and Staba, E.J. (1980) American ginseng. III. Pharmacokinetics of ginsenosides in the rabbit. *Eur. J. Drug Metab. Pharmacokinet.*, **5**, 161–168.

34 Chen, Y.J. (1987) Study on the metabolites of 20(*S*)-ginsenoside Rg_2. *J. Shenyang Coll. Pharm.*, **4**, 202.

35 Wang, B.X., Cui, J.C., and Liu, A.J. (1980) Antidiuretic effect of ginsenosides of the stems and leaves of *Panax ginseng*. *Acta Pharmacol. Sin.*, **1**, 126–130.

36 Namba, T., Yoshizaki, M., Tomimori, T., Kobashi, K., Mitsui, K., and Hase, J. (1974) Fundamental studies on the evaluation of the crude drugs. I. Hemolytic and its protective activity of ginseng saponins. *Planta Med.*, **25**, 28.

37 Ramachandran, U., Divekar, H.M., Grover, S. K., and Srivastava, K.K. (1990) New experimental model for the evaluation of adaptogenic products. *J. Ethnopharmacol.*, **29**, 275–281.

38 Yamamoto, M., Masaka, M., Yamada, K., Hayashi, Y., Hirai, A., and Kumagai, A. (1978) Stimulatory effect of ginsenosides on DNA, protein and lipid synthesis in rat bone marrow and participation of cyclic nucleotides. *Arzneim.-Forsch.*, **28**, 2238–2241.

39 Zhao, X.Z. (1990) Antisenility effect of ginseng-rhizome saponin. *Chin. J. Integr. Trad. West. Med.*, **10**, 586–589. 579.

40 Wu, C.F., Yu, Q.H., Liu, W., Guo, Y.Y., and Zhang, G.P. (1992) A study on the anti-aging effect of ginseng stem and leaf saponin in terms of the free radical theory of aging. *J. Shenyang Coll. Pharm.*, **9**, 37–40.

41 Ferrando, A., Vila, L., Voces, J.A., Cabral, A.C., Alvarez, A.I., and Prieto, J.G. (1999) Effects of a standardized *Panax ginseng* extract on the skeletal muscle of the rat: a comparative study in animals at rest and under exercise. *Planta Med.*, **65**, 239–244.

42 Jaenicke, B., Kim, E.J., Ahn, J.W., and Lee, H. S. (1991) Effect of *Panax ginseng* extract on passive avoidance retention in old rats. *Arch. Pharmacol. Res.*, **14**, 25–29.

43 Zhou, X.M., Cao, Y.L., and Dou, D.Q. (2006) Protective effect of ginsenoside-Re against cerebral ischemia/reperfusion damage in rats. *Biol. Pharm. Bull.*, **29**, 2502–2505.

44 Yuan, Q.L., Yang, C.X., Xu, P., Gao, X.Q., Deng, L., Chen, P., Sun, Z.L., and Chen, Q.Y. (2007) Neuroprotective effects of ginsenoside Rb_1 on transient cerebral ischemia in rats. *Brain Res.*, **1167**, 1–12.

45 Jie, Y.H., Cammisuli, S., and Baggiolini, M. (1984) Immunomodulatory effects of *Panax ginseng* C.A. Meyer in the mouse. *Agents Actions*, **15**, 386–391.

46 Cui, J.C., Liu, A.J., and Wang, B.X. (1982) Effect of ginsenosides from stems and leaves of ginseng on immunological functions. *J. Chin. Trad. Herbal Drugs*, **13**, 29–31.

47 Yu, J.L., Dou, D.Q., Chen, X.H., Yang, H.Z., Guo, N., and Cheng, G.F. (2004) Immunoenhancing activity of protopanaxatriol-type ginsenoside-F3 in murine spleen cells. *Acta Pharmacol. Sin.*, **25**, 1671–1676.

48 Kenarova, B., Neychev, H., Hadjiivanova, C., and Petkov, V.D. (1990) Immunomodulating activity of ginsenoside Rg1 from *Panax ginseng*. *Jpn. J. Pharmacol.*, **54**, 447–454.

49 Tian, Z.G. and Yang, G.Z. (1993) Promoting effect of panaxatriol ginsenoside on gene expression of human interleukin-1. *Acta Pharmacol. Sin.*, **14**, 159–161.

50 Yang, S.C., Wang, S.H., Zhang, D.J., and Yang, G.Z. (1994) Effects of total ginsenosides on RNA anabolism and cytokine induction of bone marrow stromal cells. *Chin. Pharmacol. Bull.*, **10**, 40–43.

51 Liu, J., Wang, S., Liu, H., Yang, L., and Nan, G. (1995) Stimulatory effect of saponin from *Panax ginseng* on immune function of lymphocytes in the elderly. *Mech. Aging Dev.*, **83**, 43–53.

52 Zhu, W.G., Guo, S.S., and Yi, Y.N. (1991) Effects of ginseng polysaccharide on immunologic functions in mice. *Bull. Hunan Med. Univ.*, **16**, 107–110.

53 Sonoda, Y., Kasahara, T., Mukaida, N., Shimizu, N., Tomoda, M., and Takeda, T. (1998) Stimulation of interleukin-8 production by acidic polysaccharides from the root of *Panax ginseng*. *Immunopharmacology*, **38**, 287–294.

54 Tomoda, M., Hirabayashi, K., Shimizu, N., Gonda, R., and Ohara, N. (1994) The core structure of ginsenan PA, a phagocytosis-activating polysaccharide from the root of *Panax ginseng*. *Biol. Pharm. Bull.*, **17**, 1287–1291.

55 Kim, K.H., Lee, Y.S., Jung, I.S., Park, S.Y., Chung, H.Y., Lee, I.R., and Yun, Y.S. (1998) Acidic polysaccharide from *Panax ginseng*, ginsan, induces Th1 cell and macrophage cytokines and generates LAK cells in synergy with rIL-2. *Planta Med.*, **64**, 110–115.

56 Hiai, S., Yokoyama, H., Oura, H., and Kawashima, Y. (1983) Evaluation of corticosterone secretion-inducing activities of ginsenosides and their prosapogenins and sapogenins. *Chem. Pharm. Bull. (Tokyo)*, **31**, 168–174.

57 Zong, R.Y., Zheng, S.Q., and Liu, J. (1985) Adrenocorticotropic hormone-like effects of ginseng saponin. *J. Bethune Univ. Med. Sci.*, **11**, 254–258.

58 Chung, E., Lee, K.Y., Lee, Y.J., Lee, Y.H., and Lee, S.K. (1998) Ginsenoside Rg_1 down-regulates glucocorticoid receptor and displays synergistic effects with cAMP. *Steroids*, **63**, 421–424.

59 Lee, Y.J., Chung, E., Lee, K.Y., Lee, Y.H., Huh, B., and Lee, S.K. (1997) Ginsenoside-Rg1, one of the major active molecules from *Panax ginseng*, is a functional ligand of glucocorticoid receptor. *Mol. Cell. Endocrinol.*, **133**, 135–140.

60 Cheng, X.J., Liu, Y.L., Deng, Y.S., Lin, G.F., and Luo, X.T. (1987) Effects of ginseng root saponins on central transmitters and plasma corticosterone in cold-stressed mice and rats. *Acta Pharmacol. Sin.*, **8**, 486–489.

61 Chen, X., Zhu, Q.Y., Liu, L.Y., and Tang, X.L. (1982) Effect of ginsenosides on cardiac performance and hemodynamics of dogs. *Acta Pharmacol. Sin.*, **3**, 236–239.

62 Zhou, X.M., Zhan, Y., Jiang, Y.P., Hu, J.G., and Li, C. (1994) Protective effects of ginsenosides on ischemic reperfused myocardium during open heart surgery: direct evidence for free radical generation by electron spin resonance. *Bull. Hunan Med. Univ.*, **18**, 378–380.

63 Chen, X., Li, Y.J., Deng, H.W., Yang, B.C., Li, D.Y., and Shen, N. (1987) Protective effects of ginsenosides on anoxia/reoxygenation of cultured rat myocytes and on reperfusion injuries against lipid peroxidation. *Biomed. Biochim. Acta*, **46**, S646–S649

64 Gao, D.W., Lou, F.Q., Jin, F., and Xia, S.Y. (1992) Protection on experimental sinus node dysfunction in rabbits with ginsenosides. *Chin. J. Cardiol.*, **20**, 38–40.

65 Ma, Z.C., Gao, Y., Wang, Y.G., Tan, H.L., Xiao, C.R., and Wang, S.Q. (2006) Ginsenoside Rg_1 inhibits proliferation of vascular smooth muscle cells stimulated by tumor necrosis factor-α. *Acta Pharmacol. Sin.*, **27**, 1000–1006.

66 Zhang, H.S. and Wang, S.Q. (2006) Ginsenoside Rg1 inhibits tumor necrosis factor-α (TNF-α)-induced human arterial smooth muscle cells (HASMCs) proliferation. *J. Cell. Biochem.*, **98**, 1471–1481.

67 Kang, S.Y., Schini-Kerth, V.B., and Kim, N.D. (1995) Ginsenosides of the protopanaxatriol group cause endothelium-dependent relaxation in the rat aorta. *Life Sci.*, **56**, 1577–1586.

68 Jiang, Y., Chen, L., Sun, C.W., Zhong, G.G., Qi, H., Ma, X.Y., and Xu, J.D. (1993) Influence of 11 ginsenoside monomers on the action

potentials of myocardiocytes. *Acta Pharmacol. Sin.*, **14** (Suppl.), S8–S12.

69 Jiang, Y., Zhong, G.G., Chen, L., and Ma, X.Y. (1992) Influences of ginsenosides Rb_1, Rb_2, and Rb_3 on electric and contractile activities of normal and damaged cultured myocardiocytes. *Acta Pharmacol. Sin.*, **13**, 403–406.

70 Chen, X. and Lee, T.J. (1995) Ginsenosides-induced nitric oxide-mediated relaxation of the rabbit corpus cavernosum. *Br. J. Pharmacol.*, **115**, 15–18.

71 Chen, X. (1996) Cardiovascular protection by ginsenosides and their nitric oxide releasing action. *Clin. Exp. Pharmacol. Physiol.*, **23**, 728–732.

72 Kim, H.J., Woo, D.S., Lee, G., and Kim, J.J. (1998) The relaxation effects of ginseng saponin in rabbit corporal smooth muscle: is it a nitric oxide donor? *Br. J. Urol.*, **82**, 744–748.

73 Gillis, C.N. (1997) *Panax ginseng* pharmacology: a nitric oxide link? *Biochem. Pharmacol.*, **54**, 1–8.

74 Kang, K.S., Kim, H.Y., Yamabe, N., Park, J.H., and Yokozawa, T. (2007) Preventive effect of 20(S)-ginsenoside Rg3 against lipopolysaccharide-induced hepatic and renal injury in rats. *Free Radic. Res.*, **41**, 1181–1188.

75 Liu, Z.Q., Luo, X.Y., Liu, G.Z., Chen, Y.P., Wang, Z.C., and Sun, Y.X. (2003) In vitro study of the relationship between the structure of ginsenoside and its antioxidative or prooxidative activity in free radical induced hemolysis of human erythrocytes. *J. Agric. Food Chem.*, **51**, 2555–2558.

76 Jeong, H.G., Pokharel, Y.R., Han, E.H., and Kang, K.W. (2007) Induction of cyclo-oxygenase-2 by ginsenoside Rd via activation of CCAAT-enhancer binding proteins and cyclic AMP response binding protein. *Biochem. Biophys. Res. Commun.*, **359**, 51–56.

77 Pan, X.X., Yan, Q.S., and Liu, T.P. (1993) Inhibitory effects of total saponins extracted from *Panax ginseng, P. quinquefolium* and *P. notoginseng* on platelet function and thrombosis in rats. *Chin. J. Pharmacol. Toxicol.*, **7**, 141–144.

78 Park, H.J., Lee, J.H., Song, Y.B., and Park, K. H. (1996) Effects of dietary supplementation of lipophilic fraction from *Panax ginseng* on GMP and cAMP in rat platelets and on blood coagulation. *Biol. Pharm. Bull.*, **19**, 1434–1439.

79 Park, H.J., Rhee, M.H., Park, K.M., Nam, K.Y., and Park, K.H. (1995) Effect of non-saponin fraction from *Panax ginseng* on cGMP and thromboxane A_2 in human platelet aggregation. *J. Ethnopharmacol.*, **49**, 157–162.

80 Zhang, Y.G. and Liu, T.P. (1994) Protective effects of total saponins of *Panax ginseng* on ischemia/reperfusion injury in rat brain. *Chin. J. Pharmacol. Toxicol.*, **8**, 7–12.

81 Zhang, Y.G. and Liu, T.P. (1996) Influences of ginsenosides Rb1 and Rg1 on reversible focal brain ischemia in rats. *Acta Pharmacol. Sin.*, **17**, 44–48.

82 Cui, X., Sakaguchi, T., Ishizuka, D., Tsukada, K., and Hatakeyama, K. (1998) Orally administered ginseng extract reduces serum total cholesterol and triglycerides that induce fatty liver in 66% hepatectomized rats. *J. Int. Med. Res.*, **26**, 181–187.

83 Yokozawa, T., Kobayashi, T., Kawai, A., Oura, H., and Kawashima, Y. (1985) Hyperlipemia - improving effects of ginsenoside-Rb_2 in cholesterol-fed rats. *Chem. Pharm. Bull. (Tokyo)*, **33**, 722–729.

84 Zheng, X.L. and Yan, Y.F. (1991) The effects of ginsenosides of ginseng stem and leaf (GSL) on the lipid regulation and lipid peroxidation in chronic hyperlipidemic rabbits. *Chin. Pharmacol. Bull.*, **7**, 110–113.

85 Kwon, B.M., Ro, S.H., Kim, M.K., Nam, J.Y., Jung, H.J., Lee, I.R., Kim, Y.K., and Bok, S.H. (1997) Polyacetylene analogs, isolated from hairy roots of *Panax ginseng*, inhibit Acyl-CoA: cholesterol acyltransferase. *Planta Med.*, **63**, 552–553.

86 Kwon, B.M., Nam, J.Y., Lee, S.H., Jeong, T.S., Kim, Y.K., and Bok, S.H. (1996) Isolation of cholesteryl ester transfer protein inhibitors from *Panax ginseng* roots. *Chem. Pharm. Bull. (Tokyo)*, **44**, 444–445.

87 Lee, H.U., Bae, E.A., Han, M.J., Kim, N.J., and Kim, D.H. (2005) Hepatoprotective effect of ginsenoside Rb1 and compound K on *tert*-butyl hydroperoxide-induced liver injury. *Liver Int.*, **25**, 1069–1073.

88 Lee, H.U., Bae, E.A., Han, M.J., and Kim, D. H. (2005) Hepatoprotective effect of 20(S)-ginsenoside Rg_3 and its metabolite 20(S)-ginsenoside Rh_2 on *tert*-butyl hydroperoxide-induced liver injury. *Biol. Pharm. Bull.*, **28**, 1992–1994.

89 Bao, T.T. (1981) Effect of total saponin of *Panax ginseng* on alloxan diabetes in mice. *Acta Pharm. Sin.*, **16**, 618–620.

90 Yokozawa, T., Kobayashi, T., Oura, H., and Kawashima, Y. (1985) Studies on the mechanism of the hypoglycemic activity of ginsenoside-Rb_2 in streptozotocin-induced rats. *Chem. Pharm. Bull. (Tokyo)*, **33**, 869–872.

91 Xie, J.T., Mehendale, S.R., Li, X., Quigg, R., Wang, X., Wang, C.Z., Wu, J.A., Aung, H.H., A. Rue, P., Bell, G.I., and Yuan, C.S. (2005) Anti-diabetic effect of ginsenoside Re in ob/ob mice. *Biochim. Biophys. Acta*, **1740**, 319–325.

92 Cho, W.C., Chung, W.S., Lee, S.K., Leung, A. W., Cheng, C.H., and Yue, K.K. (2006) Ginsenoside Re of Panax ginseng possesses significant antioxidant and antihyperlipidemic efficacies in streptozotocin-induced diabetic rats. *Eur. J. Pharmacol.*, **550**, 173–179.

93 Li, M., Wu, W., and Tang, W.Z. (1991) Comparison between the effect of ginseng and tolbutamide on free insulin in rats. *J. Lanzhou Med. Coll.*, **17**, 227–229.

94 Hasegawa, H., Matsumiya, S., Murakami, C., Kurokawa, T., Kasai, R., Ishibashi, S., and Yamasaki, K. (1994) Interactions of ginseng extract, ginseng separated fractions, and some triterpenoid saponins with glucose transporters in sheep erythrocytes. *Planta Med.*, **60**, 153–157.

95 Suzuki, Y. and Hikino, H. (1989) Antidiabetes drugs. XXXI. Mechanisms of hypoglycemic activity of panaxans A and B, glycans of *Panax ginseng* roots: effects of plasma level, secretion, sensitivity and binding of insulin in mice. *Phytother. Res.*, **3**, 20–24.

96 Wang, B.X., Yang, M., Jin, Y.L., Cui, Z.Y., and Wang, Y. (1990) Studies on the hypoglycemic effect of ginseng polypeptide. *Acta Pharm. Sin.*, **25**, 401–405.

97 Rhee, Y.H., Ahn, J.H., Choe, J., Kang, K.W., and Joe, C. (1991) Inhibition of mutagenesis and transformation by root extracts of *Panax ginseng in vitro*. *Planta Med.*, **57**, 125–128.

98 Lee, T.K., Johnke, R.M., Allison, R.R., O'Brien, K.F., and Dobbs, L.J. Jr (2005) Radioprotective potential of ginseng. *Mutagenesis*, **20**, 237–243.

99 Yun, T.K., Lee, Y.S., Kwon, H.Y., and Choi, K.J. (1996) Saponin contents and anticarcinogenic effects of ginseng depending on types and ages in mice. *Acta Pharmacol. Sin.*, **17**, 293–298.

100 Li, X. and Wu, J.G. (1992) Effect of *Panax ginseng* on liver cancer of rats induced by diethylnitrosamine. *Chin. J. Pathol.*, **21**, 113–114.

101 Chen, X.G. Li, H.Y., LI, X.H., and Han, R. (1998) Cancer chemopreventive and therapeutic activities of red ginseng. *J. Ethnopharmacol.*, **60**, 71–78.

102 Yun, Y.S., Lee, Y.S., Jo, S.K., and Jung, I.S. (1993) Inhibition of autochthonous tumor by ethanol insoluble fraction from *Panax ginseng* as an immunomodulator. *Planta Med.*, **59**, 521–524.

103 Yun, T.K. and Choi, S.Y. (1998) Non-organ specific cancer prevention of ginseng: a prospective study in Korea. *Int. J. Epidemiol.*, **27**, 359–364.

104 Kim, W., Hwang, S., Lee, H., Song, H., and Kim, S. (1999) Panax ginseng protects the testis against 2,3,7,8-tetrachlorodibenzo-p-dioxin induced testicular damage in guinea pigs. *BJU Int.*, **83**, 842–849.

105 Kim, S.H., Jeong, K.S., Ryu, S.Y., and Kim, T. H. (1998) *Panax ginseng* prevents apoptosis in hair follicles and accelerates recovery of hair medullary cells in irradiated mice. *In Vivo*, **12**, 219–222.

106 Kim, S.H., Cho, C.K., Yoo, S.Y., Koh, K.H., Yun, H.G., and Kim, T.H. (1993) *In vivo* radioprotective activity of *Panax ginseng* and diethyldithiocarbamate. *In Vivo*, **7**, 467–470.

107 Cai, R.M., Liu, C.M., Lin, M., Fan, H.X., Wei, J.J., Wang, Z.C., Xu, J.D., Wang, L.H., Ling, L., and Wang, Y.Y. (1991) Studies on the radioprotective effect of water-soluble nonsaponin components in ginseng. *China J. Chin. Mater. Med.*, **16**, 429–432.

108 Gong, S.L., Li, X.M., Lu, Z., and Liu, S.Z. (1993) Protective effect of panaxatriol on function of reproductive endocrine axis in radiation-injured rats. *Acta Pharmacol. Sin.*, **14**, 358–360.

109 Ota, T., Maeda, M., and Odashima, S. (1991) Mechanism of action of ginsenoside Rh_2: uptake and metabolism of ginsenoside Rh_2 by cultured B16 melanoma cells. *J. Pharm. Sci.*, **80**, 1141–1146.

110 Han, M.Q., Liu, J.X., and Gao, H. (1995) Effects of 24 Chinese medicinal herbs on nucleic acid, protein and cell cycle of human lung adenocarcinoma cell. *Chin. J. Integr. Trad. West. Med.*, **15**, 147–149.

111 Oh, M., Choi, Y.H., Choi, S., Chung, H., Kim, K., Kim, S.I., Kim, D.K., and Kim, N.D. (1999) Anti-proliferating effects of ginsenoside Rh_2 on MCF-7 human breast cancer cells. *Int. J. Oncol.*, **14**, 869–875.

112 Xia, L.J. and Han, R. (1996) Differentiation of B16 melanoma cells induced by ginsenoside Rh_2. *Acta Pharm. Sin.*, **31**, 742–745.

113 Jian, L.S., Wang, S.L., Li, S.L., and Wang, Y. (1992) Effect of ginsenosides (GS) from leaves and stems on gastric carcinoma cells cultured

in vitro. *J. Norman Bethune Univ. Med. Sci.*, **18**, 27–29.

114 Fujikawa-Yamamoto, K., Ota, T., Odashima, S., Abe, H., and Arichi, S. (1987) Different responses in the cell cycle of tumor cells to ginsenodide Rh_2. *Cancer J.*, **1**, 349–352.

115 Nakata, H., Kikuchi, Y., Tode, T., Hirata, J., Kita, T., Ishii, K., Kudoh, K., Nagata, I., and Shinomiya, N. (1998) Inhibitory effects of ginsenoside Rh_2 on tumor growth in nude mice bearing human ovarian cancer cells. *Jpn. J. Cancer Res.*, **89**, 733–740.

116 Kim, Y.S., Kim, D.S., and Kim, S.I. (1998) Ginsenoside Rh_2 and Rh_3 induce differentiation of HL-60 cells into granulocytes: modulation of protein kinase C isoforms during differentiation by ginsenoside Rh_2. *Int. J Biochem. Cell Biol.*, **30**, 327–338.

117 Yi, R.L., Li, W., and Hao, X.Z. (1993) Inductive differentiation effect of ginsenosides on human acute non-lymphocytic leukemic cells in 58 patients. *Chin. J. Integr. Trad. West. Med.*, **13**, 722–724.

118 Cheng, C.C., Yang, S.M., Huang, C.Y., Chen, J.C., Chang, W.M., and Hsu, S.L. (2005) Molecular mechanisms of ginsenoside Rh_2-mediated G_1 growth arrest and apoptosis in human lung adenocarcinoma A549 cells. *Cancer Chemother. Pharmacol.*, **55**, 531–540.

119 Lee, Y.N., Lee, H.Y., Chung, H.Y., Kim, S.I., Lee, S.K., Park, B.C., and Kim, K.W. (1996) *In vitro* induction of differentiation by ginsenoides in F9 teratocarcinoma cells. *Eur. J. Cancer*, **32A**, 1420–1428.

120 Zeng, X.L. and Tu, Z.G. (2004) Induction of differentiation by ginsenoside Rh_2 in hepatocarcinoma cell SMMC-7721. *Ai Zheng.*, **23**, 879–884.

121 Byun, B.H., Shin, I., Yoon, Y.S., Kim, S.I., and Joe, C.O. (1997) Modulation of protein kinase C activity in NIH 3T3 cells by plant glycosides from *Panax ginseng*. *Planta Med.*, **63**, 389–392.

122 Kim, S.E., Lee, Y.H., Park, J.H., and Lee, S.K. (1999) Ginsenoside-Rs_3, a new diol-type ginseng saponin, selectively elevates protein levels of p53 and p21WAF1 leading to induction of apoptosis in SK-HEP-1 cells. *Anticancer Res.*, **19**, 487–491.

123 Tong, C.N., Matsuda, H., and Kubo, M. (1992) Pharmacological study on *Panax ginseng* C.A. Meyer. XV. Effects of 70% methanolic extract from red and white ginseng on the antitumor activity of mitomycin C. *Yakugaku Zasshi*, **112**, 856–865.

124 Jiang, Z.M., Pang, J.N., Sun, Y.H., Chu, J.J., Zhang, S.L., Chen, Y.J., and Lin, C.J. (1991) Antitumor effect of ginsenoside Ra_1 and rhTNF-A. *Chin. J. Microbiol. Immunol.*, **11**, 238–241.

125 Wakabayashi, C., Hasegawa, H., Murata, J., and Saiki, I. (1997) *In vivo* antimetastatic action of ginseng protopanaxadiol saponins is based on their intestinal bacterial metabolites after oral administration. *Oncol. Res.*, **9**, 411–417.

126 Mochizuki, M., Yoo, Y.C., Matsuzawa, K., Sato, K., Saiki, I., Tono-oka, S., Samukawa, K., and Azuma, I. (1995) Inhibitory effect of tumor metastasis in mice by saponins, ginsenoside Rb_2, 20(*R*)- and 20(*S*)-ginsenoside Rg_3, of red ginseng. *Biol. Pharm. Bull.*, **18**, 1197–1202.

127 Sato, K., Mochizuki, M., Saiki, I., Yoo, Y.C., Samukawa, K., and Azuma, I. (1994) Inhibition of tumor angiogenesis and metastasis by a saponin of *Panax ginseng*, ginsenoside-Rb_2. *Biol. Pharm. Bull.*, **17**, 635–639.

128 Ahn, B.Z., Kim, S.I., and Lee, Y.H. (1989) Acetylpanaxydol and panaxydol chlorohydrin, two new polyynes from Korean ginseng with cytotoxic activity against L 1210 cells. *Arch. Pharm. (Weinheim)*, **322**, 223–226.

129 Sohn, J., Lee, C.H., Chung, D.J., Park, S.H., Kim, I., and Hwang, W.I. (1998) Effect of petroleum ether extract of *Panax ginseng* roots on proliferation and cell cycle progression of human renal cell carcinoma cells. *Exp. Mol. Med.*, **30**, 47–51.

130 Matsunaga, H., Saita, T., Nagumo, F., Mori, M., and Katano, M. (1995) A possible mechanism for the cytotoxicity of a polyacetylenic alcohol, panaxytriol: inhibition of mitochondrial respiration. *Cancer Chemother. Pharmacol.*, **35**, 291–296.

131 Matsunaga, H., Katano, M., Saita, T., Yamamoto, H., and Mori, M. (1994) Potentiation of cytotoxicity of mitomycin C by a polyacetylenic alcohol, panaxytriol. *Cancer Chemother. Pharmacol.*, **33**, 291–297.

132 Kim, S.I., Lee, Y.H., and Ahn, B.Z. (1999) Synthesis of ginseng diyne analogues and their antiproliferative activity against L1210 cells. *Arch. Pharm. (Weinheim)*, **332**, 133–136.

133 Wang, M.X., Li, F.W., Li, X.G., and Zhang, L.S. E.Z. (1992) Effects of ginseng volatile oil on chemical components of SGC-823 gastric carcinoma in cell culture. *China J. Chin. Mater. Med.*, **17**, 110–112.

Paris polyphylla var. *chinensis* and *Paris polyphylla* var. *yunnanensis*

Rhizoma Paridis (Chonglou) is the dried rhizome of *Paris polyphylla* Smith var. *chinensis* (Franch.) Hara or *Paris polyphylla* Smith var. *yunnanensis* (Franch.) Hand.-Mazz. (Liliaceae). It is used for the treatment of boils, carbuncles, laryngopharyngeal pain, snake bites, traumatic injuries, and convulsion.

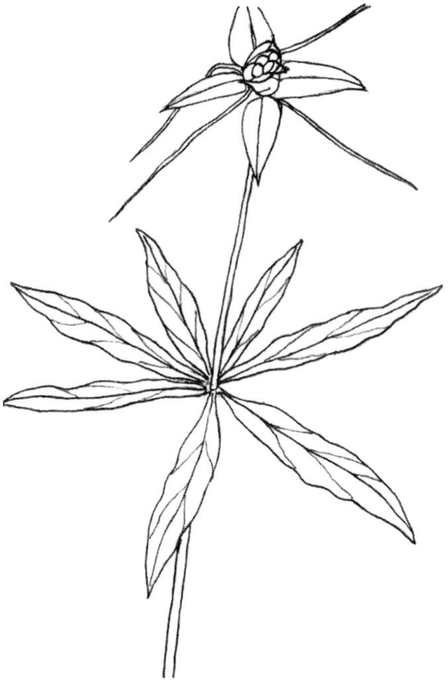

Paris polyphylla Smith var. *yunnanensis* (Franch.) Hand.-Mazz.

A galenic preparation from Rhizoma Paridis, "Gongxuening Jiaonang," is listed in the Chinese Pharmacopoeia. These capsules are prepared from the 70% ethanol extract of Rhizoma Paridis, with each capsule containing 0.13 g of the extract. Gongxuening Jiaonang is used as a hemostatic agent in the treatment of menorrhea.

Chemistry

The rhizome of *Paris* species is known to contain steroid saponins, with the diosgenin saponin polyphyllin as the major component. *Paris polyphylla* was reported to be a source of diosgenin. Minor saponins from the rhizome of *P. polyphylla* were identified as diosgenin saponins, such as polyphyllins A–H [1, 2], and pennogenin saponins [3]. Diosgenin saponins were also isolated from the rhizome of *P. polyphylla* var. *chinensis* with polyphyllin C as the major component [4]. From the rhizoma of *P. polyphylla* var. *yunnanensis*, diosgenin saponins dioscin and related saponins were isolated [5]. Saponins with 27-hydroxypennogenin, 27,23β-dihydroxypennogenin as aglycones [6], including polyphyllosides III and IV, and saponins with nuatigenin and isonuatigenin as aglycones, were isolated from the aerial parts of *P. polyphylla* var. *yunnanensis* [7].

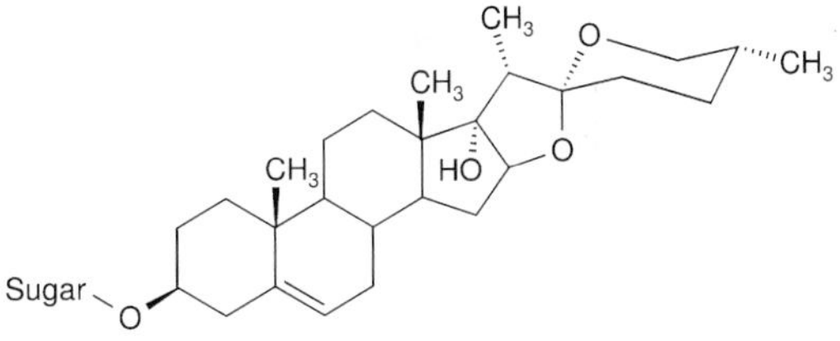

Pennogenin-Saponins (e.g. Nuatigenin, Isonuatigenin)

The Chinese Pharmacopoeia requires a quantitative determination of chonglou saponin I (polyphyllin A) and chonglou saponin II (polyphyllin B) in Rhizoma Paridis by HPLC. The total content of chonglou saponin I and chonglou saponin II in Rhizoma Paridis should be not less than 0.8%. The Chinese Pharmacopoeia requires a qualitative determination of chonglou saponin VI (polyphyllin F) and chonglou saponin VII

Handbook of Chinese Medicinal Plants: Chemistry, Pharmacology, Toxicology
Weici Tang and Gerhard Eisenbrand

ISBN: 978-3-527-32226-8

(polyphyllin G) in "Gongxuening Jiaonang" by thin-layer chromatographic comparison with reference substances, and a quantitative determination of chonglou saponin VI in "Gongxuening Jiaonang" by HPLC. The content of chonglou saponin VI in each capsule should be not less than 0.52 mg.

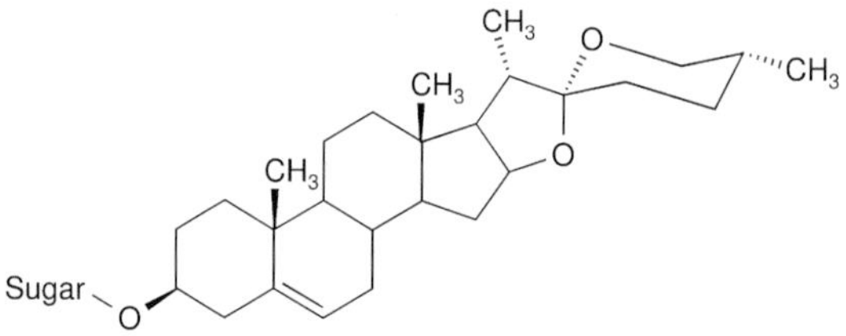

Diosgenin-Saponins (Polyphyllin A -H)

Pharmacology and Toxicology

The methanol extract of *P. polyphylla* was found to exert an antibacterial action *in vitro* and to inhibit the growth of *Shigella sonnei, Serratia marcescens, Escherichia coli,* and sensitive and resistant strains of *Staphylococcus aureus*. It also showed hemostatic action following oral administration in mice, and shortened the coagulation time of mouse blood [8]. Commonly used species and variants of Rhizoma Paridis showed sedative and analgesic activities. *Paris polyphylla* var. *chinensis* and *P. polyphylla* var. *yunnanensis* were the most potent species [9]. The rhizome of *P. polyphylla* var. *chinensis,* as an anti-snake venom herb in traditional Chinese medicine, was reported to influence the biological effects of endothelin in rats [10].

The methanol extract of the rhizome of *P. polyphylla* var. *yunnanensis* was found to potently inhibit gastric lesions in rats induced by ethanol. Four known spirostanol-type steroid saponins, including two pennogenin saponins and two diosgenin saponins, were identified as the active components. These four saponins, when given to rats at oral doses of 1.25–10 $mg\,kg^{-1}$, strongly inhibited gastric lesion formation induced by ethanol and indomethacin. With regard to the structural requirements of steroid saponins, the 3-*O*-glycoside moiety and spirostanol structure were found to be essential for the activity, while the 17-hydroxyl group in the aglycone part enhanced the protective effects against ethanol-induced gastric lesions. These compounds also weakly inhibited acid secretion in pylorus-ligated rats; this indicated that endogenous prostaglandins and sulfhydryl compounds were involved in the protective activity [11].

Polyphyllin D was reported to exert cytotoxic activity. In *in vitro* tests using MCF-7 and MDA-MB-231 mammary carcinoma cells, treatment with polyphyllin D resulted in the inhibition of viability and induction of apoptosis in a concentration-dependent manner, with IC_{50} values of 5 and 2.5 μM, respectively, after 48 h incubation. Polyphyllin D inhibited the mitochondrial membrane potential, downregulated anti-apoptotic Bcl-2 expression, upregulated proapoptotic Bax expression and activated caspase-9. Daily intravenous (i.v.) administration of polyphyllin D at a dose of 2.7 $mg\,kg^{-1}$ for 10 days to nude mice bearing MCF-7 tumors, effectively reduced tumor growth (50%). No significant toxicity was evident in the heart and liver of the mice [12].

References

1 Singh, S.B., Thakur, R.S., and Schulten, H.R. (1982) Spirostanol saponins from *Paris polyphylla*, structures of polyphyllin C, D, E, and F. *Phytochemistry*, **21**, 2925–2929.

2 Singh, S.B., Thakur, R.S. and Schulten, H.R. (1982) Plant saponins. Part V. Furostanol saponins from *Paris polyphylla*: structures of polyphyllin G and H. *Phytochemistry*, **21**, 2079–2082.

3 Wei, J.R. (1997) Study on determination of pennogenin ingredients in extract of *Paris polyphylla* Sm. by HPLC. *Chin. J. Pharm. Anal.*, **17**, 153–155.

4 Xu, X.M. and Zhong, C.C. (1988) The chemical constituents of Chinese paris (*Paris polyphylla* var. *chinensis*) I. Isolation and structural determination of sapogenins A, B, and D. *Chin. Trad. Herbal Drugs*, **19**, 194–198, 201.

5 Chen, C.X., Zhang, Y.T. and Zhou, J. (1983) Studies on the saponin components of plants in Yunnan [China]. VI. Steroid glycosides of *Paris polyphylla* Sm. var. *yunnanensis* (Fr.) H-M. 2. *Acta Bot. Yunnan*, **5**, 91–97.

6 Chen, C.X. and Zhou, J. (1992) Two new steroid sapogenins of *Paris polyphylla* var. *yunnanensis*. *Acta Bot. Yunnan*, **14**, 111–113.

7 Chen, C.X., Zhang, Y.T. and Zhou, J. (1995) The glycosides of aerial parts of *Paris polyphylla* var. *yunnanensis*. *Acta Bot. Yunnan*, **17**, 473–478.

8 Wang, Q.A., Xu, G.J. and Cheng, Y.B. (1989) Study on antibacterial and hemostatic activities of Rhizoma paridis. *J. China Pharm. Univ.*, **20**, 251–253.

9 Wang, Q.A., Xu, G.J. and Jiang, Y. (1990) Analgesic and sedative effects of Chinese drug Rhizoma Paridis. *China J. Chin. Mater. Med.*, **15**, 109–111.

10 Tian, Q., Zhao, D., Zhang, J.F., Gao, L.R., Liu, S.X., Yang, J., Su, J.Y., Zhang, Z.K., Tang, J. and Tang, C.S. (1996) Investigation on inhibition of biological effects of endothelin. *Sci. China, Ser. C, Life Sci.*, **39**, 207–216.

11 Matsuda, H., Pongpiriyadacha, Y., Morikawa, T., Kishi, A., Kataoka, S. and Yoshikawa, M. (2003) Protective effects of steroid saponins from *Paris polyphylla* var. *yunnanensis* on ethanol- or indomethacin-induced gastric mucosal lesions in rats: structural requirement for activity and mode of action. *Bioorg. Med. Chem. Lett.*, **13**, 1101–1106.

12 Lee, M.S., Yuet-Wa, J.C., Kong, S.K., Yu, B., Eng-Choon, V.O., Nai-Ching, H.W., Chung-Wai, T.M. and Fung, K.P. (2005) Effects of polyphyllin D, a steroidal saponin in *Paris polyphylla*, in growth inhibition of human breast cancer cells and in xenograft. *Cancer Biol. Ther.*, **4**, 1248–1254.

Perilla frutescens

Caulis Perillae (Zisugeng) is the dried stem of *Perilla frutescens* (L.) Britt. (Lamiaceae). It is used as an analgesic for the treatment of gastric disorders.

Perilla frutescens (L.) Britt.

Folium Perillae (Zisuye) is the dried leaf of *Perilla frutescens*. It is used as a stomachic and for the treatment of common cold, cough and nausea, vomiting in pregnancy, and fish or crab poisoning.

Fructus Perillae (Zisuzi) is the dried ripe seed of *Perilla frutescens*. It is used as an antiasthmatic, expectorant, and laxative agent.

Chemistry

The leaves, stems and fruits of *P. frutescens* are known to contain essential oil. The essential oils from leaves, stems and fruits were found to contain terpene and nonterpene compounds, with perillaldehyde and perilla alcohol (perillyl alcohol) as the major components [1]. Minor volatile components were identified as perilla ketone, caryophyllene, hexadecanoic acid, and dill apiol [2].

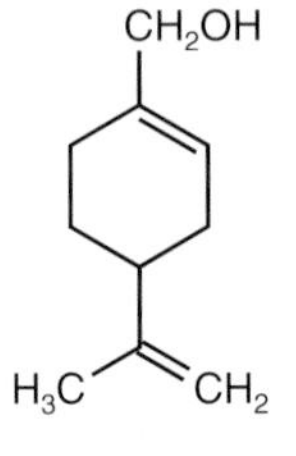

Perilla alcohol

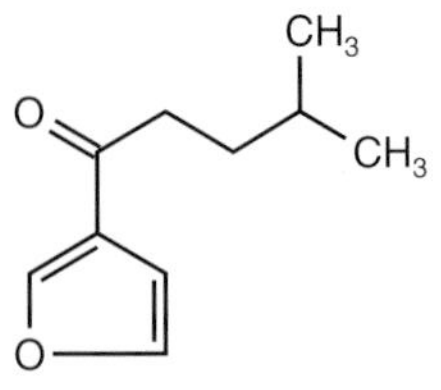

Perilla ketone

Monoterpene glucosides from the leaves of *P. frutescens* are identified as perillosides A and C [3]. Phenolic compounds from *P. frutescens* were identified as caffeic acid, methyl caffeate, rosmarinic acid, luteolin, and luteolin 7-*O*-glucuronide methyl ester [4]. Rosmarinic acid and luteolin are two major components with pharmacological significance of the leaves of *P. frutescens*. A cyanogenic glucoside from *P. frutescens* was identified as prunasin (D-mandelonitrile β-D-glucoside) [5].

Two neolignans – magnosalin and andamanicin – were isolated from the leaves of *P. frutescens*. Structurally, magnosalin is 1β,2α,3β,4α-1,2-dimethyl-3,4-bis-(2,4,5-trimethoxyphenyl)-cyclobutane, and andamanicin is the 1α,2β-isomer of magnosalin [6].

Pharmacology and Toxicology

The extract and essential oil of *P. frutescens* was shown to exert significant antipyretic effect in rabbits, and antiemetic effects in pigeons. The fatty oil extracted from the seeds had significant antitussive effects in mice and antiasthmatic effects in guinea pigs [7]. The leaf of *P. frutescens* is known to be an antiallergic agent. The intraperitoneal (i.p.) injection of the leaf extract to mice

Handbook of Chinese Medicinal Plants: Chemistry, Pharmacology, Toxicology
Weici Tang and Gerhard Eisenbrand

ISBN: 978-3-527-32226-8

immunized with anti-ovalbumin serum into the ears 15 min before ovalbumin treatment significantly suppressed the passive cutaneous anaphylaxis (PCA) reaction [8]. Rosmarinic acid and luteolin were believed to be the two active components in the leaf extract. Oral administration of the perilla leaf extract to mice inhibited inflammation, the allergic response, and the production of tumor necrosis factor-α (TNF-α). Among the isolated compounds, luteolin showed *in vivo* activity. Notably, it inhibited the production of TNF-α, as well as arachidonic acid-induced ear edema, ear edema induced by 12-*O*-tetradecanoylphorbol 13-acetate (TPA), and the allergic edema induced by oxazolone [9].

An aqueous extract of the leaves of *P. frutescens* induced nitric oxide (NO) production and mRNA expression of inducible nitric oxide synthase (iNOS) in murine cultured vascular smooth muscle cells. In contrast, it significantly inhibited NO production induced by interferon-γ (IFN-γ) combined with lipopolysaccharide (LPS). The extract also significantly inhibited the proliferation of vascular smooth muscle cells induced by platelet-derived growth factor (PDGF) or TNF-α, partially mediated by NO [10]. Rosmarinic acid was reported to exert antioxidant and free radical-scavenging activities. In RAW 264.7 murine macrophage cells, rosmarinic acid inhibited the production of NO and iNOS protein synthesis induced by LPS, and also effectively suppressed the superoxide production in macrophages activated by TPA. Moreover, rosmarinic acid was also shown to abolish the LPS-induced phosphorylation of Iκ-Bα. Rosmarinic acid might act as an effective inhibitor of superoxide and NO synthesis, based partly on its ability to inhibit the serine phosphorylation of Iκ-Bα [11]. The neolignans from the leaves of *P. frutescens*, magnosalin and andamanicin, each markedly inhibited iNOS and TNF-α levels in RAW 264.7 cells stimulated by LPS [6]. A study of the inhibitory effects of *P. frutescens* and its phenolic constituents on cytokine-induced proliferation of murine cultured mesangial cells showed that DNA synthesis of mesangial cells stimulated by PDGF at a concentration of $10\,ng\,ml^{-1}$, or by TNF-α at a concentration of $100\,U\,ml^{-1}$, was inhibited, with IC_{50} values of 3.3 and $1.4\,\mu g\,ml^{-1}$, respectively. Caffeic acid, methyl caffeate, rosmarinic acid, and luteolin 7-*O*-glucuronide methyl ester were each identified as active constituents [4].

An aqueous extract of the leaves of *P. frutescens* exhibited hepatoprotective activity against the toxicity of *tert*-butyl hydroperoxide (TBH) in rats. Treatment of hepatocytes with the extract significantly reversed the cytotoxicity and lipid peroxidation of TBH. In addition, the extract exhibited diphenyl-picrylhydrazyl free radical-scavenging activity. Pretreatment with the extract at oral doses of 1 and $3\,g\,kg^{-1}$ for five days before a single dose of TBH significantly lowered serum levels of aspartate aminotransferase and alanine aminotransferase, and increased the hepatic glutathione content and γ-glutamylcysteine synthetase activity. Histopathological examination showed that the extract reduced the incidence of liver lesions induced by TBH [12]. Rosmarinic acid was found to protect HepG2 human hepatoma-derived cells against the cytotoxicity of aflatoxin B_1 (AFB_1) and ochratoxin A (OTA). Rosmarinic acid also blocked the production of reactive oxygen species (ROS), the inhibition of protein and DNA synthesis, and the induction of apoptosis in HepG2 cells caused by the mycotoxins [13]. The leaf extract of *P. frutescens* and rosmarinic acid were each reported to inhibit the proliferation of HepG2 cells, and to induce cell apoptosis [14].

Perilla alcohol occurs widely in cherries and other fruits and edible plants. As a monoterpene compound like *d*-limonene, perilla alcohol was found to be a prenyltransferase inhibitor. Perilla alcohol and its major metabolite, perillic acid, each weakly inhibited both mammalian and yeast protein farnesyl transferase and protein geranylgeranyl transferase. In contrast, a minor metabolite of perillyl alcohol, perillic acid methyl ester, was reported to be a potent inhibitor of both enzymes [15, 16].

Perilla alcohol, as a prenyltransferase inhibitor, was found to exhibit chemopreventive efficacy against tumor induction caused by different carcinogens. For example, perilla alcohol inhibited the induction of mammary carcinomas in rats by 7,12-dimethylbenz[*a*]anthracene (DMBA) [17]; the induction of hepatocarcinoma in rats by *N*-nitroso-*N,N*-diethylamine (NDEA) [18]; the induction of lung tumors in male (C3H/HeJ×A/J) F_1 hybrid mice by the tobacco-specific nitrosamine 4-(methylnitrosamino)-1-(3-pyridyl)-1-butanone [19]; the induction of esophageal tumors in rats induced by *N*-nitroso-*N*-methyl-*N*-benzylamine (NMBA) [20]; the induction of pancreatic cancer in hamsters induced by *N*-nitroso-*N,N*-bis(2-oxopropyl)-amine [21]; the induction of colon adenocarcinomas induced by azoxymethane [22]; and the induction of nonmelanoma on BALB/c mouse skin by UVB irradiation [23]. However, it was also reported that perillyl alcohol at a daily dose of $1\,g\,kg^{-1}$, administered in the diet to rats, did not show any chemoprevention of the hepatocarcinogenesis induced by *N*-nitrosomorpholine [24].

Perillyl alcohol was also reported to be cytotoxic. The induction of cell apoptosis and cell differentiation, and the inhibition of some oncogene products, were discussed as the respective chemopreventive and cytotoxic mechanisms. Perillyl alcohol and perillic acid inhibited cell proliferation, and induced cell cycle arrest and apoptosis in NSCLC, A549, and H520 non-small-cell lung cancer cells, associated with an increasing expression of Bax, p21, and caspase-3 activity in both cell lines [25]. Perillyl alcohol was reported to inhibit the growth of hamster pancreatic tumors, to cause a complete regression [26], and to induce apoptosis in pancreatic tumor cells without affecting the rate of DNA synthesis [27]. Perillyl alcohol inhibited cell proliferation and induced apoptosis in immortalized human vascular smooth muscle cells (HVSMCs) derived from saphenous vein [28] and in human T lymphocytes [29]. It exerted a growth-inhibitory effect and induced apoptosis in human glioblastoma multiforme cells [30]. Perillyl alcohol also induced G_0/G_1 arrest and apoptosis in Bcr/Abl-transformed myeloid cells [31]. Perillyl alcohol inhibited the growth of estrogen receptor (ER)-positive KPL-1 and MCF-7 and ER-negative MKL-F and MDA-MB-231 human breast cancer cells *in vitro*, and suppressed both growth and metastasis *in vivo* [32]. The effect of perillyl alcohol on pancreatic tumor cells was significantly greater than its effect on nonmalignant pancreatic ductal cells. The perillyl alcohol-induced increase in apoptosis in pancreatic tumor cells was found to be associated with a two- to eightfold increase in the expression of the proapoptotic protein Bak, although Bak expression was not affected by perillyl alcohol in nonmalignant cells [33].

In addition to inducing apoptosis, perilla alcohol was reported to induce the differentiation of a neuroblastoma-derived cell line, Neuro-2A. Such induction was associated with several cellular effects, including a cytostatic effect and an ability to inhibit ubiquinone syntheses [34]. Perillyl alcohol inhibited the expression and function of the androgen receptor in LNCaP human prostate cancer cells [35]. In addition to inhibiting ubiquinone synthesis, perillyl alcohol also blocked the conversion of lanosterol to cholesterol in NIH 3T3 cells [36], and inhibited the incorporation of [^{14}C]mevalonolactone into proteins in NIH 3T3 cells. The inhibition of [^{14}C]mevalonolactone incorporation into proteins was associated with an inhibition of enzymes in the mevalonate pathway, and with changes in the levels of protein substrates for prenylation. Perilla alcohol inhibited the incorporation of [^{14}C]mevalonolactone into RhoA and Rab6 of NIH 3T3 cells, but not into Ras protein [37]. It was also found to be an inhibitor of angiogenesis, preventing new blood vessel growth in the chicken embryo chorioallantoic membrane and inhibiting the morphogenic differentiation of cultured endothelial cells into capillary-like networks [38].

In human myeloid THP-1 and lymphoid RPMI-8402 cells, perillyl alcohol decreased the levels of [^{35}S]methionine-labeled Ras proteins [39]. In contrast, in PANC-1 pancreas carcinoma cells carrying a K-Ras mutation, and also in 12V-H-Ras-transformed rat fibroblasts, perilla alcohol was found to efficiently inhibit cell growth, but not to induce the morphological reversion of 12V-H-Ras-transformed cells. Neither did perilla alcohol reduce MAP kinase activity or collagenase promoter activity in PANC-1 cells – two functions which are known to be downstream of Ras [40]. The cytotoxic activity of perillyl alcohol against pancreatic carcinomas might be caused by its ability to inhibit the prenylation of growth-regulatory proteins other than K-Ras, including H-ras [41]. However, in a clinical Phase I trial, high-dose perilla alcohol treatment (2800 mg m^{-2}) was unable to consistently alter the expression of p21^{ras}, Rap1, or RhoA in the peripheral blood mononuclear cells of patients. Thus, the inhibition of p21^{ras} function in humans was unlikely to occur following perilla alcohol administration at safe oral doses [42].

In rat mammary carcinomas, the gross and ultrastructural morphology of perilla alcohol-mediated tumor regression indicated that apoptosis accounted for the marked reduction in the epithelial compartment. RNA expression studies demonstrated that cell cycle- and apoptosis-related genes were differentially expressed within 48 h of perilla alcohol treatment; p21 (Cip1/WAF1), Bax, Bad, and annexin I were induced; cyclin E and cyclin-dependent kinase 2 were repressed; and bcl-2 and p53 were unchanged. A potential role for transforming growth factor-β (TGF-β) signaling in perilla alcohol-mediated carcinoma regression was explored. The nuclear localization of Smad2/Smad3 indicated that the TGF-β signaling pathway was activated in regressing carcinomas. In contrast, it was reported that none of the perilla alcohol-mediated anticancer activities was observed in normal mammary gland [43]. It was also suggested that perilla alcohol might affect c-Jun activity via the Jun *N*-terminal kinase or stress-activated protein kinase pathway, and modulate the expression of AP-1 target genes in rat mammary tumors [44]. Perilla alcohol and perillic acid were also each reported to inhibit the growth of human T-47D, MCF-7, and MDA-MB-231 breast cancer cells associated with an arrest of cells in the G_1 phase [45].

Perillyl alcohol has been developed as a clinical candidate for cancer treatment [46]. In a Phase I trial, 17 patients received perillyl alcohol orally, three times daily for 14 days of each 28-day cycle. The starting dose of perillyl alcohol (1600 mg m^{-2}) was raised to 2100 and 2800 mg m^{-2} in subsequent cohorts. Chronic nausea and fatigue proved to be the dose-limiting toxicities at 2800 mg m^{-2}. Grade 1–2 hypokalemia was common at doses of 2100 and 2800 mg m^{-2}. In this study, perilla alcohol was not detected in the plasma, but two metabolites – dihydroperillic acid and perillic acid – were identified in the plasma and urine on days 1 and 29 of perillyl alcohol treatment [47]. The mean peak plasma levels of perillic acid on days 1 and 29 were 472 and 311 μ*M* after a perilla alcohol dose of 1600 mg m^{-2}, while the mean peak plasma levels of dihydroperillic acid were 34.2 and 34.0 μ*M*, respectively. Perillyl alcohol, perillic acid and dihydroperillic acid were each detectable in the urine of all patients receiving the high perilla alcohol dose of 2800 mg m^{-2}. Approximately 9% of the total dose was recovered during the first 24 h, with perillic acid as the major component; less than 1% of the dose was recovered in the urine [48]. Perilla alcohol, as well as *cis*- and *trans*-dihydroperillic acid and perillic acid, were all detected in rat plasma following the intravenous (i.v.) injection of perilla alcohol in rats at a dose of 23 mg kg^{-1}, and in the plasma of a patient receiving oral perilla alcohol at 500 mg m^{-2}, using gas chromatography–mass spectrometry (GC-MS) [49]. In rats, approximately 70% of the perilla alcohol dose was recovered as glucuronides in the bile and urine. Perillic acid generated only the acyl glucuronide, while perilla alcohol and perillaldehyde each formed both acyl and ether glucuronides [50].

In another clinical Phase I trial involving 21 patients, a treatment schedule of perillyl alcohol, given at a starting dose of 1.6 g m^{-2} and escalating to 2.1 g m^{-2}, was recommended as "tolerable"; fatigue and low-grade nausea were dose-limiting factors. Mean peak plasma concentrations were 383 μM for perillic acid and 27 μM for dihydroperillic acid, reached at 2 h and 4 h after ingestion, respectively. Two major metabolites were detected, with half-lives of 1 h and 2.4 h, respectively. A stabilization of disease was observed in only one of the 16 patients evaluable for response [51].

A clinical Phase II study of perillyl alcohol in patients with metastatic colorectal carcinoma evaluated time to progression, objective response rate, and toxicity. A group of 27 patients received perillyl alcohol at an oral dose of 1200 mg m^{-2}, with dose escalation to 1600 mg m^{-2}. The results showed a median time to progression of 1.8 months (range: 1 to 3 months). Four patients received less than one treatment cycle and were not evaluable for response; the other 23 all had progressive disease. There were no responses, and the toxicity was relatively mild, but fatigue, nausea and anemia predominated. Three patients withdrew from therapy for reasons of toxicity. The report concluded that oral perillyl alcohol, when administered at the given dose and formulation, did not appear to possess any clinical antitumor activity in patients with advanced colorectal carcinoma [52].

In another Phase II trial, 15 patients with metastatic androgen-independent prostate cancer received perillyl alcohol orally at 1200 mg m^{-2} per dose four times daily, and continued treatment until disease progression or the development of unacceptable toxicity. Six patients received less than one cycle (4 weeks) of drug treatment; four of these withdrew because of drug intolerance. Six patients received more than two cycles of therapy and were considered evaluable for response. The main signs of toxicity were grade 1–2 gastrointestinal intolerance (nausea and vomiting in 60% of patients) and fatigue (47%). No objective responses were seen, and perillyl alcohol failed to provide any objective clinical activity among this patient population [53].

A third Phase II trial of perillyl alcohol was conducted in patients with advanced ovarian cancer. Here, perillyl alcohol was administered orally, four times daily, at a dose of 1200 mg m^{-2}. As in the other two Phase II trials, perillyl alcohol failed to show any signs of extending the time-to-progression in these patients [54].

The metabolites perillic acid and dihydroperillic acid did not affect the expression or isoprenylation of $p21^{ras}$ in MCF-7 breast or DU145 prostate carcinoma cells, at concentrations in excess of those achieved in the plasma of patients after perillyl alcohol treatment [42]. In PC12 cells, perilla alcohol was oxidized by alcohol dehydrogenase and aldehyde dehydrogenase to perillaldehyde and perillic acid. Perillaldehyde and perilla alcohol caused cell apoptosis at concentrations of 200 μM and 500 μM, but perillic acid did not induce apoptosis [49].

References

1 Pan, J.G., Xu, Z.L., Ji, L., Zhao, Z.Z., and Tang, X.J.D (1992) Constituents of essential oils from leaves, stems and fruits of *Perilla frutescens*. *China J. Chin. Mater. Med.*, **17**, 164–165.

2 Honda, G., Koezuka, Y., and Tabata, M.D (1988) Isolation of dill apiol from a chemotype of *Perilla frutescens* as an active principle for prolonging hexobarbital-induced sleep. *Chem. Pharm. Bull. (Tokyo)*, **36**, 3153–3155.

3 Fujita, T., Ohira, K., Miyatake, K., Nakano, Y., and Nakayama, M.D (1995) Inhibitory effects of perillosides A and C, and related monoterpene glucosides on aldose reductase and their structure-activity relationships. *Chem. Pharm. Bull. (Tokyo)*, **43**, 920–926.

4 Makino, T., Ono, T., Muso, E., and Honda, G.D (1998) Inhibitory effect of *Perilla frutescens* and its phenolic constituents on cultured murine mesangial cell proliferation. *Planta Med.*, **64**, 541–545.

5 Mizushina, Y., Takahashi, N., Ogawa, A., Tsurugaya, K., Koshino, H., Takemura, M., Yoshida, S., Matsukage, A., Sugawara, F., and Sakaguchi, K.D (1999) The cyanogenic glucoside, prunasin (D-mandelonitrile β-D-glucoside), is a novel inhibitor of DNA polymerase-β. *J. Biochem. (Tokyo)*, **126**, 430–436.

6 Ryu, J.H., Son, H.J., Lee, S.H., and Sohn, D.H. D (2002) Two neolignans from *Perilla frutescens* and their inhibition of nitric oxide synthase and tumor necrosis factor-α expression in murine macrophage cell line RAW 264.7. *Bioorg. Med. Chem. Lett.*, **12**, 649–651.

7 Wang, J., Tao, S., Xing, Y., and Zhu, Z.D (1997) Pharmacological effects of zisu and baisu. *China J. Chin. Mater. Med.*, **22**, 48–51, 63.

8 Makino, T., Furuta, Y., Wakushima, H., Fujii, H., Saito, K., and Kano, Y.D (2003) Anti-allergic effect of *Perilla frutescens* and its active constituents. *Phytother. Res.*, **17**, 240–243.

9 Ueda, H., Yamazaki, C., and Yamazaki, M.D (2002) Luteolin as an anti-inflammatory and anti-allergic constituent of *Perilla frutescens*. *Biol. Pharm. Bull.*, **25**, 1197–1202

10 Makino, T., Ono, T., Muso, E., and Honda, G.D (2002) Effect of *Perilla frutescens* on nitric oxide production and DNA synthesis in cultured murine vascular smooth muscle cells. *Phytother. Res.*, **16** (Suppl. 1), S19–S23

11 Qiao, S., Li, W., Tsubouchi, R., Haneda, F M., Murakami, K., Takeuchi, F., Nisimoto, Y., and Yoshino, M.D (2005) Rosmarinic acid inhibits the formation of reactive oxygen and nitrogen species in RAW264.7 macrophages. *Free Radic. Res.*, **39**, 995–1003.

12 Kim, M.K., Lee, H.S., Kim, E.J., Won, N.H., Chi, Y.M., Kim, B.C., and Lee, K.W.D (2007) Protective effect of aqueous extract of *Perilla frutescens* on *tert*-butyl hydroperoxide-induced oxidative hepatotoxicity in rats. *Food Chem. Toxicol.*, **45**, 1738–1744.

13 Renzulli, C., Galvano, F., Pierdomenico, L., Speroni, E., and Guerra, M.C.D (2004) Effects of rosmarinic acid against aflatoxin B1 and ochratoxin-A-induced cell damage in a human hepatoma cell line (HepG2). *J. Appl. Toxicol.*, **24**, 289–296.

14 Lin, C.S., Kuo, C.L., Wang, J.P., Cheng, J.S., Huang, Z.W., and Chen, C.F.D (2007) Growth inhibitory and apoptosis inducing effect of *Perilla frutescens* extract on human hepatoma HepG2 cells. *J. Ethnopharmacol.*, **112**, 557–567.

15 Gelb, M.H., Tamanoi, F., Yokoyama, K., Ghomashchi, F., Esson, K., and Gould, M.N.D (1995) The inhibition of protein prenyltransferases by oxygenated metabolites of limonene and perillyl alcohol. *Cancer Lett.*, **91**, 169–175.

16 Wiseman, D.A., Werner, S.R., and Crowell, P. L.D (2007) Cell cycle arrest by the isoprenoids perillyl alcohol, geraniol, and farnesol is mediated by p21(Cip1) and p27(Kip1) in human pancreatic adenocarcinoma cells. *J. Pharmacol. Exp. Ther.*, **320**, 1163–1170.

17 Haag, J.D. and Gould, M.N.D (1994) Mammary carcinoma regression induced by perillyl alcohol, a hydroxylated analog of limonene. *Cancer Chemother. Pharmacol.*, **34**, 477–483.

18 Mills, J.J., Chari, R.S., Boyer, I.J., Gould, M.N., and Jirtle, R.L.D (1995) Induction of apoptosis in liver tumors by the monoterpene perillyl alcohol. *Cancer Res.*, **55**, 979–983.

19 Lantry, L.E., Zhang, Z., Crist, K.A., Wang, Y., Hara, M., Zeeck, A., Lubet, F R.A., and You, M. D (2000) Chemopreventive efficacy of promising farnesyltransferase inhibitors. *Exp. Lung. Res.*, **26**, 773–790.

20 Liston, B.W., Nines, R., Carlton, P.S., Gupta, A., Aziz, R., Frankel, W., and Stoner, G.D.D (2003) Perillyl alcohol as a chemopreventive

agent in N-nitrosomethylbenzylamine-induced rat esophageal tumorigenesis. *Cancer Res.*, **63**, 2399–2403.

21 Burke, Y.D., Ayoubi, A.S., Werner, S.R., McFarland, B.C., Heilman, D.K., Ruggeri, B. A., and Crowell, P.L.D (2002) Effects of the isoprenoids perillyl alcohol and farnesol on apoptosis biomarkers in pancreatic cancer chemoprevention. *Anticancer Res.*, **22** (A), 3127–3134.

22 Reddy, B.S., Wang, C.X., Samaha, H., Lubet, R., Steele, V.E., Kelloff, G.J., and Rao, C.V.D (1997) Chemoprevention of colon carcinogenesis by dietary perillyl alcohol. *Cancer Res.*, **57**, 420–425.

23 Barthelman, M., Chen, W., Gensler, H.L., Huang, C., Dong, Z., and Bowden, G.T.D (1998) Inhibitory effects of perillyl alcohol on UVB-induced murine skin cancer and AP-1 transactivation. *Cancer Res.*, **58**, 711–716.

24 Low-Baselli, A., Huber, W.W., Kafer, M., Bukowska, K., Schulte-Hermann, R., and Grasl-Kraupp, B.D (2000) Failure to demonstrate chemoprevention by the monoterpene perillyl alcohol during early rat hepatocarcinogenesis: a cautionary note. *Carcinogenesis*, **21**, 1869–1877.

25 Yeruva, L., Pierre, K.J., Elegbede, A., Wang, R. C., and Carper, S.W.D (2007) Perillyl alcohol and perillic acid induced cell cycle arrest and apoptosis in non-small cell lung cancer cells. *Cancer Lett.*, **257**, 216–226.

26 Stark, M.J., Burke, Y.D., McKinzie, J.H., Ayoubi, A.S., and Crowell, P.L.D (1995) Chemotherapy of pancreatic cancer with the monoterpene perillyl alcohol. *Cancer Lett.*, **96**, 15–21.

27 Crowell, P.L., Siar Ayoubi, A., and Burke, F Y. D.D (1996) Antitumorigenic effects of limonene and perillyl alcohol against pancreatic and breast cancer. *Adv. Exp. Med. Biol.*, **401**, 131–136.

28 Unlu, S., Mason, C.D., Schachter, M., and Hughes, A.D.D (2000) Perillyl alcohol, an inhibitor of geranyl transferase, induces apoptosis of immortalized human vascular smooth muscle cells *in vitro*. *J. Cardiovasc. Pharmacol.*, **35**, 341–344.

29 Wei, X., Si, M.S., Imagawa, D.K., Ji, P., Tromberg, B.J., and Cahalan, M.D.D (2000) Perillyl alcohol inhibits TCR-mediated [$Ca^{(2+)}$] (i) signaling, alters cell shape and motility, and induces apoptosis in T lymphocytes. *Cell. Immunol.*, **201**, 6–13.

30 Fernandes, J., da Fonseca, C.O., Teixeira, F A., and Gattass, C.R.D (2005) Perillyl alcohol induces apoptosis in human glioblastoma multiforme cells. *Oncol. Rep.*, **13**, 943–947.

31 Sahin, M.B., Perman, S.M., Jenkins, G., and Clark, S.S.D (1999) Perillyl alcohol selectively induces G_0/G_1 arrest and apoptosis in Bcr/Abl-transformed myeloid cell lines. *Leukemia*, **13**, 1581–1591.

32 Yuri, T., Danbara, N., Tsujita-Kyutoku, F M., Kiyozuka, Y., Senzaki, H., Shikata, F N., Kanzaki, H., and Tsubura, A.D (2004) Perillyl alcohol inhibits human breast cancer cell growth in vitro and in vivo. *Breast Cancer Res. Treat.*, **84**, 251–260.

33 Stayrook, K.R., McKinzie, J.H., Burke, Y.D., Burke, Y.A., and Crowell, P.L.D (1997) Induction of the apoptosis-promoting protein Bak by perillyl alcohol in pancreatic ductal adenocarcinoma relative to untransformed ductal epithelial cells. *Carcinogenesis*, **18**, 1655–1658.

34 Shi, W. and Gould, M.N.D (1995) Induction of differentiation in neuro-2A cells by the monoterpene perillyl alcohol. *Cancer Lett.*, **95**, 1–6.

35 Chung, B.H., Lee, H.Y., Lee, J.S., and Young, C.Y.D (2006) Perillyl alcohol inhibits the expression and function of the androgen receptor in human prostate cancer cells. *Cancer Lett.*, **236**, 222–228.

36 Ren, Z. and Gould, M.N.D (1994) Inhibition of ubiquinone and cholesterol synthesis by the monoterpene perillyl alcohol. *Cancer Lett.*, **76**, 185–190.

37 Ren, Z., Elson, C.E., and Gould, M.N.D (1997) Inhibition of type I and type II geranylgeranyl-protein transferases by the monoterpene perillyl alcohol in NIH3T3 cells. *Biochem. Pharmacol.*, **54**, 113–120.

38 Loutrari, H., Hatziapostolou, M., Skouridou, V., Papadimitriou, E., Roussos, C., Kolisis, F. N., and Papapetropoulos, A.D (2004) Perillyl alcohol is an angiogenesis inhibitor. *J. Pharmacol. Exp. Ther.*, **311**, 568–575.

39 Hohl, R.J. and Lewis, K.D (1995) Differential effects of monoterpenes and lovastatin on RAS processing. *J. Biol. Chem.*, **270**, 17508–17512.

40 Karlson, J., Borg-Karlson, A.K., Unelius, F R., Shoshan, M.C., Wilking, N., Ringborg, U., and Linder, S.D (1996) Inhibition of tumor cell growth by monoterpenes *in vitro*: evidence of a Ras-independent mechanism of action. *Anticancer Drugs*, **7**, 422–429.

41 Stayrook, K.R., McKinzie, J.H., Barbhaiya, L. H., and Crowell, P.L.D (1998) Effects of the antitumor agent perillyl alcohol on H-Ras vs.

K-Ras farnesylation and signal transduction in pancreatic cells. *Anticancer Res.*, **18A**, 823–828.

42 Hudes, G.R., Szarka, C.E., Adams, A., Ranganathan, S., McCauley, R.A., Weiner, L. M., Langer, C.J., Litwin, S., Yeslow, G., Halberr, T., Qian, M., and Gallo, J.M.D (2000) Phase I pharmacokinetic trial of perillyl alcohol (NSC 641066) in patients with refractory solid malignancies. *Clin. Cancer Res.*, **6**, 3071–3080.

43 Ariazi, E.A., Satomi, Y., Ellis, M.J., Haag, J.D., Shi, W., Sattler, C.A., and Gould, M.N.D (1999) Activation of the transforming growth factor β signaling pathway and induction of cytostasis and apoptosis in mammary carcinomas treated with the anticancer agent perillyl alcohol. *Cancer Res.*, **59**, 1917–1928.

44 Satomi, Y., Miyamoto, S., and Gould, M.N.D (1999) Induction of AP-1 activity by perillyl alcohol in breast cancer cells. *Carcinogenesis*, **20**, 1957–1961.

45 Bardon, S., Picard, K., and Martel, P.D (1998) Monoterpenes inhibit cell growth, cell cycle progression, and cyclin D1 gene expression in human breast cancer cell lines. *Nutr. Cancer*, **32**, 1–7.

46 Phillips, L.R., Malspeis, L., and Supko, J.G.D (1995) Pharmacokinetics of active drug metabolites after oral administration of perillyl alcohol, an investigational antineoplastic agent, to the dog. *Drug Metab. Dispos.*, **23**, 676–680.

47 Ripple, G.H., Gould, M.N., Arzoomanian, R. Z., Alberti, D., Feierabend, C., Simon, K., Binger, K., Tutsch, K.D., Pomplun, F M., Wahamaki, A., Marnocha, R., Wilding, G., and Bailey, H.H.D (2000) Phase I clinical and pharmacokinetic study of perillyl alcohol administered four times a day. *Clin. Cancer Res.*, **6**, 390–396.

48 Ripple, G.H., Gould, M.N., Stewart, J.A., Tutsch, K.D., Arzoomanian, R.Z., Alberti, D., Feierabend, C., Pomplun, M., Wilding, G., and Bailey, H.H.D (1998) Phase I clinical trial of perillyl alcohol administered daily. *Clin. Cancer Res.*, **4**, 1159–1164.

49 Zhang, Z., Chen, H., Chan, K.K., Budd, F T. , and Ganapathi, R.D (1999) Gas chromatographic-mass spectrometric analysis of perillyl alcohol and metabolites in plasma. *J. Chromatogr. B Biomed. Sci. Appl.*, **28**, 85–95.

50 Boon, P.J., van der Boon, D., and Mulder, F G.J. D (2000) Cytotoxicity and biotransformation of the anticancer drug perillyl alcohol in PC12 cells and in the rat. *Toxicol. Appl. Pharmacol.*, **167**, 55–62.

51 Murren, J.R., Pizzorno, G., DiStasio, S.A., McKeon, A., Peccerillo, K., Gollerkari, F A., McMurray, W., Burtness, B.A., Rutherford, T., Li, X., Ho, P.T., and Sartorelli, A.D (2002) Phase I study of perillyl alcohol in patients with refractory malignancies. *Cancer Biol. Ther.*, **1**, 130–135.

52 Meadows, S.M., Mulkerin, D., Berlin, J., Bailey, H., Kolesar, J., Warren, D., and Thomas, J.P.D (2002) Phase II trial of perillyl alcohol in patients with metastatic colorectal cancer. *Int. J. Gastrointest. Cancer*, **32**, 125–128.

53 Liu, G., Oettel, K., Bailey, H., Ummersen, L.V., Tutsch, K., Staab, M.J., Horvath, D., Alberti, D., Arzoomanian, F R., Rezazadeh, H., McGovern, J., Robinson, E., DeMets, D., and Wilding, G.D (2003) Phase II trial of perillyl alcohol (NSC 641066) administered daily in patients with metastatic androgen independent prostate cancer. *Invest. New Drugs*, **21**, 367–372.

54 Bailey, H.H., Levy, D., Harris, L.S., Schink, J. C., Foss, F., Beatty, P., and Wadler, S.D (2002) A phase II trial of daily perillyl alcohol in patients with advanced ovarian cancer: Eastern Cooperative Oncology Group Study E2E96. *Gynecol. Oncol.*, **85**, 464–468.

Periploca sepium

Cortex Periplocae (Xiangjiapi) is the dried root bark of *Periploca sepium* Bge. (Asclepiadaceae). It is used as an antirheumatic and diuretic agent. The Chinese Pharmacopoeia notes that the overdosage of Cortex Periplocae should be avoided because of its toxicity.

Periploca sepium Bge.

Chemistry

The major active compound of the root bark of *P. sepium* is known to be 4-methoxysalicylaldehyde. The content of 4-methoxysalicylaldehyde in the root bark was found to be about 0.25% [1, 2]. The Chinese Pharmacopoeia requires a qualitative determination of 4-methoxysalicylaldehyde in Cortex Periplocae by thin-layer chromatographic comparison with reference substance, and a quantitative determination by HPLC. The content of 4-methoxysalicylaldehyde in Cortex Periplocae should be not less than 0.2%.

A large number of pregnane glycosides, named periplocosides A, B, C, D, E, F, J, K, L, M, N, and O [3–5], periplosides A, B, and C [6], and neridienone A [7] were isolated from the antitumor fraction of the root bark of *P. sepium*. The periplocosides are glycosides with rather complex sugar moieties. Pregn-5-ene-3β,20(*S*)-diol, pregn-5-ene-3β,16β,20(*R*)-triol, pregn-5-ene-3β,16α,20(*S*)-triol and periplocogenin were determined as aglycones [8]. Periplocoside A was found to be the major component.

The root bark of *P. sepium* was also reported to contain cardiac glycosides as a toxic component, such as periplocin, derived from cardenolide and glycosides derived from xysmalogenin [9].

Pharmacology and Toxicology

The major component periplocoside A was reported to show significant antitumor activity against S180 ascites tumors in mice at daily intraperitoneal (i.p.) doses of 10 and 20 mg kg^{-1} for five days [3]. The pregnane glycosides and three cardenolides were also reported to be differentiation inducers, inducing mouse myeloid leukemia cells into phagocytic cells. The cardenolides showed much higher activities than the pregnane glycosides. In the presence of 1 n*M* actinomycin-D, the activity of steroid glycosides was enhanced against mouse myeloid leukemia cells [10].

Periplocoside E was found to show immunosuppressive effects both *in vitro* and *in vivo*. Experimental results revealed that periplocoside E significantly inhibited the

Handbook of Chinese Medicinal Plants: Chemistry, Pharmacology, Toxicology
Weici Tang and Gerhard Eisenbrand

ISBN: 978-3-527-32226-8

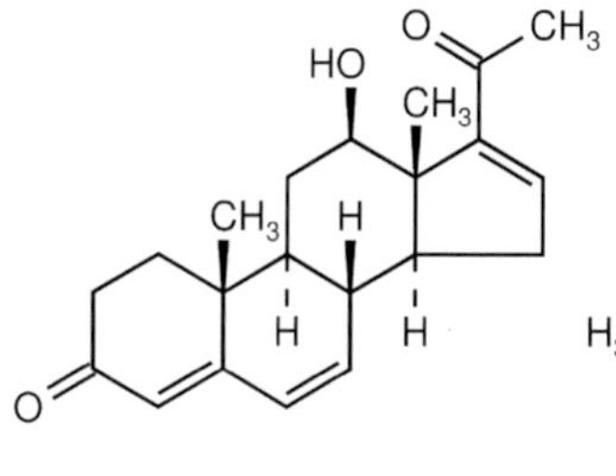

Neridienone A

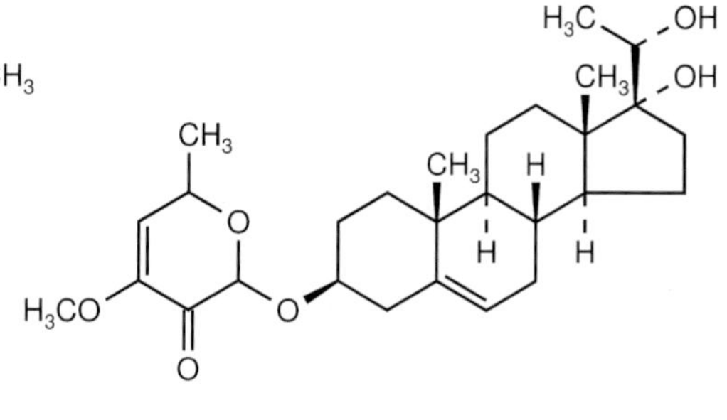

Periplocogenin

Periplocoside A: R =

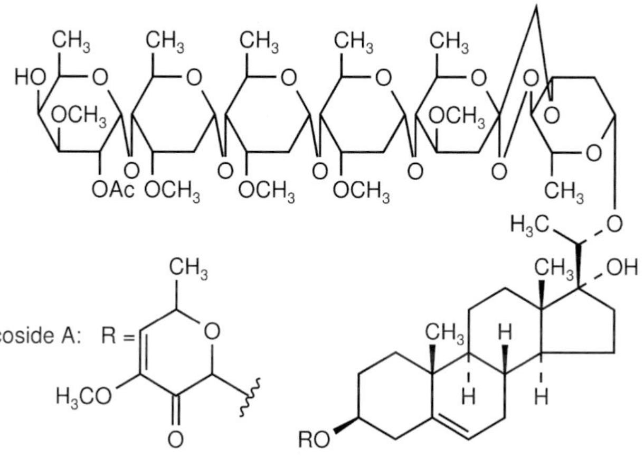

Periplocoside E: R = H

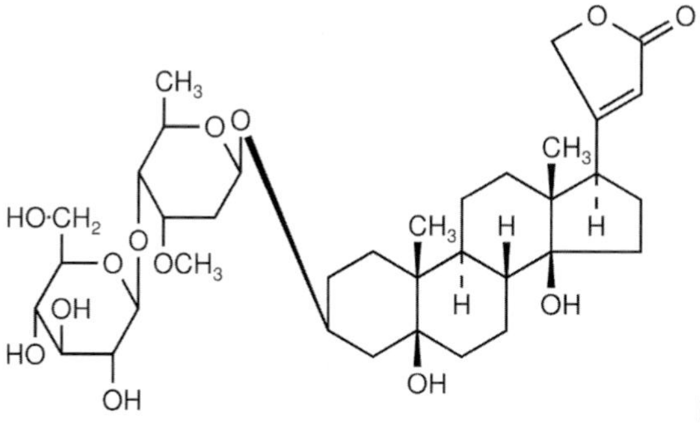

Periplocin

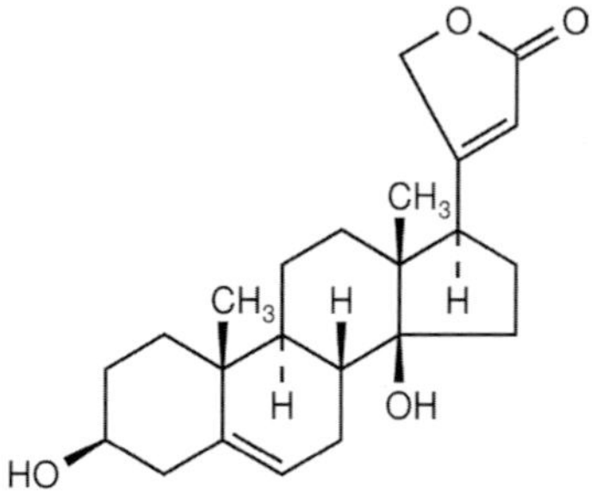

Xysmalogenin

proliferation of splenocytes induced by concanavalin A (Con A) in a mixed lymphocyte culture reaction below cytotoxic concentrations. The administration of periplocoside E suppressed a delayed-type hypersensitivity reaction and also ovalbumin-induced, antigen-specific immune responses in mice. Treatment with periplocoside E suppressed ovalbumin-induced proliferation and production of interleukin-2 (IL-2) and interferon-γ (IFN-γ) from splenocytes. Further studies showed that periplocoside E inhibited anti-CD3-induced primary T-cell proliferation, activation of IL-2Rα (CD25) expression, and the production of IFN-γ and IL-2, also at the transcriptional level. Periplocoside E was highly specific, and significantly inhibited the activation of extracellular signal-regulated kinase (ERK) and Jun N-terminal kinase (JNK). These results suggested periplocoside E to be an immunosuppressive compound, which directly inhibited T-cell activation [11].

Periplocoside E was also found to inhibit primary T-cell activation. The effect and mechanisms of periplocoside E on central nervous system (CNS) demyelination have been studied in experimental allergic encephalomyelitis in mice. C57BL/6 mice immunized with myelin oligodendrocyte glycoprotein were treated with periplocoside E following immunization and continuing throughout the study. Periplocoside E reduced the incidence and severity of allergic encephalomyelitis. The therapeutic effect of periplocoside E was found to be associated with reduced mononuclear cell infiltration and CNS inflammation [12]. Periplocoside E was further reported to inhibit the growth and IL-6 production of human rheumatoid arthritis-derived fibroblast-like cells [13].

References

1 Lu, G.B. and Zhang, M.H. (1988) Determination of 4-methoxysalicylaldehyde by HPLC in Chinese silkvine (*Periploca sepium*). *Chin. Trad. Herbal Drugs*, **19**, 16–18.

2 Huang, Q., Zhang, H.Y., Li, Y.G., Cheng, G.Y., Zhang, B., and Xi, H.S. (1995) Determination of 4-methoxysalicylic aldehyde in the *Periploca sepium* Bge. bark or root by thin layer chromatogram scanner. *Fenxi Huaxue*, **23**, 236.

3 Itokawa, H., Xu, J., Takeya, K., Watanabe, K., and Shoji, J. (1988) Studies on chemical constituents of antitumor fraction from *Periploca sepium*. II. Structures of new pregnane glycosides, periplocosides A, B and C. *Chem. Pharm. Bull. (Tokyo)*, **36**, 982–987.

4 Itokawa, H., Xu, J.P., and Takeya, K. (1988) Studies on chemical constituents of antitumor fraction from *Periploca sepium*. IV. Structures of new pregnane glycosides, periplocosides D, E, L, and M. *Chem. Pharm. Bull. (Tokyo)*, **36**, 2084–2089.

5 Itokawa, H., Xu, J.P., and Takeya, K. (1988) Studies on chemical constituents of antitumor fraction from *Periploca sepium*. V. Structures of new pregnane glycosides, periplocosides J, K, F and O. *Chem. Pharm. Bull. (Tokyo)*, **36**, 4441–4446.

6 Oshima, Y., Hirota, T., and Hikino, H. (1987) Validity of Oriental medicines. Part 125. Immunomodulating drugs. Part 1. Periplosides A, B and C, steroidal glycosides of *Periploca sepium* root bark. *Heterocycles*, **26**, 2093–2098.

7 Itokawa, H., Xu, J.P., and Takeya, K. (1987) Studies on chemical constituents of antitumor fraction from *Periploca sepium* Bge. I. *Chem. Pharm. Bull. (Tokyo)*, **35**, 4524–4529.

8 Itokawa, H., Xu, J.P., and Takeya, K. (1988) Pregnane glycosides from an antitumor fraction of *Periploca sepium*. *Phytochemistry*, **27**, 1173–1179.

9 Xu, J.P., Takeya, K., and Itokawa, H. (1990) Pregnanes and cardenolides from *Periploca sepium*. *Phytochemistry*, **29**, 344–346.

10 Umehara, K., Sumii, N., Satoh, H., Miyase, T., Kuroyanagi, M., and Ueno, A. (1995) Studies on differentiation inducers. V. Steroid glycosides from periplocae radicis cortex. *Chem. Pharm. Bull. (Tokyo)*, **43**, 1565–1568.

11 Zhu, Y.N., Zhao, W.M., Yang, Y.F., Liu, Q.F., Zhou, Y., Tian, J., Ni, J., Fu, Y.F., Zhong, X.G., Tang, W., Zhou, R., He, P.L., Li, X.Y., and Zuo, J.P. (2006) Periplocoside E, an effective compound from *Periploca sepium* Bge, inhibited T-cell activation *in vitro* and *in vivo*. *J. Pharmacol. Exp. Ther.*, **316**, 662–669.

12 Zhu, Y.N., Zhong, X.G., Feng, J.Q., Yang, Y. F., Fu, Y.F., Ni, J., Liu, Q.F., Tang, W., Zhao, W.M., and Zuo, J.P. (2006) Periplocoside E

inhibits experimental allergic encephalomyelitis by suppressing interleukin 12-dependent CCR5 expression and interferon-γ-dependent CXCR3 expression in T lymphocytes. *J. Pharmacol. Exp. Ther.*, **318**, 1153–1162.

13 Tokiwa, T., Harada, K., Matsumura, T., and Tukiyama, T. (2004) Oriental medicinal herb, *Periploca sepium*, extract inhibits growth and IL-6 production of human synovial fibroblast-like cells. *Biol. Pharm. Bull.*, **27**, 1691–1693.

Peucedanum praeruptorum

Radix Peucedani (Qianhu) is the dried root of *Peucedanum praeruptorum* Dunn (Apiaceae). It is used as an expectorant and mucolytic agent.

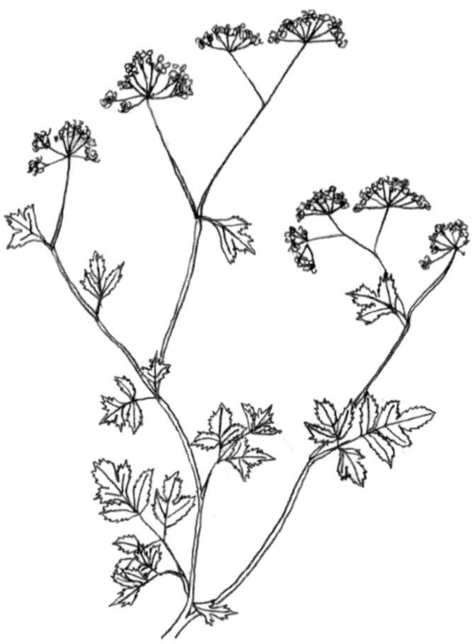

Peucedanum praeruptorum Dunn.

Chemistry

The root of *P. praeruptorum* is known to contain furocoumarins and furocoumarin glycosides, identified as psoralen, bergapten, xanthotoxin, praeroside I, isorutarin, rutarin, marmesinin, qianhucoumarin G, and apterin, a furocoumarin glycoside derived from xanthotoxin [1]. Coumarin glycosides from the root of *P. praeruptorum* were identified as scopoline, skimmin, and apiosylskimmin [1].

Pyranocoumarins from the root of *P. praeruptorum* are mainly derived from khellactone [2]. The two hydroxy groups of khellactone were either esterified or glycosylated. Khellactone esters from the root *P. praeruptorum* were identified as praeruptorins A, B, C, E, and F, pteryxin, qianhucoumarins A–D, F, H, and I [3–5]. The Chinese Pharmacopoeia requires a qualitative determination of praeruptorin A in Radix Peucedani by thin-layer chromatographic comparison with reference substance, and a quantitative determination by HPLC. The total content of praeruptorin A in Radix Peucedani should be not less than 0.9%.

Khellactone glycosides from the root of *P. praeruptorum* were identified as praerosides II, III, IV, and V. The coumarin content, including furocoumarin and pyranocoumarin in the root of *P. praeruptorum*, was determined to 0.6% [6].

In addition, a phenylpropanoid glycoside baihuaqianhuoside [7] and aliphatic dienediyne falcarindiol [8] were isolated from the root of *P. praeruptorum*. Decuroside VI and decursidate were isolated from the root of *P. decursivum*; they were structurally elucidated as 6′-*O*-crotonyl-nodakenin and 2-(4′-hydroxyphenyl)-glycol ferulate, respectively [9].

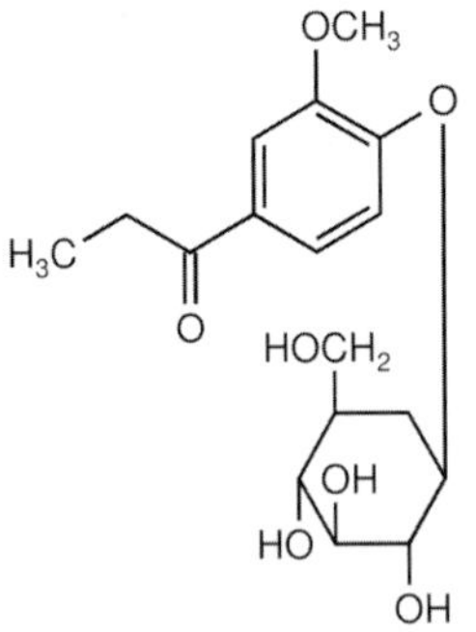

Baihuaqianhuoside

Pharmacology and Toxicology

An extract of the root of *P. praeruptorum* containing praeruptorins A and B did not

Handbook of Chinese Medicinal Plants: Chemistry, Pharmacology, Toxicology
Weici Tang and Gerhard Eisenbrand

ISBN: 978-3-527-32226-8

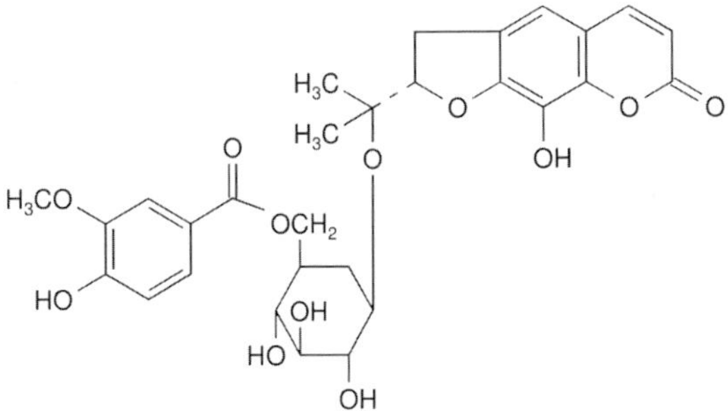

Praeroside I

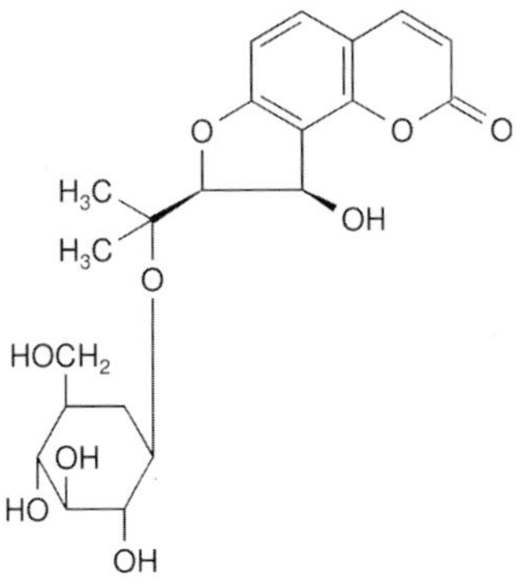

Apterin

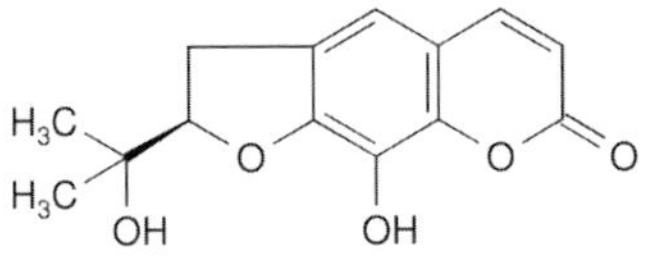

Qianhucoumarin G

show any gross behavioral effects or acute toxicity after oral administration in mice. Delayed mortality was observed with the extract and praeruptorin A only after intraperitoneal (i.p.) administration of high doses up to $1\,g\,kg^{-1}$ [10].

The ether-soluble fraction of the root of *P. praeruptorum* containing praeruptorins A, B, E, and qianhucoumarin E showed spasmolytic activity on the smooth muscle of guinea pig ileum and taenia coli contracted by acetylcholine and histamine [11]. The extract was also active in inhibiting the contractions of isolated rabbit trachea smooth muscles induced by acetylcholine and potassium chloride [12]. Praeruptorin A was the most effective compound in antagonizing acetylcholine action in the small intestine, via a noncompetitive mechanism [13]. Praeruptorin A also showed highly potent and concentration-dependent antagonistic activity against

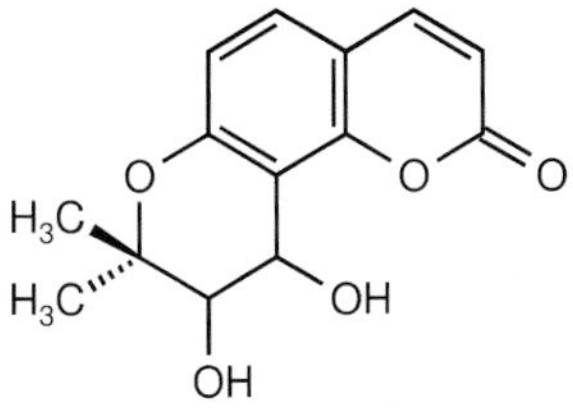

Khellactone

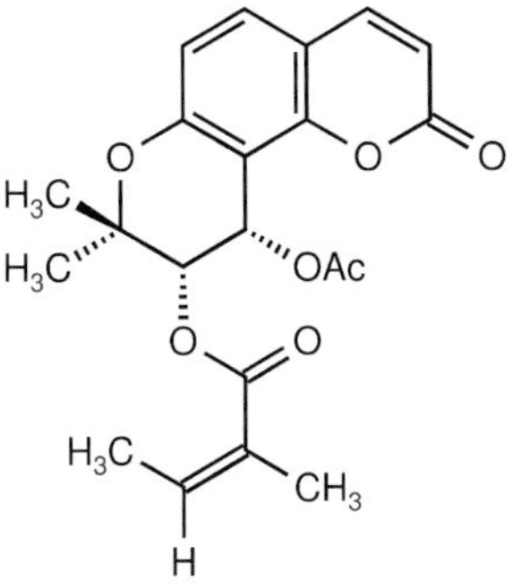

Praeruptorin A

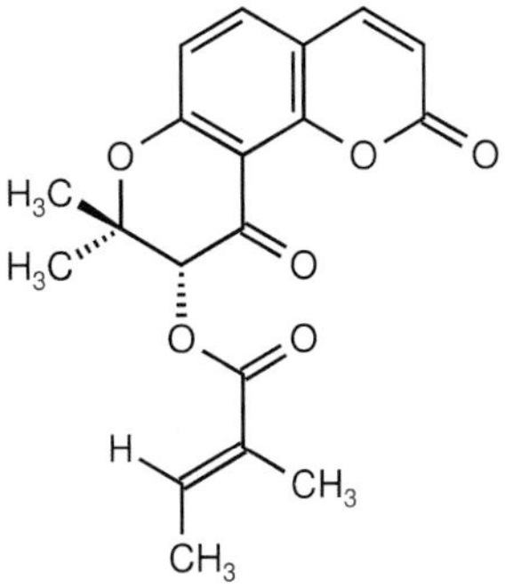

Qianhucoumarin E

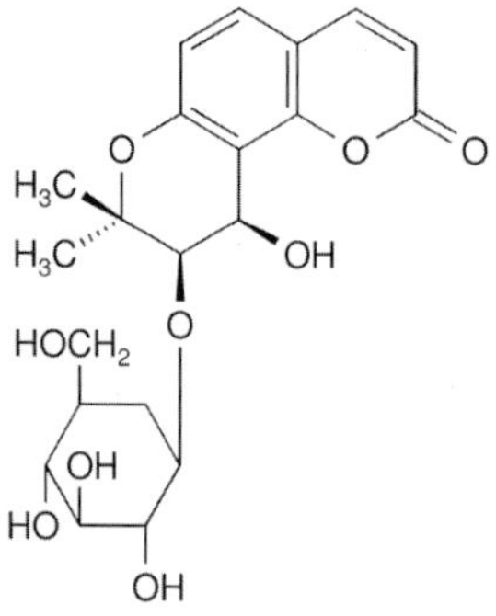

Praeroside II

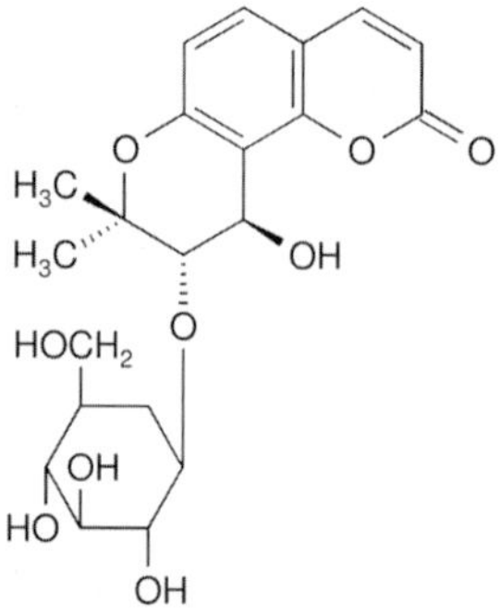

Praeroside III

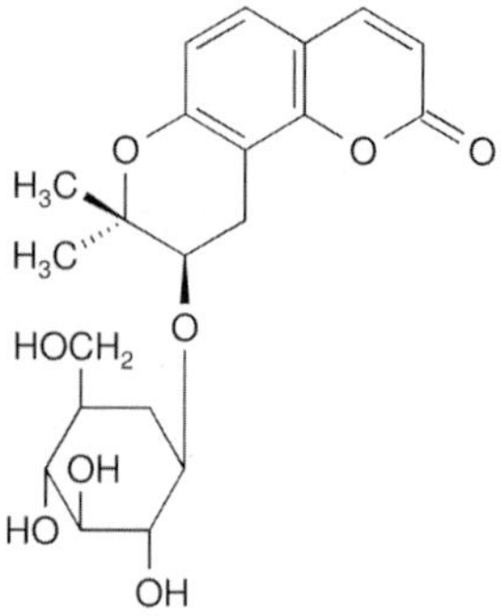

Praeroside IV

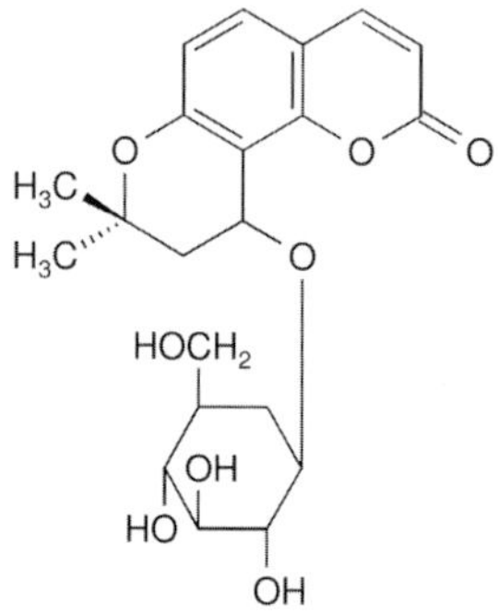

Praeroside V

histamine- and Ca^{2+}-induced contractions in isolated guinea pig smooth muscle [14]. It exhibited a Ca^{2+} channel-blocking effect in single ventricular cells of guinea pig [15].

Praeruptorins A and B inhibited isolated guinea pig ileum contractions induced by histamine or leukotriene D_4 [16]. Praeruptorin C and praeruptorin E decreased the maximum contractile effect of Ca^{2+} in potassium-depolarized swine coronary strips, in potassium-depolarized guinea pig left atria, and in cultured rat heart cells. The relaxation induced by praeruptorin C and praeruptorin E was concentration-dependent [17]. Praeruptorin C was also found to inhibit the spontaneous contraction and action potential in cultured myocardial cells of neonatal rats [18]. Long-term oral administration of the extract of *P. praeruptorum* to ventricular hypertrophic hypertensive rats resulted in a significant attenuation of the impaired left ventricular compliance, and increases in myocardial mitochondrial and arterial Ca^{2+} levels [19].

Praeruptorin A was further reported to significantly increase the coronary blood flow, to decrease the aortic pressure and systemic vascular resistance, and to increase the heart rate in anesthetized open-chest dogs [20]. Praeruptorin A also showed a cardioprotective action and improved the regional myocardial dysfunction following a brief period of ischemia in dogs [21].

Praeruptorin A and praeruptorin B each exhibited an antagonistic effect specifically on rabbit platelet aggregation induced by platelet activating factor (PAF). Other khellactone esters were also found to inhibit platelet aggregation *in vitro* [22]. Among the decurosides, decuroside III and decuroside IV showed the strongest inhibitory activities against the aggregation of human platelets [2].

In rats, an aqueous extract of *P. praeruptorum* blocked the arrhythmia induced either by $BaCl_2$ [23] or by ligation of the left coronary artery [24]. Praeruptorin C reduced the automatic rhythm and positive chronotropic effects of $CaCl_2$ in the right atrium. A modification of myocardial compliance during a 2-week period of oral praeruptorin C in patients with hypertrophic cardiomyopathy was also observed [25].

A methanol extract of *P. praeruptorum* was reported to show an antimutagenic activity against furfurylamide-induced umu gene expression of the SOS response in *Salmonella typhimurium* TA1535/pSK1002, with falcarindiol as the active component. Falcarindiol and its diacetate were also antimutagenic, counteracting the mutagenic effects of the food-borne heterocyclic amine Trp-P-1 in the Ames test using *S. typhimurium* TA100 with an S-9 metabolizing system [26]. The pyranocoumarins from *P. praeruptorum* also inhibited the promotion of two-stage skin tumor formation promoted by 12-*O*-tetradecanoyl phorbol-13-acetate (TPA) in mice after initiation with 7,12-dimethylbenz[*a*]anthracene (DMBA) [27].

Praeruptorin A, at concentrations of 10–30 $pg\,ml^{-1}$, induced apoptosis in HL-60 cells in a concentration-dependent manner, caused a loss in mitochondrial membrane potential and release of cytochrome c, increased total cellular and mitochondrial Bax protein, and stimulated an increase in caspase-dependent Bcl-2 cleavage; these findings strongly suggested an activation of the mitochondrial apoptotic pathway [28]. Praeruptorin A was further found capable of reverting P-glycoprotein-mediated multidrug resistance in cancer cells [29]. Praeruptorin A induced apoptotic cell death in drug-sensitive KB-3-1 cancer cells and multidrug resistant KB-V1 cancer cells, with IC_{50} values of approximately 42 and 17 μM, respectively. Strong synergistic interactions were demonstrated when praeruptorin A was combined with common antitumor drugs, including doxorubicin, paclitaxel, puromycin or vincristine, in the multidrug resistant KB-V1 cell line, but not in drug-sensitive KB-3-1 cells. In KB-V1 cells, praeruptorin A increased doxorubicin accumulation by about 25% after a 6 h incubation period, and after 24 h had downregulated the expression of P-glycoprotein at both protein and mRNA levels. Pyranocoumarins also transiently reduced the cellular ATP content of KB-V1 cells [30].

References

1 Takata, M., Shibata, S., and Okuyama, T. (1990) Structures of glycosides of Baihuaqianhu. *Planta Med.*, **56**, 133.

2 Ye, W.P., Guo, Z.D., and Liu, G.H. (1996) Extraction and isolation of an effective coumarin derivative from a Chinese drug *Peucedanum praeruptorum* Dunn. *J. China Med. Univ.*, **25**, 351, 354.

3 Kong, L.Y., Li, X., Pei, Y.H., and Zhu, T.R. (1994) Isolation and structural elucidation of qianhucoumarin D and qianhucoumarin E from *Peucedanum praeruptorum. Acta Pharm. Sin.*, **29**, 49–54.

4 Kong, L.Y., Li, Y., Min, Z.D., Li, X., and Zhu, T.R. (1996) Coumarins from *Peucedanum praeruptorum. Phytochemistry*, **41**, 1423–1426.

5 Kong, L.Y., Min, Z.D., Li, Y., Li, X., and Pei, Y. H. (1996) Qianhucoumarin I from *Peucedanum praeruptorum. Phytochemistry*, **42**, 1689–1691.

6 Takata, M., Okuyama, T., and Shibata, S. (1988) Studies on coumarins of a Chinese drug. Qian-Hu: VIII. Structures of new coumarin glycosides of Bai-Hua-Qian-Hu. *Planta Med.*, **54**, 323–327.

7 Kong, L.Y., Li, X., Pei, Y.H., Yu, R.M., Min, Z. D., and Zhu, T.R. (1994) Isolation and structure elucidation of baihuaqianhuoside and Pd-C-I from *Peucedanum praeruptorum*. *Acta Pharm. Sin.*, **29**, 276–280.

8 Miyazawa, M., Shimamura, H., Bhuva, R.C., Nakamura, S., and Kameoka, H. (1996) Antimutagenic activity of falcarindiol from *Peucedanum praeruptorum*. *J. Agric. Food Chem.*, **44**, 3444–3448.

9 Yao, N.H., Kong, L.Y., and Niwa, M. (2001) Two new compounds from *Peucedanum decursivum*. *J. Asian Nat. Prod. Res.*, **3**, 1–7.

10 Lu, M., Nicoletti, M., Bettinelli, L., and Mazzanti, G. (2001) Isolation of praeruptorins A and B from *Peucedanum praeruptorum* Dunn. and their general pharmacological evaluation in comparison with extracts of the drug. *Il Farmaco*, **56**, 417–420.

11 Okuyama, T. and Shibata, S. (1981) Studies on coumarins of a Chinese drug "Qian-Hu". *Planta Med.*, **42**, 89–96.

12 Jin, X., Zhang, X.H., and Zhao, N.C. (1994) Effects of petroleum ether extract of *Peucedanum praeruptorum* Dunn on rabbit trachea smooth muscles. *China J. Chin. Mater. Med.*, **19**, 365–367.

13 Kozawa, T., Sakai, K., Uchida, M., Okuyama, T., and Shibata, S. (1981) Calcium antagonistic action of a coumarin isolated from "Qian-Hu", a Chinese traditional medicine. *J. Pharm. Pharmacol.*, **33**, 317–320.

14 Takeuchi, N., Kasama, T., Aida, Y., Oki, J., Maruyama, Izumi, W., Okuyama, T., and Tobinaga, S. (1991) Pharmacological activities of the prenylcoumarins, developed from folk usage as a medicine of *Peucedanum japonicum* Thunb. *Chem. Pharm. Bull. (Tokyo)*, **39**, 1415–1421.

15 Li, J.M., Chang, T.H., Sun, X.D., Hao, L.Y., Wang, Y.P., Yu, Y.F., and Zhang, K.Y. (1994) Effect of D,L-praeruptorin A on calcium current in ventricular cells of guinea pig. *Acta Pharmacol. Sin.*, **15**, 525–527.

16 Aida, Y., Kasama, T., Takeuchi, N., and Tobinaga, S. (1995) The antagonistic effects of khellactones on platelet-activating factor, histamine, and leukotriene D_4. *Chem. Pharm. Bull. (Tokyo)*, **43**, 859–867.

17 Wu, X., Shi, C.Z., and Wu, X.D. (1993) Effects of praeruptorin C on cytosolic free calcium in cultured rat heart cells. *Acta Pharm. Sin.*, **28**, 728–731.

18 Wang, H.X., Tao, L., and Rao, M.R. (1995) Effects of praeruptorin C on spontaneous contraction and action potential in cultured myocardial cells of neonatal rats. *Acta Pharm. Sin.*, **30**, 812–817.

19 Ji, Y. and Rao, M.R. (1996) Effects of alcohol extract of white flower hogfennel (*Peucedanum praeruptorum*) on left ventricular hypertrophied hypertensive rats. *Chin. Trad. Herbal Drugs*, **27**, 413–416.

20 Chang, T.H., Adachi, H., Okuyama, T., Mori, N., Saito, I., Zhang, K.Y., Sun, X.D., and Li, J.M. (1994) Effects of 3′-angeloyloxy-4′-acetoxy-3′,4′-dihydroseselin on cardiohemodynamics in anesthetized dogs. *Acta Pharmacol. Sin.*, **15**, 507–510.

21 Chang, T.H., Adachi, H., Okuyama, T., and Zhang, K.Y. (1994) Effects of 3′-angeloyloxy-4′-acetoxy-3′,4′-dihydroseselin on myocardial dysfunction after a brief ischemia in anesthetized dogs. *Acta Pharmacol. Sin.*, **15**, 388–391.

22 Chen, I.S., Chang, C.T., Sheen, W.S., Teng, C.M., Tsai, I.L., Duh, C.Y., and Ko, F.N. (1996) Coumarins and antiplatelet aggregation constituents from Formosan *Peucedanum japonicum*. *Phytochemistry*, **41**, 525–530.

23 Chang, T.H., Wang, Y.P., Yu, Y.F., Li, J.W., Zhang, K.Y., Liu, G.H., and Ye, W.P. (1991) Antiarrhythmic effects of *Peucedanum praeruptorum*. I. Effects on arrhythmia of rats induced by barium chloride. *J. China Med. Univ.*, **20**, 337–339, 343.

24 Wang, Y.P., Chang, T.H., Yu, Y.F., Li, J.M., Zhang, K.Y., Ye, W.P., and Liu, G.H. (1991) Experimental effects of "Baihua Qianhu", a Chinese traditional herb on the prevention and treatment of arrhythmia. II. Effects on the rat arrhythmia induced by the ligation of the left coronary artery. *J. China Med. Univ.*, **20**, 420–422.

25 Wu, X. and Rao, M.R. (1990) Effects of praeruptorin C on isolated guinea pig atrium and myocardial compliance in patients. *Acta Pharmacol. Sin.*, **11**, 235–238.

26 Miyazawa, M., Shimamura, H., Bhuva, R.C., Nakamura, S., and Kameoka, H. (1996) Antimutagenic activity of falcarindiol from *Peucedanum praeruptorum*. *J. Agric. Food Chem.*, **44**, 3444–3448.

27 Nishino, H., Okuyama, T., Takata, M., Shibata, S., Tokuda, H., Takayasu, J., Hasegawa, T., Nishino, A., Ueyama, H., and Iwashima, A. (1990) Studies on the anti-tumor-promoting activity of naturally occurring substances. IV. Pd-II [(+)anomalin, (+)praeruptorin B], a seselin-type coumarin, inhibits the promotion of skin tumor formation by 12-*O*-

tetradecanoylphorbol-13-acetate in 7,12-dimethyl benz[*a*]anthracene-initiated mice. *Carcinogenesis*, **11**, 1557–1561.

28 Fong, W.F., Zhang, J.X., Wu, J.Y., Tse, K.W., Wang, C., Cheung, H.Y., and Yang, M.S. (2004) Pyranocoumarin (+/−)-4′-*O*-acetyl-3′-*O*-angeloyl-*cis*-khellactone induces mitochondrial-dependent apoptosis in HL-60 cells. *Planta Med.*, **70**, 489–495.

29 Shen, X., Chen, G., Zhu, G., and Fong, W.F. (2006) (+/−)-3′-*O*, 4′-*O*-dicinnamoyl-cis-khellactone, a derivative of (+/−)-praeruptorin A, reverses P-glycoprotein mediated multidrug resistance in cancer cells. *Bioorg. Med. Chem.*, **14**, 7138–7145.

30 Wu, J.Y., Fong, W.F., Zhang, J.X., Leung, C.H., Kwong, H.L., Yang, M.S., Li, D., and Cheung, H.Y. (2003) Reversal of multidrug resistance in cancer cells by pyranocoumarins isolated from Radix Peucedani. *Eur. J. Pharmacol.*, **473**, 9–17.

Phellodendron amurense and *Phellodendron chinense*

Cortex Phellodendri Amurensis (Guanhuangbo) is the dried stem bark of *Phellodendron amurense* Rupr. (Rutaceae). It is used as an antiphlogistic, antibacterial, anti-inflammatory agent for the treatment of diarrhea, icterus, ulcer, carbuncle, and eczema.

Phellodendron amurense Rupr.

Cortex Phellodendri Chinensis (Huangbo) is the dried stem bark of *Phellodendron chinense* Schneid. It is mainly used for the treatment of diarrhea, icterus, ulcer, and eczema.

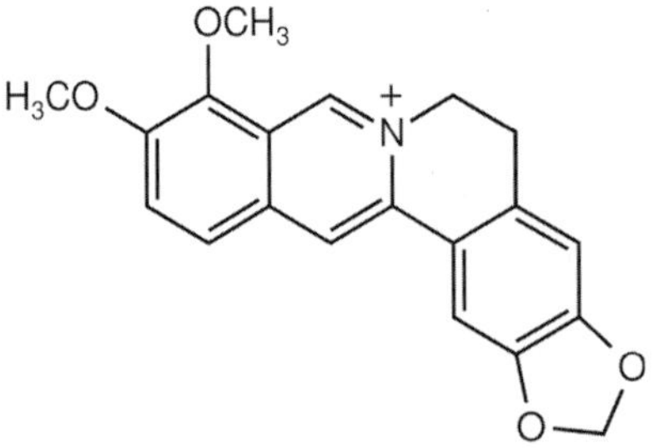

Berberine

Chemistry

The major chemical constituents in the bark of *P. amurense* and *P. chinense* are alkaloids of the protoberberine type. Protoberberine-type alkaloids isolated from the bark of *P. amurense* were identified as berberine, palmatine [1], phellodendrine, and jatrorrhizine. Other alkaloids were identified as magnoflorine and candicine [2], a quaternary ammonium compound. The Chinese Pharmacopoeia requires a qualitative determination of berberine in Cortex Phellodendri Amurensis and in Cortex Phellodendri Chinensis by thin-layer chromatographic comparison with reference substance, and a quantitative determination by HPLC. The content of berberine in Cortex Phellodendri Amurensis should be not less than 0.6%, and in Cortex Phellodendri Chinensis not less than 3.0%, calculated as berberine hydrochloride.

Berberine, jatrorrhizine, phellodendrine, and candicine were also isolated from the root bark. Fruits and seeds contained berberine and jatrorrhizine. Berberine was isolated from the wood. Cathin-6-one was also isolated from the root bark [3]. Indolopyridoquinazoline type alkaloids, 7-hydroxyrutaecarpine and 7,8-dihydroxyrutaecarpine were isolated from the callus tissue from the stems of *P. amurense* [4]. The bark of *P. chinense* contained much more berberine than that of *P. amurense*, but the latter had higher contents of palmatine and jatrorrhizine [5]. Berberine was produced from the bark of *P. amurense* with a yield of 1.7% [6]. Limonoids from the bark of *P. amurense* were identified as limonin, obakunone, and nomilin [7].

Handbook of Chinese Medicinal Plants: Chemistry, Pharmacology, Toxicology
Weici Tang and Gerhard Eisenbrand

ISBN: 978-3-527-32226-8

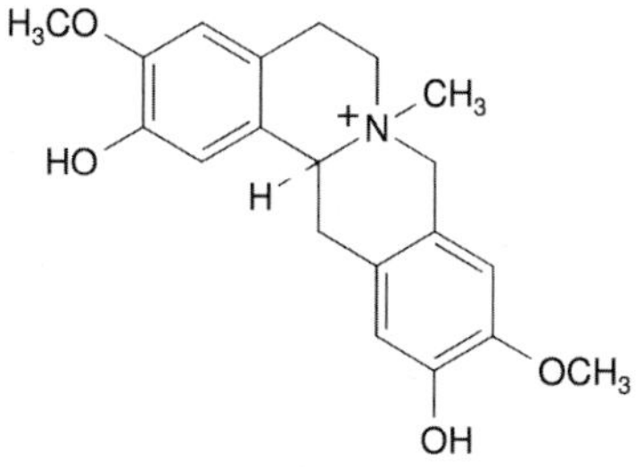

Phellodendrine

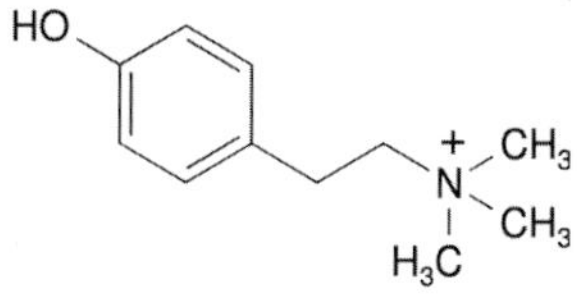

Candicine

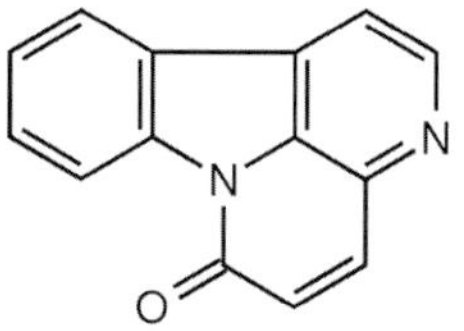

Canthin-6-one

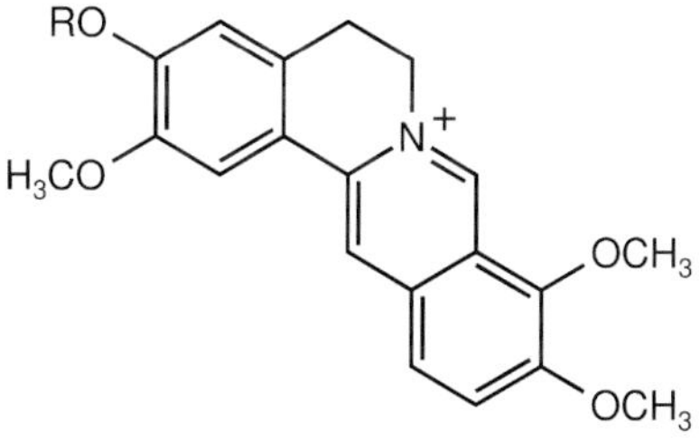

Palmatine R = CH_3
Jatrorrhizine R = H

Pharmacology and Toxicology

The bark of *P. amurense* and *P. chinense* was used in the traditional Chinese medicine for the treatment of gastric ulcer and icterus, and also as an anti-inflammatory agent. The berberine-free fraction of the extract was found to exert anti-inflammatory effects and to inhibit gastric ulcer in rats induced by ethanol, aspirin, or by pylorus-ligation and stress ulcer in mice. In addition, gastric acid secretion was found to be significantly reduced in pylorus-ligated rats by subcutaneous (s.c.) or intraduodenal administration of the fraction, but not by oral administration. These findings suggested that the suppression of ulcer formation might be due to a cytoprotective effect in combination with the reduction of gastric acid secretion [8]. A hepatoprotective effect of the bark extract against hepatotoxicity in rats induced by carbon tetrachloride (CCl_4) was also reported. Serum glutamic-pyruvic transaminase (SGPT) and serum glutamic-oxaloacetic transaminase (SGOT) activities were each significantly reduced after treatment [9].

The extract of the bark of *P. amurense* was found to inhibit the induction of various delayed-type hypersensitivities and local graft-versus-host (GVH) reactions, but did not affect the humoral immune responses or the effector phase of delayed-type hypersensitivity in mice [10]. Phellodendrine was found to be one of the active components, suppressing local semisyngeneic GVH reactions and systemic allogeneic GVH reactions in X-irradiated recipient mice. Phellodendrine also suppressed the induction phase of delayed-type hypersensitivity in mice induced by sheep red blood cells and delayed-type hypersensitivity in guinea pigs induced by tuberculin, but did not suppress the effector phase of these reactions. Moreover, phellodendrine – unlike prednisolone and cyclophosphamide – did not affect antibody production in mice against sheep red blood cells. It was considered that phellodendrine might represent a new type of immunosuppressor against the cellular immune response [11]. Phellodendrine was further reported to exhibit antinephritic efficacy in rats; the antinephritic mechanism might be due to the ability of phellodendrine to inhibit the proliferation or migration of macrophages and cytotoxic T lymphocytes in the glomeruli [12].

Aqueous and 80% alcohol extracts of the bark of *P. amurense* were reported to scavenge superoxide radicals generated through the hypoxanthine–oxidase system, hydroxyl radicals generated through the Fenton reaction, and also to inhibit lipid peroxidation induced by hydroxyl radical generation [13]. The oral or s.c. administration of canthin-6-one, at a daily dose of $5\,mg\,kg^{-1}$, to mice infected either acutely or chronically with *Trypanosoma cruzi* led to a significant trypanocidal activity. At 70 days post-infection, the serological response in the acute model was significantly improved in canthin-6-one-treated mice. Canthin-6-one treatment also exhibited a therapeutic effect in chronically infected mice, with a survival rate of 80–100% [14].

An extract of the bark of *P. amurense* suppressed the proliferation of LNCaP and PC-3 prostate cancer cells, and inhibited tumor growth in mice. The treatment of LNCaP cells with the extract markedly reduced the activity of cyclo-oxygenase-2 (COX-2) induced by tumor necrosis factor-α (TNF-α). The extract also reduced the expression and activity of COX-2 in PC-3 cells that expressed high constitutive levels of COX-2. An immunohistochemical analysis of human prostate tumors showed an increased expression of the cAMP response element (CRE) binding protein and DNA-binding activity in tumors. The activation of COX-2 mediated by the CRE binding protein as a potential signaling pathway in prostate cancer might be blocked by the bark extract [15]. The limonoids and alkaloids from the bark of *P. amurense* have been tested for cytotoxicity towards cancer cells and the reversal of multidrug resistance. Oxyberberine was found to be active in inhibiting a panel of human cancer cell lines, with ID_{50} values ranging from 0.30 to $3.0\,\mu g\,ml^{-1}$. Obacunone was found to significantly inhibit P-glycoprotein-mediated multidrug resistance in MES-SA/DX5 and HCT15 cancer cells, with ID_{50} values of 28 and $1.1\,ng\,ml^{-1}$, respectively [16]. Canthin-6-one also exhibited significant cytotoxicity against A-549 human lung cancer and MCF-7 human breast cancer cells [17].

References

1 Zhou, H. and Gu, Y. (1995) Determination of berberine in *Phellodendron chinense* Schneid and its processed products by TLC (thin layer chromatography) densitometry. *China J. Chin. Mater. Med.*, **20**, 405–407.

2 Liu, W.Z., Peng, W.B., and Yang, C.H. (1991) Berberine-electrochemical detector for the determination of berberine-type alkaloids in various Chinese patent medicines by flow injection analysis. *Acta Pharm. Sin.*, **26**, 315–319.

3 Sheridan, H. and Bhandari, P. (1992) Cathin-6-one from the root bark of *Phellodendron chinense*. *Planta Med.*, **58**, 299.

4 Ikuta, A., Urabe, H., and Nakamura, T. (1998) A new indolopyridoquinazoline-type alkaloid from *Phellodendron amurense* callus tissues. *J. Nat. Prod.*, **61**, 1012–1014.

5 Wang, Y.M., Zhao, L.B., Lin, S.L., Dong, S.S., and An, D.K. (1989) Determination of berberine and palmatine in cortex phellodendron and Chinese patent medicines by HPLC. *Acta Pharm. Sin.*, **24**, 275–279.

6 He, R. and Chen, Y. (1982) Isolation of berberine from Huang Bo (*Phellodendron amurense* Rupr) by the aqueous acid-lime method. *Chin. Trad. Herbal Drugs*, **13**, 26.

7 Min, Y.D., Kwon, H.C., Yang, M.C., Lee, K.H., Choi, S.U., and Lee, K.R. (2007) Isolation of limonoids and alkaloids from *Phellodendron amurense* and their multidrug resistance (MDR) reversal activity. *Arch. Pharm. Res.*, **30**, 58–63.

8 Uchiyama, T., Kamikawa, H., and Ogita, Z. (1989) Anti-ulcer effect of extract from

Phellodendri cortex. *Yakugaku Zasshi*, **109**, 672–676.

9 Chiu, H.F., Lin, C.C., Yang, C.C., and Yang, F. (1988) The pharmacological and pathological studies on several hepatic protective crude drugs from Taiwan. *Am. J. Chin. Med.*, **16**, 127–137.

10 Mori, H., Fuchigami, M., Inoue, N., Nagai, H., Koda, A., and Nishioka, I. (1994) Principle of the bark of *Phellodendron amurense* to suppress the cellular immune response. *Planta Med.*, **60**, 445–449.

11 Mori, H., Fuchigami, M., Inoue, N., Nagai, H., Koda, A., Nishioka, I., and Meguro, K. (1995) Principle of the bark of *Phellodendron amurense* to suppress the cellular immune response: effect of phellodendrine on cellular and humoral immune responses. *Planta Med.*, **61**, 45–49.

12 Hattori, T., Furuta, K., Hayashi, K., Nagamatsu, T., Ito, M., and Suzuki, Y. (1992) Studies on the antinephritic effects of plant components (6): antinephritic effects and mechanisms of phellodendrine (OB-5) on crescentic-type anti-GBM nephritis in rats (2). *Jpn. J. Pharmacol.*, **60**, 187–195.

13 Kong, L.D., Yang, C., Qui, X.I., Wu, H.P., and Ye, D.J. (2001) Effects of different processing products of Cortex Phellodendri on scavenging oxygen free radicals and anti-lipid peroxidation. *Zhongguo Zhong Yao Za Zhi*, **26**, 245–248.

14 Ferreira, M.E., Nakayama, H., de Arias, A.R., Schinini, A., de Bilbao, N.V., Sema, E., Lagoutte, D., Soriano-Agaton, F., Poupon, E., Hocquemiller, R., and Fournet, A. (2007) Effects of canthin-6-one alkaloids from *Zanthoxylum chiloperone* on *Trypanosoma cruzi*-infected mice. *J. Ethnopharmacol.*, **109**, 258–263.

15 Ghosh, R., Garcia, G.E., Crosby, K., Inoue, H., Thompson, I.M., Troyer, D.A., and Kumar, A. P. (2007) Regulation of Cox-2 by cyclic AMP response element binding protein in prostate cancer: potential role for nexrutine. *Neoplasia*, **9**, 893–899.

16 Min, Y.D., Kwon, H.C., Yang, M.C., Lee, K.H., Choi, S.U., and Lee, K.R. (2007) Isolation of limonoids and alkaloids from *Phellodendron amurense* and their multidrug resistance (MDR) reversal activity. *Arch. Pharm. Res.*, **30**, 58–63.

17 Kuo, P.C., Shi, L.S., Damu, A.G., Su, C.R., Huang, C.H., Ke, C.H., Wu, J.B., Lin, A.J., Bastow, K.F., Lee, K.H., and Wu, T.S. (2003) Cytotoxic and antimalarial β-carboline alkaloids from the roots of *Eurycoma longifolia*. *J. Nat. Prod.*, **66**, 1324–1327.

Phyllanthus emblica

Fructus Phyllanthi (Yuganzi) is the dried ripe fruit of *Phyllanthus emblica* L. (Euphorbiaceae). It is used as a stomachic.

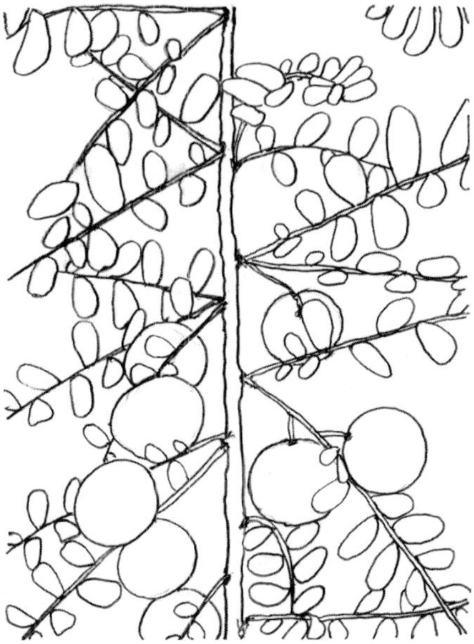

Phyllanthus emblica L.

Chemistry

The fruits of *P. Emblica* are known to be rich in tannins and gallic acid, either as the free acid, or esterified with glucose or other plant acids. The Chinese Pharmacopoeia requires a qualitative determination of gallic acid in Fructus Phyllanthi by thin-layer chromatographic comparison with a reference substance.

Galloylglucose, 1,6-di-*O*-galloyl-β-D-glucose, 1-*O*-galloyl-β-D-glucose and digallic acid; flavones and flavone glycosides kaempferol 3-*O*-β-D-glucoside, quercetin 3-*O*-β-D-glucoside [1], rutin and quercetin [2] were isolated from the fruits of *P. emblica*. The fruit of *P. emblica* is also known to be an excellent source of ascorbic acid [3]. Phenolic constituents from the fruit juice of *P. emblica* were identified as L-malic acid 2-*O*-gallate, mucic acid 2-*O*-gallate, mucic acid 1,4-lactone 2-*O*-gallate, mucic acid 1,4-lactone 5-*O*-gallate, mucic acid 1,4-lactone 3-*O*-gallate, and mucic acid 1,4-lactone 3,5-di-*O*-gallate [4].

Ellagitannins from the fruits were identified as phyllanemblinins A–F. Phyllanemblinins A and B were structurally elucidated as ellagitannins having a tetrahydroxybenzofuran dicarboxyl group and a hexahydroxydiphenoyl group, respectively. Phyllanemblinin C has a new acyl group at the glucose 2,4-positions, and is structurally related to chebulagic acid. Phyllanemblinins D–F were found to be positional isomers of neochebuloyl 1β-*O*-galloylglucose [5].

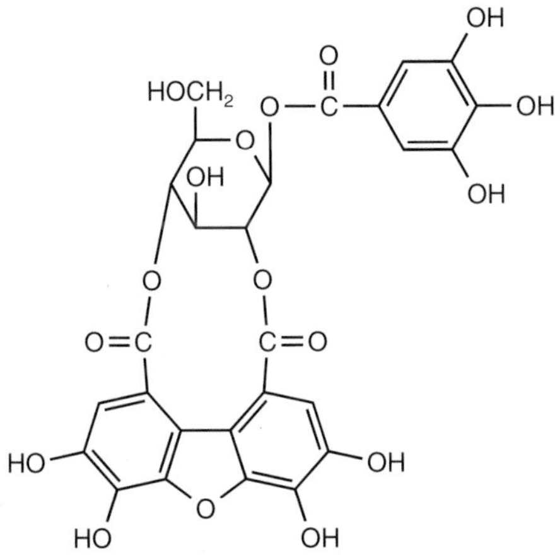

Phyllanemblinin A

Pharmacology and Toxicology

The fruits of *P. emblica* were used in the traditional Chinese medicine as an antihepatitis, antitumor agent, and for regulation of gastrointestinal function. The fruits were also regarded as a traditional immunomodulator and a natural tonic [6].

Handbook of Chinese Medicinal Plants: Chemistry, Pharmacology, Toxicology
Weici Tang and Gerhard Eisenbrand

ISBN: 978-3-527-32226-8

Oral administration of the fruit of *P. emblica* to syngeneic BALB/c mice bearing Dalton's lymphoma ascites tumor has been found to enhance natural killer cell activity and antibody-dependent cellular cytotoxicity. The fruit elicited a twofold increase in splenic natural killer cell activity on day 3 after tumor inoculation. An enhanced activity was highly significant between days 3 and 9 after tumor inoculation with respect to the untreated tumor-bearing control. A significant enhancement in antibody-dependent cellular cytotoxicity was observed on days 3 to 13 in treated mice. A 35% increase in life span was recorded in treated tumor-bearing mice. This increased survival was completely abrogated when natural killer cell and killer cell activities were depleted by cyclophosphamide [7]. A 50% ethanol extract at an oral dose of 1 g kg^{-1} significantly protected mice and rats against the hepatotoxicity of paracetamol (aetaminophen) [2]. Pretreatment of rats with an extract of the fruit of *P. emblica* at an oral dose of 75 mg kg^{-1} at 4 h before administration of ethanol significantly lowered the levels of serum aspartate transaminase and alanine transferase and the content of interleukin-1β (IL-1β). Treatment of rats with the extract for seven days after a 21-day period of ethanol intoxication improved the recovery of liver cells. Levels of transaminases and IL-1β became normal. Histopathological studies confirmed the beneficial effect of the extract against ethanol-induced liver injury in rats [8].

Pretreatment with a butanol extract of the water fraction of *P. emblica* fruits given at a dose of 100 mg kg^{-1} orally to rats for 10 consecutive days was found to enhance the secretion of gastric mucus and hexosamine in the indomethacin-induced ulceration of rats. The morphological observations also supported a protective effect to the stomach wall. Indomethacin treatment of extract premedicated animals markedly affected either malondialdehyde (MDA) or superoxide dismutase (SOD) levels in gastric tissue, while the ulcerative agent itself significantly enhanced both levels. An antioxidant property appeared to be predominantly responsible for this cytoprotective action of the drug [9].

Antimutagenic and anticarcinogenic effects of *P. emblica* have been reported. The extract of *P. emblica* fruits was able to inhibit lead and aluminum-induced sister chromatid exchanges (SCE) [10, 11] and micronuclei formation [12] in the bone marrow cells of mice. Oral pretreatment of mice intoxicated by lead or aluminum with the fruit extract significantly reduced the frequencies of SCEs induced by both metals [10]. Oral administration of an aqueous extract of *P. emblica* fruits to mice for seven days also inhibited the clastogenic effects of different doses of cesium chloride [13] and nickel chloride (10–40 mg kg^{-1}) [14] on the bone marrow cells of mice.

In addition, it was also reported that the fruit juice of *P. emblica* effectively inhibited the formation of *N*-nitrosomorpholine from morpholine and $NaNO_2$ *in vitro*, with more than 90% inhibition, while the inhibition rate of ascorbic acid at a concentration equal to that present in the fruit juice was only about 50%. In rats given $NaNO_2$ and proline (40 m*M*) orally, the coadministration of 1.3 ml of fruit juice containing 36 m*M* ascorbic acid reduced the urinary excretion of *N*-nitrosoproline, from 70 to 4 n*M*. Administration of the fruit juice (13 ml) to 12 healthy volunteers dramatically diminished the 24-h urinary excretion of *N*-nitrosoproline (from 75 to 23 n*M*) after prior ingestion of 300 mg $NaNO_3$ and 500 mg proline. The juice showed a higher inhibition in both cases than the equivalent amount of ascorbic acid [15].

References

1 el-Mekkawy, S., Meselhy, M.R., Kusumoto, I. T., Kadota, S., Hattori, M., and Namba, T. (1995) Inhibitory effects of Egyptian folk medicines on human immunodeficiency virus (HIV) reverse transcriptase. *Chem. Pharm. Bull. (Tokyo)*, **43**, 641–648.

2 Gulati, R.K., Agarwal, S., and Agrawal, S.S. (1995) Hepatoprotective studies on *Phyllanthus emblica* Linn. and quercetin. *Indian J. Exp. Biol.*, **33**, 261–268.

3 Suresh, K. and Vasudevan, D.M. (1994) Augmentation of murine natural killer cell and antibody dependent cellular cytotoxicity activities by *Phyllanthus emblica*, a new immunomodulator. *J. Ethnopharmacol.*, **44**, 55–60.

4 Zhang, Y.J., Tanaka, T., Yang, C.R., and Kouno, I. (2001) New phenolic constituents from the fruit juice of *Phyllanthus emblica*. *Chem. Pharm. Bull. (Tokyo)*, **49**, 537–540.

5 Zhang, Y.J., Abe, T., Tanaka, T., Yang, C.R., and Kouno, I. (2001) Phyllanemblinins A-F, New Ellagitannins from *Phyllanthus emblica*. *J. Nat. Prod.*, **64**, 1527–1532.

6 Xia, Q., Xiao, P., Wan, L., and Kong, J. (1997) Ethnopharmacology of *Phyllanthus emblica* L. *China J. Chin. Mater. Med.*, **22**, 515–518, 525, 574.

7 Suresh, K. and Vasudevan, D.M. (1994) Augmentation of murine natural killer cell and antibody dependent cellular cytotoxicity activities by *Phyllanthus emblica*, a new immunomodulator. *J. Ethnopharmacol.*, **44**, 55–60.

8 Pramyothin, P., Samosorn, P., Poungshompoo, S., and Chaichantipyuth, C. (2006) The protective effects of *Phyllanthus emblica* Linn. extract on ethanol induced rat hepatic injury. *J. Ethnopharmacol.*, **107**, 361–364.

9 Bandyopadhyay, S.K., Pakrashi, S.C., and Pakrashi, A. (2000) The role of antioxidant activity of *Phyllanthus emblica* fruits on prevention from indomethacin induced gastric ulcer. *J. Ethnopharmacol.*, **70**, 171–176.

10 Dhir, H., Roy, A.K., and Sharma, A. (1993) Relative efficiency of *Phyllanthus emblica* fruit extract and ascorbic acid in modifying lead and aluminium-induced sister-chromatid exchanges in mouse bone marrow. *Environ. Mol. Mutagen.*, **21**, 229–236.

11 Madhavi, D., Devil, K.R., Rao, K.K., and Reddy, P.P. (2007) Modulating effect of *Phyllanthus* fruit extract against lead genotoxicity in germ cells of mice. *J. Environ. Biol.*, **28**, 115–117.

12 Roy, A.K., Dhir, H., and Sharma, A. (1992) Modification of metal-induced micronuclei formation in mouse bone marrow erythrocytes by *Phyllanthus* fruit extract and ascorbic acid. *Toxicol. Lett.*, **62**, 9–17.

13 Ghosh, A., Sharma, A., and Talukder, G. (1992) Relative protection given by extract of *Phyllanthus emblica* fruit and an equivalent amount of vitamin C against a known clastogen: caesium chloride. *Food Chem. Toxicol.*, **30**, 865–869.

14 Dhir, H., Agarwal, K., Sharma, A., and Talukder, G. (1991) Modifying role of *Phyllanthus emblica* and ascorbic acid against nickel clastogenicity in mice. *Cancer Lett.*, **59**, 9–18.

15 Hu, J.F. (1990) Inhibitory effects of *Phyllanthus emblica* juice on formation of *N*-nitrosomorpholine *in vitro* and *N*-nitrosoproline in rat and human. *Chin. J. Prev. Med.*, **24**, 132–135.

Physalis alkekengi var. *franchetii*

Calyx seu Fructus Physalis (Jindenglong) is the dried calyx or dry calyx bearing fruit of *Physalis alkekengi* L. var. *franchetii* (Mast.) Makino (Solanaceae) collected in the fall when the fruit become ripe and the persistent calyx turns to red or orange-red. It is used as an expectorant and diuretic agent, and also for treatment of pemphigus and eczema by external application.

Physalis alkekengi L. var. *franchetii* (Mast.) Makino

Chemistry

The withanolides physalin A and physalin B [1], (25*S*)-25,27-dihydrophysalin A (physalin O) [2], and 5α,6β-dihydroxy-2,3,5,6-tetrahydrophysalin B (2,3-dihydrophysalin D or physalin T) [3], were isolated and identified from *P. alkekengi* var. *franchetii*.

As a member of the solanaceous plants, *P. alkekengi* is known to contain alkaloids. Alkaloids from the root of *P. alkekengi* were identified as 3α-tigloyloxytropane [4, 5]. Calystegins were isolated from the root of *P. alkekengi* var. *franchetii* and identified as calystegin A_3, calystegin A_5 and calystegin B_1, calystegin B_2 and calystegin B_3. Calystegin A_5 is the 2-deoxy derivative of calystegin B_2, while calystegin B_3 the 2-epimer of calystegin B_2 [6].

Citric acid was found to be the major organic acid from the fruits of *P. alkekengi* var. *franchetii* [7].

Pharmacology and Toxicology

The fruits of *P. alkekengi* were reported to exhibit antiestrogenic effects in experimental studies. Intraperitoneal (i.p.) injections of an aqueous extract of the fruits of *P. alkekengi* to female rats produced 100% diestrus, and the rats resumed their normal estrus cycle upon withdrawal of the extract. Although there was no significant decrease in the number of implantation sites, the number of pups born to rats decreased by 96% with extract administration. Treatment with the extract had no effect on body weight, uterus weight, plasma protein level or plasma total creatine kinase activity. However, the level of plasma progesterone was diminished by 44%. In addition, the uterine creatine kinase BB-isoenzyme (an estrogen-induced protein) showed a time-dependent inhibition (by 55–82%). The activity of uterine glucose 6-phosphate dehydrogenase (G6PD) activity (another estrogen-induced protein) was inhibited by 52% [8].

Daily doses of 1.88, 3.75, and 7.5 μg 17β-estradiol given i.p. to adult female rats for a period of six to eight days prolonged proestrus or estrus, and increased uterine G6PD activity by 11.5%, 26.9%, and 82.1%, respec-

Handbook of Chinese Medicinal Plants: Chemistry, Pharmacology, Toxicology
Weici Tang and Gerhard Eisenbrand

ISBN: 978-3-527-32226-8

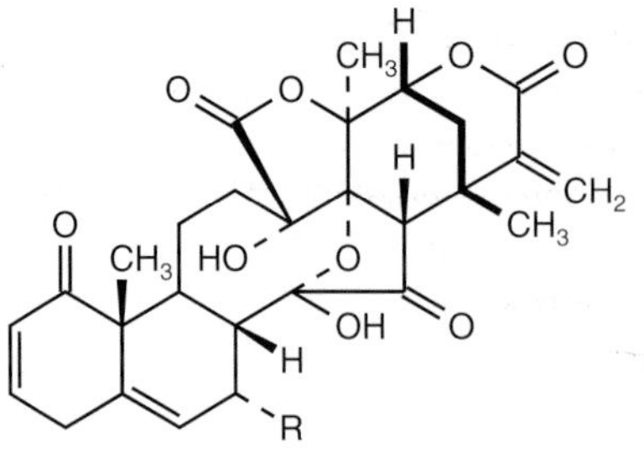

Physalin A: R = OH
Physalin C: R = H

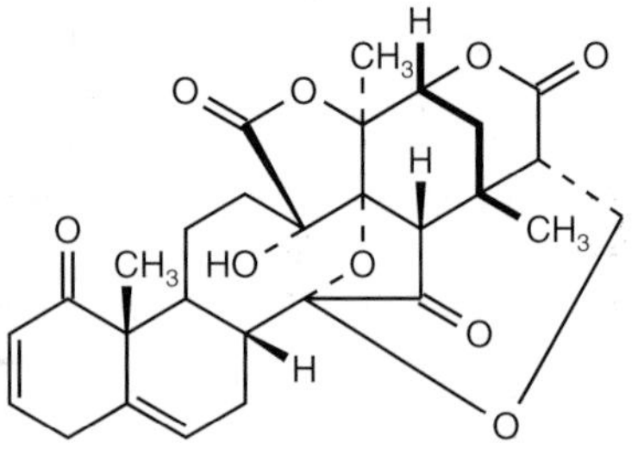

Physalin B

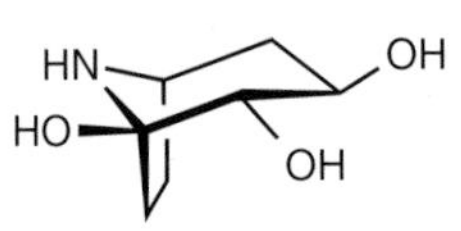

Calystegin A_3

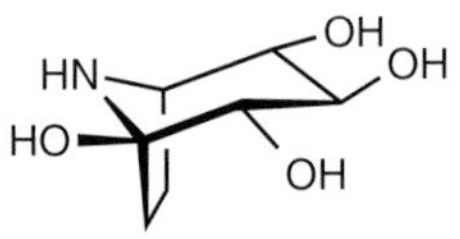

Calystegin B_2

tively. Combined i.p. injections of the extract together with 17β-estradiol for eight consecutive days shortened the time spent in diestrus in proportion to the dose employed, and proportionately reduced the uterine G6PD inhibitory activity of the extract (1.88 μg estradiol, 33.9% inhibition; 3.75 μg estradiol, 27% inhibition; and 7.5 μg estradiol, 6.0% activation). This clearly demonstrated the presence of an estrogen antagonist in the aqueous extract of *P. alkekengi* fruits [9].

The specific activities of hepatic G6PD in rats of different age and gender after i.p. treatment with the aqueous extract of *P. alkekengi* were different. The treatment of new-born and weanling female rats with the extract resulted in a 40–45% reduction in hepatic G6PD activity. However, treatment of adult females and weanling and adult males produced no significant change in the activity of this enzyme. This suggested that the capacity of rodent liver to metabolize steroidal compounds increased with age, and that low levels of circulating estradiol were necessary for enzyme induction in male rats [10].

Intraperitoneal injections of the extract to adult female estrus-cycling rats also resulted in the diminution of pituitary lysyl-aminopeptidase activity by 50%, and in the basomedial hypothalamus by 45%. Administration of daily doses of 3.75, 7.5, and 15 μg of β-estradiol for a period of five to eight days to such animals increased pituitary lysyl-aminopeptidase activity from 31% to 61.5%, and that of basomedial hypothalamus from 20% to 87%, respectively. Administration of the same doses of β-estradiol, along with a given dose of the aqueous extract for seven to eight days, diminished lysyl-aminopeptidase inhibitory effect of the extract in both the pituitary and basomedial hypothalamus. It was concluded that lysyl-aminopeptidase enzymes of both tissues, being estrogen-induced proteins, are inhibited by the estrogen antagonistic principle of *P. alkekengi*. It was further suggested that basomedial hypothalamus lysyl-aminopeptidase activity may be used as an enzyme marker for the action of β-estradiol in the hypothalamus [11].

Calystegin B_2 is known to be a potent competitive inhibitor of almond β-glucosidase, with a K_i of 1.2 μ*M*, and of coffee bean α-galactosidase with a K_i of 0.86 μ*M*. Calystegin B_1 was also found to be a potent competitive inhibitor of almond β-glucosidase, with a K_i of 1.9 μ*M*, and of bovine liver β-galactosidase with a K_i of 1.6 μ*M*, but not to be an inhibitor of α-galactosidases. Calystegin A_3 was a weaker inhibitor compared to calystegin B_2 but with the same inhibitory spectrum. Calystegin A_5 showed no inhibitory activity. Since calystegin B_3 also exhibited only a weak inhibitory activity, it was suggested that the equatorially oriented hydroxy group at C2 is the essential feature for recognition and strong binding to the active site of glycosidases [6].

References

1 Matsuura, T., Kawai, M., Makashima, R., and Butsugan, Y. (1970) Structures of physalin A and physalin B, 13,14-seco-16,24-cyclosteroids from *Physalis alkekengi* var. *franchetii*. *J. Chem. Soc.: Organic*, 664–670.

2 Kawai, M., Ogura, T., Matsumoto, A., Butsugan, Y., and Hayashi, M. (1989) Isolation of (25S)-25,27-dihydrophysalin A from *Physalis alkekengi* var. *franchetii*. *Chem. Express.*, 4, 97–100.

3 Kawai, M., Yamamoto, T., Makino, B., Yamamura, H., Araki, S., Butsugan, Y., and Saito, K. (2001) The structure of physalin T from *Physalis alkekengi* var. *franchetti*. *J. Asian Nat. Prod. Res.*, **3**, 199–205.

4 Yamaguchi, H. and Nishimoto, K. (1965) Studies on the alkaloids of the root of *Physalis alkekengi*. (1). Isolation of 3α-tigloyloxytropane. *Chem. Pharm. Bull. (Tokyo)*, **13**, 217–220.

5 Yamaguchi, H., Numata, A., and Hokimoto, K. (1974) Studies on the alkaloids of the root of *Physalis alkekengi*. II. *Yakugaku Zasshi*, **94**, 1115–1122.

6 Asano, N., Kato, A., Oseki, K., Kizu, H., and Matsui, K. (1995) Calystegins of *Physalis alkekengi* var. *franchetii* (Solanaceae). Structure determination and their glycosidase inhibitory activities. *Eur. J. Biochem.*, **229**, 369–376.

7 Dornberger, K. (1986) The potential antineoplastic acting constituents of *Physalis alkekengi* L. var. *franchetii* Mast. *Pharmazie*, **41**, 265–268.

8 Vessal, M., Mehrani, H.A., and Omrani, G.H. (1991) Effects of an aqueous extract of *Physalis alkekengi* fruit on estrus cycle, reproduction and uterine creatine kinase BB-isozyme in rats. *J. Ethnopharmacol.*, **34**, 69–78.

9 Vessal, M. and Yazdanian, M. (1995) Comparison of the effects of an aqueous extract of *Physalis alkekengi* fruits and/or various doses of 17β-estradiol on rat estrous cycle and uterine glucose 6-phosphate dehydrogenase activity. *Comp. Biochem. Physiol. C Pharmacol. Toxicol. Endocrinol.*, **112**, 229–236.

10 Vessal, M., Mostafavi-Pour, Z., and Kooshesh, F. (1995) Age and sex dependence of the effects of an aqueous extract of *Physalis alkekengi* fruits on rat hepatic glucose 6-P dehydrogenase activity. *Comp. Biochem. Physiol. B Biochem. Mol. Biol.*, **111**, 675–680.

11 Vessal, M., Rasti, M., and Kooshesh, F. (1996) Modulation of the pituitary and basomedial hypothalamic lysyl-aminopeptidase activities be β-estradiol and/or an aqueous extract of *Physalis alkekengi* fruits. *Comp. Biochem. Physiol. B Biochem. Mol. Biol.*, **115**, 267–271.

Phytolacca acinosa and *Phytolacca americana*

Radix Phytolaccae (Shanglu) is the dried root of *Phytolacca acinosa* Roxb. or *Phytolacca americana* L. (Phytolaccaceae). It is used as a diuretic and laxative agent for the treatment of edema, or for the treatment of abscess by external application. The Chinese Pharmacopoeia notes that Radix Phytolaccae is contraindicated in pregnancy.

Phytolacca acinosa Roxb.

Chemistry

The root of *P. americana* is reported to contain triterpene saponins, with phytolaccagenin as aglycone. Unusual structural features of phytolaccagenin are a high oxidation status and a carbomethoxy group, which is apparently of natural origin.

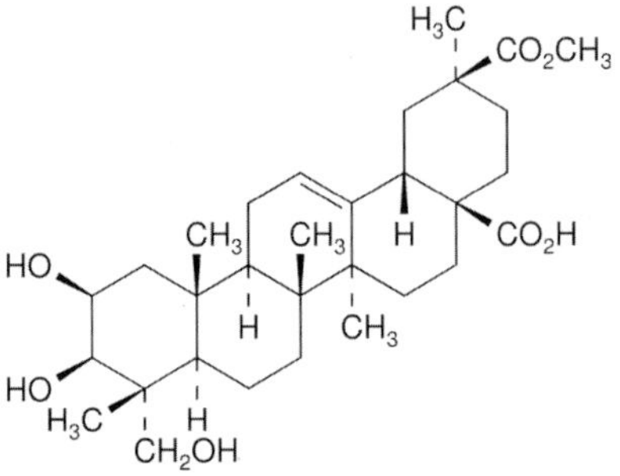

Phytolaccagenin

Saponins from the root of *P. americana* were identified as phytolaccosides A, B (phytolaccasaponin G), D, D_2, E (phytolaccasaponin E), G, I, and phytolaccasaponin B [1]. All of these saponins possess phytolaccagenin or related compounds as aglycone, including phytolaccagenic acid, jaligonic acid, and esculentic acid. The sugar moieties are all bound to the hydroxy group at position 3 of the aglycones.

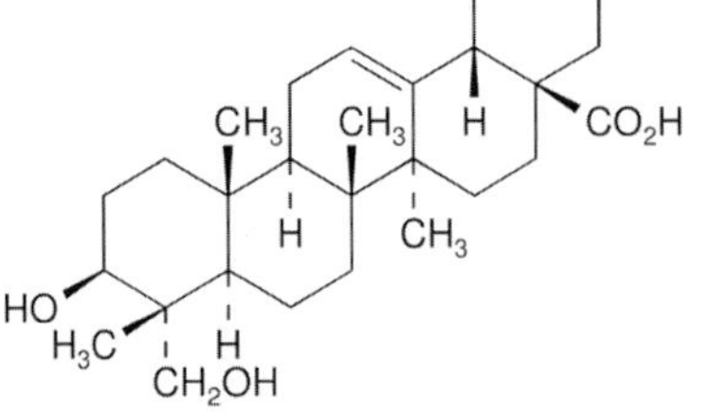

Phytolaccagenic acid

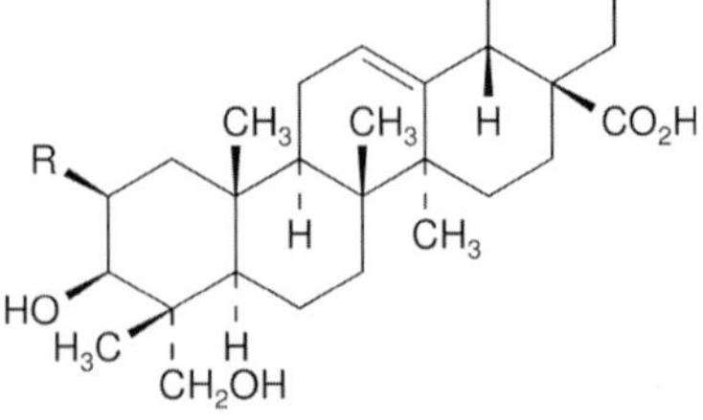

Jaligonic acid: R = OH
Esculentic acid: R = H

Handbook of Chinese Medicinal Plants: Chemistry, Pharmacology, Toxicology
Weici Tang and Gerhard Eisenbrand

ISBN: 978-3-527-32226-8

From the root of *P. acinosa* (*P. esculenta*), the triterpenes jaligonic acid, esculentagenin, esculentagenic acid were isolated [2].

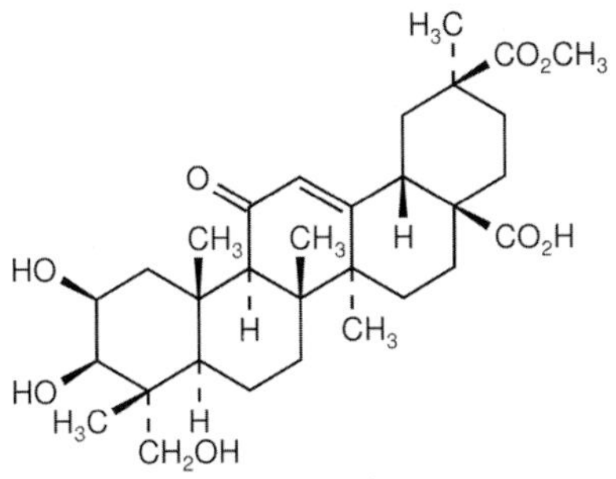

Esculentagenin

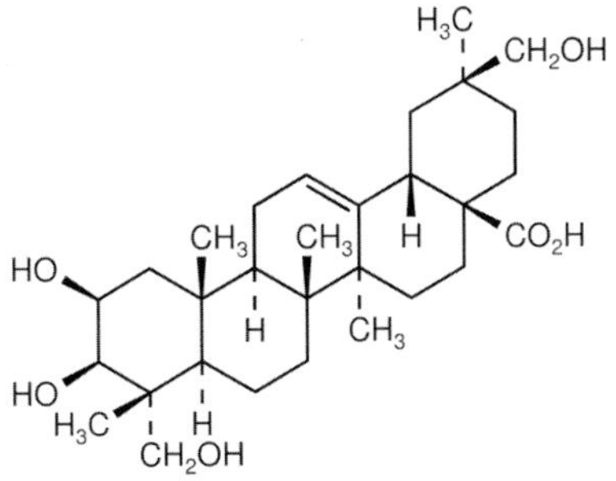

Esculentagenic acid

Saponins from the root of *P. acinosa* were identified as phytolaccosides B, D, E, G (esculentoside E), esculentosides A, F, I, J, K, L, O, P and Q, and phytolaccasaponin B [3, 4].

A number of mitogenic lectins were isolated from the root of *P. americana* (pokeweed), for example pokeweed lectins PL-A, PL-B, PL-C [5], PL-D1, and PL-D2 [6]. The amino acid sequences of PL-B around two glycosylation sites were identified as Cys-Gly-Val-Asp-Phe-Gly-Asn(CHO)-Arg [7]. PL-B is composed of seven repetitive chitin-binding domains having 48–79% sequence homology with each other. Twelve amino acid residues, including eight cysteine residues in these domains, are absolutely conserved in all other chitin-binding domains of plant lectins and class I chitinases [8]. The complete amino acid sequence of PL-C consisting of 126 residues has also been determined. PL-C is an acidic simple protein with a molecular mass of 13 747 Da, and consists of three cysteine-rich domains [9]. PL-D1 was determined to consist of 84 amino acid residues, and has a molecular mass of 9 317 Da, while PL-D2 has an identical sequence except lack of the C-terminal Leu-Thr [7]. Although all of the lectins have mitogenic activities, PL-B was reported to be a mitogenic lectin with the most potent hemagglutinating and mitogenic activities, whereas PL-C has almost no hemagglutinating activity. The extremely high hemagglutinating and mitogenic activities of PL-B may be ascribed to its seven-domain structure [8]. Polysaccharide I with a molecular mass of 10 kDa [10] and polysaccharide II composed of galactose, galacturonic acid, arabinose and rhamnose [11], were isolated from the root of *P. acinosa*.

Pharmacology and Toxicology

Esculentoside A was found to show anti-inflammatory effects. The intraperitoneal (i.p.) administration of esculentoside A markedly inhibited the elevated vascular permeability induced by acetic acid; the ear swelling induced by xylene in mice; and also inhibited the hind paw swelling induced by carrageenan in rats. Esculentoside A also suppressed the hind paw swelling in adrenalectomized rats, suggesting that its anti-inflammatory effect was not dependent on the pituitary–adrenal system [12]. It was further found that esculentoside A significantly inhibited the phagocytic activity of macrophages; inhibited the release of platelet-activating factor (PAF) from rat peritoneal macrophages stimulated by calcimycin; and significantly inhibited the intracellular and extracellular production of interleukin-1 (IL-1) from murine peritoneal macrophages induced by thioglycolate *in vitro*. *In vivo* studies showed that esculentoside A markedly decreased serum hemolysin concentration in sensitized mice challenged with sheep red blood cells. The inhibition of antibody production by B lymphocytes, phagocytosis, and the production of inflammatory mediators by macrophages, may partially explain the anti-inflammatory effect of esculentoside A [13]. Esculentoside A was further found to act on autoimmunity in mice, which might function through inhibition of expression of intercellular adhesion molecule-1 (ICAM-1) mRNA in umbilical vein endothelial cells and an acceleration of thymocyte apoptosis [14].

The i.p. treatment of mice with polysaccharide I from the root of *P. acinosa* significantly augmented the natural killer cell activity, and also lymphocyte proliferation induced by concanavalin or lipopolysaccharides seven days later. Polysaccharide I also enhanced the production of IL-2 in mice splenocytes [15]. Polysaccharide I, when given intraperitoneally at daily doses of 10 and 20 $mg\,kg^{-1}$ to mice for 10 days, was effective in inhibiting S180 growth [16]. Similar effects were observed with polysaccharide II [11].

Some ribosome-inactivating proteins named pokeweed antiviral protein were isolated from various organs, at different stages of development of *P. americana*. Three different species of the antiviral protein were isolated from spring leaves [17], summer leaves [18], and seeds [19]. cDNA clones encoding the pokeweed antiviral protein from *P. americana* have been isolated, characterized [20], and expressed in *Escherichia coli* [21], in *Saccharomyces cerevisiae* [22], and in the methylotrophic yeast *Pichia pastoris* [23]. Pokeweed antiviral protein also possessed potent antiviral activity against several viruses, including herpes simplex virus type 1 and human immunodeficiency virus (HIV) [24]. It was reported to show unique clinical potential to become the active ingredient of a nonspermicidal antiviral agent because of its potent *in vivo* anti-HIV activity, its noninterference with *in vivo* sperm functions, and a lack of cytotoxicity to genital tract epithelial cells [25]. Pokeweed antiviral protein also exhibited antiviral activity in the central nervous system of mice infected intracerebrally with an otherwise invariably fatal dose of the WE54 strain of lymphocytic choriomeningitis virus. The mice were administered pokeweed antiviral protein intraperitoneally at 24 h and 1 h prior to, and at 24, 48, 72, and 96 h after, virus inoculation [26].

Pokeweed antiviral protein is a single chain ribosome-inactivating protein with a molecular mass of 29 kDa. It inactivates both eukaryotic and prokaryotic ribosomes and inhibits translation by catalytically removing specific adenine and guanine residues from the highly conserved, α-sarcin/ricin loop in the 28S large rRNA of ribosomes, resulting in an inhibition of protein synthesis [27, 28]. The subchronic intravaginal toxicity of pokeweed antiviral protein has been studied in mice for a period of 13 weeks. Female B6C3F1 and CD-1 mice were administered intravaginally with a gel formulation containing up to 0.1% pokeweed antiviral protein, for five days each week over 13 consecutive weeks. On a molar basis, these concentrations were 500- to 2000-fold higher than the *in vitro* anti-HIV IC_{50} value. After the 13-week treatment of B6C3F1 mice, no treatment-related mortalities were observed. The mean body weight gain was not reduced by the treatment. The hemogram and blood chemistry profiles revealed a lack of systemic toxicity, and no clinically significant changes in absolute and relative organ weights were noted in the treated animals. Histopathological examinations of tissues showed no increase in treatment-related microscopic lesions.

The repeated intravaginal exposure of CD-1 mice to pokeweed antiviral protein for 13 weeks also failed to show any adverse effects on their subsequent reproductive capability, neonatal survival, or pup development [25]. The induction of irritation by pokeweed antiviral protein was studied in rabbit vagina using a gel containing up to 1.0% native pokeweed antiviral protein for 10 consecutive days. The results showed that half of the treated rabbits exhibited an acceptable range of vaginal mucosal irritation, whereas almost one-third developed moderate to marked vaginal mucosal irritation. However, no treatment-related adverse effects were seen in the hematological examinations [29].

Pokeweed antiviral protein, as well as immunotoxins containing pokeweed antiviral protein, have been tested for growth-inhibitory and antitumor activities against tumor cells in different models [30–32]. Pokeweed antiviral protein and related immunotoxins were found to induce apoptosis in a human lymphocyte cell line [33]. For example, a pokeweed antiviral protein conjugate with gonadotropin-releasing hormone (GnRH), a neuropeptide with receptor sites on several gynecological tumors, showed an IC_{50} of 3 n*M* on *in vitro* translation assays, and selectively inhibited growth of the GnRH receptor-positive Ishikawa cell line at nanomolar concentrations, whereas neither GnRH nor pokeweed antiviral protein alone had any effect [34, 35]. The cytotoxicity of the pokeweed antiviral protein conjugate with GnRH was found to depend on the number of GnRH receptors of the prostate cancer cell lines, and the duration of exposure [36].

References

1. Kang, S.S. and Woo, W.S. (1987) Two new saponins from *Phytolacca americana*. *Planta Med.*, **53**, 338–340.
2. Yi, Y.H. (1991) A triterpenoid and its saponin from *Phytolacca esculenta*. *Phytochemistry*, **30**, 4179–4181.
3. Yi, Y.H. (1992) Two new saponins from the roots of *Phytolacca esculenta*. *Planta Med.*, **58**, 99–101.
4. Yi, Y.H. (1992) A triterpenoid saponin from *Phytolacca esculenta*. *Phytochemistry*, **31**, 2552–2554.
5. Kino, M., Yamaguchi, K., Umekawa, H., and Funatsu, G. (1995) Purification and characterization of three mitogenic lectins from the roots of pokeweed (*Phytolacca americana*). *Biosci. Biotechnol. Biochem.*, **59**, 683–688.
6. Yamaguchi, K., Mori, A., and Funatsu, G. (1996) Amino acid sequence and some properties of lectin-D from the roots of pokeweed (*Phytolacca americana*). *Biosci. Biotechnol. Biochem.*, **60**, 1380–1382.
7. Kimura, Y., Yamaguchi, K., and Funatsu, G. (1996) Structural analysis of N-linked oligosaccharide of mitogenic lectin-B from the roots of pokeweed (*Phytolacca americana*). *Biosci. Biotechnol. Biochem.*, **60**, 537–540.
8. Yamaguchi, K., Yurino, N., Kino, M., Ishiguro, M., and Funatsu, G. (1997) The amino acid sequence of mitogenic lectin-B from the roots of pokeweed (*Phytolacca americana*). *Biosci. Biotechnol. Biochem.*, **61**, 690–698.
9. Yamaguchi, K., Mori, A., and Funatsu, G. (1995) The complete amino acid sequence of lectin-C from the roots of pokeweed (*Phytolacca americana*). *Biosci. Biotechnol. Biochem.*, **59**, 1384–1385.
10. Wang, H.B., Zheng, Q.Y., Shen, Y.A., Tang, H. L., and Chen, H.S. (1993) *Phytolacca acinosa* polysaccharides I antitumor activity, immune augmentation and hemopoietic protection in S180-bearing mice. *Chin. J. Pharmacol. Toxicol.*, **7**, 52–55.
11. Wang, H.B., Zheng, Q.Y., Ju, D.W., and Fang, J. (1993) Effects of *Phytolacca acinosa* polysaccharides II on lymphocyte proliferation and colony stimulating factor production from mice splenocytes in vitro. *Acta Pharm. Sin.*, **28**, 490–493.
12. Zheng, Q.Y., Mai, K.I., Pan, X.F., and Yi, Y.H. (1992) Anti-inflammatory effect of esculentoside A. *Chin. J. Pharmacol. Toxicol.*, **6**, 221–223.

13 Ju, D.W., Zheng, Q.Y., Wang, H.B., Guan, X.J., Fang, J., and Yi, Y.H. (1994) Inhibitory effects of esculentoside A on mouse macrophages and antibody production. *Acta Pharm. Sin.*, **29**, 252–255.

14 Xiao, Z.Y., Zheng, Q.Y., Zhang, J.P., Jiang, Y.Y., and Yi, Y.H. (2002) Effect of esculentoside A on autoimmunity in mice and its possible mechanisms. *Acta Pharmacol. Sin.*, **23**, 638–644.

15 Wang, H.B., Zheng, Q.Y., Qian, D.H., Fang, J., and Ju, D.W. (1993) Effects of *Phytolacca acinosa* polysaccharides I on immune function in mice. *Acta Pharmacol. Sin.*, **14**, 243–246.

16 Wang, H.B., Chen, W.Z., Bao, E.J., Zheng, Q. Y., Song, H.L., Fang, J., Xu, Y.X., and Chen, H. S. (1995) Effects of *Phytolacca acinosa* polysaccharides I combined with interleukin-2 on the cytotoxicity of murine splenocytes against tumor cells. *Acta Pharm. Sin.*, **30**, 401–407.

17 Desvoyes, B., Dulieu, P., Poyet, J.L., and Adami, P. (1995) Production and characterization of monoclonal antibodies against the ribosome-inactivating protein PAP from *Phytolacca americana. Hybridoma*, **14**, 571–575.

18 Poyet, J.L., Hoeveler, A., and Jongeneel, C.V. (1998) Analysis of active site residues of the antiviral protein from summer leaves from *Phytolacca americana* by site-directed mutagenesis. *Biochem. Biophys. Res. Commun.*, **253**, 582–587.

19 Kung, S.S., Kimura, M., and Funatsu, G. (1990) The complete amino acid sequence of antiviral protein from the seeds of pokeweed (*Phytolacca americana*). *Agric. Biol. Chem.*, **54**, 3301–3318.

20 Kataoka, J., Habuka, N., Masuta, C., Miyano, M., and Koiwai, A. (1992) Isolation and analysis of a genomic clone encoding a pokeweed antiviral protein. *Plant Mol. Biol.*, **20**, 879–886.

21 Honjo, E. and Watanabe, K. (1999) Expression of mature pokeweed antiviral protein with or without C-terminal extrapeptide in*Escherichia coli* as a fusion with maltose-binding protein. *Biosci. Biotechnol. Biochem.*, **63**, 1291–1294.

22 Hur, Y., Hwang, D.J., Zoubenko, O., Coetzer, C., Uckun, F.M., and Tumer, N.E. (1995) Isolation and characterization of pokeweed antiviral protein mutations in *Saccharomyces cerevisiae*: identification of residues important for toxicity. *Proc. Natl Acad. Sci. USA*, **92**, 8448–8452.

23 Rajamohan, F., Doumbia, S.O., Engstrom, C. R., Pendergras, S.L., Maher, D.L., and Uckun, F.M. (2000) Expression of biologically active recombinant pokeweed antiviral protein in methylotrophic yeast *Pichia pastoris. Protein Expr. Purif.*, **18**, 193–201.

24 Uckun, F.M., Rajamohan, F., Pendergrass, S., Ozer, Z., Waurzyniak, B., and Mao, C. (2003) Structure-based design and engineering of a nontoxic recombinant pokeweed antiviral protein with potent anti-human immunodeficiency virus activity. *Antimicrob. Agents Chemother.*, **47**, 1052–1061.

25 D'Cruz, O.J., Waurzyniakt, B., and Uckun, F. M. (2004) A 13-week subchronic intravaginal toxicity study of pokeweed antiviral protein in mice. *Phytomedicine*, **11**, 342–351.

26 Uckun, F.M., Rustamova, L., Vassilev, A.O., Tibbles, H.E., and Petkevich, A.S. (2005) CNS activity of Pokeweed anti-viral protein (PAP) in mice infected with lymphocytic choriomeningitis virus (LCMV). *BMC Infect. Dis.*, **5**, 9.

27 Kurinov, I.V., Rajamohan, F., Venkatachalam, T.K., and Uckun, F.M. (1999) X-ray crystallographic analysis of the structural basis for the interaction of pokeweed antiviral protein with guanine residues of ribosomal RNA. *Protein Sci.*, **8**, 2399–2405.

28 Mansouri, S., Nourollahzadeh, E., and Hudak, K.A. (2006) Pokeweed antiviral protein depurinates the sarcin/ricin loop of the rRNA prior to binding of aminoacyl-tRNA to the ribosomal A-site. *RNA*, **12**, 1683–1692.

29 D'Cruz, O.J., Waurzyniakt, B., and Uckun, F. M. (2004) Mucosal toxicity studies of a gel formulation of native pokeweed antiviral protein. *Toxicol. Pathol.*, **32**, 212–221.

30 Battelli, M.G., Polito, L., Bolognesi, A., Lafleur, L., Fradet, Y., and Stirpe, F. (1996) Toxicity of ribosome-inactivating proteins-containing immunotoxins to a human bladder carcinoma cell line. *Int. J. Cancer*, **65**, 485–490.

31 Waddick, K.G., Myers, D.E., Gunther, R., Chelstrom, L.M., Chandan-Langlie, M., Irvin, J.D., Tumer, N., and Uckun, F.M. (1995) *In vitro* and *in vivo* antileukemic activity of B43-pokeweed antiviral protein against radiation-resistant human B-cell precursor leukemia cells. *Blood*, **86**, 4228–4233.

32 Anderson, P.M., Meyers, D.E., Hasz, D.E., Covalcuic, K., Saltzman, D., Khanna, C., and Uckun, F.M. (1995) *In vitro* and *in vivo* cytotoxicity of an anti-osteosarcoma immunotoxin containing pokeweed antiviral protein. *Cancer Res.*, **55**, 1321–1327.

33 Bolognesi, A., Tazzari, P.L., Olivieri, F., Polito, L., Falini, B., and Stirpe, F. (1996) Induction of apoptosis by ribosome-inactivating proteins and related immunotoxins. *Int. J. Cancer*, **68**, 349–355.

34 Schlick, J., Dulieu, P., Desvoyes, B., Adami, P., Radom, J., and Jouvenot, M. (2000) Cytotoxic activity of a recombinant GnRH-PAP fusion toxin on human tumor cell lines. *FEBS Lett.*, **472**, 241–246.

35 Yang, W.H., Wieczorck, M., Allen, M.C. ,and Nett, T.M. (2003) Cytotoxic activity of gonadotropin-releasing hormone (GnRH)-pokeweed antiviral protein conjugates in cell lines expressing GnRH receptors. *Endocrinology*, **144**, 1456–1463.

36 Qi, L., Nett, T.M., Allen, M.C., Sha, X., Harrison, G.S., Frederick, B.A., Crawfold, E. D., and Glode, L.M. (2004) Binding and cytotoxicity of conjugated and recombinant fusion proteins targeted to the gonadotropin-releasing hormone receptor. *Cancer Res.*, **64**, 2090–2095.

Picrasma quassioides

Ramulus et Folium Picrasmae (Kumu) is the dried branch and leaf of *Picrasma quassioides* (D. Don) Benn. (Simaroubaceae). It is used as an antibacterial agent for the treatment of influenza, laryngitis, pharyngitis, enteritis, acute tonsillitis, and also for the treatment of snake bite.

Picrasma quassioides (D. Don) Benn.

Chemistry

Picrasma quassioides (*Picrasma ailanthoides* Pl.) is the third plant of the family Simaroubaceae listed in the Chinese Pharmacopoeia in addition to *Ailanthus altissima* and *Brucea javanica*. *Picrasma quassioides* was reported to contain quassinoids and β-carboline alkaloids. Quassinoids from *P. quassioides* are characterized by an oxo group at position 1 of the picrasane skeleton with quassin (nigakilactone D) and neoquassin (nigakihemiacetal B) as major components.

Minor quassinoids from *P. quassioides* were identified as nigakilactones A–F, H, K–O [1], kusulactone, picrasinols A and B, picrasins A (nigakilactone G), B (nigakilactone I), C (nigakilactone J), D–G, and nigakihemiacetals A and C [2, 3]. All of these quassinoids possessed picrasane structure, and could be divided in two groups: the δ-lactones such as quassin, and the hemiacetals such as neoquassin.

Quassinoid glycosides from *P. quassioides* were identified as picrasinosides A–H [4].

Alkaloids from the bark or wood of *P. quassioides* were mainly derived from β-carboline or β-carboline dimers [5]. Simple β-carboline derivatives were identified as kumujians A, C, and G [6, 7], picrasidines B, D, E, I, J, K, P, V, X, and Y [8, 9], kumujancine, and kumujanrine [10].

Canthin-6-one-derived alkaloids were identified as picrasidines L, O, Q [11], nigakinone, and its methyl derivative [12].

Alkaloids derived from β-carboline dimer were identified as picrasidines A, C, F–H,

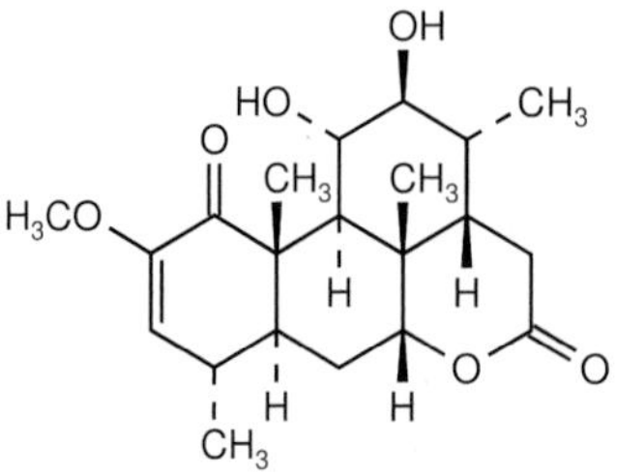

Nigakilactone A

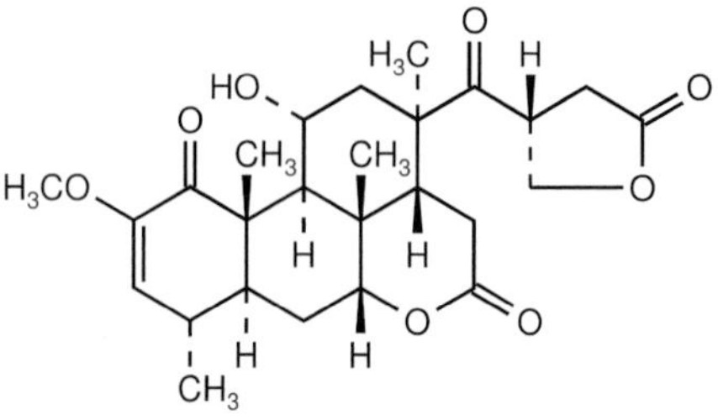

Picrasin A

Handbook of Chinese Medicinal Plants: Chemistry, Pharmacology, Toxicology
Weici Tang and Gerhard Eisenbrand

ISBN: 978-3-527-32226-8

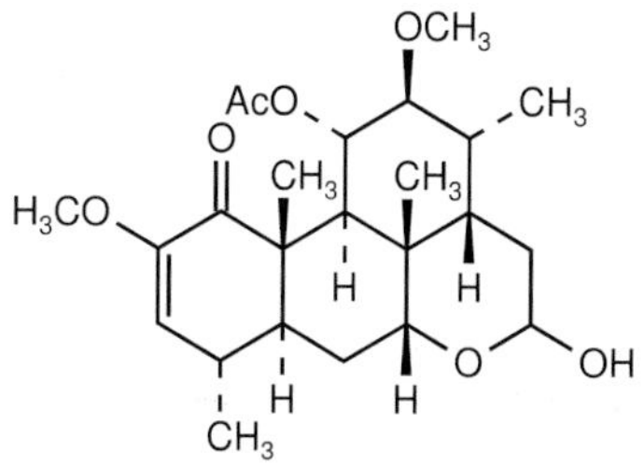

Picrasinol A

M, N, R–U [13–15], kumujansine, and kumujantine [14]. Among these, picrasidines A, C, H, and R are derived from two β-carboline moieties, picrasidines M, N and U are derived from one β-carboline moiety and one canthin-6-one moiety, while picrasidines F, G, S, T, kumujansine, and kumujantine are composed of one β-carboline and one indolo[2,3-*a*]quinolizine structure.

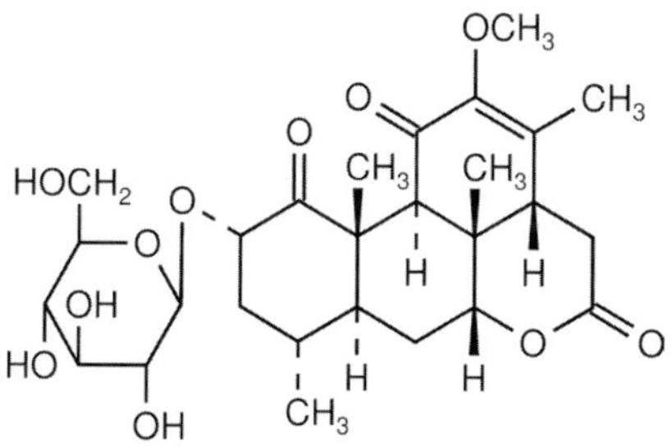

Picrasinoside A

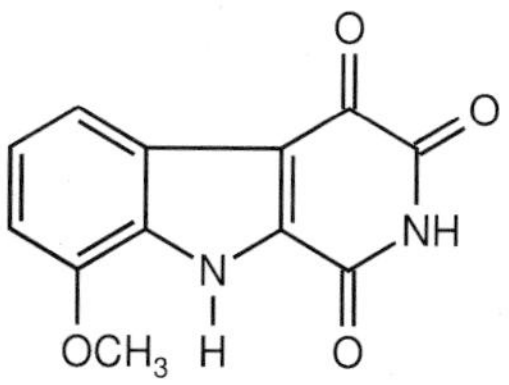

Picrasidine V

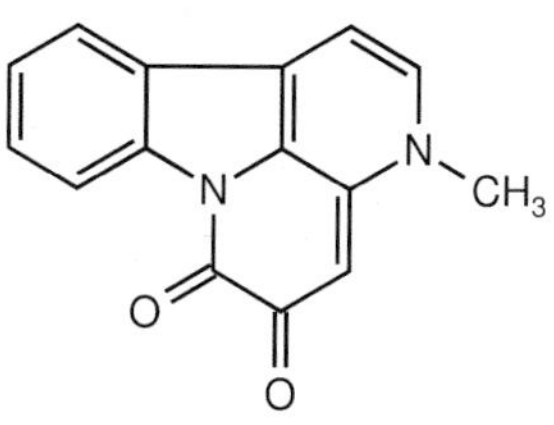

Picrasidine L

Pharmacology and Toxicology

Oral administration of quassin to rats was reported to significantly reduce the weight of the testis, epididymis and seminal vesicle; to increase the anterior pituitary gland; to decrease the epididymal sperm counts; and to reduce serum levels of testosterone, luteinizing hormone and follicle-stimulating hormone [15]. Quassin was also found to inhibit both the basal and luteinizing hormone-stimulated testosterone secretion of rat Leydig cells *in vitro* in a concentration-dependent manner [16].

The majority of the alkaloids derived from β-carboline, β-carboline dimer or canthin-6-one were found to be inhibitors of cAMP-dependent phosphodiesterase [17]. Studies on the structure–activity relationships revealed that β-carboline derivatives with a methoxycarbonyl group (e.g., β-carboline-1-carboxylic acid methyl ester) and canthin-6-one derivatives with a methoxy group (e.g., 4,5-dimethoxycanthin-6-one and 5-hydroxy-4-methoxycanthin-6-one) generally exhibited a strong inhibitory effect on cAMP-dependent phosphodiesterase [18].

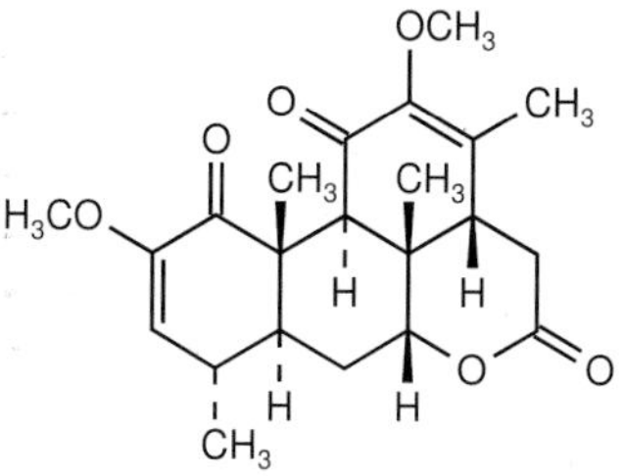

Quassin

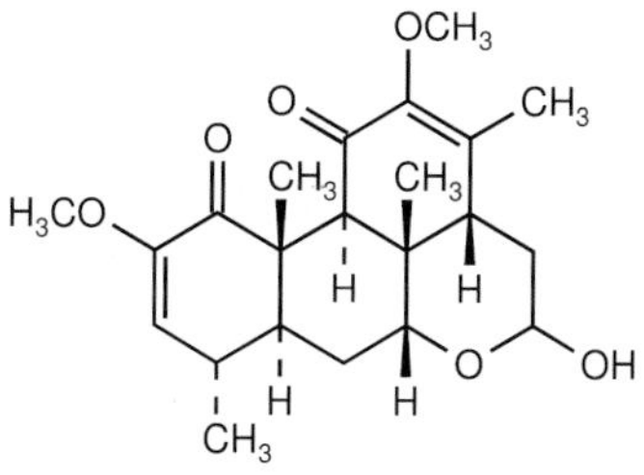

Neoquassin

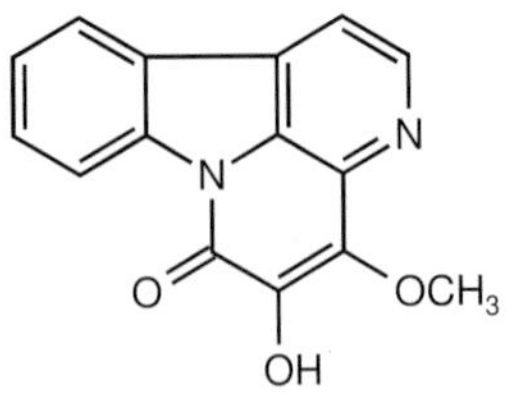

Nigakinone

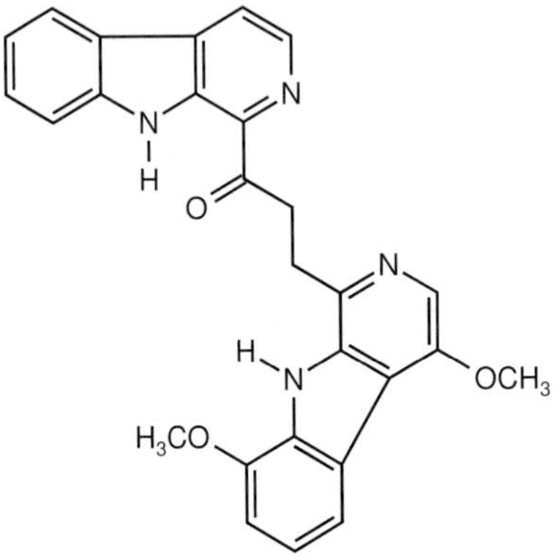

Picrasidine A

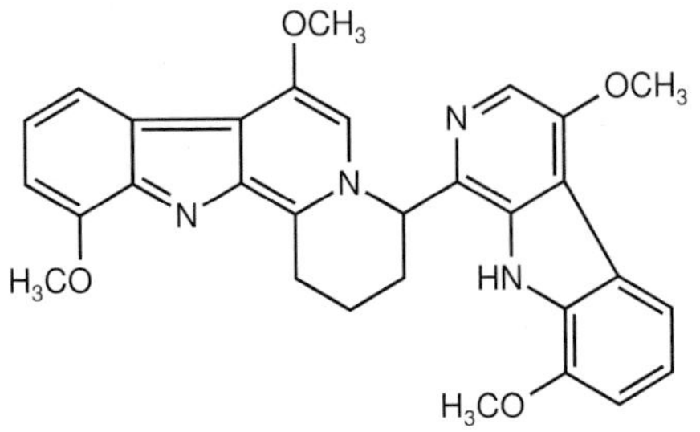

Kumujantine

The methanol extract of the wood of *P. quassioides*, when given to rats prevented the secretion of gastric juice. The extract also showed preventive effects in rats against gastric ulcer induced by aspirin. The effective components were identi-fied as nigakinone and methylnigaki-none [12].

References

1 Yang, J.S. and Gong, D. (1984) A new bitter principle, kusulactone, from Indian quassiawood (*Picrasma quassioides*). *Chin. Trad. Herbal Drugs*, **15**, 531–533.

2 Hirota, H., Yokoyama, A., Tsuyuki, T., Takahashi, T., and Waelchli, M. (1988) Structure of nigakilactone O, a new quassinoid from *Picrasma ailanthoides*. *Chem. Lett.*, 651–652.

3 Okano, M., Fujita, T., Fukamiya, N. ,and Aratani, T. (1984) New quassinoid glycosides and hemiacetals from *Picrasma ailanthoides* Planchon. Picrasinoside-B, -C, -D, -E, -F, and -G, and picrasinol-A and -B. *Chem. Lett.*, 221–224.

4 Matsuzaki, T., Fukamiya, N., Okano, M., Fujita, T., Tagahara, K. and Lee, K.H. (1991) Picrasinoside H, a new quassinoid glycoside, and related compounds from stem wood of *Picrasma ailanthoides. J. Nat. Prod.*, **54**, 844–848.

5 Ohmoto, T., and Koike, K. (1984) Studies on the constituents of *Picrasma quassioides* Bennet. III. The alkaloidal constituents. *Chem. Pharm. Bull. (Tokyo)*, **32**, 3579–3583.

6 Luo, S.R., Guo, R. ,and Yang, J.S. (1988) Deter-mination of alkaloids in *Picrasma quassioides* (D. Don) Benn. *Acta Pharm. Sin.*, **23**, 906–909.

7 Liu, J., Davidson, S.R., Van der Heijden, R., Verpoorte, R., and Howarth, O.W. (1992) Isolation of 4-hydroxy-5-methoxycanthin-6-one from *Picrasma quassioides* and revision of a previously reported structure. *Liebigs Ann. Chem.*, 987–988.

8 Koike, K., Ohmoto, T. ,and Ikeda, K. (1990) The alkaloids of *Picrasma quassioides*. XII. β-

Carboline alkaloids from *Picrasma quassioides*. *Phytochemistry*, **29**, 3060–3061.

9 Li, H.Y., Koike, K. ,and Ohmoto, T. (1993) New alkaloids, picrasidines W, X and Y, from *Picrasma quassioides* and x-ray crystallographic analysis of picrasidine Q. *Chem. Pharm. Bull. (Tokyo)*, **41**, 1807–1811.

10 Yang, J.S. and Gong, D. (1984) Kumujancine and kumujanrine, two new β-carboline alkaloids from *Picrasma quassioides* (D. Don.) Benn. *Acta Chim. Sin.*, **42**, 679–683.

11 Ohmoto, T. and Koike, K. (1985) Studies on the alkaloids from *Picrasma quassioides* Bennet. VI. Structures of picrasidines N, O, and Q. *Chem. Pharm. Bull. (Tokyo)*, **33**, 4901–4905.

12 Niiho, Y., Mitsunaga, K., Koike, K., and Ohmoto, T. (1994) Gastric antiulcer components from the woods of *Picrasma quassioides* (Simaroubaceae). *Nat. Med.*, **48**, 116–121.

13 Liu, J., Davison, R.S., and Howarth, O.W. (1993) Full assignments of proton and carbon signals of picrasidine-G by 1D and 2D NMR methods. *Magn. Reson. Chem.*, **31**, 1091–1092.

14 Yang, J.S., Yu, D.Q., and Liang, X.T. (1988) Kumujansine and kumujantine, two new dimeric β-carboline alkaloids from *Picrasma quassioides* (D. Don.) Benn. *Acta Pharm. Sin.*, **23**, 267–272.

15 Raji, Y. and Bolarinwa, A.F. (1997) Antifertility activity of *Quassia amara* in male rats: *in vivo* study. *Life Sci.*, **61**, 1067–1074.

16 Njar, V.C., Alao, T.O., Okogun, J.I., Raji, Y., Bolarinwa, A.F., and Nduka, E.U. (1995) Antifertility activity of *Quassia amara*: quassin inhibits the steroidogenesis in rat Leydig cells *in vitro*. *Planta Med.*, **61**, 180–182.

17 Ohmoto, T., Nikaido, T., Koike, K., Kohda, K., and Sankawa, U. (1988) Inhibition of cyclic AMP phosphodiesterase in medicinal plants. XV. Inhibition of adenosine 3′,5′-cyclic monophosphate phosphodiesterase by alkaloids. II. *Chem. Pharm. Bull. (Tokyo)*, **36**, 4588–4592.

18 Sung, Y.I., Koike, K., Nikaido, T., Ohmoto, T., and Sankawa, U. (1984) Inhibitors of cyclic AMP phosphodiesterase in medicinal plants. V. Inhibitors of cyclic AMP phosphodiesterase in *Picrasma quassioides* Bennet, and inhibitory activities of related β-carboline alkaloids. *Chem. Pharm. Bull. (Tokyo)*., **32**, 1872–1877.

Picria fel-terrae

The dried whole plant of *Picria fel-terrae* Lour. (Scrophulariaceae) is listed in the appendix of the Chinese Pharmacopoeia.

Picria fel-terrae Lour.

Chemistry

The whole plant of *Picria fel-terrae* was reported to contain a series of triterpenes and triterpene glycosides. The triterpenes were identified as picfeltarraegenin I, picfeltarraegenin II, picfeltarraegenin III, picfeltarraegenin IV, picfeltarraenone, picfeltarraenin IA, picfeltarraenin IB, and picfeltarraenin II. These are all tetracyclic triterpenes with a characteristic side chain of 3(2*H*)-furanone or 3(4*H*)-furanone [1–3].

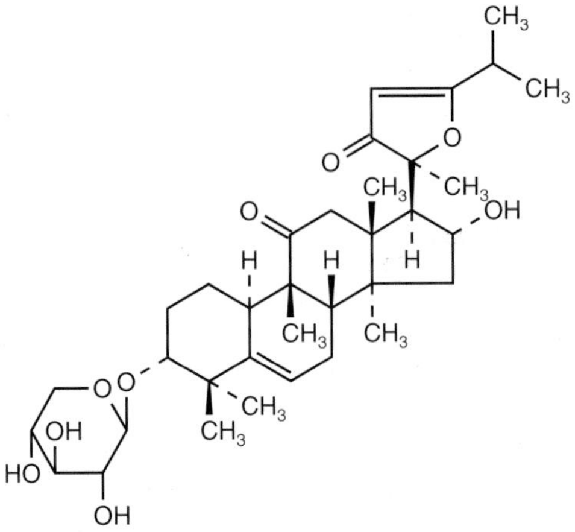

Picfeltarraenin VI

Pharmacology and Toxicology

Picfeltarraenin IA, picfeltarraenin IB, picfeltarraenin IV, and picfeltarraenin VI (picfeltarraegenin I 3-*O*-β-D-xylopyranoside) were all found to act as inhibitors of both the classical and alternative pathways of the complement system. Picfeltarraenin VI exhibited the highest inhibitory activity, with IC_{50} values of about 30 μ*M* and 20 μ*M* for the classical and alternative pathways, respectively. No antiviral, antibacterial, or antifungal activities were observed. In an *in vitro* human tumor cell line panel, picfeltarraenin IA and IB displayed no cytotoxic activity [2].

References

1 Gan, L.X., Chen, Y.Q., Zhou, W.S., Cheng, G. R., and Jin, J.L. (1986) Studies on triterpenoids and their glycosides from Chinese medicinal herb *Picria fel-terra* Lour. *Stud. Org. Chem. (Amsterdam)*, **26**, 95–108.

Handbook of Chinese Medicinal Plants: Chemistry, Pharmacology, Toxicology
Weici Tang and Gerhard Eisenbrand

ISBN: 978-3-527-32226-8

2 Huang, Y., De Bruyne, T., Apers, S., Ma, Y., Claeys, M., Vanden Berghe, D., Pieters, L., and Vlietinck, A. (1998) Complement-inhibiting cucurbitacin glycosides from *Picria fel-terrae*. *J. Nat. Prod.*, **61**, 757–761.

3 Jin, J.L., Wen, Y.X., Cheng, G.R., Gan, L.X., and Chen, Y.Q. (1987) Constituents of *Picria fel-terra* Lour. VIII. Structure of picfelterraenin II. *Acta Chim. Sin.*, **45**, 1133–1134.

Picrorhiza scrophulariiflora

Rhizoma Picrorhizae (Huhuanglian) is the dried rhizome of *Picrorhiza scrophulariiflora* Pennell (Scrophulariaceae). It is used for the treatment of dysentery and jaundice.

Picrorhiza scrophulariiflora Pennell

Chemistry

The rhizome of *P. scrophulariiflora* contains a number of organic acids, with cinnamic acid and vanillic acid as major components. The rhizome of *P. scrophulariiflora* is also known to be rich in phenylethanoid glycosides and iridoid glycosides. Phenylethanoid glycosides from the rhizome of *P. scrophulariiflora* were identified as scroside A, scroside B, and scroside C [1, 2]. Iridoid glycosides picroside I, picroside II and picroside IV have been isolated from the rhizome of *P. scrophulariiflora* [1, 2]. The Chinese Pharmacopoeia requires a qualitative determination of cinnamic acid and vanillic acid by thin-layer chromatographic comparison with reference substances, and a quantitative determination of picroside I and picroside II in Rhizoma Picrorhizae by HPLC. The total content of picroside I and picroside II in Rhizoma Picrorhizae should be not less than 9.0%.

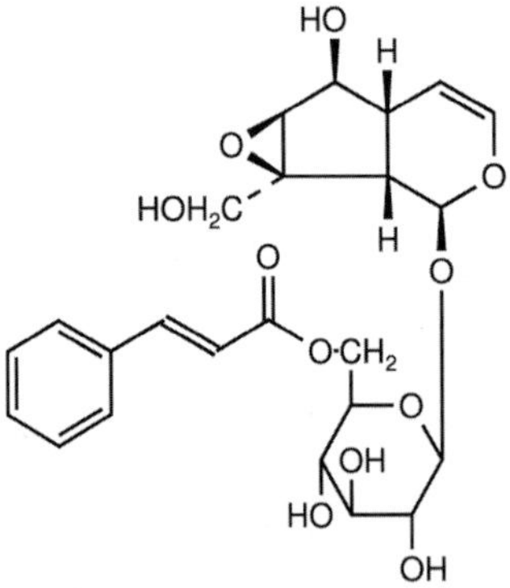

Picroside I

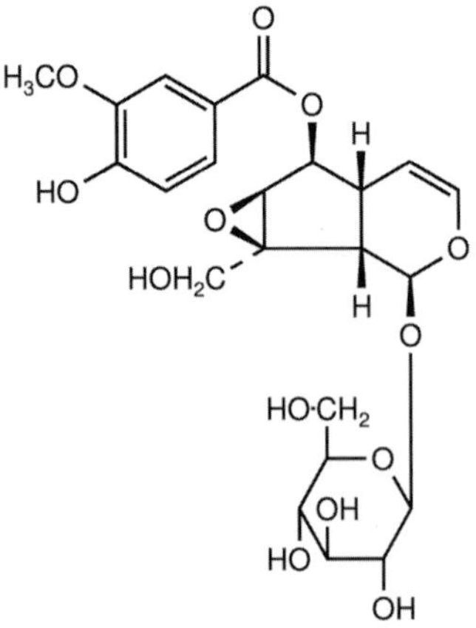

Picroside II

Handbook of Chinese Medicinal Plants: Chemistry, Pharmacology, Toxicology
Weici Tang and Gerhard Eisenbrand

ISBN: 978-3-527-32226-8

Pharmacology and Toxicology

A methanol extract of the rhizome of *P. scrophulariiflora* was found to possess nerve growth factor (NGF)-potentiating activity and to induce NGF-mediated neurite outgrowth from PC12D cells [3]. The pharmacological data suggest that picrosides I and II enhance neurite outgrowth from PC12D cells, probably by amplifying a step in the NGF receptor-mediated intracellular signaling pathway [4]. Picrosides I and II also caused a concentration-dependent enhancement of neurite outgrowth from PC12D cells induced by basic fibroblast growth factor (bFGF), staurosporine and dibutyryl cAMP, suggesting that picrosides I and II activated the mitogen activated protein (MAP) kinase-dependent signaling pathway [5]. The neuroprotective effect of picroside II was further evidenced *in vitro* using PC12 cells treated with glutamate, and in male ICR mice treated with $AlCl_3$. Pretreatment of PC12 cells with picroside II enhanced the cell viability and decreased the level of intracellular reactive oxygen species (ROS) induced by glutamate. Picrosides II also significantly prevented glutamate-induced cell apoptosis. The oral administration of picroside II to mice at daily doses of 20 and 40 mg kg^{-1} markedly ameliorated $AlCl_3$-induced learning and memory dysfunctions and attenuated $AlCl_3$-induced histological changes; the latter was associated with a significant increase in superoxide dismutase activity in mouse brain. This indicates that picroside II has therapeutic potential against neurological injuries caused by oxidative stress [6].

The diethyl ether extract of the rhizome of *P. scrophulariiflora* was found to show potent inhibitory activity towards the classical pathway of the complement system, the respiratory burst of activated polymorphonuclear leukocytes, and the mitogen-induced proliferation of T lymphocytes. In addition, it showed anti-inflammatory activity towards carrageenan-induced paw edema in mice, but no effects in experimentally induced arthritis [7]. Picroside II was reported to exhibit hepatoprotective effects. The oral administration of picroside II at doses of 5, 10, and 20 mg kg^{-1} to mice protected against liver damage induced by carbon tetrachloride (CCl_4) D-galactosamine, and acetaminophen. Picroside II significantly decreased high levels of serum alanine aminotransferase and aspartate aminotransferase levels in mice, and markedly reduced the hepatocellular damages after intoxication. Picroside II also decreased the malondialdehyde content of serum, and increased the activities of manganese-superoxide dismutase and glutathione peroxidase [8].

References

1 Li, P., Matsunaga, K., and Ohizumi, Y. (2000) Nerve growth factor-potentiating compounds from Picrorhizae Rhizoma. *Biol. Pharm. Bull.*, **23**, 890–892.

2 Li, J.X., Li, P., Tezuka, Y., Namba, T., and Kadota, S. (1998) Three phenylethanoid glycosides and an iridoid glycoside from *Picrorhiza scrophulariiflora*. *Phytochemistry*, **48**, 537–542.

3 Li, P., Matsunaga, K., and Ohizumi, Y. (1999) Enhancement of the nerve growth factor-mediated neurite outgrowth from PC12D cells by Chinese and Paraguayan medicinal plants. *Biol. Pharm. Bull.*, **22**, 752–755.

4 Li, P., Matsunaga, K., Yamakuni, T., and Ohizumi, Y. (2000) Potentiation of nerve growth factor-action by picrosides I and II, natural iridoids, in PC12D cells. *Eur. J. Pharmacol.*, **406**, 203–208.

5 Li, P., Matsunaga, K., Yamakuni, T., and Ohizumi, Y. (2002) Picrosides I and II, selective enhancers of the mitogen-activated protein kinase-dependent signaling pathway in the action of neuritogenic substances on PC12D cells. *Life Sci.*, **71**, 1821–1835.

6 Li, T., Liu, J.W., Zhang, X.D., Guo, M.C. and Ji, G. (2007) The neuroprotective effect of picroside II from hu-huang-lian against

oxidative stress. *Am. J. Chin. Med.*, **35**, 681–691.

7 Smit, H.F., Kroes, B.H., van den Berg, A.J., van der Wal, D., van den Worm, E., Beukelman, C.J., van Dijk, H., and Labadie, R.P. (2000) Immunomodulatory and anti-inflammatory activity of *Picrorhiza scrophulariiflora*. *J. Ethnopharmacol.*, **73**, 101–109.

8 Gao, H. and Zhou, Y.W. (2005) Anti-lipid peroxidation and protection of liver mitochondria against injuries by picroside II. *World J. Gastroenterol.*, **11**, 3671–3674.

Pinellia ternata

Rhizoma Pinelliae (Banxia) is the dried tuber of *Pinellia ternata* (Thunb.) Breit. (Araceae). It is used as an antiemetic, mucolytic, and antiasthmatic agent for the treatment of cough, asthma, dizziness, palpitation, vertigo, and headache. The Chinese Pharmacopoeia notes that Rhizoma Pinelliae is incompatible with Radix Aconiti and allied medicinal materials.

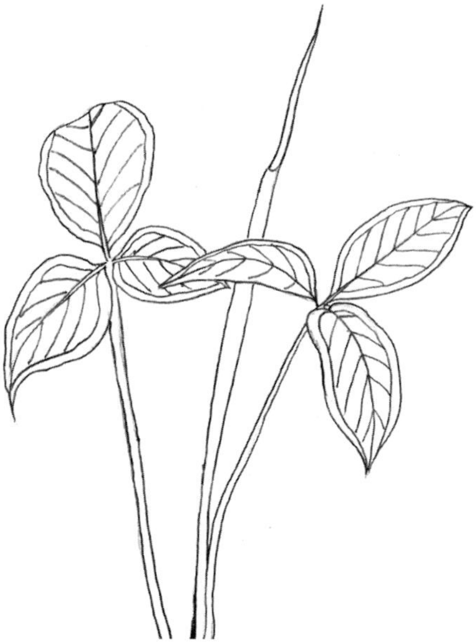

Pinellia ternata (Thunb.) Breit.

Rhizoma Pinelliae Preparatuma (Fabanxia) is the prepared tuber of *Pinellia ternata*. It is used for the same indications as Rhizoma Pinelliae, and is incompatible with Radix Aconiti and allied medicinal materials.

Chemistry

The tuber of *P. ternata* is known to have a pungent taste caused by an irritating glycoside of 3,4-dihydroxybenzaldehyde. 3,4-Dihydroxybenzaldehyde has a strong acrid taste. Further constituents from the *P. ternata* tuber were identified as ephedrine, with a content ranging from 0.002 to 0.003% [1], amino acids aspartic acid, glutamic acid, arginine, and β-aminobutyric acid, guanosine [2], β-sitosterol and β-sitosterol glucoside [3], stigmast-4-en-3-one, cycloartenol, 5α,8α-epidioxyergosta-6,22-dien-3-ol, and β-sitosterol 3-O-β-D-glucoside-6′-eicosanate [4]. Anethol was identified from the volatile oil of the *P. ternata* tuber [5]. The Chinese Pharmacopoeia requires a qualitative determination of arginine, alanine, valine and leucine in Rhizoma Pinelliae by thin-layer chromatographic (TLC) comparison with reference amino acids, and a qualitative determination of glycyrrhetic acid in Rhizoma Pinelliae Preparata by TLC comparison with reference substance, since Rhizoma Pinelliae Preparata is prepared with Radix Glycyrrhizae.

Pinellin, a crystalline plant protein isolated from the *P. ternata* tubers, was found to have a low content of cysteine, a relatively low molecular mass of 10 kDa, and did not contain hexose [6]. A trypsin inhibitor with a molecular mass of 40.8 kDa was isolated from the tubers that inhibited trypsin, but not chymotrypsin, kallikrein, or papain [7].

A polysaccharide, PT-F2-I, from *P. ternata* tuber was represented by an arabinan consisting of $\alpha(1 \rightarrow 4)$-linkages with $\alpha(1 \rightarrow 2)$-side chains as the main part, and oligosaccharide chains as the subunits involving fucose, glucose, galactose, rhamnose, ribose, and galacturonic acid [8]. An acidic polysaccharide (pinellian PA) and a glucan (pinellian G) were further isolated from *P. ternata* tuber. Pinellian PA was found to

Handbook of Chinese Medicinal Plants: Chemistry, Pharmacology, Toxicology
Weici Tang and Gerhard Eisenbrand

ISBN: 978-3-527-32226-8

be homogeneous and to be composed of L-arabinose, D-galactose, L-rhamnose, D-galacturonic acid and D-glucuronic acid, in addition to small amounts of *O*-acetyl groups and peptide moieties. The molecular mass of pinellian PA was estimated to be 118 kDa [9]. Pinellian G was also found to be homogeneous, with a molecular mass of 15 kDa. Pinellian G is a branched glucan mainly composed of α-1,4-linked D-glucopyranose residues with partially α-1,3-linked units and 4,6-branching points [10]. A glycoprotein, 6KDP, with a molecular mass of 6 kDa, was separated from the globulin fraction of *P. ternata* tubers. The content of 6KDP in the tuber was given as 6–8% [11].

Pharmacology and Toxicology

The intraduodenal infusion of a hot aqueous extract of *Pinellia* tuber in anesthetized rats resulted in a dose-related increase in efferent activity of the vagal gastric nerve. The suppressive effect on vagal gastric activity of apomorphine and copper sulfate was reported to be antagonized by prior administration of the extract [12]. *Pinellia* tuber can obviously inhibit the gastric and intestinal motility in mice and rats. It also strongly inhibits the secretion of prostaglandin E_2 (PGE_2), gastric acid, and gastric proteinase in rats [13].

The polysaccharide PT-F2-I from *Pinellia* tuber was found to be an antiemetic principle, and to inhibit the emetic action of apomorphine in frogs after oral administration [8]. The glycoprotein, 6KDP, also showed antiemetic and hemagglutination activities [11]. The acidic polysaccharide pinellian PA produced a significant stimulation of the reticuloendothelial system (RES), as shown by a carbon particle clearance test, and also exhibited potent anticomplement activity [9]. The glucan pinellian G also showed significant RES-stimulating activity, as well as pronounced anticomplement activity [10]. Amylose from *Pinellia* tuber was reported to show a marked anti-inflammatory activity in mice [14].

Lectin from the fresh juice of *Pinellia* tuber with a molecular mass of 44 kDa [15] facilitated the release of acetylcholine in the mouse motor nerve terminals [16] and formed cation channels in lipid bilayer [17]. Four tryptophan residues were identified in the *P. ternata* lectin, with one tryptophan per subunit [18]. It was further reported that the lectin-channels were voltage-independent, and had apparent subunits; the channel also exhibited a slightly higher permeability to divalent than to monovalent cations [19]. This lectin was found to raise inward Ca^{2+} concentrations by promoting Ca^{2+} influx, facilitating neurotransmitter release [20].

References

1 Zhao, L. and Su, X. (1990) Comparative analysis of alkaloid components in cultured and planted *Pinellia ternata* (Thunb.) Breit. *China J. Chin. Mater. Med.*, **15**, 146–147.

2 Wu, H., Wen, H.M., Guo, R., and Ye, D.J. (1998) Effect of ginger-processing on contents of guanosine in Pinelliae rhizoma (Banxia). *China J. Chin. Mater. Med.*, **23**, 661–663.

3 Wu, H., Su, J.Q., Cai, B.C., Zhang, L.F. and Ye, D.J. (1995) Effect of ginger processing on β-sitosterol and total alkaloid contents in Rhizome Pinelliae. *China J. Chin. Mater. Med.*, **20**, 662–664.

4 He, P., Li, S., Wang, S.J., Yang, Y.C., and Shi, J.G. (2005) Study on chemical constituents in rhizome of Pinellia ternate. *Zhongguo Zhong Yao Za Zhi*, **30**, 671–674.

5 Wang, R., Ni, J.M., and Ma, R. (1995) Volatile oils of *Pinellia ternata* (Thunb) Breit. *Chin. Pharm. J.*, **30**, 457–459.

6 Tao, Z.J., Shen, Z.M., and Yang, J.T. (1993) Conformation of the abortifacient protein

pinellin: a circular dichroic study. *J. Protein Chem.*, **12**, 387–391.

7 Wu, K.Z. and Tao, Z.J. (1981) Isolation and characterization of a trypsin inhibitor from the rhizome of *Pinellia ternata. Acta Biochim. Biophys. Sin.*, **13**, 267–274.

8 Maki, T., Takahashi, K., and Shibata, S. (1987) An antiemetic principle of *Pinellia ternata* tuber. *Planta Med.*, **53**, 410–414.

9 Gonda, R., Tomoda, M., Shimizu, N., Ohara, N., Takagi, H., and Hoshino, S. (1994) Characterization of an acidic polysaccharide with immunological activities from the tuber of *Pinellia ternata. Biol. Pharm. Bull.*, **17**, 1549–1553.

10 Tomoda, M., Gonda, R., Ohara, N., Shimizu, N., Shishido, C., and Fujiki, Y. (1994) A glucan having reticuloendothelial system-potentiating and anti-complementary activities from the tuber of *Pinellia ternata. Biol. Pharm. Bull.*, **17**, 859–861.

11 Kurata, K., Tai, T., Yang, Y., Kinoshita, K., Koyama, K., Takahashi, K., Watanabe, K., and Nunoura, Y. (1998) Quantitative analysis of anti-emetic principle in the tubers of *Pinellia ternata* by enzyme immunoassay. *Planta Med.*, **64**, 645–648.

12 Niijima, A., Okui, Y., Kubo, M., Higuchi, M., Taguchi, H., Mitsuhashi, H., and Maruno, M. (1993) Effect of *Pinellia ternata* tuber on the efferent activity of the gastric vagus nerve in the rat. *Brain Res. Bull.*, **32**, 103–106.

13 Wu, H., Cai, B.C., Rong, G.X., and Ye, D.J. (1994) The effect of *Pinellia* processed by ginger juice on gastric and intestinal function of animals. *China J. Chin. Mater. Med.*, **19**, 535–537.

14 Zhang, D.Y., Mori, M., Hall, I.H., and Lee, K. H. (1991) Anti-inflammatory agents. V. Amylose from *Pinellia ternate. Int. J. Pharmacogn.*, **29**, 29–32.

15 Wang, K.Y. and Guo, M. (1993) Purification of lectin from *Pinellia ternata. Shengwu Huaxue Zazhi*, **9**, 544–548.

16 Shi, Y.L., Xu, Y.F., Wang, W.P., Xu, K., Guo, M., and Wang, K.Y. (1992) Facilitatory effect of *Pinellia ternata* lectin on quantal release of acetylcholine from nerve terminals. *Acta Pharmacol. Sin.*, **13**, 513–516.

17 Shih, Y.L., Wang, W.P., Zhang, H., Wang, K.Y., and Guo, M. (1992) Cation channels formed in lipid bilayer by *Pinellia ternata* lectin. *Acta Physiol. Sin.*, **44**, 142–148.

18 Guo, M., Wang, K.Y., and Xu, G.J. (1992) Tryptophan residues in *Pinellia ternata* lectin. *Acta Biochim. Biophys. Sin.*, **24**, 146–151.

19 Zhang, H. and Shi, Y.L. (1994) Cation selectivity of channels formed at planar lipid bilayer by *Pinellia ternata* lectin. *Sci. China*, **37B**, 547–556.

20 Lin, J., Yao, J., Zhou, X., Sun, X., and Tang, K. (2003) Expression and purification of a novel mannose-binding lectin from *Pinellia ternata. Mol. Biotechnol.*, **25**, 215–222.

Piper kadsura

Caulis Piperis Kadsurae (Haifengteng) is the dried stem of *Piper kadsura* (Choisy) Ohwi (Piperaceae). It is used for the treatment of rheumatic and rheumatoid diseases.

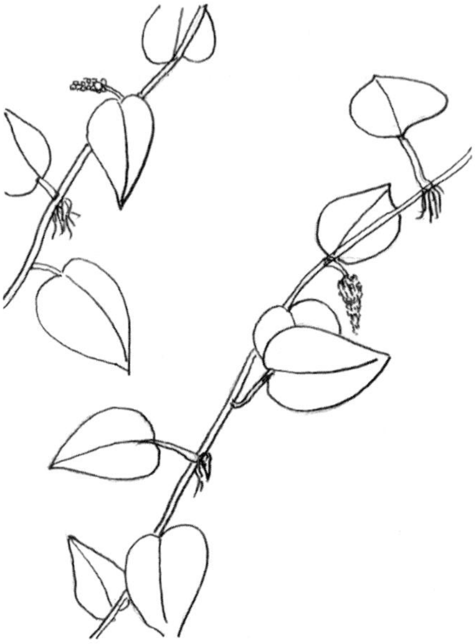

Piper kadsura (Choisy) Ohwi

The dried stem bearing the leaves of *Piper wallichii* (Miq.) Hand.-Mezz. or *Piper puberulum* (Benth.) Maxim., and the aboveground part of *Piper sarmentosum* Roxb. are listed in the appendix of the Chinese Pharmacopoeia.

Chemistry

The stem of *P. kadsura* was reported to contain a number of neolignans. These are identified as kadsurenin B, kadsurenin C, kadsurenin K, kadsurenin L [1, 2], kadsurenin M, denudatin B, kadsurenone, acuminatin, and licarin S [3]. The isolation of piperlactam S, a substituted dibenzo[*cd,f*] indol-4-one from *P. kadsura,* was also reported [4].

Pharmacology and Toxicology

Kadsurenin B, kadsurenin C, kadsurenin K, kadsurenin L [2], and denudatin B [3] were found to demonstrate significant antagonistic activity against platelet-activating factor (PAF) in the ^3H – PAF receptor-binding assay. Arterial hypotension and high-permeability pulmonary edema induced by the intravenous (i.v.) injection of PAF or endotoxin in rats were attenuated by treatment with stem extract of *P. kadsura* [5]. Among the Chinese *Piper* species, *P. kadsura* was reported to be most potent in inhibiting prostaglandin and leukotriene biosynthesis *in vitro* [6]. Piperlactam S was

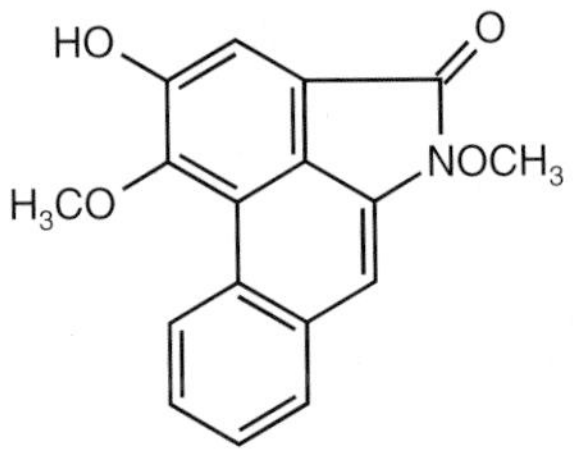

Piperlactam S

found to regulate cell proliferation, gene expression, production of cytokines, and cell-cycle progression in primary human T lymphocytes. Piperlactam S suppressed T-cell proliferation, from immediately to about 12 h after stimulation with phytohemagglutinin. The synthesis of total cellular proteins and RNA in activated cell cultures was also suppressed. The inhibitory action of piperlactam S was not via direct cytotoxicity. A cell-cycle analysis indicated that piperlactam S arrested the cell-cycle progression of activated cells from the G_1 transition to the S phase. A set of key regulatory events leading to the G_1/S

Handbook of Chinese Medicinal Plants: Chemistry, Pharmacology, Toxicology
Weici Tang and Gerhard Eisenbrand

ISBN: 978-3-527-32226-8

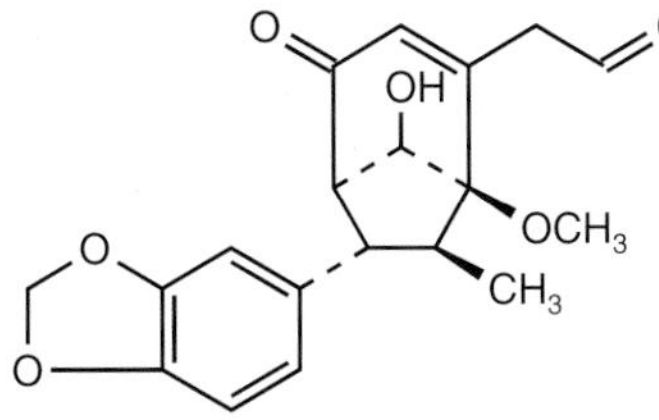

Kadsurenin B

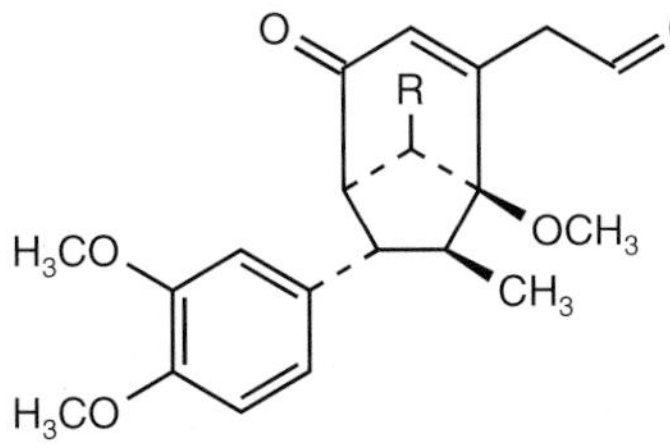

Kadsurenin C: R = OH
Kadsurenin L: R = OAc

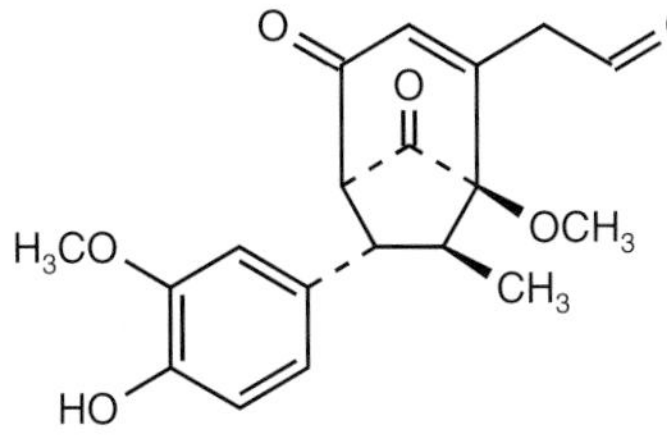

Kadsurenin K

boundary, including gene expression of cytokines and Fos protein synthesis, was examined. In activated T lymphocytes, piperlactam S suppressed the production and mRNA expression of cytokines such as interleukin-2 (IL-2), IL-4, and interferon-γ (IFN-γ) in a dose-dependent manner. In addition, Fos protein expression in activated T lymphocytes was decreased by piperlactam S. The results of a kinetic study indicated that the inhibitory effects of piperlactam S on IL-2 mRNA expressed in T cells might be related to the blockade of Fos protein synthesis. The inhibitory effects of piperlactam S on the proliferation of T cells activated by phytohemagglutinin seemed to be mediated, at least in part, through an inhibition of early transcripts of T cells (especially those of important cytokines, IL-2, IL-4), and by arresting their cell-cycle progression [4].

References

1 Ma, Y., Han, G.Q., He, C.H., and Zheng, Q.T. (1991) Neolignans from *Piper kadsura*. *Chin. Chem. Lett.*, **2**, 623–626.

2 Ma, Y., Han, G.Q., and Liu, Z.J. (1993) Two PAF-antagonistic octanoid neolignans from *Piper kadsura*. *Acta Pharm. Sin.*, **28**, 207–211.

3 Ma, Y., Han, G.Q., and Wang, Y.Y. (1993) Five PAF-antagonistic benzofuran neolignans from *Piper kadsura*. *Acta Pharm. Sin.*, **28**, 370–373.

4 Kuo, Y.C., Yang, N.S., Chou, C.J., Lin, L.C., and Tsai, W.J. (2000) Regulation of cell proliferation, gene expression, production of cytokines, and cell cycle progression in primary human T lymphocytes by piperlactam S isolated from *Piper kadsura*. *Mol. Pharmacol.*, **58**, 1057–1066.

5 Li, S.H., Fei, X., Wu, Z.L., and Cheng, S.F. (1989) Effect of an extract from the cauline of *Piper kadsura* Ohwi on endotoxin-induced hypotension and lung injury in rats. *China J. Chin. Mater. Med.*, **14**, 683–685.

6 Stoehr, J.R., Xiao, P.G., and Bauer, R. (2001) Constituents of Chinese *Piper* species and their inhibitory activity on prostaglandin and leukotriene biosynthesis *in vitro*. *J. Ethnopharmacol.*, **75**, 133–139.

Piper longum and *Piper nigrum*

Fructus Piperis Longi (Bibo) is the dried ripe or nearly ripe fruit of *Piper longum* L. (Piperaceae). It is used for the treatment of gastric diseases, epigastric pain, emesis and diarrhea; and used externally for the treatment of toothache.

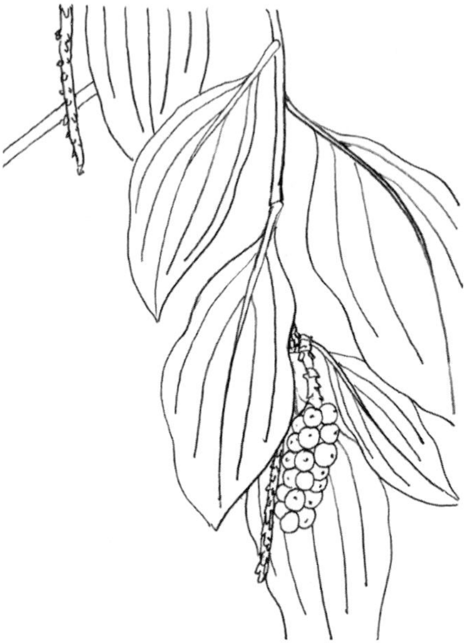

Piper nigrum L.

Fructus Piperis (Hujiao) is the dry ripe or near-ripe fruit of *Piper nigrum* L. It is used for the treatment of gastric diseases, emesis, and diarrhea. It is also used for the treatment of epilepsy. It is a well-known spice that is used worldwide.

Chemistry

The fruit of *P. nigrum*, black pepper [1], and the fruit of *P. longum*, long pepper [2], are known to contain alkaloid piperine as the major component. The Chinese Pharmacopoeia requires a qualitative determination of piperine in Fructus Piperis Longi and in Fructus Piperis by thin-layer chromatographic comparison with a reference substance, and a quantitative determination by HPLC. The content of piperine in Fructus Piperis Longi should be not less than 2.5%, and in Fructus Piperis not less than 3.0%.

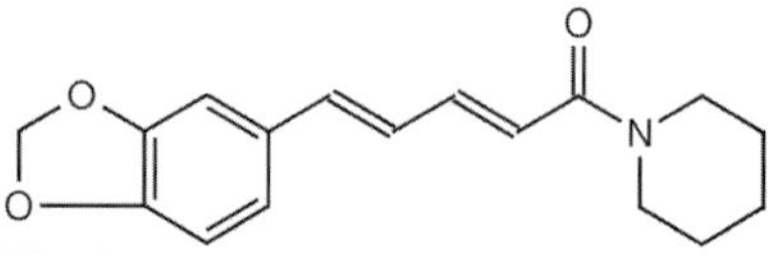

Piperine

Further alkaloids from pepper were identified as guineensine, pipericide, and *N*-feruloyltyramine, *N*-isobutyl-dodecadienamide [3]. 3′,4′,5′-Trimethoxycinnamic acid ethyl ester was isolated from the fruits of *P. longum* [4].

Pharmacology and Toxicology

About 97% of either an oral dose (170 mg kg^{-1}) or an intraperitoneal (i.p.) dose (85 mg kg^{-1}) of piperine administered to male rats was absorbed, irrespective of the administration route. Piperine was not detectable in the urine, and 3% of the dose was excreted unchanged in the feces. Only traces of piperine were detected in the serum, kidney and spleen from 30 min to 24 h. Approximately 1–2.5% of the i.p. dose of piperine was detected in the liver during 0.5–6 h after administration, in contrast to 0.1–0.25% of the oral dose. An increased excretion of conjugated uronic acids, conjugated sulfates and phenols indicated scission of the methylenedioxy group of piperine. In rats, glucuronidation and sulfation appeared to be the major steps in the disposition of piperine [5].

The LD_{50} values for single i.v., i.p., subcutaneous, intragastric and intramuscular ad-

Handbook of Chinese Medicinal Plants: Chemistry, Pharmacology, Toxicology
Weici Tang and Gerhard Eisenbrand

ISBN: 978-3-527-32226-8

ministration of piperine to male mice were 15, 43, 200, 330, and 400 mg kg^{-1}, respectively. The i.p. LD_{50} value was increased to 60 mg kg^{-1} in females, and to 132 mg kg^{-1} in weanling male mice. In female rats, the i.p. LD_{50} value was 34 mg kg^{-1}, whereas the intragastric LD_{50} value was 514 mg kg^{-1}. Most animals given a lethal dose died from respiratory paralysis within 3 to 17 min. In subacute toxicity studies, the rats died within one to three days after treatment, showing severe hemorrhagic necrosis and edema in the gastrointestinal tract, urinary bladder and adrenal glands. The death of these animals may have been attributable to multiple organ dysfunction [6].

Pepper extract was further found to possess growth-stimulatory activity towards cultured melanocytes. An aqueous extract at a concentration of 0.1 mg ml^{-1} stimulated, by 300%, the growth of a cultured mouse melanocyte line, melan-a, within eight days. Piperine also significantly stimulated melan-a cell growth. Both, pepper extract and piperine induced morphological alterations in melan-a cells, with more and longer dendrites observed. The augmentation of growth by piperine was effectively inhibited by RO-31-8220, a selective protein kinase C (PKC) inhibitor; this suggested that PKC signaling was involved in the mechanism of action [7]. The growth-stimulatory effect of the pepper extract was more potent than an equivalent concentration of pure piperine. A number of other amides with growth-stimulatory properties were detected in the extract [3].

The painting and feeding of mice with 2 mg of an extract from pepper for three days each week over a three-month period resulted in a significant increase in the number of tumor-bearing mice. The tumor incidence induced by pepper extract could be reduced by the administration of 5 or 10 mg vitamin A palmitate twice weekly for three months. Feeding of mice with pepper powder in the diet (1.7%) had no impact on carcinogenesis [8]. In V79 cells, piperine has been reported to promote DNA damage and cytotoxicity induced by benzo[*a*]pyrene (BaP). The V79 cells were treated with a nontoxic dose of piperine (1–20 μ*M*) plus 10 μ*M* BaP, or pretreated with piperine for 30 min or 2 h prior to the administration of 10 μ*M* BaP. BaP cytotoxicity was potentiated significantly by piperine under each experimental condition. The lowest activities of glutathione *S*-transferase (GST) and UDP-glucuronyltransferase in V79 cells occurred between 30 min and 1 h after piperine pretreatment. The pretreatment of V79 cells with piperine also caused an increase in the covalent binding of BaP-diolepoxide to DNA, which was 2.3-fold greater than that of cells without piperine treatment [9]. In contrast, following oral administration to mice at 25, 50, and 75 mg kg^{-1}, piperine was reported to suppress the genotoxicity of BaP and cyclophosphamide, as measured by micronucleus formation [10].

Piperine was also reported to protect against lung carcinoma induction by BaP in mice. For this, mice were administered BaP at 50 mg kg^{-1} twice weekly for four consecutive weeks to induce lung cancer by the end of the 16th week. Piperine (50 mg kg^{-1}) was administered on alternate days for 16 weeks immediately after the first dose of the carcinogen. The results showed that piperine significantly decreased not only lung carcinogenicity but also the extent of lipid peroxidation, with concomitant increases in the activities of superoxide dismutase, catalase, glutathione peroxidase, and glutathione (GSH) levels in the serum of lung cancer-bearing animals [11].

Piperine, at concentrations ranging from 0.005 to 10 μ*M* per plate, did not show any mutagenic activity in tests using *Salmonella typhimurium*, with or without metabolic activation. In the bone marrow micronucleus test in mice, piperine at doses of 10 and 20 mg kg^{-1}, was nonmutagenic. As in somatic cells, piperine at doses of 10 and

$50\,mg\,kg^{-1}$ failed to induce mutations in the male germ cells of mice, as assessed by sperm shape abnormalities and dominant lethal tests [12]. Piperine was found capable of inhibiting or reducing the oxidative changes induced by chemical carcinogens in rat intestine. Carcinogens, including 7,12-dimethylbenz[*a*]anthracene (DMBA), dimethylamino-methylazobenzene and 3-methylcholanthrene, when given to male rats induced GSH depletion, with a substantial increase in thiobarbituric-reactive substances and enzyme activities. Piperine treatment along with carcinogens resulted in an inhibition of thiobarbituric-reactive substances [13]. Piperine has also been reported to show cytotoxic activity towards several tumor cell lines. *In vivo*, piperine administered intraperitoneally at doses of 50 or $100\,mg\,kg^{-1}$, markedly inhibited tumor development in female Swiss mice transplanted with Sarcoma 180 cells when given for seven days, starting at one day after inoculation [14].

Piperine was reported to show protective effects against experimental gastric ulcer in rats or mice. The oral administration of piperine at doses of 25, 50, and $100\,mg\,kg^{-1}$ to rats or mice with experimental gastric ulceration induced by stress, indomethacin, hydrochloric acid or pyloric ligation, significantly inhibited the volume of gastric juice, gastric acidity, and pepsin A activity [15]. Studies with animals deficient in the vanilloid (capsaicin) receptor TRPV1 have confirmed the pivotal role of TRPV1 in the development of postinflammatory hyperalgesia, and an enhanced TRPV1 expression has been described in various human disorders. Piperine was found to be more efficient than capsaicin in desensitizing the human TRPV1 receptor [16]. The effect of piperine on the human vanilloid receptor TRPV1 has been demonstrated using a whole-cell electrophysiological model. Piperine produced a clear agonist activity at the human TRPV1 receptor, yielding rapidly activating whole-cell currents that were antagonized by the competitive TRPV1 antagonist and the noncompetitive TRPV1 blocker. The effect of piperine on TRPV1 receptor might also explain its activity on gastrointestinal function [17].

In rats, black pepper and piperine were reported to effectively protect against oxidative stress induced by a high-fat diet. Black pepper at doses of 0.25 g or $0.5\,g\,kg^{-1}$, and piperine at a dose of $20\,mg\,kg^{-1}$, were administered to rats fed a high-fat diet for a period of 10 weeks. This resulted in a significant lowering of the levels of thiobarbituric acid-reactive substances and conjugated diene levels, and caused elevations in the activities of superoxide dismutase, catalase, glutathione peroxidase, GST and reduced GSH in the liver, heart, kidney, intestine and aorta to near-normal values. This indicated that supplementation with black pepper or piperine might reduce high-fat diet-induced oxidative stress [18].

Piperine was reported to exert anti-inflammatory activity in rats, and to significantly inhibit carrageenin-induced rat paw edema, cotton pellet-induced granuloma, and croton oil-induced granuloma pouch. Biochemical investigations showed that piperine acted significantly on early acute changes in inflammatory processes and chronic granulative changes, and acted partially through a stimulation of pituitary adrenal axis [19]. Carrageenin-induced rat paw edema showed a direct correlation with liver lipid peroxidation. The pretreatment of rats with piperine reduced the liver lipid peroxidation, acid phosphatase activity, and paw edema. However, reduction was only observed in animals with experimental inflammation, and not in control animals after piperine treatment [20]. In B16F-10 melanoma cells, which produce very large amounts of proinflammatory cytokines such as interleukin-1β (IL-1β), IL-6, tumor necrosis factor-α (TNF-α) and granulocyte-macrophage colony-stimulating factor

(GM-CSF), treatment with piperine significantly reduced the proinflammatory cytokine production. Piperine was also found to reduce the expression of IL-1β, IL-6, TNF-α, GM-CSF and IL-12p40 genes, to inhibit collagen matrix invasion of B16F-10 melanoma cells, and to inhibit matrix metalloproteinase production. The nuclear translocation of p65, p50, Rel subunits of NF-κB and other transcription factors such as ATF-2, Fos and cAMP response element binding (CREB) were also inhibited by piperine [21]. Ethyl 3′,4′,5′-trimethoxycinnamate and piperine from *P. longum* were found to inhibit the expression of intercellular adhesion molecule 1 (ICAM-1) in primary human umbilical vein endothelial cells induced by TNF-α. Ethyl 3′,4′,5′-trimethoxycinnamate was also found to block the adhesion of neutrophils to endothelium in both time- and concentration-dependent manner, the effect being reversible. Structure–activity studies revealed that the chain length of the alcohol moiety, the substituents at the aromatic ring, and the α,β-double bond in cinnamic acid ester each have importance for the inhibition of TNF-α-induced expression of ICAM-1 in endothelial cells [4].

An ethanol extract of the fruit of *P. longum* inhibited the aggregation of rabbit platelets. The inhibitory effect of the extract seemed to be through antagonization of the thromboxane A_2 (TXA2) receptor, since the extract inhibited platelet aggregation induced by TXA2 receptor agonist in a concentration-dependent manner, but only weakly inhibited platelet aggregation induced by thrombin. In addition, the extract antagonized the TXA2 receptor in a noncompetitive manner [22]. The simultaneous administration of piperine and a high-fat diet significantly reduced plasma lipids and lipoproteins levels, as well as the plasma levels of high-density lipoprotein [23].

Piperine was found to show antihepatotoxic effects in mice, counteracting *tert*-butyl hydroperoxide (TBH) and carbon tetrachloride (CCl_4) by reducing lipid peroxidation and preventing the depletion of GSH and total thiols in the liver of intoxicated mice [24]. However, it was also reported that the hepatotoxicity of CCl_4 in rats could be potentiated by piperine. The maximum potentiation occurred when piperine was given orally at a dose of $100\,mg\,kg^{-1}$, 4 h prior to an i.p. injection of CCl_4. The activities of plasma glutamic pyruvic transaminase (GPT) and plasma glutamic oxaloacetic transaminase (GOT) were each elevated by 70–80%. Piperine pretreatment also potentiated CCl_4-induced lipid peroxidation in rat liver, the extent of potentiation correlating well with the rise in plasma enzyme activity. Piperine also exhibited a concentration-dependent potentiation of CCl_4-induced lipid peroxidation in an *in vitro* system, where rat liver tissue was preincubated with piperine and CCl_4 and subsequently added to the incubation medium [25].

Piperine was reported to be an effective anticonvulsant zing antagonizing convulsions in experimental animals induced by physical and chemical methods. Piperine also showed sedative, hypnotic, tranquilizing and muscle-relaxing actions, and an ability to intensify the depressive action of other depressants [26]. It significantly blocked convulsions of mice induced by intracerebroventricular injection of kainate, but had no (or only slight) effects on convulsions induced by L-glutamate, *N*-methyl-D-aspartate, or guanidinosuccinate. Piperine suspensions, injected intraperitoneally at 1 h before injection of the threshold intracerebroventricular dose of 1 n*M* kainate to induce clonic convulsions, blocked these convulsions, with an ED_{50} of approximately $45\,mg\,kg^{-1}$. Although piperine blocked the convulsions induced by kainate, the compound appeared not to act as a kainate receptor antagonist [27]. The i.p. administration of piperine to mice significantly increased serotonin levels in the cerebral cortex measured at 1 h after injection. On the other hand, lower levels of serotonin were observed in the hippocampus, midbrain, and cerebellum. Dopamine levels in piperine-treated mice were markedly higher in the hypothalamus, while noradrenaline levels were lower in all regions of the brain [28].

Piperine also exhibited an antidepressant-like effect in mice exposed to chronic mild stress. The repeated administration of piperine to mice for 14 days at doses of 2.5, 5, and 10 mg kg^{-1}, reversed changes in sucrose consumption, plasma corticosterone levels, and open field activity induced by chronic mild stress. The decreased proliferation of hippocampal progenitor cells was ameliorated, and the level of brain-derived neurotrophic factor in the hippocampus of stressed mice elevated by piperine treatment. In addition, piperine concentration-dependently protected primary cultured hippocampal neurons from lesions induced by corticosterone. Treatment with piperine for 72 h reversed the reduction of brain-derived neurotrophic factor mRNA expression in cultured hippocampal neurons induced by corticosterone, suggesting that an upregulation of the progenitor cell proliferation of the hippocampus, as well as a cytoprotective activity, might be involved in the antidepressant-like effect of piperine. This might be closely related to the elevation of hippocampal brain-derived neurotrophic factor levels [29].

Piperine administered orally to mature male rats at 10 mg kg^{-1} for 30 days caused a significant reduction in the weights of the testes and accessory sex organs. Histological studies showed that, at a dose of 5 mg kg^{-1}, it caused a partial degeneration of germ cell types and at a dose of 10 mg kg^{-1} severe damage to the seminiferous tubules. Correlated to the structural changes, falls in caput and cauda epididymal sperm concentrations were observed. The high dosage of piperine also caused a marked increase in serum gonadotropins and a decrease in intratesticular testosterone concentration, despite normal serum testosterone titers [30]. The body weight of piperine-treated rats remained unchanged. A significant decline in the activities of superoxide dismutase, catalase, glutathione peroxidase and glutathione reductase, along with an increase in hydrogen peroxide generation and lipid peroxidation, were observed after treatment with piperine. These results indicates that piperine caused a decrease in the activity of antioxidant enzymes and sialic acid levels in the epididymis, increasing reactive oxygen species (ROS) levels that could damage the epididymal environment and sperm function [31].

An ethanol extract of long pepper, given orally to mice at single doses of 0.5, 1.0, and 3 g kg^{-1}, and at 100 mg kg^{-1} for 90 days, did not cause any significant changes in the pre- and post-treatment body weight of the test animals, but the weight gain of control animals was significant. Long pepper caused a significant increase in the weight of the lungs and spleen of treated animals compared to controls, and a significant increase in reproductive organ weights, sperm motility, sperm count; however, the treatment failed to elicit any spermatotoxic effect [32].

References

1 Rathnawathie, M. and Buckle, K.A. (1983) Determination of piperine in pepper (*Piper nigrum*) using high-performance liquid chromatography. *J. Chromatogr.*, **264**, 316–320.

2 Li, H.S., Jia, Z.C., Zhang, M.H., and Zhou, J.Y. (1986) Determination of piperine in *Piper longum* L. by HPLC. *Chin. J. Pharm. Anal.*, **6**, 346–348.

3 Lin, Z., Liao, Y., Venkatasamy, R., Hider, R.C., and Soumyanath, A. (2007) Amides from *Piper nigrum* L. with dissimilar effects on melanocyte proliferation *in vitro*. *J. Pharm. Pharmacol.*, **59**, 529–536.

4 Kumar, S., Arya, P., Mukherjee, C., Singh, B. K., Singh, N., Parmar, V.S., Prasad, A.K., and Ghosh, B. (2005) Novel aromatic ester from *Piper longum* and its analogues inhibit expression of cell adhesion molecules on endothelial cells. *Biochemistry*, **44**, 15944–15952.

5 Bhat, B.G. and Chandrasekhara, N. (1986) Studies on the metabolism of piperine: absorption, tissue distribution and excretion of urinary conjugates in rats. *Toxicology*, **40**, 83–92.

6 Piyachaturawat, P., Glinsukon, T., and Toskulkao, C. (1983) Acute and subacute toxicity of piperine in mice, rats and hamsters. *Toxicol. Lett.*, **16**, 351–359.

7 Lin, Z., Hoult, J.R., Bennett, D.C., and Raman, A. (1999) Stimulation of mouse melanocyte proliferation by *Piper nigrum* fruit extract and its main alkaloid, piperine. *Planta Med.*, **65**, 600–603.

8 Shwaireb, M.H., Wrba, H., el-Mofty, M.M., and Dutter, A. (1990) Carcinogenesis induced by black pepper (*Piper nigrum*) and modulated by vitamin A. *Exp. Pathol.*, **40**, 233–238.

9 Chu, C.Y., Chang, J.P., and Wang, C.J. (1994) Modulatory effect of piperine on benzo[a] pyrene cytotoxicity and DNA adduct formation in V-79 lung fibroblast cells. *Food Chem. Toxicol.*, **32**, 373–377.

10 Selvendiran, K., Padmavathi, R., Magesh, V., and Sakthisekaran, D. (2005) Preliminary study on inhibition of genotoxicity by piperine in mice. *Fitoterapia*, **76**, 296–300.

11 Selvendiran, K., Banu, S.M., and Sakthisekaran, D. (2004) Protective effect of piperine on benzo[a]pyrene-induced lung carcinogenesis in Swiss albino mice. *Clin. Chim. Acta*, **350**, 73–78.

12 Karekar, V.R., Mujumdar, A.M., Joshi, S.S., Dhuley, J., Shinde, S.L., and Ghaskadbi, S. (1996) Assessment of genotoxic effect of piperine using *Salmonella typhimurium* and somatic and somatic and germ cells of Swiss albino mice. *Arzneim.-Forsch.*, **46**, 972–975.

13 Khajuria, A., Thusu, N., Zutshi, U., and Bedi, K.L. (1998) Piperine modulation of carcinogen induced oxidative stress in intestinal mucosa. *Mol. Cell. Biochem.*, **189**, 113–118.

14 Bezerra, D.P., Castro, F.O., Alves, A.P., Pessoa, C., Moraes, M.O., Silveira, E.R., Lima, M.A., Elmiro, F.J., and Costa-Lotufo, L.V. (2006) In vivo growth-inhibition of Sarcoma 180 by piplartine and piperine, two alkaloid amides from Piper. *Braz. J. Med. Biol. Res.*, **39**, 801–807.

15 Bai, Y.F. and Xu, H. (2000) Protective action of piperine against experimental gastric ulcer. *Acta Pharmacol. Sin.*, **21**, 357–359.

16 Szallasi, A. (2005) Piperine: researchers discover new flavor in an ancient spice. *Trends Pharmacol. Sci.*, **26**, 437–439.

17 McNamara, F.N., Randall, A., and Gunthorpe, M.J. (2005) Effects of piperine, the pungent component of black pepper, at the human vanilloid receptor (TRPV1). *Br. J. Pharmacol.*, **144**, 781–790.

18 Vijayakumar, R.S., Surya, D., and Nalini, N. (2004) Antioxidant efficacy of black pepper (*Piper nigrum* L.) and piperine in rats with high fat diet induced oxidative stress. *Redox Rep.*, **9**, 105–110.

19 Mujumdar, A.M., Dhuley, J.N., Deshmukh, V. K., Raman, P.H., and Naik, S.R. (1990) Anti-inflammatory activity of piperine. *Jpn. J. Med. Sci. Biol.*, **43**, 95–100.

20 Dhuley, J.N., Raman, P.H., Mujumdar, A.M., and Naik, S.R. (1993) Inhibition of lipid peroxidation by piperine during experimental inflammation in rats. *Indian J. Exp. Biol.*, **31**, 443–445.

21 Pradeep, C.R. and Kuttan, G. (2004) Piperine is a potent inhibitor of nuclear factor-κB (NF-κB), c-Fos, CREB, ATF-2 and proinflammatory cytokine gene expression in B16F-10 melanoma cells. *Int. Immunopharmacol.*, **4**, 1795–1803.

22 Iwashita, M., Saito, M., Yamaguchi, Y., Takagaki, R., and Nakahata, N. (2007) Inhibitory effect of ethanol extract of *Piper longum* L. on rabbit platelet aggregation through antagonizing thromboxane A_2 receptor. *Biol. Pharm. Bull.*, **30**, 1221–1225.

23 Vijayakumar, R.S. and Nalini, N. (2006) Piperine, an active principle from *Piper nigrum*, modulates hormonal and apo lipoprotein profiles in hyperlipidemic rats. *J. Basic Clin. Physiol. Pharmacol.*, **17**, 71–86.

24 Koul, I.B. and Kapil, A. (1993) Evaluation of the liver protective potential of piperine, an active principle of black and long peppers. *Planta Med.*, **59**, 413–417.

25 Piyachaturawat, P., Kingkaeohoi, S., and Toskulkao, C. (1995) Potentiation of carbon tetrachloride hepatotoxicity by piperine. *Drug Chem. Toxicol.*, **18**, 333–344.

26 Pei, Y.Q. (1983) A review of pharmacology and clinical use of piperine and its derivatives. *Epilepsia*, **24**, 177–182.

27 D'Hooge, R., Pei, Y.Q., Raes, A., Lebrun, P., van Bogaert, P.P., and de Deyn, P.P. (1996) Anticonvulsant activity of piperine on seizures induced by excitatory amino acid receptor agonists. *Arzneim.-Forsch.*, **46**, 557–560.

28 Mori, A., Kabuto, H., and Pei, Y.Q. (1985) Effects of piperine on convulsions and on brain serotonin and catecholamine levels in E1 mice. *Neurochem. Res.*, **10**, 1269–1275.

29 Li, S., Wang, C., Wang, M., Li, W., Matsumoto, K., and Tang, Y. (2007) Antidepressant like effects of piperine in chronic mild stress treated mice and its possible mechanisms. *Life Sci.*, **80**, 1373–1381.

30 Malini, T., Manimaran, R.R., Arunakaran, J., Aruldhas, M.M., and Govindarajulu, P. (1999) Effects of piperine on testis of albino rats. *J. Ethnopharmacol.*, **64**, 219–225.

31 D'cruz, S.C. and Mathur, P.P. (2005) Effect of piperine on the epididymis of adult male rats. *Asian J. Androl.*, **7**, 363–368.

32 Shah, A.H., Al-Shareef, A.H., Ageel, A.M., and Qureshi, S. (1998) Toxicity studies in mice of common spices, *Cinnamomum zeylanicum* bark and *Piper longum* fruits. *Plant Foods Hum. Nutr.*, **52**, 231–239.

Plantago asiatica and *Plantago depressa*

Herba Plantaginis (Cheqiancao) is the dried whole plant of *Plantago asiatica* L. and *Plantago depressa* Willd. (Plantaginaceae). It is used as a diuretic, expectorant, and hemostatic agent for the treatment of edema with oliguria, urinary infection and painful urination, cough, hemoptysis, and carbuncles.

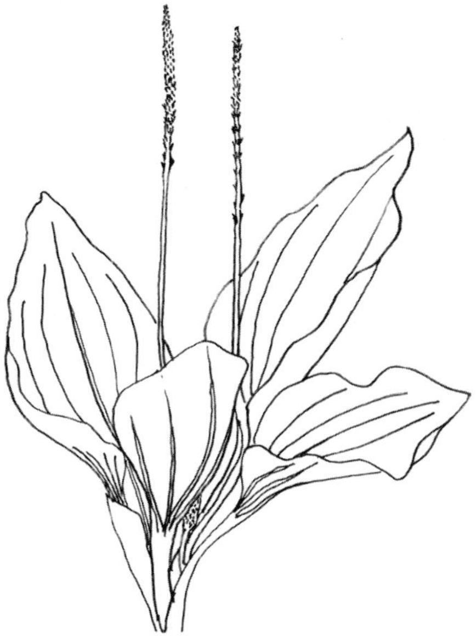

Plantago asiatica L.

Semen Plantaginis (Cheqianzi) is the dried seed of *Plantago asiatica* and *Plantago depressa*. It is used as a diuretic and expectorant agent for the treatment of edema, dysuria, painful urination, cough, and ophthalmic infections.

Chemistry

The whole plant of *P. asiatica* was reported to contain phenylethanol glycosides, with plantamajoside as the major component [1]. Structure-related phenylethanol glycosides from the whole plant of *P. asiatica* were identified as acteoside, isoacteoside, calceorioside B, leucosceptoside A, isomartynoside, martynoside, plantainosides A–F [2], and plantasioside [3]. These phenylethanol glycosides differ from each other only by the sugar moiety at position 3 of glucose, and the nature of the phenolic acid. The phenolic acids were found mainly to be either caffeic acid or ferulic acid. Related phenylethanol glycosides were also isolated from the seed of *P. depressa* [3].

The major flavanone glycoside from the seed of *P. asiatica* was identified as plantagoside [4]. A flavone glycoside from the leaves of *P. asiatica* was identified as homoplantaginin [5].

The iridoid glycoside aucubin was isolated from *P. asiatica* and *P. depressa* [6].

Pharmacology and Toxicology

The major phenylethanol glycoside of *P. asiatica*, plantamajoside, was found to show inhibitory effects on mouse ear edema induced by arachidonic acid [1]. The flavanone glycoside plantagoside was reported to inhibit the proliferation of rat lymphocytes induced by concanavalin A (Con A) [4]. Aucubin (structure, see p 511) was found to possess significant antinociceptive and anti-inflammatory activities, and also inhibited the carrageenan-induced hind paw edema and *p*-benzoquinone-induced writhings in mice [7]. Aucubin has been found to inhibit the production of tumor necrosis factor-α (TNF-α) and interleukin-6 (IL-6) in rat basophilic leukemia RBL-2H3 mast cells stimulated by antigen. Aucubin also inhibited Ag-induced nuclear translocation of the p65 subunit of nuclear factor-κB (NF-κB) and degradation of I-κ Bα [8]. In contrast, aucubigenin, the hydrolyzed product of aucubin, but not aucubin itself, suppressed the expression of TNF-α mRNA and the production of TNF-α in RAW 264.7 murine macrophage cells. The IC_{50}

Handbook of Chinese Medicinal Plants: Chemistry, Pharmacology, Toxicology
Weici Tang and Gerhard Eisenbrand

ISBN: 978-3-527-32226-8

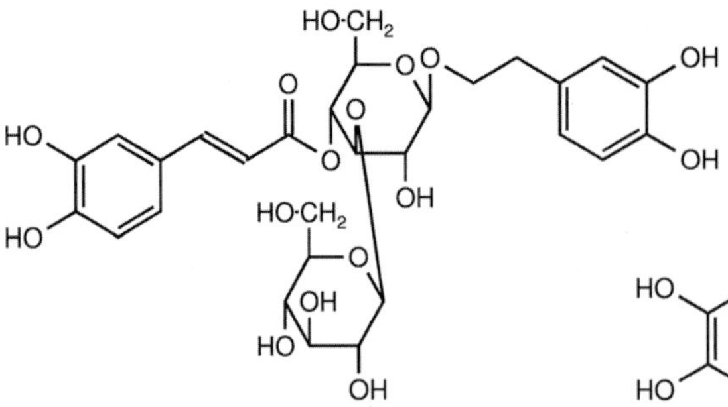

Plantamajoside

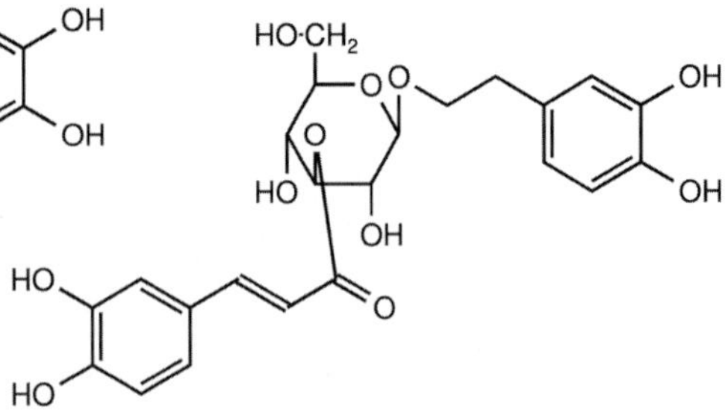

Plantainoside A

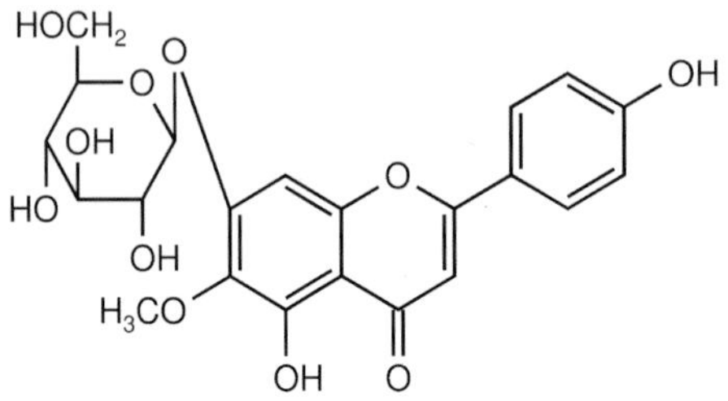

Homoplantaginin

value in inhibiting TNF-α production was reported as approximately 9 μ*M*. In addition, treatment with aucubigenin blocked both I-κ Bα degradation and the translocation of NF-κ B from the cytosol into the nuclear fraction [9].

Aucubin was reported to show cytotoxic activity. A further study revealed that aucubin was able to stabilize the covalent attachments of the topoisomerase I subunit to DNA at sites of DNA strand breaks, generating cleavage complexes that were active as poisons of topoisomerase I, but not of topoisomerase II [10].

The phenylethanol glycoside plantamajoside was reported to exhibit antispasmodic activity on isolated guinea pig ileum precontracted by acetylcholine, histamine, and potassium ions. Plantamajoside also inhibited the contraction of isolated guinea pig trachea induced by barium ions [11].

References

1 Murai, M., Tamayama, Y., and Nishibe, S. (1995) Phenylethanoids in the herb of *Plantago lanceolata* and inhibitory effect on arachidonic acid-induced mouse ear edema. *Planta Med.*, **61**, 479–480.

2 Miyase, T., Ishino, M., Akahori, C., Ueno, A., Ohkawa, Y., and Tanizawa, H. (1991) Phenylethanoid glycosides from *Plantago asiatica*. *Phytochemistry*, **30**, 2015–2018.

3 Nishibe, S., Tamayama, Y., Sasahara, M., and Andary, C. (1995) A phenylethanoid glycoside from *Plantago asiatica*. *Phytochemistry*, **38**, 741–743.

4 Yamada, H., Nagai, T., Takemoto, N., Endoh, H., Kiyohara, H., Kawamura, H., and Otsuka, Y. (1989) Plantagoside, a novel α-mannosidase inhibitor isolated from the seeds of *Plantago asiatica*, suppresses immune response. *Biochem. Biophys. Res. Commun.*, **165**, 1292–1298.

5 Aritomi, M. (1967) Homoplantaginin, a new flavonoid glycoside in leaves of *Plantago asiatica* Linnaeus. *Chem. Pharm. Bull. (Tokyo)*, **15**, 432–434.

6 Guo, Y., Cha, M., Chao, A., and Yuan, C. (1991) Determination of aucubin in *Plantago asiatica* L., *P. major* L. and *P. depressa* Willd. by HPLC. *China J. Chin. Mater. Med.*, **16**, 743–744, 763.

7 Kupali, E., Tatli, I.I., Akdemir, Z.S., and Yesilada, E. (2007) Bioassay-guided isolation of anti-inflammatory and antinociceptive glycoterpenoids from the flowers of *Verbascum lasianthum* Boiss. ex Bentham. *J. Ethnopharmacol.*, **110**, 444–450.

8 Jeong, H.J., Koo, H.N., Na, H.J., Kim, M.S., Hong, S.H., Eom, J.W., Kim, K.S., Shin, T.Y.,

and Kim, H.M. (2002) Inhibition of TNF-α and IL-6 production by aucubin through blockade of NF-κB activation RBL-2H3 mast cells. *Cytokine*, **18**, \252–259.

9 Park, K.S. and Chang, I.M. (2004) Anti-inflammatory activity of aucubin by inhibition of tumor necrosis factor-α production in RAW 264.7 cells. *Planta Med.*, **70**, 778–779.

10 Galvez, M., Martin-Cordero, C., and Ayuso, M. J. (2005) Iridoids as DNA topoisomerase I poisons. *J. Enzyme Inhib. Med. Chem.*, **20**, 389–392.

11 Fleer, H. and Verspohl, E.J. (2007) Antispasmodic activity of an extract from *Plantago lanceolata* L. and some isolated compounds. *Phytomedicine*, **14**, 409–415.

Platycladus orientalis

Cacumen Platycladi (Cebaiye) is the dried twig and leaf of *Platycladus orientalis* (L.) Franco (Cupressaceae). It is used as a hemostatic agent for the treatment of epistaxis, hematemesis, hemoptysis, hematochezia, and abnormal uterine bleeding.

Semen Platycladi (Baiziren) is the dried seed of *Platycladus orientalis*. It is used as a sedative and laxative agent for the treatment of insomnia and constipation.

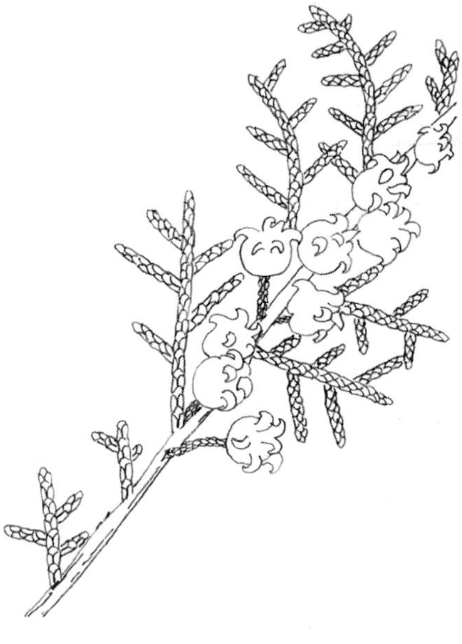

Platycladus orientalis (L.) Franco

Chemistry

The monoterpene platydiol, and the diterpenes 15-hydroxypinusolidic acid [1], platyclolactonic acid, 14,15-bisnor-8(17)-labdene-16,19-dioic acid, and 6,7-dehydrosandaracopimaric acid, were isolated from the pericarp of *P. orientalis* [2]. Cacumen Platycladi is also known to contain flavones and flavone glycosides with quercetin and quercitroside (quercitrin) as the major components. The Chinese Pharmacopoeia requires a qualitative determination of quercetin in Cacumen Platycladi by thin-layer chromatographic comparison with reference substance, and a quantitative determination of quercitroside in Cacumen Platycladi by HPLC. The content of quercitroside in Cacumen Platycladi should be not less than 0.1%.

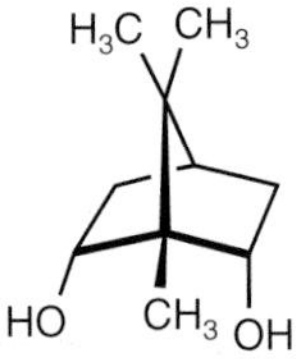

Platydiol

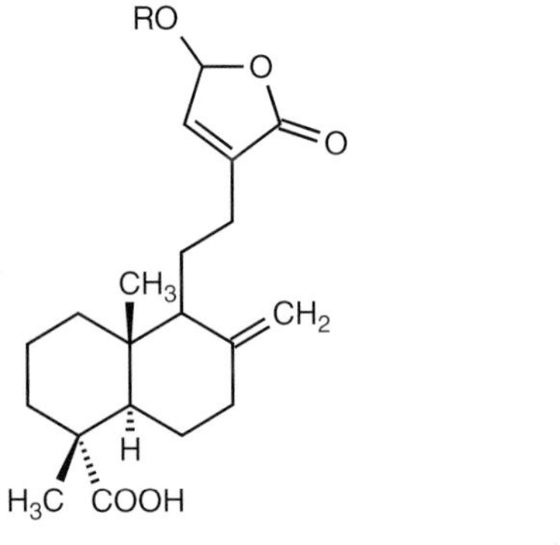

15-Hydroxypinusolidic acid: R = H
15-Methoxypinusolidic acid: R = CH_3

Further compounds from *P. orientalis* were identified as 15-methoxypinusolidic acid, *ent*-isopimara-15-en-3α,8α-diol, lambertianic acid, isopimara-8(9),15-dien-18-oic acid, isopimara-7(8),15-dien-3β,18-diol [3], and related labdane and isopimarane

Handbook of Chinese Medicinal Plants: Chemistry, Pharmacology, Toxicology
Weici Tang and Gerhard Eisenbrand

ISBN: 978-3-527-32226-8

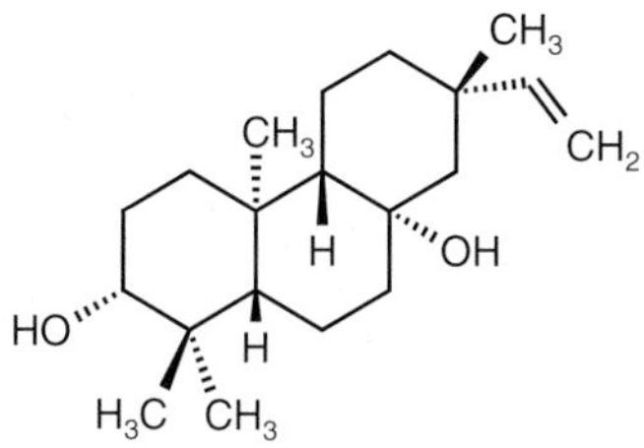

ent-Isopimara-15-en-3α,8α-diol

derivatives [4]. The seed oil from *P. orientalis* was found to contain 3% of 5,11,14-eicosatrienoic acid [5].

Pharmacology and Toxicology

The dietary administration of *P. orientalis* seed oil was reported to suppress anti-erythrocyte autoantibodies and to prolong the survival of NZB mice. These effects were paralleled by an abundance of 5,11,14-eicosatrienoic acid in serum lipids; most striking was its accumulation in serum and tissue phospholipids. 5,11,14-Eicosatrienoic acid was incorporated into all tissue phospholipids examined, except for brain phosphatidylinositol. Among the tissues, liver showed the highest incorporation of 5,11,14-eicosatrienoic acid into phosphatidylcholine, phosphatidylserine, and phosphatidylinositol; however, the spleen phospholipids contained higher quantities of 5,11,14-eicosatrienoic acid than other tissues [5].

15-Methoxypinusolidic acid was reported to show a significant protective activity against neurotoxicity in primary cultures of rat cortical cells, as induced by glutamate and other neurotoxic agents [3]. 15-Methoxypinusolidic acid exhibited a more selective protection against the neurotoxicity of *N*-methyl-D-aspartate (NMDA) than that of kainic acid, and also effectively reduced the increase of intracellular calcium in cortical cells induced by glutamate, the subsequent overproduction of nitric oxide (NO) and of cellular peroxide; it also inhibited glutathione depletion and lipid peroxidation. This suggested that 15-methoxypinusolidic acid attenuated glutamate-induced excitotoxicity via a stabilization of intracellular calcium homeostasis and suppression of oxidative stress, possibly through its action on the NMDA receptor [6].

The extract of *P. orientalis* was found to show hypouricemic activity, caused by the flavones quercetin and rutin. The oral administration of quercetin and rutin to hyperuricemic mice induced by oxonate, exerted dose-dependent hypouricemic effects. The effects of quercetin and rutin were more potent than those of the extract. Quercetin (at a dose of 50 mg kg^{-1}) or rutin (at a dose of 100 mg kg^{-1}) markedly lowered the serum urate levels of the oxonate-pretreated mice. In addition, the *P. orientalis* extract, quercetin and rutin each significantly inhibited xanthine dehydrogenase and xanthine oxidase activities in mouse liver homogenates [7].

References

1 Kuo, Y.H. and Chen, W.C. (1990) 15-Hydroxypinusolidic acid, a new diterpene from the pericarp of *Platycladus orientalis* Franco. *Heterocycles*, **31**, 1705–1709.

2 Kuo, Y.H., Chen, W.C., and Lee, C.K. (2000) Four new terpenes from the pericarp of *Platycladus orientalis*. *Chem. Pharm. Bull. (Tokyo)*, **48**, 766–778.

3 Koo, K.A., Sung, S.H., and Kim, Y.C. (2002) A new neuroprotective pinusolide derivative from the leaves of *Biota orientalis*. *Chem. Pharm. Bull. (Tokyo)*, **50**, 834–836.

4 Asili, J., Lambert, M., Ziegler, H.L., Staerk, D., Sairafianpour, M., Witt, M., Asghari, G., Ibrahimi, I.S., and Jarozewski, J.W. (2004) Labdanes and isopimaranes from *Platycladus*

orientalis and their effects on erythrocyte membrane and on *Plasmodium falciparum* growth in the erythrocyte host cells. *J. Nat. Prod.*, **67**, 631–637.

5 Lai, L.T., Naiki, M., Yoshida, S.H., German, J.B., and Gershwin, M.E. (1994) Dietary *Platycladus orientalis* seed oil suppresses anti-erythrocyte autoantibodies and prolongs survival of NZB mice. *Clin. Immunol. Immunopathol.*, **71**, 293–302.

6 Koo, K.A., Kim, S.H., Lee, M.K., and Kim, Y.C. (2006) 15-Methoxypinusolidic acid from *Biota orientalis* attenuates glutamate-induced neurotoxicity in primary cultured rat cortical cells. *Toxicol. In Vitro*, **20**, 936–941.

7 Zhu, J.X., Wang, Y., Kong, L.D., Yang, C., and Zhang, X. (2004) Effects of *Biota orientalis* extract and its flavonoid constituents, quercetin and rutin on serum uric acid levels in oxonate-induced mice and xanthine dehydrogenase and xanthine oxidase activities in mouse liver. *J. Ethnopharmacol.*, **93**, 133–140.

Platycodon grandiflorum

Radix Platycodonis (Jiegeng) is the dried root of *Platycodon grandiflorum* (Jacq.) A. DC. (Campanulaceae). It is used as an expectorant for the treatment of respiratory disorders and laryngopharyngeal pain.

Platycodon grandiflorum (Jacq.) A. DC.

Chemistry

The root of *P. grandiflorum* is known to contain saponins as major constituents. The Chinese Pharmacopoeia requires a quantitative gravimetric determination of total saponin in Radix Platycodonis after extraction with methanol. The content of total saponin in Radix Platycodonis should be not less than 6.0%. It was reported that more saponins may be obtained from the platycodon root harvested at a later nutritional stage [1]. The major sapogenins from platycodon root saponins were identified as platycodigenin and polygalacic acid; these are both polyhydroxylated oleanolic acids. Minor sapogenins were identified as platycogenic acids A, B, and C.

Saponins from platycodon root were identified as platycodins A, C, D [2], platycodin D_2, D_3, and platycosides D and E [3]. Platycodins and platycosides are saponins with rather complex sugar moieties. For example, platycodin D is platycodigenin 3-*O*-β-glucopyranoside esterified with a tetrose [4].

Polysaccharides were also isolated from the root of *P. grandiflorum*. In addition, two phenylpropanoid esters were isolated from the root and identified as coniferyl alcohol esters of palmitic acid and oleic acid [5].

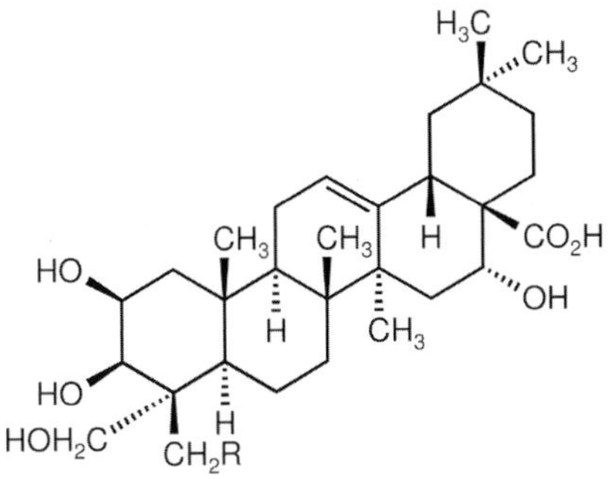

Platycodigenin: R = OH
Polygalacic acid: R = H

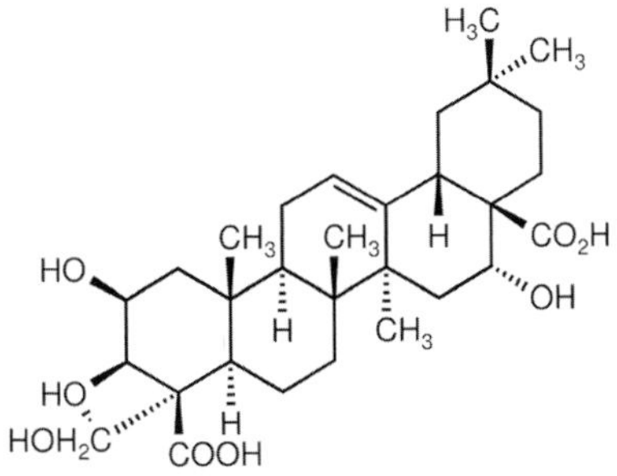

Platycogenic acid A

Handbook of Chinese Medicinal Plants: Chemistry, Pharmacology, Toxicology
Weici Tang and Gerhard Eisenbrand

ISBN: 978-3-527-32226-8

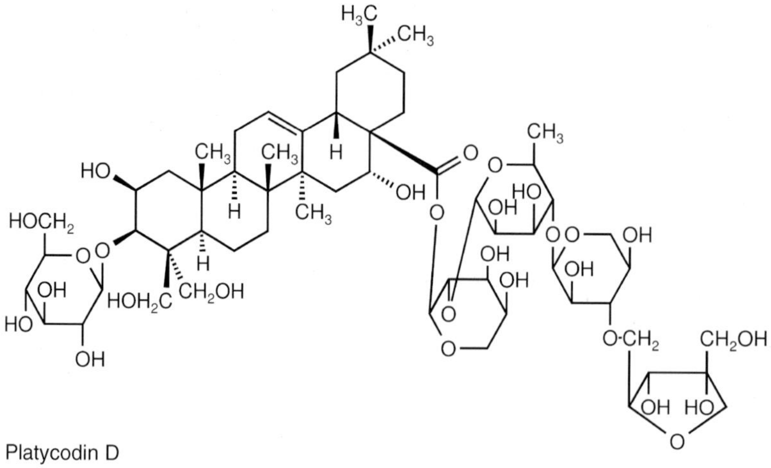

Platycodin D

Pharmacology and Toxicology

Platycodon root was mainly used in traditional Chinese medicine as an expectorant and mucolytic agent due to the saponin content. The root extract was known to be effective in expectoration, thereby improving airway respiratory function and preventing secondary airway inflammation. Saponins platycodin D and platycodin D_3 were found to increase mucin release from rat and hamster tracheal surface epithelial cell cultures, and also from intact rat trachea upon nebulization [6].

Platycodin D and platycodin D_3 were reported to show anti-inflammatory activities in rats. Treatment of RAW 264.7 macrophages with platycodon D and platycodon D_3 after activation with lipopolysaccharide (LPS) resulted in a significant inhibition of nitric oxide (NO) production and an inhibition of the expression of inducible NO synthase (iNOS). In contrast to NO production, secretion of tumor necrosis factor-α (TNF-α) as well as the expression of TNF-α mRNA, was increased by platycodon D and platycodon D_3 [7, 8]. Coniferyl alcohol esters from the platycodon root exhibited antioxidative and free radical-scavenging activities [5].

The aqueous extract of platycodon root was further shown to suppress prostaglandin E_2 (PGE_2) synthesis and NO production by inhibiting LPS-stimulated cyclooxygenase-2 (COX-2) activity and expression of iNOS mRNA, as well as reducing the LPS-induced interleukin-8 (IL-8) release in microglial cells [9]. Similar anti-inflammatory effects of the saponins from platycodon root were also observed in carrageenan-induced paw edema in rats [10]. The saponins from platycodon root were also found to exert antiatherosclerotic and anti-inflammatory activities in human endothelial cells, due partly to the reduction in cytokine-induced endothelial adhesion to monocytes by inhibiting intracellular reactive oxygen species (ROS) production, nuclear factor-κB (NF-κ B) activation, and cell adhesion molecule expression in endothelial cells [11].

A polysaccharide named PG from platycodon root was reported to activate macrophages and B cells. This polysaccharide induced the production of NO and the mRNA expression of iNOS in RAW 264.7 murine macrophage cells. PG also induced NO production in macrophages isolated from wild-type C_3H/HeN mice, but had no effect on NO production in macrophages isolated from functional

Toll-like receptor 4 (TLR4)-deficient C_3H/HeJ mice. Monoclonal antibodies directed towards TLR4 also blocked the PG-mediated induction of NO production. PG was further found to activate DNA binding of NF-κB and to cause degradation of inhibitor of nuclear factor-κB (IκB); this suggested that the PG-mediated induction of NO production and iNOS mRNA expression in macrophages was mediated, at least in part, by the TLR4/NF-κB signaling pathway [12]. Furthermore, treatment of RAW 264.7 cells with PG produced a marked induction of activator protein-1 (AP-1) DNA binding activity, and also activated mitogen-activated protein kinases (MAPKs). This indicated that macrophage activation by PG was partly mediated by MAPKs and AP-1 [13].

The intragastric administration of platycodin D at doses of 10 to 100 $mg\,kg^{-1}$ significantly stimulated pancreatic secretion in rats. This effect of platycodin D was inhibited by loxiglumide, a cholecystokinin receptor antagonist. The suppressive effect of atropine (300 $\mu g\,kg^{-1}\,h^{-1}$, intravenous) on pancreatic secretion was reduced by platycodin D (10 $mg\,kg^{-1}$, oral). No inhibition of trypsin activities *in vitro* was identified at platycodin D concentrations up to 1 m*M*. It was concluded that platycodin D causes gastrointestinal hormones, notably cholecystokinin, to be released from the duodenum [14]. The complete structure of platycodin D is required to stimulate the volumetric increase in pancreatic exocrine secretion, while the prosapogenins prepared from platycodin D increased only the protein output of pancreatic juice [15]. Aqueous extracts of the platycodon root and the saponin fraction were also found to inhibit the intestinal absorption of dietary fat by inhibiting pancreatic lipase activity. Platycodins A, C, D, and deapioplatycodin D each exhibited significant inhibitory effects on pancreatic lipase at a concentration of 500 $\mu g\,ml^{-1}$ [16]. Platycodin D was also reported to inhibit pancreatic lipase activity by a competitive mechanism [17].

The root of *P. grandiflorum* provided a preventive effect against the hepatotoxicity of carbon tetrachloride (CCl_4) in mice and rats [18, 19]. Pretreatment with platycodon root prior to CCl_4 administration led to a significant prevention of the increased serum enzymatic activities of alanine aminotransferase and aspartate aminotransferase. The root also prevented any elevation in hepatic malondialdehyde formation and depletion of reduced glutathione (GSH) content in the livers of intoxicated mice. However, the root did not affect hepatic GSH levels and glutathione *S*-transferase (GST) activities in normal mice. The hepatoprotective effect of platycodon root was also evidenced by histopathologic examination of the liver. Administration of the root to mice resulted in a significant decrease in the activity of the cytochrome P450 isoenzyme CYP2E1, the major isoenzyme involved in CCl_4 bioactivation. In addition, the root showed not only antioxidant effects on lipid peroxidation in mice liver homogenate induced by $FeCl_2$-ascorbate, but also a superoxide radical-scavenging activity [18]. The platycodon root also significantly suppressed the progress of acute hepatic fibrosis in rats, as induced by CCl_4. Blocking hepatic inflammation and activation of hepatic stellate cells were suggested to be involved [19].

The root of *P. grandiflorum* also exerted a protective action against the toxicity of *tert*-butyl hydroperoxide (TBH) in cultured rat hepatocytes. It significantly reduced the production of intracellular ROS and DNA damage in rat hepatocytes, as induced by TBH. An *in vivo* study revealed that pretreatment with the root prior to the administration of TBH significantly prevented increases in the serum levels of hepatic enzymes, and also reduced oxidative stress in the liver of treated animals [20]. The administration of platycodon root was also shown to protect against alcohol-induced fatty liver in rats, preventing elevations of serum and liver lipids. Treatment with the root also normalized the expression of hepatic liver fatty acid binding protein and CYP2E1 activity in rats intoxicated with alcohol [21].

Platycodon root was found to have a beneficial effect in preventing experimentally induced hypercholesterolemia and hyperlipidemia. Rats with dietary hyperlipidemia, fed on diets containing 5 and 10% platycodon root for three weeks, showed a marked decrease in total cholesterol and triglycerides in the serum and liver. Platycodon root also induced a reduction in low-density lipoprotein (LDL)-cholesterol, and an increase in high-density lipoprotein (HDL)-cholesterol concentrations in serum. The atherogenic index was also low in rats fed a diet containing platycodon root [22, 23].

An aqueous extract of platycodon root was reported to inhibit the hydrolysis of triolein (1,2,3-propanetriyl ester of 9-octadecenoic acid) by pancreatic lipase *in vitro*, and to reduce the elevation of rat plasma triacylglycerol level at 2–4 h after oral administration of a lipid emulsion containing corn oil. Platycodon root may inhibit the intestinal absorption of dietary fat by inhibiting its hydrolysis. The aqueous extract also protected mice against obesity induced by feeding a high-fat diet for eight weeks. The body weights at three to eight weeks, and the final parametrial adipose tissue weights, were significantly lower in mice fed a high-fat diet containing 5% aqueous extract than in animals fed only the high-fat diet. The aqueous extract also significantly reduced hepatic triacylglycerol contents that were elevated in mice fed the high-fat diet alone. The total saponin fraction of the aqueous extract inhibited pancreatic lipase activity *in vitro*; this may contribute to the antiobesity effect of platycodon root [24].

Platycodon root was also reported to show hypoglycemic effects in diabetic ICR mice, as induced by streptozotocin. A significant decrease in blood glucose levels was observed after a single administration of the root extract. However, plasma insulin levels were not increased in streptozotocin-induced diabetic mice after treatment with the root extract. This indicated that platycodon root exerted its hypoglycemic effect without stimulating insulin secretion [25].

A fraction from the extract of the platycodon root, extracted with a 7:3 petroleum ether and ethyl ether mixture, was shown to exert cytotoxic activity against human cancer cell lines, including HT-29, HRT-18, and HepG2 [26]. Among the saponins present in the root, platycodin D, platycodin D_2 and deapioplatycodin D were each found to significantly inhibit the proliferation of human tumor cell lines, including A549, SK-OV-3, SK-MEL-2, XF498 and HCT-15 *in vitro*, with ED_{50} values ranging from 4 to 18 $\mu g\,ml^{-1}$ [27]. It was also reported that an aqueous extract of platycodon root would inhibit the invasion of B16-F10 melanoma cells through a reconstituted basement membrane-coated filter, and also strongly inhibit the adhesion of B16-F10 melanoma cells to extracellular matrices. *In vivo*, the root extract inhibited experimentally induced lung cancer in rats, and prolonged the survival time of the treated animals [28]. Furthermore, the aqueous extract of the root inhibited the proliferation of A549 human lung carcinoma cells and induced cell apoptosis. The induction of apoptosis was associated with a decrease in Bcl-2 expression, an increase of Bax, and an activation of caspase-3. In addition, the root extract also markedly inhibited the activity of telomerase, and downregulated the expression of human telomerase reverse transcriptase [29]. In contrast, the aqueous extract of platycodon root was reported to be mutagenic in the chromosomal aberration and micronucleus assays in mice [30].

References

1 Zhao, G.X., Huang, Q.X., Peng, G.P., Mei, S.L., and Yu, J.M. (1987) A study on factors affecting the contents of saponins of Jiegeng (*Platycodon gradiflorum*). *Chin. Trad. Herbal Drugs*, **18**, 257–258, 254.

2 Saeki, T., Koike, K., and Nikaido, T. (1999) A comparative study on commercial, botanical gardens and wild samples of the roots of *Platycodon grandiflorum* by HPLC analysis. *Planta Med.*, **65**, 428–431.

3 Nikaido, T., Koike, K., Mitsunaga, K., and Saeki, T. (1999) Two new triterpenoid saponins from *Platycodon grandiflorum*. *Chem. Pharm. Bull. (Tokyo)*, **47**, 903–904.

4 Zhong, Y., Li, X.G., Li, S.D., and Xie, J. (1986) Isolation and identification of platycodin D from *Platycodon grandiflorum*. *Chin. Pharm. Bull.*, **21**, 71–72.

5 Lee, J.Y., Yoon, J.W., Kim, C.T., and Lim, S.T. (2004) Antioxidant activity of phenylpropanoid esters isolated and identified from *Platycodon grandiflorum* A. DC. *Phytochemistry*, **65**, 3033–3039.

6 Shin, C.Y., Lee, W.J., Lee, E.B., Choi, E.Y. and Ko, K.H. (2002) Platycodin D and D3 increase airway mucin release *in vivo* and *in vitro* in rats and hamsters. *Planta Med.*, **68**, 221–225.

7 Wang, C., Schuller Levis, G.B., Lee, E.B., Levis, W.R., Lee, D.W., Kim, B.S., Park, S.Y., and Park, E. (2004) Platycodin D and D3 isolated from the root of *Platycodon grandiflorum* modulate the production of nitric oxide and secretion of TNF-α in activated RAW 264.7 cells. *Int. Immunopharmacol.*, **4**, 1039–1049.

8 Ahn, K.S., Noh, E.J., Zhao, H.L., Jung, S.H., Kang, S.S., and Kim, Y.S. (2005) Inhibition of inducible nitric oxide synthase and cyclooxygenase II by *Platycodon grandiflorum* saponins *via* suppression of nuclear factor-κB activation in RAW 264.7 cells. *Life Sci.*, **76**, 2315–2328.

9 Jang, M.H., Kim, C.J., Kim, E.H., Kim, M.G., Leem, K.H., and Kim, J. (2006) Effects of *Platycodon grandiflorum* on lipopolysaccharide-stimulated production of prostaglandin E2, nitric oxide, and interleukin-8 in mouse microglial BV2 cells. *J. Med. Food*, **9**, 169–174.

10 Kim, J.Y., Hwang, W.I., Kim, D.H., Han, E.H., Chung, Y.C., Roh, S.H., and Jeong, H.G. (2006) Inhibitory effect of the saponins derived from roots of *Platycodon grandiflorum* on carrageenan-induced inflammation. *Biosci. Biotechnol. Biochem.*, **70**, 858–864.

11 Kim, J.Y., Kim, D.H., Kim, H.G., Song, G.Y., Chung, Y.C., Roh, S.H., and Jeong, H.G. (2006) Inhibition of tumor necrosis factor-α-induced expression of adhesion molecules in human endothelial cells by the saponins derived from roots of *Platycodon grandiflorum*. *Toxicol. Appl. Pharmacol.*, **210**, 150–156.

12 Yoon, Y.D., Han, S.B., Kang, J.S., Lee, C.W., Park, S.K., Lee, H.S., Kang, J.S., and Kim, H.M. (2003) Toll-like receptor 4-dependent activation of macrophages by polysaccharide isolated from the radix of *Platycodon grandiflorum*. *Int. Immunopharmacol.*, **3**, 1873–1882.

13 Yoon, Y.D., Kang, J.S., Han, S.B., Park, S.K., Lee, H.S., Kang, J.S., and Kim, H.M. (2004) Activation of mitogen-activated protein kinases and AP-1 by polysaccharide isolated from the radix of *Platycodon grandiflorum* in RAW 264.7 cells. *Int. Immunopharmacol.*, **4**, 1477–1487.

14 Arai, I., Komatsu, Y., Hirai, Y., Shingu, K., Ida, Y., Yamaura, H., Yamamoto, T., Kuroiwa, Y., Sasaki, K., and Taguchi, S. (1997) Stimulative effects of saponin from kikyo-to, a Japanese herbal medicine, on pancreatic exocrine secretion of conscious rats. *Planta Med.*, **63**, 419–424.

15 Ida, Y., Hirai, Y., Kajimoto, T., Shingu, K., Miura, T., Kuwahara, N., Taguchi, S., Sasaki, K., Kuroiwa, Y., Yamamoto, T., Arai, I., Amagaya, S., and Komatsu, Y. (1998) Requirement of the glycosyl parts in platycodin D to stimulate pancreatic exocrine secretion. *Bioorg. Med. Chem. Lett.*, **8**, 2209–2212.

16 Xu, B.J., Han, L.K., Zheng, Y.N., Lee, J.H., and Sung, C.K. (2005) In vitro inhibitory effect of triterpenoidal saponins from Platycodi Radix on pancreatic lipase. *Arch. Pharm. Res.*, **28**, 180–185.

17 Zhao, H.L. and Kim, Y.S. (2004) Determination of the kinetic properties of platycodin D for the inhibition of pancreatic lipase using a 1,2-diglyceride-based colorimetric assay. *Arch. Pharm. Res.*, **27**, 968–972.

18 Lee, K.J. and Jeong, H.G. (2002) Protective effect of Platycodi radix on carbon tetrachloride-induced hepatotoxicity. *Food Chem. Toxicol.*, **40**, 517–525.

19 Lee, K.J., Kim, J.Y., Jung, K.S., Choi, C.Y., Chung, Y.C., Kim, D.H., and Jeong, H.G.

(2004) Suppressive effects of *Platycodon grandiflorum* on the progress of carbon tetrachloride-induced hepatic fibrosis. *Arch. Pharm. Res.*, **27**, 1238–1244.

20 Lee, K.J., Choi, C.Y., Chung, Y.C., Kim, Y.S., Ryn, S.Y., Roh, S.H., and Jeong, H.G. (2004) Protective effect of saponins derived from roots of *Platycodon grandiflorum* on *tert*-butyl hydroperoxide-induced oxidative hepatotoxicity. *Toxicol. Lett.*, **147**, 271–282.

21 Kim, H.K., Kim, D.S., and Cho, H.Y. (2007) Protective effects of Platycodi radix on alcohol-induced fatty liver. *Biosci. Biotechnol. Biochem.*, **71**, 1550–1552.

22 Kim, K.S., Ezaki, O., Ikemoto, S., and Itakura, H. (1995) Effects of *Platycodon grandiflorum* feeding on serum and liver lipid concentrations in rats with diet-induced hyperlipidemia. *J. Nutr. Sci. Vitaminol. (Tokyo)*, **41**, 485–491.

23 Takagi, K. and Lee, E.B. (1972) Pharmacological studies on *Platycodon grandiflorum* A. DC. 3. Activities of crude platycodin on respiratory and circulatory systems and its other pharmacological activities. *Yakugaku Zasshi*, **92**, 969–973.

24 Han, L.K., Xu, B.J., Kimura, Y., Zheng, Y.n., and Okuda, H. (2000) Platycodi radix affects lipid metabolism in mice with high fat diet-induced obesity. *J. Nutr.*, **130**, 2760–2764.

25 Zheng, J., He, J., Ji, B., Li, Y., and Zhang, X. (2007) Antihyperglycemic effects of *Platycodon grandiflorum* (Jacq.) A. DC. extract on streptozotocin-induced diabetic mice. *Plant Foods Hum. Nutr.*, **62**, 7–11.

26 Lee, J.Y., Hwang, W.I., and Lim, S.T. (2004) Antioxidant and anticancer activities of organic extracts from *Platycodon grandiflorum* A. De Candolle roots. *J. Ethnopharmacol.*, **93**, 409–415.

27 Kim, Y.S., Kim, J.S., Choi, S.U., Kim, J.S., Lee, H.S., Roh, S.H., Jeong, Y.C., Kim, Y.K., and Ryn, S.Y. (2005) Isolation of a new saponin and cytotoxic effect of saponins from the root of *Platycodon grandiflorum* on human tumor cell lines. *Planta Med.*, **71**, 566–568.

28 Lee, K.J., Kim, J.Y., Choi, J.H., Kim, H.G., Chung, Y.C., Roh, S.H., and Jeong, H.G. (2006) Inhibition of tumor invasion and metastasis by aqueous extract of the radix of *Platycodon grandiflorum*. *Food Chem. Toxicol.*, **44**, 1890–1896.

29 Park, D.I., Lee, J.H., Moon, S.K., Kim, C.H., Lee, Y.T., Cheong, J., Choi, B.T., and Choi, Y.H. (2005) Induction of apoptosis and inhibition of telomerase activity by aqueous extract from *Platycodon grandiflorum* in human lung carcinoma cells. *Pharmacol. Res.*, **51**, 437–443.

30 Yin, X.J., Liu, D.X., Wang, H.C., and Zhou, Y. (1991) A study on the mutagenicity of 102 raw pharmaceuticals used in Chinese traditional medicine. *Mutat. Res.*, **260**, 73–82.

Pogostemon cablin

Herba Pogostemonis (Guanghuoxiang) is the dried, above-ground part of *Pogostemon cablin* (Blanco) Benth. (Lamiaceae). It is used for the treatment of gastrointestinal disorders, abdominal pain, vomiting and diarrhea, and for the treatment of sinusitis and headache.

Pogostemon cablin (Blanco) Benth.

Oleum Pogostemonis (Guanghuoxiang You) is the essential oil obtained by steam distillation of the above-ground part of *Pogostemon cablin*. It is used for the treatment of gastrointestinal disorders and headache.

Chemistry

The above-ground part of *P. Cablin* is known to contain essential oil, with patchouli alcohol as the major constituent. The Chinese Pharmacopoeia requires a qualitative determination of patchouli alcohol in the essential oil of Herba Pogostemonis by thin-layer chromatographic (TLC) comparison with reference substance, and a quantitative determination of patchouli alcohol in Herba Pogostemonis by gas chromatography. The content of patchouli alcohol in Herba Pogostemonis should be not less than 0.1%. Other constituents from the above-ground part of *P. cablin* were identified as pogostol, oleanolic acid, daucosterol, stigmast-4-en-3-one, β-sitosterol, friedelin, epifriedelinol, retusin, and pachypodol [1].

The Chinese Pharmacopoeia requires a qualitative determination of patchouli alcohol in Oleum Pogostemonis by TLC comparison with reference substance, and a quantitative determination by gas chromatography. The content of patchouli alcohol in Oleum Pogostemonis should be not less than 26.0%. The Chinese Pharmacopoeia gives a relative density of 0.950–0.980, a refractive index of 1.503–1.513, and a rotation of $-66°$ to $-43°$ for Oleum Pogostemonis. A series of further components were isolated from the essential oil of *P. Cablin*, including about 20% α-guaiene, 16% α-bulnesene [2], and some sesquiterpene hydroperoxides [3].

Pharmacology and Toxicology

Patchouli alcohol, pogostol, stigmast-4-en-3-one, retusin, and pachypodol were each found to exhibit antiemetic effects against copper sulfate-induced emesis in young chicks [4]. Patchouli alcohol was also found to be a Ca^{2+} antagonist, with inhibitory

Handbook of Chinese Medicinal Plants: Chemistry, Pharmacology, Toxicology
Weici Tang and Gerhard Eisenbrand

ISBN: 978-3-527-32226-8

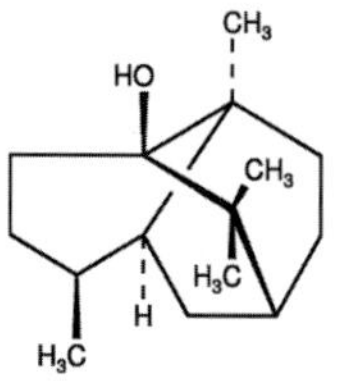

Patchouli alcohol

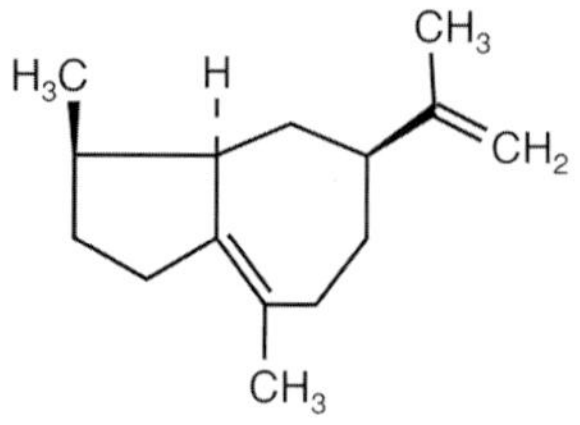

α-Bulnesene

activity on the K^+ contracture of guinea pig taenia coli [5]. Furthermore, the stem and leaf of *P. cablin* was effective in the treatment of an experimental animal model of acute otitis externa induced by *Staphylococcus aureus* [6]. The pharmacokinetics of pure patchouli alcohol and patchouli alcohol in essential oil has been studied in rat after intravenous (i.v.) administration. The pharmacokinetic parameters of patchouli alcohol were described by a two-compartment open model. The elimination half-life of patchouli alcohol in the essential oil was longer than that of pure patchouli oil, and the area under the curve (AUC) greater [7].

α-Bulnesene was reported to show a potent and concentration-dependent inhibitory effect on rabbit platelet aggregation, as induced by platelet-activating factor (PAF) and arachidonic acid. In a radioligand binding assay for the PAF receptor, α-bulnesene competitively inhibited [^{3}H]PAF binding to the PAF receptor, with an IC_{50} value of approximately 18 μ*M*. Furthermore, α-bulnesene inhibited the formation of thromboxane B_2 (TXB_2) and prostaglandin E_2 (PGE_2). These results indicated that α-bulnesene might serve as a PAF receptor antagonist and an anti-platelet-aggregation agent [8].

References

1 Yang, Y., Kinoshita, K., Koyama, K., Takahashi, K., Tai, T., Nunoura, Y., and Watanabe, K. (1999) Anti-emetic principles of *Pogostemon cablin* (Blanco) Benth. *Phytomedicine*, **6**, 89–93.

2 Tsai, Y.C., Hsu, H.C., Yang, W.C., Tsai, W.J., Chen, C.C., and Watanabe, T. (2007) α-bulnesene, a PAF inhibitor isolated from the essential oil of *Pogostemon cablin*. *Fitoterapia*, **78**, 7–11.

3 Kiuchi, F., Matsuo, K., Ito, M., Qui, T.K., and Honda, G. (2004) New sesquiterpene hydroperoxides with trypanocidal activity from *Pogostemon cablin*. *Chem. Pharm. Bull. (Tokyo)*, **52**, 1495–1496.

4 Guan, L., Quan, L.H., Xu, L.Z., and Cong, P.Z. (1994) Chemical constituents of *Pogostemon cablin* (Blanco) Benth. *China J. Chin. Mater. Med.*, **19**, 355–356.

5 Ichikawa, K., Kinoshita, T., and Sankawa, U. (1989) The screening of Chinese crude drugs for Ca^{2+} antagonist activity: identification of active principles from the aerial part of *Pogostemon cablin* and the fruits of *Prunus mume*. *Chem. Pharm. Bull. (Tokyo)*, **37**, 345–348.

6 Ho, K.Y., Juan, K.H., and Lin, C.C. (1992) The study on *Saxifraga stolonifera* and *Pogostemon cablin* on experimental animal model of acute otitis externa. *J. Chin. Med.*, **2**, 73–81.

7 Yang, F.C., Xu, L.Z., Zou, Z.M., and Yang, S.L. (2004) Pharmacokinetics of patchouli alcohol and patchouli alcohol in patchouli oil after iv administrated to rats. *Yao Xue Xue Bao*, **39**, 726–729.

8 Hsu, H.C., Yang, W.C., Tsai, W.J., Chen, C.C., Huang, H.Y., and Tsai, Y.C. (2006) α-bulnesene, a novel PAF receptor antagonist isolated from *Pogostemon cablin*. *Biochem. Biophys. Res. Commun.*, **345**, 1033–1038.

Polygala sibirica and *Polygala tenuifolia*

Radix Polygalae (Yuanzhi) is the dried root of *Polygala tenuifolia* Willd. or *Polygala sibirica* L. (Polygalaceae). It is used as a mucolytic and expectorant agent.

Two galenic preparations, Extractum Polygalae Liquidum (Yuanzhi Liujingao), and "Yuanzhi Ding," the tincture of Radix Polygalae, prepared from Extractum Polygalae Liquidum, are also officially listed in the Chinese Pharmacopoeia and are used as mucolytic and expectorant agents.

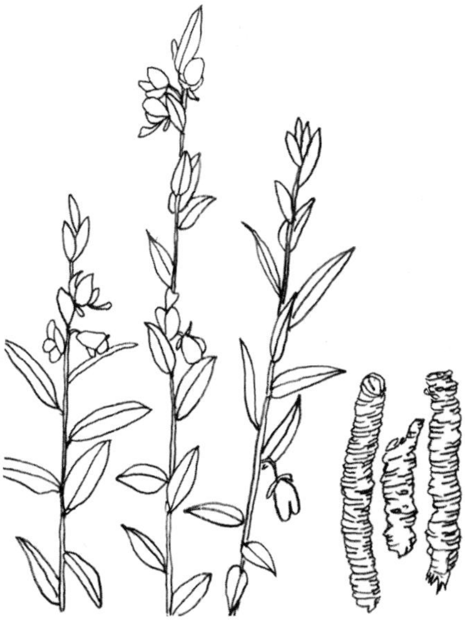

Polygala sibirica L.

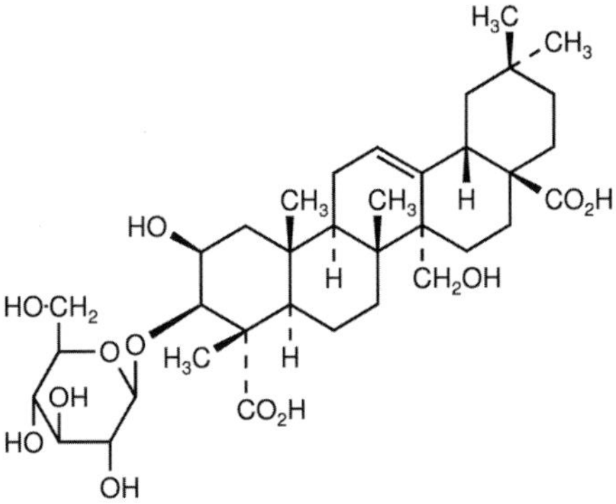

Tenuifolin

Chemistry

The root of *P. tenuifolia* is known to contain triterpene saponins with tenuifolin and triterpene sapogenins, with polygalacic acid (formula, see p. 911) as the major components. The minor saponins, named onjisaponins A, B, E, F, and G, were identified as tenuifolin 28-esters of different oligosaccharides, esterified with a substituted cinnamic acid [1]. The Chinese Pharmacopoeia requires a quantitative determination of polygalacic acid in Radix Polygalae by HPLC. The content of polygalacic acid in Radix Polygalae should be not less than 0.7%.

Xanthone derivatives substituted with hydroxy and methoxy groups were isolated from the root of *P. tenuifolia*, including onjixanthones I, II, and related xanthones. Xanthone C-glycoside from the root of *P. tenuifolia* was identified as polygalaxanthone III [2].

β-Carboline alkaloids from the root of *P. tenuifolia* were identified as harman, norharman, perlolyrine, and *N*-formylharman [3]. The contents of the β-carboline alkaloids were between 0.1 and 0.3% [4]. In addition, the isolation of tetrahydrocolumbamine from the root of *P. tenuifolia* was reported [5].

A number of mono-, di-, tri-, and tetra-acylated sucroses named tenuifolisides A–E; oligosaccharide multiesters tenuifolioses A–P and oligosaccharide derivatives, polygalatenosides A–E. were isolated from the root of *P. tenuifolia*. All of these acylated sucrose derivatives were shown to have one or two cinnamoyl ester groups on their fructofuranosyl residues. The oligosaccharide multiesters were esterified with acetic acid, benzoic acid, *p*-coumaric acid,

Handbook of Chinese Medicinal Plants: Chemistry, Pharmacology, Toxicology
Weici Tang and Gerhard Eisenbrand

ISBN: 978-3-527-32226-8

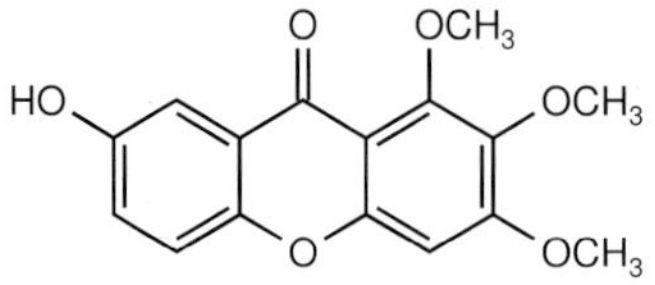

Onjixanthone I

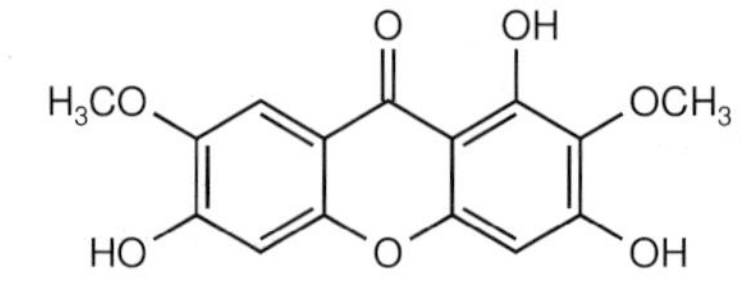

Onjixanthone II

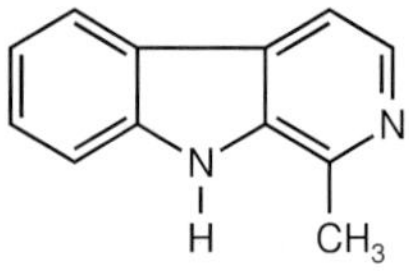

Harman

ferulic acid, and 3,4,5-trimethoxycinnamic acid [6–9].

Pharmacology and Toxicology

It is believed that 3,4,5-trimethoxycinnamic acid and related cinnamic acid derivatives belong to the active compounds of the root of *P. tenuifolia*. The sucrose derivatives possessing a 3,4,5-trimethoxycinnamoyl moiety might be considered as natural prodrugs of 3,4,5-trimethoxycinnamic acid. Rat embryo basal forebrain cells, when cultivated in the presence of a root extract for three days, showed an increased choline acetyltransferase activity. Oral administration of the extract or its hydrolyzed constituents, 3,5-dimethoxy-4-hydroxycinnamic acid, induced choline acetyltransferase activity in the cerebral cortex of basal forebrain-injured rats. The root extract further induced nerve growth factor secretion in astroglia cells [10] and inhibited the secretion of tumor necrosis factor-α (TNF-α) from primary cultures of mouse astrocytes stimulated with lipopolysaccharide (LPS). The extract may inhibit TNF-α by suppressing interleukin-1 (IL-1) secretion, and might also have an anti-inflammatory activity [11].

In rats, an extract of the root of *P. tenuifolia* was found to significantly protect against scopolamine-induced impairment of special cognition [12], and to protect primary cultured rat neurons against cell death induced by glutamate, amyloid-β protein, or the C-terminal fragment of amyloid precursor protein. In addition, the extract inhibited acetylcholinesterase activity in a concentration-dependent and noncompetitive manner, with an IC_{50} value of about 264 μg ml^{-1} [13]. Cocaine-induced behavioral effects were known to occur in parallel, with increases in Fos-related antigen-immunoreactivity and activator protein-1 DNA binding activity in the nucleus accumbens of rats. The responses induced by cocaine were consistently attenuated by concurrent treatment with the extract of the root of *P. tenuifolia* [14].

The saponins from the root of *P. tenuifolia* were found capable of treating psychosis. An *in vitro* binding study showed that the saponins possess a potential antipsychotic action, as shown by their binding affinity for both dopamine and serotonin receptors. The intraperitoneal (i.p.) or oral administration of saponins to mice reduced dose-dependently the climbing behavior, serotonin syndrome and hyperactivity induced by different agents. The saponins also inhibited hyperactivity in rats, as induced by cocaine. The dopamine and serotonin receptor antagonist properties of the saponins appeared to be one of the antipsychotic mechanisms [15]. Onjisaponins A, B, E, F, and G, as the major saponins of the root of *P. tenuifolia*, strongly enhanced the production of nerve growth factor in cultured rat astrocytes. Onjisaponin F also induced choline acetyltransferase mRNA in rat basal forebrain cells [16]. Tenuifoliside B, as one of the acylated oligosaccharides in the roots of *P. tenuifolia*, was reported to show a cerebral protective effect in mice against anoxia caused by KCN, and also to exert an ameliorative effect on the impairment of performance in a passive avoidance task in rats, as induced by scopolamine [17].

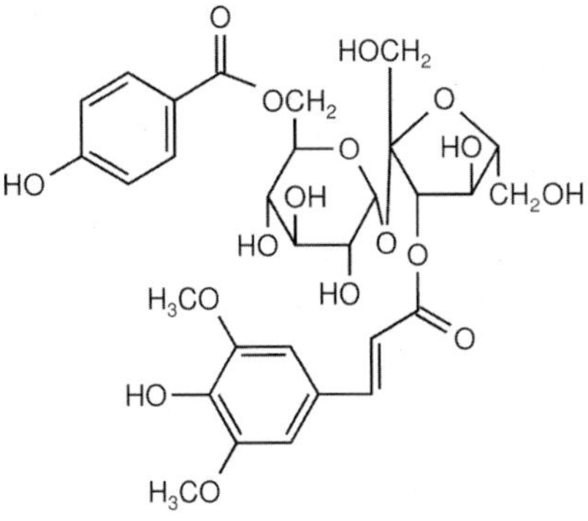

Tenuifolisids B

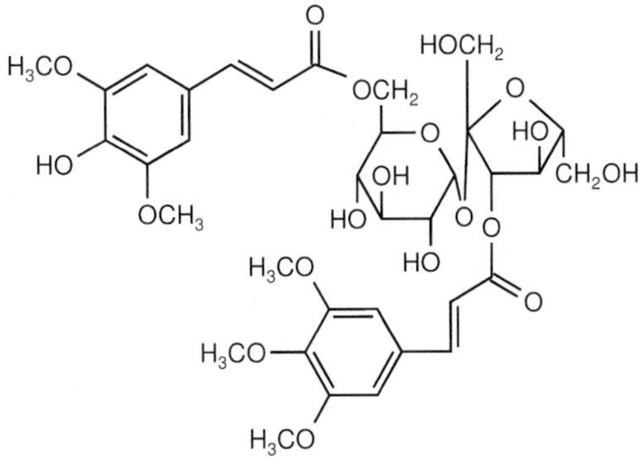

Tenuifoliside C

3,4,5-Trimethoxycinnamic acid was reported to exhibit anti-stress efficacy. Its oral administration prolonged the sleeping time induced by hexobarbital in mice, and exerted a sedative action. The i.p. injection of 3,4,5-trimethoxycinnamic acid was also found to inhibit the stress induced by repeated cold exposure, or by intracerebroventricular administration of corticotropin-releasing hormone in rats. Furthermore, the elevated content of norepinephrine (noradenaline) in the locus ceruleus of rats, as induced by the intracerebroventricular injection of corticotropin-releasing hormone, was significantly suppressed by intracerebroventricular 3,4,5-trimethoxycinnamic acid [18]. Both, tenuifoliside B and 3,6′-disinapoylsucrose from the root of *P. tenuifolia* inhibited KCN-induced hypoxia and scopolamine-induced memory impairment in mice. Sinapic acid, as a structural moiety of tenuifoliside B and 3,6′-disinapoylsucrose, exerted effects similar to the parent compounds. In addition, sinapic acid inhibited the hypoxia and mortality induced by decompression or bilateral carotid artery ligation, and memory impairment induced by CO_2 in mice. Sinapic acid further inhibited the decrease in cerebral acetylcholine concentration and choline acetyltransferase activity induced by basal forebrain lesion in rats [19]. By using a [^{125}I]RTI-55-membrane binding assay, polygalatenoside A and polygalatenoside B were both reported to inhibit norepinephrine (noradrenaline) reuptake through blocking its transport [7].

References

1. Sakuma, S. and Shoji, J. (1982) Studies on the constituents of the root of *Polygala tenuifolia* Willdenow. II. On the structures of onjisaponins A, B, and E. *Chem. Pharm. Bull. (Tokyo)*, **30**, 810–821.
2. Fujita, T., Liu, Y., Ueda, S., and Takeda, Y. (1992) Xanthones from *Polygala tenuifolia*. *Phytochemistry*, **31**, 3997–4000.
3. Jin, B.Y.and Park, J. (1993) Alkaloidal components of *Polygala tenuifolia* Willd. *China J. Chin. Mater. Med.*, **18**, 675–677.
4. Park, M.K., Park, J.H., Kim, B.Y., Kim, J.M., Liem, K.J., and Han, B.H. (1993) Analysis of alkaloids in *Polygala tenuifolia* by HPLC. *Anal. Sci. Technol.*, **6**, 255–259.
5. Shen, X.L., Witt, M.R., Dekermendjian, F K., and Nielsen, M. (1994) Isolation and identification of tetrahydrocolumbamine as a dopamine receptor ligand from *Polygala tenuifolia* Willd. *Acta Pharm. Sin.*, **29**, 887–890.
6. Ikeya, Y., Sugama, K., Okada, M., and Mitsuhashi, H. (1991) Four new phenolic glycosides from *Polygala tenuifolia*. *Chem. Pharrn. Bull.*, **39**, 2600–2605.
7. Ikeya, Y., Sugama, K., and Maruno, M. (1994) Xanthone C-glycoside and acylated sugar from

Polygala tenuifolia. Chem. Pharm. Bull. (Tokyo), **42**, 2305–2308.

8 Miyase, T., Iwata, Y., and Ueno, A. (1992) Tennuifolioses G-P, oligosaccharide multi-esters from the roots of *Polygala tenuifolia* WILLD. *Chem. Pharrn. Bull.*, **40**, 2741–2748.

9 Cheng, M.C., Li, C.Y., Ko, H.C., Ko, F.N., Lin, Y.L., and Wu, T.S. (2006) Antidepressant principles of the roots of *Polygala tenuifolia. J. Nat. Prod.*, **69**, 1305–1309.

10 Yabe, T., Iizuka, S., Komatsu, Y., and Yamada, H. (1997) Enhancements of choline acetyltransferase activity and nerve growth factor secretion by Polygalae radix-extract containing active ingredients in Kami-untan-to. *Phytomedicine*, **4**, 199–205.

11 Kim, H.M., Lee, E.H., Na, H.J., Lee, S.B., Shin, T.Y., Lyu, Y.S., Kim, N.S., and Nomura, S. (1998) Effect of *Polygala tenuifolia* root extract on the tumor necrosis factor-α secretion from mouse astrocytes. *J. Ethnopharmacol.*, **61**, 201–208.

12 Sun, X.L., Ito, H., Masuoka, T., Kamei, F C., and Hatano, T. (2007) Effect of *Polygala tenuifolia* root extract on scopolamine-induced impairment of rat spatial cognition in an eight-arm radial maze task. *Biol. Pharm. Bull.*, **30**, 1727–1731.

13 Park, C.H., Choi, S.H., Koo, J.W., Seo,J.H., Kim, H.S., Jeong, S.J., and Suh, Y.H. (2002) Novel cognitive improving and neuroprotective activities of *Polygala tenuifolia* Willdenow extract, BT-11. *J. Neurosci. Res.*, **70**, 484–492.

14 Shin, E.J., Oh, K.W., Kim, K.W., Kwon, Y.S., Jhoo, J.H., Jhoo, W.K., Cha, J.Y., Lim, Y.K., Kim, I.S., and Kim, H.C. (2004) Attenuation of cocaine-induced conditioned place preference by *Polygala tenuifolia* root extract. *Life Sci.*, **75**, 2751–2764.

15 Chung, I.W., Moore, N.A., Oh, W.K., O'Neill, M.F., Ahn, J.S., Park, J.B., Kang, U.G., and Kim, Y.S. (2002) Behavioural pharmacology of polygalasaponins indicates potential antipsychotic efficacy. *Pharmacol. Biochem. Behav.*, **71**, 191–195.

16 Yabe, T., Tuchida, H., Kiyohara, H., Takeda, T., and Yamada, H. (2003) Induction of NGF synthesis in astrocytes by onjisaponins of *Polygala tenuifolia*, constituents of kampo (Japanese herbal) medicine, Ninjin-yoei-to. *Phytomedicine*, **10**, 106–114.

17 Ikeya, Y., Takeda, S., Tunakawa, M., Karakida, H., Toda, K., Yamaguchi, T., and Aburada, M. (2004) Cognitive improving and cerebral protective effects of acylated oligosaccharides in *Polygala tenuifolia. Biol. Pharm. Bull.*, **27**, 1081–1085.

18 Kawashima, K., Miyako, D., Ishino, Y., Makino, T., Saito, K., and Kano, Y. (2004) Anti-stress effects of 3,4,5-trimethoxycinnamic acid, an active constituent of roots of *Polygala tenuifolia* (Onji). *Biol. Pharm. Bull.*, **27**, 1317–1319.

19 Karakida, F., Ikeya, Y., Tsunakawa, M., Yamaguchi, T., Ikarashi, Y., Takeda, S., and Aburada, M. (2007) Cerebral protective and cognition-improving effects of sinapic acid in rodents. *Biol. Pharm. Bull.*, **30**, 514–519.

Polygonatum cyrtonema, *Polygonatum kingianum*, *Polygonatum odoratum* and *Polygonatum sibiricum*

Rhizoma Polygonati (Huangjing) is the dried rhizome of *Polygonatum cyrtonema* Hua, *Polygonatum kingianum* Coll. et Hemsl. or *Polygonatum sibiricum* Red. (Liliaceae). It is used as a general tonic for the treatment of weakness, anorexia, cough, and diabetes.

Rhizoma Polygonati Odorati (Yuzhu) is the dried rhizome of *Polygonatum odoratum* (Mill.) Druce. It is used as a general tonic.

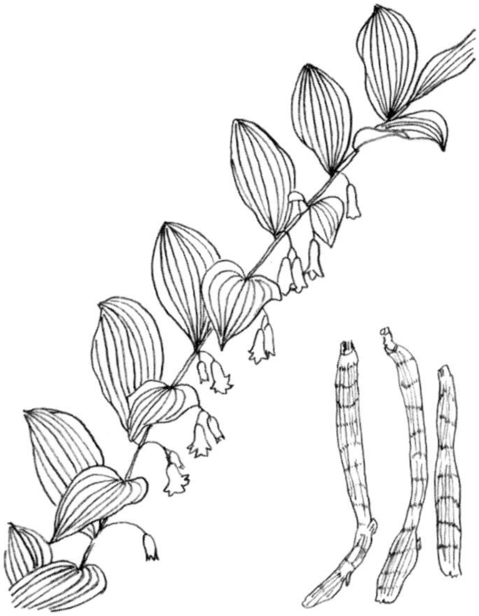

Polygonatum odoratm (Mill.) Druce

Chemistry

The rhizome of official *Polygonatum* species is known to contain polysaccharides [1, 2]. The Chinese Pharmacopoeia requires a quantitative determination of polysaccharide in Rhizoma Polygonati by ultraviolet and visible spectrophotometric analysis. The polysaccharide content in Rhizoma Polygonati should be not less than 7.0%, calculated as glucose.

Other major components were found to be steroid saponins. Steroid saponins from *P. kingianum*, named as kingianoside A, kingianoside B, kingianoside C and kingianoside D, were identified as gentrogenin glycosides with different sugar moieties. The structure of kingianoside A was elucidated as 3-*O*-β-D-glucopyranosyl-(1 → 4)-β-D-galactopyranoside; kingianoside B as gentrogenin 3-*O*-β-D-glucopyranosyl-(1 → 4)-β-D-fucopyranoside 26-*O*-β-D-glucopyranosyl-22-hydroxy-25(*R*)-furost-5-en-12-on-3β,22-diol; kingianoside C as 3-*O*-β-D-glucopyranosyl(1 → 4)-β-D-galactopyranoside 26-*O*-β-D-glucopyranosyl-22-hydroxy-25(*R*)-furost-5-en-12-on-3β,22-diol; and kingianoside D as gentrogenin 3-*O*-β-D-glucopyranosyl (1 → 4)-β-D-fucopyranoside [3]. Besides saponins, the indolizinone derivatives kinganone, 3-ethoxymethyl-5,6,7,8-tetrahydro-8-indolizinone and isomucronulatol were also isolated from the rhizome of *P. kingianum* [4].

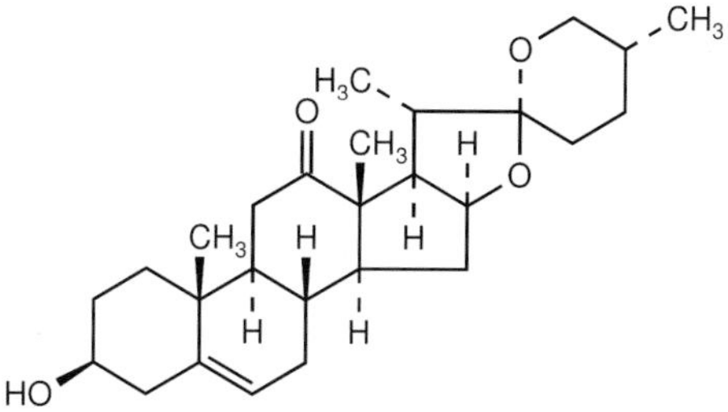

Gentrogenin

Steroidal saponins isolated from the rhizome of *P. sibiricum* were identified as neoprazerigenin A 3-*O*-lycotetraoside, its methyl prototype congeners, sibiricoside A and sibiricoside B [5], and neosibiricosides A–D. Neosibiricoside A was structurally elucidated as (23*S*,24*R*,25*R*)-1-*O*-acetylspirost-5-ene-1β,3β,23,24-tetrol 3-*O*-β-D-gluco-

Handbook of Chinese Medicinal Plants: Chemistry, Pharmacology, Toxicology
Weici Tang and Gerhard Eisenbrand

ISBN: 978-3-527-32226-8

pyranosyl-(1 → 2)-β-D-glucopyranosyl-(1 → 4)-β-D-fucopyranoside; neosibiricoside B as (25*S*)-1-*O*-acetylspirost-5-ene-1β,3β-diol 3-*O*-β-D-glucopyranosyl-(1 → 2)-[β-D-xylopyranosyl-(1 → 3)]-β-D-glucopyranosyl-(1 → 4)-β-D-galactopyranoside; neosibiricoside C as (25*S*)-spirost-5-en-3β-ol 3-*O*-β-D-glucopyranosyl-(1 → 2)-[β-D-xylopyranosyl-(1 → 3)]-β-D-glucopyranosyl-(1 → 4)-2-*O*-acetyl-β-D-galactopyranoside; and neosibiricoside D as (25*R*,*S*)-spirost-5-en-3β-ol 3-*O*-β-D-glucopyranosyl-(1 → 2)-β-D-glucopyranosyl-(1 → 4)-β-D-galactopyranoside [6].

From the ethanol extract of the rhizome of *P. odoratum*, polysaccharide, steroidal saponin POD-II, PDO-III, β-sitosterol and its glucosides were isolated [7]. The Chinese Pharmacopoeia requires a quantitative determination of polysaccharides in Rhizoma Polygonati Odorati by ultraviolet and visible spectrophotometric analysis. The content of polysaccharides in Rhizoma Polygonati Odorati should be not less than 6.0%, calculated as glucose.

Pharmacology and Toxicology

The rhizome of *P. sibiricum* was reported to show cardiotonic effects in rats. The methanol extract of the rhizome at concentrations of 1–7 mg ml^{-1} increased the developed tension of the left atrium in a concentration-dependent manner. The extract also strongly inhibited cAMP phosphodiesterase. The increase in cAMP level correlated with an increase in left atrial contraction. In contrast, the extract did not inhibit Na^+/K^+-ATPase. The cardiotonic effect of the rhizome of *P. sibiricum* was strongly

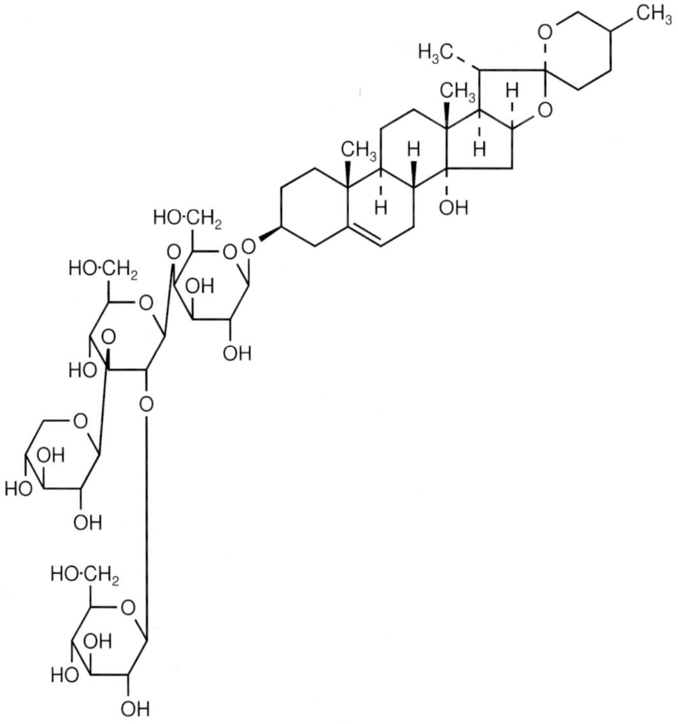

POD-II

inhibited by reserpine as a sympatholytic agent. Furthermore, the extract-treated left atria inhibited the tension produced by propranolol as a β-adrenoreceptor antagonist. It was suggested that the cardiotonic effect of the rhizome of *P. sibiricum* might be due to the stimulation of β-adrenoreceptors through activation of the sympathetic nerves [8].

The rhizome of *P. sibiricum* was further found capable of depressing the release of lambda phage from a lysogenic strain. It also showed an inhibitory effect in the SOS chromotest, and decreased the frequency of gene conversion in *Saccharomyces cerevisiae* in the presence of hydroxyurea. The effective component of the rhizome was a reductive carbohydrate with a molecular weight less than 3000 Da. The component was shown to exert an inhibitory effect on the SOS response induced at 42 °C in *E. coli* GW1060 (recA441), but had no effect on SOS network gene expression in *E. coli* GW1107 (lexA51). This suggests that the rhizome might contain an inhibitor of RecA protease [9].

Pharmacological studies revealed that POD-II was able to induce the colony-stimulating factor (CSF) in mouse serum, while POD-III was found to show synergistic activity with concanavalin A (Con A) and lipopolysaccharide (LPS) in stimulating lymphocyte proliferation [7]. An alcoholic extract from *P. odoratum* increased the index of phagocytosis of macrophages, and also stimulated the proliferation of lymphocytes up to normal levels in mice with burn damage covering 5% of the body area. This indicated that the rhizome of *P. odoratum* might serve as a immunopotentiator [10].

A mannose- and sialic acid-binding lectin from the rhizome of *P. cyrtonema* was reported to exert inhibitory activity against human immunodeficiency viruses (HIV) I and II. The anti-HIV activity of the lectin was tested in MT-4 and CEM cells, and compared with other mannose-binding lectins. The infection inhibitory activity against HIV of *P. cyrtonema* lectin was 10- to 100-fold more potent than that of other tested mannose-binding lectins, but without significant cytotoxicity towards MT-4 or CEM cells. The full-length cDNA of *P. cyrtonema* lectin contained 693 base pairs, with an open reading frame encoding a precursor protein of 160 amino acid residues, consisting of a 28-residue signal peptide, a 22-residue C-terminal cleavage peptide, and a 110-residue mature polypeptide which contained three tandemly arranged subdomains. Molecular modeling of *P. cyrtonema* lectin indicated that its three-dimensional structure resembled that of the snowdrop agglutinin. An active sialic acid-binding site was identified in the *P. cyrtonema* lectin by using docking procedures [11].

References

1. Xu, S.C., Li, S.H., Ji, Y.H., Wang, C.X., Zhang, L.Y., Liu, Y.Q., Xu, X.F., and Guan, Q. (1993) A comparative analysis of the polysaccharide contents of *Polygonatum kingianum* before and after processing. *China J. Chin. Mater. Med.*, **18**, 600–601.
2. Li, S., Guo, X., Su, S., Huang, R., Zhang, X., and Yang, Z. (1997) Culture techniques for high yield and high benefit of wild *Polygonatum sibiricum* Redoute. *China J. Chin. Mater. Med.*, **22**, 398–401, 446–447.
3. Li, X.C., Yang, C.R., Ichikawa, M., Matsuura, H., Kasai, R., and Yamasaki, K. (1992) Steroid saponins from *Polygonatum kingianum. Phytochemistry*, **31**, 3559–3563.
4. Wang, Y.F., Lu, C.H., Lai, G.F., Cao, J.X., and Luo, S.D. (2003) A new indolizinone from *Polygonatum kingianum. Planta Med.*, **69**, 1066–1068.
5. Son, K.H., Do, J.C., and Kang, S.S. (1990) Steroidal saponins from the rhizomes of *Polygonatum sibiricum. J. Nat. Prod.*, **53**, 333–339.

6 Ahn, M.J., Kim, C.Y., Yoon, K.D., Ryu, M.Y., Cheong, J.H., Chin, Y.W., and Kim, J. (2006) Steroidal saponins from the rhizomes of *Polygonatum sibiricum. J. Nat. Prod.*, **69**, 360–364.
7 Lin, H.W., Han, G.Y., and Liao, S.X. (1994) Studies on the active constituents of the Chinese traditional medicine *Polygonatum odoratum* (Mill.) Druce. *Acta Pharm. Sin.*, **29**, 215–222.
8 Hirai, N., Miura, T., Moriyasu, M., Ichimaru, M., Nishiyama, Y., Ogura, K. and Kato, A. (1997) Cardiotonic activity of the rhizome of *Polygonatum sibiricum* in rats. *Biol. Pharm. Bull.*, **20**, 1271–1273.
9 Wang, L.H., Jiang, Z.S., and Chen, Z.F. (1991) Inhibitory effects of Chinese medicines on SOS responses in*E. coli* and their mechanism. *I Chuan Hsueh Pao*, **18**, 90–96.
10 Xiao, J., Cui, F., Ning, T., and Zhao, W. (1990) Effects of alcohol extract from *Polygonatum odoratum* (Mill.) Druce and Cuscuta australis R. Br. on immunological function of mice injured by burns. *China J. Chin. Mater. Med.*, **15**, 557–559, 578.
11 An, J., Liu, J.Z., Wu, C.F., Li, J., Dai, L., Van Damme, E., Balzarini, J., De Clercq, E., Chen, F., and Bao, J.K. (2006) Anti-HIV I/II activity and molecular cloning of a novel mannose/sialic acid-binding lectin from rhizome of *Polygonatum cyrtonema* Hua. *Acta Biochem. Biophys. Sin. (Shanghai)*, **38**, 70–78.

Polygonum cuspidatum

Rhizoma et Radix Polygoni Cuspidati (Huzhang) is the dried root and rhizome of *Polygonum cuspidatum* Sieb. et Zucc. (Polygonaceae). It is used as an analgesic, antiedemic, diuretic, antitussive and mucolytic agent. It is also used for the treatment of rheumatism, icterus, menostasis, cough, burn, trauma, and ulcer. The Chinese Pharmacopoeia notes that Rhizoma Polygoni Cuspidati should be used with caution in pregnancy.

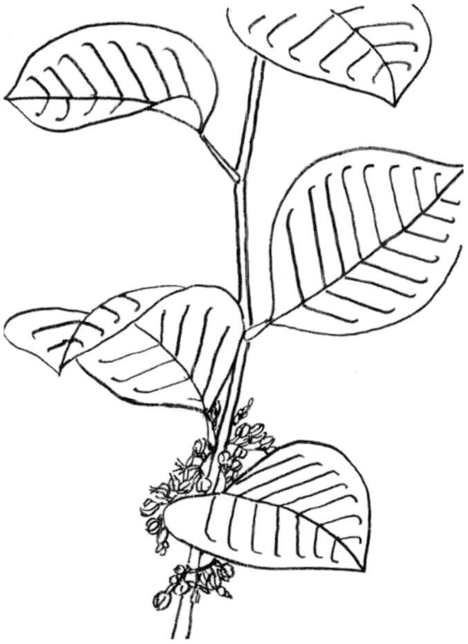

Polygonum cuspidatum Sieb. et Zucc.

Chemistry

Rhizoma et Radix Polygoni Cuspidati is known to contain a number of anthraquinone and stilbene derivatives, with emodin and physcion as the major components. Minor anthraquinones were identified as chrysophanol, fallacinol, citreorosein, questin, and questinol, as well as 8-*O*-β-D-glucopyranoside of emodin (formula, see p. 938) and physcion [1–3]. The Chinese Pharmacopoeia requires a qualitative determination of emodin and physcion in Rhizoma et Radix Polygoni Cuspidati by thin-layer chromatographic (TLC) comparison with reference substances, and a quantitative determination of emodin in Rhizoma et Radix Polygoni Cuspidati by HPLC. The content of emodin in Rhizoma et Radix Polygoni Cuspidati should be not less than 0.6%.

Stilbene derivatives from the rhizome and root of *P. cuspidatum* were identified as resveratrol and its glucoside piceid (polydatin) [4]. Both (*E*)- and (*Z*)-diastereomers of resveratrol and resveratrol glucoside were isolated from the rhizome and root of *P. cuspidatum* [5]. The Chinese Pharmacopoeia further requires a quantitative determination of polydatin in Rhizoma et Radix Polygoni Cuspidati by HPLC. The content of polydatin in Rhizoma et Radix Polygoni Cuspidati should be not less than 0.15%. Recently, the tetrahydroxystilbene piceatannol, which is structurally related to resvera-

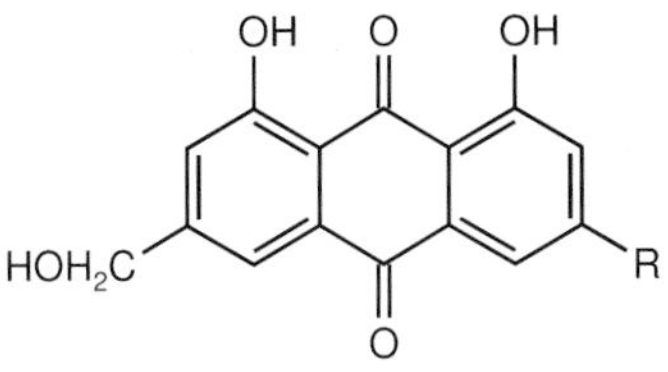

Citreorosein: R = OH
Fallacinol: R = OCH_3

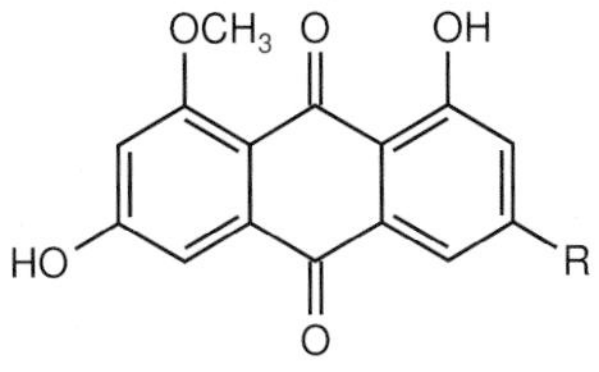

Questin: R = CH_3
Questinol: R = CH_2OH

Handbook of Chinese Medicinal Plants: Chemistry, Pharmacology, Toxicology
Weici Tang and Gerhard Eisenbrand

ISBN: 978-3-527-32226-8

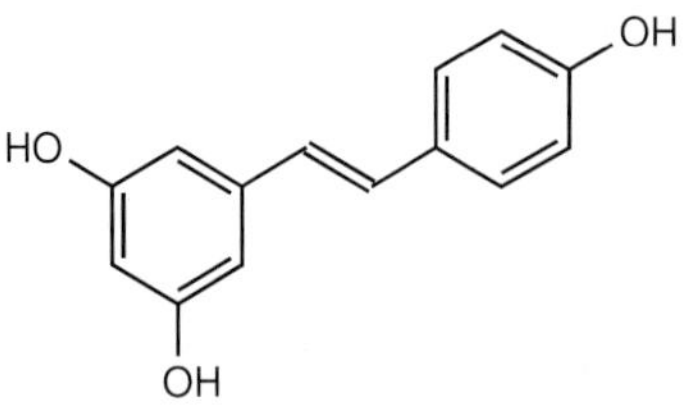
Resveratrol

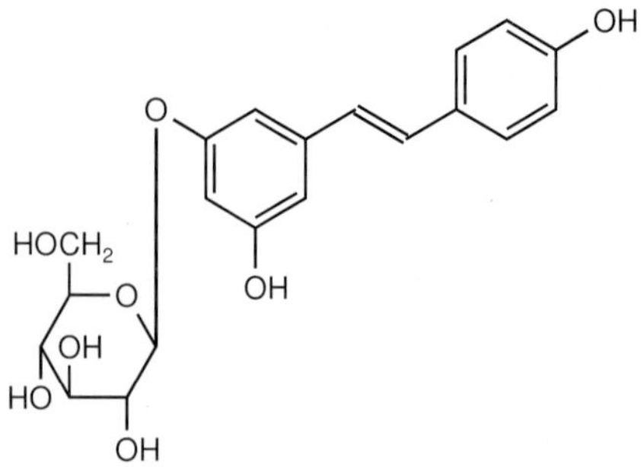
Piceid

trol, was also determined in *P. cuspidatum* [6].

Five important items with *Polygonum* species are officially listed in the Chinese Pharmacopoeia. Besides Rhizoma et Radix Polygoni Cuspidati, Caulis Polygoni Multiflori and Radix Polygoni Multiflori, the two further items are Rhizoma Bistortae and Herba Polygoni Avicularis. Rhizoma Bistortae (Quanshen) is the dried rhizome of *Polygonum bistorta* L., and is used as an antiedemic, hemostatic agent for the treatment of dysentery, gastroenteritis, acute respiratory infection, cough, carbuncle, scrofula, oral ulcer, epistaxis, hemorrhoidal bleeding, nose bleeding, and venomous snake bite. Rhizoma Bistortae is known to be rich in tannins. The Chinese Pharmacopoeia requires a qualitative determination of gallic acid in Rhizoma Bistortae by TLC comparison with a reference substance. Herba Polygoni Avicularis (Bianxu) is the dried, above-ground part of *Polygonum aviculare* L., and is used as a diuretic agent. Herba Polygoni Avicularis is known to contain flavone derivatives. The Chinese Pharmacopoeia requires a qualitative determination of quercetin in Herba Polygoni Avicularis by TLC comparison with a reference substance.

Pharmacology and Toxicology

The rhizome of *P. cuspidatum* has been used clinically for the treatment of burns. Experimental studies have shown that a *P. cuspidatum* treatment of rats with severe burns significantly increased the survival time [7]. The number of polymorphonuclear neutrophilic leucocytes adhered to the endothelium was increased after burning, but was decreased to near-normal in animals treated with *P. cuspidatum* [8]. The elevated levels of plasma tumor necrosis factor (TNF), which may be involved in the development of neutrophil adhesion, and the disturbances of microcirculation in burn shock, were almost normal after treatment; moreover, the degree of leukocyte aggregation and lung injury were attenuated [9]. In mice with thermal injury, the response to antigen signal, the proliferative capacity, the production of interleukin-2 (IL-2) and antibody by lymphocytes, were each significantly reduced and restored by treatment with *P. cuspidatum* [10].

A methanol extract from the root of *P. cuspidatum* has been reported to show broad antibacterial activity in 20 bacterial strains, including *Streptococcus mutans* and *Streptococcus sobrinus*, at minimal inhibitory concentrations (MICs) of 0.5 to 4 mg ml^{-1}. The minimal bactericidal concentration was two- to fourfold higher than the MIC. Both, *S. mutans* and *S. sobrinus* were killed after a 1-h incubation with the extract. At concentrations below the MIC, the extract also had inhibitory effects on the virulence factors of *S. mutans* and *S. sobrinus* [11]. The ethanol extract of *P. cuspidatum* also significantly inhibited hepatitis B virus (HBV) in HepG2 2.2.15 cells, a stable HBV-producing cell line, at a minimal concentration of 10 μg ml^{-1}. The expression of hepatitis B surface antigen (HbsAg) was markedly increased by the ethanol extract [12].

Resveratrol is a constituent of grapes [13] on which a number of studies have been conducted with regards to its biological activities. These studies of health effects were, in part, stimulated by previous reports of the physiological activity of resveratrol-containing plants such as *P. cuspidatum*. Epidemiological studies have suggested that the consumption of wine (notably red wine) may reduce the incidence of mortality and morbidity from coronary heart disease. This cardioprotective effect has been attributed to antioxidants, including resveratrol [14, 15].

Resveratrol was reported to inhibit platelet aggregation [16]; to block eicosanoid synthesis [17]; to protect liver peroxidation [18]; to inhibit low-density lipoprotein (LDL) oxidation [19, 20]; to inhibit the adhesion of granulocytes and monocytes to the endothelium [21]; to inhibit the expression of tissue factor in endothelial cells stimulated by interleukin-1β (IL-1β), tumor necrosis factor-α (TNF-α) and lipopolysaccharides [22]; to interfere with the release of inflammatory mediators by activated polymorphonuclear leukocytes; and to down-regulate the adhesion-dependent thrombogenic polymorphonuclear leukocyte functions [23]. In experimental studies, resveratrol also was found to attenuate hypertension in rats, as induced by ovariectomy, and to attenuate bone loss in stroke-prone, spontaneously hypertensive rats [24].

Resveratrol blocked TNF-α-induced activation of nuclear factor-κB (NF-κB) in a dose- and time-dependent manner, and also suppressed TNF-α-induced phosphorylation and nuclear translocation of the p65 subunit of NF-κB and NF-κB-dependent reporter gene transcription. NF-κB is a nuclear transcription factor that regulates the expression of various genes involved in inflammation, cytoprotection, and carcinogenesis. The suppression of TNF-α-induced NF-κB activation by resveratrol was observed in U-937 myeloid cells, in Jurkat lymphoid cells, and in HeLa and H4 epithelial cells. Resveratrol also blocked NF-κB activation induced by 12-*O*-tetradecanoylphorbol 13-acetate (TPA), lipopolysaccharides, hydrogen peroxide, okadaic acid, and ceramide. Resveratrol also inhibited the TNF-α-induced activation of mitogen-activated protein kinase (PK) and Jun *N*-terminal kinase. The generation of reactive oxygen intermediates and lipid peroxidation induced by TNF were also suppressed by resveratrol. The biological effects of resveratrol may be partially ascribed to its inhibitory effect on the activation of NF-κ B and associated kinases [25]. Resveratrol was also found to be incorporated into model membranes and to inhibit PKC activity [26]. Resveratrol showed a broad range of inhibitory potencies against purified PKC, depending on the nature of the substrate and the cofactor dependence of the phosphotransferase reaction [27].

Resveratrol reduced the proliferation of cultured smooth muscle cells induced by such diverse mitogens. The antimitogenic effects of resveratrol were not mediated by the induction of apoptosis, but rather appeared to relate to a G_1/S block in the cell cycle. Since the migration and proliferation of smooth muscle cells in the intima of susceptible vessels is widely accepted as a requisite for atherogenesis, the antiproliferation of smooth muscle cells gives further support to the hypothesis that resveratrol holds promise as an antiatherosclerotic agent [28]. Resveratrol was further found to inhibit the activity of mitogen-activated protein (MAP) kinase and nuclear translocation in porcine coronary arteries [29], and to relax isolated porcine coronary arteries [30]. Immunoblot analyses revealed a consistent reduction in the phosphorylation of extracellular regulated kinase 1 (ERK1) and ERK2, and of members of MAP kinase family, JNK-1 and p38, at active sites. Experimental studies also revealed that resveratrol reduced myocardial ischemic reperfusion injury [15]; inhibited the contractile response to noradrenaline (norepinephrine) of isolated endothelium-intact rat aorta [31]; and also scavenged free radicals [32]. Resveratrol had cardioprotective properties due to an ability to function as an *in vivo* antioxidant [33], and to its potent free radical-scavenging activity [34]. In addition to its antioxidative activity, resveratrol can also stimulate Ca^{2+}-activated K^+ channels in endothelial cells [35].

Resveratrol was also used for shock treatment. Experimental studies revealed that resveratrol inhibited the expression of intercellular adhesion molecule-1 (ICAM-1) in endothelial cells stimulated by lipopolysaccharide (LPS), attenuated the adhesion of leukocytes and endothelial cells, and increased intracellular Ca^{2+} concentration in myocardial cells with an enhancement of the contraction extent of myocardial cells. It also activated K^+ channels of vascular smooth muscle cells and decreased the intracellular pH value and intracellular Ca^{2+} concentration of vascular smooth muscle cells in shock. This indicated that resveratrol exerts multiple effects on different cells, related to the enhancement of heart function and improvement of microcirculatory perfusion in shock [36].

The neuroprotective effects of resveratrol on cerebral injury induced by ischemia–reperfusion was also reported. The intravenous (i.v.) administration of resveratrol to rats, immediately after surgery to produce middle cerebral artery occlusion, improved neurological deficits and reduced the volume of brain infarction. In addition, resveratrol decreased the levels of cell adhesion molecules, including ICAM-1, vascular cell adhesion molecule-1 (VCAM-1), E-selectin, L-selectin, and integrins. It was suggested that the protective effects of resveratrol against cerebral injury might be related to an inhibition of the expression of various cell adhesion molecules [37].

Resveratrol modulated the release of arachidonic acid, prostaglandin synthesis, and the growth of 3T6 fibroblast [38]. It exerted a strong inhibitory effect on superoxide radical and hydrogen peroxide formation by macrophages, as stimulated by LPS or TPA. It also significantly decreased arachidonic acid release and caused a significant impairment of cyclooxygenase-2 (COX-2) induction, as stimulated by LPS and TPA, or by exposure to superoxide radicals and hydrogen peroxide. The effects of resveratrol on arachidonic acid release and COX-2 overexpression in macrophages were correlated with a marked reduction of prostaglandin (PG) biosynthesis [39]. Resveratrol was further found to inhibit COX-2 transcription in human mammary epithelial cells [40]. In human erythroleukemia K562 cells, resveratrol reversed the elevation of leukotriene B_4 and prostaglandin E_2 levels induced by hydrogen peroxide. It was found to be a competitive inhibitor of purified 5-lipooxygenase and 15-lipooxygenase and prostaglandin H synthase [41].

Resveratrol inhibited platelet aggregation both *in vivo* and *in vitro*. It immediately inhibited Ca^{2+} influx in thrombin-stimulated platelets, with an IC_{50} of 0.5 μM. Resveratrol was suggested to be an inhibitor of store-operated Ca^{2+} channels in human platelets [42]. Like resveratrol, piceid also showed beneficial effects on coronary diseases, by inhibiting platelet aggregation [43] and arterial thrombosis induced by vascular endothelial injury. The selective inhibition of thromboxane A_2 (TXA_2) production rather than of prostaglandin (PGI_2) production, may be one of the mechanisms involved in the thrombosis inhibition by piceid [44]. In rats, piceid protected against acute lung injury induced by endotoxin [45], and also protected the cultured myocardial cells of newborn rats against damage by oxygen and glucose deficiencies [46].

Resveratrol and piceid partly inhibited the deposition of triglyceride and cholesterol in the liver of rats fed a mixture composed of corn oil, cholesterol, and cholic acid. Piceid reduced the serum triglyceride and low-density lipoprotein (LDL) levels of the treated rats. The intraperitoneal (i.p.) or oral administration of resveratrol or piceid reduced triglyceride synthesis from palmitate in the liver of mice [47]. The mechanism of inhibition of lipid peroxidation by resveratrol was suggested mainly to occur by the scavenging of lipid peroxyl radicals (e.g., α-tocopherol) within the membrane [48]. However, opposing results have also been reported. For example, Turrens *et al.* [49] found that resveratrol treatment has no effect on the lipoprotein profile and did not prevent the peroxidation of serum lipids in normal rats. Oxidative stress caused by phorbol esters or oxidized LDL upregulates the class A scavenger receptor in human smooth muscle cells that normally do not express this receptor. This enhancement was dependent on a concurrent upregulation of COX-2 expression in smooth muscle cells, and could be blocked by resveratrol [50]. An extract of the rhizome of *P. cuspidatum* decreased the levels of triglyceride, total cholesterol and glucose in liver, and also TNF-α mRNA levels in adipose tissue of rats with non-alcohol-induced fatty liver disease [51]. An aqueous extract of *P. cuspidatum* markedly reduced the cellular cholesterol content (notably the cholesteryl ester content), as well as the incorporation of [^{14}C]oleate into the cellular cholesteryl ester in HepG2 human hepatocytes following a 24 h incubation. The extract (with resveratrol as the active component) significantly inhibited the activity of acyl-coenzyme A-cholesterol acyltransferase, which might represent a mechanism for the inhibition of cholesterol ester formation [52].

The pharmacokinetics of resveratrol after oral administration to rats can be described by an open one- or two-compartment model. Tissue concentrations show a significant cardiac bioavailability, and a strong affinity for the liver and kidneys [53]. Up to 97% of resveratrol was absorbed by the isolated rat small intestine *in vitro* as resveratrol glucuronide. Only a small portion was absorbed across the enterocytes of the jejunum and ileum in unmetabolized form [54].

Resveratrol was found to be antimutagenic and anticarcinogenic in several test systems. It was found to block the process of multistep carcinogenesis at stages of tumor initiation, promotion, and progression [55]. Resveratrol inhibited the mutagenicity of the food-borne heterocyclic amine Trp-P-1 in the umu test and in *S. typhimurium* TA1535/pSK1002 [56]. The anti-initiation activity of resveratrol was caused by its antioxidant and antimutagenic effects; by an inhibition of the cytochrome P450 isoenzyme CYP1A1; by preventing activation of the aryl hydrocarbon receptor; and by an induction of the phase II metabolizing enzymes [57–59]. Resveratrol was reported to protect against mammary cancer formation in rodents. Following the dietary administration of resveratrol to rats, changes in cell proliferation and apoptosis in the terminal ductal structures of the mammary gland helped to explain these protective effects [60].

The antipromotion activity of resveratrol was found to be caused by its anti-inflammatory effects [61]; and by an inhibition of arachidonic acid metabolite production [62] catalyzed by either COX-1 or COX-2 [62, 63]. The antiprogression activity of resveratrol was demonstrated by its ability to induce the differentiation of human promyelocytic leukemia HL-60 cells. Moreover, the pretreatment of mouse skin with resveratrol significantly counteracted TPA-induced oxidative stress. Resveratrol reduced the generation of hydrogen peroxide and normalized levels of myeloperoxidase and glutathione reductase activities. Resveratrol selectively inhibited the TPA-induced expression of *Fos* and transforming growth factor-β_1 (TGF-β_1), but did not affect other TPA-induced gene products [64].

Resveratrol showed growth-inhibitory activity against 32Dp210, L1210 mouse leukemic and U937, HL-60 human leukemic cells, HepG2 human hepatocarcinoma, MCF-7 human mammary carcinoma and MCF-7/ADM adriamycin-resistant mammary carcinoma cells, but not against human normal hepatocytes [65]. Long-term exposure to resveratrol inhibited the clonal growth of normal hematopoietic progenitor cells at a higher IC_{50} as compared to most leukemia cell lines [66]. Resveratrol was also found to induce apoptosis in HL-60 [67]. In contrast, in normal human peripheral blood lymphocytes, resveratrol had no effect on cell survival [68]. At concentrations down to 30 μM, resveratrol caused a complete inhibition of HL-60 cell proliferation and a rapid differentiation towards a myelomonocytic phenotype. Analyses using flow cytometry demonstrated an accumulation of cells in the G_1 and S phases. A significant increase in cyclins A and E was also observed, along with an accumulation of cdc2 in the inactive phosphorylated form [69]. In THP-1 human monocytic leukemia cells, resveratrol was reported to induce *Fas* signaling-independent apoptosis [70]. Resveratrol, at a concentration of 100 μM, also inhibited the growth of SCC-25 human oral squamous carcinoma cells *in vitro* [71].

Resveratrol inhibited the tyrosine kinase activity of particulate and cytosolic fractions of human prostate adenoma [72]. A four-day treatment of rats with resveratrol reduced the levels of intracellular and secreted prostate-specific antigen (PSA) by approximately 80%, via a mechanism which was independent of changes in the androgen receptors [73]. However, resveratrol was also reported to inhibit the expression and function of androgen receptors in LNCaP prostate cancer cells [74], to induce apoptotic cell death in human prostate cancer cell lines [75], and to inhibit the cell proliferation of hormone-sensitive (MCF7, T47D) and -resistant (MDA-MB-231) breast cancer cell lines in dose- and time-dependent manner, at picomolar to nanomolar concentrations. A specific interaction of resveratrol with steroid receptors was observed [76]. Resveratrol exerted a greater growth- inhibitory effect on MDA-MB-435 mammary carcinoma cells than on minimally invasive MCF-7 cells [77]. It was also reported to strongly inhibit the proliferation of rat hepatoma Fao cells and human hepatoblastoma HepG2 cells. Resveratrol appeared to prevent or delay the entry to mitosis, since no inhibition of thymidine incorporation was observed, despite there being an increase in cell number in the S and G_2/M phases [78].

The oral administration of resveratrol to rats inoculated with Yoshida AH-130 ascites hepatoma (a fast-growing tumor) caused a significant decrease in the number of tumor cells, due to the induction of apoptosis [79]. The suppression of cell transformation and induction of apoptosis by resveratrol was only observed in cells expressing wild-type p53, but not in p53-deficient cells, which suggests that resveratrol may induce apoptosis via an increased p53 activity [80, 81].

Resveratrol was also found to be a phytoestrogen [82] and to exert estrogenic activity [83]. It inhibited the binding of estradiol to the estrogen receptor (ER); activated the transcription of estrogen-responsive reporter genes transfected into MCF-7 human breast cancer cells; and stimulated the proliferation of estrogen-dependent T47D breast cancer cells [84, 85]. However, it was also reported to have a direct antiproliferative effect on human breast epithelial cells, independent of ERs [86]. In an estrogen-responsive pituitary cell line PR1, resveratrol induced prolactin secretion in both dose- and time-dependent manner [87]. Not only case reports and cohort studies but also randomized trials have revealed that resveratrol exhibited estrogenic effects in humans [88]. As with estradiol, resveratrol stimulated the estrogen-regulated mRNA stabilizing factor in the liver of roosters, and exerted an agonistic activity with estradiol [89]. However, experimental results with rats suggested that resveratrol has little or no estrogen agonism on reproductive and nonreproductive estrogen target tissues, and may act as an estrogen antagonist [90].

The methanolic extract from the roots of *P. cuspidatum* was found to enhance cell proliferation at 30 or 100 $\mu g\,ml^{-1}$, in an estrogen-sensitive MCF-7 cell line. A bioassay-guided separation revealed emodin and emodin 8-*O*-β-D-glucopyranoside as the active principles. The methanolic extracts from *Polygonum*, *Cassia*, *Aloe*, and *Rheum* species, all of which are known to contain anthraquinones, also inhibited MCF-7 proliferation. Among various anthraquinones from plant sources and synthetic anthraquinones, aloe-emodin, chrysophanol, chrysophanol 8-*O*-β-D-glucopyranoside, and 1,8-dihydroxyanthraquinone all showed weak activity, whereas alizarin and 2,6-dihydroxyanthraquinone as well as emodin (having 2- and/or 6-hydroxyl groups) showed potent activity. Emodin and 2,6-dihydroxyanthraquinone also inhibited 17β-estradiol binding to human ERs, with K_i values of 0.77 and 0.31 μM for ER-α and 1.5 and 0.69 μM for ER-β [91]. These findings indicate that hydroxyanthraquinones such as emodin might act as phytoestrogens, with an affinity to human estrogen receptors. Mutagenicity/genotoxicity of Emodin and other hydroxy anthraquinones is discussed under *Polygonum multiflorum* (p. 938).

References

1 Chu, X., Sun, A., and Liu, R. (2005) Preparative isolation and purification of five compounds from the Chinese medicinal herb *Polygonum cuspidatum* Sieb. et Zucc by high-speed counter-current chromatography. *J. Chromatogr. A*, **1097**, 33–39.

2 Tian, K., Zhang, H., Chen, X., and Hu, Z. (2006) Determination of five anthraquinones in medicinal plants by capillary zone electrophoresis with β-cyclodextrin addition. *J. Chromatogr. A*, **1123**, 134–137.

3 Ma, Y.T., Wan, D.G., and Song, L.K. (2006) HPLC fingerprint of Rhizoma Polygoni cuspidate. *Zhongguo Zhong Yao Za Zhi*, **31**, 972–974.

4 Du, F.Y., Xiao, X.H., and Li, G.K. (2007) Application of ionic liquids in the microwave-assisted extraction of trans-resveratrol from Rhizoma Polygoni cuspidati. *J. Chromatogr. A*, **1140**, 56–62.

5 Wu, B. and Zhang, H.J. (2006) Quantitative determination of the (E)- and (Z)-diastereomers of resveratrol and resveratrol glucoside in the roots of *Polygonum cuspidatum* by HPLC and elementary study on their fluorescence. *Yao Xue Xue Bao*, **41**, 522–526.

6 Lin, L.L., Lien, C.Y., Cheng, Y.C., and Ku, K.L. (2007) An effective sample preparation approach for screening the anticancer compound piceatannol using HPLC coupled with UV and fluorescence detection. *J. Chromatogr. B*, **853**, 175–182.

7 Wang, Y., Zhao, K., and Wu, X. (1994) The prognostic implication of determination of activation of leucocyte in severe burns. *Chin. J. Plastic Surg. Burns*, **10**, 286–289.

8 Wu, X.B., Zhao, K.S., and Huang, X.L. (1994) Changes in adhesion features of leukocytes in rats with severe burns. *Natl Med. J. China*, **74**, 312–314.

9 Wu, K. and Huang, Q. (1996) Relationship between disturbances of microcirculation and TNF during burn shock. *Chin. J. Plastic Surg. Burns*, **12**, 41–44.

10 Lou, Z., Huang, W., and Liu, J. (1995) Effects of Chinese herbs on impaired lymphocyte functions after thermal injury in mice. *Chin. J. Surg.*, **33**, 571–573.

11 Song, J.H., Kim, S.K., Chang, K.W., Han, S.K., Yi, H.K., and Jeon, J.G. (2006) In vitro

inhibitory effects of *Polygonum cuspidatum* on bacterial viability and virulence factors of *Streptococcus mutans* and *Streptococcus sobrinus*. *Arch. Oral Biol.*, **51**, 1131–1140.

12 Chang, J.S., Liu, H.W., Wang, K.C., Chen, M. C., Chiang, L.C., Hua, Y.C., and Lin, C.C. (2005) Ethanol extract of *Polygonum cuspidatum* inhibits hepatitis B virus in a stable HBV-producing cell line. *Antiviral. Res.*, **66**, 29–34.

13 Mattivi, F., Reniero, F., and Korhammer, S. (1995) Isolation, characterization, and evolution in red wine vinification of resveratrol monomers. *J. Agric. Food Chem.*, **43**, 1820–1823.

14 Fremont, L. (2000) Biological effects of resveratrol. *Life Sci.*, **66**, 663–673.

15 Daniel, O., Meier, M.S., Schlatter, J., and Frischknecht, P. (1999) Selected phenolic compounds in cultivated plants: ecologic functions, health implications, and modulation by pesticides. *Environ. Health Perspect.*, **107** (Suppl. 1), 109–114.

16 Bertelli, A.A., Giovannini, L., Giannessi, D., Migliori, M., Bernini, W., Fregoni, M., and Bertelli, A. (1995) Antiplatelet activity of synthetic and natural resveratrol in red wine. *Int. J. Tissue React.*, **17**, 1–3.

17 Pace-Asciak, C.R., Hahn, S., Diamandis, E.P., Soleas, G., and Goldberg, D.M. (1995) The red wine phenolics trans-resveratrol and quercetin block human platelet aggregation and eicosanoid synthesis: implications for protection against coronary heart disease. *Clin. Chim. Acta*, **235**, 207–219.

18 Goldberg, D.M. (1996) More on antioxidant activity of resveratrol in red wine. *Clin. Chem.*, **42**, 113–114.

19 Zou, J.G., Huang, Y.Z., Chen, Q., Wei, E.H., Hsieh, T.C., and Wu, J.M. (1999) Resveratrol inhibits copper ion-induced and azo compound-initiated oxidative modification of human low density lipoprotein. *Biochem. Mol. Biol. Int.*, **47**, 1089–1096.

20 Miura, T., Muraoka, S., Ikeda, N., Watanabe, M., and Fujimoto, Y. (2000) Antioxidative and prooxidative action of stilbene derivatives. *Pharmacol. Toxicol.*, **86**, 203–208.

21 Ferrero, M.E., Bertelli, A.A., Pellegatta, F., Fulgenzi, A., Corsi, M.M., and Bertelli, A. (1998) Phytoalexin resveratrol (3,4′,5-trihydroxystilbene) modulates granulocyte and monocyte endothelial adhesion. *Transplant. Proc.*, **30**, 4191–4193.

22 Pendurthi, U.R., Williams, J.T., and Rao, L.V. (1999) Resveratrol, a polyphenolic compound found in wine, inhibits tissue factor expression in vascular cells: A possible mechanism for the cardiovascular benefits associated with moderate consumption of wine. *Arterioscler. Thromb. Vasc. Biol.*, **19**, 419–426.

23 Rotondo, S., Rajtar, G., Manarini, S., Celardo, A., Rotillo, D., de Gaetano, G., Evangelista, V., and Cerletti, C. (1998) Effect of *trans*-resveratrol, a natural polyphenolic compound, on human polymorphonuclear leukocyte function. *Br. J. Pharmacol.*, **123**, 1691–1699.

24 Mizutani, K., Ikeda, K., Kawai, Y., and Yamori, Y. (2000) Resveratrol attenuates ovariectomy-induced hypertension and bone loss in stroke-prone spontaneously hypertensive rats. *J. Nutr. Sci. Vitaminol. (Tokyo)*, **46**, 78–83.

25 Manna, S.K., Mukhopadhyay, A., and Aggarwal, B.B. (2000) Resveratrol suppresses TNF-induced activation of nuclear transcription factors NF-κB, activator protein-1, and apoptosis: potential role of reactive oxygen intermediates and lipid peroxidation. *J. Immunol.*, **164**, 6509–6519.

26 Garcia-Garcia, J., Micol, V., de Godos, A., and Gomez-Fernandez, J.C. (1999) The cancer chemopreventive agent resveratrol is incorporated into model membranes and inhibits protein kinase C activity. *Arch. Biochem. Biophys.*, **372**, 382–388.

27 Stewart, J.R., Ward, N.E., Ioannides, C.G., and O'Brian, C.A. (1999) Resveratrol preferentially inhibits protein kinase C-catalyzed phosphorylation of a cofactor-independent, arginine-rich protein substrate by a novel mechanism. *Biochemistry*, **38**, 13244–13251.

28 Zou, J., Huang, Y., Chen, Q., Wang, N., Cao, K., Hsieh, T.C., and Wu, J.M. (1999) Suppression of mitogenesis and regulation of cell cycle traverse by resveratrol in cultured smooth muscle cells. *Int. J. Oncol.*, **15**, 647–651.

29 El-Mowafy, A.M. and White, R.E. (1999) Resveratrol inhibits MAPK activity and nuclear translocation in coronary artery smooth muscle: reversal of endothelin-1 stimulatory effects. *FEBS Lett.*, **451**, 63–67.

30 Jager, U. and Nguyen-Duong, H. (1999) Relaxant effect of *trans*-resveratrol on isolated porcine coronary arteries. *Arzneim.-Forsch.*, **49**, 207–211.

31 Chen, C.K. and Pace-Asciak, C.R. (1996) Vasorelaxing activity of resveratrol and quercetin in isolated rat aorta. *Gen. Pharmacol.*, **27**, 363–366.

32 Fauconneau, B., Waffo-Teguo, P., Huguet, F., Barrier, L., Decendit, A., and Merillon, J.M.

(1997) Comparative study of radical scavenger and antioxidant properties of phenolic compounds from *Vitis vinifera* cell cultures using in vitro tests. *Life Sci.*, **61**, 2103–2110.

33 Sato, M., Ray, P.S., Maulik, G., Maulik, N., Engelman, R.M., Bertelli, A.A., Bertelli, A., and Das, D.K. (2000) Myocardial protection with red wine extract. *J. Cardiovasc. Pharmacol.*, **35**, 263–268.

34 Sato, M., Maulik, G., Bagchi, D., and Das, D.K. (2000) Myocardial protection by protykin, a novel extract of trans-resveratrol and emodin. *Free Radic. Res.*, **32**, 135–144.

35 Li, H.F., Chen, S.A., and Wu, S.N. (2000) Evidence for the stimulatory effect of resveratrol on Ca^{2+}-activated K^{+} current in vascular endothelial cells. *Cardiovasc. Res.*, **45**, 1035–1045.

36 Zhao, K.S., Jin, C., Huang, X., Liu, J., Yan, W.S., Huang, Q., and Kan, W. (2003) The mechanism of polydatin in shock treatment. *Clin. Hemorheol. Microcir.*, **29**, 211–217.

37 Cheng, Y., Zhang, H.T., Sun, L., Guo, S., Ouyang, S., Zhang, Y., and Xu, J. (2006) Involvement of cell adhesion molecules in polydatin protection of brain tissues from ischemia-reperfusion injury. *Brain Res.*, **1110**, 193–200.

38 Moreno, J.J. (2000) Resveratrol modulates arachidonic acid release, prostaglandin synthesis, and 3T6 fibroblast growth. *J. Pharmacol. Exp. Ther.*, **294**, 333–338.

39 Martinez, J. and Moreno, J.J. (2000) Effect of resveratrol, a natural polyphenolic compound, on reactive oxygen species and prostaglandin production. *Biochem. Pharmacol.*, **59**, 865–870.

40 Subbaramaiah, K., Michaluart, P., Chung, W. J., Tanabe, T., Telang, N., and Dannenberg, A.J. (1999) Resveratrol inhibits cyclooxygenase-2 transcription in human mammary epithelial cells. *Ann. N. Y. Acad. Sci.*, **889**, 214–223.

41 MacCarrone, M., Lorenzon, T., Guerrieri, P., and Agro, A.F. (1999) Resveratrol prevents apoptosis in K562 cells by inhibiting lipooxygenase and cyclooxygenase activity. *Eur. J. Biochem.*, **265**, 27–34.

42 Dobrydneva, Y., Williams, R.L., and Blackmore, P.F. (1999) *trans*-Resveratrol inhibits calcium influx in thrombin-stimulated human platelets. *Br. J. Pharmacol.*, **128**, 149–157.

43 Zhang, P.W., Yu, C.L., Wang, Y.Z., Luo, S.F., Sun, L.S., and Li, R.S. (1995) Influence of 3,4′,5-trihydroxystibene 3-β-mono-D-glucoside on vascular endothelial epoprostenol and platelet aggregation. *Acta Pharmacol. Sin.*, **16**, 265–268.

44 Wang, Y.Z., Luo, S.F., Zhang, P.W., and Yu, C. L. (1995) Reducing effect of 3,4′,5-trihydroxystibene 3-β-mono-D-glucoside on arterial thrombosis induced by vascular endothelial injury. *Acta Pharmacol. Sin.*, **16**, 159–162.

45 Mo, G.Y., Jin, L.J., and Jin, C.H. (1993) Polydatin prevents endotoxin-induced acute lung injury in rats. *Chin. J. Tubercul. Respir. Dis.*, **16**, 153–154.

46 Luo, S.F., Yu, C.L., and Zhang, P.W. (1990) Influences of 3,4′,5-trihydroxystibene 3-β-mono-D-glucoside on beat rate and injury of cultured newborn rat myocardial cells. *Acta Pharmacol. Sin.*, **11**, 147–150.

47 Kimura, Y., Ohminami, H., Okuda, H., Baba, K., Kozawa, M., and Archi, S. (1983) Effects of stilbene components of roots of *Polygonum* spp. on liver injury in peroxidized oil-fed rats. *Planta Med.*, **49**, 51–54.

48 Tadolini, B., Juliano, C., Piu, L., Franconi, F., and Cabrini, L. (2000) Resveratrol inhibition of lipid peroxidation. *Free Radic. Res.*, **33**, 105–114.

49 Turrens, J.F., Lariccia, J., and Nair, M.G. (1997) Resveratrol has no effect on lipoprotein profile and does not prevent peroxidation of serum lipids in normal rats. *Free Radic. Res.*, **27**, 557–562.

50 Mietus-Snyder, M., Gowri, M.S., and Pitas, R. E. (2000) Class A scavenger receptor up-regulation in smooth muscle cells by oxidized low density lipoprotein. Enhancement by calcium flux and concurrent cyclooxygenase-2 up-regulation. *J. Biol. Chem.*, **275**, 17661–17670.

51 Jiang, Q., Li, Y., Pan, J., Ma, J., Xu, B., Yang, H., Zhang, J., and He, M. (2005) TNF-α gene expression of NAFLD rat intervened by the extracts of Rhizoma Polygoni cuspidati. *Zhong Yao Cai*, **28**, 917–920.

52 Park, C.S., Lee, Y.C., Kim, J.D., Kim, H.M., and Kim, C.H. (2004) Inhibitory effects of *Polygonum cuspidatum* water extract (PCWE) and its component resveratrol on acyl-coenzyme A-cholesterol acyltransferase activity for cholesteryl ester synthesis in HepG2 cells. *Vasc. Pharmacol.*, **40**, 279–284.

53 Bertelli, A.A., Giovannini, L., Stradi, R., Urien, S., Tillement, J.P., and Bertelli, A. (1998) Evaluation of kinetic parameters of natural phytoalexin in resveratrol orally administered in wine to rats. *Drugs Exp. Clin. Res.*, **24**, 51–55.

54 Kuhnle, G., Spencer, J.P., Chowrimootoo, G., Schroeter, H., Debnam, E.S., Srai, S.K., Rice-Evans, C., and Hahn, U. (2000) Resveratrol is absorbed in the small intestine as resveratrol glucuronide. *Biochem. Biophys. Res. Commun.*, **272**, 212–217.

55 Lin, J.K. and Tsai, S.H. (1999) Chemoprevention of cancer and cardiovascular disease by resveratrol. *Proc. Natl Sci. Counc. Repub. China B.*, **23**, 99–106.

56 Uenobe, F., Nakamura, S., and Miyazawa, M. (1997) Antimutagenic effect of resveratrol against Trp-P-1. *Mutat. Res.*, **373**, 197–200.

57 Ciolino, H.P. and Yeh, G.C. (1999) Inhibition of aryl hydrocarbon-induced cytochrome P-450 1A1 enzyme activity and CYP1A1 expression by resveratrol. *Mol. Pharmacol.*, **56**, 760–767.

58 Casper, R.F., Quesne, M., Rogers, I.M., Shirota, T., Jolivet, A., Milgrom, E., and Savouret, J.F. (1999) Resveratrol has antagonist activity on the aryl hydrocarbon receptor: implications for prevention of dioxin toxicity. *Mol. Pharmacol.*, **56**, 784–790.

59 Chun, Y.J., Kim, M.Y., and Guengerich, F.P. (1999) Resveratrol is a selective human cytochrome P450 1A1 inhibitor. *Biochem. Biophys. Res. Commun.*, **262**, 20–24.

60 Whitsett, T.G. Jr and Lamartiniere, C.A. (2006) Genistein and resveratrol: mammary cancer chemoprevention and mechanisms of action in the rat. *Expert Rev. Anticancer Ther.*, **6**, 1699–1706.

61 Wadsworth, T.L. and Koop, D.R. (1999) Effects of the wine polyphenolics quercetin and resveratrol on pro-inflammatory cytokine expression in RAW 264.7 macrophages. *Biochem. Pharmacol.*, **57**, 941–949.

62 Shin, N.H., Ryu, S.Y., Lee, H., Min, K.R., and Kim, Y. (1998) Inhibitory effects of hydroxystilbenes on cyclooxygenase from sheep seminal vesicles. *Planta Med.*, **64**, 283–284.

63 Subbaramaiah, K., Chung, W.J., Michaluart, P., Telang, N., Tanabe, T., Inoue, H., Jang, M., Pezzuto, J.M., and Dannenberg, A.J. (1998) Resveratrol inhibits cyclooxygenase-2 transcription and activity in phorbol ester-treated human mammary epithelial cells. *J. Biol. Chem.*, **273**, 21875–21882.

64 Jang, M. and Pezzuto, J.M. (1999) Cancer chemopreventive activity of resveratrol. *Drugs Exp. Clin. Res.*, **25**, 65–77.

65 Feng, L., Zhang, L.F., Yan, T., Jin, J., and Tao, W.Y. (2006) Studies on active substance of anticancer effect in *Polygonum cuspidatum*. *Zhong Yao Cai*, **29**, 689–691.

66 Gautam, S.C., Xu, Y.X., Dumaguin, M., Janakiraman, N., and Chapman, R.A. (2000) Resveratrol selectively inhibits leukemia cells: a prospective agent for *ex vivo* bone marrow purging. *Bone Marrow Transplant.*, **25**, 639–645.

67 Surh, Y.J., Hurh, Y.J., Kang, J.Y., Lee, E., Kong, G., and Lee, S.J. (1999) Resveratrol, an antioxidant present in red wine, induces apoptosis in human promyelocytic leukemia (HL-60) cells. *Cancer Lett.*, **140**, 1–10.

68 Clement, M.V., Hirpara, J.L., Chawdhury, S. H., and Pervaiz, S. (1998) Chemopreventive agent resveratrol, a natural product derived from grapes, triggers CD_{95} signaling-dependent apoptosis in human tumor cells. *Blood*, **92**, 996–1002.

69 Ragione, F.D., Cucciolla, V., Borriello, A., Pietra, V.D., Racioppi, L., Soldati, G., Manna, C., Galletti, P., and Zappia, V. (1998) Resveratrol arrests the cell division cycle at S/G_2 phase transition. *Biochem. Biophys. Res. Commun.*, **250**, 53–58.

70 Tsan, M.F., White, J.E., Maheshwari, J.G., Bremner, T.A., and Sacco, J. (2000) Resveratrol induces Fas signalling-independent apoptosis in THP-1 human monocytic leukaemia cells. *Br. J. Haematol.*, **109**, 405–412.

71 Elattar, T.M. and Virji, A.S. (1999) The effect of red wine and its components on growth and proliferation of human oral squamous carcinoma cells. *Anticancer Res.*, **19B**, 5407–5414.

72 Palmieri, L., Mameli, M., and Ronca, G. (1999) Effect of resveratrol and some other natural compounds on tyrosine kinase activity and on cytolysis. *Drugs Exp. Clin. Res.*, **25**, 79–85.

73 Hsieh, T.C. and Wu, J.M. (2000) Grape-derived chemopreventive agent resveratrol decreases prostate-specific antigen (PSA) expression in LNCaP cells by an androgen receptor (AR)-independent mechanism. *Anticancer Res.*, **20A**, 225–228.

74 Hsieh, T.C. and Wu, J.M. (1999) Differential effects on growth, cell cycle arrest, and induction of apoptosis by resveratrol in human prostate cancer cell lines. *Exp. Cell. Res.*, **249**, 109–115.

75 Mitchell, S.H., Zhu, W., and Young, C.Y. (1999) Resveratrol inhibits the expression and function of the androgen receptor in LNCaP prostate cancer cells. *Cancer Res.*, **59**, 5892–5895.

76 Damianaki, A., Bakogeorgou, E., Kampa, M., Notas, G., Hatzoglou, A., Panagiotou, S., Gemetzi, C., Kouroumalis, E., Martin, P.M., and Castanas, E. (2000) Potent inhibitory action of red wine polyphenols on human breast cancer cells. *J. Cell. Biochem.*, **78**, 429–444

77 Hsieh, T.C., Burfeind, P., Laud, K., Backer, J.M., Traganos, F., Darzynkiewicz, Z., and Wu, J.M. (1999) Cell cycle effects and control of gene expression by resveratrol in human breast carcinoma cell lines with different metastatic potentials. *Int. J. Oncol.*, **15**, 245–252.

78 Delmas, D., Jannin, B., Malki, M.C., and Latruffe, N. (2000) Inhibitory effect of resveratrol on the proliferation of human and rat hepatic derived cell lines. *Oncol. Rep.*, **7**, 847–852.

79 Carbo, N., Costelli, P., Baccino, F.M., Lopez-Soriano, F.J., and Argiles, J.M. (1999) Resveratrol, a natural product present in wine, decreases tumour growth in a rat tumour model. *Biochem. Biophys. Res. Commun.*, **254**, 739–743.

80 Hsieh, T.C., Juan, G., Darzynkiewicz, Z., and Wu, J.M. (1999) Resveratrol increases nitric oxide synthase, induces accumulation of p53 and p21(WAF1/CIP1), and suppresses cultured bovine pulmonary artery endothelial cell proliferation by perturbing progression through S and G_2. *Cancer Res.*, **59**, 2596–2601.

81 Huang, C., Ma, W.Y., Goranson, A., and Dong, Z. (1999) Resveratrol suppresses cell transformation and induces apoptosis through a p53-dependent pathway. *Carcinogenesis*, **20**, 237–242.

82 Ashby, J., Tinwell, H., Pennie, W., Brooks, A.N., Lefevre, P.A., Beresford, N., and Sumpter, J.P. (1999) Partial and weak oestrogenicity of the red wine constituent resveratrol: consideration of its superagonist activity in MCF-7 cells and its suggested cardiovascular protective effects. *J. Appl. Toxicol.*, **19**, 39–45.

83 Basly, J.P., Marre-Fournier, F., Le Bail, J.C., Habrioux, G., and Chulia, A.J. (2000) Estrogenic/antiestrogenic and scavenging properties of (*E*)- and (*Z*)-resveratrol. *Life Sci.*, **66**, 769–777.

84 Lu, R. and Serrero, G. (1999) Resveratrol, a natural product derived from grape, exhibits antiestrogenic activity and inhibits the growth of human breast cancer cells. *J. Cell. Physiol.*, **179**, 297–304.

85 Yoon, K., Pellaroni, L., Ramamoorthy, K., Gaido, K., and Safe, S. (2000) Ligand structure-dependent differences in activation of estrogen receptor alpha in human HepG2 liver and U2 osteogenic cancer cell lines. *Mol. Cell. Endocrinol.*, **162**, 209–219.

86 Mgbonyebi, O.P., Russo, J., and Russo, I.H. (1998) Antiproliferative effect of synthetic resveratrol on human breast epithelial cells. *Int. J. Oncol.*, **12**, 865–869.

87 Stahl, S., Chun, T.Y., and Gray, W.G. (1998) Phytoestrogens act as estrogen agonists in an estrogen-responsive pituitary cell line. *Toxicol. Appl. Pharmacol.*, **152**, 41–48.

88 Calabrese, G. (1999) Nonalcoholic compounds of wine: the phytoestrogen resveratrol and moderate red wine consumption during menopause. *Drugs Exp. Clin. Res.*, **25**, 111–114.

89 Ratna, W.N. and Simonelli, J.A. (2002) The action of dietary phytochemicals quercetin, catechin, resveratrol and naringenin on estrogen-mediated gene expression. *Life Sci.*, **70**, 1577–1589.

90 Turner, R.T., Evans, G.L., Zhang, M., Maran, A., and Sibonga, J.D. (1999) Is resveratrol an estrogen agonist in growing rats? *Endocrinology*, **140**, 50–54.

91 Matsuda, H., Shimoda, H., Morikawa, T., and Yoshikawa, M. (2001) Phytoestrogens from the roots of *Polygonum cuspidatum* (Polygonaceae): structure-requirement of hydroxyanthraquinones for estrogenic activity. *Bioorg. Med. Chem. Lett.*, **11** (14), 1839–1842.

Polygonum multiflorum

Caulis Polygoni Multiflori (Shouwuteng) is the dried canes of *Polygonum multiflorum* Thunb. (Polygonaceae). It is used as a sedative and antirheumatic agent.

Polygonum multiflorum Thunb.

Radix Polygoni Multiflori (Heshouwu) is the dried tuber of *Polygonum multiflorum*. It is used as an antidote, laxative, general tonic, and for the treatment of hyperlipidemia.

Radix Polygoni Multiflori Preparata Cum Succo Glycines Sotae (Zhiheshouwu) is the processed root of *Polygonum multiflorum*. It is used for the same indications as Radix Polygoni Multiflori.

The dried stem of *Polygonum aubertii* Henry is listed in the appendix of the Chinese Pharmacopoeia.

Chemistry

The cane and root of *P. multiflorum* are known to contain anthraquinones and hydroxylated stilbenes. The major anthraquinones were identified as emodin and physcion, and minor anthraquinones as questin, questinol, 2-acetylemodin, 1-*O*-methylemodin, citreorosein, and physcion 1-*O*-β-D-glucopyranoside [1]. The Chinese Pharmacopoeia requires a qualitative determination of emodin in Caulis Polygoni Multiflori by thin-layer chromatographic comparison with a reference substance. The content of emodin in the canes of *P. multiflorum* was reported to be 0.3–0.4% [2].

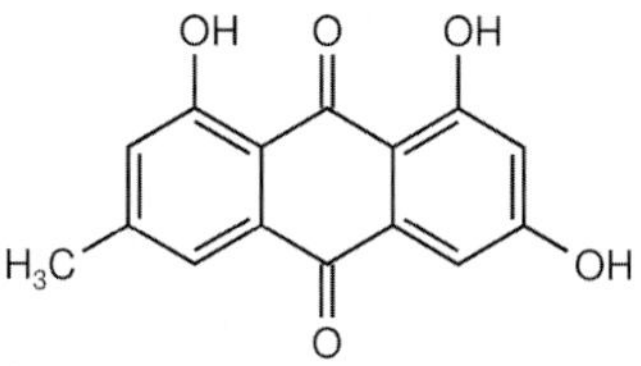

Emodin

The major stilbene glycoside in the root of *P. multiflorum* was identified as 2,3,5,4′-tetrahydroxystilbene 2-*O*-β-D-glucopyranoside, with contents ranging from 1% to 4% [3, 4]. Minor stilbene glycosides from the root were identified as 2,4,6,4′-tetrahydroxystilbene 2-*O*-β-D-glucopyranoside [5], 2,3,5,4′-tetrahydroxystilbene 2,3-bis-*O*-β-D-glucopyranoside (polygonimitin C), 2,3,5,4′-tetrahydroxystilbene 2-*O*-β-D-glucopyranoside 2″-*O*-monogalloyl ester and 3″-*O*-monogalloyl ester [6]. The Chinese Pharmacopoeia requires a quantitative determination of 2,3,5,4′-tetrahydroxystilbene 2-*O*-β-D-glucopyranoside in Radix Polygoni Multiflori and in Radix Polygoni Multiflori Preparata Cum Succo Glycines Sotae by HPLC. The content of 2,3,5,4′-tetrahydroxystilbene 2-*O*-β-D-glucopyranoside in Radix Polygoni Multiflori should be not less than 1.0%, and in Radix Polygoni

Handbook of Chinese Medicinal Plants: Chemistry, Pharmacology, Toxicology
Weici Tang and Gerhard Eisenbrand

ISBN: 978-3-527-32226-8

Multiflori Preparata Cum Succo Glycines Sotae not less than 0.7%.

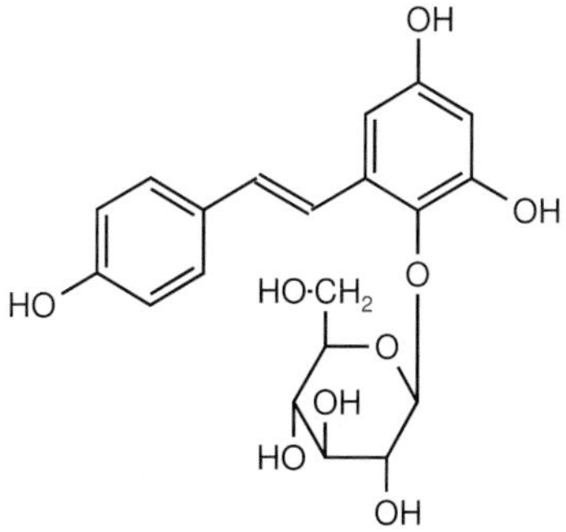

2,3,5,4′-Tetrahydroxystilbene 2-O-β-D-glucopyranoside

Pharmacology and Toxicology

The mutagenicity and carcinogenicity of the anthraquinones lucidin and rubiadin, two hydroxyanthraquinones from *Rubia* species (Rubiaceae), gave occasion for further studies of potential mutagenicity of hydroxyanthraquinones, because hydroxyanthraquinones are widely distributed not only in medicinal plants, including *P. multiflorum*, but also in several common vegetables. Physcion predominated in all vegetables tested [7]. Emodin was reported to induce mutations in *S. typhimurium* TA97 [8], TA100, TA2637 [9], and especially in *S. typhimurium* TA1537, after metabolic activation, indicating that emodin acts mainly as a frame-shift mutagen [10–13]. In addition, emodin monoglucoside showed mutagenic activity in *S. typhimurium* TA100 in the presence of a cell-free extract of rat cecal bacteria [14]. In the micronucleus test, in the HGPRT forward mutation assay, and also in the sister chromatid exchange (SCE) assay using V79 cells, emodin did not show any genotoxicity, with or without metabolic activation [15]. The inhibition of the catalytic activity of topoisomerase II should contribute to the genotoxicity and mutagenicity of emodin [16]. However, emodin was reported to induce mutations at the TK-locus and micronuclei, and was also active in the comet assay in mouse lymphoma L5178Y cells [17].

In contrast, emodin dose-dependently reduced the mutagenicity of BaP and of the food-borne mutagens IQ and Trp-P-2 in *S. typhimurium* TA98 [18], as well as the genotoxicity of 1-nitropyrene in the SOS chromotest with *Escherichia coli* PQ37 [19]. Emodin reduced the mutagenicity of IQ by a direct inhibition of the hepatic microsomal activation, and not by interaction with proximate metabolites of IQ, or by the modification of DNA repair processes in the bacterial cell [18]. Emodin was also reported to significantly inhibit the formation of 1-nitropyrene–DNA adducts in *S. typhimurium* TA98 in a ^{32}P-postlabeling study [19]. Emodin increased the unscheduled DNA synthesis (UDS) of UV-treated WI38 human fibroblast cells and reduced cisplatin-induced DNA adducts, indicating that emodin may promote nucleotide excision repair capability in cells [20].

Emodin was metabolized by rat hepatic S-9 mix to give at least five quinone metabolites with 2-hydroxyemodin as a direct mutagen in *S. typhimurium* TA1537 [21, 22]. Other metabolites, including 4-hydroxyemodin, 5-hydroxyemodin and 7-hydroxyemodin, were not mutagenic in *S. typhimurium* TA1537. A further metabolite of emodin was found to be ω-hydroxyemodin, formed from emodin in the presence of hepatic microsomes derived from various animal species. Among seven animal species, the highest activity to produce ω-hydroxyemodin was observed in the hepatic microsomes of guinea pig and rat, followed by mouse and rabbit [23]. The rates of formation of ω-hydroxyemodin from emodin were not different from microsomes of rats that had been pretreated with inducers for different cytochrome P450 enzymes, which indicated that the formation of ω-hydroxyemodin is catalyzed by several cytochrome P450 enzymes, at low rates. The formation of 2-hydroxyemodin was suggested to be catalyzed by CYP4501A2 [24].

After the administration of a single oral dose (91 mg kg^{-1}) emodin to mice, approximately 30% of the dose was excreted as unchanged emodin or its metabolites in the urine, and 21% in the feces within 24 h. The major free anthraquinone metabolites of emodin in the 24-h urine were found to be ω-hydroxyemodin, 6-hydroxyrhein, chrysophanol, and physcion [25]. Rhein, aloe emodin, and possibly 4-hydroxychrysophanol, 4-hydroxyemodin, 4-hydroxyrhein, 4,6-dihydroxyaloe emodin, and 4,6-dihydroxyemodin were also detected in the urine [26]. The urinary excretion of [^{14}C] emodin administered orally to rats at a dose of 50 mg kg^{-1} amounted to 18% and 22% after 24 and 72 h, respectively. The free anthraquinones emodin and 6-hydroxyrhein predominated in urine, compared to only 3% of conjugates. In feces, 48% (24 h) and 68% (120 h) of the dose was excreted. Biliary excretion reached a maximum at about 6 h. Although radioactivity levels in most organs decreased between days 3 and 5, they remained constant in the kidneys for five days after administration [27].

ω-Hydroxyemodin was not mutagenic in *S. typhimurium* in the absence of an S-9 mix, but exhibited mutagenicity in the presence of an activating system. The mutagenic potential in *S. typhimurium* of ω-hydroxyemodin was comparable with that of 2-hydroxyemodin [23]. However, 2-hydroxyemodin induced much higher micronucleus frequencies in mouse lymphoma L5178Y cells, compared to emodin, while ω-hydroxyemodin only induced lower micronucleus frequencies [24].

The root of *P. multiflorum* is a well-known tonic and antiaging herb in traditional Chinese medicine. It is used for the clinical treatment of hyperlipidemia [28]. The oral administration of an aqueous extract of the root of *P. multiflorum* to aged mice resulted in an increased protein content, a decreases malonaldehyde content, an increased superoxide dismutase activity, and an inhibition of monoamine oxidase in the liver and brain. The monoamine content of the brain was also increased [29]. Oral administration of the extract of *P. multiflorum* to 15-month-old mice also resulted in increases in both the size and weight of the thymus. The root of *P. multiflorum* was also found to exhibit protective effects against thymus degeneration in mice, as induced with cyclophosphamide. Mice treated with the root showed an improved thymus morphology and fewer apoptotic thymocytes [30].

Amyloid β-protein 1–40, when administered intracerebrally to rats, significantly increased the times to exert electric stimulation, and markedly reduced the expression of brain-derived neurotrophic factor in hippocampal CA1 neurons. The treatment of rats with *P. multiflorum* for 30 days reversed the downregulation of brain-derived neurotrophic factor expression in hippocampal CA1 neurons [31]. Pretreatment with an aqueous extract of the root of *P. multiflorum* for four weeks to mice significantly ameliorated the cognitive deficits caused by a single intracerebroventricular injection of amyloid β-protein 25–35. This suggested that *P. multiflorum* might exert a preventive effect against cognitive deficits induced by amyloid β-protein 25–35 accumulation in Alzheimer's disease, and that this effect might be mediated by its antioxidant properties [32]. The neuroprotective effect of the ethanol extract of *P. multiflorum* was also evidenced by experimentally induced nigrostriatal dopaminergic degeneration in mice; this suggested that the ethanol-soluble part of *P. multiflorum* might exert a beneficial effect on parkinsonism [33].

2,3,5,4′-Tetrahydroxystilbene 2-*O*-β-D-glucopyranoside was found capable of promoting the release of nitric oxide (NO) from vascular endothelial cells, and has a strong antioxidant activity. The administration of 2,3,5,4′-tetrahydroxystilbene 2-*O*-β-D-glucopyranoside to rats significantly decreased the infarct size after cardiac ischemia–reperfusion; the mechanism of action of infract reduction might be related to potassium ATP channel opening [34].

The water-soluble fraction of *P. multiflorum* containing stilbene glycosides, when administered to hypercholesterolemic rabbits (as induced by a high-cholesterol diet) at a daily dose of $100\,mg\,kg^{-1}$, led to marked reductions in plasma cholesterol, low-density lipoprotein (LDL), very low-density lipoprotein (VLDL) and triglycerides, as well as a decrease in the atherosclerotic lesion area. In U937 foam cells, the aqueous fraction decreased protein expression levels of intercellular adhesion molecule-1 (ICAM-1) and vascular endothelial growth factor (VEGF) in the medium induced by oxidized lipoprotein. These effects of the aqueous fraction on atherosclerosis might be mediated via an inhibition of the expression of ICAM-1 and VEGF in foam cells [35].

An ethyl acetate extract of *P. multiflorum*, and the anthraquinone-containing fraction of the extract, were found to protect against myocardial ischemia–reperfusion damage in the isolated perfused rat heart. This protective effect was seen to be associated with an enhancement of myocardial content of reduced glutathione (GSH) and with an inhibition of glutathione peroxidase activities [36]. The ethyl acetate-soluble fraction of *P. multiflorum* was also found to inhibit the mutagenicity of benzo[*a*]pyrene (BaP) in *Salmonella typhimurium* TA98 with an activating system. The active components were identified as tannins, and related compounds such as epigallocatechin, epigallocatechin gallate, epicatechin gallate, and tannic acid. The oral administration of an aqueous extract of *P. multiflorum* to male rats for 50 weeks significantly reduced the tumor incidence induced by the subcutaneous injection of BaP [37].

Emodin and other anthraquinone derivatives were found to exhibit antibacterial activity, caused by an inhibition of the mitochondrial respiratory chain of the microorganisms [38]. Emodin was also found to show an inhibitory effect on the pathogenicity of *Trichomonas vaginalis* in mice. The inhibition was markedly reversed by the coexistence of free radical scavengers [39]. Emodin elicited a dose-dependent growth inhibition in *Helicobacter pylori* cultures [40]. Arylamine *N*-acetyltransferase activities in *H. pylori* collected from peptic ulcer patients was found to be inhibited by emodin [41]. The decrease in *N*-acetyltransferase activity was associated with an increased emodin concentration in *H. pylori* cytosols.

As evidenced from experimental studies, emodion may partly reverse the decrease in pancreatic blood flow that occurs during the early stage of acute hemorrhagic-necrotizing pancreatits. This effect may be ascribed, at least in part, to an inhibition of abnormal synthesis of eicosanoids and improvement of the cytoprotection of acini cells [42]. Emodin also inhibited the fibronectin production of cultured rat mesangial cells, and therefore has been postulated to be effective for the treatment of mesangio-proliferative glomerulonephritis [43]. Emodin was also described as a potent inhibitor of nuclear factor-κB (NF-κB) activation in human umbilical vein endothelial cells and of the expression of adhesion molecules; thus, it may be useful for treating various inflammatory diseases. The treatment of endothelial cells with tumor necrosis factor (TNF) led to an activation of NF-κB which could be inhibited by preincubation with emodin, in dose- and time-dependent manner. Here, emodin did not modify the NF-κB subunits but rather inhibited the degradation of IκB, an inhibitory subunit of NF-κB [44].

Emodin was reported to be a protein tyrosine kinase inhibitor and to suppress HER-2/neu tyrosine kinase activity in HER-2/neu-overexpressing breast cancer cells, and also to preferentially repress the proliferation of these cells [45, 46]. Amplification and overexpression of the HER-2/neu proto-oncogene, which encodes the protein tyrosine kinase receptor $p185^{neu}$, has been observed frequently in tumors from human breast cancer patients. Emodin was found to suppress the autophosphorylation and transphosphorylation activities of HER-2/neu tyrosine kinase, resulting in the tyrosine hypophosphorylation of $p185^{neu}$ in HER-2/neu overexpressing breast cancer cells. Emodin also induced the differentiation of HER-2/neu overexpressing breast cancer cells by exhibiting a morphological maturation property [47–49]. Emodin was further reported to be a nucleotide binding-site-directed protein kinase inhibitor. The structure of a complex between the catalytic subunit of *Zea mays* protein kinase CK_2 and emodin showed that the replacement of ATP by the competitive inhibitor emodin induced conformational modifications in the catalytic site of protein kinase CK_2 [50].

Emodin selectively blocked the growth of v-*ras*-transformed human bronchial epithelial cells, with an IC_{50} of $4\,\mu g\,ml^{-1}$. In contrast, emodin, at a concentration of $100\,\mu g\,ml^{-1}$, had little effect on the growth of normal human bronchial epithelial cells. Cell-cycle analyses indicated that treatment with emodin arrested the v-*ras*-transformed cells in the G_2/M phase. Immunoblotting experiments using anti-phosphotyrosine antibodies indicated that *ras*-transformed cells, as compared to their normal counterparts, exhibited elevated levels of phosphotyrosine-containing proteins. Treatment with emodin resulted in a decrease in intracellular protein tyrosine phosphorylation [51]. In a rat glomerular mesangial cell culture, a low level of c-*Myc* mRNA was detected. The addition of lipopolysaccharide (LPS) induced a higher level expression of c-*Myc* mRNA, and this overexpressing of c-*Myc* mRNA was markedly suppressed by emodin. The downregulatory effect on c-*Myc* mRNA overexpression was suggested to contribute to its inhibitory action on glomerular mesangial cells [52]. Emodin inhibited the growth factor signaling enzyme phosphatidylinositol-3-kinase, with an IC_{50} value of $3\,\mu M$ [53].

Emodin was reported to strongly inhibit the respiration of Ehrlich ascites carcinoma cells, with an IC_{50} of $20\,\mu g\,ml^{-1}$ [54]; this effect was also confirmed in leukemia L1210 cells [55]. The anthraquinone derivatives rhein, emodin and aloe emodin, each exerted an *in vivo* inhibitory effect toward P388 leukemia in mice. The survival time of the animals tested was markedly increased, and the ascites volume and tumor cell numbers were decreased [56]. Emodin was further reported to show a growth-inhibitory activity against K562 cells, with an IC_{50} of $1.5\,\mu g\,ml^{-1}$. The IC_{50} values of emodin on Raji, HeLa, Calu-1, Wish, and Vero cells were reported to range from 3 to $9\,\mu g\,ml^{-1}$ [57]. Emodin strongly inhibited the growth of HCT-15 human colon carcinoma cells and blocked the G_1/S phase of the cell cycle [58]. It was further reported that emodin bound noncovalently to DNA and inhibited the topoisomerase II activity in L5178Y cells [59]. In FM3A mouse mammary carcinoma cells, emodin was cytotoxic at concentrations of 1 to $10\,\mu g\,ml^{-1}$, and induced HGPRT mutations [60]. Questin was reported to show a considerably high immunosuppressive activity [61].

References

1 Li, J.B. and Lin, M. (1993) Chemical constituents of tuber fleeceflower (*Polygonum multiflorum*). *Chin. Trad. Herbal Drugs*, **24**, 115–118.

2 Chen, Y.J., Shen, L.J., and Yang, Y.H. (1994) Determination of emodin in Caulis Polygoni multiflori and Huangshen infusion by TLC-scanning. *China J. Chin. Mater. Med.*, **19**, 284–285.

3 Liu, C.J., Zhang, Q.H., and Zhou, Q. (1991) Assay of stilbene glucoside in *Polygonum multiflorum* Thunb. and its processed products. *China J. Chin. Mater. Med.*, **16**, 469–472.
4 Bai, H.B., Wang, J.F., and Long, J. (2004) Study on optimizing extraction process of root of *Polygonum multiflorum. Zhongguo Zhong Yao Za Zhi*, **29**, 219–221.
5 Grech, J.N., Li, Q., Roufogalis, B.D., and Duck, C.C. (1994) Novel Ca^{2+}-ATPase inhibitors from the dried root tubers of *Polygonum multiflorum. J. Nat. Prod.*, **57**, 1682–1687.
6 Zhou, L.X., Lin, M., Li, J.B., and Li, S.Z. (1994) Chemical studies on the ethyl acetate insoluble fraction of the roots of *Polygonum multiflorum* Thunb. *Acta Pharm. Sin.*, **29**, 107–110.
7 Mueller, S.O., Schmitt, M., Dekant, W., Stopper, H., Schlatter, J., Schreier, P., and Lutz, W.K. (1999) Occurrence of emodin, chrysophanol and physcion in vegetables, herbs and liquors. Genotoxicity and anti-genotoxicity of the anthraquinones and of the whole plants. *Food Chem. Toxicol.*, **37**, 481–491.
8 Bruggeman, I.M. and van der Hoeven, J.C. (1984) Lack of activity of the bacterial mutagen emodin in HGPRT and SCE assay with V79 Chinese hamster cells. *Mutat. Res.*, **138**, 219–224.
9 Tikkanen, L., Matsushima, T., and Natori, S. (1983) Mutagenicity of anthraquinones in the *Salmonella* preincubation test. *Mutat. Res.*, **116**, 297–304.
10 Wehner, F.C., Thiel, P.G., and du Rand, M. (1979) Mutagenicity of the mycotoxin emodin in the *Salmonella*/microsome system. *Appl. Environ. Microbiol.*, **37**, 658–660.
11 Krivobok, S., Seigle-Murandi, F., Steiman, R., Marzin, D.R., and Betina, V. (1992) Mutagenicity of substituted anthraquinones in the *Salmonella*/microsome system. *Mutat. Res.*, **279**, 1–8.
12 Westendorf, J., Marquardt, H., Poginsky, B., Dominiak, M., Schmidt, J., and Marquardt, H. (1990) Genotoxicity of naturally occurring hydroxyanthraquinones. *Mutat. Res.*, **240**, 1–12.
13 Lewis, D.F., Ioannides, C., and Parke, D.V. (1996) COMPACT and molecular structure in toxicity assessment: a prospective evaluation of 30 chemicals currently being tested for rodent carcinogenicity by the NCI/NTP. *Environ. Health Perspect.*, **104** (Suppl. 5), 1011–1016.
14 Brown, J.P., and Dietrich, P.S. (1979) Mutagenicity of anthraquinone and benzanthrone derivatives in the *Salmonella*/microsome test: activation of anthraquinone glycosides by enzymic extracts of rat cecal bacteria. *Mutat. Res.*, **66**, 9–24.
15 Kersten, B., Zhang, J., Brendler-Schwaab, S.Y., Kasper, P., and Muller, L. (1999) The application of the micronucleus test in Chinese hamster V79 cells to detect drug-induced photogenotoxicity. *Mutat. Res.*, **445**, 55–71.
16 Mueller, S.O. and Stopper, H. (1999) Characterization of the genotoxicity of anthraquinones in mammalian cells. *Biochim. Biophys. Acta*, **1428**, 406–414.
17 Mueller, S.O., Eckert, I., Lutz, W.K., and Stopper, H. (1996) Genotoxicity of the laxative drug components emodin, aloe-emodin and danthron in mammalian cells: topoisomerase II mediated? *Mutat. Res*, **371**, 165–173.
18 Lee, H. and Tsai, S.J. (1991) Effect of emodin on cooked-food mutagen activation. *Food Chem. Toxicol.*, **29**, 765–770.
19 Su, H.Y., Cherng, S.H., Chen, C.C., and Lee, H. (1995) Emodin inhibits the mutagenicity and DNA adducts induced by 1-nitropyrene. *Mutat. Res.*, **329**, 205–212.
20 Chang, L.C., Sheu, H.M., Huang, Y.S., Tsai, T.R., and Kuo, K.W. (1999) A novel function of emodin: enhancement of the nucleotide excision repair of UV- and cisplatin-induced DNA damage in human cells. *Biochem. Pharmacol.*, **58**, 49–57.
21 Masuda, T. and Ueno, Y. (1984) Microsomal transformation of emodin into a direct mutagen. *Mutat. Res.*, **125**, 135–144.
22 Tanaka, H., Morooka, N., Haraikawa, K., and Ueno, Y. (1987) Metabolic activation of emodin in the reconstituted cytochrome P450 system of the hepatic microsomes of rats. *Mutat. Res.*, **176**, 165–170.
23 Murakami, H., Kobayashi, J., Masuda, T., Morooka, N., and Ueno, Y. (1987) ω-Hydroxyemodin, a major hepatic metabolite of emodin in various animals and its mutagenic activity. *Mutat. Res.*, **180**, 147–153.
24 Mueller, S.O., Stopper, H., and Dekant, W. (1998) Biotransformation of the anthraquinones emodin and chrysophanol by cytochrome P450 enzymes. Bioactivation to genotoxic metabolites. *Drug Metab. Dispos.*, **26**, 540–546.
25 Sun, Y., Li, Q., and Chen, Q.H. (1986) Excretion of emodin and its metabolites in mice. *J. Nanjing Coll. Pharm.*, **17**, 132–135.

26 Sun, Y. and Chen, Q.H. (1985) Isolation and identification of the metabolites of emodin and physcion in rat and mouse urine. *J. Nanjing Coll. Pharm.*, **16**, 72.

27 Bachmann, M. and Schlatter, C. (1981) Metabolism of ^{14}C-emodin in the rat. *Xenobiotica*, **11**, 217–225.

28 Yin, J.H., Zhou, X.Y., and Zhu, X.Q. (1992) Pharmacological and clinical studies on the processed products of radix Polygoni multiflori. *China J. Chin. Mater. Med.*, **17**, 722–724.

29 Yang, X.W. (1996) Effect of ethanolic extract from root tuber of *Polygonum multiflorum* Thunb. on liver and brain monoamine oxidase in senescence-accelerated mice *in vivo*. *China J. Chin. Mater. Med.*, **21**, 48–49.

30 Wei, X., Zhang, J., Li, J., and Chen, S. (2004) *Astragalus mongholicus* and *Polygonum multiflorum* protective function against cyclophosphamide inhibitory effect on thymus. *Am. J. Chin. Med.*, **32**, 669–680.

31 Qiu, G., Wu, X.Q., and Luo, X.G. (2006) Effect of *Polygonum multiflorum* Thunb. on BDNF expression in rat hippocampus induced by amyloid β-protein (Abeta) 1–40. *Zhong Nan Da Xue Xue Bao Yi Xue Ban*, **31**, 194–199.

32 Um, M.Y., Choi, W.H., Aan, J.Y., Kim, S.R., and Ha, T.Y. (2006) Protective effect of *Polygonum multiflorum* Thunb. on amyloid β-peptide 25–35 induced cognitive deficits in mice. *J. Ethnopharmacol.*, **104**, 144–148.

33 Li, X., Matsumoto, K., Murakami, Y., Tezuka, Y., Wu, Y., and Kadota, S. (2005) Neuroprotective effects of *Polygonum multiflorum* on nigrostriatal dopaminergic degeneration induced by paraquat and maneb in mice. *Pharmacol. Biochem. Behav.*, **82**, 345–352.

34 Ye, S., Tang, L., Xu, J., Liu, Q., and Wang, J. (2006) Postconditioning's protection of THSG on cardiac ischemia-reperfusion injury and mechanism. *J. Huazhong Univ. Sci. Technolog. Med. Sci.*, **26**, 13–16.

35 Yang, P.Y., Almofti, M.R., Lu, L., Kang, H., Zhang, J., Li, T.J., Rui, Y.C., Sun, L.N., and Chen, W.S. (2005) Reduction of atherosclerosis in cholesterol-fed rabbits and decrease of expressions of intracellular adhesion molecule-1 and vascular endothelial growth factor in foam cells by a water-soluble fraction of *Polygonum multiflorum*. *J. Pharmacol. Sci.*, **99**, 294–300.

36 Yim, T.K., Wu, W.K., Mak, D.H., and Ko, K.M. (1998) Myocardial protective effect of an anthraquinone-containing extract of *Polygonum multiflorum ex vivo*. *Planta Med.*, **64**, 607–611.

37 Horikawa, K., Mohri, T., Tanaka, Y., and Tokiwa, H. (1994) Moderate inhibition of mutagenicity and carcinogenicity of benzo[*a*] pyrene, 1,6-dinitropyrene and 3,9-dinitrofluoranthene by Chinese medicinal herbs. *Mutagenesis*, **9**, 523–526.

38 Chen, C.L. and Chen, Q.H. (1987) Biochemical study of Dahuang. XIX. Localization of inhibition of anthraquinone derivatives on the mitochondrial respiratory chain. *Acta Pharm. Sin.*, **22**, 12–18.

39 Wang, H.H. (1993) Antitrichomonal action of emodin in mice. *J. Ethnopharmacol.*, **40**, 111–116.

40 Wang, H.H. and Chung, J.G. (1997) Emodin-induced inhibition of growth and DNA damage in the *Helicobacter pylori*. *Curr. Microbiol.*, **35**, 262–266.

41 Chung, J.G., Wang, H.H., Wu, L.T., Chang, S.S. and Chang, W.C. (1997) Inhibitory actions of emodin on arylamine *N*-acetyltransferase activity in strains of *Helicobacter pylori* from peptic ulcer patients. *Food Chem. Toxicol.*, **35**, 1001–1007.

42 Wu, J.X., Xu, J.Y., and Yuan, Y.Z. (1997) Effects and mechanism of emodin and sandostatin on pancreatic ischemia in acute haemorrhagic necrotizing pancreatitis. *Chin. J. Integr. Trad. West. Med.*, **17**, 356–359.

43 Wei, J., Ni, L., and Yao, J. (1997) Experimental treatment of rhubarb on mesangio-proliferative glomerulonephritis in rats. *Chin. J. Intern. Med.*, **36**, 87–89.

44 Kumar, A., Dhawan, S., and Aggarwal, B.B. (1998) Emodin (3-methyl-1,6,8-trihydroxyanthraquinone) inhibits TNF-induced NF-κB activation, IκB degradation, and expression of cell surface adhesion proteins in human vascular endothelial cells. *Oncogene*, **17**, 913–918.

45 Zhang, L. and Hung, M.C. (1996) Sensitization of HER-2/neu-overexpressing non-small cell lung cancer cells to chemotherapeutic drugs by tyrosine kinase inhibitor emodin. *Oncogene*, **12**, 571–576.

46 Kim, J.S., Jeon, H.M., Im, S.A., and Lee, S.N. (1999) Enhanced adenoviral transduction efficiency in HER-2/neu-overexpressing human breast cancer cells not induced by an integrin pathway. *Oncol. Rep.*, **6**, 1237–1242.

47 Zhang, L., Chang, C.J., Bacus, S.S., and Hung, M.C. (1995) Suppressed transformation and induced differentiation of

HER-2/neu overexpressing breast cancer cells by emodin. *Cancer Res.*, **55**, 3890–3896.

48 Zhang, L., Lau, Y.K., Xi, L., Hong, R.L., Kim, D.S., Chen, C.F., Hortobagyi, G.N., Chang, C., and Hung, M.C. (1998) Tyrosine kinase inhibitors, emodin and its derivative repress HER-2/neu-induced cellular transformation and metastasis-associated properties. *Oncogene*, **16**, 2855–2863.

49 Zhang, L., Lau, Y.K., Xia, W., Hortobagyi, G.N., and Hung, M.C. (1999) Tyrosine kinase inhibitor emodin suppresses growth of HER-2/neu-overexpressing breast cancer cells in athymic mice and sensitizes these cells to the inhibitory effect of paclitaxel. *Clin. Cancer Res.*, **5**, 343–353.

50 Battistutta, R., Sarno, S., De Moliner, E., Papinutto, E., Zanotti, G., and Pinna, L.A. (2000) The replacement of ATP by the competitive inhibitor emodin induces conformational modifications in the catalytic site of protein kinase CK_2. *J. Biol. Chem.*, **275**, 29618–29622.

51 Chan, T.C., Chang, C.J., Koonchanok, N.M., and Geahlen, R.L. (1993) Selective inhibition of the growth of ras-transformed human bronchial epithelial cells by emodin, a protein-tyrosine kinase inhibitor. *Biochem. Biophys. Res. Commun.*, **193**, 1152–1158.

52 Liu, Z.H., Li, L.S., Hu, W.X., and Zhou, H. (1996) Effect of emodin on c-*Myc* proto-oncongene expression in cultured rat mesangial cells. *Acta Pharmacol. Sin.*, **17**, 61–63.

53 Frew, T., Powis, G., Berggren, M., Abraham, R.T., Ashendel, C.L., Zalkow, L.H., Hudson, C., Qazia, S., Gruszecka-Kowalik, E., and Merriman, R. (1994) A multiwell assay for inhibitors of phosphatidylinositol-3-kinase and the identification of natural product inhibitors. *Anticancer Res.*, **14**, 2425–2428.

54 Chen, Q.H., Liu, C.Y., and Qiu, C.H. (1980) Studies on Chinese rhubarb. XII. Effect of anthraquinone derivatives on the respiration and glycolysis of Ehrlich ascites carcinoma cells. *Acta Pharm. Sin.*, **15**, 65–70.

55 Kawai, K., Kato, T., Mori, H., Kitamura, J., and Nozawa, Y. (1984) A comparative study on cytotoxicities and biochemical properties of anthraquinone mycotoxins emodin and skyrin from *Penicillium islandicum* Sopp. *Toxicol. Lett.*, **20**, 155–160.

56 Lu, M. and Chen, Q.H. (1989) Biochemical study of Chinese rhubarb. XXIX. Inhibitory effects of anthraquinone derivatives on P 388 leukemia in mice. *J. China Pharm. Univ.*, **20**, 155–157.

57 Kuo, Y.C., Sun, C.M., Ou, J.C., and Tsai, W.J. (1997) A tumor cell growth inhibitor from *Polygonum hypoleucum* Ohwi. *Life Sci.*, **61**, 2335–2344.

58 Kamei, H., Koide, T., Kojima, T., Hashimoto, Y., and Hasegawa, M. (1998) Inhibition of cell growth in culture by quinones. *Cancer Biother. Radiopharm.*, **13**, 185–188.

59 Mueller, S.O., Lutz, W.K., and Stopper, H. (1998) Factors affecting the genotoxic potency ranking of natural anthraquinones in mammalian cell culture systems. *Mutat. Res.*, **414**, 125–129.

60 Morita, H., Umeda, M., Masuda, T., and Ueno, Y. (1988) Cytotoxic and mutagenic effects of emodin on cultured mouse carcinoma FM3A cells. *Mutat. Res.*, **204**, 329–332.

61 Fujimoto, H., Fujimaki, T., Okuyama, E., and Yamazaki, M. (1999) Immunomodulatory constituents from an ascomycete, *Microascus tardifaciens*. *Chem. Pharm. Bull. (Tokyo)*, **47**, 1426–1432.

Polyporus umbellatus

Polyporus (Zhuling) is the dried fungal body of *Polyporus umbellatus* (Pers.) Fries (Polyporaceae). It is used as a diuretic agent for the treatment of oliguria, dysuria, edema, and diarrhea.

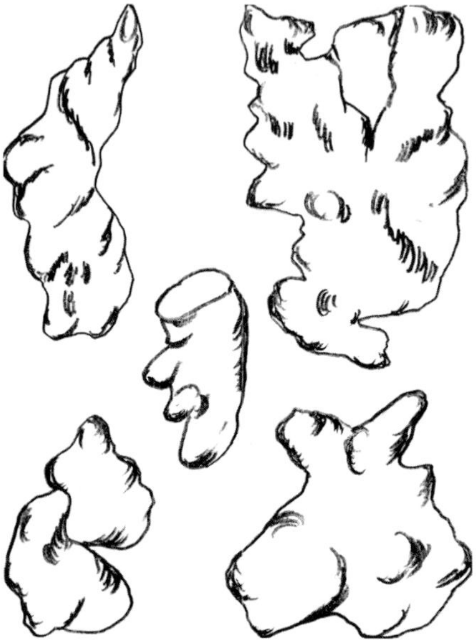

Polyporus umbellatus (Pers.) Fries

Chemistry

The fungal body of *P. umbellatus* is known to be rich in polysaccharides [1, 2]. Phytosterols polyporusterones A–G [3], acetosyringone and 3,4-dihydroxybenzaldehyde [4] were isolated from *Polyporus*. Polyporusterones are polyhydroxylated 5β-ergostan-6-ones with different substitution patterns.

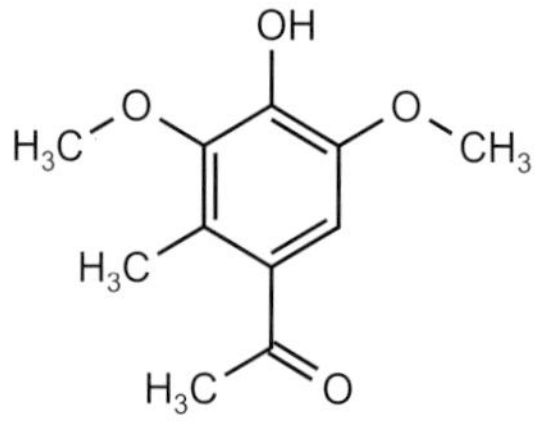

Acetosyringone

Pharmacology and Toxicology

Polyporus is mainly used as an immunostimulant, with polysaccharide being the active component. Polyporus polysaccharide not only increased the number of macrophages and the amount of H_2O_2 released in the peritoneal cavities of normal mice, but also mice with liver lesions caused by carbon tetra-chloride (CCl_4) [5]. Polyporus polysaccharide, in combination with recombinant interleukin-2 (rIL-2), significantly enhanced the lymphokine activated killer (LAK) cell activity *in vitro* by 42–57%, and reduced the required dose of rIL-2 by 50% [6].

Brucellosis in mice causes a distinct immunosuppression, potentially amenable to being counteracted by immunomodulators, including polyporus [7]. The immunosuppressive effect was seen in murine brucellosis detected by using plaque-forming cell and ^{3}H-labeled lymphocyte blastogenesis transformation assays. Polyporus was reported to reverse the suppression to normal levels [8]. Clinical and experimental studies on polyporus polysaccharide in the treatment of chronic viral hepatitis were also reported [9, 10]. Chronic hepatitis B patients with positive hepatitis B virus (HBV) replication markers and abnormalities of alanine aminotransferase (ALT) were treated with polyporus for three months. On completion of the treatment, the normalization of ALT and a negative conversion of hepatitis B e antigen (HbeAg) were seen in a proportion of the patients [11]. Polyporus polysaccharides were found capable of counteracting the immunosuppressive effect of the culture supernatant of S180 cells [12].

Polyporus was also reported to be useful in the treatment of malignant diseases. Experimental studies showed the oral administration of polyporus extract to be

Handbook of Chinese Medicinal Plants: Chemistry, Pharmacology, Toxicology
Weici Tang and Gerhard Eisenbrand

ISBN: 978-3-527-32226-8

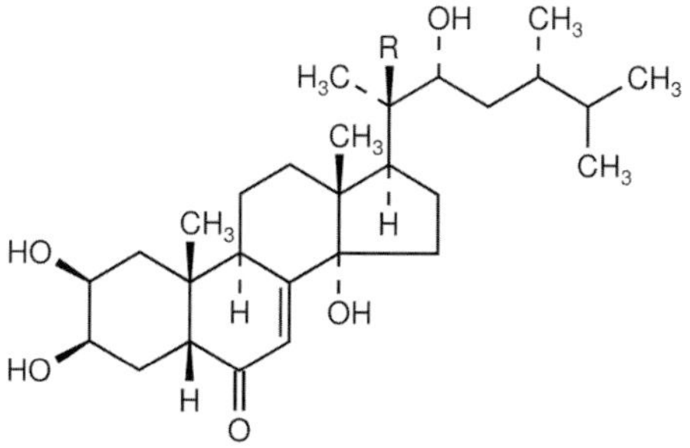

Polyporusterone A: R = OH
Polyporusterone F: R = H

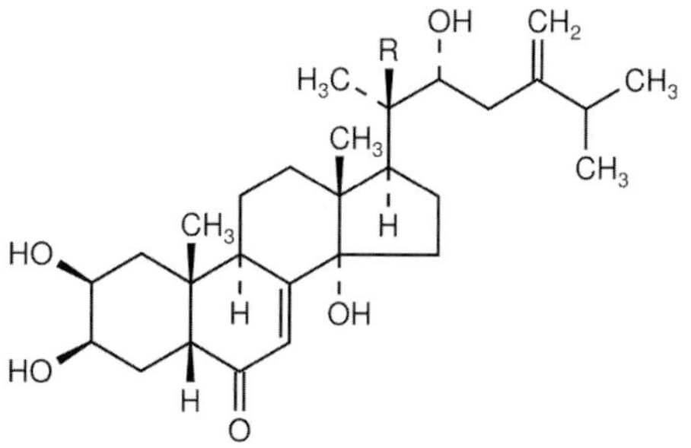

Polyporusterone B: R = OH
Polyporusterone G: R = H

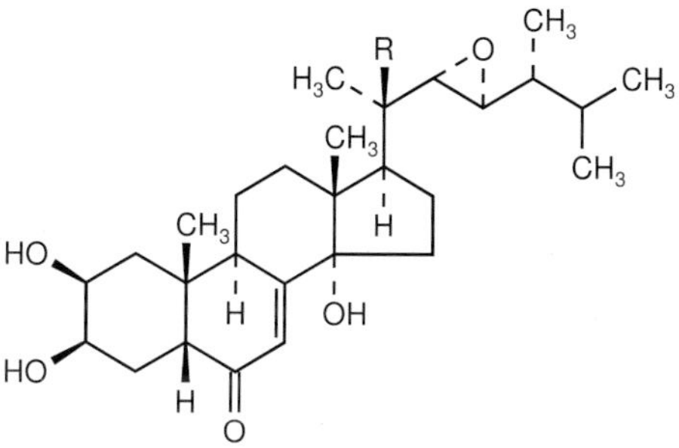

Polyporusterone C: R = OH
Polyporusterone E: R = H

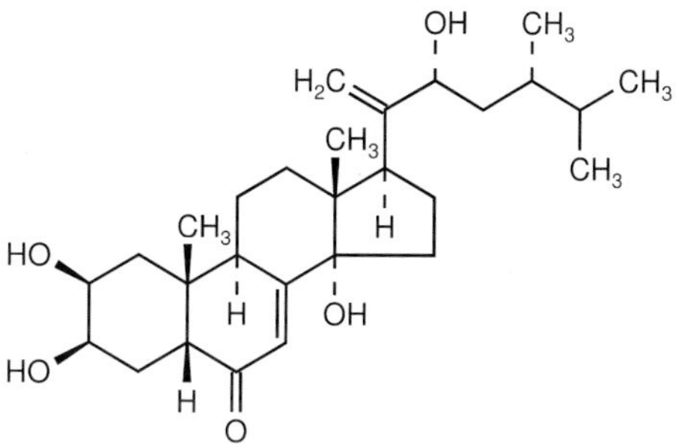

Polyporusterone D

active against the intrahepatic implantation of S180 tumor cells in mice. The 72% increase in the life span of tumor-bearing mice achieved after polyporus treatment was comparable with that observed after the intraperitoneal (i.p.) injection of mitomycin C (70%). The combined use of polyporus and mitomycin C increased the life span of tumor-bearing mice to 120%. A histopathological examination showed that the lymphocytes had infiltrated and surrounded the cancer cells, with some fibrosis being found in both normal cells and cancer cells [13].

Polyporus also inhibited the induction of bladder cancer in rats exposed to *N*-butyl-*N*-(4-hydroxybutyl)-nitrosamine, decreasing the tumor incidence from 100% (18/18) in the control group to 61% (11/18) in the treated group. Polyporus was also used, postoperatively, to treat patients with recurrent bladder cancer [14]. The polyporus polysaccharide effectively inhibited cachexia induced by toxohormone-L, which is closely related to neoplastic cachexia in rats [15]. The polyporusterones were reported to show growth-inhibitory activity against L1210 cells *in vitro* [4]. Both, polyporusterone A and polyporusterone B were reported to inhibit the lysis of erythrocytes, as induced by 2,2-azo-bis-(2-amidinopropane) dihydrochloride, though the antihemolytic effect of polyporusterone B was significantly stronger than that of polyporusterone A. The oral administration of 150 mg of polyporus to rats was also associated with a significant increase in the free radical-scavenging effect in the plasma [16].

Clinical efficiency of polyporus as a diuretic agent has been reported [17]. Polyporus was also reported to be effective with regards to hair growth in normal C3H/He mice, with acetosyringone, 3,4-dihydroxybenzaldehyde and polyporusterone A identified as the active components [1, 3].

References

1 Zhu, P. (1988) Determination of the components and molar ratio of the polysaccharide from a fermentation solution of *Polyporus umbellatus* (Pers.) Fr. *Bull. Chin. Mater. Med.*, **13**, 32–33, 62–63.

2 Ohsawa, T., Yukawa, M., Takao, C., Murayama, M., and Bando, H. (1992) Studies on constituents of fruit body of *Polyporus umbellatus* and their cytotoxic activity. *Chem. Pharm. Bull.*, **40**, 143–147.

3 Ishida, H., Inaoka, Y., Shibatani, J., Fukushima, M., and Tsuji, K. (1999) Studies of the active substances in herbs used for hair treatment. II. Isolation of hair regrowth substances, acetosyringone and polyporusterone A and B, from *Polyporus umbellatus* Fries. *Biol. Pharm. Bull.*, **22**, 1189–1192.

4 Inaoka, Y., Shakuya, A., Fukazawa, H., Ishida, H., Nukaya, H., Tsuji, K., Kuroda, H., Okada, M., Fukushima, M., and Kosuge, T. (1994) Studies on active substances in herbs used for hair treatment. I. Effects of herb extracts on hair growth and isolation of an active substance from *Polyporus umbellatus* F. *Chem. Pharm. Bull. (Tokyo)*, **42**, 530–533.

5 Zhang, Y.H., Liu, Y.L., and Yan, S.C. (1991) Effect of *Polyporus umbellatus* polysaccharide on function of macrophages in the peritoneal cavities of mice with liver lesions. *Chin. J. Integr. Trad. West. Med.*, **11**, 225–226.

6 Li, J.F., Huang, X.F., and Lin, B.Y. (1995) The effects on NK and endogenous LAK activities of splenic cells in mice by *Polyporus* polysaccharide *in vivo*. *Chin. J. Microbiol. Immunol.*, **15**, 89–91.

7 Zhang, J., Gao, B., Cun, C., Lu, X., Wang, H., Chen, X., and Tang, L. (1993) Immunosuppression in murine brucellosis. *Chin. Med. Sci.*, **8**, 134–138.

8 Zhang, J. (1992) A study on the role of immunosuppression in the pathogenesis of brucellosis. *Acta Acad. Med. Sin.*, **14**, 168–172.

9 Lin, Y.F. and Wu, G.L. (1988) Protective effect of *Polyporus umbellatus* polysaccharide on toxic hepatitis in mice. *Acta Pharmacol. Sin.*, **9**, 345–348.

10 Yan, S.C. (1988) Clinical and experimental research on *Polyporus umbellatus* polysaccharide in the treatment of chronic viral hepatitis. *Chin. J. Integr. Trad. West. Med.*, **8**, 141–143. 131.

11 Xiong, L.L. (1993) Therapeutic effect of combined therapy of *Salvia miltiorrhizae* and *Polyporus umbellatus* polysaccharide in the treatment of chronic hepatitis B. *Chin. J. Integr. Trad. West. Med.*, **13**, 533–535, 516–517.

12 Yang, L.J., Wang, R.T., Liu, J.S., Tong, H., Deng, Y.Q., and Li, Q.H. (2004) The effect of *Polyporus umbellatus* polysaccharide on the immunosuppression property of culture supernatant of S180 cells. *Xi Bao Yu Feng Zi Mian Yi Xue Za Zhi*, **20**, 234–237.

13 You, J.S., Hau, D.M., Chen, K.T., and Huang, H.F. (1994) Combined effects of chuling (*Polyporus umbellatus*) extract and mitomycin C on experimental liver cancer. *Am. J. Chin. Med.*, **22**, 19–28.

14 Yang, D.A., Li Shen, Q., Shi, B.Y., Wang, H.W., Liu, X.G., Li, Y.T., Zhou, B.M., and Wamg, X.X. (1991) Inhibitory effect of *Polyporus umbellatus* on cancer of the bladder: an experimental and clinical study. *Chin. J. Surg.*, **29**, 393–395.

15 Wu, G.S., Zhang, L.Y., and Okuda, H. (1997) Inhibitive effect of *Umbellatus polyporus* polysaccharide on cachexic manifestation induced by toxohormone-L in rats. *Chin. J. Integr. Trad. West. Med.*, **17**, 232–233.

16 Sekiya, N., Hikiami, H., Nakai, Y., Sakakibara, I., Nozaki, K., Kouta, K., Shimada, Y., and Terasawa, K. (2005) Inhibitory effects of triterpenes isolated from Chuling (*Polyporus umbellatus* Fries) on free radical-induced lysis of red blood cells. *Biol. Pharm. Bull.*, **28**, 817–821.

17 Li, Q.D., Wang, S.Q., and Sun, C.M. (1992) Clinical study of rapid bladder filling agent. *Chin. J. Integr. Trad. West. Med.*, **12**, 533–534. 517.

Poria cocos

Poria (Fuling) is the dried sclerotium of the fungus *Poria cocos* (Schw.) Wolf (Polyporaceae). It is used as a diuretic and a sedative agent.

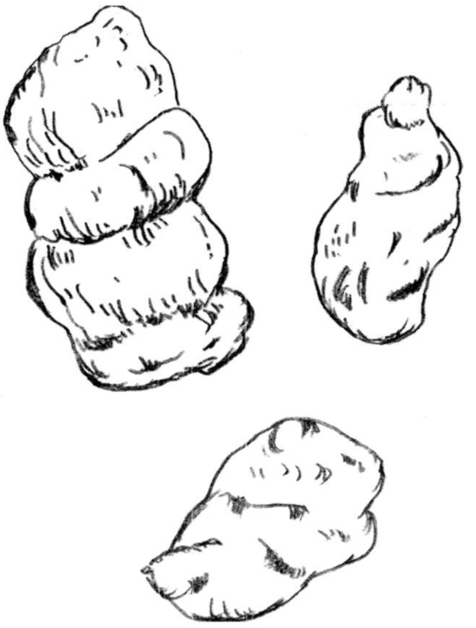

Poria cocos (Schw.) Wolf

Chemistry

The sclerotium of *P. cocos* is known to contain polysaccharides and triterpene derivatives. Triterpenes from Poria were identified as tumulosic acid (polyporenic acid B), dehydrotumulosic acid and its 3β-*p*-hydroxybenzoyl ester, pachymic acid (3-acetyltumulosic acid), dehydropachymic acid, 16α-hydroxydehydropachymic acid, *O*-acetyl-pachymic acid, 16α-hydroxytrametenolic acid, dehydrotrametenolic acid β-amyrin acetate, poricoic acid A, and poricoic acid B [1, 2].

Pharmacology and Toxicology

The sclerotium of *P. cocos* has long been used as a sedative and diuretic agent. It was also reported to exert anti-inflammatory activity and to inhibit mouse ear edema induced by 12-*O*-tetradecanoylphorbol

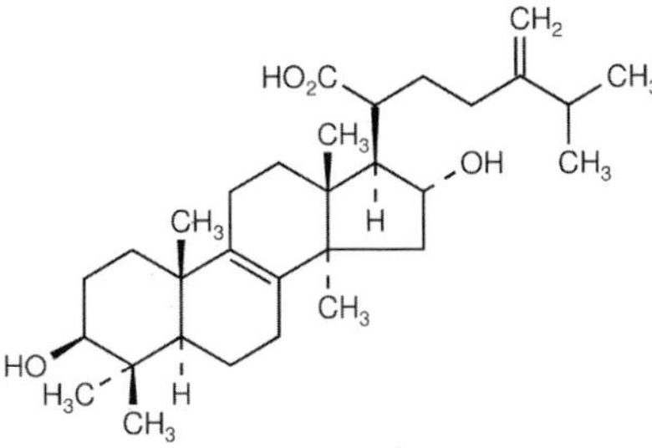

Tumulosic acid

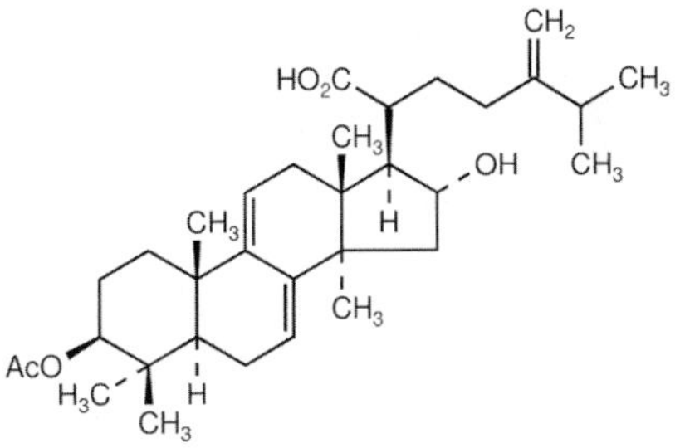

Dehydropachymic acid

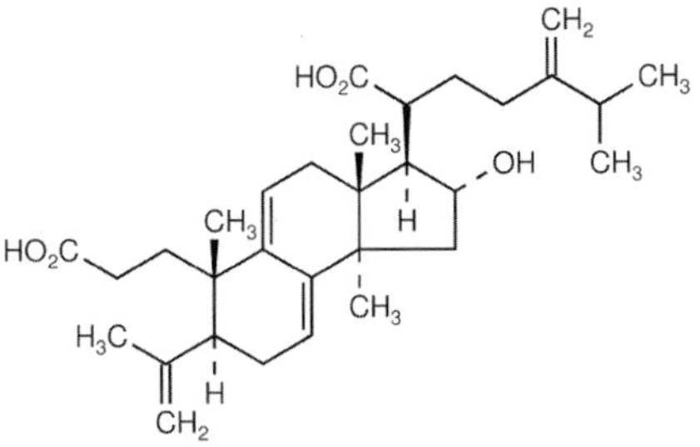

Poricoic acid A

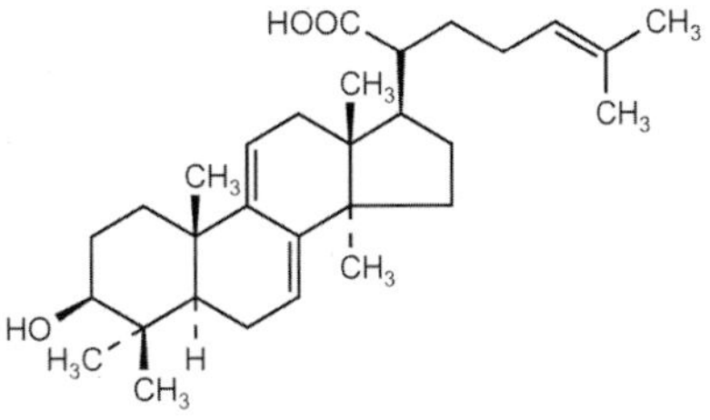

Dehydrotrametenolic acid

Handbook of Chinese Medicinal Plants: Chemistry, Pharmacology, Toxicology
Weici Tang and Gerhard Eisenbrand

ISBN: 978-3-527-32226-8

13-acetate (TPA) or arachidonic acid. Triterpenes such as dehydrotumulosic acid 3β-*p*-hydroxybenzoyl ester [1], 16α-hydroxy-dehydropachymic acid, 16α-hydroxy-trametenolic acid, dehydrotumulosic acid [2], dehydrotumulosic acid and pachymic acid [3] were found to be the active principles. The ID_{50} of dehydrotumulosic acid and pachymic acid on acute edema induced by TPA in mouse were approximately 5 and 0.7 n*M* per ear, respectively [3]. The triterpenes from *P. cocos* exhibited antiemetic activity in frogs against the emetic action of copper sulfate. Studies on the structure–activity relationship revealed that the exomethylene group at C-24 was responsible for the antiemetic effect of triterpenes [4]. Pachymic acid and dehydrotumulosic acid were each reported to be active as inhibitors of phospholipase A_2 from snake venom, with an IC_{50} of approximately 0.8 m*M* for dehydrotumulosic acid [5].

Besides the triterpenes, the polysaccharide from *P. cocos* sclerotium was also found to exert anti-inflammatory activity. A polysaccharide, PCSC, isolated from the sclerotium of *P. cocos* with 1% sodium carbonate, significantly induced the production of nitric oxide (NO) and inducible NO synthase (iNOS) transcription in RAW 264.7 murine macrophage-like cells. The effect was associated with the activation of nuclear factor-κ B/Rel (NF-κB/Rel). The DNA-binding activity of NF-κB/Rel was significantly induced by PCSC, mediated through the degradation of IκBα. An *in vivo* study revealed that the administration of PCSC induced NO production by peritoneal macrophages of B6C3F1 mice [6]. It was further reported that PCSC induced NF-κB/Rel activation and iNOS expression through the CD14, TLR4, and CR3 membrane receptor and p38 kinase, which was critically involved in the signal transduction leading to NF-κB/Rel activation in murine macrophages [7].

The polysaccharide from *P. cocos* was also reported to exhibited immunomodulating, antitumor, antinephritic, and antiviral activities. A polysaccharide fraction from *P. cocos* was reported to be active in treating chronic viral hepatitis [8]. Pachyman and other polysaccharides [9] from *P. cocos* were reported to have antitumor activity. The intraperitoneal (i.p.) injection of pachyman to mice inhibited the growth of transplanted S180 cells, but did not potentiate the effect of 5-fluorouracil and cyclophosphamide. However, pachyman prompted the recovery of the decrease of leukocytes in rats induced by cyclophosphamide and increased the phagocytic activity of macrophages in mice treated with sheep red blood cells [10]. The i.p. injection of poriatin, an active component of *P. cocos* of low molecular weight, also activated peritoneal macrophages in mice. In Vero cells, the antiviral activity of peritoneal macrophages from poriatin-treated mice was increased [11].

The lanostane-type triterpenes from *P. cocos*, such as pachymic acid and poricoic acid B, were reported to exert inhibitory efficacy against tumor promotion in two-stage carcinogenesis in mouse skin, as induced by TPA. At 0.2 μ*M* per mouse, the triterpenes markedly inhibited the promoting effect of TPA (1 μg per mouse) on skin tumor formation following initiation with 7,12-dimethylbenz[*a*]anthracene (DMBA; 50 μg per mouse) [12]. Poricoic acid G and poricoic acid H each also exhibited inhibitory activity toward tumor promotion, and potently inhibited Epstein–Barr virus early antigen (EBV-EA) activation in Raji cells, as induced by TPA [13]. 16-Deoxyporicoic acid B and poricoic acid C inhibited skin tumor promotion in two-stage carcinogenesis using DMBA as initiator and TPA as promoter [14].

Poricoic acid A and poricoic acid G were significantly cytotoxic to leukemia HL-60 cells, with an IC_{50} value of about 40 n*M* [13]. Pachymic acid inhibited the proliferation of human prostate cancer cells and induced cell apoptosis in both concentration- and time-dependent fashion, with androgen-insensitive DU145 prostate cancer cells showing a greater growth inhibition relative to androgen-responsive LNCaP cells [15]. Dehydrotrametenolic acid selectively inhibited the growth of H-Ras-transformed rat2 cells, causing cell-cycle arrest at G_2/M phase and accumulation of the sub-G_1 population, induced cell apoptosis, and activated caspase-3. Dehydrotrametenolic acid also regulated the expression of H-ras, Akt and Erk, the downstream proteins of H-ras signaling [16].

Dehydropachymic acid, pachymic acid, and tumulosic acid were each reported to exhibit cytotoxicity, with IC_{50} values of 20, 29, and 10 μM, respectively, against a human colon carcinoma cell line. They showed inhibitory activities not only on DNA topoisomerase II but also on DNA topoisomerase I [17]. Dehydroebriconic acid, one of the lanostane-type triterpene acids from *P. cocos*, potently inhibited DNA topoisomerase II activity, with an IC_{50} of 4.6 μM, while dehydrotrametenonic acid moderately inhibited DNA topoisomerase II activity, with an IC_{50} of 38 μM. Dehydroebriconic acid and dehydrotrametenonic acid both inhibited the growth of human gastric cancer cells and induced cell-cycle arrest at the G_1 phase [18]. Dehydrotrametenonic acid and dehydroeburiconic acid were further reported to inhibit eukaryotic DNA polymerases α and β [19].

A polysaccharide, PC-PS, from *P. cocos*, with a molecular weight of approximately 160 kDa, has been tested for antiproliferation and differentiation activities in U937 and HL-60 human leukemia cells. The conditioned medium prepared with human blood mononuclear cells, when stimulated with PC-PS at a concentration of 15 $\mu g\ ml^{-1}$ for five days, potently suppressed the proliferation of U937 and HL-60 cells. The medium also induced the tumor cells to differentiate into mature monocytes and macrophages, which also markedly expressed the surface antigens of CD11b, CD14, and CD 68. The differentiated U937 and HL-60 cells displayed physiological functions such as respiratory burst and phagocytosis. The growth-inhibitory and differentiation-inducing activities appeared mainly due to the elevated cytokines interferon-γ (IFN-γ) and tumor necrosis factor-α (TNF-α), indicating that PC-PS was a biological response modifier rather than a cytotoxic agent [20]. In contrast, a β-glucan from the mycelium of *P. cocos*, PCM3-II, was reported to inhibit the proliferation of human breast carcinoma MCF-7 cells. It induced cell-cycle arrest at G_1 phase, and cell apoptosis, and inhibited antiapoptotic Bcl-2 protein but not the proapoptotic Bax protein [21]. In fact, the antitumor activity of the polysaccharide from *P. cocos* mycelia was dependent upon the molecular mass [22], fungus strain, and culture media [23].

It was also reported that a decoction of *P. cocos* administered intravenously to guinea pigs inhibited the ototoxicity of kanamycin [24]. Dehydrotrametenolic acid was also found to lower plasma glucose concentrations in obese hyperglycemic db/db mice. Dehydrotrametenolic acid reduced hyperglycemia in mouse models of noninsulin-dependent diabetes mellitus and also acted as an insulin sensitizer, as indicated by the results of a glucose tolerance test [25].

References

1 Yasukawa, K., Kaminaga, T., Kitanaka, S., Tai, T., Nunoura, Y., Natori, S., and Takido, M. (1998) 3β-*p*-hydroxybenzoyldehydrotumulosic acid from *Poria cocos*, and its anti-inflammatory effect. *Phytochemistry*, **48**, 1357–1360.

2 Nukaya, H., Yamashiro, H., Fukazawa, H., Ishida, H., and Tsuji, K. (1996) Isolation of

inhibitors of TPA-induced mouse ear edema from Hoelen, *Poria cocos*. *Chem. Pharm. Bull. (Tokyo)*, **44**, 847–849.

3 Cuellar, M.J., Giner, R.M., Recio, M.C., Just, M. J., Manez, S., and Rios, J.L. (1997) Effect of the basidiomycete *Poria cocos* on experimental dermatitis and other inflammatory conditions. *Chem. Pharm. Bull. (Tokyo)*, **45**, 492–494.

4 Tai, T., Akita, Y., Kinoshita, K., Koyama, K., Takahashi, K., and Watanabe, K. (1995) Anti-emetic principles of *Poria cocos*. *Planta Med.*, **61**, 527–530.

5 Cuella, M.J.J., Giner, R.M., Recio, M.C., Just, M.J., Manez, S., and Rios, J.L. (1996) Two fungal lanostane derivatives as phospholipase A_2 inhibitors. *J. Nat. Prod.*, **59**, 977–799.

6 Lee, K.Y. and Jeon, Y.J. (2003) Polysaccharide isolated from *Poria cocos* sclerotium induces NFκB/Rel activation and iNOS expression in murine macrophages. *Int. Immunopharmacol.*, **3**, 1353–1362.

7 Lee, K.Y., You, H.J., Jeong, H.G., Kang, J.S., Kim, H.M., Rhee, S.D., and Jeon, Y.J. (2004) Polysaccharide isolated from *Poria cocos* sclerotium induces NF-κB/Rel activation and iNOS expression through the activation of p38 kinase in murine macrophages. *Int. Immunopharmacol.*, **4**, 1029–1038.

8 Guo, D.Z., Yao, P.Z., Fan, X.C., and Zhang, H.Z. (1984) Preliminary observation on carboxyl-methyl *Poria cocos* polysaccharide (CMPCP) in treating chronic viral hepatitis. *J. Trad. Chin. Med.*, **4**, 282.

9 Kanayama, H., Togami, M., Adachi, N., Fukai, Y., and Okumoto, T. (1986) Studies on the antitumor active polysaccharides from the mycelia of *Poria cocos* Wolf. III. Antitumor activity against mouse tumors. *Yakugaku Zasshi*, **106**, 307–312.

10 Chen, D.N., Fan, Y.J., Zhou, J., and Liang, Z.C. (1987) Antitumor and relevant pharmacological effects of pachyman. *Bull. Chin. Mater. Med.*, **12**, 553–555.

11 Lu, D., Xu, J. and Jiang, J.Y. (1987) Effects of poriatin on mouse peritoneal macrophages. *Acta Acad. Med. Sin.*, **9**, 433–438.

12 Kaminaga, T., Yasukawa, K., Kanno, H., Tai, T., Nunoura, Y., and Takido, M. (1996) Inhibitory effects of lanostane-type triterpene acids, the components of *Poria cocos*, on tumor promotion by 12-*O*-etradecanoylphorbol-13-acetate in two-stage carcinogenesis in mouse skin. *Oncology*, **53**, 382–385.

13 Ukiya, M., Akihisa, T., Tokuda, H., Hirano, M., Oshikubo, M., Nobukuni, Y., Kimura, Y., Tai, T., Kondo, S., and Nishino, H. (2002) Inhibition of tumor-promoting effects by poricoic acids G and H and other lanostane-type triterpenes and cytotoxic activity of poricoic acids A and G from *Poria cocos*. *J. Nat. Prod.*, **65**, 462–465.

14 Akihisa, T., Nakamura, Y., Tokuda, H., Uchiyama, E., Suzuki, T., Kimura, Y., Uchikura, K., and Nishino, H. (2007) Triterpene acids from *Poria cocos* and their anti-tumor-promoting effects. *J. Nat. Prod.*, **70**, 948–953.

15 Gapter, L., Wang, Z., Glinski, J., and Ng, K.Y. (2005) Induction of apoptosis in prostate cancer cells by pachymic acid from *Poria cocos*. *Biochem. Biophys. Res. Commun.*, **332**, 1153–1161.

16 Kang, H.M., Lee, S.K., Shin, D.S., Lee, M.Y., Han, D.C., Baek, N.I., Son, K.H., and Kwon, B.M. (2006) Dehydrotrametenolic acid selectively inhibits the growth of H-Ras transformed rat2 cells and induces apoptosis through caspase-3 pathway. *Life Sci.*, **78**, 607–613.

17 Li, G., Xu, M.L., Lee, C.S., Woo, M.H., Chang, H.W., and Son, J.K. (2004) Cytotoxicity and DNA topoisomerases inhibitory activity of constituents from the sclerotium of *Poria cocos*. *Arch. Pharm. Res.*, **27**, 829–833.

18 Mizushina, Y., Akihisa, T., Ukiya, M., Murakami, C., Kuriyama, I., Xu, X., Yoshida, H., and Sakaguchi, K. (2004) A novel DNA topoisomerase inhibitor: dehydroebriconic acid, one of the lanostane-type triterpene acids from *Poria cocos*. *Cancer Res.*, **95**, 354–360.

19 Akihisa, T., Mizushina, Y., Ukiya, M., Oshikuba, M., Kondo, S., Kimura, Y., Suzuki, T., and Tai, T. (2004) Dehydrotrametenonic acid and dehydroeburiconic acid from *Poria cocos* and their inhibitory effects on eukaryotic DNA polymerase α and β. *Biosci. Biotechnol. Biochem.*, **68**, 448–450.

20 Chen, Y.Y. and Chang, H.M. (2004) Antiproliferative and differentiating effects of polysaccharide fraction from fu-ling (*Poria cocos*) on human leukemic U937 and HL-60 cells. *Food Chem. Toxicol.*, **42**, 759–769.

21 Zhang, M., Chiu, L.C., Cheung, P.C., and Ooi, V.E. (2006) Growth-inhibitory effects of a β-glucan from the mycelium of *Poria cocos* on human breast carcinoma MCF-7 cells: cell-cycle arrest and apoptosis induction. *Oncol. Rep.*, **15**, 637–643.

22 Zhang, L., Chen, L., Xu, X., Zeng, F., and Cheung, P.C. (2005) Effect of molecular mass on antitumor activity of heteropolysaccharide from *Poria cocos*. *Biosci. Biotechnol. Biochem.*, **69**, 631–634.

23 Jin, Y., Zhang, L., Zhang, M., Chen, L., Cheung, P.C., Oi, V.E., and Lin, Y. (2003) Antitumor activities of heteropolysaccharides of *Poria cocos* mycelia from different strains and culture media. *Carbohydr. Res.*, **338**, 1517–1521.

24 Liu, Y.C., Liu, G.Y., and Liu, R.L. (1995) Effects of *Poria cocos* on ototoxicity induced by kanamycin in guinea pigs. *Chin. J. Integr. Trad. West. Med.*, **15**, 422–423.

25 Sato, M., Tai, T., Nunoura, Y., Yajima, Y., Kawashima, S., and Tanaka, K. (2002) Dehydrotrametenolic acid induces preadipocyte differentiation and sensitizes animal models of noninsulin-dependent diabetes mellitus to insulin. *Biol. Pharm. Bull.*, **25**, 81–86.

Portulaca oleracea

Herba Portulacae (Machixian) is the dried, above-ground part of *Portulaca oleracea* L. (Portulacaceae). It is used as a hemostatic agent for the treatment of hematochezia, hemorrhoidal bleeding, and abnormal uterine bleeding. It is also used for the treatment of dysentery, and for the treatment of snake and insect bites.

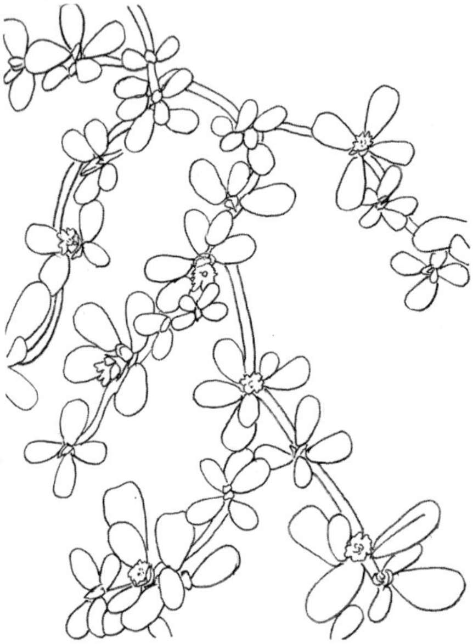

Portulaca oleracea L.

Chemistry

The above-ground part of *P. oleracea* is reported to have nutritional value for humans [1]. The leaves contain a large amount of protein, especially during the third growth stage (ca. 44 g per 100 g dry matter). The soluble carbohydrate content was significantly higher in growth stages 1 and 3. The above-ground part was found to contain phosphorus, calcium, potassium, iron, manganese, and copper. The iron content varied significantly among growth stages; the roots and leaves had the highest iron content (ca. 120 and 33 mg per 100 g dry matter, respectively). Significant accumulation of manganese was found in different growth stages. The leaves and roots had significantly higher manganese contents than did the stems [2]. The above-ground part of *P. oleracea* was reported to be a nutritious food rich in ω-3 fatty acids and antioxidants. ω-3-Fatty acids, α-tocopherol, ascorbic acid, β-carotene and glutathione (GSH) were identified in the leaves of *P. oleracea*, grown in both a controlled growth chamber and in the wild. A 100 g portion of fresh leaves contained about 300–400 mg 18: 3-fatty acid; 12.2 mg α-tocopherol; 26.6 mg ascorbic acid; 1.9 mg β-carotene; and 14.8 mg GSH [3]. The total fatty acid content was also reported to range from 1.5 to 2.5 $mg\,g^{-1}$ fresh mass in leaves, from 0.6–0.9 $mg\,g^{-1}$ in stems, and from 80–170 $mg\,g^{-1}$ in the seeds of Australian *P. oleracea*. α-Linolenic acid accounted for approximately 60% and 40% of the total fatty acid content in leaves and seeds, respectively. The β-carotene content ranged from 22 to 30 $mg\,g^{-1}$ fresh mass in leaves [4]. Flavone and polysaccharide components have also been detected in the above-ground part of *P. oleracea* [5].

Pharmacology and Toxicology

P. oleracea herb was reported to show muscle-relaxant effects on isolated nerve–muscle preparations. It was suggested that the neuromuscular activity of extracts of *P. oleracea* was partly caused by high concentrations of potassium ions [6]. An aqueous extract of *P. oleracea* leaves and stems produced a dose-dependent relaxation of

Handbook of Chinese Medicinal Plants: Chemistry, Pharmacology, Toxicology
Weici Tang and Gerhard Eisenbrand

ISBN: 978-3-527-32226-8

rabbit jejunum, and a dose-dependent contraction of the rabbit aorta. With regards to blood pressure, the extract produced a dose-dependent hypotensive effect in rats. The extract might act in part on postsynaptic α-adrenoreceptors, and/or by interference with transmembrane calcium influx [7].

The aqueous and ethanol extracts of *P. oleracea* herb exerted a protective effect in mice against gastric lesions induced by hydrochloric acid or absolute ethanol. In addition, oral and intraperitoneal administration of the extracts markedly reduced gastric acidity in pylorus-ligated mice [8]. The extract of *P. oleracea* also significantly increased the viability of adipose cells damaged by high-lipid serum, and decreased the levels of tumor necrosis factor-α (TNF-α) and interleukin-6 (IL-6) secreted by adipose cells *in vitro* [9].

After oral treatment of adult male BALB/c mice with *P. oleracea* extract for seven days, mice were adapted to a normobaric environment containing 10% oxygen and 90% nitrogen for different times. The results showed that the extract prolonged the survival time of the animals and enhanced the erythropoietin mRNA and protein expression in the mouse cortices. Mice treated with the extract at a daily dose of 1 g showed significantly higher activities of pyruvate kinase, phosphofructokinase, and lactate dehydrogenase (LDH), as well as higher levels of ATP in the cortices, especially under the hypoxic environment for 24 h. Histopathological examinations indicated that the extract had reduced inflammation damage of the mouse brain. An *in vitro* study showed that PC-12 cells and primarily cultured nerve cells had a higher viability under hypoxic conditions, and a lower level of LDH in the culture medium in the presence of the extract [10]. The flavone fraction from *P. oleracea* seemed to be the active component of the extract. Oral treatment of mice with the flavone fraction significantly prolonged the survival time of mice under hypoxic conditions [11].

The above-ground part of *P. oleracea* was also reported to show a weak to moderate inhibition of mutagenicity of the food-borne heterocyclic amine IQ in *Salmonella typhimurium* TA100 in the absence of an S-9 mix [12]. A crude sample of *P. oleracea* obtained by ethyl acetate extraction showed a specific and marked activity against dermatophytes of the genera *Trichophyton* [13].

References

1 Simopoulos, A.P., Norman, H.A., and Gillaspy, J.E. (1995) Purslane in human nutrition and its potential for world agriculture. *World Rev. Nutr. Diet.*, **77**, 47–74.

2 Mohamed, A.I. and Hussein, A.S. (1994) Chemical composition of purslane (*Portulaca oleracea*). *Plant Foods Hum. Nutr.*, **45**, 1–9.

3 Simopoulos, A.P., Norman, H.A., Gillaspy, J.E., and Duke, J.A. (1992) Common purslane: a source of ω-3 fatty acids and antioxidants. *J. Am. Coll. Nutr.*, **11**, 374–382.

4 Liu, L., Howe, P., Zhou, Y.F., Xu, Z.Q., Hocart, C., and Zhan, R. (2000) Fatty acids and β-carotene in Australian purslane (*Portulaca oleracea*) varieties. *J. Chromatogr. A*, **893**, 207–213.

5 Dong, L.W., Wang, W.Y., Yue, Y.T., and Li, M. (2005) Effects of flavones extracted from *Portulaca oleracea* on ability of hypoxia tolerance in mice and its mechanism. *Zhong Xi Yi Jie He Xue Bao*, **3**, 450–454.

6 Habtemariam, S., Harvey, A.L., and Waterman, P.G. (1993) The muscle relaxant properties of *Portulaca oleracea* are associated with high concentrations of potassium ions. *J. Ethnopharmacol.*, **40**, 195–200.

7 Parry, O., Okwuasaba, F., and Ejike, C. (1988) Effect of an aqueous extract of *Portulaca oleracea* leaves on smooth muscle and rat blood pressure. *J. Ethnopharmacol.*, **22**, 33–44.

8 Karimi, G., Hosseinzadeh, H., and Ettehad, N. (2004) Evaluation of the gastric antiulcerogenic effects of *Portulaca oleracea* L. extracts in mice. *Phytother. Res.*, **18**, 484–487.

9 Xiao, F.Y., Lu, F.E., and Xu, L.J. (2005) Effect of different parts of *Portulace oleracea* on the levels of TNF-α and IL-6 in the supernatant of cultured adipose cell. *Zhongguo Zhong Yao Za Zhi*, **30**, 1763–1766.

10 Wang, W., Gu, L., Dong, L., Wang, X., Ling, C., and Li, M. (2007) Protective effect of *Portulaca oleracea* extracts on hypoxic nerve tissue and its mechanism. *Asia Pac. J. Clin. Nutr.*, **16** (Suppl. 1), 227–233.

11 Dong, L.W., Wang, W.Y., Yue, Y.T., and Li, M. (2005) Effects of flavones extracted from *Portulaca oleracea* on ability of hypoxia tolerance in mice and its mechanism. *Zhong Xi Yi Jie He Xue Bao*, **3**, 450–454.

12 Yen, G.C., Chen, H.Y., and Peng, H.H. (2001) Evaluation of the cytotoxicity, mutagenicity and antimutagenicity of emerging edible plants. *Food Chem. Toxicol.*, **39**, 1045–1053.

13 Oh, K.B., Chang, I.M., Hwang, K.J., and Mar, W. (2000) Detection of antifungal activity in *Portulaca oleracea* by a single-cell bioassay system. *Phytother. Res.*, **14**, 329–332.

Prunella vulgaris

Spica Prunellae (Xiakucao) is the dried fruit aggregate of *Prunella vulgaris* L. (Lamiaceae). It is used for the treatment of conjunctivitis, mammitis, scrofula, and hypertension.

Prunella vulgaris L.

A galenic preparation, "Xiakucao Gao," is listed in the Chinese Pharmacopoeia. This is the aqueous extract of Spica Prunellae, and is used for the same indications as Spica Prunellae.

Chemistry

Spica Prunellae is known to be rich in ursolic acid [1]. The Chinese Pharmacopoeia requires a qualitative determination of ursolic acid in Spica Prunellae by thin-layer chromatographic comparison with reference substance, and a quantitative determination by HPLC. The content of ursolic acid in Spica Prunellae should be not less than 0.12%. The Chinese Pharmacopoeia gives a relative density of 1.40–1.46 for the galenic preparation "Xiakucao Gao."

The fruit aggregate of *P. vulgaris* was also found to contain a large amount of rosmarinic acid as major hydroxycinnamic acid derivative; the average content was reported as 6% of the dry herb [2]. Further triterpenes from the fruit aggregate were identified as betulinic acid, 2α,3α-dihydroxyurs-12-en-28-oic acid, and 2α-hydroxyursolic acid [3], the methyl esters of betulinic acid, ursolic acid, and oleanolic acid [4].

A polysaccharide named prunellin, isolated from an aqueous extract of *P. vulgaris*, was characterized as a sulfated polysaccharide composed of glucose, galactose, xylose, gluconic acid, galactonic acid and galactosamine, with a molecular weight of about 10 kDa [5]. Isolation of an anionic polysaccharide from *P. vulgaris* was also reported [6]. The major components of the essential oil from the fruit aggregate of *P. vulgaris* were identified as 1,8-cineol (45%), β-pinene (15.5%), myrcene (6%), linalyl acetate (4%), α-phellandrene (5.5%), and linalool (6.5%) [7].

Pharmacology and Toxicology

The aqueous extract of *P. vulgaris* was found to inhibit, dose-dependently, the systemic anaphylactic shock induced by compound 48/80 in rats. When the extract was given as pretreatment, at doses ranging from 5 to $100\,mg\,kg^{-1}$, the serum histamine levels induced by compound 48/80 were reduced in dose-dependent manner. The extract (1 to $100\,mg\,kg^{-1}$) also inhibited the passive cutaneous anaphylaxis activated by anti-dinitrophenyl (DNP) IgE antibody, and inhibited the histamine release induced by compound 48/80 or anti-DNP IgE from rat peritoneal mast cells. The level of cAMP in

Handbook of Chinese Medicinal Plants: Chemistry, Pharmacology, Toxicology
Weici Tang and Gerhard Eisenbrand

ISBN: 978-3-527-32226-8

rat peritoneal mast cells was increased significantly after addition of the extract. Moreover, the aqueous extract of *P. vulgaris* (0.01 and 0.1 mg ml^{-1}) exhibited a significant inhibitory effect on anti-DNP IgE-mediated tumor necrosis factor-α (TNF-α) production from rat peritoneal mast cells, indicating that the extract inhibited immediate-type allergic reactions in rats [8]. In addition, the extract inhibited the secretion of TNF-α, interleukin-6 (IL-6), and IL-8 in human mast cells stimulated by 12-*O*-tetradecanoylphorbol 13-acetate (TPA) and calcium ionophore A23187. The inhibitory effect of the extract on proinflammatory cytokines was dependent on nuclear factor-κB (NF-κ B). The extract also suppressed TPA- and A23187-induced NF-κB/DNA-binding activity [9].

The polysaccharides from *P. vulgaris* have been shown to upregulate the immune responses of monocytes/macrophages. However, the immune stimulatory effects seemed to contradict its well-known anti-inflammatory properties. One of the polysaccharide fractions, PV2IV, markedly stimulated the production of superoxide and nitric oxide (NO) from murine macrophage RAW264.7 and brain macrophage BV2 cells. The amount of NO and superoxide produced after PV2IV stimulation was as high as that stimulated by lipopolysaccharide (LPS). In addition, PV2IV increased cellular protein levels of inducible nitric oxide synthase (iNOS) and mRNA for TNF-α [10]. The immunosuppressive activity of an ethanol extract of Spica Prunellae has been studied both *in vitro* and *in vivo*. The extract significantly suppressed the splenocyte proliferation *in vitro* stimulated by concanavalin A (Con A) and LPS, in concentration-dependent manner. The ethanol extract of Spica Prunellae significantly suppressed Con A-, LPS- and OVA-induced splenocyte proliferation in mice immunized with ovalbumin, and significantly reduced total IgG, IgG1, and IgG2b levels; this suggested that Spica Prunellae might suppress the cellular and humoral response in mice [11].

Spica Prunellae was also reported to exert antihyperglycemic effects, as tested in diabetic ICR mice induced by streptozotocin. An extract of Spica Prunellae at a dose of 100 mg kg^{-1} significantly suppressed the rise in blood glucose after 30 min in an acute glucose tolerance test. A significant decrease in blood glucose levels in diabetic mice was observed after treatment with the extract. The extract enhanced and prolonged the antihyperglycemic effects of exogenous insulin on streptozotocin-induced diabetic mice. However, plasma insulin levels were not increased with the extract [12].

A hot-water extract of *P. vulgaris* was found to strongly inhibit the human immunodeficiency virus type 1 (HIV-1), with an IC_{100} value of 16 μg ml^{-1}. The active component was found to be of polar nature. In addition, the aqueous extract inhibited giant cell formation in coculture of Molt-4 cells with and without HIV-1 infection, and showed an inhibitory activity against HIV-1 reverse transcriptase [13]. The extract of the fruit aggregate of *P. vulgaris* was further reported to significantly decrease the number of copies of proviral DNA in HIV-exposed cells, in contrast to dextran sulfate. The extract and dextran sulfate suppressed HIV production to similar levels when added after HIV adsorption. However, only the extract suppressed HIV production at the same concentration when the drugs were added during HIV adsorption. Presumably, penetration of the extract into the cells was required for this activity. Moreover, the fractionated extract inhibited HIV reverse transcription in a noncompetitive manner. This fractionated extract retained its anti-HIV activity, but inhibited HIV replication and adsorption to a lesser extent compared to dextran sulfate. An active component was detected in plasma after injection into the rat intestine, demonstrating the feasibility of oral administration [14]. The aqueous extract of *P. vulgaris* was also found to moderately inhibit HIV-1 integrase with an IC_{50} of 45 μg ml^{-1} [15] and to significantly inhibit HIV-1 protease at a concentration of 200 μg ml^{-1} [16].

The polysaccharide, prunellin, also showed an anti-HIV activity [5]. The purified extract of *P. vulgaris* containing an anionic polysaccharide with a molecular weight of approximately 10 kDa as active component inhibited HIV-1 replication in the lymphoid cell line MT-4, in the monocytoid cell line U937, and in peripheral blood mononuclear cells at effective concentrations of 6–30 μg ml^{-1}. Pretreatment of uninfected cells with the extract prior to viral exposure did not prevent HIV-1 infection. By contrast, preincubation of HIV-1 with the purified extract dramatically decreased infectiousness. The purified extract was also able to block cell-to-cell transmission of HIV-1, to prevent syncytium formation, and to interfere with the ability of both HIV-1 and purified gp120 to bind to CD4. PCR analysis confirmed the absence of HIV-1 proviral DNA in cells exposed to virus in the presence of the extract; this suggested that the purified extract inhibited HIV-1 infection of susceptible cells by preventing viral attachment to the CD4 receptor [17].

The anionic polysaccharide at a concentration of 100 μg ml^{-1} was found to be active against herpes simplex virus (HSV) types 1 and 2 (HSV-1 and HSV-2), but inactive against cytomegalovirus, human influenza virus types A and B, poliovirus type 1, or vesicular stomatitis virus. The 50% plaque reduction concentration of the polysaccharide for HSV-1 and HSV-2 was 10 μg ml^{-1}. Clinical isolates and known acyclovir-resistant (TK-deficient or polymerase-defective) strains of HSV-1 and HSV-2 were similarly inhibited by the polysaccharide. The preincubation of HSV-1 with the polysaccharide at 4, 25, or 37 °C completely abrogated the infectivity of HSV-1, but pretreatment of Vero cells with the polysaccharide did not protect the cells from infection by the virus. The polysaccharide might inhibit HSV by competing for cell receptors, as well as by an unknown mechanism(s) following virus penetration of the cells. The *Prunella* polysaccharide was not cytotoxic to mammalian cells up to a concentration of 0.5 mg ml^{-1}, and did not show any anticoagulant activity [6].

Another polysaccharide fraction from *P. vulgaris* was also reported to be active against both HSV-1 and HSV-2 infections, and effectively reduced the expressions of HSV-1 and HSV-2 antigens in the host Vero cells [18]. The anti-HSV compound from *P. vulgaris* was identified as a lignin–polysaccharide complex with a molecular weight of 8500 Da. The carbohydrate moiety was composed of glucose, galactose, mannose, galacturonic acid, rhamnose, xylose, and arabinose, with glucose as the major sugar. Mechanistic studies showed that the lignin–polysaccharide complex inactivated HSV-1 directly, and also blocked HSV-1 binding to, and inhibited HSV-1 penetration into, Vero cells. The *in vivo* activities of a cream formulated with a semi-purified fraction from *P. vulgaris* were tested in HSV-1-mediated skin lesion in guinea pigs and HSV-2-infected genital lesions in BALB/c mice. Guinea pigs treated with the cream showed a significant reduction in skin lesions, while mice receiving the cream showed a significant reduction in mortality [19].

2α,3α-Dihydroxyursolic acid demonstrated a significant inhibition of the release of β-hexosaminidase by cultured RBL-2H3 cells in dose-dependent manner, with an IC_{50} value of 57 μ*M*. Ursolic acid and 2α-hydroxyursolic acid strongly each inhibited the production of nitric oxide from cultured murine macrophages, RAW 264.7 cells, with IC_{50} values of 17 and 27 μ*M*, respectively [3]. Ursolic acid was also reported to be significantly cytotoxic against leukemia P388 and L1210 cells and human lung carcinoma A-549 cells; and marginally cytotoxic against KB, HCT-8 human colon and MCF-7 human mammary carcinoma cells. Esterification of the hydroxy group at position 3, and of the carboxylic acid group at position 17, decreased the cytotoxicity against the human tumor cell lines, but increased the cytotoxicity against murine leukemia cell lines [20]. The extract of *P. vulgaris* also suppressed the proliferation of Raji cells in concentration-dependent manner, induced cell apoptosis, upregulated the expression of Bcl-1, and downregulated the expression of Bax [21].

In *Salmonella typhimurium* TA 98 in the presence of an S-9 mix, the extract of *P. vulgaris* completely inhibited the mutagenicity of benzo[*a*]pyrene (BaP), and moderately inhibited the mutagenicity of picrolonic acid [22], 1,6-dinitropyrene, and 3,9-dinitrofluoranthene [23]. The aqueous extract of *P. vulgaris* further exhibited high potency in inhibiting rat erythrocyte hemolysis and lipid peroxidation in rat kidney and brain homogenates. It also demonstrated strong superoxide- and hydroxyl radical-scavenging activity, but exerted only a slight pro-oxidant effect [24].

References

1 Lee, K.H., Lin, Y.M., Wu, T.S., Zhang, D.C., Yamagishi, T., Hayashi, T., Hall, I.H., Chang, J.J., Wu, R.Y., and Yang, T.H. (1988) The cytotoxic principles of *Prunella vulgaris*, *Psychotria serpens*, and *Hyptis capitata*: ursolic acid and related derivatives. *Planta Med.*, **54**, 308–311.

2 Lamaison, J.L., Petitjean-Freytet, C., and Carnat, A. (1991) Medicinal Lamiaceae with antioxidant properties, a potential source of rosmarinic acid. *Pharm. Acta Helv.*, **66**, 185–188.

3 Ryu, S.Y., Oak, M.H., Yoon, S.K., Cho, D.I., Yoo, G.S., Kim, T.S., and Kim, K.M. (2000) Anti-allergic and anti-inflammatory triterpenes from the herb of *Prunella vulgaris*. *Planta Med.*, **66**, 358–360.

4 Kojima, H., Tominaga, H., Sato, S., and Ogura, H. (1987) Constituents of the Labiatae plants. Part 2. Pentacyclic triterpenoids from *Prunella vulgaris*. *Phytochemistry*, **26**, 1107–1111.

5 Tabba, H.D., Chang, R.S., and Smith, K.M. (1989) Isolation, purification, and partial characterization of prunellin, an anti-HIV component from aqueous extracts of *Prunella vulgaris*. *Antiviral Res.*, **11**, 263–273.

6 Xu, H.X., Lee, S.H., Lee, S.F., White, R.L., and Blay, J. (1999) Isolation and characterization of an anti-HSV polysaccharide from *Prunella vulgaris*. *Antiviral Res.*, **44**, 43–54.

7 Yang, L.J., Li, Z.Q., Pu, F., Shi, Y., and Zhang, Z.J. (1988) The chemical composition of the essential oil of *Prunella vulgaris*. *Chin. J. Pharm. Anal.*, **8**, 264–266.

8 Shin, T.Y., Kim, Y.K., and Kim, H.M. (2001) Inhibition of immediate-type allergic reactions by *Prunella vulgaris* in a murine model. *Immunopharmacol. Immunotoxicol.*, **23**, 423–435.

9 Kim, S.Y., Kim, S.H., Shin, H.Y., Lim, J.P., Chae, B.S., Park, J.S., Hong, S.G., Kim, M.S., Jo, D.G., Park, W.H., and Shin, T.Y. (2007) Effects of *Prunella vulgaris* on mast cell-mediated allergic reaction and inflammatory cytokine production. *Exp. Biol. Med. (Maywood)*, **232**, 921–926.

10 Fang, X., Yu, M.M., Yuen, W.H., Zee, S.Y., and Chang, R.C. (2005) Immune modulatory effects of *Prunella vulgaris* L. on monocytes/macrophages. *Int. J. Mol. Med.*, **16**, 1109–1116.

11 Sun, H.X., Qin, F., and Pan, Y.J. (2005) In vitro and in vivo immunosuppressive activity of Spica Prunellae ethanol extract on the immune responses in mice. *J. Ethnopharmacol.*, **101**, 31–36.

12 Zheng, J., He, J., Ji, B., Li, Y., and Zhang, X. (2007) Antihyperglycemic activity of *Prunella vulgaris* L. in streptozotocin-induced diabetic mice. *Asia Pac. J. Clin. Nutr.*, **16** (Suppl .1), 427–431.

13 Yamasaki, K., Nakano, M., Kawahata, T., Mori, H., Otake, T., Ueba, N., Oishi, I., Inami, R., Yamane, M., Nakamura, M., Murata, H., and Nakanishi, T. (1998) Anti-HIV-1 activity of herbs in Labiatae. *Biol. Pharm. Bull.*, **21**, 829–833.

14 Kageyama, S., Kurokawa, M., and Shiraki, K. (2000) Extract of *Prunella vulgaris* spikes inhibits HIV replication at reverse

transcription *in vitro* and can be absorbed from intestine *in vivo*. *Antivir. Chem. Chemother.*, **11**, 157–164.

15 Au, T.K., Lam, T.L., Ng, T.B., Fong, W.P., and Wan, D.C. (2001) A comparison of HIV-1 integrase inhibition by aqueous and methanol extracts of Chinese medicinal herbs. *Life Sci.*, **68**, 1687–1694.

16 Lam, T.L., Lam, M.L., Au, T.K., Ip, D.T., Ng, T. B., Fong, W.P., and Wan, D.C. (2000) A comparison of human immunodeficiency virus type-1 protease inhibition activities by the aqueous and methanol extracts of Chinese medicinal herbs. *Life Sci.*, **67**, 2889–2896.

17 Yao, X.J., Wainberg, M.A., and Parniak, M.A. (1992) Mechanism of inhibition of HIV-1 infection *in vitro* by purified extract of *Prunella vulgaris*. *Virology*, **187**, 56–62.

18 Chiu, L.C., Zhu, W., and Ooi, V.E. (2004) A polysaccharide fraction from medicinal herb *Prunella vulgaris* downregulates the expression of herpes simplex virus antigen in Vero cells. *J. Ethnopharmacol.*, **93**, 63–68.

19 Zhang, Y., But, P.P., Ooi, V.E., Xu, H.X., Delaney, G.D., Lee, S.H., and Lee, S.F. (2007) Chemical properties, mode of action, and in vivo anti-herpes activities of a lignin-carbohydrate complex from *Prunella vulgaris*. *Antiviral Res.*, **75**, 242–249.

20 Lee, K.H., Lin, Y.M., Wu, T.S., Zhang, D.C., Yamagishi, T., Hayashi, T., Hall, I.H., Chang, J. J., Wu, R.Y., and Yang, T.H. (1988) Antitumor agents. LXXXVIII. The cytotoxic principles of *Prunella vulgaris*, *Psychotria serpens*, and *Hyptis capitata*: ursolic acid and related derivatives. *Planta Med.*, **54**, 308–311.

21 Zhang, K.J., Zhang, M.Z., Wang, Q.D., and Liu, W.L. (2006) The experimental research about the effect of *Prunella vulgaris* L. on Raji cells growth and expression of apoptosis related protein. *Zhong Yao Cai*, **29**, 1207–1210.

22 Lee, H. and Lin, J.Y. (1988) Antimutagenic activity of extracts from anticancer drugs in Chinese medicine. *Mutat. Res.*, **204**, 229–234.

23 Horikawa, K., Mohri, T., Tanaka, Y., and Tokiwa, H. (1994) Moderate inhibition of mutagenicity and carcinogenicity of benzo[*a*]pyrene, 1,6-dinitropyrene and 3,9-dinitrofluoranthene by Chinese medicinal herbs. *Mutagenesis*, **9**, 523–526.

24 Liu, F. and Ng, T.B. (2000) Antioxidative and free radical scavenging activities of selected medicinal herbs. *Life Sci.*, **66**, 725–735.

Prunus spp.

Semen Armeniacae Amarum (Kuxingren) is the dried ripe seed of *Prunus armeniaca* L., *Prunus armeniaca* L. var. *ansu* Maxim., *Prunus mandshurica* (Maxim.) Koehne, or *Prunus sibirica* L. (Rosaceae). It is used as an antiasthmatic, mucolytic, expectorant, and laxative agent. The Chinese Pharmacopoeia notes that overdosage of Semen Armeniacae Amarum should be avoided due to its toxicity.

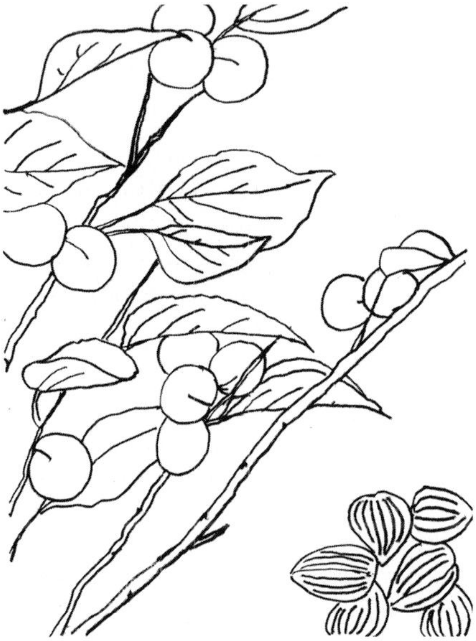

Prunus armeniaca L.

Semen Persicae (Taoren) is the dried ripe seed of *Prunus davidiana* (Carr.) Franch. or *Prunus persica* (L.) Batsch. It is used as a laxative and hemodynamic agent for the treatment of menostasia, menorrhalgia, and traumatic diseases. The Chinese Pharmacopoeia notes Semen Persicae to be used with caution in pregnancy.

Semen Pruni (Yuliren) is the dried ripe seed of *Prunus humilis* Bge., *Prunus japonica* Thunb., or *Prunus pedunculata* Maxim. It is used as a laxative and diuretic agent. The Chinese Pharmacopoeia notes Semen Pruni to be used with caution in pregnancy.

Flos Mume (Meihua) is the dried flower bud of *Prunus mume* (Sieb.) Sieb. et Zucc. It is used as an expectorant and antidotal agent.

Fructus Mume (Wumei) is the dried nearly ripe fruit of *Prunus mume*. It is used as an anthelmintic and astringent agent.

The young branch of *Prunus davidiana* and *Prunus persica* is included in the appendix of the Chinese Pharmacopoeia.

Chemistry

Semen Armeniacae Amarum and Semen Pruni are known to contain cyanogenic glycosides, with amygdalin as the major component. The Chinese Pharmacopoeia requires a qualitative determination of amygdalin in Semen Armeniacae Amarum by thin-layer chromatographic (TLC) comparison with reference substance, and a quantitative determination in Semen Armeniacae Amarum and in Semen Pruni by argentometric titration after distillation. The content of amygdalin in Semen Armeniacae Amarum should be not less than 3.0%, and in Semen Pruni not less than 1.5%. The amygdalin contents in the seeds of 11 *Prunus* species and subspecies, including *P. armeniaca*, *P. armeniaca* var. *ansu*, *P. armeniaca* var. *pendula*, *P. sibirica*, *P. manshurica*, *P. persica*, *P. davidiana*, *P. nana*, *P. tangutica*, *P. triloba*, and *P. triloba* var. *diplopetala* were measured to be 1.8–4.7%, with *P. sibirica* as the species having the highest amygdalin level [1].

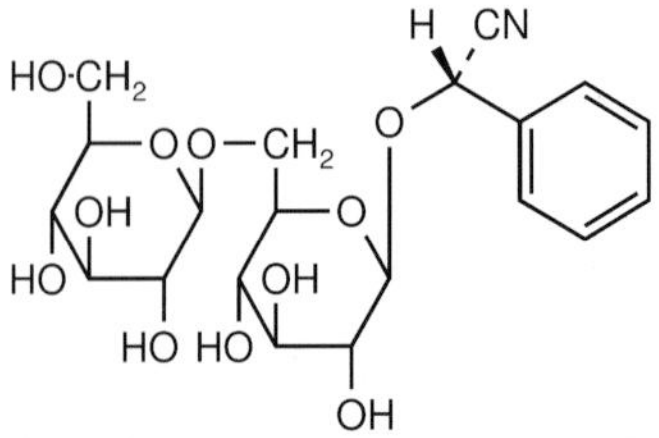

Amygdalin

Handbook of Chinese Medicinal Plants: Chemistry, Pharmacology, Toxicology
Weici Tang and Gerhard Eisenbrand

ISBN: 978-3-527-32226-8

The nearly ripe fruit of *Prunus mume* is known to contain organic acids, with citric acid as the major component. The Chinese Pharmacopoeia requires a qualitative determination of ursolic acid in Fructus Mume by TLC comparison with reference substance, and a quantitative determination of organic acids by potentiometric titration after extraction with water. The content of organic acids in Fructus Mume should be not less than 15.0%, calculated as citric acid.

Pharmacology and Toxicology

Fatal cases of cyanide poisoning caused by amygdalin have been reported [2, 3]. Cases of severe accidental cyanide poisoning following ingestion of amygdalin were also reported [4, 5]. For example, a 68-year-old patient presented to the emergency department shortly after a dose of 3 g amygdalin with a reduced Glasgow Coma Score, seizures, and severe lactic acidosis requiring intubation and ventilation [4]. The toxicological effects of amygdalin in dogs were similar to those observed in humans, including neurologic impairment, ranging from difficulty in walking to coma [6]. In rats given amygdalin intraperitoneally at doses of 250, 500, and 750 mg kg^{-1} daily for five days, mortalities of 30.8%, 44.1% and 56.8%, respectively, were caused. The mode of death and the elevated serum cyanide levels in the dying animals strongly suggested cyanide poisoning as the cause [7]. The gastrointestinal microflora played a role in amygdalin-induced cyanide toxicity [8]. Blood cyanide levels reached a maximum after amygdalin administration to mice at about 1.5 to 2 h, and were within the range of values seen after KCN administration. Ten-fold higher doses of amygdalin administered intraperitoneally produced very small increases in blood cyanide levels, and no toxic behaviour [9]. Amygdalin given intravenously at a daily dose of 4.5 g m^{-2} to humans was largely excreted unchanged in the urine, and produced no clinical or laboratory evidence of toxic reaction. Amygdalin given orally at 0.5 g three times daily produced significant blood cyanide levels of up to 2.1 g ml^{-1}. No evidence of toxic reaction was seen at this dosage. However, one patient, when challenged with a large intake of raw almonds, had transient symptoms of cyanide toxic reaction with escalating blood cyanide levels [10].

The LD_{50} of amygdalin in rats was found to be 880 mg kg^{-1} by oral administration. However, when 600 mg kg^{-1} was administered orally with β-glucosidase, all the rats died. Total and Mg-ATPase activities of the heart decreased with increasing dose. After 48 h, 2.3, 7.4 and 7.5 mg was excreted as intact amygdalin at doses of 200, 400, and 600 mg kg^{-1}, respectively. Thiocyanate excreted within the same 48 h period was 7.0, 9.1, and 9.5 mol, representing 18%, 11.2% and 7.8% of the 200, 400, and 600 mg kg^{-1} oral dosage, respectively. With 300 mg kg^{-1} amygdalin given intraperitoneally, 4.1 mg amygdalin and 3.9 mol thiocyanate was excreted [11].

The organs of 15-day-old rats had the highest capability to hydrolyze amygdalin and prunasin, and most of this activity was concentrated in the small and large intestines. The activity decreased with age. In adult rats, the ability of the organs to hydrolyze prunasin was higher than that of amygdalin, and was concentrated in the spleen, large intestine, and kidney. When 30 mg amygdalin was given orally to adult rats, its distribution after 1 h was as follows: stomach (0.89 mg), small intestine (0.78 mg), spleen (0.36 mg), large intestine (0.30 mg), kidney (0.19 mg), liver (0.10 mg), and serum (5.6 μg ml^{-1}). After 2 h, the highest amygdalin content was found in the large intestine (0.79 mg) [12]. Treatment of rats with amygdalin did not significantly affect the rate constants of D-[^{14}C]-glucose or D-[^{3}H]-mannose transport. However, treatment with 1 m*M* prunasin reduced the influx of D-[^{14}C]-glucose without affecting D-[^{3}H]-mannose values, indicating that glycoside absorption might be mediated by the epithelial sodium-dependent monosaccharide transporter SGLT1 [13].

Semen Armeniacae Amarum, the bitter apricot seed, is a widely used component in Chinese prescriptions for the treatment of common cold, fever, cough, asthma, and bronchitis. Antitussive [14] and also analgesic effects of amygdalin in mice were observed in hot plate and acetic acid-induced writhing tests. No anti-inflammatory activity was found with amygdalin [15]. Amygdalin given orally to rats at a dose of 46 mg kg^{-1} caused a small but significant increase in body temperature [16]. Amygdalin was found to exert anti-inflammatory and analgesic effects. In mouse BV2 microglial cells, amygdalin markedly suppressed prostaglandin E_2 (PGE_2) synthesis and nitric oxide (NO) production stimulated by lipopolysaccharide (LPS), by inhibiting the stimulated mRNA expressions of cyclooxygenase-2 (COX-2) and inducible NO synthase (iNOS) [17].

Amygdalin has been tested on its cytotoxicity both with and without β-glucosidase. The extract of Semen Persicae was reported to inhibit the proliferation of HL-60 human promyelocytic leukemia cells *in vitro*, with an IC_{50} of about 6 mg ml^{-1} in the presence of 250 n*M* β-glucosidase, and to induce cell apoptosis [18]. Amygdalin was also reported to inhibit the proliferation of DU145 and LNCaP human prostate cancer cells and to induce cell apoptosis. Treatment with amygdalin increased the expression of pro-apoptotic protein, Bax; decreased the expression of anti-apoptotic protein Bcl-2; and increased caspase-3 enzyme activity in prostate cancer cells. This indicated that the induction of apoptosis in cancer cells by amygdalin was caused by caspase-3 activation through downregulation of Bcl-2 and upregulation of Bax [19]. In SNU-C4 human colon cancer cells, amygdalin was also found to exert a cytotoxic activity via the downregulation of cell-cycle-related genes [20]. In murine P388 lymphocytic leukaemia and P815 mast cell leukemia, amygdalin was reported to show therapeutic effect only at high doses [21]. Amygdalin also inhibited HT1376 bladder cancer cells at high concentrations, whereas the combination of amygdalin with tumor-associated monoclonal antibody HMFG1-β-glucosidase enhanced its cytotoxic effect 36-fold [22].

Hexane extracts of Semen Armeniacae Amarum and Semen Persicae were found to inhibit the mutagenicity of benzo[*a*]pyrene (BaP). The mutagenicities of 3-amino-1,4-dimethyl-5*H*-pyrido[4,3-b]indole (Trp-P-1) and 2-(2-furyl)-3-(5-nitro-2-furyl)acrylamide (AF-2) were also inhibited by the extracts of Semen Armeniacae Amarum and Semen Persicae. The active principles in Semen Persicae were identified as oleic acid and linoleic acid. The contents of oleic acid and linoleic acid were 0.7% and 0.4%, respectively, in the hexane extract of Semen Armeniacae Amarum, and 1.5% and 0.5% in that of Semen Persicae [23]. In contrast, the extract of Semen Armeniacae Amarum was also reported to induce Epstein–Barr virus early antigen (EBV-EA) activation in Raji cells *in vitro* [24]. Processing methods influenced the special toxicity and pharmacodynamics as tested in the induction of EBV-EA activation in Raji cells [25]. The β-gentiobiosides and β-D-glucosides of mandelic acid and benzyl alcohol from the seeds of *P. persica* were also reported to exhibit anti-tumor-promoting effects. They significantly inhibited the EBV-EA activation induced by 12-*O*-tetradecanoylphorbol 13-acetate (TPA), and inhibited the promotion induced by TPA in mouse skin two-stage carcinogenesis. The anti-promotion potency of the glycosides was comparable to that of epigallocatechin gallate [26].

Amygdalin might protect from hyperoxia-induced lung injury. The action mechanism of amygdalin was investigated on type 2 alveolar epithelial cells from premature rat lungs *in vitro*. The results showed that hyperoxia inhibited cell proliferation and decreased the expression of proliferating cell nuclear antigen in epithelial cells. Amygdalin, over a concentration range of 50 to 200 μ*M*, stimulated the proliferation of epithelial cells, but at a higher concentration (400 μ*M*) it inhibited proliferation [27].

Prunasin, the monoglucoside of D-mandelonitrile, is the primary metabolite of amygdalin. It was found to be an inhibitor of DNA polymerase-β. Prunasin inhibited the activity of rat DNA polymerase-β at a concentration of 150μ*M*, but did not influence the activities of calf DNA polymerase-β and plant DNA polymerases, human immunodeficiency virus type 1 (HIV-1) reverse transcriptase, calf terminal deoxynucleotidyl transferase, or any prokaryotic DNA polymerases. The inhibition of polymerase-β by prunasin was competitive with the substrate, dTTP [28]. Prunasin was shown to be absorbed unmetabolized in the jejunum of the rat via the transport system of glucose [29].

The aqueous extract of Fructus Mume was reported to inhibit the K^+-induced contractile of guinea pig taenia coli. The active principle was identified as 5-(hydroxymethyl)-2-furaldehyde [30].

References

1 Zhu, Y.P., Su, Z.W., and Li, C.Y. (1988) Amygdalin contents in seeds of 11 kinds of plants of the genus *Prunus*. *Bull. Chin. Mater. Med.*, **13**, 356–358.

2 Humbert, J.R., Tress, J.H., and Braico, K.T. (1977) Fatal cyanide poisoning: accidental ingestion of amygdalin. *JAMA*, **238**, 482.

3 Hall, A.H., and Rumack, B.H. (1986) Clinical toxicology of cyanide. *Ann. Emerg. Med.*, **15**, 1067–1074.

4 Bromley, J., Hughes, B.G., Leong, D.C., and Buckley, N.A. (2005) Life-threatening interaction between complementary medicines: cyanide toxicity following ingestion of amygdalin and vitamin C. *Ann. Pharmacother.*, **39**, 1566–1569.

5 O'Brien, B., Quigg, C., and Leong, T. (2005) Severe cyanide toxicity from 'vitamin supplements'. *Eur. J. Emerg. Med.*, **12**, 257–258.

6 Schmidt, E.S., Newton, G.W., Sanders, S.M., Lewis, J.P., and Conn, E.E. (1978) Laetrile toxicity studies in dogs. *JAMA*, **239**, 943–947.

7 Khandekar, J.D., and Edelman, H. (1979) Studies of amygdalin (laetrile) toxicity in rodents. *JAMA*, **242**, 169–171.

8 Carter, J.H., McLafferty, M.A., and Goldman, P. (1980) Role of the gastrointestinal microflora in amygdalin (laetrile)-induced cyanide toxicity. *Biochem. Pharmacol.*, **29**, 301–304.

9 Hill, H.Z., Backer, R., and Hill, G.J. II (1980) Blood cyanide levels in mice after administration of amygdalin. *Biopharm. Drug Dispos.*, **1**, 211–220.

10 Moertel, C.G., Ames, M.M., Kovach, J.S., Moyer, T.P., Rubin, J.R., and Tinker, J.H. (1981) A pharmacologic and toxicological study of amygdalin. *JAMA*, **245**, 591–594.

11 Adewusi, S.R. and Oke, O.L. (1985) On the metabolism of amygdalin. 1. The LD_{50} and biochemical changes in rats. *Can. J. Physiol. Pharmacol.*, **63**, 1080–1083.

12 Adewusi, S.R. and Oke, O.L. (1985) On the metabolism of amygdalin. 2. The distribution of β-glucosidase activity and orally administered amygdalin in rats. *Can. J. Physiol. Pharmacol.*, **63**, 1084–1087.

13 Wagner, B. and Galey, W.R. (2003) Kinetic analysis of hexose transport to determine the mechanism of amygdalin and prunasin absorption in the intestine. *J. Appl. Toxicol.*, **23**, 371–375.

14 Miyagoshi, M., Amagaya, S., and Ogihara, Y. (1986) Antitussive effects of L-ephedrine, amygdalin, and makyokansekito (Chinese traditional medicine) using a cough model induced by sulfur dioxide gas in mice. *Planta Med.*, **52** (4), 275–278.

15 Zhu, Y.P., Su, Z.W., and Li, C.H. (1994) Analgesic effect and no physical dependence of amygdalin. *China J. Chin. Mater. Med.*, **19**, 105–107. 128.

16 Yuan, D., Komatsu, K., Cui, Z., and Kano, Y. (1999) Pharmacological properties of traditional medicines. XXV. Effects of ephedrine, amygdalin, glycyrrhizin, gypsum and their combinations on body temperature and body fluid. *Biol. Pharm. Bull.*, **22**, 165–171.

17 Yang, H.Y., Chang, H.K., Lee, J.W., Kim, Y.S., Kim, H., Lee, M.H., Shin, M.S., Ham, D.H., Park, H.K., Lee, H., and Kim, C.J. (2007) Amygdalin suppresses lipopolysaccharide-induced expressions of cyclooxygenase-2 and inducible nitric oxide synthase in mouse BV2 microglial cells. *Neurol. Res.*, **29** (Suppl. 1), S59–S64.

18 Kwon, H.Y., Hong, S.P., Hahn, D.H., and Kim, J.H. (2003) Apoptosis induction of Persicae Semen extract in human promyelocytic leukemia (HL-60) cells. *Arch. Pharm. Res.*, **26**, 157–161.

19 Chang, H.K., Shin, M.S., Yang, H.Y., Lee, J.W., Kim, Y.S., Lee, M.H., Kim, J., Kim, K.H., and Kim, C.J. (2006) Amygdalin induces apoptosis through regulation of Bax and Bcl-2 expressions in human DU145 and LNCaP prostate cancer cells. *Biol. Pharm. Bull.*, **29**, 1597–1602.

20 Park, H.J., Yoon, S.H., Han, L.S., Zheng, L.T., Jung, K.H., Uhm, Y.K., Lee, J.H., Jeong, J.S., Joo, W.S., Yim, S.V., Chung, J.H., and Hong, S. P. (2005) Amygdalin inhibits genes related to cell cycle in SNU-C4 human colon cancer cells. *World J. Gastroenterol.*, **11**, 5156–5161.

21 Chitnis, M.P., Adwankar, M.K., and Amonkar, A.J. (1985) Studies on high-dose chemotherapy of amygdalin in murine P388 lymphocytic leukaemia and P815 mast cell leukaemia. *J. Cancer Res. Clin. Oncol.*, **109**, 208–209.

22 Syrigos, K.N., Rowlinson-Busza, G., and Epenetos, A.A. (1998) *In vitro* cytotoxicity following specific activation of amygdalin by β-glucosidase conjugated to a bladder cancer-associated monoclonal antibody. *Int. J. Cancer*, **78**, 712–719.

23 Yamamoto, K., Osaki, Y., Kato, T., and Miyazaki, T. (1992) Antimutagenic substances in the Armeniacae semen and Persicae semen. *Yakugaku Zasshi*, **112**, 934–939.

24 Zeng, Y., Zhong, J.M., Ye, S.Q., Ni, Z.Y., Miao, X.Q., Mo, Y.K., and Li, Z.L. (1994) Screening of Epstein-Barr virus early antigen expression inducers from Chinese medicinal herbs and plants. *Biomed. Environ. Sci.*, **7**, 50–55.

25 Liang, A.H., Nie, S.Q., Xue, B.Y., Li, G.Q., and Li, Z.L. (1993) Effects of processing on special toxicity and pharmacodynamics in Semen Armeniacae Amarum. *China J. Chin. Mater. Med.*, **18**, 474–478, 509.

26 Fukuda, T., Ito, H., Murainaka, T., Tokuda, H., Nishino, H., and Yoshida, T. (2003) Anti-tumor promoting effect of glycosides from *Prunus persica* seeds. *Biol. Pharm. Bull.*, **26**, 271–273.

27 Zhu, H., Chang, L., Li, W., and Liu, H. (2004) Effect of amygdalin on the proliferation of hyperoxia-exposed type II alveolar epithelial cells isolated from premature rat. *J. Huazhong Univ. Sci. Technolog. Med. Sci.*, **24**, 223–225.

28 Mizushina, Y., Takahashi, N., Ogawa, A., Tsurugaya, K., Koshino, H., Takemura, M., Yoshida, S., Matsukage, A., Sugawara, F., and Sakaguchi, K. (1999) The cyanogenic glucoside, prunasin (D-mandelonitrile-β-D-glucoside), is a novel inhibitor of DNA polymerase β. *J. Biochem. (Tokyo)*, **126**, 430–436.

29 Strugala, G.J., Stahl, R., Elsenhans, B., Rauws, A.G., and Forth, W. (1995) Small-intestinal transfer mechanism of prunasin, the primary metabolite of the cyanogenic glycoside amygdalin. *Hum. Exp. Toxicol.*, **14**, 895–901.

30 Ichikawa, K., Kinoshita, T., and Sankawa, U. (1989) The screening of Chinese crude drugs for Ca^{2+} antagonist activity: identification of active principles from the aerial part of Pogostemon cablin and the fruits of *Prunus mume*. *Chem. Pharm. Bull. (Tokyo)*, **37**, 345–348.

Pseudolarix kaempferi

Cortex Pseudolaricis (Tujingpi) is the dried root bark or the stem bark near the root of *Pseudolarix kaempferi* Gord. (Pinaceae). It is used for the extradermal treatment of scabies and as an antifungal agent, due to its toxicity.

Pseudolarix kaempferi Gord.

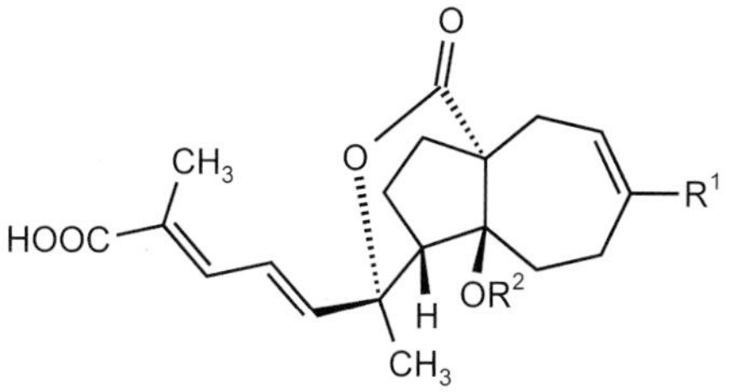

Pseudolaric acid A: $R^1 = CH_3$; $R^2 = Ac$
Pseudolaric acid B: $R^1 = COOCH_3$; $R^2 = Ac$
Pseudolaric acid C: $R^1 = COOCH_3$; $R^2 = H$

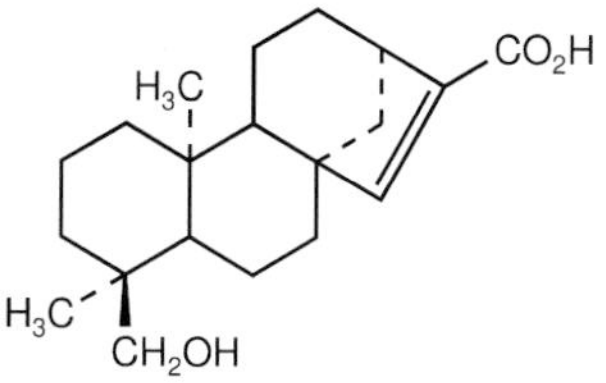

Pseudolaric acid D

Chemistry

The root bark of *P. kaempferi* was reported to contain diterpenes pseudolaric acids A, A_2, B, B_2, B_3, C, C_2, D, E, deacetylpseudolaric acid A, methyl pseudolarate A_2, and β-D-glycopyranosylesters of pseudolaric acid A and pseudolaric acid B, with pseudolaric acid B as the major component [1–3]. The Chinese Pharmacopoeia requires a qualitative determination of pseudolaric acid B in Cortex Pseudolaricis by thin-layer chromatographic comparison with reference substance, and a quantitative determination by HPLC. The content of pseudolaric acid B in Cortex Pseudolaricis should be not less than 0.25%

Triterpene compounds from the root bark of *P. kaempferi* were identified as betulinic acid and pseudolarifuroic acid [4].

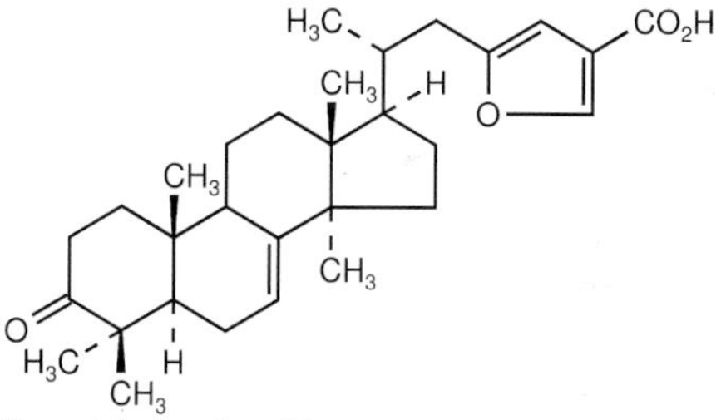

Pseudolarifuroic acid

Pharmacology and Toxicology

The root bark of *P. kaempferi* is known to be toxic, and is used only for the extradermal treatment of scabies and as an antifungal agent. The major component, pseudolaric acid B, was found to be antifungal against *Trichophyton mentagrophytes, Torulopsis petrophilum, Microsporum gypseum,* and *Candida* spp. The *in vivo* antifungal activity of pseudolaric acid B was demonstrated in a murine model of disseminated candidiasis.

Handbook of Chinese Medicinal Plants: Chemistry, Pharmacology, Toxicology
Weici Tang and Gerhard Eisenbrand

ISBN: 978-3-527-32226-8

It significantly reduced the number of recovered colony-forming units at different dosages, and prolonged the survival time of mice treated intravenously with pseudolaric acid B [3, 5].

Pseudolaric acid B was tested for its potential as an antifertility agent [6]. It was found to terminate pregnancy at an early stage in experimental animals including rats, rabbits and dogs after intraperitoneal (i.p.) or subcutaneous (s.c.) administration. The mid-term pregnancy of rats was terminated when pseudolaric acid B was given i.p. at $10\,mg\,kg^{-1}$ daily on days 10–12 after mating. However, implantation was not prevented in rats when $40\,mg\,kg^{-1}$ pseudolaric acid B was injected s.c. or i.p. daily on days 1–3 after mating [7]. The injection of pseudolaric acid B into hamster ovarian bursa at various concentrations before ovulation resulted in a significant decrease of the successful rate of fertilization of ova [8].

Pseudolaric acids A and B did not show estrogenic and antiestrogenic activities. The plasma levels of progesterone, estradiol, prostaglandins E and F, and the uterine levels of prostaglandins E and F in rats on days 8 and 12 of pregnancy were not significantly reduced, when pseudolaric acid B was given intragastrically on day 6 of pregnancy at a dose of $30\,mg\,kg^{-1}$. *In vitro*, pseudolaric acids A and B at concentration of $200\,mg\,ml^{-1}$ damaged only a part of the decidual and trophoblast cells of human uterus. In partially depolarized isolated uterine smooth muscles of early pregnant rats, pseudolaric acids A and B each caused a decline in the contractile tension. In rats, pseudolaric acid B, at a low daily dose of $2\,mg\,kg^{-1}$ given intragastrically on days 6–12 of pregnancy, caused a significant decrease in the body weight, fetal length and placental weight [9]. In addition, pseudolaric acid B was found to cause a significant decrease in the blood flow of the endometrium and myometrium in rats; this was suggested to be the most important cause of embryonal death [10].

Recently, the cytotoxicity of pseudolaric acids (especially of the major component, pseudolaric acid B) has been intensively investigated using different tumor cell lines and tumor models. Pseudolaric acids A and B were found to be cytotoxic in KB, A-549, HCT-8, P-388, and L-1210 tumor cell lines [11]. Pseudolaric acid B was reported to inhibit the proliferation of HeLa cells. It induced apoptosis in a time- and concentration-dependent manner, downregulated Bcl-2 expression, and upregulated Bax expression [12]. It was also shown that p53 was partially regulated by Jun N-terminal kinase (JNK), and that protein kinase C (PKC) participated in pseudolaric acid B-induced HeLa cell death [13]. Likewise, pseudolaric acid B inhibited the proliferation of human melanoma A375-S2 cells, induced cell apoptosis, decreased the expression of Bcl-2, Bcl-xL and ICAD, and increased the expression of Bax [14].

The growth of AGS human gastric cancer cells was also markedly suppressed by pseudolaric acid B. It induced cell-cycle arrest at G_2/M phase, accompanied by a decrease in the levels of cdc2. AGS cells treated with pseudolaric acid B showed typical characteristics of apoptosis associated with decreased levels of the Bcl-2 protein, activation of caspase-3, and proteolytic cleavage of poly(ADP-ribose)polymerase-1 [15]. In HT-29 colon cancer cells, pseudolaric acid B inhibited the cell growth, and induced cell apoptosis associated with a modulation of cyclin expression and downregulation of the protooncogene Myc. In addition, modulation of the growth-related and apoptotic factors by pseudolaric acid B was accompanied by increased protein and gene expression of the nonsteroidal anti-inflammatory drug-activated gene (NAG-1), occurring along with cyclooxygenase-2 (COX-2) inhibition [16].

In mice, pseudolaric acid B significantly inhibited the growth of transplantable tumors, such as Lewis lung cancer and hepatocarcinoma 22. The i.p. injection of pseudolaric acid B to tumor-bearing mice at daily doses of 30 and 60 mg kg^{-1} for 10 days resulted in inhibitory rates for hepatocarcinoma 22 of 14% and 40%, and for Lewis lung cancer of 39% and 47%, respectively [17]. Pseudolaric acid B was further found to be a tubulin-binding agent, significantly inhibiting the proliferation, migration, and tube formation of human microvessel enthothelial cells. Pseudolaric acid B, at noncytotoxic concentrations, induced endothelial cell retraction, intercellular gap formation, and promoted actin stress fiber formation in conjunction with a disruption of the tubulin and actin cytoskeletons. All of these effects were attributable to depolymerization of tubulin by direct interaction with a distinct binding site on tubulin [18]. It was also reported that pseudolaric acid B disrupted cellular microtubule networks and inhibited the formation of mitotic spindles. Polymerization of purified bovine brain tubulin was inhibited, concentration-dependently, by pseudolaric acid B. Moreover, pseudolaric acid B also displayed notable potency in tumor cells overexpressing P-glycoprotein [19].

After exposure to pseudolaric acid B, the proliferation of human umbilical vein endothelial cells (HUVEC) was significantly inhibited. Pseudolaric acid B potently blocked the tube formation of HUVEC induced by vascular endothelial growth factor (VEGF) in a concentration-dependent manner. In addition, pseudolaric acid B antagonized VEGF-mediated anti-apoptotic effects on serum-deprived HUVEC and increased the apoptosis of endothelial cells induced by VEGF. Moreover, pseudolaric acid B significantly inhibited VEGF-induced tyrosine phosphorylation of kinase insert domain-containing receptor/fetal liver kinase-1 (KDR/flk-1), in correlation with a marked decrease in the phosphorylation of Akt and extracellular signal-regulated kinases (ERK); this suggested that pseudolaric acid B possessed antiangiogenic activity [20].

References

1 Li, Z.L., Chen, K., Pan, D.J., and Xu, G.Y. (1989) New diterpenic constituents of Tu-Jin-Pi. IV. Isolation and identification of pseudolaric acid D and pseudolaric acid E. *Acta Chim. Sin.*, **47**, 258–261.

2 Liu, P., Guo, H., Guo, H., Sheng, Y., Wang, W., Xu, M., Feng, S., Cheng, F., and Guo, D.A. (2007) Simultaneous determination of seven major diterpenoids in *Pseudolarix kaempferi* by high-performance liquid chromatography DAD method. *J. Pharm. Biomed. Anal.*, **44**, 730–736.

3 Yang, S.P., Dong, L., Wang, Y., Wu, Y., and Yue, J.M. (2003) Antifungal diterpenoids of *Pseudolarix kaempferi*, and their structure-activity relationship study. *Bioorg. Med. Chem.*, **11**, 4577–4584.

4 Chen, K., Li, Z.L., Pan, D.J., and Xu, G.Y. (1990) Study on the triterpenic constituents of Tu-Jin-Pi. *Acta Chim. Sin.*, **48**, 591–595.

5 Li, E., Clark, A.M., and Hufford, C.D. (1995) Antifungal evaluation of pseudolaric acid B, a major constituent of *Pseudolarix kaempferi*. *J. Nat. Prod.*, **58**, 57–67.

6 Xiao, P.G. and Wang, N.G. (1991) Can ethnopharmacology contribute to the development of anti-fertility drugs? *J. Ethnopharmacol.*, **32**, 167–177.

7 Wang, W.C., Lu, R.F., Zhao, S.X., and Zhu, Y.Z. (1982) Antifertility effect of pseudolaric acid B. *Acta Pharmacol. Sin.*, **3**, 188–192.

8 Zhang, Y.L., Lu, R.Z., and Yan, A.L. (1990) Inhibition of ova fertilizability by pseudolaric acid B in hamster. *Acta Pharmacol. Sin.*, **11**, 60–62.

9 Wang, W.C., You, G.D., Jiang, X.J., Lu, R.F., and Gu, Z.P. (1991) Endocrine activity of pseudolaric acids A and B and their effects on sex hormones, prostaglandins, uteri, and fetuses. *Acta Pharmacol. Sin.*, **12**, 187–190.

10 Wang, W.C., Gu, Z.P., Koo, A., and Chen, W.S. (1991) Effects of pseudolaric acid B on blood flows of endometrium and myometrium in pregnant rats. *Acta Pharmacol. Sin.*, **12**, 423–425.

11 Pan, D.J., Li, Z.L., Hu, C.Q., Chen, K., Chang, J.J., and Lee, K.H. (1990) The cytotoxic principles of *Pseudolarix kaempferi*: pseudolaric acid-A and -B and related derivatives. *Planta Med.*, **56**, 383–385.

12 Gong, X., Wang, M., Wu, Z., Tashiro, S., Onodera, S., and Ikejima, T. (2004) Pseudolaric acid B induces apoptosis via activation of c-Jun N-terminal kinase and caspase-3 in HeLa cells. *Exp. Mol. Med.*, **36**, 551–556.

13 Gong, X., Wang, M., Tashiro, S., Onodera, S., and Ikejima, T. (2006) Involvement of JNK-initiated p53 accumulation and phosphorylation of p53 in pseudolaric acid B induced cell death. *Exp. Mol. Med.*, **38**, 428–434.

14 Gong, X.F., Wang, M.W., Tashiro, S., Onodera, S., and Ikejima, T. (2005) Pseudolaric acid B induces human melanoma A375-S2 cell apoptosis in vitro. *Zhongguo Zhong Yao Za Zhi*, **30**, 55–57.

15 Li, K.S., Gu, X.F., Li, P., Zhang, Y., Zhao, Y.S., Yao, Z.J., Qu, N.Q., and Wang, B.Y. (2005) Effect of pseudolaric acid B on gastric cancer cells: inhibition of proliferation and induction of apoptosis. *World J. Gastroenterol.*, **11**, 7555–7559.

16 Ko, J.K., Leung, W.C., Ho, W.K., and Chiu, P. (2007) Herbal diterpenoids induce growth arrest and apoptosis in colon cancer cells with increased expression of the nonsteroidal anti-inflammatory drug-activated gene. *Eur. J. Pharmacol.*, **559**, 1–13.

17 Liu, B., Chen, H., Lei, Z.Y., Yu, P.F., and Xiong, B. (2006) Studies on anti-tumour activities of pseudolaric acid-B (PLAB) and its mechanism of action. *J. Asian Nat. Prod. Res.*, **8**, 241–252.

18 Tong, Y.G., Zhang, X.W., Geng, M.Y., Yue, J.M., Xin, X.L., Tian, F., Shen, X., Tong, L.J., Li, M.H., Zhang, C., Li, W.H., Lin, L.P. and Ding, J. (2006) Pseudolaric acid B, a new tubulin-binding agent, inhibits angiogenesis by interacting with a novel binding site on tubulin. *Mol. Pharmacol.*, **69**, 1226–1233.

19 Wong, V.K., Chiu, P., Chung, S.S., Chow, L.M., Zhao, Y.Z., Yang, B.B., and Ko, B.C. (2005) Pseudolaric acid B, a novel microtubule-destabilizing agent that circumvents multidrug resistance phenotype and exhibits antitumor activity *in vivo*. *Clin. Cancer Res.*, **11**, 6002–6011.

20 Tan, W.F., Zhang, X.W., Li, M.H., Yue, J.M., Chen, Y., Lin, L.P., and Ding, J. (2004) Pseudolaric acid B inhibits angiogenesis by antagonizing the vascular endothelial growth factor-mediated anti-apoptotic effect. *Eur. J. Pharmacol.*, **499**, 219–228.

Pseudostellaria heterophylla

Radix Pseudostellariae (Taizishen) is the dried root of *Pseudostellaria heterophylla* (Miq.) Pax ex Pax et Hoffm. (Caryophyllaceae). It is used as a general tonic.

Pseudostellaria heterophylla (Miq.) Pax ex Pax et Hoffm.

Chemistry

A number of cyclopeptides were isolated from the root of *P. heterophylla* and identified as pseudostellarin D, (cyclo[Gly-Tyr-Gly-Pro-Leu-Ile-Leu]) [1], or pseudostellarin H (cyclo[Gly-Thr-Pro-Thr-Pro-Leu-Phe-Phe]) [2]. Further components in the root of *P. heterophylla* were identified as palmitic acid, behenic acid, β-sitosterol [3], and 3-furfuryl pyrrole-2-carboxylate [4]. A single-chain lectin with a molecular weight of 36 kDa and high hemagglutinating activity was also isolated from the root of *P. heterophylla* [5].

Pharmacology and Toxicology

The root of *P. heterophylla* is known to be a tonic and immunostimulatory agent. A mitogenic fraction (PH-I) from *P. heterophylla*, consisting mainly of carbohydrates (56.8%) and a small amount of proteins (7.6%), was found to exhibit both immunomodulatory and antitumor activities. Further purification of PH-1 by gel filtration chromatography resulted in the separation of three fractions (PH-IA, PH-IB and PH-IC). The fraction PH-IC was found to markedly suppress the growth of Ehrlich ascites tumor cells *in vivo*. Mechanistic studies have shown that the intraperitoneal (i.p.) injection of PH-IC into mice enhanced the phagocytic activity of thioglycolate-elicited peritoneal macrophages. Moreover, PH-IC showed a potent activating effect on the cytotoxic activity of natural killer (NK) cells and alloreactive cytotoxic T cells. PH-IC also increased the number of tumor-infiltrating lymphocytes in the tumor site of mice bearing WEHI-164 (TNF-α-sensitive) cells. The intravenous (i.v.) injection of PH-IC significantly elevated the levels of interferon-γ (IFN-γ) and interleukin-4 (IL-4) in sera of mice bearing Ehrlich ascites tumor [6].

PH-I was further found to stimulate the incorporation of [^{3}H]thymidine into the DNA of murine spleen lymphocytes in a dose-dependent manner, and to act as a priming agent for the release of tumor necrosis factor (TNF) in mice [7]. Among the subfractions of PH-I, PH-IC was the most potent priming fraction for the induction of TNF-α in serum and of TNF-α mRNA in murine macrophages. All three fractions, PH-IA, PH-IB and PH-IC, were found to increase the expression of IL-1α mRNA, while PH-IC showed the most potent activating effect on the expression of IL-1β mRNA [8]. The fraction PH-IBa was

Handbook of Chinese Medicinal Plants: Chemistry, Pharmacology, Toxicology
Weici Tang and Gerhard Eisenbrand

ISBN: 978-3-527-32226-8

reported to stimulate the incorporation of [^{3}H]thymidine into the DNA of murine bone marrow cells in a dose-dependent manner *in vitro*, and to induce the differentiation of murine bone marrow cells from pluripotent hemopoietic stem cells into macrophages-like cells *in vitro*. The autocrine or paracrine stimulation of granulocyte-macrophage colony-stimulating factor was likely to underlie the induction of differentiation. Therefore, PH-IBa was proven to be an immunostimulating agent for mouse marrow hematopoiesis [9].

PH-I was further found to exhibit potent cytotoxic activity against Ehrlich ascites tumor cells *in vivo*, but not *in vitro*. TNF could be induced in the serum of ICR mice bearing Ehrlich ascites tumor cells by i.v. administration of antitumor fractions PH-IA, PH-IB, PH-IC, PH-ICa and PH-ICb, and was also detected in the ascitic fluid of tumor-bearing mice at 2 h after i.v. injection with PH-IC, PH-ICa, or PH-ICb. The proliferation of Ehrlich ascites tumor cells was found to be significantly suppressed after the administration of the fractions. It was suggested that the antitumor activity of *P. heterophylla* might be related, at least in part, to its capacity to induce TNF-α production *in vivo*. The strongest TNF-α-eliciting and antitumor activity was found in the PH-ICb, an acidic polysaccharide or proteoglycan with a molecular weight of approximately 5000 Da [10].

The lectin from the root of *P. heterophylla* exerted some inhibitory effect on the glycohydrolases α-glucosidase, β-glucosidase and β-glucuronidase which are involved in human immunodeficiency virus-type 1 (HIV-1), infection but had no suppressive action on HIV-1 reverse transcriptase [5].

Besides the 36 kDa lectin, a 20.5 kDa trypsin inhibitor was separated from the root of *P. heterophylla*. This exhibited a trypsin-inhibitory potency similar to that of soybean trypsin inhibitor, and also exerted antifungal activity towards *Fusarium oxysporum*, whereas the 36 kDa lectin was devoid of antifungal activity. The 36 kDa lectin also showed a low thermostability and was not stable in the presence of acid and alkali [11].

References

1. Morita, H., Kayashita, T., Takeya, K., and Itokawa, H. (1996) Conformational analysis of a cyclic heptapeptide, pseudostellarin D by molecular dynamics and Monte Carlo simulations. *Chem. Pharm. Bull. (Tokyo)*, **44**, 2177–2180.
2. Morita, H., Kayashita, T., Takeya, K. and Itokawa, H. (1995) Cyclic peptides from higher plants. XV. Pseudostellarin H, a new cyclic octapeptide from *Pseudostellaria heterophylla*. *J. Nat. Prod.*, **58**, 943–947.
3. Tan, F N.H., Zhao, F S.X., Chen, F C.X., and Zhou, J. (1991) Chemical constituents from tubers of *Pseudostellaria heterophyla*. *Acta Bot. Yunnan*, **13**, 440. 431.
4. Reinecke, M.G. and Thao, Y.Y. (1988) Phytochemical studies of the Chinese herb tai-zi-shen, *Pseudostellaria heterophyla*. *J. Nat. Prod.*, **51**, 1236–1240.
5. Wang, H.X. and Ng, T.B. (2001) A novel lectin from *Pseudostellaria heterophylla* roots with sequence similarity to Kunitz-type soybean trypsin inhibitor. *Life Sci.*, **69**, 327–333.
6. Wong, C.K., Leung, K.N., Fung, K.P., and Choy, Y.M. (1994) The immunostimulating activities of anti-tumor polysaccharides from *Pseudostellaria heterophylla*. *Immunopharmacology*, **28**, 47–54.
7. Wong, C.K., Leung, K.N., Fung, K.P., Pang, P. K., and Choy, Y.M. (1992) Mitogenic and tumor necrosis factor producing activities of *Pseudostellaria heterophylla*. *Int. J. Immunopharmacol.*, **14**, 1315–1320.
8. Wong, C.K., Leung, K.N., Fung, M.C., Fung, K. P., and Choy, Y.M. (1994) The induction of cytokine gene expression in murine peritoneal macrophages by *Pseudostellaria heterophylla*. *Immunopharmacol. Immunotoxicol.*, **16**, 347–357.

9 Wong, C.K., Leung, K.N., Fung, K.P., and Choy, Y.M. (1994) Effects of *Pseudostellaria heterophylla* on proliferation and differentiation of murine bone marrow cells. *Immunopharmacol. Immunotoxicol.*, **16**, 71–84.

10 Wong, C.K., Leung, K.N., Fung, K.P., Pang, P. K., and Choy, Y.M. (1994) Tumor necrosis factor eliciting fractions separated from *Pseudostellaria heterophylla*. *Int. J. Immunopharmacol.*, **16**, 271–277.

11 Wang, H.X. and Ng, T.B. (2006) Concurrent isolation of a Kunitz-type trypsin inhibitor with antifungal activity and a novel lectin from *Pseudostellaria heterophylla* roots. *Biochem. Biophys. Res. Commun.*, **342**, 349–353.

Psoralea corylifolia

Fructus Psoraleae (Buguzhi) is the dried ripe fruit of *Psoralea corylifolia* L. (Fabaceae). It is used as an adstringent agent for the treatment of diarrhea. It is also used for treatment of vitiligo, by external application.

Psoralea corylifolia L.

Chemistry

The fruit of *P. corylifolia* is known to contain a number of furocoumarins, with psoralen and isopsoralen (angelicin) as the major components [1]. Minor furocoumarins from the fruits were identified as psoralidin and bakuchicin [2]. The Chinese Pharmacopoeia requires a qualitative determination of psoralen and isopsoralen in Fructus Psoraleae by thin-layer chromatographic comparison with reference substances, and a quantitative determination by HPLC. The total content of psoralen and isopsoralen together in Fructus Psoraleae should be not less than 0.7%.

Prenylated phenols, including prenylated flavones and isoflavones from the fruits or seeds of *P. corylifolia*, were identified as corylifolin, bavachin, corylifolinin (isobavachalcone) [1], bavachin 7-*O*-methylether (bavachinin) [3], neobavaisoflavone [4], genistein, daidzein [2], and bakuchiol [5], with bakuchiol as the major component. The contents of bakuchiol in the fruit samples were reported to range from 1% to 7% [5].

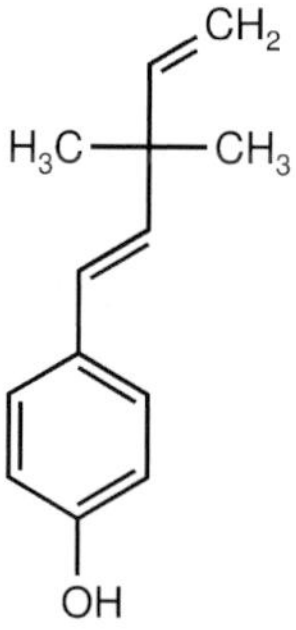

Corylifolin

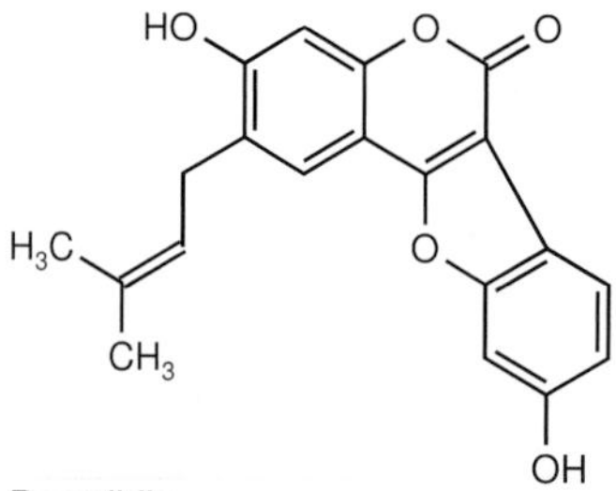

Psoralidin

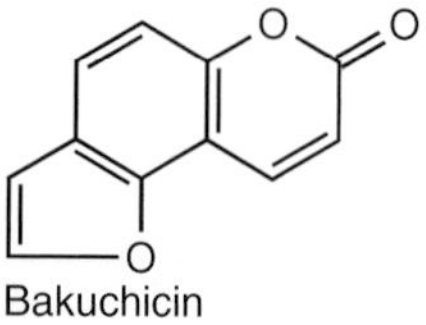

Bakuchicin

Handbook of Chinese Medicinal Plants: Chemistry, Pharmacology, Toxicology
Weici Tang and Gerhard Eisenbrand

ISBN: 978-3-527-32226-8

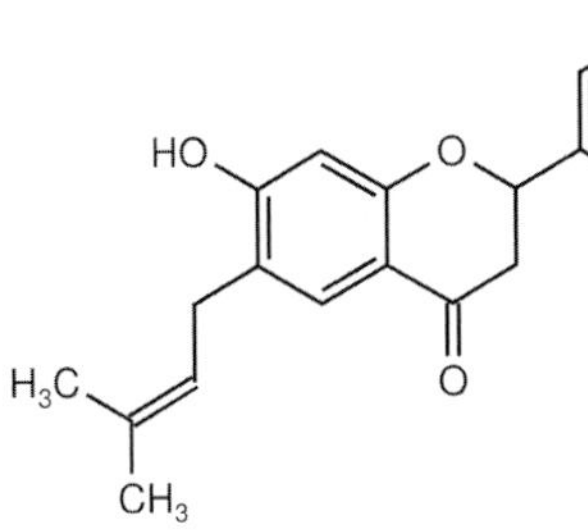

Bavachin

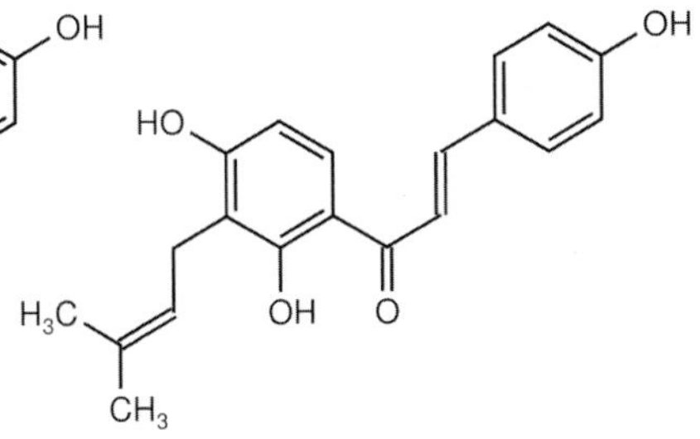

Corylifolinin

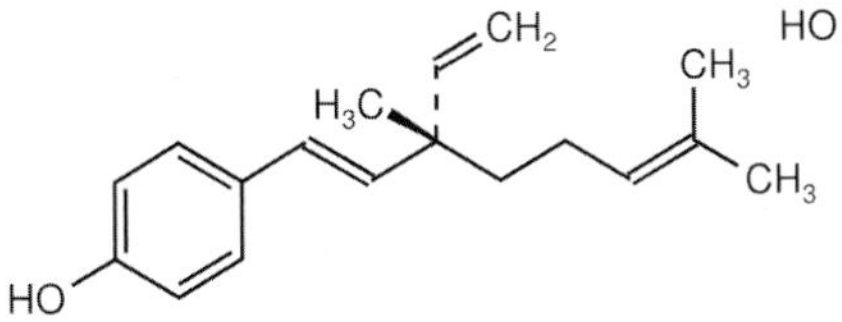

Bakuchiol

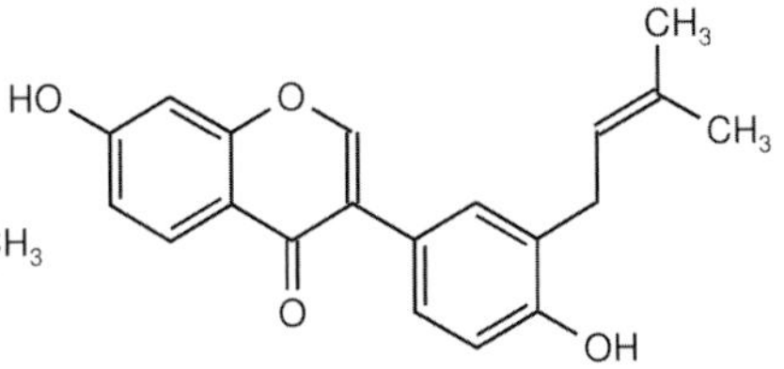

Neobavaisoflavone

Pharmacology and Toxicology

The fruit of *P. corylifolia* was reported to be nephrotoxic. Oral administration of the fruit to mice at a daily dose of 5 g kg^{-1} for 21 days caused nephrotoxicity on the basis of histology of the proximal convoluted tubule [6]. Bakuchiol was found to be the nephrotoxic component [7]. The crude fruit was significantly more nephrotoxic than processed fruit. The content of bakuchiol showed a tendency to decrease during processing [8].

The oral administration of nonpolar fractions of *P. corylifolia* seeds to experimental rachitic rats was found to exert beneficial effects on bone calcification. The fractions given to rats fed with a vitamin D-free, low-phosphorus diet not only significantly increased the concentration of inorganic phosphorus in serum, but also significantly promoted bone calcification [9]. The methanol extract of the seeds was found to inhibit the aggregation of rabbit platelet aggregation induced by arachidonic acid, collagen, and platelet-activating factor (PAF), with corylifolinin and neobavaisoflavone as the active components [4].

The ethanol extract of *P. corylifolia* was found to cause strong DNA polymerase inhibition in a whole-cell bioassay, with corylifolin, neobavaisoflavone and bakuchiol as active compounds. Bakuchicin was found to be a DNA topoisomerase II inhibitor [2]. In addition, the ethanol extract enhanced the activity of tyrosinase, indicating that the therapeutic mechanism of the fruits for the treatment of vitiligo was probably due to increased melanin synthesis [10]. A clinical trial with the fruit of *P. corylifolia* combined with the root of *Arnebia euchroma* for the treatment of 57 vitiligo cases resulted in a total effective rate of 86% after 1–6 months of treatment [11].

Bakuchiol was reported to exhibit antimicrobial activity against *Staphylococcus aureus in vitro* [12]. Bakuchiol also exerted antimicrobial activities against some oral microorganisms. It showed bactericidal effects against all bacteria strains tested, including *Streptococcus mutans, S. sanguis, S. salivarius, S. sobrinus, Enterococcus faecalis, E. faecium, Lactobacillus acidophilus, L. casei, L. plantarum, Actinomyces viscosus,* and *Porphyromonas gingivalis*. The minimal inhibitory concentrations (MICs) ranged from 1 to 4 μg ml^{-1}, and the sterilizing concentrations from 5 to 20 μg ml^{-1} for 15 min.

Furthermore, bakuchiol was also effective against adherent cells of *S. mutans* in water-insoluble glucan in the presence of sucrose, and inhibited the reduction of pH in the broth [13].

The ethanol extract of the fruits of *P. corylifolia* also showed cytotoxic activity against L929 cells *in vitro*, with bakuchiol as the active compound. The cytotoxic activity of bakuchiol in cell culture was observed after short-term incubation, and was not reversible. The mechanism of the cytotoxic activity of bakuchiol was considered to be due to damage of the cell membrane, and to hemolytic activity [14]. Structure–activity relationship studies of bakuchiol and related prenylphenols demonstrated that an alkyl group was necessary for cytotoxic activity; however, the double bonds exerted a minimal influence on the cytotoxic effect. The cytotoxic activity of bakuchiol was the strongest as compared with that of its analogues [15]. $[^{125}I]$Bakuchiol, when tested in terms of its cytotoxicity and uptake into tumor cells, showed significant uptake into both murine LS-A lymphosarcoma cells and barcl-95 radiation-induced thymic lymphoma cells. The proliferation–inhibitory activities of $[^{125}I]$bakuchiol against LS-A and barcl-95 cells were greater than those of bakuchiol [16]. Psoralidin was also found to be cytotoxic *in vitro*. The IC_{50} values of psoralidin against SNU-1 and SNU-16 stomach carcinoma cells were approximately 50 and 200 $\mu g\ ml^{-1}$ [17].

Bakuchiol was reported to be a potent antioxidant, and to inhibit lipid peroxidation in rat liver microsomes, as induced by NADPH, ascorbate, and carbon tetrachloride. Bakuchiol also prevented NADH-dependent and ascorbate-induced mitochondrial lipid peroxidation. Furthermore, bakuchiol protected human red blood cells against oxidative hemolysis [18]. Bakuchiol and its methyl ether were also found to exert scavenging activities toward a series of oxidizing radicals, and also capable of preventing lipid peroxidation in rat brain homogenate [19].

Bakuchiol was also reported to show anti-inflammatory activity and to be a weak inhibitor of secretory and intracellular phospholipase A_2. It reduced, concentration-dependently, the formation of leukotriene B_4 (LTB_4) and thromboxane B_2 (TXB_2) by human neutrophils and platelet microsomes. In addition, bakuchiol inhibited degranulation in human neutrophils, whereas superoxide generation was not affected. In mice, bakuchiol decreased cell migration, myeloperoxidase activity and eicosanoid levels in the air pouch inflammation, as induced by zymosan. After topical administration, bakuchiol was effective as an inhibitor of 12-*O*-tetradecanoylphorbol 13-acetate (TPA)-induced ear edema. It significantly reduced the prostaglandin E_2 (PGE_2) content and ear edema induced by arachidonic acid [20]. In RAW 264.7 murine peritoneal macrophage cells, bakuchiol inhibited the expression of inducible nitric oxide synthase (iNOS) via the inactivation of nuclear transcription factor-κB (NF-κ B) [21].

Corylifolinin was found to improve retinal functions and to inhibit ocular ischemia and inflammation – two major factors that induce retinal degeneration [22]. This, corylifolinin was considered to belong to the natural products used for the treatment of ischemic retinopathy [23].

Bakuchiol was further found to exert a protective activity in human liver $HepG_2$ cells against the cytotoxicity induced by 9-amino-tetrahydroacridine, with an EC_{50} value of 1.0 $\mu g\ ml^{-1}$, compared to silymarin as positive control (EC_{50} value 5.0 $\mu g\ ml^{-1}$). Bakuchicin and psoralen each also exhibited a moderate activity against 9-amino-tetrahydroacridine-induced cytotoxicity in $HepG_2$ cells, with EC_{50} values of about 50 $\mu g\ ml^{-1}$ [24]. Psoralidin and bakuchiol also exerted inhibitory activities against protein tyrosine phosphatase 1B in a concentration-dependent manner, with IC_{50} values of 9 μ*M* and 20 μ*M*, respectively [25]. Oral administration of bakuchiol to rats at doses of 25 or 50 mg kg^{-1} at 1, 24 and 48 h after subcutaneous (s.c.) injection of carbon tetrachloride (CCl_4) significantly reduced the serum levels of aspartate transaminase and alanine transaminase. Histological observations revealed that fatty acid changes, hepatocyte necrosis and inflammatory cell infiltration in CCl_4-injured liver were each improved by treatment with bakuchiol. Likewise, bakuchiol significantly reduced the serum levels of aspartate transaminase and alanine transaminase in rats intoxicated with D-galactosamine [26].

Bakuchiol was further reported to reduce activated hepatic stellate cells when given to rats during the recovery period after liver injury. It induced apoptosis in activated hepatic stellate cells and myofibroblasts, as demonstrated by DNA fragmentation, activation of caspase-3, release of cytochrome c into the cytoplasm, translocation of Bax into mitochondria, and the proteolytic cleavage of poly(ADP-ribose)polymerase *in vitro*. Bakuchiol treatment stimulated the activation of extracellular signal-regulated kinase 1/2 (ERK1/2), c-Jun NH_2-terminal protein kinase (JNK), and p38 mitogen-activated protein kinases (MAPK). It was concluded that bakuchiol induced caspase-3-dependent apoptosis in rat liver myofibroblasts through the activation of JNK, followed by Bax translocation into mitochondria [27].

Several chemical constituents from the fruits of *P. corylifolia* were found to be antimutagenic. Among these, psoralen showed a high activity in inhibiting the mutagenicity of benzo[*a*]pyrene. Bakuchiol inhibited the mutagenicity of 2-aminoanthracene in *Salmonella typhimurium* TA98 with an activating system [28, 29]. Topical application of the active fraction of *P. corylifolia* seeds at doses of 100 mg kg^{-1} inhibited the growth and delayed the onset of papilloma formation in mice, initiated with 7,12-dimethylbenz[*a*]anthracene (DMBA) and promoted with croton oil. The fraction, when administered orally at the same dose, significantly inhibited the growth of the soft tissue fibrosarcomas induced by subcutaneously injected methylcholanthrene [30].

References

1 Ji, L. and Xu, Z. (1995) Review of constituents in fruits of *Psoralea corylifolia* L. *China J. Chin. Mater. Med.*, **20**, 120–122.

2 Sun, N.J., Woo, S.H., Cassady, J.M., and Snapka, R.M. (1998) DNA polymerase and topoisomerase II inhibitors from *Psoralea corylifolia. J. Nat. Prod.*, **61**, 362–366.

3 Anand, K.K., Sharma, M.L., Singh, B., and Ghatak, B.J. (1978) Anti-inflammatory, antipyretic and analgesic properties of bavachinin, a flavanone isolated from seeds of *Psoralea corylifolia* Linn. (Babchi). *Indian J. Exp Biol.*, **16**, 1216–1217.

4 Tsai, W.J., Hsin, W.C., and Chen, C.C. (1996) Anti-platelet flavonoids from seeds of *Psoralea corylifolia. J. Nat. Prod.*, **59**, 671–672.

5 Yao, S., Yang, B., and Xu, Z. (1995) Determination of bakuchiol in the fruit of *Psoralea corylifolia* L. *China J. Chin. Mater. Med.*, **20**, 681–683.

6 Zhang, Y.S., Wu, Z.L., Hui, L.Q., Yao, X.Z., and Shen, H. (1994) Nephrotoxicity of *Psoralea corylifolia* fruits prepared by different processes. *Chin. Trad. Patent Med.*, **16**, 17–18.

7 Zhang, Y.S. (1981) A study on toxicity of bakuchiol to mice's kidney. *Bull. Chin. Mater. Med.*, **6**, 30–32.

8 Yang, B., Yao, S., and Cui, S. (1996) Comparison of ancient and modern processing methods for Fructus Psoraleae. *China J. Chin. Mater. Med.*, **21**, 537–539, 575.

9 Miura, H., Nishida, H., and Linuma, M. (1996) Effect of crude fractions of *Psoralea corylifolia* seed extract on bone calcification. *Planta Med.*, **62**, 150–153.

10 Xu, J.G. and Shang, J. (1991) Effect of *Malaytea scurfpea* (*Psoralea corylifolia*) on the activity of tyrosinase. *Chin. Trad. Herbal Drugs*, **22**, 168–169.

11 Shi, Q.L. and Pan, H.L. (1988) Treatment of 57 cases of vitiligo by integrated Chinese and western medicine. *New Chin. Med.*, **19**, 634.

12 Kaul, R. (1976) Kinetics of the anti-staphylococcal activity of bakuchiol *in vitro*. *Arzneim.-Forsch.*, **26**, 486–489.

13 Katsura, H., Tsukiyama, R.I., Suzuki, A., and Kobayashi, M. (2001) In vitro antimicrobial activities of bakuchiol against oral microorganisms. *Antimicrob. Agents Chemother.*, **45**, 3009–3013.

14 Kubo, M., Dohi, T., Odani, T., Tanaka, H., and Iwamura, J. (1989) Cytotoxicity of Corylifoliae fructus. I. Isolation of the effective compound and the cytotoxicity. *Yakugaku Zasshi*, **109**, 926–931.

15 Iwamura, J., Dohi, T., Tanaka, H., Odani, T., and Kubo, M. (1989) Cytotoxicity of corylifoliae fructus. II. Cytotoxicity of bakuchiol and the analogues. *Yakugaku Zasshi*, **109**, 962–965.

16 Bapat, K., Chintalwar, G.J., Pandey, U., Thakur, V.S., Sarma, H.D., Samuel, G., Pillai, M.R., Chattopadhyay, S., and Venkatesh, M. (2005) Preparation and *in vitro* evaluation of radioiodinated bakuchiol as an anti-tumor agent. *Appl. Radiat. Isot.*, **62**, 389–393.

17 Yang, Y.M., Hyun, J.W., Sung, M.S., Chung, H.S., Kim, B.K., Paik, W.H., Kang, S. S., and Park, J.G. (1996) The cytotoxicity of psoralidin from *Psoralea corylifolia*. *Planta Med.*, **62**, 353–354.

18 Haraguchi, H., Inoue, J., Tamura, Y., and Mizutani, K. (2002) Antioxidative components of *Psoralea corylifolia* (Leguminosae). *Phytother. Res.*, **16**, 539–544.

19 Adhikari, S., Joshi, R., Patro, B.S., Ghanty, T.K., Chintalwar, G.J., Sharma, A., Chattopadhyay, S., and Mukherjee, T. (2003) Antioxidant activity of bakuchiol: experimental evidences and theoretical treatments on the possible involvement of the terpenoid chain. *Chem. Res. Toxicol.*, **16**, 1062–1069.

20 Ferrandiz, M.L., Gil, B., Sanz, M.J., Ubeda, A., Erazo, S., Gonzalez, E., Negrete, R., Pacheco, S., Paya, M., and Alcaraz, M.J. (1996) Effect of bakuchiol on leukocyte functions and some inflammatory responses in mice. *J. Pharm. Pharmacol.*, **48**, 975–980.

21 Pae, H.O., Cho, H., Oh, G.S., Kim, N.Y., Song, E.K., Kim, Y.C., Yun, Y.G., Kang, C.L., Kim, J.D., Kim, J.M., and Chung, H.T. (2001) Bakuchiol from *Psoralea corylifolia* inhibits the expression of inducible nitric oxide synthase gene via the inactivation of nuclear transcription factor-κB in RAW 264.7 macrophages. *Int. Immunopharmacol.*, **1**, 1849–1855.

22 Liu, S.X. and Chiou, G.C. (1996) Effects of Chinese herbal products on mammalian retinal functions. *J. Ocul. Pharmacol. Ther.*, **12**, 377–386.

23 Chiou, G.C., Li, B.H., and Wang, M.S. (1994) Facilitation of retinal function recovery by natural products after temporary ischemic occlusion of central retinal artery. *J. Ocul. Pharmacol. Ther.*, **10**, 493–498.

24 Cho, H., Jun, J.Y., Song, E.K., Kang, K.H., Maek, H.Y., Ko, Y.S., and Kim, Y.C. (2001) Bakuchiol: a hepatoprotective compound of *Psoralea corylifolia* on tacrine-induced cytotoxicity in Hep G_2 cells. *Planta Med.*, **67**, 750–751.

25 Kim, Y.C., Oh, H., Kim, B.S., Kang, T.H., Ko, E. K., Han, Y.M., Kim, B.Y., and Ahn, J.S. (2005) *In vitro* protein tyrosine phosphatase 1B inhibitory phenols from the seeds of *Psoralea corylifolia*. *Planta Med.*, **71**, 87–89.

26 Park, E.J., Zhao, Y.Z., Kim, Y.C., and Sohn, D. H. (2005) Protective effect of (S)-bakuchiol from *Psoralea corylifolia* on rat liver injury in vitro and *in vivo*. *Planta Med.*, **71**, 508–513.

27 Park, E.J., Zhao, Y.Z., Kim, Y.C., and Sohn, D.H. (2007) Bakuchiol-induced caspase-3-dependent apoptosis occurs through c-Jun NH_2-terminal kinase-mediated mitochondrial translocation of Bax in rat liver myofibroblasts. *Eur. J. Pharmacol.*, **559**, 115–123.

28 Wall, M.E., Wani, M.C., Manikumar, G., Hughes, T.J., Taylor, H., McGivney, R., and Warner, J. (1988) Plant antimutagenic agents, 3. Coumarins. *J. Nat. Prod.*, **51**, 1148–1152.

29 Wall, M.E., Wani, M.C., Manikumar, G., Abraham, P., Taylor, H., Hughes, T.J., Warner, J., and McGivney, R. (1988) Plant antimutagenic agents, 2. Flavonoids. *J. Nat. Prod.*, **51**, 1084–1091.

30 Latha, P.G., and Panikkar, K.R. (1999) Inhibition of chemical carcinogenesis by *Psoralea corylifolia* seeds. *J. Ethnopharmacol.*, **68**, 295–298.

Pterocarpus santalinus

The wood of *Pterocarpus santalinus* L. (Fabaceae) is included in the appendix of the Chinese Pharmacopoeia.

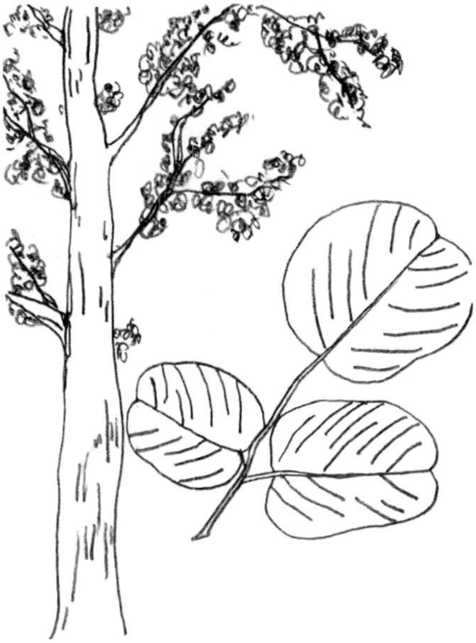

Pterocarpus santalinus L.

Chemistry

Isoflavones from the heartwood of *P. santalinus* were identified as liquiritigenin, isoliquiritigenin, 6-hydroxy,7,2′,4′,5′-tetramethoxyisoflavone [1]. Isoflavone glucosides from the heartwood of were identified as 4′,5-dihydroxy-7-*O*-methyl-isoflavone 3′-*O*-β-D-glucopyranoside [2] and its acylated derivative 4′,5-dihydroxy-7-*O*-methylisoflavone 3′-*O*-D-(3″-*E*-cinnamoyl)-β-D-glucopyranoside [3]. Dibenzyl butyrolactone-type lignan compounds having an α-arylidene γ-lactone structure from the heartwood were identified as savinin and calocedrin [4]. Further components from the heartwood were xanthen pigments, such as santalin A and santalin B [5].

Pharmacology and Toxicology

An ethanolic extract of the bark of *P. santalinus*, when administered orally at a dose of 0.25 g kg^{-1}, was reported to show antihyperglycemic activity in diabetic rats, but did not cause any hypoglycemic activity in normal rats [6]. Savinin and calocedrin significantly inhibited tumor necrosis factor-α (TNF-α) production in lipopolysaccharide (LPS)-stimulated RAW264.7 cells, and T-cell proliferation elicited by concanavalin A (Con A), without display-

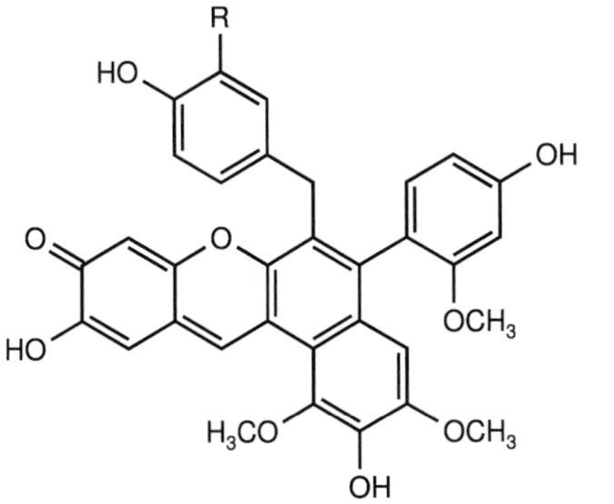

Santalin A: R = OH
Santalin B: R = OCH_3

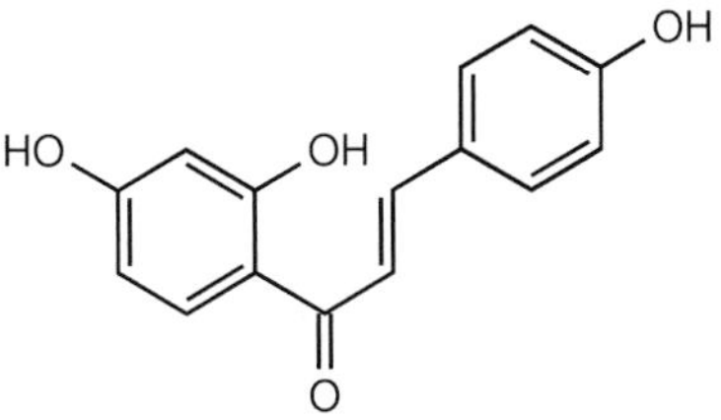

Isoliquiritigenin

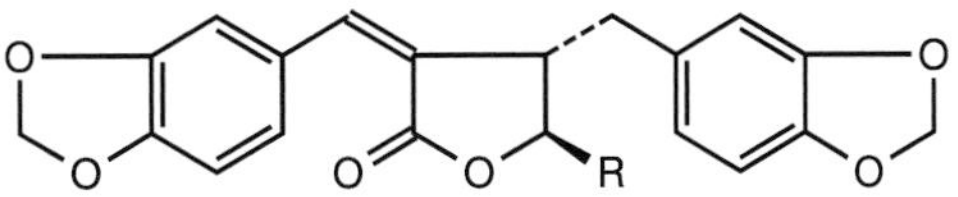

Savinin: R = H
Calocedrin: R = OH

Handbook of Chinese Medicinal Plants: Chemistry, Pharmacology, Toxicology
Weici Tang and Gerhard Eisenbrand

ISBN: 978-3-527-32226-8

ing cytotoxicity. The molecular inhibitory mechanism of savinin was confirmed to be mediated by the nonpolar butyrolactone ring, according to a structure–relationship study with structurally related and unrelated compounds, such as arctigenin (a dibenzyl butyrolactone-type lignan), eudesmin (a furofuran-type lignan), isolariciresinol (a dibenzylbutane-type lignan), and cynaropicrin (a sesquiterpene lactone). The results suggested that savinin may act as an active principle in the reported biological activities of *P. santalinus*, such as anti-inflammatory effects, by mediation of the butyrolactone ring as a pharmacophore [4].

References

1 Krishnaveni, K.S. and Rao, J.V. (2000) An isoflavone from *Pterocarpus santalinus*. *Phytochemistry*, **53**, 605–606.

2 Krishnaveni, K.S. and Srinivasa Rao, J.V. (2000) A new isoflavone glucoside from *Pterocarpus santalinus*. *J. Asian Nat. Prod. Res.*, **2**, 219–223.

3 Krishnaveni, K.S. and Srinivasa Rao, J.V. (2000) A new acylated isoflavone glucoside from *Pterocarpus santalinus*. *Chem. Pharm. Bull. (Tokyo)*, **48**, 1373–1374.

4 Cho, J.Y., Park, J., Kim, P.S., Yoo, E.S., Baik, K. U., and Park, M.H. (2001) Savinin, a lignan from *Pterocarpus santalinus* inhibits tumor necrosis factor-α production and T cell proliferation. *Biol. Pharm. Bull.*, **24**, 167–171.

5 Banerjee, A. and Mukherjee, A.K. (1981) Chemical aspects of santalin as a histological stain. *Stain. Technol.*, **56**, 83–85.

6 Kameswara Rao, B., Giri, R., Kesavulu, M.M., and Apparao, C. (2001) Effect of oral administration of bark extracts of *Pterocarpus santalinus* L. on blood glucose level in experimental animals. *J. Ethnopharmacol.*, **74**, 69–74.

Pueraria lobata and *Pueraria thomsonii*

Radix Puerariae Lobatae (Gegen) is the dried root of *Pueraria lobata* (Willd.) Ohwi (Fabaceae). It is used as muscle relaxant, antipyretic and antidysenteric agent, and also for the treatment of fever; measles, acute dysentery and diarrhea, and hypertension.

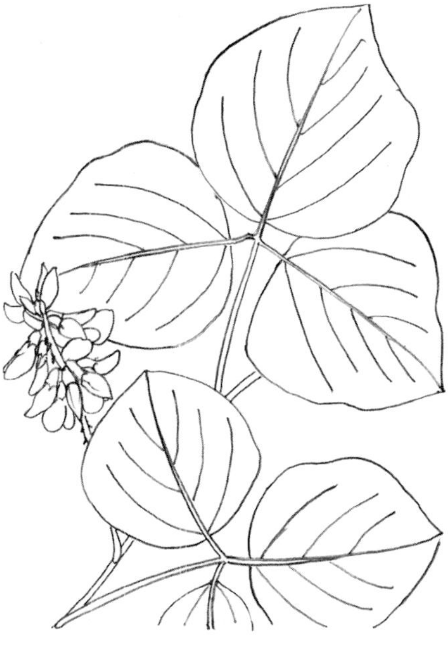

Pueraria lobata (Willd.) Ohwi

Radix Puerariae Thomsonii (Fenge) is the dried root of *Pueraria thomsonii* Benth. It is used for the same indications as Radix Puerariae Lobatae.

A galenic preparation "Yufeng Ningxin Pian," the dragees of the extract of Radix Puerariae Lobatae, is listed in the Chinese Pharmacopoeia and is used as an analgesic, spasmolytic agent for increasing cerebral and coronary blood flow; for the treatment of vertigo and headache caused by hypertension; and for coronary diseases, angina pectoris, and acute deafness.

Chemistry

The root of *P. Lobata* is known to contain isoflavones and isoflavone glycosides, with the G-glycoside puerarin as major component. Further isoflavones and isoflavone glycosides were identified as daidzein, daidzin and daidzein 7,4′-bis-*O*-β-D-glucopyranoside [1], genistein, and biochanin A [2]. The total isoflavone content in the root of *P. lobata* from various locations in China was reported to vary from 0.02% to about 5% [3]. The total flavone contents in the root of *P. lobata* were found to be higher than those in the root of *P. thomsonii* [4].

The Chinese Pharmacopoeia requires a qualitative determination of puerarin in Radix Puerariae Lobatae and in Radix Puerariae Thomsonii by thin-layer chromatographic (TLC) comparison with reference substance, and a quantitative determination by HPLC. The content of puerarin in Radix Puerariae Lobatae should be not less than 2.4%, and in Radix Puerariae Thomsonii not less than 0.3%. The Chinese Pharmacopoeia further requires a qualitative determination of puerarin in galenic preparation "Yufeng Ninxin Pian," by TLC comparison with reference substance, and a quantitative determination by HPLC. The content of puerarin in one "Yufeng Ninxin Pian" dragee should be not less than 13.0 mg.

Puerarin, as pure substance, is listed in the Chinese Pharmacopoeia Vol. II as a vasodilator. The purity of puerarin should be not less than 97%, by HPLC. Galenic preparations from puerarin listed in the Chinese Pharmacopoeia Vol. II are Puerarin Injection (50 mg per 2 ml, 100 mg per 2 ml); Puerarin and Glucose Injection (100 ml containing 0.2 g puerarin and 5.0 g glucose; 150 ml containing 0.3 g puerarin and 7.5 g glucose; 250 ml containing

Handbook of Chinese Medicinal Plants: Chemistry, Pharmacology, Toxicology
Weici Tang and Gerhard Eisenbrand

ISBN: 978-3-527-32226-8

0.3 g puerarin and 12.5 g glucose; 250 ml containing 0.5 g puerarin and 12.5 g glucose); Puerarin and Sodium Chloride Injection (100 ml containing 0.2 g puerarin and 0.9 g sodium chloride; 250 ml containing 0.4 g puerarin and 2.25 g sodium chloride).

(1 → 6)-glucosides of daidzein and genistein [5, 6]. A coumestan derivative puerarol was also isolated from the root of *P. lobata*. The isolation of 2-butenolide glycosides puerosides A and B [7] and kudzubutenolide A [5] from the root of *P. lobata* was further reported.

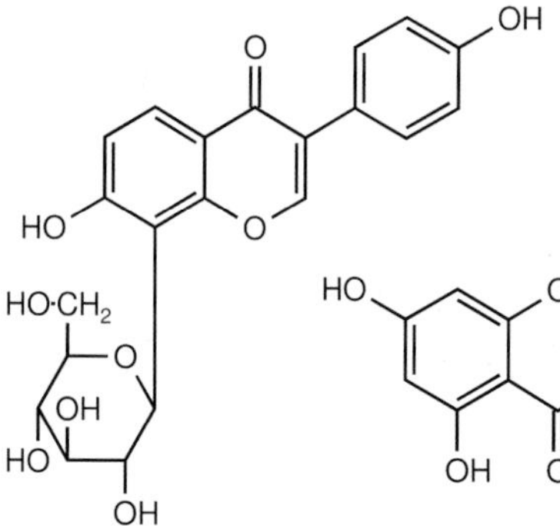

Puerarin

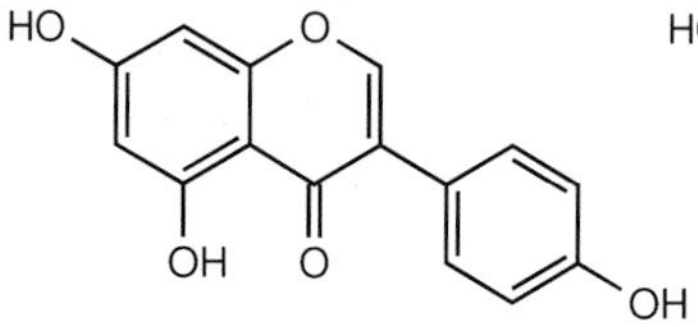

Genistein

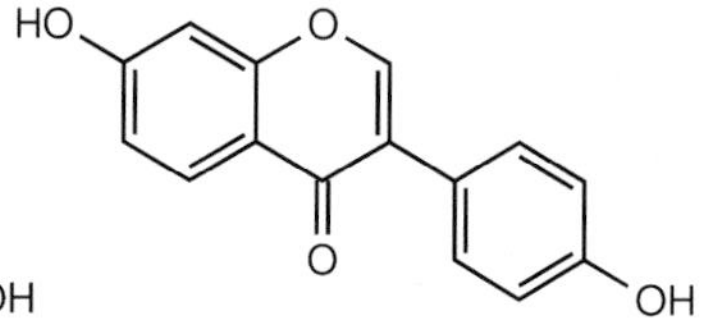

Daidzein

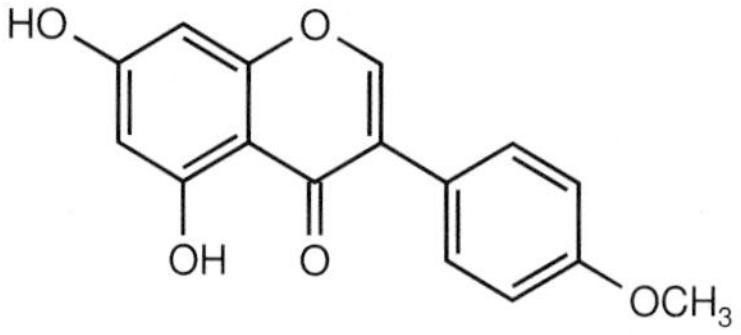

Biochanin A

Daidzein, genistein and daidzin are well-known components in soybean. Semen Sojae Preparatum (Dandouchi), the fermented seed of *Glycine max* (L.) Merr. (Fabaceae) is listed in the Chinese Pharmacopoeia. It is used for the treatment of chills, fever and headache in common cold and influenza. The seed and germinated seed of *Glycine max* are listed in the appendix of the Chinese Pharmacopoeia.

Minor isoflavones and isoflavone glycosides from the root of *P. lobata* were identified as formononetin, 3′-hydroxypuerarin, 6″-*O*-β-D-xylopyranosyl puerarin, 3′-methoxypuerarin, puerarin 4′-methylether, puerarin 4′-*O*-β-D-glucopyranoside, 3′-hydroxypuerarin 4′-*O*-β-D-glucopyranoside, 3′-methoxydaidzin, and 8-*C*-apiosyl-

Sapogenins and saponins with an oleanane skeleton from the root of *P. lobata* or *P. thomsonii* were identified as kudzusapogenols A, B, and C, sophoradiol, cantoniensistriol, soyasapogenols A and B [8], kudzusaponins A_1–A_5, B_1, C_1, SA_1–SA_4, and SB_1, acetylkaikasaponin III, soyasaponin SA_3, soyasaponin I, and acetylsoyasaponin I [9]. The root of *Pueraria* was also known to be rich in starch [10].

Pharmacology and Toxicology

The LD_{50} of puerarin by intravenous (i.v.) administration in mice was reported as 740 mg kg^{-1} [11].

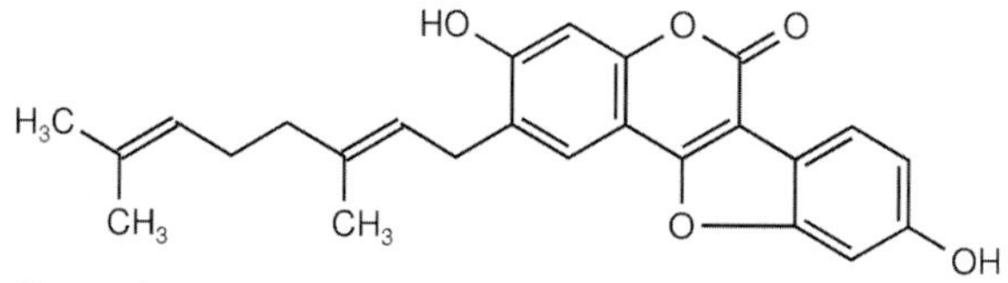

Puerarol

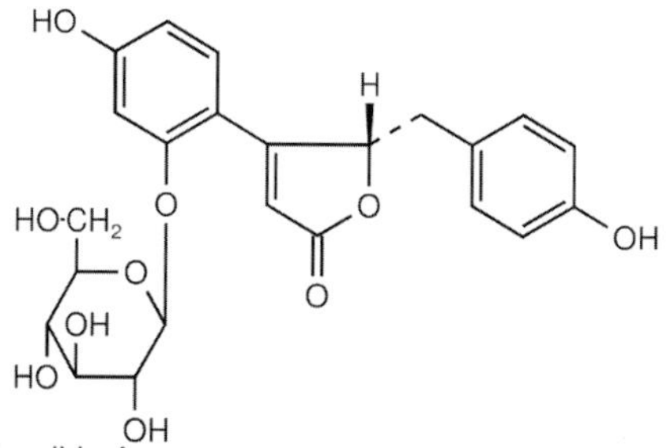

Kudzubutenolide A

The i.v. injection of puerarin was found to induce adverse reactions on multiple systems, including the immune, blood, urinary, digestive and cardiovascular systems [12]. Clinical analysis of five cases with acute intravascular hemolysis following puerarin treatment showed that all patients had a history of administering puerarin, with pre-symptoms of acute hemolysis; the clinical characteristics of acute intravascular hemolysis were observed as sudden attacks of lumbar and abdominal pain, chill, fever dyspnea, temporary consciousness loss, dark urine or hematuria, low hemoglobin, high reticular red blood cell count, and a positive Coombs' test [13].

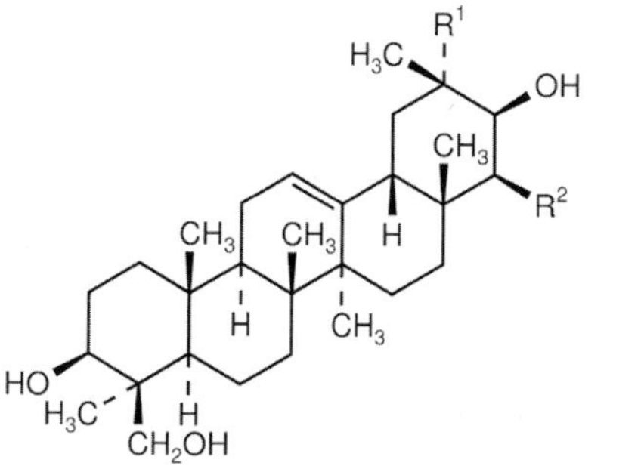

Kudzusapogenol A: $R^1 = CH_2OH$; $R^2 = OH$
Kudzusapogenol B: $R^1 = COOH$; $R^2 = OH$
Kudzusapogenol C: $R^1 = CH_3$; $R^2 = H$

The metabolism of puerarin in rats, rabbits, and dogs was reported. The plasma concentration–time course in rats and dogs after i.v. administration was best fitted to a two-compartment open model, and in rabbits to a three-compartment open model. Species specificity was found among rats, rabbits, and dogs. The elimination half-life $t_{1/2\beta}$ after i.v. administration of puerarin at a dose of $9\,mg\,kg^{-1}$ in rats, rabbits, and dogs was 11, 21, and 67 min, respectively [14]. Puerarin was widely distributed in the body, and eliminated rapidly. In rats, highest levels were found in the kidney, moderate levels in plasma, liver, and spleen, and lowest levels in the brain. Absorption of puerarin from the gastrointestinal tract following oral administration to rats was rapid, but incomplete; approximately 40% of the dose being recovered from the gastrointestinal content and feces by 24 h after administration. About 2% and 36% of the administered puerarin dose were excreted in the urine and feces within 24 h. After an i.v. administration of puerarin, about 37% of the dose was found in the urine, and 7% in feces. Puerarin was stable in the gastrointestinal tract, but appeared to be metabolized in the blood, liver, lung, and kidney [15]. After oral administration of puerarin to spontaneously hypertensive rats, the blood concentration–time profile showed a rapid initial increase, reaching a maximum and then declining within 1 h. Puerarin could not be detected in plasma after 24 h. The maximal plasma concentration of puerarin was approximately 3.5 μg ml^{-1}, and reached about 40 min after dosing [16].

After oral administration to rats, four major urinary metabolites of puerarin were identified as daidzein, daidzein 4′-*O*-sulfate, daidzein 4′,7-di-*O*-sulfate, and daidzein 7-*O*-β-D-glucuronide. The total cumulative amounts of the four metabolites excreted in the urine at 48 h following oral administration accounted for about 3.6% of the dose administered. The bile contained puerarin and two major metabolites identified as puerarin 4′-*O*-sulfate and puerarin 7-*O*-β-D-glucuronide. It was concluded that the *C*-glycoside puerarin was partially hydrolyzed to the aglycone in the body, but mainly excreted in the urine as unchanged puerarin [17].

The root of *P. lobata* was reported to have a broad spectrum of biological activities. Different *Pueraria* species showed different pharmacological activities and toxicities [18]. The total isoflavones from the root of *P. lobata* was found to show an hypotensive effect on anesthetized and unanesthetized hypertensive dogs [19]. The intraperitoneal (i.p.) administration of puerarin to spontaneously hypertensive rats decreased the blood pressure, heart rate, and plasma renin activity; however, the effects of puerarin were less marked in normal than in hypertensive rats. Puerarin, as a β-adrenoreceptor antagonist, exerted an hypotensive activity by blocking β-adrenergic receptors [20].

In dogs, the intra-arterial or i.v. injection of total isoflavones of *P. lobata* and puerarin dilated the coronary arteries to a greater extent than other arteries. Heart rate, blood pressure, and total peripheral resistance were decreased, while cardiac output remained unchanged. The i.v. administration of puerarin to dogs decreased the size of acute myocardial infarction, as demonstrated by epicardial electrocardiography, plasma creatine kinase, and radiocardiogram [21]. It was also reported that puerarin exhibited protective effects on the cardiac function after prolonged arrest and reperfusion in dogs [22]. In porcine coronary artery, puerarin potentiated endothelium-independent relaxation via the cAMP pathway [23]. Puerarin was also found to induce angiogenesis in rat myocardium with myocardial infarction, and to induce the expression of vascular endothelial growth factor (VEGF) and endothelial nitric oxide synthase (eNOS) [24]. A clinical study in patients with hypertension or angina pectoris showed that puerarin, when administered intravenously at doses of 100–200 mg, decreased blood catecholamine levels, systolic and diastolic blood pressure, heart rate and myocardial oxygen consumption [25].

A study on the protective effect of puerarin on ischemic myocardium in dogs with acute myocardial infarction (MI) revealed that puerarin might improve the opening and formation of coronary collateral circulation, and might inhibit the increase of platelet aggregation and blood viscosity during acute MI, thereby improving the microcirculation and restricting the myocardial infarct area [26]. Low doses of puerarin administered intravenously did not alter the cerebral perfusion pressure, but reduced the flow velocities of both the middle and anterior cerebral arteries. The global cerebral blood flow was enhanced due to dilatation of the intracranial arteries [27]. The oral administration of *Pueraria* isoflavones to rats also significantly antagonized the increase of water, Ca^{2+} and malondialdehyde contents in cerebral tissues. The attenuation of Ca^{2+} ATPase and superoxide dismutase activity in rat brain treated with repeated brief ischemia–reperfusion was markedly decreased [28]. Puerarin also exerted an antiarrhythmic effect in experimental studies, due partly to an inhibition of L-type calcium channels [29].

Puerarin as a β-receptor blocker was also found to show a depressive effect on intraocular pressure; in animal experiments, this was optimal when puerarin was applied in 1% solution. Clinical treatment of glaucoma patients (80 cases, 143 eyes) demonstrated a good intraocular pressure-depressive effect in primary open- or closed-angle glaucoma and secondary glaucoma. For glaucoma types not satisfactorily controlled by conventional drugs such as timolol and pilocarpine, or by operation, the addition of puerarin also yielded favorably results. The total effective rate was about 70% [30].

Interestingly, the root of *P. lobata* was reported to be effective against alcoholism or alcohol abuse, with isoflavones as the active principles [31, 32]. The extract of *P. lobata* has long been used in China to treat alcohol intoxication [33]. In an experimental study in rats, daidzin was found to decrease blood alcohol levels and sleep time induced by alcohol ingestion. However, daidzin, daidzein and puerarin were all found to be effective in suppressing voluntary alcohol consumption by the alcohol-preferring rats. The effects of these isoflavone compounds on alcohol and water intake were reversible, and appeared to be mediated via the central nervous system [34]. Puerarin and daidzein were found to inhibit [^{3}H]flunitrazepam binding to rat brain membranes *in vitro* [35]. *Pueraria* isoflavones also significantly antagonized the amnesia induced by ethanol [36]. In addition, daidzin was a potent, selective nhibitor of human mitochondrial aldehyde dehydrogenase; this might partly contribute to its effect against alcoholism [37].

Daidzein and genistein are two well-known phytoestrogens that exert a broad spectrum of pharmacological activities, including immunostimulatory effects, chemopreventive effects [38, 39], and beneficial effects for cardiovascular and cerebrovascular systems [40, 41]. Daidzein and genistein are abundant in soy products, and have been detected in human serum [42] and urine [43]. The daily dietary intake of daidzein and genistein for Japanese middle-aged women was estimated at 16 and 30 mg per person (64 and 111 μ*M*), mainly attributable to vegetable and soy product consumption. The median plasma concentrations of daidzein and genistein were 72 and 206 nmol l^{-1}. The median daily urinary excretion was about 21 μ*M* for daidzein, and 11 μ*M* for genistein [44].

The saponins from the root of *P. lobata* were found to prevent *in vitro* damage of rat primary hepatocytes induced by immunological challenge using an antiserum against rat liver plasma membranes, but with different activities. Structure–activity relationships for the sapogenol moiety suggested that the OH group at C-29 reduces the hepatoprotective activity, while the OH group at C-21 enhances it. Furthermore, structure–activity relationships for the sugar moiety suggested that the oxygen-bearing group at C-5″ contributes to an enhanced hepatoprotective activity. The configuration of the hydroxy group at C-3″ appeared of lesser importance [45]. The saponin fraction of the flowers of *P. thomsonii* and *P. lobata* also exhibited hepatoprotective activity. The saponins were identified as soyasaponin I, kaikasaponin III and kaikasaponin I [46].

References

1 Cao, X., Tian, Y., Zhang, T., Li, X., and Ito, Y. (1999) Separation and purification of isoflavones from *Pueraria lobata* by high-speed counter-current chromatography. *J. Chromatogr.*, **855A**, 709–713.
2 Wang, C.Y., Huang, H.Y., Kuo, K.L., and Hsieh, Y.Z. (1998) Analysis of Puerariae radix and its medicinal preparations by capillary electrophoresis. *J. Chromatogr.*, **802A**, 225–231.
3 Li, S.S., Liu, X., Deng, J.Z., and Zhao, S.X. (1997) Determination of flavonoid in 3 wild species of Pueraria from West-Hubei by UV-spectrophotometry. *Zhiwu Ziyuan Yu Huanjing*, **6**, 61–62.
4 Zhang, L., Zhu, R., Pan, Y., and Ma, G. (1995) Determination of puerarin in various kinds of radix Puerariae from different places. *China J. Chin. Mater. Med.*, **20**, 399–400.
5 Hirakura, K., Morita, M., Nakajima, K., Sugama, K., Takagi, K., Nitsu, K., Ikeya, Y., Maruno, M., and Okada, M. (1997) Phenolic glucosides from the root of *Pueraria lobata*. *Phytochemistry*, **46**, 921–928.
6 Rong, H.J., Stevens, J.F., Deinzer, M.L., De Cooman, L., and De Keukeleire, D. (1998) Identification of isoflavones in the roots of *Pueraria lobata*. *Planta Med.*, **64**, 620–627.
7 Nohara, T., Kinjo, J., Furusawa, J., Sakai, Y., Inoue, M., Shirataki, Y., Ishibashi, Y., Yokoe, I., and Komatsu, M. (1993) But-2-enolides from *Pueraria lobata* and revised structures of puerosides A, B and sophoroside A. *Phytochemistry*, **33**, 1207–1210.
8 Kinjo, J., Miyamoto, I., Murakami, K., Kida, K., Tomimatsu, T., Yamasaki, M., and Nohara, T. (1985) Oleanene-sapogenols from puerariae radix. *Chem. Pharm. Bull. (Tokyo)*, **33**, 1293–1296.
9 Arao, T., Kinjo, J., Nohara, T., and Isobe, R. (1997) Oleanene-type triterpene glycosides from puerariae radix. IV. Six new saponins from *Pueraria lobata*. *Chem. Pharm. Bull. (Tokyo)*, **45**, 362–366.
10 Du, X.F., Xu, S.Y., and Wang, Z. (1998) Technology of *Pueraria lobata* (Willd.) Ohwi starch production. *Zhongguo Liangyou Xuebao*, **13**, 28–32.
11 Lu, X.R., Gao, E., Xu, L.Z., Li, H.Z., Kang, B., Chen, W.N., Chen, S.M., and Chai, X.S. (1987) Puerarin β-adrenergic receptor blocking effect. *Chin. Med. J. (Engl.)*, **100**, 25–28.
12 Liu, S.D., and Mo, H.P. (2005) Occurrence and prevention of adverse reaction induced by puerarin injection. *Zhongguo Zhong Yao Za Zhi*, **25**, 852–855.
13 Wang, B., Zhu, Y.Z., Li, X.Q., Tian, Y.N., Xiong, H., and Chen, X.Y. (2004) Clinical analysis of 5 cases with acute intravascular hemolysis caused by puerarin. *Beijing Da Xue Xue Bao*, **36**, 45–46.
14 Jin, X.L. and Zhu, X.Y. (1992) Pharmaco–kinetics of puerarin in rats, rabbits, and dogs. *Acta Pharmacol. Sin.*, **13**, 284–288.
15 Zhu, X.Y., Su, G.Y., Li, Z.H., Yue, T.L., Yan, X. Z., and Wei, H.L. (1979) Metabolic fate of the effective components of puerariae. III. Metabolism of puerarin. *Acta Pharm. Sin.*, **14**, 349–355.
16 Prasain, J.K., Peng, N., Acosta, E., Moore, R., Arabshahi, A., Meezan, E., Barnes, S., and Wyss, J.M. (2007) Pharmacokinetic study of puerarin in rat serum by liquid chromatography tandem mass spectrometry. *Biomed. Chromatogr.*, **21**, 410–414.
17 Yasuda, T., Kano, Y., Saito, K., and Ohsawa, K. (1995) Urinary and biliary metabolites of puerarin in rats. *Biol. Pharm. Bull.*, **18**, 300–303.
18 Zhou, Y., Su, X., Cheng, B., Jiang, J., and Chen, H. (1995) Comparative study on pharmacological effects of various species of *Pueraria*. *China J. Chin. Mater. Med.*, **20**, 619–621, 640.
19 Yue, H.W. and Hu, X.Q. (1996) Pharmacologic value of radix Puerariae and puerarine on cardiovascular system. *Chin. J. Integr. Trad. West. Med.*, **16**, 382–384.
20 Wang, L.Y., Zhao, A.P., and Chai, X.S. (1994) Effects of puerarin on cat vascular smooth muscle *in vitro*. *Acta Pharmacol. Sin.*, **15**, 180–182.
21 Fan, L.L., O'Keefe, D.D., and Powell, W.J. Jr, (1985) Pharmacologic studies on radix puerariae. Effect of puerarin on regional myocardial blood flow and cardiac hemodynamics in dogs with acute myocardial ischemia. *Chin. Med. J. (Engl.)*, **98**, 821–832.
22 Fan, L.L., Sun, L.H., Li, J., Yue, X.H., Yu, H.X., Wang, S.Y., and Dong, S.Q. (1992) Protective effect of puerarin against myocardial reperfusion injury. Myocardial metabolism and ultrastructure. *Chin. Med. J. (Engl.)*, **105**, 451–456.
23 Yeung, D.K., Leung, S.W., Xu, Y.C., Vanhoutte, P.M., and Man, R.Y. (2006) Puerarin, an isoflavonoid derived from Radix puerariae, potentiates endothelium-independent

relaxation via the cyclic AMP pathway in porcine coronary artery. *Eur. J. Pharmcol.*, **552**, 105–111.

24 Zhang, S., Chen, S., Shen, Y., Yang, D., Liu, X., Sun-Chi, A.C., and Xu, H. (2006) Puerarin induces angiogenesis in myocardium of rat with myocardial infarction. *Biol. Pharm. Bull.*, **29**, 945–950.

25 Yang, G., Zhang, L., and Fan, L. (1990) Anti-angina effect of puerarin and its effect on plasma thromboxane A2 and prostacyclin. *Chin. J. Integr. Trad. West. Med.*, **10**, 82–84.

26 Liu, Q., Liu, Z., and Wang, L. (2000) Restrictive effect of puerarin on myocardial infarct area in dogs and its possible mechanism. *J. Tongji Med. Univ.*, **20**, 43–45.

27 Chen, L., Chai, Q., Zhao, A., and Chai, X. (1995) Effect of puerarin on cerebral blood flow in dogs. *China J. Chin. Mater. Med.*, **20**, 560–562.

28 Yu, Z.L., Zhang, G.Q., and Zhao, H.Q. (1997) Protection of *Pueraria* isoflavone against cerebral ischemia. *J. China Pharm. Univ.*, **28**, 310–312.

29 Guo, X.G., Chen, J.Z., Zhang, X., and Xia, Q. (2004) Effect of puerarin on L-type calcium channel in isolated rat ventricular myocytes. *Zhongguo Zhong Yao Za Zhi*, **29**, 248–251.

30 Kang, R.X. (1993) The intraocular pressure depressive effect of puerarin. *Chin. J. Ophthalmol.*, **29**, 336–339.

31 Lukas, S.E., Penetar, D., Berko, J., Vicens, L., Palmer, C., Mallya, G., Macklin, E.A., and Lee, D.Y. (2005) An extract of the Chinese herbal root kudzu reduces alcohol drinking by heavy drinkers in a naturalistic setting. *Alcohol Clin. Exp. Res.*, **29** (5), 756–762

32 Keung, W.M., and Vallee, B.L. (1998) Kudzu root: an ancient Chinese source of modern antidipsotropic agents. *Phytochemistry*, **47**, 499–506.

33 Xie, C.I., Lin, R.C., Antony, V., Lumeng, L., Li, T.K., Mai, K., Liu, C., Wang, Q.D., Zhao, Z. H., and Wang, G.F. (1994) Daidzin, an antioxidant isoflavonoid, decreases blood alcohol levels and shortens sleep time induced by ethanol intoxication. *Alcohol Clin. Exp. Res.*, **18**, 1443–1447.

34 Perfumi, M., Massi, M., Guerra, M.C., and Speroni, E. (1998) Effect of *Pueraria lobata* (Wild) on ethanol intake of alcohol-preferring rats. *Phytother. Res.*, **12** (Suppl. 1), S35–S38

35 Shen, X.L., Witt, M.R., Nielsen, M., and Sterner, O. (1996) Inhibition of [^{3}H]flunitrazepam binding to rat brain membranes *in vitro* by puerarin and daidzein. *Acta Pharm. Sin.*, **31**, 59–62.

36 Yu, Z.L., Zhang, G.Q., Zhao, H.Q., and Lu, J.F. (1997) Effect of Puerariae isoflavone on memory in mice. *J. China Pharm. Univ.*, **28**, 350–353.

37 Keung, W.M., Klyosov, A.A., and Vallee, B.L. (1997) Daidzin inhibits mitochondrial aldehyde dehydrogenase and suppresses ethanol intake of Syrian golden hamsters. *Proc. Natl Acad. Sci. USA*, **94**, 1675–1679.

38 Manjanatha, M.G., Shelton, S., Bishop, M.E., Lyn-Cook, L.E., and Aidoo, A. (2006) Dietary effects of soy isoflavones daidzein and genistein on 7,12-dimethylbenz[*a*]anthracene-induced mammary mutagenesis and carcinogenesis in ovariectomized Big Blue transgenic rats. *Carcinogenesis*, **27**, 2555–2564.

39 Rowell, C., Carpenter, D.M., and Lamartiniere, C.A. (2005) Chemoprevention of breast cancer, proteomic discovery of genistein action in the rat mammary gland. *J. Nutr.*, **135** (12 Suppl.), 2953S–2959S.

40 Cho, T.M., Peng, N., Clark, J.T., Novak, L., Roysommuti, S., Prasain, J., and Wyss, J.M. (2007) Genistein attenuates the hypertensive effects of dietary NaCl in hypertensive male rats. *Endocrinology*, **148**, 5396–5402.

41 Teede, H.J., Giannopoulos, D., Dalais, F.S., Hodgson, J., and McGrath, B.P. (2006) Randomised, controlled, cross-over trial of soy protein with isoflavones on blood pressure and arterial function in hypertensive subjects. *J. Am. Coll. Nutr.*, **25**, 533–540.

42 Adlercruetz, H. (1990) Western diet and Western diseases: some hormonal and biochemical mechanisms and associations. *Scand. J. Clin. Invest.*, **201** (Suppl. 50), 3–23.

43 Lampe, J.W., Gustafson, D.R., Hutchins, A. M., Martini, M.C., Li, S., Wahala, K., Grandits, G.A., Potter, J.D., and Slavin, J.L. (1999) Urinary isoflavonoid and lignan excretion on a Western diet: relation to soy, vegetable, and fruit intake. *Cancer Epidemiol. Biomarkers Prev.*, **8**, 699–707.

44 Arai, Y., Uehara, M., Sato, Y., Kimira, M., Eboshida, A., Adlercreutz, H., and Watanabe, S. (2000) Comparison of isoflavones among dietary intake, plasma concentration and urinary excretion for accurate estimation of phytoestrogen intake. *J. Epidemiol.*, **10**, 127–135.

45 Arao, T., Udayama, M., Kinjo, J., and Nohara, T. (1998) Preventive effects of saponins from the *Pueraria lobata* root on *in vitro* immunological liver injury of rat primary hepatocyte cultures. *Planta Med.*, **64**, 413–416.

46 Kinjo, J., Aoki, K., Okawa, M., Shii, Y., Hirakawa, T., Nohara, T., Nakajima, Y., Yamazaki, T., Hosono, T., Someya, M., Niiho, Y., and Kurashige, T. (1999) HPLC profile analysis of hepatoprotective oleanene-glucuronides in Puerariae Flos. *Chem. Pharm. Bull. (Tokyo)*, **47**, 708–710.

Pulsatilla chinenesis

Radix Pulsatillae (Baitouweng) is the dried root of *Pulsatilla chinensis* (Bge.) Regel (Ranunculaceae). It is used as an antidiarrheic agent for the treatment of amoebic dysentery.

Pulsatilla chinenesis (Bge.) Regel

Chemistry

The root of *P. chinensis* is known to contain lactone and triterpene compounds, with anemonin as the major component. Anemonin is a dimer of protoanemonin (5-methylene-2-furanone) [1], which is unstable and present as its glycoside ranunculin [2].

Triterpenes and triterpene glycosides from the root of *P. chinensis* were found to be structurally mainly derived from lupane, and identified as pulsatillic acid, 3-*O*-α-L-arabinopyranosyl-23-hydroxybetulinic acid, pulchinenosides A (anemoside A_3) [3], B, C (anemoside B_4) [4], and pulsatillosides A, B [5] and C [6].

Pharmacology and Toxicology

The LD_{50} of protoanemonin in male mice was reported as 190 mg kg^{-1} [7]. Anemonin was reported to be an antipyretic agent. Both, anemonin and protoanemonin exerted sedative effects [8]. Protoanemonin is known to be a skin-irritant and blistering agent [9]. It can be used as an antirheumatic and antineuralgic agent by external application [10]. Protoanemonin was found to show antifungal activity, inhibiting selected strains of dermatophytes and yeasts; the minimum inhibitory concentrations ranged from 0.2 to 0.7 m*M*. The most sensitive dermatophyte was *Epidermophyton floccosum*, and the most sensitive yeast *Rhodotorula glutinis*. The effects of protoanemonin were reversed by cysteine. RNA inhibition seemed to be the first target of protoanemonin [11]. The dermatophytes *Microsporum cookei* [12], *E. floccosum* and *Trichophyton mentagrophytes* [13], when incubated with protoanemonin at a concentration of about 0.1 m*M* for 48 h, underwent a series of ultrastructural alterations. Antibacterial activity of protoanemonine

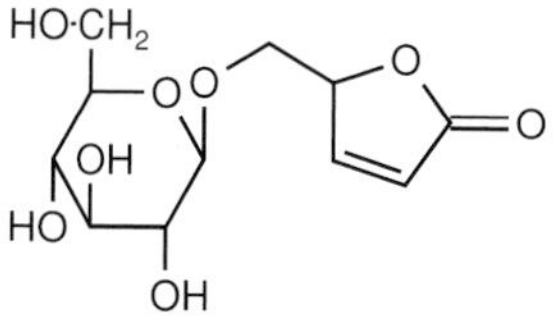

Ranunculin

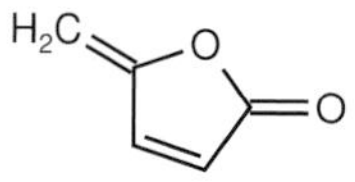

Protoanemonin

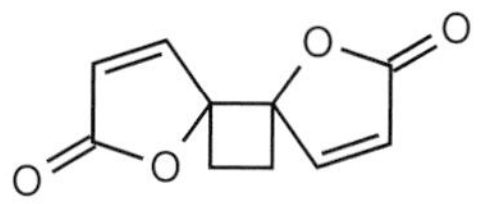

Anemonin

Handbook of Chinese Medicinal Plants: Chemistry, Pharmacology, Toxicology
Weici Tang and Gerhard Eisenbrand

ISBN: 978-3-527-32226-8

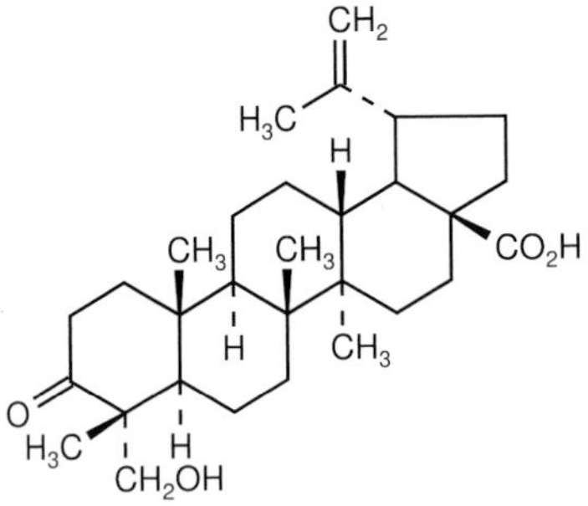

Pulsatillic acid

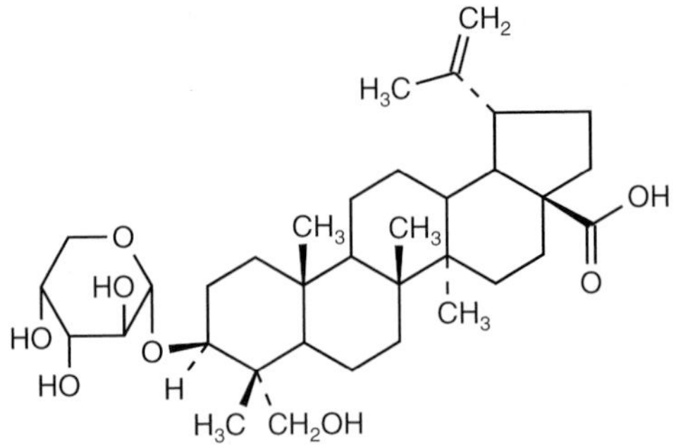

Pulsatilloside A

was also reported [14]. Protoanemonine has also been found to be generated during oxidative, DNA damage by elimination from the sugar phosphate backbone through intermediate formation of the C1′-oxidized abasic DNA lesion, 2-deoxy nibonolactone [15, 16]

Anemonin was reported to exert anti-inflammatory activity. In primary cultures of rat intestinal microvascular endothelial cells, anemonin at a concentration of 5 $\mu g\,ml^{-1}$, significantly inhibited the production of nitric oxide (NO) and endothelin-1 induced by lipopolysaccharide (LPS). At a concentration of 10 $\mu g\,ml^{-1}$, anemonin downregulated the expression of soluble intercellular adhesion molecule-1 (ICAM-1) in cells induced by LPS. It was suggested that anemonin might exert a therapeutic effect in intestinal inflammation, at least in part, by inhibiting the production of NO, endothelin-1 and ICAM-1 in rat intestinal microvascular endothelial cells, thus preventing intestinal microvascular dysfunction [17].

Protoanemonin was identified as an antimutagenic agent in *E. coli* B/rWP2trp against the mutagenicity induced by UV-light and *N*-methyl-*N*′-nitro-*N*-nitrosoguanidine [18]. Ranunculin was shown to be antimutagenic against mitomycin C or methyl methane sulfonate in *Salmonella typhimurium* TA100 and TA102. In mice, it decreased micronucleus formation in polychromatic erythrocytes induced by mitomycin C. The inhibition by ranunculin of the incorporation of [^{3}H]TdR into DNA disappeared after incubation with rat liver microsomes and cytoplasm. Ranunculin was found to be metabolized by rat liver microsomes *in vitro* [19].

Ranunculin was further found to inhibit tumor cell growth. The IC_{50}-values of ranunculin against KB and Bel7402 cells in the colony test were found to be 0.21 μM and 0.35 μM, respectively. Ranunculin inhibited the incorporation of [^{3}H]-labeled precursors into DNA and RNA of L1210 cells. At a concentration of 15 μM, ranunculin markedly decreased DNA synthesis catalyzed by DNA polymerase I. No direct reaction between ranunculin and the DNA template was observed, and no effect of ranunculin was found on DNA topoisomerase II or RNA polymerase. It was suggested that the cytotoxicity of ranunculin *in vitro* might be due to an inhibition of DNA polymerase and an increase of oxygen free radicals [20]. Pulsatillic acid was also reported to exhibit cytotoxic activities against P-388, Lewis lung carcinoma and human large-cell lung carcinoma cells [4].

References

1 Zhu, Z.Y., Yang, Y., Tao, Y., Mo, D.R., Wu, Y.H., and Wu, G.Q. (1987) Extraction, synthesis and preservation of protoanemonin. *Zhongguo Kexue Jishu Daxue Xuebao*, **17**, 336–342.

2 Zhang, X.Q., Liu, A.R., and Xu, L.X. (1990) Determination of ranunculin in *Pulsatilla chinensis* and synthetic ranunculin by reversed phase HPLC. *Acta Pharm. Sin.*, **25**, 932–935.

3 Ye, W.C., Zhao, S.X., Cai, H., and Liu, J.H. (1991) Studies on the chemical constituents of *Pulsatilla chinensis*. II. *J. China Pharm. Univ.*, **21**, 264–266.

4 Wu, Z.J., Ding, L.S., and Zhao, S.X. (1991) Glycosides from *Pulsatilla chinensis* (Bunge) Reg. *J. China Pharm. Univ.*, **22**, 265–269.

5 Ye, W.C., Ji, N.N., Zhao, S.X., Liu, J.H., Ye, T., McKervey, M.A., and Stevenson, P. (1996) Triterpenoids from *Pulsatilla chinensis*. *Phytochemistry*, **42**, 799–802.

6 Ye, W.C., He, A., Zhao, S., and Che, C.T. (1998) Pulsatilloside C from the Roots of *Pulsatilla chinensis*. *J. Nat. Prod.*, **61**, 658–659.

7 Martin, M.L., San Roman, L., and Dominguez, A. (1990) *In vitro* activity of protoanemonin, an antifungal agent. *Planta Med.*, **56**, 66–69.

8 Martin, M.L., Ortiz de Urbina, A.V., Montero, M.J., Carron, R., and San Roman, L. (1988) Pharmacologic effects of lactones isolated from *Pulsatilla alpina* subsp. *apiifolia*. *J. Ethnopharmacol.*, **24**, 185–191.

9 Kern, J.R. and Cardellina, J.H. (1983) Native American medicinal plants. Anemonin from the horse stimulant *Clematis hirsutissima*. *J. Ethnopharmacol.*, **8**, 121–123.

10 Cappelletti, E.M., Trevisan, R., and Caniato, R. (1982) External antirheumatic and antineuralgic herbal remedies in the traditional medicine of north-eastern Italy. *J. Ethnopharmacol.*, **6**, 161–190.

11 Mares, D. (1987) Antimicrobial activity of protoanemonin, a lactone from ranunculaceous plants. *Mycopathologia*, **98**, 133–140.

12 Mares, D. (1989) Electron microscopy of *Microsporum cookei* after *in vitro* treatment with protoanemonin: a combined SEM and TEM study. *Mycopathologia*, **108**, 37–46.

13 Mares, D. and Fasulo, M.P. (1990) Ultrastructural alterations in *Epidermophyton floccosum* and *Trichophyton mentagrophytes* exposed *in vitro* to protoanemonin. *Cytobios*, **61**, 89–95.

14 Didry, N., Dubreuil, L., and Pinkas, M. (1991) Antibacterial activity of vapors of protoanemonine. *Pharmazie*, **46**, 546–547.

15 Roginskaya, M., Razskazowskiy, Y., and Bernhard, W.A. (2005) 2-Deoxyribonolactone lesions in X-ray-irradiated DNA: quantitative determination by catalytic 5-methylene-2-furanone release. *Angew. Chem. Int. Ed.*, **44**, 6210–6213.

16 Bose, R.N., Fonkeng, B.S., Moghaddas, S., and Stroup, D. (1998) Mechanisms of DNA damage by chromium(V) carcinogens. *Nucleic Acids Res.*, **26**, 1588–1596.

17 Duan, H., Zhang, Y., Xu, J., Qiao, J., Suo, Z., Hu, G., and Mu, X. (2006) Effect of anemonin on NO, ET-1 and ICAM-1 production in rat intestinal microvascular endothelial cells. *J. Ethnopharmacol.*, **104**, 362–366.

18 Minakata, H., Komura, H., Nakanishi, K., and Kada, T. (1983) Protoanemonin, an antimutagen isolated from plants. *Mutat. Res.*, **116**, 317–322.

19 Li, R.Z., Pei, H.P., and Ji, X.J. (1993) Antimutagenic activity and metabolic transformation of ranunculin by rat liver microsomes. *Acta Pharm. Sin.*, **28**, 481–485.

20 Li, R.Z. and Ji, X.J. (1993) The cytotoxicity and action mechanism of ranunculin *in vitro*. *Acta Pharm. Sinica*, **28**, 326–331.

Punica granatum

Pericarpium Granati (Shiliupi) is the dried pericarp of *Punica granatum* L. (Punicaceae) which collected in the fall when the fruit has ripened. It is used as an adstringent, hemostatic, antidiarrheic and antiparasitic agent for the treatment of chronic dysentery, hematochezia, abnormal uterine bleeding, and intestinal parasitosis.

The dried fruit and seed of *Punica granatum* is included in the appendix of the Chinese Pharmacopoeia.

Punica granatum L.

Chemistry

The pericarp of *P. granatum*, pomegranate, is known to contain tannin as major component. The Chinese Pharmacopoeia requires a quantitative determination of tannin in Pericarpium Granati. The content of tannin in Pericarpium Granati should be not less than 10.0%. The pomegranate fruit was reported to contain anthocyanidins, with delphinidin, cyanidin, and pelargonidin as major components [1].

The stem bark and the root bark of *P. granatum* are known to contain simple alkaloids, mostly derived from pyridine such as pelletierine (punicine), methylisopelletierine, and pseudopelletierine. The content of total alkaloids was about 1% [2].

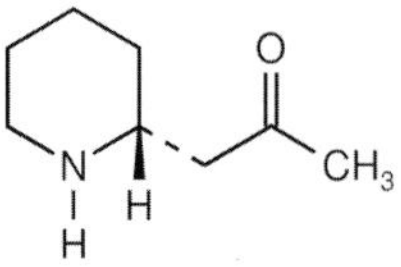

Pelletierine

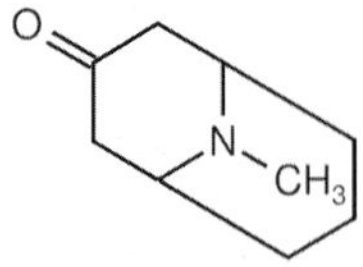

Pseudopelletierine

The root bark was also reported to contain a number of hydrolyzable tannins such as punicalin, 2-*O*-galloylpunicalin and punicalagin. Punicalin is structurally 4,6-(*S*,*S*)-gallagyl-D-glucose, whereas punicalagin 2,3-(*S*)-hexahydrodiphenoyl-4,6-(*S*,*S*)-gallagyl-D-glucose [3].

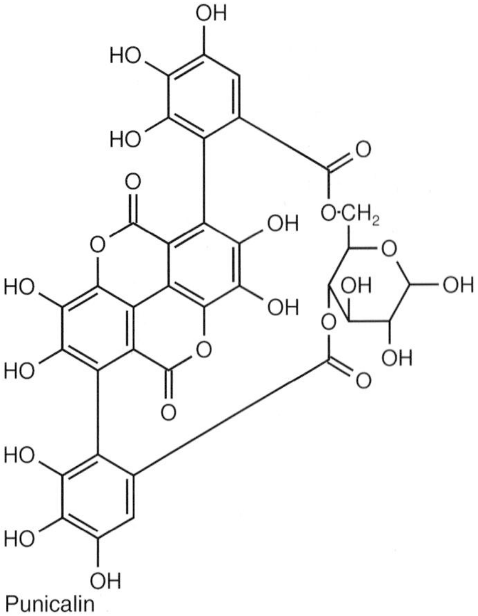

Punicalin

Handbook of Chinese Medicinal Plants: Chemistry, Pharmacology, Toxicology
Weici Tang and Gerhard Eisenbrand

ISBN: 978-3-527-32226-8

Pharmacology and Toxicology

A gastroprotective effect of the aqueous extract of the pericarp of *P. granatum* was demonstrated in the rat against ethanol-induced damage. Oral administration of the extract induced a significant decrease in gastric lesions. The observed protection was more pronounced when the test solution was given at the same time with ethanol. The acid content of the stomach was also significantly increased by *P. granatum* extract. It was suggested that monomeric and polymeric polyphenols could strengthen the gastric mucosal barrier [4].

The tannin from the pericarp of *P. granatum* was found to inhibit genital herpes virus *in vitro*. The pericarp of *P. granatum* was reported to be effective against genital herpes simplex virus (HSV-2). The active component, tannin, not only inhibited HSV-2 replication, but also showed stronger effects of killing virus and blocking its absorption to cells [5]. In addition, the pericarp, when given orally to rabbits at 100 mg kg^{-1} (as aqueous suspension) was found to stimulate the cell-mediated and humoral components of the immune system. The pericarp elicited an increase in antibody titer to typhoid-H antigen, and enhanced the inhibition of leucocyte migration [6]. The aqueous extract from the pericarp of *P. granatum* was also found to exhibit antiviral activity against human HSV type 1 [7].

The root of *P. granatum* is known to be an amebicidal agent. The aqueous extract and tannin were effective in inhibiting the growth of *Entamoeba histolytica* and *E. invadens* in cultures. The aqueous extract had a higher activity on *E. histolytica* than on *E. invadens*, producing growth inhibitions of about 100 and 40%, respectively. Alkaloid at 1 mg ml^{-1} concentration had no amebicidal activity; however, tannins at concentration of 10 μg ml^{-1} for *E. histolytica* and 100 μg ml^{-1} for *E. invadens* were sufficient to produce about 100% growth inhibition [8]. An aqueous extract of *P. granatum* was also reported to exert antibacterial [9] and antiviral [10] effects.

A 70% acetone extract of pomegranate and its three major anthocyanidins, delphinidin, cyanidin and pelargonidin, was reported to exert antioxidative activity and to scavenge hydroxyl and superoxide radicals. Inhibitory effects on lipid peroxidation were confirmed by the levels of malonaldehyde and 4-hydroxyalkenals in rat brain homogenates. The antioxidative effect of anthocyanidins might be caused by their chelating effect with ferrous ion. The IC_{50}-values of delphinidin, cyanidin, and pelargonidin were given as 2.4, 22, and 456 μ*M*, respectively. In contrast, anthocyanidins did not effectively scavenge nitric oxide (NO). Anthocyanidins also inhibited H_2O_2-induced lipid peroxidation in rat brain homogenates, with IC_{50}-values for delphinidin, cyanidin, and pelargonidin of 0.7, 3.5, and 85 μ*M*, respectively. This suggested that the anthocyanidins contribute to the antioxidant activity of pomegranate fruits [11]. Pericarpium Granati was also reported to possess anticancer activities, to inhibit tumor cell proliferation, to induce cell cycle arrest, and to suppress tumor cell invasion and angiogenesis. All of these effects seemed to be associated with the anti-inflammatory potency [12].

References

1 Noda, Y., Kaneyuki, T., Mori, A., and Packer, L. (2001) Antioxidant activities of pomegranate fruit extract and its anthocyanidins: delphinidin, cyanidin, and pelargonidin. *J. Agric. Food Chem.*, **50**, 166–171.

2 Ferrara, L., Schettino, O., Forgione, P., Rullo, V., and Di Gennaro, S. (1989)

Identification of the root of *Punica granatum* in galenic preparations using TLC. *Boll. Soc. Ital. Biol. Sper.*, **65**, 385–390.
3 Tanaka, T., Nonaka, G., and Nishioka, I. (1986) Tannins and related compounds. XL. Revision of the structures of punicalin and punicalagin, and isolation and characterization of 2-galloylpunicalin from the bark of *Punica granatum*. *Chem. Pharm. Bull. (Tokyo)*, **34**, 650–655.
4 Khennouf, S., Gharzouli, K., Amira, S., and Gharzouli, A. (1999) Effects of *Quercus ilex* L. and *Punica granatum* L. polyphenols against ethanol-induced gastric damage in rats. *Pharmazie*, **54**, 75–76.
5 Zhang, J., Zhan, B., Yao, X., Gao, Y., and Shong, J. (1995) Antiviral activity of tannin from the pericarp of *Punica granatum* L. against genital Herpes virus *in vitro*. *China J. Chin. Mater. Med.*, **20**, 556–558. 576.
6 Gracious Ross, R., Selvasubramanian, S., and Jayasundar, S. (2001) Immunomodulatory activity of *Punica granatum* in rabbits: a preliminary study. *J. Ethnopharmacol.*, **78**, 85–87.
7 Li, Y., Ooi, L.S., Wang, H., But, P.P., and Ooi, V. E. (2004) Antiviral activities of medicinal herbs traditionally used in southern mainland China. *Phytother. Res.*, **18**, 718–722.
8 Segura, J.J., Morales-Ramos, L.H., Verde-Star, J., and Guerra, D. (1990) Growth inhibition of *Entamoeba histolytica* and *E. invadens* produced by pomegranate root. *Arch. Invest. Med. (Mex)*, **21**, 235–239.
9 Prashanth, D., Asha, M.K., and Amit, A. (2001) Antibacterial activity of *Punica granatum*. *Fitoterapia*, **72**, 171–173.
10 Pendas, J., Moreira, T., Guerra, O., Pena, B.R., and Fernandez, J.A. (2001) Water relationships in *Phyllantus orbicularis* and *Punica granatum* antiviral extracts and their influence on stability after freezing and freeze-drying. *CryoLetters*, **22**, 5–12.
11 Singh, R.P., Chidambara Murthy, K.N., and Jayaprakasha, G.K. (2002) Studies on the antioxidant activity of pomegranate (*Punica granatum*) peel and seed extracts using *in vitro* models. *J. Agric. Food Chem.*, **50**, 81–86.
12 Lansky, E.P. and Newman, R.A. (2007) *Punica granatum* (pomegranate) and its potential for prevention and treatment of inflammation and cancer. *J. Ethnopharmacol.*, **109**, 177–206.

Pyrola calliantha and *Pyrola decorata*

Herba Pyrolae (Luxiancao) is the dried whole plant of *Pyrola calliantha* H. Andres or *Pyrola decorata* H. Andres (Pyrolaceae). It is used as a hemostatic and antirheumatic agent, and also for the treatment of abnormal menstrual bleeding and cough.

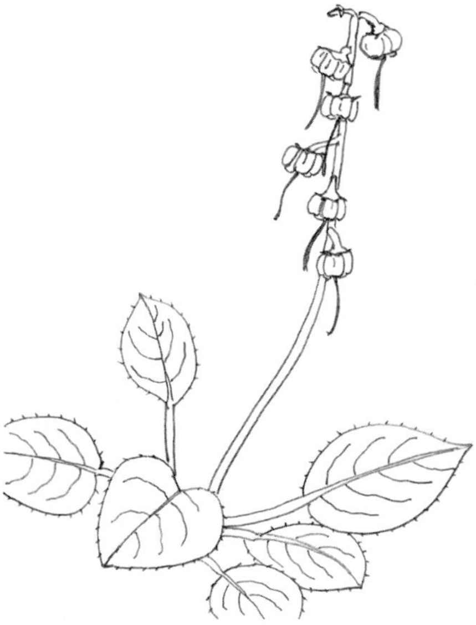

Pyrola calliantha H. Andres

Chemistry

Herba Pyrolae is known to be rich in tannins. Phenole glycosides arbutin [1], homoarbutin, 6′-*O*-galloylhomoarbutin [2]; flavone glycosides hyperin, 2″-*O*-gylloylhyperin [2, 3]; naphthalene glycoside renifolin [2]; and gallotannin [1] were isolated from the whole plant of *P. calliantha*. The content of arbutin in *P. calliantha* was reported to range from 1% to 7.5%, and the content of gallotannin from 0.2% to 1.8% [4]. In contrast, it was also reported that *P. calliantha* does not contain arbutin [5].

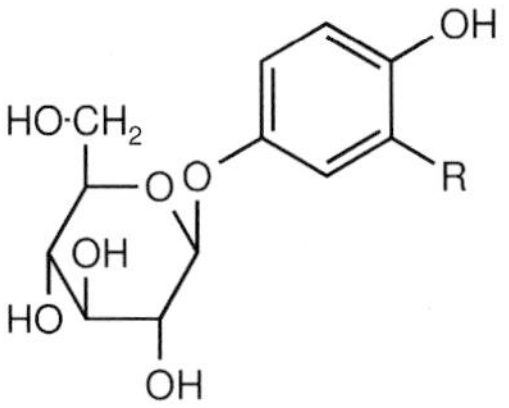

Arbutin: R = H
Homoarbutin: R = CH_3

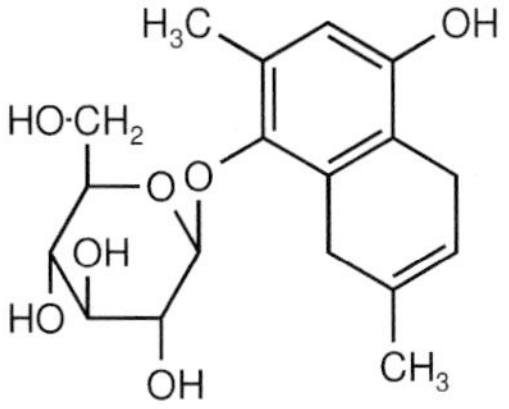

Renifolin

Pharmacology and Toxicology

Hydroquinone glucuronide and sulfate were two major metabolites in human urine after intake of arbutin. Following the administration of a single dose of a phytomedical preparation containing arbutin to 16 healthy volunteers, about 65% of the arbutin was excreted in the urine [6]. A further pharmacokinetic study with a phytomedical preparation containing arbutin, performed in three volunteers, showed that more than half of the arbutin dose was excreted within 4 h, mainly in the form of hydroquinone glucuronide and hydroquinone sulfate; in total, more than 75% of the dose was excreted within 24 h. The elimination of free hydroquinone was negligible in two out of three volunteers; the excretion of hydroquinone in the third person reached 5.6% of the total arbutin dose, indicating an individual difference [7].

Hydroquinone is classified as both mutagenic and carcinogenic. Arbutin was stable

Handbook of Chinese Medicinal Plants: Chemistry, Pharmacology, Toxicology
Weici Tang and Gerhard Eisenbrand

ISBN: 978-3-527-32226-8

in artificial gastric juice. Fecal suspensions from nine human subjects completely converted arbutin (2 m*M*) into hydroquinone. Four of nine human intestinal bacteria investigated, including *Eubacterium ramulus, Enterococcus casseliflavus, Bacteroides distasonis,* and *Bifidobacterium adolescentis,* were all found capable of hydrolyzing arbutin at different rates. In contrast, homogenates from small intestinal mucosa and cytosolic fractions from colon mucosa hydrolyzed arbutin at substantially lower rates. Arbutin, unlike hydroquinone, did not induce gene mutations in Chinese hamster V79 cells in the absence of an activating system. However, in the presence of cytosolic fractions from *E. ramulus* or *B. distasonis,* arbutin was strongly mutagenic. A cytosolic fraction from *Escherichia coli,* showing no arbutin glycosidase activity, was not able to activate arbutin. The release of the proximate mutagen hydroquinone from arbutin by intestinal bacteria in the immediate vicinity of the colon mucosa might be a genotoxic risk for humans [8].

It was also discussed, whether arbutin might be a risk factor for human leukemia, because hydroquinone as a metabolite of benzene was strongly implicated in producing leukemia associated with benzene exposure [9]. In contrast, hydroquinone has been tested for its chemopreventive activity. Male F344 rats, aged 8 weeks, were fed a diet containing either 0.05% (corresponding to 25 mg kg^{-1}) or 0.2% (100 mg kg^{-1}) hydroquinone for 13 weeks, starting one week before the administration of 2-acetylaminofluorene (AAF). AAF was given intragastrically three times weekly for 12 weeks, at a dose of 3 mg kg^{-1}. At the end of the study (13 weeks), DNA adducts, cell proliferation and preneoplastic hepatocellular altered foci were measured. The high dose of hydroquinone significantly reduced DNA adducts caused by AAF; indeed, AAF induced about a 50% increase in hepatocellular proliferation. Hydroquinone reduced the increases in preneoplastic foci per cm^2 of liver tissue by about 50%, indicating that hydroquinone in the diet inhibited AAF-induced cancer initiation in rat liver [10].

Recently, the toxicogenomics of arbutin-treated A375 human malignant melanoma cells was reported. DNA microarray analysis provided the differential gene expression pattern, showing significant changes of 324 differentially expressed genes, reflecting 88 upregulated and 236 downregulated genes. The gene ontology was classified as belonging to cellular components, molecular functions and biological processes. Four downregulated genes, *AKT1, CLECSF7, FGFR3,* and *LRP6,* correlated to suppression of the biological processes in the cell cycle during cancer progression, and to the downstream signaling pathways of melanocytic carcinogenesis [11].

Arbutin, as a hydroquinone glycoside, is present in a numerous plants, including some foods. Arbutin exhibited anti-inflammatory activity. When given at oral doses of 10 and 50 mg kg^{-1} to mice at 24 h after picryl chloride application, it speedily decreased the swelling induced by picryl chloride and sheep red cell delayed-type hypersensitivity. The combined use of arbutin with prednisolone or dexamethasone showed a stronger effect than that of prednisolone or dexamethasone alone. Prednisolone and dexamethasone decreased the weight of thymus and spleen in intact, picryl chloride and sheep red cell delayed-type hypersensitivity mice, but arbutin did not show these effects [12]. Arbutin was further found to increase the inhibitory action of indomethacin on the swelling of picryl chloride and sheep red cell delayed-type hypersensitivity, and on carrageenin-induced edema and adjuvant-induced arthritis in mice [13].

Arbutin is known to be an inhibitor of melanogenesis. It was found to be effective in the topical treatment of various cutaneous hyperpigmentations characterized by hyperactive melanocyte function. Arbutin had an inhibitory effect on tyrosinase activity, and inhibited the production of melanin from dopa by tyrosinase and from dopachrome by autoxidation in B16 melanoma cells [14]. The maximum arbutin concentration without a growth-inhibitory effect was 50 μ*M*. At this concentration, the melanin content per cell was decreased significantly to about 39%. When arbutin was added to B16 melanoma cell suspension, it was not hydrolyzed to liberate hydroquinone. It was suggested that arbutin inhibits melanogenesis by affecting not only the synthesis but also the activity of tyrosinase, rather than by killing melanocytic B16 melanoma cells. Hydroquinone, the aglycone of arbutin, was considered not to be responsible for the inhibitory effect of arbutin on the melanogenesis [15].

Arbutin also inhibited the tyrosinase activity of cultured human melanocytes at noncytotoxic concentrations. It did not affect the expression of tyrosinase mRNA. Melanin production was inhibited significantly by arbutin. Studies on the kinetics and mechanism for the inhibition of tyrosinase confirmed the reversibility of competitive inhibition. The depigmenting mechanism of arbutin in humans involved an inhibition of melanosomal tyrosinase activity, rather than a suppression of the expression and synthesis of tyrosinase [16, 17]. In contrast, it was also reported that arbutin, at concentrations in the range of 0.5–8 m*M*, increased the pigmentation of cultured melanocytes. Treatment of the cells with arbutin increased their melanin content and protein content, whereas the tyrosinase activity in the cells was reduced. Arbutin might thus promote an increase in pigmentation of cultured human melanocytes that is not mediated by augmented tyrosinase activity [18]. Arbutin inhibited the oxidation of L-tyrosine catalyzed by mushroom tyrosinase. Arbutin itself was oxidized as a monophenol substrate at an extremely slow rate; however, the oxidation was accelerated as soon as catalytic amounts (0.01 m*M*) of l-3,4-dihydroxyphenylalanine (L-DOPA) became available as a cofactor [19].

References

1 Zhang, D.K. (1987) Quantitative determination of arbutin and gallotannin in *Pyrola calliantha*. *Bull. Chin. Mater. Med.*, **12**, 45–46.
2 Chen, X.M., Li, H., Yang, L., Li, Y.F., Ren, W.X., and Thing, Z.C. (1991) Chemical constituents of *Pyrola calliantha*. *Tianran Chanwu Yanjiu Yu Kaifa*, **3**, 1–6.
3 Li, X.H., Wang, J.X., and Wang, X.Y. (1994) Determination of 2″-*O*-galloylhyperin in *Pyrola calliantha* H. Andres. *China J. Chin. Mater. Med.*, **19**, 103–104.
4 Zhang, D.K., Sha, Z.F., and Sun, W.J. (1987) Quantitative determination of arbutin and gallotannin in *Pyrola calliantha*. *Bull. Chin. Mater. Med.*, **12**, 301–302, 310.
5 Wang, J., Fu, Q., Li, X., and Liu, W. (1995) Experimental study and textual research on whether the genus of *Pyrola* contains arbutin. *China J. Chin. Mater. Med.*, **20**, 327–328.
6 Schindler, G., Patzak, U., Brinkhaus, B., von Niecieck, A., Wittig, J., Krahmer, N., Glockl, I., and Veit, M. (2002) Urinary excretion and metabolism of arbutin after oral administration of *Arctostaphylos uvae ursi* extract as film-coated tablets and aqueous solution in healthy humans. *J. Clin. Pharmacol.*, **42**, 920–927.
7 Quintus, J., Kovar, K.A., Link, P., and Hamacher, H. (2005) Urinary excretion of arbutin metabolites after oral administration of bearberry leaf extracts. *Planta Med.*, **71**, 147–152.
8 Blaut, M., Braune, A., Wunderlich, S., Sauer, P., Schneider, H., and Glatt, H. (2006) Mutagenicity of arbutin in mammalian cells after activation by human intestinal bacteria. *Food Chem. Toxicol.*, **44**, 1940–1947.
9 McDonald, T.A., Holland, N.T., Skibola, C., Duramad, P., and Smith, M.T. (2001) Hypothesis: phenol and hydroquinone derived mainly from diet and gastrointestinal flora activity are causal factors in leukemia. *Leukemia*, **15**, 10–20.

10 Williams, G.M., Iatropoulos, M.J., Jeffrey, A. M., and Duan, J.D. (2007) Inhibition by dietary hydroquinone of acetylaminofluorene induction of initiation of rat liver carcinogenesis. *Food Chem. Toxicol.*, **45**, 1620–1625.

11 Cheng, S.L., Liu, R.H., Sheu, J.N., Chen, S.T., Sinchaikui, S., and Tsay, G.J. (2007) Toxicogenomics of A375 human malignant melanoma cells treated with arbutin. *J. Biomed. Sci.*, **14**, 87–105.

12 Matsuda, H., Nakata, H., Tanaka, T., and Kubo, M. (1990) Pharmacological study on *Arctostaphylos uva-ursi* (L.) Spreng. II. Combined effects of arbutin and prednisolone or dexamethazone on immuno-inflammation. *Yakugaku Zasshi*, **110**, 68–76.

13 Matsuda, H., Tanaka, T., and Kubo, M. (1991) Pharmacological studies on leaf of *Arctostaphylos uva-ursi* (L.) Spreng. III. Combined effect of arbutin and indomethacin on immuno-inflammation. *Yakugaku Zasshi*, **111**, 253–258.

14 Matsuda, H., Nakamura, S., Shiomoto, H., Tanaka, T., and Kubo, M. (1992) Pharmacological studies on leaf of *Arctostaphylos uva-ursi* (L.) Spreng. IV. Effect of 50% methanolic extract from *Arctostaphylos uva-ursi* (L.) Spreng. (bearberry leaf) on melanin synthesis. *Yakugaku Zasshi*, **112**, 276–282.

15 Akiu, S., Suzuki, Y., Asahara, T., Fujinuma, Y., and Fukuda, M. (1991) Inhibitory effect of arbutin on melanogenesis: biochemical study using cultured B16 melanoma cells. *Nippon Hifuka Gakkai Zasshi*, **101**, 609–913.

16 Parvez, S., Kang, M., Chung, H.S., Cho, C., Hong, M.C., Shin, M.K., and Bae, H. (2006) Survey and mechanism of skin depigmenting and lightening agents. *Phytother. Res.*, **20**, 921–934.

17 Chakraborty, A.K., Funasaka, Y., Komoto, M., and Ichihashi, M. (1998) Effect of arbutin on melanogenic proteins in human melanocytes. *Pigment Cell Res.*, **11**, 206–212.

18 Nakajima, M., Shinoda, I., Fukuwatari, Y., and Hayasawa, H. (1998) Arbutin increases the pigmentation of cultured human melanocytes through mechanisms other than the induction of tyrosinase activity. *Pigment Cell Res.*, **11**, 12–17.

19 Hori, I., Nihei, K., and Kubo, I. (2004) Structural criteria for depigmenting mechanism of arbutin. *Phytother. Res.*, **18**, 475–479.

Pyrrosia lingua, *Pyrrosia petiolosa* and *Pyrrosia sheareri*

Folium Pyrrosiae (Shiwei) is the dried leaf of *Pyrrosia lingua* (Thunb.) Farwell, *Pyrrosia petiolosa* (Christ) Ching or *Pyrrosia sheareri* (Bak.) Ching (Polypodiaceae). It is mainly used as a diuretic, antiasthmatic and hemostatic agent for the treatment of urinary infection, urolithiasis, hematuria, abnormal uterine bleeding, cough, and asthma.

Pyrrosia lingua (Thunb.) Farwell

Chemistry

The leaf of the official *Pyrrosia* species was reported to contain compounds of different types, with mangiferin, isomangiferin, and chlorogenic acid as the major compounds [1]. Mangiferin and isomangiferin are xanthone C-glycosides. The Chinese Pharmacopoeia requires a quantitative determination of chlorogenic acid in Folium Pyrrosiae by HPLC. The content of chlorogenic acid should be not less than 0.3%.

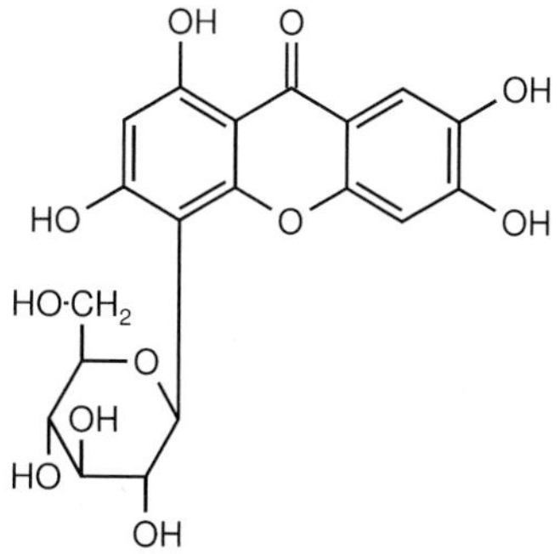

Isomangiferin

Other components from *P. lingua* were identified as astragalin, liquiritin, quercetin, isoquercetin, kaempferol, trifolin, and β-sitosterol [3]. The major common compounds detected in the essential oil of *P. lingua* were identified as hexanal, vanillin, 4-hexen-1-ol, 1-hepten-3-ol, and 3-hydroxy-2,2,4-trimethylpentyl isobutyrate [4].

Hopane (A′-neogammacerane) derivative diploptene, β-sitosterol, vanillic acid, protocatechuic acid, and fumaric acid were isolated from *P. sheareri* [5]. From the rhizome of *P. Lingua* hopane derivatives were identified as cyclohopenol, cyclohopanediol, hop-22(29)-en-28-al [6], 22,28-epoxyhopane, 22,28-epoxyhopan-30-ol, hopane-22,30-diol, hop-22(29)-en-30-ol, and hop-22(29)-en-28-ol [7]. Stigmasterol, ursolic acid, mangiferin and sucrose were also isolated from *P. gralla* [8].

Pharmacology and Toxicology

The methanol insoluble fraction of an aqueous extract of *P. lingua* was reported to inhibit platelet aggregation induced by ADP and collagen in a dose-dependent manner. The methanol soluble fraction of the aqueous extract had high inhibitory effect on ADP- but not collagen-induced

Handbook of Chinese Medicinal Plants: Chemistry, Pharmacology, Toxicology
Weici Tang and Gerhard Eisenbrand

ISBN: 978-3-527-32226-8

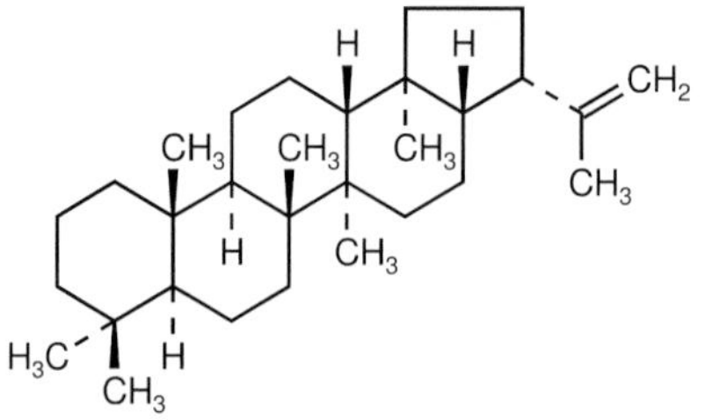

Diploptene

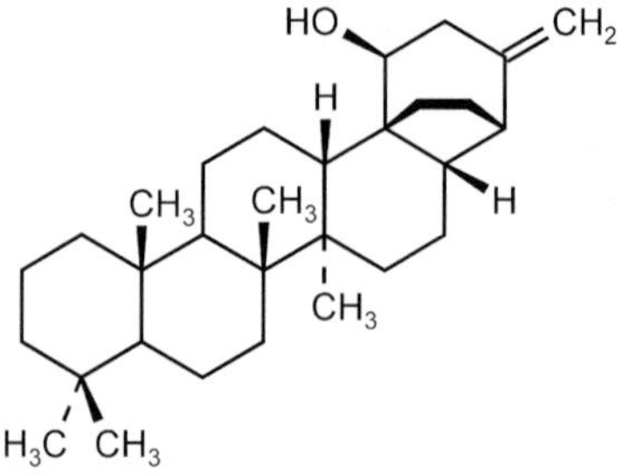

Cyclohopenol

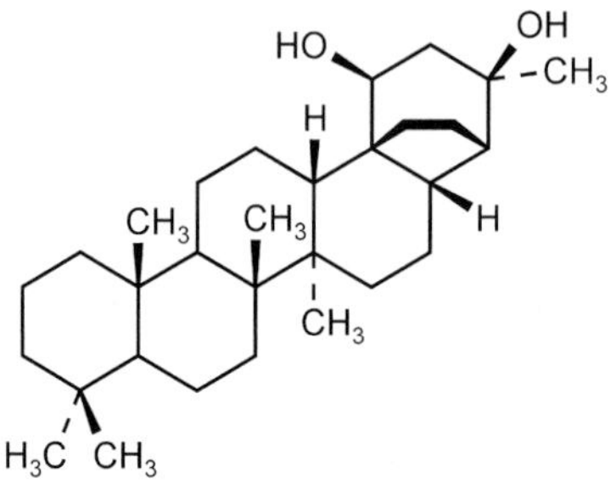

Cyclohopanediol

platelet aggregation [9]. Therapeutic efficacy of *P. sheareri* for clinical treatment of a great number of bacillary dysentery was reported [10].

References

1 Li, J. and Tong, Y.Y. (1992) Determination of active constituents in "Shiwei" (plants from *Pyrrosia*) by high performance liquid chromatography. *Acta Pharm. Sin.*, **27**, 153–156.

2 Do, J.C., Jung, K.Y., and Son, K.H. (1992) Flavonoid glycosides from the fronds of *Pyrrosia lingua*. *Saengyak Hakhoechi*, **23**, 276–279 (CA: 119: 188377).

3 Mizuno, M., Iinuma, M., Imai, T., Tanaka, T., and Min, Z.D. (1986) The chemical constituents of *Pyrrosia lingua*. *Acta Bot Sin.*, **28**, 339–340.

4 Okuno, M., Kameoka, H., Yamashita, M., and Miyazawa, M. (1993) Components of volatile oil from plants of Polypodiaceae. *Yukagaku*, **42**, 44–48 (CA 118: 154259).

5 Han, G.S. and Wang, M.S. (1984) Chemical constituents of *Pyrrosia sheareri* (Bak.) Ching. *J. Nanjing Coll. Pharm.*, **15**, 40–44.

6 Yamashita, H., Masuda, K., Ageta, H., and Shiojima, K. (1998) Fern constituents: cyclohopenol and cyclohopanediol, novel skeletal triterpenoids from rhizomes of *Pyrrosia lingua*. *Chem. Pharm. Bull. (Tokyo)*, **46**, 730–732.

7 Masuda, K., Yamashita, H., Shiojima, K., Itoh, T., and Ageta, H. (1997) Fern constituents: triterpenoids isolated from rhizomes of *Pyrrosia lingua*. I. *Chem. Pharm. Bull. (Tokyo)*, **45**, 590–594.

8 Zheng, X., Xu, Y., and Xu, J. (1998) Chemical studies on *Pyrrosia gralla* (Gies.) Ching. *China J. Chin. Mater. Med.*, **23**, 98–99, 128–129.

9 Sawabe, Y., Iwagami, S., Suzuki, S., and Nakazawa, H. (1991) Inhibitory effect of *Pyrrosia lingua* on platelet aggregation. *Osaka-furitsu Koshu Eisei Kenkyusho Kenkyu Hokoku, Yakuji Shido-hen*, **25**, 39–40 (CA 116: 227914).

10 Zhao, Y.H. (1985) Analysis of the therapeutic effect of 1,148 cases of bacillary dysentery treated with *Pyrrosia sheareri*. *Chin. J. Integr. Trad. West. Med.*, **5**, 530–533.

Rabdosia rubescens

Donglingcao, the dried above-ground part of *Rabdosia rebescens* (Hemsl.) Hara (Lamiaceae) is not listed in the Chinese Pharmacopoeia. It is used in Chinese folk medicine as an antitumor, analgesic and anti-inflammatory agent. In fact, a great number of *Rabdosia* species have been used in Chinese folk medicine.

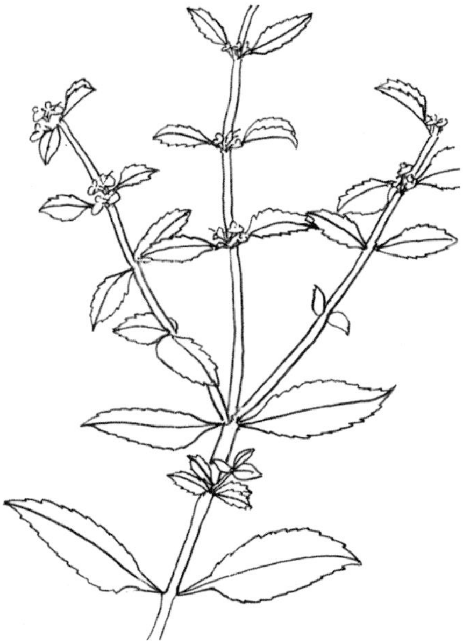

Rabdosia rubescens (Hemsl.) Hara

Chemistry

The above-ground part of *R. rubescens* was reported to contain diterpenes, with oridonin and ponicidin as major components [1].

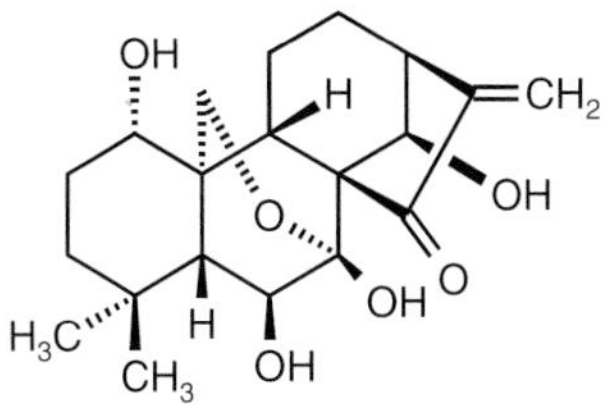

Oridonin

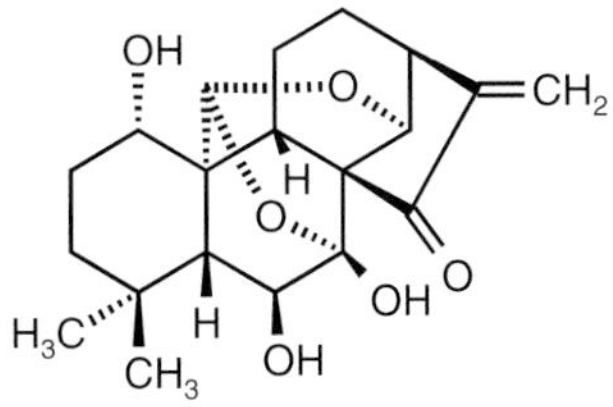

Ponicidin

Related minor diterpenes from *R. rubescens* were identified as rubescensins A–E [2], lushanrubescensin A, ludongnin, xindongnins A and B, guidongnin, and suimiyain A [3].

Pharmacology and Toxicology

Rabdosia rubescens was found to exert growth-inhibitory activity against a series of tumor cell lines, and to show antitumor activity in animal models [4]. In a clinical trial, 115 patients with inoperable esophageal carcinoma were treated by either chemotherapy alone (group A) or chemotherapy plus *R. rubescens* (group B). In group A, out of 31 patients treated with bleomycin and nitrocaphane, 10 (32.3%) responded to the treatment, with two partial remissions (>50% tumor regression) and eight minimal responses. In group B, out of 84 patients treated with bleomycin, nitrocaphane and *R. rubescens*, 59 (70.2%) responded, with 10 complete remissions (100% tumor regression), 16 partial responses and 33 minimal responses. The one-year survival rates of groups A and B were 13.6% and 41.3%, respectively. Statistical significance was present in these two groups, both in the response rate and one-year survival rate. As regards the drug toxicity, there was no

Handbook of Chinese Medicinal Plants: Chemistry, Pharmacology, Toxicology
Weici Tang and Gerhard Eisenbrand

ISBN: 978-3-527-32226-8

significant difference between the two groups. Alopecia, anorexia, nausea and hyperpyrexia occurred in more than 30% of patients. Mild leucopenia with thrombocytopenia and interstitial pneumonia were noted in some patients, and two patients died of toxicity in the lungs [5].

Between August 1974 and January 1987, 650 cases of moderately and advanced esophageal carcinoma were treated with a combination of chemotherapy plus *R. rubescens*, or *R. rubescens* plus a traditional Chinese medicinal prescription. After treatment, 40 patients (6.15%) survived for over five years. Among these 40 patients, 32 survived for over six years, 23 for more than 10 years, five for more than 15 years, and 20 died from the tumor ($n = 16$) or other diseases ($n = 4$). It was believed that the patient age, state of activity, duration of illness, efficacy of primary treatment, multicourse extensive therapy, and long-term maintenance treatment were all important factors affecting the results of drug treatment [6].

A nonrandomized contemporary controlled clinical study to compare the action of local thermotherapy with *Rabdosia* liquid and repeated perfusion of mitomycin C in postoperative prophylaxis of superficial urinary bladder carcinoma has been reported. The study involved 123 patients divided into two groups. Patients in group A received local thermotherapy with *Rabdosia* liquid, starting from one to two months after operation, once tri-monthly for one year. Patients in group B received intravesical perfusion of mitomycin C, starting from two weeks after operation, once weekly (six times in total), and thereafter once monthly for one year. A significant difference was seen in the recurrence rates of groups A and B, at 5.0% and 14.3%, respectively. The occurrence rates of cystitis, hematuria, vesical contracture, and urethral stricture in the two groups did not show any significant difference [7].

Oridonin was reported to be one of the major active principles of *Rabdosia* species. Oridonin, given intraperitoneally to mice at a dose of 15 mg kg^{-1} on days 5 or 8 after implantation of L1210 leukemia cells, exhibited 73% and 39% cell killing, respectively. The G_2 and S phases of L1210 cells were prolonged, but the G_1 phase was unchanged [8]. The intraperitoneal (i.p.) injection of oridonin to mice inoculated with Ehrlich ascites carcinoma also showed significant antitumor activity [9]. In rabbits given i.p. injections of L1210 cells, the i.p. administration of oridonin (15 mg kg^{-1}) on days 5 or 8 killed 73% and 39% of the cells, respectively [10]. Studies on the growth-inhibitory effects of oridonin on MGc80-3 human gastric adenocarcinoma cells in different proliferation stages showed there to be no differences in the sensitivities of cells in the different phases of the cell cycle [11].

In Ehrlich ascites tumor cells, oridonin inhibited, concentration-dependently, the incorporation of [^{3}H]thymidine into the acid-insoluble fraction, but increased [^{3}H] thymine nucleotide uptake in the acid-soluble fraction. Apparently, oridonin blocked DNA polymerization but did not affect the phosphorylation of thymidine [12]. A decreased incorporation of [^{3}H]TdR, [^{3}H]UR, and [^{3}H]leucine into Ehrlich carcinoma cells showed that oridonin inhibits synthesis of DNA, RNA, and protein in a concentration-dependent manner. The inhibition of DNA and RNA synthesis was fast, but reversible upon the removal of oridonin; the inhibition of protein synthesis, however, was strong and longlasting [13]. Oridonin at a concentration of 40 μg ml^{-1} also inhibited DNA synthesis in a cell-free system by 62% after 60 min. This inhibition was not reversed by increasing the concentration of DNA, but was changed by preincubation with partially purified DNA polymerase II [14].

The i.p. administration of oridonin at a dose of 15 mg kg^{-1} to mice bearing L1210 cells inhibited the incorporation of [^{3}H] TdR, [^{3}H]UR, and [^{3}H]leucine into DNA, RNA and protein in leukemia cells. The inhibition of DNA synthesis by oridonin preceded that of RNA and protein synthesis [15]. The α-methylenecyclopentanone moiety in oridonin was believed to be the structural part responsible for anticancer activity [16]. The cytotoxicity of oridonin was compared with that of methotrexate and 5-fluorouracil, using human Mgc803 cells. Tumor cells, when exposed for 24 h to increasing concentrations of methotrexate (0–1.6 μg ml^{-1}), 5-fluorouracil (0–12 μg ml^{-1}) or oridonin (0–25 μg ml^{-1}), showed different types of dose–survival curve in the colony-forming assay. Different modes of action of these compounds have been suggested [17]. Oridonin, at a concentration of 2.0 μg ml^{-1}, inhibited human esophageal cancer CaEs-17 cells by 35–52% [18].

Recent studies on the mechanism of action of oridonin revealed that it inhibited the proliferation of tumor cells by inducing cell-cycle arrest and apoptosis, a decrease in the mitochondrial membrane potential, downregulation of Bcl-2, and upregulation of Bax expression. For example, oridonin inhibited the growth of HT29 human colorectal carcinoma cells, induced cell-cycle arrest at G_2/M phase and induced cell apoptosis. At doses of 10, 15, and 20 mg kg^{-1}, it also exhibited antitumor activity against HT29 cells in mice [19]. In BEL-7402 human hepatocelluar carcinoma cells, oridonin inhibited cell proliferation, induced cell apoptosis, downregulated Bcl-2, upregulated Bax expression, and decreased the expression of hTERT mRNA and telomerase activity [20]. Oridonin was also found to induce apoptosis in human epidermoid carcinoma A431 cells. Total tyrosine kinase activity was inhibited and the protein expression of epidermal growth factor receptor (EGFR) and phosphorylated EGFR were decreased. The expression of EGFR downstream effector proteins, Grb2, Ras, Raf-1 and extracellular signal-regulated kinase (ERK) was also downregulated; this suggested that the inactivation of Ras, Raf, or ERK participated in oridonin-induced apoptosis in A431 cells [21].

Oridonin was also found to be cytotoxic against NB4 cells and fresh leukemia cells obtained from acute promyelocytic leukemia patients [22]. It inhibited U266 and RPMI8226 multiple myeloma, Jurkat acute lymphoblastic T-cell leukemia and MT-1 adult T-cell leukemia cells, with IC_{50}-values ranging from 0.75 to 2.7 μg ml^{-1}. Further studies revealed that oridonin inhibited nuclear factor-κB (NF-κB) DNA-binding activity in these tumor cells. It also blocked NF-κB activity in Jurkat cells and in RAW264.7 murine macrophages stimulated by tumor necrosis factor-α (TNF-α) and lipopolysaccharide (LPS). Oridonin decreased the survival of freshly isolated adult T-cell leukemia, acute lymphoblastic leukemia, chronic lymphocytic leukemia, nonHodgkin's lymphoma and multiple myeloma cells from patients, in association with an inhibition of NF-κB DNA-binding activity. However, it did not affect the survival of normal lymphoid cells from healthy volunteers [23].

The plasma concentrations of oridonin after an intravenous (i.v.) injection to rats decreased exponentially. The pharmacokinetic parameters of oridonin were dose-independent within the dose range of 5, 10, and 15 mg kg^{-1}. After oral administration, oridonin was absorbed rapidly, with a half-life of less than 15 min. The bioavailability of oridonin following oral administration at doses of 20, 40, and 80 mg kg^{-1} was between 4.3 to 10.8%. Following an i.p. dose of 10 mg kg^{-1} first-order rate pharmacokinetics where observed [24]. The pharmacokinetics of oridonin-loaded poly(D,L-lactic acid) nanoparticles in rats has also been reported. After an i.v. administration of oridonin-loaded poly(D,L-lactic acid) nanoparticles, a stable and high concentration was found in the liver, lung and spleen. Distribution to the heart and kidney decreased [25].

Like oridonin, ponicidin was also found to exhibit cytotoxic effects against a large variety of cancer cell lines. Ponicidin inhibited the proliferation of QGY-7701 and HepG-2 hepatocellular carcinoma cells, induced cell apoptosis, downregulated the expression of both survivin and Bcl-2, and upregulated Bax expression [26]. In lung cancer A549 and GLC-82 cells, ponicidin was found to inhibit cell growth, to induce apoptosis, to disrupt the mitochondrial membrane potential, and to trigger the activation of caspase-3, caspase-8, and caspase-9 in cancer cells. Ponicidin also downregulated the antiapoptotic proteins Bcl-2 and survivin, and upregulated the proapoptotic protein Bax in lung carcinoma cells [27].

References

1 Lu, D., Zheng, Q.T., and Zhou, T.H. (1995) Crystal structure of oridonin. *Jiegou Huaxue*, **14**, 237–239.

2 Lin, C. and Zhao, Z. (1997) Structure of rubescensin E. *China J. Chin. Mater. Med.*, **22**, 612–613, 640.

3 Sun, X.P. and Yue, S.J. (1992) Chemical constituents of blushed rabdosia (*Rabdosia rubescens*). *Chin. Trad. Herbal Drugs*, **23**, 59–60.

4 Tang, W. (2002) Recent advances in antineoplastic principles of Traditional Chinese Medicine. *Pharmazie*, **57**, 223–232.

5 Wang, R.L., Gao, B.L., Xiong, M.L., Mei, Q.D., Fan, K.S., Zuo, Z.K., Lang, T.L., Gao, G.Q., Ji, Z.C., and Wei, D.C. (1986) Potentiation by *Rabdosia rubescens* on chemotherapy of advanced esophageal carcinoma. *Chin. J. Oncol.*, **8**, 297–299.

6 Wang, R.L. (1993) A report of 40 cases of esophageal carcinoma surviving for more than 5 years after treatment with drugs. *Chin. J. Oncol.*, **15**, 300–302.

7 Xu, P.Y., Zhao, G.X., and Chang, L.S. (2005) Local thermotherapy with rabdosia liquid as prophylactic measure for recurrence of superficial urinary bladder carcinoma: a non-randomized contemporary controlled study. *Zhongguo Zhong Xi Yi Jie He Za Zhi*, **25**, 1115–1117.

8 Wang, M.Y., Lin, C., and Zhang, T.M. (1985) Cytokinetic effects of oridonin on leukemia L1210 cells. *Acta Pharmacol. Sin.*, **6**, 195–196.

9 Fujita, E., Nagao, Y., Node, M., Kaneko, K., Nakazawa, S., and Kuroda, H. (1976) Antitumor activity of the *Isodon* diterpenoids: structural requirements for the activity. *Experientia*, **32**, 203–205.

10 Wang, M.Y., Lin, C., and Zhang, T.M. (1985) Cytokinetic effects of oridonin on leukemia L1210 cells. *Acta Pharmacol. Sin.*, **6**, 195–198.

11 Li, X.T., Lin, C., and Li, P.Y. (1986) Characteristics of the cytostatic effects of oridonin *in vitro*. *Acta Pharmacol. Sin.*, **7**, 361–363.

12 Li, Y., and Zhang, T.M. (1987) Effect of oridonin on metabolism of thymine nucleotides in mouse tumor cells. *Acta Pharmacol. Sin.*, **8**, 271–274.

13 Wang, M.Y., Shou, M.G., Wang, Q.D., Wang, Q., and Zhang, T.M. (1985) Effect of oridonin on DNA, RNA and protein syntheses *in vitro*. *Acta Acad. Med. Henan.*, **20**, 15–18.

14 Li, Y., and Zhang, T.M. (1988) Effect of oridonin on cell-free DNA synthesis *in vitro*. *Acta Pharmacol. Sin.*, **9**, 465–467.

15 Wang, M.Y., Lin, C., and Zhang, T.M. (1987) Autoradiographic study on the effects of oridonin on DNA, RNA and protein synthesis of leukemia L 1210 cells. *Acta Pharmacol. Sin.*, **8**, 164–165.

16 Zhai, J.K., Han, W.C., and Ju, X.H. (1993) Crystal structure of nervosin and the electronic structure of its anticancer-active zone. *Acta Chim. Sin.*, **51**, 854–859.

17 Li, X.T., Lin, C., and Li, P.Y. (1986) Comparison of *in vitro* assays for the cytotoxic effect of anticancer drugs. *Chin. J. Oncol.*, **8**, 184–186.

18 Yung, K.H., Ku, Y.C., Wang, S.H., and Chu, Y. M. (1981) Preliminary observation on using CaEs-17 cell line for *in vitro* screening of antitumor drugs. *Chin. J. Oncol.*, **3**, 77.

19 Zhu, Y., Xie, L., Chen, G., Wang, H., and Zhang, R. (2007) Effects of oridonin on proliferation of HT29 human colon carcinoma cell lines both *in vitro* and *in vivo* in mice. *Pharmazie*, **62**, 439–444.

20 Zhang, J.F., Chen, G.H., Lu, M.Q., Li, H., Cai, C.J., and Yang, Y. (2006) Change of Bcl-2 expression and telomerase during apoptosis induced by oridonin on human hepatocelluar

carcinoma cells. *Zhongguo Zhong Yao Za Zhi*, **31**, 1811–1814.

21 Li, D., Wu, L.J., Tashiro, S., Onodera, S., and Ikejima, T. (2007) Oridonin-induced A431 cell apoptosis partially through blockage of the Ras/Raf/ERK signal pathway. *J. Pharmacol. Sci.*, **103**, 56–66.

22 Liu, J., Huang, R., Lin, D., Wu, X., Peng, J., Lin, Q., Pan, X., Zhang, M., Hou, M., and Chen, F. (2005) Apoptotic effect of oridonin on NB4 cells and its mechanism. *Leuk. Lymphoma*, **46**, 593–597.

23 Ikezoe, T., Yang, Y., Bandobashi, K., Saito, T., Takemoto, S., Machida, H., Togitani, K., Koeffler, H.P., and Taguchi, H. (2005) Oridonin, a diterpenoid purified from *Rabdosia rubescens*, inhibits the proliferation of cells from lymphoid malignancies in association with blockade of the NF-κB signal pathways. *Mol. Cancer Ther.*, **4**, 578–586.

24 Xu, W., Sun, J., Zhang, T.T., Ma, B., Cui, S.M., Chen, D.W., and He, Z.G. (2006) Pharmacokinetic behaviors and oral bioavailability of oridonin in rat plasma. *Acta Pharm. Sin.*, **27**, 1642–1646.

25 Xing, J., Zhang, D., and Tan, T. (2007) Studies on the oridonin-loaded poly(D,L-lactic acid) nanoparticles *in vitro* and *in vivo*. *Int. J. Biol. Macromol.*, **40**, 153–158.

26 Zhang, J.F., Liu, P.Q., Chen, G.H., Lu, M.Q., Cai, C.J., Yang, Y., and Li, H. (2007) Ponicidin inhibits cell growth on hepatocellular carcinoma cells by induction of apoptosis. *Dig. Liver Dis.*, **39**, 160–166.

27 Liu, J.J., Huang, R.W., Lin, D.J., Peng, J., Zhang, M., Pan, X., Hou, M., Wu, X.Y., Lin, Q., and Chen, F. (2006) Ponicidin, an ent-kaurane diterpenoid derived from a constituent of the herbal supplement PC-SPES, *Rabdosia rubescens*, induces apoptosis by activation of caspase-3 and mitochondrial events in lung cancer cells *in vitro*. *Cancer Invest.*, **24**, 136–148.

Raphanus sativus

Semen Raphani (Laifuzi) is the dried ripe seed of *Raphanus sativus* L. (Brassicaceae) collected in the summer. It is used as a stomachic and laxative agent, and as an expectorant for the treatment of abdominal distension and pain, constipation, and cough.

Raphanus sativus L.

Chemistry

Antifungal proteins were reported to be isolated from the seed of *R. sativus*, the radish seeds. One class of the antifungal proteins consists of two homologous proteins, designated *Raphanus sativus*-antifungal protein Rs-AFP1, Rs-AFP2, Rs-AFP3, and Rs-AFP4. These are highly basic oligomeric proteins composed of small (5 kDa) polypeptides rich in cysteine [1]. These proteins were found to be located in the cell wall, and to occur predominantly in the outer cell layers lining the different seed organs. Rs-AFP1 and Rs-AFP2 are released preferentially during seed germination after disruption of the seed coat. Rs-AFP1 and Rs-AFP2 are both composed of 51 amino acids. A cDNA clone encoding the Rs-AFP2 was modified by recombinant DNA methods to allow expression in the yeast *Saccharomyces cerevisiae* [2]. The three-dimensional structure of Rs-AFP1 was reported [3].

An isothiocyanate derivative, sulforaphene, was isolated from the seeds of *R. sativa* [4].

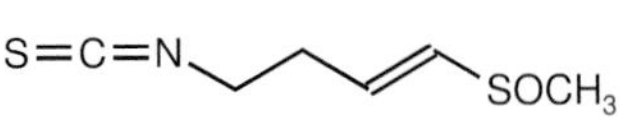

Sulforaphene

Pharmacology and Toxicology

Both proteins Rs-AFP1 and Rs-AFP2 showed a broad antifungal spectrum, and are among the most potent antifungal proteins. In comparison with many other plant antifungal proteins, the activities of Rs-AFP1 and Rs-AFP2 were less sensitive to the presence of cations. Moreover, their antibiotic activity had a high degree of specificity to filamentous fungi [2]. Rs-AFP1 and Rs-AFP2 were believed to play a role in host defense [5]. Rs-AFP2 was found to interact with glucosylceramides in the membranes of susceptible yeast and fungi, and to induce membrane permeabilization and fungal cell death. It was shown that Rs-AFP2 induced the production of reactive oxygen species (ROS) in *Candida albicans* wild-type, in a concentration-dependent manner [6].

A crude aqueous extract of the seeds of *R. sativa* exhibited a dose-dependent hypotensive effect and delayed the heart rate of anaesthetized normotensive rats. In isolated guinea-pig atria, the extract inhibited the contractile force. In endothelium-intact,

Handbook of Chinese Medicinal Plants: Chemistry, Pharmacology, Toxicology
Weici Tang and Gerhard Eisenbrand

ISBN: 978-3-527-32226-8

but not endothelium-denuded, rat aorta preparations, the extract inhibited contractions induced by phenylephrine [7]. Sulforaphene was found to exhibit antimutagenic activity, by inhibiting the mutagenicity of a series of food-borne mutagens in *Salmonella typhimurium* TA98 and TA100 in the presence of a rat liver S-9 fraction [4].

References

1 Terras, F.R., Schoofs, H.M., De Bolle, M.F., Van Leuven, F., Rees, S.B., Vanderleyden, J., Cammue, B.P., and Broekaert, W.F. (1992) Analysis of two novel classes of plant antifungal proteins from radish (*Raphanus sativus* L.) seeds. *J. Biol. Chem.*, **267**, 15301–15309.

2 Alves, A.L., De Samblanx, G.W., Terras, F.R., Cammue, B.P., and Broekaert, W.F. (1994) Expression of functional *Raphanus sativus* antifungal protein in yeast. *FEBS Lett.*, **348**, 228–232.

3 Fant, F., Vranken, W., Broekaert, W., and Borremans, F. (1998) Determination of the three-dimensional solution structure of *Raphanus sativus* antifungal protein 1 by 1H NMR. *J. Mol. Biol.*, **279**, 257–270.

4 Shishu, I., Singla, A.K., and Kaur, I.P. (2003) Inhibition of mutagenicity of food-derived heterocyclic amines by sulforaphene: an isothiocyanate isolated from radish. *Planta Med.*, **69**, 184–186.

5 Terras, F.R., Eggermont, K., Kovaleva, V., Raikhel, N.V., Osborn, R.W., Kester, A., Rees, S.B., Torrekens, S., Van Leuven, F., and Vanderleyden, J. (1995) Small cysteine-rich antifungal proteins from radish: their role in host defense. *Plant Cell*, **7**, 573–588.

6 Aerts, A.M., Francois, I.E., Meert, E.M., Li, Q.T., Cammue, B.P., and Thevissen, K. (2007) The antifungal activity of RsAFP2, a plant defensin from *Raphanus sativus*, involves the induction of reactive oxygen species in *Candida albicans*. *J. Mol. Microbiol. Biotechnol.*, **13**, 243–247.

7 Ghayur, M.N. and Gilani, A.H. (2006) Radish seed extract mediates its cardiovascular inhibitory effects via muscarinic receptor activation. *Fundam. Clin. Pharmacol.*, **20**, 57–63.

Rehmannia glutinosa

Radix Rehmanniae (Dihuang) is the fresh or dried root of *Rhemannia glutinosa* Libosch. (Scrophulariaceae). It is used in fresh or dried form as an antipyretic and hemostatic agent for the treatment of febrile diseases, diabetes, hemostypsis, epistaxis, skin eruption, and maculation.

Rehmannia glutinosa Libosch.

Radix Rehmanniae Preparata (Shudihuang) is the processed root of *Rehmannia glutinosa*. It is used as a tonic, hemostatic and sedative agent for the treatment of consumptive fever, night sweating, diabetes, cardiac palpitation, menstrual disorders, abnormal uterine bleeding, dizziness, and tinnitus.

Chemistry

The root of *R. glutinosa* is one of the traditional Chinese medicines containing iridoid glycosides. The major iridoid glycoside was identified as catalpol [1]. The Chinese Pharmacopoeia requires a qualitative determination of catalpol in Radix Rehmanniae by thin-layer chromatographic (TLC) comparison with a reference substance, and a quantitative determination by HPLC. The content of catalpol in Radix Rehmanniae should be not less than 0.2%. In addition, the Chinese Pharmacopoeia requires a qualitative determination of 5-hydroxymethylfurfural in Radix Rehmanniae Preparata by TLC comparison with reference substance. 5-Hydroxymethylfurfural is not a natural component, and is formed from fructose and molasses by thermic treatment of the root, because the root is rich in sugars. Minor iridoid glycosides in the root of *R. glutinosa* were identified as 6′-*O*-acetylcatalpol [2], rehmanniosides A–D, ajugol, aucubin, melittoside, ajugol esters, and jioglutosides A and B [2–4].

Simple iridoids from *Rehmannia* root were identified as rehmaglutins A–D, jioglutins A–E, jioglutolide, and jiofuran, some of which contain chlorine [5–7]. A bis-furan derivative and two new natural furan derivatives in the dried roots were identified as 1,5-bis(5-methoxymethyl)furan-2-yl-penta-1,4-dien-3-one, (*E*)-4-(5-(methoxymethyl)furan-2-yl)but-3-en-2-one, and (*E*)-4-(5-(hydroxymethyl)furan-2-yl)but-3-en-2-one [8].

Ionone glycosides from the root of *R. glutinosa* were identified as rehmaionosides A, B, and C [9]. Norcarotenoid glycosides from the root were identified as jiocarotenoside A_1, jiocarotenoside A_2, and 6-*O*-*sec*-hydroxyarginetoyl ajugol [10]. 2-Phenylethyl glycosides including salidroside have also been isolated from the root of *R. glutinosa* [11].

The monosaccharide contents of raw root and processed root were found to be quite different. Raw root contained about 16% monosaccharide processed root about 52% [12]. Saccharide [13] and a number of polysaccharides were reported to be

Handbook of Chinese Medicinal Plants: Chemistry, Pharmacology, Toxicology
Weici Tang and Gerhard Eisenbrand

ISBN: 978-3-527-32226-8

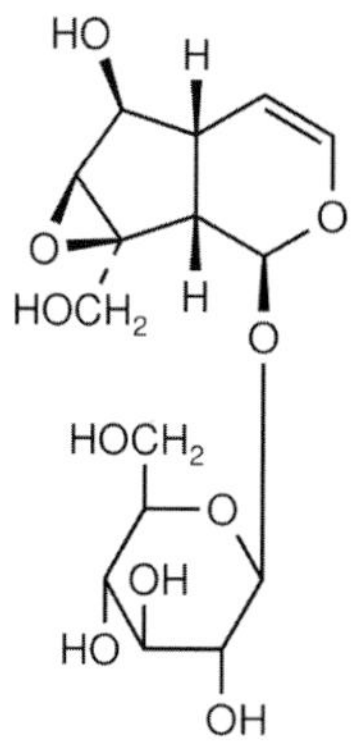

Catalpol

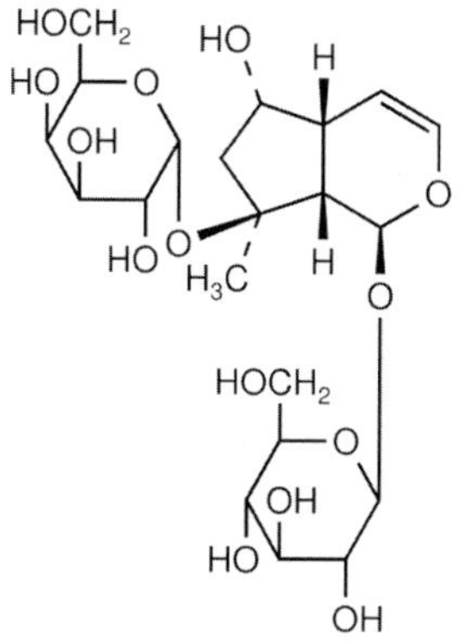

Rehmannioside C

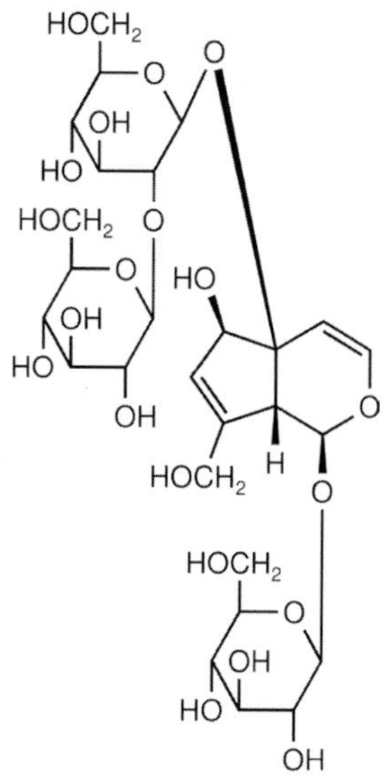

Rehmannioside D

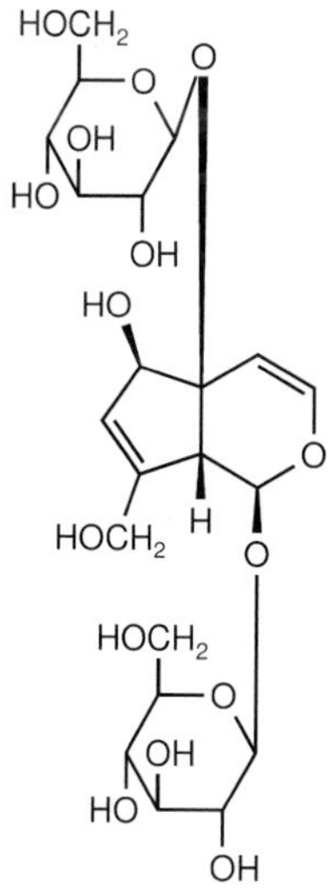

Melittoside

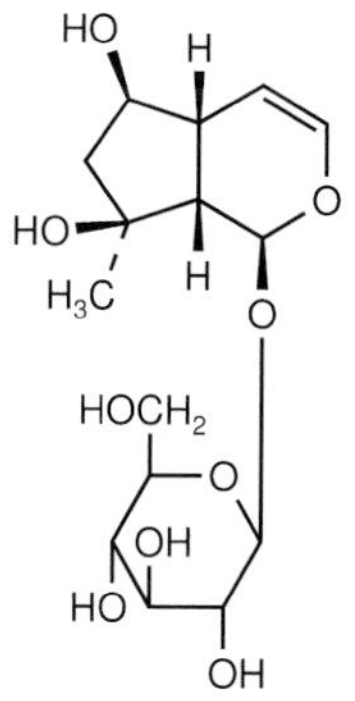

Ajugol

isolated from the root of *R. glutinosa*. Polysaccharide B had an average molecular mass of 160 kDa [14]. Acidic polysaccharides, rehmannans FS-I, FS-II [15], SA and SB [16] from the raw root of *R. glutinosa* were found to be homogeneous, with molecular masses of 580, 660, 640, and 790 kDa, respectively. Rehamannan FS-I is composed of L-arabinose, D-galactose, L-rhamnose, D-galacturonic acid, and D-glucuronic acid; rehmannan FS-II of L-arabinose, D-galactose, L-rhamnose, D-galacturonic acid, in addition to small amounts of *O*-acetyl groups [15]. Rehmannan SA and rehmannan SB were commonly composed of L-arabinose, D-galactose, L-

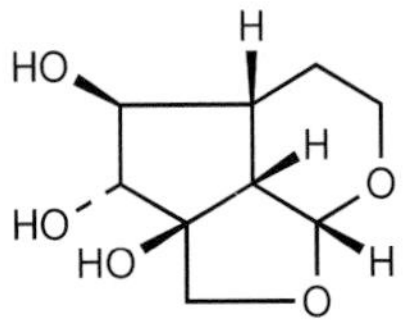

Rehmaglutin A

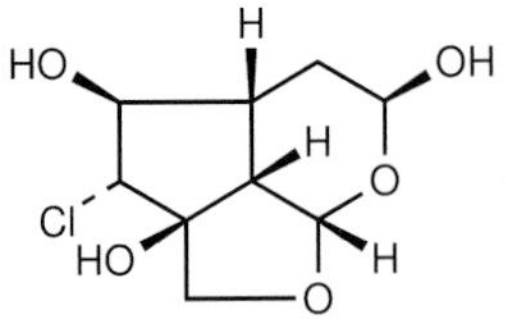

Rehmaglutin B

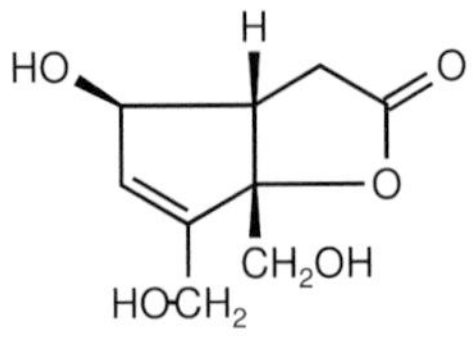

Rehmaglutin C

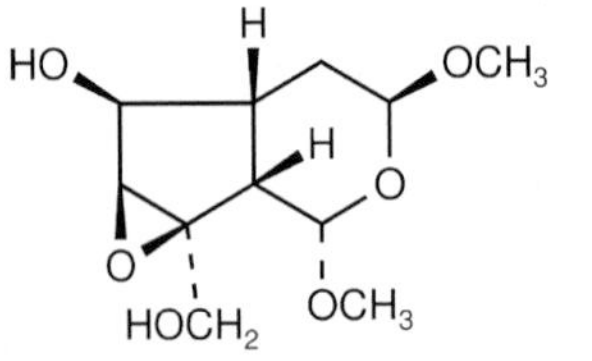

Jioglutin D

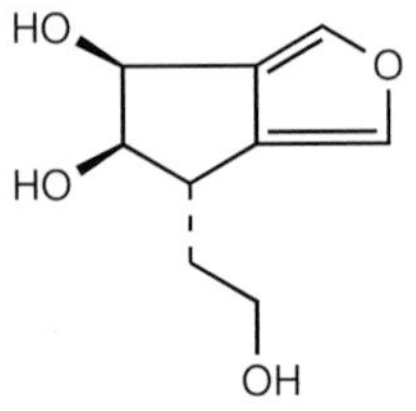

Jiofuran

rhamnose and D-galacturonic acid in different molar ratios, in addition to small amounts of peptide moieties [16, 17].

Pharmacology and Toxicology

The root of *R. glutinosa* is known to have hypoglycemic activity, and is used clinically for the treatment of diabetes. The antidiabetic effect of the root was demonstrated in KK-Ay mice, one of the non-insulin-dependent diabetes mellitus types. It decreased the blood glucose levels and improved glucose tolerance for up to five weeks after repeated administration in KK-Ay mice [18]. The polysaccharide fraction from the root of *R. glutinosa* exhibited hypoglycemic activity in normal mice and in mice made diabetic by the intraperitoneal (i.p.) administration of streptozotocin. Administration of the polysaccharide fraction to normal mice significantly increased the activities of glucokinase and glucose-6-phosphatase dehydrogenase, but decreased the activities of glucose-6-phosphatase and phosphofructokinase in the liver. It stimulated the secretion of insulin and reduced the glycogen content in the liver of normal mice [19]. The oligosaccharides from the root of *R. glutinosa* were reported to exert hypoglycemic activity in diabetic rats, as induced by alloxan. The i.p. pretreatment of rats with the oligosaccharide at a dose of 100 $mg\,kg^{-1}$ for three days partially prevented hyperglycemia caused by the i.p.

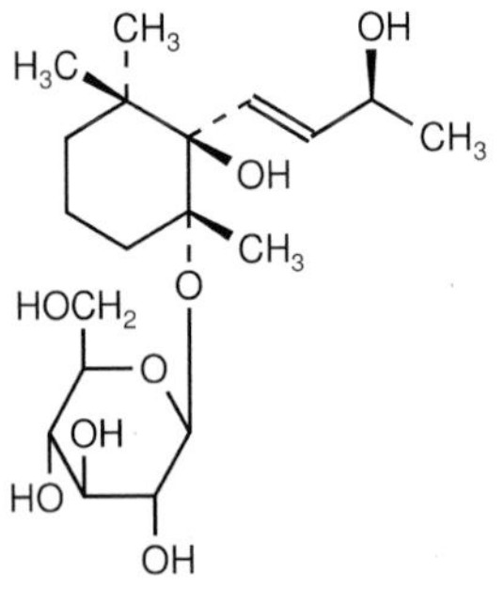

Rehmaionoside A

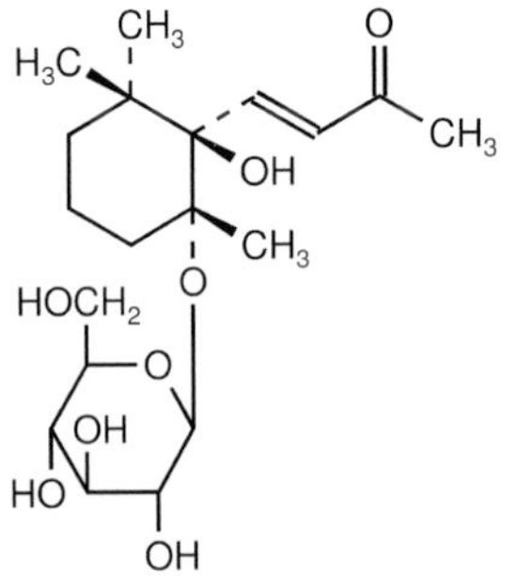

Rehmaionoside C

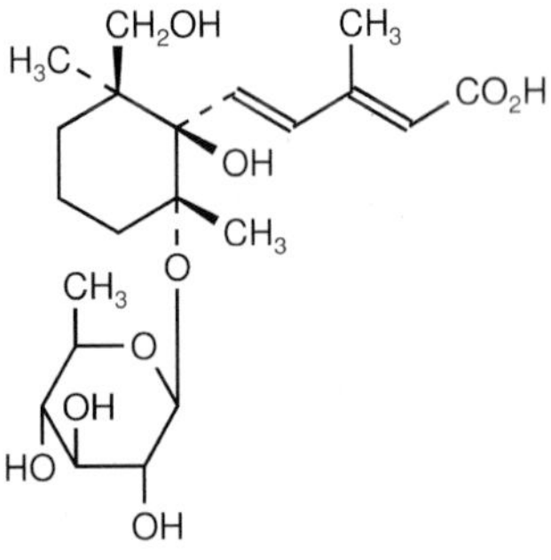

Jiocarotenoside A_1

administration of glucose (2 g kg^{-1}) in normal rats, but not in adrenalectomized rats. The i.p. treatment of 15-day alloxan-induced diabetic rats with the oligosaccharides significantly decreased blood glucose level and hepatic glucose-6-phosphatase activity, with an increase in hepatic glycogen content. The hypoglycemic effect of the oligosaccharides seemed to be adrenal-dependent, and related to the neuroendocrine system [20]. The oligosaccharides were also reported to significantly increase glucose consumption in 3T3-L1 preadipocytes and adipocytes in culture, in a concentration-dependent manner. They also promoted the proliferation of 3T3-L1 preadipocytes, inhibited the proliferation of 3T3-L1 adipocytes, and improved the sensitivity of 3T3-L1 adipocytes to insulin [21].The effects of the ethanolic extract of the processed root of *R. glutinosa* on the hemorheology of inflammatory, thrombosic and normal animals have been studied. Oral administration of the extract to rats with adjuvant-induced arthritis inhibited the reduction of fibrinolytic activity; erythrocyte deformability; the decrease in erythrocyte counts; and the increase in connective tissue of the thoracic artery. However, the extract was found to be ineffective on the development of edema in arthritic rats, and on acute and chronic inflammation [22]. The extract from raw root had only weak or no anti-inflammatory effect [23]. The bis-furan derivative 1,5-bis(5-methoxymethyl)furan-2-yl-penta-1,4-dien-3-one was found to inhibit blood platelet aggregation [8].

An aqueous extract of processed root of *R. glutinosa* was found to inhibit the secretion of tumor necrosis factor-α (TNF-α) from primary cultures of mouse astrocytes stimulated with substance P and lipopolysaccharide (LPS). The inhibition of TNF-α secretion from mouse astrocytes was caused by inhibiting interleukin-1 (IL-1) secretion [24]. In rats, the aqueous extract protected against the systemic anaphylactic reaction induced by compound 48/80, and reduced plasma histamine levels in dose-dependent manner. It also dose-dependently inhibited histamine release from rat peritoneal mast cells by compound 48/80 or anti-dinitrophenyl IgE, and significantly inhibited anti-dinitrophenyl IgE-induced TNF-α production by rat peritoneal mast cells [25].

The root of *R. glutinosa* was also used in traditional Chinese medicine for the treatment of dementia. Radix Rehmanniae Preparata was reported to improve the function of learning and memory. The effect of the prepared root was tested with injured thalamic arcuate nucleus rats, as induced by sodium glutamate. Improving functions of the root on learning and memory were observed by the step-down task and Morris water maze task. Moreover, the expression of c-Fos and nerve growth factor in rat hippocampus was increased after treatment. It was concluded that the root of *R. glutinosa* could improve the function of learning and memory of injured rats, the mechanism possibly being related to an increase of the expression of hippocampal c-Fos and nerve growth factor [26].

The root extract induced expression of glial cell line-derived neurotrophic factor in C6 glioblastoma cells and primary cultured astrocytes. The upregulation of glial cell line-derived neurotrophic factor mRNA in C6 glioblastoma cells was completely inhibited by protein kinase C (PKC) inhibitors, and by a MAPK/ERK kinase 1 (MEK1) inhibitor. On the other hand, the root was found to stimulate the phosphorylation of extracellular signal-regulated kinase 1 and 2 (ERK1/2), which preceded glial cell line-derived neurotrophic factor mRNA induction in C6 glioblastoma cells. However, none of the PKC inhibitors significantly changed ERK1/2 phosphorylation stimulated by the root of *R. glutinosa*. The results showed that glial cell line-derived neurotrophic factor gene expression, stimulated by the root of *R. glutinosa*, is independently upregulated through PKC and ERK 1/2 pathways in C6 glioblastoma cells [27].

A hot-water extract of the root of *R. glutinosa* was reported to inhibit the proliferation of H-4-II-E rat hepatocellular carcinoma and HA22T/VGH human hepatocellular carcinoma cells in a concentration-dependent manner. Apoptotic cell numbers and the expression of p53 protein in H-4-II-E cells were each significantly increased after 24 h treatment with the extract; this indicated that the extract might inhibit proliferation and stimulate p53-mediated apoptosis in hepatocellular carcinoma cells [28].

The prepared root of *R. glutinosa* was also used to treat inner ear diseases, such as tinnitus and hearing loss. An experimental investigation showed that an ethanol extract of the prepared root of *R. glutinosa* protected HEI-OC1 auditory cells from cytotoxicity caused by cisplatin, in concentration-dependent fashion at concentrations of 5 to 100 $\mu g\,ml^{-1}$. In addition, the root extract reduced lipid peroxidation, showed strong scavenging activity against superoxide and hydroxyl radicals, and hydrogen peroxide and diphenylpicrylhydrazyl radicals. It was suggested that the root of *R. glutinosa* protected against cisplatin-induced HEI-OC1 cell damage via an inhibition of lipid peroxidation and by the scavenging of free radicals [29].

The iridoid glycoside catalpol was found to exert neuroprotection in gerbils subjected to transient global cerebral ischemia. Catalpol, at an i.p. dose of 1 $mg\,kg^{-1}$ injected immediately after reperfusion and repeatedly at 12, 24, 48 and 72 h, significantly rescued neurons in the hippocampal CA1 subfield and reduced working errors during behavioral testing. The neuroprotective effect of catalpol became more evident when the doses were increased to 5 and 10 $mg\,kg^{-1}$. A significant neuroprotective effect of catalpol was also observed when catalpol was administered up to 3 h after ischemia. The neuroprotective effect of catalpol was seen not only during a short post-ischemic period (12 days) but also over a long period (35 days) [30]. Catalpol was further reported to protect against neurodegeneration mediated by inflammation. Microglia, the resident immune cells in the central nervous system, are pivotal in the inflammatory reaction. Activated microglia can induce the expression of inducible nitric-oxide synthase (iNOS) and release significant amounts of nitric oxide (NO) and TNF-α, which can damage the dopaminergic neurons. Catalpol was found to significantly reduce the release of reactive oxygen species (ROS), TNF-α and NO in microglia in mesencephalic neuron-glia cultures after activation induced by LPS. Catalpol also suppressed the expression of iNOS induced by LPS. These results suggested that catalpol exerted a protective effect on dopaminergic neurons by inhibiting microglial activation and reducing the production of proinflammatory factors [31].

The polysaccharide B from the root of *R. glutinosa* was reported to inhibit the growth of S180, Lewis lung, B16, and H22 tumor cells in mice at daily i.p. doses of 20–40 $mg\,kg^{-1}$ for eight days. It was also effective against S180 by oral treatment, but ineffective in an *in vitro* experiment. Polysaccharide B showed a positive immunomodulatory activity by enhancing the proliferation of splenic T lymphocytes and partly blocking the inhibition of natural killer cell activity caused by tumor cell growth [14, 32]. The low-molecular-weight polysaccharides from the root were found to markedly increase p53 gene expression in Lewis lung cancer tissue in mice at doses of 20 and 40 $mg\,kg^{-1}$, which were the effective doses in inhibiting the growth of tumor cells *in vivo* [33]. The root of *R. glutinosa*, and patent preparations of the root as a major component, were used to reduce the side effects of chemotherapeutic treatments [34–36]. In hemorrhagic anemic mice, treatment with the root of *R. glutinosa* var. *hueichingensis* caused a stimulation of the bone marrow hematopoietic cells CFU-S, CFU-E [37].

All of the acidic polysaccharides rehmannan FS-I, rehamannan FS-II rehamannan SA, and rehamannan SB showed remarkable reticuloendothelial system-potentiating activity in a carbon clearance test [15, 16]. Both, rehmannan SA and rehmannan SB showed pronounced anti-complementary activity [17].

References

1 Kitagawa, I., Fukuda, Y., Taniyama, T., and Yoshikawa, M. (1995) Chemical studies on crude drug processing. X. On the constituents of rehmanniae radix (4): comparison of the constituents of various rehmanniae radixes originating in China, Korea, and Japan. *Yakugaku Zasshi*, **115**, 992–1003.

2 Nishimura, H., Sasaki, H., Morota, T., Chin, M., and Mitsuhashi, H. (1989) Six iridoid glycosides from *Rehmannia glutinosa*. *Phytochemistry*, **28**, 2705–2709.

3 Morota, T., Sasaki, H., Nishimura, H., Sugama, K., Chin, M., and Mitsuhashi, H. (1989) Two iridoid glycosides from *Rehmannia glutinosa*. *Phytochemistry*, **28**, 2149–2153.

4 Oshio, H., and Inouye, H. (1981) Iridoid glycosides of *Rehmannia glutinosa*. *Phytochem.*, **21**, 133–138.

5 Kitagawa, I., Fukuda, Y., Taniyama, T., and Yoshikawa, M. (1986) Absolute stereostructures of rehmaglutins A,B, and D, three new iridoids isolated from Chinese rehmanniae radix. *Chem. Pharm. Bull. (Tokyo)*, **34**, 1399–1402.

6 Morota, T., Nishimura, H., Sasaki, H., Chin, M., Sugama, K., Katsuhara, T., and Mitsuhashi, H. (1989) Chemical and biological studies on rehmanniae radix. Part 5. Five cyclopentanoid monoterpenes from *Rhemmania glutinosa*. *Phytochemistry*, **28**, 2385–2391.

7 Morota, T., Sasaki, H., Sugama, K., Nishimura, H., Chin, M., and Mitsuhashi, H. (1990) Chemical and biological studies on rehmanniae radix. Part 6. Two nonglycosidic iridoids from *Rhemmania glutinosa*. *Phytochemistry*, **29**, 523–526.

8 Li, Y.S., Chen, Z.J., and Zhu, D.Y. (2005) A novel bis-furan derivative, two new natural furan derivatives from *Rehmannia glutinosa* and their bioactivity. *Nat. Prod. Res.*, **19**, 165–170.

9 Yoshikawa, M., Fukuda, Y., Taniyama, T., Cha, B.C., and Kitagawa, I. (1986) Absolute configurations of rehmaionosides A,B, and C and rehmapicroside three new ionone glucosides and a new monoterpene glucoside from Rehmanniae radix. *Chem. Pharm. Bull. (Tokyo)*, **34**, 2294–2297.

10 Sasaki, H., Nishimura, H., Morota, T., Chin, M., Mitsuhashi, H., and Katsuhara, T. (1991) Chemical and biological studies on rehmanniae radix. Part 9. Norcarotenoid glycosides from *Rhemmania glutinosa* var. *purpurea*. *Phytochemistry*, **30**, 1639–1644.

11 Nishimura, H., Sasaki, H., Morota, T., Chin, M., and Mitsuhashi, H. (1990) Chemical and biological studies on rehmanniae radix. Part 7. Six glycosides from *Rhemmania glutinosa* var. *purpurea*. *Phytochemistry*, **29**, 3303–3306.

12 Liu, Z.Y. (1984) Monosaccharide content in raw and processed roots of *Rhemannia glutinosa*. *Bull. Chin. Mater. Med.*, **9**, 17–18.

13 Bian, B., Wang, H., and Ni, M. (1995) Determination of total saccharide and several main saccharides in *Rehmannia glutinosa* Libosh, and its processed products. *China J. Chin. Mater. Med.*, **20**, 469–471.

14 Chen, L.Z., Feng, X.W., and Zhou, J.H. (1995) Effects of *Rehmannia glutinosa* polysaccharide b on T-lymphocytes in mice bearing sarcoma 180. *Acta Pharmacol. Sin.*, **16**, 337–340.

15 Tomoda, M., Miyamoto, H., Shimizu, N., Gonda, R., and Ohara, N. (1994) Two acidic polysaccharides having reticuloendothelial system-potentiating activity from the raw root of *Rehmannia glutinosa*. *Biol. Pharm. Bull.*, **17**, 1456–1459.

16 Tomoda, M., Miyamoto, H., Shimizu, N., Gonda, R., and Ohara, N. (1994) Characterization of two polysaccharides having activity on the reticuloendothelial system from the root of *Rehmannia glutinosa*. *Chem. Pharm. Bull. (Tokyo)*, **42**, 625–629.

17 Tomoda, M., Miyamoto, H., and Shimizu, N. (1994) Structural features and anti-complementary activity of rehmannan SA, a polysaccharide from the root of *Rehmannia glutinosa*. *Chem. Pharm. Bull. (Tokyo)*, **42**, 1666–1668.

18 Miura, T., Kako, M., Ishihara, E., Usami, M., Yano, H., Tanigawa, K., Sudo, K., and Seino, Y. (1997) Antidiabetic effect of seishin-kanro-to in KK-Ay mice. *Planta Med.*, **63**, 320–322.

19 Kiho, T., Watanabe, T., Nagai, K., and Ukai, S. (1992) Hypoglycemic activity of polysaccharide fraction from rhizome of *Rehmannia glutinosa* Libosch. f. *hueichingensis* Hsiao and the effect on carbohydrate metabolism in normal mouse liver. *Yakugaku Zasshi*, **112**, 393–400.

20 Guo, X.N., Zhang, R.X., Jia, Z.P., Li, M.X., and Wang, J. (2006) Effects of *Rehmannia glutinosa* oligosaccharides on proliferation of 3T3-L1 adipocytes and insulin resistance. *Zhongguo Zhong Yao Za Zhi*, **31**, 403–407.

21 Zhang, R., Zhou, J., Jia, Z., Zhang, Y., and Gu, G. (2004) Hypoglycemic effect of *Rehmannia glutinosa* oligosaccharide in hyperglycemic and alloxan-induced diabetic rats and its mechanism. *J. Ethnopharmacol.*, **90**, 39–43.

22 Kubo, M., Asano, T., Shiomoto, H., and Matsuda, H. (1994) Studies on rehmanniae radix. I. Effect of 50% ethanolic extract from steamed and dried rehmanniae radix on hemorheology in arthritic and thrombosic rats. *Biol. Pharm. Bull.*, **17**, 1282–1286.

23 Kubo, M., Asano, T., Matsuda, H., Yutani, S., and Honda, S. (1996) Studies on Rehmanniae radix. III. The relation between changes of constituents and improvable effects on hemorheology with the processing of roots of *Rehmannia glutinosa*. *Yakugaku Zasshi*, **116**, 158–168.

24 Kim, H.M., An, C.S., Jung, K.Y., Choo, Y.K., Park, J.K., and Nam, S.Y. (1999) *Rehmannia glutinosa* inhibits tumor necrosis factor-α and interleukin-1 secretion from mouse astrocytes. *Pharmacol. Res.*, **40**, 171–176.

25 Kim, H., Lee, E., Lee, S., Shin, T., Kim, Y., and Kim, J. (1998) Effect of *Rehmannia glutinosa* on immediate type allergic reaction. *Int. J. Immunopharmacol.*, **20**, 231–240.

26 Cui, Y., Yan, Z., Hou, S., and Chang, Z. (2004) Effect of radix Rehmanniae preparata on the expression of c-fos and NGF in hippocampi and learning and memory in rats with damaged thalamic arcuate nucleus. *Zhong Yao Cai*, **27**, 589–592.

27 Yu, H., Oh-Hashi, K., Tanaka, T., Sai, A., Inoue, M., Hirata, Y., and Kiuchi, K. (2006) *Rehmannia glutinosa* induces glial cell line-derived neurotrophic factor gene expression in astroglial cells via cPKC and ERK1/2 pathways independently. *Pharmacol. Res.*, **54**, 39–45.

28 Chao, J.C., Chiang, S.W., Wang, C.C., Tsai, Y.H., and Wu, M.S. (2006) Hot water-extracted Lycium barbarum and *Rehmannia glutinosa* inhibit proliferation and induce apoptosis of hepatocellular carcinoma cells. *World J. Gastroenterol.*, **12**, 4478–4484.

29 Yu, H.H., Seo, S.J., Kim, Y.H., Lee, H.Y., Park, R.K., So, H.S., Jang, S.L., and You, Y.O. (2006) Protective effect of *Rehmannia glutinosa* on the cisplatin-induced damage of HEI-OC1 auditory cells through scavenging free radicals. *J. Ethnopharmacol.*, **107**, 383–388.

30 Li, D.Q., Li, Y., Liu, Y., Bao, Y.M., Hu, B., and An, L.J. (2005) Catalpol prevents the loss of CA1 hippocampal neurons and reduces working errors in gerbils after ischemia-reperfusion injury. *Toxicon*, **46**, 845–851.

31 Tian, Y.Y., An, L.J., Jiang, L., Duan, Y.L., Chen, J., and Jiang, B. (2006) Catalpol protects dopaminergic neurons from LPS-induced neurotoxicity in mesencephalic neuron-glia cultures. *Life Sci.*, **80**, 193–199.

32 Chen, L.Z., Feng, X.W., Zhou, J.H., and Tang, J.F. (1993) Immuno-tumoricidal effect of *Rhemmania glutinosa* polysaccharide B and its mechanism. *Chin. J. Pharmacol. Toxicol.*, **7**, 153–156.

33 Wei, X.L. and Ru, X.B. (1997) Effects of low-molecular-weight *Rehmannia glutinosa* polysaccharides on p53 gene expression. *Acta Pharmacol. Sin.*, **18**, 471–474.

34 Xu, J.P. (1992) Research on liu wei *Rehmannia* oral liquid against side-effect of drugs of anti-tumor chemotherapy. *Chin. J. Integr. Trad. West. Med.*, **12**, 734–737.

35 Zee-Cheng, R.K. (1992) Shi-quan-da-bu-tang (ten significant tonic decoction), SQT. A potent Chinese biological response modifier in cancer immunotherapy, potentiation and detoxification of anticancer drugs. *Methods Find. Exp. Clin. Pharmacol.*, **14**, 725–736.

36 Zhuang, J., Zhang, M., Zeng, Z., Xu, F., Han, T., Hu, S., and Sun, Y. (1992) The use of 6-flavor *Rehmannia* decoction with additives in the prevention of ototoxic deafness induced by gentamicin in guinea pigs. *China J. Chin. Mater. Med.*, **17**, 496–499.

37 Yuan, Y., Hou, S., Lian, T., and Han, Y. (1992) Studies of *Rehmannia glutinosa* Libosch. f. *hueichingensis* as a blood tonic. *China J. Chin. Mater. Med.*, **17**, 366–368.

Rhaponticum uniflorum

Radix Rhapontici (Loulu) is the dried root of *Rhaponticum uniflorum* (L.) DC. (Asteraceae). It is used for the treatment of mastitis, galactostasis, scrofula and ulcers, arthritis, and carbuncle. The Chinese Pharmacopoeia notes Radix Rhapontici to be used with caution in pregnancy.

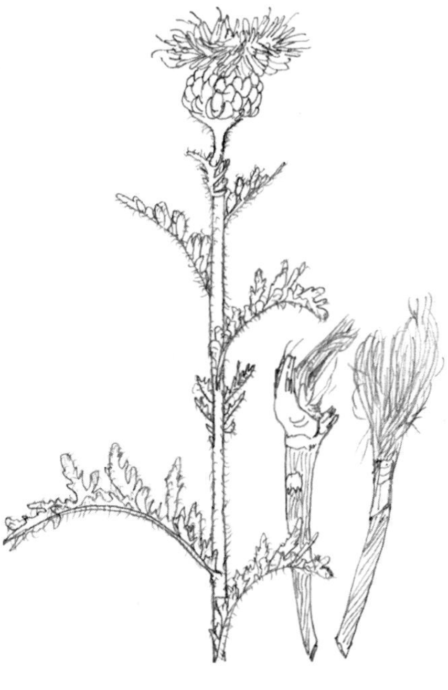

Rhaponticum uniflorum (L.) DC.

Chemistry

Three phytoecdysterols were isolated from the root of *R. uniflorum* and identified as rhapontisterone, ecdysterone, and turkesterone [1]. Further phytoecdysones from the roots of *R. uniflorum* were identified as ajugasterone C, ajugasterone C-20,22-monoacetonide, ajugasterone C-2,3,20,22-diacetonide and 5-deoxykaladasterone-20,22-monoacetonide [2], and rhapontisterone R1 (2,3,11,14,20,22,25-heptahydroxy-6-oxo-stigma-7,24(28)-dein-29-oic acid γ-lactone) [3].

Through a critical study of Loulu in ancient Chinese herbals, it was concluded that the botanical origin of Loulu should be the dried root of *R. uniflorum*, while the dried root of *Echinops latifolius*, which is also specified in the Chinese Pharmacopoeia, should not be used as Loulu, because of its totally different chemical constituents from those of the former [4].

Pharmacology and Toxicology

The roots of *R. uniflorum* were found to exhibit antioxidative activity [5, 6], to inhibit lipid peroxide production, and to improve membrane fluidity in smooth muscle cells *in vitro* [7].

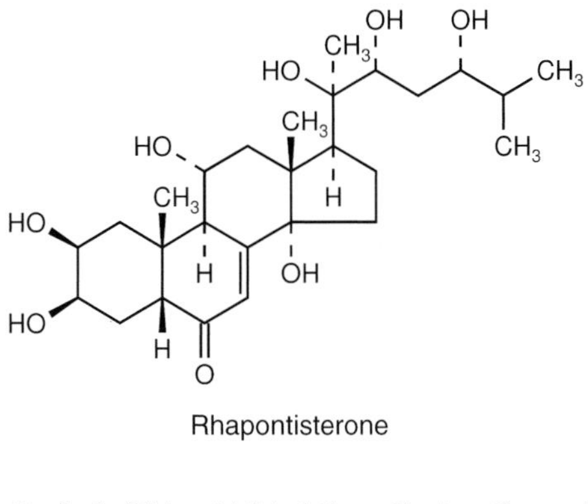

Rhapontisterone

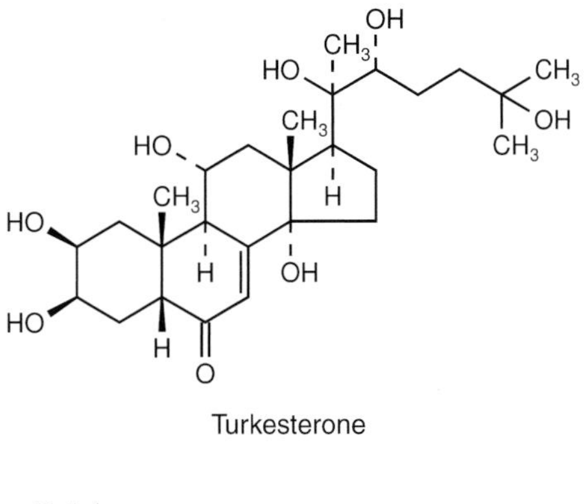

Turkesterone

Handbook of Chinese Medicinal Plants: Chemistry, Pharmacology, Toxicology
Weici Tang and Gerhard Eisenbrand

ISBN: 978-3-527-32226-8

References

1 Guo, D.A., Lou, Z.C., Gao, C.Y., Qiao, L., and Peng, J.R. (1991) Phytoecdysteroids of *Rhaponticum uniflorum* root. *Acta Pharm. Sin.*, **26**, 442–446.

2 Zhang, Y.H. and Wang, H.Q. (2001) Ecdysteroids from *Rhaponticum uniflorum. Pharmazie*, **56**, 828–829.

3 Li, X.Q., Wang, J.H., Wang, S.X., and Li, X. (2000) A new phytoecdysone from the roots of *Rhaponticum uniflorum. J. Asian Nat. Prod. Res.*, **2**, 225–229.

4 Guo, D. and Lou, Z. (1992) Textual study of Chinese drug loulu. *Zhongguo Zhong Yao Za Zhi*, **17**, 579–581, 638.

5 Lu, Y.C. (1985) Experimental study on the anti-atherosclerotic action of the Chinese medicinal *Rhaponticum uniflorum. Natl Med. J. China*, **65**, 750–753.

6 Fu, N.W. (1988) Antioxidation effect of *Rhaponticum uniflorum* against hematoporphyrin derivative photooxidation. *Acta Acad. Med. Sci.*, **10**, 95–99.

7 Lu, Y.C. (1993) Lipid peroxide, membrane fluidity of smooth muscle cells and atherosclerosis. *Chin. J. Pathol.*, **22**, 42–45.

Rheum officinale, *Rheum palmatum* and *Rheum tanguticum*

Radix et Rhizoma Rhei (Dahuang) is the dried root and rhizome of *Rheum palmatum* L., *Rheum tanguticum* Maxim. ex Balf., or *Rheum officinale* Baill. (Polygonaceae). It is used as a laxative, antiphlogistic, and hemostatic agent for the treatment of obstipation, gastrointestinal indigestion, diarrhea, and jaundice. It is also used for the treatment of gastrointestinal bleeding, menstrual disorders, conjunctivitis, traumatic diseases, carbuncle, and ulcer. By external application, it can also be used for the treatment of thermal burn. It is utilized either as the raw material, or after processing. The Chinese Pharmacopoeia notes that Radix et Rhizoma Rhei should be used with caution in pregnancy.

Rheum palmatum L.

Two galenic preparations from Radix et Rhizoma Rhei are officially listed in the Chinese Pharmacopoeia: Extractum Rhei Liquidum and "Xinqingning Pian." Extractum Rhei Liquidum (Dahuang Liujingao) is the fluid extract of Radix et Rhizoma Rhei; it is used as a laxative and bitter stomachic agent for the treatment of constipation and anorexia. "Xinqingning Pian" is the dragees prepared from Radix et Rhizoma Rhei; it is used for the treatment of obstipation and other symptoms caused by obstipation, and for diarrhea, fever, and toothache.

Chemistry

Rheum root and rhizome, rhubarb, are known to be one of the herbal medicines containing anthraquinones and anthraquinone glycosides and are used as a laxative agent. Other herbal medicines containing anthraquinones listed in the Chinese Pharmacopoeia are Aloe, the dried matter of the leaf exudate from *Aloe barbadensis* (Curacao Aloe) and *A. ferox* (Cape Aloe); Folium Sennae, the dried leaf of *Cassia acutifolia* and *C. angustifolia*; Rhizoma et Radix Polygoni Cuspidati, the dried root and rhizome of *Polygonum cuspidatum*, Caulis Polygoni Multiflori and Radix Polygoni Multiflori, the dried canes and the dried tuber of *Polygonum multiflorum*; Semen Cassiae; the dried ripe seed of *C. obtusifolia* and *C. tora*; as well as Radix et Rhizoma Rubiae, the dried root and rhizome of *Rubia cordifolia*.

Anthraquinone from rhubarb were identified as chrysophanol, aloe-emodin, emodin, physcion, and rhein [1]. All five anthraquinones possess hydroxy substituents at positions 1 and 8. The anthraquinone derivatives occur in the rhubarb root mainly as glycosides. The Chinese Pharmacopoeia requires a qualitative determination of rhein in Radix et Rhizoma Rhei and in Extractum Rhei Liquidum by thin-layer chromatographic comparison with reference substance, and a quantitative determi-

Handbook of Chinese Medicinal Plants: Chemistry, Pharmacology, Toxicology
Weici Tang and Gerhard Eisenbrand

ISBN: 978-3-527-32226-8

nation of aloe-emodin, rhein, emodin, chrysophanol and physcion in Radix et Rhizoma Rhei by HPLC. The total content of aloe-emodin, rhein, emodin, chrysophanol and physcion in Radix et Rhizoma Rhei should be not less than 1.5%. The Chinese Pharmacopoeia requires a quantitative determination of emodin and chrysophanol in Extractum Rhei Liquidum by HPLC. The total content of emodin and chrysophanol in Extractum Rhei Liquidum should be not less than 0.45%. The Chinese Pharmacopoeia also requires a quantitative determination of emodin and chrysophanol in "Xinqingning Pian" dragees by HPLC. The total content of emodin and chrysophanol in each "Xinqingning Pian" dragee should be not less than 1.8 mg.

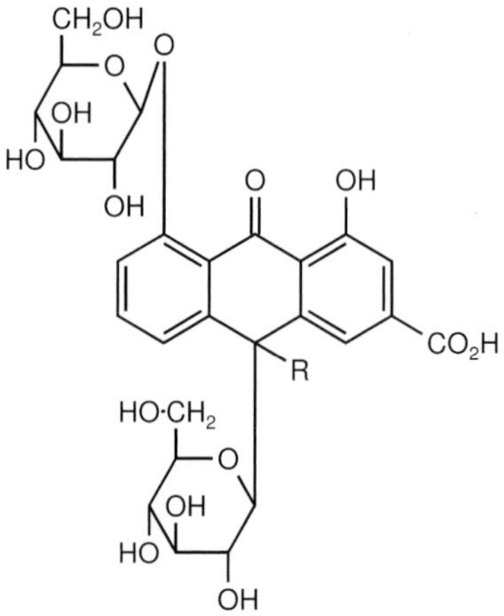

Rheinoside A: R = OH
Rheinoside C: R = H

Bianthrones (dianthrones) from rhubarb are derived from the five major anthraquinone derivatives physcion, rhein, emodin, aloe-emodin and chrysophanol, including chrysophanol bianthrones together with the heterobianthrones palmidins, rheidins and sennidins. Bianthrone glycosides from rhubarb root were identified as stereoisomeric pair sennosides A and B; sennosides C and D; and sennosides E and F. Anthrone *C,O*-diglycosides, rheinosides A–D were isolated from the rhubarb root. Rheinosides A and B are stereoisomeric 5-*O*-β-D-glucopyranosyl-9-*C*-β-D-glucopyranosyl 9-hydroxyrheinanthrones. Rheinosides C and rheinoside D differ from rheinosides A and B by the absence of the 9-hydroxyl group [2].

Rhubarb is known to contain tannins. Samples of *R. palmatum* were found to contain about 11% tannin, whereas those of *R. officinale* and *R. tanguticum* contained 4–7% [3]. Tannins from rhubarb root belong to both the condensed [4] and the hydrolyzable tannins [5]. Chromones such as 2,5-dimethyl-7-hydroxychromone, 2-methyl-5-carboxymethyl-7-hydroxychromone, 2-(2-hydroxypropyl)-5-methyl-7-hydroxychromone [6]; and stilbene glycosides such as piceid, 3,5,4′-trihydroxystilbene 4′-*O*-β-D-glucopyranoside, 3,5,4′-trihydroxystilbene 4′-*O*-β-D-(6″-galloyl)-glucopyranoside [7] were also isolated from rhubarb. The Chinese Pharmacopoeia further requires a differential determination of rhaponticin in Radix et Rhizoma Rhei and in Extractum Rhei Liquidum. Rhaponticin is a stilbene glycoside present in the root of *Rheum rhaponticum* L. as a major component. The root of *R. rhaponticum* should

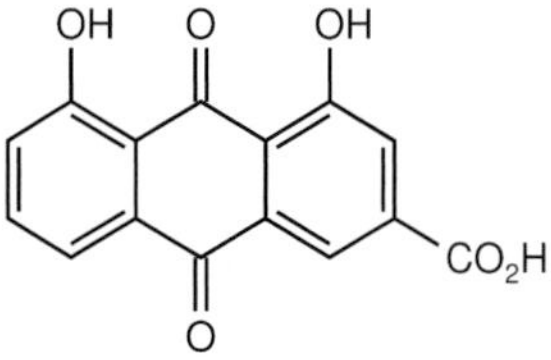

Rhein

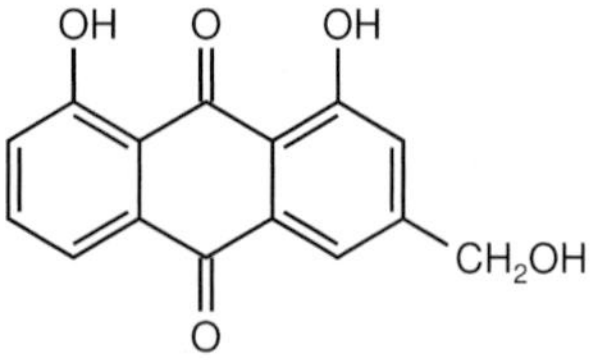

Aloe-emodin

not be used as a substitute of Radix et Rhizoma Rhei.

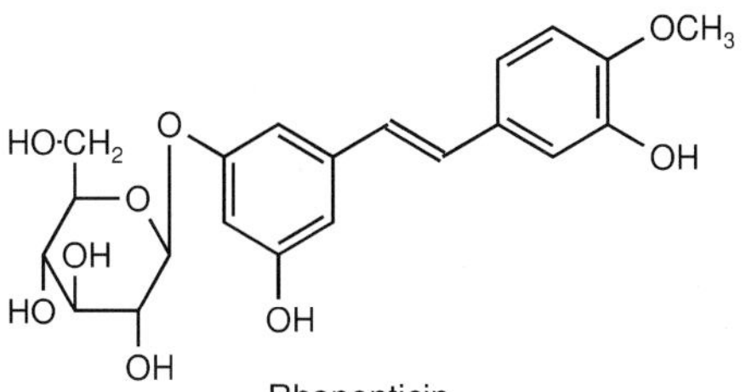

Rhaponticin

Two homogeneous acidic heteroglycans, DHP-1 and DHP-2, were obtained from the root of *R. palmatum*, the mean molecular weights being 110 and 25 kDa, respectively. Both heteroglycans contained the same major residues of glucose, galactose, arabinose, rhamnose, lyxose, xylose, glucuronic acid, and galacturonic acid [8].

Pharmacology and Toxicology

Rhubarb was reported to exert antimicrobial [9] and antiparasitic [10] activities. An activity of *Rheum* extract against herpes viruse (HSV) was also observed in cell culture, with a minimum inhibitory concentration (MIC) of 100 μg ml^{-1}; moreover, the extract showed no cytotoxicity even at a concentration of 20 mg ml^{-1} [11]. Emodin, rhein, and emodin bianthrone each showed antiviral activity against a normal laboratory human cytomegalovirus strain, AD-169 [12]. Aloe-emodin inactivated various viruses, including HSV types 1 and 2, varicella zoster virus and influenza virus, but not adenovirus and rhinovirus. Electron microscopic examination of aloe-emodin-treated HSV showed the envelopes to be partially disrupted; this suggested that aloe-emodin and some related anthraquinones were directly virucidal to enveloped viruses [13]. Interestingly, emodin was recently reported to exert antiviral activity against severe acute respiratory syndrome (SARS), an emerging infectious disease caused by a novel coronavirus. SARS coronavirus spike protein, a type I membrane-bound protein, was found to be essential for the viral attachment to the host cell receptor for angiotensin-converting enzyme 2. Emodin significantly blocked the SARS spike protein and angiotensin-converting enzyme 2 interaction, in a concentration-dependent manner [14].

Rhein and sennosides in *R. palmatum* are believed to be responsible for the laxative activity [15]. Sennoside content in rhubarb correlated highly with its laxative activity. In addition to laxative activity, the oral administration of emodin and rhein provoked marked diuretic, natriuretic, and kaliuretic effects in rabbits [16]. Sennoside A and sennoside C, when injected directly into the cecum of mice, showed equal purgative activities. Intracecal administration reduced the time to onset of stool, as induced by sennoside C, from about 3 h after oral administration to about 24 min. At 2.3 h after oral administration of sennoside C, near-equimolar amounts of aloe-emodin anthrone and rhein anthrone were detected in the large intestine of mice. It was concluded that aloe-emodin anthrone and rhein anthrone, formed by intraluminal bacteria, were the true laxative metabolites of sennoside C in mice and that both anthrones exert their purgative effects synergistically [17]. An *in vitro* study using isolated gastric muscle strips of guinea pigs revealed that rhubarb exerted excitatory actions on gastric smooth muscle strips, partly mediated via the cholinergic M and N receptor, and L-type calcium channel [18]. A new possibility for the treatment of postprandial hyperlipidemia has been demonstrated in alloxan-induced diabetic rats with gastroparesis by improving gastrointestinal transit with orally administered rhubarb [19].

Rhubarb was also used as a hemostatic agent in traditional Chinese medicine. The hemostatic activity has been demonstrated experimentally and clinically. Significant therapeutic effects of powdered rhubarb were reported in the treatment of gastric and duodenal ulcer bleeding. Emodin, at an intraperitoneal (i.p.) dose of $15\,mg\,kg^{-1}$, exhibited antiulcer activity against pylorus-ligated, aspirin and immobilization stress-induced gastric ulcers [20]. Aloe-emodin and emodin each elicited a dose-dependent growth inhibition of *Helicobacter pylori* in culture [21]. Arylamine *N*-acetyltransferase activities in *H. pylori* collected from peptic ulcer patients were found to be inhibited by aloe-emodin and emodin [22, 23]. The decreased *N*-acetyltransferase activity was associated with increased aloe-emodin and emodin in *H. pylori* cytosols.

Among the anthraquinones from rhubarb, emodin and rhein were found to significantly inhibit nitric oxide (NO) production from RAW 264.7 murine macrophage cells, as stimulated by lipopolysaccharide (LPS). The inhibitory activity of emodin and rhein was not due to a direct inhibition of inducible NO synthase (iNOS) activity, but to result from an inhibited expression of iNOS and the cyclooxygenase-2 (COX-2) protein. The production of prostaglandin E_2 (PGE_2) was also reduced by emodin in LPS-activated RAW 264.7 cells. Rhein also inhibited LPS-induced iNOS protein expression, but not COX-2 or PGE_2 production. This suggested that emodin and rhein were major iNOS inhibitors in rhubarb, and might exert anti-inflammatory effects [24].

Rhubarb has been used for wound healing. Emodin was shown capable of enhancing cutaneous wound healing in rats. The topical administration of an emodin solution significantly improved wound healing over five days after injury. Emodin treatment led to a markedly higher hydroxyproline content in day 7 wounds and tensile strength in day 14 wounds than in control, untreated wounds. Levels of transforming growth factor-β_1 (TGF-β_1) in the wound tissues showed a dose-dependent increase in emodin-treated wounds. The results showed that emodin promoted wound repair in rats via a complex mechanism that involved the stimulation of tissue regeneration and regulation of TGF-β_1 signaling pathways [25].

Rhubarb was reported to be effective in the treatment of chronic renal failure [26]. A rhubarb extract was active in reducing proteinuria and severity of glomerulosclerosis in rats with remnant kidneys [27]. The clinical treatment of chronic renal failure with rhubarb was also reported; it reduced urea nitrogen, adjusted the anemia, and regulated the membrane function [28]. The addition of serum obtained from rats fed with rhubarb significantly inhibited the proliferation of renal tubular cells in culture in a dose-dependent manner, as measured by the uptake of [^{3}H]thymidine [29]. Renal hypertrophy and an elevated glomerular filtration rate in the early stage of nephropathy in streptozotocin-induced diabetic rats were effectively inhibited by rhubarb [30]. A study on the effects of rhubarb on arachidonic acid metabolism in the renal medulla of rabbits indicated that rhubarb blocked the biosynthesis of PGE_2 and $PGF_{2\alpha}$ in a dose-dependent manner. Rhubarb also inhibited the biosynthesis of PGA_2 and thromboxane B_2 [31]. In contrast, a study on the nephrotoxicity of the total rhubarb anthraquinones in Sprague-Dawley rats has been reported recently, using DNA microarrays. The mitogen-activated protein kinase (MAPK) kinase 6 seemed to be a target gene, resulting in cell-cycle arrest and contributing to nephrotoxicity in rats [32].

Among anthraquinones and stilbenes from rhubarb, emodin and aloe-emodin showed higher cytotoxic activities against human oral squamous cell carcinoma HSC-2 and salivary gland tumor HSG cells, as compared to normal human gingival fibroblasts. Chrysophanol 8-*O*-β-(6′-acetyl)-glucopyranoside, 4-(4′-hydroxyphenyl)-2-butanone 4′-*O*-β-D-(2″-*O*-galloyl-6″-*O*-cinnamoyl)-glucopyranoside, and 6″-*O*-(4‴-hydroxybenzoyl)-resveratroloside exhibited relatively higher cytotoxic activities against tumor and normal cells [33]. In CH27 human lung squamous carcinoma cells, emodin exerted cytotoxicity, induced cell-cycle arrest at sub-G_1 phase and cell apoptosis, upregulated Bak and Bax expression, and activated caspase-3, caspase-9, and caspase-8 [34]. Aloe-emodin also induced cell death and cell apoptosis in CH27 cells. Apoptosis involved a modulation of the expression of Bcl-2 family proteins; this suggested that aloe-emodin induced cell death by the Bax and Fas death pathway [35].

Emodin exhibited significant cytotoxicity against human myeloma cells and induced cell apoptosis. It selectively inhibited the interleukin-6 (IL-6)-induced activation of Janus-activated kinase 2 and phosphorylation of signal transducer and activator of transcription 3, inducing apoptosis in myeloma cells via downregulation of myeloid cell leukemia 1 [36]. Rhein was reported to inhibit growth and to induce apoptosis, and to block cell-cycle progression in the G_1 phase of HepG2 human hepatoblastoma cells [37].

Aloe-emodin was found to induce cell apoptosis in H460 human non-small-cell lung carcinoma cells. The release of nucleophosmin from the nucleus into cytosol, and the degradation of nucleophosmin, were each associated with H460 cell apoptosis induced by aloe-emodin [38]. It also inhibited the proliferation of MGC-803 human gastric cancer cells with an increase in S phase in cell cycle and decreased alkaline phosphatase activity as an indicator of cell differentiation [39]. It was also reported that the aloe-emodin-induced apoptosis of H460 cells involved a modulation of protein expression of cAMP-dependent protein kinase, protein kinase C, Bcl-2, caspase-3 and p38. The expression of p38 seemed to be an important determinant of apoptotic death induced by aloe-emodin [40].

Emodin was further reported to inhibit angiogenesis induced by vascular endothelial growth factor-A (VEGF-A), both *in vitro* and *in vivo*. *In vitro*, it inhibited concentration-dependently the proliferation, migration into the denuded area, invasion through a layer of matrigel and tube formation of human umbilical vein endothelial cells (HUVEC) stimulated with VEGF-A. Emodin induced cell-cycle arrest of HUVEC at G_0/G_1 phase by blocking the expression of cyclin D1 and E and the phosphorylation of Rb protein. Emodin also suppressed matrigel invasion by inhibiting the basal secretion of matrix metalloproteinase-2 and the expression of urokinase plasminogen activator receptor in HUVEC stimulated by VEGF-A. *In vivo*, emodin strongly suppressed neovessel formation in the chorioallantoic membrane of chicks, and VEGF-A-induced angiogenesis of the matrigel plug in mice [41].

References

1 He, L.Y. (1985) Determination of anthraquinone derivatives in rhubarb for export. *Bull. Chin. Mater. Med.*, **10**, 33–35.

2 Yamagishi, T., Nishizawa, M., Ikura, M., Hikichi, K., Nonaka, G., and Nishioka, I. (1987) Studies on rhubarb (Rhei rhizoma). Part XI. New laxative constituents of rhubarb, isolation and characterization of rheinosides A, B, C and D. *Chem. Pharm. Bull. (Tokyo)*, **35**, 3132–3138.

3 Luo, W.Y. and Zhang, Y.Z. (1986) Studies on the determination of tannin in Dahuang. I. Casein method. *Chin. J. Pharm. Anal.*, **6**, 15–18.

4 Kashiwada, Y., Nonaka, G., and Nishioka, I. (1984) Tannins and related compounds XXIII. Rhubarb. 4. Isolation and structures of new classes of gallotannins. *Chem. Pharm. Bull. (Tokyo)*, **32**, 3461–3470.

5 Nonaka, G. and Nishioka, I. (1983) Tannins and related compounds X. Rhubarb. 2.

Isolation and structures of a glycerol gallate, gallic acid glucoside gallates, galloylglucose and isolinallyin. *Chem. Pharm. Bull. (Tokyo)*, **31**, 1652–1658.

6 Kashiwada, Y., Nonaka, G., and Nishioka, I. (1984) Studies on rhubarb (rhei rhizoma). V. Isolation and characterization of chromone and chromanone derivatives. *Chem. Pharm. Bull. (Tokyo)*, **32**, 3493–3500.

7 Nonaka, G., Minami, M., and Nishioka, I. (1977) Studies on Rhubarb (Rhei rhizoma). III. Stilbene glycosides. *Chem. Pharm. Bull. (Tokyo)*, **25**, 2300–2305.

8 Zhang, S.J., Zhang, S.Y., Wang, L., and Zhu, B. (1993) Studies on polysaccharide of *Rheum palmatum* L. *China J. Chin. Mater. Med.*, **18**, 679–681.

9 Cyong, J.C., Matsumoto, T., Arakawa, K., Kiyohara, H., Yamada, H., and Otsuka, Y. (1987) Anti-*Bacteroides fragilis* substance from rhubarb. *J. Ethnopharmacol.*, **19**, 279–283.

10 Wang, H.H. (1993) Antitrichomonal action of emodin in mice. *J. Ethnopharmacol.*, **40**, 111–116.

11 Wang, Z., Wang, G., Xu, H., and Wang, P. (1996) Anti-herpes virus action of ethanol-extract from the root and rhizome of *Rheum officinale* Baill. *China J. Chin. Mater. Med.*, **21**, 364–366.

12 Barnard, D.L., Huffman, J.H., Morris, J.L., Wood, S.G., Hughes, B.G., and Sidwell, R.W. (1992) Evaluation of the antiviral activity of anthraquinones, anthrones and anthraquinone derivatives against human cytomegalovirus. *Antiviral Res.*, **17**, 63–77.

13 Sydiskis, R.J., Owen, D.G., Lohr, J.L., Rosler, K. H., and Blomster, R.N. (1991) Inactivation of enveloped viruses by anthraquinones extracted from plants. *Antimicrob. Agents Chemother.*, **35**, 2463–2466.

14 Ho TYWu, S.L., Chen, J.C., Li, C.C., and Hsiang, C.Y. (2007) Emodin blocks the SARS coronavirus spike protein and angiotensin-converting enzyme 2 interaction. *Antiviral Res.*, **74**, 92–101.

15 Liu, X.X. (1992) Pharmacological action and clinical application of *Rheum palmatum*. *Chin. J. Integr. Trad. West. Med.*, **12**, 571–573.

16 Zhou, X.M. and Chen, Q.H. (1988) Biochemical study of Chinese rhubarb XXII. Inhibitory effect of anthraquinone derivatives on sodium-potassium-ATPase of rabbit renal medulla and their diuretic action. *Acta Pharm. Sin.*, **23**, 17–20.

17 Yamauchi, K., Shinano, K., Nakajima, K., Yagi, T., and Kuwano, S. (1992) Metabolic activation of sennoside C in mice: synergistic action of anthrones. *J. Pharm. Pharmacol.*, **44**, 973–976.

18 Yu, M., Luo, Y.L., Zheng, J.W., Ding, Y.H., Li, W., Zheng, T.Z., and Qu, S.Y. (2005) Effects of rhubarb on isolated gastric muscle strips of guinea pigs. *World J. Gastroenterol.*, **11**, 2670–2673.

19 Xie, W., Xing, D., Zhao, Y., Su, H., Meng, Z., Chen, Y., and Du, L. (2005) A new tactic to treat postprandial hyperlipidemia in diabetic rats with gastroparesis by improving gastrointestinal transit. *Eur. J. Pharmacol.*, **510**, 113–120.

20 Goel, R.K., Das Gupta, G., Ram, S.N., and Pandey, V.B. (1991) Antiulcerogenic and anti-inflammatory effects of emodin, isolated from *Rhamnus triquerta* Wall. *Indian J. Exp. Biol.*, **29**, 230–232.

21 Wang, H.H. and Chung, J.G. (1997) Emodin-induced inhibition of growth and DNA damage in the *Helicobacter pylori*. *Curr. Microbiol.*, **35**, 262–266.

22 Wang, H.H., Chung, J.G., Ho, C.C., Wu, L.T., and Chang, S.H. (1998) Aloe-emodin effects on arylamine N-acetyltransferase activity in the bacterium *Helicobacter pylori*. *Planta Med.*, **64**, 176–178.

23 Chung, J.G., Wang, H.H., Wu, L.T., Chang, S.S., and Chang, W.C. (1997) Inhibitory actions of emodin on arylamine N-acetyltransferase activity in strains of *Helicobacter pylori* from peptic ulcer patients. *Food Chem. Toxicol.*, **35**, 1001–1007.

24 Wang, C.C., Huang, Y.J., Chen, L.G., Lee, L.T., and Yang, L.L. (2002) Inducible nitric oxide synthase inhibitors of Chinese herbs III. *Rheum palmatum*. *Planta Med.*, **68**, 869–874.

25 Tang, T., Yin, L., Yang, J., and Shan, G. (2007) Emodin, an anthraquinone derivative from *Rheum officinale* Baill, enhances cutaneous wound healing in rats. *Eur. J. Pharmacol.*, **567**, 177–185.

26 Huang, C.L., Li, C., and Deng, Y.B. (1995) Progress on research of mechanism of *Rheum palmatum* in delaying the chronic renal failure. *Chin. J. Integr. Trad. West. Med.*, **15**, 506–508.

27 Li, L. (1996) *Rheum officinale*: a new lead in preventing progression of chronic renal failure. *Chin. Med. J. (Engl.)*, **109**, 35–37.

28 Kang, Z., Bi, Z., Ji, W., Zhao, C., and Xie, Y. (1993) Observation of therapeutic effect in 50 cases of chronic renal failure treated with rhubarb and adjuvant drugs. *J. Trad. Chin. Med.*, **13**, 249–252.

29 Zheng, F. (1993) Effect of *Rheum officinal* on the proliferation of renal tubular cells *in vitro*. *Natl Med. J. China*, **73**, 343–345.

30 Yang, J.W. and Li, L.S. (1993) Effects of *Rheum* on renal hypertrophy and hyperfiltration of experimental diabetes in rat. *Chin. J. Integr. Trad. West. Med.*, **13**, 286–288.

31 Guo, C.Y., Zhao, S.Y., and Lin, C.R. (1989) Effects of rhubarb on arachidonic acid metabolism of the renal medulla in the rabbit. *Chin. J. Integr. Trad. West. Med.*, **9**, 161–163.

32 Yan, M., Zhang, L.Y., Sun, L.X., Jiang, Z.Z., and Xiao, X.H. (2006) Nephrotoxicity study of total rhubarb anthraquinones on Sprague Dawley rats using DNA microarrays. *J. Ethnopharmacol.*, **107**, 308–311.

33 Shi, Y.Q., Fukai, T., Sakagami, H., Kuroda, J., Miyaoka, R., Tamura, M., Yoshida, N., and Nomura, T. (2001) Cytotoxic and DNA damage-inducing activities of low molecular weight phenols from rhubarb. *Anticancer Res.*, **21** (A), 2847–2853.

34 Lee, H.Z. (2001) Effects and mechanisms of emodin on cell death in human lung squamous cell carcinoma. *Br. J. Pharmacol.*, **134**, 11–20.

35 Lee, H.Z., Hsu, S.L., Liu, M.C., and Wu, C.H. (2001) Effects and mechanisms of aloe-emodin on cell death in human lung squamous cell carcinoma. *Eur. J. Pharmacol.*, **431**, 287–295.

36 Muto, A., Hori, M., Sasaki, Y., Saitoh, A., Yasuda, I., Maekawa, T., Uchida, T., Asakura, K., Nakazato, T., Kaneda, T., Kizaki, M., Ikeda, Y., and Yoshida, T. (2007) Emodin has a cytotoxic activity against human multiple myeloma as a Janus-activated kinase 2 inhibitor. *Mol. Cancer Ther.*, **6**, 987–994.

37 Kuo, P.L., Hsu, Y.L., Ng, L.T., and Lin, C.C. (2004) Rhein inhibits the growth and induces the apoptosis of Hep G2 cells. *Planta Med.*, **70**, 12–16.

38 Lee, H.Z., Wu, C.H., and Chang, S.P. (2005) Release of nucleophosmin from the nucleus: Involvement in aloe-emodin-induced human lung non small carcinoma cell apoptosis. *Int. J. Cancer*, **113**, 971–976.

39 Guo, J., Xiao, B., Zhang, S., Liu, D., Liao, Y., and Sun, Q. (2007) Growth inhibitory effects of gastric cancer cells with an increase in S phase and alkaline phosphatase activity repression by aloe-emodin. *Cancer Biol. Ther.*, **6**, 85–88.

40 Yeh, F.T., Wu, C.H., and Lee, H.Z. (2003) Signaling pathway for aloe-emodin-induced apoptosis in human H460 lung non-small carcinoma cell. *Int. J. Cancer*, **106**, 26–33.

41 Kwak, H.J., Park, M.J., Park, C.M., Moon, S.I., Yoo, D.H., Lee, H.C., Lee, S.H., Kim, M.S., Lee, H.W., Shin, W.S., Park, I.C., Rhee, C.H., and Hong, S.I. (2006) Emodin inhibits vascular endothelial growth factor-A-induced angiogenesis by blocking receptor-2 (KDR/Flk-1) phosphorylation. *Int. J. Cancer*, **118**, 2711–2720.

Rhododendron dauricum and *Rhododendron molle*

Folium Rhododendri Daurici (Manshanhong) is the dried leaf of *Rhododendron dauricum* L. (Ericaceae). It is used as an expectorant and mucolytic for the treatment of acute and chronic bronchitis.

Rhododendron dauricum L.

Oleum Rhododendri Daurici (Manshanhongyou) is the essential oil obtained by steam distillation of the leaves of *R. dauricum*. It is used for the treatment of acute and chronic bronchitis. Oleum Rhododendri Daurici is used mainly in form of its galenic preparation, "Manshanhongyou Jiaowan".

Flos Rhododendri Mollis (Naoyanghua) is the dried flower of *Rhododendron molle* G. Don. It is used as an analgesic agent for the treatment of traumatic and rheumatic diseases. The Chinese Pharmacopoeia notes that Flos Rhododendri Mollis is toxic; overdosage or long-term administration should be avoided. It is also contraindicated in debility and in pregnancy.

Two galenic preparation from Folium Rhododendri Daurici, "Xiaokechuan Tangjiang" and "Manshanhongyou Jiaowan," are officially listed in the Chinese Pharmacopoeia. "Xiaokechuan Tangjiang" is a syrup of Folium Rhododendri Daurici, prepared by extraction of the leaf with 40% ethanol. "Manshanhongyou Jiaowan" is capsules of Oleum Rhododendri Daurici (0.05 or 0.1 g essential oil per capsule). These are both used as an expectorant, mucolytic, or antiasthmatic agent for the treatment of chronic bronchitis.

Chemistry

The leaf of *R. dauricum* is known to contain flavones and flavone glycosides, with farrerol and quercetin glycosides as the major components. The content of farrerol in the dry leaves of *R. dauricum* was reported to range from 0.07 to 0.15% [1]. The Chinese Pharmacopoeia requires a qualitative determination of farrerol in Folium Rhododendri Daurici by thin-layer chromatographic (TLC) comparison with a reference substance, and a quantitative determination by HPLC. The content of farrerol in Folium Rhododendri Daurici should be not less than 0.08%.

The Chinese Pharmacopoeia requires a quantitative determination of total flavonoids in the syrup "Xiaokechuan Tangjiang" by spectrophotometric analysis, and of farrerol in the syrup by HPLC. The content of total flavonoids in the syrup "Xiaokechuan Tangjiang" should be not less than 2.0 mg ml^{-1}, calculated as rutin, and the content of farrerol in the syrup should be not less than 50 $\mu g\,ml^{-1}$. The Chinese Pharmacopoeia also requires a qualitative determination of farrerol in "Manshanhongyou Jiaowan" by TLC comparison with reference substance.

Handbook of Chinese Medicinal Plants: Chemistry, Pharmacology, Toxicology
Weici Tang and Gerhard Eisenbrand

ISBN: 978-3-527-32226-8

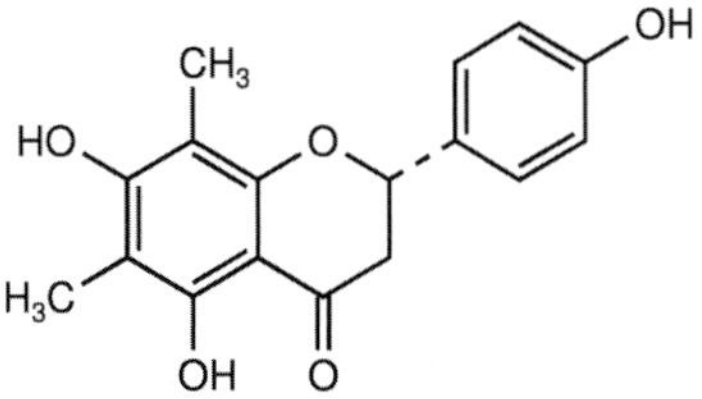

Farrerol

A toxic diterpene derivative andromedotoxin (grayanotoxin I) was also isolated from the leaves of *R. dauricum* [2]. The toxic diterpene derivatives related to andromedotoxin are the major components in the flower of *R. molle*. Diterpenes from the flower of *R. molle* were identified as rhodojaponins II, III and IV, grayanotoxin II, rhodomolleins I, II, III, IX, X, XI, XII, XIII, XIV, and XIX [3], and rhodomolins A and B [4]. All of these diterpenes possess a grayanotoxane skeleton except for rhodomollein XIV, which is a kalmane-type diterpene [3].

The essential oil of *R. dauricum* was detected to contain germacrone, menthol, juniper camphor, and α-eudesmol, β-eudesmol, and γ-eudesmol [5]. The Chinese Pharmacopoeia requires a qualitative determination of germacrone in Oleum Rhododendri Daurici by TLC comparison with a reference substance. The Chinese Pharmacopoeia gives the following physico-chemical parameters for Oleum Rhododendri Daurici: relative density 0.935–0.950 at 25 °C; and refractive index 1.50–1.52.

Pharmacology and Toxicology

Andromedotoxin and related toxins are known as sodium channel blockers; they exert selective effects on voltage-dependent sodium channels by eliminating fast sodium inactivation, and causing a hyperpolarizing shift in the voltage dependence of channel activation [6, 7]. Rhodojaponin III exhibited a positive inotropic effect on isolated guinea pig papillary muscles and decreased the resting membrane potential and amplitude of the action potential. It acts selectively on myocardial sodium channels, causing an increase in sodium influx; hence, it is recommended as a tool for studying cardiac sodium channels [8]. Intoxication cases caused by the consumption of honey produced from the rhododendron plant flowers containing andromedotoxin and related toxins were

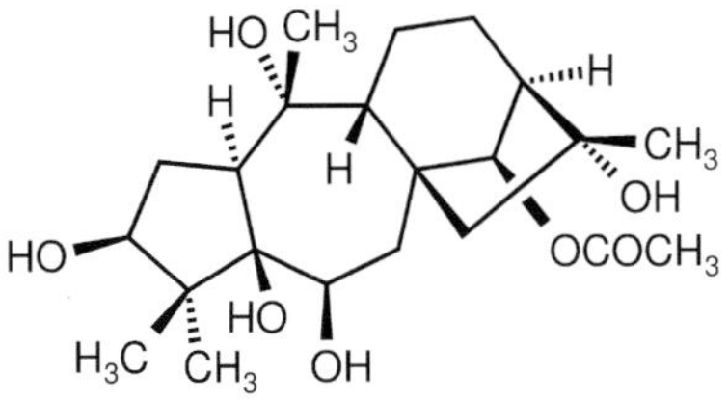

Grayanotoxin I

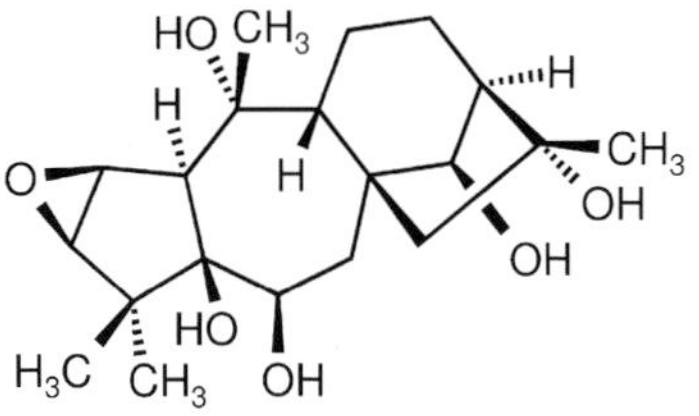

Rhodojaponin III

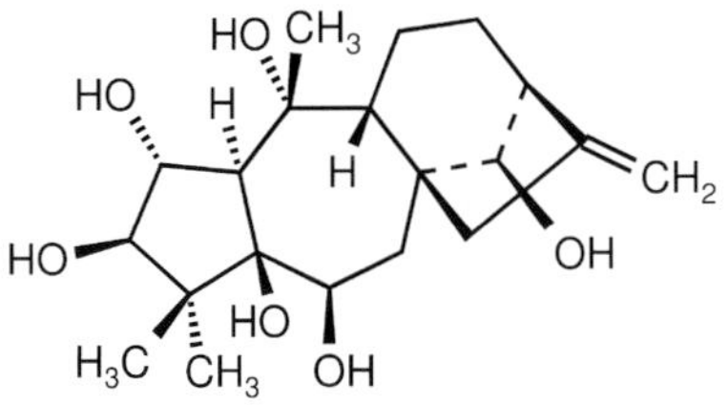

Rhodomollein IX

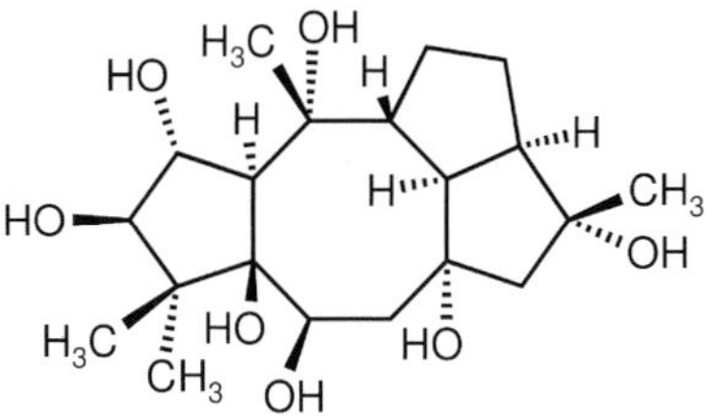

Rhodomollein XIV

reported. Symptoms of intoxication at low doses were dizziness, hypotension, and bradycardia. At high doses impaired consciousness, seizures, and atrioventricular block were reported [9, 10].

Farrerol was reported to exhibit an expectorant effect. The oral or intraperitoneal (i.p.) administration of farrerol to rats increased the output of phenol red from the respiratory tract. By repeated oral administration, farrerol decreased the protein content of respiratory tract washings in mice and rats [11]. Farrerol was found to be well absorbed from the gastrointestinal tract of rats, with 70–80% of an oral dose of 200 mg kg^{-1} in rats being cleared from the gastrointestinal tract within 6–12 h. Approximately 30% of the dose was excreted in the feces. At 1 h after intravenous (i.v.) administration, the highest tissue level of farrerol was found to be in the lungs [12].

References

1 Zhang, G.D., Wang, M.Z., and Zhang, S.R. (1980) Studies on the determination of farrerol content in *Rhododendron dauricum* leaves. *Acta Pharm. Sin.*, **15**, 736–740.

2 Fu, F.Y., Liu, Y.L., Liang, H.T., King, P.Y., Chen, Y.Y., and Yu, C.K. (1976) Studies on the constituents of Man-Shang-Hong (*Rhododendron dauricum* L.). II. *Acta Pharm. Sin.*, **34**, 223–227.

3 Chen, S.N., Zhang, H.P., Wang, L.Q., Bao, G. H., and Qin, G.W. (2004) Diterpenoids from the flowers of *Rhododendron molle*. *J. Nat. Prod.*, **67**, 1903–1906.

4 Zhong, G., Hu, M., Wei, X., Weng, Q., Xie, J., Liu, J., and Wang, W. (2005) Grayanane diterpenoids from the flowers of *Rhododendron molle* with cytotoxic activity against a *Spodoptera frugiperda* cell line. *J. Nat. Prod.*, **68**, 924–926.

5 Hsu, C.C. and Yu, T.C. (1976) Studies on the chemical constituents of Man-Shang Hong (*Rhododendron dauricum* L.) III. *Acta Chim. Sin.*, **34**, 275–281.

6 Yuki, T., Yamaoka, K., Yakehiro, M., and Seyama, I. (2001) State-dependent action of grayanotoxin I on Na^+ channels in frog ventricular myocytes. *J. Physiol.*, **534** (3), 777–790.

7 Maejima, H., Kinoshita, E., Seyema, I., and Yamaoka, K. (2003) Distinct sites regulating grayanotoxin binding and unbinding to D4S6 of $Na_v1.4$ sodium channel as revealed by improved estimation of toxin sensitivity. *J. Biol. Chem.*, **278**, 9464–9471.

8 Jin, M.W., Zong, X.G., Fang, D.C., and Jiang, M.X. (1985) Effect of rhomotoxin on the membrane potential and contractile force of guinea pig papillary muscles. *Acta Pharm. Sin.*, **20**, 481–484.

9 Koca, I. and Koca, A.F. (2007) Poisoning by mad honey: a brief review. *Food Chem. Toxicol.*, **45**, 1315–1318.

10 Gunduz, A., Turedi, S., Uzun, H., and Topbas, M. (2006) Mad honey poisoning. *Am. J. Emerg. Med.*, **24**, 595–598.

11 Chinese Academy of Medical Sciences (1977) Expectorant action of farrerol. *Chin. Med. J. (Engl.)*, **3**, 259–265.

12 Feng, Y.S. and Zhu, D.Z. (1979) Metabolism of farrerol. *Acta Pharm. Sin.*, **14**, 149–155.

Ricinus communis

Semen Ricini (Bimazi) is the dried ripe seed of *Ricinus communis* L. (Euphorbiaceae) collected in the fall. It is used a laxative, and for the treatment of carbuncle and scrofula.

Ricinus communis L.

Oleum Ricini (Bima You) is the fatty oil from the seed of *R. communis* produced by pressing and purification. It is used as an irritant laxative. The Chinese Pharmacopoeia notes that Oleum Ricini is incompatible with lipophilic anthelmintics, and should not be used by pregnant women.

Chemistry

The seed of *R. communis* is known to contain fatty oil. The Chinese Pharmacopoeia provides the following physico-chemical parameters for *Ricinus* oil: relative density 0.956–0.969 at 25 °C; refractive index 1.478–1.480; acid value not more than 2.0; saponification value 176–186; and iodine value 82–90. *Ricinus* oil is composed of 90% ricinoleate (12-hydroxyoleate), which is responsible for its laxative activity [1]. Besides fatty oil, a *N*-methyl pyridone alkaloid bearing a nitrile group, ricinine, was isolated from *Ricinus* seed [2]. A ribosome-inactivating protein, ricin, was isolated from the seeds of *R. communis* [3].

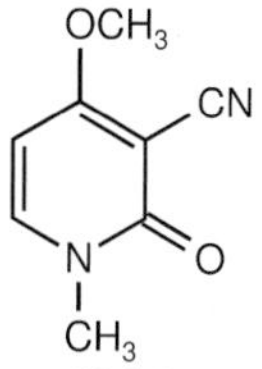

Ricinine

Pharmacology and Toxicology

The mechanism of sodium ricinoleate-induced secretion was investigated in perfused hamster small intestine *in vivo*. Sodium ricinoleate at a concentration of 8 m*M* resulted not only in the secretion of water and sodium, but also in an increase of the intestinal clearance of inulin and dextran. Sodium ricinoleate at a concentration of 2 m*M* had not effect on water transport or intestinal permeability. Intestinal secretion induced by sodium ricinoleate was also accompanied by increased mucosal cell exfoliation. Both, light and electron microscopy studies demonstrated substantial mucosal architectural changes with 8 m*M* sodium ricinoleate, with villus shortening and injury to epithelial cells at the villus tips [4]. Mucosal alterations were also observed in rabbit ileum perfused with isotonic buffer containing 10 m*M* sodium ricinoleate [5].

Fluid secretion induced by sodium ricinoleate in the rat small intestine was markedly inhibited by hexamethonium and lidocaine, but not by atropine; this suggested that the intestinal fluid secretion

Handbook of Chinese Medicinal Plants: Chemistry, Pharmacology, Toxicology
Weici Tang and Gerhard Eisenbrand

ISBN: 978-3-527-32226-8

might be mainly caused by the stimulation of an active secretory process via an activation of enteric nerves [6]. Sodium ricinoleate similarly stimulated muscle contractions of the isolated circular muscle of the cat colon. It was also reported that sodium ricinoleate might act as a calcium ionophore in jejunal brush border vesicles, and promoted passive uptake and efflux of Ca^{2+} across brush border vesicles in a concentration-dependent manner [7].

The intraduodenal application of sodium ricinoleate to unanesthetized cats at doses of 0.1–10 $g\,kg^{-1}$ briefly increased, and then decreased, spike activity in the small intestine. Sodium ricinoleate also initiated spike complexes of 10 to 215 s duration. Transit from the right to left colon was sometimes accelerated and sometimes delayed. The number of uncoupled slow waves was increased by about 280% of control [8]. In addition, sodium ricinoleate was found to inhibit alanine absorption in the rabbit intestine, both *in vitro* and *in vivo* [9].

A clinical study on the efficacy of *Ricinus* oil to initiate labor in pregnant women with intact membranes at weeks 40–42 of pregnancy was reported. Fifty-two women received a single oral dose of 60 ml *Ricinus* oil, and 48 women rested without treatment. Following the administration of *Ricinus* oil, 30 of 52 women (57.7%) began active labor compared to 2 of 48 (4.2%) without treatment. It was concluded that women receiving *Ricinus* oil have an increased likelihood of initiation of labor within 24 h, compared to women without treatment [10].

Ricinine was reported to be a central nervous system (CNS) stimulant. An extract of the pericarp of *Ricinus* seed showed some typical CNS stimulant effects in mice; typically, the animals became exophthalmic, presented tremors and clonic seizures, and died a few minutes after receiving larger doses of the extract. At lower doses, the extract improved memory consolidation and showed some neuroleptic-like properties, such as a decrease in exploratory behavior and catalepsy. The memory-improving effect and seizure-eliciting properties of the extract were also observed with ricinine; however, the neuroleptic-like properties of the extract were not observed with ricinine [11].

Ricinine induced seizures when administered to mice at doses higher than 20 $mg\,kg^{-1}$. Animals presenting seizures showed a marked preconvulsive phase, followed by short-duration hind limb myoclonus, respiratory spasms, and death. The lethal nature of ricinine seizures was also noted as a good model to study the events causing death in clonic seizures, particularly those related to respiratory spasms, which were also observed in some types of human epilepsy. The behavioral signs of ricinine-elicited seizures are accompanied by electrographic alterations that were more evident during the preconvulsive phase in the cerebral cortex and more intense during the ictal phase, both in the cortex and hippocampus. The ricinine-elicited seizures could be inhibited by diazepam, but not by phenobarbital. Micromolar concentrations of ricinine caused a small decrease in the binding of $[^3H]$flunitrazepam to cerebral cortex membranes. Although ricinine possesses a cyano substituent, only higher concentrations of ricinine (4 m*M*) caused a minor impairment of mitochondrial respiration. It was suggested that the mechanism of action of ricinine might involve the benzodiazepine site in the γ-aminobutyric acid ($GABA_A$) receptor [12].

The ribosome-inactivating protein ricin [13] was used for the preparation of immunotoxins coupled to a monoclonal antibodies. Ricin immunotoxins have been entered into clinical trials, mainly for the treatment of lymphomas [14–19]. For example, 82 adult patients with acute lymphoblastic leukemia were treated with anti-B4-blocked ricin, an immunotoxin composed of an anti-CD19 murine monoclonal antibody (B4) conjugated to blocked ricin. Patients with CD19-positive acute lymphoblastic leukemia were given anti-B4-blocked ricin as two seven-day continuous infusions, one week apart. Among the 82 patients, 78 were eligible. Subsequently, 66 patients (85%) achieved complete remission. Forty-six patients received the anti-B4-blocked ricin, which generally was well tolerated; of these patients, 80% were able to receive both courses. The most common toxicity was asymptomatic transient elevation of liver enzymes in 72% of cases. Lymphopenia occurred in 46% of the patients, and two developed antibodies to the anti-B4-blocked ricin [19]. In an *in vitro* study or the cytotoxicity of ricin towards cancer cells and normal cells, ricin at high concentrations (5×0.1 to $5\times10\,nM$) did not show any difference between normal cell and SW480 colon cancer cell killing. However, ricin was capable of exerting a selective cytotoxicity towards tumor cells at low concentrations (5×0.1 to $5\times10\,pM$) [20]. In HeLa human cervical cancer cells, ricin was found to induce apoptosis, to generate reactive oxygen species (ROS), and to deplete intracellular glutathione levels. Ricin also activated caspase-3 and increased protease activity. This suggested that it might induce cell death by ROS generation and subsequent activation of the caspase-3 cascade, followed by downstream events leading to apoptotic cell death [21].

References

1 McKeon, T.A., Lin, J.T., and Stafford, A.E. (1999) Biosynthesis of ricinoleate in castor oil. *Adv. Exp. Med. Biol.*, **464**, 37–47.

2 Waller, G.R., Ryhage, R., and Meyerson, S. (1966) Mass spectrometry of biosynthetically labeled ricinine. *Anal. Biochem.*, **16**, 277–286.

3 Despeyroux, D., Walker, N., Pearce, M., Fisher, M., McDonnell, M., Bailey, S.C., Griffiths, G.D., and Watts, P. (2000) Characterization of ricin heterogeneity by electrospray mass spectrometry, capillary electrophoresis, and resonant mirror. *Anal. Biochem.*, **279**, 23–36.

4 Cline, W.S., Lorenzsonn, V., Benz, L., Bass, P., and Olsen, W.A. (1976) The effects of sodium ricinoleate on small intestinal function and structure. *J. Clin. Invest.*, **58**, 380–390.

5 Gaginella, T.S. and Phillips, S.F. (1976) Ricinoleic acid (castor oil) alters intestinal surface structure. A scanning electron microscopic study. *Mayo Clin. Proc.*, **51**, 6–12.

6 Karlstrom, L., Cassuto, J., Jodal, M., and Lundgren, O. (1986) Involvement of the enteric nervous system in the intestinal secretion induced by sodium deoxycholate and sodium ricinoleate. *Scand. J. Gastroenterol.*, **21**, 331–340.

7 Maenz, D.D. and Forsyth, G.W. (1982) Ricinoleate and deoxycholate are calcium ionophores in jejunal brush border vesicles. *J. Membr. Biol.*, **70**, 125–133.

8 Wienbeck, M., Wallenfels, M., and Kortenhaus, E. (1987) Ricinoleic acid and loperamide have opposite motor effects in the small and large intestine of the cat. *Z. Gastroenterol.*, **25**, 355–363.

9 Hajjar, J.J., Murphy, D.M., and Scheig, R.L. (1979) Mechanism of inhibition of alanine absorption by Na ricinoleate. *Am. J. Physiol.*, **236**, E534–E538.

10 Garry, D., Figueroa, R., Guillaume, J., and Cucco, V. (2000) Use of castor oil in pregnancies at term. *Altern. Ther. Health Med.*, **6**, 77–79.

11 Ferraz, A.C., Angelucci, M.E., Da Costa, M.L., Batista, I.R., De Oliveira, B.H., and Da Cunha, C. (1999) Pharmacological evaluation of ricinine, a central nervous system stimulant isolated from *Ricinus communis*. *Pharmacol. Biochem. Behav.*, **63**, 367–375.

12 Ferraz, A.C., Pereira, L.F., Ribeiro, R.L., Wolfman, C., Medina, J.H., Scorza, F.A., Santos, N.F., Cavalheiro, E.A., and Da Cunha, C. (2000) Ricinine-elicited seizures. A novel chemical model of convulsive seizures. *Pharmacol. Biochem. Behav.*, **65**, 577–583.

13 Zhong, R.K., van De Winkel, J.G., Thepen, T., Schultz, L.D., and Ball, E.D. (2001) Cytotoxicity of anti-cd64-ricin, a chain immunotoxin against human acute myeloid leukemia cells *in vitro* and in SCID mice. *J. Hematother. Stem Cell Res.*, **10**, 95–105.

14 van Oosterhout, Y.V., van Emst, J.L., Bakker, H.H., Preijers, F.W., Schattenberg, A.V., Ruiter, D.J., Evers, S., Koopman, J.P., and de Witte, T. (2001) Production of anti-CD3 and anti-CD7 ricin A-immunotoxins for a clinical pilot study. *Int. J. Pharm.*, **221**, 175–186.

15 Schindler, J., Sausville, E., Messmann, R., Uhr, J.W., and Vitetta, E.S. (2001) The toxicity of deglycosylated ricin A chain-containing immunotoxins in patients with non-Hodgkin's lymphoma is exacerbated by prior radiotherapy: a retrospective analysis of patients in five clinical trials. *Clin. Cancer Res.*, **7**, 255–258.

16 Longo, D.L., Duffey, P.L., Gribben, J.G., Jaffe, E.S., Curti, B.D., Gause, B.L., Janik, J.E., Braman, V.M., Esseltine, D., Wilson, W.H., Kaufman, D., Wittes, R.E., Nadler, L.M., and Urba, W.J. (2000) Combination chemotherapy followed by an immunotoxin (anti-B4-blocked ricin) in patients with indolent lymphoma: results of a phase II study. *Cancer J. Sci. Am.*, **6**, 146–150.

17 Messmann, R.A., Vitetta, E.S., Headlee, D., Senderowicz, A.M., Figg, W.D., Schindler, J., Michiel, D.F., Creekmore, S., Steinberg, S.M., Kohler, D., Jaffe, E.S., Stetler-Stevenson, M., Chen, H., Ghetie, V., and Sausville, E.A. (2000) A phase I study of combination therapy with immunotoxins IgG-HD37-deglycosylated ricin A chain (dgA) and IgG-RFB4-dgA (Combotox) in patients with refractory CD19 (+), CD22(+) B cell lymphoma. *Clin. Cancer Res.*, **6**, 1302–1313.

18 Schnell, R., Vitetta, E., Schindler, J., Borchmann, P., Barth, S., Ghetie, V., Hell, K., Drillich, S., Diehl, V., and Engert, A. (2000) Treatment of refractory Hodgkin's lymphoma patients with an anti-CD25 ricin A-chain immunotoxin. *Leukemia*, **14**, 129–135.

19 Szatrowski, T.P., Dodge, R.K., Reynolds, C., Westbrook, C.A., Frankel, S.R., Sklar, J., Stewart, C.C., Hurd, D.D., Kolitz, J.E., Velez-Garcia, E., Stone, R.M., Bloomfield, C.D., Schiffer, C.A., and Larson, R.A. (2003) Lineage specific treatment of adult patients with acute lymphoblastic leukemia in first remission with anti-B4-blocked ricin or high-dose cytarabine: Cancer and Leukemia Group B Study 9311. *Cancer*, **97**, 1471–1480.

20 Zou, L.B., and Zhan, J.B. (2005) Purification and anti-cancer activity of ricin. *Zhejiang Da Xue Xue Bao Yi Xue Ban*, **34**, 217–219.

21 Rao, P.V., Jayaraj, R., Bhaskar, A.S., Kumar, O., Bhattacharya, R., Saxena, P., Dash, P.K., and Vijayarahavan, R. (2005) Mechanism of ricin-induced apoptosis in human cervical cancer cells. *Biochem. Pharmacol.*, **69**, 855–865.

Rosa chinensis, *Rosa laevigata* and *Rosa rugosa*

Flos Rosae Chinensis (Yuejihua) is the dried flower of *Rosa chinensis* Jacq. (Rosaceae). It is used for the treatment of menstrual disorders.

Flos Rosae Rugosae (Meiguihua) is the dried flower bud of *Rosa rugosa* Thunb. It is used for the treatment of menstrual disorders and traumatic diseases.

Rosa laevigata Michx.

Fructus Rosae Laevigatae (Jinyingzi) is the dried ripe fruit of *Rosa laevigata* Michx. It is used as an astringent and antidiarrheic agent.

The root of *Rosa cymosa* Tratt., *Rosa laevigata* Michx., and *Rosa multiflora* var. *cathayensis* Rehd. et Wils. is listed in the appendix of the Chinese Pharmacopoeia.

Chemistry

From the fruits of *R. laevigata* triterpenes and sterols were isolated and identified as euscaphic acid, 2α,19α,23-trihydroxyursolic acid and its 3-epimer, β-sitosterol, daucosterol and 2α,3β-diacetyloxy-lup-12-en-28-oic acid methyl ester [1]. The fruits are known to be rich in polysaccharides. The Chinese Pharmacopoeia requires a quantitative determination of polysaccharides in Fructus Rosae Laevigatae by spectrophotometry. The content of polysaccharides in the fruit flesh of Fructus Rosae Laevigatae should be not less than 25.0%, calculated as glucose.

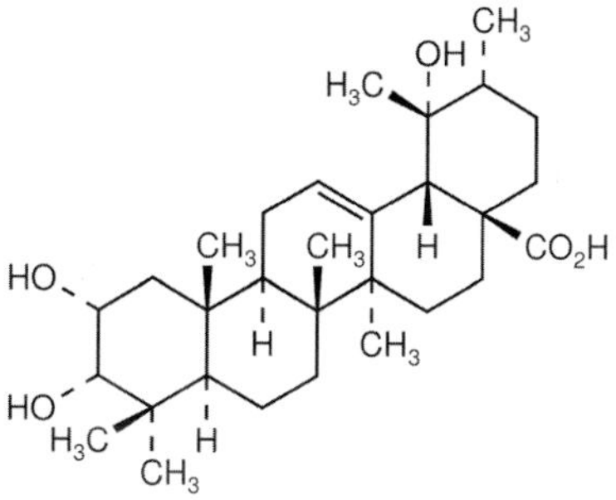

Euscaphic acid

A number of polysaccharides and other compounds were isolated from the dried flowers of *R. rugosa*. An aqueous extract of the flowers of *R. rugosa* was chromatographed on CM-cellulose in ammonium acetate buffer to yield three fractions: F1, F2, and F3. F1a, derived from F1 by gel filtration, is mainly a polysaccharide–peptide complex, while F2 is a polysaccharide and F3 a gallic acid derivative [2]. The fraction P from an aqueous extract from dried flowers of *R. rugosa* was obtained by precipitation with ethanol. By chromatography on DEAE-cellulose of fraction P, the components retained were eluted with a linear gradient of 0 to 2 *M* NaCl solution, providing two fractions, P1 and P2, which eluted at concentrations of 0.5 and 1 *M* NaCl, respectively. Two subfractions, P1a and P1b, were produced from fraction P1 by gel filtration on Sephadex G-200. P1a was identified as a polysaccharide–peptide

Handbook of Chinese Medicinal Plants: Chemistry, Pharmacology, Toxicology
Weici Tang and Gerhard Eisenbrand

ISBN: 978-3-527-32226-8

complex with a molecular mass of 150 kDa, and P1b as a polymer consisting of acteoside and acteoside derivatives with a molecular mass of 8 kDa [3].

Pharmacology and Toxicology

The fruits of *R. laevigata* were used mainly as an astrigent and as a tonic [4]. Fruit juice prepared from the fruits of *R. rugosa* was found to strongly inhibit the proliferation of cancer cell lines, but was substantially less cytotoxic towards normal human cell lines [5]. It also strongly induced the differentiation of HL-60 cells to monocyte/macrophage characteristics, in a concentration-dependent manner [6].

An extract of the flower of *R. rugosa* was reported to exert an antioxidant activity. In senescence-accelerated mice, treatment with the extract markedly increased catalase and glutathione peroxidase activities in the whole blood and liver. The gene expression level of catalase and glutathione peroxidase was also upregulated in the liver, while the malondialdehyde content of the liver and brain was decreased. Male senescence-accelerated mice were more sensitive to treatment with the extract than females. Both, the mean and longest lifespan of senescence-accelerated mice were extended after treatment with the extract [7]. The gallic acid derivative F3 fraction was found to exhibit high antioxidative potency [2]. Both subfractions of the aqueous extract of the flower of *R. rugosa*, P1a and P1b, possessed antioxidant activities [3]. The subfractions P1a and P1b were further reported to inhibit the activity of human immunodeficiency virus type-1 reverse transcriptase *in vitro*, with IC_{50}-values of 158 n*M* and 148 $\mu g\,ml^{-1}$ (18.5 μ*M*), respectively [8].

References

1 Gao, Y., Cheng, W.M., and Li, G.Y. (1993) Chemical constituents of *Rosa laevigata* Michx. *China J. Chin. Mater. Med.*, **18**, 426–428, 447.

2 Ng, T.B., He, J.S., Niu, S.M., Zhao, L., Pi, Z.F., Shao, W., and Liu, F. (2004) A gallic acid derivative and polysaccharides with antioxidative activity from rose (*Rosa rugosa*) flowers. *J. Pharm. Pharmacol.*, **56**, 537–545.

3 Ng, T.B., Pi, Z.F., Yue, H., Zhao, L., Fu, M., Li, L., Hou, J., Shi, L.S., Chen, R.R., Jiang, Y., and Liu, F. (2006) A polysaccharopeptide complex and a condensed tannin with antioxidant activity from dried rose (*Rosa rugosa*) flowers. *J. Pharm. Pharmacol.*, **58**, 529–534.

4 Wang, Q. and Chen, Y. (1997) Advances in the study of the fruits of *Rosa davurica* Pall. *China J. Chin. Mater. Med.*, **22**, 131–134.

5 Yoshizawa, Y., Kawaii, S., Urashima, M., Fukase, T., Sato, T., Tanaka, R., Murofushi, N., and Nishimura, H. (2000) Antiproliferative effects of small fruit juices on several cancer cell lines. *Anticancer Res.*, **20B**, 4285–4289.

6 Yoshizawa, Y., Kawaii, S., Urashima, M., Fukase, T., Sato, T., Murofushi, N., and Nishimura, H. (2000) Differentiation-inducing effects of small fruit juices on HL-60 leukemic cells. *J. Agric. Food Chem.*, **48**, 3177–3182.

7 Ng, T.B., Gao, W., Li, L., Niu, S.M., Zhao, L., Liu, J., Shi, L.S., Fu, M., and Liu, F. (2005) Rose (*Rosa rugosa*)-flower extract increases the activities of antioxidant enzymes and their gene expression and reduces lipid peroxidation. *Biochem. Cell Biol.*, **83**, 78–85.

8 Fu, M., Ng, T.B., Jiang, Y., Pi, Z.F., Liu, Z.K., Li, L., and Liu, F. (2006) Compounds from rose (*Rosa rugosa*) flowers with human immunodeficiency virus type 1 reverse transcriptase inhibitory activity. *J. Pharm. Pharmacol.*, **58**, 1275–1280.

Rubia cordifolia

Radix et Rhizoma Rubiae (Qiancao) is the dried root and rhizome of *Rubia cordifolia* L. (Rubiaceae). It is mainly used as a hemostatic agent for the treatment of hematorrhea, hematemesis, nose bleeding, traumatic bleeding, dysmenorrhea, and for the treatment of arthritis.

Rubia cordifolia L.

Chemistry

The root of *R. cordifolia* madder root is known to contain anthraquinones, naphthoquinones, and triterpenes. The isolation of a number of cyclic peptides with cytotoxic activity has also been described. Anthraquinones from the root of *R. cordifolia* have mainly substitutions in only one of the aromatic rings, including purpurin, munjistin, pseudopurpurin, alizarin, nordamnacanthal, rubiadin and lucidin, with rubiadin and lucidin being the major components [1].

Anthraquinone glycosides were also isolated from the root of *R. cordifolia* such as lucidin primveroside and ruberythric acid [2].

Naphthoquinones and their glycosides were isolated from the root of *R. cordifolia*. These were identified as mollugin (rubimaillin), hydroxymollugin, methoxymollugin, dihydromollugin, dihydroxydihydromollugin, rubidatum, rubilactone, and furomollugin [3]. The Chinese Pharmacopoeia requires a qualitative determination of mollugin in Radix et Rhizoma Rubiae by thin-layer chromatographic comparison with reference substance, and a quantitative determination by HPLC. The content of mollugin in Radix et Rhizoma Rubiae should be not less than 0.4%. Naphthoquinone dimers were also isolated from *R. cordifolia* [4].

An iridoid glycoside from the root of *R. cordifolia* was identified as 6-methoxygeniposidic acid [5].

Cyclic hexapeptides named RA-I, RA-II, RA-III, RA-IV, RA-V, RA-VI, RA-VII, RA-VIII, RA-IX, RA-X, RA-XI, RA-XII, RA-XIII, RA-XIV, RA-XV, and RA-XVI [6, 7] were isolated from the roots of *R. cordifolia*. RA-V and RA-VII were found to be the major components, with an average content

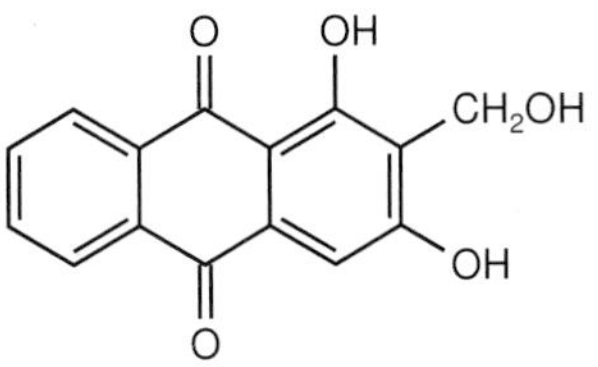

Lucidin

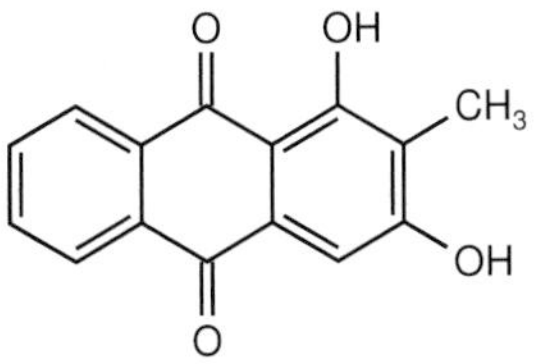

Rubiadin

Handbook of Chinese Medicinal Plants: Chemistry, Pharmacology, Toxicology
Weici Tang and Gerhard Eisenbrand

ISBN: 978-3-527-32226-8

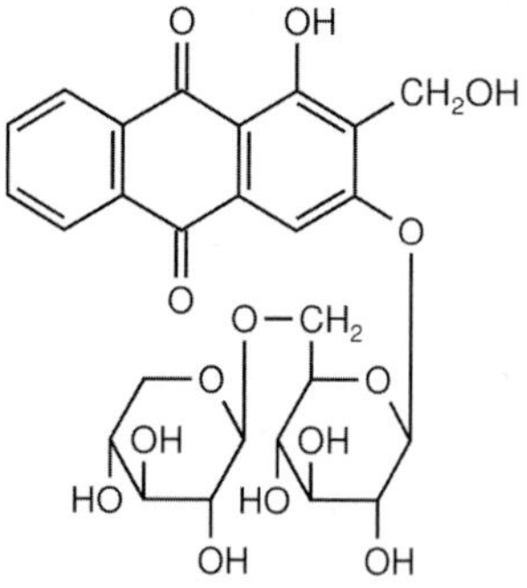

Lucidin primveroside

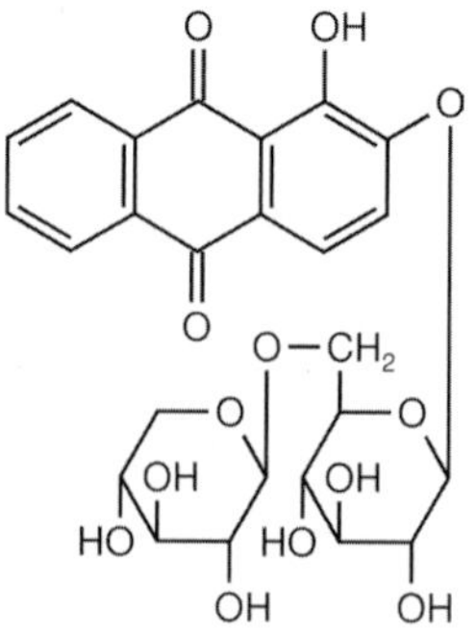

Ruberythric acid

of 94 and 36 mg kg^{-1}, respectively [8]. They are cyclic hexapeptides composed of three tyrosine residues, one D-alanine, one alanine, and one further amino acid. The hydroxy group of the tyrosine residue 5 is etherified with the tyrosine residue 6 at position 3. Compounds RA-XII to RA-XVI are glycosides of the cyclic hexapeptides. Conformational isomers of RA-III and RA-VI (RAI-III and RAI-VI) were also isolated from the root of *R. cordifolia* [9]. Cytotoxic cyclic hexapeptides were first isolated from *Bouvardia ternifolia* (Rubiaceae). Therefore, the basic structure was designated as bouvardin. Bouvardin is characterized by a hydroxy group at the α-position of the tyrosine residue 5. RA compounds from *R. cordifolia* are derived from deoxybouvardin (RA-V) [10].

Pharmacology and Toxicology

Lucidin and lucidin 3-*O*-primveroside were found to be mutagenic in *Salmonella typhimurium*, with or without metabolic activation, but the mutagenic activity was increased after addition of the activation system [11]. Rubiadin was highly mutagenic in *S. typhimurium*; however, in contrast to lucidin, it required metabolic activation. The ethyl ether of lucidin showed a weak mutagenic activity *in vitro* with activating systems. Lucidin 2-ethylether, formed

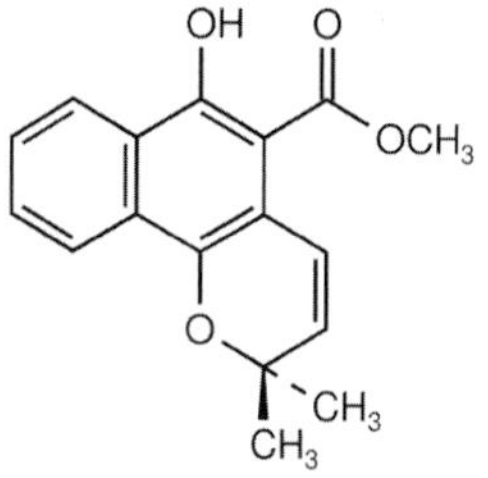

Mollugin

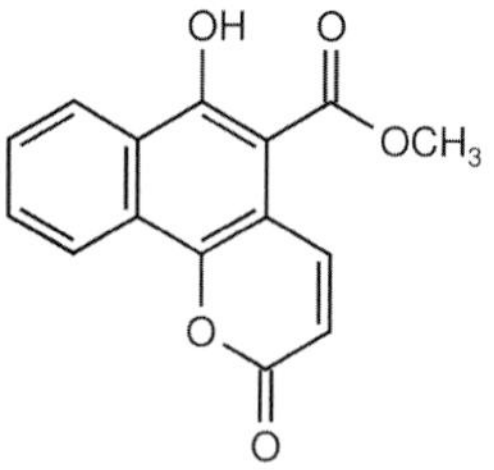

Rubilactone

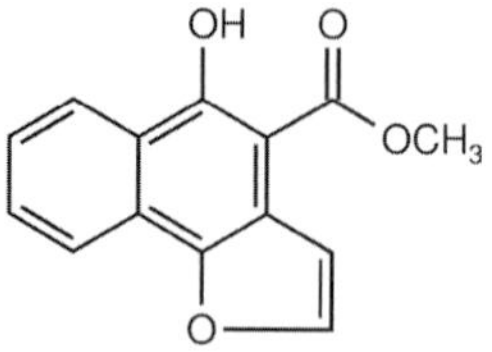

Furomollugin

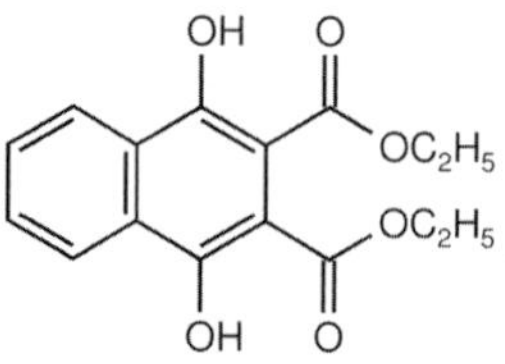

Rubidatum

from lucidin by extraction of the root with boiling ethanol, was also mutagenic in *S. typhimurium*, but only after addition of a rat liver S-9 mix [12], inducing mainly frameshift mutations [13]. Studies on the structure–mutagenicity relationship of the anthraquinones revealed that the greatest activity is exhibited by 1,3-dihydroxyanthraquinones possessing a methyl or hydroxymethyl group at position 2, such as rubiadin, lucidin and lucidin 2-ethyl ether. was even more potent than lucidin and equal to the positive control 7,12-dimethylbenzo[*a*]anthracene (DMBA) [14].

It was demonstrated that the uptake of lucidin 3-*O*-primveroside led to the formation of the rodent carcinogen lucidin, and to the highly genotoxic compound rubiadin. Therefore, it was suggested that the therapeutic use of *R. tinctorum* might involve a carcinogenic risk. The oral administration of lucidin 3-*O*-primveroside to rats resulted in the excretion of lucidin and rubiadin. Treatment of lucidin 3-*O*-primveroside with rat liver extract and NADPH led to the conversion to rubiadin 3-*O*-primveroside. Rubiadin was formed from rubiadin 3-*O*-primveroside after hydrolysis [15].

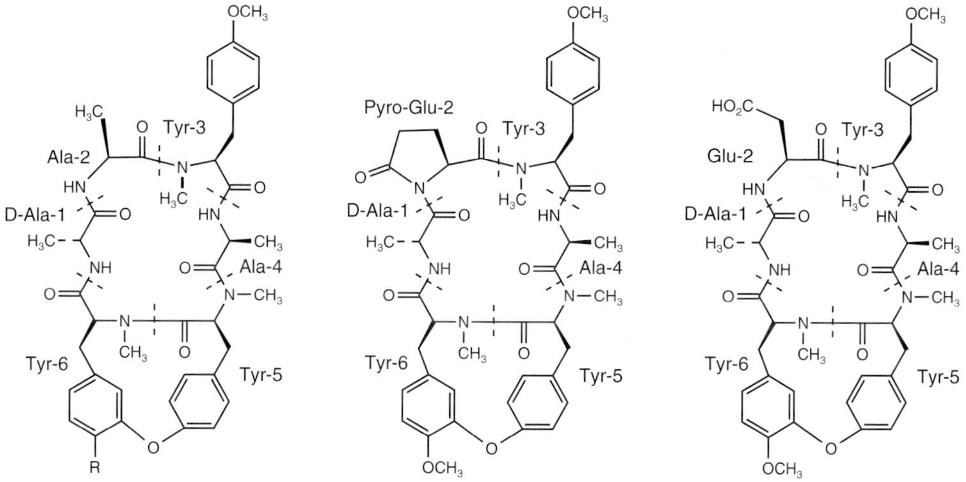

RA-V: R = OH
RA-VII: R = OCH_3

RA-IX

RA-X

Evidence for the mutagenicity of hydroxy-substituted anthraquinones from *Rubia* root was also observed in mammalian test systems such as the V79 mutation assay at the hypoxanthine guanine phosphoribosyl transferase (HGPRT) gene locus, the DNA repair test in primary rat hepatocytes, and the transformation assay in C3H/M2 mouse fibroblasts. Lucidin induced DNA single-strand breaks and DNA–protein crosslinks in V79 cells, as determined by DNA alkaline filter elution. Lucidin 2-ethylether was weakly mutagenic to V79 cells cocultivated with rat hepatocytes [12]. In the unscheduled DNA synthesis (UDS) assay in primary rat hepatocytes, rubiadin

The mechanism of the genotoxic activity of lucidin and rubiadin has been studied. Up to five different DNA adducts in the range from about one to three adducts per 10^8 nucleotides were formed by incubation of mouse DNA with lucidin in the presence of a S-9 mix using the ^{32}P-post-labeling technique. A similar adduct pattern was observed by incubation of polydGpolydC with lucidin in the presence of S-9 mix.

Lucidin reacted with nucleic bases under physiological conditions in the order of the reactivity as follows: adenine > guanine ≫ pyrimidine bases. The isolated purine base adducts were identified as condensed reactants at the benzylic position 1 with a nitrogen atom of a purine base. This indicated a strong possibility of the formation of an exomethylenic compound as an electrophilic intermediate. A ^{32}P-post-labeling analysis showed an increase in the overall level of DNA adducts observed in the liver, kidney and colon of rats treated with 10% madder root in the diet for two weeks. HPLC analysis of ^{32}P-labeled DNA adducts revealed a peak that comigrated with an adduct obtained after the *in vitro* treatment of deoxyguanosine-3′-phosphate with lucidin [14]. The DNA adduct formation by lucidin strongly suggested that the genotoxic effects of lucidin observed in *in vitro* tests resulted from a covalent interaction between lucidin and the cellular DNA [16, 17].

The enzymatic activation of lucidin and formation of electrophilic intermediates from benzylic alcohol was postulated to proceed via elimination of water or via formation of a sulfuric acid ester (Figure 192.1).

The root of *R. cordifolia* was used in traditional Chinese medicine mainly as a hemostatic agent in combination preparations. It also used for the treatment of γ-hyperglobulinemia in chronic hepatitis [18]. In an experimental study, an alcoholic extract of *R. cordifolia* exerted antiperoxidative properties [19]. Rubiadin was found to possess a potent antioxidant activity, preventing lipid peroxidation in rat liver homogenate induced by $FeSO_4$ and *t*-butyl hydroperoxide, in a concentration-dependent manner [20]. Rubiadin was also reported to exert hepatoprotective effects in rats against the hepatotoxicity of carbon tetrachloride (CCl_4). The oral treatment of rats with rubiadin at daily doses of 50, 100, and 200 mg kg^{-1} for 14 days significantly reduced the elevated serum levels of glutamic oxaloacetic transaminase (GOT), glutamate pyruvate transaminase (GPT), alkaline phosphatase (ALP) and γ-glutamyltransferase. In addition, rubiadin significantly prevented the elevation of hepatic malondialdehyde formation and depletion of reduced glutathione (GSH) content in the liver of CCl_4-intoxicated rats. The biochemical results were supported by a histopathological examination of rat liver [21]. The extract of *R. cordifolia* reduced Fe^{3+} ions, but had no radical-scavenging activity. It maintained the glutathione content and inhibited GSH depletion in the presence of $FeSO_4$ [22].

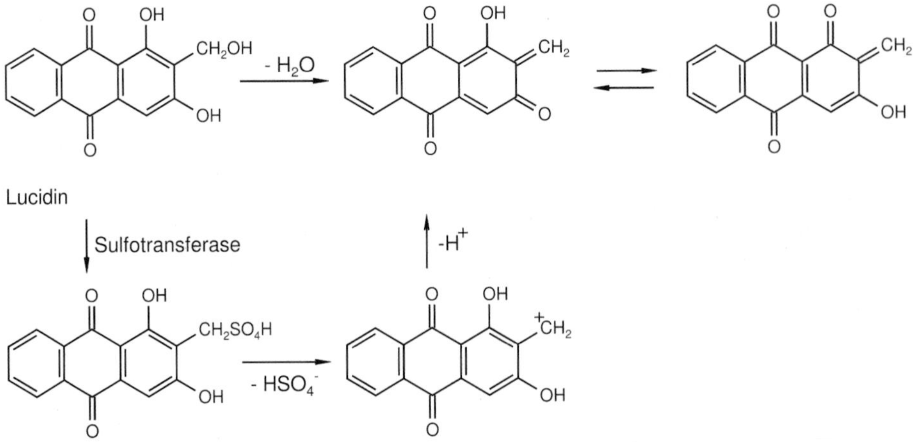

Figure 192.1 Possible toxification pathways to reactive electrophilic intermediates of lucidin.

An alcoholic extract of the root of *R. cordifolia* was reported to exert radioprotective effects in mice. The survival of animals after irradiation was significantly prolonged by intraperitoneal (i.p.) treatment with the extract at a dose of 460 mg kg^{-1} 90 min before the radiation exposure. The extract also inhibited radiation-induced lipid peroxidation, as measured by the inhibition of thiobarbituric acid reactive substance (TBARS) in the liver, and protected against radiation-induced suppression of endogenous colony-forming units in the spleen. A significant inhibition of radiation-induced micronuclei formation was also observed. Thus, the root extract of *R. cordifolia* exhibited significant protection against radiation-induced lipid peroxidation, hemopoietic damage and genotoxicity, possibly through its antioxidant, metal chelation, and anti-inflammatory properties [23].

The extract from Radix Rubiae was found to be effective on mast cell-mediated allergic reactions. An oral dose of 300 mg kg^{-1} potently inhibited systemic anaphylaxis in mice induced by compound 48/80, and also exerted inhibitory activity in cells, with an IC_{50} value of approximately 35 μg ml^{-1}. The extract also inhibited expression of the proinflammatory mediator tumor necrosis factor-α (TNF-α). A mechanistic study revealed that it suppressed the activating phosphorylation of Syk, a key enzyme in mast cell-signalling processes, and that of Akt. It also inhibited the mitogen-activated protein (MAP) kinase, which is critical for the production of inflammatory cytokines in mast cells, as indicated by the suppression of the activating phosphorylation of extracellular signal-regulated kinases 1 and 2 (ERK1/2) [24].

Rubidatum, 1,4-dihydroxy-naphthalene-2,3-dicarboxylic acid diethyl ester, was reported to stimulate hematopoiesis in mice and dogs. It increased peripheral leukocyte counts and stimulated the proliferation and differentiation of hematopoietic stem cells. Rubidatum prevented leucopenia induced by cyclophosphamide in mice by i.p. administration [25]. Rubidatum was also found to be effective in treating abnormal uterine bleeding by oral administration to the patient [26]. It was able to scavenge oxygen radicals in human polymorphonuclear leukocytes [27]. Mollugin and furomollugin have been found to show antiviral activity, and to strongly suppress the secretion of hepatitis B surface antigen in human hepatoma Hep3B cells, with an IC_{50} of 2 μg ml^{-1} [28].

The hypoglycemic activity of an ethanol extract of the root of *R. cordifolia* was demonstrated by decreasing the elevated blood glucose level in diabetic mice, as induced by alloxan. The extract was also found to be effective against stress induced by cold restraint, and on memory impairment induced by scopolamine. The extract enhanced γ-aminobutyric acid (GABA) levels and decreased dopamine levels in the brain. It also decreased plasma levels of corticosterone. Gastric acidity and ulcers caused by cold restraint were inhibited by the extract [29].

The cyclic hexapeptides were reported to exert growth inhibitory activity on P388 lymphocytic leukemia and human nasopharynx carcinoma (KB) cells *in vitro*, and to exhibit significant antitumor activity in mice transplanted with leukemias P388 and L1210, melanoma B16, solid tumor colon 38, Lewis lung carcinoma, and Ehrlich carcinoma. RA-V was also found to be highly effective against MM2 mammary carcinoma in mice [30]. The acute LD_{50} values of RA-VII were estimated as 10 mg kg^{-1} by i.p. dosing; 16.5 mg kg^{-1} by intravenous dosing; and 63 mg kg^{-1} by oral administration. The optimal dose of RA-VII against P388 leukemia in mice was 4 mg kg^{-1} [31]. The IC_{50} values of RA-VII against KB and P388 cells *in vitro* were given as 14 and 18 ng ml^{-1}. The treated/control (T/C) value against P388 in mice at the optimal dose by i.p. administration was 174% [32]. The cytotoxic mechanism of the cyclic hexapeptides was postulated to be due to an inhibition of protein synthesis by binding to the eukaryotic 80S ribosome, and subsequently inhibiting elongation factor 1 (EF1)-dependent binding of aminoacyl-tRNA and EF2-dependent translocation of peptide-tRNA [33].

The biotransformation of RA-VII by rat hepatic microsomes resulted in *N*-demethylation of tyrosine residue 3, and *O*-demethylation and hydroxylation of the aromatic rings of tyrosine residue 3 and tyrosine residue 5. In rabbits given RA-VII, a similar biotransformation pattern was observed in the bile, whereas RA-X was mostly excreted unchanged in the bile [34]. Treatment of RA-VII under basic conditions led to isomerization. Two of the resulting isomers showed antileukemia activity [35].

References

1 Wang, S.X., Hua, H.M., Wu, L.J., Li, X., and Zhu, T.R. (1992) Anthraquinones from the roots of *Rubia cordifolia* L. *Acta Pharm. Sin.*, **27**, 743–747.

2 Itokawa, H., Qiao, Y.F., and Takeya, K. (1989) Anthraquinones and naphthohydroquinones from *Rubia cordifolia*. *Phytochemistry*, **28**, 3465–3468.

3 Hua, H.M., Wang, S.X., Wu, L.J., Li, X., and Zhu, T.R. (1992) Naphthoic acid esters from *Rubia cordifolia* L. *Acta Pharm. Sin.*, **27**, 279–282.

4 Hassanean, H.A., Ibraheim, Z.Z., Takeya, K., and Itorawa, H. (2000) Further quinoidal derivatives from *Rubia cordifolia* L. *Pharmazie*, **55**, 317–319.

5 Wu, L.J., Wang, S.X., Hua, H.M., Li, X., Zhu, T.R., and Miyase, T. (1991) 6-Methoxygeniposidic acid, and iridoid glycoside from *Rubia cordifolia*. *Phytochemistry*, **30**, 1710–1711.

6 Morita, H., Yamamiya, T., Takeya, K., and Itokawa, H. (1992) New antitumor bicyclic hexapeptides, RA-XI, -XII, -XIII and -XIV from *Rubia cordifolia*. *Chem. Pharm. Bull. (Tokyo)*, **40**, 1352–1354.

7 Takeya, K., Yamamiya, T., Morita, H., and Itokawa, H. (1993) Two antitumor bicyclic hexapeptides from *Rubia cordifolia*. *Phytochemistry*, **33**, 613–615.

8 Itokawa, H., Takeya, K., Mori, N., Takanashi, M., Yamamoto, H., Sonobe, T., and Kidokoro, S. (1984) Cell growth-inhibitory effects of derivatives of antitumor cyclic hexapeptide RA-V obtained from Rubiae radix V. *Gann (Jpn. J. Cancer Res.)*, **75**, 929–936.

9 Itokawa, H., Morita, H., Takeya, K., Nomioka, N., and Itai, A. (1991) RAI-III and VI, conformational isomers of antitumor cyclic hexapeptides, RA-III and VI from *Rubia cordifolia*. *Chem Lett.*, 2217–2220.

10 Boger, D.L., Patane, M.A., Jin, Q., and Kitos, P. A. (1994) Design, synthesis and evaluation of bouvardin, deoxybouvardin and RA-I-XIV pharmacophore analogs. *Bioorg. Med. Chem.*, **2**, 85–100.

11 Chulasiri, M., Matsushima, T., and Yoshihira, K. (1995) Activation of lucidin 3-*O*-primveroside mutagenicity by hesperidinase. *Phytother. Res.*, **9**, 421–424.

12 Westendorf, J., Poginsky, B., Marquardt, H., Groth, G., and Marquardt, H. (1988) The genotoxicity of lucidin, a natural component of *Rubia tinctorum* L., and lucidinethylether, a component of ethanolic *Rubia* extracts. *Cell Biol. Toxicol.*, **4**, 225–239.

13 Brown, J.P. and Dietrich, P.S. (1979) Mutagenicity of anthraquinone and benzanthrone derivatives in the Salmonella/microsome test: activation of anthraquinone glycosides by enzymic extracts of rat cecal bacteria. *Mutat. Res.*, **66**, 9–24.

14 Bloemeke, B., Poginsky, B., Schmutte, C., Marquardt, H., and Westendorf, J. (1992) Formation of genotoxic metabolites from anthraquinone glycosides, present in *Rubia tinctorium* L. *Mutat. Res.*, **265**, 263–272.

15 Bertram, B., Hemm, I., and Tang, W. (2001) Mutagenic and carcinogenic constituents of medicinal herbs used in Europe or in the USA. *Pharmazie*, **56**, 99–120.

16 Poginsky, B., Westendorf, J., Blomeke, B., Marquardt, H., Hewer, A., Grover, P.L., and Phillips, D.H. (1991) Evaluation of DNA-binding activity of hydroxyanthraquinones occurring in *Rubia tinctorum* L. *Carcinogenesis*, **12**, 1265–1271.

17 Kawasaki, Y., Goda, Y., Noguchi, H., and Yamada, T. (1994) Identification of adducts formed by reaction of purine bases with a mutagenic anthraquinone, lucidin: mechanism of mutagenicity by anthraquinones

occurring in rubiaceae plants. *Chem. Pharm. Bull. (Tokyo)*, **42**, 1971–1973.

18 Li, Y. (1993) Gamma-hyperglobulinemia in chronic hepatitis treated with massive dosage of *Rubia cordifolia* and *Siegesbeckia orientalis*. *J. Trad. Chin. Med.*, **34**, 603–604.

19 Pandey, S., Sharma, M., Chaturvedi, P., and Tripathi, Y.B. (1994) Protective effect of *Rubia cordifolia* on lipid peroxide formation in isolated rat liver homogenate. *Indian J. Exp. Biol.*, **32**, 180–183.

20 Tripathi, Y.B., Sharma, M., and Manickam, M. (1997) Rubiadin, a new antioxidant from *Rubia cordifolia*. *Indian J. Biochem. Biophys.*, **34**, 302–306.

21 Rao, G.M., Rao, C.V., Pushpangadan, P., and Shirwaikar, A. (2006) Hepatoprotective effects of rubiadin, a major constituent of *Rubia cordifolia* Linn. *J. Ethnopharmacol.*, **103**, 484–490.

22 Tripathi, Y.B. and Sharma, M. (1999) The interaction of *Rubia cordifolia* with iron redox status: a mechanistic aspect in free radical reactions. *Phytomedicine*, **6**, 51–57.

23 Tripathi, Y.B., and Singh, A.V. (2007) Role of *Rubia cordifolia* Linn. in radiation protection. *Indian J. Exp. Biol.*, **45**, 620–625.

24 Lee, J.H., Kim, N.W., Her, E., Kim, B.K., Hwang, K.H., Choi, D.K., Lim, B.O., Han, J.W., Kim, Y.M., and Choi, W.S. (2006) Rubiae Radix suppresses the activation of mast cells through the inhibition of Syk kinase for anti-allergic activity. *J. Pharm. Pharmacol.*, **58**, 503–512.

25 Song, S.Y., Ding, L.M., Chen, Y., and Bai, Y.Z. (1985) Studies on the hematopoietic effect and toxicity of rubidate. *Chin. J. Integr. Trad. West. Med.*, **5**, 625–626.

26 Wu, X.Z., Shao, J.Y., and Liu, Y.X. (1992) Rubidatum in treatment of abnormal uterine bleeding after intra-uterine device insertion. *Reprod. Contracept.*, **12**, 33–36.

27 Kang, X., Fan, Y.Z., Li, X.J., Zhao, B.L., Hou, J. W., and Xin, W.J. (1990) Effects of rubidatum and aminopropylaminoethylthiophosphoric acid on chemiluminescence and ESR of human polymorphonuclear leukocytes. *Chin. J. Pharmacol. Toxicol.*, **4**, 251–254.

28 Ho, L.K., Don, M.J., Chen, H.C., Yeh, S.F., and Chen, J.M. (1996) Inhibition of hepatitis B surface antigen secretion on human hepatoma cells. Components from *Rubia cordifolia*. *J. Nat. Prod.*, **59**, 330–333.

29 Patil, R.A., Jagdale, S.C., and Kasture, S.B. (2006) Antihyperglycemic, antistress and nootropic activity of roots of *Rubia cordifolia* Linn. *Indian J. Exp. Biol.*, **44**, 987–992.

30 Itokawa, H., Takeya, K., Mori, N., Hamanaka, T., Sonobe, T., and Mihara, K. (1984) Isolation and antitumor activity of cyclic hexapeptides isolated from rubiae radix. *Chem. Pharm. Bull. (Tokyo)*, **32**, 284–290.

31 Itokawa, H., Takeya, K., Mori, N., Hamanaka, T., Sonobe, T., Mihara, K., and Tsukagoshi, S. (1983) Studies on the antineoplastic cyclic hexapeptides obtained from Rubiae Radix. II. Biological activities of RA-VII, RA-V, KA-IV and RA-III. in Proceedings of the 13th International Congress on Chemotherapy. Egermann, Vienna (eds K.G. Spitzy and K. Karrer), pp. 284/114–284/116.

32 Itokawa, H., Takeya, K., Mori, N., Kidokoro, S., and Yamamoto, H. (1984) Studies on antitumor cyclic hexapeptides RA obtained from Rubiae Radix, Rubiaceae. IV. quantitative determination of RA-VII and RA-V in commercial Rubiae Radix and collected plants. *Planta Med.*, **50**, 313–316.

33 Morita, H., Yamamiya, T., Takeya, K., Itokawa, H., Sakuma, C., Yamada, J.J., and Suga, T. (1993) Conformational recognition of RA-XII by 80S ribosomes: a differential line broadening study in proton NMR spectroscopy. *Chem. Pharm. Bull. (Tokyo)*, **41**, 781–783.

34 Itokawa, H., Saitou, K., Morita, H., Takeya, K., and Yamada, K. (1992) Structures and conformations of metabolites of antitumor cyclic hexapeptides, RA-VII and RA-X. *Chem. Pharm. Bull. (Tokyo)*, **40**, 2984–2989.

35 Itokawa, H., Morita, H., Kondo, K., Hitotsuyanagi, Y., Takeya, K., and Itaka, Y. (1992) Isomerization of antitumor bicyclic hexapeptide, RA-VII from *Rubia cordifolia*. IV. Conformation-antitumor activity relationship. *J. Chem. Soc., Perkin Trans.*, **2**, 1635–1642.

Salvia miltiorrhiza

Radix et Rhizoma Salviae Miltiorrhizae (Danshen) is the dried root and rhizome of *Salvia miltiorrhiza* Bge. (Lamiaceae). It is used for the treatment of menstrual disorder, menostasis, menorrhalgia, insomnia, vascular diseases and angina pectoris. It is also used as an anti-inflammatory and sedative agent. The Chinese Pharmacopoeia notes that Radix Salviae Miltiorrhizae is incompatible with Rhizome et Radix Veratri.

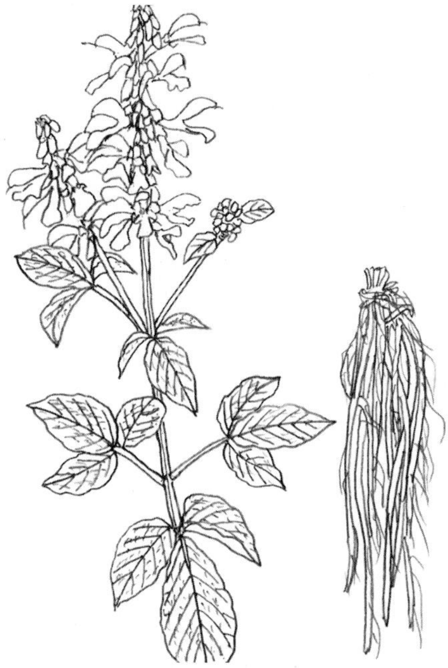

Salvia miltiorrhiza Bge.

A galenic preparation of Radix Salviae Miltiorrhizae, "Danshen Pian," listed in the Chinese Pharmacopoeia, is the tablets of Radix et Rhizoma Salviae Miltiorrhizae. It is used for the treatment of vascular diseases, thoracalgia, and angina pectoris.

Chemistry

The root of *S. miltiorrhiza*, the danshen root, is known to contain diterpene pigments with a phenanthrenequinone structure, especially phenanthrofurane quinone derivatives. Tanshinone I, tanshinone II (tanshinone IIA), and cryptotanshinone have a phenanthro[1,2-*b*]furan-10,11-dione structure, whereas isotanshinone I, isotanshinone II and isocryptotanshinone are derived from phenanthro[3,2-*b*]furan-7,11-dione. Tanshinones I, II, and cryptotanshinone were reported to be the major pigments [1]. The Chinese Pharmacopoeia requires a qualitative determination of tanshinone IIA in Radix et Rhizoma Salviae Miltiorrhizae by thin-layer chromatographic (TLC) comparison with reference substance, and a quantitative determination by HPLC. The content of tanshinone IIA in Radix et Rhizoma Salviae Miltiorrhizae should be not less than 0.2%.

Minor diterpene pigments derived from phenanthro[1,2-*b*]furan-10,11-dione in danshen root were identified as tanshinone IIB,

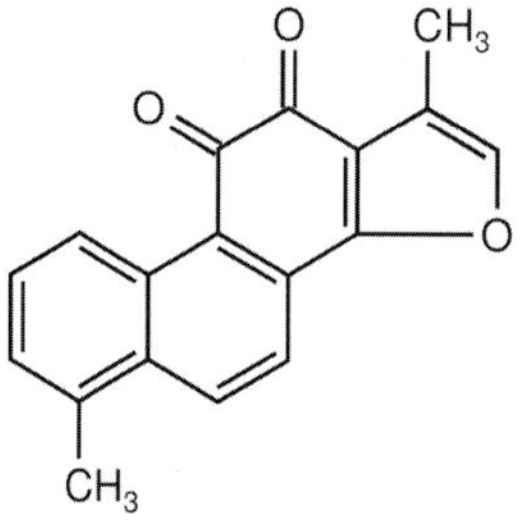

Tanshinone I

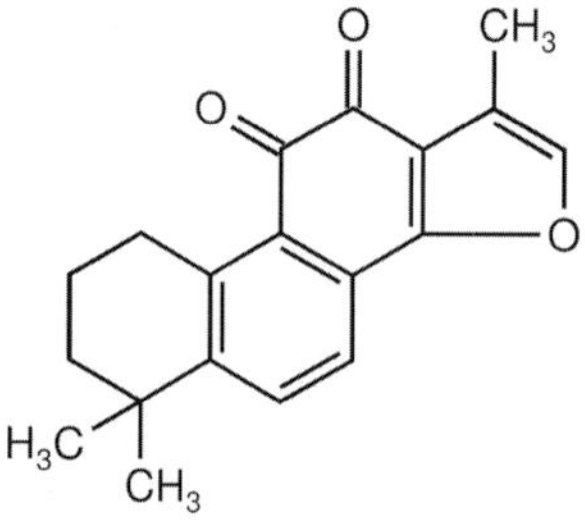

Tanshinone II

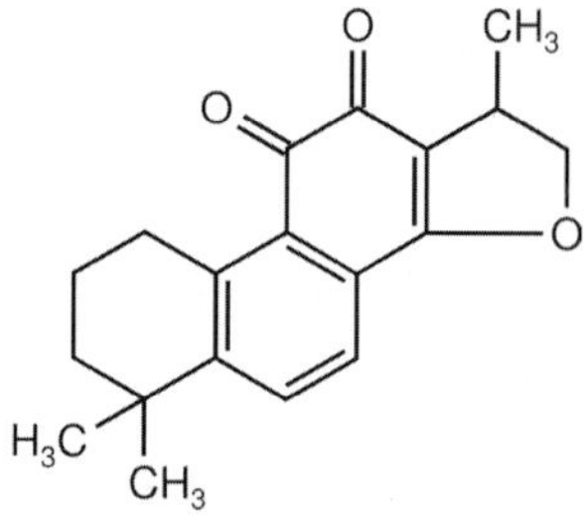

Cryptotanshinone

Handbook of Chinese Medicinal Plants: Chemistry, Pharmacology, Toxicology
Weici Tang and Gerhard Eisenbrand

ISBN: 978-3-527-32226-8

hydroxytanshinone II, tanshinonic acid and its methyl ester, methylene tanshinquinone, tanshindiols A, B and C, tanshinonal (tanshinaldehyde), nortanshinone, dihydrotanshinone I, dihydrotanshinquinone, didehydrotanshinone II, przewaquinones A, B and C, and neocryptotanshinone II [2].

Diterpene pigments derived from phenanthro[3,2-*b*]furan-7,11-dione from danshen root were identified as isotanshinones

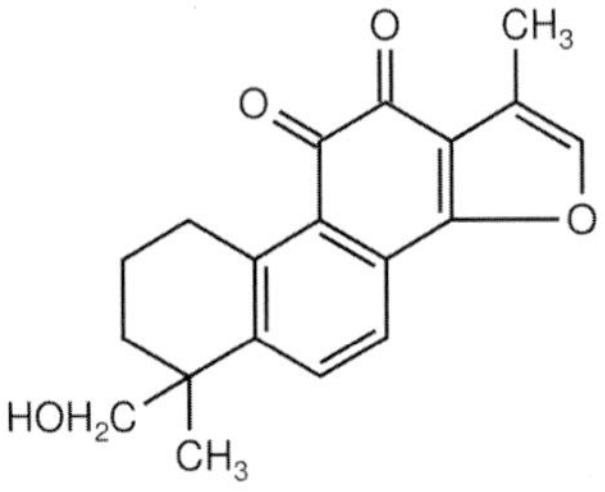

Tanshinone IIB

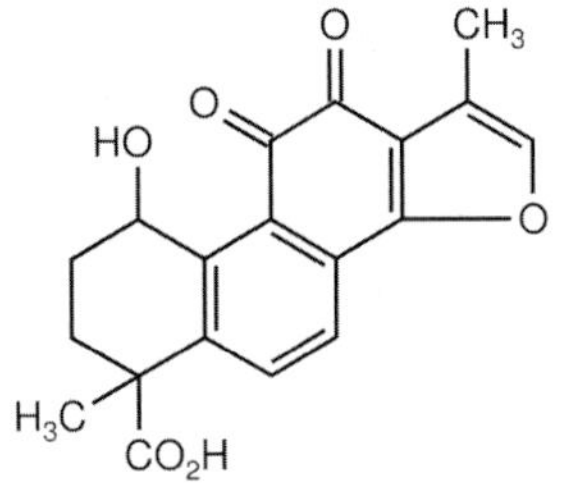

Tanshinonic acid

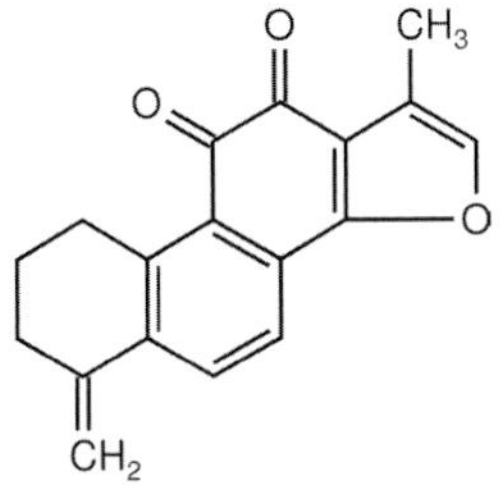

Methylene tanshinquinone

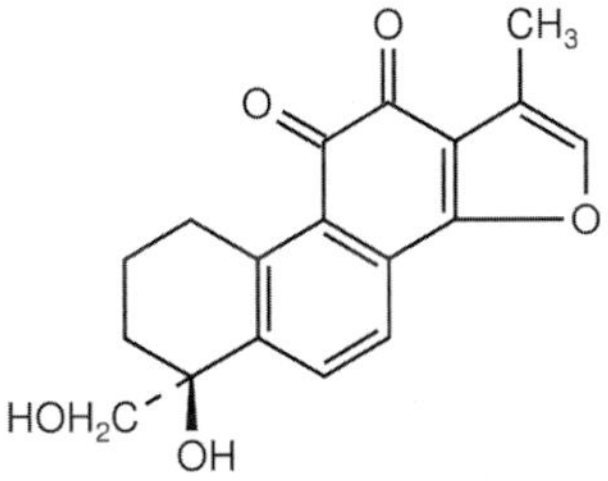

Tanshindiol A

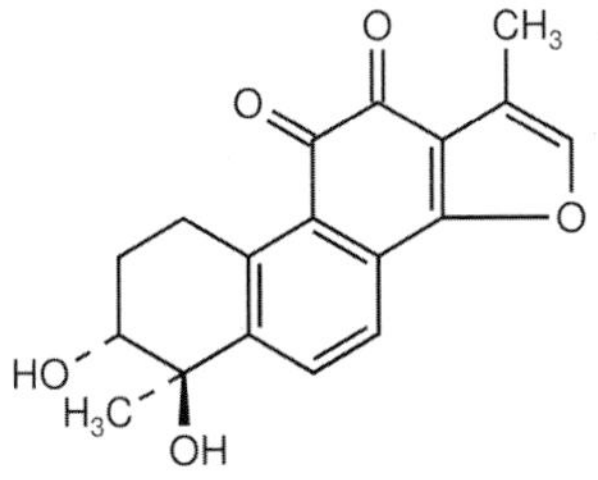

Tanshindiol B

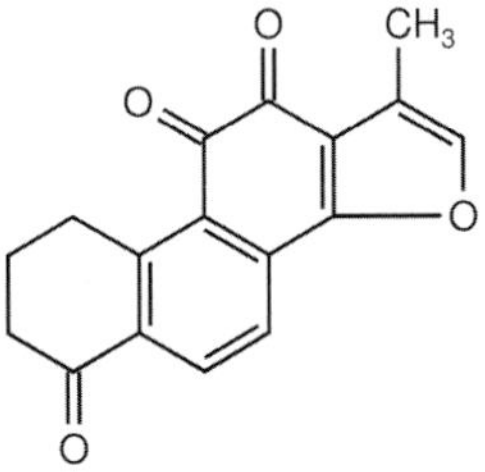

Nortanshinone

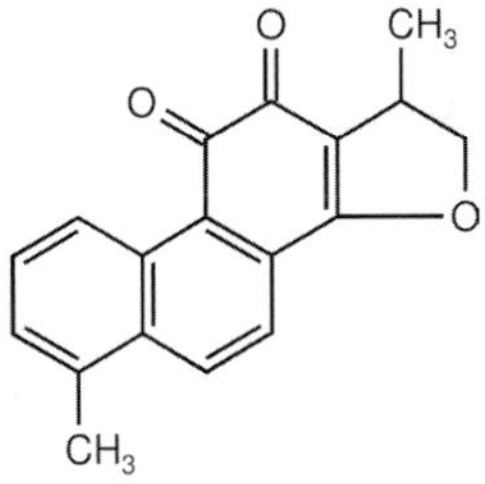

Dihydrotanshinone I

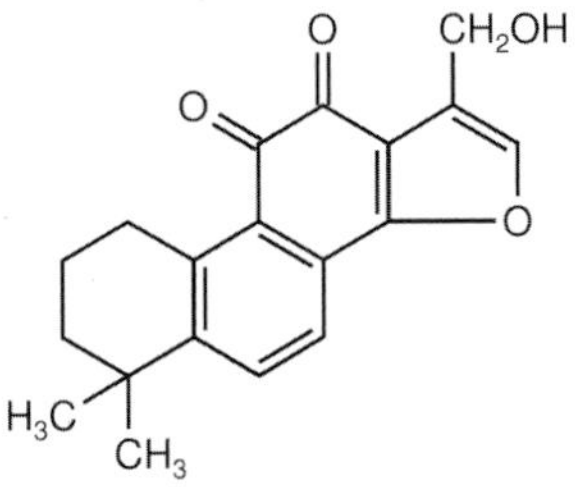

Przewaquinone A

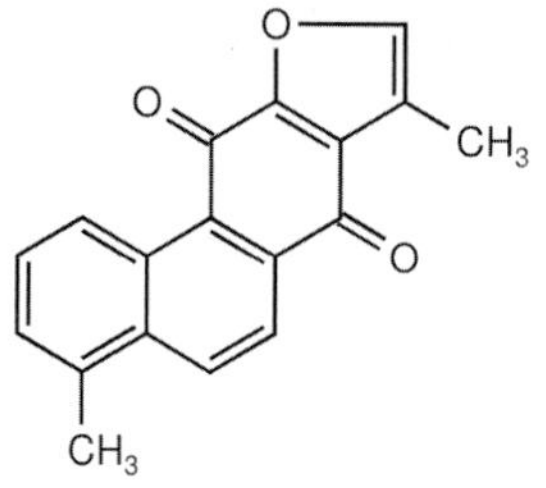
Isotanshinone I

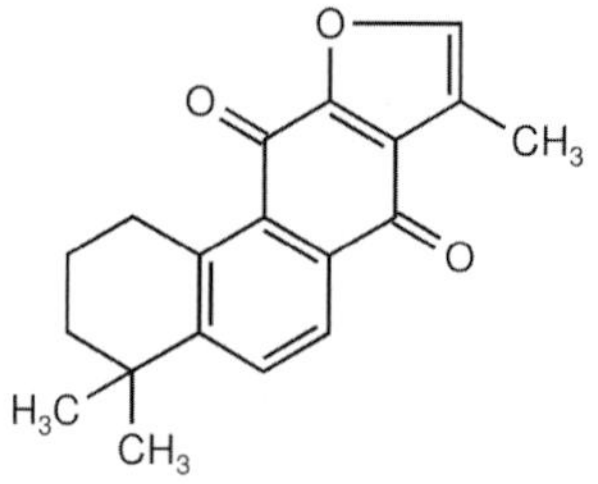
Isotanshinone II

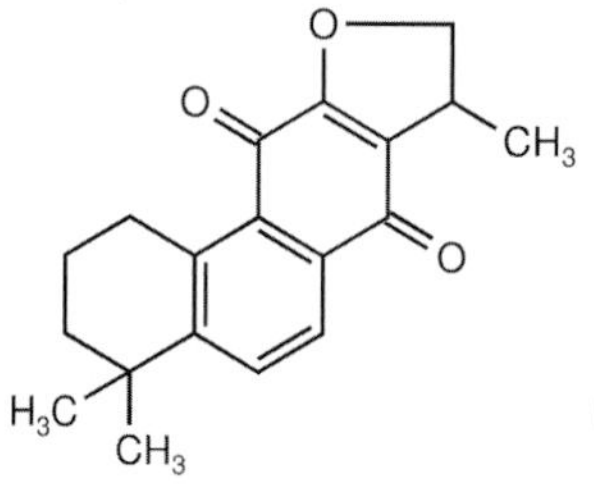
Isocryptotanshinone

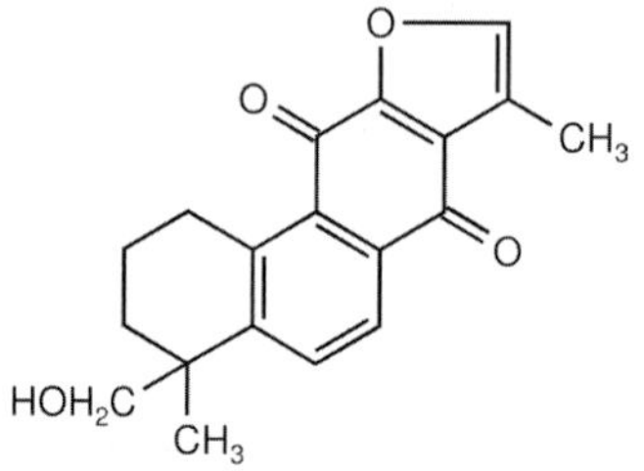
Isotanshinone IIB

I and II and isocryptotanshinone, dihydroisotanshinone I and isotanshinone IIB [3].

3,4-Phenanthrenediones from danshen root were identified as tanshinones V and VI, miltirone, and dehydromiltirone [4].

1,4-Phenanthrenedione from danshen root were identified as danshenxinkuns A,

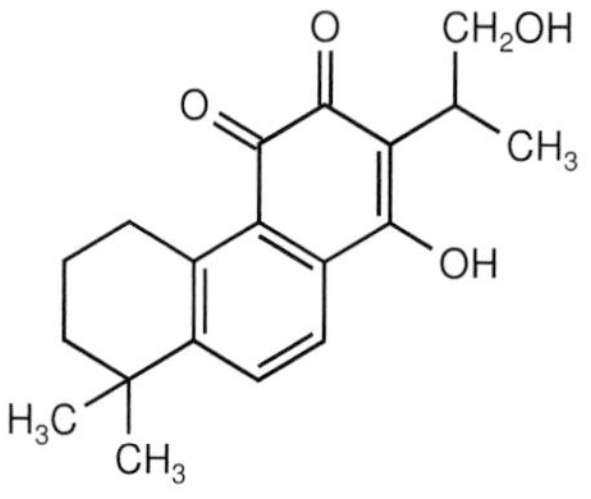
Tanshinone V

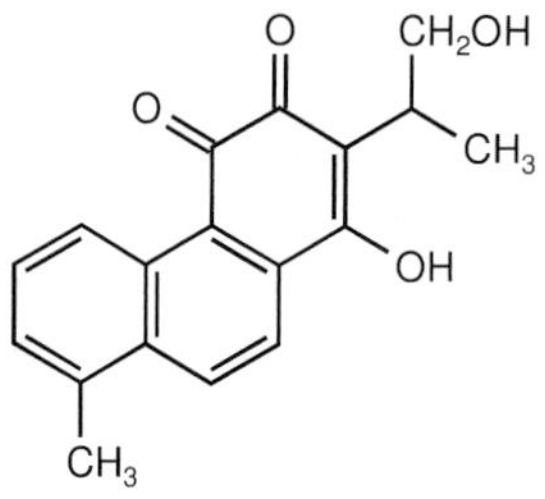
Tanshinone VI

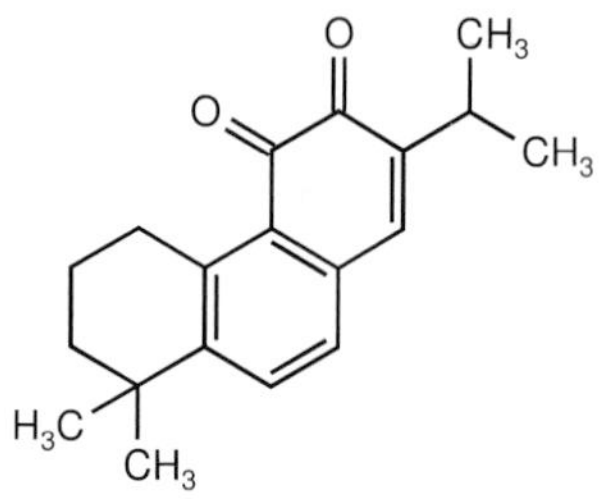
Miltirone

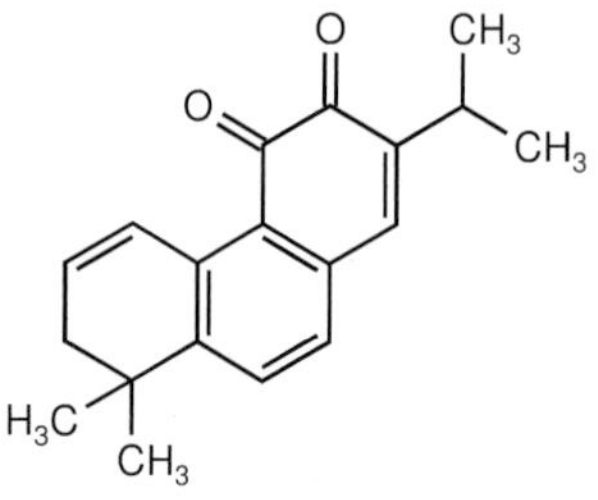
Dehydromiltirone

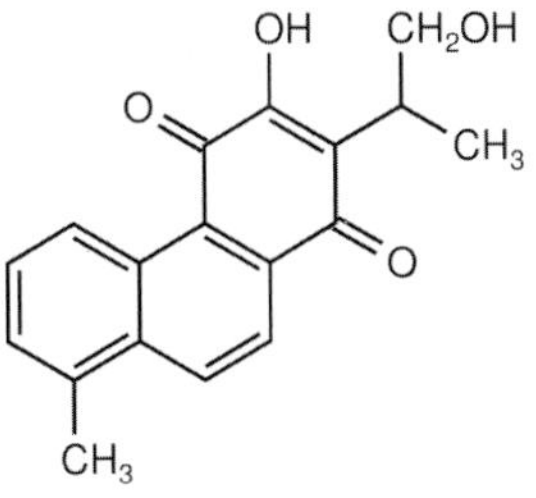

Danshenxinkun A

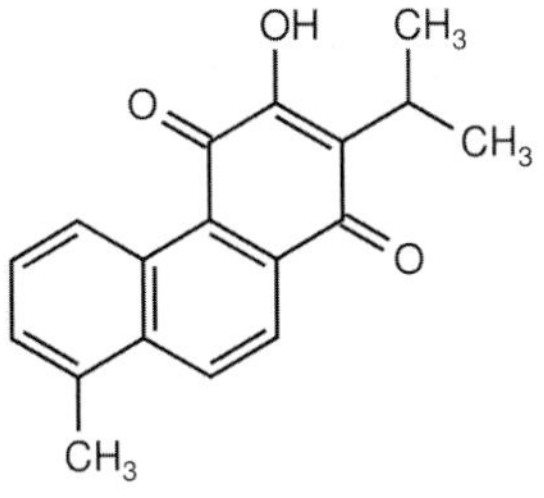

Danshenxinkun B

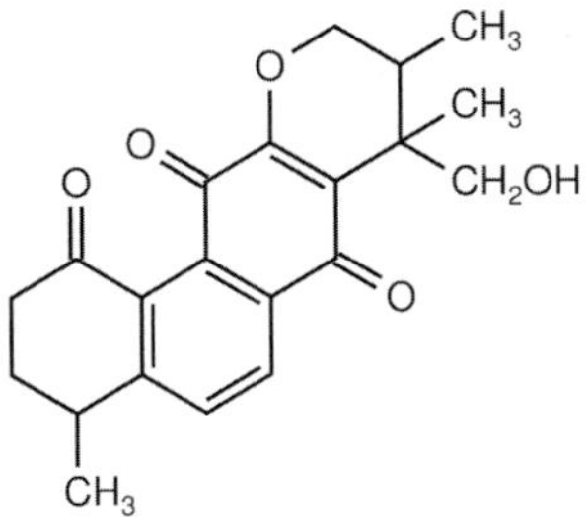

Danshenxinkun D

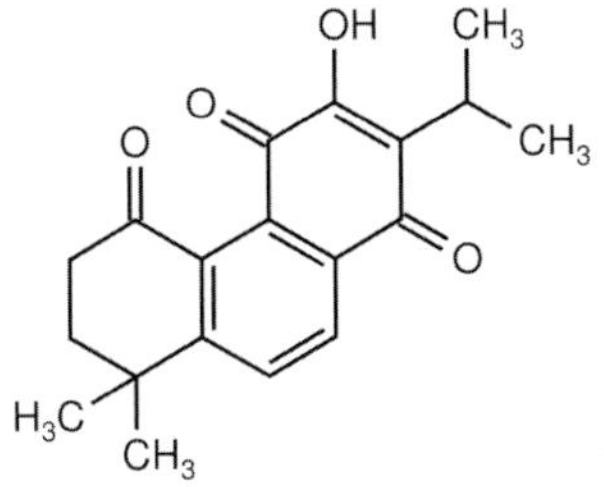

Miltionone I

B, C and D, miltionone I, neocryptotanshinone and ketoisocryptotanshinone [5].

Further constituents of different types from danshen root were reported to be miltionone II, ferruginol, salviol, miltiodiol, norsalvioxide, salviolone, salvilenone, danshenspiroketallactone, danshenspiroketallactone II, cryptoacetalide, tanshinlactone, dihydrotanshinlactone, danshenols A and B, tanshinketolactone, and miltipolone [6].

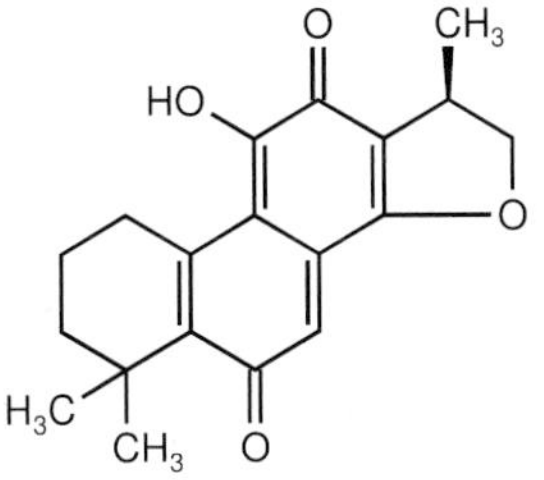

Miltionone II

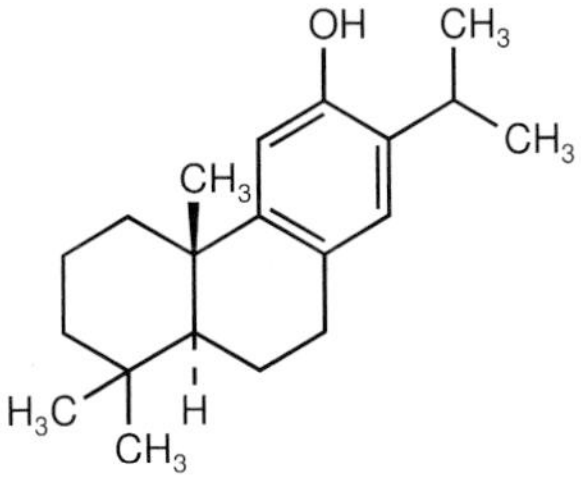

Ferruginol

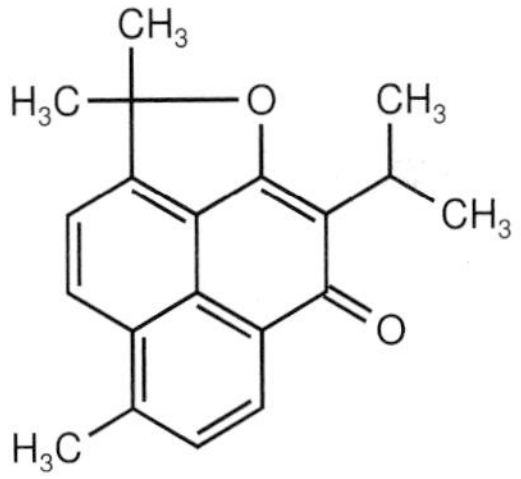

Salvilenone

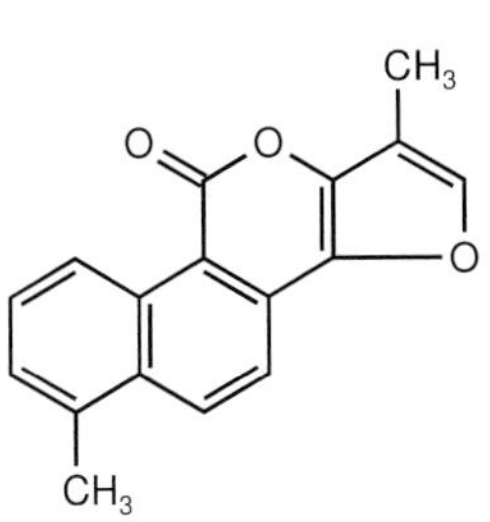

Tanshinlactone

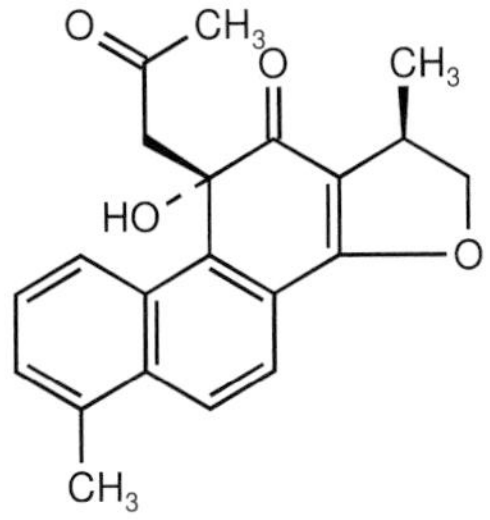

Danshenol A

Hydroxycarboxylic acids and related depsides as condensation products of the hydroxycarboxylic acids from danshen root were identified as protocatechualdehyde, protocatechuic acid, isoferulic acid, danshensu, salvianolic acids A–C, lithospermic acid, rosmarinic acid, and lithospermic acid B [7]. The salvianolic acids are considered to be derived from caffeic acid. The Chinese Pharmacopoeia requires a qualitative determination of salvianolic acid B in Radix et Rhizoma Salviae Miltiorrhizae by TLC comparison with reference substance, and a quantitative determination by HPLC. The content of salvianolic acid B in Radix Salviae Miltiorrhizae should be not less than 3.0%.

The Chinese Pharmacopoeia requires a qualitative determination of tanshinone IIA in the tablets of Radix et Rhizoma Salviae Miltiorrhizae by TLC comparison with reference substances, and a quantitative determination of salvianolic acid B in the tablets of Radix et Rhizoma Salviae Miltiorrhizae by HPLC. The total content of salvianolic acid B in each tablet of Radix et Rhizoma Salviae Miltiorrhizae should be not less than 11 mg.

Pharmacology and Toxicology

Danshen root is one of the most widely used traditional Chinese medicines for the treatment of cardiovascular diseases. A number of results from experimental studies on the effects of danshen and its constituents have been reported. The major active principles of danshen root appeared to be the diterpene pigments and the phenolic acids including water-soluble phenolic acids and lipophilic tanshinones. Phenolic acids

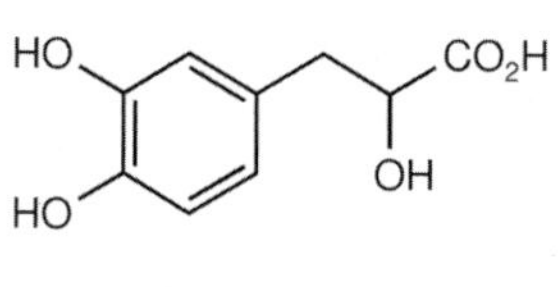

Danshensu

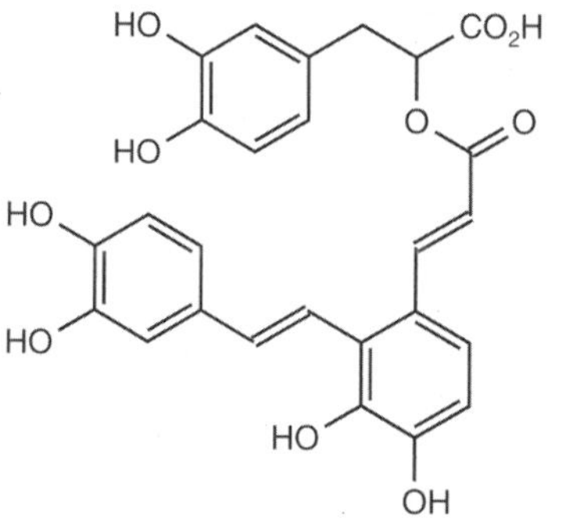

Salvianolic acid A

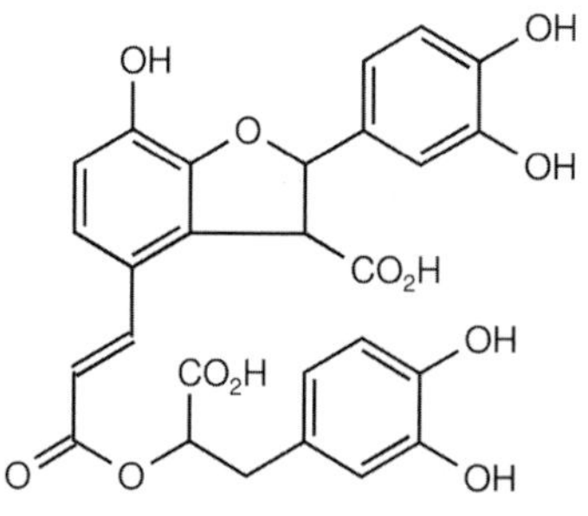

Lithospermic acid

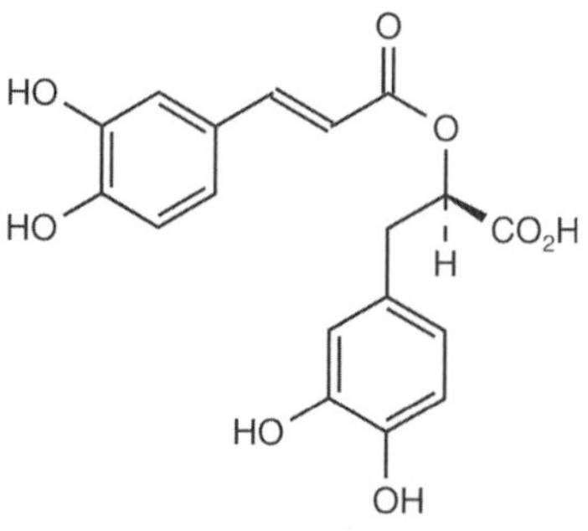

Rosmarinic acid

possess antioxidant and anticoagulant activities and exhibit effects on the cardiovascular system, whereas tanshinones show antibacterial, antioxidant, antineoplastic activities, and also exert effects on the cerebrovascular and cardiovascular system [7, 8].

Tanshinone IIA showed anti-inflammatory activity. It reduced the production of proinflammatory mediators in RAW 264.7 cells stimulated with lipopolysaccharide (LPS). It inhibited NF-κB binding activity, and the phosphorylation of IκBα in a concentration-dependent manner. Tanshinone IIA also inhibited the translocation of NF-κ B from cytosol to nucleus and suppressed the phosphorylation of NF-κB-inducing kinase (NIK)-IκBα kinase (IKK) as well as the phosphorylation of p38, extracellular signal-regulated kinases 1/2 (ERK1/2), and c-Jun N-terminal kinase (JNK) in LPS-stimulated RAW 264.7 cells [9].

Danshen extract was reported to be effective against microcirculation disturbances in experimental animals by increasing the arteriolar diameter and microcirculation rate. Similar effects were also observed in the venous and capillary microcirculation [10]. The aqueous extract of danshen was found to protect the acute myocardial ischemia and arrhythmia of rats induced by isoproterenol or $BaCl_2$. The intraperitoneal (i.p.) or intravenous (i.v.) administration of the extract significantly reduced ventricular contraction, ventricular fibrillation, bradycardia and mortality rate as induced by a bolus i.v. injection of $BaCl_2$ [11]. Acute fatal ventricular fibrillation in rats induced by the subcutaneous (s.c.) injection of isoproterenol was significantly decreased by the aqueous extract of *S. miltiorrhiza*, leading to a reduced mortality rate and prolonged survival time [12]. A study using ^{131}Cs distribution in experimental animals with acute myocardial infarction (MI) and coronary insufficiency confirmed the vasodilatory effect of danshen root [13]. Both, danshen and nitroglycerin showed similar vasodilatory effects, reduced the filling pressure of the left ventricle and increased the cardiac output, but with differences in the time of appearance and duration. The effect of danshen was superior to nitroglycerin; it was more persistent, and the improvement of cardiac functions was better than that of nitroglycerin [14]. Danshen also efficiently inhibited cellular cholesterol biosynthesis [15].

Tanshinone I, tanshinone II, cryptotanshinone, dihydrotanshinone I, dihydrotanshinquinone, and danshenxinkun A all exhibited coronary dilatory activity. It was postulated that the prophylactic effects of tanshinone derivatives on MI may partly result from the inhibition of circulating neutrophil functions. The oral administration of tanshinone derivatives to rabbits suppressed the neutrophil functions dose-dependently, and reduced myocardial necrosis [16]. Tanshinone IIA also protected cardiac myocytes against oxidative stress-triggered damage and apoptosis. Treatment with tanshinone IIA prior to H_2O_2 exposure significantly increased cell viability of neonatal rat ventricular myocytes, and markedly inhibited H_2O_2-induced cardiomyocyte apoptosis. Tanshinone IIA significantly inhibited ischemia–reperfusion-induced cardiomyocyte apoptosis, associated with an increased ratio of Bcl-2 to Bax protein in cardiomyocytes, an elevation of serum superoxide dismutase (SOD) activity and a decrease in serum malondialdehyde (MDA) levels in adult rats [17].

Tanshinone IIA, as one of the major lipophilic components of danshen, exhibited antiatherosclerotic activity. In atherosclerotic rabbits induced by feeding a high-fat diet, tanshinone IIA significantly reduced the formation of atherosclerotic lesions in the aorta, downregulating protein expression and activities of matrix metalloproteinase-2 (MMP-2) and MMP-9 as well as the serum levels of vascular adhesion molecule-1 (VCAM-1) and interleukin-1β (IL-1β) [18]. Tanshinone IIA also attenuated atherosclerotic calcification in rats by inhibition of oxidative stress [19].

Prophylactic effects of tanshinone derivatives against experimental cerebral infarction in rats were also reported. Pretreatment with tanshinone derivatives inhibited the chemoluminescence of peripheral neutrophils; reduced leukocyte infiltration in ischemic brain; inhibited peroxidation-mediated brain damage; and reduced the size of the cerebral infarction [20]. In mice, oral treatment with tanshinone I, tanshinone IIA, cryptotanshinone, and 15,16-dihydrotanshinone I significantly reversed cognitive impairments induced by scopolamine. Tanshinone I and tanshinone IIA also reversed diazepam-induced cognitive dysfunctions. Cryptotanshinone and 15,16-dihydrotanshinone I inhibited acetylcholinesterase *in vitro* [21]. Neuroprotective effects of tanshinone IIA have also been demonstrated in PC12 cells against the toxicity of ethanol. The formation of reactive oxygen species (ROS) and lactate dehydrogenase (LDH) release in culture medium, as induced by ethanol, were significantly reversed by preincubation with tanshinone IIA. The antiapoptotic effects of tanshinone IIA on ethanol-induced toxicity were accompanied by the downregulation of proapoptotic p53 protein expression [22].

Tanshinone II sodium sulfonate, a water-soluble derivative, was found to show beneficial effects in the cardiovascular and cerebrovascular system. It was found to reduce the MI size in ischemic rabbits and in dogs. Its beneficial effects on the ischemic heart may be related to its dilatory activity on the coronary collateral vessels [23]. Tanshinone II sodium sulfonate was also reported to inhibit thrombus formation and platelet aggregation in rats and mice. A clinical investigation of tanshinone II sodium sulfonate in 180 patients with coronary heart disease showed a significant effect, as evidenced by electrocardiograms, and improvements in angina pectoris and chest oppression [24]. The number and intensity of adhesion of erythrocytes from cerebral thrombosis patients to cultured human umbilical vein endothelial cells (HUVEC) was decreased after the treatment with danshen [25]. Tanshinone II sodium sulfonate was also found to protect against lipid peroxidation. This may be related not only to its antioxidant activity but also to its regulation of antioxidant enzyme activities in the heart [26].

Tanshinones scavenged free radicals generated from lipid peroxidation of myocardial mitochondrial membranes, and protected membranes against ischemia–reperfusion injury [27]. Studies on the effects of tanshinone II on DNA damage by lipid peroxidation in hepatocytes showed that tanshinone II inhibited the lipid peroxidation. The protective effect of tanshinone II may be through breaking the chain reactions of peroxidation by scavenging free radicals [28]. The chemoluminescence of human neutrophils induced by 12-*O*-tetradecanoylphorbol 13-acetate (TPA) was markedly depressed by incubation with tanshinone II, in a concentration-dependent manner. Since protein kinase C (PKC) is a receptor for TPA, it was suggested that tanshinones may competitively inhibit PKC activation [29].

Tanshinone II A was also found to protect against lung injury in rats, as induced by LPS. The treatment of LPS-exposed rats with tanshinone IIA significantly lowered the expression of adhesion molecule CD18 on the surface of polymorphonuclear neutrophils, and MDA content, thus ameliorating the coagulation abnormality. Histopathological examinations showed a decrease in the polymorphonuclear neutrophil sequestration and also in the width of the alveolar septa in lung tissue; this indicated that tanshinone IIA exerted a protective effect in LPS-induced lung injury in rats [30].

The clinical treatment of patients with coronary heart disease with danshen significantly decreased levels of serum lipid peroxides and increased superoxide dismutase activity. The mechanism might be correlated with the effect of danshen in inhibiting platelet aggregation, reducing blood viscosity, improving myocardial ischemia, and protecting cytomembranes [31]. Danshen was further reported to inhibit the growth of fibroblasts; such inhibition was shown to be caused by a suppression of cell mitosis and cell arrest at the G_2/M stage [32].

Danshensu was found to dilate isolated pig coronary artery and to antagonize the constriction induced by morphine and propranolol. On the other hand, danshensu showed no antagonistic effect on the contracting response of coronary artery caused by a high potassium medium [33]. An aqueous danshen extract and danshensu also exhibited a vasorelaxant effect on isolated rat coronary artery rings precontracted with serotonin. The vasodilatory activity of danshen extract and danshensu were produced by the inhibition of Ca^{2+} influx into vascular smooth muscle cells [34]. Danshensu significantly inhibited the increased expression of atrial natriuretic peptide (ANP) and β-actin mRNA in myocardial cells induced by angiotensin II; this apparently is the mechanism for its protective effect against myocardial hypertrophy [35]. Danshensu and protocatechualdehyde, as calcium-antagonist compounds, markedly decreased cytosolic free calcium concentrations in human erythrocytes in a concentration-dependent manner [36].

The aqueous extract of danshen root, danshensu, protocatechuic acid, catechin and protocatechualdehyde were each found to suppress endothelial dysfunction, as induced by homocysteine. The latter, as a byproduct of methionine metabolism, may be one of the risk factors for the development of several vascular diseases, including thromboembolism, atherosclerosis, stroke, vascular diseases, and dementia. The protective effect of the aqueous extract and its active compounds on endothelial cell function was demonstrated through an *in vitro* tube formation assay, which mimics new blood vessel formation [37].

Salvianolic acid B was reported to activate the opening of the iberiotoxin-sensitive Ca^{2+}-activated K^+ channels of porcine coronary artery smooth muscle cells through the activation of guanylate cyclase, without the involvement of nitric oxide synthase (NOS) activation [38]. Salvianolic acid B, at a concentration of $10\,\mu g\,ml^{-1}$, was shown to remarkably inhibit the increased expression of vascular endothelial growth factor (VEGF) in U937 foam cells induced by incubation with oxidized low-density lipoprotein (LDL) [39]. Salvianolic acid A strongly inhibited the oxidation of human LDL. It was suggested that salvianolic acid A exerted an inhibitory activity on Cu^{2+}-mediated LDL oxidation through chelating Cu^{2+} and scavenging free radicals [40].

A protective effect of salvianolic acid A on cerebral cell injuries induced by oxidative stress was also reported [41]. Salvianolic acids significantly inhibited the platelet aggregation induced by collagen or ADP. Given at an i.v. dose of $10\,mg\,kg^{-1}$ to rats they significantly inhibited the platelet aggregation induced by collagen or by ADP, and improved the regional cerebral blood flow in rats after ischemia; however, no effect on normal regional cerebral blood flow was observed [42].

Beneficial effects of acetylsalvianolic acid A on cerebral ischemic rats subjected to middle cerebral artery thrombosis were reported. Intravenous treatment with acetylsalvianolic acid A significantly reduced the cerebral infarction [43]. The i.v. administration of rosmarinic acid to rats significantly inhibited venous thrombosis formation and platelet aggregation, as induced by collagen. Rosmarinic acid showed a mild antithrombotic effect that potentially was due to an inhibition of platelet aggregation and promotion of fibrinolytic activity [44]. Lithospermic acid B infused into the postischemic rabbit heart reduced the myocardial damage [45].

Danshensu also scavenged superoxide anion free radicals generated from the reaction system of xanthine and xanthine oxidase in the myocardium, and protected the mitochondrial membrane from ischemia–reperfusion injury and lipid peroxidation [27]. Among the phenolic carboxylic acids and related depsides, danshensu, rosmarinic acid, lithospermic acid, and salvianolic acids all exhibited significant antioxidative and free radical-scavenging activities [46]. Salvianolic acids A, B and C, danshensu, protocatechuic aldehyde, and rosmarinic acid each prevented the hemolysis of mouse erythrocytes, as induced by H_2O_2. The protection of salvianolic acid A against myocardial anoxia in experimental animals may be mainly ascribed to its oxygen free radical-scavenging activity [47]. Danshensu was also found to be the active principle of *S. miltiorrhiza* against aflatoxin B_1 (AFB_1)-induced cytotoxicity in cultured primary rat hepatocytes. It suppressed lactate dehydrogenase leakage, inhibited intracellular ROS formation caused by AFB_1, and increased the intracellular glutathione (GSH) content [48]. Salvianolic acid A, salvianolic acid B and rosmarinic acid were each found to inhibit both NADPH-ascorbic acid and Fe^{2+}-cysteine-induced lipid peroxidation in rat brain, liver and kidney microsomes suppressing the production of superoxide anion radical in the xanthine–xanthine oxidase system [49]. Salvianolic acid A also significantly scavenged oxygen radicals released by activated neutrophils, without affecting their functional ability [50].

The simultaneous administration of an aqueous danshen extract and a high-cholesterol diet to rabbits resulted in a significant reduction in plasma cholesterol; of the atherosclerotic area in the abdominal aorta; and of cholesterol deposition in the thoracic aorta [51]. Salvianolic acid A may reverse the inhibition of the potassium channel in rat cardiac myocytes caused by oxygen free radicals. Danshensu was also found to protect from decrease of H^+-ATPase activity in rat myocardial mitochondria induced by free radicals [52]. The antioxidative effect of danshen extract was also observed in rats with acute pancreatitis. It inhibited the generation of oxygen-derived free radicals, and increased their elimination [53].

A danshen decoction was further found to exert hepatoprotective effects and to reduce the enhanced serum glutamic-pyruvic transaminase (SGPT) and pathological changes in rabbits with acute liver damage caused by carbon tetrachloride (CCl_4) [54]. It also protected against acute liver damage in rats, as induced by D-galactosamine [55] or by CCl_4 [56]. Salvianolic acid A prevented CCl_4-induced liver damage and liver fibrosis in rats due to its effect against lipid peroxidation [57]. The danshen extract showed preventive and therapeutic effects in rats against acute renal failure, as induced by an intramuscular injection of glycerol. The treatment of rabbits with acute renal failure with danshen extract resulted in preventive and therapeutic effects, as evidenced by measuring renal function and renal morphology [58]. The protective effect of danshen on acute renal failure in experimental studies, and in clinical observations, was suggested to be caused by an increased volume of renal blood flow, an increased clearance rate of creatinine, a decreased nitrogen content in blood urea, and diuresis [59]. It was also postulated that the effect of danshen against acute renal failure may be the result of adenosine antagonism [60].

The i.p. injection of danshensu at a daily dose of 300 mg kg^{-1} to Wistar rats with liver fibrosis, as induced by pig serum, resulted in a decline of serum hyaluronic acid level and a significant reduction of liver fibrosis. Moreover, *in vitro* danshensu inhibited the proliferation of hepatic stellate cells at concentrations of 50–200 μg ml^{-1}; this indicated that its therapeutic effect on liver fibrosis in rats might be related to an inhibition of the proliferation of hepatic stellate cells [61].

Salvianolic acid B was also reported to inhibit intermediate hepatic stellate cell proliferation, to decrease transforming growth factor-β1 (TGF-β1) -stimulated hepatic stellate cell activation as well as matrix protein and gene expression, and to inhibit stimulated hepatic stellate cell Smad2 and 3 protein expression, phosphorylation, and nuclear translocation. The inhibition of TGF-β1 signaling in hepatic stellate cells and its biological responses might represent an important mechanism of salvianolic acid B activity against hepatic fibrosis [62]. The proliferation inhibitory activity of salvianolic acid B on hepatic stellate cells was also evidenced by the measurement of [^{3}H]TdR uptake. Salvianolic acid B inhibited soluble type I collagen secretion and decreased matrix collagen deposition, also inhibiting mitogen-activated protein kinase (MAPK) activity. The inhibition of hepatic stellate cell proliferation and collagen production, as well as the decrease in TGF-β1 and MAPK activity, might contribute to the mechanism of salvianolic acid B action against hepatic fibrosis [63]. A clinical trial with salvianolic acid B in 60 patients with definite diagnosis of liver fibrosis with hepatitis B revealed that salvianolic acid B effectively reversed liver fibrosis in chronic hepatitis. It did better than interferon-γ (IFN-γ) in reducing serum hyaluronic acid levels. Salvianolic acid B overall decreased four serum fibrotic markers, also decreasing the ultrasound imaging score. Hepatic fibrosis in chronic hepatitis B with slight liver injury was more suitable to salvianolic acid B treatment. Salvianolic acid B showed no obvious side effects [64].

The i.v. application of an extract of Radix Salviae Miltiorrhizae to rats with experimental pulmonary fibrosis, as induced by bleomycin for 14 days, starting from the 15th day of bleomycin application, was also reported to exert a significant therapeutic effect against pulmonary fibrosis, as evidenced by pathological examination [65]. In human cutaneous fibroblasts, danshensu effectively inhibited the nuclear transcription factor NF-κB activity, inducing apoptosis. Furthermore, danshensu also inhibited nuclear transcription factor NF-1 activity in fibroblasts and modulated the synthesis and secretion of collagen [66].

Tanshinone derivatives were found to be cytotoxic *in vitro* against human tumor cell lines A549 (non-small-cell lung), SK-OV-3 (ovary), SK-MEL-2 (melanoma), XF498 (tumor of CNS), HCT-15 (colon), KB (nasopharyngeal carcinoma), HeLa (cervical carcinoma), Colo 205 (colon adenocarcinoma) and Hep-2 (laryngeal epidermal carcinoma) [67], with IC_{50}-values ranging from 0.2 to 8 μg ml^{-1}. Dihydroisotanshinone, with a *para*-quinone moiety, showed significant cytotoxicity against HeLa, Colon 205, and Hep-2 cells [68]. The growth-inhibitory activity of tanshinone II was postulated to be associated with inhibiting DNA synthesis, proliferating cell nuclear antigen expression, and the DNA polymerase activity of tumor cells [69]. It was suggested that the planar phenanthrene ring of the tanshinones may be essential for interaction with DNA, whereas the furanoquinone moiety may be responsible for the production of reactive free radicals in close vicinity of the bases so as to cause DNA damage [70]. Tanshinone II effectively inhibited growth of the human cervical carcinoma cell line, ME180, and reduced its tumorigenicity in nude mice. The tumor cells became well differentiated *in vitro* after treatment with tanshinone II [71].

Tanshinone IIA exhibited cytotoxic activities against multiple human cancer cells. It inhibited the growth of human glioma cells in a concentration-dependent manner, with an IC_{50} of 100 ng ml^{-1}. It significantly inhibited colony formation and BrdU incorporation, induced cell-cycle arrest at G_0/G_1 phase and apoptosis, also inducing differentiation of human glioma cells [72]. In NB4 acute promyelocytic leukemia cells, tanshinone IIA inhibited proliferation and induced apoptosis by the activation of caspase-3, the downregulation of anti-apoptotic protein Bcl-2 and Bcl-XL, and the upregulation of proapoptotic protein Bax, as well as by disruption of the mitochondrial membrane potential [73].

In $HepG_2$ human hepatoma cells, tanshinone IIA also exerted strong inhibition of cell proliferation and induced apoptosis [74]. Induction of apoptosis by tanshinone IIA was also observed in HL-60 human promyelocytic leukemic cells and K562 human erythroleukemic cells. The apoptosis induction in HL60 and K562 cells by tanshinone II may be associated with the activation of selective members of the caspase family [75]. Danshen extract also enhanced the antineoplastic effect of cyclophosphamide against S180 in mice, and was found to be more effective than urokinase. The serum level of fructose-1,6-diphosphate of tumor-bearing mice was higher than that in normal mice, while that in treated mice was lower than in untreated mice [76]. Tanshinone II sodium sulfonate prevented and inhibited bleomycin-induced pulmonary fibrosis in rats, and significantly reduced the levels of hydroxyproline in rat lung tissue homogenate [77].

Tanshinone IIA also exerted both concentration- and time-dependent inhibitory effects on cell growth of human breast cancer cells, with an IC_{50} of 0.25 μg ml^{-1}. It also significantly inhibited colony formation and BrdU incorporation of cancer cells. Oligonucleotide microarray analysis identified 41 upregulated (1.22%) and 24 downregulated (0.71%) genes in cancer cells after tanshinone IIA treatment. The upregulated genes were involved predominantly in cycle regulation, cell proliferation, apoptosis, signal transduction and transcriptional regulation, while the downregulated genes were associated mainly with apoptosis and extracellular matrix and adhesion molecules. A 45% tumor mass volume reduction and significant increase in caspase-3 protein expression were observed in nude mice bearing human breast infiltrating duct carcinoma when treated with a s.c. injection of tanshinone IIA at a dose of 30 mg kg^{-1} three times weekly for 10 weeks [78].

The danshen extract was reported to enhance the mutagenicity of the food-borne mutagens Trp-P-1 and benzo[*a*]pyrene (BaP) in *Salmonella typhimurium* TA98 at low concentrations, but to suppress them at high concentrations [79]. Danshen extract remarkably inhibited the X-linked lethal mutation and the somatic chromosome mutations during the spermatogenesis of *Drosophila melanogaster*, as induced by *N*-methyl-*N*′-nitro-*N*-nitroso-guanidine (MNNG) [80]. It was also reported that danshen prevented pulmonary (but not thymus) damage in mice after X-ray irradiation of the right thorax [81]. Danshen also exhibited significant protective effects against multiple organ injuries caused by high-energy wounding in animals, through a free radical mechanism. The injuries were characterized by hemorrhage, tissue rupture, hematoma, local edema, and necrosis [82].

When tanshinone IIA was given orally to rats at 60 mg kg^{-1}, tissue concentrations decreased in the order of stomach, small intestine, lung, liver, fat, muscle, kidneys, spleen, heart, plasma, brain, and testes; however, tanshinone IIA was still detected in most tissues at 20 h after administration [83]. Following an i.v. injection of [^{3}H] tanshinone II sodium sulfonate to rats, the highest levels of radioactivity were found in the liver, followed by spleen, kidney, and lung. Peak levels of radioactivity in organs were reached at 2 h after injection. The half-lives $t_{1/2\alpha}$ and $t_{1/2\beta}$ were reported to be 27 and 200 min. Within 72 h, 75% of the administered activity was excreted in the feces, and 18% in the urine [84]. The biotransformation of cryptotanshinone into tanshinone II by hydrogenation in the liver was observed after oral administration to rats [85]. In mice, the pharmacokinetics of sodium tanshinone IIA sulfonate were characterized with a distribution half-life ($t_{1/2\alpha}$) of 1.2 min, a terminal half-life ($t_{1/2\beta}$) of 21.6 min, and an area under the curve (AUC) of 58.4 $\mu g \cdot h\, ml^{-1}$ [86]. Metabolite identification *in vitro* and *in vivo* showed that tanshinone IIA and tanshinone I were the major metabolites of cryptotanshinone and dihydrotanshinone I, respectively [87]. After the i.v. injection of tanshinone IIA, tanshinone IIB, hydroxytanshinone IIA, przewaquinone A and dehydrotanshinone IIA and glucuronide conjugates of two different semiquinones, derived from hydrogenation products of dehydrotanshinone IIA and tanshinone IIA, were identified as metabolites in rat bile, urine, and feces [88].

After the i.v. injection of danshensu to rabbits, high concentrations were found in the kidney, liver, lung, and heart, whereas concentrations in brain and spleen were relatively low [89]. The plasma half-life of danshensu after i.v. administration to rabbits was 32 min [90]. The pharmacokinetics of danshensu has also been studied in healthy human volunteers after the oral administration of two types of danshen preparations containing determined amounts of danshensu. The elimination half-life of danshensu after oral administration of a compound granule or a danshen decoction was about 0.9 h in both cases. This resulted in 6% and 14% of the dose being excreted in the urine, respectively, within 8 h [91]. The pharmacokinetics of danshensu and protocatechuic aldehyde has been studied in rats after the oral administration of an extract of the root of *S. miltiorrhiza* [92].

References

1 Zhang, H., Yu, C., Jia, J.Y., Leung, S.W., Siow, Y. L., Man, R.Y., and Zhu, D.Y. (2002) Contents of four active components in different commercial crude drugs and preparations of danshen (*Salvia miltiorrhiza*). *Acta Pharmacol. Sin.*, **23**, 1163–1168.

2 Lin, H.C. and Chang, W.L. (2000) Diterpenoids from *Salvia miltiorrhiza. Phytochemistry*, **53**, 951–953.

3 Lee, A.R., Wu, W.L., Chang, W.L., Lin, H.C., and King, M.L. (1987) Isolation and bioactivity of new tanshinones. *J. Nat. Prod.*, **50**, 157–160.

4 Lin, L.Z., Wang, X.M., Huang, X.L., Huang, Y., and Xang, B.J. (1988) A new diterpenoid quinone dehydromiltirone. *Acta Pharm. Sin.*, **23**, 273–275.

5 Lin, H.C. and Chang, W.L. (1993) A new tanshinone from *Salvia miltiorrhiza. Chin. Pharm J.*, **45**, 615–618.

6 Zhou, L., Zuo, Z., and Chow, M.S. (2005) Danshen: an overview of its chemistry, pharmacology, pharmacokinetics, and clinical use. *J. Clin. Pharmacol.*, **45**, 1345–1359.

7 Wang, X., Morris-Natschke, S.L., and Lee, K.H. (2007) New developments in the chemistry and biology of the bioactive constituents of Tanshen. *Med. Res. Rev.*, **27**, 133–148.

8 Chen, L.N. and Zhu, X.X. (2005) Advances in study of the pharmacological effects of danshen on hemorheology. *Zhongguo Zhong Yao Za Zhi*, **30**, 630–633. 640.

9 Jang, S.I., Kim, H.J., Kim, Y.J., Jeong, S.I., and You, Y.O. (2006) Tanshinone IIA inhibits LPS-induced NF-kappaB activation in RAW 264.7 cells: possible involvement of the NIK-IKK, ERK1/2, p38 and JNK pathways. *Eur. J. Pharmacol.*, **542**, 1–7.

10 Yu, G.R. (1988) Clinical and experimental study on the effect of *Salvia miltiorrhiza* on microcirculation and 2,3 diphosphoglyceric acid in patients with coronary heart disease. *Chin. J. Integr. Trad. West. Med.*, **8**, 596–598.

11 Cheng, Y.Y., Fong, S.M., and Hon, P.M. (1992) Effect of *Salvia miltiorrhiza* on the cardial ischemia in rats induced by ligation. *Chin. J. Integr. Trad. West. Med.*, **12**, 424–426.

12 Cheng, Y.Y., Fong, S.M., Hon, P.M., Li, C.M., and Chang, H.M. (1991) Prevention and treatment of isoproterenol induced ventricular fibrillation in rats by aqueous extract of *Salvia miltiorrhiza*. *Chin. J. Integr. Trad. West. Med.*, **11**, 543–546.

13 Chen, K.Y., Wen, S.F., Zhi, Z.J., and Shao, J.S. (1983) Preliminary observation of ^{131}Cs distribution in experimental acute myocardial infarction and coronary insufficiency treated with root of *Salvia miltiorrhiza*, flower of *Chrysanthemum morifolium* and *Chrysanthemum indicum*. *J. Trad. Chin. Med.*, **3**, 265–270.

14 Bai, Y.R. and Wang, S.Z. (1994) Hemodynamic study on nitroglycerin compared with *Salvia miltiorrhiza*. *Chin. J. Integr. Trad. West. Med.*, **14**, 24–25.

15 Sun, X.M. and Cai, H.J. (1989) Use of amphotericin B-cells to detect inhibitors of cellular cholesterol biosynthesis. *Chin. J. Integr. Trad. West. Med.*, **9**, 604–606.

16 Li, X.H. and Tang, R.Y. (1991) Relationship between inhibitory action of tanshinone on neutrophil function and its prophylactic effects on myocardial infarction. *Acta Pharmacol. Sin.*, **12**, 269–272.

17 Fu, J., Huang, H., Liu, J., Pi, R., Chen, J., and Liu, P. (2007) Tanshinone IIA protects cardiac myocytes against oxidative stress-triggered damage and apoptosis. *Eur. J. Pharmacol.*, **568**, 213–221.

18 Fang, Z.Y., Lin, R., Yuan, B.X., Liu, Y., and Zhang, H. (2007) Tanshinone IIA inhibits atherosclerotic plaque formation by down-regulating MMP-2 and MMP-9 expression in rabbits fed a high-fat diet. *Life Sci.*, **81**, 1339–1345.

19 Tang, F., Wu, X., Wang, T., Wang, P., Li, R., Zhang, H., Gao, J., Chen, S., Bao, L., Huang, H., and Liu, P. (2007) Tanshinone IIA attenuates atherosclerotic calcification in rat model by inhibition of oxidative stress. *Vasc. Pharmacol.*, **46**, 427–438.

20 Liu, J.D. and Tang, R.Y. (1993) The prophylactic effects of tanshinone against experimental cerebral infarction in rats. *Chin. J. Pathophysiol.*, **9**, 369–372.

21 Kim, D.H., Jeon, S.J., Jung, J.W., Lee, S., Yoon, B.H., Shin, B.Y., Son, K.H., Cheong, J.H., Kim, Y.S., Kang, S.S., Ko, K.H., and Ryu, J.H. (2007) Tanshinone congeners improve memory impairments induced by scopolamine on passive avoidance tasks in mice. *Eur. J. Pharmacol.*, **574**, 140–147.

22 Meng, X.F., Zou, X.J., Peng, B., Shi, J., Guan, X.M., and Zhang, C. (2006) Inhibition of ethanol-induced toxicity by tanshinone IIA in PC12 cells. *Acta Pharmacol. Sin.*, **27**, 659–664.

23 Hu, G.J., Zhang, J.G., Jiang, W.D., and Wei, P. J. (1981) Effects of intracoronary injections of sodium tanshinone II-A sulfonate and dipyridamole on myocardial infarct size in acute ischemic dogs. *Acta Pharmacol Sin.*, **2**, 34–35.

24 Li, C.Z., Yang, S.C., and Zhao, F.D. (1984) Effects of tanshinone II-A sulfonate on thrombus formation, platelet and blood coagulation in rats and mice. *Acta Pharmacol. Sin.*, **5**, 39–42.

25 Jiang, K.Y., Ruan, C.G., Gu, Z.L., Zhou, W.Y., and Guo, C.Y. (1998) Effects of tanshinone II-A sulfonate on adhesion molecule expression of endothelial cells and platelets *in vitro*. *Acta Pharmacol. Sin.*, **19**, 47–50.

26 Zhou, G.Y., Zhao, B.L., Hou, J.W., Ma, G.E., and Xin, W.J. (1999) Protective effects of sodium tanshinone IIA sulphonate against adriamycin-induced lipid peroxidation in mice hearts *in vivo* and *in vitro*. *Pharmacol. Res.*, **40**, 487–491.

27 Zhao, B.L., Jiang, W., Zhao, Y., Hou, J.W., and Xin, W.J. (1996) Scavenging effects of *Salvia miltiorrhiza* on free radicals and its protection for myocardial mitochondrial membranes from ischemia-reperfusion injury. *Biochem. Mol. Biol. Int.*, **38**, 1171–1182.

28 Cao, E.H., Liu, X.Q., Wang, J.J., and Xu, N.F. (1996) Effect of natural antioxidant tanshinone II-A on DNA damage by lipid peroxidation in liver cells. *Free Radic. Biol. Med.*, **20**, 801–806.

29 Xu, L.L., Wu, Q., Wang, B.Y., Yang, X.C., and Wu, L.P. (1994) Inhibitory effects of tanshinone on oxygen free radical production by human neutrophils. *Chin. J. Pathophysiol.*, **10**, 635–638.

30 Shi, X.M., Huang, L., Xiong, S.D., and Zhong, X.Y. (2007) Protective effect of tanshinone II A on lipopolysaccharide-induced lung injury in rats. *Chin. J. Integr. Med.*, **13**, 137–140.

31 Xing, Z.Q., Zeng, X.C., and Yi, C.T. (1996) Effect of *Salvia miltiorrhiza* on serum lipid peroxide, superoxide dismutase of the patients with coronary heart disease. *Chin. J. Integr. Trad. West. Med.*, **16**, 287–288.

32 Shang, Q., Zhang, D., and Guan, W. (1999) The inhibitory effect *in vitro* of *Salvia miltiorrhiza* and tetramethyl pyrazine on the growth of fibroblasts. *Chin. J. Repar. Reconstr. Surg.*, **12**, 321–324.

33 Dong, Z.T. and Jiang, W.D. (1982) Effect of Dan Shen Su on isolated swine coronary artery perfusion preparation. *Acta Pharm. Sin.*, **17**, 226–228.

34 Lam, F.F., Yeung, J.H., Chan, K.M., and Or, P. M. (2007) Relaxant effects of danshen aqueous extract and its constituent danshensu on rat coronary artery are mediated by inhibition of calcium channels. *Vasc. Pharmacol.*, **46**, 271–277.

35 Guo, Z.Q., Wang, S.R., and Zhu, L.Q. (2005) Effect of danshensu and ligustrazine on related genes of myocardial hypertrophy induced by angiotensin II. *Zhongguo Zhong Xi Yi Jie He Za Zhi*, **25**, 342–344.

36 Shen, L.H., Wang, B.Y., Wang, C.Q., Xie, X.L., Yu, G.R., Yao, Z.Y., Zhu, Y.L., and Yang, B.J. (2004) Effect of danshensu, protocatechualdehyde and danshen injection on calcium ion concentration in cytoplasm of human erythrocytes. *Zhongguo Zhong Yao Za Zhi*, **29**, 984–988.

37 Chan, K., Chui, S.H., Wong, D.Y., Ha, W.Y., Chan, C.L., and Wong, R.N. (2004) Protective effects of Danshensu from the aqueous extract of *Salvia miltiorrhiza* (Danshen) against homocysteine-induced endothelial dysfunction. *Life Sci.*, **75**, 3157–3171.

38 Lam, F.F., Seto, S.W., Kwan, Y.W., Yeung, J.H., and Chen, P. (2006) Activation of the iberiotoxin-sensitive BKCa channels by salvianolic acid B of the porcine coronary artery smooth muscle cells. *Eur. J. Pharmacol.*, **546**, 28–35.

39 Yang, P.Y., Rui, Y.C., Zhang, L., Li, T.J., Qiu, Y., Wang, J.S., and Zhang, W.D. (2002) Expression of vascular endothelial growth factor in U937 foam cells and the inhibitory effect of drugs. *Yao Xue Xue Bao*, **37**, 86–89.

40 Liu, Y.L. and Liu, G.T. (2002) Inhibition of human low-density lipoprotein oxidation by salvianolic acid-A. *Yao Xue Xue Bao*, **37**, 81–85.

41 Li, L. (1998) Protective effects of schisanhenol, salvianolic acid A and SY-L on oxidative stress induced injuries of cerebral cells and their mechanisms. *Sheng Li Ke Xue Jin Zhan*, **29**, 35–38.

42 Tang, M.K., Ren, D.C., Zhang, J.T., and Du, G. H. (2002) Effect of salvianolic acids from Radix Salviae miltiorrhizae on regional cerebral blood flow and platelet aggregation in rats. *Phytomedicine*, **9**, 405–409.

43 Dong, J.C. and Xu, L.N. (1996) Beneficial effects of acetylsalvianolic acid A on focal cerebral ischemic rats subjected to middle cerebral artery thrombosis. *Acta Pharm. Sin.*, **31**, 6–9.

44 Zou, Z.W., Xu, L.N., and Tian, J.Y. (1993) Antithrombotic and antiplatelet effects of rosmarinic acid, a water-soluble component isolated from radix salviae miltiorrhizae (Danshen). *Acta Pharm. Sin.*, **28**, 241–245.

45 Fung, K.P., Zeng, L.H., Wu, J., Wong, H.N.C., Lee, C.M., Hon, P.M., Chang, H.M., and Wu, T.S. (1993) Demonstration of the myocardial salvage effect of lithospermic acid B isolated from the aqueous extract of *Salvia miltiorrhiza*. *Life Sci.*, **52**, PL239–PL244.

46 Kang, H.S., Chung, H.Y., Jung, J.H., Kang, S. S., and Choi, J.S. (1997) Antioxidant effect of *Salvia miltiorrhiza*. *Arch Pharmacol. Res.*, **20**, 496–500.

47 Li, D.Y., Xu, L.N., and Liu, X.G. (1995) Effects of water-soluble components isolated from *Salvia miltiorrhiza* on oxygen free radical generation and lipid peroxidation. *J. Chin. Pharm. Sci.*, **4**, 107–112.

48 Liu, J., Yang, C.F., Lee, B.L., Shen, H.M., Ang, S.G., and Ong, C.N. (1999) Effect of *Salvia miltiorrhiza* on aflatoxin B1-induced oxidative stress in cultured rat hepatocytes. *Free Radic. Res.*, **31**, 559–568.

49 Liu, Y. and Zhang, J.T. (1994) Hydroxyl radical scavenging effect of salvianolic acids. *J. Chin. Pharm. Sci.*, **3**, 43–50.

50 Lin, T.J., Zhang, K.J., and Liu, G.T. (1996) Effects of salvianolic acid A on oxygen radicals released by rat neutrophils and on neutrophil function. *Biochem. Pharmacol.*, **51**, 1237–1241.

51 Wu, Y.J., Hong, C.Y., Lin, S.J., Wu, P., and Shiao, M.S. (1998) Increase of vitamin E content in LDL and reduction of atherosclerosis in cholesterol-fed rabbits by a water-soluble antioxidant-rich fraction of *Salvia miltiorrhiza*. *Arterioscler. Thromb. Vasc. Biol.*, **18**, 481–486.

52 Bao, G.H., Yu, D.J., Qu, J.H., Zheng, Y.F., and Xu, L.N. (1993) Oxygen free radical inhibition of potassium channel activity in cardiac myocytes and antagonistic effect of salvianolic acid A. *Acta Acad. Med. Sin.*, **15**, 320–324.

53 Zhu, N. and Du, J.H. (1991) The effect of the anti-oxidation of *Salvia miltiorrhiza* on rats

with acute pancreatitis. *Chin. J. Pathophysiol.*, 7, 449–452.
54 Cheng, Y.X., Gan, S.X., Shu, S.F., Wang, Q.Q., Guo, X., and Jing, F.Y. (1985) Treatment for acute toxipathic hepatitis in rabbits with variously processed danshen. *Bull. Chin. Mater. Med.*, **10**, 161–162.
55 Qi, X.G. (1991) Protective mechanism of *Salvia miltiorrhiza* and *Paeonia lactiflora* for experimental liver damage. *Chin. J. Integr. Trad. West. Med.*, **11**, 102–104.
56 Wasser, S., Ho, J.M., Ang, H.K., and Tan, C.E. (1998) *Salvia miltiorrhiza* reduces experimentally-induced hepatic fibrosis in rats. *J. Hepatol.*, **29**, 760–771.
57 Hu, Y.Y., Liu, P., Liu, C., Xu, L.M., Liu, C.H., Zhu, D.Y., and Huang, M.F. (1997) Actions of salvianolic acid A on CCl_4-induced liver injury and fibrosis in rats. *Acta Pharmacol. Sin.*, **18**, 478–480.
58 Hu, L., Yu, T., and Jia, Z. (1996) Experimental study of the protective effects of *Astragalus* and *Salvia miltiorrhiza* Bunge on glycerol induced acute renal failure in rabbits. *Chin. J. Surg.*, **34**, 311–314.
59 Zhang, B.Z., Huang, L.N., and Yuan, L.G. (1991) The therapeutic observation and mechanism research on treatment of acute renal failure with Danshen. *Acta Med. Sin.*, **6**, 90–92.
60 Jin, H.J., Wang, A.M., and Wang, Y.K. (1997) Preventive and therapeutic effects of radix *Salvia miltiorrhiza* on Glycerol-induced acute renal failure in rats. *China J. Chin. Mater. Med.*, **22**, 236–238.
61 Zheng, Y.Y., Dai, L.L., Wang, W.B., Jia, L.P., and Zhou, X. (2003) Effect and mechanism of Tanshensu on fibrotic rats. *Zhonghua Gan Zang Bing Za Zhi*, **11**, 288–290.
62 Liu, C., Liu, P., Hu, Y., and Zhu, D. (2002) Effects of salvianolic acid-B on TGF-β1 stimulated hepatic stellate cell activation and its intracellular signaling. *Zhonghua Yi Xue Za Zhi*, **82**, 1267–1272.
63 Liu, P., Liu, C.H., Wang, H.N., Hu, Y.Y., and Liu, C. (2002) Effect of salvianolic acid B on collagen production and mitogen-activated protein kinase activity in rat hepatic stellate cells. *Acta Pharmacol. Sin.*, **23**, 733–738.
64 Liu, P., Hu, Y.Y., Liu, C., Zhu, D.Y., Xue, H.M., Xu, Z.Q., Xu, L.M., Liu, C.H., Gu, H.T., and Zhang, Z.Q. (2002) Clinical observation of salvianolic acid B in treatment of liver fibrosis in chronic hepatitis B. *World J. Gastroenterol.*, **8**, 679–685.
65 Dai, L.J., Hou, J., and Cai, H.R. (2004) Experimental study on treatment of pulmonary fibrosis by Chinese drugs and integrative Chinese and Western medicine. *Zhongguo Zhong Xi Yi Jie He Za Zhi*, **24**, 130–132.
66 Jiang, H., Ha, T., and Wei, D. (2001) A study on the mechanism of the biological roles of danshensu on fibroblast. *Zhonghua Shao Shang Za Zhi*, **17**, 36–38.
67 Ryu, S.Y., Lee, C.O., and Choi, S.U. (1997) *In vitro* cytotoxicity of tanshinones from *Salvia miltiorrhiza. Planta Med.*, **63**, 339–342.
68 Lin, H.C., Chang, W.L., and Chen, C.F. (1995) Phytochemical and pharmacological study on *Salvia miltiorrhiza*. VI. Cytotoxic activity of tanshinones. *Chin. Pharm. J. (Taipei)*, **47**, 77–80.
69 Wang, X., Yuan, S., and Wang, C. (1996) A preliminary study of the anti-cancer effect of tanshinone on hepatic carcinoma and its mechanism of action in mice. *Chin. J. Oncol.*, **18**, 412–414.
70 Wu, W.L., Chang, W.L., and Chen, C.F. (1991) Cytotoxic activities of tanshinones against human carcinoma cell lines. *Am. J. Chin. Med.*, **19**, 207–216.
71 Yuan, S., Huang, G., and Wang, X. (1995) The differentiation-inducing effect of tanshinone and retinoic acid on human cervical carcinoma cell line *in vitro*. *Chin. J. Oncol.*, **17**, 422–424.
72 Wang, J., Wang, X., Jiang, S., Yuan, S., Lin, P., Zhang, J., Lu, Y., Wang, Q., Xiong, Z., Wu, Y., Ren, J., and Yang, H. (2007) Growth inhibition and induction of apoptosis and differentiation of tanshinone IIA in human glioma cells. *J. Neurooncol.*, **82**, 11–21.
73 Liu, J.J., Lin, D.J., Liu, P.Q., Huang, M., Li, X. D., and Huang, R.W. (2006) Induction of apoptosis and inhibition of cell adhesive and invasive effects by tanshinone IIA in acute promyelocytic leukemia cells *in vitro*. *J. Biomed. Sci.*, **13**, 813–823.
74 Yuan, S.L., Wei, Y.Q., Wang, X.J., Xiao, F., Li, S. F., and Zhang, J. (2004) Growth inhibition and apoptosis induction of tanshinone IIA on human hepatocellular carcinoma cells. *World J. Gastroenterol.*, **10**, 2024–2028.
75 Song, Y., Yuan, S.L., Yang, Y.M., Wang, X.J., and Huang, G.Q. (2005) Alteration of activities of telomerase in tanshinone IIA inducing apoptosis of the leukemia cells. *Zhongguo Zhong Yao Za Zhi*, **30**, 207–211.
76 Zheng, Y.W., Tie, Y., Jiao, Z.K., Zhang, J.W., Xue, Y., and Sun, B.Z. (1991) Comparison of

the effect of increase in cyclophosphamide antitumor activity between *Salvia miltiorrhiza* and urokinase. *J. Xian Med. Univ.*, **12**, 139–141.

77 Wang, C.M., He, Q.Z., and Zhang, R.X. (1994) Effects of tanshinone to bleomycin induced pulmonary fibrosis of rats on histological changes and production of lipid peroxides and hydroxyproline. *Chin. J. Tubercul. Respir. Dis.*, **17**, 308–310.

78 Wang, X., Wei, Y., Yuan, S., Liu, G., Lu, Y., Zhang, J., and Wang, W. (2005) Potential anticancer activity of tanshinone IIA against human breast cancer. *Int. J. Cancer*, **116**, 799–807.

79 Sato, M., Sato, T., Ose, Y., Nagase, H., Kito, H., and Sakai, Y. (1992) Modulating effect of tanshinones on mutagenic activity of Trp-P-1 and benzo[a]pyrene in *Salmonella typhimurium*. *Mutat. Res.*, **265**, 149–154.

80 Choi, Y.H., Chung, H.Y., Yoo, M.A., and Lee, W.H. (1994) Effects of ginseng and *Salvia miltiorrhiza* extracts on the mutagenicity of MNNG in Drosophila. *Yakhak Hoechi*, **38**, 332–337 (CA 121: 172831.

81 Du, H., Qian, Z., and Wang, Z. (1990) Prevention of radiation injury of the lungs by *Salvia miltiorrhiza* in mice. *Chin. J. Integr. Trad. West. Med.*, **10**, 230–231.

82 Fu, X., Tian, H., Sheng, Z., and Wang, D. (1992) Multiple organ injuries after abdominal high energy wounding in animals and the protective effect of antioxidants. *Chin. Med. Sci. J.*, **7**, 86–91.

83 Bi, H.C., Law, F.C., Zhong, G.P., Xu, C.S., Pan, Y., Ding, L., Chen, X., Zhao, L.Z., Xu, Q., and Huang, M. (2007) Study of tanshinone IIA tissue distribution in rat by liquid chromatography-tandem mass spectrometry method. *Biomed. Chromatogr.*, **21**, 473–479.

84 Shao, H.S., Jing, X.A., Ying, L.Q., Zhao, M.M., and Gu, J.E. (1981) Distribution and metabolism of ^{3}H-tanshinone II A sulfonate in rats. *Nucl. Tech.*, 55–57.

85 Luo, H.S., Shen, L.S., Zhang, S.Q., Xu, L.F., and Wei, P. (1983) Tanshinones: antimicrobacterial agents - bile excretion and biotransformation in rat liver. *Acta Pharm. Sin.*, **18**, 1–6.

86 Mao, S., Jin, H., Bi, Y., Liang, Z., Li, H., and Hou, S. (2007) Ion-pair reversed-phase HPLC method for determination of sodium tanshinone IIA sulfonate in biological samples and its pharmacokinetics and biodistribution in mice. *Chem. Pharm. Bull. (Tokyo)*, **55**, 753–756.

87 Liu, J., Wu, J., Wang, X., and Cai, Z. (2007) Study of the phase I and phase II metabolism of a mixture containing multiple tanshinones using liquid chromatography/tandem mass spectrometry. *Rapid Commun. Mass Spectrom.*, **21**, 2992–2998.

88 Li, P., Wang, G.J., Li, J., Hao, H.P., and Zheng, C.N. (2006) Characterization of metabolites of tanshinone IIA in rats by liquid chromatography/tandem mass spectrometry. *J. Mass Spectrom.*, **41**, 670–684.

89 Zhao, F.Q., Wang, A.L., and Wang, L. (1994) The distribution of danshensu in rabbits. *Chin. Pharm. J.*, **29**, 291–293.

90 Zhao, F.Q., Zheng, N.X., Sato, H., Adachi, I., and Horikoshi, I. (1997) Pharmacokinetics of a Chinese traditional medicine, danshensu (3,4-dihydroxyphenyllactic acid), in rabbits using high-performance liquid chromatography. *Biol. Pharm. Bull.*, **20**, 285–287.

91 Liu, Q. and Chao, R.B. (2003) Determination of danshensu in urine and its pharmacokinetics in human. *Yao Xue Xue Bao*, **38**, 771–774.

92 Ye, G., Wang, C.S., Li, Y.Y., Ren, H., and Guo, D.A. (2003) Simultaneous determination and pharmacokinetic studies on (3,4-Dihydroxyphenyl)-lactic acid and protocatechuic aldehyde in rat serum after oral administration of Radix Salviae Miltiorrhizae extract. *J. Chromatogr. Sci.*, **41**, 327–330.

Sanguisorba officinalis and *Sanguisorba officinalis* var. *longifolia*

Radix Sanguisorbae (Diyu) is the dried root of *Sanguisorba officinalis* L. or *Sanguisorba officinalis* L. var. *longifolia* (Bert.) Yü et Li (Rosaceae). It is mainly used as a hemostatic agent.

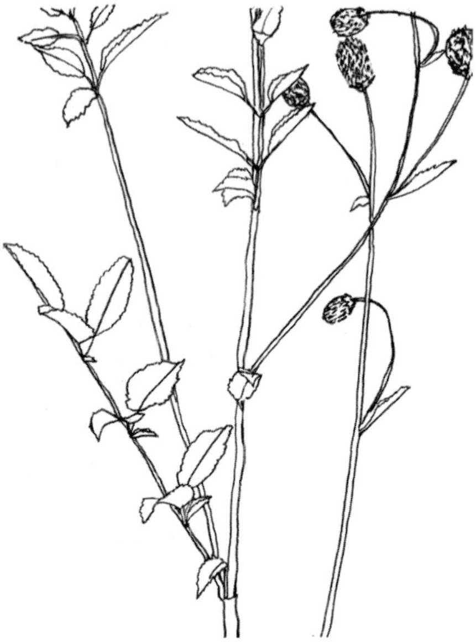

Sanguisorba officinalis L.

Chemistry

The root of *S. officinalis* and *S. officinalis* var. *longifolia* is known to be rich in tannin. The Chinese Pharmacopoeia requires a qualitative determination of gallic acid in Radix Sanguisorbae by thin-layer chromatographic comparison with reference substance, and a quantitative determination of tannin in Radix Sanguisorbae. The content of tannin in Radix Sanguisorbae should be not less than 10%. The major flavone compound from *S. officinalis* was identified as hyperin [1]. Also isolated from the root of *S. officinalis* were triterpenes and triterpene glycosides, such as sanguisorbin with sanguisorbigenin as aglycone [2] and 3β-[(α-L-arabinopyranosyl)oxy]-19β-hydroxyurs-12,20(30)-dien-28-oic acid, 3β-[(α-L-arabinopyranosyl)oxy]-urs-11,13(18)-dien-28-oic acid β-D-glucopyranosyl ester, 2α, 3α, 23-trihydroxyurs-12-en-24,28-dioic acid 28-β-D-glucopyranosyl ester, 3β-[(α-L-arabinopyranosyl)oxy]-urs-12,19(20)-dien-28-oic acid, 3β-[(α-L-arabinopyranosyl)oxy]-urs-12,19(29) -dien-28-oic acid, 3β-[(α-L-arabinopyranosyl)oxy]-19α-hydroxyolean-12-en-28-oic acid, 2α,3β-dihydroxy-28-norurs-12,17,19(20),21-tetraen-23-oic acid [3].

Pharmacology and Toxicology

The ethanol extract of *S. officinalis* was found to exhibit antimicrobial activity against *Bacillus cereus, Escherichia coli, Staphylococcus aureus, Pseudomonas aeruginosa,* and *Candida albicans* [4]. The aqueous extract of *S. officinalis* was found to exert antiviral activity against hepatitis B virus (HBV). The extract inhibited HBV multiplication as to measured by HBV DNA and surface antigen (HBsAg) levels in the extracellular medium of a HBV-producing cell line, HepG2 2.2.15. The extract also decreased the levels of extracellular HBV virion DNA and inhibited the secretion of HBsAg, in concentration-dependent manner [5].

The aqueous extract of the root of *S. officinalis* L. exhibited antiallergic activity, both *in vivo* and *in vitro*. Administration of the extract to rats at doses of 0.01 to 1 g kg^{-1} inhibited the systemic allergic reaction induced by compound 48/80, and reduced the plasma histamine levels in a dose-dependent manner. The extract also dose-dependently inhibited passive cutaneous

Handbook of Chinese Medicinal Plants: Chemistry, Pharmacology, Toxicology
Weici Tang and Gerhard Eisenbrand

ISBN: 978-3-527-32226-8

anaphylaxis activated by anti-dinitrophenyl (DNP) IgE. The extract, at concentrations of 0.001 to 1 mg ml^{-1}, also inhibited the histamine release from rat peritoneal mast cells activated by compound 48/80 or anti-DNP IgE in a concentration-dependent manner. The cAMP level in rat peritoneal mast cells was increased significantly after addition of the extract. Moreover, the extract also inhibited the production of tumor necrosis factor-α (TNF-α), as induced by anti-DNP IgE [6]. The disaccharide 5-*O*-α-D-(3-C-hydroxymethyl)lyxofuranosyl-β-D-(2-C-hydroxymethyl)arabinofuranose from *S. officinalis* was found to exert an antiallergic activity. The intravenous (i.v.) administration of the disaccharide to rats significantly inhibited the passive cutaneous anaphylaxis response in rats in a dose-dependent manner, with an ED_{50} of 9.6 mg kg^{-1}. The disaccharide also inhibited histamine release in rat peritoneal mast cells induced by both compound 48/80 and calcium ionophore A23187. In passively sensitized isolated guinea-pig hearts, the disaccharide markedly diminished both coronary flow reduction and histamine release; this indicated that the mechanism of the disaccharide's antiallergic activity might be the inhibition of mast cell mediator release [7].The topical application of an extract of *S. officinalis* on the hind limb skins of Sprague-Dawley rats inhibited wrinkle formation, maintained skin elasticity, and inhibited the decrease of dermal elastic fiber linearity after chronic ultraviolet-B (UVB) irradiation. This suggested that the extract might prevent chronic dermal photodamage following UVB irradiation [8]. After administration of the extract to human keratinocytes following UVB irradiation, the secretion of endothelin-1 was reduced, accompanied by a concomitant increase in the secretion of the inactive precursor, big endothelin-1. The treatment of hairless mice with the extract after exposure to UVB light resulted in an inhibition of endothelin-1 induction in the UVB-irradiated epidermis. Moreover, the extract significantly diminished pigmentation in UVB-exposed areas of guinea pig skin [9].

Sanguiin H-6, a dimeric ellagitannin from *S. officinalis*, was found to be an inhibitor of DNA topoisomerases. It inhibited topoisomerase I and topoisomerase II *in vitro*, with IC_{50}-values of 1 and 0.01 μ*M*, respectively. By comparison, the IC_{50} of camptothecin for the inhibition of topoisomerase I was 0.02 μ*M*, while the IC_{50} of VP-16 for the inhibition of topoisomerase II was 0.16 μ*M*. The inhibition of topoisomerases by sanguiin H-6 was not associated with the stabilization of covalent enzyme–DNA complexes, but rather by a mechanism which prevented the formation of covalent intermediates. The *in vitro* inhibitory effects of sanguiin H-6 on topoisomerases were irreversible. Sanguiin H-6 was also reported to exert proliferation inhibition against HeLa cells, with an ED_{50} of 12 μ*M*, and also interfered with intracellular topoisomerase activities in a concentration-dependent fashion [10].

References

1 Sha, M., Cao, A., Wang, B., Liu, C., Geng, J., and Liu, W. (1998) Determination of hyperin in *Sanguisorba officinalis* L. by high performance liquid chromatography. *Se Pu*, **16**, 226–228.

2 Mimaki, Y., Fukushima, M., Yokosuka, F A., Sashida, Y., Furuya, S., and Sakagami, H. (2001) Triterpene glycosides from the roots of *Sanguisorba officinalis*. *Phytochemistry*, **57**, 773–779.

3 Liu, X., Cui, Y., Yu, Q., and Yu, B. (2005) Triterpenoids from *Sanguisorba officinalis*. *Phytochemistry*, **66**, 1671–1679.

4 Kokoska, L., Polesny, Z., Rada, V., Nepovim, A., and Vanek, T. (2002) Screening of some Siberian medicinal plants for antimicrobial activity. *J. Ethnopharmacol.*, **82**, 51–53.

5 Kim, T.G., Kang, S.Y., Jung, K.K., Kang, F J.H., Lee, E., Han, H.M., and Kim, S.H. (2001)

Antiviral activities of extracts isolated from *Terminalis chebula* Retz., *Sanguisorba officinalis* L., *Rubus coreanus* Miq. and *Rheum palmatum* L. against hepatitis B virus. *Phytother. Res.*, **15**, 718–720.

6 Shin, T.Y., Lee, K.B., and Kim, S.H. (2002) Antiallergic effects of *Sanguisorba officinalis* on animal models of allergic reactions. *Immunopharmacol. Immunotoxicol.*, **24**, 455–468.

7 Park, K.H., Koh, D., Kim, K., Park, J., and Lim, Y. (2004) Antiallergic activity of a disaccharide isolated from *Sanguisorba officinalis*. *Phytother. Res.*, **18**, 658–662.

8 Tsukahara, K., Moriwaki, S., Fujimura, T., and Takema, Y. (2001) Inhibitory effect of an extract of *Sanguisorba officinalis* L. on ultraviolet-B-induced photodamage of rat skin. *Biol. Pharm. Bull.*, **24**, 998–1003.

9 Hachiya, A., Kobayashi, A., Ohuchi, A., Kitahara, T., and Takema, Y. (2001) The inhibitory effect of an extract of *Sanguisorba officinalis* L. on ultraviolet B-induced pigmentation *via* the suppression of endothelin-converting enzyme-1α. *Biol. Pharm. Bull.*, **24**, 688–692.

10 Bastow, K.F., Bori, I.D., Fukushima, Y., Kashiwada, Y., Tanaka, T., Nonaka, G., Nishioka, I., and Lee, K.H. (1993) Inhibition of DNA topoisomerases by sanguiin H-6, a cytotoxic dimeric ellagitannin from *Sanguisorba officinalis*. *Planta Med.*, **59**, 240–245.

Santalum album

Lignum Santali Albi (Tanxiang) is the dried heartwood of *Santalum album* L. (Santalaceae). It is used as an analgesic and stomachic agent for the treatment of abdominal pain and angina pectoris.

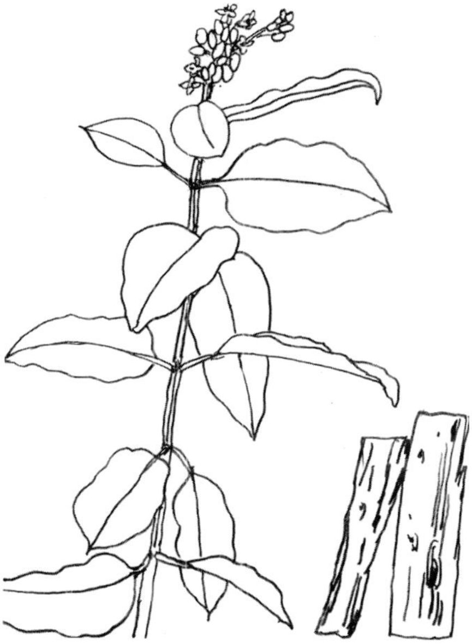

Santalum album L.

macopoeia requires a qualitative determination of α-santalol in Lignum Santali Albi by thin-layer chromatographic comparison with reference substance; and a quantitative determination of essential oil in Lignum Santali Albi. The content of essential oil in Lignum Santali Albi should be not less than 3.0% (ml g^{-1}). Further sesquiterpenes in the essential oil were identified as β-santalol, α-santalene, β-santalene, α-santalal, β-santalal, α-curcumone, nuciferol, tricyclokasantolol [2], *trans*-α-bergamotenol [3], 2β-hydroxy-14-hydro-β-santalol, 2α-hydroxy-albumol, campherene-2,13-diol, campherene-2β,13-diol, 7-hydroxynuciferol, 1β-hydroxy-2-hydrolanceol [4], and bisabolane sesquiterpenes bisabolenols A, B, C, D, and E [2].

Pharmacology and Toxicology

Sandal oil was reported to increase glutathione *S*-transferase (GST) activity in the liver of mice given orally at a daily dose of 5 and

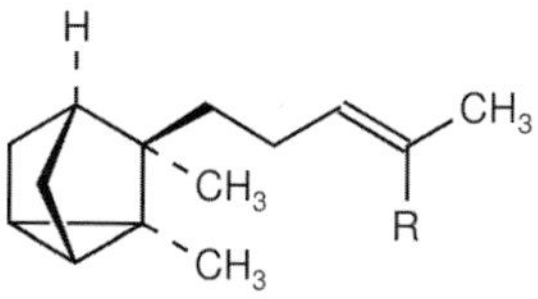

α-Santalene: R = CH_3
α-Santalol: R = CH_2OH
α-Santalol: R = CHO

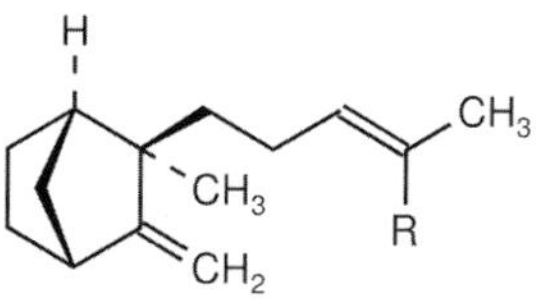

β-Santalene: R = CH_3
β-Santalol: R = CH_2OH
β-Santalol: R = CHO

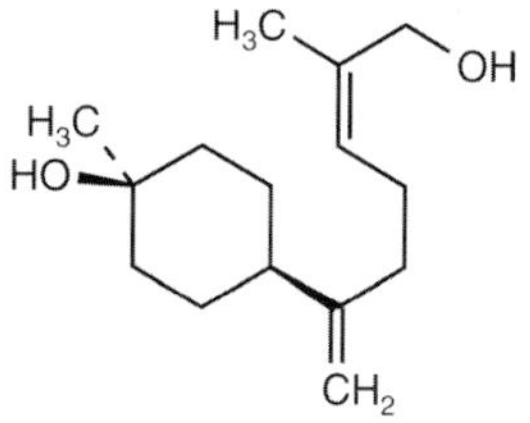

1β-Hydroxy-2-hydrolanceol

Chemistry

The heartwood of *Santalum album*, the sandalwood, is known to contain essential oil with the sesquiterpene α-santalol as the major component [1]. The Chinese Phar-

15 μl per animal for 10 and 20 days in both time- and dose-dependent manner. Feeding a dose of 5 μl sandalwood oil per animal for 10 and 20 days caused 1.8- to 1.9-fold increases in GST activity, whilst a dose of 15 μl oil per animal for 10 and 20 days induced

Handbook of Chinese Medicinal Plants: Chemistry, Pharmacology, Toxicology
Weici Tang and Gerhard Eisenbrand

ISBN: 978-3-527-32226-8

4.7- to 6.1-fold increases in enzyme activity. In addition, there were significant increases in acid-soluble thiol levels in the hepatic tissue of the mice following feeding of the oil [5]. α-Santalol and β-santalol were found to exert strong antibacterial activities against a series of *Helicobacter pylori* strains, including a strain (TS281) that is resistant to clarithromycin [4].

The chemopreventive effect of sandal oil was also observed in a mouse skin model initiated by 7,12-dimethylbenz[*a*]anthracene (DMBA) and promoted by 12-*O*-tetradecanoylphorbol 13-acetate (TPA). The formation of skin papilloma and TPA-induced ornithine decarboxylase activity were significantly decreased after topical treatment with sandal oil in acetone at different concentrations (1.25, 2.5, 3.75, and 5%) and at different times (0.5, 1, and 2 h) before TPA application [6, 7]. The sandalwood was also reported to possess antioxidant activity [8]. *In vitro*, the extract of sandalwood exhibited a direct scavenging activity towards nitric oxide (NO), when using sodium nitroprusside as an NO donor [9].

Sandal oil was further reported to possess *in vitro* antiviral activity against herpes simplex virus-1 (HSV-1) and HSV-2. Replication of the viruses was inhibited in the presence of the oil in a concentration-dependent manner; the antiviral effect of sandal oil was more pronounced against HSV-1. The oil was not virucidal and showed no cytotoxicity [10]. Both, α-santalol and β-santalol were found to inhibit acetic acid-induced writhing in mice. Inhibitory activities of α-santalol on opioid receptors were shown only by the δ-antagonist, but not by the μ- and κ-antagonists. The mechanism of inhibitory activity on the opioid receptor by α-santalol was different from that of morphine. α-Santalol was shown to be highly potent as an antagonist of dopamine D_2 and serotonin 5-HT_{2A} receptor binding. The effect of α-santalol as an antipsychotic agent was similar to that of chlorpromazine, but less potent [11]. After absorption through the skin of healthy human subjects, α-santalol was found to cause physiological changes, which were interpreted in terms of relaxing and sedative effects. In contrast, sandalwood oil provoked physiological deactivation but behavioral activation [12]. α-Santalol at a concentration of 6.1 ng ml^{-1}, β-santalol at 5.3 ng ml^{-1}, and α-santalene at 0.5 ng ml^{-1}, were each found in the serum of mice given sandal oil by inhalation [13].

References

1 Wang, Z. and Hong, X. (1991) Comparative GC analysis of essential oil in imported sandalwood. *China J. Chin. Mater. Med.*, **16**, 40–43, 64.

2 Yu, J.G., Cong, P.Z., Lin, J.T., and Fang, H.J. (1988) Studies on the chemical constituents of Chinese sandalwood oil and preliminary structures of five novel compounds. *Acta Pharm. Sin.*, **23**, 868–872.

3 Yu, J.G., Cong, P.Z., Lin, J.T., Zhang, Y.J., Hong, S.L., and Tu, G.Z. (1993) Studies on the structure of α-*trans*-bergamotenol from Chinese sandalwood oil. *Acta Pharm. Sin.*, **28**, 840–844.

4 Ochi, T., Shibata, H., Higuti, T., Kodama, K.H., Kusumi, T., and Takaishi, Y. (2005) Anti-*Helicobacter pylori* compounds from *Santalum album*. *J. Nat. Prod.*, **68**, 819–824.

5 Banerjee, S., Ecavade, A., and Rao, A.R. (1993) Modulatory influence of sandalwood oil on mouse hepatic glutathione *S*-transferase activity and acid soluble sulphydryl level. *Cancer Lett.*, **68**, 105–109.

6 Dwivedi, C. and Abu-Ghazaleh, A. (1997) Chemopreventive effects of sandalwood oil on skin papillomas in mice. *Eur. J. Cancer Prev.*, **6**, 399–401.

7 Dwivedi, C. and Zhang, Y. (1999) Sandalwood oil prevent skin tumour development in CD1 mice. *Eur. J. Cancer Prev.*, **8**, 449–455.

8 Scartezzini, P. and Speroni, E. (2000) Review on some plants of Indian traditional medicine

with antioxidant activity. *J. Ethnopharmacol.*, **71**, 23–43.

9 Jagetia, G.C. and Baliga, M.S. (2004) The evaluation of nitric oxide scavenging activity of certain Indian medicinal plants *in vitro*: a preliminary study. *J. Med. Food*, **7**, 343–348.

10 Benencia, F. and Courreges, M.C. (1999) Antiviral activity of sandalwood oil against herpes simplex viruses-1 and -2. *Phytomedicine*, **6**, 119–123.

11 Okugawa, H., Ueda, R., Matsumoto, K., Kawanishi, K., and Kato, K. (2000) Effects of sesquiterpenoids from "Oriental incenses" on acetic acid-induced writhing and D2 and 5-HT_{2A} receptors in rat brain. *Phytomedicine*, **7**, 417–422.

12 Hongratanaworakit, T., Heuberger, E., and Buchbauer, G. (2004) Evaluation of the effects of East Indian sandalwood oil and α-santalol on humans after transdermal absorption. *Planta Med.*, **70**, 3–7.

13 Jirovetz, L., Buchbauer, G., Jager, W., Woidich, A., and Nikiforov, A. (1992) Analysis of fragrance compounds in blood samples of mice by gas chromatography, mass spectrometry, GC/FTIR and GC/AES after inhalation of sandalwood oil. *Biomed. Chromatogr.*, **6**, 133–134.

Saposhnikovia divaricata

Radix Saposhnikoviae (Fangfeng) is the dried root of *Saposhnikovia divaricata* (Turcz.) Schischk. (Apiaceae). It is used as a spasmolytic agent for the treatment of cold, rheumatic diseases, and tetanus.

Saposhnikovia divaricata (Turcz.) Schischk.

Chemistry

The root of *S. divaricata* was reported to contain furocoumarins and furochromones khellin, 5-*O*-methylvisamminol, psoralen, xanthotoxin, bergapten and imperatorin [1], cimifugin and its primary *O*-β-D-glucopyranoside (prim-*O*-glucosylcimifugin), 5-*O*-methylvisamminol 4′-*O*-β-D-glucopyranoside (5-*O*-methylvisamminoside), divaricatol, ledebouriellol, hamaudol, *sec*-*O*-glucosylhamaudol [2], angeloylhamaudol [3], 3′-hydroxydeltoin, and 1-acylglycerols [4]. The Chinese Pharmacopoeia requires a qualitative determination of prim-*O*-glucosylcimifugin and 5-*O*-methylvisamminoside in Radix Saposhnikoviae by thin-layer chromatographic comparison with reference substances, and a quantitative determination by HPLC. The total content of prim-*O*-glucosylcimifugin and 5-*O*-methylvisamminoside together in Radix Saposhnikoviae should be not less than 0.24%.

In addition to furocoumarins and furochromones, polyacetylenes falcarinone, panaxynol, falcarindiol, panaxydol, and pa-

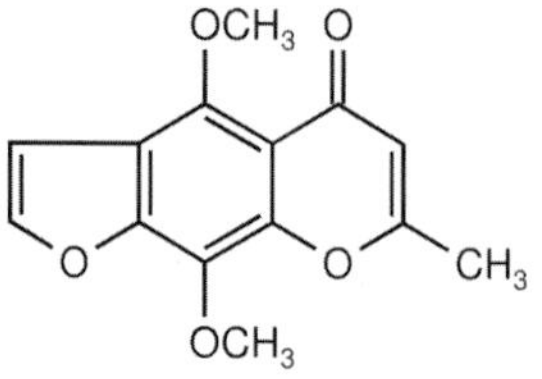

Khellin

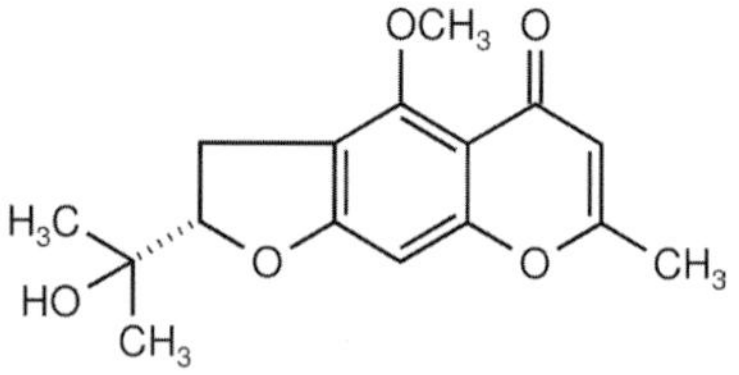

5-*O*-Methylvisamminol

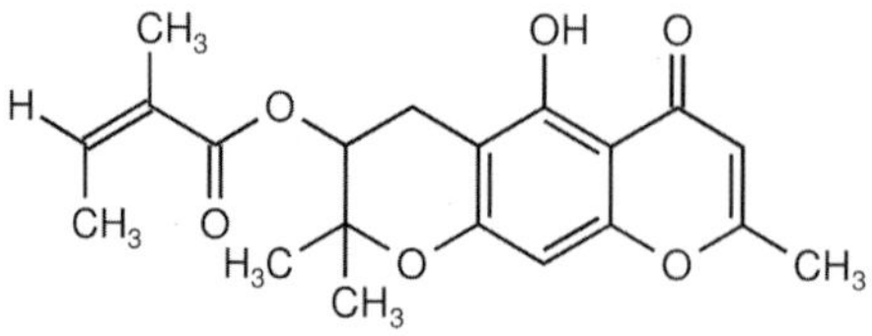

Angeloylhamaudol

Handbook of Chinese Medicinal Plants: Chemistry, Pharmacology, Toxicology
Weici Tang and Gerhard Eisenbrand

ISBN: 978-3-527-32226-8

naxytriol were isolated from the root of *S. divaricata* [5].

An acidic polysaccharide having activity on the reticuloendothelial system was isolated from the root of *S. divaricata*. The polysaccharide was homogeneous as judged by electrophoresis and gel chromatography, and was composed of D-galacturonic acid:L-rhamnose:L-arabinose:D-galactose in a molar ratio of 27: 7: 8: 8, and its molecular weight was estimated to be 132 000 Da. Approximately 30% of the D-galacturonic acid residues exist as the methyl esters. *O*-Acetyl groups were identified, and the content amounted to 3.3%. The polysaccharide has a pectin-like rhamnogalacturonan backbone with branched arabinan and galactan side chains [6, 7]. The major components from the essential oil of the root of *S. divaricata* were identified as caryophyllene oxide, sabinene, β-pinene, myrtenal, myrtenol, α-terpineol, p-cymene, α-pinene, and nonanoic acid [8].

Pharmacology and Toxicology

Biological effects of furocoumarins are discussed under *Angelica dahurica*. The extract of the root of *S. divaricata* was reported to exhibit antiproliferative, antioxidant and anti-inflammatory activities *in vitro*. It inhibited the proliferation of K562, HL-60, MCF7 and MDA-MB-468 tumor cells. The extract also induced differentiation in HL-60 cells along the granulocyte lineage, but not along the monocyte and macrophage lineages. At noncytotoxic concentrations, the extract significantly reduced nitric oxide (NO) production and inducible NO synthase (iNOS) levels, and inhibited iNOS mRNA expression in RAW 264.7 murine macrophage cells stimulated by lipopolysaccharide (LPS) [9]. Imperatorin was found to inhibit the synthesis of NO induced by LPS in the macrophage cell line RAW 264.7, with an IC_{50} of about 17 μg ml^{-1} [10].

The polyacetylenes falcarinone, panaxynol, falcarindiol, panaxydol and panaxytriol were identified as active principles inhibiting nitrite production by iNOS. Treatment with 10 μ*M* of panaxynol, falcarindiol, panaxydol and panaxytriol decreased the LPS/interferon-γ (IFN-γ)-stimulated accumulation of nitrite by about 72, 70, 45, and 37%, respectively. The IC_{50} values of falcarinone, panaxynol, falcarindiol, panaxydol and panaxytriol were >20, 2.2, 2.0, 6.6, and 9.9 μ*M*, respectively [5]. A decoction of the root of *S. divaricata* also inhibited human cytochrome P450 isoenzymes CYPIIIA, as determined by microsomal testosterone 6β-hydroxylation [11].

The chromones divaricatol, ledebouriellol and hamaudol were found to be analgesic principles, which inhibited writhing induction at an oral dose of 1 mg kg^{-1} in mice. Acylglycerols also showed inhibition significantly at a dose of 5 mg kg^{-1}. In some pharmacological tests using *sec-O*-glucosylhamaudol, the compound showed analgesia, as monitored by the tail pressure method [4].

As aqueous extract of the root of both cultivated and wild *S. divaricata* was found to exhibit similar antipyretic, analgesic, and anticonvulsive activities [12]. The polysaccharide was found to show remarkable potentiating activity on the reticuloendothelial system in a carbon-clearance test [6, 7].

References

1 Kang, J., Sun, J.-H., Zhou, L., Ye, M., Han, J., Wang, B.-R., and Guo, D.-A. (2008) Characterization af compounds form the roots of *Saposhnikovia divaricata* by highperformance liquid chromatography coupled with electrospray ionIation tandem mass spectrometry. *Rapid Commun. Mass Spectrom.*, **22**, 1899–1911.

2 Wang, J.H., Tian, Z., and Lou, Z.C. (1988) Determination of four chromones in Fangfeng, root of *Saposhnikovia divaricata* (Turcz.) Schischk., by high performance liquid chromatography. *Chin. J. Pharm. Anal.*, **8**, 325–328.

3 Ding, A.R., Wang, Q.Z., Li, S.L., and Kiao, K.F. (1987) Chemical constituents of Quanfangfeng (*Saposhnikovia divaricata*). *Chin. Trad. Herbal Drugs*, **18**, 247–249.

4 Okuyama, E., Hasegawa, T., Matsushita, T., Fujimoto, H., Ishibashi, M., and Yamazaki, M. (2001) Analgesic components of saposhnikovia root (*Saposhnikovia divaricata*). *Chem. Pharm. Bull. (Tokyo)*, **49**, 154–160.

5 Wang, C.N., Shiao, Y.J., Kuo, Y.H., Chen, C.C., and Lin, Y.L. (2000) Inducible nitric oxide synthase inhibitors from *Saposhnikovia divaricata* and *Panax quinquefolium*. *Planta Med.*, **66**, 644–647.

6 Shimizu, N., Tomoda, M., Gonda, R., Kanari, M., Takanashi, N., and Takahashi, N. (1989) The major pectic arabinogalactan having activity on the reticuloendothelial system from the roots and rhizomes of *Saposhnikovia divaricata*. *Chem. Pharm. Bull. (Tokyo)*, **37**, 1329–1332.

7 Shimizu, N., Tomoda, M., Gonda, R., Kanari, M., Kubota, A., and Kubota, A. (1989) An acidic polysaccharide having activity on the reticuloendothelial system from the roots and rhizomes of *Saposhnikovia divaricata*. *Chem. Pharm. Bull. (Tokyo)*, **37**, 3054–3057.

8 Wang, J.H., Li, S., and Luo, Z.C. (1991) Volatile constituents of *Saposhnikovia divaricata* fruit oil. *Chin. Pharm. J.*, **26**, 465–467.

9 Tai, J., and Cheung, S. (2007) Anti-proliferative and antioxidant activities of *Saposhnikovia divaricata*. *Oncol. Rep.*, **18**, 227–234.

10 Wang, C.C., Chen, L.G., and Yang, L.L. (1999) Inducible nitric oxide synthase inhibitor of the Chinese herb I. *Saposhnikovia divaricata* (Turcz.) Schischk. *Cancer Lett.*, **145**, 151–157.

11 Guo, L.Q., Taniguchi, M., Chen, Q.Y., Baba, K., and Yamazoe, Y. (2001) Inhibitory potential of herbal medicines on human cytochrome P450-mediated oxidation: properties of umbelliferous or citrus crude drugs and their relative prescriptions. *Jpn. J. Pharmacol.*, **85**, 399–408.

12 Wang, F.R., Xu, Q.P., and Li, P. (1991) Comparative studies on the febrifugal analgesic and anticonvulsive activities of water extracts from cultivated and wild *Saposhnikovia divaricata*. *Chin. J. Integr. Trad. West. Med.*, **11**, 730–732, 710.

Sarcandra glabra

Herba Sarcandrae (Zhongjiefeng) is the dried whole herb of *Sarcandra glabra* (Thunb.) Nakai (Chloranthaceae). It is used for the treatment of rheumatic pain and traumatic injuries.

Sarcandra glabra (Thunb.) Nakai

Two galenic preparations from Herba Sarcandrae are listed in the Chinese Pharmacopoeia: "Xuekang Koufuye", the oral ampoules of Herba Sarcandrae, and "Zhongjiefeng Pian", the dragees of Herba Sarcandrae. "Xuekang Koufuye" is used for the treatment of primary and secondary thrombopenic purpura. "Zhongjiefeng Pian" is used as an antipyretic and antiphlogistic agent for the treatment of pneumonitis, appendicitis, phlegmona, and as adjuvant in cancer therapy.

Chinese Pharmacopoeia requires a qualitative determination of isofraxidin in Herba Sarcandrae by thin-layer chromatographic (TLC) comparison with a reference substance, and a quantitative determination by HPLC. The content of isofraxidin in Herba Sarcandrae should be not less than 0.02%. The Chinese Pharmacopoeia also requires a qualitative determination of isofraxidin in the galenic preparations "Xuekang Koufuye" and "Zhongjiefeng Pian" by TLC comparison with a reference substance; and a quantitative determination by HPLC. The content of isofraxidin in "Xuekang Koufuye" should be not less than 0.9 mg per 10 ml oral ampoule, and in "Zhongjiefeng Pian" not less than 0.5 mg per dragee.

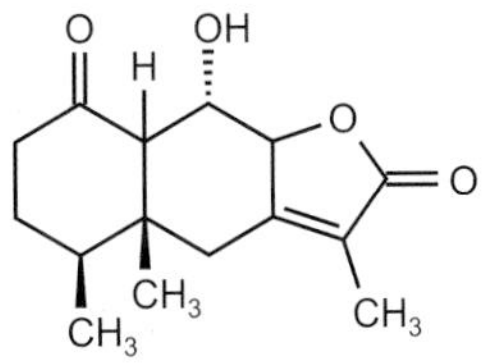

Istanbulin A

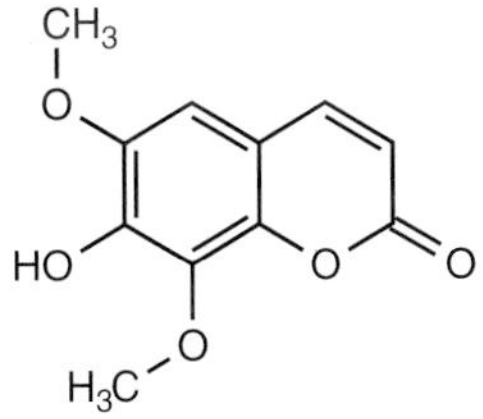

Isofraxidin

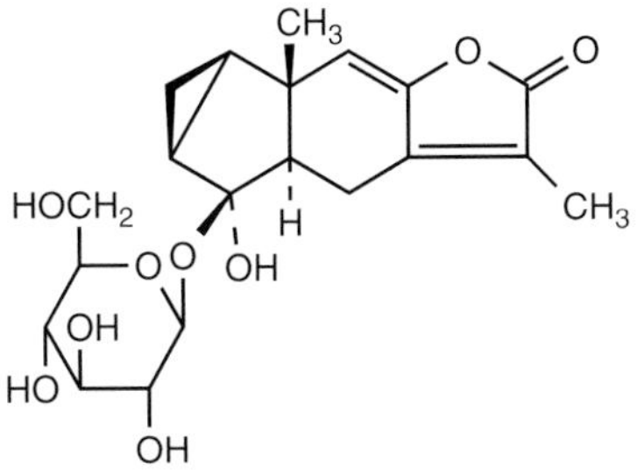

Chloranoside A

Chemistry

Herba Sarcandrae is known to contain isofraxidin as the major constituent [1, 2]. The

Fumaric acid is the major organic acid present in Herba Sarcandrae [3]. Istanbulin A from Herba Sarcandrae is a perhydro naphthofuran derivative [4]. Sesquiterpene

Handbook of Chinese Medicinal Plants: Chemistry, Pharmacology, Toxicology
Weici Tang and Gerhard Eisenbrand

ISBN: 978-3-527-32226-8

glycosides with eudesmanolide, elemanolide, lindenane, and germacranolide as the aglycones, along with chloranoside A, have been isolated from the whole plant of *S. glabra* [5].

Sarcandrosides A and B are two triterpene saponins from *S. glabra*. The structure of sarcandroside A and sarcandroside B were established as 3β,19α,20β-trihydroxy-urs-11,13(18)-diene-28,20β-lactone 3-*O*-β-D-glucopyranosyl-(1 → 3)-[α-L-rhamnopyranosyl(1 → 2)]-β-D-xylopyranoside and 3-*O*-β-D-glucopyranosyl-(1 → 3)-[α-L-rhamnopyranosyl(1 → 2)]-β-D-xylopyranosyl-pomolic acid 28-*O*-β-D-glucopyranosyl ester, respectively [6].

Pharmacology and Toxicology

Preventive effects of *S. glabra* on thrombocytopenia after chemotherapy have been studies in Babl/c mice. Thrombocytopenia was induced by the intraperitoneal (i.p.) injection of 5-FU at high dosage. The oral administration of an extract of *S. glabra*, without chemotherapy significantly elevated platelet counts. After chemotherapy with 5-FU, the platelet counts were not significantly decreased in mice treated with the extract. Moreover, treatment with the extract caused bone marrow hyperplasia with some megakaryocytes in the medullary cavity of bone, while bone marrow hyperplasia was low in control animals with no megakaryocytes in the bone medullary cavity [7].

The ethyl acetate extract of *S. glabra* was reported to inhibit the proliferation of HL60 human leukemic cells, with an IC_{50} of $58\,\mu g\,ml^{-1}$. Flow cytometric studies showed that the extract might interfere with DNA replication and induce cell-cycle arrest at S phase in leukemic cells, followed by DNA fragmentation and a loss of phospholipid asymmetry in the plasma membrane. The ratio of proapoptotic proteins Bax/Bcl-2 was also markedly upregulated, suggesting that the extract had induced cell apoptosis [8].

The sesquiterpene glycosides from *S. glabra* were found to exhibit pronounced hepatoprotective activities against D-galactosamine-induced toxicity in rat hepatic epithelial stem-like cells [5].

References

1 Zhou, G., Liu, H., Wang, H., and Kuang, P. (1999) Determination of isofraxidin in *Sarcandra glabra* (Thunb.) Nakai by HPLC. *Zhongguo Zhong Yao Za Zhi*, **24**, 481–502.

2 Luo, Q.Z., Dai, K.J., and Ma, A.D. (2006) High-performance liquid chromatography-electrospray ionization/tandem mass spectrometry of isofraxidin in *Sarcandra glabra*. *Nan Fang Yi Ke Da Xue Xue Bao*, **26**, 1821–1823.

3 You, Y. and Cheng, G. (1997) Determination of fumaric acid in Sarcandra glabra (Thunb.) Nakai by HPLC. *Zhongguo Zhong Yao Za Zhi*, **22**, 554, 576.

4 Zeng, A.H., Luo, Y.M., and Liu, N. (2006) Determination of istanbulin A in *Sarcandra glabra*. *Zhong Yao Cai*, **29**, 443–444.

5 Li, Y., Zhang, D.M., Li, J.B., Yu, S.S., Li, Y., and Luo, Y.M. (2006) Hepatoprotective sesquiterpene glycosides from *Sarcandra glabra*. *J. Nat. Prod.*, **69**, 616–620.

6 Luo, Y.M., Liu, A.H., Zhang, D.M., and Huang, L.Q. (2005) Two new triterpenoid saponins from *Sarcandra glabra*. *J. Asian Nat. Prod. Res.*, **7**, 829–834.

7 Zhong, L., Liu, T., Chen, Y., Zhong, X., Du, X., Lu, Z., Weng, J., Wu, S., and Lin, W. (2005) The study on effect of *Sarcandra glabra* on prevention and treatment of thrombocytopenia by chemotherapy. *Zhong Yao Cai*, **28**, 35–38.

8 Li, W.Y., Chiu, L.C., Lam, W.S., Wong, W.Y., Chan, Y.T., Ho, Y.P., Wong, E.Y., Wong, Y.S., and Ooi, V.E. (2007) Ethyl acetate extract of Chinese medicinal herb *Sarcandra glabra* induces growth inhibition on human leukemic HL-60 cells, associated with cell cycle arrest and up-regulation of pro-apoptotic Bax/Bcl-2 ratio. *Oncol. Rep.*, **17**, 425–431.

Saururus chinensis

Herba Saururi (Sanbaicao) is the dried above-ground part of *Saururus chinensis* (Lour.) Baill. (Saururaceae). It is used as a diuretic and antiedemic agent for the treatment of dysuria, urinary infection, nephritic edema. It is also used externally for the treatment of boils, sores, and eczema.

Saururus chinensis (Lour.) Baill.

Chemistry

The above-ground part of *S. chinensis* is known to contain flavones, with hyperin, quercetin, and isoquercetin as the major components [1]. Lignans and phenylpropanoids from the *n*-hexane fraction of *S. chinensis* were identified as sauchinone, sarisan, galbacin, saucernetin, sauchinone A, and 1′-*epi*-sauchinone [2]. Tetrahydrofuran-type sesquilignans, saucerneols A, B and C and the diarylbutane lignan saururin A [3], were isolated from the underground parts of *S. chinensis*, together with di-*O*-methyltetrahydrofuriguaiacin B, machilin D_2 and virolin (machilin D_4-methyl ether) [4].

Acyclic furanoditerpene compounds from the root of *S. chinenesis* were identified as saurufuran A and saurufuran B [5]. Dineolignan derivatives from the underground part of *S. chinenesis* were identified as manassantin A (saucernitin-8) and manassantin B (saucernitin-7), while sesquineolignans and lignans from the root were identified as saucerneol D and saucerneol E, (−)-saucerneol methyl ether, and (+)-saucernetin [6, 7].

Aristolactam-type alkaloids from the aerial part of *S. chinensis* were identified as aristololactam BII and saurolactam, which differ from the aristololactams from aristolochiaceous plants only in the absence of the methylenedioxy group [8].

Pharmacology and Toxicology

The aerial part of *S. chinensis* was reported to exert anti-inflammatory activity. A methanol extract of the aerial part of *S. chinensis* potently inhibited the production of nitric oxide (NO), the release of prostaglandin E_2 (PGE_2), reducing the protein level of inducible NO synthase (iNOS), and mRNA expression of iNOS and cyclooxygenase-2 (COX-2) in RAW264.7 murine macrophage cells stimulated by lipopolysaccharide (LPS). In addition, the extract also suppressed the LPS-induced DNA binding activity of nuclear factor-κB (NF-κB) associated with decreased p65 protein levels in the nucleus. This suggested that the extract of *S. chinensis* inhibited LPS-induced iNOS and COX-2 expression by blocking NF-κB activation [9]. Sauchinone seemed to be the

Handbook of Chinese Medicinal Plants: Chemistry, Pharmacology, Toxicology
Weici Tang and Gerhard Eisenbrand

ISBN: 978-3-527-32226-8

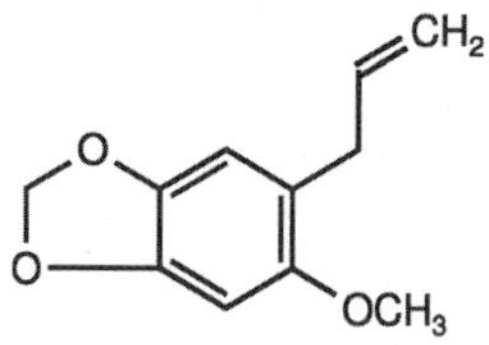

Sarisan

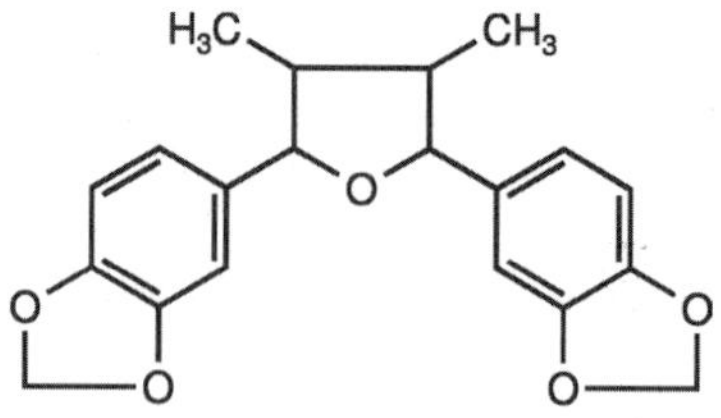

Galbacin

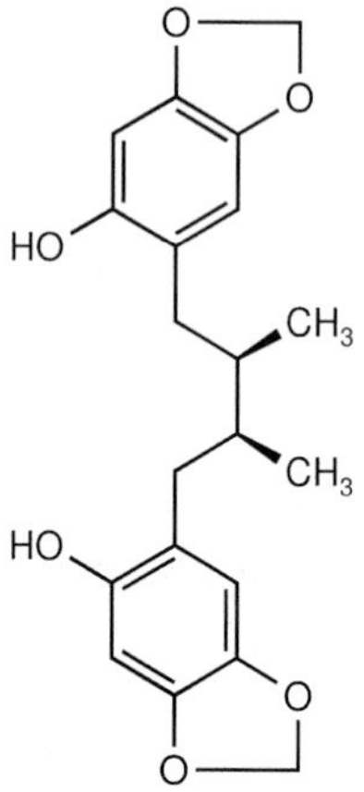

Saururin A

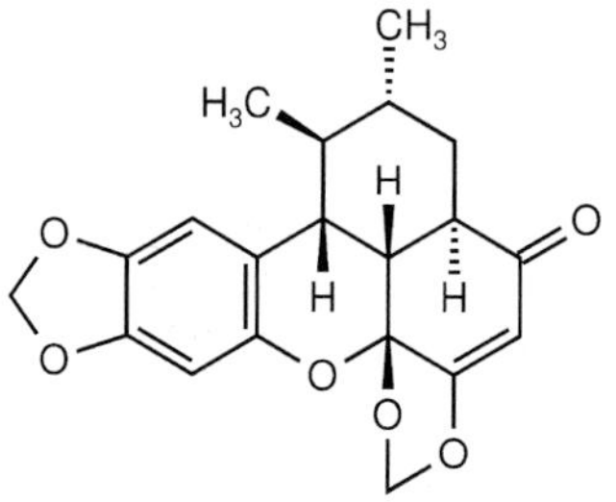

Sauchinone

active principle of *S. chinensis.* It inhibited the induction of iNOS, tumor necrosis factor-α (TNF-α) and COX-2 in RAW 264.7 cells activated by LPS, with suppression of the mRNA expression [10].

Saucernetin-7 and saucernetin-8 from the underground part of *S. chinensis* also exerted potent inhibitory effects on LPS-induced NO and PGE_2 production in RAW 264.7 cells [6]. The lignans manassantin A and manassantin B, as well as the sesquineolignans saucerneol D, saucerneol E, and saucerneol methyl ether, effectively inhibited NF-κB-dependent reporter gene expression in HeLa cells transfected with a NF-κB reporter construct [7]. An ethanol extract of *S. chinensis* exhibited antiasthmatic activity. The extract inhibited the generation of COX-2-dependent phases of PGD_2 in bone marrow-derived mast cells

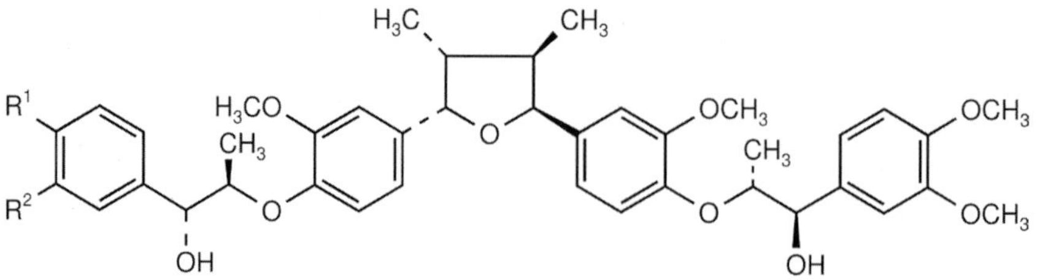

Manassantin A: $R^1 = R^2 = OCH_3$
Manassantin B: $R^1, R^2 = \text{-}OCH_2O\text{-}$

in a concentration-dependent manner, with an IC_{50} of about 14 μg ml^{-1}, and inhibited leukotriene C_4 (LTC_4) production, with an IC_{50} of 0.3 μg ml^{-1}. An *in vivo* study showed that the oral administration of the extract to mice at doses of 50–200 mg kg^{-1} reduced the number of infiltrating eosinophils in a bronchoalveolar lavage fluid after treatment with ovalbumin [11]. Manassantin A and manassantin B were further reported to inhibit TNF-α-induced cell adhesion molecule expression of human umbilical vein endothelial cells (HUVEC). The pretreatment of HUVEC with manassantin A and manassantin B, followed by stimulation with TNF-α, decreased the adhesion of THP-1 cells to HUVEC in a concentration-dependent manner (IC_{50} values 5–7 ng ml^{-1}), without cytotoxicity. Manassantin A and manassantin B also inhibited TNF-α-induced upregulation of intercellular adhesion molecule-1 (ICAM-1) and vascular cell adhesion molecule-1 (VCAM-1) [12].

The lignan derivative sauchinone, at a concentration of 50 μ*M*, was found to be cytoprotective in rat hepatocytes, significantly reducing the levels of glutamic-pyruvic transaminase released after carbon tetrachloride (CCl_4) induced cell damage. Sauchinone protected the hepatocytes against the cytotoxicity of CCl_4 by preventing the decrease of glutathione, superoxide dismutase and glutathione peroxidase as important factors in the cellular defense against oxidative stress. Sauchinone also ameliorated lipid peroxidation, as demonstrated by the reduction in malondialdehyde content; this suggested that sauchinone might exert its cytoprotective activity through antioxidant activity [13]. Saururin A, machilin D_2 and virolin exhibited significant low-density lipoprotein (LDL)-antioxidant activity in the thiobarbituric acid-reactive substance (T-BAR) assay, with IC_{50} values of 8.5, 2.9, and 4.3 μ*M*, respectively [4]. An aqueous extract of *S. chinensis* exhibited a hypolipidemic effect in rats fed a high-fat diet, significantly reducing triglyceride and cholesterol levels in the plasma and liver [14]. Sauchinone was also reported to exert neuroprotective effects. Sauchinone effectively inhibited apoptosis in C6 rat glioma cells induced by staurosporine, associated with an upregulation of antiapoptotic protein Bcl-2 significantly decreasing the activity of caspase-3 [15]. The cytotoxicity of saucernetin-7 has been demonstrated in HL-60 human promyelocytic leukemia cells. Saucernetin-7 inhibited the proliferation of HL-60 cells, induced cell apoptosis, activated caspases-3, -8 and -9, and induced Bid cleavage, the mitochondrial translocation of Bax from the cytosol, and cytochrome c release from mitochondria [16]. Saucernetin-8 also inhibited the proliferation of HL-60 cells and induced cell differentiation [17]. The neolignans, especially threo,erythro-manassantin A, strongly inhibited the proliferation of PC-3 prostate cancer cells, induced cell-cycle arrest in G_1 phase and apoptosis, decreasing hyperphosphorylated Rb and increasing hypophosphorylated Rb. The induction of cell apoptosis was accompanied by the cleavage of caspases-3, -8 and -9, as well as by the downregulation of Bcl-2 and the upregulation of Bax [18]. The neolignans manassantin A and its erythro,erythro- and threo,erythro-epimers from *S. chinensis* were further reported to be cytotoxic in a series of human cancer cell lines, including SK-Hep-1, PC-3, DU-145, BT-20, SK-BR-3, T-47D, HeLa, T98G, and SK-MEL-28. These compounds were more active than cisplatin and doxorubicin [19].

References

1 Xu, L.X., Zhang, X.Q., and Liu, A.R. (1988) Differential pulsed polarographic determination of flavonoids in *Saururus chinensis* (Lour.) Baill. *Chin. J. Pharm. Anal.*, 8, 223–225.

2 Sung, S.H. and Kim, Y.C. (2000) Hepatoprotective diastereomeric lignans from *Saururus chinensis* herbs. *J. Nat. Prod.*, **63**, 1019–1021.
3 Ahn, B.T., Lee, S., Lee, S.B., Lee, E.S., Kim, J.G., Bok, S.H., and Jeong, T.S. (2001) Low-density lipoprotein-antioxidant constituents of *Saururus chinensis*. *J. Nat. Prod.*, **64**, 1562–1564.
4 Sung, S.H., Huh, M.S., and Kim, Y.C. (2001) New tetrahydrofuran-type sesquilignans of *Saururus chinensis* root. *Chem. Pharm. Bull. (Tokyo)*, **49**, 1192–1194.
5 Hwang, B.Y., Lee, J.H., Nam, J.B., Kim, H.S., Hong, Y.S., and Lee, J.J. (2002) Two new furanoditerpenes from *Saururus chinenesis* and their effects on the activation of peroxisome proliferator-activated receptor-γ. *J. Nat. Prod.*, **65**, 616–617.
6 Park, H.J., Kim, R.G., Seo, B.R., Ha, J., Ahn, B.T., Bok, S.H., Lee, Y.S., Kim, H.J., and Lee, K.T. (2003) Saucernetin-7 and saucernetin-8 isolated from *Saururus chinensis* inhibit the LPS-induced production of nitric oxide and prostaglandin E_2 in macrophage RAW264.7 cells. *Planta Med.*, **69**, 947–950.
7 Hwang, B.Y., Lee, J.H., Nam, J.B., Hong, Y.S., and Lee, J.J. (2003) Lignans from *Saururus chinensis* inhibiting the transcription factor NF-κB. *Phytochemistry*, **64**, 765–771.
8 Kim, S.R., Sung, S.H., Kang, S.Y., Koo, K.A., Kim, S.H., Ma, C.J., Lee, H.S., Park, M.J., and Kim, Y.C. (2004) Aristolactam BII of *Saururus chinensis* attenuates glutamate-induced neurotoxicity in rat cortical cultures probably by inhibiting nitric oxide production. *Planta Med.*, **70**, 391–396.
9 Cho, H.Y., Cho, C.W., and Song, Y.S. (2005) Antioxidative and anti-inflammatory effects of *Saururus chinensis* methanol extract in RAW 264.7 macrophages. *J. Med. Food.*, **8**, 190–197.
10 Lee, A.K., Sung, S.H., Kim, Y.C., and Kim, S.G. (2003) Inhibition of lipopolysaccharide-inducible nitric oxide synthase, TNF-α and COX-2 expression by sauchinone effects on IκBα phosphorylation, C/EBP and AP-1 activation. *Br. J. Pharmacol.*, **139**, 11–20.
11 Lee, E., Haa, K., Yook, J.M., Jin, M.H., Seo, C.S., Son, K.H., Kim, H.P., Bae, K.H., Kang, S.S., Son, J.K., and Chang, H.W. (2006) Anti-asthmatic activity of an ethanol extract from *Saururus chinensis*. *Biol. Pharm. Bull.*, **29**, 211–215.
12 Kwon, O.E., Lee, H.S., Lee, S.W., Chung, M.Y., Bae, K.H., Rho, M.C., and Kim, Y.K. (2005) Manassantin A and B isolated from *Saururus chinensis* inhibit TNF-α-induced cell adhesion molecule expression of human umbilical vein endothelial cells. *Arch. Pharm. Res.*, **28**, 55–60.
13 Sung, S.H., Lee, E.J., Cho, J.H., Kim, H.S., and Kim, Y.C. (2000) Sauchinone, a lignan from *Saururus chinensis*, attenuates CCl_4-induced toxicity in primary cultures of rat hepatocytes. *Biol. Pharm. Bull.*, **23**, 666–668.
14 Yun, Y.R., Kim, M.J., Kwon, M.J., Kim, H.J., Song, Y.B., Song, K.B., and Song, Y.O. (2007) Lipid-lowering effect of hot water-soluble extracts of *Saururus chinensis* Bail on rats fed high fat diets. *J. Med. Food*, **102**, 316–322.
15 Song, H., Kim, Y.C., and Moon, A. (2003) Sauchinone, a lignan from *Saururus chinensis*, inhibits staurosporine-induced apoptosis in C6 rat glioma cells. *Biol. Pharm. Bull.*, **26**, 1428–1430.
16 Choi, S.K., Seo, B.R., Lee, K.W., Cho, W., Jeong, S.H., and Lee, K.T. (2007) Saucernetin-7 isolated from *Saururus chinensis* induces caspase-dependent apoptosis in human promyelocytic leukemia HL-60 cells. *Biol. Pharm. Bull.*, **30**, 1516–1522.
17 Seo, B.R., Yoo, C.B., Park, H.J., Choi, J.W., Seo, K., Choi, S.K., and Lee, K.T. (2004) Saucernetin-8 isolated from *Saururus chinensis* induced the differentiation of human acute promyelocytic leukemia HL-60 cells. *Biol. Pharm. Bull.*, **27**, 1594–1598.
18 Song, S.Y., Lee, I., Park, C., Lee, H., Hahm, J. C., and Kang, W.K. (2005) Neolignans from *Saururus chinensis* inhibit PC-3 prostate cancer cell growth via apoptosis and senescence-like mechanisms. *Int. J. Mol. Med.*, **16**, 517–523.
19 Hahm, J.C., Lee, I.K., Kang, W.K., Kim, S.U., and Ahn, Y.J. (2005) Cytotoxicity of neolignans identified in *Saururus chinensis* towards human cancer cell lines. *Planta Med.*, **71**, 464–469.

Schisandra chinensis and *Schisandra sphenanthera*

Fructus Schisandrae Chinensis (Wuweizi) is the dried fruit of *Schisandra chinensis* (Turcz.) Baill. (Magnoliaceae). It is used as a tonic and sedative and for the treatment of chronic cough and asthma; frequent urination; diarrhea; night sweating; and palpitation and insomnia.

Schisandra chinensis (Turcz.) Baill.

Fructus Schisandrae Sphenantherae (Nanwuweizi) is the dried fruit of *Schisandra sphenanthera* Rehd. et Wils. It is used for the same indications as Fructus Schisandrae Chinensis.

Chemistry

The fruits, especially the seeds of the official *Schisandra* species, are known to contain lignans of dibenzo[*a*,*c*]cyclooctadiene structure as the active principles. The lignan content of *Schisandra* seeds were reported to vary between 7% and 19%, according to the species and geographical origin and to be highest from May to July. The stems were also reported to have a lignan content between 1% and 10% [1].

A great number of tetrahydro-dibenzo[*a*, *c*]cyclooctadiene derivatives and related lignans have been isolated from the seeds of *S. chinensis* and *S. sphenanthera*. The dibenzo[*a*,*c*]cyclooctadiene derivatives were found to possess different conformations. The majority of the lignans possess a dibenzo[*a*,*c*]cyclooctadiene skeleton in *R* biphenyl configuration twist-boat-chair cyclooctadiene conformation, *S* biphenyl configuration twist-boat-chair cyclooctadiene conformation, or *S* biphenyl configuration twist-boat cyclooctadiene conformation. These stereocharacteristics have been confirmed by X-ray analysis [2]. For example, schisandrin A (deoxyschizandrin, wuweizisu A), schisandrin B (wuweizisu B, γ-schizandrin), schisandrol A (schizandrin, wuweizi alcohol A) and schisandrol B (wuweizi alcohol B, gomisin A) are lignans with dibenzo[*a*,*c*]cyclooctadiene *R* biphenyl configuration and twist-boat-chair cyclooctadiene conformation; schisantherin A (wuweizi ester A, gomisin C) and gomisin G belong to lignans with dibenzo[*a*,*c*]cyclooctadiene *S* biphenyl configuration and twist-boat-chair cyclooctadiene conformation; and gomisin O is a lignan with dibenzo[*a*,*c*]cyclooctadiene *S* biphenyl configuration and twist-boat cyclooctadiene conformation.

The Chinese Pharmacopoeia requires a qualitative determination of deoxyschizandrin in Fructus Schisandrae Chinensis and in Fructus Schisandrae Sphenantherae by thin-layer chromatographic comparison with reference substance; and a quantitative determination of schizandrin in Fructus Schisandrae Chinensis and schisantherin

Handbook of Chinese Medicinal Plants: Chemistry, Pharmacology, Toxicology
Weici Tang and Gerhard Eisenbrand

ISBN: 978-3-527-32226-8

A in Fructus Schisandrae Sphenantherae by HPLC. The content of schizandrin in Fructus Schisandrae Chinensis should be not less than 0.4%, and the content of schisantherin A in Fructus Schisandrae Sphenantherae not less than 0.12%.

fraction containing an NADPH-generating system, and *in vivo* in rats and dogs. They were 7,8-dihydroxy-schisandrin A, 7,8-dihydroxy-2-demethyl-schisandrin A, and 7,8-dihydroxy-3-demethyl-schisandrin A [4]. The maximum plasma concentration

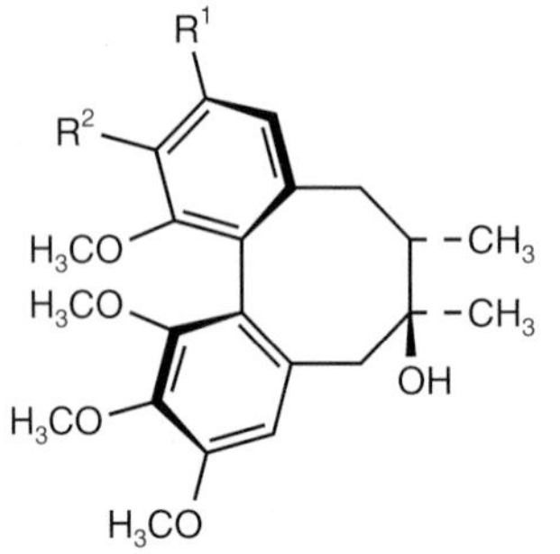

Schisandrol A: $R^1 = R^2 = -OCH_3$
Schisandrol B: $R^1, R^2 = -OCH_2O-$

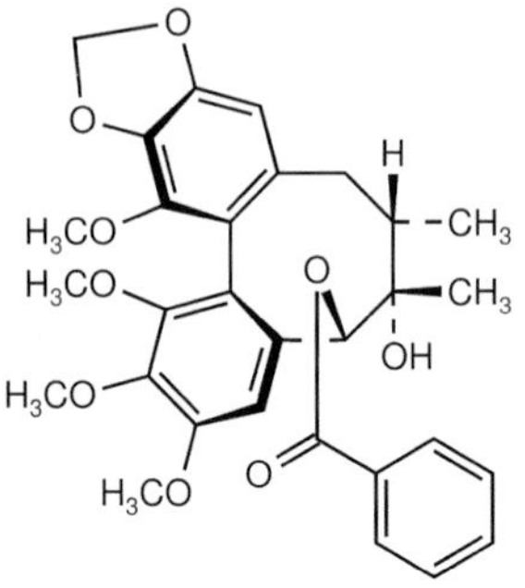

Schisantherin A

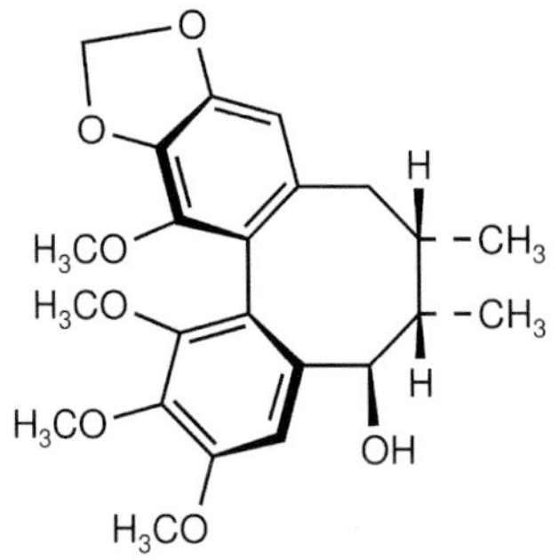

Gomisin O

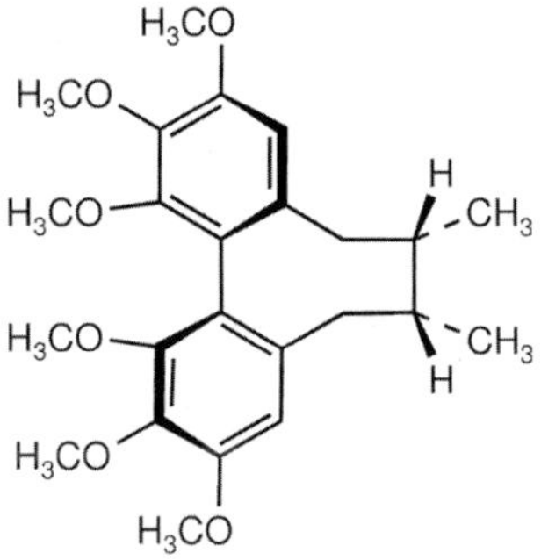

Deoxyschizandrin

In addition, lignan derivatives without dibenzo[*a*,*c*]cyclooctadiene structure such as pregomisin, anwulignan, epigalbacin, henricine, chicanine, schisandrone and ganschisandrin were also isolated from the seeds of *S. chinensis* and *S. sphenanthera* [2].

A plant steroid derived from 9,19-cyclo-3,4-secolanostan from *S. sphaerandra* was identified as nigranoic acid [3].

Pharmacology and Toxicology

Major metabolites of schisandrin A were identified *in vitro* using rat liver microsomal

of schisandrin A after oral administration at a dose of 15 mg per person to healthy men was about 96 ng ml^{-1} [5].

After oral administration to rats, schisandrol A was absorbed from the gastrointestinal tract with a half-life $t_{1/2\alpha}$ of about 1 h. Following intravenous (i.v.) injection, the blood level showed a biphasic decline. The half-life of the distribution phase was 1.4 min, and of the elimination phase 42 min. Schisandrol A was detectable in urine at 1 h after oral administration. At 5 min after i.v. injection, high levels of schisandrol A were found in the lungs, moderate amounts in the liver, heart, brain, kid-

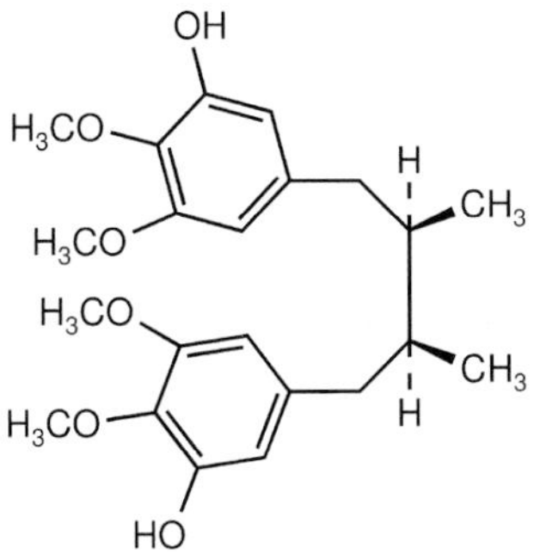

Pregomisin

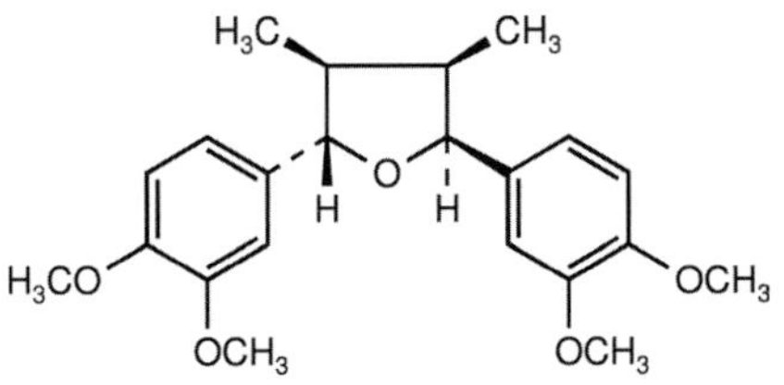

Ganschisandrin

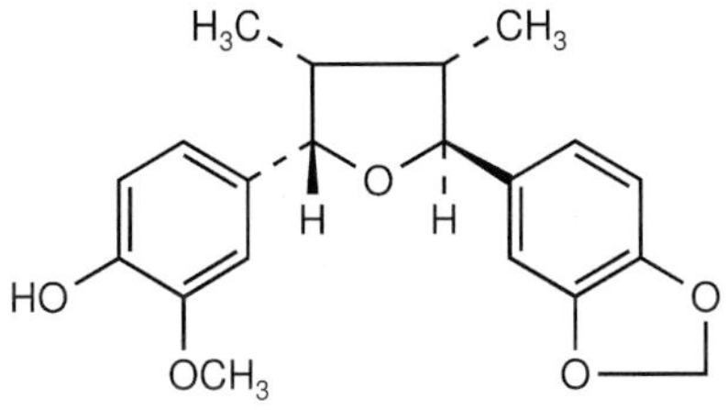

Chicanine

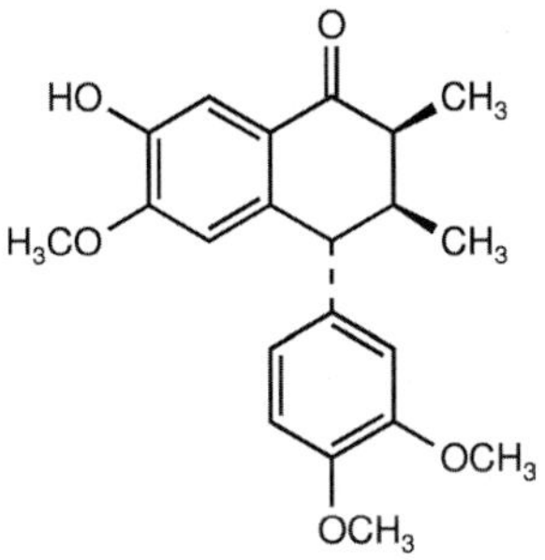

Schisandrone

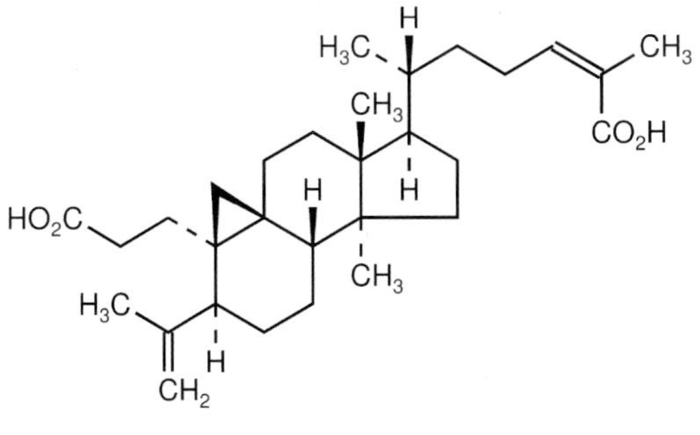

Nigranoic acid

neys, and low amounts in the ileum and spleen. Schisandrol A was eliminated rather rapidly. In the brain, the highest amounts were found in the hypothalamus, striatum, and hippocampus, and moderate amounts in the cerebral cortex and cerebellum. These differences may be relevant to the neuroleptic and anticonvulsant properties of schisandrol A [6].

After i.v. administration of schisandrol B, the serum concentration decreased biphasically; the elimination half-life $t_{1/2\beta}$ was about 70 min. Dose-dependency was observed for the area under the concentration–time curve (AUC). The serum concentration of schisandrol B increased rapidly, reaching a maximum within 15–30 min after oral administration. The biotransformation of schisandrol B to its demethylenated derivative as one of the major metabolites, was very rapid after both i.v. and oral administration [7]. More than 80% of schisandrol B was bound with rat serum protein *in vitro* and *in vivo* [8]. The urinary excretion of schisandrol B or its demethylenated metabolite was lower in carbon tetrachloride (CCl_4) pretreated rats than in normal rats. Similar results were also observed for excretion in the feces [9].

The fruit of *Schisandra* has been used in traditional Chinese medicine as a tonic and a sedative agent since ancient times. In experimental studies, sedative and hypnotic activities of the ethanol fraction of Fructus Schisandrae were demonstrated in mice and rats [10]. Schisandrol A was found to exert inhibitory effects on the central nervous system (CNS). Schisandrol A prolonged the sleeping time of mice induced by pentobarbital; decreased the spontaneous activity; antagonized the stimulatory effect of amphetamine and caffeine on spontaneous activity; and protected the animals against tonic convulsion induced by electric shock [11]. The mechanism of the CNS inhibition by schisandrol A may be related to the dopamine system [12]. Schisandrol B was also found to effectively prolong hexobarbital-induced sleeping time in mice [13].

In M146L cells, which can produce considerable amyloid β-protein 42 *in vitro*, schisandrin B was found to be able to suppress the production of amyloid β-protein 42 through an inhibition of γ-secretase [14]. A methanolic extract of the dried fruit of *S. chinensis* significantly attenuated the neurotoxicity induced by glutamate in primary cultures of rat cortical cells. Deoxyschisandrin, gomisin N, and schisandrin C from the extract significantly attenuated glutamate-induced neurotoxicity, as measured by inhibition of the increase in intracellular Ca^{2+}, by improvement of the glutathione (GSH) defense system, increasing the level of GSH, and the activity of glutathione peroxidase, and by inhibiting the formation of cellular peroxide [15].

The neuroprotective effect of schisandrin B has been demonstrated in mice against *tert*-butylhydroperoxide (t-BHP)-induced cerebral toxicity. The t-BHP toxicity was associated with an increase in the extent of cerebral lipid peroxidation and an impairment in cerebral GSH antioxidant status, as evidenced by a steep decrease in GSH level and by the inhibition of Se-GSH peroxidase activity at 5 min following the t-BHP challenge. Schisandrin B pretreatment at a daily dose of $2\,mM\,kg^{-1}$ for three days produced a protection against t-BHP induced mortality. Such protection was associated with a decrease in the extent of lipid peroxidation and an enhancement in the GSH antioxidant status of the brain tissue. These biochemical parameters returned to normal values at 60 min. This suggested that the antioxidant potential of schisandrin B was responsible for the protection against cerebral oxidative stress [16]. The oral treatment of mice with gomisin A at a dose of $5\,mg\,kg^{-1}$ significantly reversed cognitive impairment induced by scopolamine. In an *in vitro* study, gomisin A was found to inhibit acetylcholinesterase activity in a concentration-dependent manner, with an IC_{50} of 15.5 μ*M* [17].

The seed of *Schisandra* was shown to exert hepatoprotective effects, and has been introduced clinically for the treatment of hepatitis [18]. An ethanol extract of *Schisandra* seeds effectively reduced the elevated serum glutamic-pyruvic transaminase (GPT) in mice, rats and rabbits, as induced by different hepatotoxic agents such as CCl_4 and acetaminophen (paracetamol) [18]. The pretreatment of rats with a lignan-enriched extract of *S. chinensis* also protected against aflatoxin B_1 (AFB_1)-induced hepatotoxicity [19]. The mechanism of hepatoprotection of *Schisandra* seeds was believed to involve both antioxidative and detoxification processes in the liver, as indicated by increases in hepatic GSH levels as well as hepatic glutathione reductase and glutathione *S*-transferase (GST) activities [20].

Pure lignan derivatives, such as schisandrin B, also showed hepatoprotective activities. The intragastric administration of schisandrin B to mice protected against CCl_4-induced liver damage, and decreased serum GPT and enhanced hepatic GSH levels. The hepatoprotective action of schisandrin B against CCl_4 toxicity was mediated by both an enhancement of mitochondrial GSH status and an induction of heat shock proteins [21]. In rats, schisandrin B also exhibited a protective effect against oxygen free radical-induced lipoperoxidative damage of the hepatocyte plasma membrane [22]. The possible inhibition of CCl_4 metabolism by schisandrin B, and the inhibition of lipid peroxidation, may contribute to the hepatoprotective action [21]. The pretreatment of mice with schisandrin B or schisandrin C produced almost complete protection, with significant increases in hepatic mitochondrial GSH level and glutathione reductase activity. The methylenedioxy substituents might represent an important structural determinant in the stimulation of hepatic mitochondrial GSH, particularly under conditions of CCl_4 intoxication [23]. The hepatoprotective effect of schisandrin B against CCl_4 was also observed in streptozotocin-induced diabetic rats, and was associated with an improvement in hepatic GSH redox status, as well as increased hepatic ascorbic acid and microsomal GST activities. Schisandrin B may be a useful hepatoprotective agent under diabetic conditions [24].

The pretreatment of rats with schisandrol B decreased the activities of serum glutamic oxalacetic transaminase (GOT), GPT, lactate dehydrogenase, total bilirubin, and total cholesterol, as elevated by CCl_4. Hepatocellular necrosis caused by CCl_4 was inhibited by pretreatment with schisandrol B, the action of which was reported to be even stronger than that of schisandrol A [25]. The oral administration of schisandrol B also exhibited protective activity against hepatotoxicity of acetaminophen in rats [26]. Schisandrol B was also found to suppress fibrosis associated proliferation and to accelerate both liver regeneration and recovery of liver function in rats after partial hepatectomy. The mitotic index and the increase of DNA synthesis after partial hepatectomy were significantly enhanced by schisandrol B. The increase in ornithine decarboxylase (ODC) activity in the early stages of liver regeneration was also enhanced by schisandrol B. It was suggested that schisandrol B stimulated liver regeneration after partial hepatectomy by enhancing ODC activity as an important biochemical event in the early stages of liver regeneration [27]. The effect of schisandrol B on immunologic liver injury in mice revealed that the hepatoprotective effect of schisandrol B could be related to the protective effect on the hepatocyte plasma membrane, rather than to inhibitory effects on antibody formation and complement activity [28]. Schisandrol B was also found to induce hepatocyte growth factor mRNA through different mechanisms in rats [29].

Schisandrin B, when applied to normal and CCl_4-intoxicated mice at a daily oral dose of 1 mmol kg^{-1} for three days, resulted in a significant decrease in plasma alanine aminotransferase in CCl_4-intoxicated mice. However, a significant decrease in plasma sorbital dehydrogenase was only observed after schisandrin B pretreatment. The lowering of plasma sorbital dehydrogenase activity, indicative of hepatoprotection by schisandrin B pretreatment, was found to be associated with an enhancement in hepatic mitochondrial GSH redox status, as well as an increase in mitochondrial GSH reductase activity in mice [30]. In CCl_4-intoxicated mice, both the methylenedioxy group and the cyclooctadiene ring are important structural determinants in the enhancement of liver mitochondrial GSH status [31].

Dibenzo[*a,c*]cyclooctadiene lignans were reported to inhibit lipid peroxidation in rat liver microsomes induced by iron/cysteine, as well as superoxide anion production in the xanthine/xanthine oxidase system. They were much more potent than vitamin E at the same concentration [31]. Schisandrin B, schisandrin C, and schisandrol A each exhibited scavenging effects on hydroxyl radicals in cell systems that were superior to vitamin E and vitamin C [32]. The oxygen radical-scavenging effects of dibenzo[*a,c*]cyclooctadiene lignans of *S* biphenyl configuration were found to be stronger than those of *R* biphenyl configuration [33]. Protective effects of schizandrin A on the cardiotoxic activity of adriamycin by antioxidation were reported [34]. Radioprotective effects of schisandrin B have been suggested due to its antioxidant activity [33].

An aqueous extract of the fruits of *S. chinensis* was found to cause vasorelaxation. In isolated rat thoracic aorta preconstricted with norepinephrine (noradrenaline), the extract exhibited an endothelium-dependent relaxation, especially at lower concentrations (0.1 and 0.3 $mg\,ml^{-1}$). At higher concentrations (0.3 and 1.0 $mg\,ml^{-1}$) the extract was also involved in inhibition of the extracellular calcium influx to vascular smooth muscle, and caused endothelium-independent vasorelaxation [35]. In contrast, schisandrin B was reported to exert an acute hyperlipidemic effect in mice. At 24 h after schisandrin B administration to mice, the serum triglyceride level was increased in a dose-dependent manner, whereas the serum low-density lipoprotein (LDL) cholesterol level was significantly decreased. Schisandrin B, when given once daily for four days, also dose-dependently elevated the serum triglyceride level [36].

The cardioprotective effect of schisandrin B has been demonstrated by modulation of ischemia–reperfusion injury in the isolated rat heart. The pretreatment of rats with schisandrin B at a daily dose of 1.2 $mM\,kg^{-1}$ for three days protected against ischemia–reperfusion injury in isolated rat heart, and improved the recovery of contractile force. The protection against ischemia–reperfusion injury was associated with significant increases in myocardial α-tocopherol levels in pretreated heart. This suggested that the cardioprotection afforded by schisandrin B pretreatment might be attributed in part to a modulation of the status of nonenzymatic antioxidants [37]. A cardioprotective effect of schisandrin B against ischemia–reperfusion injury has also been reported in rats. The induction of expression of heat shock protein 25 and heat shock protein 70 might contribute to the cardioprotection by schisandrin B pretreatment against ischemia–reperfusion injury [38].

Gomisin G was found to exhibit inhibitory activity against human immunodeficiency virus (HIV), with an IC_{50} of 6 ng ml^{-1}. Schisantherin D and schisandrin C showed IC_{50} values of 0.5 and 1.2 $\mu g\,ml^{-1}$. The cyclooctadiene ring of the natural lignans, and the position and substitution of hydroxy groups, are important for an enhanced anti-HIV activity [39]. Some halogenated dibenzo[*a,c*]cyclooctadiene derivatives, such as halogen-substituted gomisin J, were synthesized and found to be inhibitors of HIV-1 reverse transcriptase [40]. Nigranoic acid was reported to show activity in several anti-HIV reverse transcriptase and polymerase assays [3].

The lignans from *Schisandra* species were also found to be platelet-activating factor (PAF) antagonists. Strong activity was shown in lignans without an ester group at position 7, a hydroxyl group at position 6, or a methylene dioxy moiety and with an *R* biphenyl configuration. Dehydroschisandrol A showed the highest activity, with an IC_{50} of about 2 μ*M* [41].

Dibenzocyclooctadiene lignans were also reported to be a novel class of P-glycoprotein inhibitors. The P-glycoprotein inhibitory activities of five dibenzocyclooctadiene lignans, including schisandrin A, schisandrin B, schisantherin A, schisandrol A, and schisandrol B, were demonstrated to reverse drug resistance mediated by multidrug resistance-associated protein-1 in HL-60/adriamycin (HL-60/ADR) and HL-60/multidrug resistance-associated protein (HL-60/MRP), two human promyelocytic leukemia cell lines overexpressing of multidrug resistance-associated protein-1, but not P-glycoprotein. The five lignans could effectively reverse drug resistance of the two cell lines to vincristine, daunorubicin, and VP-16. This indicated that dibenzocyclooctadiene lignans possessed multiple activities against cancer multidrug resistance, including an inhibition of P-glycoprotein and multidrug resistance-associated protein-1 [42]. Schisandrin B was found to be a dual inhibitor of P-glycoprotein and multidrug resistance-associated protein 1, as demonstrated in HL-60/ADR and HL-60/MRP cells [43]. Schisandrol A was found to reverse multidrug resistance mediated by P-glycoprotein. In HepG2-DR cells, schisandrol A enhanced the cytotoxicity of cancer drugs that were P-glycoprotein substrates, and restored vinblastine-induced G_2/M arrest without lowering P-glycoprotein expression. It was suggested that schisandrol A might affect P-glycoprotein–substrate complexes [44].

The reversal of P-glycoprotein-mediated multidrug resistance in HepG2-DR cells by schisandrol B has also been reported. Schisandrol B may bind to P-glycoprotein simultaneously with substrates, and thus alter P-glycoprotein–substrate interaction [45]. Schisandrin B was also found capable of enhancing doxorubicin-induced apoptosis in SMMC7721 human hepatic carcinoma and in MCF-7 human breast cancer cells. However, whilst the enhancement was irrelevant to the action of schisandrin B on P-glycoprotein or other drug-transporters, it was associated with the activation of caspase-9. On the other hand, schisandrin B did not enhance the doxorubicin-induced apoptosis of primary rat cardiomyocytes and primary human fibroblasts [46].

Schisandrin B was also reported to inhibit cell proliferation and to induce cell apoptosis in SMMC-7721 human hepatoma cells. In addition, treatment with schisandrin B resulted in a downregulation of heat shock protein 70 and an upregulation of caspase-3, but not of caspase-9. This indicated that the cell growth inhibition and cell apoptosis induction of human hepatoma SMMC-7721 cells by schisandrin B was via caspase-3-dependent, caspase-9-independent pathways accompanied by a downregulation of heat shock protein 70 protein expression at an early stage [47].

In vitro, schisandrin B and schisandrol B have been found to decrease the mutagenicity of benzo[*a*]pyrene (BaP) in *Salmonella typhimurium* [48]. In a two-stage carcinogenesis model of mouse skin, schisandrol B was reported to inhibit the tumor-promoting activity of 12-*O*-tetradecanoylphorbol 13-acetate (TPA) subsequent to initiation with 7,12-dimethylbenz[*a*]anthracene (DMBA). It was postulated that the inhibition of tumor promotion by schisandrol B was due to its anti-inflammatory activity [49]. Schisandrol B also showed a weak suppressive effect on tumor promotion in the liver after short-term feeding of 3′-methyl-4-dimethylamino-azobenzene to rats [50]. The simultaneous oral administration of schisandrol B to rats significantly decreased the number and area of GST placental form-positive foci in the liver, as induced by 3′-methyl-4-dimethylamino-azobenzene, and the concentrations of serum bile acids (especially deoxycholic acid) as a potential endogenous risk factor for hepatocarcinogenesis. The anti-promotion effect of schisandrol B was also suggested to be partly caused by a metabolic modification of bile acids, including deoxycholic acid [51].

References

1 Wang, K., Tong, Y.Y., and Song, W.Z. (1990) Determination of the active ingredients in Chinese drug Wuweizi (*Schisandra chinensis*) by TLC-densitometry. *Acta Pharm. Sin.*, **25**, 49–53.

2 Li, L.N. (1989) Biologically active constituents of *Schisandra chinensis* and related plants. *Abstr. Chin. Med. (Hong Kong)*, **3**, 414–428.

3 Sun, H.D., Qiu, S.X., Lin, L.Z., Wang, Z.Y., Lin, Z.W., Pengsuparp, T., Pezzuto, J.M., Fong, H.H., Cordell, G.A., and Farnsworth, N. R. (1996) Nigranoic acid, a triterpenoid from *Schisandra sphaerandra* that inhibits HIV-1 reverse transcriptase. *J. Nat. Prod.*, **59**, 525–527.

4 Ikeya, Y., Sugama, K., Tanaka, M., Wakamatsu, T., Ono, H., Takeda, S., Oyama, T., and Maruno, M. (1995) Structure determination of biliary metabolites of schizandrin in rat and dog. *Chem. Pharm. Bull. (Tokyo)*, **43**, 121–129.

5 Ono, H., Matsuzaki, Y., Wakui, Y., Takeda, S., Ikeya, Y., Amagaya, S., and Maruno, M. (1995) Determination of schizandrin in human plasma by gas chromatography-mass spectrometry. *J. Chromatogr. B Biomed.*, **674**, 293–297.

6 Niu, X.Y., Bian, Z.J., and Ren, Z.H. (1983) Metabolism of schisandrol A in rats and its distribution in brain determined by TLC-UV. *Acta Pharm. Sin.*, **18**, 491–495.

7 Matsuzaki, Y., Ishibashi, E., Koguchi, S., Wakui, Y., Takeda, S., Aburada, M., and Oyama, T. (1991) Determination of gomisin A (TJN-101) and its metabolite in rat serum by gas chromatography-mass spectrometry. *Yakugaku Zasshi*, **111**, 617–620.

8 Matsuzaki, Y., Matsuzaki, T., Takeda, S., Koguchi, S., Ikeya, Y., Mitsuhashi, H., Sasaki, H., Aburada, M., Hosoya, E., and Oyama, T. (1991) Studies on the metabolic fate of gomisin A (TJN-101). I. Absorption in rats. *Yakugaku Zasshi*, **111**, 524–530.

9 Matsuzaki, Y., Matsuzaki, T., Ono, H., Koguchi, S., Takeda, S., Takeda, S., Funo, S., Aburada, M., Hosoya, E., and Oyama, T. (1991) Studies on the metabolic fate of gomisin A (TJN-101). II. Absorption and excretion in CCl_4 treated rats. *Yakugaku Zasshi*, **111**, 531–537.

10 Huang, F., Xiong, Y., Xu, L., Ma, S., and Dou, C. (2007) Sedative and hypnotic activities of the ethanol fraction from Fructus Schisandrae in mice and rats. *J. Ethnopharmacol.*, **110**, 471–475.

11 Niu, X.Y., Wang, W.J., Bian, Z.J., and Ren, Z.H. (1983) Effects of schizandrol on the central nervous system. *Acta Pharm. Sin.*, **18**, 416–421.

12 Zhang, L. and Niu, X. (1991) Effects of schizandrol A on monoamine neurotransmitters in the central nervous system. *Acta Acad. Med. Sin.*, **13**, 13–16.

13 Maeda, S., Sudo, K., Aburada, M., Ikeya, Y., Taguchi, H., Yosioka, I., and Harada, M. (1981) Pharmacological studies on schizandra fruit. I. General pharmacological effects of gomisin A and schisandrin. *Yakugaku Zasshi*, **101**, 1030–1041.

14 Liu, W., Yu, R., Wu, J.H., and Luo, H.M. (2006) γ-Schisandrin inhibits production of amyloid β-protein 42 in M146L cells. *Yao Xue Xue Bao*, **41**, 1136–1140.

15 Kim, S.R., Lee, M.K., Koo, K.A., Kim, S.H., Sung, S.H., Lee, N.G., Markelonis, G.J., Oh, T.H., Yang, J.H., and Kim, Y.C. (2004) Dibenzocyclooctadiene lignans from *Schisandra chinensis* protect primary cultures of rat cortical cells from glutamate-induced toxicity. *J. Neurosci. Res.*, **76**, 397–405.

16 Ko, K.M. and Lam, B.Y. (2002) Schisandrin B protects against *tert*-butylhydroperoxide induced cerebral toxicity by enhancing glutathione antioxidant status in mouse brain. *Mol. Cell. Biochem.*, **238**, 181–186.

17 Kim, D.H., Hung, T.M., Bae, K.H., Jung, J.W., Lee, S., Yoon, B.H., Cheong, J.H., Ko, K.H., and Ryu, J.H. (2006) Gomisin A improves scopolamine-induced memory impairment in mice. *Eur. J. Pharmacol.*, **542**, 129–135.

18 Zhu, M., Lin, K.F., Yeung, R.Y., and Li, R.C. (1999) Evaluation of the protective effects of *Schisandra chinensis* on Phase I drug metabolism using a CCl_4 intoxication model. *J. Ethnopharmacol.*, **67**, 61–68.

19 Ip, S.P., Mak, D.H., Li, P.C., Poon, M.K., and Ko, K.M. (1996) Effect of a lignan-enriched extract of *Schisandra chinensis* on aflatoxin B_1 and cadmium chloride-induced hepatotoxicity in rats. *Pharmacol. Toxicol.*, **78**, 413–416.

20 Ko, K.M., Ip, S.P., Poon, M.K., Wu, S.S., Che, C.T., Ng, K.H., and Kong, Y.C. (1995) Effect of a lignan-enriched fructus schisandrae extract on hepatic glutathione status in rats: protection against carbon tetrachloride toxicity. *Planta Med.*, **61**, 134–137.

21 Tang, M.H., Chiu, P.Y., and Ko, K.M. (2003) Hepatoprotective action of schisandrin B against carbon tetrachloride toxicity was mediated by both enhancement of mitochondrial glutathione status and induction of heat shock proteins in mice. *Biofactors*, **19**, 33–42.

22 Zhang, T.M., Wang, B.E., and Liu, G.T. (1992) Effect of schisandrin B on lipoperoxidative damage to plasma membrane of rat liver *in vitro*. *Acta Pharmacol. Sin.*, **13**, 255–258.

23 Ip, S.P., Ma, C.Y., Che, C.T., and Ko, K.M. (1997) Methylenedioxy group as determinant of schisandrin in enhancing hepatic mitochondrial glutathione in carbon tetrachloride-intoxicated mice. *Biochem. Pharmacol.*, **54**, 317–319.

24 Mak, D.H. and Ko, K.M. (1997) Alterations in susceptibility to carbon tetrachloride toxicity and hepatic antioxidant/detoxification system in streptozotocin-induced short-term diabetic rats: effects of insulin and schisandrin B treatment. *Mol. Cell. Biochem.*, **175**, 225–232.

25 Maeda, S., Sudo, K., Miyamoto, Y., Takeda, S., Shinbo, M., Aburada, M., Ikeya, Y., Taguchi, H., and Harada, M. (1982) Pharmacological studies on schizandra fruits. II. Effects of constituents of schizandra fruits on chemical-induced hepatic damage in rats. *Yakugaku Zasshi*, **102**, 579–588.

26 Yamada, S., Murawaki, Y., and Kawasaki, H. (1993) Preventive effect of gomisin A, a lignan component of shizandra fruits, on acetaminophen-induced hepatotoxicity in rats. *Biochem. Pharmacol.*, **46**, 1081–1085.

27 Kubo, S., Ohkura, Y., Mizoguchi, Y., Matsui-Yuasa, I., Otani, S., Morisawa, S., Kinoshita, H., Takeda, S., Aburada, M., and Hosoya, E. (1992) Effect of gomisin A (TJN-101) on liver regeneration. *Planta Med.*, **58**, 489–492.

28 Nagai, H., Yakuo, I., Aoki, M., Teshima, K., Ono, Y., Sengoku, T., Shimazawa, T., Aburada, M., and Koda, A. (1989) The effect of gomisin A on immunologic liver injury in mice. *Planta Med.*, **55**, 13–17.

29 Shiota, G., Yamada, S., and Kawasaki, H. (1996) Rapid induction of hepatocyte growth factor mRNA after administration of gomisin A, a lignan component of shizandra fruits. *Res. Commun. Mol. Pathol. Pharmacol.*, **94**, 141–146.

30 Ip, S.P., Yiu, H.Y., and Ko, K.M. (2000) Differential effect of schisandrin B and dimethyl diphenyl bicarboxylate (DDB) on hepatic mitochondrial glutathione redox status in carbon tetrachloride intoxicated mice. *Mol. Cell. Biochem.*, **205**, 111–114.

31 Ip, S.P., Che, C.T., and Ko, K.M. (1998) Structure-activity relationship of schisandrins in enhancing liver mitochondrial glutathione status in CCl_4-poisoned mice. *Acta Pharmacol. Sin.*, **19**, 313–316.

32 Li, X.J., Zhao, B.L., Liu, G.T., and Xin, W.J. (1990) Scavenging effects on active oxygen radicals by schizandrins with different structures and configurations. *Free Radic. Biol. Med.*, **9**, 99–104.

33 Lin, T.J. (1991) Antioxidation mechanism of schizandrin and tanshinonatic acid A and their effects on the protection of cardiotoxic action of adriamycin. *Prog. Physiol. Sci.*, **22**, 342–345.

34 Liu, L., Zheng, R., Zhang, H., Zhang, X., Liu, Y., and Liu, M. (1983) Radiation protection mechanism of -schizandrin. Part VI. Effect of ?-schizandrin on the formation of lipid peroxide in homogenates of liver and brain. *J. Lanzhou Univ. [Nat. Sci.]*, **19**, 99–104.

35 Rhyu, M.R., Kim, E.Y., Yoon, B.K., Lee, Y.J., and Chen, S.N. (2006) Aqueous extract of *Schizandra chinensis* fruit causes endothelium-dependent and -independent relaxation of isolated rat thoracic aorta. *Phytomedicine*, **13**, 651–657.

36 Pan, S.Y., Dong, H., Han, Y.F., Li, W.Y., Zhao, X.Y., and Ko, K.M. (2006) A novel experimental model of acute hypertriglyceridemia induced by schisandrin B. *Eur. J. Pharmacol.*, **537**, 200–204.

37 Ko, K.M. and Yiu, H.Y. (2001) Schisandrin B modulates the ischemia-reperfusion induced changes in non-enzymatic antioxidant levels in isolated-perfused rat hearts. *Mol. Cell. Biochem.*, **220**, 141–147.

38 Chiu, P.Y. and Ko, K.M. (2004) Schisandrin B protects myocardial ischemia-reperfusion injury partly by inducing Hsp25 and Hsp70 expression in rats. *Mol. Cell. Biochem.*, **266**, 139–144.

39 Chen, D.F., Zhang, S.X., Xie, L., Xie, J.X., Chen, K., Kashiwada, Y., Zhou, B.N., Wang, P., Cosentino, L.M., and Lee, K.H. (1997) Anti-AIDS agents. XXVI. Structure-activity correlations of gomisin-G-related anti-HIV lignans from *Kadsura interior* and of related synthetic analogues. *Bioorg. Med. Chem.*, **5**, 1715–1723.

40 Fujihashi, T., Hara, H., Sakata, T., Mori, K., Higuchi, H., Tanaka, A., Kaji, H., and Kaji, A. (1995) Anti-human immunodeficiency virus (HIV) activities of halogenated gomisin J derivatives, new nonnucleoside inhibitors of HIV type 1 reverse transcriptase. *Antimicrob. Agents Chemother.*, **39**, 2000–2007.

41 Lee, I.S., Jung, K.Y., Oh, S.R., Park, S.H., Ahn, K.S., and Lee, H.K. (1999) Structure-activity relationships of lignans from *Schisandra chinensis* as platelet activating factor antagonists. *Biol. Pharm. Bull.*, **22**, 265–267.

42 Li, L., Pan, Q., Sun, M., Lu, Q., and Hu, X. (2007) Dibenzocyclooctadiene lignans: a class of novel inhibitors of multidrug resistance-associated protein 1. *Life Sci.*, **80**, 741–748.

43 Sun, M., Xu, X., Lu, Q., Pan, Q., and Hu, X. (2007) Schisandrin B: a dual inhibitor of P-glycoprotein and multidrug resistance-associated protein 1. *Cancer Lett.*, **246**, 300–307.

44 Fong, W.F., Wan, C.K., Zhu, G.Y., Chattopadhyay, A., Dey, S., Zhao, Z., and Shen, X.L. (2007) Schisandrol A from *Schisandra chinensis* reverses P-glycoprotein-mediated multidrug resistance by affecting Pgp-substrate complexes. *Planta Med.*, **73**, 212–220.

45 Wan, C.K., Zhu, G.Y., Shen, X.L., Chattopadhyay, A., Dey, S., and Fong, W.F. (2006) Gomisin A alters substrate interaction and reverses P-glycoprotein-mediated multidrug resistance in HepG2-DR cells. *Biochem. Pharmacol.*, **72**, 824–837.

46 Li, L., Lu, Q., Shen, Y., and Hu, X. (2006) Schisandrin B enhances doxorubicin-induced apoptosis of cancer cells but not normal cells. *Biochem. Pharmacol.*, **71**, 584–595.

47 Wu, Y.F., Cao, M.F., Gao, Y.P., Chen, F., Wang, T., Zumbika, E.P., and Qian, K.X. (2004) Down-modulation of heat shock protein 70 and up-modulation of caspase-3 during schisandrin B-induced apoptosis in human hepatoma SMMC-7721 cells. *World J. Gastroenterol.*, **10**, 2944–2948.

48 Liu, K.T. and Lesca, P. (1982) Pharmacological properties of dibenzo[*a,c*]cyclooctene derivatives isolated from Fructus Schisandrae chinensis. I. Interaction with rat liver cytochrome P450 and inhibition of xenobiotic metabolism and mutagenicity. *Chem. Biol. Interact.*, **39**, 301–314.

49 Yasukawa, K., Ikeya, Y., Mitsuhashi, H., Iwasaki, M., Aburada, M., Nakagawa, S., Takeuchi, M., and Takido, M. (1992) Gomisin A inhibits tumor promotion by 12-*O*-tetradecanoylphorbol 13-acetate in two-stage carcinogenesis in mouse skin. *Oncology*, **49**, 68–71.

50 Miyamoto, K., Wakusawa, S., Nomura, M., Sanae, F., Sakai, R., Sudo, K., Ohtaki, Y., Takeda, S., and Fujii, Y. (1991) Effects of gomisin A on hepatocarcinogenesis by 3′-methyl-4-dimethylaminoazobenzene in rats. *Jpn. J. Pharmacol.*, **57**, 71–77.

51 Ohtaki, Y., Hida, T., Hiramatsu, K., Kanitani, M., Ohshima, T., Nomura, M., Wakita, H., Aburada, M., and Miyamoto, K.I. (1996) Deoxycholic acid as an endogenous risk factor for hepatocarcinogenesis and effects of gomisin A, a lignan component of *Schizandra* fruits. *Anticancer Res.*, **16**, 751–755.

Schizonepeta tenuifolia

Herba Schizonepetae (Jingjie) is the dried above-ground part of *Schizonepeta tenuifolia* Briq. (Lamiaceae). It is used for the treatment of common cold, headache, measles and rubella.

Schizonepeta tenuifolia Briq.

Herba Schizonepetae Carbonisatum (Jingjietan) is the carbonized (charcoaled), above-ground part of *Schizonepeta tenuifolia*. It is mainly used as a hemostatic agent.

Spica Schizonepetae (Jingjiesui) is the dried spikes of *Schizonepeta tenuifolia*. It is used for the treatment of cold, headache, measles and rubella.

Spica Schizonepetae Carbonisata (Jingjiesuitan) is the carbonized spikes of *Schizonepeta tenuifolia*. It is used as a hemostatic agent.

Chemistry

The above-ground part of *S. tenuifolia* was reported to contain essential oil, flavones, and monoterpenes [1]. The Chinese Pharmacopoeia requires a quantitative determination of the essential oil in Herba Schizonepetae and in Spica Schizonepetae; a qualitative determination of pulegone in Spica Schizonepetae by thin-layer chromatographic comparison with reference substance; and a quantitative determination of pulegone in Herba Schizonepetae and in Spica Schizonepetae by HPLC. The content of essential oil in Herba Schizonepetae should be not less than 0.6%, in Spica Schizonepetae not less than 0.4% (ml g^{-1}), the respective contents of pulegone not less than 0.02%, and 0.08%. After preparation, the content of essential oil in prepared samples should be not less than 0.3%, of pulegone not less than 0.02%.

From the spikes of *S. tenuifolia* flavones diosmetin, hesperetin and luteolin were isolated. Monoterpenes from the spikes were identified as schizonol and schizonodiol [2].

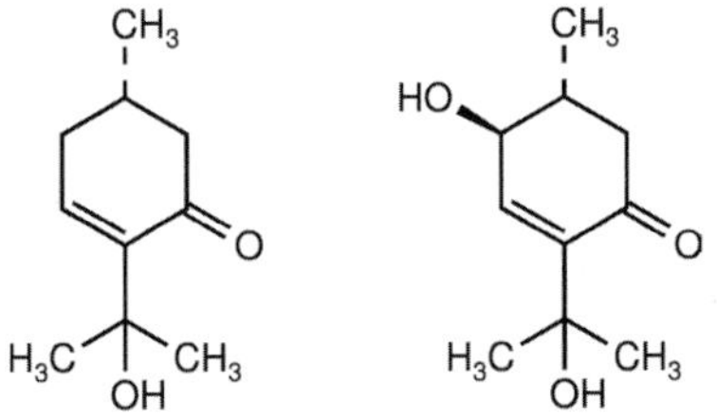

Schizonol Schizonodiol

Monoterpene glycosides from the spikes were identified as schizonepetoside A, schizonepetoside B, schizonepetoside C, schizonepetoside D, and schizonepetoside E [2].

Pharmacology and Toxicology

The extract of *S. tenuifolia* was reported to inhibit the systemic allergic reaction

Handbook of Chinese Medicinal Plants: Chemistry, Pharmacology, Toxicology
Weici Tang and Gerhard Eisenbrand

ISBN: 978-3-527-32226-8

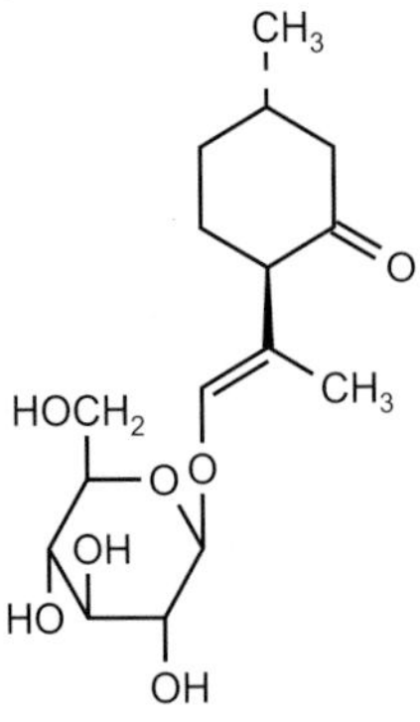

Schizonepetoside A

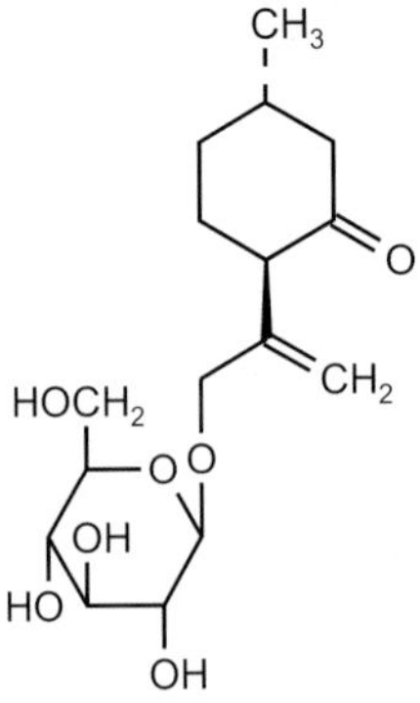

Schizonepetoside D

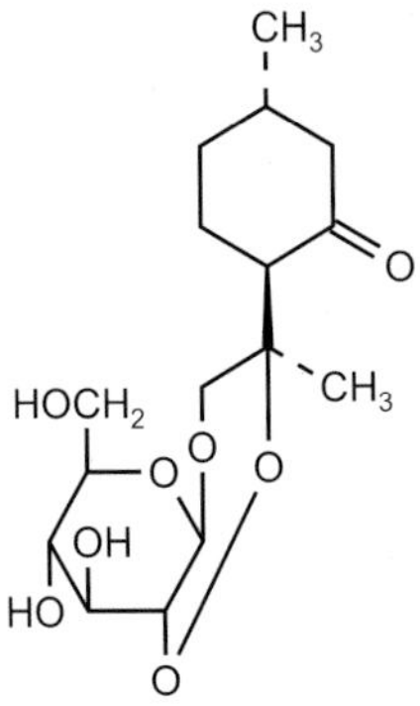

Schizonepetoside B

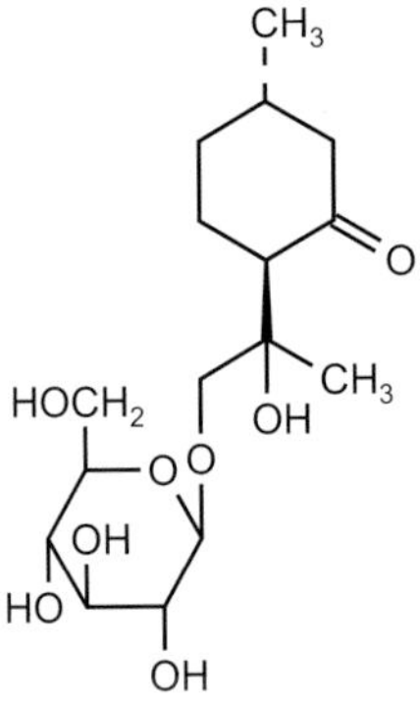

Schizonepetoside E

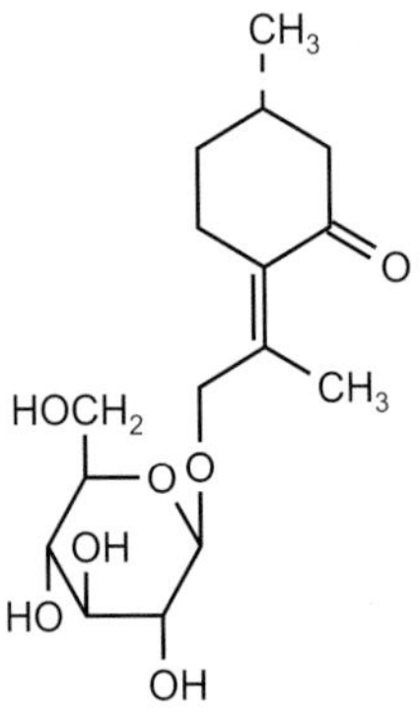

Schizonepetoside C

induced by compound 48/80 in rats, in a dose-dependent manner. The extract also decreased plasma histamine levels induced by compound 48/80, and local allergic reactions activated by anti-dinitrophenyl (DNP) IgE. In addition, the extract dose-dependently inhibited histamine release from rat peritoneal mast cells activated by compound 48/80 or anti-DNP IgE. However, *S. tenuifolia* extract had a significant enhancing effect on anti-DNP IgE-induced tumor necrosis factor-α (TNF-α) production from rat peritoneal mast cells. This indicated that *S. tenuifolia* inhibits immediate-type hypersensitivity, and suggested that *S. tenuifolia* may selectively activate the

TNF-α production from rat peritoneal mast cells [3]. The methanol extract of the aerial part of *S. tenuifolia* inhibited the substance P-induced itch–scratch response in mice at a dose of 200 mg kg^{-1}, without affecting locomotor activity [4].

The aerial part of *S. tenuifolia* is also widely used as a hemostatic agent, especially after its carbonization [5, 6]. The hemostatic mechanism of the extract of carbonized *S. tenuifolia* was found to occur through promoting coagulation and inhibiting fibrinolysis [7]. Hemostatic action was observed after mice had been administered intraperitoneally and orally for 0.5 and 1 h, respectively. The hemostatic time of the former was 6 h and the latter 12 h. The LD_{50} of the fat-soluble extract was 2.6 g kg^{-1} after oral dosing, and 1.9 g kg^{-1} after intraperitoneal dosing [8].

References

1 Ye, D.J. (1985) Components of essential oils from different parts and the charcoal of *Schizonepeta tenuifolia. Bull. Chin. Mater. Med.*, **10**, 19–21.

2 Oshima, Y., Takata, S., and Hikino, H. (1989) Validity of the oriental medicines. Part 137. Schizonodiol, schizonol, schizonepetoside D and schizonepetoside E, monoterpenoids of *Schizonepeta tenuifolia spikes. Planta Med.*, **55**, 179–180.

3 Shin, T.Y., Jeong, H.J., Jun, S.M., Chae, H.J., Kim, H.R., Baek, S.H., and Kim, H.M. (1999) Effect of *Schizonepeta tenuifolia* extract on mast cell-mediated immediate-type hypersensitivity in rats. *Immunopharmacol. Immunotoxicol.*, **21**, 705–715.

4 Tohda, C., Kakihara, Y., Komatsu, K., and Kuraishi, Y. (2000) Inhibitory effects of methanol extracts of herbal medicines on substance P-induced itch-scratch response. *Biol. Pharm. Bull.*, **23**, 599–601.

5 Ding, A.W. (1986) Hemostatic effect of *Schizonepeta tenuifolia* before and after carbonization. *Bull. Chin. Mater. Med.*, **11**, 23–25.

6 Fung, D. and Lau, C.B. (2002) *Schizonepeta tenuifolia*: chemistry, pharmacology, and clinical applications. *J. Clin. Pharmacol.*, **42**, 30–36.

7 Ding, A.W., Wu, H., Kong, L.D., Wang, S.L., Gao, Z.Z., Zhao, M.X., and Tan, M. (1993) Research on hemostatic mechanism of extracts from carbonized *Schizonepeta tenuifolia. China J. Chin. Mater. Med.*, **18**, 598–600.

8 Ding, N.W., Kong, L.D., Wu, H., Wang, S.L., Long, Q.J., Yao, Z., and Chen, J. (1993) Research on hemostatic constituents in carbonized *Schizonepeta tenuifolia* Brig. *China J. Chin. Mater. Med.*, **18**, 535–538.

Scrophularia ningpoensis

Radix Scrophulariae (Xuanshen) is the dried root of *Scrophularia ningpoensis* Hemsl. (Scrophulariaceae). It is used for the treatment of constipation, tuberculosis, cough, conjunctivitis, scrofula, diphtheria, boils and sores. The Chinese Pharmacopoeia notes Radix Scrophulariae to be incompatible with Radix et Rhizoma Veratri.

Scrophularia ningpoensis Hemsl.

Chemistry

The root of *S. ningpoensis* is reported to contain a number of phenylpropanoids and iridoids and their glycosides, including phenylpropanoid glycosides named ningposides A, B and C, harpagide, harpagoside, sibirioside A, angoroside C, acteoside, decaffeoylacteoside and cistanoside D and F [1, 2], and scrophuloside B4. Scrophuloside B4 is derived from catalpol [3]. The Chinese Pharmacopoeia requires a qualitative determination of harpagide and harpagoside in Radix Scrophulariae by thin-layer chromatographic comparison with reference substances, and a quantitative determination of harpagoside in Radix Scrophulariae by HPLC. The content of harpagoside in Radix Scrophulariae should be not less than 0.05%. Further compounds from the root of *S. ningpoensis* were identified as oleanonic acid, ursolonic acid, cinnamic acid, 3-hydroxy-4-methoxy-benzoic acid, 5-(hydroxymethyl)-2-furfural and β-sitosterol [3]

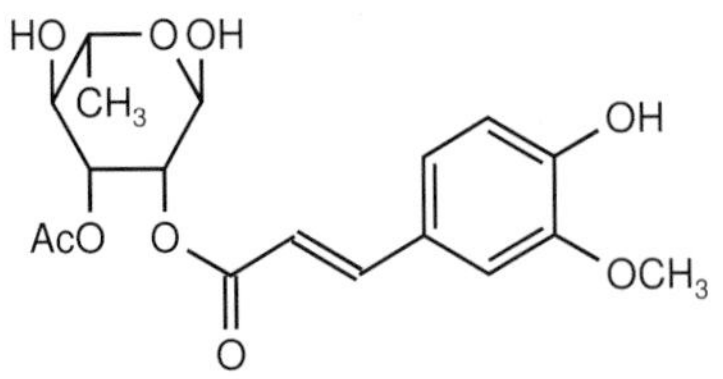

Ningposide A

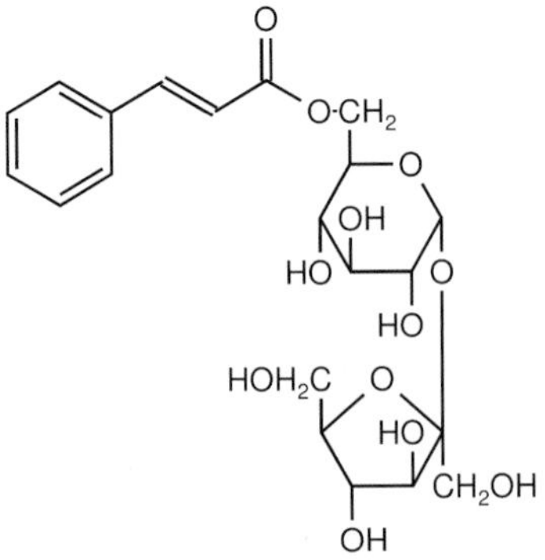

Sibirioside A

Handbook of Chinese Medicinal Plants: Chemistry, Pharmacology, Toxicology
Weici Tang and Gerhard Eisenbrand

ISBN: 978-3-527-32226-8

Pharmacology and Toxicology

The iridoid glycosides harpagoside and harpagide, and phenylpropanoid glycosides angoroside C and acteoside from *S. ningpoensis*, were found to significantly "repair" the oxidized OH radical adducts of dAMP or dGMP, indicating the aerial part of *S. tenuifolia* to exert an antioxidative/reductive activity [4].

References

1 Kajimoto, T., Hidaka, M., Shoyama, K., and Nohara, T. (1989) Iridoids from *Scrophularia ningpoensis*. *Phytochemistry*, **28**, 2701–2704.

2 Li, Y.M., Jiang, S.H., Gao, W.Y., and Zhu, D.Y. (2000) Phenylpropanoid glycosides from *Scrophularia ningpoensis*. *Phytochemistry*, **54**, 923–925.

3 Nguyen, A.T., Fontaine, J., Malonne, H., Claeys, M., Luhmer, M., and Duez, P. (2005) A sugar ester and an iridoid glycoside from *Scrophularia ningpoensis*. *Phytochemistry*, **66**, 1186–1191.

4 Li, Y.M., Han, Z.H., Jiang, S.H., Jiang, Y., Yao, S.D., and Zhu, D.Y. (2000) Fast repairing of oxidized OH radical adducts of dAMP and dGMP by phenylpropanoid glycosides from *Scrophularia ningpoensis* Hemsl. *Acta Pharmacol. Sin.*, **21**, 1125–1128.

Scutellaria baicalensis

Radix Scutellariae (Huangqin) is the dried root of *Scutellaria baicalensis* Georgi (Lamiaceae). It is used for the treatment of nausea and vomiting in febrile diseases; acute dysentery; epistaxis; carbuncle and sores; threatening abortion; and bacterial infections of the respiratory and gastrointestinal tracts.

Scutellaria baicalensis Georgi.

Extractum Scutellariae Siccus (Huangqin Tiquwu) is the extract of Radix Scutellariae. It is used for the same indications as Radix Scutellariae.

Chemistry

The root of *S. baicalensis* is known to contain a number of flavones, with baicalein, wogonin and the corresponding 7-*O*-β-D-glucopyranosiduronide baicalin and wogonoside as the major components. Baicalin and wogonoside contents in root sample of *S. baicalensis* from China were found to vary from 12–17% and 3–4%, respectively [1]. The Chinese Pharmacopoeia requires a qualitative determination of baicalin, baicalein and wogonin in Radix Scutellariae by thin-layer chromatographic (TLC) comparison with reference substances, and a quantitative determination of baicalin in Radix Scutellariae by HPLC. The content of baicalin in Radix Scutellariae should be not less than 9.0%, and in the prepared root not less than 8.0%.

The Chinese Pharmacopoeia requires a qualitative determination of baicalin in Extractum Scutellariae Siccus by TLC comparison with reference substance, and a quantitative determination by HPLC. The content of baicalin in Extractum Scutellariae Siccus should be not less than 85.0%.

Another item officially listed in the Chinese Pharmacopoeia containing baicalein and baicalin as major flavone constituents is Semen Oroxyli. Semen Oroxyli (Muhudie) is the dried ripe seed of *Oroxylum indicum* (L.) Vent. (Bignoniaceae). It is used for the treatment of respiratory diseases. The Chinese Pharmacopoeia requires a qualitative determination of baicalin in Semen Oroxyli by TLC comparison with reference substance.

Minor flavone derivatives from the root of *S. baicalensis* were identified as chrysin, ganhuangenin, hispidulin, norwogonin, oroxylin A, scutellarein, skullcapflavone I, skullcapflavone II (neobaicalein), tenaxin I, viscidulin I, viscidulin II, dihydrobaicalein, dihydrooroxylin A, and some of their glycosides [2].

Pharmacology and Toxicology

The absorption of baicalin by the rat small intestine *in situ* was reported as 20–40%, with $t_{1/2\alpha}$ of 11–16 h [3]. In fact, baicalin was only poorly absorbed from the rat

Handbook of Chinese Medicinal Plants: Chemistry, Pharmacology, Toxicology
Weici Tang and Gerhard Eisenbrand

ISBN: 978-3-527-32226-8

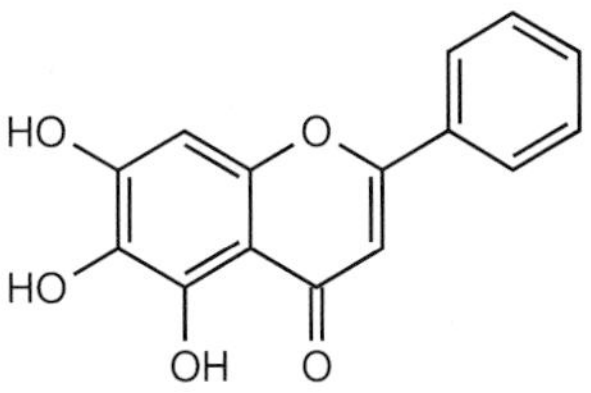

Baicalein

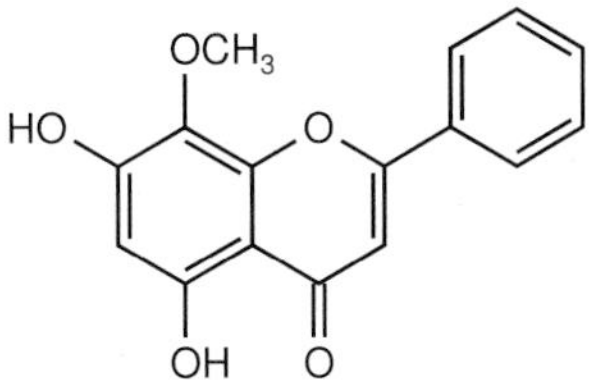

Wogonin

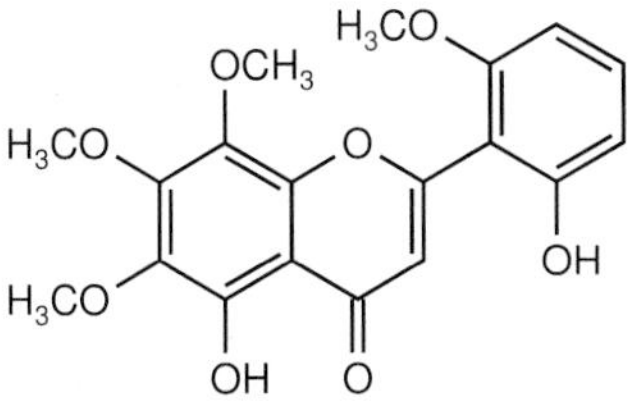

Skullcapflavone II

intestine, but was hydrolyzed to baicalein by intestinal bacteria. A large proportion of baicalein absorbed was transformed into baicalin within the intestinal mucosal cells, and excreted through multidrug resistance-associated protein 2 into the intestinal lumen [4]. After intramuscular (i.m.) injection to humans, baicalin appeared in the urine after 24 min. Bioavailability was assessed to be about 90%, the mean half-life $t_{1/2\beta}$ was 36 min, indicating a rapid elimination [5]. The metabolites excreted in the bile after oral administration of baicalin or baicalein to rats were identified as baicalein 6-*O*-β-glucuronopyranoside, baicalin, 6-*O*-methyl-baicalin, 6-*O*-β-glucuronopyranosyl-baicalein 7-*O*-sulfate, and baicalein 6,7-di-*O*-β-glucuronopyranoside. A slower biliary excretion of the metabolites after baicalin administration suggested that it was absorbed as baicalein after hydrolysis in the gastrointestinal tract [6]. After oral administration of an extract of the root of *S. baicalensis,* the major biliary metabolites were identified as baicalein 6,7-di-*O*-β-glucuronopyranoside, baicalein 6-β-glucuronopyranoside, 6-*O*-methyl-baicalein 7-*O*-β-glucuronopyranoside, and baicalein 6-*O*-β-glucopyranoside-7-*O*-β-glucuronopyranoside. Glucuronide, glucoside, and methylated products were also found in rat urine after administration of the extract [7]. In contrast, baicalein was extensively metabolized to baicalin, that was detected in rat small intestine after oral administration [8]. Baicalin was also found to be the major metabolite of baicalein by incubation with liver and intestine microsomal, cytosol or S-9 fractions from human, rat, and various human recombinant UDP-glucuronosyltransferase isoenzymes. Baicalin was extensively generated in liver and jejunum microsomes in both humans and rats, mainly catalyzed by UDP-glucuronosyltransferase isoenzyme 1A9 and also mediated by UDP-glucuronosyltransferase isoenzymes 1A1, 1A3, 1A8, 1A7 and 2B15, with different kinetic profiles [9]. The intravenous (i.v.) administration of wogonin resulted in a rapid distribution phase, followed by a slower elimination phase, as observed from the plasma concentration–time profile [10].

The pharmacokinetics of baicalein, wogonin and oroxylin A in Sprague-Dawley rats after an oral dose of a purified extract of *S. baicalensis* has also been studied. Following oral administration of the extract at doses of 10, 20, and 40 mg kg^{-1}, equivalent to 4.5, 9.0, and 18 mg kg^{-1} baicalein, baicalein and its metabolite baicalin showed dose-linear pharmacokinetics. The extract doses were equivalent to 0.4, 0.8, and 1.6 mg kg^{-1} wogonin. The area under the curve (AUC) could be determined for wogonin at extract doses of 20 and 40 mg kg^{-1}, but not at 10 mg kg^{-1}. The oral doses of the extract were equivalent to 1.5, 3.0, and 6.0 mg kg^{-1} oroxylin A, but concentrations of oroxylin A in plasma, urine and gastrointestine samples were below the detection limit. The pharmacokinetics of baicalein and baicalin were different after oral application of the extract, as compared to pure baicalein [11]. After oral administration of an extract at a dose corresponding to 9 g Radix Scutellariae to healthy male human volunteers, the cumulative renal excretion of baicalein glucuronides and sulfates was 43 μmol (2.9% of dose) and 65 μmol (4.3% of dose), respectively, whereas wogonin glucuronides and sulfates were 22 μmol (5.9% of dose) and 21 μmol (5.7% of dose), respectively. This indicated that the renal excretion of wogonin metabolites was greater than that of baicalein [12].

The root of *S. baicalensis* has been widely used as an antibacterial, antiviral, anti-inflammatory, neuroprotective, hepatopreventive, and antitumor agent, as well as for the prevention and treatment of cardiovascular diseases. No doubt, the flavones and flavone glycosides, especially the major flavones, baicalein and its glucuromide, baicalin, are the most important active ingredients as they possess a broad spectrum of pharmacological properties. Baicalein markedly reduced the minimum inhibitory concentration (MIC) of tetracycline against a clinically isolated methicillin-resistant *Staphylococcus aureus* strain which possessed a *tetK* gene encoding the efflux pump for tetracycline. Using *Escherichia coli* strain KAM32/pTZ1252 carrying the *tetK* gene, baicalein strongly inhibited the transport of tetracycline. Baicalein also showed a synergistic effect with tetracycline, or with β-lactam antibiotics, against a methicillin-resistant *S. aureus* strain without the *tetK* gene [13].

The flavones from *S. baicalensis* were reported to inhibit lipid peroxidation in rat liver homogenate, as stimulated by a mixture of $FeCl_2$ and ascorbic acid or by a mixture of NADPH and ADP [14]. The extract of *S. baicalensis* and the flavones baicalein, baicalin and wogonin, each significantly inhibited lipid peroxidation in lecithin liposomal membranes, as induced by UV light [15], and inhibited lipid peroxidation in rat liver microsomes induced by ascorbic acid. Iron release induced by ascorbic acid from liver microsomes of baicalein-treated rats was markedly lower than from liver microsomes of control rats. It was suggested that baicalein might form an inert iron complex [16]. Quantitative structure–activity relationships studies showed that the presence of the 1,4- and 1,2-hydroquinone and the hydrophobicity of the molecule were responsible for the inhibitory activity on iron-induced lipid peroxidation in rat liver mitochondria [17]. It was also reported that the extract of *S. baicalensis* and baicalein might attenuate oxidant stress and protect cultured embryonic cardiomyocytes against lethal oxidant damage [18]. Baicalein was found to be a potent radical scavenger in different *in vitro* and *in vivo* models [19]. Baicalein, baicalin, and wogonin have been reported to strongly inhibit xanthine oxidase, with baicalein as the most active compound. Baicalein and baicalin also exhibited a strong eliminating activity on the superoxide radical [20]. In human umbilical vein endothelial cells (HUVEC), baicalein inhibited oxidative stress-induced endothelin-1 secretion, increased the formation of reactive oxygen species (ROS), elevating the phosphorylation of extracellular signal-regulated kinases (ERK), and stress-stimulated activator protein-1 reporter activity [21]. Baicalein also lowered the blood pressure in renin-dependent hypertension. The *in vivo* hypotensive effect of baicalein might partly be attributed to its inhibition of lipooxygenase, resulting in a reduced biosynthesis and release of arachidonic acid-derived vasoconstrictive products. Furthermore, baicalein inhibited the expression of endothelial leukocyte adhesion molecule-1 in HUVEC, as induced by thrombin and thrombin receptor agonist peptides, and also inhibited the expression of intercellular adhesion molecule-1 (ICAM-1), as induced by thrombin or by 12-*O*-tetradecanoylphorbol 13-acetate (TPA) [22].

The exposure of rat isolated aortic rings to baicalein completely abolished endothelium-dependent relaxation as induced by acetylcholine, and significantly attenuated the endothelium-independent relaxation as induced by sodium nitroprusside. Baicalein inhibited endothelial nitric oxide (NO) synthase and endothelium-derived NO-mediated vascular tone in rat aortas through inhibiting endothelium-derived NO bioavailability, concomitant with a reduced bioactivity of endothelium-derived NO and cyclooxygenase (COX)-mediated release of superoxide anions [23]. Baicalein was also found to scavenge ROS generated in chick cardiomyocytes, and to protect against cell death caused by ischemia–reperfusion [24]. The pretreatment of neonatal rat cardiomyocytes with baicalein significantly reduced the release of lactate dehydrogenase, while baicalin was ineffective. Hydrogen peroxide might be involved in the cardioprotective effect of baicalein [25]. In human aortic smooth muscle cells, wogonin, but not baicalin or baicalein, significantly and selectively suppressed matrix metalloproteinase-9 (MMP-9) expression, as induced by tumor necrosis factor-α (TNF-α), and reduced the TNF-α-induced migration of human aortic smooth muscle cells. This suggested that wogonin might effectively suppress TNF-α-induced human aortic smooth muscle cell migration through a selective inhibition of MMP-9 expression [26].

Baicalin, baicalein, and wogonin were each reported to show anti-inflammatory activity, notably against carrageenan-induced rat paw edema [27]. The anti-inflammatory activity of baicalein was greater in the chronic inflammation model than in the rat carrageenan-induced paw edema. It was suggested that inhibition of the 5-lipooxygenase pathway of arachidonic acid metabolism might be one mechanism of anti-inflammatory activity of baicalein [28]. Baicalein inhibited both the lucigenin-dependent chemiluminescence and leukotriene (LTB_4) synthesis in human alveolar macrophages obtained by bronchoalveolar lavage from patients with various respiratory diseases [29]. Baicalein exerted a marked inhibition of the release of LTB_4 and LTC_4 from human polymorphonuclear leukocytes stimulated with the Ca^{2+} ionophore A23,187 at micromolar concentrations [30]. The anti-inflammatory mechanism of baicalin was also thought to occur by binding to chemokines. Baicalin did not compete directly with chemokines for binding to receptors, but rather acted through its selective binding to chemokine ligands [31].

Baicalein, but not baicalin, inhibited COX-2 gene expression in RAW 264.7 cells induced by lipopolysaccharide (LPS). However, both baicalein and baicalin inhibited LPS-induced inducible NO synthase (iNOS) protein expression, iNOS mRNA expression, and NO production. The inhibition of COX-2 gene expression by baicalein might be mediated through inhibition of the DNA-binding activity of CCAAT/enhancer binding protein-β [32]. The root of *S. baicalensis* and the flavone components baicalin, baicalein and chrysin, but not wogonin, also upregulated the expression of transforming growth factor-β1 (TGF-β1) gene and protein levels in RAW 264.7 cells in a concentration-dependent manner. The upregulation of TGF-β1 gene expression seemed to be through nuclear factor-κB (NF-κB) and protein kinase C (PKC) pathways [33]. The root of *S. baicalensis* and baicalin were found to inhibit cAMP phosphodiesterase from beef heart *in vitro* [34]. The most active inhibitor was 2′-hydroxywogonin [35]. Baicalin, baicalein were also found to markedly reduce the increase in cGMP levels, as stimulated by acetylcholine, in rat endothelium-intact rings, and by sodium nitroprusside in endothelium-denuded rings [36]. Baicalein inhibited the release of slow-reacting substance of anaphylaxis and reduced the release of LTC_4 and LTD_4 from sensitized guinea pig lung after antigen challenge. Baicalin relaxed isolated guinea pig tracheal smooth muscle contracted by LTD_4. It was suggested that baicalein exerted its action via two different mechanisms: the inhibition of slow-reacting substance of anaphylaxis; and direct relaxing effects on the trachea [37]. Baicalein was further reported to exert contractile effects at low concentrations, and relaxant effects at high concentrations, in endothelium-intact arteries precontracted by phenylephrine or by high concentrations of K^+. In endothelium-denuded arteries, the contractile response to baicalein was absent while the relaxant response remained. At low concentrations, baicalein caused a contractile response and inhibited endothelium-dependent relaxation, probably through an inhibition of endothelial formation and release of NO. At higher concentrations, baicalein relaxed the arterial smooth muscle, partially through an inhibition of the PKC-mediated contractile mechanism [38].

Baicalein strongly counteracted the reduced production of tissue-type plasminogen activator (t-PA) and elevated the production of plasminogen activator inhibitor 1 in HUVEC, as induced by trypsin [39] or thrombin and a thrombin receptor agonist peptide [30]. Other flavones from *S. baicalensis*, including wogonin, oroxylin A and skullcapflavone II, showed similar effects, though less strongly. It was suggested that baicalein might prevent the thrombotic tendency induced by trypsin [39]. Baicalein reduced the migration of vascular smooth muscle cells of bovine carotid artery induced by platelet-derived growth factor-BB [41].

Baicalein also inhibited the adipogenesis of 3T3-L1 preadipocytes and triglyceride accumulation during adipogenesis, and significantly decreased the mRNA expression of fatty acid-binding protein as a marker of adipogenesis. Microarray analysis revealed that several genes were modulated. The upregulation of the expression of fatty acid-binding protein, apolipoprotein D, and insulin-like growth factor 2 during adipogenesis was normalized by baicalein. The decrease in COX-2 mRNA expression during adipogenesis was alleviated by baicalein. The antiadipogenic effect of baicalein might be mediated by its ability to enhance the expression of COX-2 [42].

Flavones from *S. baicalensis* were reported to be antivirally active. Wogonin was found to suppress hepatitis B virus (HBV) surface antigen production in an HBV-producing cell line (MS-G2) *in vitro*, without any evidence of cytotoxicity. The relaxed circular and the linear forms of HBV DNA were significantly reduced after wogonin treatment [43]. Baicalein was found to be a potent inhibitor of reverse transcriptases from murine leukemia viruses and human immunodeficiency virus (HIV). The enzyme activities were inhibited by more than 90% in the presence of baicalein at a concentration of $2\,\mu g\,ml^{-1}$ [44]. Baicalin inhibited HIV-1 infection and replication, as measured by focal syncytium formation of human T-lymphoid leukemia cells and HIV-1 specific core antigen p24 expression, as well as retroviral reverse transcriptase activity in the HIV-1-infected H9 cells. Baicalin also inhibited the purified recombinant HIV-1 reverse transcriptase activity. The anti-HIV-1 activity of baicalin was also observed in cultures of primary human peripheral blood mononuclear cells infected with HIV-1, without any cytotoxicity on the indicator cells [45, 46]. Baicalin was reported to be a noncompetitive inhibitor of HIV reverse transcriptase *in vitro* [47], but was also found to inhibit HIV-1 infection by interfering with viral entry, a process known to involve interaction between HIV-1 envelope proteins and the cellular CD_4 and chemokine receptors [48]. Baicalein and baicalin were reported to be therapeutically effective in treating HIV-infected patients. Baicalin might also selectively induce the apoptosis of HIV-infected human T-lymphoid leukemia cells with a high virus-releasing capacity, and stimulate the proliferation of HIV-infected human T-lymphoid leukemia cells which have a relatively lower capacity of HIV-production [49]. Moreover, baicalin treatment selectively reduced the detectable levels of human T-cell lymphotropic virus HTLV-I p19 Gag protein in infected cells at concentrations that produced insignificant effects on total cellular protein and DNA synthesis [50].

The hepatoprotective activity of baicalein has been studied *in vitro* using rat hepatocytes. Baicalein exerted an inhibitory effect on cytotoxicity and oxidative damage in primary rat hepatocytes induced by *tert*-butyl hydroperoxide (TBH). It showed effective free radical-scavenging properties toward 1,1-diphenyl-2-picrylhydrazyl radicals, and decreased lactate dehydrogenase and alanine aminotransferase, as well as the formation of malondialdehyde induced by TBH in hepatocytes. Baicalein also inhibited the mitochondrial depolarization and DNA repair synthesis caused by TBH. In addition, baicalein decreased the 8-hydroxy-2′-deoxyguanosine level in hepatocytes as a marker for DNA damage [51]. Baicalein, baicalin, and wogonin were each also found to inhibit protein nitration and lipid peroxidation in liver homogenate, as well as in $HepG_2$ human hepatoblastoma cells induced by hemin-nitrite-H_2O_2 [52]. Skullcapflavone I was reported to induce apoptosis in activated rat hepatic stellate cells, which play a central role in liver fibrogenesis. Skullcapflavone I increased caspase-3 and caspase-9 activities, accompanied by the proteolytic cleavage of poly (ADP-ribose)polymerase [53].

The neuroprotective activity of baicalein and baicalin has been intensively investigated. On incubation of human neuroblastoma SH-SY5Y cells in the presence of 400 μ*M* H_2O_2 for 2 h, the viability of cells was decreased remarkably, while the cell lipid peroxidation and percentage of lactate dehydrogenase released into the culture medium was significantly increased. The addition of 10 μ*M* of baicalein and baicalin significantly attenuated the cellular damage induced by H_2O_2, while the effect of wogonin was marginal, and wogonoside showed no effect at the tested concentration. Baicalein and baicalin also antagonized the elevated intracellular free-calcium concentration caused by H_2O_2 [54]. Baicalein was further found to protect SH-SY5Y cells from damages induced by 6-hydroxydopamine. The mechanism was believed to be via an inhibition of ROS formation, the inhibition of caspase-9 and caspase-3 activation, and the inhibition of Jnk phosphorylation, which was known as an apoptotic mediator in 6-hydroxydopamine-induced neuronal cell death [55]. The protective effect of baicalein against the neurotoxicity of 6-hydroxydopamine was also seen in young mice [56].

An inhibition of oxidative stress-induced apoptosis by baicalein was observed in C6 rat glioma cells, suggested to be caused by the ROS-scavenging activity of baicalein and by ERK modulation [57]. Pretreatment of neurons from rat embryo with baicalein resulted in an inhibition of the decreased uptake of [^{3}H]dopamine induced by LPS. Excessive production of TNF-α, NO and superoxide stimulated by LPS were also attenuated by baicalein, indicating that baicalein exerted potent neuroprotective effects on LPS-induced injury of dopaminergic neurons. The inhibition of LPS-induced production of NO and free radicals from microglia might underlie the mechanisms of the neuroprotection [58]. Baicalein was also reported to inhibit NO production and the apoptosis of LPS-activated, but not interferon-γ (IFN-γ)-activated, BV-2 mouse microglial as well as rat primary microglial cells. The inhibition of NO production by baicalein was due to a suppression of iNOS induction. Moreover, baicalein inhibited LPS-induced NF-κB activity in BV-2 cells, without affecting caspase-11 activation [59]. In human glioma cells, baicalein was found to prevent a loss of cell viability and apoptosis induced by cisplatin. Baicalein also prevented Bax expression, mitochondrial depolarization, cytochrome c release from mitochondria, and caspase activation in human glioma cells induced by cisplatin [60].

The i.p. injection of low doses of baicalein (0.3–1.0 mg kg^{-1}) to mice significantly attenuated the striatal dopamine transporter loss induced by methamphetamine in a dose-dependent manner, and diminished methamphetamine-induced increase in striatal malondialdehyde content and myeloperoxidase activity. Moreover, baicalein effectively diminished the ROS production by leukocytes stimulated with methamphetamine or TPA. This suggested that baicalein might inhibit methamphetamine-induced dopamine transporter loss by inhibiting the neutrophil increase and the lipid peroxidation caused by neutrophil-derived ROS in striatum [61]. Baicalein significantly improved cognitive deficits and neuropathological changes in rats caused by a permanent occlusion of the bilateral common carotid arteries. It reduced the increased activity of superoxide dismutase and malondialdehyde content, and attenuated the decreased activity of glutathione peroxidase and catalase in treated rats. The therapeutic potential of baicalein was partly due to its antioxidant effect [62]. The neuroprotective effect of baicalein was also evidenced in mice with transient middle cerebral artery occlusion. Baicalein protected against ischemia–reperfusion injury by inhibiting the 12-lipooxygenase and 15-lipooxygenase pathway to neuronal cell death [63].

Baicalein was also found to inhibit both apoptosis and Jun protein overexpression in rat cortical cells, as induced by β-amyloid peptide [64, 65]. Likewise, the preventive effect of baicalein and baicalin against apoptosis and cell death induced by β-amyloid peptide was also observed in PC12 nerve cells [66]. Baicalein was reported to act on the benzodiazepine-binding sites and to exert an anxiolytic-like effect in mice. Pretreatment with a single dose of baicalein, or post-treatment for 7 or 13 days with multiple doses of baicalein, attenuated β-amyloid peptide-induced amnesia [67].

Baicalein was reported to inhibit the mutagenicity of aflatoxin B_1 (AFB_1) and *N*-methyl-*N*′-nitro-*N*-nitrosoguanidine (MNNG) in *Salmonella typhimurium* [68], and to show a marked inhibition of TPA-induced epidermal ornithine decarboxylase activity, also inhibiting tumor promotion by TPA after initiation with benzo[*a*]pyrene (BaP) in a mouse skin model [69]. Baicalein and wogonin both reduced the genotoxicity of BaP and AFB_1 using the *umuC* gene expression response in *S. typhimurium* TA1535/pSK1002. Treatment of mice with a diet containing 5 m*M* baicalein and wogonin resulted in decreased liver microsomal BaP hydroxylation. Baicalein treatment also reduced the formation of AFQ_1 and AFB_1-epoxide formation in mouse liver microsomes. Pretreatment of mice with wogonin for one week significantly decreased hepatic DNA adduct formation induced by 200 mg kg^{-1} of BaP [70]. 5,7,2′-Trihydroxyflavone and 5,7,2′,3′-tetrahydroxyflavone both showed remarkable inhibitory effects on the Epstein–Barr virus early antigen activation, and markedly inhibited mouse skin tumor promotion in a two-stage carcinogenesis test *in vivo* [71].

Baicalein and chrysin markedly inhibited the cytochrome P450-dependent aryl hydrocarbon hydroxylase in rat liver microsomes [72]. Baicalein, chrysin, and related flavones were found to bind to the active site of aromatase, competitively inhibiting cytochrome P450 aromatase [73]. The induction of quinone reductase with a methanol extract of *S. baicalensis* and its flavone components, baicalin, baicalein and wogonin, in murine Hepa 1c1c7 cells was also reported [74]. The inhibition of DNA adduct formation from 7,12-dimethylbenz[*a*]anthracene (DMBA) in MCF-7 cells by baicalein was found to be associated with a modulation of the activity of cytochrome P450 isoenzymes CYP1A1 and CYP1B1 [75]. Baicalein and 2′,5,6′,7-tetrahydroxyflavone also inhibited CYP3A4 activity, whereas oroxylin A inhibited CYP2C9 activity. All flavones from the root of *S. baicalensis* were found to significantly inhibit CYP1A2 activity [76]. In HL-60 human promyelocytic leukemia cells, the *N*-acetylation of 2-aminofluorene and DNA adduct formation with 2-aminofluorene in HL-60 cells were each inhibited by baicalein. Baicalein also inhibited *N*-acetyltransferase-1 mRNA gene expression in HL-60 cells [77].

Skullcapflavone II was reported to be cytotoxic against L1210 leukemia cells *in vitro*, with an IC_{50} of 1.5 μg ml^{-1} [78]. Baicalin and baicalein exhibited growth inhibition in human hepatoma cell lines PLC/PRF/5 and Hep-G_2, and a human pancreatic cancer cell line BxPC-3 with IC_{50} of 20 μg ml^{-1} for baicalin and 50 μg ml^{-1} for baicalein [79]. Cytotoxicity of baicalein, baicalin and wogonin was also observed on human bladder cancer cell lines KU-1 and EJ-1, and on a murine bladder cancer cell line MBT-2. All three flavones inhibited cell proliferation *in vitro* in a concentration-dependent manner, with baicalin as the most active compound. The IC_{50} of baicalin was 3.4 μg ml^{-1} for KU-1; 4.4 μg ml^{-1} for EJ-1; and 0.9 μg ml^{-1} for MBT-2 cells [80]. The root of *S. baicalensis,* baicalin and baicalein exerted a cytotoxic effect on HL-60 cells *in vitro* comparable to that of cisplatin and doxorubicin [81]. Baicalein was reported to induce apoptosis in HL-60 cells through a ROS-mediated mitochondrial dysfunction pathway [82, 83]. It was also found to significantly inhibit the proliferation of K562 human leukemia cells, and to selectively induce apoptosis. Flow cytometric analysis showed that baicalein caused cell-cycle arrest of K562 cells in the S-phase, and increased the protein expression of FAS and caspase-3, but not of BCL-2 [84]. The methanol extract of cultured *S. baicalensis* cells, containing baicalin, baicalein and wogonin, was found to inhibit the proliferation of a human monocytic leukemia cell line THP-1 and human osteogenic sarcoma cell line HOS. It has been noted that wogonin did not show any inhibitory effect on the human fetal lung normal diploid cell line TIG-1, as compared to cancer cells. Wogonin induced cancer cell-specific apoptosis, and caused cell-cycle arrest at G_2/M phase and apoptosis in THP-1 cells [85]. Baicalein also inhibited the growth of CH27 human lung squamous carcinoma cells, induced cell-cycle arrest at the S-phase and apoptosis, accompanied by a decrease in BCL-2 and caspase-3 [86]. The root of *S. Baicalensis,* given at an oral daily dose of 10 mg per mouse for 10 days, significantly inhibited the growth of MBT-2 cells implanted in C3H/HeN mice [80].

Baicalein strongly inhibited DNA topoisomerase II and the proliferation of three human hepatocellular carcinoma cell lines. Among the three cell lines, baicalein induced apoptosis in a concentration-dependent manner in only one cell line, and caused cell death via necrosis in the other two lines [87]. Baicalein induced mammalian topoisomerase II-dependent DNA-cleavage *in vitro* and caused an unwinding of duplex DNA. The cleavage specificity was the same as that for the known intercalator 4′-(acridin-9-ylamino)methanesulfon-*m*-anisidide. An analysis of the structure–activity relationships for topoisomerase II-mediated DNA cleavage revealed that the flavones with hydroxy groups at the 5, 7, 3′ and 4′ positions were favorable for efficient cleavage. Similar requirements have been reported for flavones to inhibit PKC that act competitively with ATP. Formation of the cleavable complex might contribute to the growth-inhibitory activity of flavones observed for some human tumor cell lines [88].

Baicalin and baicalein were also found to inhibit cell proliferation and to induce apoptosis in several human prostate cancer cell lines, including DU145, PC-3, LNCaP and CA-HPV-10. The proliferation inhibition of prostate cancer cells after a short period of exposure to baicalin was associated with induction by apoptosis [89]. Baicalein, wogonin, neobaicalein, and skullcapflavone I each inhibited the proliferation of prostate cancer cells, and exhibited antiandrogenic activities with a reduced expression of the androgen receptor and androgen-regulated genes in cancer cells. At a daily oral dose of $20\,mg\,kg^{-1}$, it significantly reduced the growth of prostate cancer xenografts in nude mice by 55% at two weeks, and increased the average time for tumors to reach a volume of approximately $1000\,mm^3$ from 16 to 47 days [90]. The oral treatment of SCID mice bearing DU-145 human prostate cancer cells with baicalein for 28 days at daily doses of 10, 20, and $40\,mg\,kg^{-1}$, significantly and dose-dependently reduced the tumor volume [91]. At intragastric doses of 130 and $260\,mg\,kg^{-1}$, it significantly inhibited prostate hyperplasia in castrated rats, as induced by testosterone propionate. Baicalein at doses of 260 and $520\,mg\,kg^{-1}$ also significantly inhibited prostate hyperplasia induced by transplantation of homologous strain fetal mouse urogenital sinus, and by testosterone propionate in mice [92]. Baicalein strongly inhibited human T-lymphoid leukemia cell proliferation, with an IC_{50} of about $5\,\mu M$. The protein tyrosine kinase activity in the human T-lymphoid leukemia cells was significantly reduced by baicalein. On the other hand, the PKC activity stimulated by TPA was reduced by direct incubation with baicalein. PCR analysis of platelet-derived growth factor (PDGF)-A and transforming growth factor-β1 messenger RNA levels demonstrated that baicalein reduced the PDGF-A mRNA level, but less affected the transforming growth factor-β 1 mRNA [93].

Baicalein potently inhibited the growth of the human breast carcinoma cell line, MDA-MB-435 (IC_{50} ca. $6\,\mu g\,ml^{-1}$), being more active than some flavones isolated from citrus fruits, including hesperetin and naringenin [94]. Baicalein and related flavones also inhibited the proliferation of estrogen receptor-positive MCF-7 human breast cancer cells; however, the inhibition was not reversed with the addition of estrogen [95]. Baicalein as a α-glucosidase inhibitor suppressed the *in vitro* invasion and *in vivo* metastasis of mouse melanoma cells [96]. In addition, baicalein and baicalin were reported to be potent inhibitors of angiogenesis [97].

References

1 Wang, J.Z., Chen, D.Y., and Su, Y.Y. (1994) Analytical study on processing of *Scutellaria baicalensis* Georyl by HPLC. *China J. Chin. Mater. Med.*, **19**, 340–341.

2 Ishimaru, K., Nishikawa, K., Omoto, T., Asai, I., Yoshihira, K., and Shimomura, K. (1995) Two flavone 2′-glucosides from *Scutellaria baicalensis. Phytochemistry*, **40**, 279–281.

3 Lai, M.Y., Hsiu, S.L., Tsai, S.Y., Hou, Y.C. and Chao, P.D. (2003) Comparison of metabolic pharmacokinetics of baicalin and baicalein in rats. *J. Pharm. Pharmacol.*, **55**, 205–209.

4 Akao, T., Sakashita, Y., Hanada, M., Goto, H., Shimada, Y., and Terasawa, K. (2004) Enteric excretion of baicalein, a flavone of Scutellariae Radix, via glucuronidation in rat: involvement of multidrug resistance-associated protein 2. *Pharm. Res.*, **21**, 2120–2126.

5 Mao, F.F., Tu, X.D., Zhu, J.B., Zhao, L.H., Shen, G.M., Philippe, C.C., and Fan, B.T. (1983) Absorption of baicalin in humans. *J. Nanjing Pharm. Coll.*, **14**, 61–66

6 Abe, K., Inoue, O., and Yumioka, E. (1990) Biliary excretion of metabolites of baicalin and baicalein in rats. *Chem. Pharm. Bull. (Tokyo)*, **38**, 208–211.

7 Feng, N.P., Di, B., and Liu, W.Y. (2005) Comparison of the metabolism of baicalin in rats orally administered with Radix scutellariae extract and Shuang-Huang-Lian extract. *Chem. Pharm. Bull. (Tokyo)*, **53**, 978–983.

8 Zhang, L., Lin, G., Chang, Q., and Zuo, Z. (2005) Role of intestinal first-pass metabolism of baicalein in its absorption process. *Pharm. Res.*, **22**, 1050–1058.

9 Zhang, L., Lin, G., and Zuo, Z. (2007) Involvement of UDP-glucuronosyl-transferases in the extensive liver and intestinal first-pass metabolism of flavonoid baicalein. *Pharm. Res.*, **24**, 81–89.

10 Tsai, T.H., Chou, C.J., Tsai, T.R., and Chen, C. F. (1996) Determination of wogonin in rat plasma by liquid chromatography and its pharmacokinetic application. *Planta Med.*, **62**, 263–266.

11 Kim, Y.H., Jeong, D.W., Kim, Y.C., Sohn, D.H., Park, E.S., and Lee, H.S. (2007) Pharmacokinetics of baicalein, baicalin and wogonin after oral administration of a standardized extract of *Scutellaria baicalensis,* PF-2405 in rats. *Arch. Pharm. Res.*, **30**, 260–265.

12 Lai, M.Y., Hsiu, S.L., Chen, C.C., Hou, Y.C. and Chao, P.D. (2003) Urinary pharmacokinetics of baicalein, wogonin and their glycosides after oral administration of Scutellariae Radix in humans. *Biol. Pharm. Bull.*, **26**, 79–83.

13 Fujita, M., Shiota, S., Kuroda, T., Hatano, T., Yoshida, T., Mizushima, T., and Tsuchiya, T. (2005) Remarkable synergies between baicalein and tetracycline, and baicalein and β-lactams against methicillin-resistant *Staphylococcus aureus. Microbiol. Immunol.*, **49**, 391–436

14 Lim, B.O., Yu, B.P., Kim, S.C., and Park, D.K. (1999) The antioxidative effect of ganhuangenin against lipid peroxidation. *Phytother. Res.*, **13**, 479–483.

15 Gabrielska, J., Oszmianski, J., Zylka, R., and Komorowska, M. (1997) Antioxidant activity of flavones from *Scutellaria baicalensis* in lecithin liposomes. *Z. Naturforsch.*, **52C**, 817–823.

16 Gao, D., Sakurai, K., Katoh, M., Chen, J., and Ogiso, T. (1996) Inhibition of microsomal lipid peroxidation by baicalein: a possible formation of an iron-baicalein complex. *Biochem. Mol. Biol. Int.*, **39**, 215–225.

17 Shimizu, I., Ma, Y.R., Mizobuchi, Y., Liu, F., Miura, T., Nakai, Y., Yasuda, M., Shiba, M., Horie, T., Amagaya, S., Kawada, N., Hori, H., and Ito, S. (1999) Effects of Sho-saiko-to, a Japanese herbal medicine, on hepatic fibrosis in rats. *Hepatology*, **29**, 149–160.

18 Shao, Z.H., Li, C.Q., Vanden Hoek, T.L., Becker, L.B., Schumacker, P.T., Wu, J.A., Attele, A.S., and Yuan, C.S. (1999) Extract from *Scutellaria baicalensis* Georgi attenuates oxidant stress in cardiomyocytes. *J. Mol. Cell. Cardiol.*, **31**, 1885–1895.

19 Gao, Z., Huang, K., Yang, X., and Xu, H. (1999) Free radical scavenging and antioxidant activities of flavonoids extracted from the radix of *Scutellaria baicalensis* Georgi. *Biochim. Biophys. Acta*, **1472**, 643–650.

20 Shieh, D.E., Liu, L.T., and Lin, C.C. (2000) Antioxidant and free radical scavenging effects of baicalein, baicalin and wogonin. *Anticancer Res.*, **20** (A), 2861–2865.

21 Chen, H.H., Hong, H.J., Chou, Y.H., Cheng, T. H., Chen, J.J., and Lin, H. (2006) Inhibition of cyclic strain-induced endothelin-1 secretion by baicalein in human umbilical vein endothelial cells. *Planta Med.*, **72**, 1318–1321.

22 Kimura, Y., Matsushita, N., Yokoi-Hayashi, K., and Okuda, H. (2001) Effects of baicalein isolated from *Scutellaria baicalensis* Radix on adhesion molecule expression induced by thrombin and thrombin receptor agonist peptide in cultured human umbilical vein endothelial cells. *Planta Med.*, **67**, 331–334.

23 Machha, A., Achike, F.I., Mohd, M.A., and Mustafa, M.R. (2007) Baicalein impairs vascular tone in normal rat aortas: role of superoxide anions. *Eur. J. Pharmacol.*, **565**, 144–150.

24 Shao, Z.H., Vanden Hoek, T.L., Qin, Y., Becker, L.B., Schumacker, P.T., Li, C.Q., Dey, L., Barth, E., Halpern, H., Rosen, G.M., and Yuan, C.S. (2002) Baicalein attenuates oxidant stress in cardiomyocytes. *Am. J. Physiol. Heart Circ. Physiol.*, **282**, H999–H1006

25 Woo, A.Y., Cheng, C.H., and Waye, M.M. (2005) Baicalein protects rat cardiomyocytes from hypoxia/reoxygenation damage via a prooxidant mechanism. *Cardiovasc. Res.*, **65**, 244–253.

26 Lee, S.O., Jeong, Y.J., Yu, M.H., Lee, J.W., Hwangbo, M.H., Kim, C.H., and Lee, I.S. (2006) Wogonin suppresses TNF-α-induced MMP-9 expression by blocking the NF-κB activation via MAPK signaling pathways in human aortic smooth muscle cells. *Biochem. Biophys. Res. Commun.*, **351**, 118–125.

27 Huang, W.H., Lee, A.R., and Yang, C.H. (2006) Antioxidative and anti-inflammatory activities of polyhydroxyflavonoids of *Scutellaria baicalensis* Georgi. *Biosci. Biotechnol. Biochem.*, **70**, 2371–2380.

28 Butenko, I.G., Gladtchenko, S.V., and Galushko, S.V. (1993) Anti-inflammatory properties and inhibition of leukotriene C4 biosynthesis *in vitro* by flavonoid baicalein from *Scutellaria baicalensis* Georgy roots. *Agents Actions*, **39**, C49–C51

29 Tanno, Y., Kakuta, Y., Aikawa, T., Shindoh, Y., Ohno, I., and Takishima, T. (1988) Effects of qing-fei-tang (seihai-to) and baicalein, its main component flavonoid, on lucigenin-dependent chemiluminescence and leukotriene B_4 synthesis of human alveolar macrophages. *Am. J. Chin. Med.*, **16**, 145–154.

30 Homma, M., Minami, M., Taniguchi, C., Oka, K., Morita, S., Niitsuma, T., and Hayashi, T. (2000) Inhibitory effects of lignans and flavonoids in saiboku-to, a herbal medicine for bronchial asthma, on the release of leukotrienes from human polymorphonuclear leukocytes. *Planta Med.*, **66**, 88–91.

31 Li, B.Q., Fu, T., Gong, W.H., Dunlop, N., Kung, H., Yan, Y., Kang, J., and Wang, J.M. (2000) The flavonoid baicalin exhibits anti-inflammatory activity by binding to chemokines. *Immunopharmacology*, **49**, 295–306.

32 Woo, K.J., Lim, J.H., Suh, S.I., Kwon, Y.K., Shin, S.W., Kim, S.C., Choi, Y.H., Park, J.W., and Kwon, T.K. (2006) Differential inhibitory effects of baicalein and baicalin on LPS-induced cyclooxygenase-2 expression through inhibition of C/EBPβ DNA-binding activity. *Immunobiology*, **211**, 359–368.

33 Chuang, H.N., Wang, J.Y., Chiu, J.H., Tsai, T. H., Yeh, S.F., Fu, S.L., Lui, W.Y., and Wu, C.W. (2005) Enhancing effects of *Scutellaria baicalensis* and some of its constituents on TGF-β1 gene expression in RAW 264.7 murine macrophage cell line. *Planta Med.*, **71**, 440–445.

34 Nikaido, T., Ohmoto, T., Kinoshita, T., Sankawa, U., Delle Monache, F., Botta, B., Tomimori, T., Miyaichi, Y., and Shirataki, Y. (1989) Inhibition of cAMP phosphodiesterase in medicinal plants. Part XVI. Inhibition of adenosine 3′,5′-cyclic monophosphate phosphodiesterase by flavonoids. III. *Chem. Pharm. Bull. (Tokyo)*, **37**, 1392–1395.

35 Nikaido, T., Ohmoto, T., Sankawa, U., Tomimori, T., Miyaichi, Y., and Imoto, Y. (1988) Inhibitors of cAMP phosphodiesterase in medicinal plants. Part XIII. Inhibition of adenosine 3′,5′-cyclic monophosphate phosphodiesterase by flavonoids. II. *Chem. Pharm. Bull. (Tokyo)*, **36**, 654–661.

36 Huang, Y., Wong, C.M., Lau, C.W., Yao, X., Tsang, S.Y., Su, Y.L., and Chen, Z.Y. (2004) Inhibition of nitric oxide/cyclic GMP-mediated relaxation by purified flavonoids, baicalin and baicalein, in rat aortic rings. *Biochem. Pharmacol.*, **67**, 787–794.

37 Miyamoto, K., Katsuragi, T., Abdu, P., and Furukawa, T. (1997) Effects of baicalein on prostanoid generation from the lung and contractile responses of the trachea in guinea pig. *Am. J. Chin. Med.*, **25**, 37–50.

38 Chen, Z.Y., Su, Y.L., Lau, C.W., Law, W.I., and Huang, Y. (1999) Endothelium-dependent contraction and direct relaxation induced by baicalein in rat mesenteric artery. *Eur. J. Pharmacol.*, **374**, 41–47.

39 Kimura, Y., Okuda, H., and Ogita, Z. (1997) Effects of flavonoids isolated from scutellariae radix on fibrinolytic system induced by trypsin in human umbilical vein endothelial cells. *J. Nat. Prod.*, **60**, 598–601.

40 Kimura, Y., Yokoi, K., Matsushita, N., and Okuda, H. (1997) Effects of flavonoids isolated from scutellariae radix on the production of tissue-type plasminogen activator and plasminogen activator inhibitor-1 induced by thrombin and thrombin receptor agonist peptide in cultured human umbilical vein endothelial cells. *J. Pharm. Pharmacol.*, **49**, 816–822.

41 Kanayasu-Toyoda, T., Morita, I., and Murota, S. (1998) Arachidonic acid pretreatment enhances smooth muscle cell migration via increased Ca^{2+} influx. *Prostaglandins Leukot. Essent. Fatty Acids*, **58**, 25–31.

42 Cha, M.H., Kim, I.C., Lee, B.H., and Yoon, Y. (2006) Baicalein inhibits adipocyte differentiation by enhancing COX-2 expression. *J. Med. Food*, **9**, 145–153.

43 Huang, R.L., Chen, C.C., Huang, H.L., Chang, C.G., Chen, C.F., Chang, C., and Hsieh, M.T. (2000) Anti-hepatitis B virus effects of wogonin isolated from *Scutellaria baicalensis*. *Planta Med.*, **66**, 694–698.

44 Ono, K., Nakane, H., Fukushima, M., Chermann, J.C., and Barre-Sinoussi, F. (1989) Inhibition of reverse transcriptase activity by a flavonoid compound, 5,6,7-trihydroxyflavone. *Biochem. Biophys. Res. Commun.*, **160**, 982–987.

45 Li, B.Q., Fu, T., Yan, Y.D., Baylor, N.W., Ruscetti, F.W., and Kung, H.F. (1993) Inhibition of HIV infection by baicalin: a flavonoid compound purified from Chinese herbal medicine. *Cell. Mol. Biol. Res.*, **39**, 119–124.

46 Zhang, X., Tang, X., and Chen, H. (1991) Inhibition of HIV replication by baicalin and *S. baicalensis* extracts in H9 cell culture. *Chin. Med. Sci. J.*, **6**, 230–232.

47 Tang, X., Chen, H., and Zhang, X. (1990) Inhibition of human immunodeficiency virus reverse transcriptase by Chinese medicines *in vitro*. *Proc. Chin. Acad. Med. Sci. Peking Union Med. Coll.*, **5**, 140–144.

48 Li, B.Q., Fu, T., Dongyan, Y., Mikovits, J.A., Ruscetti, F.W., and Wang, J.M. (2000) Flavonoid baicalin inhibits HIV-1 infection at the level of viral entry. *Biochem. Biophys. Res. Commun.*, **276**, 534–538.

49 Wu, X., Akatsu, H., and Okada, H. (1995) Apoptosis of HIV-infected cells following treatment with Sho-Saiko-to and its components. *Jpn. J. Med. Sci. Biol.*, **48**, 79–87.

50 Baylor, N.W., Fu, T., Yan, Y.D., and Ruscetti, F. W. (1992) Inhibition of human T cell leukemia virus by the plant flavonoid baicalin (7-glucuronic acid, 5,6-dihydroxyflavone). *J. Infect. Dis.*, **165**, 433–437.

51 Hwang, J.M., Tseng, T.H., Tsai, Y.Y., Lee, H.J., Chou, F.P., Wang, C.J., and Chu, C.Y. (2005) Protective effects of baicalein on tert-butyl hydroperoxide-induced hepatic toxicity in rat hepatocytes. *J. Biomed. Sci.*, **12**, 389–397.

52 Zhao, Y., Li, H., Gao, Z., Gong, Y., and Xu, H. (2006) Effects of flavonoids extracted from *Scutellaria baicalensis* Georgi on hemin-nitrite-H_2O_2 induced liver injury. *Eur. J. Pharmacol.*, **536**, 192–199.

53 Park, E.J., Zhao, Y.Z., Lian, L., Kim, Y.C., and Sohn, D.H. (2005) Skullcapflavone I from *Scutellaria baicalensis* induces apoptosis in activated rat hepatic stellate cells. *Planta Med.*, **71**, 885–887.

54 Gao, Z., Huang, K., and Xu, H. (2001) Protective effects of flavonoids in the roots of *Scutellaria baicalensis* Georgi against hydrogen peroxide-induced oxidative stress in HS-SY5Y cells. *Pharmacol. Res.*, **43**, 173–178.

55 Lee, H.J., Noh, Y.H., Lee, D.Y., Kim, Y.S., Kim, K.Y., Chung, Y.H., Lee, W.B., and Kim, S. S. (2005) Baicalein attenuates 6-hydroxydopamine-induced neurotoxicity in SH-SY5Y cells. *Eur. J. Cell Biol.*, **84**, 897–905.

56 Im, H.I., Joo, W.S., Nam, E., Lee, E.S., Hwang, Y.J., and Kim, Y.S. (2005) Baicalein prevents 6-hydroxydopamine-induced dopaminergic dysfunction and lipid peroxidation in mice. *J. Pharmacol. Sci.*, **98**, 185–189.

57 Chen, Y.C., Chow, J.M., Lin, C.W., Wu, C.Y., and Shen, S.C. (2006) Baicalein inhibition of oxidative-stress-induced apoptosis via modulation of ERKs activation and induction of HO-1 gene expression in rat glioma cells C6. *Toxicol. Appl. Pharmacol.*, **216**, 263–273.

58 Li, F.Q., Wang, T., Pei, Z., Liu, B., and Hong, J. S. (2005) Inhibition of microglial activation by the herbal flavonoid baicalein attenuates inflammation-mediated degeneration of dopaminergic neurons. *J. Neural Transm.*, **112**, 331–347.

59 Suk, K., Lee, H., Kang, S.S., Cho, G.J., and Choi, W.S. (2003) Flavonoid baicalein attenuates activation-induced cell death of brain microglia. *J. Pharmacol. Exp. Ther.*, **305**, 638–645.

60 Lee, S.W., Song, G.S., Kwon, C.H., and Kim, Y. K. (2005) Beneficial effect of flavonoid baicalein in cisplatin-induced cell death of human glioma cells. *Neurosci. Lett.*, **382**, 71–75.

61 Wu, P.H., Shen, Y.C., Wang, Y.H., Chi, C.W., and Yen, J.C. (2006) Baicalein attenuates methamphetamine-induced loss of dopamine transporter in mouse striatum. *Toxicology*, **226**, 238–245.

62 Liu, C., Wu, J., Gu, J., Xiong, Z., Wang, F., Wang, J., Wang, W., and Chen, J. (2007) Baicalein improves cognitive deficits induced by chronic cerebral hypoperfusion in rats. *Pharmacol. Biochem. Behav.*, **86**, 423–430.

63 van Leyen, K., Kim, H.Y., Lee, S.R., Jin, G., Arai, K., and Lo, E.H. (2006) Baicalein and 12/15-lipoxygenase in the ischemic brain. *Stroke*, **37**, 3014–3018.

64 Lebeau, A., Esclaire, F., Rostene, W., and Pelaprat, D. (2001) Baicalein protects cortical neurons from β-amyloid (25–35) induced toxicity. *NeuroReport*, **12**, 2199–2202.

65 Zhu, J.T., Choi, R.C., Chu, G.K., Cheung, A.W., Gao, Q.T., Li, J., Jiang, Z.Y., Dong, T.T., and Tsim, K.W. (2007) Flavonoids possess neuroprotective effects on cultured pheochromocytoma PC12 cells: a comparison of different flavonoids in activating estrogenic effect and in preventing β-amyloid-induced cell death. *J. Agric. Food Chem.*, **55**, 2438–2445.

66 Heo, H.J., Kim, D.O., Choi, S.J., Shin, D.H., and Lee, C.Y. (2004) Potent Inhibitory effect of flavonoids in *Scutellaria baicalensis* on amyloid-β protein-induced neurotoxicity. *J. Agric. Food Chem.*, **52**, 4128–4132.

67 Wang, S.Y., Wang, H.H., Chi, C.W., Chen, C. F., and Liao, J.F. (2004) Effects of baicalein on β-amyloid peptide-(25–35)-induced amnesia in mice. *Eur. J. Pharmacol.*, **506**, 55–61.

68 Lee, B.H., Lee, S.J., Kang, T.H., Kim, D.H., Sohn, D.H., Ko, G.I., and Kim, Y.C. (2000) Baicalein: an *in vitro* antigenotoxic compound from *Scutellaria baicalensis*. *Planta Med.*, **66**, 70–71.

69 Lee, M.J., Wang, C.J., Tsai, Y.Y., Hwang, J.M., Lin, W.L., Tseng, T.H., and Chu, C.Y. (1999) Inhibitory effect of 12-O-tetradecanoylphorbol 13-acetate-caused tumor promotion in benzo [*a*]pyrene-initiated CD-1 mouse skin by baicalein. *Nutr. Cancer*, **34**, 185–191.

70 Ueng, Y.F., Shyu, C.C., Liu, T.Y., Oda, Y., Lin, Y. L., Liao, J.F., and Chen, C.F. (2001) Protective effects of baicalein and wogonin against benzo [*a*]pyrene- and aflatoxin B_1-induced genotoxicities. *Biochem. Pharmacol.*, **62**, 1653–1660.

71 Konoshima, T., Kokumai, M., Kozuka, M., Iinuma, M., Mizuno, M., Tanaka, T., Tokuda, H., Hishino, H., and Iwashima, A. (1992) Studies on inhibitors of skin tumor promotion. XI. Inhibitory effects of flavonoids from *Scutellaria baicalensis* on Epstein-Barr virus activation and their antitumor promotion activity. *Chem. Pharm. Bull. (Tokyo)*, **40**, 531–533.

72 Friedman, F.K., West, D., Sugimura, T., and Gelboin, H.V. (1985) Flavone modulators of rat hepatic aryl hydrocarbon hydroxylase. *Pharmacology*, **31**, 203–207.

73 Kao, Y.C., Zhou, C., Sherman, M., Laughton, C.A., and Chen, S. (1998) Molecular basis of the inhibition of human aromatase (estrogen synthetase) by flavone and isoflavone phytoestrogens: a site-directed mutagenesis study. *Environ. Health Perspect.*, **106**, 85–92.

74 Park, H.J., Lee, Y.W., Park, H.H., Lee, Y.S., Kwon, I.B., and Yu, J.H. (1999) Induction of quinone reductase by a methanol extract of *Scutellaria baicalensis* and its flavonoids in murine Hepa 1c1c7 cells. *Eur. J. Cancer Prev.*, **7**, 465–471.

75 Chan, H.Y., Chen, Z.Y., Tsang, D.S., and Leung, L.K. (2002) Baicalein inhibits DMBA-DNA adduct formation by modulating CYP1A1 and CYP1B1 activities. *Biomed. Pharmacother.*, **56**, 269–275.

76 Kim, J.Y., Lee, S., Kim, D.H., Kim, B.R., Park, R., and Lee, B.M. (2002) Effects of flavonoids isolated from Scutellariae radix on cytochrome P-450 activities in human liver microsomes. *J. Toxicol. Environ. Health A*, **65**, 373–381.

77 Li, Y.C., Tyan, Y.S., Lee, Y.M., Tsao, T.Y., Chuang, J.Y., Kuo, H.M., Hsia, T.C., Yang, J.H., and Chung, J.G. (2005) N-acetyltransferase is involved in baicalein-induced N-acetylation of 2-aminofluorene and DNA-2-aminofluorene adduct formation in human leukemia HL-60 cells. *In Vivo*, **19**, 399–405.

78 Ryu, S.H., Ahn, B.Z., and Pack, M.Y. (1985) The cytotoxic principle of Scutellariae radix against L 1210 cell. *Planta Med.*, **51**, 462.

79 Motoo, Y., and Sawabu, N. (1994) Antitumor effects of saikosaponins, baicalin and baicalein on human hepatoma cell lines. *Cancer Lett.*, **86**, 91–95.

80 Ikemoto, S., Sugimura, K., Yoshida, N., Yasumoto, R., Wada, S., Yamamoto, K., and Kishimoto, T. (2000) Antitumor effects of scutellariae radix and its components baicalein, baicalin, and wogonin on bladder cancer cell lines. *Urology*, **55**, 951–955.

81 Ciesielska, E., Wolszczak, M., Gulanowski, B., Szulawska, A., Kochman, A., and Metodiewa, D. (2004) In vitro antileukemic, antioxidant and prooxidant activities of Antoksyd S (C/E/XXI): a comparison with baicalin and baicalein. *In Vivo*, **18**, 497–503.

82 Roy, M.K., Nakahara, K., Na, T.V., Trakoontivakorn, G., Takenaka, M., Isobe, S., and Tsushida, T. (2007) Baicalein, a flavonoid extracted from a methanolic extract of Oroxylum indicum inhibits proliferation of a cancer cell line in vitro via induction of apoptosis. *Pharmazie*, **62**, 149–153.

83 Wang, J., Yu, Y., Hashimoto, F., Sakata, Y., Fujii, M., and Hou, D.X. (2004) Baicalein induces apoptosis through ROS-mediated mitochondrial dysfunction pathway in HL-60 cells. *Int. J. Mol. Med.*, **14**, 627–632.

84 Dong, Q.H., Zheng, S., Xu, R.Z., and Lu, Q.H. (2003) Baicalein selectively induce apoptosis in human leukemia K562 cells. *Yao Xue Xue Bao*, **38**, 817–820.

85 Himeji, M., Ohtsuki, T., Fukazawa, H., Tanaka, M., Yazaki, S., Ui, S., Nishio, K., Yamamoto, H., Tasaka, K., and Mimura, A. (2007) Difference of growth-inhibitory effect of *Scutellaria baicalensis* producing flavonoid wogonin among human cancer cells and normal diploid cell. *Cancer Lett.*, **245**, 269–274.

86 Lee, H.Z., Leung, H.W., Lai, M.Y., and Wu, C. H. (2005) Baicalein induced cell cycle arrest and apoptosis in human lung squamous carcinoma CH27 cells. *Anticancer Res.*, **25** (2A), 959–964.

87 Matsuzaki, Y., Kurokawa, N., Terai, S., Matsumura, Y., Kobayashi, N., and Okita, K. (1996) Cell death induced by baicalein in human hepatocellular carcinoma cell lines. *Jpn. J. Cancer Res.*, **87**, 170–177.

88 Austin, C.A., Patel, S., Ono, K., Nakane, H. and Fisher, L.M. (1992) Site-specific DNA cleavage by mammalian DNA topoisomerase II induced by novel flavone and catechin derivatives. *Biochem. J.*, **282**, 883–889.

89 Chen, S., Ruan, Q., Bedner, E., Deptala, A., Wang, X., Hsieh, T.C., Traganos, F., and Darzynkiewicz, Z. (2001) Effects of the flavonoid baicalin and its metabolite baicalein on androgen receptor expression, cell cycle progression and apoptosis of prostate cancer cell lines. *Cell Prolif.*, **34**, 293–304.

90 Bonham, M., Posakony, J., Coleman, I., Montgomery, B., Simon, J., and Nelson, P.S. (2005) Characterization of chemical constituents in *Scutellaria baicalensis* with antiandrogenic and growth-inhibitory activities toward prostate carcinoma. *Clin. Cancer Res.*, **11**, 3905–3914.

91 Miocinovic, R., McCabe, N.P., Keck, R.W., Jankun, J., Hampton, J.A., and Selman, S.H. (2005) *In vivo* and *in vitro* effect of baicalein on human prostate cancer cells. *Int. J. Oncol.*, **26**, 241–246.

92 Guo, Q.L., Ding, Q.L., and Wu, Z.Q. (2004) Effect of baicalein on experimental prostatic hyperplasia in rats and mice. *Biol. Pharm. Bull.*, **27**, 333–337.

93 Huang, H.C., Hsieh, L.M., Chen, H.W., Lin, Y. S., and Chen, J.S. (1994) Effects of baicalein and esculetin on transduction signals and growth factors expression in T-lymphoid leukemia cells. *Eur. J. Pharmacol.*, **268**, 73–78.

94 So, F.V., Guthrie, N., Chambers, A.F., Moussa, M., and Carroll, K.K. (1996) Inhibition of human breast cancer cell proliferation and delay of mammary tumorigenesis by flavonoids and citrus juices. *Nutr. Cancer.*, **26**, 167–181.

95 So, F.V., Guthrie, N., Chambers, A.F., and Carroll, K.K. (1997) Inhibition of proliferation of estrogen receptor-positive MCF-7 human breast cancer cells by flavonoids in the presence and absence of excess estrogen. *Cancer Lett.*, **112**, 127–133.

96 Umezawa, K. (1996) Inhibition of experimental metastasis by enzyme inhibitors from microorganisms and plants. *Adv. Enzyme Regul.*, **36**, 267–281.

97 Liu, J.J., Huang, T.S., Cheng, W.F., and Lu, F.J. (2003) Baicalein and baicalin are potent inhibitors of angiogenesis: Inhibition of endothelial cell proliferation, migration and differentiation. *Int. J. Cancer*, **106**, 559–565.

Scutellaria barbata

Herba Scutellariae Barbatae (Banzhilian) is the dried whole plant of *Scutellaria barbata* D. Don (Lamiaceae). It is used as an antiphlogistic and diuretic agent for the treatment of pharyngitis, laryngitis, edema, jaundice, traumatic pain, and snake bite.

Scutellaria barbata D. Don

Chemistry

The whole plant of *S. barbata* was reported to contain flavones and flavone glycosides, with scutellarin as the major component and a small amount of baicalin [1]. The Chinese Pharmacopoeia requires a quantitative determination of the total flavone by spectrophotometry, and a quantitative determination of scutellarin by HPLC. The content of total flavone in Herba Scutellariae Barbatae should be not less than 1.5%, calculated as scutellarin. The content of scutellarin in Herba Scutellariae Barbatae should be not less than 0.2%.

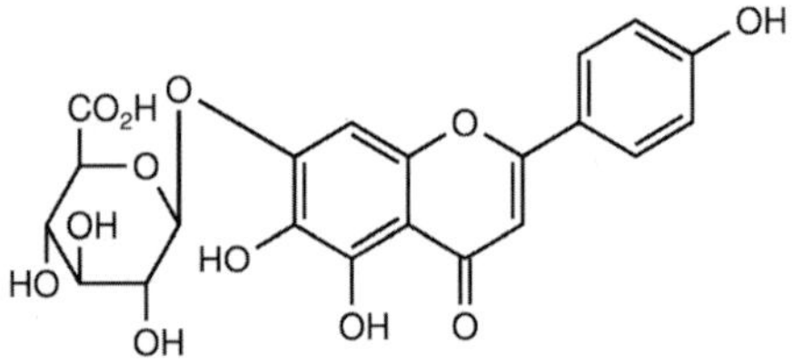

Scutellarin

In addition, carthamidin [2], isocarthamidin, wogonin [3], baicalin, resveratrol, apigenin, luteolin [4] and 1-(4-hydroxyphenyl)-but-1-en-3-one [5] were isolated from *S. barbata*. A polysaccharide SPS_4 from *S. barbata* was found to have an average molecular weight of 10 kDa, and was composed of rhamnose, fucose, arabinose, xylose, glucose, galactose, and mannose [6]. Cytotoxic compounds, including three neo-clerodane diterpenoids, barbatins A–C, a neo-clerodane diterpenoid nicotinyl ester, scutebarbatine B [7], and pheophorbide a [8] were isolated from *S. barbata* herb.

From the essential oil of *S. barbata*, hexahydrofarnesylacetone (11.0%), 3,7,11,15-tetramethyl-2-hexadecen-1-ol (7.8%), menthol (7.7%) and 1-octen-3-ol (7.1%) were identified as the major components [9].

Another item officially listed in the Chinese Pharmacopoeia with scutellarin as major ingredient is Herba Erigerontis. Herba Erigerontis (Dengzhanxixin) is the dried whole plant of *Erigeron breviscapus* (Vant.) Hand.-Mazz. (Asteraceae). It is used as an analgesic for the treatment of cold, toothache, and rheumatic pain.

Handbook of Chinese Medicinal Plants: Chemistry, Pharmacology, Toxicology
Weici Tang and Gerhard Eisenbrand

ISBN: 978-3-527-32226-8

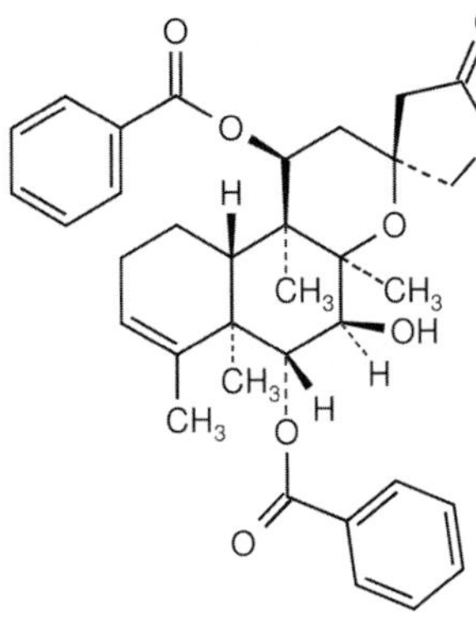

Barbatin A

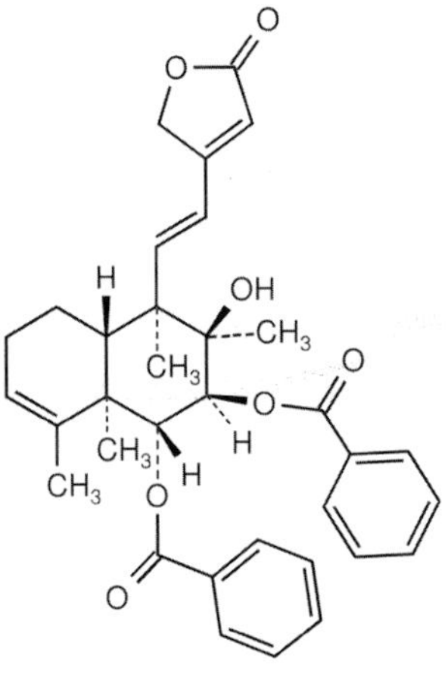

Scutebarbatine B

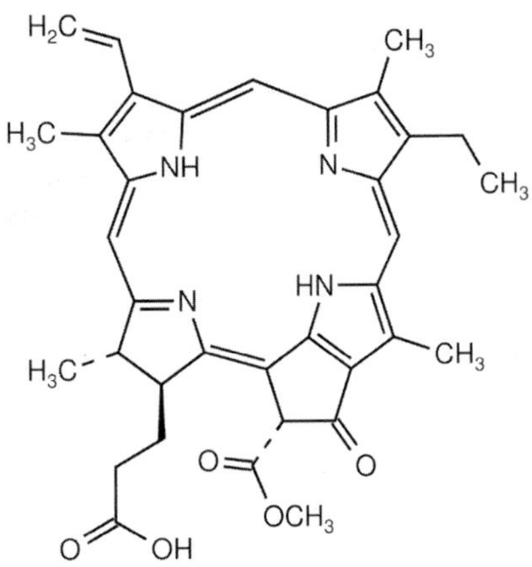

Pheophorbide a

Pharmacology and Toxicology

The whole plant of *S. barbata* has been used in traditional Chinese medicine for the treatment of liver, lung and rectal tumors, mostly in a combination with other herbal medicines. For example, *S. barbata* was reported to be used in combination preparation for the treatment of cervical cancer [10], hepatocarcinoma [11], and esophageal cancer [12]. *Scutellaria barbata* was reported to augment, dose-dependently, the oxidative burst as an indicator of phagocytic function in J774 murine macrophage cell line. An antitumor effect of *S. barbata* was observed in Balb/c mice transplanted subcutaneously with Renca murine renal cell carcinoma cells. Oral treatment with *S. barbata* significantly inhibited the growth of Renca cells in mice [13]. The polysaccharide SPS_4 showed significant *in vitro* growth inhibition against sarcoma S180 and hepatoma ascites cells [6].

The essential oil of *S. barbata* exerted antimicrobial activity against several Gram-positive bacteria, including methicillin-resistant *Staphylococcus aureus* [9]. The flavones apigenin and luteolin were found to be selectively toxic to *S. aureus*, including both methicillin-resistant and methicillin-sensitive strains [4].

A standardized extract from *S. barbata* induced cell death and exerted a significant cell-cycle arrest of LoVo human colon cancer cells in the sub-G_1 phase [3]. The fresh juice of *S. barbata* herb exerted a growth-inhibitory effect and induced apoptosis in a series of cancer cell lines, including HepG2 hepatoblastoma, Hep3B hepatocellular carcinoma, MDA-MB231 mammary carcinoma, A549 lung cancer, and KG-1 acute myelogenous leukemia *in vitro* [14]. The methylene chloride fraction of *S. barbata* inhibited the proliferation of U937 human leukemia cells, with an IC_{50} of about $10\,\mu g\,ml^{-1}$, increased the cell-cycle arrest at sub-G_1 stage, induced apoptosis, and activated caspase-3 and caspase-9. In addition, the methylene chloride fraction also effectively cleaved poly(ADP-ribose) polymerase, increased the ratio of Bax/Bcl-2, and released cytochrome c from mitochondria during apoptosis in U937 cells; this suggested that the fraction might induce apoptosis via the mitochondria-mediated signaling pathway [15]. The ethanol extract of *S. barbata* was reported to inhibit cell growth and to induce cell apoptosis of

A549 human lung cancer cells. Several genes involved in DNA damage, cell cycle control, nucleic acid binding and protein phosphorylation were affected. In particular, CD209, which is related to dendritic cell function, was dramatically downregulated [16].

The antiproliferative activity of the aqueous extract of *S. barbata* has also been tested in myometrial smooth muscle cells and smooth muscle cells of uterine leiomyoma as the most common benign smooth muscle cell tumor of the myometrium. The extract exerted growth inhibition and induced apoptosis in both cell lines, inducing arrest in the G_1 phase. The markers of uterine smooth muscle cell differentiation, such as α-smooth muscle actin, calponin h1 and cyclin-dependent kinase inhibitor p27, were induced by the extract in both myometrial and leiomyomal smooth muscle cells [17]. Although the expression of insulin-like growth factor-I (IGF-I) in myometrial and leiomyomal cells after treatment with the extract was similar, the IGF-I protein was more abundant in leiomyomal cells than in myometrium before treatment. This indicated that IGF-I expression might be associated with a proliferation of leiomyomal cells. Two flavones, apigenin and luteolin, were found to be responsible for the antiproliferative activity of *S. barbata* [18]. The extract was also found to reduce the proliferation of myometrial and leiomyomal cells promoted by human chorionic gonadotropin [19]. It was also reported that β_2-adrenergic receptors were expressed at higher levels in uterine leiomyomal cells than in myometrial smooth muscle cells. *Scutellaria barbata* was suggested to induce the expression of c-fos gene in uterine leiomyomal cells by increasing cyclic adenosine monophosphate (cAMP), which in turn activated the cAMP-dependent protein kinase A (PKA) pathway, resulting in leiomyoma regression [20].

Furthermore, barbatins A–C and scutebarbatine B were found to exhibit significant cytotoxic activities against HONE-1 human nasopharyngeal cancer, KB human oral epidermoid carcinoma, and HT29 human colorectal carcinoma cells, with IC_{50} values in the range of 3.5 to 8.1 μ*M* [7]. Pheophorbide a was found to show an antiproliferative effect against HepG2 and Hep3B human hepatoma cells. Mechanistic studies revealed that pheophorbide a induced apoptosis in Hep3B cells, a viral-induced hepatoma cell line, but was nontoxic in normal WRL-68 human liver cells. At a concentration of 40 μg ml^{-1}, pheophorbide a caused DNA fragmentation, cell-cycle arrest at sub-G_1 stage, suppression of the antiapoptotic protein Bcl-2, release of cytochrome c to the cytosol, and activation of pro-caspase 3 and pro-caspase 9 [8].

The whole plant of *S. barbata* was also found to modulate the mutagenesis of some known mutagens. The aqueous extract of *S. barbata* exerted a significant concentration-dependent inhibition of mutagenicity of benzo[*a*]pyrene (BaP) 7,8-dihydrodiol and BaP 7,8-dihydrodiol-9,10-epoxide in *Salmonella typhimurium* TA100 in the presence of a rat liver S-9 fraction. It also significantly inhibited BaP 7,8-dihydrodiol and BaP 7,8-dihydrodiol-9,10-epoxide binding to calf thymus DNA. It was suggested that *S. barbata* acted as a blocking agent through a scavenging mechanism [21], and may be an inhibitor of the cytochrome P450 isoenzyme CYP1A1 [22]. A similar antimutagenic effect of *S. barbata* was observed at a concentration of 1.5 mg per plate against the mutagenicity of aflatoxin B_1 (AFB_1) in *S. typhimurium* TA 100 in the presence of an activation system. *Scutellaria barbata* significantly inhibited AFB_1 binding to DNA, reducing AFB_1–DNA adduct formation [23]. Inhibition of the cytochrome P450 isoenzyme may also be a mechanism of *S. barbata* against the mutagenicity of AFB_1 [24]. An antimutagenic effect of *S. barbata* was further observed in the unscheduled DNA synthesis (UDS) test in human peripheral lymphocytes against cigarette tar at a concentration of 125 mg ml^{-1} [25].

Scutellarin was found inhibit human immunodeficiency virus (HIV). It inhibited the HIV-1IIIB laboratory-derived virus direct infection in C8166 cells, with an EC_{50} of 26 μ*M*. By comparing the inhibitory effects on p24 antigen, scutellarin was also found to be active against HIV-1(74V) drug-resistant virus (EC_{50} 253 μ*M*) and HIV-1KM018 low-passage clinical isolated virus (EC_{50} 136 μ*M*). At a concentration of 433 μ*M*, scutellarin inhibited 48% of the cell free recombinant HIV-1 RT activity; at a concentration of 54 μ*M*, it inhibited 82% of HIV-1 particle attachment and 45% of cell fusion. It was suggested that scutellarin inhibited the essential events for viral transmission and replication [26].

Scutellarin was also reported to exert neuroprotective effects against neuronal damage caused by cerebral ischemia–reperfusion in rats. Scutellarin was given intragastrically to rats at doses of 50 or 75 mg kg^{-1} for seven days before middle cerebral artery occlusion. Pretreatment with scutellarin significantly reduced the infarct volume, ameliorated the neurological deficit, and reduced the permeability of the blood–-brain barrier. It increased the expression of nitric oxide synthase isoform eNOS, and inhibited the expression of vascular endothelial growth factor (VEGF) and basic fibroblast growth factor (bFGF) and iNOS [27]. Scutellarin was also found to protect rat retinal neurons *in vitro* [28].

The pharmacokinetic results indicated that scutellarin underwent rapid and extensive biotransformation in rats after i.p. and oral administration. After i.p. injection, scutellarin was absorbed rapidly. The profiles of scutellarin and scutellarein conjugates were fitted to a two-compartment open model. Scutellarin and scutellarein conjugates showed a similar time course, without significant difference in the half-lives $t_{1/2\alpha}$ and $t_{1/2\beta}$. After oral administration, fluctuations were observed in the concentration–time profiles of both scutellarin and scutellarein conjugates. The pharmacokinetics could not be explained by a classical compartment model. The bioavailability of oral administration was about 10.7% for scutellarin and about 7.9% for scutellarein conjugates [29]. After oral administration of scutellarin, two major metabolites were isolated from the urine and identified as scutellarein 6,7-di-*O*-β-D-glucuronide and scutellarein [30]. In addition, 6-methyl- scutellarin, 6-methyl-scutellarein, and two scutellarin disulfates, were identified [31].

References

1 Wang, Y.Q., Lu, L., Li, D.D., and Yang, M. (1990) Comparison of baicalin content in eight *Scutellaria* species by RP-HPLC and determination of their blood coagulation and fibrinolytic activity. *Chin. Trad. Herbal Drugs*, **21**, 538–540.

2 Zhao, H.R., Hu, S.Q., Ren, X.H., and Xiang, R.D. (1993) Inhibitory effect of carthamidin on lens aldose reductase. *Chin. J. Pharmacol. Toxicol.*, **7**, 159–160.

3 Goh, D., Lee, Y.H., and Ong, E.S. (2005) Inhibitory effects of a chemically standardized extract from *Scutellaria barbata* in human colon cancer cell lines, LoVo. *J. Agric. Food Chem.*, **53**, 8197–8204.

4 Sato, Y., Suzaki, S., Nishikawa, T., Kihara, M., Shibata, H., and Higuti, T. (2000) Phytochemical flavones isolated from *Scutellaria barbata* and antibacterial activity against methicillin-resistant *Staphylococcus aureus*. *J. Ethnopharmacol.*, **72**, 483–488.

5 Ducki, S., Hadfield, J.A., Lawrence, N.J., Liu, C.Y., McGown, A.T., and Zhang, X. (1996) Isolation of E-1-(4'-hydroxyphenyl)-but-1-en-3-one from *Scutellaria barbata*. *Planta Med.*, **62**, 185–186.

6 Meng, Y.F., Li, Z.X., Zhang, L., and Chen, Y.Z. (1993) Isolation, purification and analysis of polysaccharide SPS_4 from *Scutellaria barbata*. *Chin. Biochem. J.*, **9**, 224–228.

7 Dai, S.J., Tao, J.Y., Liu, K., Jiang, Y.T., and Shen, L. (2006) neo-Clerodane diterpenoids from *Scutellaria barbata* with cytotoxic activities. *Phytochemistry*, **67**, 1326–1330.

8 Chan, J.Y., Tang, P.M., Hon, P.M., Au, S.W., Tsui, S.K., Waye, M.M., Kong, S.K., Mak, T.C., and Fung, K.P. (2006) Pheophorbide a, a major antitumor component purified from *Scutellaria barbata*, induces apoptosis in human hepatocellular carcinoma cells. *Planta Med.*, **72**, 28–33.

9 Yu, J., Lei, J., Yu, H., Cai, X., and Zou, G. (2004) Chemical composition and antimicrobial activity of the essential oil of *Scutellaria barbata*. *Phytochemistry*, **65**, 881–884.

10 Tian, J.F. (1989) Ten year follow up of 30 cases of stages II and III cervix cancer treated by Chinese medicine. *J. Trad. Chin. Med.*, **30**, 542–543.

11 Zheng, R.Q. (1987) Treatment of 47 cases of hepatocarcinoma with "Lianhua Tablet". *Guangzhou Med. J.*, **18**, 48–50.

12 Ma, J.F. (1985) 178 cases of esophagus-cardia cancer treated with compound "Bajiao Jinpan Decoction". *Liaoning J. Trad. Chin. Med.*, **9**, 23.

13 Wong, B.Y., Lau, B.H., Jia, T.Y., and Wan, C.P. (1996) *Oldenlandia diffusa* and *Scutellaria barbata* augment macrophage oxidative burst and inhibit tumor growth. *Cancer Biother. Radiopharm.*, **11**, 51–56.

14 Chui, C.H., Lau, F.Y., Tang, J.C., Kan, K.L., Cheng, G.Y., Wong, R.S., Kok, S.H., Lai, P.B., Ho, R., Gambari, R., and Chan, A.S. (2005) Activities of fresh juice of *Scutellaria barbata* and warmed water extract of Radix Sophorae tonkinensis on anti-proliferation and apoptosis of human cancer cell lines. *Int. J. Mol. Med.*, **16**, 337–341.

15 Cha, Y.Y., Lee, E.O., Lee, H.J., Park, Y.D., Ko, S.G., Kim, D.H., Kim, H.M., Kang, I.C., and Kim, S.H. (2004) Methylene chloride fraction of *Scutellaria barbata* induces apoptosis in human U937 leukemia cells via the mitochondrial signaling pathway. *Clin. Chim. Acta*, **348**, 41–48.

16 Yin, X., Zhou, J., Jie, C., Xing, D., and Zhang, Y. (2004) Anticancer activity and mechanism of *Scutellaria barbata* extract on human lung cancer cell line A549. *Life Sci.*, **75**, 2233–2244.

17 Lee, T.K., Lee, D.K., Kim, D.I., Lee, Y.C., Chang, Y.C., and Kim, C.H. (2004) Inhibitory effects of *Scutellaria barbata* D. Don on human uterine leiomyomal smooth muscle cell proliferation through cell cycle analysis. *Int. Immunopharmacol.*, **4**, 447–454.

18 Kim, D.I., Lee, T.K., Lim, I.S., Kim, H., Lee, Y.C., and Kim, C.H. (2005) Regulation of IGF-I production and proliferation of human leiomyomal smooth muscle cells by Scutellaria barbata D. Don in vitro: isolation of flavonoids of apigenin and luteolin as acting compounds. *Toxicol. Appl. Pharmacol.*, **205**, 213–224.

19 Lee, T.K., Kim, D.I., Song, Y.L., Lee, Y.C., Kim, H.M., and Kim, C.H. (2004) Differential inhibition of *Scutellaria barbata* D. Don (Lamiaceae) on HCG-promoted proliferation of cultured uterine leiomyomal and myometrial smooth muscle cells. *Immunopharmacol. Immunotoxicol.*, **26**, 329–342.

20 Lee, T.K., Cho, H.L., Kim, D.I., Lee, Y.C., and Kim, C.H. (2004) *Scutellaria barbata* D. Don induces c-fos gene expression in human uterine leiomyomal cells by activating β_2-adrenergic receptors. *Int. J. Gynecol. Cancer*, **14**, 526–531.

21 Wong, B.Y., Lau, B.H., and Teel, R.W. (1992) Chinese medicinal herbs modulate mutagenesis, DNA binding and metabolism of benzo[*a*]pyrene 7,8-dihydrodiol and benzo[*a*]pyrene 7,8-dihydrodiol-9,10-epoxide. *Cancer Lett.*, **62**, 123–131.

22 Wong, B.Y., Lau, B.H., Yamasaki, T., and Teel, R.W. (1993) Modulation of cytochrome P-450IA1-mediated mutagenicity, DNA binding and metabolism of benzo[a]pyrene by Chinese medicinal herbs. *Cancer Lett.*, **68**, 75–82.

23 Wong, B.Y., Lau, B.H., Tadi, P.P., and Teel, R. W. (1992) Chinese medicinal herbs modulate mutagenesis, DNA binding and metabolism of aflatoxin B_1. *Mutat. Res.*, **279**, 209–216.

24 Wong, B.Y., Lau, B.H., Yamasaki, T., and Teel, R.W. (1993) Inhibition of dexamethasone-induced cytochrome P450-mediated mutagenicity and metabolism of aflatoxin B_1 by Chinese medicinal herbs. *Eur. J. Cancer Prev.*, **2**, 351–356.

25 Han, F., Hu, J., and Xu, H. (1997) Effects of some Chinese herbal medicine and green tea antagonizing mutagenesis caused by cigarette tar. *Chin. J. Prev. Med.*, **31**, 71–74.

26 Zhang, G.H., Wang, Q., Chen, J.J., Zhang, X.M., Tam, S.C., and Zheng, Y.T. (2005) The anti-HIV-1 effect of scutellarin. *Biochem. Biophys. Res. Commun.*, **334**, 812–816.

27 Hu, X.M., Zhou, M.M., Hu, X.M., and Zeng, F.D. (2005) Neuroprotective effects of scutellarin on rat neuronal damage induced by cerebral ischemia/reperfusion. *Acta Pharmacol. Sin.*, **26**, 1454–1459.

28 Zhang, Y., Sheng, Y.M., Meng, X.L., and Long, Y. (2005) Effect of caffeic acid, scopoletin and scutellarin on rat retinal neurons in vitro. *Zhongguo Zhong Yao Za Zhi*, **30**, 907–909.

29 Huang, J.M., Weng, W.Y., Huang, X.B., Ji, Y.H., and Chen, E. (2005) Pharmacokinetics of scutellarin and its aglycone conjugated metabolites in rats. *Eur. J. Drug Metab. Pharmacokinet.*, **30**, 165–170.

30 Qiu, F., Xia, H., Zhang, T., Di, X., Qu, G., and Yao, X. (2007) Two major urinary metabolites of scutellarin in rats. *Planta Med.*, **73**, 363–365.

31 Zhang, J.L., Che, Q.M., Li, S.Z., and Zhou, T. H. (2003) Study on metabolism of scutellarin in rats by HPLC-MS and HPLC-NMR. *J. Asian Nat. Prod. Res.*, **5**, 249–256.

Sedum sarmentosum

Herba Sedi (Chuipencao) is the fresh or dried whole plant of *Sedum sarmentosum* Bunge (Crassulaceae). It is used as a diuretic agent, and also for the treatment of jaundice and acute and chronic hepatitis.

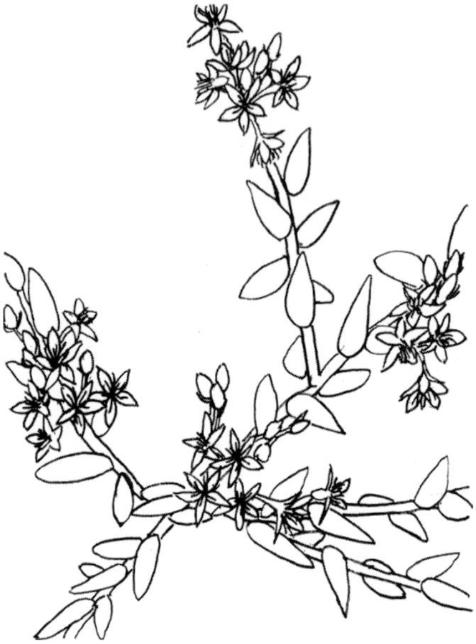

Sedum sarmentosum Bunge

Chemistry

The whole plant of *S. sarmentosum* was reported to contain sterols identified as sarmentosterol, 3β,6β-stigma-4-en-diol, β-sitosterol and daucosterol [1, 2]. In addition, alkaloids were reported to be present in the plant [3]. A series of megastigmanes and megastigmane glycosides, including sedumoside A_4, sedumoside A_5, sedumoside A_6, sedumoside H, sedumoside I, myrsinionoside A and myrsinionoside D, were isolated from *Sedum sarmentosum* [4].

The cyano-ethylene glucoside sarmentosin was isloated from whole herbs of Sedum sarmentosum [5].

Pharmacology and Toxicology

The whole plant of *S. sarmentosum* has been traditionally used for the treatment of chronic viral hepatitis in China. By incubation of murine hepatoma (BNL CL. 2) and human hepatoma (HepG2) cell lines with the crude alkaloid fraction of *S. sarmentosum* at concentrations of 50–150 μg ml^{-1} for 24 or 48 h, a concentration-dependent inhi-

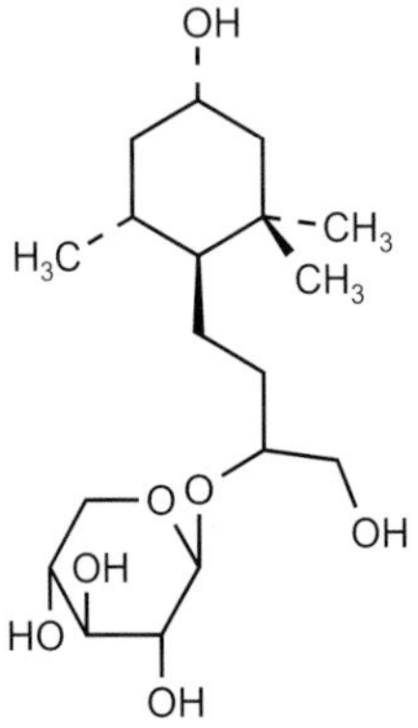

Sedumoside A_4

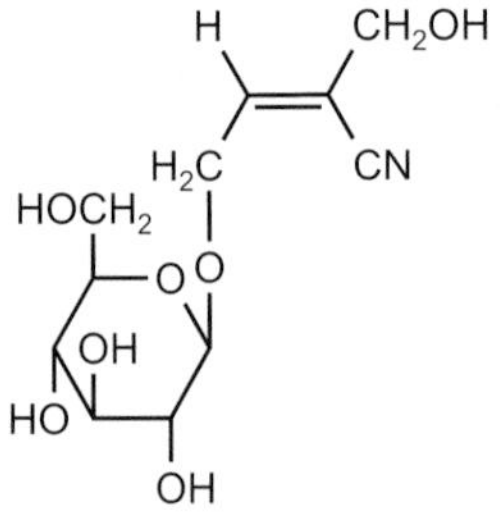

Sarmentosin

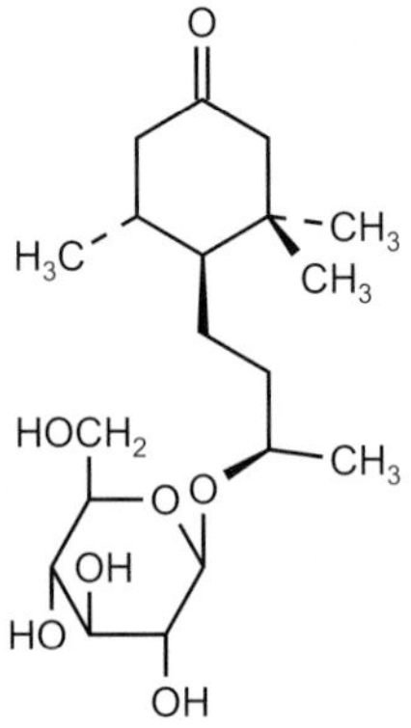

Myrsinionoside A

Handbook of Chinese Medicinal Plants: Chemistry, Pharmacology, Toxicology
Weici Tang and Gerhard Eisenbrand

ISBN: 978-3-527-32226-8

bition of cell proliferation without necrosis or apoptosis was observed. The antiproliferative effects of the crude alkaloid fraction were associated with an increase in the number of cells in the G_1 phase of the cell cycle [3]. Among the megastigmane derivatives, sedumoside F, megastigmane-3,9-diol, myrsinionoside A, and myrsinionoside D were found to show strong cytoprotective activity in primary cultured mouse hepatocytes against the cytotoxicity induced by D-galactosamine [4].

S. sarmentosum has been studied for estrogenic effects in ovariectomized rats. From day 2 until day 37 after ovariectomy, female Sprague-Dawley rats were given orally either a diethyl ether fraction or an ethyl acetate fraction of *S. sarmentosum*, at a daily dose of 10 mg kg^{-1}. The treatment of ovariectomized rats with the diethyl ether and ethyl acetate fractions increased 17β-estradiol-related transcriptional activity, indicating that *S. sarmentosum* exerted an estrogenic activity. The fractions also decreased the serum triglyceride levels and prevented the decrease of collagen in bone and cartilage tissues in ovariectomized rats [6].

References

1 Lu, X.L., Cao, X.L., Zhang, S.L., Hu, Y.C., Bao, X.S., and Wang, Y.X. (1984) Analytical studies of *Sedum sarmentosum* Bunge and its preparations. *Acta Pharm. Sin.*, **19**, 914–920.

2 He, A.M., Hao, H.Y., Wang, M.S., and Zhang, D.C. (1997) Sterols from *Sedum sarmentosum*. *J. China Pharm. Univ.*, **28**, 271–274.

3 Kang, T.H., Pae, H.O., Yoo, J.C., Kim, N.Y., Kim, Y.C., Ko, G.I., and Chung, H.T. (2000) Antiproliferative effects of alkaloids from *Sedum sarmentosum* on murine and human hepatoma cell lines. *J. Ethnopharmacol.*, **70**, 177–182.

4 Ninomiya, K., Morikawa, T., Zhang, Y., Nakamura, S., Matsuda, H., Muraoka, O., and Yoshikawa, M. (2007) Bioactive constituents from Chinese natural medicines. XXIII. Absolute structures of new megastigmane glycosides, sedumosides A(4), A(5), A(6), H, and I, and hepatoprotective megastigmanes from *Sedum sarmentosum*. *Chem. Pharm. Bull. (Tokyo)*, **55**, 1185–1191.

5 Xu, R.-S., Zhu, Q.-Z., and Xie, Y.-Y. (1985) Recent advances in studies on Chinese medicinal herbs with physiological activity. *J. Ethnopharm.*, **14**, 223–253

6 Kim, W.H., Park, Y.J., Park, M.R., Ha, T.Y., Lee, S.H., Bae, S.J., and Kim, M. (2004) Estrogenic effects of *Sedum sarmentosum* Bunge in ovariectomized rats. *J. Nutr. Sci. Vitaminol. (Tokyo)*, **50**, 100–105.

Selaginella pulvinata and *Selaginella tamariscina*

Herba Selaginellae (Juanbai) is the dried whole herb of *Selaginella pulvinata* (Hook. et Grev.) Maxim. or *Selaginella tamariscina* (Beauv.) Spring (Selaginellaceae). It is used for the treatment of menstrual disorders, amenorrhea, dysmenorrhea and traumatic diseases. The Chinese Pharmacopoeia notes Herba Selaginellae to be used with caution in pregnancy.

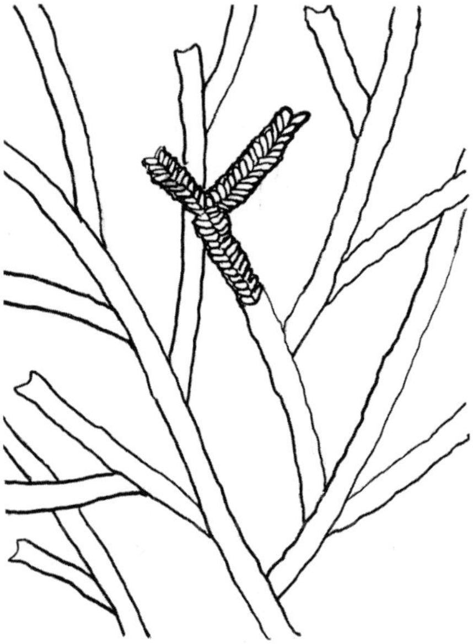

Selaginella tamariscina (Beauv.) Spring.

Chemistry

The whole plant of the official *Selaginella* species was reported to contain amentoflavone and related biflavones, isocryptomerin, cryptomerin B, sumaflavone and robustaflavone [1]. Amentoflavone is also present in the leaf of *Ginkgo biloba* L. (Ginkgoaceae) as one of the biflavones.

Recently, some steroid derivatives were isolated from Herba Selaginellae and identified as 3β,16α-dihydroxy-5α,17β-cholestan-21-carboxylic acid, 3β-acetoxy-16α-hydroxy-5α,17β-cholestan-21-carboxylic acid and 3β-(3-hydroxybutyroxy)-16α-hydroxy-5α,17β-cholestan-21-carboxylic acid [2]

Pharmacology and Toxicology

Amentoflavone is known to possess various biological activities including antitumor, neuroprotective and anti-inflammatory activities. Amentoflavone from *S. tamariscina* inhibited human pathogenic fungi without hemolysis of human erythrocytes. The antifungal mechanism of amentoflavone on human pathogenic yeast *Candida albicans* was associated with a cell-cycle arrest at S-phase [3]. Moreover, amentoflavone particularly induced the accumulation of intracellular trehalose in *C. albicans* as a stress response to the compound [4]. Amentoflavone also showed potent antiviral activity against respiratory syncytial virus, with an IC_{50} of 5.5 μg ml^{-1} [5].

Amentoflavone inhibited the production of nitric oxide (NO) in RAW 264.7 murine macrophages induced by lipopolysaccharide (LPS) via nuclear factor-κB (NF-κB) inactivation. Sumaflavone also inhibited NO production in macrophages induced by LPS and blocked the LPS-induced expression of inducible NO synthase (iNOS). In contrast, robustaflavone only marginally affected iNOS gene expression and NO production [1]. Amentoflavone downregulated cyclooxygenase-2 (COX-2) expression in A549 human lung adenocarcinoma cells activated by tumor necrosis factor-α (TNF-α), with concomitant inhibition of the NF-κB mediated signaling cascades. It inhibited NF-κB DNA-binding activity in TNF-α-activated A549 cells potently, along with an inhibition of the degradation of IκBα, an inhibitor of NF-κB, and NF-κB

Handbook of Chinese Medicinal Plants: Chemistry, Pharmacology, Toxicology
Weici Tang and Gerhard Eisenbrand

ISBN: 978-3-527-32226-8

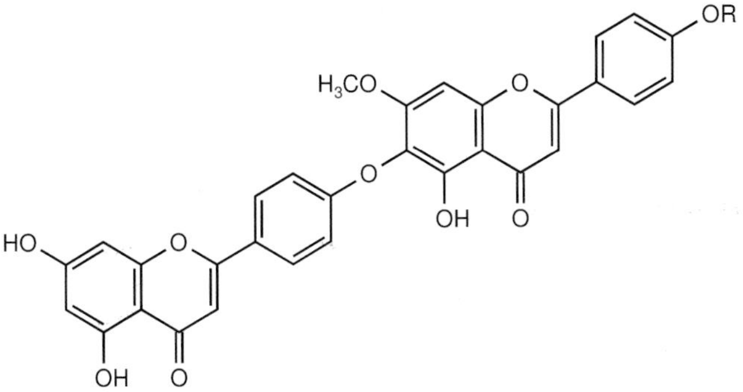

Isocryptomerin: R=H
Cryptomerin B: R=CH_3

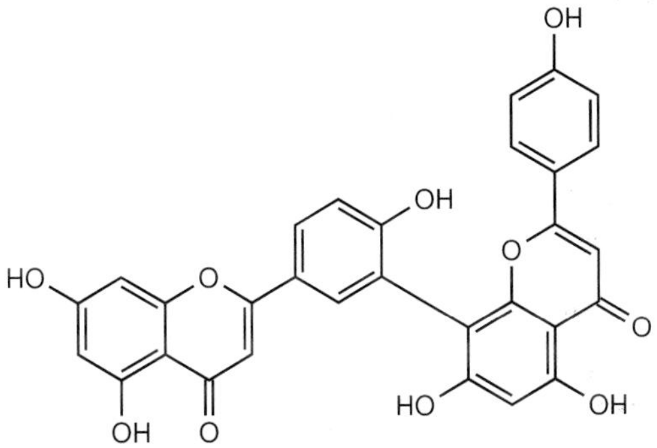

Amentoflavone

translocation into nucleus [6, 7]. Amentoflavone also strongly inhibited prostaglandin E_2 (PGE_2) biosynthesis in A549 cells [8].

The ethyl acetate extract of *S. tamariscina* was found to exhibit a marked vasorelaxant activity, with amentoflavone as the active principle. Amentoflavone induced a concentration-dependent relaxation of rat aorta rings precontracted with phenylephrine; this disappeared after removal of the functional endothelium, indicating the relaxation to be endothelium-dependent. Amentoflavone also inhibited cAMP phosphodiesterase activity, resulting in vasorelaxation. It exhibited a relaxation of rat aorta rings via an enhanced generation of NO, leading to elevated cGMP levels. It also inhibited cGMP-specific phosphodiesterase-5, which suggested that amentoflavone relaxed vascular smooth muscle via endothelium-dependent cGMP signaling, with the possible involvement of nonspecific K^+ and Ca^{2+} channels [9, 10]. The topical application of amentoflavone to the mouse earprovoked a dose-dependent reduction of ear edema induced by croton oil (ID_{50} $0.16\,\mu M\,cm^{-2}$), comparable to that of indomethacin (ID_{50} $0.26\,\mu M\,cm^{-2}$), indicating anti-inflammatory activity [11].

The systemic administration of amentoflavone to post-natal day 7 rats at a dose of $30\,mg\,kg^{-1}$ markedly reduced the hypoxic–ischemic brain damages induced by unilateral carotid ligation and hypoxia, including neuronal cell death caused by both apoptosis and necrosis. Amentoflavone blocked the activation of caspase-3, which is characteristic of apoptosis, and the proteolytic cleavage of its substrates following hypoxic–ischemic injury. The treatment of mouse microglial cells with amentoflavone resulted in a significant decrease in NO production as induced by LPS, and in an induction of iNOS and COX-2 [12]. Amentoflavone also exhibited strong neuroprotective effects against cytotoxicity in neuronal cells, as induced by oxidative stress or amyloid-β protein [13].

Radioligand binding studies revealed that amentoflavone recognition paralleled that of the classical benzodiazepine derivative, diazepam [14]. Amentoflavone was fitted into a pharmacophore model for ligands binding to the $GABA_A$-receptor benzodiazepine site with an IC_{50} of approximately 15 n*M* [15]. In an *in vitro* study using primary cell cultures of porcine brain capillary endothelial cells, the concentration-dependent uptake of amentoflavone over a concentration range of 37 to 2000 n*M* was neither saturable nor temperature-sensitive. The results revealed that amentoflavone was capable of passing the blood–brain barrier *in vitro*, and that passive diffusion might be considered as the major mechanism for the uptake. Experiments with the P-glycoprotein overexpressing cell line P388-MDR cells showed that amentoflavone uptake was significantly enhanced by addition of the P-glycoprotein inhibitor verapamil; this suggested a P-glycoprotein mediated back-transport out of the cells [16]. Amentoflavone, at a dose of 0.1 mg kg^{-1}, was found to markedly decrease stress-induced hyperthermia in mice [17].

The intraperitoneal (i.p.) injection of an extract of *S. tamariscina* to diabetic rats, as induced by alloxan, resulted in a decrease in blood sugar and serum lipid peroxide levels, as well as in an increase in serum insulin concentrations. Histological observations have shown that the injection could repair the structure of pancreatic inlet B cells injured by alloxan [18]. Amentoflavone also inhibited protein tyrosine phosphatase 1B as a target for the treatment of type 2 diabetes and obesity. The IC_{50} value of amentoflavone for inhibiting protein tyrosine phosphatase 1B *in vitro* was given as about 7 μ*M*. A kinetic study revealed amentoflavone to be a noncompetitive inhibitor of the enzyme. The treatment of 32D cells overexpressing the insulin receptor with amentoflavone resulted in a concentration-dependent increase in tyrosine phosphorylation of the insulin receptor; this suggested that amentoflavone might enhance insulin-induced intracellular signaling, possibly through an inhibition of protein tyrosine phosphatase 1B activity [19].

Extracts of *S. tamariscina* with different solvents showed significant growth-inhibitory activities against human leukemia cells, but not against normal human lymphocytes, *in vitro*. The aqueous extract efficiently increased p53 gene expression and induced G_1 arrest in leukemic cells. The extract of *S. tamariscina* induced apoptotic cell death in HL-60 leukemia cells, accompanied by activation of caspase-3 and proteolytic cleavage of poly(ADP-ribose)polymerase. In addition, treatment of HL-60 cells with the extract resulted in an increase of proapoptotic Bax, while Bcl-2 expression was decreased [20]. *Selaginella tamariscina* was further reported to exert antimetastatic activity on lung cancer cells, *in vitro* and *in vivo*. The extract of *S. tamariscina* significantly decreased the expression of matrix metalloproteinase-2 (MMP-2), MMP-9 and urokinase plasminogen activator (u-PA) in a concentration-dependent manner in the highly metastatic A549 lung cancer cells, and also in Lewis lung carcinoma cells. The antimetastatic effect of the extract was also demonstrated by *in vivo* experiments [21].

The administration of 150 mg kg^{-1} *N*-methyl-*N*′-nitro-*N*-nitrosoguanidine (MNNG) to rats by gavage, and subsequent feeding a diet containing 1% *S. tamariscina*, resulted in a significant reduction in the proliferating cell nuclear antigen (PCNA) labeling index of the glandular stomach epithelium, as compared to rats treated with MNNG alone; this indicated that the extract might exert a chemopreventive effect [22]. The extract of *S. tamariscina* was also reported to inhibit human mesangial cell proliferation activated by interleukin-1β (IL-1β) and IL-6, with a median inhibitory concentration of 56 μg ml^{-1} [23]. Amentoflavone was also found to exhibit mutagenicity in *Salmonella typhimurium* TA98, with or without an S-9 fraction as activating system [24]. Amentoflavone was further found to degrade calf thymus DNA in the presence of Cu^{2+}, reducing Cu^{2+} to Cu^{+} and generating hydroxyl radicals. In the presence of Cu^{2+} ions, the absorption spectrum of amentoflavone underwent a shift and a quenching effect, indicating that amentoflavone was capable of binding copper ions [25].

Amentoflavone was reported to be an inhibitor of a series of enzymes. It was revealed to inhibit CYP2C9 with an IC_{50} of 19 ng ml^{-1} (35 n*M*) [26, 27]. It was also found to strongly inhibit human cathepsin B, with an IC_{50} of 1.75 μ*M* [28]. Amentoflavone was further found to be an inhibitor of phospholipase Cγ1, and to inhibit phosphoinositide turnover in phospholipase Cγ1-overexpressing NIH-3T3 fibroblasts. The IC_{50} of phospholipase Cγ1 inhibition was reported as 29 μ*M*, while the IC_{50} for inhibiting the formation of total inositol phosphates in platelet-derived growth factor (PDGF)-stimulated NIH 3T3 cells was 9.2 μ*M*. Amentoflavone did not show any inhibitory activity against protein kinase C (PKC) [29].

References

1 Yang, J.W., Pokharel, Y.R., Kim, M.R., Woo, E.R., Choi, H.K., and Kang, K.W. (2006) Inhibition of inducible nitric oxide synthase by sumaflavone isolated from *Selaginella tamariscina*. *J. Ethnopharmacol.*, **105**, 107–113.

2 Gao, L.L., Yin, S.L., Li, Z.L., Sha, Y., Pei, Y.H., Shi, G., Jing, Y.K., and Hua, H.M. (2007) Three novel sterols isolated from *Selaginella tamariscina* with antiproliferative activity in leukemia cells. *Planta Med.*, **73**, 1112–1115.

3 Jung, H.J., Park, K., Lee, I.S., Kim, H.S., Yeo, S.H., Woo, E.R., and Lee, D.G. (2007) S-phase accumulation of *Candida albicans* by anticandidal effect of amentoflavone isolated from *Selaginella tamariscina*. *Biol. Pharm. Bull.*, **30**, 1969–1971.

4 Jung, H.J., Sung, W.S., Yeo, S.H., Kim, H.S., Lee, I.S., Woo, E.R., and Lee, D.G. (2006) Antifungal effect of amentoflavone derived from *Selaginella tamariscina*. *Arch. Pharm. Res.*, **29**, 746–751.

5 Ma, S.C., But, P.P., Ooi, V.E., He, Y.H., Lee, S.H., Lee, S.F., and Lin, R.C. (2001) Antiviral amentoflavone from *Selaginella sinensis*. *Biol. Pharm. Bull.*, **24**, 311–312.

6 Banerjee, T., Valacchi, G., Ziboh, V.A., and van der Vliet, A. (2002) Inhibition of TNFα-induced cyclooxygenase-2 expression by amentoflavone through suppression of NF-κB activation in A549 cells. *Mol. Cell. Biochem.*, **238**, 105–110.

7 Woo, E.R., Lee, J.Y., Cho, I.J., Kim, S.G., and Kang, K.W. (2005) Amentoflavone inhibits the induction of nitric oxide synthase by inhibiting NF-κB activation in macrophages. *Pharmacol. Res.*, **51**, 539–546.

8 Banerjee, T., van der Vliet, A., and Ziboh, V.A. (2002) Downregulation of COX-2 and iNOS by amentoflavone and quercetin in A549 human lung adenocarcinoma cell line. *Prostaglandins Leukot. Essent. Fatty Acids*, **66**, 485–492.

9 Kang, D.G., Yin, M.H., Oh, H., Lee, D.H., and Lee, H.S. (2004) Vasorelaxation by amentoflavone isolated from *Selaginella tamariscina*. *Planta Med.*, **70**, 718–722.

10 Dell'Agli, M., Galli, G.V., and Bosisio, E. (2006) Inhibition of cGMP-phosphodiesterase-5 by biflavones of *Ginkgo biloba*. *Planta Med.*, **72**, 468–470.

11 Sosa, S., Pace, R., Bornancin, A., Morazzoni, P., Riva, A., Tubaro, A., and Della Loggia, R. (2007) Topical anti-inflammatory activity of extracts and

compounds from *Hypericum perforatum* L. *J. Pharm. Pharmacol.*, **59**, 703–709.

12 Shin, D.H., Bae, Y.C., Kim-Han, J.S., Lee, J.H., Choi, I.Y., Son, K.H., Kang, S.S., Kim, W.K., and Han, B.H. (2006) Polyphenol amentoflavone affords neuroprotection against neonatal hypoxic-ischemic brain damage via multiple mechanisms. *J. Neurochem.*, **96**, 561–572.

13 Kang, S.S., Lee, J.Y., Choi, Y.K., Song, S.S., Kim, J.S., Jeon, S.J., Han, Y.N., Son, K.H., and Han, B.H. (2005) Neuroprotective effects of naturally occurring biflavonoids. *Bioorg. Med. Chem. Lett.*, **15**, 3588–3591.

14 Hansen, R.S., Paulsen, I., and Davies, M. (2005) Determinants of amentoflavone interaction at the $GABA_A$ receptor. *Eur. J. Pharmacol.*, **519**, 199–207.

15 Svenningsen, A.B., Madsen, K.D., Liljefors, T., Stafford, G.I., van Staden, J., and Jager, A.K. (2006) Biflavones from *Rhus* species with affinity for the $GABA_A$/benzodiazepine receptor. *J. Ethnopharmacol.*, **103**, 276–280.

16 Gutmann, H., Bruggisser, R., Schaffner, W., Bogman, K., Botomino, A., and Drewe, J. (2002) Transport of amentoflavone across the blood-brain barrier *in vitro*. *Planta Med.*, **68**, 804–807.

17 Grundmann, O., Kelber, O., and Butterweck, V. (2005) Effects of St. John's wort extract and single constituents on stress-induced hyperthermia in mice. *Planta Med.*, **72**, 1366–1371.

18 Miao, N., Tao, H., Tong, C., Xuan, H., and Zhamg, G. (1996) The *Selaginella tamariscina* (Beauv.) Spring complex in the treatment of experimental diabetes and its effect on blood rheology. *China J. Chin. Mater. Med.*, **21**, 493–495. 512.

19 Na, M., Kim, K.A., Oh, H., Kim, B.Y., Oh, W.K., and Ahn, J.S. (2007) Protein tyrosine phosphatase 1B inhibitory activity of amentoflavone and its cellular effect on tyrosine phosphorylation of insulin receptors. *Biol. Pharm. Bull.*, **30**, 379–381.

20 Ahn, S.H., Mun, Y.J., Lee, S.W., Kwak, S., Choi, M.K., Baik, S.K., Kim, Y.M., and Woo, W.H. (2006) *Selaginella tamariscina* induces apoptosis via a caspase-3-mediated mechanism in human promyelocytic leukemia cells. *J. Med. Food*, **9**, 138–144.

21 Yang, S.F., Chu, S.C., Liu, S.J., Chen, Y.C., Chang, Y.Z., and Hsieh, Y.S. (2007) Antimetastatic activities of *Selaginella tamariscina* (Beauv.) on lung cancer cells *in vitro* and *in vivo*. *J. Ethnopharmacol.*, **110**, 483–489.

22 Lee, I.S., Nishikawa, A., Furukawa, F., Kasahara, K., and Kim, S.U. (1999) Effects of *Selaginella tamariscina* on *in vitro* tumor cell growth, p53 expression, G1 arrest and *in vivo* gastric cell proliferation. *Cancer Lett.*, **144**, 93–99.

23 Kuo, Y.C., Sun, C.M., Tsai, W.J., Ou, J.C., Chen, W.P., and Lin, C.Y. (1998) Chinese herbs as modulators of human mesangial cell proliferation: preliminary studies. *J. Lab. Clin. Med.*, **132**, 76–85.

24 Cardoso, C.R., de Syllos Colus, I.M., Bernardi, C.C., Sannomiya, M., Vilegas, W., and Varanda, E.A. (2006) Mutagenic activity promoted by amentoflavone and methanolic extract of *Byrsonima crassa* Niedenzu. *Toxicology*, **225**, 55–63.

25 Uddin, Q., Malik, A., Azam, S., Hadi, N., Azmi, A.S., Parveen, N., Khan, N.U., and Hadi, S.M. (2004) The biflavonoid, amentoflavone degrades DNA in the presence of copper ions. *Toxicol. In Vitro*, **18**, 435–440.

26 von Moltke, L.L., Weemhoff, J.L., Bedir, E., Khan, I.A., Harmatz, J.S., Goldman, P., and Greenblatt, D.J. (2004) Inhibition of human cytochromes P450 by components of *Ginkgo biloba*. *J. Pharm. Pharmacol.*, **56**, 1039–1044.

27 Greenblatt, D.J., von Moltke, L.L., Luo, Y., Perloff, E.S., Horan, K.A., Bruce, A., Reynolds, R.C., Harmatz, J.S., Avula, B., Khan, I.A., and Goldman, P. (2006) *Ginkgo biloba* does not alter clearance of flurbiprofen, a cytochrome P450-2C9 substrate. *J. Clin. Pharmacol.*, **46**, 214–221.

28 Pan, X., Tan, N., Zeng, G., Zhang, Y., and Jia, R. (2005) Amentoflavone and its derivatives as novel natural inhibitors of human cathepsin B. *Bioorg. Med. Chem.*, **13**, 5819–5825.

29 Lee, H.S., Oh, W.K., Kim, B.Y., Ahn, S.C., Kang, D.O., Shin, D.I., Kim, J., Mheen, T.I., and Ahn, J.S. (1996) Inhibition of phospholipase Cγ1 activity by amentoflavone isolated from *Selaginella tamariscina*. *Planta Med.*, **62**, 293–296.

Senecio scandens

The dried, above-ground part of *Senecio scandens* Buch-Ham. (Asteraceae) is listed in the appendix of the Chinese Pharmacopoeia. It is the major component of a Chinese patent medicine "Qianbai Biyan Pian (Senecio-Selaginella Antirhinitis Dragees)." "Qianbai Biyan Pian" is composed of seven medicinal materials, including the above-ground part of *Senecio scandens* (2424 g), Herba Selaginellae (404 g), Rhizoma et Radix Notopterygii (16 g), Semen Cassiae (242 g), Herba Ephedrae (81 g), Rhizoma Chuanxiong (8 g), and Radix Angelicae Dahuricae (8 g). The total weight of the herbal medicines for 1000 dragees is 3183 g; therefore, the above-ground part of *Senecio scandens* amounts to more than 76% of the total weight. One dragee corresponds about 2.4 g of the above-ground part of *S. scandens*.

Senecio scandens Buch-Ham.

Chemistry

The above-ground part *S. scandens* was reported to contain the pyrrolizidine alkaloids senecionine and seneciphylline [1]. In addition, hydroquinone, *p*-hydroxyphenylacetic acid, vanillic acid, salicylic acid and 2-furancarboxylic acid were isolated from *S. scandens* [2].

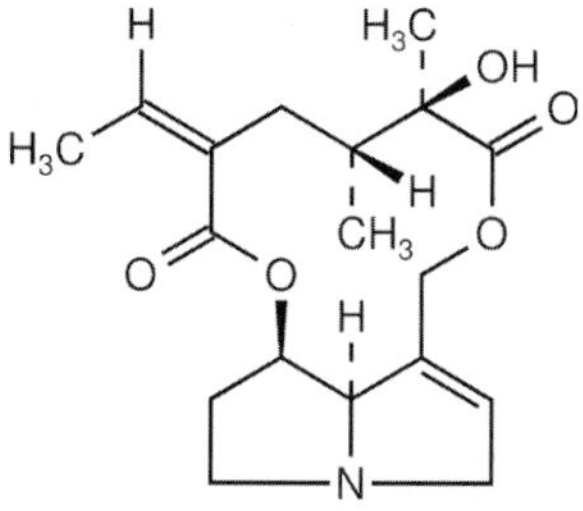

Senecionine

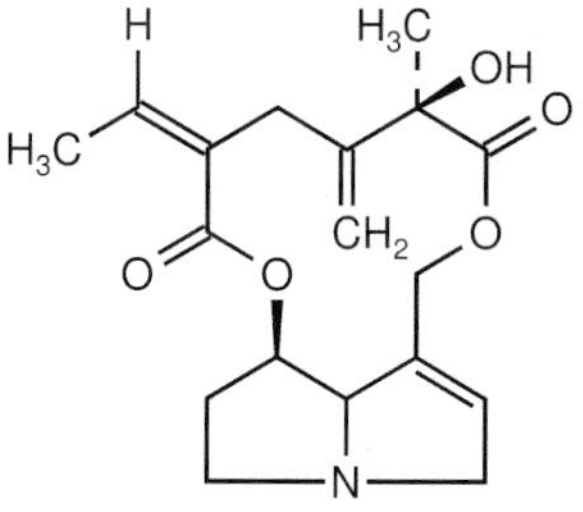

Seneciphylline

Pharmacology and Toxicology

Senecio scandens, and tablets prepared from its extract, were officially listed in the Chinese Pharmacopoeia edition 1977, with an indication for the treatment of bacterial diarrhea, enteritis, conjunctivitis, and respiratory tract infections. It was eliminated from later editions of the Chinese Pharmacopoeia, but reintroduced in 1995 as a component of a patent medicine. In contrast, the whole plant of *S. scandens* is still used in compositions for the clinical treatment of

Handbook of Chinese Medicinal Plants: Chemistry, Pharmacology, Toxicology
Weici Tang and Gerhard Eisenbrand

ISBN: 978-3-527-32226-8

rhinitis [3], ulcerative colitis [4], pneumonia in children [5], and burns [6]. The dragees "Qianbai Biyan Pian," as an official Chinese patent medicine, were reported to be effective in the treatment of rhinitis. Patients were treated with the dragees at a daily dose of nine dragees (3×3) for two weeks. Among 413 cases of rhinitis and sinusitis (229 cases of chronic rhinitis, 150 of allergic rhinitis, and 34 of chronic sinusitis), the total effective rate was reported as 88% [3].

The aqueous extract of *S. scandens* was found to have high potency in inhibiting rat erythrocyte hemolysis and lipid peroxidation in rat kidney and brain homogenates. It also demonstrated strong superoxide- and hydroxyl radical-scavenging activity [7].

Pyrrolizidine alkaloids are widely distributed in plant species such as Apocynaceae, Asteraceae, Fabaceae, and Boraginaceae, especially in the genus *Senecio* and *Crotalaria* (Fabaceae). The pyrrolizidine alkaloids are therefore also called senecio alkaloids. A large number of pyrrolizidine alkaloids were reported to be hepatotoxic, pneumotoxic, genotoxic, neurotoxic, and cytotoxic [8]. Chemically, the majority of pyrrolizidine alkaloids represents esters of retronecine or its 7-*cis* analogue, heliotridine. Senecionine and seneciphylline are macrocyclic diesters of retronecine. The hepatotoxicity of pyrrolizidine alkaloids in humans has long been known. Seneciphylline, as a major alkaloid in *Senecio* species, was a grain contaminant and has thus caused "bread poisoning" of humans in South Africa. Over 80 cases of *Senecio* poisoning, mainly in young people and mostly fatal, were observed in the George and Mossel Bay districts some 80–90 years ago; 12 cases were hospitalized in one area in 1931–1941. The most common symptoms were severe abdominal pain, rapidly developing ascites, and hepatomegaly [9, 10].

Senecionine and seneciphylline were both reported to be mutagenic and genotoxic. Seneciphylline showed a weak mutagenic activity in *Salmonella typhimurium* in the presence of an activating system [11], and significantly induced sister-chromatid exchange in V79 cells in the presence of primary chick embryo hepatocytes [12]. Both, senecionine and seneciphylline were reported to be mutagenic in V79 cells in the presence of an S-9 mix [13], and to induce DNA repair synthesis in primary rat hepatocytes in culture. Seneciphylline was also positive in the DNA repair test with hamster or mouse hepatocytes [14].

Pyrrolizidine alkaloids of macrocyclic diester type with a double bond in the acid moiety, such as seneciphylline and senecionine, produced a concentration-dependent inhibition of colony formation at 50, 100, and 300 μM, and induction of megalocytosis at 500 μM [15].

DNA and proteins from rat liver, lung and kidney of both genders, when treated with [^{3}H]senecionine and [^{3}H]seneciphylline, showed covalent binding of the alkaloids to DNA. A covalent binding index (CBI), expressed as μmol bound alkaloid per mol nucleotides per dose (in mmol kg^{-1}) of 210 was found in the liver of female animals treated with [^{3}H]senecionine, whereas the CBI of binding to liver DNA in males was about 50. The DNA damage 6 h after treatment persisted during the following four days. The administration of [^{3}H]seneciphylline to female and male rats resulted in CBIs of 69 and 73–92, respectively, for the liver DNA. Binding of both alkaloids to lung and kidney DNA in male and female rats was also observed [16]. Crosslinking of DNA, including intra- and intermolecular crosslinking, was reported with pBR322 plasmid DNA and M13 viral DNA [17] and with synthetic oligonucleotides [18]. The desoxyguanosine group of 5′-d(CG) was shown to be the predominant site for crosslinking by activated pyrrolizidine diesters [18].

A structure–activity study concerning DNA crosslinking by pyrrolizidine alkaloids

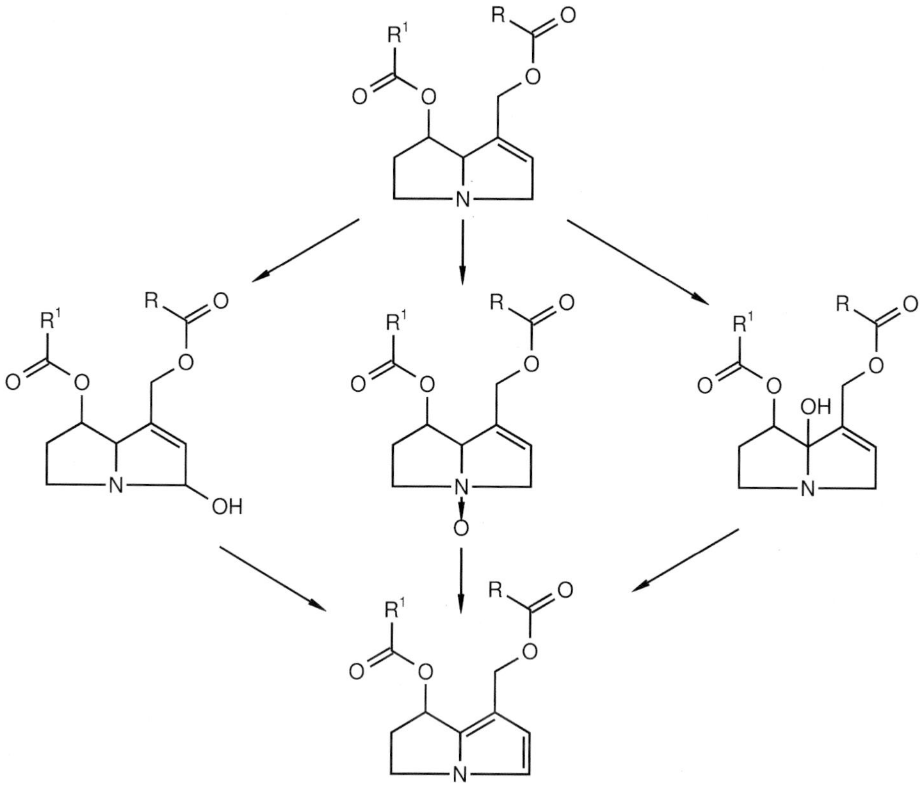

Figure 206.1 Metabolic activation of pyrrolizidine alkaloids derived from 1,2-didehydropyrrolizidine [20].

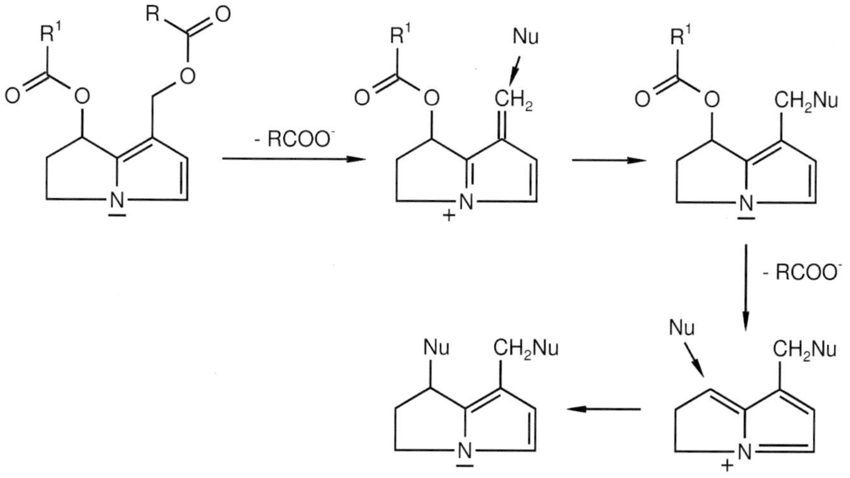

Figure 206.2 Alkylation and crosslinking reactions by activated pyrrolizidine alkaloids [20].

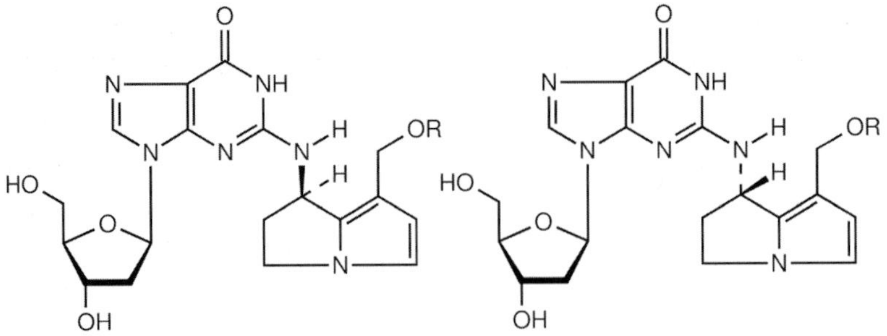

Diastereomers of N^2-(dehydroretronecin-7-yl)-deoxyguanosine

in mammalian cells revealed that the presence of both a macrocyclic diester and an unsaturated ester function was necessary. Thus, the crosslinking ability of seneciphylline and senecionine by far exceeded that of macrocyclic type pyrrolizidine alkaloids without an unsaturated ester function, and that of retronecine. The stereochemical orientation of the ester linkage was found to have no effect on biological activity [19].

The carcinogenic effects of pyrrolizidine alkaloids were found to also depend on their structure. The pyrrolizidine alkaloids of diester type were highly carcinogenic. Studies of structure–activity relationships revealed that pyrrolizidine alkaloids derived from 1,2-didehydropyrrolizidine, esterified at O^7 or O^9 position with a branched carboxylic acid containing more than five carbon atoms were toxic, carcinogenic and mutagenic [8, 20].

Pyrrolizidine alkaloids derived from 1,2-didehydropyrrolizidine were found to be mainly activated by cytochrome P450 isoenzymes [21], including CYPIIIA4 [22]. This resulted in 1,2,3,4-tetradehydropyrrolizidine metabolites, consisting of a pyrrole and a pyrrolidine ring, which are much more electrophilic than the parent compounds and represent the ultimate reactive species. They are able to react with alcohol, amine, and thiol groups of biological macromolecules [23]. The formation of *N*-oxides, 3- or 8-hydroxy derivatives as intermediates has also been discussed [24] (Figure 206.1).

Pyrrolizidine alkaloids of the diester type derived from 1,2-didehydropyrrolizidine may act as bifunctional alkylating agents, causing DNA alkylation and DNA–DNA or DNA–protein crosslinking [25]. The DNA-binding of pyrrolizidine alkaloids has been analyzed using ^{14}C- or ^{3}H-labeled compounds and alkaline elution [26]. The reactivity of the positions 7 and 9 depended on the steric hindrance exerted by the ester function (Figure 206.2).

In a model study, two diastereomers of N^2-(dehydroretronecin-7-yl)-deoxyguanosine were analyzed after reacting dehydroretronecine with deoxyguanosine. The adducts were formed by a nucleophilic attack of the exocyclic amino group of deoxyguanosine at the electrophilic center at position 7 of dehydroretronecine [27].

According to the activating mechanism, the parent compound retronecine practically did not possess any alkylating potential. The hydrolysis of pyrrolizidine alkaloids by either base-catalyzed or enzymatic hydrolysis of the ester linkages should be considered as a detoxification process [28].

A further detoxification mechanism is the conjugation of the active pyrrolic metabolites with glutathione (GSH) to form 7-glutathionyl-1-hydroxymethyl-1,2,3,8-tetradehydropyrrolizidine [29]. Senecionine was metabolized primarily to putative pyrrolic metabolites, and to its *N*-oxide by rat liver microsomal monooxygenases. The pyrrolic metabolites were highly reactive, and either bound covalently to nucleophiles or were hydrolyzed to the more stable 1-hydroxymethyl-7-hydroxy-1,2,3,8-tetradehydropyrrolizidine and the corresponding necic acid. The addition of GSH to incubation mixtures containing rat liver microsomes and senecionine, resulted in the formation of a GSH conjugate with 1-hydroxymethyl-7-hydroxy-1,2,3,8-tetradehydropyrrolizidine. However, only negligible amounts of conjugate were formed without rat liver microsomes, which indicated that the microsomal conversion of senecionine to the highly reactive metabolite was followed by conjugation with GSH [30].

Sulfur-bound pyrrolic metabolites were identified in the blood and liver tissue from rats after a single intraperitoneal (i.p.) injection of senecionine and some related pyrrolizidine alkaloids [31]. An *N*-acetylcysteine conjugate with pyrrolic metabolites of senecionine has been identified in rat urine following the administration of senecionine [32].

Following an intravenous (i.v.) injection of [^{14}C]senecionine at a dose of 60 mg kg^{-1} to rats, 44% and 43% of the total administered radioactivity was excreted in the bile and urine, respectively. Senecionine *N*-oxide was identified as the major metabolite. Less than 5% in bile and 18% in urine was excreted as the parent alkaloid. The plasma concentration of senecionine decreased from 107 to 12 nmol g^{-1} over a 7 h period, while red blood cell concentrations declined from 109 to 26 nmol g^{-1} [33]. [^{14}C]Senecionine and [^{14}C]seneciphylline, when administered to mice, were rapidly excreted, with radioactivity in the urine and feces accounting for 84% of the injected dose within 16 h. The liver contained over 1.5% of the dose at 16 h and a small amount (0.04%) was secreted into the milk over 16 h [34].

An interspecies difference of the urinary retrorsine metabolites has been observed. Guinea pigs are generally resistant to the toxicity of pyrrolizidine alkaloids. The urinary *N*-oxide levels of administered pyrrolizidine alkaloids were higher in guinea pigs relative to mice, hamsters, and rats. It has been suggested that the resistance of guinea pigs to pyrrolizidine alkaloids is attributed to a preferential metabolism of pyrrolizidine alkaloids to *N*-oxides, combined with a correspondingly low conversion to pyrrolic metabolites [35]. A flavin-containing monooxygenase was found to be the major detoxifying enzyme for the pyrrolizidine alkaloid senecionine in guinea pig tissues. In guinea pig lung and kidney microsomes, the *N*-oxide was the major metabolite formed from senecionine, with little or no production of pyrrolic metabolites. The high rate of detoxification, coupled with the low level of activation of senecionine in liver, lung, and kidney, may help to explain the apparent resistance of the guinea pig to intoxication by senecionine and other pyrrolizidine alkaloids [36]. Carboxylesterase GPH1 from guinea pig liver microsomes was found capable of hydrolyzing seneciphylline, senecionine and related pyrrolizidine alkaloids, whereas carboxylesterase GPL1 was much less active [37].

The clinical use of *S. scandens* might represent a carcinogenic risk for humans. For example, if Qianbai Biyan Pian were used to treat rhinitis at the recommended dose, the daily intake of pyrrolizidine alkaloids would be approximately 4 mg.

References

1 Batra, V. and Rajagopalan, T.R. (1977) Alkaloidal constituents of *Senecio scandens*. *Curr. Sci.*, **46**, 141.

2 Wang, X.F. and Tu, D.J. (1980) Studies on the chemical constituents of *Senecio scandens* Buch-Ham. *Acta Pharm. Sin.*, **15**, 503–505.

3 Guo, Z.G. and Zeng, X.C. (1987) Clinical application of "Qianbai rhinitis tablet". *Guanzhou Med. J.*, **18**, 50–52.
4 Guo, Q., Wan, S.Q., and Guo, Z. (1989) Treatment of ulcerative colitis by traditional Chinese medicine. *Chin. J. Integr. Trad. West. Med.*, **9**, 474–474, 452.
5 Wu, K.H., Hu, T.C., and Liu, X.F. (1994) "Jieredu" injection: treatment of 119 children with pneumonia. *Chin. J. Integr. Trad. West. Med.*, **14**, 106–107.
6 Xiao, J.X. and Liang, G.C. (1989) Clinical report of 1000 cases of burns treated with Chinese drugs. *Beijing J. Trad. Chin. Med.*, 30–31.
7 Liu, F. and Ng, T.B. (2000) Antioxidative and free radical scavenging activities of selected medicinal herbs. *Life Sci.*, **66**, 725–735.
8 Roeder, E. (2000) Medicinal plants in China containing pyrrolizidine alkaloids. *Pharmazie*, **55**, 711–726.
9 Habermehl, G.G., Martz, W., Tokarnia, C.H., Dobereiner, J., and Mendez, M.C. (1988) Livestock poisoning in South America by species of the *Senecio* plant. *Toxicon*, **26**, 275–286.
10 International Agency for Research on Cancer (1976) IARC monographs on the evaluation of carcinogenic risk of chemicals to man, *Some Naturally Occurring Substances, vol.* **10**, International Agency for Research on Cancer, Lyon.
11 Rubiolo, P., Pieters, L., Calomme, M., Bicchi, C., Vlietinck, A., and Vanden Berghe, D. (1992) Mutagenicity of pyrrolizidine alkaloids in the *Salmonella typhimurium*/mammalian microsome system. *Mutat. Res.*, **281**, 143–147.
12 Bruggeman, I.M. and van der Hoeven, J.C. (1985) Induction of SCEs by some pyrrolizidine alkaloids in V79 Chinese hamster cells co-cultured with chick embryo hepatocytes. *Mutat. Res.*, **142**, 209–212.
13 Berry, D.L., Schoofs, G.M., Schwass, D.E., and Molyneux, R.J. (1996) Genotoxic activity of a series of pyrrolizidine alkaloids in primary hepatocyte-mediated V79 cell mutagenesis and DNA repair assay. *J. Nat. Toxins*, **5**, 7–24.
14 Mori, H., Sugie, S., Yoshimi, N., Asada, Y., Furuya, T., and Williams, G.M. (1985) Genotoxicity of a variety of pyrrolizidine alkaloids in the hepatocyte primary culture-DNA repair test using rat, mouse, and hamster hepatocytes. *Cancer Res.*, **45**, 3125–3129.
15 Kim, H.Y., Stermitz, F.R., Molyneux, R.J., Wilson, D.W., Taylor, D., and Coulombe, R.A. Jr (1993) Structural influences on pyrrolizidine alkaloid-induced cytopathology. *Toxicol. Appl. Pharmacol*, **122**, 61–69.
16 Candrian, U., Lüthy, J., and Schatter, C. (1985) *In vivo* covalent binding to retronecine labelled [^{3}H]seneciphylline and [^{3}H]senecionine to DNA of rat liver, lung and kidney. *Chem.-Biol. Interact.*, **54**, 57–69.
17 Reed, R.L., Ahern, K.G., Pearson, G.D., and Buhler, D.R. (1988) Crosslinking of DNA by dehydroretronecine, a metabolite of pyrrolizidine alkaloids. *Carcinogenesis*, **9**, 1355–1361.
18 Weidner, M.F., Millard, J.T., and Hopkins, P.B. (1989) Determination of single-nucleotide resolution of the sequence specificity of DNA interstrand cross linking agents in DNA fragments. *J. Am. Chem. Soc.*, **111**, 9270–9272.
19 Hincks, J.R., Kim, H.Y., Segall, H.J., Molyneux, R.J., Stermitz, F.R., and Coulombe, R.A. Jr (1991) DNA cross-linking in mammalian cells by pyrrolizidine alkaloids: structure-activity relationships. *Toxicol. Appl. Pharmacol*, **111**, 90–98.
20 Roeder, E. (1995) Medicinal plants in Europe containing pyrrolizidine alkaloids. *Pharmazie*, **50**, 83–98.
21 Buhler, D.R., Miranda, C.L., Kedzierski, B., and Reed, R.L. (1991) Mechanisms for pyrrolizidine alkaloid activation and detoxification. *Adv. Exp. Med. Biol.*, **283**, 597–603.
22 Miranda, C.L., Reed, R.L., Guengerich, F.P., and Buhler, D.R. (1991) Role of cytochrome P450IIIA4 in the metabolism of the pyrrolizidine alkaloid senecionine in human liver. *Carcinogenesis*, **12**, 515–519.
23 Winter, C.K., Segall, H.J., and Jones, A.D. (1988) Determination of pyrrolizidine alkaloid metabolites from mouse liver microsomes using tandem mass spectrometry and gas chromatography/mass spectrometry. *Biomed. Environ. Mass Spectrom.*, **15**, 265–273.
24 Bertram, B., Hemm, I., and Tang, W. (2001) Mutagenic and carcinogenic constituents of medicinal herbs used in Europe or in the USA. *Pharmazie*, **56**, 99–120.
25 Mattocks, A.R. and Bird, I. (1983) Pyrrolic and N-oxide metabolites from pyrrolizidine alkaloids by hepatic microsomes *in vitro*: relevance to *in vivo* hepatotoxicity. *Chem.-Biol. Interact.*, **53**, 209–222.
26 Couet, C.E., Hopley, J., and Hanley, A.B. (1996) Metabolic activation of pyrrolizidine alkaloids by human, rat and avocado microsomes. *Toxicon*, **34**, 1058–1061.

27 Robertson, K.A. (1982) Alkylation of N^2 in deoxyguanosine by dehydroretronecine, a carcinogenic metabolite of the pyrrolizidine alkaloid monocrotaline. *Cancer Res.*, **42**, 8–14.

28 Dueker, S.R., Lame, M.W., and Segall, H.J. (1995) Hydrolysis rates of pyrrolizidine alkaloids derived from *Senecio jacobaea*. *Arch. Toxicol.*, **69**, 725–728.

29 Yan, C.C., Cooper, R.A., and Huxtable, R.J. (1995) The comparative metabolism of the four pyrrolizidine alkaloids, seneciphylline, retrorsine, monocrotaline, and trichodesmine in the isolated, perfused rat liver. *Toxicol. Appl. Pharmacol.*, **133**, 277–284.

30 Reed, R.L., Miranda, C.L., Kedzierski, B., Henderson, M.C., and Buhler, D.R. (1992) Microsomal formation of a pyrrolic alcohol glutathione conjugate of the pyrrolizidine alkaloid senecionine. *Xenobiotica*, **22**, 1321–1327.

31 Mattocks, A.R. and Jukes, R. (1992) Detection of sulphur-conjugated pyrrolic metabolites in blood and fresh or fixed liver tissue from rats given a variety of toxic pyrrolizidine alkaloids. *Toxicol. Lett.*, **63**, 47–55.

32 Estep, J.E., Lame, M.W., Jones, A.D., and Segall, H.J. (1990) N-acetylcysteine-conjugated pyrrole identified in rat urine following administration of two pyrrolizidine alkaloids, monocrotaline and senecionine. *Toxicol. Lett.*, **54**, 61–69.

33 Estep, J.E., Lame, M.W., and Segall, H.J. (1990) Excretion and blood radioactivity levels following [^{14}C]senecionine administration in the rat. *Toxicology*, **64**, 179–189.

34 Eastman, D.F., Dimenna, G.P., and Segall, H.J. (1982) Covalent binding of two pyrrolizidine alkaloids, senecionine and seneciphylline, to hepatic macromolecules and their distribution, excretion, and transfer into milk of lactating mice. *Drug Metab. Dispos.*, **10**, 236–240.

35 Chu, P.S. and Segall, H.J. (1991) Species difference in the urinary excretion of isatinecic acid from the pyrrolizidine alkaloid retrorsine. *Comp. Biochem. Physiol. C*, **100**, 683–686.

36 Miranda, C.L., Chung, W., Reed, R.E., Zhao, X., Henderson, M.C., Wang, J.L., Williams, D. E., and Buhler, D.R. (1991) Flavin-containing monooxygenase: a major detoxifying enzyme for the pyrrolizidine alkaloid senecionine in guinea pig tissues. *Biochem. Biophys. Res. Commun.*, **178**, 546–552.

37 Dueker, S.R., Lame, M.W., and Segall, H.J. (1992) Hydrolysis of pyrrolizidine alkaloids by guinea pig hepatic carboxylesterases. *Toxicol. Appl. Pharmacol.*, **117**, 116–121.

Sesamum indicum

Semen Sesami Nigrum (Heizhima) is the dried ripe seed of *Sesamum indicum* L. (Pedaliaceae). It is used as a general tonic and laxative for the treatment dizziness, tinnitus, impaired hearing, and constipation.

Sesamum indicum L.

Oleum Sesami (Ma You) is the fatty oil from the seed of *Sesamum indicum*. It is used as laxative and as base material for ointments. It is also a known food oil.

Chemistry

Semen Sesami Nigrum, the black sesame, is a common food, especially in Asian countries, and is known to contain fatty oil. Sesame oil is mainly composed of glycerol ester of oleic acid and linoleic acid [1]. The major sterols in sesame oil were identified as β-sitosterol and spinasterol, with a total content of about 600 mg kg^{-1}. The unsaponifiable part of sesame oil contains considerable amounts of lignan derivatives sesamin, episesamin and sesamolin. Sesamolin produces sesamol by acidic hydrolysis [2]. Episesamin is not a naturally occuring compound, but is formed during the refining process of non-roasted sesame seed oil [3]. Other lignan derivatives from sesame seed were identified as sesamolinol, sesaminol, and *epi*-sesaminol. The Chinese Pharmacopoeia requires a qualitative determination of β-sitosterol and sesamin in Semen Sesami Nigrum by thin-layer chromatographic comparison with reference substances. The Chinese Pharmacopoeia gives a relative density of 0.917–0.923, a refractive index of 1.471–1.475, an acid value of less than 2.5, a saponification value of 188–195, and an iodine value of 103–116 for Oleum Sesami.

The major protein component of sesame seed was reported to be α-globulin, which was resistant to proteases [4] but hydrolyzed by α-chymotrypsin [5].

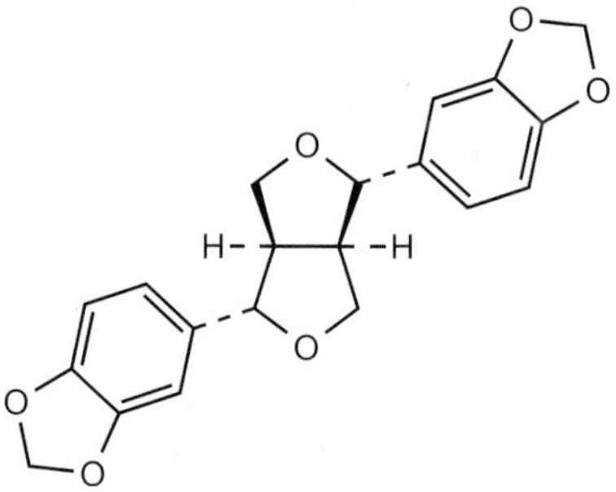

Sesamin

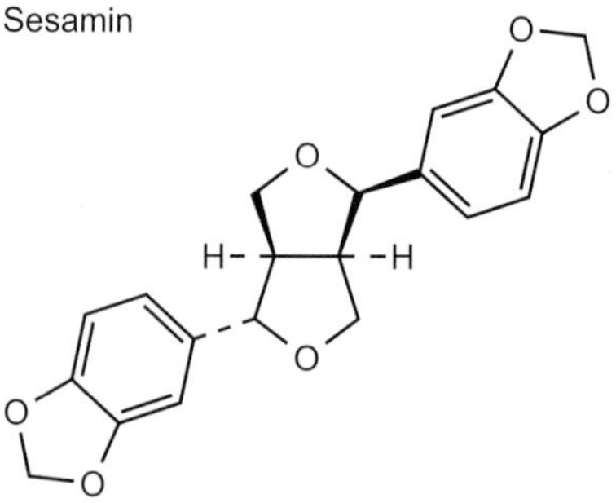

Episesamin

Handbook of Chinese Medicinal Plants: Chemistry, Pharmacology, Toxicology
Weici Tang and Gerhard Eisenbrand

ISBN: 978-3-527-32226-8

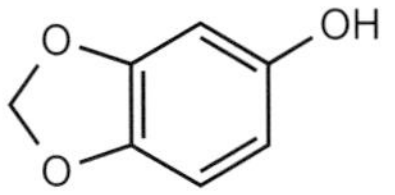

Sesamol

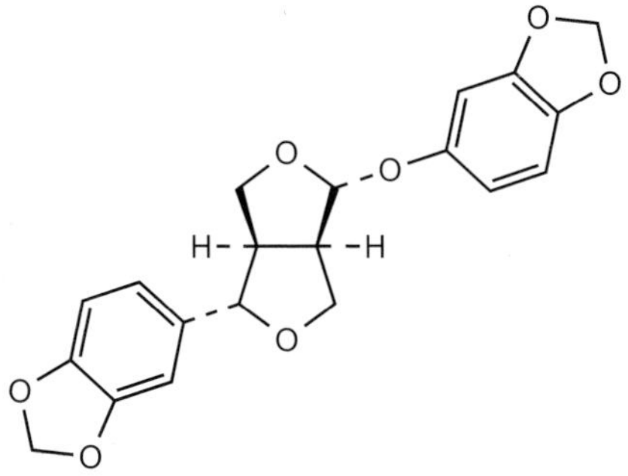

Sesamolin

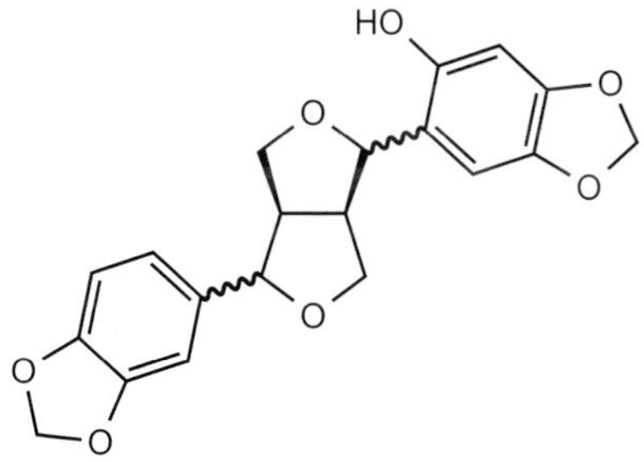

Sesaminol

Pharmacology and Toxicology

Sesamin was found to inhibit cholesterol absorption and synthesis in rats. When given at a dietary level of 0.5% for four weeks, sesamin significantly reduced the concentration of serum and liver cholesterol, irrespective of the presence or absence of cholesterol in the diet. On feeding sesamin, there was a decrease in lymphatic absorption of cholesterol, accompanyed by an increase in the fecal excretion of neutral steroids. Only a marginal proportion (ca. 0.15%) of sesamin administered intragastrically was recovered in the lymph. Although the weight and phospholipid concentration of the liver increased unequivocally on feeding sesamin, the histological examination showed no abnormality, and the activities of serum glutamic oxalacetic transaminase (GOT) and glutamic pyruvic transaminase (GPT) remained unchanged [6].

Feeding sesamin at the dietary level of 0.2% for 14–16 days resulted in an enlargement of liver weight. Sesamin feeding caused a stimulation of ketone body production, especially when exogenous oleic acid was provided. The ratio of β-hydroxybutyrate to acetoacetate (an index of mitochondrial redox potential) was consistently lowered by dietary sesamin. The cumulative secretion of triacylglycerol (but not of cholesterol) in the livers of sesamin-fed rats was decreased markedly, suggesting an inverse relationship between the rates of ketogenesis and triacylglycerol secretion. Dietary sesamin may exert its hypolipidemic effect through an enhanced metabolism of exogenous fatty acid in rat liver [7].

It was reported that α-tocopherol could enhance the hypocholesterolemic action of sesamin in rats. Rats were fed different levels of sesamin in the diet (0.05 and 0.2%). Supplementation with 1% α-tocopherol significantly increased the hypocholesterolemic action of sesamin, particularly at a higher sesamin level, although α-tocopherol alone did not affect the concentration of serum cholesterol [8]. A hypocholesterolemic effect of sesame lignan in humans was also reported [9]. Sesamin-feeding to rats increased γ-tocopherol levels in the plasma, liver, and lung; however, the increase was nonsignificant for α-tocopherol [10]. Sesamin given at a concentration of 0.4% in a diet containing 0.2% cholesterol to rats for four weeks induced substantially higher γ-tocopherol levels in plasma, liver, and lung. Sesamin tended to lower the plasma total, very low-density lipoprotein (VLDL) and low-density lipoprotein (LDL) cholesterol concentrations [11].

Sesame lignans also modulated cholesterol metabolism in normocholesterolemic and hypercholesterolemic spontaneously hypertensive rats. In normocholesterolemic spontaneously hypertensive rats fed a regular diet, both sesamin and episesamin significantly increased the concentration of serum total cholesterol, which was due to an increase of high-density lipoprotein (HDL) subfraction rich in apoE (apoE-HDL). In addition, both substances effectively decreased serum VLDL levels. In the liver, only episesamin significantly decreased the activity of microsomal acyl-CoA:cholesterol acyltransferase. In hypercholesterolemic spontaneously hypertensive rats fed a high-fat and high-cholesterol diet, only episesamin improved serum lipoprotein metabolism, with an increase in apoA-I and a decrease in apoB. In the liver, both sesamin and episesamin significantly suppressed cholesterol accumulation [12].

γ-Tocopherol in sesame seed exerted vitamin E activity in rats equal to that of α-tocopherol through a synergistic interaction with sesame seed lignans [13]. Sesame seed and its lignans also caused a significant enhancement of vitamin E activity in rats fed a low α-tocopherol diet [14]. Sesamin was also reported to suppress lipid peroxides in rats fed a high-docosahexaenoic acid diet synergistically with α-tocopherol [15]. It was suggested that sesamin increases tissue tocopherol concentration by inhibiting tocopherol catabolism, as evidenced by inhibition of tocopherol metabolism in HepG2/C3A cells. The result supported a CYP3A-dependent mechanism of the side-chain metabolism of tocopherols to water-soluble carboxychromans, and provided the first evidence of a specific enzyme involved in vitamin E metabolism [16].

Feeding sesamin and α-tocopherol in combination at 0.5% dietary level to rats for three weeks resulted in an interference with the metabolism of linoleic acid. This dietary manipulation significantly reduced the production of leukotriene C_4 (LTC_4) in the lung, the splenic production of LTB_4, and a reduction of plasma histamine levels. Simultaneous administration of sesamin and α-tocopherol significantly increased the production of immunoglobulin A (IgA), IgG, and IgM by mesenteric lymph node lymphocytes, while the IgE level tended to be reduced. Sesamin and α-tocopherol in combination appear to be effective for regulating the eicosanoid production and modifying the immune function [17].

Hypertension in rats induced by deoxycorticosterone acetate and salt was markedly suppressed by feeding a diet containing 1% sesamin. In rats, treatment with sesamin for five weeks significantly inhibited the increase in left ventricle weight induced by deoxycorticosterone acetate and salt [18]. Dietary sesamin supplementation efficiently improved the abnormal vasodilator and vasoconstrictor responses in hypertensive rats caused by deoxycorticosterone acetate and salt [19]. The antihypertensive effect of sesamin was also observed in renal hypertensive [20] and in spontaneously hypertensive rats [21]. Sesamin increased the nitric oxide (NO) concentration in the medium of human umbilical vein endothelial cells (HUVEC) in a concentration-dependent manner after 24 h incubation, increased NO synthase, and induced NO synthase mRNA expressions. The content of cGMP was induced by sesamin through NO signaling. On the other hand, the endothelin-1 concentration in the medium was suppressed, and the endothelin converting enzyme-1 protein and mRNA expressions were also inhibited by sesamin. These results suggested that sesamin might improve hypertension by its ability to induce NO and inhibit endothelin-1 production from endothelial cells [22].

Dietary sesamin dose-dependently increased both mitochondrial and peroxisomal palmitoyl-coenzyme A oxidation rates. Mitochondrial activity almost doubled in rats fed a 0.5% sesamin diet. Peroxisomal activity increased more than 10-fold in rats fed a 0.5% sesamin diet in relation to rats fed a sesamin-free diet. Dietary sesamin greatly increased the hepatic activity of fatty acid oxidation enzymes, including carnitine palmitoyltransferase, acyl-CoA dehydrogenase, acyl-coenzyme A oxidase, 3-hydroxyacyl-coenzyme A dehydrogenase, enoyl-coenzyme A hydratase, and 3-ketoacyl-coenzyme A thiolase. Dietary sesamin also increased the activity of 2,4-dienoyl-coenzyme A reductase and Δ3,Δ2-enoyl-coenzyme A isomerase; these enzymes are involved in the auxiliary pathway for β-oxidation of unsaturated fatty acids. Sesamin also induced increased the gene expression of mitochondrial and peroxisomal fatty acid oxidation enzymes. Among these various enzymes, peroxisomal acyl-coenzyme A oxidase and bifunctional enzyme gene expression were affected most by dietary sesamin. An alteration in hepatic fatty acid metabolism may therefore account for the serum lipid-lowering effect of sesamin in the rat [23]. In rats fed on a diet containing 0.2% sesamin or episesamin, both compounds increased mitochondrial and peroxisomal palmitoyl-CoA oxidation rates in parallel to enhanced expression and activity of various fatty acid oxidation enzymes with higher increase observed for episesamin. In contrast, activity and gene expression of hepatic lipogenic enzymes was lowered with no difference being observed between sesamin and episesamin. This indicates that responses of hepatic lipogenesis are markedly different from those observed for fatty acid oxidation [3].

The effects of sesamin and sesamolin on hepatic fatty acid metabolism were compared in rats. Sesamin and sesamolin dose-dependently increased the activity of various enzymes involved in hepatic fatty acid oxidation. The increase was much greater with sesamolin than with sesamin. Sesamolin accumulated in serum at 33- and 46-fold the level of sesamin at dietary concentrations of 0.6 and 2 g kg^{-1}, respectively. The amount of sesamolin accumulated in liver was ten- and sevenfold that of sesamin at the respective dietary levels. Sesamolin, rather than sesamin, could account for the potent physiological effect of sesame seeds in increasing hepatic fatty acid oxidation. Differences in bioavailability might contribute to the different effects of sesamin and sesamolin. Sesamin, compared to sesamolin, was more effective in reducing serum and liver lipid levels, although sesamolin more strongly increased hepatic fatty acid oxidation [24].

A mixture of sesamin and episesamin given to rats at a dietary level of 0.5% for 13 days significantly increased the proportions of dihomo-γ-linolenic acid not only in the liver but also in the plasma and hemocytes. The mixture at a dietary level of 1% improved changes in various blood parameters of mice, such as aspartate aminotransferase and alanine aminotransferase activities, and the concentrations of total cholesterol, triglyceride and total bilirubin, caused by the continuous inhalation of ethanol. In addition, sesamin showed a significant protective effect against the accumulation of fat droplets and vacuolar degeneration in the mouse liver, as confirmed by histological examination. In mice, sesamin at a dose of 100 mg kg^{-1} tended to prevent liver lipid accumulation caused carbon tetrachloride [25]. A study in rats suggested that sesamin ingestion might regulate the transcription levels of hepatic metabolizing enzymes for alcohol and lipids. The changes of gene expression were investigated in rats given 250 mg kg^{-1} sesamin for three days, and the profiles of gene expression in rat livers were determined at 4 h after the final ingestion. The results showed that 38 transcripts were upregulated and eight transcripts downregulated. The gene expression levels of the early-stage enzymes of β-oxidation were not changed; however, those of the late-stage enzymes were significantly increased by sesamin ingestion. Also, in rats given sesamin, the gene expression of aldehyde dehydrogenase was increased, whereas alcohol dehydrogenase, liver catalase and cytochrome P450 isoenzyme CYP2E1 were not changed [26].

A hot-water extract from defatted sesame seeds was reported to exert hypoglycemic activity in genetically type II diabetic KK-Ay mice. The hypoglycemic effect of sesame seeds was suggested to be caused by delayed glucose absorption [27]. A neuroprotective effect of sesamin and sesamolin has been demonstrated in murine BV-2 microglia cell line under hypoxia. Sesamin and sesamolin protected BV-2 cells from hypoxia-induced cell death, and reduced lactate dehydrogenase (LDH) release and the production of reactive oxygen species (ROS). Sesamin and sesamolin also inhibited the activation of caspase-3 and mitogen-activated protein kinases (MAPK) in BV-2 cells caused by hypoxia. This indicated that the antioxidant activity of sesamin and sesamolin involved an inhibition of MAPK pathways and of apoptosis through scavenging of ROS in hypoxia-stressed BV-2 cells [28]. A similar effect of sesamin and sesamolin was observed in PC12 cells against hypoxia or H_2O_2-induced cell injury [29].

Dietary supplementation of 0.2% sesamin significantly reduced by 36% the cumulative number of palpable mammary cancers in female rats induced by 7,12-dimethylbenz[*a*]anthracene (DMBA) at 12 weeks after administration of the carcinogen. Concentrations of lipid peroxides in plasma, liver and tumors were all decreased in sesamin-treated rats. The activity of peripheral blood mononuclear cells was increased in rats fed sesamin. The fatty acid compositions of plasma, liver and tumor phosphatidylcholine showed a decreased tendency of the metabolism of linoleic acid to arachidonic acid, and hence of the plasma concentration of prostaglandin E_2 in the sesamin-treated animals. The inhibitory effect of sesamin on DMBA-induced mammary carcinogenesis may be ascribed, at least in part, to immunopotentiation and increased antioxidative activity [30]. In contrast, sesamin exerted no significant influence on pancreatic carcinogenesis in female Syrian golden hamsters induced by *N*-nitroso-*N*,*N*-bis(2-oxopropyl)amine, at least within the four-month period after carcinogen treatment [31]. Sesamin inhibited the growth of human stomach cancer KATO III cells and induced apoptosis [32]. Sesamin and episesamin were also reported to induce apoptosis in human lymphoid leukemia Molt 4B cells [33].

The concentrations of sesamin and episesamin in rat liver after administration for eight weeks of a diet containing 0.5% sesame lignans were very low (both <0.5 $\mu g\,g^{-1}$ tissue). Sesamin and episesamin may be incorporated into the liver and then transported to the other tissues, including the lungs, heart, kidneys, and brain. Both sesamin and episesamin were eliminated within 24 h. There was no significant difference in lymphatic absorption between sesamin and episesamin, but levels of sesamin were significantly lower than those of episesamin in all tissues and serum. It was suggested that sesamin is absorbed into the lymph similar to episesamin, but that sesamin is subsequently metabolized faster by the liver [34]. The major metabolite of sesamin in human urine of healthy volunteers after a single dose of sesame oil containing about 500 μM sesamin was sesamin catechol, structurally elucidated as (1*R*,2*S*,5*R*,6*S*)-6-(3,4-dihydroxyphenyl)-2-(3,4-methylenedioxyphenyl)-3,7-dioxabicyclo-[3,3,0]octane. The excretion of the sesamin catechol metabolite ranged from 22 to 39% of the ingested dose within 12 h after ingestion [35]. Dietary ingested sesamin was found to convert into enterolactone in humans. Four healthy volunteers were given a single dose of sesame seeds after having been on a low-lignan diet for one week. The results showed that enterolactone was the major metabolite of sesamin, both *in vivo* and *in vitro*. The abundance of sesamin in sesame seed indicated that the seeds might be a major food source of enterolactone precursors [36].

Feeding rabbits with diets containing 1% cholesterol and 10% defatted sesame flour did not protect against cholesterol-induced hypercholesterolemia. Susceptibility to oxidative stress was found decreased in rabbits perhaps due to the antioxidative activity of sesaminol, which was present in defatted sesame flour at a concentration of 1% as sesaminol glucosides [37].

Sesamol was reported to be an antioxidant [38]. In rats fed a diet containing 1% sesamolin for two weeks, about 75% of the ingested sesamolin was excreted unmetabolized in the feces, but was not detected in the urine. Sesamolin and its metabolites, sesamol and sesamolinol, were excreted primarily as sulfates and glucuronides. The amount of sesamolin and its metabolites was lower in the plasma than in the liver or kidneys. Lipid peroxidation activity, measured as 2-thiobarbituric acid reactive substances, was significantly lower in the kidneys and liver of sesamolin-fed rats than in controls. The amount of 8-hydroxy-2′-deoxyguanosine excreted in the urine was also significantly lower in sesamolin-fed rats. Sesamolin and its metabolites may contribute to the antioxidative properties of sesame seeds and sesame oil, supporting the hypothesis that sesame lignans reduce susceptibility to oxidative stress [39]. Sesame oil and sesamol each inhibited C6 glial cell monoamine oxidase activity and scavenged peroxides [40]. Sesamol also exhibited a dose-dependent suppressive effect of NO production in C6 astrocyte cells stimulated by lipopolysaccharide (LPS) and interferon-γ (IFN-γ), with an IC_{50} of less than 1 m*M* [41]. Human erythrocytes incubated with sesamol at various concentrations up to 10 m*M* at 37 °C for 1 h resulted in increased amounts of hemoglobin being bound to the membranes, and the presence of detergent C12E8-insoluble membrane protein aggregates, in a concentration-dependent manner [42]. Sesamol as an antioxidant exerted a protective effect against CCl_4-induced liver injury in rats [43].

Sesamol was shown to exhibit strong antimutagenic effects towards mutagens generating oxygen radicals, such as *tert*-butylhydroperoxide (TBH) or hydrogen peroxide in *Salmonella typhimurium* TA100 and TA102. The antimutagenic property of sesamol was attributed to its antioxidant properties. Sesamol also inhibited the mutagenicity of sodium azide in *S. typhimurium* TA100, but had no effect on nitroquinoline *N*-oxide mutagenesis in *S. typhimurium* TA98 [44]. Generation of the imidazoquinoxaline-type heterocyclic amines in a heated model system composed of glucose, glycine and creatinine, was effectively prevented by sesamol in a dose-dependent manner. Electron spin resonance (ESR) studies showed that the heated model mixture of glucose and glycine generated the unstable pyrazine cation radical, and its formation was inhibited by sesamol [45]. The continuous UV irradiation of sesamol in benzene produced two types of radical: the neutral sesamolyl radical, and the dimer radical. The 3,4-dimethoxyphenoxyl radical had a shorter lifetime than the neutral sesamolyl radical [46]. Sesamol fed to 30 male F344 rats at a dietary level of 2% for two years induced a retardation of body weight and an elevation of relative liver weights. The numbers and areas of glutathione *S*-transferase (placental form)-positive foci per unit area of liver section were reduced. This suggested that, in rats, the long-term feeding of sesamol inhibited the development of naturally occurring preneoplastic hepatocytic foci [47].

In contrast, sesamol was found to be carcinogenic in rodents, targeting the forestomach or glandular stomach [48]. In a carcinogenicity study, groups of 30 male F344 rats were treated with 0.4% sesamol for up to 104 weeks. All surviving animals were killed at the end of week 28, and the major organs examined histopathologically. A slightly increased incidence of forestomach papillomas was found in animals given sesamol (15.8%) as compared with basal diet (0%) [49]. Male rats given a diet containing sesamol at a level of 2% for four weeks showed an induction of large ulcers and hyperplasia in the central region of the forestomach [50]. Sesamol given at a dietary level of 2% to 30 male and female F344/DuCrj rats for 104 weeks induced squamous cell carcinomas in the forestomach in nine of 29 (31%) male rats, and in three of 30 (10%) female rats. Papillomas developed in ten of 29 (34%) male rats and fourteen of 30 (47%) female rats. Sesamol given at the same level in the diet to B6C3F1 mice for 96 weeks induced forestomach cancer in eleven of 29 (38%) male mice, and in five of 30 (17%) female mice. No papilloma was observed in any of the control mice. Hyperplasias developed in almost all rats and mice of both genders [51]. Sesamol may act primarily as mitogen in the rat forestomach epithelium, with regeneration due to toxicity further enhancing cell proliferation [52].

References

1 Sengupta, A. and Roychoudhury, S.K. (1976) Triglyceride composition of *Sesamum indicum* seed oil. *J. Sci. Food Agric.*, **27**, 165–169.

2 Kato, M.J., Chu, A., Davin, L.B., and Lewis, N.G. (1998) Biosynthesis of antioxidant lignans in *Sesamum indicum* seeds. *Phytochem.*, **47**, 583–591.

3 Kushiro, M., Masaoka, T., Hageshita, S., Takahashi, Y., Ide, T., Sugano, M. (2002) Comparative effect of sesamin and episesamin on the activity and gene expression of enzymes in fatty acid oxidation and synthesis in rat liver. *J. Nutr. Biochem.*, **13**, 289–295.

4 Tasneem, R. and Prakash, V. (1989) Resistance of α-globulin from *Sesamum indicum* L. to proteases in relationship to its structure. *J. Protein Chem.*, **8**, 251–261.

5 Tasneem, R. and Prakash, V. (1992) The nature of the unhydrolysed fraction of α-globulin, the major protein component of *Sesamum indicum* L. hydrolysed by α-chymotrypsin. *Indian J. Biochem. Biophys.*, **29**, 160–167.

6 Hirose, N., Inoue, T., Nishihara, K., Sugano, M., Akimoto, K., Shimizu, S., and Yamada, H. (1991) Inhibition of cholesterol absorption and synthesis in rats by sesamin. *J. Lipid Res.*, **32**, 629–638.

7 Fukuda, N., Miyagi, C., Zhang, L., Jayasooriya, A.P., Sakono, M., Yamamoto, K., Ide, T., and Sugano, M. (1998) Reciprocal effects of dietary sesamin on ketogenesis and triacylglycerol secretion by the rat liver. *J. Nutr. Sci. Vitaminol. (Tokyo)*, **44**, 715–722.

8 Nakabayashi, A., Kitagawa, Y., Suwa, Y., Akimoto, K., Asami, S., Shimizu, S., Hirose, N., Sugano, M., and Yamada, H. (1995) α-Tocopherol enhances the hypocholesterolemic action of sesamin in rats. *Int. J. Vitam. Nutr. Res.*, **65**, 162–168.

9 Hirata, F., Fujita, K., Ishikura, Y., Hosoda, K., Ishikawa, T., and Nakamura, H. (1996) Hypocholesterolemic effect of sesame lignan in humans. *Atherosclerosis*, **122**, 135–136.

10 Kamal-Eldin, A., Pettersson, D., and Appelqvist, L.A. (1995) Sesamin (a compound from sesame oil) increases tocopherol levels in rats fed ad libitum. *Lipids*, **30**, 499–505.

11 Kamal-Eldin, A., Frank, J., Razdan, A., Tengblad, S., Basu, S., and Vessby, B. (2000) Effects of dietary phenolic compounds on tocopherol, cholesterol, and fatty acids in rats. *Lipids*, **35**, 427–435.

12 Ogawa, H., Sasagawa, S., Murakami, T., and Yoshizumi, H. (1995) Sesame lignans modulate cholesterol metabolism in the stroke-prone spontaneously hypertensive rat. *Clin. Exp. Pharmacol. Physiol.*, **22** (Suppl. 1), S310–S312.

13 Parker, R.S., Sontag, T.J., and Swanson, J.E. (2000) Cytochrome P4503A-dependent metabolism of tocopherols and inhibition by sesamin. *Biochem. Biophys. Res. Commun.*, **277**, 531–534.

14 Yamashita, K., Nohara, Y., Katayama, K., and Namiki, M. (1992) Sesame seed lignans and γ-tocopherol act synergistically to produce vitamin E activity in rats. *J. Nutr.*, **122**, 2440–2446.

15 Yamashita, K., Iizuka, Y., Imai, T., and Namiki, M. (1995) Sesame seed and its lignans produce marked enhancement of vitamin E activity in rats fed a low α-tocopherol diet. *Lipids*, **30**, 1019–1028.

16 Yamashita, K., Kagaya, M., Higuti, N., and Kiso, Y. (2000) Sesamin and α-tocopherol synergistically suppress lipid-peroxide in rats fed a high docosahexaenoic acid diet. *Biofactors*, **11**, 11–13.

17 Gu, J.Y., Wakizono, Y., Dohi, A., Nonaka, M., Sugano, M., and Yamada, K. (1998) Effect of dietary fats and sesamin on the lipid metabolism and immune function of Sprague-Dawley rats. *Biosci. Biotechnol. Biochem.*, **62**, 1917–1924.

18 Matsumura, Y., Kita, S., Morimoto, S., Akimoto, K., Furuya, M., Oka, N., and Tanaka, T. (1995) Antihypertensive effect of sesamin. I. Protection against deoxycorticosterone acetate-salt-induced hypertension and cardiovascular hypertrophy. *Biol. Pharm. Bull.*, **18**, 1016–1019.

19 Matsumura, Y., Kita, S., Ohgushi, R., and Okui, T. (2000) Effects of sesamin on altered vascular reactivity in aortic rings of deoxycorticosterone acetate-salt-induced hypertensive rat. *Biol. Pharm. Bull.*, **23**, 1041–1045.

20 Kita, S., Matsumura, Y., Morimoto, S., Akimoto, K., Furuya, M., Oka, N., and Tanaka, T. (1995) Antihypertensive effect of sesamin. II. Protection against two-kidney, one-clip renal hypertension and cardiovascular hypertrophy. *Biol. Pharm. Bull.*, **18**, 1283–1285.

21 Matsumura, Y., Kita, S., Tanida, Y., Taguchi, Y., Morimoto, S., Akimoto, K., and Tanaka, T. (1998) Antihypertensive effect of sesamin. III. Protection against development and maintenance of hypertension in stroke-prone spontaneously hypertensive rats. *Biol. Pharm. Bull.*, **21**, 469–473.

22 Lee, C.C., Chen, P.R., Lin, S., Tsai, S.C., Wang, B.W., Chen, W.W., Tsai, C.E., and Shyu, K.G. (2004) Sesamin induces nitric oxide and decreases endothelin-1 production in HUVECs: possible implications for its antihypertensive effect. *J. Hypertens.*, **22**, 2329–2338.

23 Ashakumary, L., Rouyer, I., Takahashi, Y., Ide, T., Fukuda, N., Aoyama, T., Hashimoto, T., Mizugaki, M., and Sugano, M. (1999) Sesamin, a sesame lignan, is a potent inducer of hepatic fatty acid oxidation in the rat. *Metabolism*, **48**, 1303–1313.

24 Lim, J.S., Adachi, Y., Takahashi, Y., and Ide, T. (2007) Comparative analysis of sesame lignans (sesamin and sesamolin) in affecting hepatic fatty acid metabolism in rats. *Br. J. Nutr.*, **97**, 85–95.

25 Akimoto, K., Kitagawa, Y., Akamatsu, T., Hirose, N., Sugano, M., Shimizu, S., and Yamada, H. (1993) Protective effects of sesamin against liver damage caused by alcohol or carbon tetrachloride in rodents. *Ann. Nutr. Metab.*, **37**, 218–224.

26 Kiso, Y., Tsuruoka, N., Kidokoro, A., Matsumoto, I., and Abe, K. (2005) Sesamin ingestion regulates the transcription levels of hepatic metabolizing enzymes for alcohol and lipids in rats. *Alcohol Clin. Exp. Res.*, **29** (11 Suppl.), 116S–120S.

27 Takeuchi, H., Mooi, L.Y., Inagaki, Y., and He, P. (2001) Hypoglycemic effect of a hot-water extract from defatted sesame (*Sesamum indicum* L.) seed on the blood glucose level in genetically diabetic KK-Ay mice. *Biosci. Biotechnol. Biochem.*, **65**, 2318–2321.

28 Hou, R.C., Wu, C.C., Yang, C.H., and Jeng, K. C. (2004) Protective effects of sesamin and sesamolin on murine BV-2 microglia cell line under hypoxia. *Neurosci. Lett.*, **367**, 10–13.

29 Hou, R.C., Huang, H.M., Tzen, J.T., and Jeng, K.C. (2003) Protective effects of sesamin and sesamolin on hypoxic neuronal and PC12 cells. *J. Neurosci. Res.*, **74**, 123–133.

30 Hirose, N., Doi, F., Ueki, T., Akazawa, K., Chijiiwa, K., Sugano, M., Akimoto, K., Shimizu, S., and Yamada, H. (1992) Suppressive effect of sesamin against 7,12-dimethylbenz[a]-anthracene induced rat mammary carcinogenesis. *Anticancer Res.*, **12**, 1259–1265.

31 Ogawa, T., Makino, T., Hirose, N., and Sugano, M. (1994) Lack of influence of low blood cholesterol levels on pancreatic carcinogenesis after initiation with *N*-nitrosobis(2-oxopropyl) amine in Syrian golden hamsters. *Carcinogenesis*, **15**, 1663–1666.

32 Hibasami, H., Fujikawa, T., Takeda, H., Nishibe, S., Satoh, T., Fujisawa, T., and Nakashima, K. (2000) Induction of apoptosis by *Acanthopanax senticosus* Harms and its component, sesamin in human stomach cancer KATO III cells. *Oncol. Rep.*, **7**, 1213–1216.

33 Miyahara, Y., Komiya, T., Katsuzaki, H., Imai, K., Nakagawa, M., Ishi, Y., and Hibasami, H. (2000) Sesamin and episesamin induce apoptosis in human lymphoid leukemia Molt 4B cells. *Int. J. Mol. Med.*, **6**, 43–46.

34 Umeda-Sawada, R., Ogawa, M., and Igarashi, O. (1999) The metabolism and distribution of sesame lignans (sesamin and episesamin) in rats. *Lipids*, **34**, 633–637.

35 Moazzami, A.A., Andersson, R.E., and Kamal-Eldin, A. (2007) Quantitative NMR analysis of a sesamin catechol metabolite in human urine. *J. Nutr.*, **137**, 940–944.

36 Penalvo, J.L., Heinonen, S.M., Aura, A.M., and Adlercreutz, H. (2005) Dietary sesamin is converted to enterolactone in humans. *J. Nutr.*, **135**, 1056–1062.

37 Kang, M.H., Kawai, Y., Naito, M., and Osawa, T. (1999) Dietary defatted sesame flour decreases susceptibility to oxidative stress in hypercholesterolemic rabbits. *J. Nutr.*, **129**, 1885–1890.

38 Uchida, M., Nakajin, S., Toyoshima, S., and Shinoda, M. (1996) Antioxidative effect of sesamol and related compounds on lipid peroxidation. *Biol. Pharm. Bull.*, **19**, 623–626.

39 Kang, M.H., Naito, M., Tsujihara, N., and Osawa, T. (1998) Sesamolin inhibits lipid peroxidation in rat liver and kidney. *J. Nutr.*, **128**, 1018–1022.

40 Mazzio, E.A., Harris, N., and Soliman, K.F. (1998) Food constituents attenuate monoamine oxidase activity and peroxide levels in C6 astrocyte cells. *Planta Med.*, **64**, 603–606.

41 Soliman, K.F. and Mazzio, E.A. (1998) *In vitro* attenuation of nitric oxide production in C6 astrocyte cell culture by various dietary compounds. *Proc. Soc. Exp. Biol. Med.*, **218**, 390–397.

42 Ando, K., Sako, K., Takahashi, M., Beppu, M., and Kikugawa, K. (2000) Increased band 3 protein aggregation and anti-band 3 binding of erythrocyte membranes on treatment with sesamol. *Biol. Pharm. Bull.*, **23**, 159–164.

43 Ohta, S., Suzuki, M., Sato, N., Kamogawa, A., and Shinoda, M. (1994) Protective effects of sesamol and its related compounds on carbon tetrachloride induced liver injury in rats. *Yakugaku Zasshi*, **114**, 901–910.

44 Kaur, I.P. and Saini, A. (2000) Sesamol exhibits antimutagenic activity against oxygen species mediated mutagenicity. *Mutat. Res.*, **470**, 71–76.

45 Kato, T., Harashima, T., Moriya, N., Kikugawa, K., and Hiramoto, K. (1996) Formation of the mutagenic/carcinogenic imidazoquinoxaline-type heterocyclic amines through the unstable free radical Maillard intermediates and its inhibition by phenolic antioxidants. *Carcinogenesis*, **17**, 2469–2476.

46 Nakagawa, K., Tero-Kubota, S., Ikegami, Y., and Tsuchihashi, N. (1994) EPR and TREPR spectroscopic studies of antioxidant sesamolyl and related phenoxyl radicals. *Photochem. Photobiol.*, **60**, 199–204.

47 Hagiwara, A., Kokubo, Y., Takesada, Y., Tanaka, H., Tamano, S., Hirose, M., Shirai, T., and Ito, N. (1996) Inhibitory effects of phenolic compounds on development of naturally occurring preneoplastic hepatocytic foci in long-term feeding studies using male F344 rats. *Teratog. Carcinog. Mutagen.*, **16**, 317–325.

48 Hirose, M., Shirai, T., Takahashi, S., Ogawa, K., and Ito, N. (1993) Organ-specific modification of carcinogenesis by antioxidants in rats. *Basic Life Sci.*, **61**, 181–188.

49 Hirose, M., Takesada, Y., Tanaka, H., Tamano, S., Kato, T., and Shirai, T. (1998) Carcinogenicity of antioxidants BHA, caffeic acid, sesamol, 4-methoxyphenol and catechol at low doses, either alone or in combination, and modulation of their effects in a rat medium-term multi-organ carcinogenesis model. *Carcinogenesis*, **19**, 207–212.

50 Ito, N. and Hirose, M. (1989) Antioxidants: carcinogenic and chemopreventive properties. *Adv. Cancer Res.*, **53**, 247–302.

51 Tamano, S., Hirose, M., Tanaka, H., Asakawa, E., Ogawa, K., and Ito, N. (1992) Forestomach neoplasm induction in F344/DuCrj rats and B6C3F1 mice exposed to sesamol. *Jpn. J. Cancer Res.*, **83**, 1279–1285.

52 Ito, N., Hirose, M., and Takahashi, S. (1993) Cell proliferation and forestomach carcinogenesis. *Environ. Health Perspect.*, **101** (Suppl. 5), 107–110.

Siegesbeckia glabrescens, *Siegesbeckia orientalis* and *Siegesbeckia pubescens*

Herba Siegesbeckiae (Xixiancao) is the dried, above-ground part of *Siegesbeckia glabrescens* Makino, *Siegesbeckia orientalis* L. or *Siegesbeckia pubescens* Makino (Asteraceae). It is used as an antirheumatic agent for the treatment of rheumatic arthralgia and paralysis.

Siegesbeckia orientalis L.

The pills prepared from Herba Siegesbeckiae, "Xixian Wan," are also listed in the Chinese Pharmacopoeia with the same indications as described under Herba Siegesbeckiae.

Chemistry

The whole plant of *S. glabrescens* was reported to contain diterpenes and diterpene glycosides identified as darutigenol, darutoside, neodarutoside [1], kirenol, acetylkirenol, and isopropylidenkirenol [2]. The Chinese Pharmacopoeia requires a qualitative determination of kirenol in Herba Siegesbeckiae by thin-layer chromatographic comparison with reference substance, and a quantitative determination by HPLC. The content of kirenol in Herba Siegesbeckiae should be not less than 0.05%.

Diterpenes of *ent*-kaurane type from the whole plant of *S. pubescens* were identified as siegesbeckioside, siegesbeckiol, siegesbeckic acid, and grandifloric acid [3]. Diterpene kirenol was also isolated from the whole plant of *S. pubescens* [3]. An *ent*-kaurane type diterpene, siegeskaurolic acid, was isolated from the root of *S. pubescens*. The structure of siegeskaurolic acid was elucidated as *ent*-16αH,17-hydroxy-kauran-19-oic acid [4]. Related *ent*-kaurane type diterpenes from the aerial part of *S. glabrescens* were identified as *ent*-16βH,17-isobutyryloxy-kauran-19-oic acid, ent-16βH, 17-acetoxy-18-isobutyryloxy-kauran-19-oic acid, and *ent*-16βH,17-hydroxykauran-19-oic acid [5].

Pharmacology and Toxicology

The oral administration of an aqueous extract of *S. pubescens* to rats at a dose of $100\,mg\,kg^{-1}$ was reported to significantly inhibit passive cutaneous anaphylaxis and to inhibit histamine release from rat peritoneal mast cells induced by anti-dinitrophenyl (DNP) immunoglobulin E (IgE) or DNP-treated human serum albumin. This suggested that *S. pubescens* exerted antiallergic activity, and that this action might be due to an inhibition of histamine release from mast cells [6]. Inhibition of

Handbook of Chinese Medicinal Plants: Chemistry, Pharmacology, Toxicology
Weici Tang and Gerhard Eisenbrand

ISBN: 978-3-527-32226-8

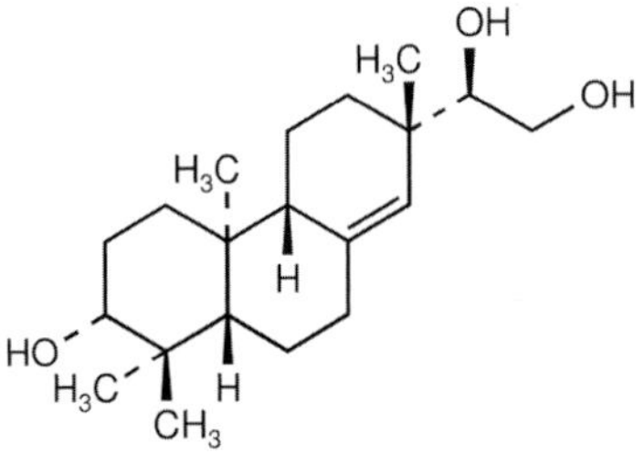

Darutigenol

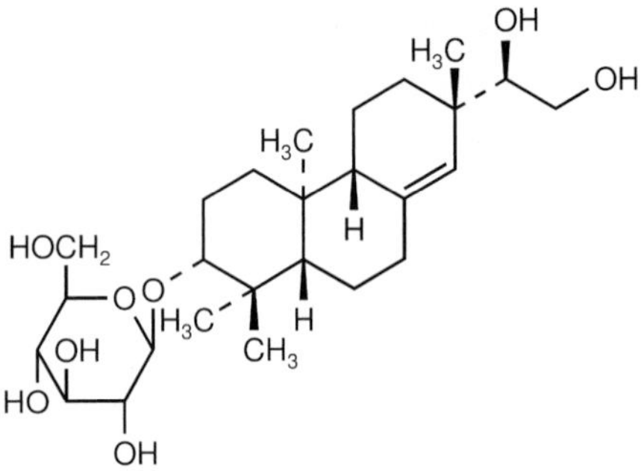

Darutoside

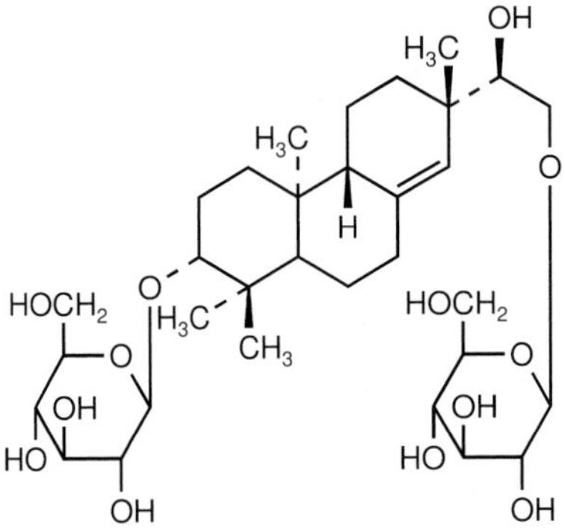

Neodarutoside

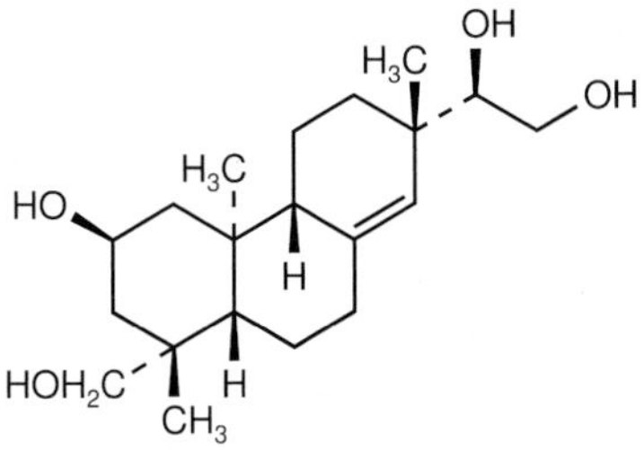

Kirenol

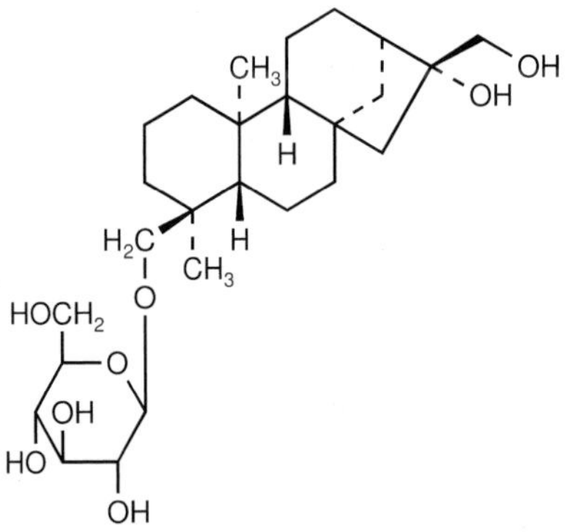

Siegesbeckioside

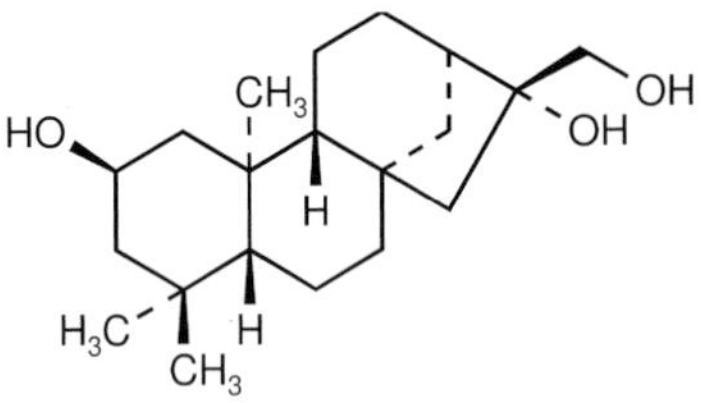

Siegesbeckiol

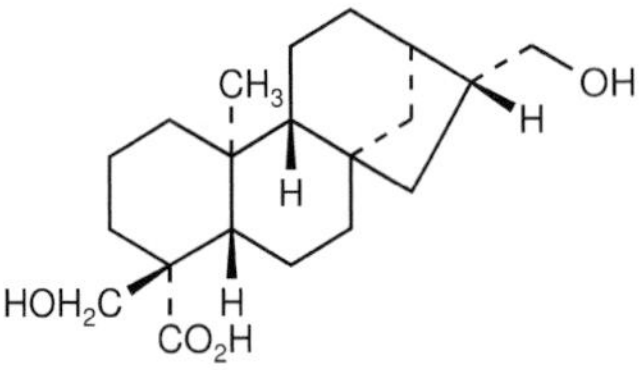

Siegesbeckic acid

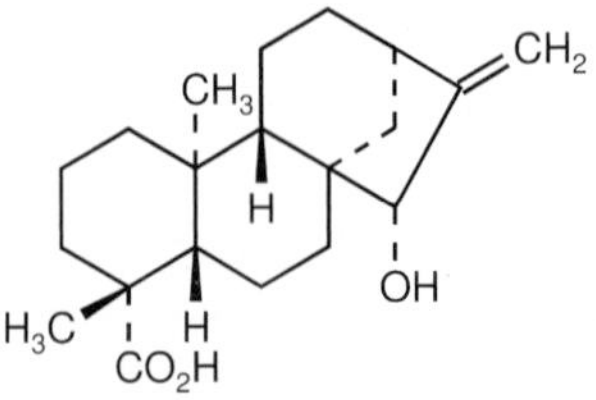

Grandifloric acid

histamine release from mast cells *in vivo* and *in vitro* was also described for the aqueous extract of *S. glabrescens* [7]. The aqueous extract of *S. glabrescens* was also found to inhibit the production of IgE both *in vivo* and *in vitro*. It dose-dependently inhibited the active systemic anaphylaxis and serum IgE production in mice induced by immunization with ovalbumin and *Bordetella pertussis* toxin adsorbed to an aluminum hydroxide gel. The extract also inhibited interleukin-4 (IL-4)-dependent IgE production in murine spleen cells stimulated by lipopolysaccharide (LPS) [8]. The extract of *S. orientalis* also inhibited the production of IgE *in vivo* and *in vitro*. In addition, using U266B I human IgE-bearing B cells, it also inhibited the production of IgE activated by LPS plus IL-4 [9]. The ethanol extract of *S. orientalis* was further reported to significantly suppress mouse splenocyte proliferation stimulated by concanavalin A (Con A) and LPS, in a concentration-dependent manner. ICR mice were immunized subcutaneously with ovalbumin on days 0 and 14, and administered intraperitoneally with the ethanol extract of *S. orientalis* on the day of immunization and at intervals of seven days. The extract significantly suppressed splenocyte proliferation in the ovalbumin-immunized mice induced by Con A, LPS and ovalbumin in a dose-dependent manner. The ovalbumin-specific serum IgG, IgG_1, and IgG_{2b} levels in the ovalbumin-immunized mice were also significantly reduced by the extract [10].

Siegeskaurolic acid was reported to exhibit anti-inflammatory activity. The pretreatment with siegeskaurolic acid of experimental animals at daily oral doses of 20 and 30 $mg\,kg^{-1}$ exhibited an anti-inflammatory effect against paw edema in rats induced by carrageenan, and an antinociceptive effect against abdominal constriction induced by acetic acid in mice. *In vitro*, siegeskaurolic acid significantly inhibited the production of nitric oxide (NO), prostaglandin E_2 (PGE_2), and tumor necrosis factor-α (TNF-α) in RAW 264.7 murine macrophage cells stimulated by LPS. In addition, inducible nitric oxide synthase (iNOS) and cyclooxygenase-2 (COX-2) proteins, as well as iNOS, COX-2, and TNF-α mRNA expression, were found to be inhibited by siegeskaurolic acid. Furthermore, siegeskaurolic acid inhibited nuclear factor-κB (NF-κB) activation in RAW 264.7 cells induced by LPS. This was found to be associated with the prevention of inhibitor κB degradation (IκB), and with decreased nuclear p65 and p50 protein levels [4].

A methanol extract of the aerial part of *S. glabrescens* was found to inhibit the activity of protein tyrosine phosphatase 1B as a therapeutic target for the treatment of diabetes and obesity, at 30 $\mu g\,ml^{-1}$. The active principles were found to be ent-16βH,17-isobutyryloxy-kauran-19-oic acid and ent-16βH,17-acetoxy-18-isobutyryloxy-kauran-19-oic acid, with IC_{50} values of about 9 and 31 μM, respectively. Kinetic studies suggested that both compounds were non-competitive inhibitors of protein tyrosine phosphatase 1B [5].

The extract of *S. glabrescens* was further found to inhibit the proliferation of estrogen receptor-positive MCF-7 and -negative MDA-MB-231 human mammary carcinoma cells. The extract also induced apoptosis in human mammary carcinoma cells, accompanied by the cleavage of procaspases, and poly(ADP-ribose)polymerase in MCF-7 and in MDA-MB-231 cells. In addition, apoptosis was associated with a decrease in Bcl-2 mRNA expression and an increase in Bax mRNA expression in MCF-7 cells. There was no detectable change in the MDA-MB-231 cells, which suggested that *S. glabrescens* might exert an antiproliferative action in human mammary carcinoma cells via two different apoptotic pathways: an intrinsic signal in MCF-7 cells; and an extrinsic signal in MDA-MB-231 cells [11].

Darutoside was reported to capable of terminating early pregnancy in rats over a dose range of 20–40 $mg\,kg^{-1}$ [1].

References

1 Dong, X.Y., Chen, M., Jin, W., Huang, D.X., Shen, S.M., and Li, H.T. (1989) Studies on antifertility constituents of *Siegesbeckia glabrescens* Mak. *Acta Pharm. Sin.*, **24**, 833–836.

2 Cheng, Z.H., Chou, G.X., and Wang, Z.T. (2005) A study on quality standard for Herba Siegesbeckidae. *Zhongguo Zhong Yao Za Zhi*, **30**, 257–259.

3 Xiong, J., Ma, Y.B., and Xu, Y.L. (1992) Diterpenoids from *Siegesbeckia pubescens*. *Phytochemistry*, **31**, 917–921.

4 Park, H.J., Kim, I.T., Won, J.H., Jeong, S.H., Park, E.Y., Nam, J.H., Choi, J., and Lee, K.T. (2007) Anti-inflammatory activities of ent-16αH,17-hydroxy-kauran-19-oic acid isolated from the roots of *Siegesbeckia pubescens* are due to the inhibition of iNOS and COX-2 expression in RAW 264.7 macrophages via NF-κB inactivation. *Eur. J. Pharmacol.*, **558**, 185–193.

5 Kim, S., Na, M., Oh, H., Jang, J., Sohn, C.B., Kim, B.Y., Oh, W.K., and Ahn, J.S. (2006) PTP1B inhibitory activity of kaurane diterpenes isolated from *Siegesbeckia glabrescens*. *J. Enzyme Inhib. Med. Chem.*, **21**, 379–383.

6 Kim, H.M. (1997) Effects of *Siegesbeckia pubescens* on immediate hypersensitivity reaction. *Am. J. Chin. Med.*, **25**, 163–167.

7 Kang, B.K., Lee, E.H., and Kim, H.M. (1997) Inhibitory effects of Korean folk medicine "Hi-Chum" on histamine release from mast cells *in vivo* and *in vitro*. *J. Ethnopharmacol.*, **57**, 73–79.

8 Kim, H.M., Lee, J.H., Won, J.H., Park, E.J., Chae, H.J., Kim, H.R., Kim, C.H., and Baek, S.H. (2001) Inhibitory effect on immunoglobulin E production *in vivo* and *in vitro* by *Siegesbeckia glabrescens*. *Phytother. Res.*, **15**, 572–576.

9 Hwang, W.J., Park, E.J., Jang, C.H., Han, S.W., Oh, G.J., Kim, N.S., and Kim, H.M. (2001) Inhibitory effect of immunoglobulin E production by jin-deuk-chal (*Siegesbeckia orientalis*). *Immunopharmacol. Immunotoxicol.*, **23**, 555–563.

10 Sun, H.X. and Wang, H. (2006) Immunosuppressive activity of the ethanol extract of *Siegesbeckia orientalis* on the immune responses to ovalbumin in mice. *Chem. Biodivers.*, **3**, 754–761.

11 Jun, S.Y., Choi, Y.H., and Shin, H.M. (2006) *Siegesbeckia glabrescens* induces apoptosis with different pathways in human MCF-7 and MDA-MB-231 breast carcinoma cells. *Oncol. Rep.*, **15**, 1461–1467.

Sinomenium acutum and *Sinomenium acutum* var. *cinereum*

Caulis Sinomenii (Qingfengteng) is the dried stem of *Sinomenium acutum* (Thunb.) Rehd. et Wils. or *Sinomenium acutum* (Thunb.) Rehd. et Wils. var. *cinereum* Rehd. et Wils. (Menispermaceae). It is used as a diuretic and antirheumatic agent.

Sinomenium acutum (Thunb.) Rehd. et Wils.

Chemistry

Sinomenium acutum belongs to the family of Menispermaceae. Other items of menispermaceous plants officially listed in the Chinese Pharmacopoeia are Herba Cissampelotis (*Cissampelos pareira* var. *hirsuta*), Rhizoma Menispermi (*Menispermum dauricum*), Radix Stephaniae Tetrandrae (*Stephania tetrandra*), and Radix Tinosporae (*Tinospora sagittata* or *T. capillipes*). The stem of *S. acutum* is known to contain alkaloids with sinomenine, an alkaloid with a morphinan skeleton, as the major component. The contents of sinomenine in different samples of *S. acutum* and *S. acutum* var. *cinereum* were found to range from 0.1% to 1.6%. Plant samples collected in June showed the highest content of sinomenine [1]. The Chinese Pharmacopoeia requires a qualitative determination of sinomenine in Caulis Sinomenii by thin-layer chromatographic comparison with reference substance, and a quantitative determination by HPLC. The content of sinomenine in Caulis Sinomenii should be not less than 0.5%.

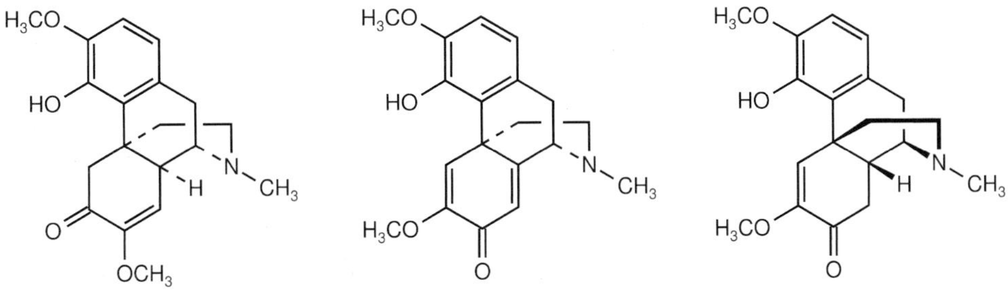

Sinomenine   Sinoacutine   Dihydrosalutaridine

Handbook of Chinese Medicinal Plants: Chemistry, Pharmacology, Toxicology
Weici Tang and Gerhard Eisenbrand

ISBN: 978-3-527-32226-8

Minor alkaloids isolated from *S. acutum* were identified as reticuline, sinoacutine, acutumine, acutumidine, stepholidine, liriodenine, magnoflorine, stepharanine, bianfugenine, dihydrosalutaridine, menisperine, laurifoline, dehydrodiscretine, epiberberine, palmatine, and sinomenidine [2]. These belong to different structural types. Only sinoacutine and dihydrosalutaridine are alkaloids related to sinomenine with a morphinan skeleton. Acutumine and acutumidine are alkaloids containing chlorine.

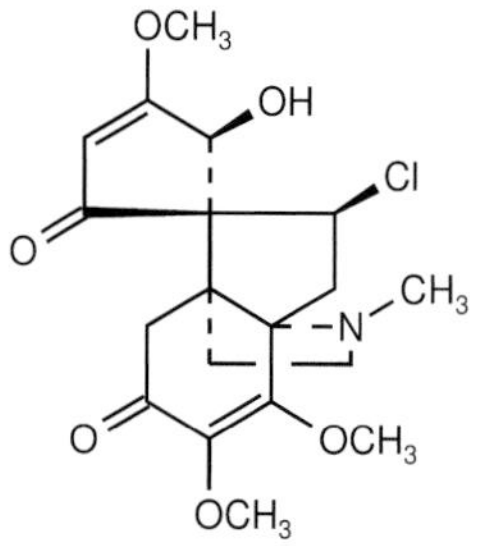

Acutumine

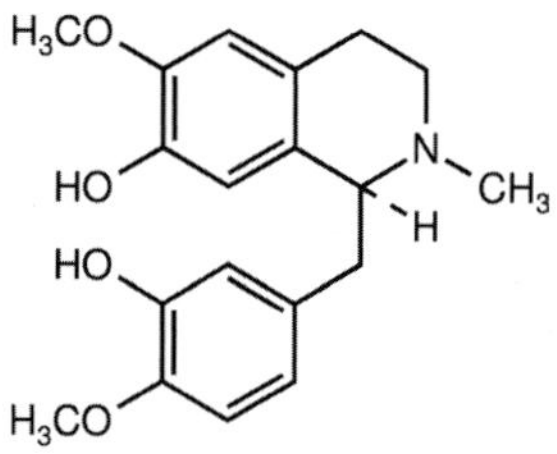

Reticuline

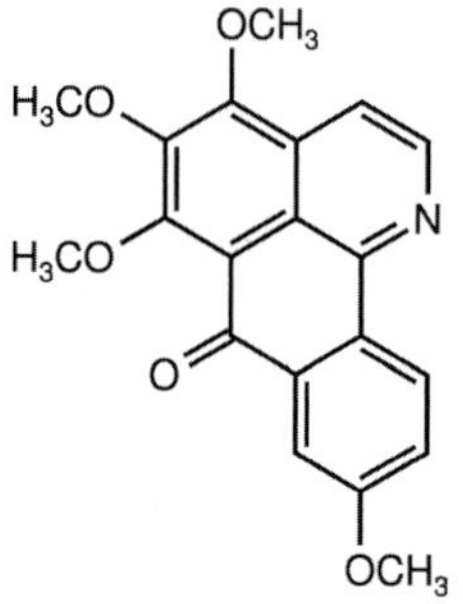

Bianfugenine

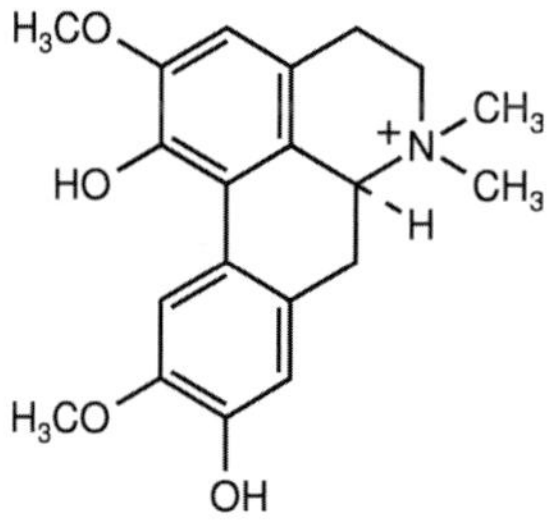

Laurifoline

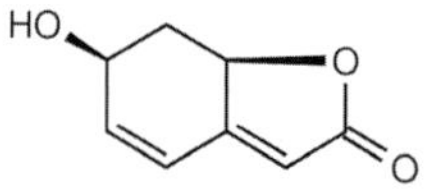

Menisdaurilide

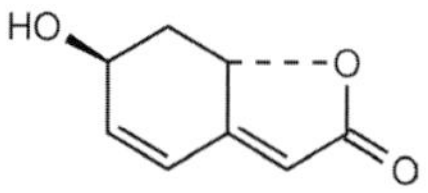

Aquilegiolide

Caffeine, 1,7-dimethylxanthine [3] and butenolides menisdaurilide and aquilegiolide [4] were also isolated from the stem of *S. acutum*.

Pharmacology and Toxicology

Pharmacokinetics of sinomenine in rabbits after intravenous (i.v.) administration followed a two-compartment open model [5]. Results from a pharmacokinetic study of sinomenine in rats showed that disposition of sinomenine in the bile represented a slow elimination phase, reaching a peak concentration at 20–40 min after an i.v. dose [6]. When a single oral dose of sinomenine was given to rats (90 mg kg^{-1}), about 80% was absorbed and distributed widely among the organs, with tissue concentrations at 45 min after dosing being highest in the kidneys, followed by the liver, lungs, spleen and heart, brain, and testicles. At 90 min after dosing, tissue concentrations in the organs were markedly decreased. The liver and kidneys had high tissue concentrations, reflecting metabolism and elimination of sinomenine. At 4 μg ml^{-1} in the plasma of rats and rabbits, or in albumin solution, sinomenine achieved a protein binding rate >60% [7]. Most pharmacokinetic parameters after oral dosing of sinomenine were comparable with those after i.v. injection.

The stem of *S. acutum* has been used in traditional Chinese medicine for the treatment of rheumatic diseases since ancient times. The major alkaloid component, sinomenine, was reported to be effective against inflammation, neuralgia, and rheumatoid arthritis. Experimental studies confirmed the antiarthritis effect of sinomenine in rats. It was reported that long-term intraperitoneal (i.p.) treatment of rats with sinomenine caused significant improvements in the arthritic score, hind paw swelling, and erythrocyte sedimentation rate (ESR). Short- and middle-term treatment with sinomenine in acute antigen-induced arthritis of rats resulted in a dose-dependent decrease of both joint swelling and ESR and a significant reduction in joint destruction. The long-term treatment of chronic antigen-induced arthritis with sinomenine partially ameliorated the symptoms, and significantly counteracted joint destruction. *In vitro*, sinomenine markedly inhibited the proliferation of synovial fibroblasts from rats with antigen-induced arthritis or from normal rats, both at rest and following activation with either transforming growth factor β_2 (TGF-β_2) or interleukin-1β (IL-1β) [8].

Sinomenine, at concentrations of 0.5 and 1.0 *M*, significantly inhibited the expression of vascular adhesion molecule-1 (VCAM-1) in human umbilical vein endothelial cells (HUVEC) stimulated by TNF-α [9]. As an active compound for the treatment of rheumatoid arthritis, it was found to possess antiangiogenic properties. It inhibited the proliferation of HUVEC induced by fibroblast growth factor, and induced cell-cycle arrest in G_1 phase. Sinomenine disrupted tube formation of HUVEC on Matrigel, and suppressed chemotaxis of the cells. In addition, it reduced neovascularization and microvascular outgrowth in isolated rat aorta rings. Furthermore, it also reduced the transmigration of granulocytic differentiated HL-60 cells across HUVEC monolayers activated by IL-1β. The inhibition of leukocyte migration across blood vessel walls and the antiangiogenic effect of sinomenine might contribute to its therapeutic mechanisms in alleviating the pathogenesis of rheumatoid arthritis [10].

The treatment of peritoneal macrophages and synoviocytes from rats with adjuvant arthritis resulted in a significant inhibition of nuclear factor-κB (NF-κB) activity, TNF-α and IL-1β mRNA expression in a concentration-dependent manner, at concentrations from 30 to 120 $\mu g\,ml^{-1}$. In addition, the protein level of IκBα, the inhibitor of NF-κB was increased by treatment with sinomenine. This indicated that sinomenine decreased the mRNA expression of TNF-α and IL-1β by inhibiting the NF-κB DNA binding activity as a consequence of upregulating IκBα expression of peritoneal macrophages and synoviocytes from adjuvant arthritis rats [11]. Sinomenine also inhibited the proliferation of RAW264.7 murine macrophages by inducing apoptosis. Sinomenine-induced apoptosis in RAW264.7 cells required the activation of extracellular signal-regulated protein kinase (ERK). In addition, it increased the expression of p27/KIP1 and of the proapoptotic factor Bax, and decreased the expression of Bcl-2 [12]. Sinomenine was also found to inhibit the production of prostaglandin E_2 (PGE_2) and to selectively suppress the expression of cyclooxygenase-2 (COX-2) mRNA in human peripheral monocytes stimulated by lipopolysaccharide (LPS) [13].

The oral administration of an aqueous extract of *S. acutum* to mice inhibited, dose-dependently, the systemic anaphylactic reaction induced by compound 48/80, with an EC_{50} of 1 $g\,kg^{-1}$, significantly reducing the plasma histamine levels. The extract at concentrations of 1 to 1000 $\mu g\,ml^{-1}$ dose-dependently inhibited histamine release from rat peritoneal mast cells activated by compound 48/80, and significantly inhibited tumor necrosis factor-α

(TNF-α) production [14]. Mice, immunized with ovalbumin and complete Freund's adjuvant at day 0, were treated orally by sinomenine over a period of 21 days. On day 21, anti-ovalbumin immunoglobulin G_{2a} (IgG_{2a}) and interferon-γ (IFN-γ) were used as indicators of Th1 immune response. Anti-ovalbumin IgG_1, IgE, and IL-5 served as indicators of Th2 responses, and TGF-β of Th3 immune responses. The results showed that treatment with sinomenine was followed by decreases in anti-ovalbumin IgG and antigen-specific splenocyte proliferation. Sinomenine also suppressed the production of anti-ovalbumin IgG_{2a}, IgG_1 and IgE, as well as the secretion of IFN-γ and IL-5. In addition, sinomenine enhanced the secretion of TGF-β, which suggested that it appeared to have suppressive effects on both Th1 and Th2 immune responses, and that Th1 responses might be more preferentially suppressed by sinomenine [15].

An inhibitory activity of sinomenine on lymphocyte proliferation was demonstrated *in vitro*, using mouse spleen cells and human peripheral blood mononuclear cells. It was found that sinomenine markedly inhibited [^{3}H]thymidine incorporation in mouse spleen cells activated with concanavalin A (Con A), and also in human peripheral blood mononuclear cells activated with phytohemagglutinin, 12-*O*-tetradecanoylphorbol-13-acetate (TPA) plus ionomycin, or mixed lymphocyte culture. The inhibition of lymphocyte proliferation was reversible. Sinomenine showed no direct cytotoxicity, and had no inhibitory effect on cell proliferation linked to cytokine-independent growth [16]. It markedly decreased prostaglandin E_2 (PGE_2) and leukotriene C_4 (LTC_4) synthesis of mouse macrophages *in vitro* stimulated by zymosan or calcium ionophore, and also significantly inhibited the nitric oxide production of macrophage-like RAW 264.7 cells activated by IFN-γ and LPS [17].

Sinomenine was also reported to exert anti-inflammatory and neuroprotective effects. It reduced the production of cytokines from the rat retinal microglia, as activated by advanced glycation end products (AGE) at both gene and protein levels. In addition, it concentration-dependently attenuated reactive oxygen species (ROS) production and reduced the nuclear translocation of NF-κB p65 in retinal microglia stimulated by advanced glycation end products [18]. Sinomenine also effectively protected dopaminergic neuron death in rat midbrain neuron-glial cultures, associated with an inhibition of microglial activation. It significantly decreased TNF-α, PGE_2 and ROS production by microglia [19]. It also down-regulated the levels of COX-2 and the expression of COX-2 mRNA, and reduced the production of PGE_2 in PC12 cells stimulated by LPS. Sinomenine also significantly suppressed NF-κB activity in LPS-stimulated PC12 cells. No inhibition of proliferation of PC12 cells by sinomenine treatment was observed [20].

The immunosuppressive effects of Sinomenine were further investigated. It decreased the thymus and spleen weights of mice; inhibited the antibody production and delayed hypersensitivity induced by sheep erythrocytes; and prolonged the survival of myocardial allografts. Sinomenine was also found to inhibit the proliferation of mouse spleen cells *in vitro* [21]. Sinomenine and cyclosporine were reported to show a synergistic effect in the high responder ACI-to-Lewis cardiac allograft model, with allograft survival being prolonged from 5 to 42 days [22].

In experimental studies, sinomenine was also found to show a hypotensive effect in spontaneously hypertensive rats. The i.p. injection of sinomenine at doses of 2.5–10 mg kg^{-1} decreased the systolic blood pressure in spontaneously hypertensive rats in dose-dependent manner, but had no effect in normotensive rats. In isolated rat aortic rings precontracted with phenylephrine or KCl, sinomenine produced concentration-dependent relaxation and decreased the elevated intracellular Ca^{2+} concentration elicited by phenylephrine or KCl. It was suggested that sinomenine caused vascular relaxation by opening ATP-sensitive K^+ channels, thus decreasing intracellular Ca^{2+} concentrations [23]. Sinomenine also dilated rat aorta ring strips contracted by norepinephrine (noradrenaline), KCl, and TPA [24].

Clinical side effects encountered with high doses of sinomenine were reported to be pruritus in the head and upper part of the body, edema around the lips and eyelids, and temporary cephalalgia [25].

Sinomenine was also reported to decrease the contraction amplitude of isolated guinea pig left auricle; to inhibit myocardial excitability induced by epinephrine (adrenaline); and to prolong the functional refractory period; this suggested that sinomenine may inhibit the influx of Na^+ and Ca^{2+} in the myocardium [26]. Inhibition of the slow response action potential [27] and contraction force of guinea pig heart muscle [28] by sinomenine were described. It has also been found to protect against acute myocardial ischemia–reperfusion damage in rats, as produced by temporary ligation of the left coronary artery [29]. The pretreatment of rat polymorphonuclear neutrophils with sinomenine inhibited morphological changes and lysosome enzyme of the cells [30].

Sinomenine was also found to protect against hepatic damage in galactosamine-sensitized mice, as induced by LPS. It was proposed that sinomenine prevented hepatic failure by suppressing TNF production and ROS generation. It significantly suppressed the *in vitro* production of superoxide anion and hydrogen peroxide in macrophages stimulated by TPA [31]. A number of alkaloids from the stems of *S. acutum* such as acutumine, sinomenine and magnoflorine, were found to be able to reverse the multidrug resistance of cancer cells mediated by P-glycoprotein [32].

References

1 Li, A.J., Wang, X.M., and Feng, H.L. (1987) Quantitative determination of sinomenine in qingfengteng (*Sinomenium acutum* and *S. acutum* var. *cinereum*) by TLC densitometry and relation between the content and the seasons. *Bull. Chin. Mater. Med.*, **12**, 332–334.

2 Moriyasu, M., Ichimaru, M., Nishiyama, Y., and Kato, A. (1994) Isolation of alkaloids from plant materials by the combination of ion-pair extraction and preparative ion-pair chromatography using sodium perchlorate. II Sinomeni Caulis et Rhizoma. *Nat. Med.*, **48**, 287–290.

3 Jiang, M., Kameda, K., Han, L.K., Kimura, Y., and Okuda, H. (1998) Isolation of lipolytic substances caffeine and 1,7-dimethylxanthine from the stem and rhizome of *Sinomenium acutum*. *Planta Med.*, **64**, 375–377.

4 Otsuka, H., Ito, A., Fujioka, N., Kawamata, K., Kasai, R., Yamasaki, K., and Satoh, T. (1993) Butenolides from *Sinomenium acutum*. *Phytochemistry*, **33**, 389–392.

5 Lou, S., Cai, H., Zhang, X., and Pan, X. (1992) Pharmacokinetics of sinomenine by HPLC. *China J. Chin. Mater. Med.*, **17**, 424–426.

6 Tsai, T.H. and Wu, J.W. (2003) Regulation of hepatobiliary excretion of sinomenine by P-glycoprotein in Sprague-Dawley rats. *Life Sci.*, **72**, 2413–2426.

7 Liu, Z.Q., Chan, K., Zhou, H., Jiang, Z.H., Wong, Y.F., Xu, H.X., and Liu, L. (2005) The pharmacokinetics and tissue distribution of sinomenine in rats and its protein binding ability in vitro. *Life Sci.*, **77**, 3197–3209.

8 Liu, L., Buchner, E., Beitze, D., Schmidt-Weber, C.B., Kaever, V., Emmrich, F., and Kinne, R.W. (1996) Amelioration of rat experimental arthritis by treatment with the

alkaloid sinomenine. *Int. J. Immunopharmacol.*, **18**, 529–543.

9 Huang, J., Lin, Z., Luo, M., Lu, C., Kim, M.H., Yu, B., and Gu, J. (2007) Sinomenine suppresses TNF-α-induced VCAM-1 expression in human umbilical vein endothelial cells. *J. Ethnopharmacol.*, **114**, 180–185.

10 Kok, T.W., Yue, P.Y., Mak, N.K., Fan, T.P., Liu, L., and Wong, R.N. (2005) The anti-angiogenic effect of sinomenine. *Angiogenesis*, **8**, 3–12.

11 Wang, Y., Fang, Y., Huang, W., Zhou, X., Wang, M., Zhong, B., and Peng, D. (2005) Effect of sinomenine on cytokine expression of macrophages and synoviocytes in adjuvant arthritis rats. *J. Ethnopharmacol.*, **98**, 37–43.

12 He, X., Wang, J., Guo, Z., Liu, Q., Chen, T., Wang, X., and Cao, X. (2005) Requirement for ERK activation in sinomenine-induced apoptosis of macrophages. *Immunol. Lett.*, **98**, 91–96.

13 Wang, W.J., Wang, P.X., and Li, X.J. (2003) The effect of sinomenine on cyclooxygenase activity and the expression of COX-1 and COX-2 mRNA in human peripheral monocytes. *Zhongguo Zhong Yao Za Zhi*, **28**, 352–355.

14 Kim, H.M., Moon, P.D., Chae, H.J., Kim, H.R., Chung, J.G., Kim, J.J., and Lee, E.J. (2000) The stem of *Sinomenium acutum* inhibits mast cell-mediated anaphylactic reactions and tumor necrosis factor-alpha production from rat peritoneal mast cells. *J. Ethnopharmacol.*, **70**, 135–141.

15 Feng, H., Yamaki, K., Takano, H., Inoue, K., Yanagisawa, R., and Yoshino, S. (2006) Suppression of Th1 and Th2 immune responses in mice by Sinomenine, an alkaloid extracted from the Chinese medicinal plant *Sinomenium acutum*. *Planta Med.*, **72**, 1383–1388.

16 Liu, L., Resch, K., and Kaever, V. (1994) Inhibition of lymphocyte proliferation by the anti-arthritic drug sinomenine. *Int. J. Immunopharmacol.*, **16**, 685–691.

17 Liu, L., Riese, J., Resch, K., and Kaever, V. (1994) Impairment of macrophage eicosanoid and nitric oxide production by an alkaloid from *Sinomenium acutum*. *Arzneim.-Forsch.*, **44**, 1223–1226.

18 Wang, A.L., Li, Z., Yuan, M., Yu, A.C., Zhu, X., and Tso, M.O. (2007) Sinomenine inhibits activation of rat retinal microglia induced by advanced glycation end products. *Int. Immunopharmacol.*, **7**, 1552–1558.

19 Qian, L., Xu, Z., Zhang, W., Wilson, B., Hong, J.S., and Flood, P.M. (2007) Sinomenine, a natural dextrorotatory morphinan analog, is anti-inflammatory and neuroprotective through inhibition of microglial NADPH oxidase. *J. Neuroinflammation.*, **4**, 23.

20 Chen, W., Shen, Y.D., Zhao, G.S., and Yao, H. P. (2004) Inhibitory effect of sinomenine on expression of cyclooxygenase-2 in lipopolysaccharide-induced PC-12 cells. *Zhongguo Zhong Yao Za Zhi*, **29**, 900–903.

21 Peng, H.M., Ding, X.Y., Liu, X.S., and Liu, Z.G. (1988) Immunosuppressive effects of sinomenine. *Acta Pharmacol. Sin.*, **9**, 377–380.

22 Vieregge, B., Resch, K., and Kaever, V. (1999) Synergistic effects of the alkaloid sinomenine in combination with the immunosuppressive drugs tacrolimus and mycophenolic acid. *Planta Med.*, **65**, 80–82.

23 Lee, P.Y., Chen, W., Liu, I.M., and Cheng, J.T. (2007) Vasodilatation induced by sinomenine lowers blood pressure in spontaneously hypertensive rats. *Clin. Exp. Pharmacol. Physiol.*, **34**, 979–984.

24 Nishida, S. and Satoh, H. (2006) *In vitro* pharmacological actions of sinomenine on the smooth muscle and the endothelial cell activity in rat aorta. *Life Sci.*, **79**, 1203–1206.

25 Okuda, T., Umezawa, Y., Ichikawa, M., Hirata, M., Oh-i, T., and Koga, M. (1995) A case of drug eruption caused by the crude drug Boi (*Sinomenium* stem/Sinomeni caulis et Rhizoma). *J. Dermatol.*, **22**, 795–800.

26 Li, C.X. and Zhao, G.S. (1986) Effects of sinomenine on isolated guinea pig auricle. *J. Xian Med. Univ.*, **7**, 24–26.

27 Li, C.X., Zhao, G.S., and Li, X.G. (1987) Effects of sinomenine on slow response action potential in guinea pig papillary muscle. *Acta Pharm. Sin.*, **22**, 566–569.

28 Li, C.X., Zhao, G.S., and Li, X.G. (1987) Effects of sinomenine on action potential and force of contraction in guinea pig heart muscle. *Acta Pharm. Sin.*, **22**, 561–565.

29 Xie, S.X. and Jin, Q.Q. (1993) Prevention of sinomenine on isolated rat myocardial reperfusion injury. *Acta Pharmacol. Sin.*, **14** (Suppl.), S12–S15.

30 Wang, J.Z. and Jin, L.J. (1992) The inhibitory effects of sinomenine on complement-induced shape change and degranulation in rat neutrophils. *Chin. Pharmacol. Bull.*, **8**, 288–291.

31 Kondo, Y., Takano, F., Yoshida, K., and Hojo, H. (1994) Protection by sinomenine against endotoxin-induced fulminant hepatitis in galactosamine-sensitized mice. *Biochem. Pharmacol.*, **48**, 1050–1052.

32 Min, Y.D., Choi, S.U., and Lee, K.R. (2006) Aporphine alkaloids and their reversal activity of multidrug resistance (MDR) from the stems and rhizomes of *Sinomenium acutum*. *Arch. Pharm. Res.*, **29**, 627–632.

Smilax glabra

Rhizoma Smilacis Glabrae (Tufuling) is the dried rhizome of *Smilax glabra* Roxb. (Liliaceae). It is used as a diuretic and antirheumatic agent.

The whole plant of *Smilax riparia* A DC. is listed in the appendix of the Chinese Pharmacopoeia.

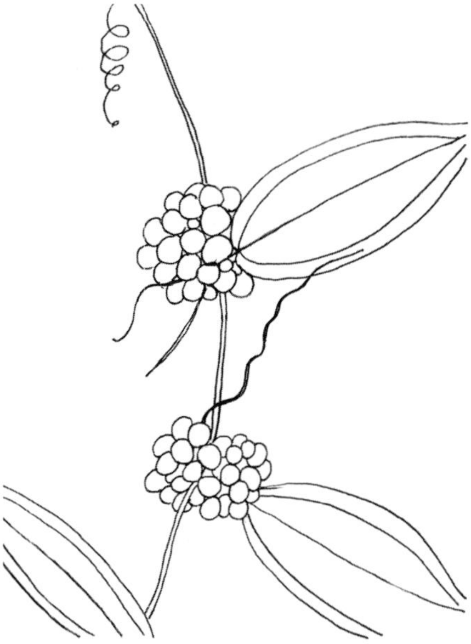

Smilax glabra Roxb.

Chemistry

The rhizome of *S. glabra* was reported to be rich in flavanones and flavanone glycosides, with the flavanone glycoside astilbin as the major component [1]. Minor flavone derivatives were identified as engeletin and its *cis*-isomer isoengeletin, isostilbin, dihydroquercetin, 7,6′-dihydroxy-3′-methoxyisoflavone, quercetin, and kaempferol [2, 3]. A chromone glycoside from the rhizome of *S. glabra* was identified as smiglanin [4, 5].

The stilbene derivative resveratrol, phytosterols diosgenin, stigmasterol and stigmasterol 3-*O*-β-D-glucopyranoside, β-sitosterol and β-sitosterol 3-*O*-β-D-glucopyranoside and daucosterol were isolated from the rhizome of *S. glabra*. In addition, shikimic acid, 3-*O*-caffeoylshikimic acid, and ferulic acid were also isolated from the rhizome [1–5]. The main components of the essential oil of the rhizome were found to be long-chain aliphatic esters [6].

Structurally related phenylpropane esters of sucrose glycosides from the rhizome of *S. glabra* were identified as smiglasides A–E, and helonioside [7].

Another item with *Smilax* species listed in the Chinese Pharmacopoeia is Rhizoma Smilacis Chinae. Rhizoma Smilacis Chinae (Bagia) is the dried rhizome of *Smilax China* L. It is used as a diuretic agent. The rhizome of *S. China* is known to contain saponins and sapogenins with diogenin as the major component. The Chinese Pharmacopoeia requires a qualitative determination of diosgenin in Rhizoma Smilacis Chinae by thin-layer chromatographic comparison with reference substance and a quantitative determination of diosgenin by HPLC. The content of diosgenin in Rhizoma Smilacis Chinae should be not less than 0.04%.

Pharmacology and Toxicology

The flavanone glycosides, including astilbin and engeletin and the stilbene resveratrol, were found to protect hepatocytes from damage in mice with an immunological liver injury. The saponin fraction from the rhizome of *S. glabra* was found to protect atherosclerosis in quails, as induced by a diet containing high cholesterol concentration. Both, the serum cholesterol levels and incidence of atherosclerosis in treated animals were significantly reduced [8].

Handbook of Chinese Medicinal Plants: Chemistry, Pharmacology, Toxicology
Weici Tang and Gerhard Eisenbrand

ISBN: 978-3-527-32226-8

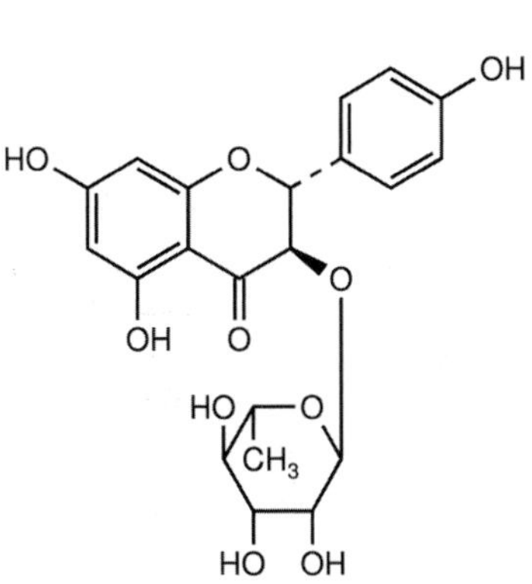

Engeletin

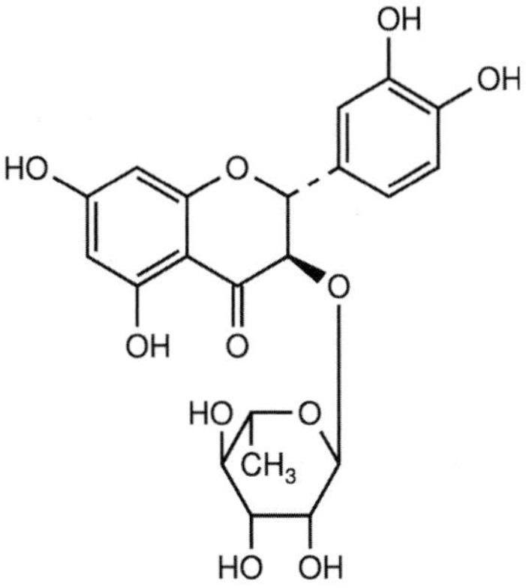

Astilbin

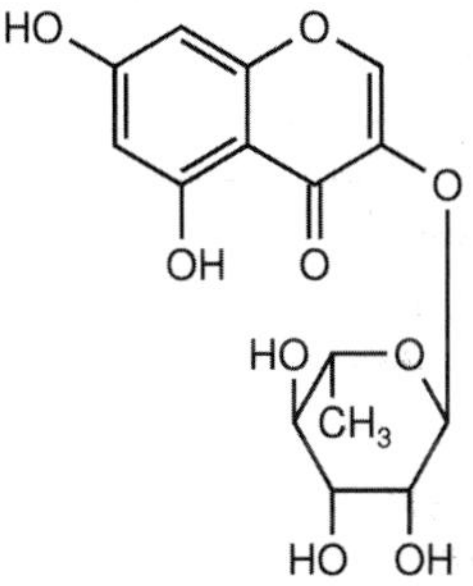

Smiglanin

Smiglanin given orally to mice was found to prevent acute myocardial ischemia, as induced by noradrenaline. The protective activity was mediated by an inhibition of lipid peroxidation [9].

A hypoglycemic effect of *S. glabra* was observed in normal and KK-Ay mice, an animal model of non-insulin-dependent diabetes with hyperinsulinemia. The methanol extract of the rhizome at a dose of 100 mg kg^{-1} significantly reduced the blood glucose of normal mice at 4 h after intraperitoneal (i.p.) administration, and also significantly lowered the blood glucose of KK-Ay mice under similar conditions; however, the extract did not affect the blood glucose in streptozotocin-induced diabetic mice as a model of insulin-dependent

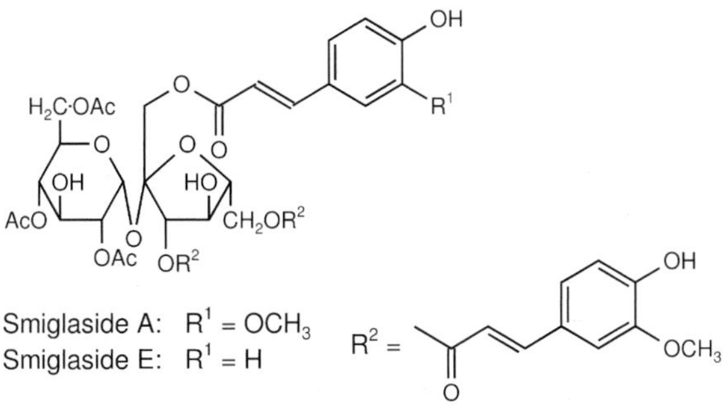

Smiglaside A: $R^1 = OCH_3$
Smiglaside E: $R^1 = H$

diabetes with hypoinsulinemia. The rhizome extract also inhibited epinephrine (adrenaline)-induced hyperglycemia in mice [10].

The aqueous extract of the rhizome exerted a remarkable anti-inflammatory effect, inhibiting the contact dermatitis in mice induced by picryl chloride and the footpad reaction induced by sheep red blood cells. In mice, the aqueous extract also showed a significant anti-inflammatory activity against xylene-induced ear edema and egg white-induced footpad edema. However, it did not show any notable effect on immunoglobulin IgM and IgG levels in mice against sheep red blood cells. This suggested that the extract exhibits a selective activity to inhibit the cellular immune response, without inhibiting the humoral immune response [11]. Clinical trials with an extract of *S. glabra* for the treatment of 164 psoriasis patients were reported. Among patients treated, 79 were cured in 10–100 days (average 55 days), with a total effective rate of 77%. Treatment was most effective for simple acute punctata psoriasis [12].

The aqueous extract of the rhizome of *S. glabra* was reported to remarkably inhibit the primary inflammation of adjuvant arthritis in rats. Oral administration of the extract to rats at doses of 400 and 800 $mg\,kg^{-1}$ significantly inhibited the swelling of the adjuvant-non-injected footpad of adjuvant arthritis rats. The extract also significantly reduced the production of interleukin-1 (IL-1), tumor necrosis factor (TNF) and nitric oxide (NO) by peritoneal macrophages, as stimulated by lipopolysaccharide (LPS). It significantly attenuated the decrease in weight gain of adjuvant arthritis rats, as well as T-lymphocyte proliferation and IL-2 production by splenocytes induced with concanavalin A (Con A). Furthermore, the extract significantly attenuated delayed-type hypersensitivity induced by picryl chloride to almost normal levels from levels induced by different treatments of cyclophosphamide, with a normalization of the CD4/CD8 ratio. This suggested that the extract of the rhizome of *S. glabra* improved adjuvant arthritis via a downregulation of overactivated macrophages and an upregulation of the dysfunctional T lymphocytes during the later phase of arthritis [13].

Shikimic acid was reported to protect from focal cerebral ischemia injury in rats subjected to middle cerebral artery thrombosis. Shikimic acid, at i.p. doses of 25 and 50 $mg\,kg^{-1}$ for three days before middle cerebral artery thrombosis, significantly prevented the neurologic deficit and reduced the infarct size [14].

Astilbin was reported to show a unique immunosuppressive activity through a selective inhibition of activated T lymphocytes, which might be beneficial for the treatment of human immune diseases. The incubation of astilbin with rat liver microsomal and cytosolic fractions resulted in the isolation of 3′-*O*-methylastilbin as a metabolite. After oral administration of astilbin to rats at a dose of 0.22 $mM\,kg^{-1}$, 3′-*O*-methylastilbin was detected both in blood and urine. 3′-*O*-Methylastilbin was found to inhibit picryl chloride-induced ear swelling in mice, and to suppress the expression of TNF-α and interferon-γ (INF-γ), similarly to astilbin [15].

References

1 Chen, G., Shen, L., and Jiang, P. (1996) Flavanonol glucosides of *Smilax glabra* Roxb. *China J. Chin. Mater. Med.*, **21**, 355–357.

2 Chen, T., Li, J., Cao, J., Xu, Q., Komatsu, K., and Namba, T. (1999) A new flavanone isolated from rhizoma smilacis glabrae and the structural requirements of its derivatives for

preventing immunological hepatocyte damage. *Planta Med.*, **65**, 56–59.

3 Yi, Y.J., Cao, Z.Z., Yang, D.L., Cao, Y., Wu, Y.P., and Zhao, S.X. (1998) Studies on chemical constituents of *Smilax glabra* Roxb. IV. *Acta Pharm. Sin.*, **33**, 873–875.

4 Li, Z., Li, D., Owen, N.L., Len, Z.K., Cao, Z.Z., and Yi, Y.J. (1996) Structure determination of a new chromone glycoside by 2D inadequate NMR and molecular modeling. *Magn. Reson. Chem.*, **34**, 512–517.

5 Cao, Z.Z., Yi, Y.J., Cao, Y., and Leng, Z.K. (1995) A new chromone glycoside from *Smilax glabra* Roxb. *Chin. Chem. Lett.*, **6**, 587–588.

6 Cao, Z.Z., Yi, Y.J., Cao, Y., and Yang, D.L. (1994) Chemical constituents of essential oil from *Smilax glabra* Roxb. *Tianran Chanwu Yanjiu Yu Kaifa*, **6**, 33–36.

7 Chen, T., Li, J.X., and Xu, Q. (2000) Phenylpropanoid glycosides from *Smilax glabra*. *Phytochemistry*, **53**, 1051–1055.

8 Zhang, K.J., Zou, Y.L., and Zhou, C.M. (1991) Preventive effects of crude saponins of glabrous greenbrier (*Smilax glabra*) on the cholesterol-fed atherosclerosis of quails. *Chin. Trad. Herbal Drugs*, **22**, 411–412. 418.

9 Li, Y.Q., Zhou, C.M., Wang, X.W., Li, Z.L., Zhou, K., and Zhang, K.J. (1996) Protective effect of smiglabrin from *Smilax glabra* on isopropyl noradrenaline induced mouse myocardial ischemia. *Chin. Trad. Herbal Drugs*, **27**, 418–420.

10 Fukunaga, T., Miura, T., Furuta, K., and Kato, A. (1997) Hypoglycemic effect of the rhizomes of *Smilax glabra* in normal and diabetic mice. *Biol. Pharm. Bull.*, **20**, 44–46.

11 Xu, Q., Wang, R., Xu, L.H., and Jiang, J.Y. (1993) Effects of *Smilax glabra* on cellular and humoral immune responses. *Chin. J. Immunol.*, **9**, 39–42.

12 Sun, Y.Q. (1986) Treatment of 164 cases of psoriasis with Yinxieling. *Chin. J. Dermatol.*, **19**, 30–31.

13 Jiang, J. and Xu, Q. (2003) Immunomodulatory activity of the aqueous extract from rhizome of *Smilax glabra* in the later phase of adjuvant-induced arthritis in rats. *J. Ethnopharmacol.*, **85**, 53–59.

14 Ma, Y., Xu, Q.P., Sun, J.N., Bai, L.M., Guo, Y.J., and Niu, J.Z. (1999) Antagonistic effects of shikimic acid against focal cerebral ischemia injury in rats subjected to middle cerebral artery thrombosis. *Acta Pharmacol. Sin.*, **20**, 701–704.

15 Guo, J., Qian, F., Li, J., Xu, Q., and Chen, T. (2007) Identification of a new metabolite of astilbin, 3′-*O*-methylastilbin, and its immunosuppressive activity against contact dermatitis. *Clin. Chem.*, **53**, 465–471.

Sophora flavescens

Radix Sophorae Flavescentis (Kushen) is the dried root of *Sophora flavescens* Ait. (Fabaceae). It is used as a diuretic and for the treatment of diarrhea, gastrointestinal hemorrhage, jaundice with oliguria, and eczema. The Chinese Pharmacopoeia notes that Radix Sophorae Flavescentis is incompatible with Rhizoma et Radix Veratri.

Sophora flavescens Ait.

Chemistry

The root of *S. flavescens* is known to contain a number of quinolizidine alkaloids, with matrine, oxymatrine, and sophoridine as the major components. The Chinese Pharmacopoeia requires a qualitative determination of oxymatrine, matrine and sophoridine in Radix Sophorae Flavescentis by thin-layer chromatographic (TLC) comparison with reference substances, and a quantitative determination of matrine and oxymatrine in Radix Sophorae Flavescentis by HPLC. The total content of matrine and oxymatrine in Radix Sophorae Flavescentis should be not less than 1.2%. Minor alkaloids from the root of *S. flavescens* were identified as stereoisomers and dehydroanalogues of matrine, including sophoranol, sophocarpine, sophoramine, sophocarpine *N*-oxide, 5- episophocarpine, isomatrine, oxysophocarpine, 9α-hydroxymatrine, sophoranol *N*-oxide, 7,8-dehydrosophoramine [1, 2], 12α-hydroxysophocarpine, oxysophocarpine, 9α-hydroxysophocarpine, lehmannine, and 13,14-dehydrosophoridine [3].

Another item listed in the Chinese Pharmacopoeia with matrine and oxymatrine as major components is Radix et Rhizoma Sophorae Tonkinensis (Shandougen), the dried root of *Sophora tonkinensis* Gapnep. This is used as an antipyretic and anti-inflammatory agent for the treatment of painful swelling of the throat and gums, gingivitis, and gingivalgia. The Chinese Pharmacopoeia requires a qualitative determination of matrine and oxymatrine in Radix et Rhizoma Sophorae Tokinensis by TLC comparison with the reference substances, and a quantitative determination of oxymatrine in Radix et Rhizoma Sophorae Tonkinensis by TLC-densitometry. The content of oxymatrine in Radix Sophorae

Handbook of Chinese Medicinal Plants: Chemistry, Pharmacology, Toxicology
Weici Tang and Gerhard Eisenbrand

ISBN: 978-3-527-32226-8

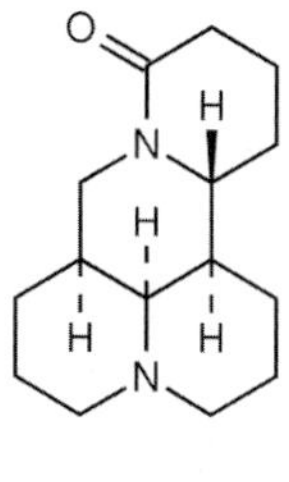
Matrine

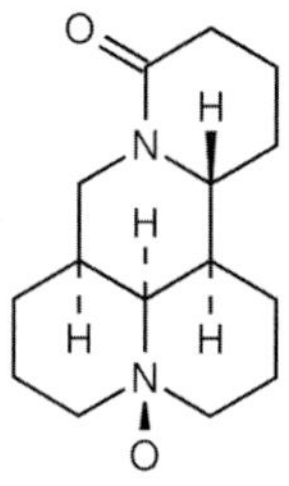
Oxymatrine

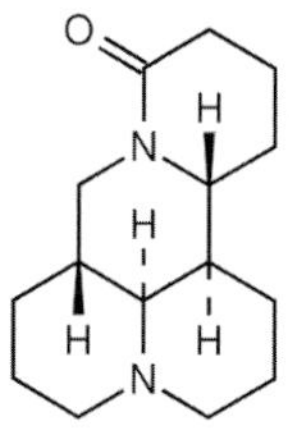
Sophoridine

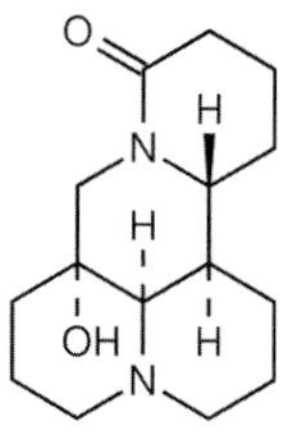
Sophoranol

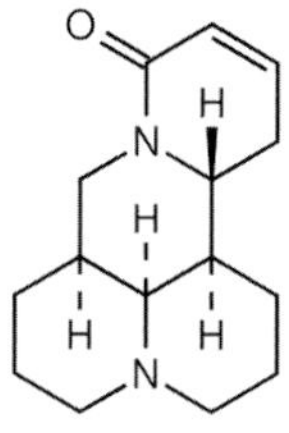
Sophocarpine

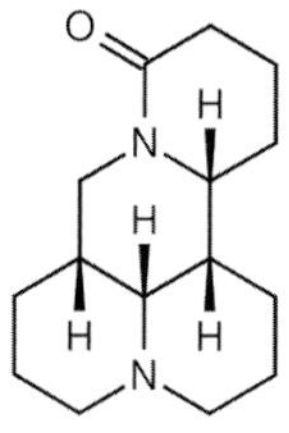
Isomatrine

Tonkinensis should be not less than 0.4%. Other types of quinolizidine alkaloids without matridine skeleton from the roots of *S. flavescens* were identified as methylcytisine, anagyrine, and baptifoline [4].

Flavone and related compounds, especially prenylated derivatives from the root 7-*O*-β-D-xylopyranosyl-(1 → 6)-β-D-glucopyranoside of formononetin. Kuraridinol, kushenol D and kuraridin are prenylchalcones. Kurarinol, neokurarinol, norkurarinol, isokurarinone, kushenols A, B, E, F, H, I, K–N, kurarinone, norkurarinone, and sophoraflavanone B all represent prenylfla-

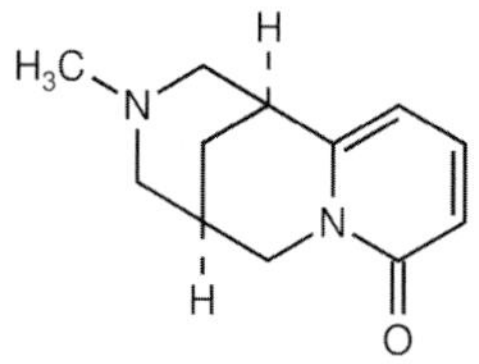
Methylcytisine

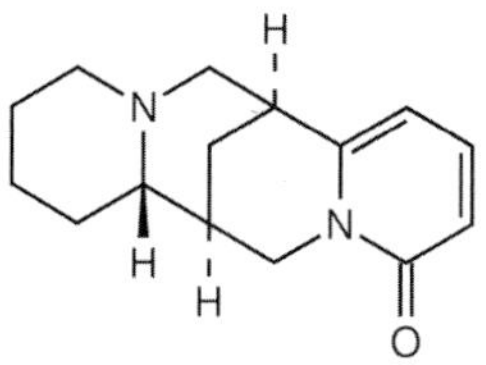
Anagyrine

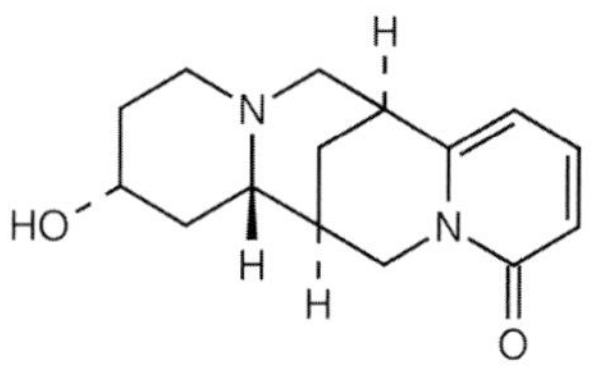
Baptifoline

of *S. flavescens*, were identified as kuraridinol, kurarinol, neokurarinol, norkurarinol, isokurarinone, formononetin, kushenols A–O [5], kushequinone A, kushenin, kuraridin, kurarinone, norkurarinone, noranhydroicaritin, sophoraflavanone B, maackiain, 4-methoxy-maackiain, and trifolirhizin [6]. Among these compounds, formononetin is an isoflavone and kushenol O is the vanones. Kushenol J is 7-(6-*O*-β-D-xylopyranosyl-β-D-glucopyranosyloxy-4′-methoxyflvanonol. Kushenols C and G, and noranhydroicaritin are three prenylflavones. Kushequinone A is a quinone derivative with prenyl side chain, whereas kushenin and maackiain are compounds related to coumestrol. Trifolirhizin is the β-D-glucopyranoside of maackiain.

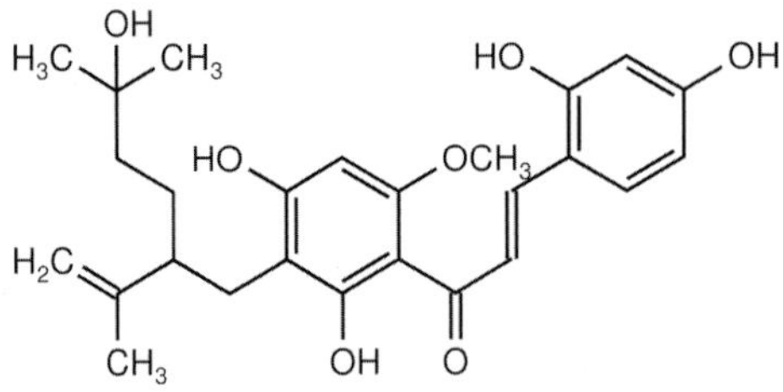

Kuraridinol

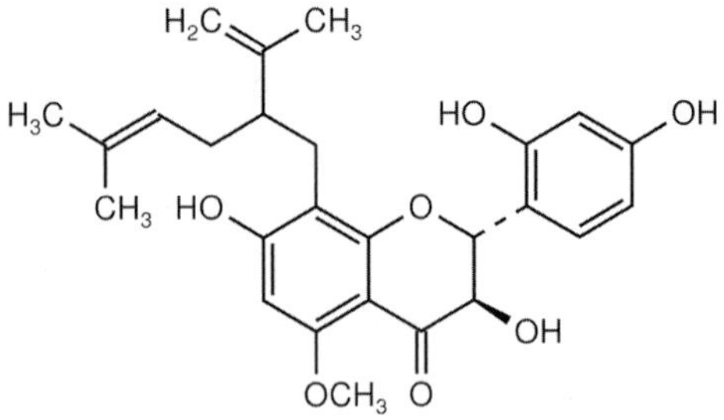

Kushenol I

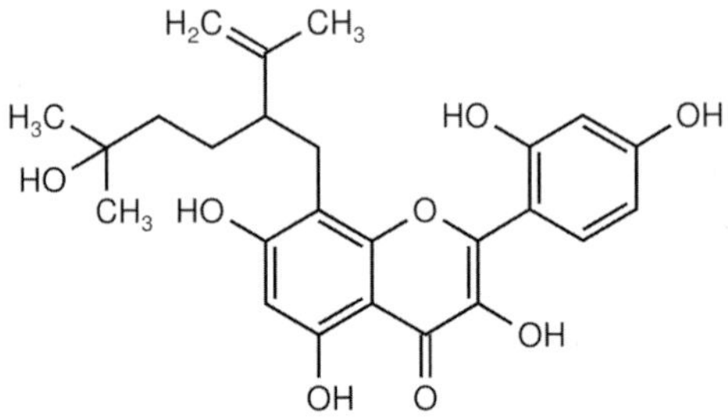

Kushenol G

Saponins sophoraflavoside I with soyasapogenol B as aglycone, soyasaponin I, oxytrogenin and its glycosides were isolated from the root of *S. flavescens* [7].

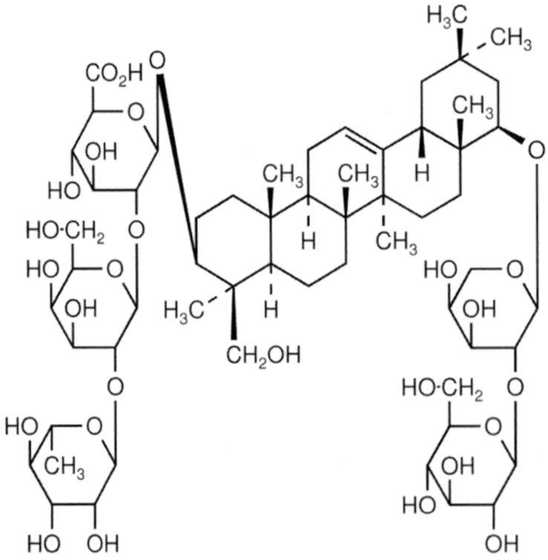

Sophoraflavoside I

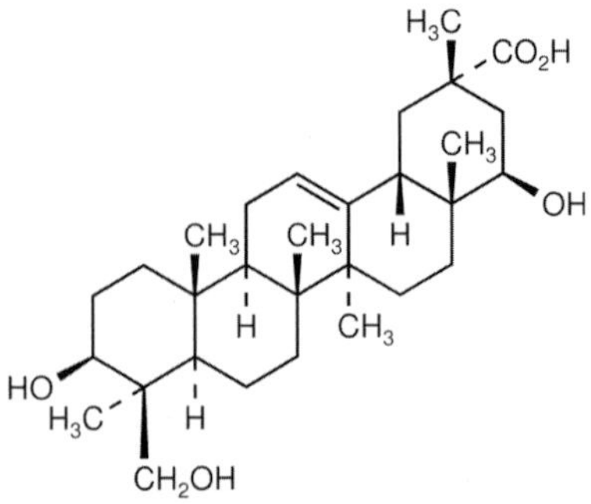

Oxytrogenin

Pharmacology and Toxicology

The pharmacokinetic data of matrine in rabbits after intravenous (i.v.) administration fitted an open two-compartment model, with $t_{1/2\alpha}$ of 4.4 min and $t_{1/2\beta}$ of 79 min [8]. Similar pharmacokinetic parameters were observed in healthy volunteers after i.v. infusion. The excretion of unchanged matrine in urine of the tested persons was 53% of the dose after 32 h [9]. The intraperitoneal (i.p.) LD_{50} of oxymatrine in mice was reported as 520 mg kg^{-1}. The LD_{50} of matrine by i.v. injection to mice was about 70 mg kg^{-1} [10].

The pharmacokinetics of oxymatrine was also described by an open, two-compartment model in mice and rats following i.v. injection, with $t_{1/2\alpha}$ of 5 min and $t_{1/2\beta}$ of 2 h. In mice, following the intramuscular (i.m.) injection of [^{3}H]oxymatrine, highest levels of radioactivity were found in the gallbladder, kidneys, liver and intestine, followed by the heart, lungs, spleen, stomach, bone marrow, and muscle; low radioactivity levels were found in the brain, testes, bone, and fat. About 83% of the total radioactivity was excreted in the urine within 48 h, but only 6% in the feces. Pharmacokinetic studies in rats also showed that oxymatrine, after i.m. injection, appeared at high concentrations in the tissues, bile, and urine. In contrast, when oxymatrine was given orally, concentrations of matrine exceeded those of oxymatrine in all samples, indicating that oxymatrine was transformed to matrine in the gastrointestinal tract and liver [11]. Matrine is, therefore, probably the pharmacologically active metabolite for asthma treatment arising from oxymatrine. In healthy volunteers given 100 mg oxymatrine orally, about 40% of the dose was excreted in the urine, with 13–33% representing oxymatrine [12]. Marked differences in pharmacokinetic and pharmacodynamic parameters between matrine and oxymatrine after i.v. injection to rabbits have been observed; however, their pharmacokinetic and pharmacodynamic properties were dose-independent [13].

The pharmacokinetic parameters of oxymatrine and matrine as the metabolite of oxymatrine in dogs following a single oral dose of 300 mg oxymatrine, were also reported. The disposition conformed with a two-compartment model. The $t_{1/2\alpha}$, t_{max}, C_{max} and area under the curve (AUC) for oxymatrine were 5.5 h, 1.0 h, 2.4 μg ml^{-1}, and 5.8 μg·h ml^{-1}, respectively, whereas the $t_{1/2\alpha}$, t_{max}, C_{max} and AUC for matrine were 9.8 h, 1.9 h, 1.5 μg ml^{-1} and 5.5 μg·h ml^{-1} [14]. The pharmacokinetics of oxymatrine and its metabolite matrine were also studied in healthy Chinese men after an i.v. infusion of 600 mg oxymatrine in 100 ml 5% glucose over 0.5 h. The plasma concentration–time profiles of oxymatrine and matrine were best fitted with two-compartment and one-compartment models, respectively. The $t_{1/2\alpha}$ of oxymatrine and matrine were 2.2 and 9.4 h, the C_{max} were 20.5 and 0.25 μg ml^{-1}, the t_{max} were 0.5 and 5.6 h, and the AUC were 20.4 and 3.8 μg·h ml^{-1}, respectively, indicating only a small amount of oxymatrine was reduced to matrine following i.v. administration [15].

Studies on the pharmacokinetic and pharmacodynamic profiles of intravenously administered sophocarpine and oxysophocarpine in rabbits showed that the plasma concentration–time profiles of these two compounds can as well be described by an open, two-compartment model. Although marked differences in pharmacokinetic and pharmacodynamic parameters were observed between the two compounds, their pharmacokinetic and pharmacodynamic properties were dose-independent [16]. Sophocarpine was found to be well absorbed from the gastrointestinal tract. The distribution of sophocarpine in different organs after oral administration was similar to that after i.v. or i.m. administration, with some delay in the appearance of maxima. Sophocarpine was eliminated mainly via the renal route [17].

The root of *S. flavescens*, its alkaloid fraction and matrine, were found to exhibit antiarrhythmic activity. They are used clinically for the treatment of arrhythmia. The i.v. administration of matrine to mice, rats and rabbits exhibited significant effects against arrhythmia induced by aconitine, $BaCl_2$ or coronary ligation [10, 18]. The effects of matrine on myocardial contraction and arrhythmia was also demonstrated in isolated heart atria [19]. Sophocarpine was reported to show antiarrhythmic effects against $CaCl_2$-induced ventricular arrhythmia in mice, aconitine-induced arrhythmia in rats, ouabain-induced arrhythmia in rabbits, and arrhythmia in dogs, as induced by coronary artery occlusion and reperfusion [20].

Antiarrhythmic activities of sophoridine [21] and sophoramine [22] were also reported. Sophoramine exerted an inhibitory effect on arrhythmia in rats induced by aconitine and coronary artery ligation. It elevated the ventricular fibrillation threshold to electrical stimulation in rabbits [23] and in cats [24]. Sophoridine inhibited the electrophysiological changes in dogs with ischemic ventricular tachyarrhythmia and spontaneous ventricular fibrillation after myocardial infarction [25]. The total flavone from *S. flavescens* was also found to have antiarrhythmic activity [26].

Positive inotropic effects in guinea pig papillary muscle stimulated electrically, and in isolated perfused frog heart, were observed by treatment with matrine, sophoridine, or sophoramine, but not by oxymatrine. However, these alkaloids showed only little or no inotropic effect in guinea pig papillary muscle stimulated by isoproterenol or Ca^{2+} [27]. Sophoramine enhanced the contractile force of isolated rabbit atria, but had no effect on the contraction of rabbit aorta strips induced by high K^+ concentration and norepinephrine (noradrenaline). In isolated guinea pig ventricular papillary muscle, sophoramine greatly prolonged the action potential duration and effective refractory period. Sophoramine was also found to prolong the action potential duration of rabbit sinoatrial node, cells and to reduce the spontaneous excitatory frequency [23]. Sophramine produced a concentration-related relaxation of isolated dog mesenteric, coronary and cerebral arteries and mesenteric veins contracted with prostaglandin $F_{2\alpha}$ ($PGF_{2\alpha}$). The relaxation seemed to be nonspecific [28]. It was further reported that oxymatrine, sophoridine, sophocarpine, sophoramine, matrine and cytisine each increased the amplitude of isolated papillary muscle contractions of guinea pigs; high concentrations of these alkaloids induced spontaneous contractions and depressed muscle excitability [29]. Matrine and oxymatrine were found to inhibit Na^+/K^+-ATPase of rabbit erythrocytes, in a concentration-dependent manner [30].

The root of *S. flavescens* was reported to exhibit a potent antiulcer effect. The intraduodenal administration of oxymatrine decreased acid secretion in rats and inhibited gastric motility, as induced by experimental stress. The protective effect of oxymatrine on stress ulcer was postulated to occur by the inhibition of acid secretion and gastric motility. On the other hand, matrine showed only a weak inhibition of gastric acid secretion, but was fairly effective for preventing stress ulcer after i.v. administration [31]. Sophocarpine was reported to inhibit the gastric ulcer induced by water-immersion stress, and also that induced by indomethacin or alcohol in mice [32]. Flavones and related compounds from *S. flavescens*, including kushenol A, kurarinone and kuraridin, showed inhibitory activity on cAMP phosphodiesterase. The prenyl group in the structure has been found to be important for the high inhibition of cAMP phosphodiesterase, in a noncompetitive manner [33].

Matrine was found to be an anti-inflammatory agent. The i.m. administration of matrine at dose of 25 mg kg^{-1} to rats markedly decreased carrageenan-induced hind paw inflammation [34]. The anti-inflammatory effect was also observed after i.p. administration, but not after oral administration. The anti-inflammatory effect was dose-dependent, and the edema formation reduced in the initial and later stages [35]. Adrenalectomy did not influence the anti-inflammatory effect of matrine in mice. The anti-inflammatory activity of matrine was similar to that of nonsteroidal anti-inflammatory agents, and its effect was probably not related to the hypothalamic-adrenal axis [34]. The i.p. administration of oxymatrine was found to significantly inhibit the passive cutaneous anaphylaxis in rats and delayed-type hypersensitivity to sheep erythrocytes in mice. *In vitro*, oxymatrine decreased the degranulation rate of mast cells mediated by homocytotropic antibody [36]. An anti-inflammatory and antiallergic action of aloperine was also reported [37].

Matrine, given intraperitoneally or orally at doses of 20 or 30 mg kg^{-1}, inhibited the yeast-provoked elevation of body temperature in rats. The antipyretic effect of matrine has been considered to be mediated by dopamine release or through dopaminergic receptor blockade [38]. The subcutaneous injection of matrine produced a marked and dose-dependent inhibition of acetic acid-induced abdominal contractions and the tail flick response in mice. The mechanism of the antinociceptive effect of matrine was postulated to occur mainly through an activation of opioid receptors [39]. The prenylflavone derivatives from the root of *S. flavescens* were reported to relatively strongly inhibit phospholipase $C_{1\alpha}$ [40].

Matrine was also found to exhibit antitrichomonal activity. Although it cured the intravaginal infection of trichomonads in mice, the action was observed only by combined intravaginal and s.c. administration, and it failed to produce the same activity through oral administration [41]. It was also reported that a crude extract of *S. flavescens* inhibited coxsackie group B virus, which lost its toxicity towards HeLa cells. The survival time of mice infected with coxsackie group B virus was significantly prolonged after treatment with the extract [42]. Sophoridine was found to exert an antiviral effect in primary cultured myocardial cells and in BALB/c mice against coxsackievirus B3 as a major pathogen for acute and chronic viral myocarditis. Serum samples obtained from rats after oral administration of sophoridine reduced the virus titers in infected myocardial cells. Moreover, sophoridine significantly enhanced the mRNA expression of interleukin-10 (IL-10) and interferon-γ (IFN-γ), but decreased the mRNA expression of tumor necrosis factor-α (TNF-α). This suggested that sophoridine exerted its antiviral activity against coxsackievirus B3 by regulating cytokine expression. A pharmacokinetic study revealed that sophoridine itself, but not its metabolites, might be mainly responsible for the antiviral activities, because the serum concentration–time profile of sophoridine correlated closely with its antiviral activity profile [43].

Matrine was shown to inhibit hepatitis B virus (HBV) in 2.2.15 cultured cells; however, the effect of liposome-encapsulated matrine was better than that of matrine. The median toxic concentration (TC_{50}) of liposome-encapsulated matrine and matrine was 7.3 mg ml^{-1} and 1.3 mg ml^{-1}, respectively. Hepatitis B surface antigen (HBsAg) and hepatitis B e antigen (HBeAg) expression were found to be inhibited. The antiviral activity of matrine and liposome-encapsulated matrine was also confirmed in ducks infected with duck hepatitis B virus. The gastric perfusion of matrine and liposome-encapsulated matrine at a dose of 20 mg kg^{-1}, starting at 13 days after infection, resulted in a significant reduction of

DNA levels of duck hepatitis B virus in treated animals. The effect of liposome-encapsulated matrine was better (though not significantly) than that of matrine [44].

A clinical trial on matrine treatment of chronic hepatitis B has been reported. A group of 120 patients with chronic hepatitis B were allocated at random to a matrine treatment group ($n = 60$) or a control group ($n = 60$). Patients in the treated group received matrine at a daily dose of 100 mg i.m. for 90 days, in addition to conventional liver-protective drugs, while the control patients received only conventional liver-protective drugs. The results showed that matrine caused a significant improvement in clinical symptoms, recovery of liver functions, and serum conversion from HBeAg to antibody, and from positive to negative serum HBV DNA. No serious side effects were observed, except for mild pain at the injection site in some patients [45]. A review of 22 randomized trials with Radix Sophorae Flavescentis for the treatment of chronic hepatitis B included 2409 patients. According to this review, matrine (an aqueous extract of Radix Sophorae Flavescentis) had antiviral activity, positive liver biochemical effects, and improved the symptoms and signs. A combination of matrine and IFN-α, thymosin, or basic treatment showed better effects on viral and liver biochemical responses [46]. Sophocarpine, oxysophocarpine, lehmannine, 13,14-dehydrosophoridine [3], sophoranol and cytisine [47] were also found to inhibit HBV *in vitro*, with inhibitory potency against HBsAg and HBeAg expression.

Recently, matrine was found to exert cytotoxic and antitumor effects. Treatment of K562 leukemia cells with matrine resulted in a greater inhibition of cell survival than that of non-cancer fibroblast NIH-3T3 cells [48]. Matrine at a concentration of 0.2 mg ml^{-1} significantly inhibited the proliferation of C6 glioma cells, and suppressed the expression of Myc proto-oncogene [49]. Matrine also exerted growth inhibition of JM cells, causing cell-cycle arrest at the G_1 stage, and inducing apoptosis [50].

Matrine was found to markedly induce apoptosis in tumor cells. The incubation of K562 leukemia cells with matrine resulted in growth inhibition, and an increased number of apoptotic cells with activated poly(ADP-ribose)polymerase and caspase-3. The proapoptotic protein Bax was enhanced, causing a release of cytochrome c [48]. In MKN45 human gastric cancer cells, matrine inhibited cell proliferation, induced apoptosis, activated caspase-3 and caspase-7, and upregulated the pro-apoptotic proteins Bok, Bak, Bax, Puma, and Bim; this suggested that matrine induced the apoptosis of MKN45 gastric cancer cells by increasing pro-apoptotic molecules of the Bcl-2 family [51]. Likewise, matrine induced apoptosis in U937 leukemia cells via a cytochrome c-triggered caspase activation pathway [52].

In HXO-Rb44 retinoblastoma cells, matrine inhibited growth, induced apoptosis and downregulated the telomerase activity [53]. In HT-29 colon cancer cells, matrine selectively inhibited the mRNA and protein expression of cyclooxygenase-2 (COX-2) and the synthesis of prostaglandin E_2 (PGE_2), but did not inhibit the expression of COX-1 [54]. Matrine also inhibited the expression of P170 expressed by the multidrug resistance gene. The activity of topoisomerase II in mouse S180 tumor cells with acquired multidrug resistance induced by chemotherapy was also inhibited [55]. The reduction of gene expression and inhibition of topoisomerase II correlated well with the multidrug resistance of tumor cells [56]. Furthermore, matrine was reported to be active against tumor invasion and tumor metastasis [57]. At concentrations of 0.8 mg ml^{-1} and 1.0 mg ml^{-1}, it not only markedly inhibited the proliferation of SMMC-7721 human hepatoma cells and induced differentiation but also significantly inhibited the invasion and metastasis potential of tumor cells through several steps, involving cell adhesion, migration, and invasion [58].

Matrine and 5-fluorouracil exhibited a synergistic effect on tumor growth of the implanted SGC-7901 human gastric adenocarcinoma in nude mice [59]. A clinical trial also revealed that matrine infusion at a daily dose of 150 mg markedly protected the liver functions of patients with primary hepatic carcinoma after trans-artery chemo-embolization for two weeks, improving their tolerance to treatment [60].

References

1 Jin, L.X., Cui, Y.Y., and Zhang, G.D. (1993) HPLC analysis of alkaloids in *Sophora flavescens* Ait. *Acta Pharm. Sin.*, **28**, 136–139.

2 Yamamoto, K., Arimoto, K., Iwata, Y., Ujita, K., Sakurai, K., Shimaoka, Y., Shimada, Y., Takagi, A., and Taniyama, T. (1994) Analysis of matrine, matrine *N*-oxide and sophocarpine *N*-oxide in *Sophora* root "Kushen". *Nat. Med.*, **48**, 181–184.

3 Ding, P.L., Liao, Z.X., Huang, H., Zhou, P., and Chen, D.F. (2006) (+)-12α-Hydroxysophocarpine, a new quinolizidine alkaloid and related anti-HBV alkaloids from *Sophora flavescens*. *Bioorg. Med. Chem. Lett.*, **16**, 1231–1235.

4 Okuda, S., Murakoshi, I., Kamata, H., Kashida, Y., Haginiwa, J., and Tsuda, K. (1965) Lupine alkaloids. I. Minor alkaloids of Japanese *Sophora flavescens*. *Chem. Pharm. Bull. (Tokyo)*, **13**, 482–487.

5 Wu, L.J., Miyase, T., Ueno, A., Kuroyanagi, M., Noro, T., and Fukushima, S. (1985) Studies on the constituents of *Sophora flavescens* Aiton. II. *Chem. Pharm. Bull. (Tokyo)*, **33**, 3231–3236.

6 Zhang, Y.Y., Wang, B., Lei, L.M., Guo, M.J., Zhang, R.Y., Huang, L.R., Yi, Y.Y., and Lou, Z. C. (1993) Studies on the constituents of the flavonoids from the roots of *Sophora flavescens*. *Acta Bot. Sin.*, **35**, 304–306.

7 Ding, Y., Tian, R.H., Kinjo, J., Nohara, T., and Kitagawa, I. (1992) Constituents of leguminous plants. XXXIII. Three new oleanene glycosides from *Sophora flavescens*. *Chem. Pharm. Bull. (Tokyo)*, **40**, 2990–2994.

8 Liu, X.D., Yuan, H.N., and Huan, S.K. (1986) Pharmacokinetics of matrine in rabbits. *J. Nanjing Pharm. Coll.*, **17**, 309–311.

9 Wang, P.Q., Lu, G.H., Zhou, X.B., Shen, J.F., Chen, S.X., Mei, S.W., and Chen, M.F. (1994) Pharmacokinetics of matrine in healthy volunteers. *Acta Pharm. Sin.*, **29**, 326–329.

10 Zhang, B.H., Wang, N.S., Li, X.J., Kong, X.J., and Cai, Y.L. (1990) Anti-arrhythmic effects of matrine. *Acta Pharmacol. Sin.*, **11**, 253–257.

11 Song, Y.L., Dong, Q.L., Chen, H.Y., and Jin, Y.Y. (1986) The metabolism of ^{3}H-oxymatrine in mice and rats. *Acta Acad. Med. Sin.*, **8**, 261–265.

12 Xie, M.Z., Zhou, W.Z., and Zhang, Y. (1983) Oxymatrine metabolic fate. *Chin. Med. J. [Engl.]*, **96**, 145–150.

13 Wang, X.H. and Huang, S.K. (1992) Pharmacokinetics and pharmacodynamics of matrine and oxymatrine. *Acta Pharm. Sin.*, **27**, 572–576.

14 Wang, S.J., Wang, G.J., Li, X.T., Ma, R.L., Sun, J.G., and Sheng, L.S. (2005) Pharmacokinetics of oxymatrine and its metabolite in beagle dogs by LC-MS. *Zhongguo Zhong Yao Za Zhi*, **30**, 133–136.

15 Wu, X.L., Hang, T.J., Shen, J.P., and Zhang, Y. D. (2006) Determination and pharmacokinetic study of oxymatrine and its metabolite matrine in human plasma by liquid chromatography tandem mass spectrometry. *J. Pharm. Biomed. Anal.*, **41**, 918–924.

16 Wang, X.H., Huang, S.K., and Liu, R. (1992) Pharmacokinetics and pharmacodynamics of sophocarpine and oxysophocarpine. *J. China Pharm. Univ.*, **23**, 161–164.

17 Li, Y.Q., Yao, D.F., Yu, S.L., Wang, J.X., Liu, C. R., and Cheng, Y. (1982) Neural mechanism of bronchospasmolytic effect of sophocarpine hydrobromide. *Acta Pharmacol. Sin.*, **3**, 169–171.

18 Xu, C.Q., Dong, D.L., Du, Z.M., Chen, Q.W., Gong, D.M., and Yang, B.F. (2004) Comparison of the anti-arrhythmic effects of matrine and berbamine with amiodarone and RP58866. *Yao Xue Xue Bao*, **39**, 691–694.

19 Xin, H.B. and Liu, S.F. (1987) Effects of matrine on myocardial contraction and arrhythmia in isolated heart atria. *Acta Pharmacol. Sin.*, **8**, 501–505.

20 Zhao, Z.Y., Li, Y.Q., and Lin, Z.Y. (1983) Antiarrhythmic effect of sophocarpine

hydrobromide. *Acta Pharmacol. Sin.*, **4**, 173–176.

21 Cui, L.H. and Zhang, B.H. (1986) The antiarrhythmic effects and mechanism of sophoridine. *Chin. J. Pharmacol. Toxicol.*, **1**, 3–10.

22 Wu, Q.H., Long, Y.D., Yao, L.Y., Luo, W., Xiong, H.F., Zhang, J.M., and Zeng, J.Y. (1988) Effects of sophoramine on ventricular fibrillation threshold in dogs. *Acta Pharmacol. Sin.*, **9**, 137–139.

23 Yao, J.A. and Zhang, B.H. (1989) Anti-arrhythmic mechanisms of sophoramine. *Acta Pharmacol. Sin.*, **10**, 315–319.

24 Chen, X.S., Long, Y.D., Yao, L.Y., Luo, W., Wu, Q.H., Xiao, L.M., and Zhang, J.M. (1988) Effects of sophoramine on ventricular fibrillation threshold and dispersion of refractoriness in cats. *Acta Pharmacol. Sin.*, **9**, 551–554.

25 Guo, Z.B., Cao, H.Y., Xu, Z., and Li, Q. (1991) Effects of sophoridine on the electrophysiological changes in dogs with ischemic ventricular tachyarrhythmias and spontaneous ventricular fibrillation after myocardial infarction. *Chin. J. Pharmacol. Toxicol.*, **5**, 104–107.

26 Zhang, B.H., Su, Y., Ma, L., and Li, Q.H. (1979) Experimental study on antiarrhythmic activity of total flavones from "Ku Shen" (*Sophora flavescens* Ait.). *Acta Pharm. Sin.*, **14**, 449–454.

27 Kimura, M., Kimura, I., Chui, L.H., and Okuda, S. (1989) Positive inotropic action and conformation difference of lupine alkaloids in isolated cardiac muscle of guinea pig and bullfrog. *Phytother. Res.*, **3**, 101–105.

28 Bian, K. and Toda, N. (1988) Effects of sophoramine, an alkaloid from *Sophora alopecuroides* on isolated dog blood vessels. *J. Ethnopharmacol.*, **24**, 167–178.

29 Li, R.S. and Chen, S.Y. (1986) Effects of seven alkaloids of *Sophora alopecuroides* on contractility of papillary muscles of guinea pigs. *Acta Pharmacol. Sin.*, **7**, 219–221.

30 Liu, X.T., Tang, H.F., and Xu, Y.F. (1993) Effects of paeonoflorine, matrine and oxymatrine on membrane enzymes *in vitro*. *Chin. Pharm. J.*, **28**, 658–660.

31 Yamazuki, M., Arai, A., Suzuki, S., and Takeuchi, T. (1984) Protective effects of matrine and oxymatrine on stress ulcer in relation to their effects on the central nervous system. *Yakugaku Zasshi*, **104**, 293–301.

32 Zhang, M.F. and Zhu, Z.P. (1992) Anti-ulcerous actions of *Sophora viciifolia* alkaloids and sophocarpine. *Tianran Chanwu Yanjiu Yu Kaifa*, **4**, 71–73.

33 Ohmoto, T., Aikawa, R., Nikaido, T., Sankawa, U., and Wu, L.J. (1986) Inhibition of cyclic AMP phosphodiesterase in medicinal plants. Part XI. Inhibition of adenosine 3′,5′-cyclic monophosphate phosphodiesterase by components of *Sophora flavescens* Aiton. *Chem. Pharm. Bull. (Tokyo)*, **34**, 2094–2099.

34 Tan, H.R. and Zhang, B.H. (1985) Antiinflammatory effects of matrine. *Chin. J. Integr. Trad. West. Med.*, **5**, 108–110.

35 Cho, C.H. and Chuang, C.Y. (1986) Study of the anti-inflammatory action of matrine: an alkaloid isolated from *Sophora subprostrata*. *IRCS Med. Sci.*, **14**, 441–442.

36 Ma, J.J., Si, L.F., Ding, Y., Lin, Z.B., Chen, X.R., Han, C.L., Peng, Q.S., and Fang, Y. (1991) Inhibition of type I-IV allergic reactions by oxymatrine. *J. Beijing Med. Univ.*, **23**, 445–447.

37 Zhou, C.C., Gao, H.B., Sun, X.B., Shi, H.B., Lin, W., Yuan, H.N., and Wang, Z.X. (1989) Anti-inflammatory and antiallergic action of aloperine. *Acta Pharmacol. Sin.*, **10**, 360–365.

38 Cho, C.H., Chuang, C.Y., and Chen, C.F. (1986) Study of the antipyretic activity of matrine. A lupine alkaloid isolated from *Sophora subprostrata*. *Planta Med.*, **52**, 343–345.

39 Kamei, J., Xiao, P., Ohsawa, M., Kubo, H., Higashiyama, K., Takahashi, H., Li, J., Nagase, H., and Ohmiya, S. (1997) Antinociceptive effects of (+)-matrine in mice. *Eur. J. Pharmacol.*, **337**, 223–226.

40 Lee, H.S., Ko, H.R., Ryu, S.Y., Oh, W.K., Kim, B.Y., Ahn, S.C., Mheen, T.I., and Ahn, J.S. (1997) Inhibition of phospholipase $C_{?1}$ by the prenylated flavonoids from *Sophora flavescens*. *Planta Med.*, **63**, 266–268.

41 Wang, H.H. and Cheng, J.T. (1994) Antitrichomonal activity of matrine, an active substance from *Sophora flavescens*. *Phytother. Res.*, **8**, 70–73.

42 Liu, J.X., Lu, D.Y., Yang, Z.M., Chen, S.X., and Qian, F.R. (1991) Preliminary studies on the role of anti-coxsackie group B virus of Chinese medicine *Sophora flavescens*. *Acta Univ. Med. Sec. Shanghai*, **11**, 140–142.

43 Zhang, Y., Zhu, H., Ye, G., Huang, C., Yang, Y., Chen, R., Yu, Y., and Cui, X. (2006) Antiviral effects of sophoridine against coxsackievirus B3 and its pharmacokinetics in rats. *Life Sci.*, **78**, 1998–2005.

44 Li, C.Q., Zhu, Y.T., Zhang, F.X., Fu, L.C., Li, X. H., Cheng, Y., and Li, X.Y. (2005) Anti-HBV effect of liposome-encapsulated matrine *in vitro* and *in vivo*. *World J. Gastroenterol.*, **11**, 426–428.

45 Long, Y., Lin, X.T., Zeng, K.L., and Zhang, L. (2004) Efficacy of intramuscular matrine in the treatment of chronic hepatitis B. *Hepatobiliary Pancreat. Dis. Int.*, **3**, 69–72.

46 Liu, J., Zhu, M., Shi, R., and Yang, M. (2003) Radix Sophorae flavescentis for chronic hepatitis B: a systematic review of randomized trials. *Am. J. Chin. Med.*, **31**, 337–354.

47 Ding, P.L., Huang, H., Zhou, P., and Chen, D. F. (2006) Quinolizidine alkaloids with anti-HBV activity from *Sophora tonkinensis*. *Planta Med.*, **72**, 854–856.

48 Liu, X.S. and Jiang, J. (2006) Molecular mechanism of matrine-induced apoptosis in leukemia K562 cells. *Am. J. Chin. Med.*, **34**, 1095–1103.

49 Deng, H., Luo, H., Huang, F., Li, X., and Gao, Q. (2004) Inhibition of proliferation and influence of proto-oncogenes expression by matrine in C6 cell. *Zhong Yao Cai*, **27**, 416–419.

50 Feng, J.L., Huang, G.S., Zhang, Y.Q., Wang, Z., Zhang, X.H., Guo, Y., and Yan, G.Q. (2003) Matrine effects on JM cells by inhibiting proliferation and inducing apoptosis. *Zhongguo Zhong Yao Za Zhi*, **28**, 437–442.

51 Luo, C., Zhu, Y., Jiang, T., Lu, X., Zhang, W., Jing, Q., Li, J., Pang, L., Chen, K., Qiu, F., Yu, X., Yang, J., and Huang, J. (2007) Matrine induced gastric cancer MKN45 cells apoptosis *via* increasing pro-apoptotic molecules of Bcl-2 family. *Toxicology*, **229**, 245–252.

52 Liu, X.S., Jiang, J., Jiao, X.Y., Wu, Y.E., and Lin, J.H. (2006) Matrine-induced apoptosis in leukemia U937 cells: involvement of caspases activation and MAPK-independent pathways. *Planta Med.*, **72**, 501–506.

53 Yu, W., Li, B., Ren, R.J., Gao, F., Li, L.Q., Liu, X.C., and Wang, Y. (2006) The effects of matrine on cell proliferation and telomerase activity in retinoblastoma cells *in vitro*. *Zhonghua Yan Ke Za Zhi*, **42**, 594–599.

54 Huang, J., Zhang, M.J., and Qiu, F.M. (2005) Study on inhibitory effect of matrine on cyclooxygenase-2 expression in colon cancer HT-29 cell line. *Zhongguo Zhong Xi Yi Jie He Za Zhi*, **25**, 240–243.

55 Sun, F., Wang, N., Li, G., Wang, X., Li, X., and Yin, G. (2004) The effect of matrine on the expression of P170 LRP and TOPO II of obtained multi-drug resistance of mouse S180's tumour cell. *Zhong Yao Cai*, **27**, 838–840.

56 Li, G.H., Wang, M., Sun, F.J., Wang, X.R., Li, X.J., and Yin, G.P. (2006) Study of matrine's use on the reversion of obtained multi-drug resistance of mice S180 tumour cell. *Zhong Yao Cai*, **29**, 40–42.

57 Wang, X. and Lin, C.Q. (2004) Progress of the study on effective integredients from Chinese herbal medicine in anti-tumor metastasis. *Zhongguo Zhong Xi Yi Jie He Za Zhi*, **24**, 178–181.

58 Wang, Y., Peng, C., Zhang, G., Liu, Y., Li, H., and Shan, J. (2003) Study on invasion and metastasis related factors in differentiation of SMMC-7721 cells induced by matrine. *Zhong Yao Cai*, **26**, 566–569.

59 Hu, M.J., Zeng, H., Wu, Y.L., Zhang, Y.P., Zhang, S., Qiao, M.M., and Fu, H. (2005) Synergistic effects of matrine and 5-fluorouracil on tumor growth of the implanted gastric cancer in nude mice. *Chin. J. Dig. Dis.*, **6**, 68–71.

60 Lao, Y. (2005) Clinical study on effect of matrine injection to protect the liver function for patients with primary hepatic carcinoma after trans-artery chemo-embolization (TAE). *Zhong Yao Cai*, **28**, 637–638.

Sophora japonica

Flos Sophorae (Huaihua) is the dried flower and flower bud of *Sophora japonica* L. (Fabaceae). It is used as a hemostatic agent for treatment of different hemorrhagic diseases such as hematochezia, hemorrhoidal bleeding, dysentery with bloody stools, and abnormal uterine bleeding.

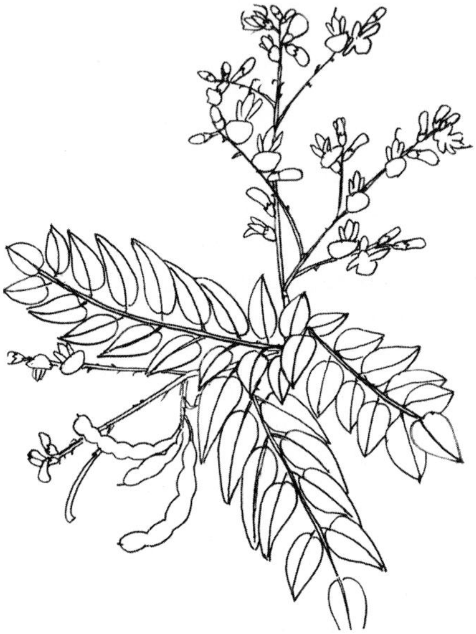

Sophora japonica L.

Fructus Sophorae (Huaijiao) is the dried ripe fruit of *Sophora japonica*. It is used for the treatment of intestinal hemorrhage.

The dried young branch of *Sophora japonica* is listed in the appendix of the Chinese Pharmacopoeia.

Chemistry

The flower of *S. japonica* is one of the richest sources of rutin [1]. In addition to rutin, minor flavones and isoflavones from the fruits of *S. japonica* were identified as quercetin, genistein and its glycosides sophoricoside, sophorabioside, and isorhamnetin, kaempferol and its glycosides [2]. The Chinese Pharmacopoeia requires a qualitative determination of rutin in Flos Sophorae by thin-layer chromatographic comparison with reference substance, and a quantitative determination of total flavone glycosides in Flos Sophorae after extraction by spectrophotometry and a quantitative determination of rutin by HPLC. The content of total flavone glycosides in the buds of Flos Sophorae should be not less than 20.0%, and in the flowers of Flos Sophorae not less than 8.0%, calculated as rutin. The content of rutin in the buds of Flos Sophorae should be not less than 15.0%, and in the flowers of Flos Sophorae not less than 6.0%.

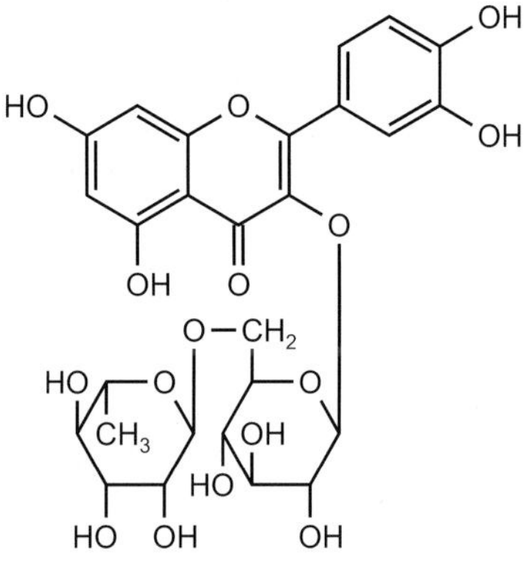

Rutin

The fruit of *S. japonica* is also known to contain flavones and flavone glycosides. The Chinese Pharmacopoeia requires a qualitative and a quantitative determination of sophoricoside in Fructus Sophorae by HPLC. The content of sophoricoside in Fructus Sophorae should be not less than 4.0%.

Triterpene glycosides from the flower buds of *S. japonica* were identified as glycosides of betulin and sophoradiol, including soyasaponins I and III, azukisaponins I, II, and V, kaisasaponins I, II, and III [3]. Saponins

Handbook of Chinese Medicinal Plants: Chemistry, Pharmacology, Toxicology
Weici Tang and Gerhard Eisenbrand

ISBN: 978-3-527-32226-8

from the seeds of *S. japonica* were identified as glycosides of soyasapogenol B [4]. For *Sophora japonica* as a member of the Fabaceae, the seeds and other parts of the plant contained lectins. Lectin from the seeds was a glycoprotein and *N*-acetylgalactosamine-specific, had a molecular mass of about 130 kDa [5]. Molecular cloning of the seed lectin genes revealed that *Sophora* seeds contain only a *N*-acetylgalactosamine-specific lectin which was highly homologous to, but not identical with, the *N*-acetylgalactosamine-specific lectin from the bark [6]. Two major lectins with molecular weights of 40 and 34 kDa were isolated from the flowers; both are also *N*-acetylgalactosamine-specific [7].

Pharmacology and Toxicology

In view of the ubiquitous occurrence of quercetin in vegetables and fruits, reports concerning the mutagenic activity of quercetin have attracted great attention. The daily intake of quercetin per person was estimated as between 20 and 300 mg, and the intake of total flavone derivatives to range from 50 to 1000 mg.

Quercetin was reported to be mutagenic in *Salmonella typhimurium* TA98, TA100, TA1535, TA1537 and TA 1538 without metabolic activation; however, its mutagenicity is increased by the addition of an S-9 mix. Rutin requires metabolic activation for its mutagenic activity [8]. The mutagenicity of crude extracts from nutritional items containing rutin was found to correlate with the flavone content. Among the flavones, quercetin showed the highest mutagenic activity. *Salmonella typhimurium* TA97 was found to be far more susceptible to quercetin than TA1537 [9]. A double bond between positions 2 and 3, and a hydroxyl group at position 3, were required for strong mutagenicity; here, wogonin was an exception, as it does not possess a hydroxyl group at position 3, although it was mutagenic in the presence of an S-9 mix [10].

Mutagenic and genotoxic activities of quercetin were also detected by the induction of DNA strand breakage in isolated plasmids, in L5178 mouse lymphoma cells, and in human cells [11]. It was also found to induce forward mutations at the thymidine kinase (TK) locus in L5178Y and Chinese hamster ovary (CHO) cells [12]. It also induce chromosome aberrations and sister chromatid exchanges (SCE) in CHO cells, in V79 cells and in human fibroblasts and lymphocytes. It was found to induce cell transformation in Syrian hamster embryonal (SHE) cells. However, quercetin was not found to be mutagenic in *Drosophila melanogaster* [13]. Likewise, quercetin was found not to be toxic *in vivo*, and to lack genotoxic and carcinogenic properties [14].

In vivo studies demonstrated that quercetin and rutin, when administered intraperitoneally to normal mice at daily doses of 50, 100 or 150 mg kg^{-1} for seven days, did not show any effect on frequency of micronuclei in bone-marrow polychromatic erythrocytes. Quercetin also showed no mutagenic activity when given orally to mice at concentrations that were approximately 1000-fold greater than the estimated average human intake of total flavones, using the micronucleus test and the host-mediated Ames test [15]. It was also discussed whether catechol *O*-methyltransferase (COMT)-catalyzed rapid *O*-methylation might be responsible for the lack of mutagenicity *in vivo* [16]. The formation of covalent quercetin adducts with DNA or proteins has been studied. The DNA adducts formed were of a transient nature. If this transient nature represented a chemical reversibility of adduct formation, it would provide a possible explanation for the apparent lack of *in vivo* carcinogenicity of this *in vitro* mutagen [17].

In contrast, quercetin was reported to be antimutagenic against a number of known mutagens in various test systems. The mutagenic activity of heterocyclic amines Trp-P-1, IQ [18], PhIP [19], and other food-borne mutagens [20] was reduced by quercetin and related compounds. Administered intraperitoneally at daily doses of 100 and 150 mg kg^{-1} for seven to ten days to mice pretreated with cyclophosphamide, it decreased micronuclei frequency [21]. The mutagenicity of benzo[*a*]pyrene (BaP) and its ultimate carcinogenic metabolite, diolepoxide, were inhibited by quercetin, most likely by either affecting the metabolic activation of BaP [22] or by directly interacting with the active diolepoxide. However, quercetin enhanced the mutagenicity of 2-acetylaminofluorene (AAF). It has been suggested that the effect of quercetin on AAF may be caused by an inhibition of aryl-hydroxylation, and the promotion of *N*-hydroxylation and deacetylation in S-9 mix [23]. In addition, quercetin increased the severity of estradiol-induced carcinogenesis in hamster kidney [24].

On the basis of the reported mutagenicities of quercetin in a variety of experimental models, a number of studies were focused on the potential carcinogenic activity of rutin and quercetin, as well as the related flavones. Positive results on the induction of intestinal and bladder tumors in rats by quercetin were reported twice by the same group [25]. All other studies showed negative results regarding the carcinogenicity of quercetin. For example, ACI rats fed a diet containing 10% rutin or 10% quercetin for 850 days showed no significant difference in the incidence of tumors between the experimental and control groups [26]. In mice, when 2% quercetin was administered in the diet throughout the life span, no difference was observed in tumor incidence between the test and control groups [27]. Likewise, no carcinogenic activity was observed in an experiment with golden hamsters fed with 10% rutin or 10% quercetin in the diet [28]. No evidence was obtained that quercetin was carcinogenic in Fischer rats [29]. The sodium salt of the rutin sulfate ester was devoid of carcinogenic activity in Sprague-Dawley rats [30]. In a NTP-study, dietary quercetin (4%), when fed to Fischer 344 rats throughout their life span, caused a slight increase in hyperplasia and adenomas of the glomeruli, but only in males. The increased induction of glomerular hyperplasia and adenomas seemed to be associated with the species and gender-specific high expression of α_2-microglobulin. The comment was made that the carcinogenic risk of quercetin in men would be negligible [31].

Quercetin did not show any tumor-promoting activity on rat liver preneoplastic foci induced by aflatoxin B_1 (AFB_1) *in vivo*, or on gap junctional intercellular communication *in vitro* [32]. On the other hand, quercetin was found to inhibit tumor promotion by 12-*O*-tetradecanoylphorbol 13-acetate (TPA). For example, it suppressed the effect of TPA on skin tumor formation in CD-1 mice initiated by 7,12-dimethyl-benz[*a*]anthracene (DMBA). However, quercetin did not inhibit the stimulation of epidermal DNA synthesis by TPA. The inhibition of lipoxygenase by quercetin has been proposed as one of the major mechanisms of its inhibitory effect on TPA-induced tumor promotion and ornithine decarboxylase activity [33].

Quercetin was found to interfere with signal transduction cascades, especially those stimulated by the tumor promoter TPA, and to be an inhibitor of tyrosine phosphorylation by transforming protein kinase $pp60^{Src}$. Quercetin competed with ATP for binding to protein kinases [34]. Quercetin and related flavones commonly and potently induced tyrosine dephosphorylation, and concurrently inactivated oncogenic proline-directed protein kinase FA in human prostate carcinoma cells, in a concentration-dependent manner [35]. Quercetin, as a modulator of protein kinases, has undergone Phase I clinical evaluation. The major dose-limiting factor of quercetin was nephrotoxicity [36]. Quercetin was a modulator of the cellular neoplastic phenotype, and inhibited the expression of mutated H-*Ras* and *p53* in rodent and human cells [37]. Quercetin was able to inhibit PI and PIP kinase activities, and to reduce IP_3 concentrations both *in vivo* and in tissue culture systems [38]. It downregulated the expression of *Myc* and K-*Ras* oncogenes, and led to induced differentiation and apoptosis in K562 cells.

Quercetin and rutin were further reported to inhibit colonic neoplasia induced by azoxymethane [39]. The dietary administration of quercetin to rats caused a significant reduction in the frequency of tongue carcinoma induced by 4-nitroquinoline-1-oxide during the initiation and postinitiation phases, and significantly decreased the bromodeoxyuridine labeling index of the tongue squamous epithelium [40]. Quercetin also inhibited tumor formation in the hamster buccal pouch, as induced by DMBA [41]. Rutin significantly inhibited the lung metastasis induced by B16F10 melanoma cells in mice, and increased the life span of the animals [42].

Quercetin and related flavones were found to inhibit topoisomerase I-catalyzed DNA religation. In contrast to the well-known topoisomerase I inhibitor, camptothecin, these flavones did not act directly on the catalytic intermediate; neither did they interfere with DNA cleavage. However, the formation of a ternary complex with topoisomerase I and DNA during the cleavage reaction inhibited the following DNA religation step [43]. 3,7,3′,4′-Tetrahydroxy-substituted flavones stabilize the covalent topoisomerase I–DNA intermediate most efficiently. An enhanced formation of the covalent topoisomerase I–DNA complex was also seen in human HL-60 cells [44]. Rutin was found to be a selective inhibitor of topoisomerase IV of *Escherichia coli* [45].

Other mechanisms of the anticarcinogenic or cytotoxic activities of quercetin were suggested to include protection of cells by suppression of the tyrosine kinase-Jun/activator protein-1 [46]; interaction with nuclear and cytosolic type II estrogen binding sites present in primary lung cancers [47]; inhibition of heat shock protein induction, but not heat shock factor DNA-binding in human breast carcinoma cells [48]; and a decrease in lactate release from tumor cells and an elevation of intratumoral acidity [49].

Following the oral administration of quercetin to rats, 20% was absorbed from the digestive tract and about 30% was excreted unchanged in the feces. The absorbed quercetin was rapidly excreted into the bile and urine as the glucuronide and sulfate, and as quercetin 3′-methylether and quercetin 4′-methylether [50]. It was reported that humans absorb appreciable amounts of quercetin, and that absorption is enhanced by glycosidation. In healthy subjects, 24% of an oral dose of quercetin was absorbed, whereas for quercetin glycosides about 50% of the dose was absorbed [51]. A pharmacokinetic study on quercetin in men showed a biphasic elimination, with $t_{1/2\alpha}$ of 8.8 min and $t_{1/2\beta}$ of 2.4 h after a single i.v. dose. Protein binding was more than 98%. About 7% of the i.v. dose was excreted in the urine in conjugated form, while about 0.6% was excreted unchanged. After oral administration of quercitin, no measurable plasma concentrations could be detected [52]. After oral administration of quercetin glycosides to men, peak plasma levels of quercetin were reached after 3 h, with a half-life of absorption of 0.9 h. The half-life of the distribution phase was about 4 h, and of the elimination phase about 17 h [53]. It was also found that the bioavailability of dietary quercetin or quercetin glycosides depended on their origin [54]. Both, plasma concentrations of quercitin and the amount excreted in the urine can be used as biomarkers for the intake [55].

Rutin is absorbed more slowly than quercetin because it must be hydrolyzed by the cecal microflora, whereas quercetin is absorbed from the small intestine [59]. The urinary metabolites in rat urine of orally administered rutin were identified as 3-hydroxyphenylacetic acid, homovanillic acid, homoprotocatechuic acid, 3,4-dihydroxytoluene, and 3-(3-hydroxyphenyl)- propionic acid. Unchanged rutin and quercetin were not detected in urine [56]. The metabolites of quercetin in human urine were identified as homoprotocatechuic acid, 3-hydroxyphenylacetic acid, and homovanillic acid [57]. After i.p. administration of rutin to bile duct-cannulated rats, three conjugates were detected as biliary metabolites; one of these was identified as quercetin 3′-methylether [58].

There is no doubt that quercetin is the most important component in the flowers of *S. japonica*. Rutin exerts its biological effects after hydrolysis to quercetin by the cecal microflora [59]. Quercetin is known to have a broad spectrum of pharmacological activities, including cardiovascular and chemopreventive actions.

Quercetin and related flavones were found to show protective effects against cardiovascular diseases. Quercetin was reported to significantly inhibit platelet aggregation in human platelets *in vitro*, as induced by collagen and ADP [60]; and to suppress serotonin release in parallel to an inhibition of thromboxane (TX) synthesis induced by arachidonic acid [61]. A study of the structure–activity relationship showed that the inhibitory effect of flavones on human platelet functions was diminished by saturation of the double bond, by lack of the carbonyl function, and by glycosylation of the 3-hydroxy group. Quercetin inhibited platelet aggregation by blocking cAMP-phosphodiesterase and increasing cAMP levels [62]. Quercetin was one of the compounds with a maximal inhibitory effect on both cAMP and cGMP phosphodiesterases. In general, flavones inhibited the phosphodiesterases more potently than did the corresponding flavone glycosides. The inhibition of rabbit heart cAMP phosphodiesterase by quercetin and related flavones was competitive [63]. The ingestion by human subjects of either 150 or 300 mg of quercetin 4′-*O*-β-D-glucopyranoside supplement resulted in peak plasma concentrations of about 5 and 10 μM after 30 min, respectively. These results showed that quercetin was bioavailable, with plasma concentrations achieved within the range known to affect platelet function *in vitro*. Platelet aggregation was inhibited at 30 and 120 min after the ingestion of both doses of quercetin 4′-*O*-β-D-glucopyranoside [64].

Quercetin also inhibited the rise in intraplatelet free calcium concentration, as induced by thrombin. The inhibitory effects of quercetin on the aggregation and rise of intraplatelet free calcium concentration were mainly due to an inhibition of Ca^{2+} influx [65]. Quercetin decreased the activities of the Na^{+}/K^{+}-ATPase of rat brain plasma membranes and Ca^{2+}/Mg^{2+} ATPase of heart sarcolemmal membrane after oral administration. In rats, the Na^{+}/K^{+}-ATPase of the myocardium was seen to be more sensitive than that of the brain [66].

Oral administration of quercetin has been found to exhibit hypolipidemic activity in rats fed 1% cholesterol in the diet for six and 12 weeks. The serum triglyceride levels in both mice and rats were depressed by quercetin [67]. Quercetin and its glycosides were capable of inhibiting lipoxygenase-induced low-density lipoprotein (LDL) oxidation [68]. Some conjugated metabolites of quercetin were found to act as effective antioxidants when plasma was subject to metal ion-induced lipid peroxidation [69]. Dogs or rats given rutin or quercetin showed decreased total cholesterol levels in their arterial and venous blood, increased thoracic lymph cholesterol, and decreased hepatic cholesterol [70]. Epidemiological studies suggested that people with very low intakes of flavonoids have higher risks of coronary disease [71]. Quercetin also inhibited arrhythmias in rats, as induced by platelet reperfusion. It markedly improved ultrastructural platelet deviations, inhibited platelet aggregation and TXA_2 formation, and increased prostacyclin generation [72].

Vascular and nonvascular inflammatory diseases were believed to be associated with the presence of leukotrienes in the organ systems. The presence of cysteinyl leukotrienes as slow-reacting substances of anaphylaxis was found in inflammatory diseases such as asthma, anaphylactic rhinitis, and neonatal pulmonary hypertension. A study on the effect of quercetin on anaphylactic smooth muscle contraction of ileum from guinea pigs sensitized to egg albumin showed that quercetin inhibited the contraction in a concentration-dependent manner, with an IC_{50} of about 10 μ*M* [73]. *In vitro*, quercetin and related flavones were reported to affect various pathways involved in the metabolism of arachidonic acid. In intact human platelets, quercetin was effective as an inhibitor of both cyclooxygenase and lipoxygenase [74].

Quercetin and related flavones showed a palliative effect on inflammation by inhibiting leukotriene (LT) synthesis, histamine release, and by antioxidative [75] and superoxide-scavenging activities [76]. In rats, quercetin and rutin also prevented gastric ulcer as induced by ethanol [77] and prevented colitis induced by acetic acid and by trinitrobenzenesulfonic acid [78]. Quercetin also prevented the oxidative stress in liver cirrhotic rats induced by carbon tetrachloride (CCl_4). At a daily dose of 50 mg kg^{-1} it successfully attenuated the hepatotoxic effects of CCl_4, significantly suppressing lipid peroxidation levels [79]. Oral pretreatment of mice with quercetin at a dose of 200 mg kg^{-1} also significantly protected against the toxicity of microcystin, a cyanobacterial heptapeptide. Quercetin treatment caused decreases in the levels of serum alanine amino transferase (ALT) and aspartate aminotransaminase (AST), indicating that quercetin could reverse the hepatotoxic effects of microcystin in mice [80]. Furthermore, the antioxidant activity of quercetin was also evidenced against galactose-induced hyperglycemic oxidative stress in rats [81].

In fact, quercetin possessed both antioxidant and pro-oxidant properties. In A549 cells, quercetin at low concentrations stimulated cell proliferation and increased the total antioxidant capacity. At higher concentrations however, it decreased cell survival and viability, thiol content and total antioxidant capacity, and the activities of superoxide dismutase, catalase and glutathione *S*-transferase (GST). Quercetin decreased the production of reactive oxygen species (ROS) in the cells, but produced peroxides in the medium [82]. However, it was suggested that quercetin exerted antioxidant, rather than pro-oxidant, activities in normal cells. In mouse thymocytes, quercetin did not induce oxidative damage, but protected the cells from apoptosis mediated by glucose oxidase. It also suppressed the glucose oxidase-mediated DNA binding activity of redox state-sensitive transcription factors, such as nuclear factor-κB (NF-κB), AP-1, and p53, suggesting antioxidative effects in thymocytes [83]. Both, antioxidant and pro-oxidant effects of chronic quercetin administration have also been observed in rats. The oral administration of quercetin at doses of 2 or 20 mg kg^{-1} for four weeks significantly decreased the concentration of malondialdehyde in rat liver, as an indicator of lipid peroxidation. In contrast, as signs of pro-oxidant activity, quercetin treatment decreased the glutathione concentration and glutathione reductase activity in the liver [84].

References

1 Xu, L.X., Liu, A.R., and Zhang, X.Q. (1989) Coulometric titration of total flavonoids in *Sophora japonica* L. *Acta Pharm. Sin.*, **24**, 755–758.

2 Ishida, H., Umino, T., Tsuji, K., and Kosuge, T. (1989) Studies on the antihemostatic substances in herbs classified as hemostatics in traditional Chinese medicine. I. On the antihemostatic principles in *Sophora japonica* L. *Chem. Pharm. Bull. (Tokyo)*, **37**, 1616–1618.

3 Kitagawa, I., Taniyama, T., Hong, W.W., Hori, K., and Yoshikawa, M. (1988) Saponin and sapogenol XLV. Structures of kaisasaponins I, II, and III from Sophorae flos, the bids of *Sophora japonica*. *Yakugaku Zasshi*, **108**, 538–546.

4 Gorbacheva, L.A., Grishkovets, V.I., Drozd, G.A., and Chirva, V.Ya. (1996) Triterpene glycosides from *Sophora japonica* L. seeds. *Adv. Exp. Med. Biol.*, **404**, 501–504.

5 McPherson, A., Hankins, C.N., and Shannon, L. (1987) Preliminary X-ray diffraction analysis of crystalline lectins from the seeds and leaves of *Sophora japonica*. *J. Biol. Chem.*, **262**, 1791–1794.

6 Van Damme, E.J., Barre, A., Rouge, P., and Peumans, W.J. (1997) Molecular cloning of the bark and seed lectins from the Japanese pagoda tree (*Sophora japonica*). *Plant Mol. Biol.*, **33**, 523–536.

7 Ito, Y. (1986) Occurrence of lectins in leaves and flowers of *Sophora japonica*. *Plant Sci.*, **47**, 77–82.

8 Brown, J.P. and Dietrich, P.S. (1979) Mutagenicity of plant flavonols in the *Salmonella*/mammalian microsome test. Activation of flavonol glycosides by mixed glycosidase from rat cecal bacteria and other sources. *Mutat. Res.*, **66**, 223–240.

9 Busch, D.B., Hatcher, J.F., and Bryan, G.T. (1986) Urine recovery experiments with quercetin and other mutagens using the Ames test. *Environ. Mutagen.*, **8**, 393–399.

10 Nagao, M., Marita, N., Yahagi, T., Shimizu, M., Kuroyanagi, M., Fukuoka, M., Yoshihara, K., Natori, S., Fujino, T., and Sugimura, T. (1981) Mutagenicities of 61 flavonoids and 11 related compounds. *Environ. Mutagen.*, **3**, 401–419.

11 Duthie, S.J., Johnson, W., and Dobson, V.L. (1997) The effect of dietary flavonoids on DNA damage (strand breaks and oxidised pyrimidines) and growth in human cells. *Mutat. Res.*, **390**, 141–151.

12 Van der Hoeven, J.C.M., Bruggeman, I.M., and Debets, F.M.H. (1984) Genotoxicity of quercetin in cultured mammalian cells. *Mutat. Res.*, **136**, 9–21.

13 Schramm, D.D., Collins, H.E., Hawley, R.S., and German, J.B. (1998) Unaltered meiotic chromosome segregation in *Drosophila melanogaster* raised on a 5% quercetin diet. *Food Chem. Toxicol.*, **36**, 585–589.

14 Harwood, M., Danielewska-Nikiel, B., Borzelleca, J.F., Flamm, G.W., Williams, G.M., and Lines, T.C. (2007) A critical review of the data related to the safety of quercetin and lack of evidence of in vivo toxicity, including lack of genotoxic/carcinogenic properties. *Food Chem. Toxicol.*, **45**, 2179–2205.

15 Aeschbacher, H.U., Meier, H., and Ruch, E. (1982) Nonmutagenicity in vivo of the food flavonol quercetin. *Nutr. Cancer*, **4**, 90–98.

16 Zhu, B.T., Ezell, E.L., and Liehr, J.G. (1994) Catechol O-methyltransferase catalyzed rapid *O*-methylation of mutagenic flavonoids. Metabolic inactivation as a possible reason for their lack of carcinogenicity *in vivo*. *J. Biol. Chem.*, **269**, 292–299.

17 Van der Woude, H., Alink, G.M., van Rossum, B.E., Walle, K., van Steeg, H., Walle, T., and Rietjens, I.M. (2005) Formation of transient covalent protein and DNA adducts by quercetin in cells with and without oxidative enzyme activity. *Chem. Res. Toxicol.*, **18**, 1907–1916.

18 Anderson, D., Basaran, N., Dobrzynska, M.M., Basaran, A.A., and Yu, T.W. (1997) Modulating effects of flavonoids on food mutagens in human blood and sperm samples in the comet assay. *Teratog. Carcinog. Mutagen.*, **17**, 45–58.

19 Bacon, J.R., Williamson, G., Garner, R.C., Lappin, G., Langouet, S., and Bao, Y. (2003) Sulforaphane and quercetin modulate PhIP-DNA adduct formation in human HepG2 cells and hepatocytes. *Carcinogenesis*, **24**, 1903–1911.

20 Taj, S. and Nagarajan, B. (1996) Inhibition by quercetin and luteolin of chromosomal alterations induced by salted, deep-fried fish and mutton in rats. *Mutat. Res.*, **369**, 97–106.

21 Hang, B.Q., Wu, Y., Hang, S., Yang, Y., and Wang, M.S. (1985) Effect of quercetin and rutin on the occurrence of micronuclei in mouse bone-marrow polychromatic erythrocytes. *J. Nanjing Coll. Pharm.*, **16**, 52–55.

22 Huang, M.T., Wood, A.W., Newmark, H.L., Sayer, J.M., Yagi, H., Jerina, D.M., and Conney, A.H. (1983) Inhibition of the mutagenicity of bay-region diol-epoxides of polycyclic aromatic hydrocarbons by phenolic plant flavonoids. *Carcinogenesis*, **4**, 1631–1637.

23 Ogawa, S., Hirayama, T., Tokuda, M., Hirai, K., and Fukui, S. (1986) The effect of quercetin, a mutagenicity-enhancing agent, on the metabolism of 2-acetylaminofluorene with mammalian metabolic activation systems. *Mutat. Res.*, **162**, 179–186.

24 Zhu, B.T. and Liehr, J.G. (1994) Quercetin increases the severity of estradiol-induced tumorigenesis in hamster kidney. *Toxicol. Appl. Pharmacol.*, **125**, 149–158.

25 Ertürk, J.F., Hatcher, J.F., and Pamukcu, A.M. (1985) Bracken fern carcinogens and quercetin. *Fed. Proc.*, **43**, 2344.

26 Hirono, I., Ueno, I., Hosaka, S., Takanashi, H., Matsushima, T., Sugimura, T., and Natori, S. (1981) Carcinogenicity examination of quercetin and rutin in ACI rats. *Cancer Lett.*, **13**, 15–21.

27 Saito, D., Shirai, A., Matsushima, T., Sugimura, T., and Hirono, I. (1980) Test of carcinogenicity of quercetin, a widely distributed mutagen in food. *Teratogen. Carcinog. Mutagen.*, **1**, 213–221.

28 Morino, K., Matsukura, N., Kawachi, T., Ohgaki, H., Sugimura, T., and Hirono, I. (1982) Carcinogenicity test of quercetin and rutin in golden hamsters by oral administration. *Carcinogenesis*, **3**, 93–97.

29 Ito, N., Hagiwara, A., Tamano, S., Kagawa, M., Shibata, M., Kurata, Y., and Fukushima, S. (1989) Lack of carcinogenicity of quercetin in F344/DuCrj rats. *Jpn. J. Cancer Res.*, **80**, 317–325.

30 Habs, M., Habs, H., Berger, M.R., and Schmaehl, D. (1984) Negative dose- response study for carcinogenicity of orally administered rutin sulfate in Sprague-Dawley rats. *Cancer Lett.*, **23**, 103–108.

31 Hirono, I. (1992) Is quercetin carcinogenic? *Jpn. J. Cancer Res.*, **83**, 313–314.

32 Chaumontet, C., Suschetet, M., Honikman-Leban, E., Krutovskikh, V.A., Berges, R., Le Bon, A.M., Heberden, C., Shahin, M.M., Yamasaki, H., and Martel, P. (1996) Lack of tumor-promoting effects of flavonoids: studies on rat liver preneoplastic foci and on *in vivo* and *in vitro* gap junctional intercellular communication. *Nutr. Cancer*, **26**, 251–263.

33 Kato, R., Nakadate, T., Yamamoto, S., and Sugimura, T. (1983) Inhibition of 12-*O*-tetradecanoylphorbol 13-acetate-induced tumor promotion and ornithine decarboxylase activity by quercetin: possible involvement of lipoxygenase inhibition. *Carcinogenesis*, **4**, 1301–1305.

34 Kang, T.B. and Liang, N.C. (1997) Studies on the inhibitory effects of quercetin on the growth of HL-60 leukemia cells. *Biochem. Pharmacol.*, **54**, 1013–1018.

35 Lee, S.C., Kuan, C.Y., Yang, C.C., and Yang, S. D. (1998) Bioflavonoids commonly and potently induce tyrosine dephosphorylation/inactivation of oncogenic proline-directed protein kinase FA in human prostate carcinoma cells. *Anticancer Res.*, **18**, 1117–1121.

36 Gescher, A. (1995) Modulators of signal transduction as cancer chemotherapeutic agents - novel mechanisms and toxicities. *Toxicol. Lett.*, **82–83**, 159–165.

37 Avila, M.A., Cansado, J., Harter, K.W., Velasco, J.A., and Notario, V. (1996) Quercetin as a modulator of the cellular neoplastic phenotype. Effects on the expression of mutated H-ras and p53 in rodent and human cells. *Adv. Exp. Med. Biol.*, **401**, 101–110.

38 Agullo, G., Gamet-Payrastre, L., Manenti, S., Viala, C., Remesy, C., Chap, H., and Payrastre, B. (1997) Relationship between flavonoid structure and inhibition of phosphatidylinositol 3-kinase: a comparison with tyrosine kinase and protein kinase C inhibition. *Biochem. Pharmacol.*, **53**, 1649–1657.

39 Femia, A.P., Cademi, G., Ianni, M., Salvadori, M., Schijlen, E., Collins, G., Bovy, A., and Dolara, P. (2003) Effect of diets fortified with tomatoes or onions with variable quercetin-glycoside content on azoxymethane-induced aberrant crypt foci in the colon of rats. *Eur. J. Nutr.*, **42**, 346–352.

40 Makita, H., Tanaka, T., Fujitsuka, H., Tatematsu, N., Satoh, K., Hara, A., and Mori, H. (1996) Chemoprevention of 4-nitroquinoline 1-oxide-induced rat oral carcinogenesis by the dietary flavonoids chalcone, 2-hydroxychalcone, and quercetin. *Cancer Res.*, **56**, 4904–4909.

41 Balasubramanian, S. and Govindasamy, S. (1996) Inhibitory effect of dietary flavonol quercetin on 7,12-dimethylbenzaanthracene-induced hamster buccal pouch carcinogenesis. *Carcinogenesis*, **17**, 877–879.

42 Menon, L.G., Kuttan, R., and Kuttan, G. (1995) Inhibition of lung metastasis in mice induced

by B16F10 melanoma cells by polyphenolic compounds. *Cancer Lett.*, **95**, 221–225.

43 Boege, F., Straub, T., Kehr, A., Boesenberg, C., Christiansen, K., Andersen, A., Jakob, F., and Kohrle, J. (1996) Selected novel flavones inhibit the DNA binding or the DNA religation step of eukaryotic topoisomerase I. *J. Biol. Chem.*, **271**, 2262–2270.

44 Constantinou, A., Mehta, R., Runyan, C., Rao, K., Vaughan, A., and Moon, R. (1995) Flavonoids as DNA topoisomerase antagonists and poisons: structure-activity relationships. *J. Nat. Prod.*, **58**, 217–225.

45 Bernard, F.X., Sable, S., Cameron, B., Provost, J., Desnottes, J.F., Crouzet, J., and Blanche, F. (1997) Glycosylated flavones as selective inhibitors of topoisomerase IV. *Antimicrob. Agents Chemother.*, **41**, 992–998.

46 Yokoo, T. and Kitamura, M. (1997) Unexpected protection of glomerular mesangial cells from oxidant-triggered apoptosis by bioflavonoid quercetin. *Am. J. Physiol.*, **273** (2 Pt 2), F206–F212

47 Caltagirone, S., Ranelletti, F.O., Rinelli, A., Maggiano, N., Colasante, A., Musiani, P., Aiello, F.B., and Piantelli, M. (1997) Interaction with type II estrogen binding sites and antiproliferative activity of tamoxifen and quercetin in human non-small-cell lung cancer. *Am. J. Respir. Cell. Mol. Biol.*, **17**, 51–59.

48 Hansen, R.K., Oesterreich, S., Lemieux, P., Sarge, K.D., and Fuqua, S.A. (1997) Quercetin inhibits heat shock protein induction but not heat shock factor DNA-binding in human breast carcinoma cells. *Biochem. Biophys. Res. Commun.*, **239**, 851–856.

49 Kaplan, A.E. and Bunow, M.R. (1986) Spectrophotometric determination of intracellular pH with cultured rat liver cells. *J. Histochem. Cytochem.*, **34**, 749–752.

50 Ueno, I., Nakano, N., and Hirono, I. (1983) Metabolic fate of ^{14}C-quercetin in the ACI rat. *Jpn. J. Exp. Med.*, **53**, 41–50.

51 Hollman, P.C., de Vries, J.H., van Leeuwen, S.D., Mengelers, M.J., and Katan, M.B. (1995) Absorption of dietary quercetin glycosides and quercetin in healthy ileostomy volunteers. *Am. J. Clin. Nutr.*, **62**, 1276–1282.

52 Gugler, R., Leschik, M., and Dengler, H.J. (1975) Disposition of quercetin in man after single oral and intravenous doses. *Eur. J. Clin. Pharmacol.*, **9**, 229–234.

53 Hollman, P.C., van der Gaag, M., Mengelers, M.J., van Trijp, J.M., de Vries, J.H., and Katan, M.B. (1996) Absorption and disposition kinetics of the dietary antioxidant quercetin in man. *Free Radic. Biol. Med.*, **21**, 703–707.

54 Hollman, P.C. and Katan, M.B. (1998) Bioavailability and health effects of dietary flavonols in man. *Arch. Toxicol. Suppl.*, **20**, 237–248.

55 de Vries, J.H., Hollman, P.C., Meyboom, S., Buysman, M.N., Zock, P.L., van Staveren, W. A., and Katan, M.B. (1998) Plasma concentrations and urinary excretion of the antioxidant flavonols quercetin and kaempferol as biomarkers for dietary intake. *Am. J. Clin. Nutr.*, **68**, 60–65.

56 Baba, S., Futura, T., Fujioka, M., and Goromaru, T. (1983) Studies on drug metabolism by use of isotopes. XXVII: Urinary metabolites of rutin in rats and the role of intestinal microflora in the metabolism of rutin. *J. Pharm. Sci.*, **72**, 1155–1158.

57 Gross, M., Pfeiffer, M., Martini, M., Campbell, D., Slavin, J., and Potter, J. (1996) The quantitation of metabolites of quercetin flavonols in human urine. *Cancer Epidemiol. Biomarkers Prev.*, **5**, 711–720.

58 Griffiths, L.A. and Brown, S. (1983) New metabolites of naturally occurring flavonols; identification by E.I. mass spectrometry and chromatographic techniques. *Anal. Chem. Symp. Ser.*, **14**, 121–124.

59 Manach, C., Morand, C., Demigne, C., Texier, O., Regerat, F., and Remesy, C. (1997) Bioavailability of rutin and quercetin in rats. *FEBS Lett.*, **409**, 12–16.

60 Janssen, K., Mensink, R.P., Cox, F.J., Harryvan, J.L., Hovenier, R., Hollman, P.C., and Katan, M.B. (1998) Effects of the flavonoids quercetin and apigenin on hemostasis in healthy volunteers: results from an *in vitro* and a dietary supplement study. *Am. J. Clin. Nutr.*, **67**, 255–262.

61 Corvazier, E. and Maclouf, J. (1985) Interference of some flavonoids and nonsteroidal anti-inflammatory drug with oxidative metabolism of arachidonic acid by human platelets and neutrophils. *Biochim. Biophys. Acta*, **835**, 315–321.

62 Landolfi, R., Mower, R.L., and Steiner, M. (1984) Modification of platelet function and arachidonic acid metabolism by bioflavonoids. Structure-activity relationship. *Biochem. Pharmacol.*, **33**, 1525–1530.

63 He, S.P. and Qiao, J. (1985) The inhibition kinetic study of flavonoids on cAMP phosphodiesterase. *Shenwu huaxue Zazhi*, **1**, 55–58.

64 Hubbard, G.P., Wolffram, S., Lovegrove, J.A., and Gibbins, J.M. (2004) Ingestion of quercetin inhibits platelet aggregation and essential components of the collagen-stimulated platelet activation pathway in humans. *J. Thromb. Haemost.*, **2**, 2138–2145.

65 Xiao, D., Gu, Z.L., Bai, J.P., and Wang, Z. (1995) Effects of quercetin on aggregation and intracellular free calcium of platelets. *Acta Pharmacol. Sin.*, **16**, 223–226.

66 Gu, Z.L., Xiao, D., Jin, L.Q., Fan, P.S., and Qian, Z.N. (1994) Effects of quercetin on Na^+/K^+-exchanging ATPase and Ca^{2+}/Mg^{2+}-ATPase in rats. *Acta Pharmacol. Sin.*, **15**, 414–416.

67 Kato, N., Tosa, N., Doudou, T., and Imamura, T. (1983) Effects of dietary quercetin on serum lipids. *Agric. Biol. Chem.*, **47**, 2119–2120.

68 Terao, J., Yamaguchi, S., Shirai, M., Miyoshi, M., Moon, J.H., Oshima, S., Inakuma, T., Tsushida, T., and Kato, Y. (2001) Protection by quercetin and quercetin 3-O-β-D-glucuronide of peroxynitrite-induced antioxidant consumption in human plasma low-density lipoprotein. *Free Radic. Res.*, **35**, 925–931.

69 Manach, C., Morand, C., Crespy, V., Demigne, C., Texier, O., Regerat, F., and Remesy, C. (1998) Quercetin is recovered in human plasma as conjugated derivatives which retain antioxidant properties. *FEBS Lett.*, **426**, 331–336.

70 Igarashi, K. and Ohmuma, M. (1995) Effects of isorhamnetin, rhamnetin, and quercetin on the concentrations of cholesterol and lipoperoxide in the serum and liver and on the blood and liver antioxidative enzyme activities of rats. *Biosci. Biotechnol. Biochem.*, **59**, 595–601.

71 Knekt, P., Jarvinen, R., Reunanen, A., and Maatela, J. (1996) Flavonoid intake and coronary mortality in Finland: a cohort study. *Br. Med. J.*, **312**, 478–481.

72 Xiao, D., Gu, Z.L., and Qian, Z.N. (1993) Effects of quercetin on platelet and reperfusion-induced arrhythmias in rats. *Acta Pharmacol. Sin.*, **14**, 505–508.

73 Fanning, M.J., Macander, P., Drzewiecki, G., and Middleton, E. Jr (1983) Quercetin inhibits anaphylactic contraction of guinea pig ileum smooth muscle. *Int. Arch. Allergy Appl. Immunol.*, **71**, 371–373.

74 Shirai, M., Moon, J.H., Tsushida, T., and Terao, J. (2001) Inhibitory effect of a quercetin metabolite, quercetin 3-*O*-β-D-glucuronide, on lipid peroxidation in liposomal membranes. *J. Agric. Food Chem.*, **49**, 5602–5608.

75 Grinberg, L.N., Rachmilewitz, E.A., and Newmark, H. (1994) Protective effects of rutin against hemoglobin oxidation. *Biochem. Pharmacol.*, **48**, 643–649.

76 Kostyuk, V.A. and Potapovich, A.I. (1998) Antiradical and chelating effects in flavonoid protection against silica-induced cell injury. *Arch. Biochem. Biophys.*, **355**, 43–48.

77 Martin, M.J., La-Casa, C., Alarcon-de-la-Lastra, C., Cabeza, J., Villegas, I., and Motilva, V. (1998) Anti-oxidant mechanisms involved in gastroprotective effects of quercetin. *Z. Naturforsch [C]*, **53**, 82–88.

78 Cruz, T., Galvez, J., Ocete, M.A., Crespo, M.E., Sanchez de Medina, L., and Zarzuelo, A. (1998) Oral administration of rutoside can ameliorate inflammatory bowel disease in rats. *Life Sci.*, **62**, 687–695.

79 Amalia, P.M., Possa, M.N., Augusto, M.C., and Francisca, L.S. (2007) Quercetin prevents oxidative stress in cirrhotic rats. *Dig. Dis. Sci.*, **52**, 2616–2621.

80 Jayaraj, R., Deb, U., Bhaskar, A.S., Prasad, G.B., and Rao, P.V. (2007) Hepatoprotective efficacy of certain flavonoids against microcystin induced toxicity in mice. *Environ. Toxicol.*, **22**, 472–479.

81 Ramana, B.V., Kumar, W., Krishna, P.N., Kumar, C.S., Reddy, P.U., and Raju, T.N. (2006) Effect of quercetin on galactose-induced hyperglycaemic oxidative stress in hepatic and neuronal tissues of Wistar rats. *Acta Diabetol.*, **43**, 135–141.

82 Robaszkiewicz, A., Balcerczyk, A., and Bartosz, G. (2007) Antioxidative and prooxidative effects of quercetin on A549 cells. *Cell Biol. Int.*, **31**, 1245–1250.

83 Lee, J.C., Kim, J., Park, J.K., Chung, G.H., and Jang, Y.S. (2003) The antioxidant, rather than prooxidant, activities of quercetin on normal cells: quercetin protects mouse thymocytes from glucose oxidase-mediated apoptosis. *Exp. Cell Res.*, **291**, 386–397.

84 Choi, E.J., Chee, K.M., and Lee, B.H. (2003) Anti- and prooxidant effects of chronic quercetin administration in rats. *Eur. J. Pharmacol.*, **482**, 281–285.

Spatholobus suberectus

Caulis Spatholobi (Jixueteng) is the dried cane of *Spatholobus suberectus* Dunn (Fabaceae). It is used for the treatment of menstrual disorders, anemia, numbness and paralysis, and rheumatic arthralgia.

Spatholobus suberectus Dunn.

Chemistry

Caulis Spatholobi is known to contain a number of different types of compounds. Catechin and its analogues, epicatechin and gallocatechin, are the major components in the stem of *S. suberectus* [1]. A number of flavones, flavanones, isoflavones and chalcones was also identified, including 3′,4′,7-trihydroxyflavone, eriodictyol, plathymenin, dihydroquercetin, butin, liquiritigenin, neoisoliquiritigenin, dihydrokaempferol, 6-methoxyeriodictyol, ononin, prunetin, pseudobaptigenin, genistein, sativan, 7-hydroxy-6-methoxy-flavanone, and biochanin B. Pterocarpans (6a*R*,11a*R*)-maackiain, (6a*R*,11a*R*)-medicarpin, as well as betulinic acid and β-sitosterol, were also isolated from Caulis Spatholobi [2–5]. The Chinese Pharmacopoeia requires a qualitative determination of biochanin B in Caulis Spatholobi by thin-layer chromatographic comparison with reference substance.

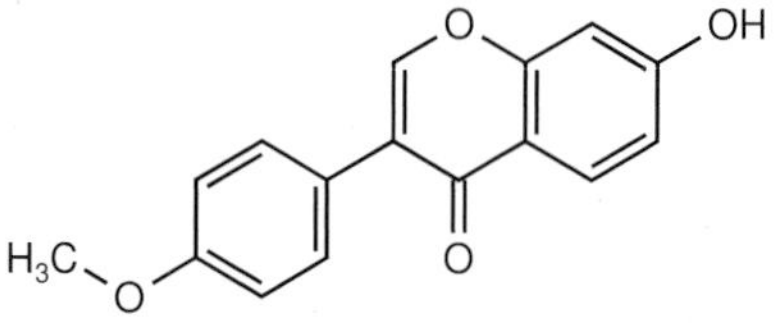

Biochanin B

Pharmacology and Toxicology

The aqueous extract of *S. suberectus* was reported to exert antiviral activity against enteroviruses, including coxsackievirus B3, coxsackievirus B5, poliovirus I, ECHO virus 9, and ECHO virus 29. During infection with coxsackievirus B3, three steps, including inactivation, adsorption and replication, were inhibited. The inhibitory concentration of the extract was less than that causing cytotoxicity [6]. The aqueous and methanol extracts of *S. suberectus* were also found to significantly inhibit human immunodeficiency virus type-1 protease; a concentration of $0.2\,mg\,ml^{-1}$ led to an inhibitory activity >90% [7].

The ethanol extract of the stem of *S. suberectus* was reported to exhibit anti-inflammatory activity, and to inhibit a panel of key enzymes relating to inflammation, including cyclooxygenase-1, phospholipase A_2, 5-lipoxygenase and 12-lipoxygenase, with IC_{50} values of 158, 54, 31 and 35 $\mu g\,ml^{-1}$, respectively [8]. The extract also exhibited a potent inhibitory effect on tyrosinase, with an IC_{50} of $84\,\mu g\,ml^{-1}$, but cytotoxicity on human epidermal melanocytes was

Handbook of Chinese Medicinal Plants: Chemistry, Pharmacology, Toxicology
Weici Tang and Gerhard Eisenbrand

ISBN: 978-3-527-32226-8

rather low. Furthermore, the extract exerted a free radical-scavenging activity, tested using 1,1-diphenyl-2-picrylhydrazyl and hydroxyl radicals, with IC_{50} values of about 10 and 4 $\mu g\, ml^{-1}$, respectively. The phenolic content of the extract was determined as 189 $mg\, g^{-1}$ (estimated as gallic acid), which suggested that the extract might be a good candidate for cosmetic application [9]. Butin was found to inhibit the expression of tyrosinase and tyrosinase-related proteins 1 and 2 in human epidermal melanocytes, in a concentration-dependent manner [10].

An extract fraction, SS8, from *S. suberectus* markedly stimulated the proliferation of hematopoietic progenitor cells in mice with bone marrow depression, in both time- and dose-dependent manner [11]. The intraperitoneal administration of catechin to mice immediately after irradiation for seven consecutive days was shown to promote the expression of IL-6 mRNA and GM-CSF mRNA in the spleen cells of mice. Catechin stimulated the bone marrow cells of normal mice into cell cycle, and helped those of bone marrow-depressed mice to enter the cell cycle. Catechin also accelerated the proliferation and differentiation of hematopoietic stem cells and hematopoietic progenitor cells [4, 12]. The extract of *S. suberectus* also markedly promoted the proliferation of bone marrow cells in healthy and anemic mice [13]. A clinical trial using *S. suberectus* composita to treat aplastic anemia revealed that the preparation not only stimulated hematopoiesis, but also exerted a reconstructive activity to the marrow microenvironment [14].

References

1 Liu, C., Ma, L., Chen, R.Y., and Liu, P. (2005) Determination of catechin and its analogues in *Spatholobus suberectus* by RP-HPLC. *Zhongguo Zhong Yao Za Zhi*, **30**, 1433–1435.

2 Cui, Y.J., Liu, P., and Chen, R.Y. (2005) Studies on the active constituents in vine stem of *Spatholobus suberectus*. *Zhongguo Zhong Yao Za Zhi*, **30**, 121–123.

3 Cui, Y.J., Liu, P., and Chen, R.Y. (2002) Studies on the chemical constituents of *Spatholobus suberectus* Dunn. *Yao Xue Xue Bao*, **37**, 784–787.

4 Yoon, J.S., Sung, S.H., Park, J.H., and Kim, Y. C. (2004) Flavonoids from *Spatholobus suberectus*. *Arch. Pharm. Res.*, **27**, 589–592.

5 Cheng, J., Liang, H., Wang, Y., and Zhao, Y.Y. (2003) Studies on the constituents from the stems of *Spatholobus suberectus*. *Zhongguo Zhong Yao Za Zhi*, **28**, 1153–1155.

6 Guo, J.P., Pang, J., Wang, X.W., Shen, Z.Q., Jin, M., and Li, J.W. (2006) *In vitro* screening of traditionally used medicinal plants in China against enteroviruses. *World J. Gastroenterol.*, **12**, 4078–4081.

7 Lam, T.L., Lam, M.L., Au, T.K., Ip, D.T., Ng, T.B., Fong, W.P., and Wan, D.C. (2000) A comparison of human immunodeficiency virus type-1 protease inhibition activities by the aqueous and methanol extracts of Chinese medicinal herbs. *Life Sci.*, **67**, 2889–2896.

8 Li, R.W., David Lin, G., Myers, S.P., and Leach, D.N. (2003) Anti-inflammatory activity of Chinese medicinal vine plants. *J. Ethnopharmacol.*, **85**, 61–67.

9 Wang, K.H., Lin, R.D., Hsu, F.L., Huang, Y.H., Chang, H.C., Huang, C.Y., and Lee, M.H. (2006) Cosmetic applications of selected traditional Chinese herbal medicines. *J. Ethnoparmacol.*, **106**, 353–359.

10 Lee, M.H., Lin, Y.P., Hsu, F.L., Zhan, G.R., and Yen, K.Y. (2006) Bioactive constituents of *Spatholobus suberectus* in regulating tyrosinase-related proteins and mRNA in HEMn cells. *Phytochemistry*, **67**, 1262–1270.

11 Wang, D.X., Chen, M.L., Yin, J.F., and Liu, P. (2003) Effect of SS8, the active part of *Spatholobus suberectus* Dunn, on proliferation of hematopoietic progenitor cells in mice with bone marrow depression. *Zhongguo Zhong Yao Za Zhi*, **28**, 152–155.

12 Liu, P., Wang, D.X., Chen, R.Y., Chen, M.L., Yin, J.F., and Chen, G.Y. (2004) Effect of catechin on bone marrow cell cycle and gene expression of hematopoietic growth factors. *Yao Xue Xue Bao*, **39**, 424–428.

13 Chen, D.H., Luo, X., Yu, M.Y., Zhao, Y.Q., Cheng, Y.F., and Yang, Z.R. (2004) Effect of *Spatholobus suberectus* on the bone marrow cells and related cytokines of mice. *Zhongguo Zhong Yao Za Zhi*, **29**, 352–355.

14 Su, E.Y. and Chen, H.S. (1997) Clinical observation on aplastic anemia treated by *Spatholobus suberectus* composita. *Zhongguo Zhong Xi Yi Jie He Za Zhi*, **17**, 213–215.

Stemona japonica, *Stemona sessilifolia* and *Stemona tuberosa*

Radix Stemonae (Baibu) is the dried bulb of *Stemona sessilifolia* (Miq.) Miq., *Stemona japonica* (Bl.) Miq. or *Stemona tuberosa* Lour. (Stemonaceae). It is used for the treatment of acute and chronic cough and whooping cough.

Stemona sessilifolia (Miq.) Miq.

Chemistry

The root of *Stemona* is known to contain alkaloids as major components [1]. Alkaloids from the root of *S. sessilifolia* were identified as tuberostemonine, oxotuberostemonine, and protostemotinine [2].

Alkaloids from the root of *S. tuberosa* were identified as tuberostemonine, stenine, oxotuberostemonine, stemonine, the stereoisomer of tuberostemonine, stemotinine, its stereoisomer isostemotinine, neotuberostemonine, bisdehydroneotuberostemonine, tuberostemoninol, tuberostemoamide, tuberostemonone, tuberostemonol, stemoamide, tuberostemospironine, bisdehydrotuberostemonine, bisdehydrostemoninine, isobisdehydrostemoninine, bisdehydroneostemoninine, bisdehydros-

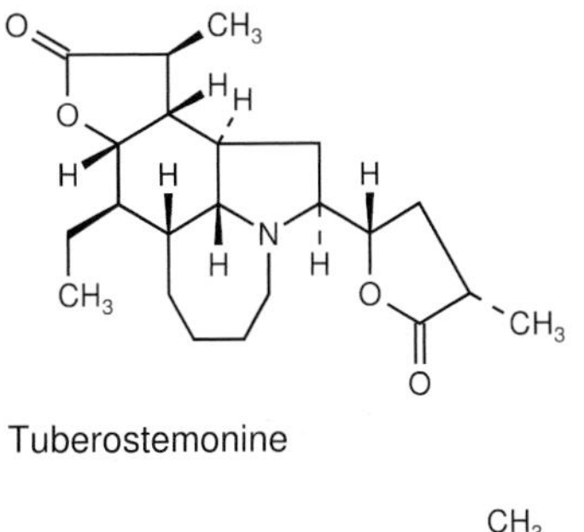

Tuberostemonine

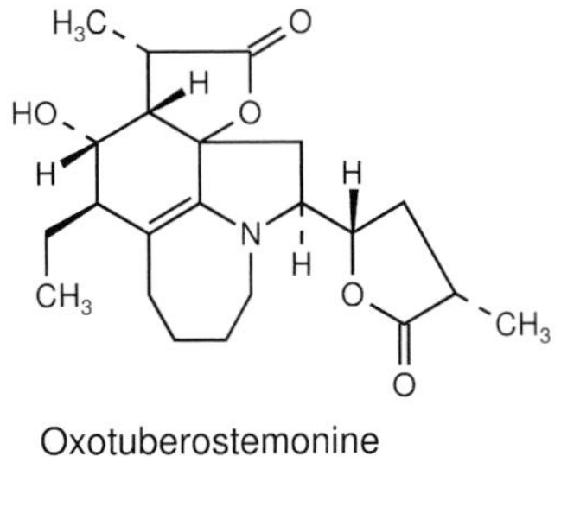

Oxotuberostemonine

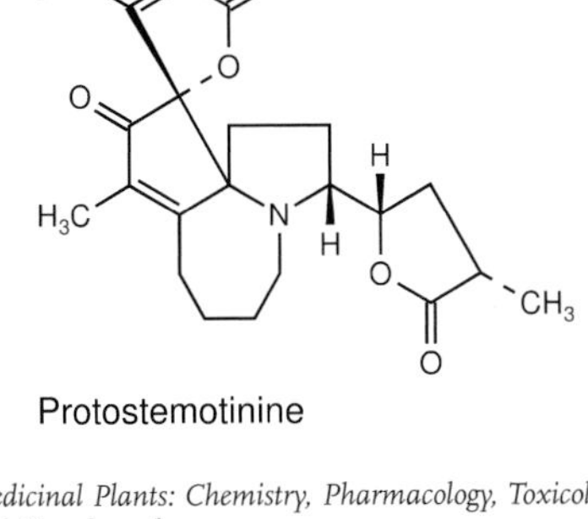

Protostemotinine

Handbook of Chinese Medicinal Plants: Chemistry, Pharmacology, Toxicology
Weici Tang and Gerhard Eisenbrand

ISBN: 978-3-527-32226-8

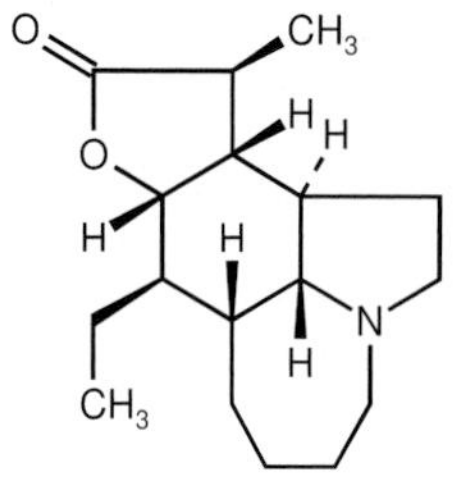

Stenine

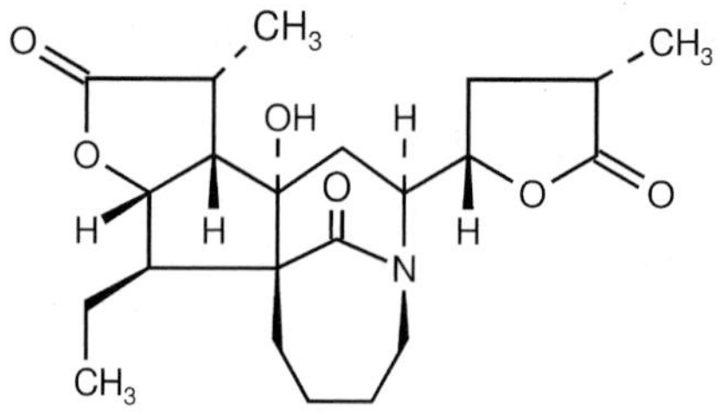

Tuberostemoninol

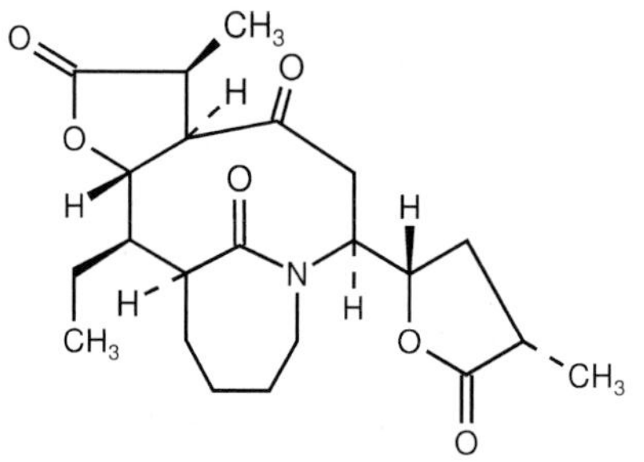

Tuberostemonone

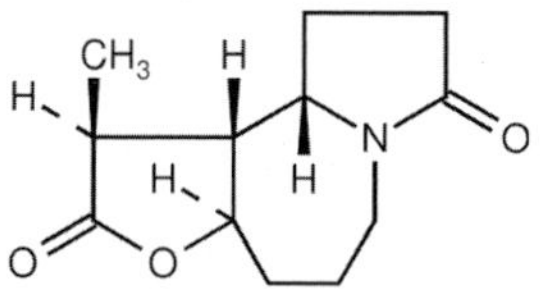

Stemoamide

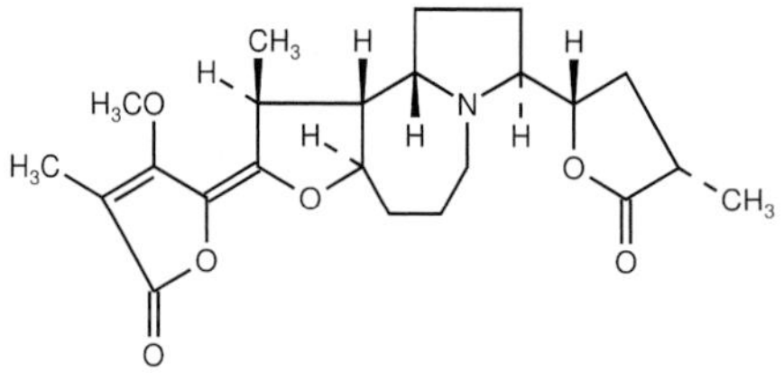

Protostemonine

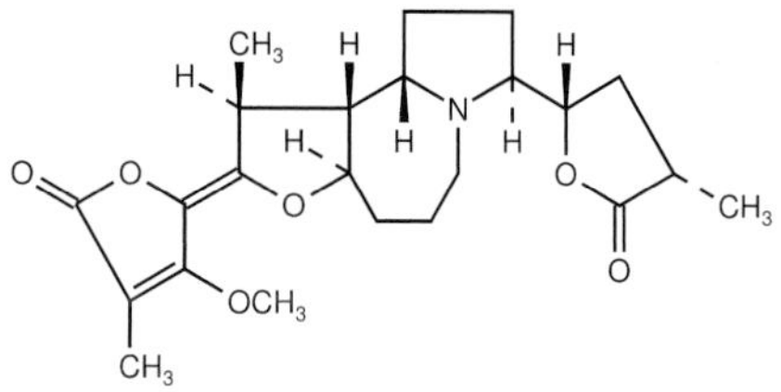

Isoprotostemonine

temoninine A, bisdehydrostemoninine B, tuberostemonine J, tuberostemonine H, epi-bisdehydrotuberostemonine J and neostenine [3–6].

Alkaloids from the root of *S. japonica* were identified as protostemonine, stemonamine and its stereoisomer isostemonamine, neostemonine, bisdehydroneostemonine, bisdehydroprotostemonine, stemonamide, isostemonamide and isoprotostemonine [7].

The content of total alkaloids was determined to range from 0.3% to 3% in the root of *S. sessilifolia*; from 0.8% to 1.4% in the root of *S. japonica*; and from 0.5% to 3% in the root of *S. tuberosa* [8]. Benzyl derivatives from the root of *S. tuberosa* were identified as 3,5-dihydroxy-4-methylbibenzyl, 3,5-dihydroxy-2′-methoxy-4-methylbibenzyl and 3-hydroxy-2′,5-dimethoxy-2-methylbibenzyl [9]. Bibenzyl glycosides stilbostemin B 3′-β-D-glucopyranoside, stilbostemin H 3′-β-D-glucopyranoside, and stilbostemin I 2″-β-D-glucopyranoside were also isolated from the root of *S. tuberosa* [10].

Pharmacology and Toxicology

Only limited studies on the biological activity of the root of *Stemona* species and its constituents have been reported [11]. Neotuberostemonine, tuberostemonine J, tuberostemonine H, *epi*-bisdehydrotuberostemonine J and neostenine were reported to exert antitussive activity in guinea pig after cough induction by citric acid aerosol. A study on the structure–activity relationship revealed that the saturated tricyclic pyrrolo[3,2,1-jk] benzazepine nucleus was the primary key structure contributing to the antitussive activity, and all *cis* configurations at the three ring junctions were the optimal structure for the activity of stenine-type alkaloids [12]. Bibenzyl glycosides stilbostemin B 3′-β-D-glucopyranoside, stilbostemin H 3′-β-D-glucopyranoside, and stilbostemin I 2″-β-D-glucopyranoside from the roots of *S. tuberosa* significantly protected human neuroblastoma SH-SY5Y cells from 6-hydroxydopamine-induced neurotoxicity [10].

References

1 Pilli, R.A. and Ferreira de Oliveira, M.C. (2000) Recent progress in the chemistry of the *Stemona* alkaloids. *Nat. Prod. Rep.*, **17**, 117–127.

2 Cong, X.D., Zhao, H.R., Guillaume, D., Xu, G.J., Lu, Y., and Zheng, Q.T. (1995) Crystal structure and NMR analysis of the alkaloid protostemotinine. *Phytochemistry*, **40**, 615–617.

3 Ye, Y., Qin, G.W., and Xu, R.S. (1994) Alkaloids from *Stemona tuberosa*. *Phytochemistry*, **37**, 1201–1203.

4 Lin, W.H., Ma, L., Cai, M.S., and Barnes, R.A. (1994) Two minor alkaloids from roots of *Stemona tuberosa*. *Phytochemistry*, **36**, 1333–1335.

5 Lin, L.G., Zhong, Q.X., Cheng, T.Y., Tang, C.P., Ke, C.Q., Lin, G., and Ye, Y. (2006) Stemoninines from the roots of *Stemona tuberosa*. *J. Nat. Prod.*, **69**, 1051–1054.

6 Chung, H.S., Hon, P.M., Lin, G., But, P.P., and Dong, H. (2003) Antitussive activity of Stemona alkaloids from *Stemona tuberosa*. *Planta Med.*, **69**, 914–920.

7 Ye, Y., Qin, G.W., and Xu, R.S. (1994) Alkaloids of *Stemona japonica*. *Phytochemistry*, **37**, 1205–1208.

8 Cong, X.D., Xu, G.J., Jin, R.L., and Zhi, H.J. (1992) Pharmacognostical studies on Baibu radix stemonae and allied drugs. IX. Determination and evaluation of total alkaloid content in the roots of Chinese *Stemona* spp. *Acta Pharm. Sin.*, **27**, 556–560.

9 Zhao, W., Qin, G., Ye, Y., Xu, R., and Le, X. (1995) Bibenzyls from *Stemona tuberosa*. *Phytochemistry*, **38**, 711–713.

10 Lee, K.Y., Sung, S.H., and Kim, Y.C. (2006) Neuroprotective bibenzyl glycosides of *Stemona tuberosa* roots. *J. Nat. Prod.*, **69**, 679–681.

11 Qin, G.W. and Xu, R.S. (1998) Recent advances on bioactive natural products from Chinese medicinal plants. *Med. Res. Rev.*, **18**, 375–382.

12 Xu, Y.T., Hon, P.M., Jiang, R.W., Cheng, L., Li, S.H., Chan, Y.P., Xu, H.X., Shaw, P.C., and But, P.P. (2006) Antitussive effects of *Stemona tuberosa* with different chemical profiles. *J. Ethnopharmacol.*, **108**, 46–53.

Stephania cepharantha

The dried root of *Stephania cepharantha* Hayata (Menispermaceae) is a Chinese folk medicine used as an analgesic, diuretic, and tuberculostatic agent. It is not listed in the Chinese Pharmacopoeia.

Stephania cepharantha Hayata

Chemistry

The tuber of *S. cepharantha* was known to contain alkaloids with cepharanthine and isotetrandrine as the major components [1]. They are both alkaloids of the bisbenzylisoquinoline type.

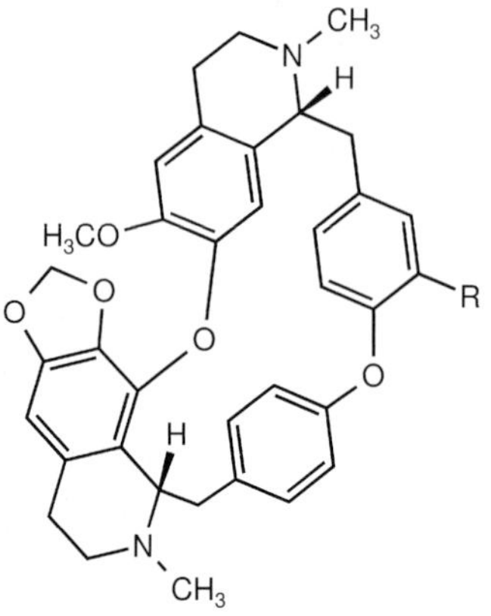

Cepharanthine: R = OCH_3
Cepharanoline: R = OH

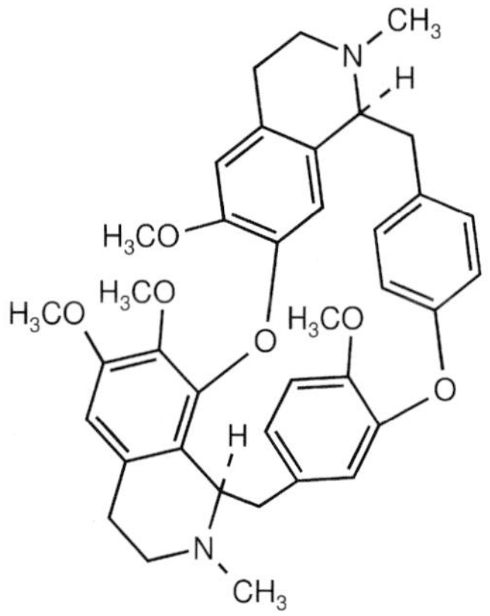

Isotetrandrine

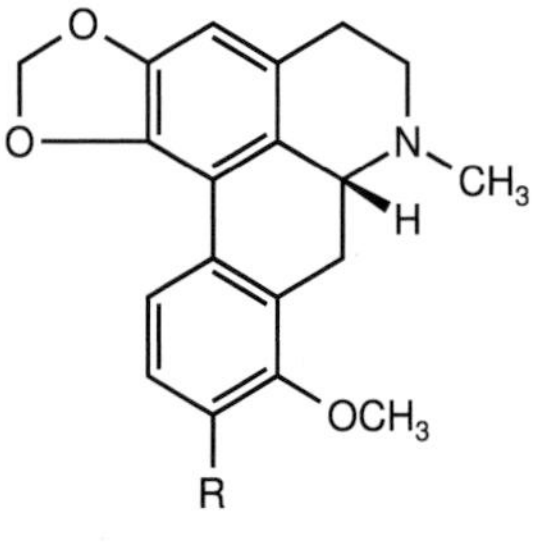

Stephanine: R = H
Stesakine: R = OH

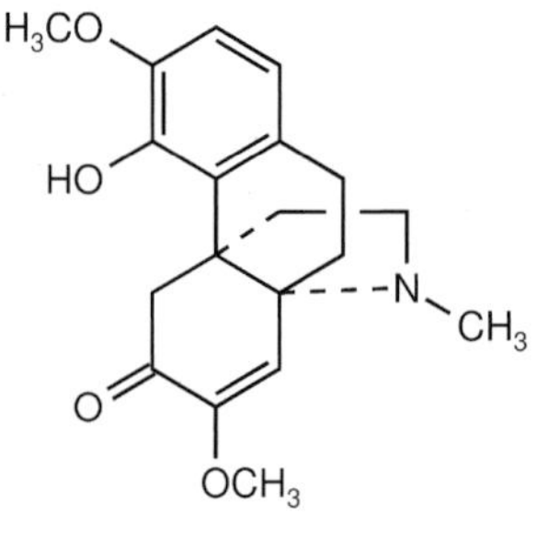

Cepharamine

Handbook of Chinese Medicinal Plants: Chemistry, Pharmacology, Toxicology
Weici Tang and Gerhard Eisenbrand

ISBN: 978-3-527-32226-8

Minor alkaloids of different structural types were identified as berbamine, cepharanoline, cycleanine, stephanine, crebanine, dehydrocrebanine, stesakine, magnoflorine, menisperine palmatine, steponine, cyclanoline, cepharamine, stecepharine, and tetradehydroreticuline [2].

Pharmacology and Toxicology

The extract of *S. cepharantha* contains biscoclaurine alkaloids, with cepharanthine as the major component. It has been used for the treatment of different diseases. Cepharanthine exerted diverse pharmacological effects, including anti-inflammatory, antiallergic, free radical-scavenging, immunomodulatory, anti-HIV-1, membrane-stabilizing, antitumor, apoptosis-inducing, and multidrug resistance-reversing effects [3]. The methanol extract of the tuber of *S. cepharantha* and a number of its alkaloid components were found to exert antiviral activity against herpes simplex virus type 1 (HSV-1) [4]. Cepharanthine also dose-dependently inhibited human immunodeficiency virus type 1 (HIV-1) replication in monocytic U1 cells stimulated with tumor necrosis factor-α (TNF-α) or 12-*O*-tetradecanoylphorbol 13-acetate (TPA) with IC_{50} values of about 0.02 and 2 $\mu g\,ml^{-1}$, but not in T-lymphocytic ACH-2 cells. Inhibitory effects of cepharanthine on HIV-1 long terminal repeat (LTR)-driven gene expression through the inhibition of nuclear transcription factor κB (NF-κB) activation were also observed [5].

Cepharanthine was reported to exert vasodilatory effects in rabbits after intravenous (i.v.) administration at doses of 1 and 3 $mg\,kg^{-1}$. The microvascular dilator effect of cepharanthine appeared to have no direct association with systemic hemodynamics [6]. Intraperitoneal (i.p.) treatment with cepharanthine to mice 10 min before an i.v. challenge by lipopolysaccharide (LPS) resulted in protection against LPS-mediated lethality in the D-galactosamine-sensitized mouse strain BALB/c, but not in C57BL/6 and C57BL/10ScSn mouse strains. Treatment with cepharanthine before the LPS challenge significantly reduced serum TNF-α levels in a dose-dependent manner. Cepharanthine also significantly inhibited the production of TNF-α by LPS-stimulated monocytes *in vitro* [7]. Cepharanthine was further found to effectively inhibit pulmonary vascular injury in rats induced by LPS. It was suggested that cepharanthine prevents LPS-induced pulmonary vascular injury by inhibiting leukocyte activation [8].

Cepharanthine has been considered to exhibit antiperoxidation activity due to its membrane-stabilizing effect. It was found to scavenge free radicals in solution, and to inhibit lipid peroxidation in mitochondria and liposomes. The antiperoxidation activity of cepharanthine in rat liver mitochondria initiated by Fe^{2+}/ADP at pH 7.4 was much greater than that of α-tocopherol, with an IC_{50} of about 23 μM. However, cepharanthine was effective only at neutral pH values, but not in a moderately acidic pH region below pH 6.5. The amine moiety was suggested to be responsible for this pH-dependent radical-scavenging activity [9]. The inhibitory effect of cepharanthine on mitochondrial membrane permeability transition was found to be caused by its inhibitory action on Ca^{2+} release and antioxidant activity [10]. The anti-lipid peroxidation by cepharanthine was believed to be caused by its direct radical-scavenging activity [11]. Cepharanthine inhibited platelet aggregation induced by thrombin. The inhibition of thrombin-induced platelet aggregation by cepharanthine correlated with the inhibition of arachidonate release from the phospholipids of the membrane, notably from phosphatidylcholine [12].

Cepharanthine was also reported to be active in the treatment of experimental acute lung injury and septic shock in sheep.

In a post-smoke inhalation model of sepsis in female sheep, an infusion of cepharanthine at a dose of 1.3 mg $kg^{-1} h^{-1}$ significantly attenuated changes in lung histology as well as in the lung wet/dry weight ratio. An *in vitro* experiment revealed that cepharanthin inhibited the release of neutrophil elastase from isolated neutrophils stimulated with either formyl-methyl-leucyl-phenylalanine (fMLP) or TPA with an IC_{50} of 60 μ*M*; this indicated that cepharanthin inhibited protein kinase C (PKC) or a more downstream signaling pathway in neutrophil activation [13]. In mice, cepharanthine clearly prevented the lethal shock induced by endotoxin or recombinant human TNF-α combined with endotoxin [14].

Cepharanthine suppressed two-stage skin tumor promotion by TPA and mezerein in mice initiated with 7,12-dimethylbenz[*a*]anthracene (DMBA) by topical application. The oral administration of cepharanthine also was inhibitory. Furthermore, it was found that topical application of berbamine and isotetrandrine markedly suppressed the tumor-promoting effect of TPA [15]. Pretreatment of platelet-derived growth factor (PDGF)-stimulated rat mesangial cells with cepharanthine at concentrations of 0.1–2 μ*M* suppressed the PDGF-stimulated increase in fibronectin, in concentration-dependent manner. Cepharanthine inhibited tyrosine phosphorylation of several proteins, including the PDGF β receptor in PDGF-stimulated cells, and also tyrosine kinase activity. This suggested that it inhibits fibronectin production induced by growth factors, probably through a suppression of receptor autophosphorylation [16].

Cepharanthine exhibited cytotoxicity towards murine P388 cell lines both sensitive (P388/S) and resistant (P388/DOX) to doxorubicin. At a concentration of 10 μg ml^{-1}, it markedly induced apoptosis in resistant cells after incubation for 6 h and 24 h. Furthermore, it was found that the inhibition of DNA and protein synthesis caused by cepharanthine was more significant in resistant cells than in sensitive cells [17]. Cepharanthine was also reported to exert growth-inhibitory activity against a number of human oral squamous cell carcinoma cell lines, especially B88 cells. The treatment of B88 cancer cells with cepharanthine resulted in cell-cycle arrest at G_1 stage and DNA fragmentation. The subcutaneous injection of cepharanthine at a daily dose of 40 mg kg^{-1} to nude mice bearing B88 tumors significantly inhibited tumor growth, induced p27Kip1, and reduced cyclin E and Skp2. Thus, cepharanthine might inhibit the growth of B88 human oral squamous cell carcinoma cells by downregulating cyclin E to induce G_1 arrest through a pathway of p27Kip1 induction [18].

Cepharanthine has been shown to reverse multidrug resistance in P-glycoprotein-expressing cell lines. The augmentation of doxorubicin sensitivity by cepharanthine, and the correlation to the decreased expression of P-glycoprotein, were demonstrated in fresh human gastrointestinal tumor cells from 73 cancer patients [19]. Cepharanthine enhanced doxorubicin accumulation in cancer cells [20], and synergistically accelerated apoptosis in resistant cancer cells [21, 22], but lowered doxorubicin accumulation in normal liver cells and did not affect that in spleen cells [23]. It was also suggested that the mechanism of cepharanthine to reverse drug resistance might be related to suppressing the constitutive activity and activation of NF-κB [24].

Cepharanthine also showed additive or synergistic effects in combination with vincristine, vinblastine and vindesine against the RPMI 4788 human colon cancer cell line, and against HeLa cells. Cepharanthine significantly enhanced the antiproliferative activity of vincristine against RPMI 4788 cells transplanted subcutaneously or intraperitoneally into nude mice [25]. A reversal of multidrug-resistance in 50MT-1 cells towards irinotecan, doxorubicin and VP-16 by cepharanthine treatment was also reported [26]. Synergistic effects of tamoxifen and cepharanthine for circumventing the multidrug resistance of tumor cell lines towards doxorubicin were observed [27]. It was also postulated that increased doxorubicin accumulation caused by cepharanthine may be due to an increased influx of Ca^{2+} [28]. Cepharanthine also enhanced the thermosensitivity of murine mammary carcinoma cells not only *in vitro* but also *in vivo* when transplanted in mice [29]. Similar results were obtained with FSa-II mouse fibrosarcoma. Cepharanthine as a thermosensitizer was found to increase the apoptosis of tumor cells [30]. A retrospective clinical study revealed that cepharanthine was able to lower the degree of xerostomia in 37 cases of head and neck cancer after radiotherapy [31].

Cepharanthine was found not only to be active against the multidrug resistance of tumor cells, but also against the drug-resistance of human malaria parasites, *Plasmodium falciparum*. Like chloroquine, cepharanthine inhibited the trophozoite stage of parasite growth, exerted synergistic activity with chloroquine, and lowered the half-maximal inhibitory concentration of chloroquine from 148.5 to 37.8 n*M* [32].

The minor alkaloid berbamine from the root of *S. cepharantha* was found to significantly inhibit NB4 leukemia cells with an IC_{50} value of about 4 $\mu g\,ml^{-1}$ at 48 h. Apoptosis and the expression of caspase 3 were induced at a berbamine concentration of 12 $\mu g\,ml^{-1}$, in time-dependent manner [33].

It was suggested that the suppression of matrix metalloproteinase 9 (MMP-9), as induced by TNF-α, could prevent the destruction of acinar tissue in the salivary glands of patients with Sjögren's syndrome. Cepharanthine was found to markedly inhibit the TNF-α-induced production of MMP-9 in NS-SV-AC, an SV40-immortalized normal human acinar cell clone. In addition, cepharanthine suppressed the TNF-α-stimulated NF-κB activity by partly preventing the degradation of IκBα protein in NS-SV-AC cells. When NS-SV-AC cells were seeded on type IV collagen-coated dishes in the presence of both TNF-α and plasmin, type IV collagen interaction with the cells was lost and the cells entered apoptosis. However, pretreatment with cepharanthine restored the aberrant *in vitro* morphogenesis of the NS-SV-AC cells [34]. A preventive effect on the destruction of acinar tissues by cepharanthin was also observed in a murine model. The i.p. administration of cepharanthin to thymectomized female NFS/sld mice resulted in a significant inhibition of mononuclear cell infiltration and the destruction of acinar tissue in salivary and lacrimal glands. An immunohistochemical analysis revealed that p65, phosphorylated IκBα, and MMP-9 were less stained in the acinar cells after cepharanthin treatment. Destruction of acinar tissues was attributed to the induction of apoptosis, suggesting that cepharanthin inhibits apoptosis by suppressing phosphorylation of IκBα, followed by prevention of MMP-9 activation [35].

References

1 Xia, G.C., Qin, Y.Q., Yang, Q.Z., Zhang, Y.S., and Liu, X.M. (1985) Palmitine from 16 Menispermaceae species. *Chin. Trad. Herbal Drugs*, **16**, 266–267.

2 Tanahashi, T., Su, U., Nagakura, N., and Nayeshiro, H. (2000) Quaternary isoquinoline alkaloids from *Stephania cepharantha*. *Chem. Pharm. Bull. (Tokyo)*, **48**, 370–373.

3 Furusawa, S. and Wu, J. (2007) The effects of biscoclaurine alkaloid cepharanthine on mammalian cells: implications for cancer, shock, and inflammatory diseases. *Life Sci.*, **80**, 1073–1079.

4 Nawawi, A., Ma, C., Nakamura, N., Hattori, M., Kurokawa, M., Shiraki, K., Kashiwaba, N., and Ono, M. (1999) Anti-herpes simplex

virus activity of alkaloids isolated from *Stephania cepharantha*. *Biol. Pharm. Bull.*, **22**, 268–274.

5 Okamoto, M., Okamoto, T., and Baba, M. (1999) Inhibition of human immunodeficiency virus type 1 replication by combination of transcription inhibitor K-12 and other antiretroviral agents in acutely and chronically infected cells. *Antimicrob. Agents Chemother.*, **43**, 492–497.

6 Kamiya, T., Sugimoto, Y., and Yamada, Y. (1993) Vasodilator effects of bisbenzylisoquinoline alkaloids from *Stephania cepharantha*. *Planta Med.*, **59**, 475–476.

7 Maruyama, H., Kikuchi, S., Kawaguchi, K., Hasunuma, R., Ono, M., Ohbu, M., and Kumazawa, Y. (2000) Suppression of lethal toxicity of endotoxin by biscoclaurine alkaloid cepharanthine. *Shock*, **13**, 160–165.

8 Murakami, K., Okajima, K., and Uchiba, M. (2000) The prevention of lipopolysaccharide-induced pulmonary vascular injury by pretreatment with cepharanthine in rats. *Am. J. Respir. Crit. Care Med.*, **161**, 57–63.

9 Kogure, K., Goto, S., Abe, K., Ohiwa, C., Akasu, M., and Terada, H. (1999) Potent antiperoxidation activity of the bisbenzylisoquinoline alkaloid cepharanthine: the amine moiety is responsible for its pH-dependent radical scavenge activity. *Biochim. Biophys. Acta*, **1426**, 133–142.

10 Nagano, M., Kanno, T., Fujita, H., Muranaka, S., Fujiwara, T., and Utaumi, K. (2003) Cepharanthine, an anti-inflammatory drug, suppresses mitochondrial membrane permeability transition. *Physiol. Chem. Phys. Med. NMR*, **35**, 131–143.

11 Kogure, K., Tsuchya, K., Abe, K., Akasu, M., Tamaki, T., Fukuzawa, K., and Terada, H. (2003) Direct radical scavenging by the bisbenzylisoquinoline alkaloid cepharanthine. *Biochim. Biophys. Acta*, **1622**, 1–5.

12 Kanaho, Y., Sato, T., and Fujii, T. (1982) Mechanism of the inhibitory effect of cepharanthine on the aggregation of platelets. *Cell Struct. Funct.*, **7**, 39–48.

13 Murakami, K., Cox, R.A., Hawkins, H.K., Schmalstieg, F.C., McGuire, R.W., Jodoin, J. M., Traber, L.D., and Traber, D.L. (2003) Cepharanthin, an alkaloid from *Stephania cepharantha*, inhibits increased pulmonary vascular permeability in an ovine model of sepsis. *Shock*, **20**, 46–51.

14 Sakaguchi, S., Furusawa, S., Wu, J., and Nagata, K. (2007) Preventive effects of a biscoclaurine alkaloid, cepharanthine, on endotoxin or tumor necrosis factor-α-induced septic shock symptoms: involvement of from cell death in L929 cells and nitric oxide production in raw 264.7 cells. *Int. Immunopharmacol.*, **7**, 191–197.

15 Yasukawa, K., Akasu, M., Takeuchi, M., and Takido, M. (1993) Bisbenzylisoquinoline alkaloids inhibit tumor promotion by 12-*O*-tetradecanoylphorbol 13-acetate in two-stage carcinogenesis in mouse skin. *Oncology*, **50**, 137–140.

16 Hayama, M., Inoue, R., Akiba, S., and Sato, T. (2000) Inhibitory effect of cepharanthine on fibronectin production in growth factor-stimulated rat mesangial cells. *Eur. J. Pharmacol.*, **390**, 37–42.

17 Furusawa, S., Wu, J., Fujimura, T., Nakano, S., Nemoto, S., Takayanagi, M., Sasaki, K., and Takayanagi, Y. (1998) Cepharanthine inhibits proliferation of cancer cells by inducing apoptosis. *Methods Find. Exp. Clin. Pharmacol.*, **20**, 87–97.

18 Harada, K., Yamamoto, S., Kawaguchi, S., Yoshida, H., and Sato, M. (2003) Cepharanthine exerts antitumor activity on oral squamous cell carcinoma cell lines by induction of p27Kip1. *Anticancer Res.*, **23**, 1441–1448.

19 Hotta, T., Tanimura, H., Yamaue, H., Iwahashi, M., Tani, M., Tsunoda, T., Tamai, M., Noguchi, K., Mizobata, S., Arii, K., and Terasawa, H. (1997) Modulation of multidrug resistance by cepharanthine in fresh human gastrointestinal tumor cells. *Oncology*, **54**, 153–157.

20 Wada, H., Saikawa, Y., Niida, Y., Nishimura, R., Noguchi, T., Matsukawa, H., Ichihara, T., and Koizumi, S. (1999) Selectively induced high MRP gene expression in multidrug-resistant human HL60 leukemia cells. *Exp. Hematol.*, **27**, 99–109.

21 Katsui, K., Kuroda, M., Wang, Y., Komatsu, M., Himei, K., Takemoto, M., Akaki, S., Asaumi, J., Kanazawa, S., and Hiraki, Y. (2004) Cepharanthin enhances adriamycin sensitivity by synergistically accelerating apoptosis for adriamycin-resistant osteosarcoma cell lines, SaOS2-AR and SaOS2 F-AR. *Int. J. Oncol.*, **25**, 47–56.

22 Nakajima, A., Yamamoto, Y., Taura, K., Hata, K., Fukumoto, M., Uchinami, H., Yonezawa, K., and Yamaoka, Y. (2004) Beneficial effect of cepharanthine on overcoming drug-resistance of hepatocellular carcinoma. *Int. J. Oncol.*, **24**, 635–645.

23 Nishikawa, K., Asaumi, J., Kawasaki, S., Kuroda, M., Takeda, Y., and Hiraki, Y. (1997) Influence of cepharanthin on the intracellular accumulation of adriamycin in normal liver cells and spleen cells of mice *in vitro* and *in vivo*. *Anticancer Res.*, **17A**, 3617–3621.

24 Song, Y.C., Xia, W., Jiang, J.H., and Wang, Q.D. (2005) Reversal of multidrug resistance in drug-resistant cell line EAC/ADR by cepharanthine hydrochloride and its mechanism. *Yao Xue Xue Bao*, **40**, 204–207.

25 Ono, M. and Tanaka, N. (1997) Positive interaction of bisbenzylisoquinoline alkaloid, cepharanthin, with vinca alkaloid agents against human tumors. *In Vivo*, **11**, 233–241.

26 Aogi, K., Nishiyama, M., Kim, R., Hirabayashi, N., Toge, T., Mizutani, A., Okada, K., Sumiyoshi, H., Fujiwara, Y., Yamakido, M., Kusano, T., and Andoh, T. (1997) Overcoming CPT-11 resistance by using a biscoclaurine alkaloid, cepharanthine, to modulate plasma trans-membrane potential. *Int. J. Cancer*, **72**, 295–300.

27 Ikeda, R., Che, X.F., Yamaguchi, T., Ushiyama, M., Zheng, C.L., Okumura, H., Takeda, Y., Shibayama, Y., Nakamura, K., Jeung, H.C., Furukawa, T., Sumizawa, T., Haraguchi, M., Akiyama, S., and Yamada, K. (2005) Cepharanthine potently enhances the sensitivity of anticancer agents in K562 cells. *Cancer Sci.*, **96**, 372–376.

28 Nishikawa, K., Asaumi, J., Kawasaki, S., Shibuya, K., Kuroda, M., Takeda, Y., and Hiraki, Y. (1998) Influence of cepharanthin and hyperthermia on the intracellular accumulation of adriamycin and Fluo3, an indicator of Ca^{2+}. *Anticancer Res.*, **18A**, 1649–1654.

29 Yamamoto, M., Kuroda, M., Honda, O., Ono, E., Asaumi, J.I., Shibuya, K., Kawasaki, S., Joja, I., Takemoto, M., Kanazawa, S., and Hiraki, Y. (1999) Cepharanthin enhances thermosensitivity without a resultant reduction in the thermotolerance of a murine mammary carcinoma. *Int. J. Oncol.*, **15**, 95–99.

30 Wang, Y., Kuroda, M., Gao, X.S., Akaki, S., Asaumi, J., Okumura, Y., Shibuya, K., Kawasaki, S., Joja, I., Kato, H., Himei, K., Dendo, S., Kanazawa, S., and Hiraki, Y. (2004) Cepharanthine enhances *in vitro* and *in vivo* thermosensitivity of a mouse fibrosarcoma, FSa-II, based on increased apoptosis. *Int. J. Mol. Med.*, **13**, 405–411.

31 Imada, H., Nomoto, S., Ohguri, T., Yahara, K., Kato, F., Morioka, T., and Korogi, Y. (2004) Effect of Cepharanthin to prevent radiation induced xerostomia in head and neck cancer. *Gan To Kagaku Ryoho*, **31**, 1041–1045.

32 Tamez, P.A., Lantvit, D., Lim, E., and Pezzuto, J.M. (2005) Chemosensitizing action of cepharanthine against drug-resistant human malaria, *Plasmodium falciparum*. *J. Ethnopharmacol.*, **98**, 137–142.

33 He, Z.W., Zhao, X.Y., Xu, R.Z., and Wu, D. (2006) Effects of berbamine on growth of leukemia cell line NB4 and its mechanism. *Zhejiang Da Xue Xue Bao Yi Xue Ban*, **35**, 209–214.

34 Azuma, M., Aota, K., Tamatani, T., Motegi, K., Yamashita, T., Ashida, Y., Hayashi, Y., and Sato, M. (2002) Suppression of tumor necrosis factor α-induced matrix metalloproteinase 9 production in human salivary gland acinar cells by cepharanthine occurs *via* down-regulation of nuclear factor κB: a possible therapeutic agent for preventing the destruction of the acinar structure in the salivary glands of Sjögren's syndrome patients. *Arthritis Rheum.*, **46**, 1585–1594.

35 Azuma, M., Ashida, Y., Tamatani, T., Motegi, K., Takamaru, N., Ishimaru, N., Hayashi, Y., and Sato, M. (2006) Cepharanthin, a biscoclaurine alkaloid, prevents destruction of acinar tissues in murine Sjögren's syndrome. *J. Rheumatol.*, **33**, 912–920.

Stephania tetrandra

Radix Stephaniae Tetrandrae (Fangji) is the dried root of *Stephania tetrandra* S. Moore (Menispermaceae). It is used as an analgesic, antirheumatic, diuretic, and antiedemic agent for the treatment of edema, oliguria, rheumatic arthritis, and hypertension.

Stephania tetrandra S. Moore

Chemistry

The root of *S. tetrandra* is known to contain tetrandrine and fangchinoline as major alkaloids. Tetrandrine and fangchinoline are both alkaloids of the bisbenzylisoquinoline type, in which the two benzylisoquinoline moieties are connected via two ether bridges. The tetrandrine content in the root is reported to range from 0.7 to 1.3% [1]. The Chinese Pharmacopoeia requires a qualitative determination of tetrandrine and fangchinoline in Radix Stephaniae Tetrandrae by thin-layer chromatographic comparison with reference substances, and a quantitative determination by HPLC. The total content of tetrandrine and fangchinoline in Radix Stephaniae Tetrandrae should be not less than 1.6%. The total content of tetrandrine and fangchinoline in prepared Radix Stephaniae Tetrandrae should be not less than 1.4%.

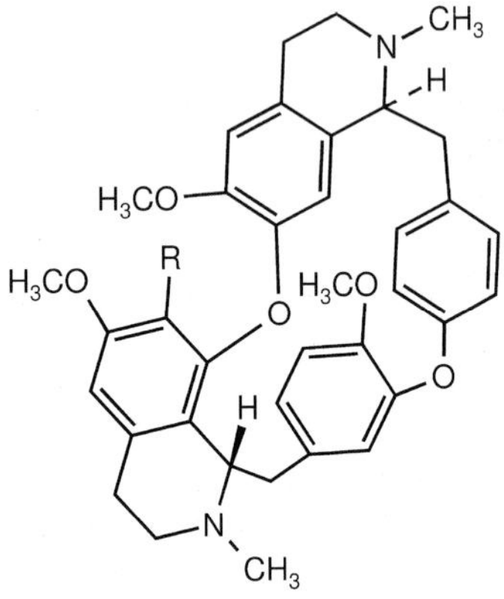

Tetrandrine: R = OCH_3
Fangchinoline: R = OH

Minor bisbenzylisoquinoline alkaloids from the roots were identified as berbamine, oxofangchirine, stereoisomers of *N*- or *N*′-oxides of tetrandrine and fangchinoline, fenfangjine D, *N*′-methyltetrandrine, *N,N*′-dimethyltetrandrine, cycleanine, and *N*-methylfangchinoline [2].

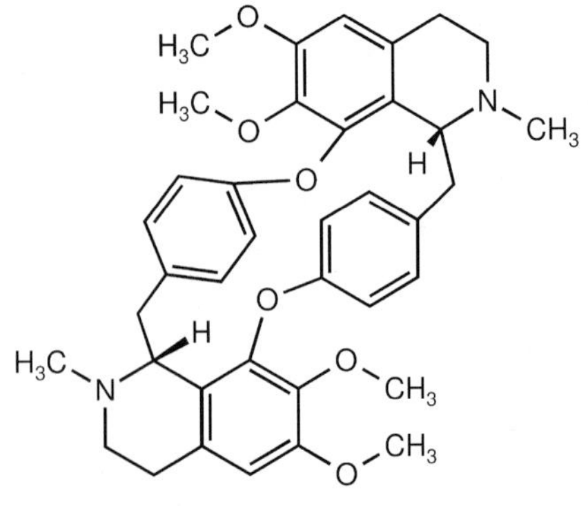

Cycleanine

Handbook of Chinese Medicinal Plants: Chemistry, Pharmacology, Toxicology
Weici Tang and Gerhard Eisenbrand

ISBN: 978-3-527-32226-8

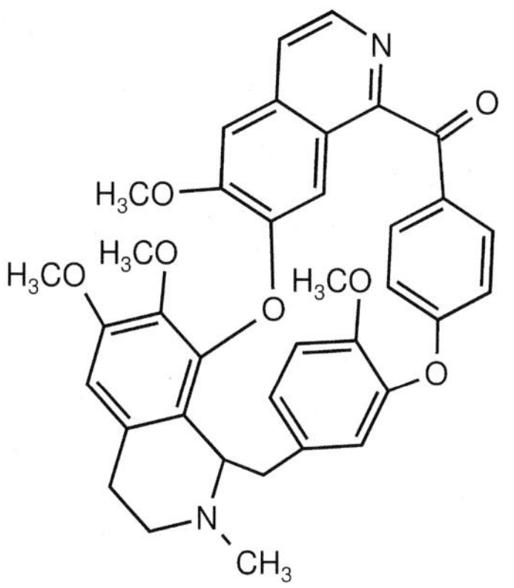

Oxofangchirine

In addition to the bisbenzylisoquinoline alkaloids, cyclanoline (a quaternary alkaloid of protoberberine type) and stephenanthrine (an alkaloid with a phenanthrene skeleton and a tertiary amine side chain [2]) were isolated from the root of *S. tetrandra*. Tetrandrine has been found to be cleaved by treatment with Na/NH_3 into two components, *O*-methylarmepavine and *N*-methylcoclaurine [3].

Pharmacology and Toxicology

The biological activities of *S. tetrandra* are known to depend on the presence of the alkaloid constituents, especially of tetrandine and fangchinoline. Tetrandrine and some other related bisbenzylisoquinoline alkaloids were found to exhibit antimalarial activity *in vitro* against both chloroquine-sensitive and -resistant strains of *Plasmodium falciparum* [4]. Tetrandrine did not reverse the chloroquine resistance, but it potentiated the antimalarial activity of chloroquine and artemisinin against chloroquine-sensitive and -resistant *P. falciparum* [5].

Tetrandrine is known to block voltage-dependent Ca^{2+} and Ca^{2+}-activated K^+ channels, and to inhibit the contraction of smooth muscles [6]. It inhibited the contraction of isolated pig coronary artery strips induced by ouabain [7], and of the rat aorta in the absence of endothelium, as induced by ATP and H_2O_2 [8]. Tetrandrine given before myocardial ischemia to rats might reduce the microvascular permeability in myocardium, in association with its

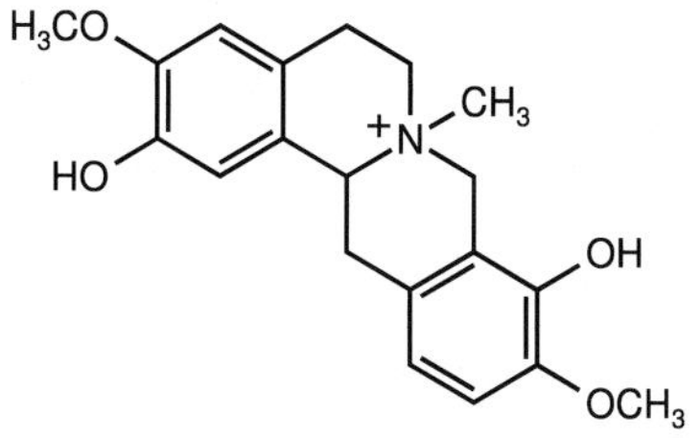

Cyclanoline

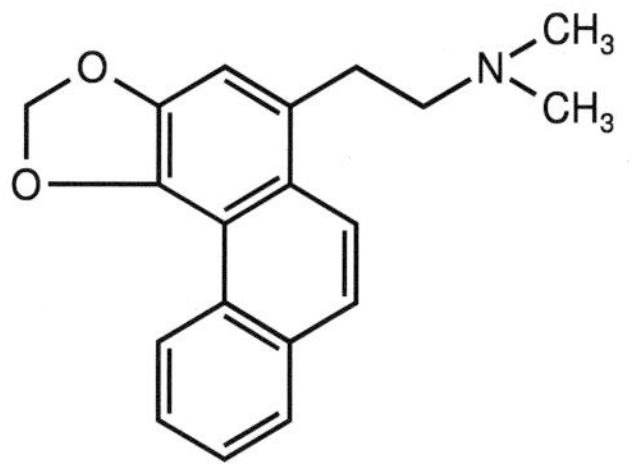

Stephenanthrine

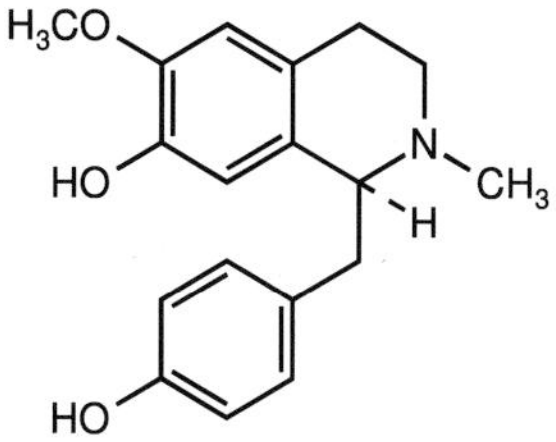

N-Methylcoclaurine

calcium channel-blocking effect [9]. Tetrandrine was found to inhibit both Ca^{2+} entry from the extracellular medium and Ca^{2+} release from intracellular stores in rat glioma C6 cells [10] and in bovine chromaffin cells [11]. It was postulated that tetrandrine directly inhibits the Ca^{2+} current in vascular smooth muscle cells, which may predominantly contribute to its vasodilatory actions [12]. In addition to the contraction inhibitory effects on vascular smooth muscle, tetrandrine was also found to inhibit the production of nitric oxide (NO) [13].

In an endothelial cell line originally derived from human umbilical vein, tetrandrine reversibly decreased the amplitude of K^+ outward currents, with an IC_{50} value of 5 μM. The tetrandrine-sensitive component of the outward current was believed to be a Ca^{2+}-activated K^+ current. Tetrandrine was capable of suppressing the activity of Ca^{2+}-activated K^+ channels in endothelial cells; the direct inhibition of these channels by tetrandrine may contribute to its effect on the functional activities of endothelial cells [14]. The synergistic effect of ouabain and Ca^{2+} on contraction might be antagonized by tetrandrine. All actions of tetrandrine might be reversed by an excess of Ca^{2+}. The effects of tetrandrine were similar to those of the clinically used drug verapamil, but less potent [15]. Structure–activity relationship studies revealed that the methoxy group of one benzene ring of tetrandrine partly contributed to the pharmacological actions of tetrandrine [16]. The protective effect of tetrandrine against myocardial ischemia–reperfusion injury in rats was suggested to be due to inhibition of neutrophil priming and activation, and to the abolition of subsequent infiltration and formation of reactive oxygen species [17].

Tetrandrin did not inhibit the Na^+/K^+-ATPase activity of rat myocardial microsomes *in vitro*. It increased the affinity of Na^+/K^+-ATPase to ATP, but inhibited Mg^{2+}-ATPase activity. The reduced myocardial Na^+/K^+-ATPase activity of renovascular hypertensive rats can be increased by treatment with tetrandrine, both *in vitro* and *in vivo*. This effect of tetrandrine was believed not only secondary to the calcium channel-blocking or hypotensive effect but also due to its direct action on the Na^+/K^+-ATPase and Mg^{2+}-ATPase [18]. Similar effects of tetrandrine were also observed in rats with spontaneous hypertension [19].

Tetrandrine at a concentration of 10 μM was reported to significantly inhibit the proliferation and DNA synthesis of rat pulmonary artery smooth muscle cells induced by fetal bovine serum (10%) alone, or by fetal bovine serum (1%) in combination with platelet-derived growth factor (PDGF, 50 ng ml^{-1}), fibroblast growth factor (50 ng ml^{-1}), or interleukin-1α (IL-1α, 100 pg ml^{-1}). The antiproliferative effects of tetrandrine counteracting mitogenic stimuli for vascular smooth muscle may contribute to the inhibition of pulmonary vascular remodeling associated with pulmonary hypertension [20]. The pathological changes, such as a narrowed diameter of the small pulmonary arteria in rats with hypoxia and pulmonary hypertension, decreased significantly after treatment with tetrandrine. It inhibited the hypoxia-induced thickening and muscularization of the small pulmonary arteria by inhibiting the proliferation of collagenous fibers [21].

Tetrandrine, as a calcium blocker, is known to exhibit hypotensive activity in spontaneous hypertensive rats [22] and in hypertensive dogs [23]. Tetrandrine has been used clinically in China for the treatment of cardiovascular diseases, mainly due to its vasodilatory and hypotensive actions [24]. Tetrandrine was reported to reduce portal venous pressure, mean arterial pressure and total peripheral resistance in cirrhotic rats [25]. It also inhibited the opening of *N*-methyl-D-asparate receptor channels in cortical neurons of rats, as induced by anoxia [26].

Tetrandrine was also reported to have antiarrhythmic action and to inhibit platelet aggregation [27]. It inhibited rabbit platelet aggregation *in vitro* induced by ADP or collagen; these effects may be caused by a reduced generation of endogenous platelet-activating factor (PAF) [28]. Tetrandrine and fangchinoline each inhibited human platelet aggregation, as induced by PAF, with IC_{50} values of 22–29 μ*M*. Neither tetrandrine nor fangchinoline showed any inhibitory effects on the specific binding of PAF to its receptor. However, they inhibited thromboxane B_2 formation in washed human platelets, as induced by PAF, thrombin, and arachidonic acid. This indicates that tetrandrine or fangchinoline inhibits platelet aggregation partly by interfering with the intracellular messengers system, but not by inhibiting the binding of PAF to the PAF-receptor [29]. Tetrandrine was also found to inhibit PAF activity in cultured bovine cerebral microvascular endothelial cells, and to markedly inhibit platelet adhesion and thrombosis [30].

Some bisbenzylisoquinoline alkaloids, including tetrandrine, fangchinoline and berbamine, are known to affect both immune and inflammatory responses. They have been used for the treatment of inflammatory diseases. For example, tetrandrine and berbamine inhibited acute paw edema and adjuvant-induced polyarthritis in rats [31]; fangchinoline selectively inhibited cytokine transcription in human peripheral blood mononuclear cells [32]. Tetrandrine was also reported to protect human mononuclear cells *in vitro* against damage caused by a single high dose of ionizing irradiation, and to inhibit the inflammatory responses induced by irradiation, including the release of superoxide and phagocytic activity [33]. Fangchinoline and tetrandrine were found both to exert anti-inflammatory effects on mouse ear edema induced by croton oil [34]. In human monocytic cells, tetrandrine markedly suppressed NO release and prostaglandin E_2 (PGE_2) generation induced by lipopolysaccharide (LPS). It also significantly attenuated the LPS-induced transcription of proinflammatory cytokines, including tumor necrosis factor-α (TNF-α), interleukin-4 (IL-4) and IL-8, in a concentration-dependent manner. Furthermore, tetrandrine at a concentration of 100 μ*M* significantly blocked the LPS-mediated induction of inducible NO synthase (iNOS) and cyclooxygenase-2 (COX-2) expression, but not of COX-1 [35].

Tetrandrine has been used clinically for the treatment of rheumatic diseases, such as rheumatoid arthritis and systemic lupus erythematosus. It was reported that tetrandrine significantly inhibited the secretion of IL-1, TNF-α, IL-6, and IL-8 from monocytes, IgG secretion from B cells, and phagocytic activity of neutrophils [36]. The effect of tetrandrine was also described to involve nuclear factor-κB (NF-κB) transcription factors, and to downregulate inhibitor IκBα kinases in human peripheral blood T cells [37]. These effects might account for the therapeutic efficacy in rheumatic diseases [38]. Tetrandrine and berbamine were also found to inhibit leukotriene and prostaglandin generation by human monocytes and neutrophils. Tetrandrine had a much greater effect than berbamine on leukotriene generation; however, the compounds were equally potent in the suppression of prostaglandin generation [39]. In patients with essential hypertension, oral treatment with tetrandrine at doses of 100–200 mg, three times daily for three months, significantly decreased the levels of thromboxane B_2 (TXB_2) and $apoB_{100}$ in serum, and increased the levels of $apoA_2$ and 6-keto-prostaglandin F_1 (6-keto-PGF_1). It was suggested that tetrandrine stimulates the synthesis of PGI_2 and inhibits the release of TXA_2, thus normalizing the metabolism of lipids in patients with essential hypertension [40].

Studies on the antiallergic activities of tetrandrine indicated that it not only antagonized allergenic activities, but also blocked the release of allergens. Tetrandrine suppressed both the passive cutaneous anaphylaxis reaction in rats and the allergic contraction of isolated ileum from sensitized guinea pigs. The asthmatic response and the contraction of ileum in guinea pigs induced by histamine or acetylcholine, as well as the increase in cutaneous vessel permeabilitiy induced by serotonin, were antagonized by tetrandrine. The release of the slow-reacting substance of anaphylaxis from the lungs of sensitized guinea pigs, and the release of histamine from mast cells of dextran-treated rats, were inhibited by tetrandrine [41]. A structure–activity relationship study on the inhibition of histamine release from guinea pig ileum by bisbenzylisoquinoline alkaloids *in vitro* demonstrated that the inhibitory activity of fangchinoline is higher than that of tetrandrine. *N*-Methylcoclaurine, the half-molecule type, showed an inhibitory effect comparable to that of fangchinoline [42].

Tetrandrine was reported to effectively inhibit the development of experimental silicosis in rats when given to animals immediately after dusting, or after the formation of silicotic nodules. These effects are characterized by a lowering of the total weight and total collagen content of the lungs, as compared with that of silicotic control animals [43]. The antisilicotic effect was also observed in rabbits. Silicotic rabbits induced by powdered quartz in saline and detected by X-ray four months later were effectively treated by the intramuscular administration of tetrandrine at a dose of 40 mg per animal, three times weekly over a three-month period. After treatment, the symptoms had almost disappeared in silicotic rabbits [44].

One of the antisilicotic mechanisms of tetrandrine was postulated to be the inhibition of particle-induced inflammation and the secretion of reactive compounds from alveolar phagocytes [45]. A positive correlation was observed between binding affinity to alveolar macrophages and the antifibrotic potency of bisbenzylisoquinoline alkaloids. This suggested that the ability of the alkaloids to interact with alveolar macrophages may be one of the key steps in the inhibition of silicosis [46]. It was also found that tetrandrine effectively blocks the generation of oxygen radicals from pulmonary phagocytes stimulated by quartz [47], and exerts scavenging effects on active oxygen radicals *in vitro* [48]. Free radical generation on the silica surface was also found to play an important role in the silica-induced activation of nuclear transcription NF-κB [49]. It was further postulated that tetrandrine exhibits a cytotoxic effect on macrophages, possibly by interfering with calcium homeostasis, leading to an overproduction of prostaglandin and a protection against silicosis [50].

The antifibrotic activity was proposed to be a further mechanism of the anti-silicotic activity of tetrandrine, which may be partly mediated by a direct inhibition of the fibroblast proliferation normally associated with the development and progression of silicosis [51]. The wet weight and lipid contents of lung tissues of tetrandrine-treated rats were higher than those of normal rats, but lower than those of silicotic rats [52]. Tetrandrine inhibited the biosynthesis of collagen and glycosaminoglycan in cultured rat peritoneal macrophages and fibroblasts [53]. The combined use of tetrandrine and polyvinylpyridine *N*-oxide was found to stimulate the degradation of collagen in silicotic rats [54], and to inhibit the gene expression of type I and type III collagen during experimental silicosis in rats [55]. In addition, tetrandrine was reported to inhibit hepatic fibrosis in rats induced by carbon tetrachloride (CCl_4) [56].

Tetrandrine was reported to be cytotoxic. *In vitro*, it was found to inhibit the proliferation of human leukemia cell lines, including U937 and HL-60 [57]. The growth inhibition was both dose- and time-dependent, and accompanied by evidence of apoptosis [58]. Tetrandrine was also found to inhibit cell proliferation, IL-2 secretion and expression of the T-cell activation antigen, CD71, of human peripheral blood T cells, as induced by 12-*O*-tetradecanoylphorbol 13-acetate (TPA). A further investigation demonstrated that tetrandrine downregulates PKC-dependent signaling pathway in T cells [59]. An inhibitory effect of tetrandrine on expression of the protooncogene *Fos* in rat cerebrum was also reported [60]. It also inhibited the cell proliferation of human lung fibroblasts and rat liver cells, as induced by PDGF [61].

Tetrandrine was also reported to reverse the resistance of human cancer cell lines to doxorubicin, harringtonine, and vincristine *in vitro* [62, 63]. The mechanism of tetrandrine to reverse multidrug resistance was due to a modulation of P-glycoprotein [64]. Tetrandrine induced resistance to apoptosis in harringtonine-resistant HL-60 cells associated with high protein phosphorylation, and reduced the expression of *Myc* mRNA [65]. Tetrandrine was also reported to reverse multidrug resistance of human multidrug-resistant KBV200 cells. It increased vincristine cytotoxicity towards KBV200 cells in a concentration-dependent manner, and was more effective than verapamil [66]. Tetrandrine also enhanced the radiosensitivity of human glioblastoma U138MG cells [67].

Tetrandrine has been shown to be weakly mutagenic in *Salmonella typhimurium* TA98 with metabolic activation, but not in the SOS response. In addition, tetrandrine was found to significantly increase the mutagenic activity of benzo[*a*]pyrene (BaP), trinitrofluorenone, 2-aminoanthracene, diesel emission particles, airborne particles, cigarette smoke condensate, aflatoxin B_1 (AFB_1), and fried beef in *S. typhimurium* TA98, but not in *S. typhimurium* TA1538 [68]. Therefore, the enhancement of genotoxicity by tetrandrine may result from an increase in error-prone DNA repair [69]. Furthermore, tetrandrine was also found to induce micronuclei and sister-chromatid exchange (SCE) in V79 cells *in vitro,* as well as to induce micronuclei in mouse bone marrow cells and SCE in mouse spleen cells *in vivo* [70]. The incubation of human lymphoid T cells (Jurkat cells) with tetrandrine prior to cell stimulation inhibited TPA-induced NF-κB activation completely at a concentration of 50 μ*M*. It was reported that the free radical-scavenging activity of tetrandrine is responsible for its inhibition of TPA-induced NF-κB activation [71].

Tetrandrine was administered orally to rats and also to patients. It was recovered predominantly unchanged from the rat liver, lung, and urine and also from human urine. In addition to tetrandrine, tetrandrine 2′-oxide, and 2′-nortetrandrine were detected in human and rat urine and in rat organs [72].

References

1 Yang, Y.F. (1985) Thin layer chromatography-UV-spectrophotometry for tetrandrine in *Stephania tetrandra*. *Chin. Trad. Herbal Drugs*, **16**, 281.

2 Hu, T.M. and Zhao, S.X. (1986) The structure of oxofangchirine and stephenanthrine isolated from *Stephania tetrandra* S. Moore. *Acta Pharm. Sin.*, **21**, 29–34.

3 Huang, W.L., Zhang, H.B., and Peng, S.X. (1993) Isolation and identification of phenolic by-products obtained from cleavage of tetrandrine. *J. China Pharm. Univ.*, **24**, 269–271.

4 Lin, L.Z., Shieh, H.L., Angerhofer, C.K., Pezzuto, J.M., Cordell, G.A., Xue, L., Johnson, M.E., and Ruangrungsi, N. (1993) Cytotoxic and antimalarial bisbenzylisoquinoline alkaloids from *Cyclea barbata*. *J. Nat. Prod.*, **56**, 22–29.

5 Ye, Z.G., Van Dyke, K., and Castranova, V. (1989) The potentiating action of tetrandrine in combination with chloroquine or qinghaosu against chloroquine-sensitive and resistant falciparum malaria. *Biochem. Biophys. Res. Commun.*, **165**, 758–765.

6 Liu, J.H., Chen, J., Wang, T., Liu, B., Yang, J., Chen, X.W., Wang, S.G., Yin, C.P., and Ye,

Z.Q. (2006) Effects of tetrandrine on cytosolic free calcium concentration in corpus cavernosum smooth muscle cells of rabbits. *Asian J. Androl.*, **8**, 405–409.

7 Achike, F.I. and Kwan, C.Y. (2002) Characterization of a novel tetrandrine-induced contraction in rat tail artery. *Acta Pharmacol. Sin.*, **23**, 698–704.

8 Shen, J.Z., Zheng, X.F., and Kwan, C.Y. (2000) Evidence for P_2-purinoceptors contribution in H_2O_2-induced contraction of rat aorta in the absence of endothelium. *Cardiovasc. Res.*, **47**, 574–585.

9 Wong, T.M., Wu, S., Yu, X.C., and Li, H.Y. (2000) Cardiovascular actions of Radix Stephaniae Tetrandrae: a comparison with its main component, tetrandrine. *Acta Pharmacol. Sin.*, **21**, 1083–1088.

10 Takemura, H., Imoto, K., Ohshika, H., and Kwan, C.Y. (1996) Tetrandrine as a calcium antagonist. *Clin. Exp. Pharmacol. Physiol.*, **23**, 751–753.

11 Bickmeyer, U., Weinsberg, F., Muller, E., and Wiegand, H. (1998) Blockade of voltage-operated calcium channels, increase in spontaneous catecholamine release and elevation of intracellular calcium levels in bovine chromaffin cells by the plant alkaloid tetrandrine. *Naunyn Schmiedeberg's Arch. Pharmacol.*, **357**, 441–445.

12 Kwan, C.Y. and Achike, F.I. (2002) Tetrandrine and related bis-benzylisoquinoline alkaloids from medicinal herbs: cardiovascular effects and mechanisms of action. *Acta Pharmacol. Sin.*, **23**, 1057–1068.

13 Kwan, C.Y., Ma, F.M., and Hui, S.C. (1999) Inhibition of endothelium-dependent vascular relaxation by tetrandrine. *Life Sci.*, **64**, 2391–2400.

14 Wu, S.N., Li, H.F., and Lo, Y.C. (2000) Characterization of tetrandrine-induced inhibition of large-conductance calcium-activated potassium channels in a human endothelial cell line (HUV-EC-C). *J. Pharmacol. Exp. Ther.*, **292**, 188–195.

15 Shen, Y.C., Chou, C.J., Chiou, W.F., and Chen, C.F. (2001) Anti-inflammatory effects of the partially purified extract of radix *Stephaniae tetrandrae*: comparative studies of its active principles tetrandrine and fangchinoline on human polymorphonuclear leukocyte functions. *Mol. Pharmacol.*, **60**, 1083–1090.

16 Leung, Y.M., Berdik, M., Kwan, C.Y., and Loh, T.T. (1996) Effects of tetrandrine and closely related bis-benzylisoquinoline derivatives on cytosolic Ca^{2+} in human leukaemic HL-60 cells: a structure-activity relationship study. *Clin. Exp. Pharmacol. Physiol.*, **23**, 653–659.

17 Shen, Y.C., Chen, C.F., and Sung, Y.J. (1999) Tetrandrine ameliorates ischaemia-reperfusion injury of rat myocardium through inhibition of neutrophil priming and activation. *Br. J. Pharmacol.*, **128**, 1593–1601.

18 Chen, N.H., Wang, Y.L., and Ding, J.H. (1991) Effect of tetrandrine on myocardial Na^+, K^+-ATPase in renovascular rats. *Acta Pharmacol. Sin.*, **12**, 488–493.

19 Wang, H.L., Zhang, X.H., and Chang, T.H. (2002) Effects of tetrandrine on smooth muscle contraction induced by mediators in pulmonary hypertension. *Acta Pharmacol. Sin.*, **23**, 1114–1120.

20 Wang, H.L., Kilfeather, S.A., Martin, G.R., and Page, C.P. (2000) Effects of tetrandrine on growth factor-induced DNA synthesis and proliferative response of rat pulmonary artery smooth muscle cells. *Pulm. Pharmacol. Ther.*, **13**, 53–60.

21 Li, W., Chen, W., Zhang, S., and Lei, S. (1997) Effect of tetrandrine on pulmonary vascular morphology in rats with hypoxia pulmonary hypertension. *J. West. China Univ. Med. Sci.*, **28**, 388–391.

22 Liu, T.B., Lin, H.C., Huang, Y.T., Sun, C.M., and Hong, C.Y. (1997) Portal hypotensive effects of tetrandrine and verapamil in portal hypertensive rats. *J. Pharm. Pharmacol.*, **49**, 85–88.

23 Wang, H.L., Liu, K.Y., Jin, X., and Zhang, X.H. (1994) Effects of tetrandrine on acute hypoxic pulmonary hypertension in dogs. *Chin. J. Pharmacol. Toxicol.*, **8**, 246–249.

24 Kang, J. (1993) Effects of tetrandrine on pulmonary hypertension in patients with chronic obstructive pulmonary disease. *Chin. J. Tubercul. Respir. Dis.*, **16**, 93–94.

25 Wang, H. and Chen, X. (2004) Tetrandrine ameliorates cirrhosis and portal hypertension by inhibiting nitric oxide in cirrhotic rats. *J. Huazhong Univ. Sci. Technolog. Med. Sci.*, **24**, 385–388. 395.

26 Wang, Z.F., Xue, C.S., Zhou, Q.X., Wan, Z.B., and Luo, Q.S. (1999) Effects of tetrandrine on changes of NMDA receptor channel in cortical neurons of rat induced by anoxia. *Acta Pharmacol. Sin.*, **20**, 729–732.

27 Xie, Q.M., Tang, H.F., Chen, J.Q., and Bian, R.L. (2002) Pharmacological actions of tetrandrine in inflammatory pulmonary diseases. *Acta Pharmacol. Sin.*, **23**, 1107–1113.

28 Zhang, M., Zhang, L.Z., and Lu, J.S. (1995) Effects of tetrandrine on rabbit platelet aggregation and platelet activating factor generation. *Acta Pharmacol. Sin.*, **16**, 209–212.

29 Kim, H.S., Zhang, Y.H., Fang, L.H., Yun, Y.P., and Lee, H.K. (1999) Effects of tetrandrine and fangchinoline on human platelet aggregation and thromboxane B_2 formation. *J. Ethnopharmacol.*, **66**, 241–246.

30 Hu, J.H., Sun, D.X., Zeng, G.Q., Lin, A.Y., and Rui, Y.C. (1995) Platelet adhesion to cultured bovine cerebral microvascular endothelial cells by stimulation of platelet activating factor and antagonism of drugs. *Acta Pharmacol. Sin.*, **16**, 318–321.

31 Whitehouse, M.W., Fairlie, D.P., and Thong, Y. H. (1994) Anti-inflammatory activity of the isoquinoline alkaloid, tetrandrine, against established adjuvant arthritis in rats. *Agents Actions*, **42**, 123–127.

32 Onai, N., Tsunokawa, Y., Suda, M., Watanabe, N., Nakamura, K., Sugimoto, Y., and Kobayashi, Y. (1995) Inhibitory effects of bisbenzylisoquinoline alkaloids on induction of proinflammatory cytokines, interleukin-1 and tumor necrosis factor-?. *Planta Med.*, **61**, 497–501.

33 Chen, Y.J., Tu, M.L., Kuo, H.C., Chang, K.H., Lai, Y.L., Chung, C.H., and Chen, M.L. (1997) Protective effect of tetrandrine on normal human mononuclear cells against ionizing irradiation. *Biol. Pharm. Bull.*, **20**, 1160–1164.

34 Choi, H.S., Kim, H.S., Min, K.R., Kim, Y., Lim, H.K., Chang, Y.K., and Chung, M.W. (2000) Anti-inflammatory effects of fangchinoline and tetrandrine. *J. Ethnopharmacol.*, **69**, 173–179.

35 Wu, S.J. and Ng, L.T. (2007) Tetrandrine inhibits proinflammatory cytokines, iNOS and COX-2 expression in human monocytic cells. *Biol. Pharm. Bull.*, **30**, 59–62.

36 Wong, C.W., Seow, W.K., O'Callaghan, J.W., and Thong, Y.H. (1992) Comparative effects of tetrandrine and berbamine on subcutaneous air pouch inflammation induced by interleukin-1, tumour necrosis factor and platelet-activating factor. *Agents Actions*, **36**, 112–118.

37 Ho, L.J., Juan, T.Y., Chao, P., Wu, W.L., Chang, D.M., Chang, S.Y., and Lai, J.H. (2004) Plant alkaloid tetrandrine downregulates IκBα kinases-IκBα-NF-κB signaling pathway in human peripheral blood T cell. *Br. J. Pharmacol.*, **143**, 919–927.

38 Chang, D.M., Chang, W.Y., Kuo, S.Y., and Chang, M.L. (1997) The effects of traditional antirheumatic herbal medicines on immune response cells. *J. Rheumatol.*, **24**, 436–441.

39 Teh, B.S., Seow, W.K., Li, S.Y., and Thong, Y.H. (1990) Inhibition of prostaglandin and leukotriene generation by the plant alkaloids tetrandrine and berbamine. *Int. J. Immunopharmacol.*, **12**, 321–326.

40 Chen, S.Y. and Ye, S.B. (1992) The effects of tetrandrine on metabolism of prostaglandin and lipids in patients with essential hypertension. *Chin. Pharmacol. Bull.*, **8**, 436–439.

41 Bian, R.L., Zhou, H.L., Xie, Q.M., Tong, F.D., Yang, W., and Wang, Y. (1984) Observation on antiallergic effect of tetrandrine. *Chin. Trad. Herbal Drugs*, **15**, 262–264.

42 Nakamura, K., Tsuchiya, S., Sugimoto, Y., Sugimura, Y., and Yamada, Y. (1992) Histamine release inhibition activity of bisbenzylisoquinoline alkaloids. *Planta Med.*, **58**, 505–508.

43 Yu, X.F., Zou, C.Q., and Lin, M.B. (1983) Observation of the effect of tetrandrine on experimental silicosis of rats. *Ecotoxicol. Environ. Safety*, **7**, 306–312.

44 Jiang, H.X., Hu, T.X., Peng, B.X., and Shi, D.Z. (1983) Preliminary study on the mechanism of therapeutic effect of tetrandrine on silicosis. *Chin. J. Tuberc. Resp. Dis.*, **6**, 92–94.

45 Kang, J.H., Lewis, D.M., Castranova, V., Rojanasakul, Y., Banks, D.E., Ma, J.Y., and Ma, J.K. (1992) Inhibitory action of tetrandrine on macrophage production of interleukin-1 (IL-1)-like activity and thymocyte proliferation. *Exp. Lung Res.*, **18**, 715–729.

46 Castranova, V., Kang, J.H., Ma, J.K., Mo, C.G., Malanga, C.J., Moore, M.D., Schwegler-Berry, D., and Ma, J.Y. (1991) Effects of bisbenzylisoquinoline alkaloids on alveolar macrophages: correlation between binding affinity, inhibitory potency, and antifibrotic potential. *Toxicol. Appl. Pharmacol.*, **108**, 242–252.

47 Shi, X., Mao, Y., Saffiotti, U., Wang, L., Rojanasakul, Y., Leonard, S.S., and Vallyathan, V. (1995) Antioxidant activity of tetrandrine and its inhibition of quartz-induced lipid peroxidation. *J. Toxicol. Environ. Health*, **46**, 233–248.

48 Cao, Z.F. (1996) Scavenging effect of tetrandrine of active oxygen radicals. *Planta Med.*, **62**, 413–414.

49 Chen, F., Sun, S., Kuhn, D.C., Lu, Y., Gaydos, L.J., Shi, X., and Demers, L.M. (1997) Tetrandrine inhibits signal-induced NF-κB activation in rat alveolar macrophages. *Biochem. Biophys. Res. Commun.*, **231**, 99–102.

50 Pang, L. and Hoult, J.R. (1997) Cytotoxicity to macrophages of tetrandrine, an antisilicosis alkaloid, accompanied by an overproduction of prostaglandins. *Biochem. Pharmacol.*, **53**, 773–782.

51 Reist, R.H., Dey, R.D., Durham, J.P., Rojanasakul, Y., and Castranova, V. (1993) Inhibition of proliferative activity of pulmonary fibroblasts by tetrandrine. *Toxicol. Appl. Pharmacol.*, **122**, 70–76.

52 San, S., Kang, A., and Liu, Z. (1992) Effects of tetrandrine on lung-lipid contents of rats with experimental silicosis. *Biomed. Environ. Sci.*, **5**, 362–371.

53 Li, Y.R., Cai, G.P., Liu, L.F., Chen, N.M., and Li, C.L. (1988) Studies on the inhibitory effect of tetrandrine on silica-induced fibrogenesis *in vitro*. *Biomed. Environ. Sci.*, **1**, 125–129.

54 Yu, L., Zhou, C.Q., Li, Y.R., Qu, L., Xing, K.J., and Du, Q.C. (1995) A biochemical study on combined treatment of experimental silicosis with tetradrine-PVNO and tetradrine-QOHP in rats. *Biomed. Environ. Sci.*, **8**, 265–268.

55 He, Y., Liu, B., and Miao, Q. (1995) Inhibition of mRNA expression of silicotic collagen gene by tetrandrine. *Chin. J. Prev. Med.*, **29**, 18–20.

56 Li, D.G., Liu, Y.L., Lu, H.M., Jiang, Z.M., and Xu, Q.F. (1994) Effects of tetrandrine on mitochondria of hepatic fibrosis in rats. *Chin. J. Digestion*, **14**, 339–342.

57 Lai, Y.L., Chen, Y.J., Wu, T.Y., Wang, S.Y., Chang, K.H., Chung, C.H., and Chen, M.L. (1998) Induction of apoptosis in human leukemic U937 cells by tetrandrine. *Anticancer Drugs*, **9**, 77–81.

58 Teh, B.S., Chen, P., Lavin, M.F., Seow, W.K., and Thong, Y.H. (1991) Demonstration of the induction of apoptosis (programmed cell death) by tetrandrine, a novel anti-inflammatory agent. *Int. J. Immunopharmacol.*, **13**, 1117–1126.

59 Ho, L.J., Chang, D.M., Lee, T.C., Chang, M.L., and Lai, J.H. (1999) Plant alkaloid tetrandrine downregulates protein kinase C-dependent signaling pathway in T cells. *Eur. J. Pharmacol.*, **367**, 389–398.

60 Che, J.T., Zhang, J.T., Qu, Z.W., and Wei, G. (1997) Effect of tetrandrine on proto-oncogene c-*Fos* expression in rat cerebrum. *Acta Pharmacol. Sin.*, **18**, 371–373.

61 Tian, Z., Liu, S., and Li, D. (1997) Blocking action of tetrandrine on the cell proliferation induced by PDGF in human lung fibroblasts and liver Ito cell of rats. *Natl Med. J. China*, **77**, 50–53.

62 Ye, Z., Sun, A., Li, L., Cao, X., and Ye, W. (1996) Reversal of adriamycin or vincristine resistance by tetrandrine in human cancer cells *in vitro*. *China J. Chin. Mater. Med.*, **21**, 368–371.

63 He, Q.Y., Zhou, W.D., Ji, L., Zhang, H.Q., He, N.G., and Xue, S.B. (1996) Characteristics of harringtonine-resistant human leukemia HL60 cell. *Acta Pharmacol. Sin.*, **17**, 463–467.

64 Choi, S.U., Park, S.H., Kim, K.H., Choi, E.J., Kim, S., Park, W.K., Zhang, Y.H., Kim, H.S., Jung, N.P., and Lee, C.O. (1998) The bisbenzylisoquinoline alkaloids, tetrandine and fangchinoline, enhance the cytotoxicity of multidrug resistance-related drugs via modulation of P-glycoprotein. *Anticancer Drugs*, **9**, 255–261.

65 He, Q.Y., Zhang, H.Q., Pang, D.B., Chi, X.S., and Xue, S.B. (1996) Resistance to apoptosis of harringtonine-resistant HL60 cells induced by tetrandrine. *Acta Pharmacol. Sin.*, **17**, 545–549.

66 Xu, J.Y., Zhou, Q., Shen, P., and Tang, W. (1999) Reversal effect of TTD on human multidrug resistant KBV200 cell line. *J. Exp. Clin. Cancer Res.*, **18**, 549–552.

67 Chang, K.H., Chen, M.L., Chen, H.C., Huang, Y.W., Wu, T.Y., and Chen, Y.J. (1999) Enhancement of radiosensitivity in human glioblastoma U138MG cells by tetrandrine. *Neoplasma*, **46**, 196–200.

68 Rosenkranz, H.S. and Klopman, G. (1990) Novel structural concepts in elucidating the potential genotoxicity and carcinogenicity of tetrandrine, a traditional herbal drug. *Mutat. Res.*, **244**, 265–271.

69 Whong, W.Z., Lu, C.H., Stewart, J.D., Jiang, H. X., and Ong, T. (1989) Genotoxicity and genotoxic enhancing effect of tetrandrine in *Salmonella typhimurium*. *Mutat. Res.*, **222**, 237–244.

70 Xing, S.G., Shi, X.C., Wu, Z.L., Whong, W.Z., and Ong, T. (1989) Effect of tetrandrine on micronucleus formation and sister-chromatid exchange in both *in vitro* and *in vivo* assays. *Mutat. Res.*, **224**, 5–10.

71 Ye, J., Ding, M., Zhang, X., Rojanasakul, Y., and Shi, X. (2000) On the role of hydroxyl radical and the effect of tetrandrine on nuclear factor ? B activation by phorbol 12-myristate 13-acetate. *Ann. Clin. Lab. Sci.*, **30**, 65–71.

72 Lin, M.B., Zhang, W., Zhao, X.W., Lu, J.X., Wang, M., and Chen, L.G. (1982) Biotransformation of tetrandrine in rats and in men. *Acta Pharm. Sin.*, **17**, 728–735.

Tribulus terrestris

Fructus Tribuli (Jili) is the dried fruits of *Tribulus terrestris* L. (Zygophyllaceae). It is used as an analgesic for the treatment of headache and dizziness.

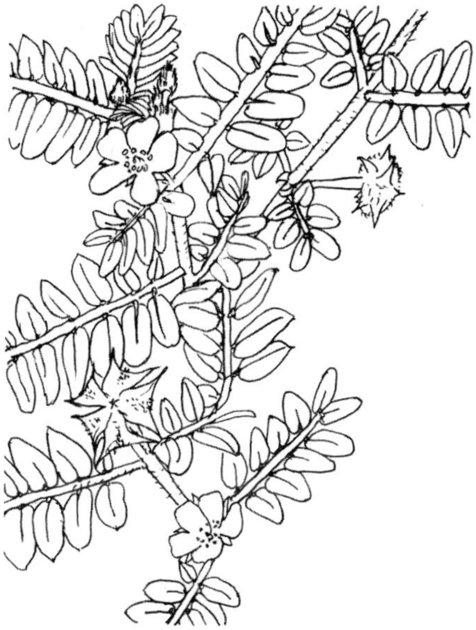

Tribulus terrestris L.

Chemistry

Fructus Tribuli is known to contain steroidal saponins with tigogenin, hecogenin, and gitogenin as aglycones. The analysis of market products showed considerable variations of 0.17 to 6.49% in the protodioscin content [1]. Diosgenin, ruscogenin and ruscogenin diacetate were detected as terminal metabolites in the urine of experimental animals given saponins of *T. terrestris* [2].

A number of steroidal saponins were isolated from the fruits of *T. terrestris* and identified as gitonin, F-gitonin, terrestrosins A–E [3] and further structurally related steroid saponins, such as 26-*O*-β-D-glucopyranosyl-(25*S*)-5β-furost-20(22)-en-3β,26-diol 3-*O*-α-L-rhamnopyranosyl-(1 → 2)-[α-L-rhamnopyranosyl-(1 → 4)]-β-D-glucopyranoside, 26-*O*-β-D-glucopyranosyl-(25*S*)- 5β-furost-20(22)-en-3β,26-diol 3-*O*-α-L-rhamnopyranosyl-(1 → 2)-[β-D-glucopyranosyl-(1 → 4)]-β-D-galactopyranoside, and 25(*S*)-5β-spirostan-3β-ol 3-*O*-α-L-rhamnopyranosyl-(1 → 2)-[β-D-glucopyranosyl-(1 → 4)]-β-D-galactopyranoside [4–7].

Flavones from the fruits of *T. terrestris* were identified as quercetin and kaempferol [8]. Further components from the fruits were identified as tribulusamides A and B, lignanamides, terrestriamide, *N-trans*-feruloyltyramine, *N-trans*-coumaroyltyramine, and β-sitosterol [9]. From the overground part of the plant, β-sitosterol-β-D-glucoside, dioscin, neohecogenin-3-O-β-D-glucopyranoside and tribulosin, a neotigogenin glucoside, were isolated [10]. The polysaccharide from the stem and leaf of *T. terristris* were found to be composed of arabinose, rhamnose, xylose, galacturonic acid, galactose, glucose, and mannose. [11].

Pharmacology and Toxicology

The saponins tigogenin 3-*O*-β-D-xylopyranosyl-(1 → 2)-[β-D-xylopyranosyl-(1 → 3)]-β-D-glucopyranosyl-(1 → 4)-[α-L-rhamnopyranosyl-(1 → 2)]-β-D-galactopyranoside and tigogenin 3-*O*-β-D-glucopyranosyl-(1 → 2)-[β-D-xylopyranosyl-(1 → 3)]-β-D-glucopyranosyl-(1 → 4)-β-D-galactopyranoside from the fruits of *T. terrestris* were reported to exert high antifungal activity against a number of pathogenic *Candida* spp. and *Cryptococcus neoformans in vitro*. The IC_{80} values of the two saponins to inhibit *Candida albicans* were 10 and 2.3 μg ml^{-1}, respectively, whereas the IC_{80} values of the two saponins to inhibit *C. neoformans* were 1.7 and 6.7 μg ml^{-1}, respectively. The first saponin inhibited hyphal formation and destroyed the cell

Handbook of Chinese Medicinal Plants: Chemistry, Pharmacology, Toxicology
Weici Tang and Gerhard Eisenbrand

ISBN: 978-3-527-32226-8

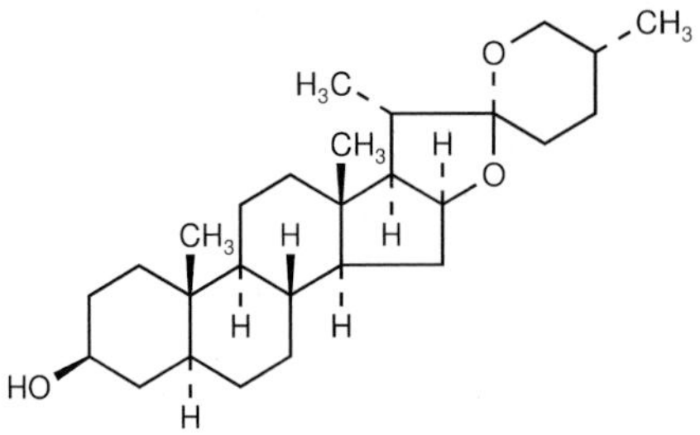

Tigogenin

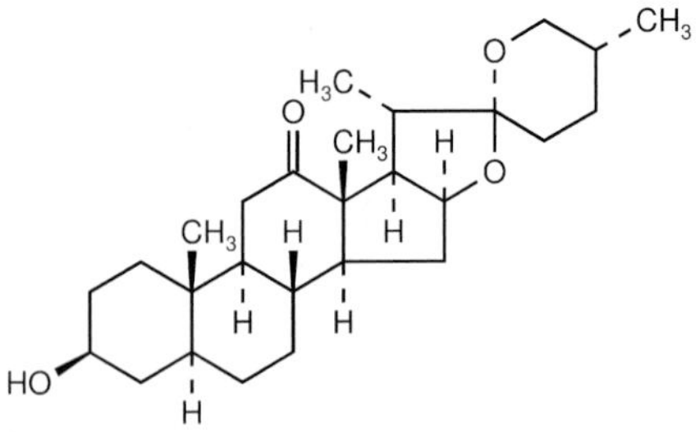

Hecogenin

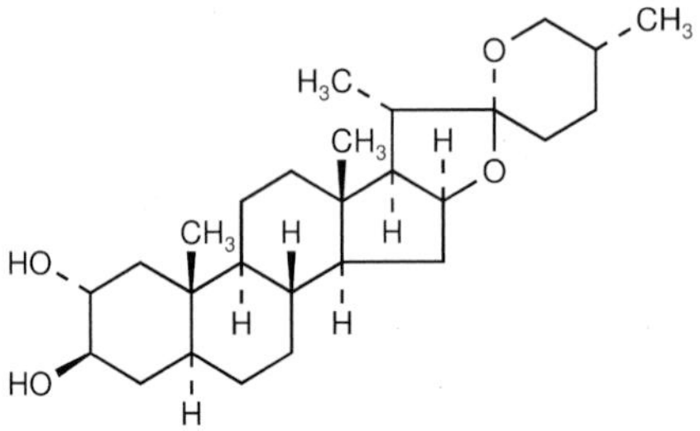

Gitogenin

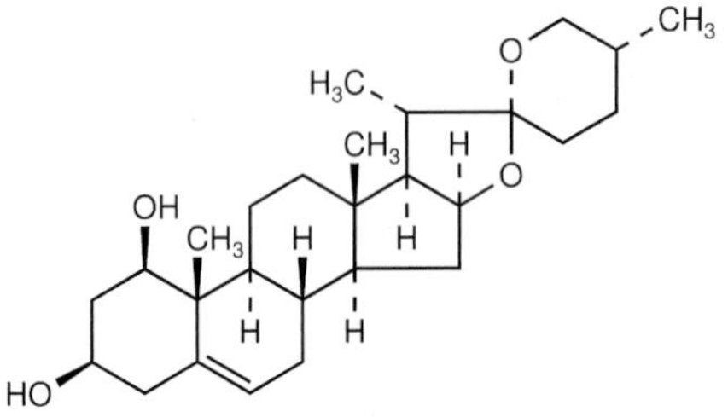

Ruscogenin

membrane of *C. albicans*. The antifungal activity of the first saponin was also evidenced in a vaginal infection model in mice [12]. The saponins were also reported to inhibit fluconazole-resistant *C. albicans* and *C. neoformans* [13]. Tribulosin and β-sitosterol D-glucoside were found to be anthelmintic principles of *T. terrestris* [14].

The methanol extract of the fruits of *T. terrestris* exerted anti-inflammatory activity and markedly inhibited the cyclooxygenase-2 (COX-2) activity in mouse macrophages

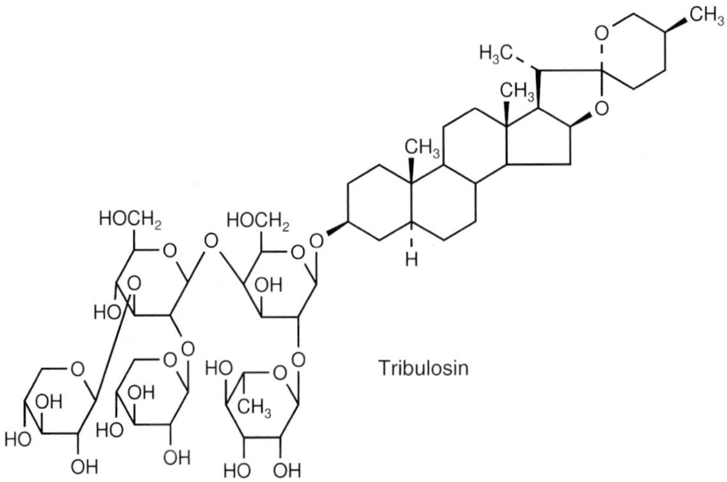

Tribulosin

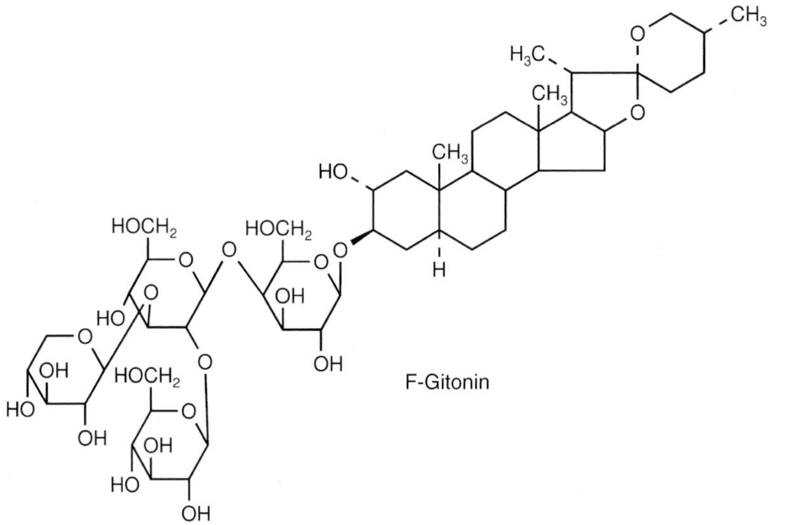

F-Gitonin

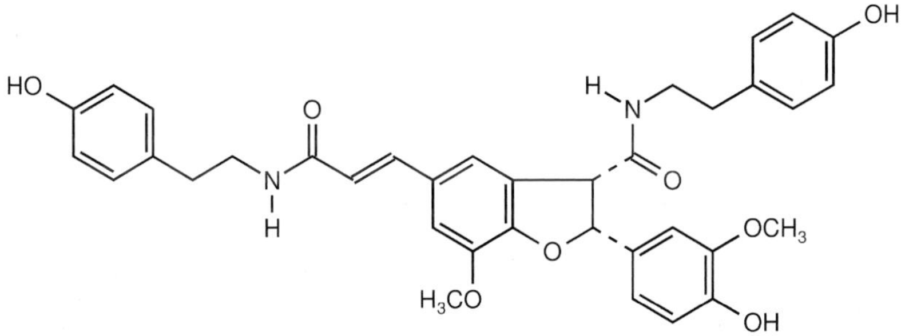

Tribulusamide A

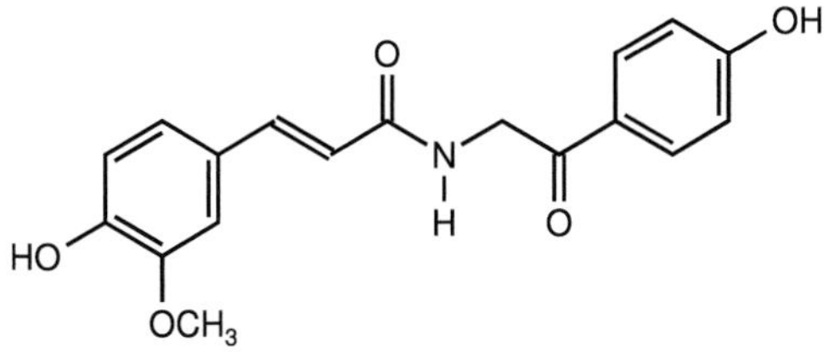

Terrestriamide

RAW264.7 cells, as induced by lipopolysaccharide (LPS) [15].

The aqueous extract of the fruits of *T. terrestris* was reported to be capable of stimulating melanocyte proliferation [16]. Tribulusamides A and B significantly prevented cell death of primary cultured mouse hepatocytes caused by D-galactosamine and tumor necrosis factor-α (TNF-α) [9]. A decoction of the fruit of *T. terrestris* can significantly inhibit the gluconeogenesis and influence glycometabolism in normal mice. The decoction can also reduce the level of triglyceride and the content of cholesterol in the plasma [17]. It was further reported that the methanol extract of the fruit of *T. terrestris* exhibited an antidote activity in mice intoxicated with a median-lethal dose of mercuric chloride [18].

An ethanolic extract of the fruits of *T. terrestris* showed significant dose-dependent protection against urolithiasis, as induced by glass bead implantation in rats. It provided significant protection against the deposition of calculogenic material around the glass bead, and also protected against leucocytosis and elevation in serum urea levels [19]. The aqueous extract of *T. terrestris*, when given to Wistar rats at an oral dose of $5\,g\,kg^{-1}$, exerted a diuretic effect, while urinary concentrations of sodium, potassium and chlorine were markedly increased. In addition, *T. terrestris* exhibited a contractile activity on guinea pig ileum, indicating that the extract might propel urinary stones [20]. The administration of *T. terrestris* to sodium glycolate-fed rats produced a significant decrease in urinary oxalate excretion, and a significant increase in urinary glyoxylate excretion. The supplementation of *T. terrestris* with sodium glycolate also caused a reduction in the activities of liver glycolate oxidase and glycolate dehydrogenase as oxalate-synthesizing enzymes, whereas liver lactate dehydrogenase (LDH) activity remained unaltered. The isoenzyme pattern of kidney LDH revealed that the normalization of kidney LDH by *T. terrestris* feeding was mainly due to an increase in the LDH 5 fraction. The LDH 1 isoenzyme remained unchanged [21].

The fruit of *T. terrestris* was also found to show cardiac effects [22, 23]. In a clinical trial, the total saponin of *T. terrestris* was highly active against angina pectoris, with a total effective rate of 82% compared to a total effective rate of 67% when treated with a common Chinese patent medicine for treatment of angina pectoris. The total effective rate of electrocardiogram (ECG) improvement (53%) was even higher than that of the control group (34%). It was shown that the saponins of *T. terrestris* had the action of dilating coronary artery and improving coronary circulation and ECG of myocardial ischemia. It showed no adverse reaction on the blood system and hepatic and renal functions, nor had any other side effects in long-term use [24].

The hypotensive and vasodilatory effects of the aqueous extract of *T. terrestris* have been demonstrated in spontaneously hypertensive rats. *In vitro* experiments showed that the aqueous extract produced a concentration-dependent decrease in the perfusion pressure of the mesenteric vascular vessel, as induced by phenylephrine [25]. The saponin fraction from *T. tribulus*, when given to hyperlipidemic rats, was found to reduce the left ventricular remodeling after acute myocardial infarction, and to improve cardiac function in the early phase after acute myocardial infarction [26].

The effect of the lyophilized saponin mixture of the plant on several smooth muscle preparations *in vitro* was reported. The LD_{50} of an intraperitoneally administered saponin mixture on mice was calculated to be about $0.8\,g\,kg^{-1}$. The saponin mixture caused a significant decrease in the peristaltic movements of the isolated sheep ureter and rabbit jejunum preparations, in dose-dependent manner. However, no significant effect on isolated rabbit aorta and its contractile response to KCl or noradrenaline (norepinephrine) was observed [27].

The saponin fraction from *T. terrestris* exhibited a hypoglycemic effect in diabetic mice, as induced by alloxan. The serum glucose level was significantly reduced after treatment with the saponin fraction, with reduction rates of about 26% and 41% for normal and diabetic mice, respectively. The serum triglyceride level was reduced by 23% after treatment. The saponin fraction also decreased the serum cholesterol level, and increased the activity of superoxide dismutase (SOD) [28].

The fruit of *T. Terrestris* was further found to exert anti-anaphylactic activity and to inhibit delayed-type hypersensitivity reactions [29]. When rabbits were treated orally with the fruit extract at different dose levels (2.5, 5, and 10 mg kg^{-1}) for eight weeks, the isolated penile tissue of the rabbits was found to show an increase in relaxation to electrical field stimulation, acetylcholine and nitroglycerin in noradrenaline-precontracted tissues. The enhanced relaxant effect observed was probably due to an increase in the release of nitric oxide (NO) from the endothelium and nitrergic nerve endings, which may account for its claims as an aphrodisiac [30].

T. terrestris has been used as an aphrodisiac agent. The administration of an extract of *T. terrestris extract* was reported to increase sexual behavior in both normal and castrated rats. The aphrodisiac effect of *T. terrestris* was speculated to be due to its androgen-increasing activity. The oral treatment of adult male Sprague-Dawley rats with an extract of *T. tribulus* at a dose of 5 mg kg^{-1} resulted in a significant increase in both nicotinamide adenine dinucleotide phosphate-diaphorase (NADPH-d) activity and androgen receptor immunoreactivity. Moreover, chronic treatment of rats with the extract increased the NADPH-d positive neurons and androgen receptor immunoreactivity in the paraventricular area of the hypothalamus. The mechanism for the observed increase in androgen receptor and NADPH-d positive neurons was probably due to the androgen-increasing property of the extract [31]. The active compound for the aphrodisiac properties of *T. terrestris* extract might be protodioscin [32, 33]. However, a study in healthy men aged 20–36 years revealed that *T. terrestris* did not influence androgen production [34]. Several herbal preparations were believed capable of enhancing physical performance [35]. A study with *T. terrestris* in men showed that *Tribulus* supplementation did not enhance body composition or exercise performance in resistance-trained male subjects [36].

The steroidal saponins derived from spirostanol, but not from furostanol of *T. terrestris*, were found to exhibit remarkable cytotoxicity against human cancer cell lines, including SK-MEL malignant melanoma, KB oral epidermoid carcinoma, BT-549 mammary carcinoma, and SK-OV-3 ovary carcinoma [37]. The saponins from *T. terrestris* also significantly inhibited the proliferation of Bcap37 breast cancer cells [38], BEL-7402 hepatoma cells, and 786-0 renal carcinoma cells [39]. The molecular mechanisms of the cytotoxicity of the saponins from *T. terrestris* involved the upregulation and downregulation of polyamine homeostasis, the suppression of proliferation, and the induction of apoptosis. The saponins were found to be less toxic for normal human fibroblasts than for many cancer lines [40]. The polysaccharide fraction from the fruit of *T. terrestris* was reported to show protective effects against genetic damage [41].

References

1 Ganzera, M., Bedir, E., and Khan, I.A. (2001) Determination of steroidal saponins in *Tribulus terrestris* by reversed-phase high-performance liquid chromatography and evaporative light scattering detection. *J. Pharm. Sci.*, **90**, 1752–1758.

2 Iskenderov, G.B. (1991) Terminal metabolites of saponins of *Tribulus terrestris*. *Azerb. Med. Zh.*, 28–31 (CA 116: 75670).

3 Yan, W., Ohtani, K., Kasai, R., and Yamasaki, K. (1996) Steroidal saponins from fruits of *Tribulus terrestris*. *Phytochemistry*, **42**, 1417–1422.

4 Bedir, E. and Khan, I.A. (2000) New steroidal glycosides from the fruits of *Tribulus terrestris*. *J. Nat. Prod.*, **63**, 1699–1701.

5 Xu, Y.X., Chen, H.S., Liang, H.Q., Gu, Z.B., Liu, W.Y., Leung, W.N., and Li, T.J. (2000)

Three new saponins from *Tribulus terrestris*. *Planta Med.*, **66**, 545–550.
6 Cai, L., Wu, Y., Zhang, J., Pei, F., Xu, Y., Xie, S., and Xu, D. (2001) Steroidal saponins from *Tribulus terrestris*. *Planta Med.*, **67**, 196–198.
7 Bedir, E. and Khan, I.A. (2000) New steroidal glycosides from the fruits of *Tribulus terrestris*. *J. Nat. Prod.*, **63**, 1699–1701.
8 Zafar, R. and Nasa, A.K. (1987) Quercetin and kaempferol from the fruits and stem of *Tribulus terrestris* Linn. *Indian J. Nat. Prod.*, **3**, 17–18.
9 Li, J.X., Shi, Q., Xiong, Q.B., Prasain, J.K., Tezuka, Y., Hareyama, T., Wang, Z.T., Tanaka, K., Namba, T., and Kadota, S. (1998) Tribulusamide A and B, new hepatoprotective lignanamides from the fruits of *Tribulus terrestris*: indications of cytoprotective activity in murine hepatocyte culture. *Planta Med.*, **64**, 628–631.
10 Mahato, S.B., Sahu, N.P., and Ganguly, A.N. (1981) Steroidal glycosides of *Tribulus terrestris* Linn. *J. Chem. Soc.* Perkin **1**, 2405–2410.
11 Huang, X.L., Zhang, Y.S., and Liang, Z.Y. (1991) Studies on water soluble polysaccharides isolated from *Tribulus terrestris* L.: purification and preliminary structural determination of heteropolysaccharide H. *Acta Pharm. Sin.*, **26**, 578–583.
12 Zhang, J.D., Xu, Z., Cao, Y.B., Chen, H.S., Yan, L., An, M.M., Gao, P.H., Wang, Y., Jia, X.M., and Jiang, Y.Y. (2006) Antifungal activities and action mechanisms of compounds from *Tribulus terrestris* L. *J. Ethnopharmacol.*, **103**, 76–84.
13 Zhang, J.D., Cao, Y.B., Xu, Z., Sun, H.H., An, M.M., Yan, L., Chen, H.S., Gao, P.H., Wang, Y., Jia, X.M., and Jiang, Y.Y. (2005) *In vitro* and *in vivo* antifungal activities of the eight steroid saponins from *Tribulus terrestris* L. with potent activity against fluconazole-resistant fungal pathogens. *Biol. Pharm. Bull.*, **28**, 2211–2215.
14 Deepak, M., Dipankar, G., Prashanth, D., Asha, M.K., Amit, A., and Ventakaraman, B.V. (2002) Tribulosin and β-sitosterol D-glucoside, the anthelmintic principles of *Tribulus terrestris*. *Phytomedicine*, **9**, 753–756.
15 Nong, C.H., Hur, S.K., Oh, O.J., Kim, S.S., Nam, K.A., and Lee, S.K. (2002) Evaluation of natural products on inhibition of inducible cyclooxygenase (COX-2) and nitric oxide synthase (iNOS) in cultured mouse macrophage cells. *J. Ethnopharmacol.*, **83**, 153–159.
16 Lin, Z.X., Hoult, J.R., and Raman, A. (1999) Sulphorhodamine B assay for measuring proliferation of a pigmented melanocyte cell line and its application to the evaluation of crude drugs used in the treatment of vitiligo. *J. Ethnopharmacol.*, **66**, 141–150.
17 Li, M., Qu, W., Chu, S., Wang, H., Tian, C., and Tu, M. (2001) Effect of the decoction of *Tribulus terrestris* on mice gluconeogenesis. *Zhong Yao Cai*, **24**, 586–588.
18 Jagadeesan, G., Kavitha, A.V., and Subashini, J. (2005) FT-IR Study of the influence of *Tribulus terrestris* on mercury intoxicated mice, *Mus musculus* liver. *Trop. Biomed.*, **22**, 15–22.
19 Anand, R., Patnaik, G.K., Kulshreshtha, D.K., and Dhawan, B.N. (1994) Activity of certain fractions of *Tribulus terrestris* fruits against experimentally induced urolithiasis in rats. *Indian J. Exp. Biol.*, **32**, 548–552.
20 Al-Ali, M., Wahbi, S., Twaij, H., and Al-Badr, A. (2003) *Tribulus terrestris*: preliminary study of its diuretic and contractile effects and comparison with *Zea mays*. *J. Ethnopharmacol.*, **85**, 257–260.
21 Sangeeta, D., Sidhu, H., Thind, S.K., and Nath, R. (1994) Effect of *Tribulus terrestris* on oxalate metabolism in rats. *J. Ethnopharmacol.*, **44**, 61–66.
22 Seth, S.D. and Jagadeesh, G. (1976) Cardiac action of *Tribulus terrestris*. *Indian J. Med. Res.*, **64**, 1821–1825.
23 Seth, S.D. and Prabhakar, M.C. (1974) Preliminary pharmacological investigations of *Tribulus terrestris* Linn. (Gokhru) part 1. *Indian J. Med. Sci.*, **28**, 377–380.
24 Wang, B., Ma, L., and Liu, T. (1990) 406 cases of angina pectoris in coronary heart disease treated with saponin of *Tribulus terrestris*. *Chin. J. Integr. Trad. West. Med.*, **10**, 85–87.
25 Phillips, O.A., Mathew, K.T., and Oriowo, M.A. (2006) Antihypertensive and vasodilator effects of methanolic and aqueous extracts of *Tribulus terrestris* in rats. *J. Ethnopharmacol.*, **104**, 351–355.
26 Guo, Y., Yin, H.J., Shi, D.Z., and Chen, K.J. (2005) Effects of Tribuli saponins on left ventricular remodeling after acute myocardial infarction in rats with hyperlipidemia. *Chin. J. Integr. Med.*, **11**, 142–146.
27 Arcasoy, H.B., Erenmemisoglu, A., Tekol, Y., Kurucu, S., and Kartal, M. (1998) Effect of *Tribulus terrestris* L. saponin mixture on some smooth muscle preparations: a preliminary study. *Boll. Chim. Farm.*, **137**, 473–475.
28 Li, M., Qu, W., Wang, Y., Wan, H., and Tian, C. (2002) Hypoglycemic effect of saponin

from *Tribulus terrestris*. *Zhong Yao Cai*, **25**, 420–422.

29 Xu, Q., Zhao, H., and Hang, B.Q. (1991) Inhibition of delayed type hypersensitivity reactions by Fructus Tribuli. *J. China Pharm. Univ.*, **22**, 12–16.

30 Adaikan, P.G., Gauthaman, K., Prasad, R.N., and Ng, S.C. (2000) Proerectile pharmacological effects of *Tribulus terrestris* extract on the rabbit corpus cavernosum. *Ann. Acad. Med. Singapore*, **29**, 22–26.

31 Gauthaman, K. and Adaikan, P.G. (2005) Effect of *Tribulus terrestris* on nicotinamide adenine dinucleotide phosphate-diaphorase activity and androgen receptors in rat brain. *J. Ethnopharmacol.*, **96**, 127–132.

32 Gauthaman, K., Adaikan, P.G., and Prasad, R.N. (2002) Aphrodisiac properties of *Tribulus terrestris* extract (protodioscin) in normal and castrated rats. *Life Sci.*, **71**, 1385–1396.

33 Gauthaman, K., Ganesan, A.P., and Prasad, R. N. (2003) Sexual effects of puncturevine (*Tribulus terrestris*) extract (protodioscin): an evaluation using a rat model. *J. Altern. Complement. Med.*, **9**, 257–265.

34 Neychev, V.K. and Mitev, V.I. (2005) The aphrodisiac herb *Tribulus terrestris* does not influence the androgen production in young men. *J. Ethnopharmacol.*, **101**, 319–323.

35 Bucci, L.R. (2000) Selected herbals and human exercise performance. *Am. J. Clin. Nutr.*, **72** (2 Suppl.), 624S–636S.

36 Antonio, J., Uelmen, J., Rodriguez, R., and Earnest, C. (2000) The effects of *Tribulus terrestris* on body composition and exercise performance in resistance-trained males. *Int. J. Sport Nutr. Exerc. Metab.*, **10**, 208–215.

37 Bedir, E., Khan, I.A., and Walker, L.A. (2002) Biologically active steroidal glycosides from *Tribulus terrestris*. *Pharmazie*, **57**, 491–493.

38 Sun, B., Qu, W., and Bai, Z. (2003) The inhibitory effect of saponins from *Tribulus terrestris* on Bcap-37 breast cancer cell line *in vitro*. *Zhong Yao Cai*, **26**, 104–106.

39 Yang, H.J., Qu, W.J., and Sun, B. (2005) Experimental study of saponins from *Tribulus terrestris* on renal carcinoma cell line. *Zhongguo Zhong Yao Za Zhi*, **30**, 1271–1274.

40 Neychev, V.K., Nikolova, E., Zhelev, N., and Mitev, V.I. (2007) Saponins from *Tribulus terrestris* L. are less toxic for normal human fibroblasts than for many cancer lines: influence on apoptosis and proliferation. *Exp. Biol. Med. (Maywood)*, **232**, 126–133.

41 Liu, Q., Chen, Y., Wang, J., Chen, X., and Han, Y. (1995) Protective effects of *Tribulus terrestris* L. polysaccharide on genetic damage. *China J. Chin. Mater. Med.*, **20**, 427–429. 449.

Trichosanthes kirilowii and *Trichosanthes rosthornii*

Fructus Trichosanthis (Gualou) is the dried ripe fruit of *Trichosanthes kirilowii* Maxim. or *Trichosanthes rosthornii* Harms (Cucurbitaceae). It is used as an antipyretic and expectorant, and in the treatment of constipation.

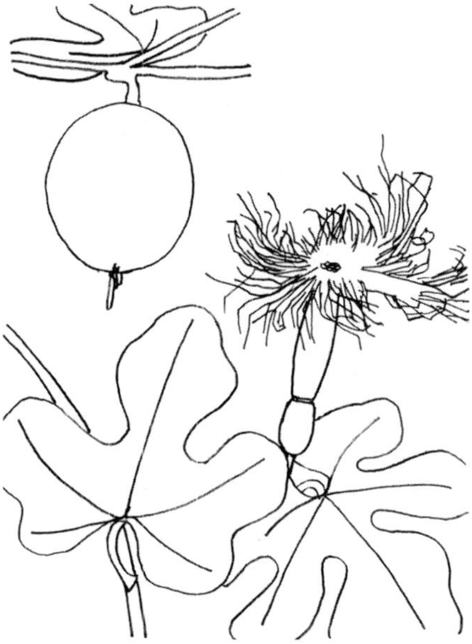

Trichosanthes kirilowii Maxim.

Pericarpum Trichosanthis (Gualoupi) is the dried pericarp of the ripe fruits of *Trichosanthes kirilowii* or *Trichosanthes rosthornii*. It is used as an expectorant.

Radix Trichosanthis (Tianhuafen) is the dried root of *Trichosanthes kirilowii* or *Trichosanthes rosthornii*. It is used in as an antiphlogistic and antiedemic agent.

Semen Trichosanthis (Gualouzi) is the dried ripe seed of *Trichosanthes kirilowii* or *Trichosanthes rosthornii*. It is used as an expectorant and in the treatment of constipation.

Semen Trichosanthis Tostum (Chaogualouzi) is the roasted seed of *Trichosanthes kirilowii* or *Trichosanthes rosthornii*. It is used as a laxative and as an expectorant, and for the treatment of constipation and cough.

The Chinese Pharmacopoeia notes that the medicinal materials derived from *Trichosanthes* species are incompatible with Radix Aconiti and allied materials.

Chemistry

The root of *T. kirilowii* was reported to contain the basic protein trichosanthin as the active principle, inducing abortion and terminating early pregnancy. The molecular mass of trichosanthin was estimated to be about 27 kDa. Trichosanthin is a relatively simple linear polypeptide composed of 246 (7) amino acid residues with a C-terminus of Asn-Asn-Met and Asn-Asn-Met-Ala [1]. X-ray diffraction and Raman spectroscopy of trichosanthin revealed eight segments of α-helix with about 85 amino acids, and 13 strands of β-sheet structure with about 70 amino acids, as well as some extended chains. The α-helices are in the center of the molecule and are surrounded by a β-sheet [2].

A rapid and simple one-step purification procedure for trichosanthin from dried root tubers has been described by using cation-exchange chromatography, with a yield of 0.16% of electrophoretically pure trichosanthin [3]. It was also reported that trichosanthin was highly expressed in the root tuber of *T. kirilowii* when grown under normal greenhouse conditions. The expression level of trichosanthin was significantly reduced when the seeds were germinated and subsequently grown in a sterile environment. However, cocultivation of the sterile plant with microorganisms resulted in an accumulation of trichosanthin, suggesting a possible role of trichosanthin in defense against pathogens [4].

Handbook of Chinese Medicinal Plants: Chemistry, Pharmacology, Toxicology
Weici Tang and Gerhard Eisenbrand

ISBN: 978-3-527-32226-8

The modification of arginine residues in trichosanthin had little or no effect on antigen–antibody interaction, whereas modification of lysine residues significantly reduced the antigen–antibody binding [5]. Trichosanthin was identified as a ribosome-inactivating protein [6]. Ribosome-inactivating proteins, which are widespread throughout the plant kingdom, are a specific type of RNA N-glycosidase, which inactivates ribosome 60S subunits [7].

α-Kirilowin and β-kirilowin, two ribosome-inactivating proteins, were isolated from the seed of *T. kirilowii*. β-Kirilowin, with an apparent molecular mass of 27.5 kDa, exhibited a strong abortifacient activity in pregnant mice. No cross-reactivity with trichosanthin could be detected by immunodiffusion. A sequence comparison of the first 10 residues of β-kirilowin with ribosome-inactivating proteins such as trichosanthin, trichokirin and karasurin, indicated 60–70% identity [8]. α-Kirilowin was reported to have a molecular mass of 28.8 kDa. The amino acid composition of α-kirilowin grossly resembled β-kirilowin and other ribosome-inactivating proteins. The N-terminal sequence of α-kirilowin was identical to that of β-kirilowin, at least in the first ten residues. It was proposed that the size difference between α-kirilowin and β-kirilowin is either due to a C-terminal extension in α-kirilowin or differences in glycosylation, or a combination of both [9].

Trichokirin was also reported to be isolated from the seed of *T. kirilowii* [10]. It inhibited protein synthesis in rabbit reticulocyte lysate, but showed a relatively low toxicity to intact cells. Trichokirin was also conjugated to monoclonal antibodies H65 that recognize human T-lymphocyte CD5 antigen. The resultant immunotoxin showed potent cytotoxicity to the target cell molt-4, but not to nontarget cell Daudi [11]. A further protein, TAP 29 (Trichosanthes Anti-HIV Protein, 29 kDa), from *T. kirilowii* differed from trichosanthin in size, *N*-terminal amino acid sequence, and cytotoxicity. In addition to three conservative substitutions, Arg-29 → Lys, Ile-37 → Val, and Pro-42 → Ser, a total difference of residues 12–16 was found. The sequence of amino acids 12–16 in TAP 29 was Lys-Lys-Lys-Val-Tyr, whereas in trichosanthin it was Ser-Ser-Tyr-Gly-Val [12].

Further cucurbitaceous plants containing ribosome-inactivating proteins listed in the Chinese Pharmacopoeia are Semen Momordicae (Mubiezi), the dried ripe seed of *Momordica cochinchinensis* and Retinervus Luffae Fructus (Sigualuo), the fibrovascular bundle of the dried ripe fruits of *Luffa cylindrica* (L.) Roem.

Triterpenes karounidiol, 7-oxodihydrokarounidiol, 5-dehydrokarounidiol, isokarounidiol 3,29-dibenzoylkarounitriol, 3-*epi*-karounidiol, 7-oxoisomultiflorenol, 3-*epi*-bryonolol and bryonolol were isolated from the seeds of *T. kirilowii*. They are all derived from a D:C-friedooleanane skeleton [13]. The Chinese Pharmacopoeia requires a qualitative determination of 3,29-dibenzoylkarounitriol in Semen Trichosanthis and in Semen Trichosanthis Tostum by thin-layer chromatographic (TLC) comparison with reference substance. In addition, the Chinese Pharmacopoeia requires a quantitative determination of 3,29-dibenzoylkarounitriol in Semen Trichosanthis Tostum by HPLC. The content of 3,29-dibenzoylkarounitriol in Semen Trichosanthis Tostum should be not less than 0.06%.

The cucurbitane triterpenes cucurbitacin B, isocucurbitacin B, cucurbitacin D, isocucurbitacin D, 3-*epi*-isocucurbitacin B, dihydrocucurbitacin B, dihydroisocucurbitacin B and dihydrocucurbitacin E were isolated from the root of *T. kirilowii* [14]. The phytosterols stigmastane-3β,6α-diol, poriferastane-3β,6α-diol, stigmast-5-ene-3β,4β-diol, poriferast-5-ene-3β,4β-diol, and poriferasta-5,25-diene-3β,4β-diol were isolated from the seed of *T. kirilowii* [15].

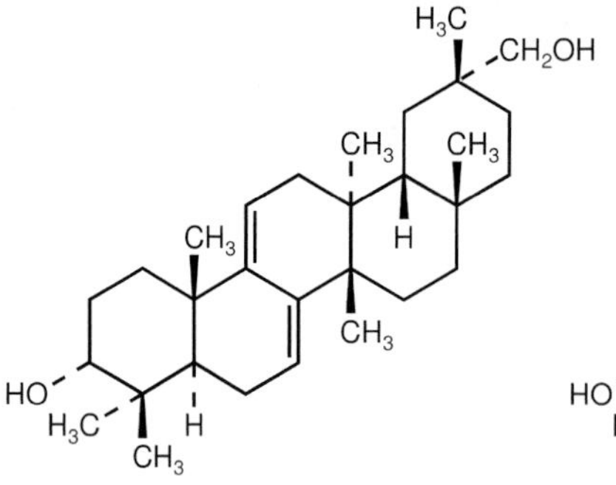

Karounidiol

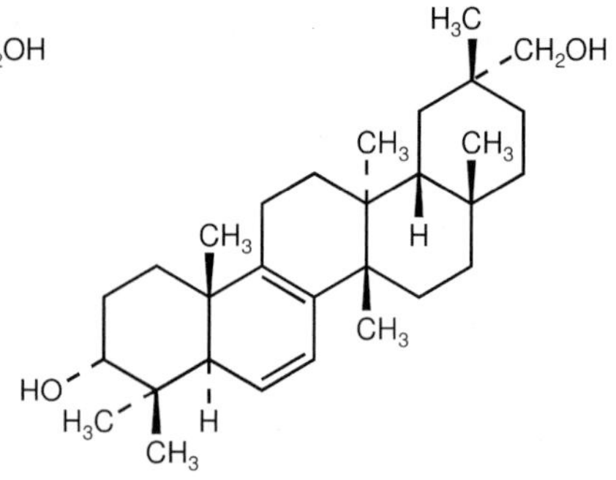

Isokarounidiol

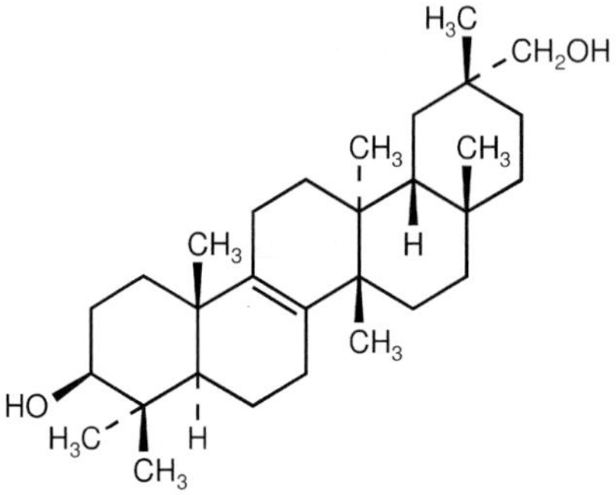

Bryonolol

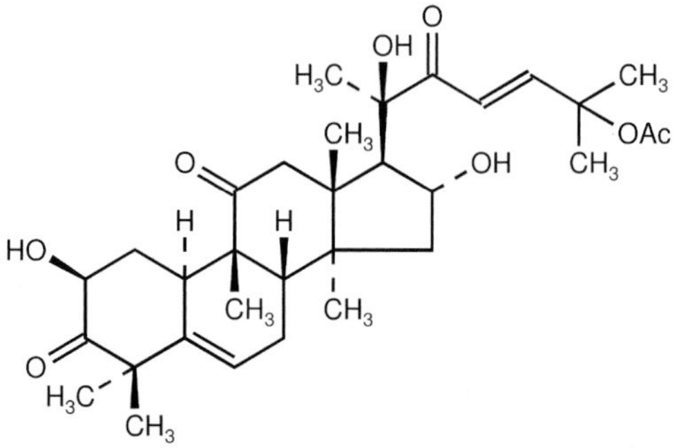

Cucurbitacin B

A polysaccharide mainly composed of glucose, galactose, fructose, mannose, xylose, and a small amount of protein from the root of *T. kirilowii* [16], and a polysaccharide composed of arabinose and galactose in a molar ratio of 1: 0.06 with a molecular mass of 5.2 Da from the fresh juice of the root [17], were reported.

The major components of the essential oil from the pericarp of *T. rosthornii* were reported to be dibutyl phthalate, methyl hexadecanoate, phenanthrene, fluranthene, and 3-methyl phenanthrene [18]. *Trichosanthes kirilowii* seeds were found to contain fatty oil which makes up for about 6% of the dry matter. The oil was composed of 95% triglycerides, 2% glycolipids and 1.3% phospholipids, with linoleic acid, oleic acid and palmitic acid as the major fatty acids [19]. Long-chain fatty acids, such as palmitic acid, linolenic acid, linoleic acid, lauric acid, and myristic acid, were isolated from the pericarp of *T. kirilowii* and *T. rosthornii*, with palmitic acid being the major organic acid [20]. The seeds of *T. kirilowii* and *T. rosthornii* were reported to contain 10–19% amino acids, including citrulline [21]. The Chinese Pharmacopoeia requires a qualitative determination of L-citrulline in Radix Trichosanthis by TLC comparison with reference substance.

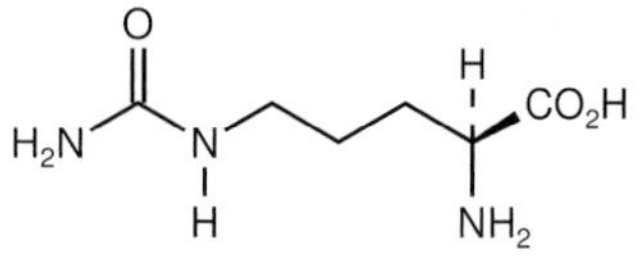

Citrulline

Pharmacology and Toxicology

Renal dysfunction in experimental rats has reported to be induced by a single dose of trichosanthin. Trichosanthin induced proximal tubular toxicity, resulting in a reduction in glomerular filtration rate and in tubular proteinuria, in a dose-dependent manner. Both, necrotic cell death and apoptosis participated in the loss of cells from the proximal tubules [22]. The embryotoxic and teratogenic effects of trichosanthin were studied *in vivo* and *in vitro*. When given at a high intraperitoneal (i.p.) dose ($7.5\,mg\,kg^{-1}$) to mice, trichosanthin decreased the viability of the fetuses; increased the number of resorbed fetuses; and reduced the crown–rump length of the surviving fetuses. At doses of 5.0 and $7.5\,mg\,kg^{-1}$, 2.3% and 9.0% respectively of the surviving fetuses were found to be abnormal; the abnormalities observed included exencephaly, micromelia, and short tail. When mouse embryos at the early organogenesis stage were cultured with trichosanthin at concentrations higher than $200\,\mu g\,ml^{-1}$, a significantly larger number of embryos was abnormal; the abnormalities were found in the head, trunk, and limb regions [23].

Trichosanthin was reported to be capable of inducing abortion in both experimental studies and clinical trials. It also possessed immunomodulatory, antitumor and anti-human immunodeficiency virus (HIV) activities. Trichosanthin belongs to the single chain ribosome-inactivating proteins, and inactivated eukaryotic ribosomes by its N-glycosidase activity. Recently, trichosanthin has been found to induce apoptosis, to enhance the action of chemokines, and to inhibit HIV-1 integrase [24].

Trichosanthin was found to be effective in inducing abortion in several mammalian models, including mice, rats, hamsters, and rabbits. The i.p. dose effective for the induction of abortion in the 10- or 11-day pregnant mouse was $50\,\mu g$ per animal. In rabbits, a response which depended on the dose and state of pregnancy was observed. Trichosanthin, when given to 19-day pregnant mice or to 28-day pregnant rabbits, resulted in a premature delivery comparable to the administration of prostaglandin $F_{2\alpha}$ ($PGF_{2\alpha}$) [25]. Trichosanthin injection terminated late pregnancy in rats, increased spontaneous uterine contraction, and significantly elevated the uterine $PGF_{2\alpha}$ content, without affecting the plasma and uterine progesterone contents. The elevated $PGF_{2\alpha}$ was postulated to be the key factor in enforcing uterine spontaneous contraction [26].

Trichosanthin has been used clinically to terminate mid-term pregnancy. Success rates of usually greater than 90% have been reported. A preliminary intradermal hypersensivity test of trichosanthin should be carried out before treatment. Trichosanthin was reported occasionally to induce anaphylaxis in some patients, which limits its clinical applications. In cases of negative skin test, 0.05 mg trichosanthin was given intramuscularly 20 min later. When there were no significant changes in blood pressure, pulse rate, and other signs within 2 h of observation, a therapeutic dose of trichosanthin was given by intra-amniotic administration. Trichosanthin induced an extensive coagulative necrosis of the trophoblastic tissues of the placental villi [27]. The use of trichosanthin either intra-amniotically or intramuscularly was reported to be very effective in the induction of expulsion of trophoblastic tissues, without the need for operative intervention and with less blood loss, especially in the case of hydatidiform mole; usually, repeated curettage caused extensive uterine bleeding. However, no significant differences in serum IgE levels were observed between women after trichosanthin treatment compared with untreated controls [28].

The combined use of trichosanthin, reserpine and testosterone propionate for the termination of early and mid-term pregnancy in 7754 cases resulted in an effective rate of 95.4%, and an incidence of serious hypersensitivity of 0.13%. The effective rate in patients treated with trichosanthin combined with dexamethasone was 96.4%, and the incidence of hypersensitivity 0.07%. No death, sequela, or complications were observed [29]. The side effect was slightest by intra-amniotic injection, followed by intrauterine and then intracervical administration. The intramuscular (i.m.) injection had the most severe side effects and local reaction at the injection site [30].

Trichosanthin, as a ribosome-inactivating protein, inactivated ribosomes and arrested protein synthesis by removing a specific adenine from 28S rRNA. The binding modes of trichosanthin were studied with oligonucleotides GAG, GAGA, and CGAGAG as substrates. All of the oligoribonucleotides could dock into the active cleft of trichosanthin, without unfavorable contacts [31]. Positions 120–123 of the native trichosanthin molecule may play a critical role in maintaining its inhibitory activity on protein biosynthesis [32]. Fragments corresponding to amino acids 1–72 and 153–246 might be the antigenic sites [33].

Trichosanthin was found to inhibit the growth of trichoblastic cells, and has been used for the treatment of abnormal growth of trichoblastic cells, such as hydatidiform and malignant moles and choriocarcinomas [34]. Studies on the *in vitro* cytotoxicity of trichosanthin showed that it selectively damaged choriocarcinoma and melanoma cells. Hepatoma cells represented the most resistant cell line among the cell lines tested. The cytotoxicity profiles of trichosanthin differed from those of anticancer drugs which interfere with DNA metabolism, such as cisplatin, methotrexate and 5-fluorouracil [35]. Trichosanthin induced apoptosis of HeLa cells [36]. Apoptosis induction in HeLa cells by trichosanthin was accompanied by decrease of Bcl-2 and phosphorylated cAMP response element-binding protein levels [37]. It was further reported that apoptosis induction in HeLa cells by trichosanthin was also accompanied by specific changes of the cytoskeleton configuration, and by decreasing the expression level of actin and tubulin subunit genes in different stages [38]. The N-terminal peptides composed of amino acid residues 1–15 and 16–30 caused increases in the concanavalin A (ConA)-stimulated incorporation of [^{3}H]thymidine into normal spleen cells at a concentration of 5 $\mu g\,ml^{-1}$. These peptides also showed growth-inhibitory effects against L1210 leukemic cells *in vivo* [39]. In HL-60 leukemia cells, trichosanthin induced cell apoptosis and activated caspase-8, -9, and -3 [40].

Clinical trials using trichosanthin for the treatment of hydatidiform mole resulted in a success in 44 of 52 patients (85%), including 38 cases (73%) of complete remission and six cases (12%) of incomplete remission [41]. *In vitro* studies revealed that trichosanthin induced the generation of reactive oxygen species (ROS) in human choriocarcinoma cell line (JAR cells). The formation of ROS induced by trichosanthin was dependent on the presence of extracellular Ca^{2+}, and was involved in the apoptosis of JAR cells [42].

Various immune functions were determined in mice bearing Ehrlich ascites carcinoma treated and untreated with the root of *T. kirilowii*, and compared with those of normal mice. In treated mice, the rosette formation of red cell immune complex was enhanced, and the superoxide dismutase activity was higher than in untreated mice bearing Ehrlich ascites carcinoma, and almost the same as in normal mice [43]. Trichosanthin inhibited surface molecule B7-1 expression of antigen-processing cells and T-cell proliferation at a concentration of 10 $\mu g\,ml^{-1}$, and therefore exerted multiple immunoinhibitory effects [44].

Trichosanthin was also reported to selectively inhibit the replication of HIV *in vitro*. It was specifically cytotoxic for HIV-infected macrophages and lymphocytes. The anti-HIV activity of trichosanthin was observed in both acutely and chronically infected cells of both lymphoid and mononuclear phagocytic lineage [45]. Under normal enzymatic digestion conditions, trichosanthin was reported to cleave the supercoiled double-stranded DNA to produce nicked circular and linear DNA [46]. Trichosanthin inhibited syncytium formation between infected H9 cells and uninfected Sup-T1 cells, and inhibited HIV replication in H9 and CEM-SS cells [47]. Although both proteins, TAP 29 and trichosanthin, exhibited similar inhibitory activity against HIV, they differed significantly in their cytotoxicity. Trichosanthin was more toxic than TAP 29 [12].

Trichosanthin and related ribosome-inactivating proteins exhibited a very weak suppressive effect on HIV-1 reverse transcriptase and on HIV-1 protease, but strongly inhibited HIV-1 integrase [48]. Studies with trichosanthin mutants revealed that the activity of trichosanthin against HIV-1 correlated well with the ribosome-inactivating property [49]. On the other hand, it was also reported that ribosome inactivation alone might not suffice to explain the anti-HIV action of trichosanthin [50]. Cell viability was always lower in HIV-1-infected cells after trichosanthin treatment. A DNA fragmentation study confirmed more laddering in infected cells, demonstrating that trichosanthin was more effective in inducing apoptosis in HIV-1 infected cells. This might explain in part the antiviral action of trichosanthin [51].

Trichosanthin was reported to greatly enhance chemokine-stimulated chemotaxis, with an EC_{50} of approximately 1 n*M* in leukocytes, and to enhance pertussis toxin-sensitive G protein activation, with an EC_{50} of approximately 20 n*M*. These effects were found to be caused by the interaction of trichosanthin chemokine receptors [52]. Human peripheral monocytes and macrophages were found to be highly sensitive to trichosanthin *in vitro*, with an IC_{50} of 1.7 μg ml^{-1}. Trichosanthin suppressed lymphocyte proliferation stimulated by Con A or lipopolysaccharide (LPS). Human T-cell lines and macrophage cell lines were more sensitive to trichosanthin than B-cell lines and myeloid cells; this suggested that a selective cytotoxicity of trichosanthin towards human macrophages and monocytes may be implicated in its anti-HIV activity [53].

Clinical trials have also been carried out to evaluate the effectiveness of trichosanthin in patients with AIDS and AIDS-related complications [54]. In a Phase I study of trichosanthin for the treatment of AIDS and AIDS-related complex, no consistent or sustained changes in CD4+ lymphocyte populations or HIV antigen levels were observed. A single intravenous (i.v.) infusion of trichosanthin was not associated with notable toxicity, with the exception of one subject among 18 patients who experienced a severe neurological adverse reaction [55]. A Phase I/II study with trichosanthin treatment of 51 patients with advanced HIV disease was reported. The patients received three i.v. infusions over a 9- to 21-day period in a dose range of 10–30 μg kg^{-1}. The maximum tolerated dose was estimated to be 30 μg kg^{-1}. Reversible but severe fatigue and myalgias were the major dose-limiting effects; mild leucocytosis and elevations in serum transaminases were noted and were reversible. Non-dose-related reversible mental status changes were seen in six patients and were considered to be associated with the drug. Decreases in serum p24 antigen levels were noted at one month after the first infusion in 10 of 18 patients who entered the study with elevated levels; one patient converted to negative. In patients with CD4+ cell levels greater than 50×10^6

cells l^{-1}, significant decreases in sedimentation rate and increases in CD4+ cell numbers were also noted [56].

Two HIV-infected patients treated with trichosanthin were reported to develop coma and multifocal neurological deficits. Neuropathological examination revealed regions of severe, multifocal necrosis with histiocytic infiltrates. These reactions to trichosanthin may be mediated by soluble factors released by HIV-infected macrophages [57]. In a Phase II study, treatment with trichosanthin combined with zidovudine of patients with HIV disease resulted in a significant increase in CD4+ cell numbers in all 85 patients [58]. Another reported clinical trial involved 22 patients with AIDS or AIDS-related complex. Trichosanthin, at doses of 8, 16, 24, 36, or 50 μg kg^{-1}, was administered by constant i.v. infusion over 3 h to achieve a serum concentration of 50 ng ml^{-1}; this concentration is known to be associated with anti-HIV effects *in vitro*. For patients who received 36 and 50 μg kg^{-1}, target serum concentrations were achieved and increases in CD4+ and CD8+ T cells observed [59]. Trichosanthin seemed to reduce viral activity and to improve certain symptoms in healthy AIDS-related complex patients [60].

Triterpenes with a D: C-friedooleanane skeleton from *Trichosanthes* seed were reported to show marked inhibitory activity against 12-*O*-tetradecanoylphorbol 13-acetate (TPA)-induced ear inflammation in mice [13]. Karounidiol was also found to markedly inhibit the promoting effect of TPA on skin tumor formation in mice following initiation with 7,12-dimethylbenz[*a*]anthracene (DMBA) [61]. Bryonolic acid exhibited cytotoxic activity towards a number of tumor cells *in vitro* by induction of apoptosis, whereas normal cells such as rat hepatocytes were less sensitive [13]. Cucurbitacin B, isocucurbitacin B, cucurbitacin D, isocucurbitacin D all exhibited high cytotoxicity against human tumor cells in culture [62].

References

1 Wang, Y., Gu, Z.W., Ye, G.J., Sun, X.J., Wang, Q.H., and Jin, S.W. (1993) Revision of the primary structure of trichosanthin and study on the trichosanthin from different places of origin. *Acta Chim. Sin.*, **51**, 1023–1029.

2 Zhou, K.J., Fu, Z.J., Chen, M.H., Lin, Y.J., and Pan, K.Z. (1992) Structure of the orthorhombic crystal of trichosanthin at 2.4 ANG. resolution. *Jiegou Huaxue*, **11**, 401–406.

3 Bhatia, N., McDonald, K.A., Jackman, A.P., and Dandekar, A.M. (1996) A simplified procedure for the purification of trichosanthin (A type 1 ribosome inactivating protein) from *Trichosanthes kirilowii* root tubers. *Protein Expr. Purif.*, **7**, 143–146.

4 Wong, R.N.S., Mak, N.K., Choi, W.T., and Law, P.T.W. (1995) Increased accumulation of trichosanthin in *Trichosanthes kirilowii* induced by microorganisms. *J. Exp. Bot.*, **46**, 355–358.

5 Keung, W.M., Yeung, H.W., Feng, Z., and Ng, T.B. (1993) Importance of lysine and arginine residues to the biological activity of trichosanthin, a ribosome-inactivating protein from *Trichosanthes kirilowii* tubers. *Int. J. Pept. Protein Res.*, **42**, 504–508.

6 Yeung, H.W., Li, W.W., Feng, Z., Barbieri, L., and Stirpe, F. (1988) Trichosanthin α-momocharin and β-momocharin: identity of abortifacient and ribosome-inactivating proteins. *Int. J. Pept. Protein Res.*, **31**, 265–268.

7 Wang, R.H., Zheng, S., Chen, X., and Shen, B.F. (1992) Inhibition of protein synthesis in cell-free system by single chain ribosome-inactivating proteins. *Chin. Biochem. J.*, **8**, 395–399.

8 Dong, T.X., Ng, T.B., Yeung, H.W., and Wong, R.N.S. (1994) Isolation and characterization of a novel ribosome-inactivating protein, β-kirilowin, from the seeds of *Trichosanthes kirilowii*. *Biochem. Biophys. Res. Commun.*, **199**, 387–393.

9 Wong, R.N.S., Dong, T.X., Ng, T.B., Choi, W.T., and Yeung, H.W. (1996) α-Kirilowin, a novel

ribosome-inactivating protein from seeds of *Trichosanthes kirilowii* (family Cucurbitaceae): a comparison with ß-kirilowin and other related proteins. *Int. J. Pept. Protein Res.*, **47**, 103–109.

10 Barbieri, L., Ferreras, J.M., Barraco, A., Ricci, P., and Stirpe, F. (1992) Some ribosome-inactivating proteins depurinate ribosomal RNA at multiple sites. *Biochem. J.*, **286**, 1–4.

11 Wang, R.H., Zheng, S., and Shen, B.F. (1993) Purification of trichokirin, a ribosome-inactivating protein from the seeds of *Trichosanthes kirilowii* Maximowicz, and preparation of immunotoxins. *Chin. Biochem. J.*, **9**, 586–590.

12 Lee-Huang, S., Huang, P.L., Chen, H.C., Kung, H.F., Li, B.Q., Huang, P.L., Huang, P., Huang, H.I., and Chen, H.C. (1991) TAP: 29 an anti-human immunodeficiency virus protein from *Trichosanthes kirilowii* that is nontoxic to intact cells. *Proc. Natl Acad. Sci. USA*, **88**, 6570–6574.

13 Akihisa, T., Yasukawa, K., Kimura, Y., Takido, M., Kokke, W.C.M.C., and Tamura, T. (1994) Five D:C-Friedo-oleanane triterpenes from the seeds of *Trichosanthes kirilowii* Maxim. and their anti-inflammatory effects. *Chem. Pharm. Bull. (Tokyo)*, **42**, 1101–1105.

14 Ryu, S.Y., Lee, S.H., Choi, S.U., Lee, C.O., No, Z., and Ahn, J.W. (1994) Antitumor activity of *Trichosanthes kirilowii*. *Arch. Pharmacol. Res.*, **17**, 348–353.

15 Kimura, Y., Akihisa, T., Yasukawa, K., Takido, M., and Tamura, Y. (1995) Structures of five hydroxylated sterols from the seeds of *Trichosanthes kirilowii* Maxim. *Chem. Pharm. Bull. (Tokyo)*, **43**, 1813–1817.

16 Chung, Y.B., Lee, C.C., Park, S.W., and Lee, C. K. (1990) Studies on antitumor and immunopotentiating activities of polysaccharides from *Trichosanthes* rhizome. *Arch. Pharmacol. Res.*, **13**, 285–288.

17 Tian, G.Y., Li, S.T., Tang, T.B., and Wang, D.C. (1985) *Trichosanthes* polysaccharide. I. Isolation and physical and chemical properties. *Acta Biochim. Biophys. Sin.*, **17**, 582–586.

18 Chao, Z., and Liu, J. (1996) Chemical constituents of the essential oil from the pericarp of *Trichosanthes rosthornii* Harms. *China J. Chin. Mater. Med.*, **21**, 357–359, 384.

19 Huang, Y., He, P., Bader, K.P., Radunz, A., and Schmid, G.H. (2000) Seeds of *Trichosanthes kirilowii*, an energy-rich diet. *Z. Naturforsch. [C]*, **55**, 189–194.

20 Chao, Z.M., Liu, J.M., Wang, F.H., Liu, D., and Yang, L.X. (1992) Analysis of the volatile organic acids in five kinds of Pericarpium trichosanthes. *China J. Chin. Mater. Med.*, **17**, 673–674.

21 Chao, Z.M., Liu, J.M., and Wang, F.H. (1992) Analysis on the amino acids in the seeds of *Trichosanthes* (Gualouzi). *Tianran Chanwu Yanjiu Yu Kaifa*, **4**, 31–34.

22 Chan, S.H., Shaw, P.C., Mulot, S.F., Xu, L.H., Chan, W.L., Tam, S.C., and Wong, K.B. (2000) Engineering of a mini-trichosanthin that has lower antigenicity by deleting its C-terminal amino acid residues. *Biochem. Biophys. Res. Commun.*, **270**, 279–285.

23 Chan, W.Y., Ng, T.B., Wu, P.J., and Yeung, H. W. (1993) Developmental toxicity and teratogenicity of trichosanthin, a ribosome-inactivating protein, in mice. *Teratog. Carcinog. Mutagen.*, **13**, 47–57.

24 Shaw, P.C., Lee, K.M., and Wong, K.B. (2005) Recent advances in trichosanthin, a ribosome-inactivating protein with multiple pharmacological properties. *Toxicon.*, **45**, 683–689.

25 Zhou, M.H., LI, Q., Shu, H.D., Bao, Y.M., and Chu, Y.H. (1982) Pharmacological study of the effect of Radix Trichosanthis on terminating early pregnancy. *Acta Pharm. Sin.*, **17**, 176–181.

26 Chen, M.X., and Chu, Y.H. (1992) Effect of trichosanthin injection on uterine contraction in rats in late pregnancy. *Chin. J. Pharm.*, **23**, 168–170.

27 Lu, P.X., and Jin, Y.C. (1989) Ectopic pregnancy treated with trichosanthin. Clinical analysis of 71 patients. *Chin. Med. J. (Engl.)*, **102**, 365–367.

28 Liu, F.Y., Chen, S.Y., Li, Y.J., Liu, G.W., Zhou, Z.R., Lu, C.L., and Zhao, M.L. (1986) Allergic reaction to trichosanthin-assays of serum IgE and specific anti-trichosanthin IgE levels. *Chin. J. Obstet. Gynecol.*, **21**, 165–167.

29 Liu, S.L., Wu, X.D., Liu, G.W., Li, Y.J., Huang, J.X., Jia, F.H., Wu, S.Y., Sun, J.Z., Zhang, Y.Z., Duan, L.N., Wang, A.M., Yu, S.H., Fan, L.X., and Shi, X.N. (1991) Primary study on safety of trichosanthin combined with reserpine and testosterone propionate for termination of early and middle stage pregnancy: a clinical report of 7754 cases. *Reprod. Contracept.*, **11**, 46–51.

30 Xu, M.F. and Jin, Y.C. (1991) Clinical trial of trichosanthin with or without dexamethasone in induction of abortion by four different routes of administration. *Reprod. Contracept.*, **11**, 47–50.

31 Gu, Y.J. and Xia, Z.X. (2000) Crystal structures of the complexes of trichosanthin with four substrate analogs and catalytic mechanism of RNA N-glycosidase. *Proteins*, **39**, 37–46.

32 Nie, H., Cai, X., He, X., Xu, L., Ke, X., Ke, Y., and Tam, S.C. (1998) Position 120-123, a potential active site of trichosanthin. *Life Sci.*, **62**, 491–500.

33 Mulot, S., Chung, K.K., Li, X.B., Wong, C.C., Ng, T.B., and Shaw, P.C. (1997) The antigenic sites of trichosanthin, a ribosome-inactivating protein with multiple pharmacological properties. *Life Sci.*, **61**, 2291–2303.

34 Xia, X., Hou, F., Li, J., Ke, Y., and Nie, H. (2006) Two novel proteins bind specifically to trichosanthin on choriocarcinoma cell membrane. *J. Biochem.*, **139**, 725–731.

35 Tsao, S.W., Ng, T.B., and Yeung, H.W. (1990) Toxicities of trichosanthin and ?-momorcharin, abortifacient proteins from Chinese medicinal plants, on cultured tumor cell lines. *Toxicon*, **28**, 1183–1192.

36 Ru, Q.H., Luo, G.A., Liao, J.J., and Liu, Y. (2000) Capillary electrophoretic determination of apoptosis of HeLa cells induced by trichosanthin. *J. Chromatogr. A*, **894**, 165–170.

37 Wang, P., Yan, H., and Li, J.C. (2007) CREB-mediated Bcl-2 expression in trichosanthin-induced HeLa cell apoptosis. *Biochem. Biophys. Res. Commun.*, **363**, 101–105.

38 Wang, P. and Li, J.C. (2007) Trichosanthin-induced specific changes of cytoskeleton configuration were associated with the decreased expression level of actin and tubulin genes in apoptotic HeLa cells. *Life Sci.*, **81**, 1130–1140.

39 Takemoto, D.J. (1998) Effect of trichosanthin, an anti-leukemia protein on normal mouse spleen cells. *Anticancer Res.*, **18** (1A), 357–361.

40 Li, J., Xia, X., Ke, Y., Nie, H., Smith, M.A., and Zhu, X. (2007) Trichosanthin induced apoptosis in HL-60 cells via mitochondrial and endoplasmic reticulum stress signaling pathways. *Biochim. Biophys. Acta*, **1770**, 1169–1180.

41 Lu, P.X. and Jin, Y.C. (1990) Trichosanthin in the treatment of hydatidiform mole. Clinical analysis of 52 cases. *Chin. Med. J. (Engl.)*, **103**, 183–185.

42 Zhang, C.Y., Gong, Y.X., Ma, H., An, C.C., and Chen, D.Y. (2000) Trichosanthin induced calcium-dependent generation of reactive oxygen species in human choriocarcinoma cells. *Analyst*, **125**, 1539–1542.

43 Guo, F. (1989) Enhancement of the therapeutic effect and red cell immune function by Radix Trichosanthis in mice bearing Ehrlich ascites carcinoma. *Chin. J. Integr. Trad. West. Med.*, **9**, 418–420.

44 Fan, Z.S. and Ma, B.L. (1999) IL-10 and trichosanthin inhibited surface molecule expression of antigen processing cells and T-cell proliferation. *Acta Pharmacol. Sin.*, **20**, 353–357.

45 McGrath, M.S., Santulli, S., and Gaston, I. (1990) Effects of GLQ223 on HIV replication in human monocyte/macrophages chronically infected *in vitro* with HIV. *AIDS Res. Hum. Retroviruses*, **6**, 1039–1043.

46 Li, M.X., Yeung, H.W., Pan, L.P., and Chan, S.I. (1991) Trichosanthin, a potent HIV-1 inhibitor, can cleave supercoiled DNA *in vitro*. *Nucleic Acids Res.*, **19**, 6309–6312.

47 Ferrari, P., Trabaud, M.A., Rommain, M., Mandine, E., Zalisz, R., Desgranges, C., and Smets, P. (1991) Toxicity and activity of purified trichosanthin. *AIDS (London)*, **5**, 865–870.

48 Au, T.K., Collins, R.A., Lam, T.L., Ng, T.B., Fong, W.P., and Wan, D.C. (2000) The plant ribosome inactivating proteins luffin and saporin are potent inhibitors of HIV-1 integrase. *FEBS Lett.*, **471**, 169–172.

49 Wang, J.H., Nie, H.L., Tam, S.C., Huang, H., and Zheng, Y.T. (2002) Anti-HIV-1 property of trichosanthin correlates with its ribosome inactivating activity. *FEBS Lett.*, **531**, 295–298.

50 Wang, J.H., Nie, H.L., Huang, H., Tam, S.C., and Zheng, Y.T. (2003) Independency of anti-HIV-1 activity from ribosome-inactivating activity of trichosanthin. *Biochem. Biophys. Res. Commun.*, **302**, 89–94.

51 Wang, Y.Y., Ouyang, D.Y., Huang, H., Chan, H., Tam, S.C., and Zheng, Y.T. (2005) Enhanced apoptotic action of trichosanthin in HIV-1 infected cells. *Biochem. Biophys. Res. Commun.*, **331**, 1075–1080.

52 Zhao, J., Ben, L.H., Wu, Y.L., Hu, W., Ling, K., Xin, S.M., Nie, H.L., Ma, L., and Pei, G. (1999) Anti-HIV agent trichosanthin enhances the capabilities of chemokines to stimulate chemotaxis and G protein activation, and this is mediated through interaction of trichosanthin and chemokine receptors. *J. Exp. Med.*, **190**, 101–111.

53 Zheng, Y.T., Zhang, W.F., Ben, K.L., and Wang, J.H. (1995) *In vitro* immunotoxicity and cytotoxicity of trichosanthin against human normal immunocytes and leukemia-lymphoma cells. *Immunopharmacol. Immunotoxicol.*, **17**, 69–79.

54 Byers, V.S. and Baldwin, P.W. (1991) Trichosanthin treatment of HIV disease. *AIDS (London)*, **5**, 1150–1151.

55 Kahn, J.O., Kaplan, L.D., Gambertoglio, J.G., Bredesen, D., Arri, C.J., Turin, L., Kibort, T., Williams, R.L., Lifson, J.D., and Volberding, P.A. (1990) The safety and pharmacokinetics of GLQ223 in subjects with AIDS and AIDS-related complex: a phase I study. *AIDS (London)*, **4**, 1197–1204.

56 Byers, V.S., Levin, A.S., Waites, L.A., Starrett, B.A., Mayer, R.A., Clegg, J.A., Price, M.R., Robins, R.A., Delaney, M., and Baldwin, R.W. (1990) A phase I/II study of trichosanthin treatment of HIV disease. *AIDS (London)*, **4**, 1189–1196.

57 Garcia, P.A., Bredesen, D.E., Vinters, H.V., Graefin von Einsiedel, R., Williams, R.L., Kahn, J.O., Byers, V.S., Levin, A.S., Waites, L. A., and Messing, R.O. (1993) Neurological reactions in HIV-infected patients treated with trichosanthin. *Neuropathol. Appl. Neurobiol.*, **19**, 402–405.

58 Byers, V.S., Levin, A.S., Malvino, A., Waites, L., Robins, R.A., and Baldwin, R.W. (1994) A phase II study of effect of addition of trichosanthin to zidovudine in patients with HIV disease and failing antiretroviral agents. *AIDS Res. Hum. Retroviruses*, **10**, 413–420.

59 Kahn, J.O., Gorelick, K.J., Gatti, G., Arri, C.J., Lifson, J.D., Gambertoglio, J.G., Bostrom, A., and Williams, R. (1994) Safety, activity, and pharmacokinetics of GLQ223 in patients with AIDS and AIDS-related complex. *Antimicrob. Agents Chemother.*, **38**, 260–267.

60 Mayer, R.A., Sergios, P.A., Coonan, K., and O'Brien, L. (1992) Trichosanthin treatment of HIV-induced immune dysregulation. *Eur. J. Clin. Invest.*, **22**, 113–122.

61 Yasukawa, K., Akihisa, T., Tamura, T., and Takido, M. (1994) Inhibitory effect of karounidiol on 12-*O*-tetradecanoylphorbol-13-acetate-induced tumor promotion. *Biol. Pharm. Bull.*, **17**, 460–462.

62 Ryu, S.Y., Choi, S.U., Lee, S.H., Lee, C.O., No, Z., and Ahn, J.W. (1995) Cytotoxicity of cucurbitacins in vitro. *Arch. Pharmacol. Res.*, **18**, 60–61.

Trigonella foenum-graecum

Semen Trigonellae (Huluba) is the dried ripe seed of *Trigonella foenum-graecum* L. (Fabaceae). It is used as an analgesic and diuretic agent.

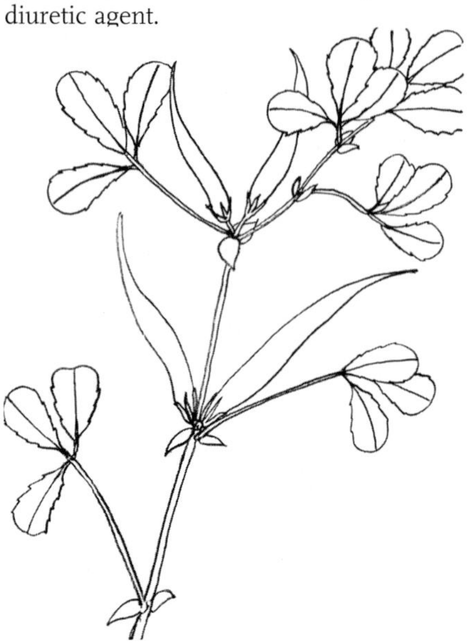

Trigonella foenum-graecum L.

Chemistry

The seed of *T. foenum-graecum* was reported to contain saponins. All these saponins possess sapogenins derived from furostan and were identified as trigoneosides Ia, Ib, IIa, IIb, IIIa, IIIb [1] and diosgenin [2]. Trigoneosides Ia, Ib, Va, Xa, Xb, Xlb, XIIa, Xllb, XIIIa were reported to be isolated from the seeds of Egyptian *T. foenum-graecum* L [3].

The alkaloid trigonelline isolated from the seeds of *T. foenum-graecum* is the inner salt of 1-methyl-3-carboxypridinium. The Chinese Pharmacopoeia requires a qualitative determination of trigonelline in Semen Trigonellae by thin-layer chromatographic comparison with reference substance, and a quantitative determination by HPLC. The content of trigonelline in Semen Trigonellae should be not less than 0.45%.

Pharmacology and Toxicology

The seed of *T. foenum-graecum* was reported to show hypoglycemic activity. Oral administration of the seed of *T. foenum-graecum* at doses of 2 and $8\,g\,kg^{-1}$ to normal and diabetic rats, as induced by alloxan or streptozotocin, produced a significant decrease of blood glucose levels in a dose-related manner [4, 5]. The seed powder of *T. foenum-graecum* was found to improve glucose homeostasis in alloxan-induced diabetic rat tissues by reversing the altered glycolytic, gluconeogenic and lipogenic enzymes [6]. The hypoglycemic effect of the seeds of *T. foenum-graecum* was also observed in normal mice [7].

It was also found that a subfraction of the seeds composed of testa and endosperm and rich in fibers (about 80%) exhibited a hypoglycemic effect in alloxan-induced diabetic dogs. The subfraction was given orally to dogs in addition to insulin treatment for a period of 21 days. The addition of subfraction to insulin treatment resulted in a clear decrease of hyperglycemia and glycosuria, accompanied by a reduction of the high plasma glucagon and somatostatin levels in diabetic dogs [8]. The soluble dietary fiber fraction with a galactomannan as major constituent, when fed simultaneously with glucose, was reported to show a significant hypoglycemic effect in diabetic rats [9].

A hypoglycemic effect of the seeds in alloxan-induced diabetic and normal rats was also reported [10]. Oral treatment of alloxan-induced diabetic rats with seed pow-

Handbook of Chinese Medicinal Plants: Chemistry, Pharmacology, Toxicology
Weici Tang and Gerhard Eisenbrand

ISBN: 978-3-527-32226-8

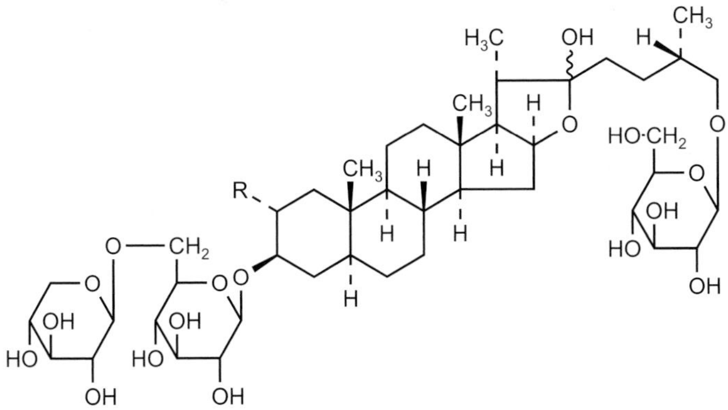

Trigoneoside Ia: R = OH
Trigoneoside IIa: R = H

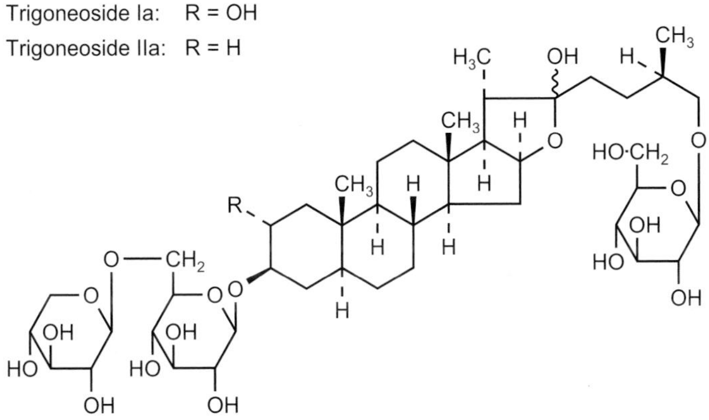

Trigoneoside Ib: R = OH
Trigoneoside IIb: R = H

der of *T. foenum-graecum* resulted in a marked decrease in the plasma glucose levels. The treatment partially restored the altered expression of pyruvate kinase and phosphoenolpyruvate carboxykinase in rat liver as two key enzymes of glycolysis and gluconeogenesis. Treatment of the rats with the seed powder also exerted beneficial effects on the alterations in the distribution of glucose transporter (GLUT4) in the skeletal muscle [11]. The seeds of *T. foenum-graecum* significantly decreased the elevated creatine kinase activities in heart, skeletal muscle and liver [12], the elevated activities of glucose-6-phosphatase and fructose-1,6-bisphosphatase in liver and kidney [13], and the elevated glyoxalase I activity in liver [14] of experimentally induced diabetic rats to almost control values. An increase in plasma insulin in rats given the seed extract at daily doses of 10 and 100 mg per rat was also observed [15].

In a clinical study, isocaloric diets with and without seeds of *T. foenum-graecum* were each given to insulin-dependent diabetic patients randomly for 10 days. Diet with seeds of *T. foenum-graecum* significantly reduced fasting blood sugar and improved the glucose tolerance. There was a 54% reduction in 24 h urinary glucose excretion. In diabetic patients, a diet with seeds of *T. foenum-graecum* also significantly reduced serum total cholesterol, low-density lipoprotein (LDL) and very low-density lipoprotein (VLDL) cholesterol and triglycerides; however, serum levels of high-density lipoprotein (HDL) cholesterol remained unchanged [16].

In experimental studies, supplementation of the diet with seeds of *T. foenum-graecum* given to alloxan-induced diabetic rats lowered blood lipid peroxidation, increased the contents of glutathione and β-carotene, and decreased the α-tocopherol content of serum [17]. Administration of the total steroid saponins in diet to streptozotocin-induced diabetic rats, corresponding to a daily dose of 12.5 mg per rat, resulted in a significant increase in food intake, and a decrease in total plasma cholesterol, without any change in triglycerides [18]. Hypocholesterolemic properties of an ethanol extract from defatted seeds were also observed in hypercholesterolemic rats fed 30 or 50 $g\,kg^{-1}$ for four weeks. Saponins appeared to be the hypocholesterolemic components, interacting with bile salts in the digestive tract [19].

Furthermore, the seed extract of *T. foenum-graecum* was found to be antineoplastic. The intraperitoneal administration of an alcohol extract of the seed, both before and after inoculation of Ehrlich ascites carcinoma in Balb-C mice, produced more than 70% inhibition of tumor cell growth with respect to controls. Treatment with the extract was found to enhance both the peritoneal exudate cell and macrophage cell counts [20].

The seed of *T. foenum-graecum* was found to contain proteinase inhibitors, inhibiting human trypsin, bovine trypsin, human chymotrypsin, and bovine chymotrypsin. Twenty-seven inhibitors were active in inhibiting all the four enzymes, including a group of acid inhibitors (TFI-A1 to TFI-A10), a group of neutral inhibitors (TFI-N1 to TFI-N6), and a group of basic inhibitors (TFI-B1 to TFI-B7). The purified inhibitor TFI-A8 exhibited potent inhibition of trypsin and a very low inhibition of chymotrypsin. TFI-N2 inhibited the four enzymes to about the same extent. TFI-B2 displayed a high inhibition of trypsin and a low inhibition of chymotrypsin. The human enzymes were inhibited better than the bovine enzymes by the purified inhibitors. The inhibitors contained high amounts of cystine (five or six disulfide bridges per molecule), aspartic acid, threonine, serine and proline, no valine and methionine; two also contained no tryptophan. Their molecular weights were given to be about 6 kDa. The reactive sites and the C-terminal sequences of TFI-B2, TFI-N2, and TFI-A8 were determined. TFI-B2 contained lysine and leucine in the trypsin- and chymotrypsin-reactive sites, and -(Lys)-Phe-Leu-Ile was the C-terminal sequence. TFI-N2 possessed arginine and leucine in the trypsin- and chymotrypsin-reactive sites, and -(Tyr)-Lys-Ile-Leu at the C-terminus. TFI-A8 contained two arginines, one in each of the two reactive sites, and -(Leu)-Phe-Ile-Arg at the C-terminus in TFI-A8. All three inhibitors belong to the Bowman–Birk proteinase inhibitor family [21].

References

1 Yoshikawa, M., Murakami, T., Komatsu, H., Murakami, N., Yamahara, J., and Matsuda, H. (1997) Medicinal foodstuffs. IV. Fenugreek seed. 1: structures of trigoneosides Ia, Ib, IIa, IIb, IIIa, and IIIb, new furostanol saponins from the seeds of Indian *Trigonella foenum-graecum* L. *Chem. Pharm. Bull. (Tokyo)*, **4**, 81–87.

2 Taylor, W.G., Elder, J.L., Chang, P.R., and Richards, K.W. (2000) Microdetermination of diosgenin from fenugreek (*Trigonella foenum-graecum*) seeds. *J. Agric. Food Chem.*, **48**, 5206–5210.

3 Murakami, T., Kishi, A., Matsuda, H., and Yoshikawa, M. (2000) Medicinal foodstuffs. XVII. Fenugreek seed. 3: structures of new

furostanol-type steroid saponins, trigoneosides Xa, Xb, XIb, XIIa, XIIb, and XIIIa, from the seeds of Egyptian *Trigonella foenum-graecum* L. *Chem. Pharm. Bull. (Tokyo)*, **48**, 994–1000.

4 Xue, W.L., Li, X.S., Zhang, J., Liu, Y.H., Wang, Z.L., and Zhang, R.J. (2007) Effect of *Trigonella foenum-graecum* (fenugreek) extract on blood glucose, blood lipid and hemorheological properties in streptozotocin-induced diabetic rats. *Asia Pac. J. Clin. Nutr.*, **16** (Suppl. 1), 422–426.

5 Vats, V., Grover, J.K., and Rathi, S.S. (2002) Evaluation of anti-hyperglycemic and hypoglycemic effect of *Trigonella foenum-graecum* Linn, *Ocimum sanctum* Linn and *Pterocarpus marsupium* Linn in normal and alloxanized diabetic rats. *J. Ethnopharmacol.*, **79**, 95–100.

6 Raju, J., Gupta, D., Rao, A.R., Yadava, P.K., and Baquer, N.Z. (2001) *Trigonella foenum-graecum* (fenugreek) seed powder improves glucose homeostasis in alloxan diabetic rat tissues by reversing the altered glycolytic, gluconeogenic and lipogenic enzymes. *Mol. Cell. Biochem.*, **224**, 45–51.

7 Zia, T., Hasnain, S.N., and Hasan, S.K. (2001) Evaluation of the oral hypoglycaemic effect of *Trigonella foenum-graecum* L. (methi) in normal mice. *J. Ethnopharmacol.*, **75**, 191–195.

8 Hannan, J.M., Rokeya, B., Faruque, O., Nahar, N., Mosihuzzaman, M., Azad Khan, A. K., and Ali, L. (2003) Effect of soluble dietary fibre fraction of *Trigonella foenum graecum* on glycemic, insulinemic, lipidemic and platelet aggregation status of Type 2 diabetic model rats. *J. Ethnopharmacol.*, **88**, 73–77.

9 Ali, L., Azad Khan, A.K., Hassan, Z., Mosihuzzaman, M., Nahar, N., Nasreen, T., Nur-e-Alam, M., and Rokeya, B. (1995) Characterization of the hypoglycemic effects of *Trigonella foenum-graecum* seed. *Planta Med.*, **61**, 358–360.

10 Preet, A., Siddiqui, M.R., Taha, A., Badhai, J., Hussain, M.E., Yadava, P.K., and Baquer, N.Z. (2006) Long-term effect of *Trigonella foenum graecum* and its combination with sodium orthovanadate in preventing histopathological and biochemical abnormalities in diabetic rat ocular tissues. *Mol. Cell. Biochem.*, **289**, 137–147.

11 Mohammad, S., Taha, A., Akhtar, K., Bamezai, R.N., and Baquer, N.Z. (2006) In vivo effect of Trigonella foenum graecum on the expression of pyruvate kinase, phosphoenolpyruvate carboxykinase, and distribution of glucose transporter (GLUT4) in alloxan-diabetic rats. *Can. J. Physiol. Pharmacol.*, **84**, 647–654.

12 Siddiqui, M.R., Moorthy, K., Taha, A., Hussain, M.E., and Baquer, N.Z. (2006) Low doses of vanadate and Trigonella synergistically regulate Na^+/K^+-ATPase activity and GLUT4 translocation in alloxan-diabetic rats. *Mol. Cell. Biochem.*, **285**, 17–27.

13 Gupta, D., Raju, J., and Baquer, N.Z. (1999) Modulation of some gluconeogenic enzyme activities in diabetic rat liver and kidney: effect of antidiabetic compounds. *Indian J. Exp. Biol.*, **37**, 196–199.

14 Raju, J., Gupta, D., Rao, A.R., and Baquer, N.Z. (1999) Effect of antidiabetic compounds on glyoxalase I activity in experimental diabetic rat liver. *Indian J. Exp. Biol.*, **37**, 193–195.

15 Petit, P., Sauvaire, Y., Ponsin, G., Manteghetti, M., Fave, A., and Ribes, G. (1993) Effects of a fenugreek seed extract on feeding behaviour in the rat: metabolic-endocrine correlates. *Pharmacol. Biochem. Behav.*, **45** (2), 369–374.

16 Sharma, R.D., Raghuram, T.C., and Rao, N.S. (1990) Effect of fenugreek seeds on blood glucose and serum lipids in type I diabetes. *Eur. J. Clin. Nutr.*, **44**, 301–306.

17 Ravikumar, P., and Anuradha, C.V. (1999) Effect of fenugreek seeds on blood lipid peroxidation and antioxidants in diabetic rats. *Phytother. Res.*, **13**, 197–201.

18 Petit, P.R., Sauvaire, Y.D., Hillaire-Buys, D.M., Leconte, O.M., Baissac, Y.G., Ponsin, G.R., and Ribes, G.R. (1995) Steroid saponins from fenugreek seeds: extraction, purification, and pharmacological investigation on feeding behavior and plasma cholesterol. *Steroids*, **60**, 674–680.

19 Stark, A. and Madar, Z. (1993) The effect of an ethanol extract derived from fenugreek (*Trigonella foenum-graecum*) on bile acid absorption and cholesterol levels in rats. *Br. J. Nutr.*, **69**, 277–287.

20 Sur, P., Das, M., Gomes, A., Vedasiromoni, J. R., Sahu, N.P., Banerjee, S., Sharma, R.M., and Ganguly, D.K. (2001) *Trigonella foenum-graecum* (fenugreek) seed extract as an antineoplastic agent. *Phytother. Res.*, **15**, 257–259.

21 Weder, J.K. and Haussner, K. (1991) Inhibitors of human and bovine trypsin and chymotrypsin in fenugreek (*Trigonella foenum-graecum* L.) seeds. Reaction with the human and bovine proteinases. *Z. Lebensm. Unters. Forsch.*, **193**, 321–325.

Tussilago farfara

Flos Farfarae (Kuandonghua) is the dried flower bud of *Tussilago farfara* L. (Asteraceae). It is used as an expectorant and mucolytic agent.

Tussilago farfara L.

Chemistry

The flower bud of *T. farfara* is known to be rich in mucopolysaccharides as the expectorant principle. Sesquiterpenes from the flower bud were identified as tussilagone [1], tussilagonone I [2] and related sesquiterpenes with different acid moieties [3].

Structurally related sesquiterpene lactones from the flower buds of *T. farfara* were identified as tussilagolactone and neotussilagolactone I [2, 3]. A bisabolene epoxide from *T. farfara* was structurally elucidated as 1α,5α-bisacetoxy-8-angeloyloxy-3β,4β-epoxy-bisabola-7(14),10-dien-2-one [4].

Triterpenes from the flower buds of *T. farfara* were identified as bauerenol, isobauerenol, and 16-hyroxybauerenol [5].

A number of pyrrolizidine alkaloids were reported to be isolated from the flower of *T. farfara*, with senkirkine and senecionine as the major components. Minor pyrrolizidine alkaloids were identified as integerrimine, seneciphylline, tussilagine and isotussilagine, and the 1-epimers of tussilagine and isotussilagine [6]. The alkaloid content in the flower buds of *T. farfara* was reported as 0.01%. Pyrrolizidine alkaloids were also isolated from the leaves of *T. farfara* [7]. 2-Pyrrolidine acetic acid was further isolated from the flower buds [8]. Senecionine represents a macrocyclic diester of retronecine, whereas senkirkine is derived from 4,8-secosenecionan, having a methyl substituent at nitrogen and a keto function at position 8. Tussilagine and isotussilagine are derived from pyrrolizidine-7-carboxylic acid methyl ester, and are not toxic [6].

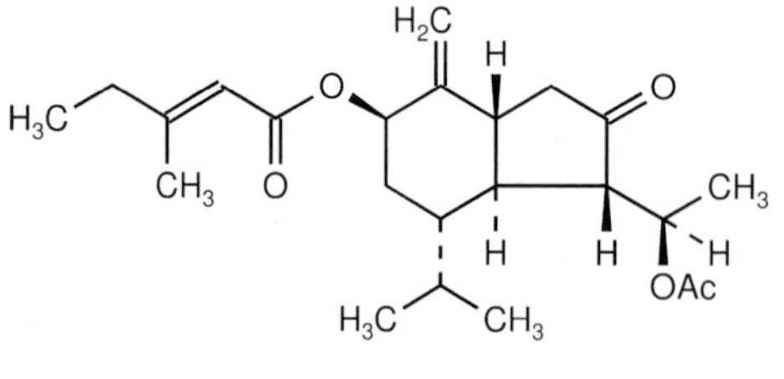

Tussilagone

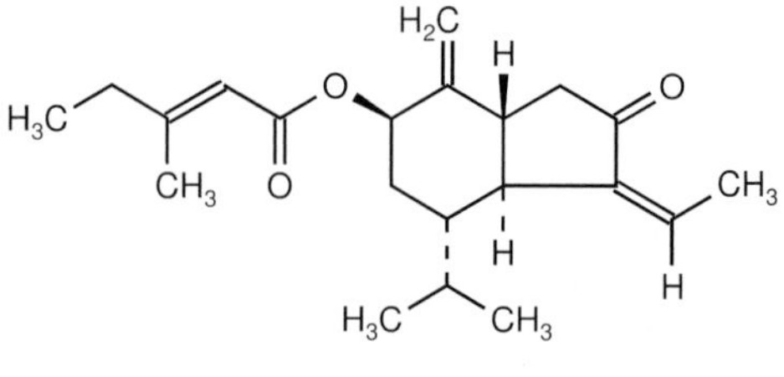

Tussilagonone I

Handbook of Chinese Medicinal Plants: Chemistry, Pharmacology, Toxicology
Weici Tang and Gerhard Eisenbrand

ISBN: 978-3-527-32226-8

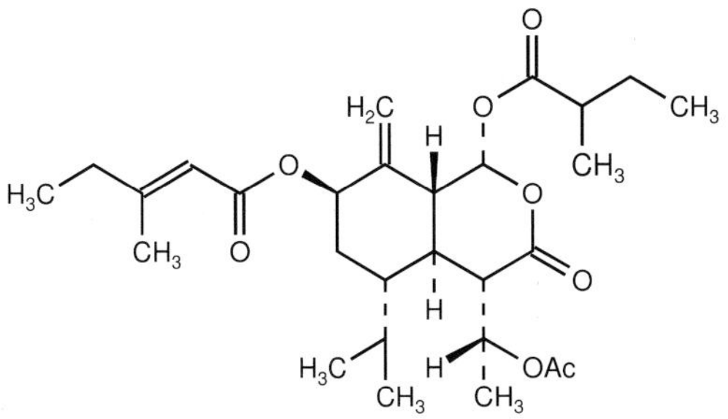

Tussilagolactone

Pharmacology and Toxicology

Senkirkine significantly induced sister-chromatid exchange (SCE) in V79 cells in the presence of an activating system [9]. Senkirkine was also reported to induce chromosome aberrations and 8-azaguanine-resistant mutation in V79 cells [10]. Senkirkine was further reported to be positive in the DNA repair test with hamster or mouse hepatocytes [11]. Senkirkine was carcinogenic in rats after intraperitoneal (i.p.) administration at 22 mg kg^{-1} twice weekly for four weeks, and then once weekly for 52 weeks. All 20 rats in the experimental group survived for more than 290 days after the start of treatment, and nine of these developed liver cell adenomas [12].

Carcinogenicity of the flowers of *T. farfara* has been studied in rats. Rats were administered the dried flowers of *T. farfara* in the diet. In the high-dose group, the diet contained 16–32% flowers, in the middle-dose group 8%, and in the low-dose group 4%. In the high-dose group, eight of 12 rats developed hemangioendothelial sarcomas in the liver, and three of these developed additional tumors, including one hepatocellular carcinoma, one hepatocellular adenoma and one papilloma of the urinary bladder. In the middle-dose group, one rat developed hemangioendothelial sarcoma in the liver. No tumors were observed in the low-dose group [13]. A case of reversible hepatic veno-occlusive disease in an infant after consumption of herbal tea containing pyrrolizidine alkaloids was reported. The condition was diagnosed in an 18-month-old boy who had regularly consumed a herbal tea mixture since the third month of life. A pharmacological analysis of the tea compounds revealed high levels of pyrrolizidine alkaloids. Seneciphylline and the corresponding *N*-oxide were identified as the major components. It was calculated that the child had consumed at least 60 μg kg^{-1}

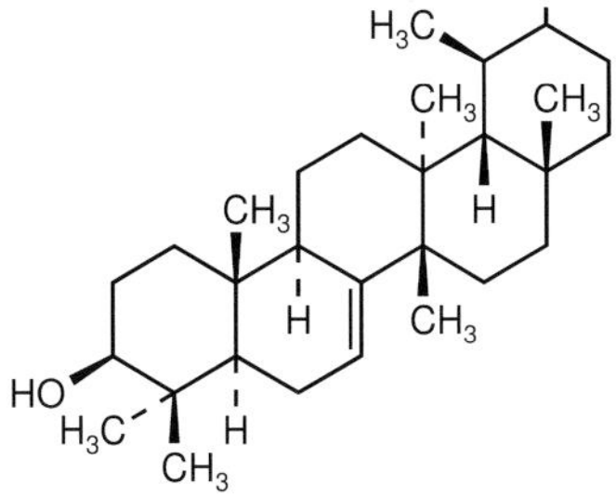

Bauerenol

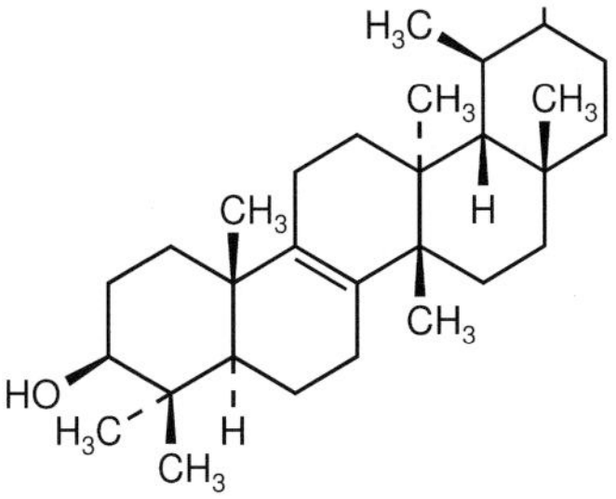

Isobauerenol

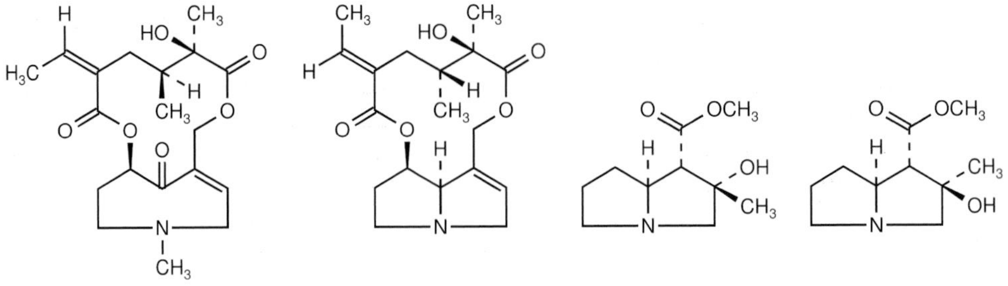

Senkirkine Integerrimine Tussilagine Isotussilagine

per day of the toxic pyrrolizidine alkaloid mixture over 15 months [14].

The mutagenicity, genotoxicity, metabolic activation and DNA adduct formation of senecionine and seneciphylline have been discussed under *Senecio scandens*. Senkirkine, as a 4,8-secosenecionan derivative, can be metabolically activated to the corresponding 8-hydroxy-retronecine ester via *N*-demethylsenkirkine. 8-Hydroxy-retronecine ester has been postulated as an intermediate in the activation of pyrrolizidine alkaloids to dehydropyrrolizidine alkaloids (Figure 220.1) [6].

The sesquiterpene tussilagone and some related sesquiterpenes were found to show moderate activity in inhibiting platelet aggregation induced by platelet-activating factor (PAF) [15], and to be dual receptor antagonists of PAF. They also competitively inhibited the specific binding of Ca^{2+} channel blockers in cardiac sarcolemmal vesicles. At a concentration of $10\,\mu M$, tussilagone caused a 60% relaxation of rat thoracic aorta strips contracted by Ca^{2+}. According to the dual antagonistic activities, tussilagone potently inhibited the gel-filtered rabbit platelet aggregation. In *in vivo* studies, tussilagone was reported to be orally active in inhibiting the PAF-induced rat foot edema and the first phase of carrageenan-induced rat hind paw edema [16]. The intravenous (i.v.) administration of tussilagone dose-dependently increased the peripheral resistance in anesthetized dogs, cats and rats, with a potency similar to that of dopamine [17]. In addition to the cardiovascular effect, tussilagone stimulated respiration in experimental animals. The acute i.v. LD_{50} value of tussilagone in mice was $29\,mg\,kg^{-1}$ [17].

The ethyl acetate fraction prepared from the flower buds of *T. farfara* was reported to exert neuroprotective and antioxidant effects. It exhibited anti-inflammatory activity by inhibiting arachidonic acid metabolism and nitric oxide (NO) production in macrophages activated by lipopolysaccharide (LPS). In primary cultured rat cortical cells,

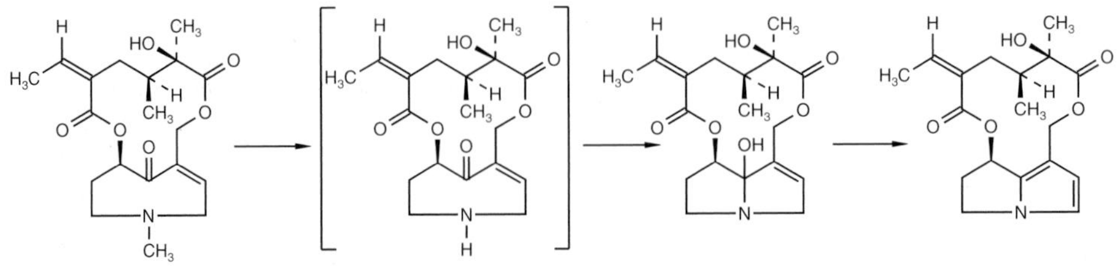

Figure 220.1 Postulated metabolic activation of senkirkine [6].

the ethyl acetate fraction potently inhibited the neuronal damage induced by arachidonic acid. In addition, it inhibited the neurotoxicity of β-amyloid-25-35 and excitotoxicity induced by glutamate or *N*-methyl-D-aspartic acid. The fraction also inhibited the oxidative neuronal damage in rat brain homogenates induced by H_2O_2, xanthine and xanthine oxidase, or Fe^{2+} and ascorbic acid [18]

References

1 Wang, C.D., Takayanagi, H., Mi, C.F., Qiao, B. L., Yang, J., Zheng, Q.T., and He, C.H. (1989) Chemical studies of flower buds of *Tussilago farfara* L. *Acta Pharm. Sin.*, **24**, 913–916.

2 Shi, W., and Han, G.Q. (1996) Chemical constituents of *Tussilago farfara* L. *J. Chin. Pharm. Sci.*, **5**, 63–67.

3 Kikuchi, M. and Suzuki, N. (1992) Studies on the constituents of *Tussilago farfara* L. II. Structures of new sesquiterpenoids isolated from the flower buds. *Chem. Pharm. Bull. (Tokyo)*, **40**, 2753–2755.

4 Ryu, J.H., Jeong, Y.S., and Sohn, D.H. (1999) A new bisabolene epoxide from *Tussilago farfara*, and inhibition of nitric oxide synthesis in LPS-activated macrophages. *J. Nat. Prod.*, **62**, 1437–1438.

5 Yaoita, Y. and Kikuchi, M. (1998) Triterpenoids from flower buds of *Tussilago farfara* L. *Nat. Med. (Tokyo)*, **52**, 273–275.

6 Roeder, E. (2000) Medicinal plants in China containing pyrrolizidine alkaloids. *Pharmazie*, **55**, 711–726.

7 Bartkowski, J.B., Wiedenfeld, H., and Roeder, E. (1997) Quantitative photometric determination of senkirkine in farfara folium. *Phytochem. Anal.*, **8**, 1–4.

8 Passreiter, C.M. (1992) Co-occurrence of 2-pyrrolidine acetic acid with the pyrrolizidines tussilagic acid and isotussilagic acid and their 1-epimers in *Arnica* species and *Tussilago farfara*. *Phytochemistry*, **31**, 4135–4137.

9 Bruggeman, I.M., and van der Hoeven, J.C. (1985) Induction of SCEs by some pyrrolizidine alkaloids in V79 Chinese hamster cells co-cultured with chick embryo hepatocytes. *Mutat. Res.*, **142**, 209–212.

10 Takanashi, H., Umeda, M., and Hirono, I. (1980) Chromosomal aberrations and mutation in cultured mammalian cells induced by pyrrolizidine alkaloids. *Mutat. Res.*, **78**, 67–77.

11 Mori, H., Sugie, S., Yoshimi, N., Asada, Y., Furuya, T., and Williams, G.M. (1985) Genotoxicity of a variety of pyrrolizidine alkaloids in the hepatocyte primary culture-DNA repair test using rat, mouse, and hamster hepatocytes. *Cancer Res.*, **45**, 3125–3129.

12 Hirono, I., Haga, M., Fujii, M., Matsuura, S., Matsubara, N., Nakayama, M., Furuya, T., Hikichi, M., Takanashi, H., Uchida, E., Hosaka, S., and Ueno, I. (1979) Induction of hepatic tumors in rats by senkirkine and symphytine. *J. Natl Cancer Inst.*, **63**, 469–472.

13 Hirono, I., Mori, H., and Culvenor, C.C. (1976) Carcinogenic activity of coltsfoot, *Tussilago farfara* l. *Gann*, **67**, 125–129.

14 Sperl, W., Stuppner, H., Gassner, I., Judmaier, W., Dietze, O., and Vogel, W. (1995) Reversible hepatic veno-occlusive disease in an infant after consumption of pyrrolizidine-containing herbal tea. *Eur. J. Pediatr.*, **154**, 112–116.

15 Han, G.Q., Yang, Y.J., and Li, C.L. (1987) The investigation of principles against platelet activating factor (PAF) from *Tussilago farfara* L. *J. Beijing Med. Univ.*, **19**, 33–35.

16 Hwang, S.B., Chang, M.N., Garcia, M.L., Han, Q.Q., Huang, L., King, V.F., Kaczorowski, G.J., and Winquist, R.J. (1987) L-652,469: a dual receptor antagonist of platelet activating factor and dihydropyridines from *Tussilago farfara* L. *Eur. J. Pharmacol.*, **141**, 269–281.

17 Li, Y.P., and Wang, Y.M. (1988) Evaluation of tussilagone: a cardiovascular-respiratory stimulant isolated from Chinese herbal medicine. *Gen. Pharmacol.*, **19**, 261–263.

18 Cho, J., Kim, H.M., Ryu, J.H., Jeong, Y.S., Lee, Y.S., and Jin, C. (2005) Neuroprotective and antioxidant effects of the ethyl acetate fraction prepared from *Tussilago farfara* L. *Biol. Pharm. Bull.*, **28**, 455–460.

Typha angustifolia and *Typha orientalis*

Pollen Typhae (Puhuang) is the dried pollen of *Typha angustifolia* L., *Typha orientalis* Presl, or other *Typha* species (Typhaceae). It is mainly used as a hemostatic agent for the treatment of different bleeding, and also for traumatic diseases. The Chinese Pharmacopoeia notes Pollen Typhae to be used with caution in pregnancy.

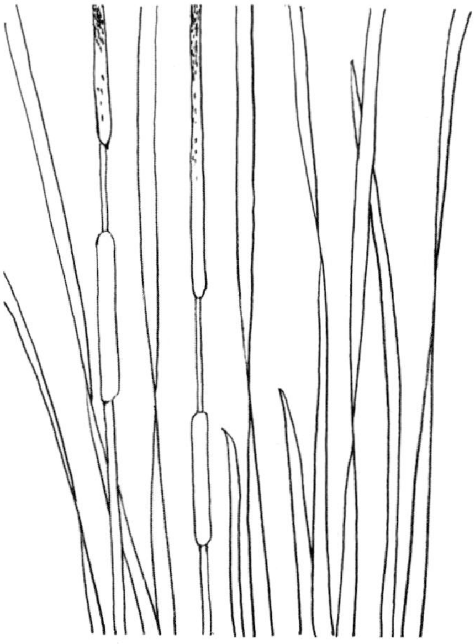

Typha angustifolia L.

Chemistry

The pollen of *Typha* species is known to contain flavone derivatives, with isorhamnetin 3-*O*-neohesperidoside and typhaneoside as the major components [1]. The Chinese Pharmacopoeia requires a qualitative determination of isorhamnetin 3-*O*-neohesperidoside and typhaneoside in Pollen Typhae by thin-layer chromatographic comparison with reference substances, and a quantitative determination of isorhamnetin 3-*O*-neohesperidoside in Pollen Typhae by HPLC. The content of isorhamnetin 3-*O*-neohesperidoside in Pollen Typhae should be not less than 0.1%. The contents of isorhamnetin 3-*O*-neohesperidoside and typhaneoside in Pollen Typhae were found to be reduced after processing [1]. Besides isorhamnetin 3-*O*-neohesperidoside, flavones and flavone glycosides in Pollen Typhae were identified as isorhamnetin, kaempferol, quercetin, naringetin, isorhamnetin 3-*O*-α-L-rhamnosyl-(1 → 2)-β-D-glucoside, kaempferol, and quercetin 3-*O*-α-rhamnosyl-β-glucoside [2]. Further constituents from Pollen Typhae were identified as β-sitosterol, β-sitosterol palmitate, palmitic acid and β-sitosterol 3-*O*-β-D-glucopyranoside [3].

Pharmacology and Toxicology

The pollen of *T. angustifolia* is used for the treatment of angina pectoris. It was also reported that the pollen is able to prevent and treat atherosclerosis [3]. In experimental studies, the pollen of *Typha* species was found to inhibit the proliferation of rat aortic smooth muscle cells in culture [4]; to influence 6-keto-prostaglandin $F_{1\alpha}$ (6-keto-$PGF_{1\alpha}$) and thromboxane B_2 (TXB_2) in chronic hyperlipemic rabbits [5]; to increase macrophage activity and the regression of cholesterol granuloma of rats [6]; to prevent ventricular fibrillation and sudden death caused by isoproterenol and also arrhythmia induced by the infusion of $BaCl_2$ [7]; and to prevent acute experimental myocardial infarction in rabbits [8]. In addition, the pollen of *Typha* was found to be capable of enhancing the osteoinductive potential of demineralized bone matrix [9]. The ethanol extract of Pollen Typhae significantly suppressed splenocyte proliferation stimulated by concanavalin A

Handbook of Chinese Medicinal Plants: Chemistry, Pharmacology, Toxicology
Weici Tang and Gerhard Eisenbrand

ISBN: 978-3-527-32226-8

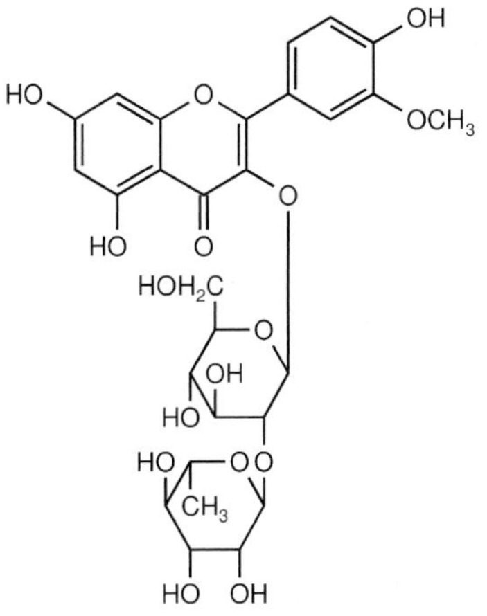

Isorhamnetin 3-*O*-neohesperidoside

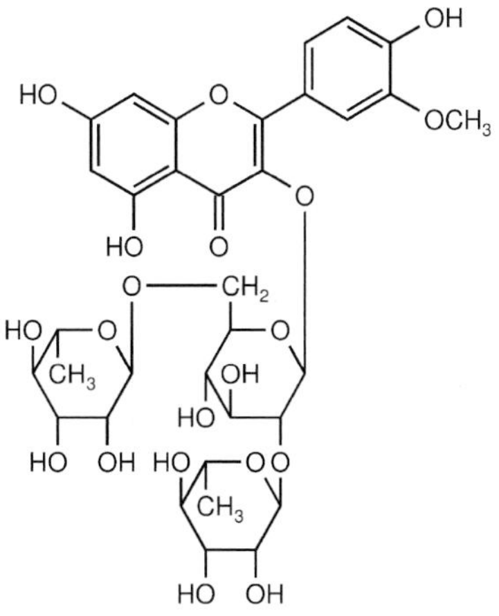

Typhaneoside

(Con A) and lipopolysaccharide (LPS) in a concentration-dependent manner. The intraperitoneal injection of the extract in ICR mice immunized subcutaneously with ovalbumin significantly suppressed splenocyte proliferation in ovalbumin-immunized mice stimulated by Con A, LPS or ovalbumin in a dose-dependent manner; this suggested that the extract elicited cellular and humoral immune response in mice [10].

The flavone fraction from the pollen was able to raise tissue plasminogen activator (tPA) and PGI_2 production in cultured porcine aortic endothelial cells [11]. Isorhamnetin 3-*O*-rhamnosyl-glucoside stimulated endothelial cells to produce tPA and PGI_2; quercetin 3-*O*-neohesperidose was found to protect endothelial cells from injury by fibrin, as well as to raise tPA activity; β-sitosterol 3-*O*-palmitate could inhibit smooth muscle cell proliferation and β-sitosterol glucoside showed an inhibitory effect on platelet aggregation [12]. Polysaccharides from *T. augustata* with a molecular weight of approximately 30 kDa were also found to show an anticoagulant effect in normal human plasma. The polysaccharide accelerated the recalcified plasma times at concentrations below $100\,\mu g\,ml^{-1}$, while inhibiting them at higher concentrations. The anticoagulant effect was mainly directed towards fibrinogen [13].

In clinical studies, the pollen of *Typha* was reported to be hypolipidemic [14]; to enhance the immune state of nonspecific ulcerative colitis [15]; to show labor-inducing activity [16]; to enhance the tolerance to hypoxia; and to promote adaptation to living at high altitude [17].

References

1 Liu, B. and Lu, Y. (1998) HPLC determination of two flavonoids in pollen Typhae (puhuang) and its different processed products. *China J. Chin. Mater. Med.*, **23**, 402–404, 447.

2 Chen, Y., Fang, S.D., Gu, Y.L., Zhang, C.Y., and Zhao, J. (1990) Active principles of the pollen of narrowleaf cattail (*Typha angustifolia*). *Chin. Trad. Herbal Drugs*, **21**, 50–53.

3 Yin, Y. and Yu, T. (1992) Advances in the pharmacological study of Pollen Typhae. *China J. Chin. Mater. Med.*, **17**, 374–377.

4 Huang, G.Q., Xu, D.M., Zhang, C.Y., Xu, Y.L., Zhao, J., and Wang, Z.Y. (1991) Protection of the monomers of the pollen of typhae on cultured endothelial cells. *Acta Univ. Med. Sec. Shanghai*, **11**, 185–188.

5 Zhang, L. (1986) Effect of Pollen Typhae on 6-keto-PGF1α, TXB2, TC and HDL-C in chronic hyperlipemic rabbits. *Chin. J. Cardiovasc. Dis.*, **14**, 291–293, 319–320.

6 Qin, F. and Sun, H.X. (2005) Immunosuppressive activity of Pollen Typhae ethanol extract on the immune responses in mice. *J. Ethnopharmacol.*, **102**, 424–429.

7 Ren, W.H. (1984) Effect of Pollen Typhae on macrophage activity and regression of cholesterol granuloma of rats. *Chin. J. Integr. Trad. West. Med.*, **4**, 176–178.

8 Zheng, R.X., Fang, S.M., Li, Z.M., and Zhang, X.M. (1993) Prevention of arrhythmia in rats by puhuang. *China J. Chin. Mater. Med.*, **18**, 108–110.

9 Huang, S.Y. and Wang, Y.J. (1985) Effect of *Typha angustata* on acute experimental myocardial infarction in rabbits. *Chin. J. Integr. Trad. West. Med.*, **5**, 297–298.

10 Yan, S.Q., Wang, G.J., and Shen, T.Y. (1994) Effects of pollen from *Typha angustata* on the osteoinductive potential of demineralized bone matrix in rat calvarial defects. *Clin. Orthop.*, **306**, 239–246.

11 Zhao, J., Zhang, C.Y., Xu, D.M., Huang, G.Q., Xu, Y.L., Wang, Z.Y., Fang, S.D., Chen, Y., and Gu, Y.L. (1989) Further study of pollen typhae's effects on the production of tPA and PGI, by cultured endothelial cells. *Thromb. Res.*, **56**, 677–685.

12 Zhao, J., Zhang, C.Y., Xu, D.M., Huang, G.Q., Xu, Y.L., Wang, Z.Y., Fang, S.D., Chen, Y., and Gu, Y.L. (1990) The antiatherogenic effects of components isolated from pollen typhae. *Thromb. Res.*, **57**, 957–966.

13 Gibbs, A., Green, C., and Doctor, V.M. (1983) Isolation and anticoagulant properties of polysaccharides of *Typha Augustata* and *Daemonorops* species. *Thromb. Res.*, **32**, 97–108.

14 Zhang, B.Z. (1985) Clinical and experimental study of *Typha angustata* on hyperlipoidemia. *Chin. J. Integr. Trad. West. Med.*, **5**, 141–144.

15 Yang, X.H., Zhang, J.G., and Peng, Z.H. (1987) Effect of extracts of Pollen Typhae on the immune state of nonspecific ulcerative colitis and clinical analysis. *Bull. Chin. Mater. Med.*, **12**, 48–50, 65.

16 Geng, Q.M. (1985) Labor-inducing effect of *Typha angustata* in injectable form and its pharmacological properties. *Chin. J. Integr. Trad. West. Med.*, **5**, 299–300.

17 Peng, H. (1990) The effect of pollen in enhancing tolerance to hypoxia and promoting adaptation to highlands. *Natl. Med. J. China*, **70** (2), 77–81, 6.

Uncaria hirsuta, Uncaria macrophylla, Uncaria rhynchophylla, Uncaria sessilifructus and *Uncaria sinensis*

Ramulus Uncariae cum Uncis (Gouteng) is the dried branch-bearing hooks of *Uncaria rhynchophylla* (Miq.) Jacks., *Uncaria macrophylla* Wall., *Uncaria hirsuta* Havil., *Uncaria sinensis* (Oliv.) Havil., or *Uncaria sessilifructus* Roxb. (Rubiaceae). It is used as an antipyretic, and as an anticonvulsant agent for the treatment of headache, hypertension, vertigo, and epilepsy.

Uncaria rhynchophylla (Miq.) Jacks.

Chemistry

The stem and hook of the official *Uncaria* species were reported to contain alkaloids, with the corynoxan-type alkaloid rhynchophylline as the major component. Rhynchophylline possesses a tetracyclic system composed of an indole moiety and an indolizidine moiety. Related alkaloids were identified as isorhynchophylline, the 7-epimer of rhynchophylline, the isomers corynoxeine and isocorynoxeine. It was reported that approximately 97% of total alkaloids detected in the hook, small stem, and leaf represented the corynoxan-type alkaloids rhynchophylline, isorhynchophylline, corynoxine, corynoxine B, corynoxeine, and isocorynoxeine [1].

Indole alkaloids isolated from *U. rhynchophylla* and other *Uncaria* species were identified as hirsuteine, hirsutine, corynantheine, dihydrocorynantheine, geissoschizine, vallesiachotamine, and akuammigine [1]. Whereas, hirsuteine, hirsutine, corynantheine, dihydrocorynantheine, geissoschizine, vallesiachotamine all represent tetracyclic indole alkaloids, akuammigine is a pentacyclic indole alkaloid.

Indole alkaloid glycosides from different *Uncaria* species were identified as cadambine, 3α-dihydrocadambine, 3β-isodihydrocadambine, strictosamide, and its 3-epimer vincoside lactam together with 6′-feruloylvincoside lactam (rhynchophine). Geissoschizine methylether, corynantheine and dihydrocorynantheine were isolated from the branches and hooks of *Uncaria sinensis* [2, 3].

Triterpenes from *U. rhynchophylla* were identified as uncarinic acids A, B, C, D, and E [4, 5].

Pharmacology and Toxicology

Uncaria species and the alkaloid constituents such as rhynchophylline, isorhynchophylline and hirsutine were found to show activities on cardiovascular and central nervous system, including hypotension, vasodilation, brachycardia, antiarrhythmia,

Handbook of Chinese Medicinal Plants: Chemistry, Pharmacology, Toxicology
Weici Tang and Gerhard Eisenbrand

ISBN: 978-3-527-32226-8

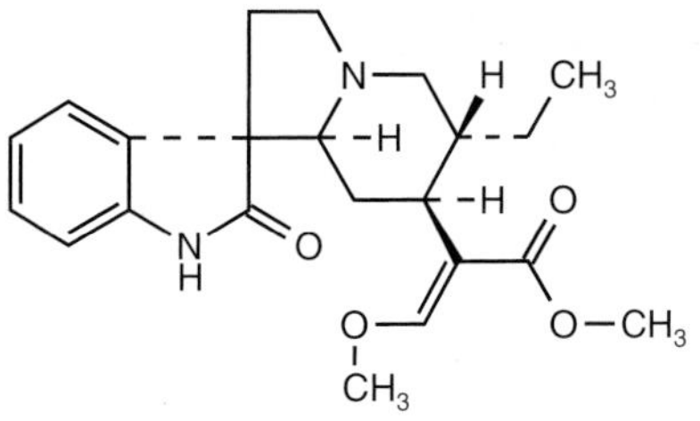

Rhynchophylline

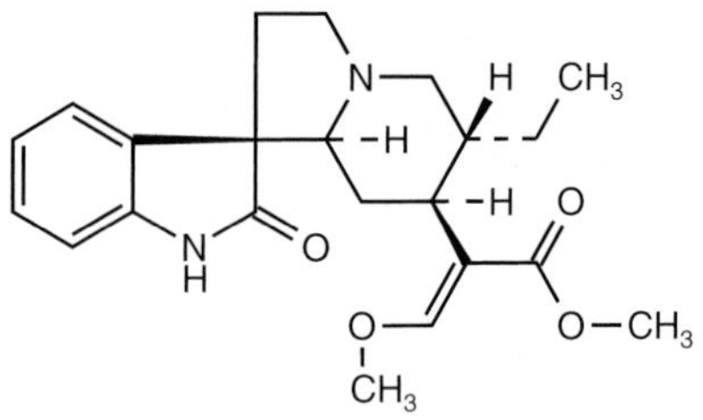

Isorhynchophylline

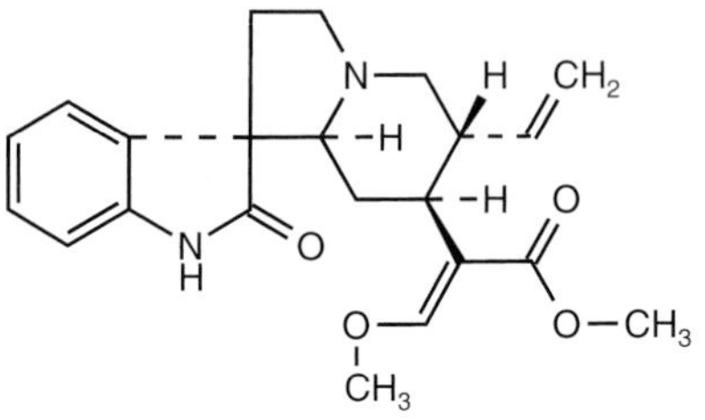

Corynoxeine

protection from cerebral ischemia, and sedation. The active mechanisms were related to the blocking and opening of potassium channels, and the regulation of nerve transmitter transport and metabolism [6].

The total alkaloids from *U. rhynchophylla* and rhynchophylline were reported to exert a hypotensive effect in cats after intravenous (i.v.) administration. Marked hypotensive effects of total alkaloids and rhynchophylline were also observed in rats [7]. The i.v. infusion of total alkaloids of *U. macrophylla* in anesthetized dogs caused moderate hypotension, marked bradycardia, a rise in

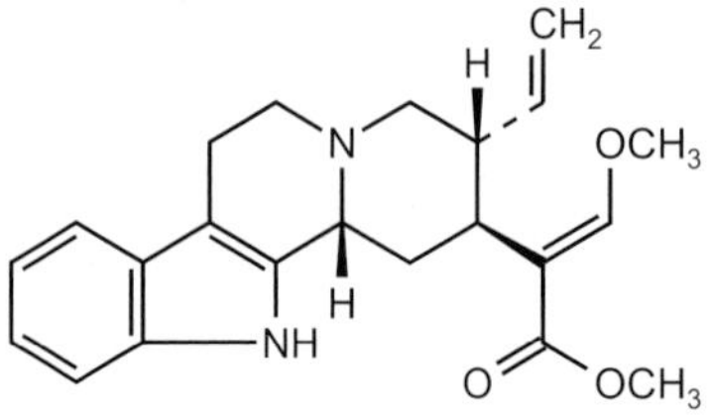

Hirsuteine

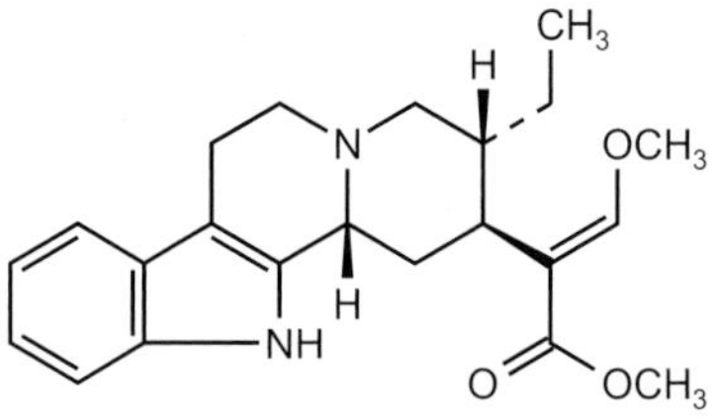

Hirsutine

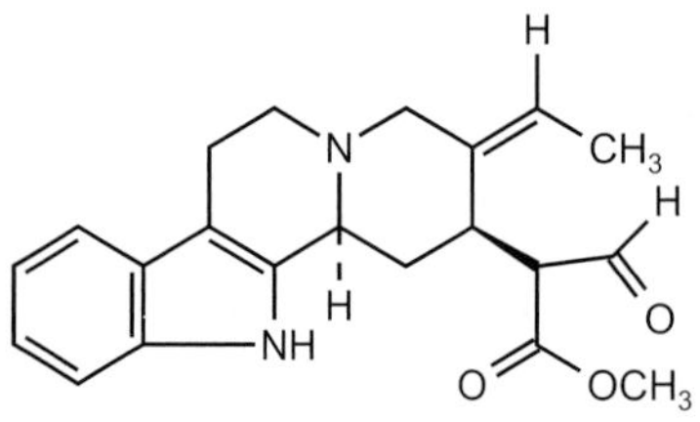

Geissoschizine

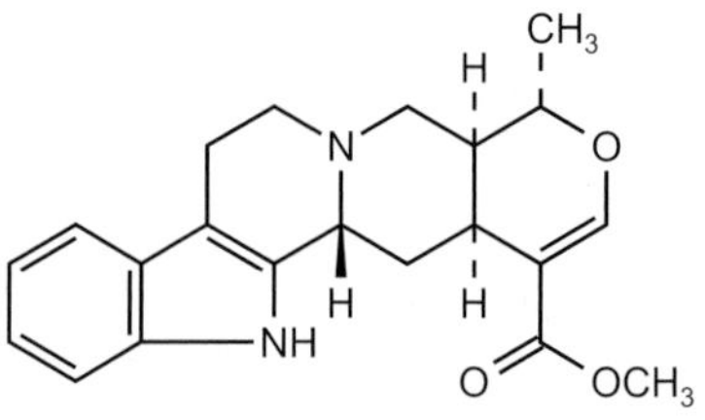

Akuammigine

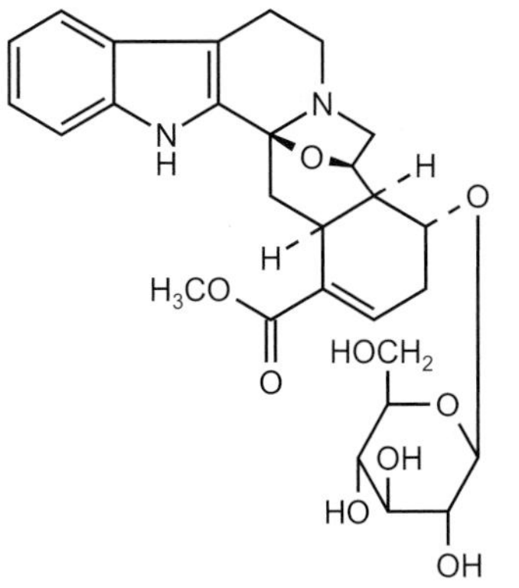

Cadambine

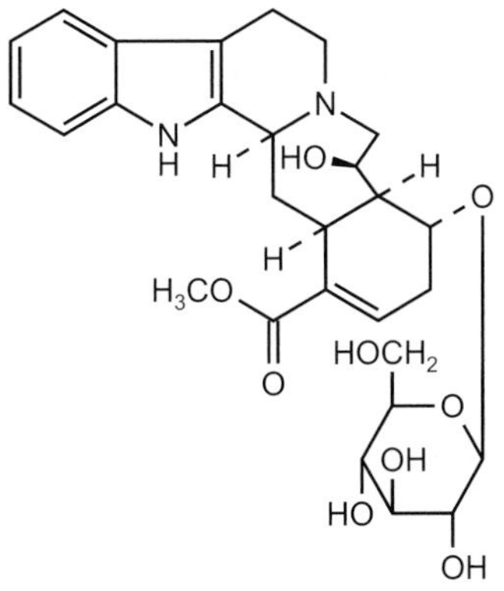

3α-Dihydrocadambine

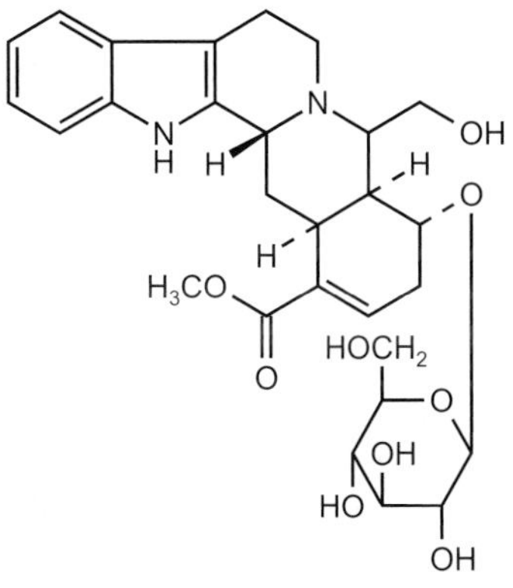

3β-Isodihydrocadambine

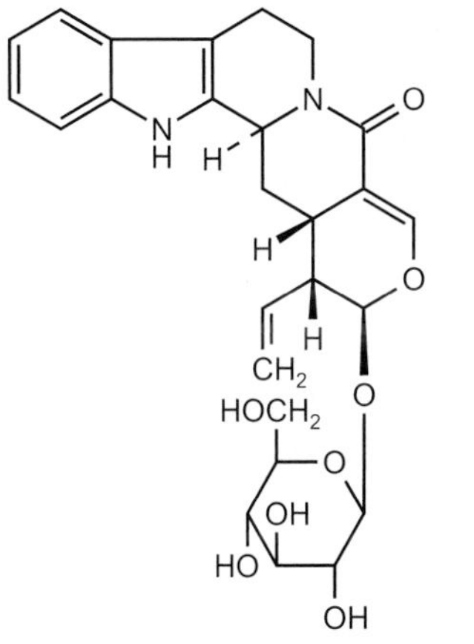

Strictosamide

stroke volume, and a decrease in total peripheral resistance. Hypotensive activity of the total alkaloids from *U. macrophylla* was suggested to arise from decreases in cardiac output and peripheral resistance [8]. An extract of *U. rhynchophylla* relaxed rat aortic ring preparations with and without intact endothelium contracted by norepinephrine (noradrenaline). The rat aortic ring preparation with intact endothelium was significantly more sensitive than that without intact endothelium. It was concluded that the extract of *U. rhynchophylla* relaxed the norepinephrine-precontracted rat aorta through endothelium-dependent and, to lesser extent, through endothelium-independent mechanisms [9].

The hypotensive effect of isorhynchophylline was demonstrated in anesthetized dogs. Isorhynchophylline reduced the mean arterial pressure, heart rate, and coronary blood flow comparable to rhynchophylline [10]. The extract of *U. sinensis* also exhibited a hypotensive effect on normal and renal-type hypertensive rats and spasmolytic and sedative effects in mice [11]. Dihydrocadambine and isodihydrocadambine were also reported to be hypotensive in rats [2]. Intra-arterial administration of hirsutine and hirsuteine into the hind-limb artery of anesthetized dogs caused a vasodilatation. Vasodilatation was also observed in the coronary and cerebral arteries [12]. Similar effects of dihydrocorynantheine in were also reported in rats [13].

Spasmolytic effects of rhynchophylline on the contraction of rabbit aorta *in vitro* [14], and on myocardial contraction in anesthetized dogs and cats [15], were reported. Isorhynchophylline inhibited the contraction of isolated guinea pig atrium in a concentration-dependent manner [16].

Rhynchophylline, corynoxeine and their isomers were further reported to show calcium antagonistic activity, and to inhibit the contraction of rat and rabbit thoracic aorta preparations as induced by calcium ions [17]. The blocking effect of rhynchophylline on calcium channels in isolated rat ventricular myocytes was postulated to be voltage-independent [18]. The i.v. administration of isorhynchophylline to rats produced a negative chronotropic effect, without influencing the blood pressure. This negative chronotropic effect may be related to the blockade of calcium channels, and was not dependent upon interaction with cardiac receptors [19]. In addition, rhynchophylline was also found to block potassium channels in isolated rat or guinea pig ventricular myocytes. The antiarrhythmic action of rhynchophylline was suggested partially to be due to the potassium channel-blocking effect [20].

Hirsutine markedly decreased the enhanced cytosolic calcium levels in the smooth muscle of the isolated rat aorta, as induced by noradrenaline and high potassium concentration. This result indicated that hirsutine inhibited calcium influx mainly through a voltage-dependent calcium channel. A study on the effect of hirsutine on intracellular calcium store revealed that it inhibited calcium release and increased calcium uptake [21]. Hirsutine produced a dose-dependent relaxation of the isolated rat aorta with or without the endothelium contracted by norepinephrine and high potassium concentration. It also inhibited the contractions induced by serotonin and calcium channel activator YC-170, but not by calcium ionophore A23187 [22]. Hirsutine and dihydrocorynantheine were also shown to have direct effects on the action potential of cardiac muscle through inhibition of multiple ion channels, which may explain their negative chronotropic and antiarrhythmic activities [23].

Rhynchophylline inhibited rabbit platelet aggregation induced by arachidonic acid, collagen and ADP, and suppressed malondialdehyde formation in a platelet suspension stimulated by thrombin. However, it did not alter intraplatelet cAMP concentrations. The i.v. administration of rhynchophylline to rats resulted in a significant inhibition of venous and cerebral thrombosis [24]. Rhynchopylline also significantly reduced the number of mice that died due to thrombosis by platelet aggregates [25].

Isorhynchophylline dose-dependently inhibited 5-HT_{2A} receptor-mediated head-twitch, but not 5-HT_{1A} receptor-mediated head-weaving, responses in mice, as evoked by 5-methoxy-*N,N*-dimethyltryptamine. Isocorynoxeine also reduced the head-twitch response in reserpinized mice over the same dose range as isorhynchophylline, whereas both rhynchophylline and corynoxeine (the stereoisomers of isorhynchophylline and isocorynoxeine) did not show any effect on behavioral responses mediated by serotonin receptors. *In vitro* experiments revealed that isorhynchophylline and isocorynoxeine concentration-dependently and competitively inhibited 5-HT-evoked currents in *Xenopus* oocytes expressing the 5-HT_{2A} receptor. This indicates that isorhynchophylline and isocorynoxeine preferentially suppress 5-HT_{2A} receptors in the brain, probably via a competitive antagonistic mechanism, and that the configuration of the oxindole moiety of the alkaloids seemed to be essential for their antagonistic activity [26]. In contrast to serotonin receptors, rhynchophylline and isorhynchophylline were found to reversibly reduce *N*-methyl-D-aspartate (NMDA)-induced current in *Xenopus* oocytes expressing NMDA receptors in a concentration-dependent, but voltage-independent, manner [27].

The oral administration of the extract of *Uncaria* hooks to mice showed significant inhibitory activity against glutamate-induced convulsion, in a dose-dependent manner. It was not effective against the picrotoxine-induced, strychnine-induced, and electroshock convulsions. Geissoschizine methylether, hirsuteine, hirsutine, and isocorynoxeine were identified as the active components. Rhynchophylline was also found to reduce the spontaneous motor activity and to enhance the sedative and hypnotic effects of pentobarbital in mice [1]. The i.p. administration of an extract of *U. rhynchophylla* at doses of 0.5 and 1 g kg^{-1} to rats 30 min prior to an i.p. administration of kainic acid (12 mg kg^{-1}) significantly inhibited the induced epileptic seizure. Treatment with *U. rhynchophylla* in a dose range of 0.25–1 g kg^{-1} significantly decreased kainic acid-induced lipid peroxide levels in the cerebral cortex. The anticonvulsant effect of *U. rhynchophylla* possibly resulted from its suppressive effect on lipid peroxidation in the brain [28]. The extract of *U. rhynchophylla* significantly inhibited the increase of lipid peroxide levels in the ipsilateral cortex of the rat, and induced an early increase of the activity of superoxide dismutase in the mitochondrial fraction. Using electron spin resonance spectroscopy, the extract of *U. rhynchophylla* was shown to exhibit significant dose-dependent scavenging effects on free radicals. It was proposed that the antiepileptic effect of *U. rhynchophylla* might be attributed to its antioxidant activity [29].

Neuroprotective effects of the oxyindole alkaloids corynoxeine, rhynchophylline, isorhynchophylline and isocorynoxeine and the indole alkaloids geissoschizine methylether, hirsuteine and hirsutine on glutamate-induced neuronal death in cultured cerebellar granule cells from rats were reported. Cell viability was significantly increased by treatment with rhynchophylline, isorhynchophylline, isocorynoxeine, hirsuteine or hirsutine at concentrations of 0.1 to 1 m*M*, with isorhynchophylline being the most potent compound. The increased $^{45}Ca^{2+}$ influx into cells induced by glutamate was significantly inhibited by the alkaloids [30]. Clinical trials also showed that a combination preparation with *U. sinensis* as the major component was effective in the treatment of patients with vascular dementia. The preparation was superior in terms of global improvement rating, utility rating and improvement of subjective symptoms, psychiatric symptoms and disturbance in daily living activities, both in a well-controlled study and in a further double-blind study [31].

The phenolic compounds, including epicatechin, catechin, procyanidin B-1, procyanidin B-2, hyperin and caffeic acid, from the hooks and stems of *U. sinensis* were also reported to protect cultured rat cerebellar granule cells against neuronal death, as induced by glutamate. The mechanism of the protective effect seemed to be an inhibition of Ca^{2+} influx [32]. In NT2 neurons, rhynchophylline also exhibited a protective effect against cell apoptosis induced by dopamine [33].

The aqueous extracts of *U. macrophylla* and *U. sinensis*, as well as indole alkaloids from *Uncaria* species including corynoxine, corynoxine B, isorhynchophylline and geissoschizine methylether, were reported to significantly decrease locomotor activity in mice after oral administration. The depression of locomotor activity upon administration of the alkaloids appeared to be mediated by the central dopaminergic system [34]. Hirsuteine at concentrations of 0.3–10 μ*M* was also reported to inhibit dopamine release in rat pheochromocytoma PC12 cells, as evoked by 100 μ*M* nicotine. Dopamine release caused by 60 and 155 m*M* KCl was also inhibited by hirsuteine, but only at concentrations higher than 10 μ*M*. Hirsuteine was suggested to noncompetitively antagonize the nicotine-evoked release of dopamine by blocking ion permeation through nicotinic receptor channel complexes [35].

The incubation of HT29 (epithelial) and RAW264.7 (macrophage) cells with an extract of *U. tomentosa* at a concentration of 100 μg ml^{-1} attenuated apoptosis induced by peroxynitrite; inhibited nitric oxide synthase gene expression and nitrite formation induced by lipopolysaccharide (LPS); and inhibited the activation of NF-κB [36]. *Uncaria tomentosa* was further reported to greatly stimulate interleukin-1 (IL-1) and IL-6 production by rat macrophages in a dose-dependent manner. *Uncaria tomentosa* was also able to enhance IL-1 and IL-6 in macrophages stimulated by LPS [37]. An aqueous extract of *U. tomentosa* was found to be an effective scavenger of the free radical 1-diphenyl-2-picrilhydrazyl (DPPH), and was fully protective against DPPH and UV irradiation-induced cytotoxicity in murine macrophages RAW264.7 cells at a concentration of 10 μg ml^{-1}. The level of tumor necrosis factor-α (TNF-α) in culture media induced by LPS was significantly decreased by adding the aqueous extract. *Uncaria tomentosa* suppressed TNF-α production at concentrations considerably lower than its antioxidant activity. The primary mechanism for the anti-inflammatory actions of *U. tomentosa* appeared to be immunomodulation via the suppression of TNF-α synthesis [38].

The aqueous extract of *U. tomentosa* did not show toxicity *in vitro* in several test systems using Chinese hamster ovary (CHO) cells and the bacterium *Photobacterium phosphoreum* [39]. However, the aqueous extract of *U. tomentosa* was reported to significantly inhibit the proliferation of the human leukemic cell line (HL60) and Epstein-Barr virus (EBV)-transformed human B lymphoma cell line (Raji), while K562 was more resistant to the extract. The suppressive effect of *U. tomentosa* on tumor cell growth appeared to be mediated through an induction of apoptosis, and was demonstrated by characteristic morphological changes, internucleosomal DNA fragmentation after agarose gel electrophoresis, and DNA fragmentation. Both, DNA single and double strand breaks were increased at 24 h after treatment with the extract [40].

The aqueous extract of *U. tomentosa* depleted of indole alkaloids and given orally to female rats, resulted in a significant increase in lymphocyte proliferation. The leukocyte counts from rats treated with the aqueous extract at doses of 40 and 80 mg kg^{-1} for eight weeks, or at a dose of 160 mg kg^{-1} for four weeks, were significantly elevated. The LD_{50} value of a single oral dose of the extract in the rat was greater than 8 g kg^{-1}. Rats treated with *U. tomentosa* extract at daily doses of 10–80 mg kg^{-1} for eight weeks, or at a daily dose of 160 mg kg^{-1} for four weeks, did not show any signs of acute or chronic toxicity. The repair of DNA single- and double-strand breaks at 3 h after 12 Gy whole-body irradiation of rats was also significantly improved in treated rats [41]. The aqueous extract of *U. tomentosa* was also found to suppress leucopenia in rats induced by i.p. injection of doxorubicin, when given 24 h after doxorubicin treatment for 16 consecutive days [42]. In a human study, an aqueous extract of *U. tomentosa* was given daily at 5 mg kg^{-1} for six consecutive weeks to four healthy adult males, without any toxicity but with a significant elevation of the leukocyte count [41].

The pentacyclic oxindole alkaloids from *U. tomentosa* were reported to induce human endothelial cells to release lymphocyte proliferation regulating factor. This was shown to significantly enhance the proliferation of normal human resting or weakly activated B and T lymphocytes. In contrast, the proliferation of normal human lymphoblasts, and of both the human lymphoblastoid B cell line Raji and the human lymphoblastoid T cell line Jurkat, was inhibited significantly by the proliferation-regulating factor. Tetracyclic oxindole alkaloids dose-dependently reduced the activity of pentacyclic oxindole alkaloids on human endothelial cells [43]. Triterpenes from the hooks of *U. rhynchophylla* showed dose-dependent inhibition of phospholipase Cγ1 *in vitro*, with IC_{50} values of 9–45 μ*M*, and inhibited the proliferation of human cancer cells with IC_{50} values of 0.5–6.5 $\mu g\, ml^{-1}$ [5]. The methanol extract from *U. sinensis* was found to show a suppressive effect on umu gene expression of the SOS response in *Salmonella typhimurium* TA1535/pSK1002 against mutagenicity of the food-borne mutagen Trp-P-1 in the presence of an activating system. The active compounds were identified as ursolic acid and its derivatives, methyl ursolate, acetylursolic acid, and methyl acetylursolate [44].

References

1 Shi, J.S., Huang, B., Wu, Q., Ren, R.X., and Xie, X.L. (1993) Effects of rhynchophylline on motor activity of mice and serotonin and dopamine in rat brain. *Acta Pharmacol. Sin.*, **14**, 114–117.

2 Aisaka, K., Hattori, Y., Kihara, T., Ishihara, T., Endo, K., and Hikino, H. (1985) Hypotensive action of 3α-dihydrocadambine, an indole alkaloid glycoside of *Uncaria* hooks. *Planta Med.*, **51**, 424–427.

3 Kanatani, H., Kohda, H., Yamasaki, K., Hotta, I., Nakata, Y., Segawa, T., Yamanaka, E., Aimi, N., and Sakai, S. (1985) The active principles of the branchlet and hook of *Uncaria sinensis* Oliv. examined with a 5-hydroxytryptamine receptor binding assay. *J. Pharm. Pharmacol.*, **37**, 401–404.

4 Lee, J.S., Yang, M.Y., Yeo, H., Kim, J., Lee, H.S., and Ahn, J.S. (1999) Uncarinic acids: phospholipase C1 inhibitors from hooks of *Uncaria rhynchophylla. Bioorg. Med. Chem. Lett.*, **9**, 1429–1432.

5 Lee, J.S., Kim, J., Kim, B.Y., Lee, H.S., Ahn, J. S., and Chang, Y.S. (2000) Inhibition of phospholipase Cγ1 and cancer cell proliferation by triterpene esters from *Uncaria rhynchophylla. J. Nat. Prod.*, **63**, 753–756.

6 Shi, J.S., Yu, J.X., Chen, X.P., and Xu, R.X. (2003) Pharmacological actions of *Uncaria* alkaloids, rhynchophylline and isorhynchophylline. *Acta Pharmacol. Sin.*, **24**, 97–101.

7 Kuramochi, T., Chu, J., and Suga, T. (1994) Gou-teng (from *Uncaria rhynchophylla* Miquel)-induced endothelium-dependent and -independent relaxations in the isolated rat aorta. *Life Sci.*, **54**, 2061–2069.

8 Liu, G.X., Huang, X.N., and Peng, Y. (1983) Hemodynamic effects of the total alkaloids of *Uncaria macrophylla* in anesthetized dogs. *Acta Pharmacol. Sin.*, **4**, 114–116.

9 Kuramochi, T., Chu, J., and Suga, T. (1994) Gou-teng (from *Uncaria rhynchophylla* Miquel)-induced endothelium-dependent and -independent relaxations in the isolated rat aorta. *Life Sci.*, **54**, 2061–2069.

10 Shi, J.S., Liu, G.X., Wu, Q., Huang, Y.P. and Zhang, X.D. (1992) Effects of rhynchophylline and isorhynchophylline on blood pressure and blood flow of organs in anesthetized dogs. *Acta Pharmacol. Sin.*, **13**, 35–38.

11 Qin, C.L., Liu, J.Y., Cheng, Z.M., and Jiao, Y. (1994) Experimental studies on *Uncaria sinensis* (Oliv.) Havil and *Achyranthes bidentata* Blume and their compatibility. *China J. Chin. Mater. Med.*, **19**, 371–373.

12 Ozaki, Y. (1990) Vasodilative effects of indole alkaloids obtained from domestic plants, *Uncaria rhynchophylla* Miq. and *Amsonia elliptica* Roem. et Schult. *Nippon Yakurigaku Zasshi*, **95**, 47–54.

13 Chang, P., Koh, Y.K., Geh, S.L., Soepadmo, E., Goh, S.H., and Wong, A.K. (1989) Cardiovascular effects in the rat of dihydrocorynantheine isolated from *Uncaria callophylla. J. Ethnopharmacol.*, **25**, 213–215.

14 Zhang, W., Liu, G.X., and Huang, X.N. (1987) Effect of rhynchophylline on the contraction

of rabbit aorta. *Acta Pharmacol. Sin.*, **8**, 425–429.

15 Zhang, W., and Liu, G.X. (1986) Effects of rhynchophylline on myocardial contractility in anesthetized dogs and cats. *Acta Pharmacol. Sin.*, **7**, 426–428.

16 Zhu, Y., Huang, X.N., and Liu, G.X. (1995) Effects of isorhynchophylline on the physiological characteristics of isolated guinea pig atrium. *China J. Chin. Mater. Med.*, **20**, 112–114.

17 Yamahara, J., Miki, S., Matsuda, H., Kobayashi, G., and Fujimura, H. (1987) Screening for calcium antagonists in natural products. The active principles of uncariae ramulus et uncus. *Yakurigaku Zasshi*, **90**, 133–140.

18 Wang, X.L., Zhang, L.M., and Hua, Z. (1994) Blocking effect of rhynchophylline on calcium channels in isolated rat ventricular myocytes. *Acta Pharmacol. Sin.*, **15**, 115–118.

19 Zhu, Y. (1993) The negative chronotropic effect of isorhynchophylline and its mechanism. *China J. Chin. Mater. Med.*, **18**, 745–747.

20 Wang, X.L., Zhang, L.M., and Hua, Z. (1994) Effect of rhyncophylline on potassium channels in isolated rat or guinea pig ventricular myocytes. *Acta Pharm. Sin.*, **29**, 9–14.

21 Horie, S., Yano, S., Aimi, N., Sakai, S., and Watanabe, K. (1992) Effects of hirsutine, an antihypertensive indole alkaloid from *Uncaria rhynchophylla*, on intracellular calcium in rat thoracic aorta. *Life Sci.*, **50**, 491–498.

22 Yano, S., Horiuchi, H., Horie, S., Aimi, N., Sakai, S., and Watanabe, K. (1991) Ca^{2+} channel blocking effects of hirsutine, an indole alkaloid from *Uncaria* genus, in the isolated rat aorta. *Planta Med.*, **57**, 403–405.

23 Masumiya, H., Saitoh, T., Tanaka, Y., Horie, S., Aimi, N., Takayama, H., Tanaka, H., and Shigenobu, K. (1999) Effects of hirsutine and dihydrocorynantheine on the action potentials of sino-atrial node, atrium and ventricle. *Life Sci.*, **65**, 2333–2341.

24 Chen, C.X., Jin, R.M., Li, Y.K., Zhong, J., Yue, L., Chen, S.C., and Zhou, J.Y. (1992) Inhibitory effects of rhynchophylline on platelet aggregation and thrombosis. *Acta Pharmacol. Sin.*, **13**, 126–130.

25 Jin, R.M., Chen, C.X., Li, Y.K., and Xu, P.K. (1991) Effect of rhyncophylline on platelet aggregation and experimental thrombosis. *Acta Pharm. Sin.*, **26**, 246–249.

26 Matsumoto, K., Morishige, R., Murakami, Y., Tohda, M., Takayama, H., Sakakibara, I., and Watanabe, H. (2005) Suppressive effects of isorhynchophylline on $5\text{-}HT_{2A}$ receptor function in the brain: behavioral and electrophysiological studies. *Eur. J. Pharmacol.*, **517**, 191–199.

27 Kang, T.H., Murakami, Y., Matsumoto, K., Takayama, H., Kitajima, M., Aimi, N., and Watanabe, H. (2002) Rhynchophylline and isorhynchophylline inhibit NMDA receptors expressed in *Xenopus* oocytes. *Eur. J. Pharmacol.*, **455**, 27–34.

28 Hsieh, C.L., Tang, N.Y., Chiang, S.Y., Hsieh, C.T., and Lin, J.G. (1999) Anticonvulsive and free radical scavenging actions of two herbs, *Uncaria rhynchophylla* (MIQ) Jack and *Gastrodia elata* Bl, in kainic acid-treated rats. *Life Sci.*, **65**, 2071–2082.

29 Watano, T., Nakazawa, K., Obama, T., Mori, M., Inoue, K., Fujimori, K., and Takanaka, A. (1993) Non-competitive antagonism by hirsuteine of nicotinic receptor-mediated dopamine release from rat pheochromocytoma cells. *Jpn. J. Pharmacol.*, **61**, 351–356.

30 Shimada, Y., Goto, H., Itoh, T., Sakakibara, I., Kubo, M., Sasaki, H., and Terasawa, K. (1999) Evaluation of the protective effects of alkaloids isolated from the hooks and stems of *Uncaria sinensis* on glutamate-induced neuronal death in cultured cerebellar granule cells from rats. *J. Pharm. Pharmacol.*, **51**, 715–722.

31 Itoh, T., Shimada, Y., and Terasawa, K. (1999) Efficacy of Choto-san on vascular dementia and the protective effect of the hooks and stems of *Uncaria sinensis* on glutamate-induced neuronal death. *Mech. Ageing Dev.*, **111**, 155–173.

32 Shimada, Y., Goto, H., Kogure, T., Shibahara, N., Sakakibara, I., Sasaki, H., and Terasawa, K. (2001) Protective effect of phenolic compounds isolated from the hooks and stems of *Uncaria sinensis* on glutamate-induced neuronal death. *Am. J. Chin. Med.*, **29**, 173–180.

33 Shi, J.S. and Kenneth, H.G. (2002) Effect of rhynchophylline on apoptosis induced by dopamine in NT2 cells. *Acta Pharmacol. Sin.*, **23**, 445–449.

34 Sakakibara, I., Terabayashi, S., Kubo, M., Higuchi, M., Komatsu, Y., Okada, M., Taki, K., and Kamei, J. (1999) Effect on locomotion of indole alkaloids from the hooks of *Uncaria* plants. *Phytomedicine*, **6**, 163–168.

35 Liu, J. and Mori, A. (1992) Antioxidant and free radical scavenging activities of *Gastrodia elata* Bl. and *Uncaria rhynchophylla* (Miq.) Jacks. *Neuropharmacology*, **31**, 1287–1298.

36 Sandoval-Chacon, M., Thompson, J.H., Zhang, X.J., Liu, X., Mannick, E.E., Sadowska-Krowicka, H., Charbonnet, R.M., Clark, D.A., and Miller, M.J. (1998) Antiinflammatory actions of cat's claw: the role of NF-kappaB. *Aliment. Pharmacol. Ther.*, **12**, 1279–1289.

37 Lemaire, I., Assinewe, V., Cano, P., Awang, D.V., and Arnason, J.T. (1999) Stimulation of interleukin-1 and -6 production in alveolar macrophages by the neotropical liana, *Uncaria tomentosa*. *J. Ethnopharmacol.*, **64**, 109–115.

38 Sandoval, M., Charbonnet, R.M., Okuhama, N.N., Roberts, J., Krenova, Z., Trentacosti, A.M., and Miller, M.J. (2000) Cat's claw inhibits TNFα production and scavenges free radicals: role in cytoprotection. *Free Radic. Biol. Med.*, **29**, 71–78.

39 Santa Maria, A., Lopez, A., Diaz, M.M., Alban, J., Galan de Mera, A., Vicente Orellana, J.A., and Pozuelo, J.M. (1997) Evaluation of the toxicity of *Uncaria tomentosa* by bioassays *in vitro*. *J. Ethnopharmacol.*, **57**, 183–187.

40 Sheng, Y., Pero, R.W., Amiri, A., and Bryngelsson, C. (1998) Induction of apoptosis and inhibition of proliferation in human tumor cells treated with extracts of *Uncaria tomentosa*. *Anticancer Res.*, **18A**, 3363–3368.

41 Sheng, Y., Bryngelsson, C., and Pero, R.W. (2000) Enhanced DNA repair, immune function and reduced toxicity of C-MED-100, a novel aqueous extract from *Uncaria tomentosa*. *J. Ethnopharmacol.*, **69**, 115–126.

42 Sheng, Y., Pero, R.W., and Wagner, H. (2000) Treatment of chemotherapy-induced leukopenia in a rat model with aqueous extract from *Uncaria tomentosa*. *Phytomedicine*, **7**, 137–143.

43 Wurm, M., Kacani, L., Laus, G., Keplinger, K., and Dierich, M.P. (1998) Pentacyclic oxindole alkaloids from *Uncaria tomentosa* induce human endothelial cells to release a lymphocyte-proliferation-regulating factor. *Planta Med.*, **64**, 701–704.

44 Miyazawa, M., Okuno, Y., and Imanishi, K. (2005) Suppression of the SOS-inducing activity of mutagenic heterocyclic amine, Trp-P-1, by triterpenoid from *Uncaria sinensis* in the *Salmonella typhimurium* TA1535/pSK1002 umu test. *J. Agric. Food Chem.*, **53**, 2312–2315.

Vaccaria segetalis

Semen Vaccariae (Wangbuliuxing) is the dried ripe seed of *Vaccaria segetalis* (Neck.) Garcke (Caryophyllaceae). It is used for the treatment of menstrual disorders, and as a lactigenous agent. The Chinese Pharmacopoeia notes Semen Vaccariae to be used with caution in pregnancy.

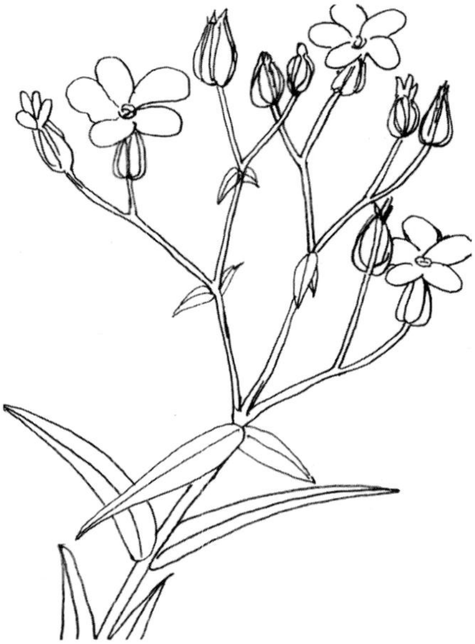

Vaccaria segetalis (Neck.) Garcke.

Chemistry

The seeds of *V. segetalis* were reported to contain triterpenes and triterpene saponins as well as cyclic peptides. Triterpene sapogenins were identified as segetalic acid, vaccaric acid, quillaic acid, gypsogenin [1], gypsogenic acid, and 3,4-secogypsogenic acid [2].

Saponins from the seeds were identified as vaccarosides A–H [1, 2], vaccaroid B [3], dianoside G, segetoside C [4], and segetosides G, H, and I [5].

Cyclic peptides from the seeds of *V. segitalis* were identified as cyclic pentapeptides segetalins A, B (cyclo(-L-Ala-Gly-L-Val-L-Ala-L-Try-) [6], G (cyclo(-Gly-Val-Lys-Tyr-Ala-)) and H (cyclo(-Gly-Tyr-Arg-Phe-Ser-)) [7], and cyclic heptapeptide (cyclo(-Gly-Tyr-Val-Pro-Leu-Trp-Pro-)) [8],

Pharmacology and Toxicology

The cyclic pentapeptides segetalins A, B, G and H were shown to have an estrogen-like activity [9]. The segetalins were also reported to exhibit a vasorelaxant activity against contractions of rat aorta, as induced by norepinephrine (noradrenaline) [10].

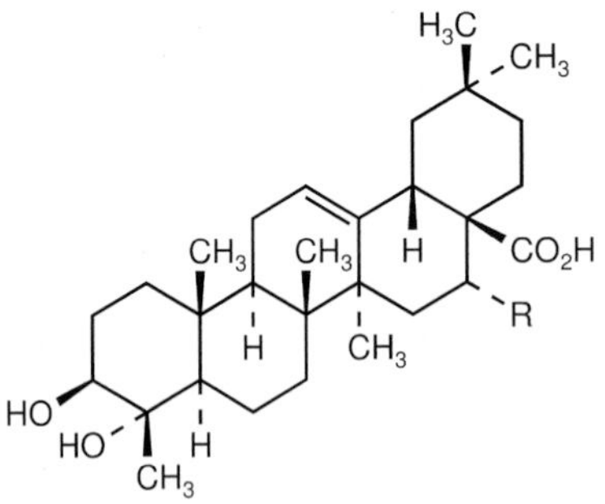

Segetalic acid: R = OH
Vaccaric acid: R = H

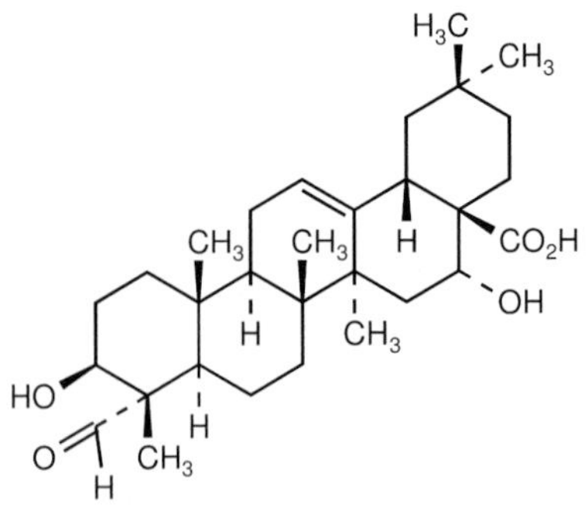

Quillaic acid

Handbook of Chinese Medicinal Plants: Chemistry, Pharmacology, Toxicology
Weici Tang and Gerhard Eisenbrand

ISBN: 978-3-527-32226-8

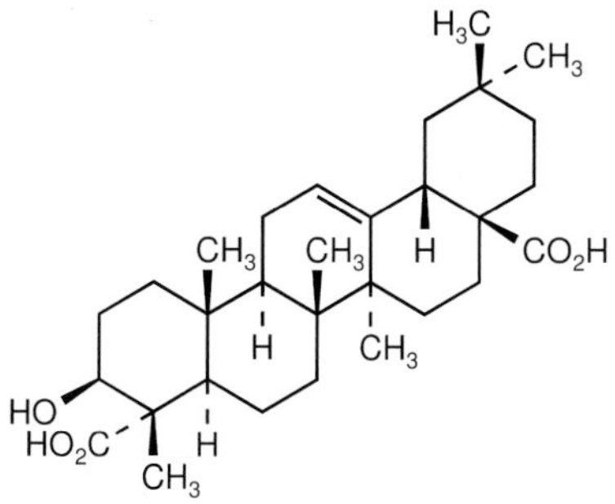

Gypsogenic acid

References

1 Jia, Z., Koike, K., Kudo, M., Li, H., and Nikaido, T. (1998) Triterpenoid saponins and sapogenins from *Vaccaria segetalis. Phytochemistry*, **48**, 529–536.

2 Koike, K., Jia, Z., and Nikaido, T. (1998) Triterpenoid saponins from *Vaccaria segetalis. Phytochemistry*, **47**, 1343–1349.

3 Yun, Y.S., Shimizu, K., Morita, H., Takeya, K., Itokawa, H., and Shirota, O. (1998) Triterpenoid saponin from *Vaccaria segetalis. Phytochemistry*, **47**, 143–144.

4 Sang, S., Lao, A., Wang, H., Chen, Z., Uzawa, J., and Fujimoto, Y. (1999) Triterpenoid saponins from *Vaccaria segetalis. J. Asian Nat. Prod. Res.*, **1**, 199–205.

5 Sang, S.M., Lao, A.N., Chen, Z.L., Uzawa, J., and Fujimoto, Y. (2000) Three new triterpenoid saponins from the seeds of *Vaccaria segetalis. J. Asian Nat. Prod. Res.*, **2**, 187–193.

6 Itokawa, H., Yun, Y., Morita, H., Takeya, K., and Yamada, K. (1995) Estrogen-like activity of cyclic peptides from *Vaccaria segetalis* extracts. *Planta Med.*, **61**, 561–562.

7 Morita, H., Yun, Y.S., Takeya, K., Itokawa, H., and Shirota, O. (1996) A cyclic heptapeptide from *Vaccaria segetalis. Phytochemistry*, **42**, 439–441.

8 Yun, Y.S., Morita, H., Takeya, K., and Itokawa, H. (1997) Cyclic peptides from higher plants. 34. Segetalins G and H, structures and estrogen-like activity of cyclic pentapeptides from *Vaccaria segetalis. J. Nat. Prod.*, **60**, 216–218.

9 Morita, H., Yun, Y.S., Takeya, K., and Itokawa, H. (1997) Conformational preference for segetalins G and H, cyclic peptides with estrogen-like activity from seeds of *Vaccaria segetalis. Bioorg. Med. Chem.*, **5**, 2063–2067.

10 Morita, H., Eda, M., Iizuka, T., Hirasawa, Y., Sekiguchi, M., Yun, Y.S., Itokawa, H., and Takeya, K. (2006) Structure of a new cyclic nonapeptide, segetalin F, and vasorelaxant activity of segetalins from *Vaccaria segetalis. Bioorg. Med. Chem. Lett.*, **16**, 4458–4461.

Verbena officinalis

Herba Verbenae (Mabiancao) is the dried aerial part of *Verbena officinalis* L. (Verbenaceae). It is used as a hemostatic, antimalarial, and diuretic agent.

Verbena officinalis L.

Chemistry

The aerial part of *V. officinalis* was reported to contain iridoid glycosides verbenalin (cornin) and aucubin (hastatoside) [1].

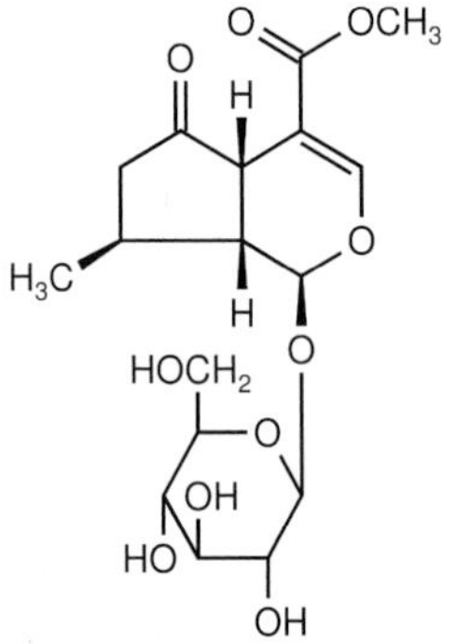

Verbenalin

Flavones from the aerial part of *V. officinalis* were identified as artemetin, nepetin, pedalitin, sorbifolin; and flavone glycoside as luteolin 7-diglucuronide [2]. Triterpenes and phytosterol from *V. officinalis* were identified as oleanolic acid, lupeol, ursolic acid, and β-sitosterol [2]. The Chinese Pharmacopoeia requires a qualitative determination of ursolic acid in Herba Verbenae by thin-layer chromatographic (TLC) comparison with reference substance, and a quantitative determination by TLC-densitometry. The content of ursolic acid in Herba Verbenae should be not less than 0.36%.

Phenylethanoid glycosides from *V. officinalis* were identified as verbascoside and eukovoside [3], with a verbascoside content of 0.8% [4].

Pharmacology and Toxicology

The ethanol extract of *V. officinalis* was reported to exhibit anti-inflammatory activity against carrageenan-induced paw edema in mice [5]. The anti-inflammatory and analgesic effects of the extract of *V. officinalis* were also demonstrated in mice by topical application [6]. Verbenalin showed significant anti-inflammatory activity against mouse paw edema induced by carrageenan, and against the mouse ear edema induced by 12-*O*-tetradecanoylphorbol 13-acetate (TPA) [7]. The antitussive action of *V. officinalis* was found to be related to the content of verbenalin and β-sitosterol [2]. In a study on the interaction between *Verbena* constituents and prostaglandin, synergistic effects were reported on the contraction of isolated rat uterus preparations with prostaglandin E_2 (PGE_2) and additive effects with PGF_2. The extract of *V. officinalis* was found to effectively prevent and to treat urolithiasis in rats [8].

Handbook of Chinese Medicinal Plants: Chemistry, Pharmacology, Toxicology
Weici Tang and Gerhard Eisenbrand

ISBN: 978-3-527-32226-8

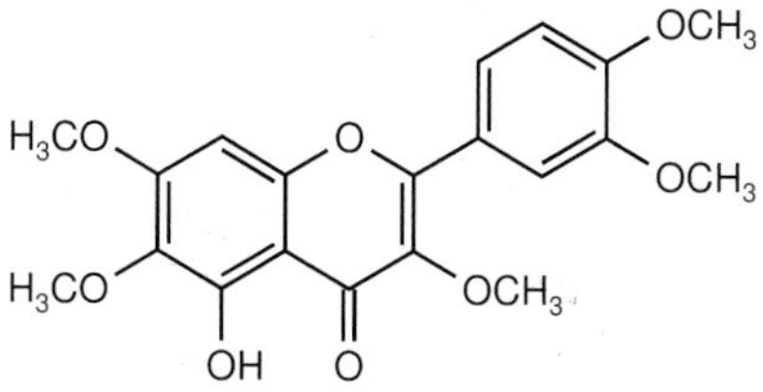

Artemetin

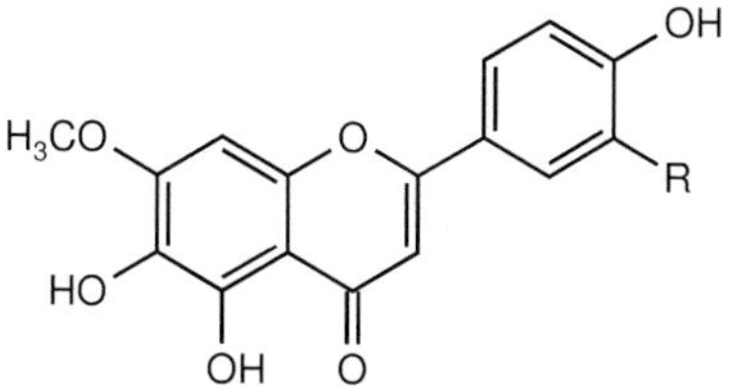

Pedalitin: R = OH
Sorbifolin: R = H

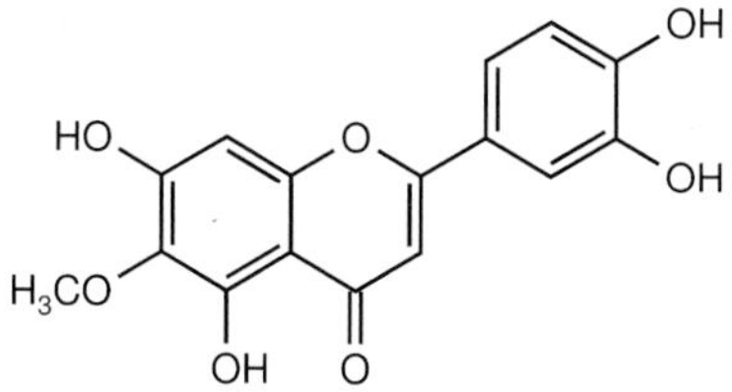

Nepetin

Verbascoside was found to show hepatoprotective activity *in vitro.* It significantly inhibited lipid peroxidation in rat liver microsomes induced by NADPH and carbon tetrachloride (CCl_4); efficiently prevented cell damage induced by CCl_4 or D-galactosamine; and exhibited hepatoprotective activity against CCl_4 in rats [9]. The scavenging activity of verbascoside on superoxide anion and hydroxyl radicals was demonstrated using the spin trapping method [10]. The repair activity of verbascoside towards hydroxyl radical oxidized deoxyguanosine monophosphate (dGMP) was demonstrated with pulse radiolysis [11]. The antioxidative effect of verbascoside was also demonstrated with linoleic acid in cetyl trimethylammonium bromide micelles [12]. The radical-scavenging action of verbascoside was postulated to occur both at the stage of superoxide anion formation and at the stage of lipid peroxide production [13].

Heroin was shown to induce reactive oxygen species (ROS) formation in several cells, to decrease enzyme activity of the antioxidant defense system including

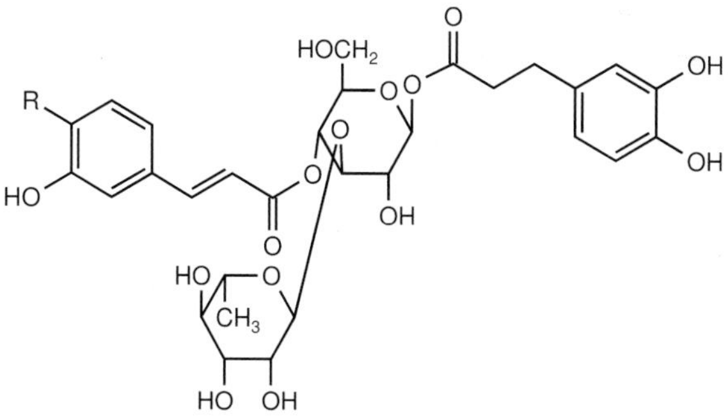

Verbascoside: R = OH
Eukovoside: R = OCH_3

superoxide dismutase (SOD), catalase and glutathione peroxidase, and to deplete antioxidants, glutathione (GSH), Se, and vitamins. Mice administered intraperitoneally with heroin showed the indices of oxidative damage, such as significantly increased contents of 8-hydroxy-2′-deoxyguanosine (8-OHdG), protein carbonyl groups and malondialdehyde (MDA) in the brains, markedly decreased ratio of reduced to oxidized GSH and decreased total antioxidant capacity in the serum. After treatment of heroin-dependent mice with verbascoside, the oxidative stress was found attenuated [14].

Protective effects of verbascoside against the Fenton reaction on plasmid pBR322 DNA have been studied. The pBR322 plasmid DNA was damaged by hydroxyl radicals generated from the Fenton reaction with H_2O_2 and Fe^{3+} or Fe^{2+} ions, as characterized by a diminution of supercoiled DNA forms or by an increase of relaxed or linear DNA forms after oxidative attack. Verbascoside was found to form complexes with Fe^{3+} or Fe^{2+} ions and to inhibit the Fenton reaction [15]. Verbascoside was also reported to exert photoprotective activity against UV-B induced cellular death in *Escherichia coli* [16].

Verbascoside was further reported to exhibit neuroprotective effects on apoptosis and oxidative stress in PC12 neuronal cells induced by 1-methyl-4-phenylpyridinium ion (MPP^+). Simultaneous treatment of PC12 cells with MPP^+ and verbascoside markedly reduced MPP^+-induced apoptotic death, increased extracellular hydrogen peroxide level, the activation of caspase-3, and the collapse of mitochondrial membrane potential. This indicated that verbascoside might be useful for the treatment of oxidative stress-induced neurodegenerative disease [17].

Verbascoside induced cell death in promyelocytic leukemia HL-60 cells with an IC_{50} of 27 μ*M*. The mechanism of the cytotoxic effect was the induction of apoptosis [18]. In addition, verbascoside was reported to exhibit a growth-inhibitory activity against human gastric adenocarcinoma MGc80-3 cells inoculated in nude mice. It was postulated that verbascoside might reverse malignant phenotypic characteristics and induce the differentiation of MGc80-3 cells [19]. The cytotoxic activity of verbascoside was also demonstrated by the inhibition of murine P 388 lymphocytic leukemia [20] and of a series of other tumor cell lines *in vitro* [21]. Verbascoside interacted with the catalytic domain of protein kinase C (PKC) from rat brain, acting a competitive inhibitor with respect to ATP and as a noncompetitive inhibitor with respect to the phosphate acceptor. Therefore, the cytotoxic activity of verbascoside was further postulated to be due, at least in part, to an inhibition of PKC [22].

The treatment of MKN45 human gastric carcinoma cells with verbascoside resulted in a significant inhibition of telomerase activity in the cells, but not in the cellular supernatant. The average telomere length became remarkably short. The cell number in sub-G_0/G_1 and G_2/M phases was increased, which suggested that verbascoside-mediated cell differentiation and apoptosis may be affected by a telomere–telomerase cell-cycle-dependent modulation. The cytotoxic mechanism of verbascoside was demonstrated to be due to its inhibiting effect on telomerase activity in tumor cells [23].

The oral administration of verbascoside to rats significantly inhibited the elevation of protein excretion in urine of rat with crescentic-type, anti-glomerular basement membrane nephritis. This effect indicated that verbascoside may be a useful compound against rapidly progressive glomerulonephritis characterized by severe glomerular lesions with diffuse crescents. The cholesterol and creatinine levels and antibody production against rabbit α-globulin in the plasma of rats treated with verbascoside were lower than those of the nephritic controls [24]. Verbascoside markedly inhibited leukocyte accumulation in the glomeruli of nephritic rats [25]. It also markedly inhibited the overexpression of intercellular adhesion molecule-1 (ICAM-1) in nephritic glomeruli *in vivo* [26]. The inhibition of leukocyte accumulation in the glomeruli and inhibition of the upregulation of ICAM-1 in nephritic glomeruli may be the potential mechanisms of the antinephritic effect of verbascoside.

References

1 Makboul, A.M. (1986) Chemical constituents of *Verbena officinalis. Fitoterapia*, **57**, 50–51.

2 Kui, C.H. and Tang, R.J. (1985) Studies on the antitussive constituents of *Verbena officinalis. Bull. Chin. Mater. Med.*, **10**, 467.

3 Bianco, A., Guiso, M., and Passacantilli, P. (1984) Iridoid and phenylpropanoid glycosides from new sources. *J. Nat. Prod.*, **47**, 901–902.

4 Haensel, R., and Kallmann, S. (1986) Verbascoside, a main constituent of *Verbena officinalis. Arch. Pharm.*, **319**, 227–230.

5 Deepak, M. and Handa, S.S. (2000) Antiinflammatory activity and chemical composition of extracts of *Verbena officinalis. Phytother. Res.*, **14**, 463–465.

6 Calvo, M.I. (2006) Anti-inflammatory and analgesic activity of the topical preparation of *Verbena officinalis* L. *J. Ethnopharmacol.*, **107**, 380–382.

7 Recio, M.C., Giner, R.M., Manez, S., and Rios, J.L. (1994) Structural considerations on the iridoids as anti-inflammatory agents. *Planta Med.*, **60**, 232–234.

8 Grases, F., Melero, G., Costa-Bauza, A., Prieto, R., and March, J.G. (1994) Urolithiasis and phytotherapy. *Int. Urol. Nephrol.*, **26**, 507–511.

9 Xiong, Q., Hase, K., Tezuka, Y., Tani, T., Namba, T., and Kadota, S. (1998) Hepatoprotective activity of phenylethanoids from *Cistanche deserticola. Planta Med.*, **64**, 120–125.

10 Wang, P., Kang, J., Zheng, R., Yang, Z., Lu, J., Gao, J., and Jia, Z. (1996) Scavenging effects of phenylpropanoid glycosides from *Pedicularis* on superoxide anion and hydroxyl radical by the spin trapping method. *Biochem. Pharmacol.*, **51**, 687–691.

11 Li, W., Zheng, R., Su, B., Jia, Z., Li, H., Jiang, Y., Yao, S., and Lin, N. (1996) Repair of dGMP hydroxyl radical adducts by verbascoside via electron transfer: a pulse radiolysis study. *Int. J. Radiat. Biol.*, **69**, 481–485.

12 Zheng, R.L., Wang, P.F., Li, J., Liu, Z.M., and Jia, Z.J. (1993) Inhibition of the autoxidation of linoleic acid by phenylpropanoid glycosides from *Pedicularis* in micelles. *Chem. Phys. Lipids*, **65**, 151–154.

13 Zhou, Y.C. and Zheng, R.L. (1991) Phenolic compounds and an analog as superoxide anion scavengers and antioxidants. *Biochem. Pharmacol.*, **42**, 1177–1179.

14 Qiusheng, Z., Yuntao, Z., Zheng, R.L., Rongliang, Z., and Changling, L. (2005) Effects of verbascoside and luteolin on oxidative damage in brain of heroin treated mice. *Pharmazie*, **60**, 539–543.

15 Zhao, C., Dodin, G., Yuan, C., Chen, H., Zheng, R., Jia, Z., and Fan, B.T. (2005) "*In vitro*" protection of DNA from Fenton reaction by plant polyphenol verbascoside. *Biochim. Biophys. Acta*, **1723**, 114–123.

16 Avila Acevedo, J.G., Castaneda, C.M., Benitez, F.J., Duran, D.A., Barroso, V.R., Martinez, C. G., Munoz, L.J., Martinez, C.A., and Romo de Vivar, A. (2005) Photoprotective activity of *Buddleja scordioides. Fitoterapia*, **76**, 301–309.

17 Sheng, G.Q., Zhang, J.R., Pu, X.P., Ma, J., and Li, C.L. (2002) Protective effect of verbascoside on 1-methyl-4-phenylpyridinium ion-induced neurotoxicity in PC12 cells. *Eur. J. Pharmacol.*, **451**, 119–124.

18 Inoue, M., Sakuma, Z., Ogihara, Y., and Saracoglu, I. (1998) Induction of apoptotic cell death in HL-60 cells by acteoside, a phenylpropanoid glycoside. *Biol. Pharm. Bull.*, **21**, 81–83.

19 Li, J., Zheng, Y., Zhou, H., Su, B., and Zheng, R. (1997) Differentiation of human gastric adenocarcinoma cell line MGc80-3 induced by verbascoside. *Planta Med.*, **63**, 499–502.

20 Pettit, G.R., Numata, A., Takemura, T., Ode, R. H., Narula, A.S., Schmidt, J.M., Cragg, G.M., and Pase, C.P. (1990) Antineoplastic agents, 107. Isolation of acteoside and isoacteoside from *Castilleja linariaefolia. J. Nat. Prod.*, **53**, 456–458.

21 Saracoglu, I., Inoue, M., Calis, I., and Ogihara, Y. (1995) Studies on constituents with cytotoxic and cytostatic activity of two Turkish medicinal plants *Phlomis armeniaca* and *Scutellaria salviifolia*. *Biol. Pharm. Bull.*, **18**, 1396–1400.

22 Herbert, J.M., Maffrand, J.P., Taoubi, K., Augereau, J.M., Fouraste, I., and Gleye, J. (1991) Verbascoside isolated from *Lantana camara*, an inhibitor of protein kinase C. *J. Nat. Prod.*, **54**, 1595–1600.

23 Zhang, F., Jia, Z., Deng, Z., Wei, Y., Zheng, R., and Yu, L. (2002) *In vitro* modulation of telomerase activity, telomere length and cell cycle in MKN45 cells by verbascoside. *Planta Med.*, **68**, 115–118.

24 Hayashi, K., Nagamatsu, T., Ito, M., Hattori, T., and Suzuki, Y. (1994) Acteoside, a component of *Stachys sieboldii* MIQ, may be a promising antinephritic agent: effect of acteoside on crescentic-type anti-GBM nephritis in rats. *Jpn. J. Pharmacol.*, **65**, 143–151.

25 Hayashi, K., Nagamatsu, T., Ito, M., Hattori, T., and Suzuki, Y. (1994) Acteoside, a component of *Stachys sieboldii* MIQ, may be a promising antinephritic agent (2): Effect of acteoside on leukocyte accumulation in the glomeruli of nephritic rats. *Jpn. J. Pharmacol.*, **66**, 47–52.

26 Hayashi, K., Nagamatsu, T., Ito, M., Yagita, H., and Suzuki, Y. (1996) Acteoside, a component of *Stachys sieboldii* MIQ, may be a promising antinephritic agent (3): effect of acteoside on expression of intercellular adhesion molecule-1 in experimental nephritic glomeruli in rats and cultured endothelial cells. *Jpn. J. Pharmacol.*, **70**, 157–168.

Viscum coloratum

Herba Visci (Hujisheng) is the dried stem and leaf of *Viscum coloratum* (Komar.) Nakai (Loranthaceae). It is used for the treatment of rheumatic pain.

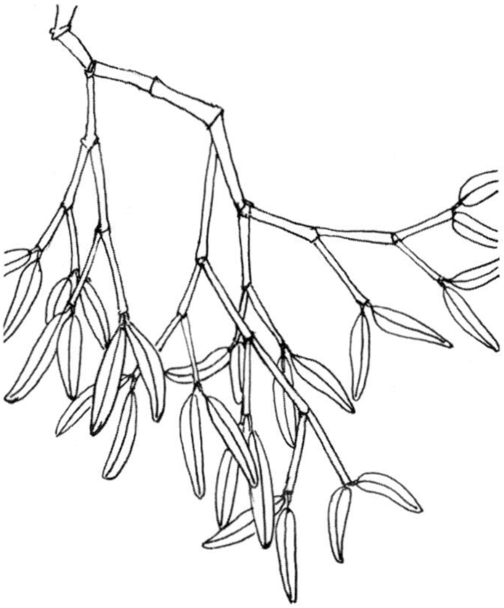

Viscum coloratum (Komar.) Nakai

Chemistry

The stem and leaf of *V. coloratum* is reported to contain flavones and flavone glycosides. Several flavone glycosides were designated as viscumneosides I, II [1], III [2], IV [3], V, VI [4], and VII [5]. Viscumneosides II, IV and VII are flavone glycosides with rhamnazin as aglycone, while viscumneosides I, III, V and VI are flavanone glycosides with homoeriodictyol as aglycone. Recently, a series of further flavanone glycosides were isolated from *V. coloratum*, for example, (2*S*)-7,4′-dihydroxy-5,3′-dimethoxyflavanone [6], (2*S*)-homoeriodictyol 7,4′-di-*O*-β-D-glucopyranoside, (2*R*)-eriodictyol 7,4′-di-*O*-β-D-glucopyranoside, (2*S*)-homoeriodictyol 7-*O*-β-D-glucopyranoside, (2*S*)-eriodictyol 7-*O*-β-D-glucopyranoside (2), and (2S)-naringenin 7-*O*-β-D-glucopyranoside [7]. In addition, a 1,3-diphenylpropane derivative, viscolin, has been isolated and structurally elucidated as 4′,4″-dihydroxy-2′,3′,6′,3″-tetramethoxy-1,3-diphenylpropane [6]. The total synthesis of viscolin was also described [8].

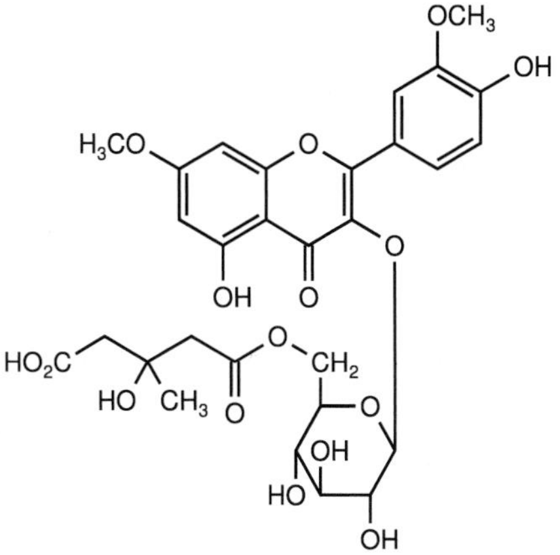

Viscumneoside IV

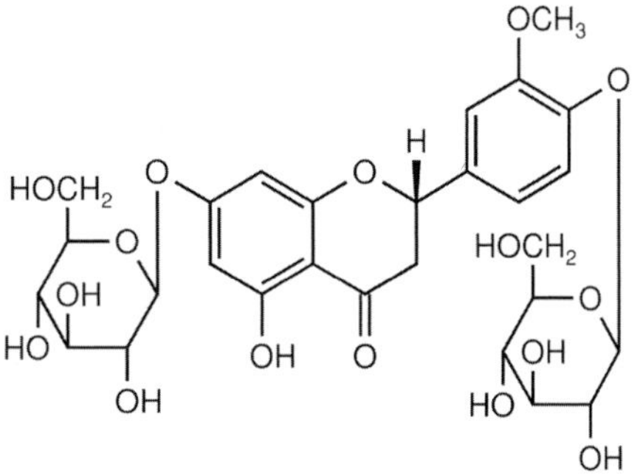

(2*S*)-Homoeriodictyol 7,4'-di-*O*-β-D-glucopyranoside

Handbook of Chinese Medicinal Plants: Chemistry, Pharmacology, Toxicology
Weici Tang and Gerhard Eisenbrand

ISBN: 978-3-527-32226-8

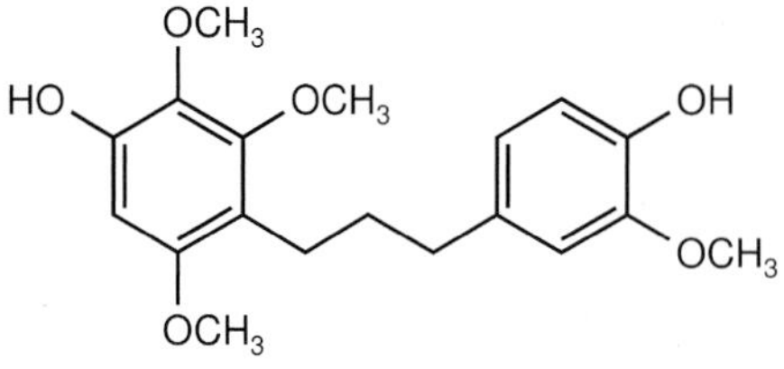

Viscolin

Other constituents from *V. coloratum* were identified as 2-β-D-glucosyl-3-methylpropanol, syringin, eleutheroside E, syringenin 4′-*O*-D-apiosylglucoside [9], β-amyrin, β-acetylamyrin, lupeol, oleanolic acid, β-sitosterol, betulinic acid, daucosterol [10], and *epi*-oleanolic acid [11]. The Chinese Pharmacopoeia requires a qualitative determination of oleanolic acid in Herba Visci by thin-layer chromatographic (TLC) comparison with a reference substance; and a quantitative determination by TLC-densitometry. The content of oleanolic acid in Herba Visci should be not less than 0.17%.

Herba Visci is known to be rich in lectins. The lectins from *V. coloratum* show a high degree of sequence and secondary structure homology with those from *V. album*, which have been used in European countries as biological response modifiers for the adjuvant therapy of cancer. The lectins from *V. album* are classified as ribosome-inactivating proteins (RIPs). The glycoprotein heterodimers contain the A chain with catalytic activity, and the B chain with sugar-binding properties [12]. The alignment of the A chain of lectin III from *V. album* with the A chains of three lectins from *V. coloratum* demonstrates the rigid conservation of all amino acid residues, responsible for the RNA-*N*-glycosidase activity [13]. Besides lectins, toxins were also isolated from *V. coloratum*. Viscotoxin B2 from *V. coloratum* has a high similarity with viscotoxins from *V. album*. The primary structure of the peptide viscotoxin B2 was elucidated [14]. The three-dimensional structure of viscotoxin C1 from *V. coloratum* has also been reported [15].

A further item with loranthaceous plant listed in the Chinese Pharmacopoeia is Herba Taxilli (Sangjisheng), the dried stem and leaf of *Taxillus chinensis* (DC.) Danser. It is used for the treatment of rheumatic and rheumatoid arthralgia, abnormal uterine bleeding, excessive menstrual flow, threatened abortion, and hypertension. The Chinese Pharmacopoeia requires a qualitative determination of quercetin in Herba Taxilli by TLC comparison with reference substance. In addition, the Chinese Pharmacopoeia notes that Herba Taxilli should not contain cardiac glycosides.

Pharmacology and Toxicology

The flavanone glucosides, homoeriodictyol 7,4′-di-*O*-β-D-glucopyranoside, eriodictyol 7,4′-di-*O*-β-D-glucopyranoside, homoeriodictyol 7-*O*-β-D-glucopyranoside, eriodictyol 7-*O*-β-D-glucopyranoside, and naringenin 7-*O*-β-D-glucopyranoside from *V. coloratum* were reported to exert antioxidative activity and scavenging effects on hydroxyl and superoxide anion radicals [16]. The partially purified extract of *V. coloratum* was found to significantly inhibit superoxide anion generation and elastase release in human neutrophils induced by formyl-L-methionyl-L-leucyl-L-phenylalanine (fMLP), with IC_{50} values of 0.58 and 4.9 $\mu g\, ml^{-1}$, respectively. Viscolin as an active component inhibited superoxide anion generation and elastase release in human neutrophils activated by fMLP. The inhibitory effect of viscolin was reversed by an inhibitor of protein kinase A (PKA), which suggested that the inhibition was mediated by PKA. Moreover, viscolin induced a substantial increase in cAMP levels through the inhibition of phosphodiesterase (PDE) activity, but not an increase in adenylate cyclase function. This indicated that the inhibition of inflammatory responses in human neutrophils by viscolin was associated with an elevation of cellular cAMP through an inhibition of PDE [17].

Viscum coloratum was reported to exhibit preventive and curative effects on acute myocardial infarction in experimental animals. The mechanism of action was suggested to be through an improvement of myocardial oxygen consumption [18]. The production of prostaglandin I_2 (PGI_2) in cultured endothelial cells originated from human umbilical veins morphologically damaged to various degrees was significantly inhibited by the extract of *V. coloratum*. However, after pretreatment of the endothelial cells with thrombin, neither the synthesis nor release of PGI_2 was affected by the extract of *V. coloratum* [19].

The total flavone from *V. coloratum* was found to exert anti-tachyarrhythmic effects *in vitro*. A study on the effects of the total flavonoids of *V. coloratum* on the fast response action potentials of canine Purkinje fibers and guinea pig ventricular papillary muscles demonstrated that the total flavones, at a concentration of $100\,\mu g\,ml^{-1}$, accelerated the repolarization of the 2nd and 3rd phases of fast response action potentials. The cellular electrophysiologic mechanisms of the total flavones on anti-tachyarrhythmias appear responsible to prolong the effective refractory period, and to abolish the reentry [20].

The lectins from *V. coloratum* were reported to enhance humoral and cellular immune responses, and therefore might be of use as adjuvants in cancer therapy. *Viscum coloratum* lectins significantly enhanced antibody titers against keyhole limpet hemocyanine in mice when immunized subcutaneously with hemocyanine combined with lectins, and augmented specific antibody titers of IgG_1, IgG_{2a} and IgG_{2b}. The culture supernatants from the splenocytes of immunized mice showed a high level of the specific Th1 cytokines interleukin-2 (IL-2) and interferon-γ (IFN-γ), as well as Th2 type cytokine IL-4 [21]. It was suggested that lectins from *V. coloratum* might be able to modulate murine splenocyte proliferation and act on the balance of Th1/Th2 cellular immune responses [22].

The lectins from *V. coloratum* inhibited the proliferation of B16-BL6 melanoma cells, and caused cell-cycle arrest at the sub-G_1 stage consistent with apoptosis. An *in vivo* treatment with lectins of C57BL6 mice inoculated with B16-BL6 melanoma cells resulted in an increased survival time and an inhibition of lung metastasis. The inhibition of tumor growth and metastasis by the lectins was suggested to be associated with induction of apoptosis and with inhibition of angiogenesis [23]. The intravenous (i.v.) administration of the lectins at doses of 20–50 ng per mouse two days before tumor inoculation also significantly inhibited lung metastases of B16-BL6 melanoma and 26-M3.1 colon carcinoma cells, as well as liver and spleen metastasis of L5178Y-ML25 lymphoma cells in mice. The lectins significantly augmented natural killer (NK) cell activity against Yac-1 tumor cells 2 days after i.v. injection. In addition, the lectins induced tumoricidal activity of peritoneal macrophages against B16-BL6 and 3LL cells. This suggested that they exerted an immunomodulating effect, enhancing the host defense system against tumors. The prophylactic and therapeutic action on tumor metastasis thus appeared to be associated with the activation of NK cells and macrophages [24].

Mechanistic studies further revealed that the lectins from *V. coloratum* induced apoptosis in both p53-positive SK-Hep-1 and p53-negative Hep 3B hepatocarcinoma cells through p53- and p21-independent pathways. The lectins induced apoptosis by the downregulation of Bcl-2 and by the upregulation of Bax, resulting in an activation of caspase-3 in both cell lines. In addition, the downregulation of telomerase activity in both hepatocarcinoma cell lines after treatment with the lectins was observed, indicating that they exhibited antitumor potential by an inhibition of telomerase and consequent apoptosis induction [25]. The induction of apoptotic cell death through the activation of caspase-3, and the inhibition of telomerase activity through transcriptional downregulation of hTERT, were also observed in A253 cells treated with the lectins from *V. coloratum* [26].

The lectin-II from *V. coloratum* also induced the apoptotic death of U937 human myeloleukemic cells, and was a strong inducer of pro-oxidant generation, such as H_2O_2. This is thought to mediate the activation of Jun N-terminal kinase (JNK) and stress-activated protein kinase (SAPK), the release of cytosolic cytochrome c, activation of caspase-9 and caspase 3-like protease, and poly(ADP-ribose) polymerase (PARP) cleavage in U937 cells [27]. Likewise, the induction of apoptosis by lectins from *V. coloratum* in Hep3B cells was also postulated by induction of reactive oxygen species (ROS) production and by a loss of mitochondrial membrane potential, in which JNK phosphorylation plays a critical role [28]. Viscotoxin exerted growth inhibition against 17/2.8 rat osteoblast-like sarcoma cells with an IC_{50} value of 1.6 μg ml^{-1} [14]. *epi*-Oleanolic acid was also found to inhibit the proliferation of various human and murine tumor cell lines, and to induce apoptotic cell death [11].

References

1 Kong, D.Y., Luo, S.G., Li, H.T., and Lei, X.H. (1987) Chemical components of *Viscum cloratum*. I. *Yiyao Gongye*, **18**, 123–127.

2 Kong, D.Y., Luo, S.Q., Li, H.T., and Lei, X.H. (1988) Studies on the chemical components of *Viscum coloratum*. III. Structure of viscumneoside III, V and VI. *Acta Pharm. Sin.*, **23**, 593–600.

3 Kong, D.Y., Li, H.T., Luo, S.Q., and Lei, X.H. (1990) Studies on chemical components of *Viscum coloratum*. VI. Chirality of the acyl group of viscumneoside IV. *Acta Pharm. Sin.*, **25**, 349–352.

4 Kong, D.Y., Dong, Y.Y., Luo, S.Q., Li, H.T., and Lei, X.B. (1988) Determination of new flavone glycosides in colored mistletoe (*Viscum coloratum*]) by HPLC. *Chin. Trad. Herbal Drugs*, **19**, 495–496, 498.

5 Kong, D.Y., Li, H.T., and Luo, S.Q. (1990) Studies on the chemical components of *Viscum coloratum* VII. isolation and structure of viscumneoside VII. *Acta Pharm. Sin.*, **25**, 608–611.

6 Leu, Y.L., Hwang, T.L., Chung, Y.M., and Hong, P.Y. (2006) The inhibition of superoxide anion generation in human neutrophils by *Viscum coloratum*. *Chem. Pharm. Bull. (Tokyo)*, **54**, 1063–1066.

7 Yao, H., Liao, Z.X., Wu, Q., Lei, G.Q., Liu, Z.J., Chen, D.F., Chen, J.K., and Zhou, T.S. (2006) Antioxidative flavanone glycosides from the branches and leaves of *Viscum coloratum*. *Chem. Pharm. Bull. (Tokyo)*, **54**, 133–135.

8 Su, C.R., Shen, Y.C., Kuo, P.C., Leu, Y.L., Damu, Z.G., Wang, Y.H., and Wu, T.S. (2006) Total synthesis and biological evaluation of viscolin, a 1,3-diphenylpropane as a novel potent anti-inflammatory agent. *Bioorg. Med. Chem. Lett.*, **16**, 6155–6160.

9 Kong, D.Y., Li, H.T., and Luo, S.Q. (1992) Studies on the chemical components of *Viscum coloratum*. VIII. Isolation and structure of 3-β-D-glucopyranosyloxy-butanol-2. *Acta Pharm. Sin.*, **27**, 792–795.

10 Kong, D.Y., Luo, S.G., Li, H.T., and Lei, X.H. (1987) Chemical components of *Viscum cloratum*. II. *Yiyao Gongye*, **18**, 445–447.

11 Jung, M.J., Yoo, Y.C., Lee, K.B., Kim, J.B., and Song, K.S. (2004) Isolation of *epi*-oleanolic acid from Korean mistletoe and its apoptosis-inducing activity in tumor cells. *Arch. Pharm. Res.*, **27**, 840–844.

12 Ye, W., Nanga, R.P., Kang, C.B., Song, J.H., Song, S.K., and Yoon, H.S. (2006) Molecular characterization of the recombinant A-chain of a type II ribosome-inactivating protein (RIP) from *Viscum album coloratum* and structural basis on its ribosome-inactivating activity and the sugar-binding properties of the B-chain. *J. Biochem. Mol. Biol.*, **39**, 560–570.

13 Wacker, R., Stoeva, S., Pfuller, K., Pfuller, U., and Voelter, W. (2004) Complete structure determination of the A chain of mistletoe lectin III from *Viscum album* L. ssp. *album*. *J. Pept. Sci.*, **10**, 138–148.

14 Kong, J.L., Du, X.B., Fan, C.X., Xu, J.F., and Zheng, X.J. (2004) Determination of primary structure of a novel peptide from mistletoe and its antitumor activity. *Yao Xue Xue Bao*, **39**, 813–817.

15 Romagnoli, S., Fogolari, F., Catalano, M., Zetta, L., Schaller, G., Urech, K., Giannattasio, M., Ragona, L., and Molinari, H. (2003) NMR solution structure of viscotoxin C1 from *Viscum album* subspecies *coloratum* Ohwi: toward a structure-function analysis of viscotoxins. *Biochemistry*, **42**, 12503–12510.

16 Yao, H., Liao, Z.X., Wu, Q., Lei, G.Q., Liu, Z. J., Chen, D.F., Chen, J.K., and Zhou, T.S. (2006) Antioxidative flavanone glycosides from the branches and leaves of *Viscum coloratum*. *Chem. Pharm. Bull. (Tokyo)*, **54**, 133–135.

17 Hwang, T.L., Leu, Y.L., Kao, S.H., Tang, M.C., and Chang, H.L. (2006) Viscolin, a new chalcone from *Viscum coloratum*, inhibits human neutrophil superoxide anion and elastase release via a cAMP-dependent pathway. *Free Radic. Biol. Med.*, **41**, 1433–1441.

18 Cheng, B.H., and Zhu, S.H. (1985) An experimental observation on the preventive and curative effect of *Viscum coloratum* on acute myocardial infarction through improvement of the myocardial oxygen consumption. *Chin. J. Integr. Trad. West. Med.*, **5**, 565–566.

19 Cai, W.J., Gong, L.S., Ding, H.Y., Hou, G.Y., Wu, C.F., and Chen, F.Y. (1991) Effects of *Viscum coloratum* and tetramethylpyrazine on cultured endothelial cells of human umbilical veins. *Shanghai Med. J.*, **14**, 462–464.

20 Wu, J.X., Yu, G.R., Wang, B.Y., Zhong, D.S., and Huang, D.J. (1994) Effects of *Viscum coloratum* flavonoids on fast response action potentials of hearts. *Acta Pharmacol. Sin.*, **15**, 169–172.

21 Yoon, T.J., Yoo, Y.C., Kang, T.B., Her, E., Kim, Sh., Kim, K., Azuma, I., and Kim, J.B. (2001) Cellular and humoral adjuvant activity of lectins isolated from Korean mistletoe (*Viscum album coloratum*). *Int. Immunopharmacol*, **1**, 881–889.

22 Lyu, S.Y. and Park, W.B. (2006) Mistletoe lectin (*Viscum album coloratum*) modulates proliferation and cytokine expressions in murine splenocytes. *J. Biochem. Mol. Biol.*, **39**, 662–670.

23 Park, W.B., Lyu, S.Y., Kim, J.H., Choi, S.H., Chung, H.K., Ahn, S.H., Hong, S.Y., Yoon, T.J., and Choi, M.J. (2001) Inhibition of tumor growth and metastasis by Korean mistletoe lectin is associated with apoptosis and antiangiogenesis. *Cancer Biother. Radiopharm.*, **16**, 439–447.

24 Yoon, T.J., Yoo, Y.C., Kang, T.B., Song, S.K., Lee, K.B., Her, E., Song, K.S., and Kim, J.B. (2003) Antitumor activity of the Korean mistletoe lectin is attributed to activation of macrophages and NK cells. *Arch. Pharm. Res.*, **26**, 861–867.

25 Lyu, S.Y., Choi, S.H., and Park, W.B. (2002) Korean mistletoe lectin-induced apoptosis in hepatocarcinoma cells is associated with inhibition of telomerase *via* mitochondrial controlled pathway independent of p53. *Arch. Pharm. Res.*, **25**, 93–101.

26 Choi, S.H., Lyu, S.Y., and Park, W.B. (2004) Mistletoe lectin induces apoptosis and telomerase inhibition in human A253 cancer cells through dephosphorylation of Akt. *Arch. Pharm. Res.*, **27**, 68–76.

27 Kim, M.S., Lee, J., Lee, K.M., Yang, S.H., Choi, S., Chung, S.Y., Kim, T.Y., Jeong, W.H., and Park, R. (2003) Involvement of hydrogen peroxide in mistletoe lectin-II-induced apoptosis of myeloleukemic U937 cells. *Life Sci.*, **73**, 1231–1243.

28 Kim, W.H., Park, W.B., Gao, B., and Jung, M. H. (2004) Critical role of reactive oxygen species and mitochondrial membrane potential in Korean mistletoe lectin-induced apoptosis in human hepatocarcinoma cells. *Mol. Pharmacol.*, **66**, 1383–1396.

Vitex negundo var. *cannabifolia*, *Vitex trifolia*, and *Vitex trifolia* var. *simplicifolia*

Folium Viticis Negundo (Mujingye) is the dried leaf of *Vitex negundo* L. var. *cannabifolia* (Sieb. et Zucc.) Hand.-Mazz. (Verbenaceae).

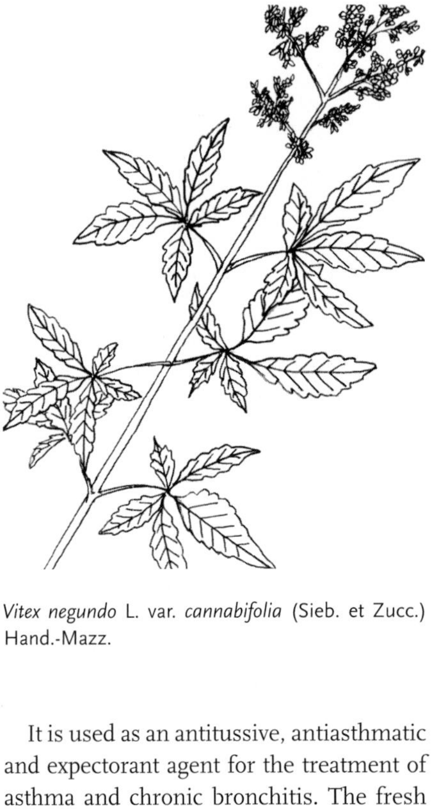

Vitex negundo L. var. *cannabifolia* (Sieb. et Zucc.) Hand.-Mazz.

It is used as an antitussive, antiasthmatic and expectorant agent for the treatment of asthma and chronic bronchitis. The fresh material of Folium Viticis Negundo is used for preparation of the essential oil, Oleum Viticis Negundo.

Oleum Viticis Negundo (Mujingyou) is the essential oil of *Vitex negundo* var. *cannabifolia*, obtained by steam distillation of fresh leaves. It is used as an antiasthmatic, antitussive and expectorant agent in the treatment of chronic bronchitis and asthma. Oleum Viticis Negundo is used mainly in form of its galenic preparation, "Mujingyou Jiaowan," the capsules of Oleum Viticis Negundo, which is also officially listed in the Chinese Pharmacopoeia. Each capsule contains 20 mg of essential oil.

Fructus Viticis (Manjingzi) is the dried ripe seed of *Vitex trifolia* L. var. *simplicifolia* Cham. or *Vitex trifolia* L. It is used for treatment of cold, headache, toothache, and vertigo.

The dried root of *Vitex trifolia* or *Vitex trifolia* var. *simplicifolia*, is listed in the appendix of the Chinese Pharmacopoeia.

Chemistry

The essential oil from the leaf of *V. negundo* var. *cannabifolia* is reported to contain β-caryophyllene, caryophyllene oxide, 1,8-cineole, *p*-cymene, β-elemene, linalool, α-phellandrene, α-pinene, β-pinene, sabinene, and γ-terpinene, with β-caryophyllene as the major component. The content of β-caryophyllene in the essential oil was found to be about 50% [1]. The composition of the essential oil from the leaf of *V. trifolia* was similar to that of *V. negundo* var. *cannabifolia*, however, with α-pinene as the major component [1]. From the leaf of *V. negundo* iridoid glycosides aucubin, agnuside, mussaenosidic acid 2′-*p*-hydroxybenzoate, 6′-*p*-hydroxybenzoate and nishindaside were isolated [2, 3].

Phytosterols from the fruit of *V. trifolia* were identified as β-sitosterol and its 3-*O*-glucoside. Flavones from the fruit of *V. trifolia* were identified as vitexicarpin (casticin) and the corresponding 4′-demethyl analogue [4]. The fatty oil from the seeds con-

Handbook of Chinese Medicinal Plants: Chemistry, Pharmacology, Toxicology
Weici Tang and Gerhard Eisenbrand

ISBN: 978-3-527-32226-8

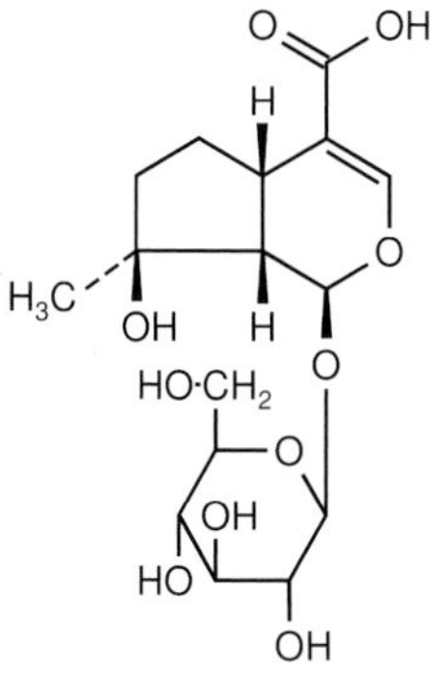

Mussaenosidic acid

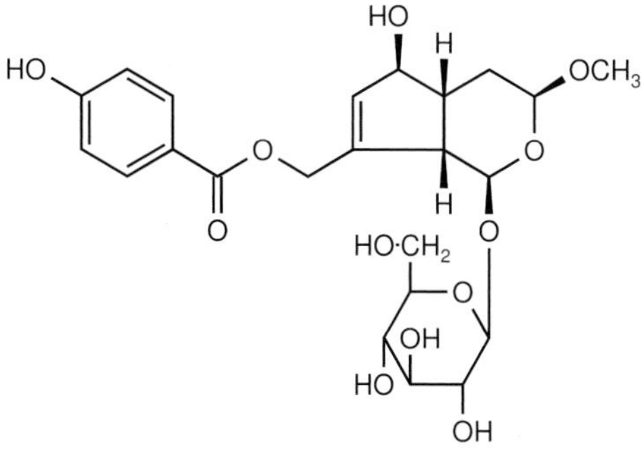

Nishindaside

tained linoleic, myristic, oleic, palmitic, palmitoleic, and stearic acids, with linoleic (20%), oleic (28%), palmitic (35%), and stearic (15%) acids as the major components [5]. The Chinese Pharmacopoeia requires a qualitative determination of vitexicarpin in Fructus Viticis by thin-layer chromatographic comparison with reference substance, and a quantitative determination by HPLC. The content of vitexicarpin in Fructus Viticis should be not less than 0.03%. The Chinese Pharmacopoeia gives physico-chemical parameters for Oleum Viticis Negundo: relative density 0.890–0.910 at 25 °C, and refractive index 1.485–1.500.

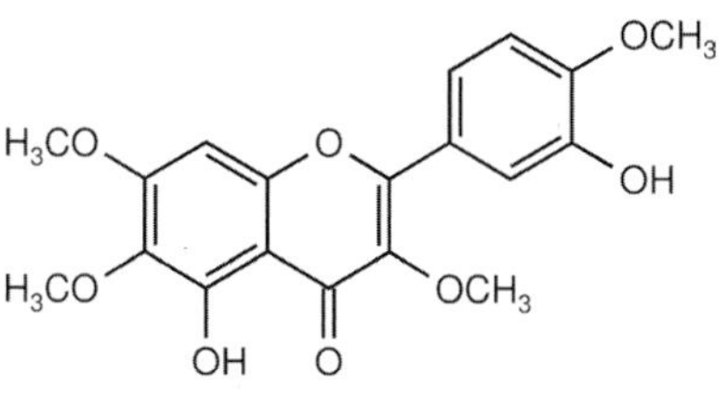

Vitexicarpin

Pharmacology and Toxicology

The essential oil of *V. negundo* var. *cannabifolia* was found to have a therapeutic effect on bronchitis and asthma, similar to that found in *Rhododendron racemosum* [6]. A high degree of similarity in chemical composition of the essential oil from *R. racemosum* and from some *Vitex* species was reported [8]. The lipid fraction of *V. negundo* var. *cannabifolia* exhibited spasmolytic activity in isolated perfused mouse lung in a concentration-dependent manner. Contraction of the isolated mouse trachea smooth muscle induced by acetylcholine was relaxed by the lipid fraction. The spasmolytic effect of the lipid fraction was also observed in guinea pigs with experimental asthma induced by acetylcholine and histamine [7]. The methanol extract of the leaves of *V. negundo* was found to significantly potentiate the sleeping time induced by pentobarbitone, diazepam and chlorpromazine in mice, to possess analgesic properties, and to potentiate analgesia induced by morphine, as well as to show significant protection against strychnine-induced convulsions. This suggested that the extract exhibits depressant activity on the central nervous system in a dose-dependent manner [8].

An orally administered extract of the seeds of *V. trifolia* showed a significant antipyretic effect in rabbits given yeast by subcutaneous injection [9].

References

1 Fan, J.F., Pan, J.Q., Sun, Y.F., Ciu, M.S., He, H.J., Duan, S.M., Wang, X.Z., Liu, S.M., and Sun, Y.R. (1981) Studies on the chemical constituents of Chinese *Vitex*. I. Chemical constituents of essential oils of certain Vitex species. *Chin. Trad. Herbal Drugs*, **12**, 393–396.

2 Sehgal, C.K., Taneja, S.C., Dhar, K.L., and Atal, C.K. (1983) 6′-*p*-Hydroxybenzoylmussaenosidic acid, an iridoid glycoside from *Vitex negundo*. *Phytochemistry*, **22**, 1036–1038.

3 Dutta, P.K., Chowdhury, U.S., Chakravarty, A.K., Achari, B., and Pakrashi, S.C. (1983) Studies on Indian medicinal plants. LXXV. Nishindaside, a novel iridoid glycoside from *Vitex negundo*. *Tetrahedron*, **39**, 3067–3072.

4 Zeng, X., Fang, Z., Wu, Y., and Zhang, H. (1996) Chemical constituents of the fruits of *Vitex trifolia* L. *China J. Chin. Mater. Med.*, **21**, 167–168.

5 Prasad, Y.R. and Nigam, S.S. (1982) Detailed chemical investigation of the seed oil of *Vitex trifolia* Linn. *Proc. Natl Acad. Sci. India [A]*, **52**, 336–339.

6 Fang, H.J., Chen, L.S., and Zhou, T.H. (1980) Studies on the components of the essential oils. III. Studies of chemical constituents of the essential oil from *Rhododendron racemosum* Franch. Comparison of the constituents of *Vitex negundo* L. var. *cannabifolia* (Sieb. et Zucc.) Hand. -Mazz. and *V. negundo* L. var. heterophylla (Franch.) Rehd. *Acta Pharm. Sin.*, **15**, 284–287.

7 Liu, M.S., Liu, C.L., Gu, G.M., and Xu, X.H. (1993) Effects of lipids from Mujingzi on smooth muscles of trachea in animals. *Chin. Pharmacol. Bull.*, **9**, 307–310.

8 Gupta, M., Mazumder, U.K., and Bhawal, S.R. (1999) CNS activity of *Vitex negundo* Linn. in mice. *Indian J. Exp. Biol.*, **37**, 143–146.

9 Ikram, M., Khattak, S.G., and Gilani, S.N. (1987) Antipyretic studies on some indigenous Pakistani medicinal plants: II. *J. Ethnopharmacol.*, **19**, 185–192.

Zanthoxylum bungeanum and *Zanthoxylum schinifolium*

Pericarpium Zanthoxyli (Huajiao) is the dried pericarp of the ripe fruits of *Zanthoxylum schinifolium* Sieb. et Zucc. or *Zanthoxylum bungeanum* Maxim. (Rutaceae). It is used for the treatment of gastrointestinal disorders such as vomiting, diarrhea and abdominal pain and abdominal parasitosis, as well as for the treatment of eczema by external application. It is also a well-known spice.

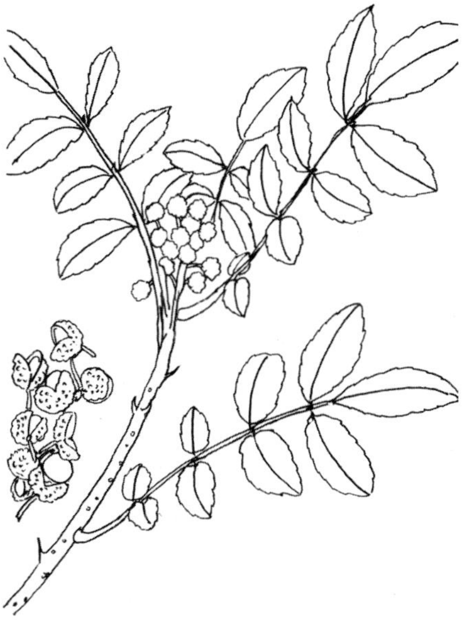

Zanthoxylum bungeanum Maxim.

Chemistry

The pericarp of *Z. bungeanum* was reported to contain essential oil with limonene (27%) as the major component, along with β-myrcene (17%) and β-phellandrene (6%). Minor components of the essential oil were identified as isopulegol, linalool, α-pinene, sabinene, terpinen-4-ol, α-terpineol, piperitone, and β-ocimene [1]. The fruit secretory glands of *Z. bungeanum* were found to contain β-phellandrene (37%), piperitone (9%) and β-pinene (9%) as major components in the essential oil [1]. The content and composition of the essential oil from the pericarp of *Z. schinifolium* were similar to that from *Z. bungeanum* [2]. The Chinese Pharmacopoeia requires a quantitative determination of essential oil in Pericarpium Zanthoxyli. The content of essential oil should be not less than 1.5%.

Pungent principles of the pericarp were identified as aliphatic acid amides, including α-sanshool, hydroxy-α-sanshool, hydroxy-β-sanshool, γ-sanshool, and hydroxy-γ-sanshool. The fruit secretory glands of *Z. bungeanum* were found to contain about 20% hydroxy-α-sanshool [1].

Further constituents from the pericarp of *Z. bungeanum* were identified as quinoline and furoquinoline alkaloids kokusaginine, haplopine, and skimmianine; coumarin and furocoumarin derivatives herniarin (7-methoxycoumarin) and bergapten; quercetin glycosides hyperin and quercitrin; and xanthoxylin [3, 4]. Skimmianine, schinifoline, umbelliferone (7-hydroxycoumarin), bergapten and xanthoxylin were also isolated from the pericarp of *Z. schinifolium* [5]. Skimmianine and related furoquinoline alkaloids occur widely in rutaceous plants. A further item listed in the Chinese Pharmacopoeia containing skimmianine is Cortex Dictamni (Baixianpi), the dried root bark of *Dictamnus dasycarpus* Turcz.

Pharmacology and Toxicology

The aqueous and ether extracts of the pericarp of *Z. bungeanum* were reported to be active against gastric ulcer, diarrhea, and other gastrointestinal disorders [6]. The aqueous and methanol extracts of the fruit

Handbook of Chinese Medicinal Plants: Chemistry, Pharmacology, Toxicology
Weici Tang and Gerhard Eisenbrand

ISBN: 978-3-527-32226-8

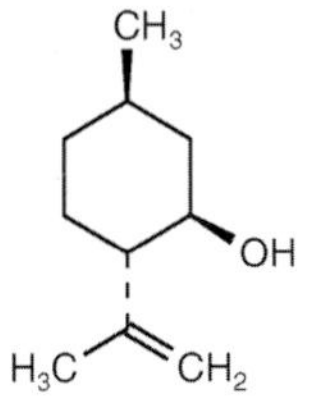

Isopulegol

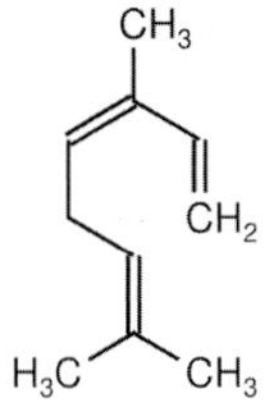

β-Ocimene

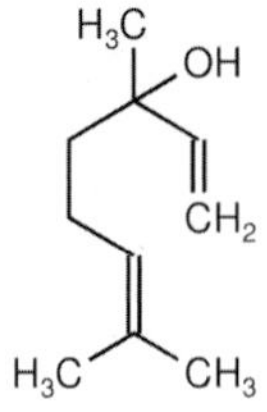

Linalool

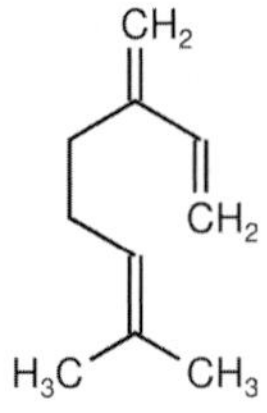

β-Myrcene

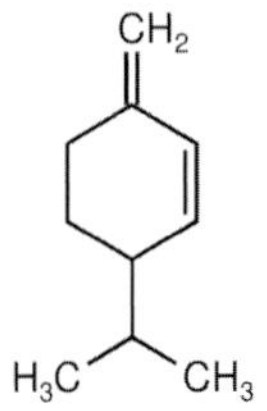

β-Phellandrene

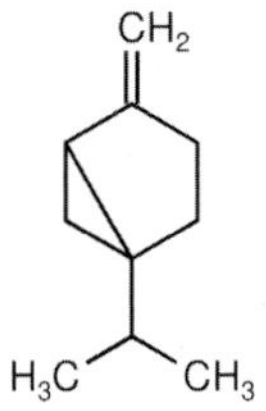

Sabinene

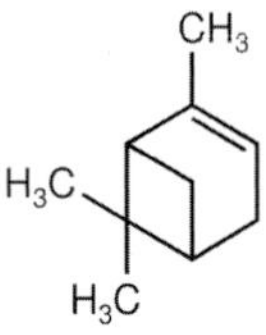

α-Pinene

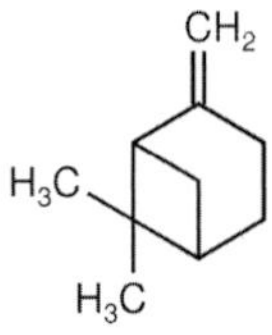

β-Pinene

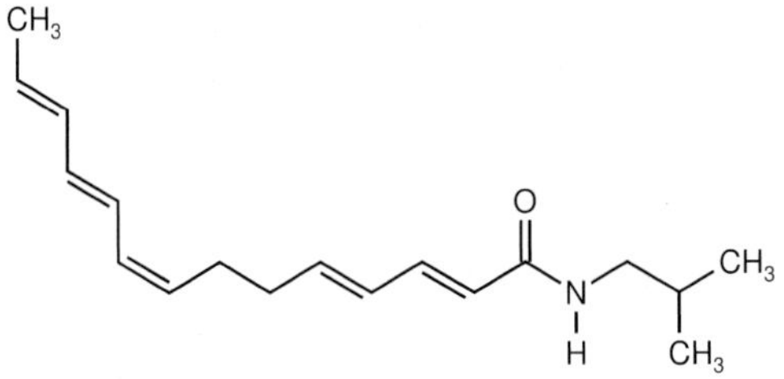

γ-Sanshool

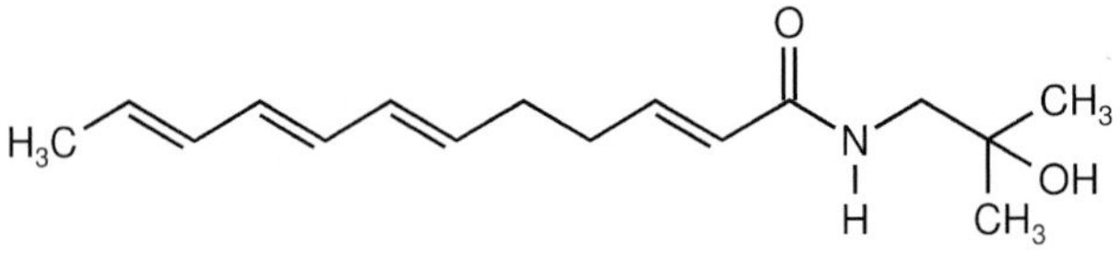

Hydroxy-β-sanshool

of *Z. bungeanum* significantly increased the spontaneous beating rate of embryonic mouse myocardial cell sheets in culture. Hydroxy-β-sanshool, xanthoxylin and quercetin glycosides were found to be the active principles. They increased the spontaneous beating rate in a standard medium with 2.1 m*M* Ca^{2+}, and suppressed the decrease of the spontaneous beating rate induced by low Ca^{2+} concentration (0.5 m*M*). Hydroxy-β-sanshool and xanthoxylin stimulated the calcium uptake of the cultured myocardial cells, but hyperin and quercitrin were inactive [3].

α-Sanshool was reported to be antiparasitically active against *Ascaris suum in vitro*, with a LC_{50} of 0.8 m*M*. On the other hand, α-sanshool induced tonic–clonic seizures

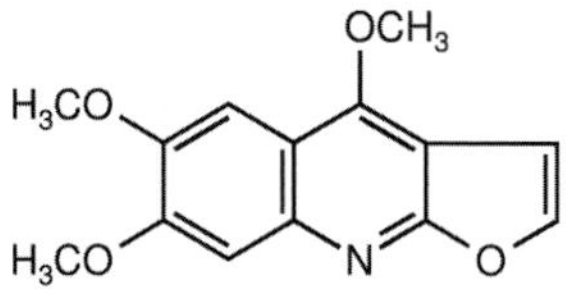

Kokusaginine

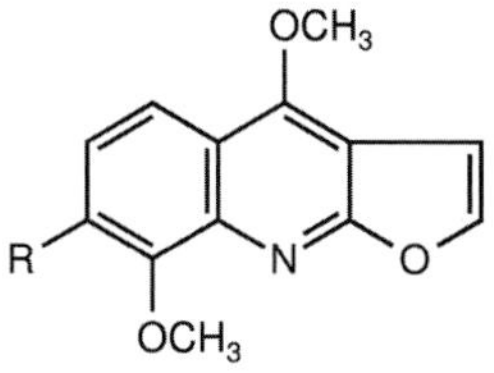

Haplopine: R = OH
Skimmianine: R = OCH_3

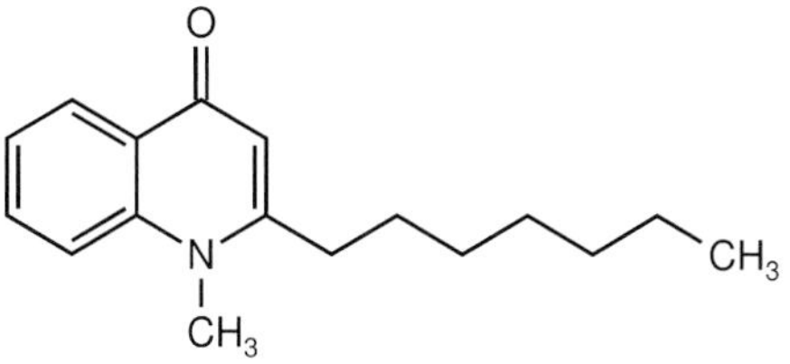

Schinifoline

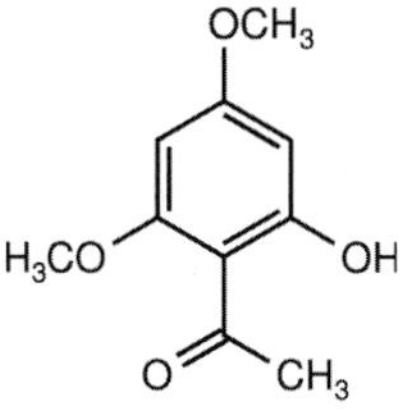

Xanthoxylin

when injected intraperitoneally into mice [7]. Hydroxy-α-sanshool was found to alter the levels of spontaneous activity in cool-sensitive fibers obtained from the lingual nerve of rats, and also to induce activity in tactile fibers [8]. β-Sanshool and γ-sanshool relaxed the circular muscle of the gastric body and contracted the longitudinal muscle of the ileum and distal colon of guinea pig [9].

Skimmianine showed analgesic, spasmolytic and sedative effects *in vivo*, which might contribute to relieve abdominal pain [10]. Skimmianine was found to exhibit a selective inhibitory effect on the serotonin (5-HT)-induced vasopressor responses of rats. A similar inhibitory action was observed in isolated atria. At high concentrations, skimmianine produced a nonspecific blockade of the cardiovascular functions [11]. Skimmianine and kokusaginine displaced [^{3}H]5-HT and [^{3}H]ketanserin as radioligands to receptors in isolated membranes of rat cerebrocortex, in a concentration-dependent manner. They inhibited 5-HT_2-receptor-mediated contractions, but were inactive against 5-HT_1-receptor- and 5-HT_3-receptor-mediated processes. Studies in isolated preparations indicated that furoquinoline alkaloids might act on 5-HT receptors in animals, with a preferential selectively towards the 5-HT_2 subtype [12].

It was reported that both the aqueous and ether extract of *Z. schinifolium* showed significant inhibition on thrombus formation in rats [13]. Some coumarins were found to be the active components with antiplatelet aggregation activity of the bark of *Z. schinifolium* [14]. Skimmianine and kokusaginine were also found to inhibit platelet aggregation [15].

Skimmianine was found to show significant inhibitory effects on spontaneous motor activity, exploratory behavior, cataleptogenic activity, conditioned avoidance response, and long-term isolation-induced fighting of animals [16]. No physical dependence of skimmianine in mice, rats and monkeys was observed [17].

Skimmianine and kokusaginine were reported to be cytotoxic against the KB-V1$^+$ human tumor cell line [18]. Skimmianine exhibited strong mutagenicity in *Salmonella typhimurium* TA98 and TA100 in the presence of metabolic systems, but had only weak or no activity in *S. typhimurium* TA1538 and TA1535, indicating that it acted primarily as a frameshift mutagen [19]. It was suggested that skimmianine and related furoquinoline alkaloids were activated to mutagenic metabolites by cytochrome P450, and possibly by the flavin-containing monooxygenase [20]. Skimmianine was further found to be genotoxic in the SOS-chromotest in the presence of a metabolic activation system [21]. Furoquinoline alkaloids, including skimmianine, were also found to be photomutagenic in an arginine-requiring mutant strain of the green alga *Chlamydomonas reinhardtii*. However, the mutagenic activity of furoquinoline alkaloids was much less as compared to that of furocoumarins, such as bergapten. The lower phototoxicity and photomutagenicity of the furoquinoline alkaloids might be due to the fact that furoquinolines form only monoadducts with DNA in the presence of UV-A; this is in contrast to furocoumarins, which also form biadducts [22].

The essential oils from *Z. schinifolium* pericarp, mainly composed of geranyl acetate, citronella, and sabinene, was found to inhibit the proliferation of HepG2 human hepatoma cells and to induce cell apoptosis, but not to change the activity of caspase-3. In addition, the essential oil increased the production of reactive oxygen species (ROS). In nude mice inoculated with Huh-7 human hepatoma cells, the essential oil also significantly inhibited tumor growth. It was suggested that ROS were the key signaling molecules in the cytotoxic activity of the essential oil inducing cell death in HepG2 cells [23].

Linalool was reported to exert anti-inflammatory activity. The systemic administration of linalool at a dose of 25 mg kg^{-1} to rats inhibited the paw edema induced by carrageenin [24]. Linalool was further reported to exert antiproliferative activities against a broad spectrum of carcinoma cells. The IC_{50} values of linalool against tumor cells of cervix carcinoma, stomach carcinoma, skin carcinoma, lung carcinoma, and osteosarcoma were given as 0.37, 14.1, 14.9, 21.5, and 21.7 μg ml^{-1}, respectively [25].

References

1 Tirillini, B. and Stoppini, A.M. (1994) Volatile constituents of the fruit secretory glands of *Zanthoxylum bungeanum* Maxim. *J. Essent. Oil Res.*, **6**, 249–252.

2 Xiong, Q.B. and Shi, D.W. (1992) Analysis of the volatile oils of "Huajiao" (Pericarpium zanthoxyli) and its allied drugs. *Acta Univ. Med. Shanghai*, **19**, 301–306.

3 Huang, X.L., Kakiuchi, N., Che, Q.M., Huang, S.L., Hattori, M., and Namba, T. (1993) Effects of extracts of *Zanthoxylum* fruit and their constituents on spontaneous beating rate of myocardial cell sheets in culture. *Phytother. Res.*, **7**, 41–48.

4 Lin, S.L. and Wei, L.X. (1991) Determination of chemical constituents in *Zanthoxylum schinifolium* and *Z. bungeanum*. *Chin. Trad. Herbal Drugs*, **22**, 16–18.

5 Liu, S.L., Wei, L.X., Wang, D., and Gao, C.Y. (1991) Chemical constituents from the peel of *Zanthoxylum schinifolium* Sieb et Zucc. *Acta Pharm. Sin.*, **26**, 836–840.

6 Zhang, M.F., Shen, Y.Q., Zhu, Z.P., and Chen, G.J. (1991) Pharmacological studies on warming the middle-jiao to alleviate pain by Pericarpium Zanthoxyli. *China J. Chin. Mater. Med.*, **16**, 493–497.

7 Navarrete, A., and Hong, E. (1996) Anthelmintic properties of α-sanshool from *Zanthoxylum liebmannianum*. *Planta Med.*, **62**, 250–251.

8 Bryant, B.P. and Mezine, I. (1999) Alkylamides that produce tingling paresthesia activate tactile and thermal trigeminal neurons. *Brain Res.*, **842**, 452–460.

9 Hashimoto, K., Satoh, K., Kase, Y., Ishige, A., Kubo, M., Sasaki, H., Nishikawa, S., Kurosawa, S., Yakabi, K., and Nakamura, T. (2001) Modulatory effect of aliphatic acid amides from *Zanthoxylum piperitum* on isolated gastrointestinal tract. *Planta Med.*, **67**, 179–181.

10 Chang, Z.Q., Wang, S.L., Hao, C.Y., Liu, F., Bian, C.F., and Chen, J.M. (1982) Analgesic, antispastic and sedative effects of

skimmianine. *Acta Pharmacol. Sin.*, **3**, 163–165.

11 Cheng, J.T., Chang, S.S., and Chen, I.S. (1990) Cardiovascular effect of skimmianine in rats. *Arch. Int. Pharmacodyn. Ther.*, **306**, 65–74.

12 Cheng, J.T., Chang, T.K., and Chen, I.S. (1994) Skimmianine and related furoquinolines function as antagonists of 5-hydroxytryptamine receptors in animals. *J. Auton. Pharmacol.*, **14**, 365–374.

13 Xu, Q.Y., Yu, L.S., Zhang, X.L., Chen, R.M., and Chen, C.M. (1990) Effects of crude extract of *Zanthoxylum schinifolium* on experimental thrombus formation and coagulation system. *Chin. Trad. Herbal Drugs*, **21**, 545–546.

14 Chen, I.S., Lin, Y.C., Tsai, I.L., Teng, C.M., Ko, F.N., Ishikawa, T., and Ishii, H. (1995) Coumarins and anti-platelet aggregation constituents from *Zanthoxylum schinifolium*. *Phytochemistry*, **39**, 1091–1097.

15 Chen, K.S., Chang, Y.L., Teng, C.M., Chen, C.F., and Wu, Y.C. (2000) Furoquinolines with antiplatelet aggregation activity from leaves of *Melicope confusa*. *Planta Med.*, **66**, 80–81.

16 Cheng, J.T. (1986) Effect of skimmianine on animal behavior. *Arch. Int. Pharmacodyn. Ther.*, **281**, 35–43.

17 Chang, Z.Q., Wang, S.L., Liu, F., Zhu, M.Y., and Tang, X.C. (1982) No physical dependence of skimmianine in mice, rats and monkeys. *Acta Pharmacol. Sin.*, **3**, 223–226.

18 Cui, B., Chai, H., Dong, Y., Horgen, F.D., Hansen, B., Madulid, D.A., Soejarto, D.D., Farnsworth, N.R., Cordell, G.A., Pezzuto, J.M., and Kinghorn, A.D. (1999) Quinoline alkaloids from *Acronychia laurifolia*. *Phytochemistry*, **52**, 95–98.

19 Paulini, H., Eilert, U., and Schimmer, O. (1987) Mutagenic compounds in an extract from rutae herba (*Ruta graveolens* L.). I. Mutagenicity is partially caused by furoquinoline alkaloids. *Mutagenesis*, **2**, 271–273.

20 Hafele, F. and Schimmer, O. (1988) Mutagenicity of furoquinoline alkaloids in the *Salmonella*/microsome assay. Mutagenicity of dictamnine is modified by various enzyme inducers and inhibitors. *Mutagenesis*, **3**, 349–353.

21 Henriques, J.A., Moreno, P.R., Von Poser, G.L., Querol, C.C., and Henriques, A.T. (1991) Genotoxic effect of alkaloids. *Mem. Inst. Oswaldo Cruz.*, **86** (Suppl. 2), 71–74.

22 Schimmer, O., and Kuhne, I. (1990) Mutagenic compounds in an extract from Rutae Herba (*Ruta graveolens* L.). II. UV-A mediated mutagenicity in the green alga *Chlamydomonas reinhardtii* by furoquinoline alkaloids and furocoumarins present in a commercial tincture from Rutae Herba. *Mutat. Res.*, **243**, 57–62.

23 Paik, S.Y., Koh, K.H., Beak, S.M., Paek, S.H., and Kim, J.A. (2005) The essential oils from *Zanthoxylum schinifolium* pericarp induce apoptosis of HepG2 human hepatoma cells through increased production of reactive oxygen species. *Biol. Pharm. Bull.*, **28**, 802–807.

24 Peana, A.T., D'Aquila, P.S., Panin, F., Serra, G., Pippia, P., and Moretti, M.D. (2002) Anti-inflammatory activity of linalool and linalyl acetate constituents of essential oils. *Phytomedicine*, **9**, 721–726.

25 Cherng, J.M., Shieh, D.E., Chiang, W., Chang, M.Y., and Chiang, L.C. (2007) Chemopreventive effects of minor dietary constituents in common foods on human cancer cells. *Biosci. Biotechnol. Biochem.*, **71**, 1500–1504.

Zanthoxylum nitidum

Radix Zanthoxyli (Liangmianzhen) is the dried root of *Zanthoxylum nitidum* (Roxb.) DC. (Rutaceae). It is used as an analgesic agent for treatment of traumatic injuries; rheumatic arthralgia; abdominal pain; toothache; snake bite; and for the treatment of scald by external use. The Chinese Pharmacopoeia notes that overdosing of Radix Zanthoxyli should be avoided, and that Radix Zanthoxyli is incompatible with sour food.

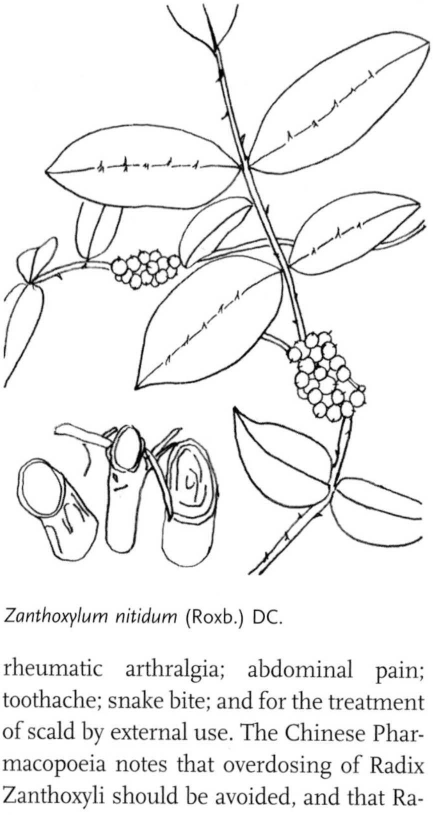

Zanthoxylum nitidum (Roxb.) DC.

The dried root and rhizome of *Zanthoxylum dissitum* Hemsl. is listed in the appendix of the Chinese Pharmacopoeia.

Chemistry

The root of *Z. nitidum* is known to contain a number of alkaloids, with the benzo[*c*]phenanthridine alkaloids nitidine, fagaronine and 6-ethoxychelerythrine as the major components. However, 6-ethoxychelerythrine has been discovered to be formed as an artefact during ethanol extraction [1]. Minor benzo[*c*]phenanthridine alkaloids from the root were identified as oxynitidine, dihydronitidine, oxyterihanine, bocconoline, chelerythrine, oxychelerythrine, isofararidine, demethylchelerythrine, decarine [2–4], methoxy- and ethoxydihydrochelerythrine [4], hydroxyethyldihydrochelerythrine and methoxynorchelerythrine [5]. The contents of the quaternary benzo[*c*]phenanthridine alkaloids in the roots showed a seasonal variation [2, 6]. The Chinese Pharmacopoeia requires a qualitative determination of ethoxychelerythrine in Radix Zanthoxyli by thin-layer chromatographic (TLC) comparison with reference substance, and a quantitative determination of nitidine in Radix Zanthoxyli by TLC-densitometry. The content of nitidine in Radix Zanthoxyli should be not less than 0.25%. The Chinese Pharmacopoeia further requires a differential determination of toddalolactone in Radix Zanthoxyli by TLC comparison with reference substance. Radix Zanthoxyli should not contain toddalolactone.

Alkaloids of other structural types from the root of *Z. nitidum* were identified as skimmianine [1]; magnoflorine [7]; allocryptopine, arnottianamide, liriodenine, integriamide, and isoarnottianamide [2].

The lignans asarinin, sesamin, and syringaresinol; the coumarin aesculetin dimethylether; and β-sitosterol were also isolated from the roots of *Z. nitidum* [2].

Handbook of Chinese Medicinal Plants: Chemistry, Pharmacology, Toxicology
Weici Tang and Gerhard Eisenbrand

ISBN: 978-3-527-32226-8

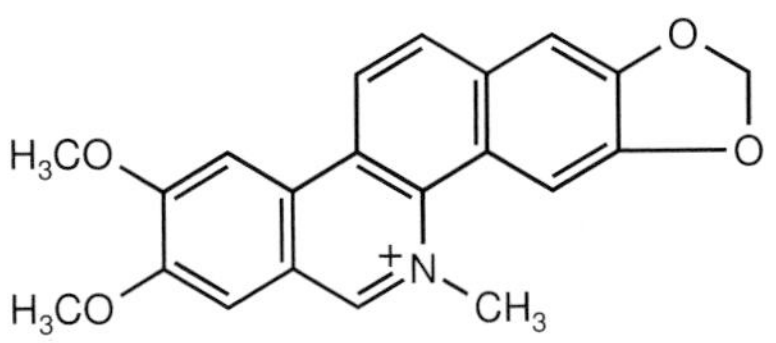

Nitidine

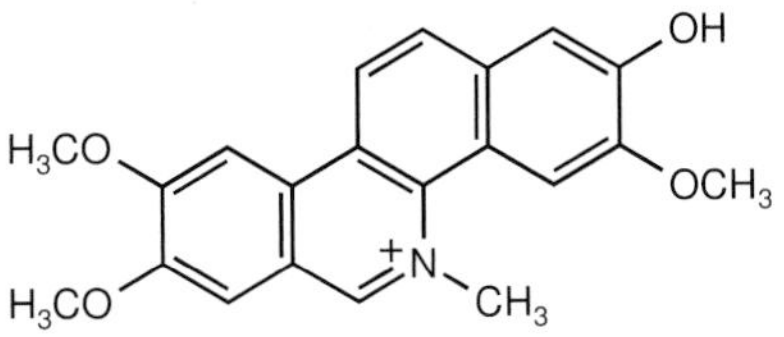

Fagaronine

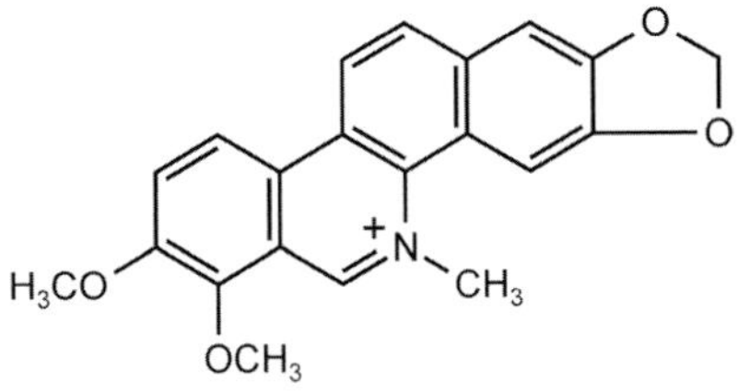

Chelerythrine

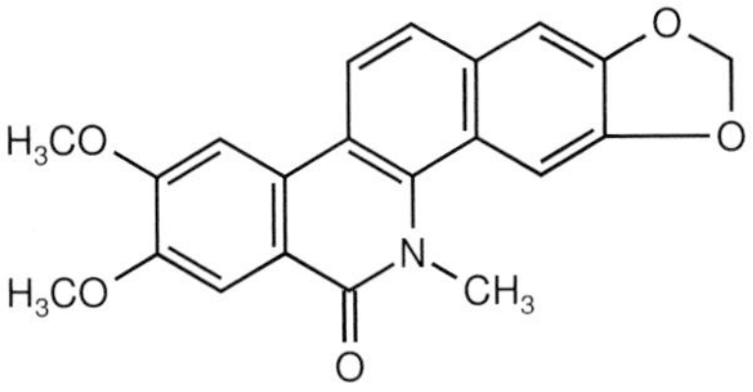

Oxynitidine

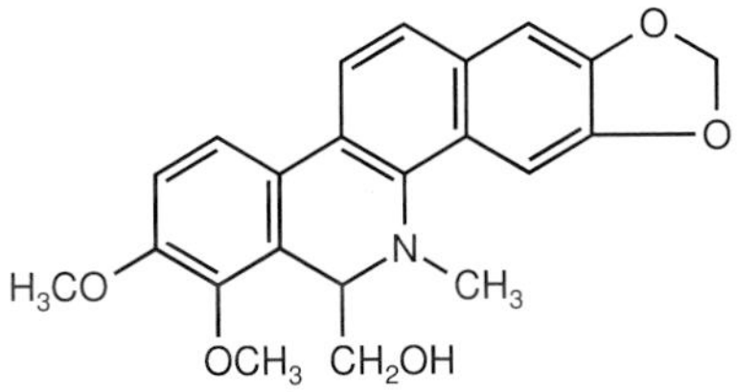

Bocconoline

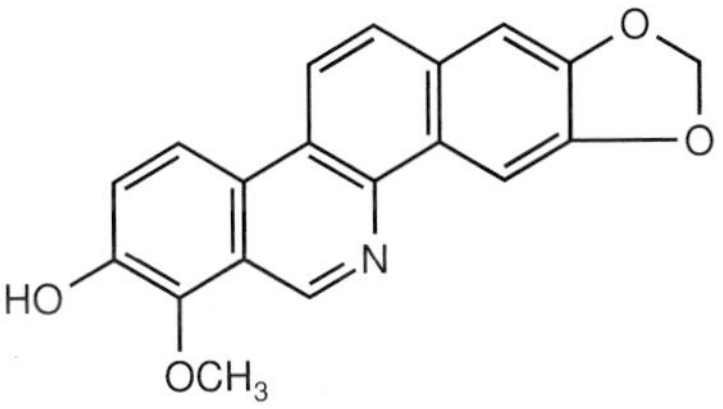

Decarine

Pharmacology and Toxicology

Nitidine and related alkaloids have been reported to possess antileukemic activity against both L1210 and P388 in mice, and some were curative against Lewis lung carcinoma [8]. Nitidine was marginally active against B16 melanoma *in vivo* [9]. Nitidine chloride and 6-methoxy-5,6-dihydrochelerythrine were each found capable of increasing the life span of mice inoculated with Ehrlich ascites tumor, and caused a decrease in the mitotic index and size of the tumor cells. An inhibition of DNA and RNA formation in tumors *in vivo* by nitidine was also observed [10]. 6-Ethoxychelerythrine also showed inhibitory activity against Ehrlich ascites carcinoma cells. Nitidine chloride was reported to be used effectively in the clinical treatment of chronic myelocytic leukemia [11].

The methanol extract of the roots of *Z. nitidum* and the alkaloids nitidine, chelerythrine, and isofararidine, were found to inhibit topoisomerase I (topo I)-mediated DNA relaxation, and to stabilize the covalent binary complex formed between topo I and DNA. While all three alkaloids inhibited the relaxation of supercoiled plasmid pSP64 DNA by calf thymus topo I, only nitidine exhibited a strong stabilization of the enzyme–DNA covalent binary complex [4]. Nitidine was found to bind to calf thymus DNA by an intercalative mode [12]. Yeast cells expressing human DNA topo I were shown to be specifically sensitive to niti-

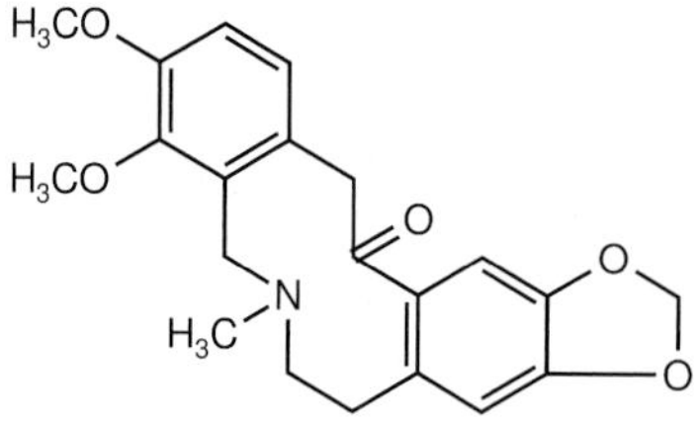

Allocryptopine

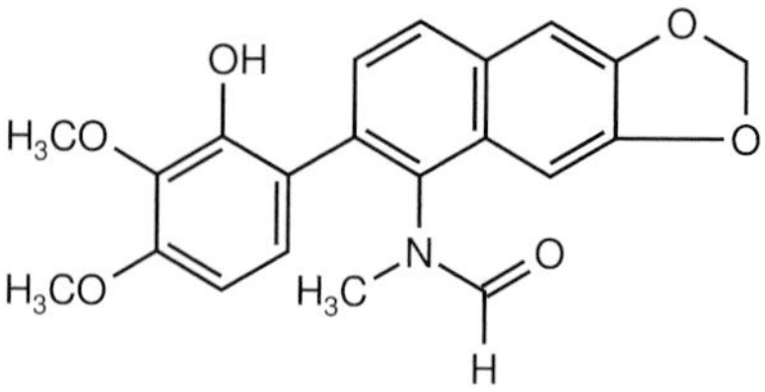

Arnottianamide

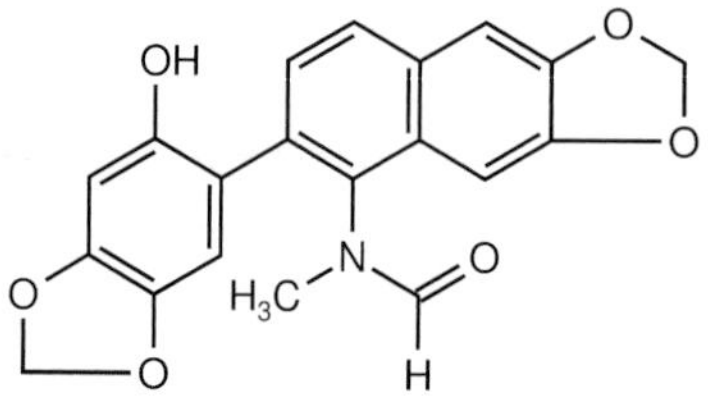

Integriamide

dine, which suggested that cellular DNA topo I was the cytotoxic target.

Two human camptothecin-resistant cell lines, CPT-K5 and A2780/CPT-2000, which were known to highly express topo I resistance to camptothecin, were only marginally resistant to nitidine. Purified human topo I demonstrated similar marginal cross-resistance to nitidine [13, 14]. Unlike camptothecin as a reference topo I inhibitor, the benzo[*c*]phenanthridine-type alkaloids nitidine and fagaronine bound directly to, and mediated the unwinding of, the B-form DNA. A moderate topo II inhibition of nitidine was observed, but only at high concentrations [15]. However, cross-resistance of vinblastine- and taxol-resistant mutants of Chinese hamster ovary (CHO) cells and puromycin-resistant mutants of HeLa cells to nitidine was reported [16, 17]. The anticancer mechanism of nitidine has also been proposed via the crystal structure of dihydronitidine [18]. A study on the intercalative binding of nitidine with calf thymus DNA, poly(dG-dC).poly(dG-dC), poly(dA-dT).poly(dA-dT), and sequence-designed double-stranded oligodeoxynucleotides, revealed that nitidine binds preferentially to DNA containing alternating GC base pairs [d(TGCGCA)$_2$] [19].

The interaction of fagaronine with nucleic acids was also described. The addition of calf thymus DNA to fagaronine produced a hypsochromic shift in the absorption spectrum of fagaronine, with a remarkable increase in intensity in the range of 400 nm. The extent of the spectral alterations was dependent on the concentration of DNA. The maximal extent of binding was one fagaronine molecule per 4.5 base pairs. Fagaronine binding was not limited to regions of DNA containing adenine and thymine. Fagaronine showed bactericidal activity in *Salmonella typhimurium* strain TM677. In contrast, *N*-demethylfagaronine was neither bactericidal nor cytotoxic, nor did it effectively interact with DNA, which suggested that the quaternary nitrogen atom was of importance. The key mechanism of cytotoxic activity may be an interaction with cellular nucleic acids [20]. The antitumor mechanism of fagaronine was believed to be due to the inhibition of topoisomerases [21]. In the K562 erythroleukemic cell line, fagaronine was found to induce cell differentiation. It was further found that fagaronine exerts a differentiating activity by a specific activation of the transcription factor GATA-1, regulating regions of genes involved in the erythroid phenotype expression [22].

Chelerythrine has also been found to show antitumor activity and to be a potent, selective antagonist of protein kinase C (PKC) from rat brain. Chelerythrine interacted with the catalytic domain of PKC as a competitive inhibitor with respect to the phosphate acceptor, and as a noncompetitive inhibitor with respect to ATP. This effect was further evidenced by the fact that chelerythrine inhibited native PKC and its catalytic fragment identically, but did not affect [^{3}H]-phorbol 12,13-dibutyrate binding to PKC. The potent antitumor activity of chelerythrine measured *in vitro* might thus be due, at least in part, to an inhibition of PKC [23]. In contrast to nitidine, chelerythrine preferentially binds to DNA containing contiguous GC base pairs [5′-TGGGGA-3′/3′-ACCCCT-5′] [19]. Toxic effects of chelerythrine on human and porcine hepatocytes in primary cultures were also demonstrated. Chelerythrine showed a dose- and time-dependent toxicity within the range of 25 to 100 μM [24].

After incubation of calf thymus DNA with chelerythrine or fagaronine in the presence of rat hepatic microsomes as metabolic activator and NADPH as cofactor, chelerythrine, but not fagaronine, was found to form DNA adducts, as detected by the ^{32}P-postlabeling technique. DNA adduct formation by chelerythrine was concentration-dependent. It was postulated that chelerythrine may be metabolized by hepatic microsomes to species which form adducts with DNA [25]. Studies on structure–activity relationship of benzo[*c*]phenanthridine alkaloids showed that analogues possessing a methylenedioxy moiety and at least one (preferably two) methoxy group were active as topoisomerase I inhibitors [26, 27].

Nitidine chloride and fagaronine chloride were reported to be inhibitors of avian myeloblastosis virus reverse transcriptase. They also demonstrated potent activity in the human immunodeficiency virus (HIV) type 1 reverse transcriptase system [28]. Nitidine and some related alkaloids have also been shown to posses strong inhibitory activity against reverse transcriptase of RNA tumor viruses [29]. The strong inhibitory activities on nucleic acid polymerizing enzyme activities, such as reverse transcriptase or DNA polymerase, were found to be template-directed. It was postulated that they inhibited DNA polymerase activity by interaction with A: T base pairs of the template primer [30, 31].

Nitidine showed a weak inhibitory effect on transcription activity in cell nucleus transcription systems from mouse liver and mouse HepA ascites hepatoma [32], and was found to inhibit calmodulin-dependent cyclic nucleotide phosphodiesterase [33, 34]. The inhibition of catechol *O*-methyltransferase and transfer RNA methyltransferases by nitidine and related compounds was also reported [35].

Besides the antitumor activity, nitidine was found to show an antimalarial effect. Nitidine inhibited the growth of *Plasmodium falciparum*, with an IC_{50} less than 270 ng ml^{-1}, and showed no cross-resistance with chloroquine [36, 37]. Acute cardiovascular effects of nitidine chloride following continuous intravenous infusion to anesthetized dogs was reported [38]. Fagaronine was also found to be an antimalarial agent against *P. falciparum*; the IC_{50} of fagaronine was reported as only 18 ng ml^{-1} [39].

Neither nitidine nor fagaronine showed any mutagenic activity in the somatic cells of *Drosophila melanogaster*. It was postulated that the alkoxy groups and the quaternary nitrogen of the benzo[*c*]phenanthridine might reduce or eliminate their mutagenicity [40].

References

1 Huang, Z.X. and Li, Z.H. (1980) Studies on the antitumor constituents of *Zanthoxylum nitidum* (Roxb.) DC. *Acta Chim. Sin.*, **38**, 535–542.

2 Ishii, H., Ishikawa, T., Akaike, M., Tohjoh, T., Toyoki, M., Ishikawa, M., Chen, I.S., and Lu, S. T. (1984) Studies on the chemical constituents of rutaceous plants. LIX. The chemical constituents of *Xanthoxylum nitidum* (Roxb.) D.C. (*Fagara nitida* Roxb.). 1. Examination of the alkaloidal fraction of the bark. *Yakugaku Zasshi*, **104**, 1030–1042.

3 Fang, S.D., Wang, L.K., and Hecht, S.M. (1993) Inhibitors of DNA topoisomerase I isolated from the roots of *Zanthoxylum nitidum*. *J. Org. Chem.*, **58**, 5025–5027.

4 Chen, Y.Z., Yang, L.L., Xu, B.J., and Huang, Z. X. (1989) Crystal structure of 7-demethyl-6-methoxy-5,6-dihydrochelerythrine, a new alkaloid from *Zanthoxylum nitidum*. *Acta Chim. Sin.*, **47**, 1048–1051.

5 Hu, J., Zhang, W.D., Liu, R.H., Zhang, C., Shen, Y.H., Li, H.L., Liang, M.J., and Xu, X.K. (2006) Benzophenanthridine alkaloids from *Zanthoxylum nitidum* (Roxb.) DC, and their analgesic and anti-inflammatory activities. *Chem. Biodivers.*, **3**, 990–995.

6 Deyun, K., Gray, A.I., Hartley, T.G., and Waterman, P.G. (1996) Alkaloids from an Australian accession of *Zanthoxylum nitidum* (Rutaceae). *Biochem. Syst. Ecol.*, **24**, 87–88.

7 Ishii, H., Imai, M., Johji, S., Tan, S., Chen, I.S., and Ishikawa, T. (1994) Studies on the chemical constituents of *Xanthoxylum nitidum* (Roxb.) D. C. (*Fagara nitida* Roxb.). II. Examination of the chemical constituents by membrane filtration: identification of magnoflorine, a water-soluble quaternary aporphine alkaloid. *Chem. Pharm. Bull. (Tokyo)*, **42**, 108–111.

8 Zee-Cheng, R.K.Y., and Cheng, C.C. (1975) Preparation and antileukemic activity of some alkoxybenzo[*c*]phenanthridinium salts and corresponding dihydro derivatives. *J. Med. Chem.*, **18**, 66–71.

9 Stermitz, F.R., Gillespie, J.P., Amoros, L.G., Romero, R., Stermitz, T.A., Larson, K.A., Earl, S., and Ogg, J.E. (1975) Synthesis and biological activity of some antitumor benzophenanthridinium salts. *J. Med. Chem.*, **18**, 708–713.

10 Fan, Y.J., Zhou, J., and Li, M. (1981) Effect of nitidine chloride on the life cycle of Ehrlich ascites carcinoma cells in mice. *Acta Pharmacol. Sin.*, **2**, 46–49.

11 Tang, W. (2002) Recent advances in antineoplastic principles of Traditional Chinese Medicines. *Pharmazie*, **57**, 223–232.

12 Kubova, N., Smekal, E., Kleinwachter, V., and Cushman, M. (1986) Binding properties of nitidine and its indenoisoquinoline analog with DNA. *Stud. Biophys.*, **114**, 251–256.

13 Gatto, B., Sanders, M.M., Yu, C., Wu, H.Y., Makhey, D., LaVoie, E.J., and Liu, L.F. (1996) Identification of topoisomerase I as the cytotoxic target of the protoberberine alkaloid coralyne. *Cancer Res.*, **56**, 2795–2800.

14 Makhey, D., Gatto, B., Yu, C., Liu, A., Liu, L.F., and LaVoie, E.J. (1996) Coralyne and related compounds as mammalian topoisomerase I and topoisomerase II poisons. *Bioorg. Med. Chem.*, **4**, 781–791.

15 Wang, L.K., Johnson, R.K., and Hecht, S.M. (1993) Inhibition of topoisomerase I function by nitidine and fagaronine. *Chem. Res. Toxicol.*, **6**, 813–818.

16 Gupta, R.S. (1985) Cross-resistance of vinblastine- and taxol-resistant mutants of Chinese hamster ovary cells to other anticancer drugs. *Cancer Treat. Rep.*, **69**, 515–521.

17 Gupta, R.S., Murray, W., and Gupta, R. (1988) Cross resistance pattern towards anticancer drugs of a human carcinoma multidrug-resistant cell line. *Br. J. Cancer*, **58**, 441–447.

18 Chen, Y.Z., Tang, G.Y., Xu, B.J., Wu, Q.J., Lu, C. Z., Li, J.Q., and Huang, Z.X. (1992) The formation and crystal structure of dihydronitidine and discussion of anticancer

mechanism of nitidine cation. *Sci. China Ser. B*, **35**, 1101–1109.
19 Bai, L.P., Zhao, Z.Z., Cai, Z., and Jiang, Z.H. (2006) DNA-binding affinities and sequence selectivity of quaternary benzophenanthridine alkaloids sanguinarine, chelerythrine, and nitidine. *Bioorg. Med. Chem.*, **14**, 5439–5445.
20 Pezzuto, J.M., Antosiak, S.K., Messmer, W.M., Slaytor, M.B., and Honig, G.R. (1983) Interaction of the antileukemic alkaloid, 2-hydroxy-3,8,9-trimethoxy-5-methylbenzo[*c*] phenanthridine (fagaronine), with nucleic acids. *Chem.-Biol. Interact.*, **43**, 323–339.
21 Larsen, A.K., Grondard, L., Couprie, J., Desoize, B., Comoe, L., Jardillier, J.C., and Riou, J.F. (1993) The antileukemic alkaloid fagaronine is an inhibitor of DNA topoisomerases I and II. *Biochem. Pharmacol.*, **46**, 1403–1412.
22 Dupont, C., Couillerot, E., Gillet, R., Caron, C., Zeches-Hanrot, M., Riou, J.F., and Trentesaux, C. (2005) The benzophenanthridine alkaloid fagaronine induces erythroleukemic cell differentiation by gene activation. *Planta Med.*, **71**, 489–494.
23 Herbert, J.M., Augereau, J.M., Gleye, J., and Maffrand, J.P. (1990) Chelerythrine is a potent and specific inhibitor of protein kinase C. *Biochem. Biophys. Res. Commun.*, **172**, 993–999.
24 Ulrichova, J., Dvorak, Z., Vicar, J., Lata, J., Smrzova, J., Sedo, A., and Simanek, V. (2001) Cytotoxicity of natural compounds in hepatocyte cell culture models. The case of quaternary benzo[*c*]phenanthridine alkaloids. *Toxicol. Lett.*, **125**, 125–132.
25 Stiborova, M., Simanek, V., Frei, E., Hobza, P., and Ulrichova, J. (2002) DNA adduct formation from quaternary benzo[c] phenanthridine alkaloids sanguinarine and chelerythrine as revealed by the ^{32}P-postlabeling technique. *Chem.- Biol. Interact.*, **140**, 231–242.
26 Li, D., Zhao, B., Sim, S.P., Li, T.K., Liu, A., Liu, L.F., and LaVoie, E.J. (2003) 2, 3-Dimethoxybenzo[*i*]phenanthridines: topoisomerase I-targeting anticancer agents. *Bioorg. Med. Chem.*, **11**, 521–528.
27 Prabo, S., Michel, S., Tillequin, F., Koch, M., Pfeiffer, B., Pierre, A., Leonce, S., Colson, P., Balseyrou, B., Lansiaux, A., and Bailly, C. (2004) Synthesis and cytotoxic activity of benzo [*c*][1,7] and [1,8]phenanthrolines analogues of nitidine and fagaronine. *Bioorg. Med. Chem.*, **12**, 3943–3953.
28 Tan, G.T., Pezzuto, J.M., Kinghorn, A.D., and Hughes, S.H. (1991) Evaluation of natural products as inhibitors of human immunodeficiency virus type 1 (HIV-1) reverse transcriptase. *J. Nat. Prod.*, **54**, 143–154.
29 Sethi, M.L. (1979) Inhibition of reverse transcriptase activity by benzophenanthridine alkaloids. *J. Nat. Prod.*, **42**, 187–196.
30 Sethi, V.S. (1976) Inhibition of mammalian and oncornavirus nucleic acid polymerase activities by alkoxybenzophenanthridine alkaloids. *Cancer Res.*, **36**, 2390–2395.
31 Sethi, V.S. (1977) Base specificity in the inhibition of oncornavirus reverse transcriptase and cellular nucleic acid polymerases by antitumor drugs. *Ann. N. Y. Acad. Sci.*, **284**, 508–524.
32 Zhao, Q., Li, M.S., and He, K.L. (1982) Effects of some anticancer drugs on transcription activity in nuclei. *Acta Acad. Med. Primae. Shanghai*, **9**, 171–177.
33 Yang, D.L., Gu, X.F., and Ma, J.C. (1995) Factors affecting calmodulin-dependent cyclic nucleotide phosphodiesterase activity observed by orthogonal design. *Chin. Biochem. J.*, **11**, 490–492.
34 Yang, D.L., Chen, Y., Gu, X.F., Ma, R.Q., Nie, L., and Ren, H. (1995) Effects of six effective components of Chinese Herbal drugs on calmodulin-dependent cyclic nucleotide phosphodiesterase. *Chin. Trad. Herbal Drugs*, **26**, 582–584.
35 Lee, J.W., MacFarlane, J.O., Zee-Cheng, R.K., and Cheng, C.C. (1977) Inhibition of catechol *O*-methyltransferase and transfer RNA methyltransferases by coralyne, nitidine, and related compounds. *J. Pharm. Sci.*, **66**, 986–989.
36 Jullian, V., Bourdy, G., Georges, S., Maurel, S., and Sauvain, M. (2006) Validation of use of a traditional antimalarial remedy from French Guiana, Zanthoxylum rhoifolium Lam. *J. Ethnopharmacol.*, **106**, 348–352.
37 Gakunju, D.M., Mberu, E.K., Dossaji, S.F., Gray, A.I., Waigh, R.D., Waterman, P.G., and Watkins, W.M. (1995) Potent antimalarial activity of the alkaloid nitidine, isolated from a Kenyan herbal remedy. *Antimicrob. Agents Chemother.*, **39**, 2606–2609.
38 Hamlin, R.L., Pipers, F.S., Mguyen, K., Milhalko, P., and Folk, R.M. (1976) Acute cardiovascular effects of nitidine chloride dihydrate (NSC-146397) following continuous intravenous infusion to anesthetized beagle hounds. US NTIS PB Rep PB-261267, p. 72 (CA 87: 145670).

39 Kassim, O.O., Loyevsky, M., Elliott, B., Geall, A., Amonoo, H., and Gordeuk, V.R. (2005) Effects of root extracts of *Fagara zanthoxyloides* on the *in vitro* growth and stage distribution of *Plasmodium falciparum*. *Antimicrob. Agents Chemother.*, **49**, 264–268.

40 Cheng, C.C., Engle, R.R., Hodgson, J.R., Ing, R.B., Wood, H.B. Jr, Yan, S.J., and Zee-Cheng, R.K. (1977) Absence of mutagenicity of coralyne and related antileukemic agents: structural comparison with the potent carcinogen 7,12-dimethylbenz[*a*]anthracene. *J. Pharm. Sci.*, **66**, 1781–1783.

Zingiber officinale

Rhizoma Zinigiberis (Ganjiang) is the dried rhizome of *Zingiber officinale* Rosc. (Zingiberaceae). It is used as a stomachic, antiemetic, antidiarrheal, expectorant, antiasthmatic, hemostatic, and cardiotonic agent for the treatment of several gastrointestinal and respiratory diseases.

Zingiber officinale Rosc.

Rhizoma Zinigiberis Praeparatum (Paojiang) is the prepared dried rhizome of *Zingiber officinale*. It is used as a stomachic, antiemetic, antidiarrheal, and hemostatic agent.

Rhizoma Zingiberis Recens (Shengjiang) is the fresh rhizome of *Zingiber officinale*. It is used for the treatment of gastrointestinal and respiratory disorders, and is also a well-known spice.

Two galenic preparations of the ginger rhizome are officially listed in the Chinese Pharmacopoeia: Extractum Zingiberis Liquidum and "Jiang Ding." Extractum Zingiberis Liquidum (Jiang Liujingao) is the fluid extract of Rhizoma Zingiberis and is used as a stomachic and carminative agent. "Jiang Ding" is the tincture of Rhizoma Zingiberis. It is prepared from Extractum Zingiberis Liquidum; it is also used as carminative and stomachic agent.

Chemistry

The rhizome of *Z. officinale*, ginger, is known to contain pungent components and flavoring principles. The major pungent principles of ginger were identified as zingerone, shogaols, and gingerols. Gingerols are 1-(4-hydroxy-3-methoxyphenyl)-5-hydroxyalkan-3-ones with an *S*(+)-configuration, and are designated as [3]-gingerol, [4]-gingerol, [5]-gingerol, [6]-gingerol, [8]-gingerol, and [10]-gingerol, according to their side chains of different length [1]. Zingerone and shogaols are reported to be formed during isolation of gingerols [2].

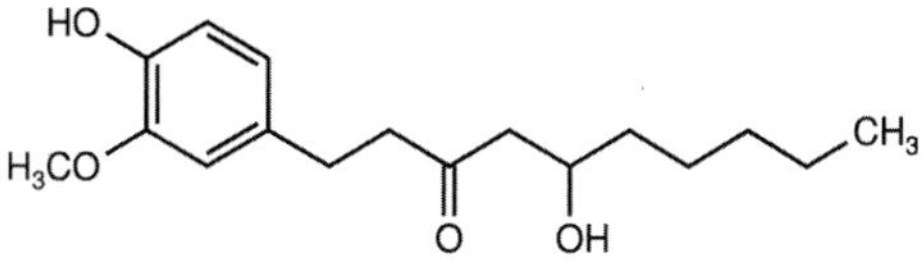

[6]-Gingerol

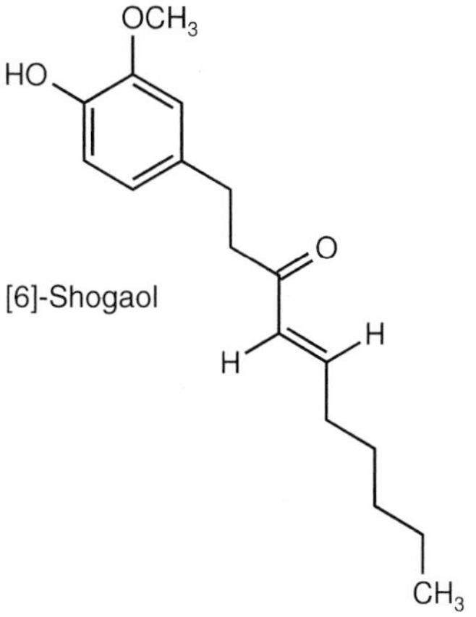

[6]-Shogaol

Handbook of Chinese Medicinal Plants: Chemistry, Pharmacology, Toxicology
Weici Tang and Gerhard Eisenbrand

ISBN: 978-3-527-32226-8

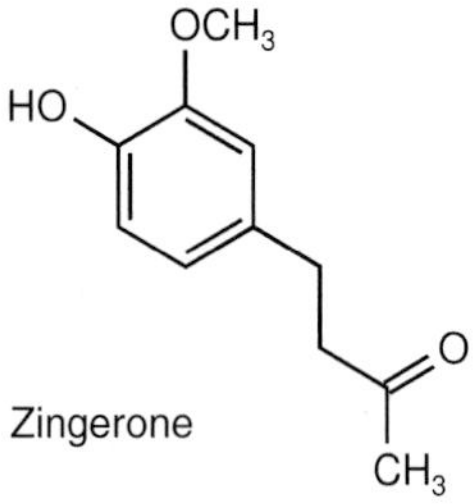

Zingerone

Minor constituents related to gingerol were identified as gingediols, methylgingediols, and their diacetates, gingerdiones, dehydrogingerdiones [1] and [6]-gingesulfonic acid [3]. Gingerdiones might be intermediates in gingerol biosynthesis. Glucosides of [6]-gingerdiol were isolated from fresh ginger [4].

Diarylheptane derivatives from ginger were identified as gingerenones A, B, C, isogingerenone B, and related heptane or heptanone derivatives [5]. The diarylheptane derivatives are known to be the major constituents in the root and rhizome of several *Curcuma* species.

Glycosides monoacyldigalactosylglycerols, gingerglycolipids A, B, and C, and angelicoidenol 2-*O*-β-D-glucopyranoside were also isolated from ginger rhizome [3]. Gingerglycolipid A is 1-(9,12,15-octadecatrienoyl)-glycerol 3-digalactoside, whereas gingerglycolipid B and gingerglycolipid C represent the corresponding 9,12-octadecadienoyl and 9-octadecenoyl analogues.

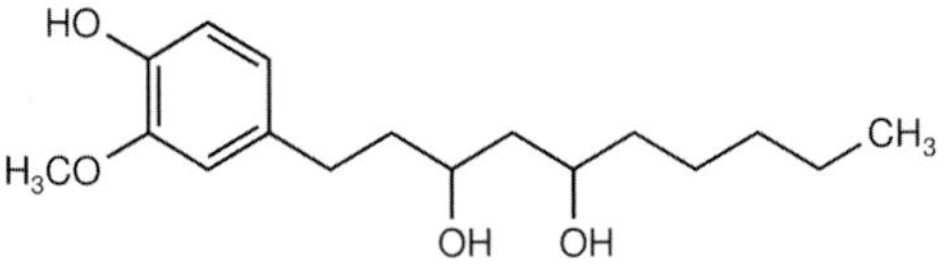

[6]-Gingerdiol

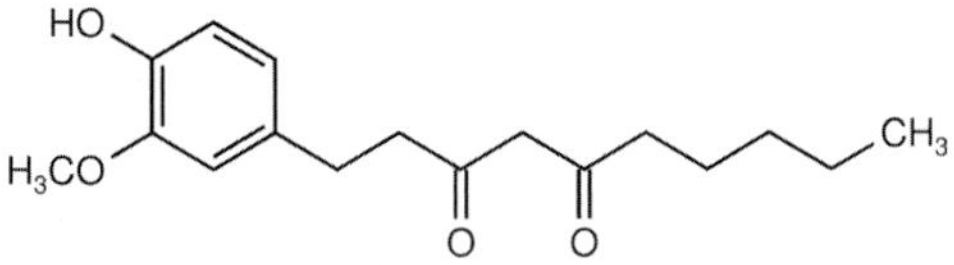

[6]-Gingerdione

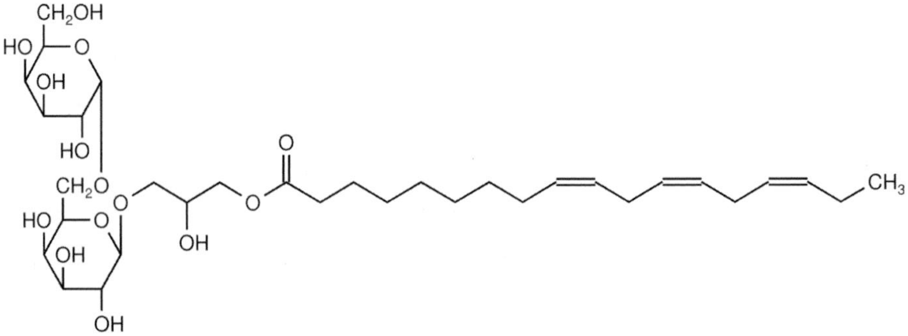

Gingerglycolipid A

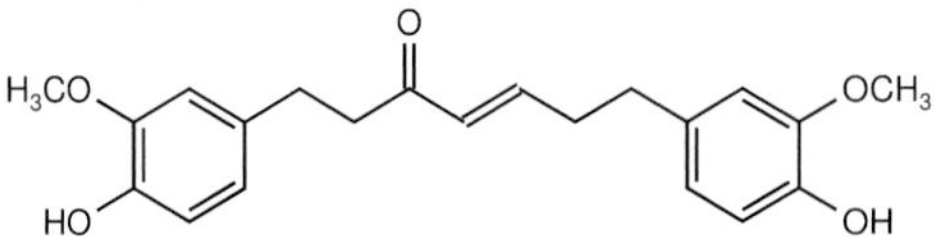

Gingerenone A

In addition, [E]-8β,17-epoxylabd-12-ene-15,16-dial, was isolated from the rhizome [6].

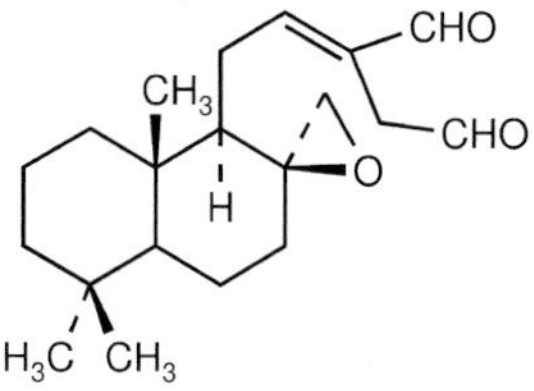

[E]-8ß,17-Epoxylabd-12-ene-15,16-dial

The purification and characterization of ginger proteases has been described. Ginger proteases GP-I and GP-II were found to cleave peptides and proteins. The enzyme GP-II is a glycoprotein composed of 221 amino acids possessing two N-linked oligosaccharide chains at Asn99 and Asn156. The three-dimensional structure of GP-II has been determined using X-ray crystallography [7]. Both proteases, GP-I and GP-II, which are 82% similar, have a cysteine residue at position 27 and histidine at position 161, corresponding to the essential cysteine–histidine diads found in the papain family, and six cysteine residues that form the three invariant disulfide linkages seen in this family [8]. Some divalent metal ions, such as Hg^{2+}, Cu^{2+}, Cd^{2+}, and Zn^{2+} strongly inhibited these enzymes.

A number of flavoring components were isolated and identified from the essential oil of *Z. officinalis*, with zingiberene as the major component [9]. Other volatile components include β-bisabolene, borneol, bornyl acetate, camphene, cineole, citral, cumene, ar-curcumene, *p*-cymene, β-elemene, farnesene, geraniol, limonene, linalool, myrcene, β-phellandrene, α-pinene, β-pinene, sabinene, γ-selinene, *cis*- and *trans*-sesquiphellandrol, β-sesquiphellandrene, sesquithujene, sesquisabinene hydrate, and zingiberenol [10, 11]. The Chinese Pharmacopoeia requires a quantitative determination of essential oil in Rhizoma Zinigiberis. The content of essential oil in Rhizoma Zinigiberis should be not less than 0.8%.

Pharmacology and Toxicology

The rhizome of *Z. officinale*, ginger, is a well-known spice, exhibiting a wide pharmacological spectrum, including antiemetic, anti-inflammatory, analgesic, antipyretic, antimicrobial, hepatoprotective, antioxidative,

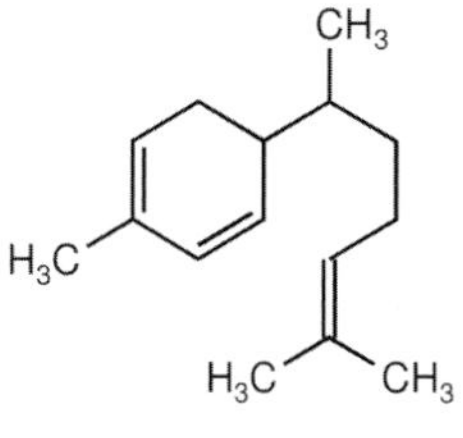

Zingiberene

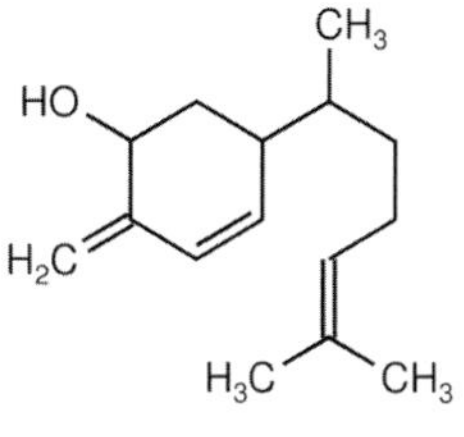

Sesquiphellandrol

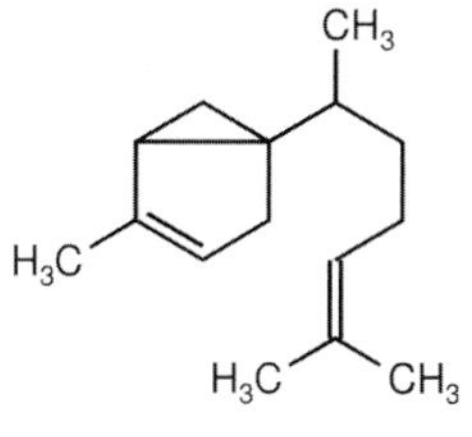

Sesquithujene

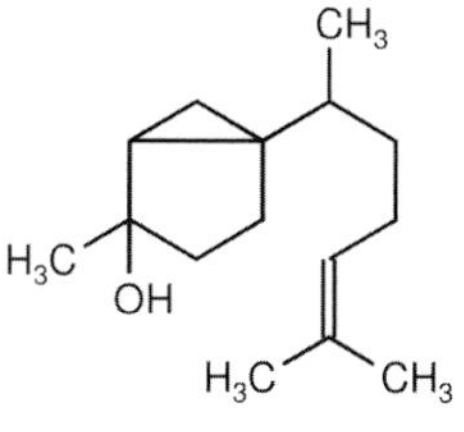

Sesquisabinene hydrate

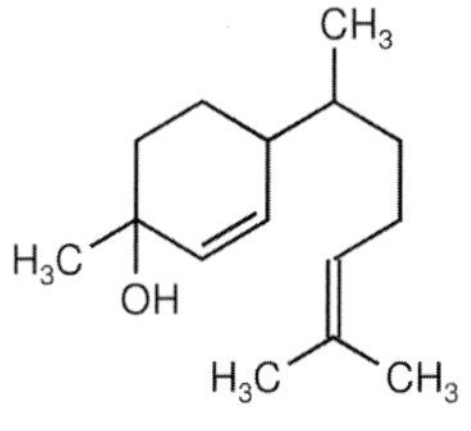

Zingiberenol

hypoglycemic, and hypolipidemic activities [12]. Ginger extract significantly inhibited Gram-positive and Gram-negative bacteria [12]. Ginger and gingerols were also reported to inhibit the growth of *Helicobacter pylori* [13]. β-Sesquiphellandrene from ginger was found to be the most active compound, showing antirhinoviral activity with an IC_{50} of about 0.4 μ*M* against rhinovirus IB *in vitro* [14]. Molluscicidal and antischistosomal activities of ginger were also described. [7]-Gingerol was reported to be the major active principle in ginger responsible for its molluscicidal activity against *Biomphalaria glabrata*, and its antischistosomal activity on different stages of *Schistosoma mansonii* [15].

The oral administration of [6]-gingerol to rats at doses of 70–140 mg kg^{-1}, and the intravenous (i.v.) administration of [6]-gingerol at a dose of 1.75–3.5 mg kg^{-1}, produced an inhibition of spontaneous motor activity, showed antipyretic and analgesic effects, and prolonged hexabarbital-induced sleeping times [16]. Ginger was also shown to be effective in the treatment and prophylaxis of migraine [17]. Clinical studies showed that ginger exhibited anti-motion sickness activity [18]. The mechanism of the anti-motion sickness actions of ginger was suggested to be caused by the central and peripheral anticholinergic and antihistaminic effects [19]. Ginger also reduced vertigo in clinical studies [20].

Ginger has been used clinically for the treatment of nausea and vomiting, especially in gynecology and oncology [21]. Double-blind randomized clinical trials revealed that ginger was effective for the treatment of hyperemesis gravidarum in pregnant women [22]. It was also reported that ginger can be used for the treatment of postoperative nausea and vomiting after major gynecological surgery [23]. In cancer therapy, ginger was reported to relieve vomiting induced by cytotoxic agents. The antiemetic efficacy of ginger against cisplatin-induced emesis was also observed in dogs [24]. Acetone and ethanol extracts of ginger, given orally at doses of 100, 200 and 500 mg kg^{-1}, and ginger juice at doses of 2 and 4 ml kg^{-1}, significantly reversed the cisplatin-induced delay in gastric emptying in rats [25]. The antiemetic effect of ginger was not associated with an effect on gastric emptying [26]. In contrast, in a randomized, double-blind crossover trial in 48 gynecologic cancer patients receiving cisplatin-based chemotherapy, the antiemetic activity of ginger could not be confirmed. In this clinical trial, ginger was given orally to patients in the form of capsules containing ginger powder at a daily dose of 1 g per patient for five days, starting on the first day of chemotherapy. Among 43 evaluable patients who received both cycles of treatment, no advantage in reducing nausea or vomiting in the acute phase of cisplatin-induced emesis could be observed [27].

In rats, the oral administration of ginger extract at a dose of 500 mg kg^{-1} resulted in highly significant cytoprotection against gastric lesions induced by 80% ethanol, 0.6 *M* HCl, 0.2 *M* NaOH, or 25% NaCl. The extract also prevented the occurrence of gastric ulcers induced by nonsteroidal anti-inflammatory drugs and hypothermic restraint stress [28]. The antiulcer principles in ginger were identified as [6]-gingerol, zingiberene, β-sesquiphellandrene, β-bisabolene, ar-curcumene, and 6-gingesulfonic acid. The oral administration to rats of the acetone extract of ginger composed of volatile oil and bitter substances at 75 mg kg^{-1}, or gingerols at 5 mg kg^{-1}, stimulated gastrointestinal motility and enhanced charcoal transport [29]. In a clinical study, ginger extract (2 × 100 mg) significantly increased the interdigestive antral motility during phase III of the migrating motor complex in healthy volunteers. The oral administration of ginger improved gastroduodenal motility in the fasting state, and also after a standard test meal [30].

Ginger was reported to exert hepatoprotective effects. An ethanol extract of ginger significantly inhibited the hepatotoxicity in rats induced by carbon tetrachloride (CCl_4) and acetaminophen (paracetamol), and decreased the serum activities of alanine aminotransferase (ALT), aspartate aminotransferase (AST), lactate dehydrogenase (LDH), alkaline phosphatase (ALP) as well as sorbitol dehydrogenase and glutamate dehydrogenase of intoxicated rats [31]. The antioxidant status in liver such as activities of superoxide dismutase (SOD), catalase (CAT), glutathione peroxidase and glutathione-*S*-transferase (GST), and levels of reduced glutathione (GSH) diminished by acetaminophen pretreatment were normalized by treatment with ginger extract [32]. The hepatoprotective effect of ginger extract was also confirmed by a histopathological examination of rat liver.

Ginger is used in traditional Chinese medicine as a cardiotonic agent. The methanolic extract of ginger and gingerols have been found to possess potent positive inotropic effects on isolated guinea pig left atria. Cardiotonic activity decreased in the order [8]-gingerol, [10]-gingerol, and [6]-gingerol. [8]-Gingerol was reported to activate the Ca^{2+}-pumping ATPase, which is rather specific to sarcoplasmic reticulum Ca^{2+}-ATPase activity [33]. The stimulation of sarcoplasmic reticulum Ca^{2+}-ATPase was also observed by [6]-gingerol and [10]-gingerol [34]. [6]-Gingerol at a concentration of 1 μM exerted significant positive inotropic effects in failing human myocardium under basal experimental conditions, and also normalized post-rest behavior. In addition, [6]-gingerol increased sarcoplasmic reticulum Ca^{2+}-uptake significantly in myocardial homogenates [35]. [6]-Gingerol and [8]-gingerol were each found to potentiate the contraction of isolated mouse and rat blood vessels induced by a series of prostaglandins (PG) such as $PGF_{2\alpha}$, PGI_2, PGE_2, and PGI_2, and inhibited the contraction produced by PGD_2, thromboxane A_2 (TXA_2), leukotriene C_4 (LTC_4) and LTD_4. This suggested that [6]-gingerol or [8]-gingerol affect the responses induced by eicosanoids. The aliphatic hydroxyl group present in gingerols is necessary for the potentiation of $PGF_{2\alpha}$-induced contraction [36].

A significant decrease in blood glucose, serum total cholesterol and serum alkaline phosphatase, as well as a significant increase in serum high-density lipoprotein (HDL), were found in rats fed diets containing 0.5% ginger [37]. The oral administration of 8β,17-epoxylabd-12-ene-15,16-dial from ginger exerted an inhibitory effect on the cholesterol biosynthesis in livers of treated mice and rats [6]. In apolipoprotein E-deficient mice, the oral administration of ginger extract in the diet at a daily dose of 250 μg per mouse for 10 weeks resulted in significant reductions of aortic atherosclerotic lesion areas, in plasma triglycerides, in cholesterol, in very low-density lipoprotein (VLDL), and in LDL. These results were associated with a marked reduction in cellular cholesterol biosynthesis in peritoneal macrophages derived from treated mice. Treatment with 250 μg ginger extract also reduced the basal level of LDL-associated lipid peroxides. It was suggested that the dietary consumption of ginger extract by apolipoprotein E-deficient mice would significantly attenuate the development of atherosclerotic lesions associated with a significant reduction in plasma cholesterol and and LDL levels, and a significant reduction in the LDL basal oxidative state [38]. A hypolipidemic effect was also observed in cholesterol-fed rabbits, when ginger was given orally as an ethanol extract at a daily dose of 200 mg kg^{-1} for 10 weeks [39].

Ginger was found to exert a hypoglycemic effect in type I diabetic rats, as induced by streptozotocin. In normoglycemic rats, an intraperitoneal (i.p.) injection of serotonin at a dose of 1 mg kg^{-1}, produced hyperglycemia and hypoinsulinemia, which was significantly prevented by ginger juice. Diabetic rats, as induced by streptozotocin, showed a significant increase in fasting glucose levels, associated with a significant decrease in serum insulin levels. Treatment of the diabetic rats with ginger significantly increased the insulin levels and decreased the fasting glucose levels. In an oral glucose tolerance test, treatment with ginger significantly decreased the area under the curve (AUC) value of glucose, and increased the AUC value of insulin in diabetic rats. It was suggested that ginger exhibited a hypoglycemic effect in type I diabetic rats, possibly involving serotonin receptors [40].

[8]-Gingerol was also reported to inhibit platelet aggregation induced by arachidonic acid and collagen, in a concentration-dependent manner. It did not inhibit platelet aggregation induced by platelet-activating factor (PAF) and thrombin. The maximal antiplatelet effect was obtained when platelets were incubated with [8]-gingerol for 30 min. The inhibition was reversible. It was suggested that the antiplatelet action of [8]-gingerol is mainly due to the inhibition of TX formation [41]. The oral administration of ginger to patients with coronary artery disease at a dose of 10 g resulted in a significant reduction in platelet aggregation *ex vivo*, as induced by ADP or epinephrine (adrenaline) [42].

Ginger extract was further found to exhibit a strong antioxidative effect against the oxidation of linoleic acid [43]. The antioxidative effect was also observed for [6]-gingerol [44]. The efficacy of zingerone in scavenging superoxide anions was demonstrated by nitroblue tetrazolium reduction in the xanthine–xanthine oxidase system [45]. Ginger significantly lowered lipid peroxidation in rats by maintaining the activities of the antioxidant enzymes, including superoxide dismutase, catalase and glutathione peroxidase. The serum GSH content was significantly increased in ginger-fed rats [46].

Ginger also exhibited anti-inflammatory activity. [6]-Gingerol, [6]-gingerdione, [8]-gingerdione, [6]-dehydrogingerdione and [8]-dehydrogingerdione were all potent inhibitors of prostaglandin biosynthesis. [6]-Gingerol, [6]-gingerdione and related compounds were found to inhibit 5-lipoxygenase activity and PGE_2 formation in intact human neutrophils [47]. The inhibition of arachidonic acid metabolism by [6]-gingerol and related compounds could account in part for their anti-inflammatory properties [48]. Ginger was also reported to be useful in the treatment of rheumatic disorders, with patients with rheumatic and musculoskeletal disorders reporting relief from pain and associated symptoms on ginger administration [49].

The anti-inflammatory activity of ginger rhizome has been demonstrated in U937 cells stimulated by lipopolysaccharide (LPS). The crude extract of ginger potently inhibited the LPS-induced production of prostaglandin E_2 (PGE_2) in U937 cells, with an IC_{50} value below 0.1 $\mu g\ ml^{-1}$, but did not effectively suppress the production of tumor necrosis factor-α (TNF-α). Furthermore, extracts containing either predominantly gingerols or shogaols were both highly active at inhibiting LPS-induced PGE_2 production. On the other hand, an extract containing predominantly gingerols was capable of inhibiting LPS-induced cyclooxygenase-2 (COX-2) expression, whereas an extract containing predominantly shogaol had no effect on COX-2 expression [50]

Ginger extract and [6]-gingerol were shown to be mutagenic in *Salmonella typhimurium* TA 100 and TA 1535 with metabolic activation [51]. When ginger juice was added to a solution of *N*-methyl-*N*′-nitro-*N*-nitrosoguanidine (MNNG), it significantly increased its mutagenic activity [52]. An aqueous extract of ginger applied at doses corresponding to 0.5, 1, 2, 5, and 10 $g\ kg^{-1}$ to male mice suppressed the clastogenic effects in bone marrow cells following colchicine injection [53].

The chemopreventive effect of ginger seemed to be due to the presence of gingerols, shogaols, and related compounds. A number of mechanisms that might be involved in the chemopreventive effects of ginger have been reported from experimental studies [54]. In anticarcinogenicity studies, [6]-gingerol was found to significantly reduce the multiplicity of intestinal neoplasms in rats, as induced by azoxymethane [55]. The pre-application of an ethanol extract of ginger onto the skin of mice resulted in a significant inhibition of 12-*O*-tetradecanoylphorbol 13-acetate (TPA)-mediated induction of epidermal ornithine decarboxylase (ODC), cyclooxygenase and lipoxygenase activities and ODC mRNA expression, in a dose-dependent manner. The pre-application of ginger extract to mouse skin also afforded significant inhibition of TPA-induced epidermal edema and hyperplasia. In long-term studies, topical application of ginger extract 30 min prior to each TPA application to 7,12-dimethylbenz[*a*]anthracene (DMBA)-initiated mice resulted in a highly significant protection against skin tumor incidence and multiplicity. These results provide clear evidence that ginger extract possesses antitumor-promoting effects. The antipromotion mechanism of ginger may involve the inhibition of cellular, biochemical, and molecular changes in mouse skin caused by tumor promoters [56]. Ginger extract also inhibited (TPA)-induced Epstein–Barr virus (EBV) early antigen expression in Raji cells [57].

[6]-Gingerol was also found to exert inhibitory effects on the viability and DNA synthesis of human promyelocytic leukemia HL-60 cells. The cytotoxic and antiproliferative effects of [6]-gingerol were associated with an induction of apoptotic cell death [58]. A structure–activity relationship study on the cytotoxic and apoptosis-inducing activities of diarylheptanes from ginger in HL-60 cells indicated that the appropriate alkyl side-chain length, the *ortho*-diphenoxyl function in the aromatic ring, the α,β-unsaturated ketone moiety in the side chain, and the acetoxyl groups at the 3- and 5-positions of the side chain were structural determinants for the cytotoxicity [59].

A patented standardized ginger extract given to rats had no significant effect on blood glucose levels, and no significant effects on coagulation parameters. It neither decreased systolic blood pressure nor increased heart rate in rats [60]. Pregnant female rats fed a ginger tea instead of drinking water corresponding to a concentration of up to 50 $g\,l^{-1}$ from gestation day 6 to day 15 did not show any maternal toxicity. However, the embryonic loss in the treatment groups was double that of the controls. No gross morphologic malformations were seen in the fetuses of treated animals. Fetuses exposed to ginger tea were found to be significantly heavier than controls, an effect that was greater in female fetuses and was not correlated with increased placental size. Treated fetuses also had more advanced skeletal development. It was suggested that *in utero* exposure to ginger tea might result in an increased early embryo loss, with increased growth in surviving fetuses [61].

The pharmacokinetics of [6]-gingerol after i.v. administration to rats was reported. After i.v. administration at a dose of 3 $mg\,kg^{-1}$, the plasma concentration–time curve of [6]-gingerol was described by a two-compartment open model. [6]-Gingerol was rapidly cleared from plasma with a terminal half-life of about 7 min, and a total body clearance of about 17 $ml\,min^{-1}\,kg^{-1}$. The serum protein binding of [6]-gingerol was 92% [62]. The pharmacokinetics of [6]-gingerol after i.v. administration to rats with acute renal failure induced by bilateral nephrectomy were found to be similar to those of normal rats. In contrast, the pharmacokinetics of [6]-gingerol after i.v. administration to rats with acute hepatic failure induced by a single oral administration of carbon tetrachloride (CCl_4) were found to cause elevated plasma concentrations of [6]-gingerol at the terminal phase and to increase the elimination half-life [63]. Two diastereomers of [6]-gingerdiol were formed by the enzymatic metabolism of [6]-gingerol using rat liver postmitochondrial supernatant fraction *in vitro*. The ratio of the two isomers formed was about 1:5, suggesting a stereospecific reduction of [6]-gingerol by carbonyl reductase [64].

References

1 Kikuzaki, H., Tsai, S.M., and Nakatani, N. (1992) Constituents of Zingiberaceae. Part 5. Gingerdiol related compounds from the rhizomes of *Zingiber officinale*. *Phytochemistry*, **31**, 1783–1786.

2 Connell, D.W. and Sutherland, M.D. (1969) A re-examination of gingerol, shogaol, and zingerone, the pungent principles of ginger (*Zingiber officinale* Roscoe). *Austr. J. Chem.*, **22**, 1033–1043.

3 Yoshikawa, M., Yamaguchi, S., Kunimi, K., Matsuda, H., Okuno, Y., Yamahara, J., and Murakami, N. (1994) Stomachic principles in ginger. III. An anti-ulcer principle, 6-gingesulfonic acid, and three monoacyldigalactosylglycerols, gingerglycolipids A, B, and C, from Zingiberis Rhizoma originating in Taiwan. *Chem. Pharm. Bull. (Tokyo)*, **42**, 1226–1230.

4 Sekiwa, Y., Kubota, K., and Kobayashi, A. (2000) Isolation of novel glucosides related to gingerdiol from ginger and their antioxidative activities. *J. Agric. Food Chem.*, **48**, 373–377.

5 Kikuzaki, H., Kobayashi, M., and Nakatani, N. (1991) Constituents of Zingiberaceae. Part 4. Diarylheptanoids from rhizomes of *Zingiber officinale*. *Phytochemistry*, **30**, 3647–3651.

6 Tanabe, M., Chen, Y.D., Saito, K., and Kano, Y. (1993) Cholesterol biosynthesis inhibitory component from *Zingiber officinale* Roscoe. *Chem. Pharm. Bull. (Tokyo)*, **41**, 710–713.

7 Choi, K.H., Laursen, R.A., and Allen, K.N. (1999) The 2.1 A structure of a cysteine protease with proline specificity from ginger rhizome, *Zingiber officinale*. *Biochemistry*, **38**, 11624–11633.

8 Choi, K.H. and Laursen, R.A. (2000) Amino-acid sequence and glycan structures of cysteine proteases with proline specificity from ginger rhizome *Zingiber officinale*. *Eur. J. Biochem.*, **267**, 1516–1526.

9 Millar, J.G. (1998) Rapid and simple isolation of zingiberene from ginger essential oil. *J. Nat. Prod.*, **61**, 1025–1026.

10 Denyer, C.V., Jackson, P., Loakes, D.M., Ellis, M.R., and Young, D.A. (1994) Isolation of antirhinoviral sesquiterpenes from ginger (*Zingiber officinale*). *J. Nat. Prod.*, **57**, 658–662.

11 Terhune, S.J., Hogg, J.W., Bromstein, A.C., and Lawrence, B.M. (1975) Four new sesquiterpene analogs of common monoterpenes. *Can. J. Chem.*, **53**, 3285–3293.

12 Ojewole, J.A. (2006) Analgesic, antiinflammatory and hypoglycaemic effects of ethanol extract of *Zingiber officinale* Roscoe rhizomes (Zingiberaceae) in mice and rats. *Phytother. Res.*, **20**, 764–772.

13 Mahady, G.B., Pendland, S.L., Yun, G.S., Lu, Z. Z., and Stoia, A. (2003) Ginger (Zingiber officinale Roscoe) and the gingerols inhibit the growth of Cag A + strains of *Helicobacter pylori*. *Anticancer Res.*, **23** (A), 3699–3702.

14 Denyer, C.V., Jackson, P., Loakes, D.M., Ellis, M.R., and Young, D.A. (1994) Isolation of antirhinoviral sesquiterpenes from ginger (*Zingiber officinale*). *J. Nat. Prod.*, **57**, 658–662.

15 Adewunmi, C.O., Oguntimein, B.O., and Furu, P. (1990) Molluscicidal and antischistosomal activities of *Zingiber officinale*. *Planta Med.*, **56**, 374–376.

16 Suekawa, M., Ishige, A., Yuasa, K., Sudo, K., Aburada, M., and Hosoya, E. (1984) Pharmacological studies on ginger. I. Pharmacological actions of pungent constituents, (6)-gingerol and (6)-shogaol. *J. Pharmacobio-Dyn.*, **7**, 836–848.

17 Mustafa, T. and Srivastava, K.C. (1990) Ginger (*Zingiber officinale*) in migraine headache. *J. Ethnopharmacol.*, **29**, 267–273.

18 Holtmann, S., Clarke, A.H., Scherer, H., and Hohn, M. (1989) The anti-motion sickness mechanism of ginger. A comparative study with placebo and dimenhydrinate. *Acta Otolaryngol. (Stockh.)*, **108**, 168–174.

19 Qian, D.S. and Liu, Z.S. (1992) Pharmacologic studies of anti-motion sickness actions of ginger. *Chin. J. Integr. Trad. West. Med.*, **12**, 95–98.

20 Grontved, A. and Hentzer, E. (1986) Vertigo-reducing effect of ginger root. A controlled clinical study. *ORL J. Otorhinolaryngol. Relat. Spec.*, **48**, 282–286.

21 Ernst, E., and Pittler, M.H. (2000) Efficacy of ginger for nausea and vomiting: a systematic review of randomized clinical trials. *Br. J. Anaesth.*, **84**, 367–371.

22 Aikins Murphy, P. (1998) Alternative therapies for nausea and vomiting of pregnancy. *Obstet. Gynecol.*, **91**, 149–155.

23 Phillips, S., Ruggier, R., and Hutchinson, S.E. (1993) *Zingiber officinale* (ginger) - an antiemetic for day case surgery. *Anaesthesia*, **48**, 715–717.

24 Sharma, S.S., Kochupillai, V., Gupta, S.K., Seth, S.D., and Gupta, Y.K. (1997) Antiemetic efficacy of ginger (*Zingiber officinale*) against cisplatin-induced emesis in dogs. *J. Ethnopharmacol.*, **57**, 93–96.

25 Sharma, S.S. and Gupta, Y.K. (1998) Reversal of cisplatin-induced delay in gastric emptying in rats by ginger (*Zingiber officinale*). *J. Ethnopharmacol.*, **62**, 49–55.

26 Phillips, S., Hutchinson, S., and Ruggier, R. (1993) *Zingiber officinale* does not affect gastric emptying rate. A randomised, placebo-controlled, crossover trial. *Anaesthesia*, **48**, 393–395.

27 Manusirivithaya, S., Sripramote, M., Tangjitgamol, S., Sheanakul, C., Leelahakom, S., Thavaramara, T., and Tangcharoenpanich, K. (2004) Antiemetic effect of ginger in gynecologic oncology patients receiving cisplatin. *Int. J. Gynecol. Cancer*, **14**, 1063–1069.

28 al-Yahya, M.A., Rafatullah, S., Mossa, J.S., Ageel, A.M., Parmar, N.S., and Tariq, M. (1989) Gastroprotective activity of ginger (*Zingiber officinale* rocs.) in albino rats. *Am. J. Chin. Med.*, **17**, 51–56.

29 Yamahara, J., Huang, Q.R., Li, Y.H., Xu, L., and Fujimura, H. (1990) Gastrointestinal motility enhancing effect of ginger and its active constituents. *Chem. Pharm. Bull. (Tokyo)*, **38**, 430–431.

30 Micklefield, G.H., Redeker, Y., Meister, V., Jung, O., Greving, I., and May, B. (1999) Effects of ginger on gastroduodenal motility. *Int. J. Clin. Pharmacol. Ther.*, **37**, 341–346.

31 Yemitan, O.K., and Izegbu, M.C. (2006) Protective effects of *Zingiber officinale* (Zingiberaceae) against carbon tetrachloride and acetaminophen-induced hepatotoxicity in rats. *Phytother. Res.*, **20**, 997–1002.

32 Ajith, T.A., Hema, U., and Aswathy, M.S. (2007) *Zingiber officinale* Roscoe prevents acetaminophen-induced acute hepatotoxicity by enhancing hepatic antioxidant status. *Food Chem. Toxicol.*, **45**, 2267–2272.

33 Kobayashi, M., Shoji, N., and Ohizumi, Y. (1987) Gingerol, a novel cardiotonic agent, activates the $Ca2^+$-pumping ATPase in skeletal and cardiac sarcoplasmic reticulum. *Biochim. Biophys. Acta*, **903**, 96–102.

34 Ohizumi, Y., Sasaki, S., Shibusawa, K., Ishikawa, K., and Ikemoto, F. (1996) Stimulation of sarcoplasmic reticulum Ca (2 +)-ATPase by gingerol analogues. *Biol. Pharm. Bull.*, **19**, 1377–1379.

35 Maier, L.S., Schwan, C., Schillinger, W., Minami, K., Schutt, U., and Pieske, B. (2000) Gingerol, isoproterenol and ouabain normalize impaired post-rest behavior but not force-frequency relation in failing human myocardium. *Cardiovasc. Res.*, **45**, 913–924.

36 Kimura, I., Kimura, M., and Pancho, L.R. (1989) Modulation of eicosanoid-induced contraction of mouse and rat blood vessels by gingerols. *Jpn. J. Pharmacol.*, **50**, 253–261.

37 Ahmed, R.S. and Sharma, S.B. (1997) Biochemical studies on combined effects of garlic (*Allium sativum* Linn) and ginger (*Zingiber officinale* Rosc.) in albino rats. *Indian J. Exp. Biol.*, **35**, 841–843.

38 Fuhrman, B., Rosenblat, M., Hayek, T., Coleman, R., and Aviram, M. (2000) Ginger extract consumption reduces plasma cholesterol, inhibits LDL oxidation and attenuates development of atherosclerosis in atherosclerotic, apolipoprotein E-deficient mice. *J. Nutr.*, **130**, 1124–1131.

39 Bhandari, U., Sharma, J.N., and Zafar, R. (1998) The protective action of ethanolic ginger (*Zingiber officinale*) extract in cholesterol fed rabbits. *J. Ethnopharmacol.*, **61**, 167–171.

40 Akhani, S.P., Vishwakarma, S.L., and Goyal, R. K. (2004) Anti-diabetic activity of *Zingiber officinale* in streptozotocin-induced type I diabetic rats. *J. Pharm. Pharmacol.*, **56**, 101–105.

41 Guh, J.H., Ko, F.N., Jong, T.T., and Teng, C.M. (1995) Antiplatelet effect of gingerol isolated from *Zingiber officinale*. *J. Pharm. Pharmacol.*, **47**, 329–332.

42 Bordia, A., Verma, S.K., and Srivastava, K.C. (1997) Effect of ginger (*Zingiber officinale* Rosc.) and fenugreek (*Trigonella foenumgraecum* L.) on blood lipids, blood sugar and platelet aggregation in patients with coronary artery disease. *Prostaglandins Leukot. Essent. Fatty Acids*, **56**, 379–384.

43 Shobana, S., and Naidu, K.A. (2000) Antioxidant activity of selected Indian spices. *Prostaglandins Leukot. Essent. Fatty Acids*, **62**, 107–110.

44 Aeschbach, R., Loliger, J., Scott, B.C., Murcia, A., Butler, J., Halliwell, B., and Aruoma, O.I. (1994) Antioxidant actions of thymol, carvacrol, 6-gingerol, zingerone and hydroxytyrosol. *Food Chem. Toxicol.*, **32**, 31–36.

45 Krishnakantha, T.P., and Lokesh, B.R. (1993) Scavenging of superoxide anions by spice principles. *Indian J. Biochem. Biophys.*, **30**, 133–134.

46 Ahmed, R.S., Seth, V., and Banerjee, B.D. (2000) Influence of dietary ginger (*Zingiber officinales* Rosc) on antioxidant defense system in rat: comparison with ascorbic acid. *Indian J. Exp. Biol.*, **38**, 604–606.

47 Suekawa, M., Yuasa, K., Isono, M., Sone, H., Ikeya, Y., Sakakibara, I., Aburada, M., and Hosoya, E. (1986) Pharmacological studies on ginger. IV. Effect of [6]-shogaol on the arachidonic cascade. *Nippon Yakurigaku Zasshi*, **88**, 263–269.

48 Srivastava, K.C. and Mustafa, T. (1989) Ginger (*Zingiber officinale*) and rheumatic disorders. *Med. Hypotheses*, **29**, 25–28.

49 Srivastava, K.C., and Mustafa, T. (1992) Ginger (*Zingiber officinale*) in rheumatism and musculoskeletal disorders. *Med. Hypotheses*, **39**, 342–348.

50 Lantz, R.C., Chen, G.J., Sarihan, M., Solyom, A.M., Jolad, S.D., and Timmermann, B.N. (2007) The effect of extracts from ginger rhizome on inflammatory mediator production. *Phytomedicine*, **14**, 123–128.

51 Nagabhushan, M., Amonkar, A.J., and Bhide, S.V. (1987) Mutagenicity of gingerol and shogaol and antimutagenicity of zingerone in *Salmonella*/microsome assay. *Cancer Lett.*, **36**, 221–233.

52 Nakamura, H. and Yamamoto, T. (1982) Mutagen and anti-mutagen in ginger, *Zingiber officinale. Mutat. Res.*, **103**, 119–126.

53 Mukhopadhyay, M.J., and Mukherjee, A. (2000) Clastogenic effect of ginger rhizome in mice. *Phytother. Res.*, **14**, 555–557.

54 Shukla, Y. and Singh, M. (2007) Cancer preventive properties of ginger: a brief review. *Food Chem. Toxicol.*, **45**, 683–690.

55 Yoshimi, N., Wang, A., Morishita, Y., Tanaka, T., Sugie, S., Kawai, K., Yamahara, J., and Mori, H. (1992) Modifying effects of fungal and herb metabolites on azoxymethane-induced intestinal carcinogenesis in rats. *Jpn. J. Cancer Res.*, **83**, 1273–1278.

56 Park, K.K., Chun, K.S., Lee, J.M., Lee, S.S., and Surh, Y.J. (1998) Inhibitory effects of [6]-gingerol, a major pungent principle of ginger, on phorbol ester-induced inflammation, epidermal ornithine decarboxylase activity and skin tumor promotion in ICR mice. *Cancer Lett.*, **129**, 139–144.

57 Vimala, S., Norhanom, A.W., and Yadav, M. (1999) Anti-tumour promoter activity in Malaysian ginger rhizobia used in traditional medicine. *Br. J. Cancer*, **80**, 110–116.

58 Surh, Y. (1999) Molecular mechanisms of chemopreventive effects of selected dietary and medicinal phenolic substances. *Mutat. Res.*, **428**, 305–327.

59 Wei, Q.Y., Ma, J.P., Cai, Y.J., Yang, L., and Liu, Z.L. (2005) Cytotoxic and apoptotic activities of diarylheptanoids and gingerol-related compounds from the rhizome of Chinese ginger. *J. Ethnopharmacol.*, **102**, 177–184.

60 Weidner, M.S., and Sigwart, K. (2000) The safety of a ginger extract in the rat. *J. Ethnopharmacol.*, **73**, 513–520.

61 Wilkinson, J.M. (2000) Effect of ginger tea on the fetal development of Sprague-Dawley rats. *Reprod. Toxicol.*, **14**, 507–512.

62 Ding, G.H., Naora, K., Hayashibara, M., Katagiri, Y., Kano, Y., and Iwamoto, K. (1991) Pharmacokinetics of 6-gingerol after intravenous administration in rats. *Chem. Pharm. Bull. (Tokyo)*, **39**, 1612–1614.

63 Naora, K., Ding, G., Hayashibara, M., Katagiri, Y., Kano, Y., and Iwamoto, K. (1992) Pharmacokinetics of 6-gingerol after intravenous administration in rats with acute renal or hepatic failure. *Chem. Pharm. Bull. (Tokyo)*, **40**, 1295–1298.

64 Surh, Y.J. and Lee, S.S. (1994) Enzymic reduction of 6-gingerol, a major pungent principle of ginger, in the cell-free preparation of rat liver. *Life. Sci*, **54**, PL321–PL326.

Ziziphus jujuba and *Ziziphus jujuba* var. *spinosa*

Fructus Jujubae (Dazao) is the dried ripe fruit of *Ziziphus jujuba* Mill. (Rhamnaceae). It is mainly used as a general tonic and sedative, and also for the treatment of anorexia and hysteria in women. The fresh fruits of *Ziziphus jujuba* are also used as a common fruit. The dried fruits are used as a food supplement.

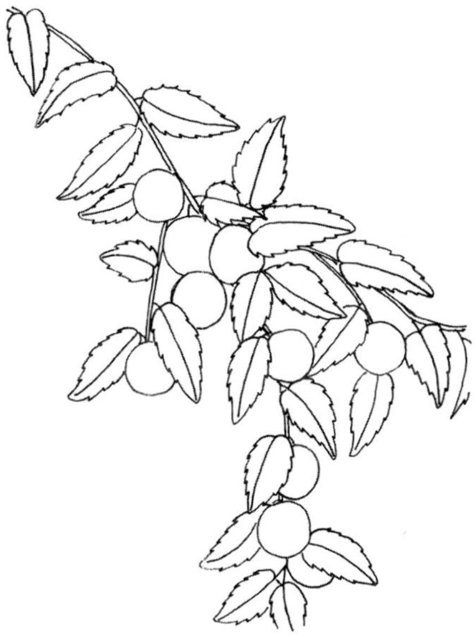

Ziziphus jujuba Mill.

Semen Ziziphi Spinosae (Suanzaoren) is the dried ripe seed of *Ziziphus jujuba* Mill. var. *spinosa* (Bunge) Hu ex H. F. Chou. It is used as a sedative and tonic agent for the treatment of insomnia and dream-disturbed sleep.

Chemistry

The fruit of *Z. jujuba* [1] and the semen of *Z. jujuba* var. *spinosa* [2] were reported to contain triterpene saponins, jujuboside A and jujuboside B with jujubogenin as sapogenin. The Chinese Pharmacopoeia requires a qualitative determination of jujuboside A and jujuboside B in Semen Ziziphi Spinosae by thin-layer chromatographic (TLC) comparison with reference substances.

Triterpenes from the fruits of *Z. jujuba* were identified as betulinic acid, alphitolic acid, dihydroalphitolic acid methyl ester, betulonic acid, oleanonic acid, maslinic acid, oleanolic acid, and ursolic acid [3]. Triterpenes from the seeds of *Z. jujube* var. *spinosa* were identified as betulin, betulinic acid, ceanothic acid, alphitolic acid, and daucosterol [4]. The Chinese Pharmacopoeia requires a qualitative determination of oleanolic acid in Fructus Jujubae by TLC comparison with reference substance.

In addition to triterpenes and triterpene saponins, alkaloids and cyclopeptides were isolated from the fruits and seeds of *Z. jujuba* and *Z. jujuba* var. *spinosa*. Zizyphusine, a quaternary aporphine alkaloid, nornuciferine, lysicamine and the cyclopeptide daechucyclopeptide-1 were isolated from the fruits of *Z. jujube* [5]. Zizyphusine, nuciferine, nornuciferine, *N*-methylasimilobine, norisocorydine, caaverine, and coclaurine were isolated from the seeds of *Z. jujuba* var. *spinosa* [6]. A cyclopeptide isolated from the seeds of *Z. jujuba* var. *spinosa* was identified as frangufoline [7]. Triterpenes from the fruits of *Z. jujuba* were identified as colubrinic acid, alphitolic acid, 3-*O*-*cis*-*p*-coumaroylalphitolic acid, 3-*O*-*trans*-*p*-coumaroylalphitolic acid, 3-*O*-*cis*-*p*-coumaroylmaslinic acid, 3-*O*-*trans*-*p*-coumaroylmaslinic acid, betulinic acid,

Handbook of Chinese Medicinal Plants: Chemistry, Pharmacology, Toxicology
Weici Tang and Gerhard Eisenbrand

ISBN: 978-3-527-32226-8

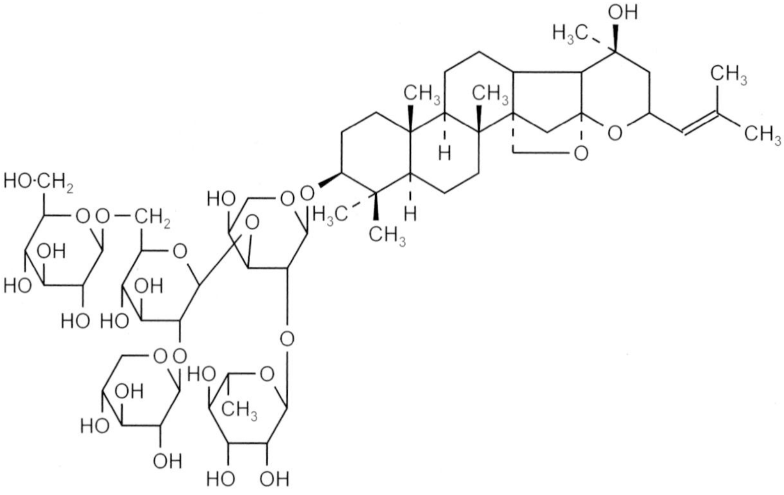

Jujuboside A

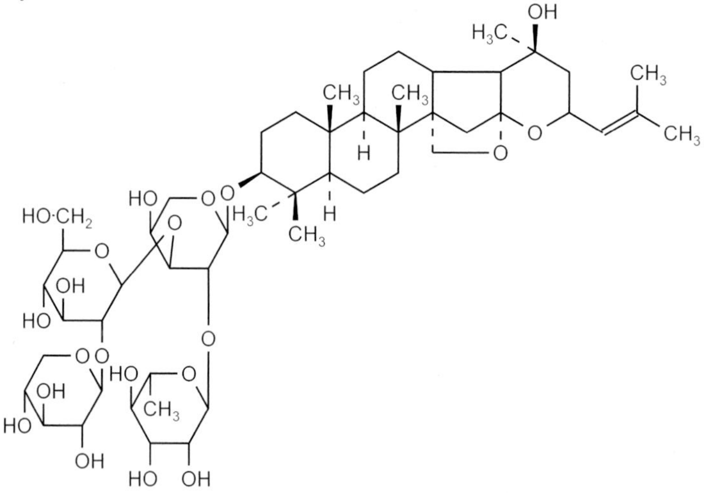

Jujuboside B

oleanolic acid, betulonic acid, oleanonic acid, and zizyberenalic acid [8].

Flavone *C*-glycosides swertisin, spinosin and the acylated derivatives of spinosin were isolated from the seeds of *Z. jujuba* [9]. Flavone *C*-glycosides from the seeds of *Z. jujuba* var. *spinosa* were identified as zivulgarin and spinosin. Spinosin is swertisin 2′-*O*-β-D-glucopyranoside, whereas zivulgarin is swertisin 4′-*O*-β-D-glucopyranoside [2].

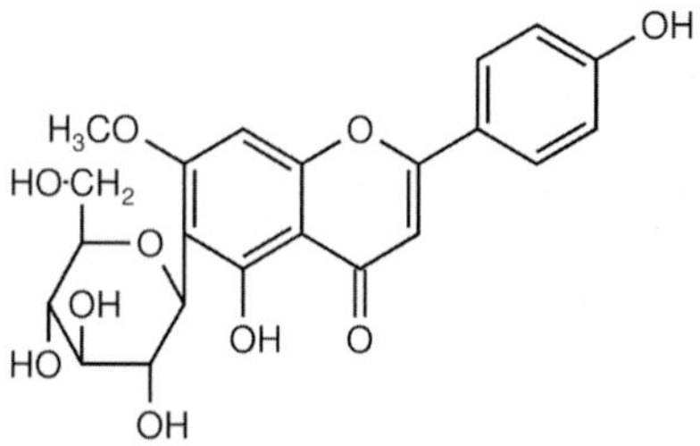

Swertisin

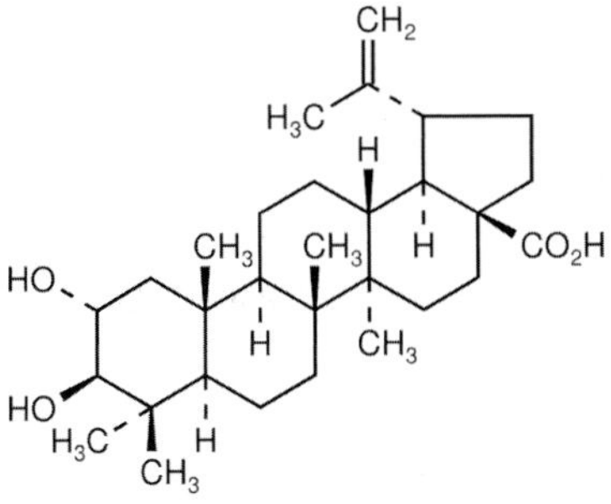

Alphitolic acid

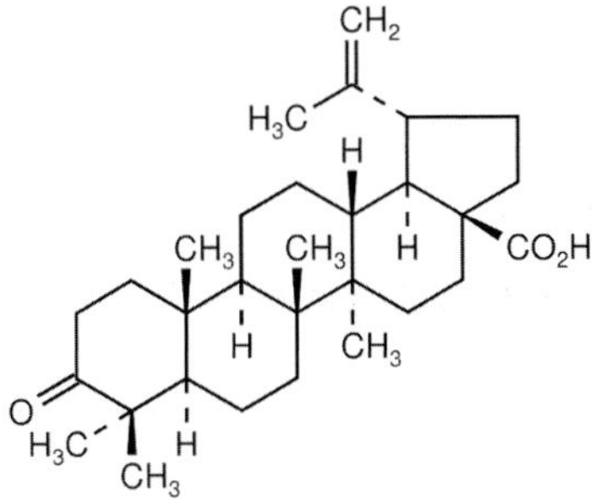

Betulonic acid

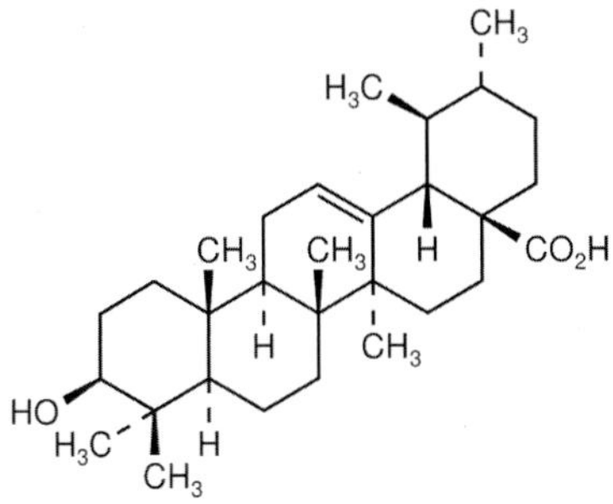

Ursolic acid

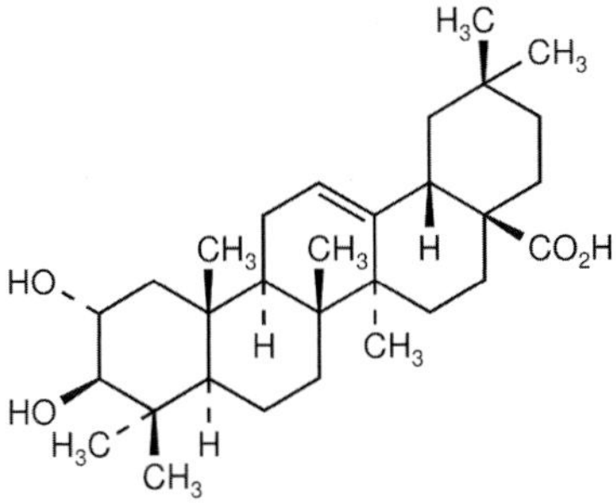

Maslinic acid

The fruits of *Z. jujuba* are rich in ascorbic acid. Immature fruits were found to contain as much as 0.6–0.8% of free vitamin C, together with about 0.3% in bound form. The free form of vitamin C decreased when the fruits reached complete ripeness, and decreased further during storage [10]. This loss was ascribed to oxidation by ascorbic acid oxidase. Fruit products were high in vitamin C. The fruits of *Z. jujuba* also contained esters, carboxylates, ketones, alcohols, and aldehydes, amino acids [11], and polysaccharide [12].

Interestingly, relatively high amounts of cAMP and cGMP were detected in the fruits of *Z. jujuba*. The cAMP and cGMP contents in dried fruits were reported to be 100–500 and 30–50 $nmol\,g^{-1}$, respectively [13]. In another study, however, only 3 $nmol\,g^{-1}$ cAMP and cGMP, respectively, were found in the dry fruits of *Z. jujuba*, by radioimmunoassay [14].

Pharmacology and Toxicology

The seed of *Z. jujuba* var. *spinosa* was reported to exert inhibitory effects on the central nervous system (CNS). The fruits exhibited a synergistic effect with pentobarbital and thiopental. Jujuboside A had no inhibiting effect, but exerted a synergism with phenylalanine on functions of the CNS [15]. The flavone *C*-glycosides swertsin, spinosin and the acylspinosins from *Zizyphus* seeds were found to exhibit mild sedative activity in animal experiments, with swertisin being the most effective compound [16]. The alkaloids nornuciferine, lysicamine, daechucyclopeptide-1 and

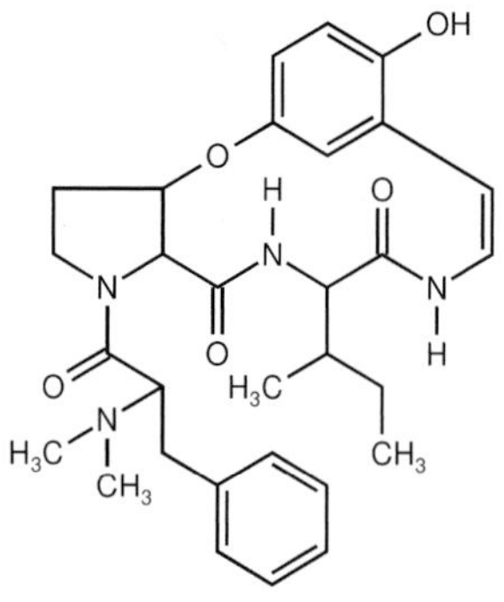

Daechucyclopeptide-1

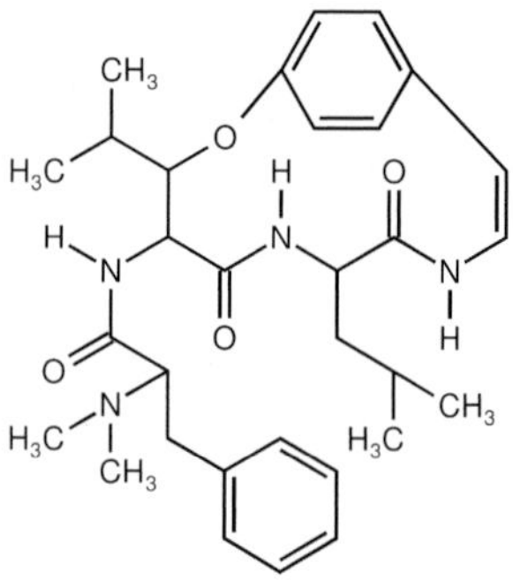

Frangufoline

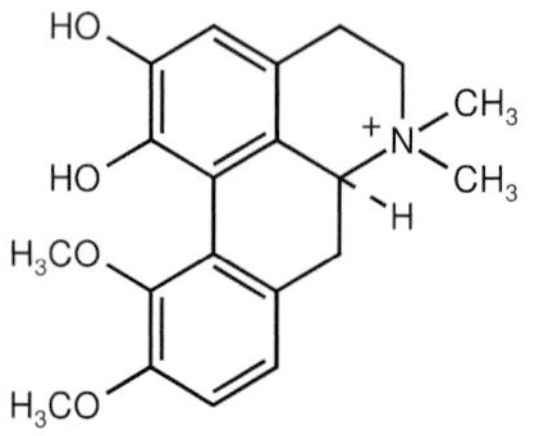

Zizyphusine

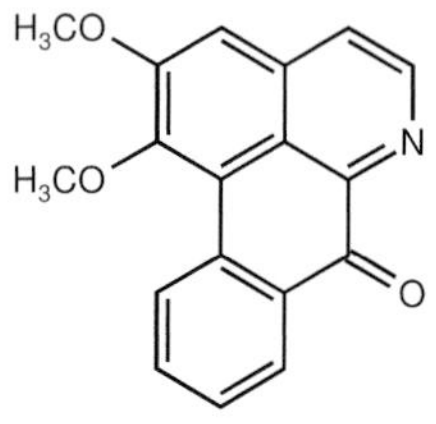

Lysicamine

frangufoline, were all reported to exhibit sedative effects and to prolong the sleeping time of mice, as induced by hexobarbital [5, 10].

A methanol extract of the seeds of *Z. jujuba* var. *spinosa* was reported to protect cultured rat cerebellar granule neurons from cell death induced by *N*-methyl-D-aspartate (NMDA). Pretreatment with the extract inhibited NMDA (1 m*M*)-induced elevation of cytosolic calcium concentration and the generation of reactive oxygen species (ROS) in rat cerebellar granule neurons caused by NMDA. This suggested that the extract of the seeds might prevent neuronal cell damage *in vitro* induced by NMDA [17]. The inhibitory effect of jujuboside A on the CNS was also demonstrated in rats [18]. In addition, jujuboside A significantly blocked glutamate release in rat hippocampus, as induced by penicillin. Moreover, jujuboside A significantly inhibited the intracellular Ca^{2+} increase in cultured hippocampal neurons induced by glutamate. Trifluoperazine, as a calmodulin antagonist, exerted a similar inhibitory effect as jujuboside A. This indicates jujuboside A to have inhibitory effects on the glutamate-mediated excitatory signal pathway in the hippocampus, and to probably act through an anti-calmodulin action [19].

The total saponin from the seeds of *Z. jujuba* had a marked hypotensive effect in rats and cats. It did not affect the hypertensive effect of noradrenaline in rats, but markedly inhibited the pressor reflex in cats, as induced by occlusion of the common carotid artery. In rats, total saponin also inhibited the spontaneous discharges of the preganglionic fibers of the greater splanchnic nerve, and decreased the plasma renin activity. Therefore, the hypotensive effect of the total saponin was probably central in origin [20]. The seed of *Z. jujuba* was shown to exert an anxiolytic effect in mice, after oral administration [21].

3-*O*-*cis*-*p*-coumaroylalphitolic acid, 3-*O*-*trans*-*p*-coumaroylalphitolic acid, betulinic acid and betulonic acid were found to show cytotoxic activity against a series of tumor cell lines, including K562, B16(F-10), SK-MEL-2, PC-3, LOX-IMVI, and A549 [8]. In HepG2 human hepatoma cells, the extract of the fruit of *Z. jujuba* inhibited proliferation, induced cell-cycle arrest in G_1 (100 μg ml^{-1}) and in G_2/M phase (200 μg ml^{-1}), and also induced cell apoptosis. In addition, an increase in intracellular ROS levels, a decline in mitochondrial membrane potential at low concentrations, and a ROS-independent mitochondrial dysfunction pathway at high concentrations of the extract, were observed in tumor cells. The cell-cycle arrest at G_1 in HepG2 cells was associated with a decrease in phosphorylated Rb, whereas the cell-cycle arrest at G_2/M correlated with a decrease in the p27(Kip1) levels. This indicated that the extract of *Z. jujuba* exhibited a concentration-dependent effect on apoptosis, and a differential cell-cycle arrest in HepG2 cells [22].

Polysaccharides from the seeds of *Z. jujuba* var. *spinosa* enhanced both cellular and humoral immune function in mice, and also protected mice against radiation injury [23]. The fruits of *Z. jujuba* were reported to be active against lipid peroxidation in mice. The oral administration of a fruit extract of *Z. jujuba* to mice significantly enhanced the erythrocyte superoxide dismutase activity, and significantly decreased the malondialdehyde content [24]. The fruit extract of *Z. jujuba* was also found to scavenge oxygen free radicals and to inhibit the lipid peroxidation of mouse liver homogenate; to decrease hyaluronic acid depolymerization induced by oxygen; and to inhibit the adenosine deaminase activity of mice liver homogenate [25]. Oral administration of the seed oil or seed extract of *Z. jujuba* var. *spinosa* to rats and rabbits significantly reduced serum total cholesterol, low-density lipoprotein (LDL), and triglycerides; and significantly increased high-density lipoprotein (HDL) [26]. The interaction of calmodulin with jujuboside A was studied. ^1H–NMR experiments showed that jujuboside A has two types of binding site on calmodulin: one site locates in the N-terminus, the other locates in the C-terminal region of the polypeptide chain. One calmodulin molecule can bind at least two jujuboside A molecules. The binding of jujuboside A affected the environments of lysine residues, especially Lys-74 and Lys-94, suggesting that jujuboside A binds to calmodulin through hydrophobic interaction [27].

Ziziphin was found to suppress sweetness taste in human [28]. The sweetness-inhibitory mechanism of ziziphin was identified as a taste modification [29].

References

1 Li, S.Z. and Zhang, B. (1983) Pharmacological and chemical studies on Da Zao (*Ziziphus jujuba*). *Chin. Trad. Herbal Drugs*, **14**, 471–475.

2 Zeng, L., Zhang, R.Y. and Wang, X. (1986) Constituents of the Chinese traditional drug - Suanzaoren. *Acta Bot. Sin.*, **28**, 517–522.

3 Bai, G., Ren, Y.L., Zhang, B., and Zhang, H.X. (1992) Studies on chemical constituents of *Zizyphus jujuba* in Hebei China. *Chem. Res. Chin. Univ.*, **8**, 177–179.

4 Zeng, L., Zhang, R.Y., and Wang, X. (1987) Constituents of *Ziziphus spinosus*. *Acta Pharm. Sin.*, **22**, 114–120.

5 Han, B.H. and Park, M.H. (1987) Sedative activity and the active components of ziziphi fruits. *Arch. Pharmacol. Res.*, **10**, 208–211.

6 Han, B.H., Park, M.H., and Han, Y.N. (1989) Aporphine and tetrahydrobenzylisoquinoline alkaloids from the seeds of *Zizyphus vulgaris* var. *spinosus*. *Arch. Pharmacol. Res.*, **12**, 263–268.

7 Han, B.H., and Park, M.H. (1987) Alkaloids are the sedative principles of the seeds of *Zizyphus vulgaris* var. *spinosus*. *Arch. Pharmacol. Res.*, **10**, 203–207.

8 Lee, S.M., Min, B.S., Lee, C.G., Kim, K.S., and Kho, Y.H. (2003) Cytotoxic triterpenoids from the fruits of *Zizyphus jujuba*. *Planta Med.*, **69**, 1051–1054.

9 Woo, W.S., Kang, S.S., Wagner, H., Seligmann, O., and Chari, V.M. (1980) Acylated flavone-C-glycosides from seeds of *Zizyphus jujuba*. *Phytochemistry*, **19**, 2791–2793.

10 Yao, G.S., Li, Y.J., Chang, X.Q., and Lu, J.H. (1983) Vitamin C content in vegetables and fruits in Shenyang (China) market during four seasons. *Acta Nutr. Sin.*, **5**, 373–379.

11 Wang, H., Zhang, H., and Sun, Y. (1988) Amino acids and chemical constituents of *Zizyphus jujube* Mill in Liaoxi. *Jilin Daxue Ziran Kexue Xuebao*, Issue 2, 89–92.

12 Lang, X.C., Li, M.X., Jia, B.Y., Wu, S.X., Li, L.F., Zhao, S.Y., Shi, S.L., and Zhang, Z.Z. (1991) Experimental research of the enhancing effect of polysaccharides in semen of *Ziziphus spinosa* on the immune function of mice injured by radiation. *China J. Chin. Mater. Med.*, **16**, 366–368.

13 Cyong, J.C. and Takahashi, M. (1982) Guanosine 3′,5′-monophosphate activity in fruits of *Zizyphus jujube*. *Chem. Pharm. Bull. (Tokyo)*, **30**, 1081–1083.

14 Liu, J.R., and Liu, S.T. (1983) Determination of cAMP- and cGMP-like substances in traditional Chinese medicines and their preparations. *Shanxi Med. J.*, **12**, 131–132.

15 Wu, S.X., Zhang, J.X., Xu, T., Li, L.F., Zhao, S.Y., and Lan, M.Y. (1993) Effects of seeds, leaves and fruits of *Ziziphus spinosa* and jujuboside A on central nervous system function. *China J. Chin. Mater. Med.*, **18**, 685–687.

16 Yuan, C.L., Wang, Z.B., Jiao, Y., Cao, A.M., Huo, Y.L., and Cui, C.X. (1987) Sedative and hypnotic constituents of flavonoids in seeds of Ziziphi spinosae. *Bull. Chin. Mater. Med.*, **12**, 546–548.

17 Park, J.H., Lee, H.J., Koh, S.B., Ban, J.Y., and Seong, Y.H. (2004) Protection of NMDA-induced neuronal cell damage by methanol extract of zizyphi spinosi semen in cultured rat cerebellar granule cells. *J. Ethnopharmacol.*, **95**, 39–45.

18 Shou, C., Feng, Z., Wang, J., and Zheng, X. (2002) The inhibitory effects of jujuboside A on rat hippocampus *in vivo* and *in vitro*. *Planta Med.*, **68**, 799–803.

19 Zhang, M., Ning, G., Shou, C., Lu, Y., Hong, D., and Zheng, X. (2003) Inhibitory effect of jujuboside A on glutamate-mediated excitatory signal pathway in hippocampus. *Planta Med.*, **69**, 692–695.

20 Gu, W.X., Liu, J.F., Zhang, J.X., Liu, X.M., Liu, S.J., and Chen, Y.R. (1987) A study of the hypotensive action of total saponin of *Ziziphus jujuba* Mill. and its mechanism. *J. Med. Coll. PLA*, **2**, 315–318.

21 Peng, W., Hsieh, M., Lee, Y., Lin, Y., and Liao, J. (2000) Anxiolytic effect of seed of *Ziziphus jujuba* in mouse models of anxiety. *J. Ethnopharmacol.*, **72**, 435–441.

22 Huang, X., Kojima-Yuasa, A., Norikura, T., Kennedy, D.O., Hasuma, T., and Matsui-Yuasa, I. (2007) Mechanism of the anti-cancer activity of *Zizyphus jujuba* in HepG2 cells. *Am. J. Chin. Med.*, **35**, 517–532.

23 Lang, X.C., Li, M.X., Jia, B.Y., Wu, S.X., Li, L.F., and Zhao, S.Y. (1988) Effects of the seeds of *Ziziphus spinosa* Hu on the immune function of mice. *Bull. Chin. Mater. Med.*, **13**, 43–45.

24 Liu, X.F., Gao, W.H., Zhao, Z.L., Dai, F.Q., Li, X.D., and Gao, C.L. (1994) Extracts of the fruits of *Ziziphus jujuba* and *Lycium barbarum*: actions against lipid peroxidation in mice. *Chin. J. Prev. Med.*, **28**, 254.

25 Wang, W. and Chen, W.W. (1991) Antioxidative activity studies on the meaning of same original of herbal drug and food. *Chin. J. Integr. Trad. West. Med.*, **11**, 159–161.

26 Wu, S.X., Li, L.F., Lan, X.C., Zhao, S.Y., Qi, H., Liu, C.P., Lan, M.Y., and Zhang, J.Y. (1991) Effects of semen Ziziphus spinosae and Ziziphus spinosae extract on decrease of serum lipoprotein and inhibition of platelet aggregation. *China J. Chin. Mater. Med.*, **16**, 435–437.

27 Zhou, Y., Li, Y., Wang, Z., Ou, Y., and Zhou, X. (1994) ^{1}H-NMR and spin-labeled EPR studies on the interaction of calmodulin with jujuboside A. *Biochem. Biophys. Res. Commun.*, **202**, 148–154.

28 Suttisri, R., Lee, I.S., and Kinghorn, A.D. (1995) Plant-derived triterpenoid sweetness inhibitors. *J. Ethnopharmacol.*, **47**, 9–26.

29 Smith, V.V. and Halpern, B.P. (1983) Selective suppression of judged sweetness by ziziphin. *Physiol. Behav.*, **30**, 867–874.

Index

a

Abrisapogenol 1
Abrus cantoniensis 1
Acacetin 268
Acacic acid lactone 53
Acacigenin 53
Acanthopanax gracilistylus 5
Acanthopanax senticosus 9
Acemannan 83
Acerinol 320
Acetoxychavicol acetate 88
Acetoxyeugenol acetate 88
Acetylacteol 318
Acetylangelicin 371
Acetylindirubin 661
Acetyloxy-epoxylathyrol phenylacetate 532
Acetylrinderine 527
Acetyl-toosendantriol 753
Achyranthes bidentata 16
Achyranthoside 16
Aconitum carmichaeli 23
Aconitum kusnezoffii 23
Acorus calamus 31
Acorus tatarinowii 31
Acteoside 334
Actinidine 788
Acutumine 1134
Adlumidine 412
Afzelin 629
Agrimonia pilosa 36
Ailanthone 45
Ailanthus altissima 45
Ailantinol 45
Ajoene 63
Ajugol 1009
Ajugoside 511
Akebia quinata 50
Akebia trifoliata 50
Akebia trifoliata var. *australis* 50
Akuammigine 1214
Alaternin 292
Albiflorin 811
Albizia julibrissin 53
Aliphitolic acid 1261
Alisma orientalis 57
Alismol 58
Alismoxide 58
Alisol 57
Alkannin 186
Allicin 63
Alliin 63
Allinase 63
Allium chinensis 64
Allium macrostemon 64
Allium sativum 63
Allium tuberosum 64
Allocryptopine 1244
Alloimperatorin 127
Alloisoimperatorin 127
Aloctin 81
Aloe 80
Aloe barbadensis 80
Aloe ferox 80
Aloe resin 80
Aloe-emodin 1018
Aloe-emodin anthrone 288
Aloesin 80
Aloin 80
Alpinia galanga 88
Alpinia katsumadai 93
Alpinia officinarum 96
Alpinia oxyphylla 102
Amarogentin 588
Amarolide 45
Amentoflavone 1109
Amomum compactum 106

Handbook of Chinese Medicinal Plants: Chemistry, Pharmacology, Toxicology
Weici Tang and Gerhard Eisenbrand

ISBN: 978-3-527-32226-8

Amomum kravanh 106
Amomum tsao-ko 106
Amygdalin 11, 962
Anagyrine 1145
Andrographis paniculata 110
Andrographolide 110
Andromedotoxin 1025
Anemarrhena asphodeloides 121
Anemone raddeana 17
Anemonin 989
Anethole 545, 645
Angelica dahurica 127
Angelica dahurica var. *formosana* 127
Angelica pubescens biserrata 139
Angelica sinensis 143
Angelicin 127
Angelol 139
Angeloylhamaudol 1062
Angeloylretronecine 186
Anisaldehyde 545, 645
Anisatin 646
Anisodamine 465
Anisodine 465
Annuadiepoxide 192
Annulide 192
Anthraquinones 2
Apigenin 268
Apioglycyrrhizin 603
Apocynum venetum 153
Apterin 861
Aquilegiolide 1134
Arbutin 995
Arctigenin 158
Arctiin 158
Arctium lappa 158
Ardisia crenata 163
Ardisia japonica 163
Ardisinol 163
Areca catechu 169
Areca nut 169
Arecaine 169
Arecoline 169
Aristolochia contorta 176
Aristolochia debilis 176
Aristolochic acid 214
Arjunolic acid 50
Arnebia euchroma 186
Arnebia guttata 186
Arnebinol 186
Arnottianamide 1244
Aromadendrene 829
Arteannuin 192
Artemetin 1224
Artemisia annua 192
Artemisia argyi 202
Artemisia capillaris 207
Artemisia scoparia 207
Artemisic acid 192
Artemisinin 192
Artemisinol 192
Artemisitene 192
Artepillin 208
Asarone 31
Asarum heterotropoides var. *mandshuricum* 214
Asarum sieboldii 214
Asarum sieboldii var. *seoulense* 214
Aschantin 738
Asiaticoside 299
Asparacoside 222
Asparacosin 222
Asparagus cochinchinensis 222
Asperuloside 774
Aster tataricus 225
Asterin 227
Asterinin 227
Asterogenic acid 225
Astersaponin 226
Astertarones 226
Astilbin 1142
Astin 227
Astragaloside 230
Astragalus membranaceus 229
Astragalus membranaceus var. *mongholicus* 229
Atanine 535
Atractylenolactam 240
Atractylenolide 240
Atractylodes chinensis 240
Atractylodes lancea 240
Atractylodes macrocephala 240
Atractyloside 242
Atractylodin 241
Atropa belladonna 464
Aucklandia lappa 652
Aucubin 511
Auranetin 343
Aurantiamide acetate 227
Aurantioobtusin 292

b
Baicalein 1087
Baihuaqianhuoside 861
Bakuchicin 975
Bakuchiol 975
Baphicacanthus cusia 658
Baptifoline 1145
Barbatin 1101

Bauerenol 1207
Belamcanda chinensis 246
Belamcandal 247
Belamcanidin 247
Benzoylpaeoniflorin 811
Berberastine 385
Berberine 384
Berberrubine 385
Bergamottin 799
Bergapten 127, 139
Bergaptol 127
Bergenin 163
Betulinic acid 11, 735, 1261
Bianfugecine 758
Bianfugenine 1134
Biatractylolide 240
Bigelovin 651
Bilobalide 592
Bilobetin 592
Bilobol 593
Biochanin 982, 1164
Bocconoline 1243
Bolbostemma paniculatum 250
Borneol 107
Bornyl acetate 107
Bornylene 526
Bornylmagnolol 741
Bowdichione 454
Brassica juncea 253
Brazilin 278
Britannin 651
Brucea javanica 45, 258
Bruceaketolic acid 261
Bruceantin 259
Bruceene 260
Bruceine 259
Bruceolide 259
Bruceoside 259, 261
Brusatol 259
Bryonolol 1193
Buddleja officinalis 267
Buddleoside 268
Bulbocapnine 412
Bulbus Fritillariae Cirrhosae 561
Bulbus Fritillariae Hupehensis 562
Bulbus Fritillariae Pallidiflorae 562
Bulbus Fritillariae Thunbergii 563
Bulbus Fritillariae Ussuriensis 563
Bupleurotoxin 273
Bupleurum chinense 271
Bupleurum scorzonerifolium 271
Bupleurynol 273
Butein 454
Byakangelicin 139

c

Cacumen Platycladi 908
Cadambine 1215
Caesalpinia sappan 278
Caffeic acid 11
Calocedrin 979
Calycosin 231
Calystegine 728
Calyx seu Fructus Physalis 874
Campestanol trans-ferulate 380
Camphor 107
Candicine 868
Canthin-6-one 47
Canthine 47
Cantleyine 486
Cantleyoside 486
Cantoniensistriol 1
Cape Aloe 80
Capillanol 207
Capillaridin 208
Capillarin 208
Capillarisin 208
Capillene 207
Capillin 207
Capillone 207
Capsaicin 900
Carboxymethylindirubin 661
Carthamin 283
Carthamus tinctorius 283
Carveol 202
Caryophyllene oxide 88
Caryophyllenol 88
Cassia acutifolia 287
Cassia angustifolia 287
Cassia obtusifolia 292
Cassia tora 292
Cassiaside 292
Catalpol 1009
Caulis Dendrobii 468
Caulis Erycibes 465
Caulis Lonicerae Japonicae 715
Caulis Perillae 850
Caulis Piperis Kadsurae 896
Caulis Polygoni Multiflori 938
Caulis Sinomenii 1133
Caulis Spathilobi 1164
Cedrelopsin 127
Celogenamide 297
Celogentin 297
Celosia argentea 296
Celosia cristata 296
Centella asiatica 299
Centipeda minima 305
Cepharanoline 1171

Cepharanthine 1170
Chaihunaphthone 272
Chaparrinone 45
Chaparrolide 45
Cheilanthifoline 758
Chicanine 1072
Chikusetsusaponin 17
Chloranoside 1065
Chlorochrymorin 314
Chlorogenic acid 11, 208, 716
Chrysanthetriol 311, 312
Chrysandiol 314
Chrysanthemol 312
Chrysanthemum indicum 311
Chrysanthemum morifolium 314
Chrysanthenol 306
Chrysartemin 203
Chrysoobtusin 292
Chrysophanol 2
Chuanbeinone 562
Chuanxiongol 693
Cimicidanol 319
Cimicidol 319
Cimicifuga dahurica 318
Cimicifuga foetida 318
Cimicifuga heracleifolia 318
Cimicifugamide 320
Cimifugin 320
Cimigenol 318
Cinchonain 509
Cineole 97, 106, 214
Cinnamaldehyde 323
Cinnamic acid 323
Cinnamomum cassia 323
Cinnamtannin 324
Cirsilineol 207
Cissamine 331
Cissampareine 330
Cissampelos pareira var. *hirsuta* 330
Cistanche deserticola 334
Cistanche tubulosa 334
Cistanoside 334
Citreorosein 927
Citromitin 362
Citrulline 1194
Citrus aurantium 342
Citrus grandis 342
Citrus reticulata 361
Citrus sinensis 342
Clematichinenoside 369
Clematis chinensis 368
Clematis hexapetala 368
Clematis manshurica 368
Cnidilide 693
Cnidilin 127
Cnidimol 372
Cnidium monnieri 371
Cniforin 371
Codonopsis pilosula 376
Codonopsis pilosula var. *modesta* 376
Codonopsis tangshen 376
Coix lacryma-jobi var. *ma-yuen* 380
Columbianetin 139
Coniferin 9
Convolvulinolic acid 448
Coptis chinensis 384
Coptis deltoidea 384
Coptis teeta 384
Coptisine 385
Corchioside 432
Coriolus 397
Coriolus versicolor 397
Cornoside 551
Cornus officinalis 404
Cornuside 404
Cortex Acanthopanacis 5
Cortex Ailanthi 45
Cortex Albiziae 53
Cortex Cinnamomi 323
Cortex Dictamni 475
Cortex Eucommiae 511
Cortex Fraxini 555
Cortex Lycii 727
Cortex Magnoliae Officinalis 741
Cortex Meliae 752
Cortex Mori 777
Cortex Moutan 820
Cortex Periplocae 858
Cortex Phellodendri 868
Cortex Pseudolaricis 967
Corybulbine 417
Corydaline 416
Corydalis bungeana 411
Corydalis decumbens 411
Corydalis yanhusuo 416
Corylifolin 975
Corylifolinin 975
Corynoline 412
Corynoxeine 1214
Coumarin 127
Coumestrol 488
Coumurrayin 139
Crocin 421
Crocus sativus 421
Croton tiglium 428
Cryptomerin 1109
Cryptotanshinone 1041
Cucurbitacin 1193

Cumambrin 311
Curacao Aloe 80
Curculigo orchioides 432
Curculigenin 433
Curculigine 433
Curculigol 433
Curculigoside 432
Curcuma kwangsiensis 435
Curcuma longa 435
Curcuma phaeocaulis 435
Curcuma wenyujin 435
Curcumin 435
Curcumol 437
Curlone 436
Cuscuta chinensis 448
Cyclamiretin 164
Cyclanoline 1177
Cycleanine 330
Cyclohopanediol 1000
Cyclohopenol 1000
Cyclomulberrin 777
Cyclomulberrochromene 778
Cymene 526
Cyperene 451
Cyperolone 451
Cyperone 451
Cyperus rotundus 451

d

Daechucyclopeptide 1262
Dahurinol 318
Daidzein 982
Dalbergia odorifera 454
Dalbergin 454
Dammarane 57
Danshenol 1043
Danshensu 1044
Danshenxinkun 1043
Daphne genkwa 460
Darutigenol 1130
Darutoside 1130
Datura metel 464
Dauricine 757
Daurisoline 757
Decarine 1243
Decumbensine 412
Dehydroandrographolide 110
Dehydrobruceine 261
Dehydroeffusal 666
Dehydrofalcarinol 207
Dehydrofalcarinone 207
Dehydroglaucarubinone 45
Dehydroglaucarubolone 45
Dehydromiltirone 1042
Dehydropachymic acid 949
Dehydroretronecin-7-yl-adenosine 1116
Dehydroretronecin-7-yl-guanosine 1116
Dehydroretronecin-7-yl-thymidine 1116
Demethylflavasperone gentiobioside 292
Demethyllasiodiplodin 186
Denbinobin 469
Dendrine 468
Dendrobine 468
Dendrobium candidum 468
Dendrobium fimbriatum var. *oculatum* 468
Dendrobium nobile 468
Dendroxine 468
Deoxyalisol 57
Deoxyartemisinin 192
Deoxyshikonin 186
Desmodilactone 471
Desmodimine 471
Desmodium styracifolium 471
Dianchinenoside 473
Dianthus chinensis 473
Dianthus superbus 473
Diarylheptanoids 96
Dicaffeoyl quinic acid 11, 505
Dictamnine 475
Dictamnoside 475
Dictamnus dasycarpus 475
Dihydrobruceine 260
Dihydrocadambine 1215
Dihydrosalutaridine 1133
Dihydrosanguinarine 416
Dihydrotanshinone 1041
Diligustilide 693
Dimethylacrylalkannin 186
Dioscin 222, 480
Dioscorea futschauensis 479
Dioscorea hypoglauca 479
Dioscorea nipponica 479
Dioscorea opposita 479
Dioscorea septembloba 479
Diosgenin 480
Diploptene 1000
Dipsacus asperoides 485
Dipsacus saponin 485
Drynaria fortunei 342

e

Echinacoside 334
Echinatin 603
Echinocystic acid 225, 226
Echinops latifolius 1015
Ecklonia kurome 675
Eckol 675
Eclipta prostrata 488

Effusol 666
Elemene 97
Elemol 437
Eleutheroside 5, 9
Embelin 163
Emodin 227, 938
Enanthotoxin 273
Engeletin 1141
Enterodiol 706
Enterolactone 706
Ephedra equisetina 491
Ephedra intermedia 491
Ephedra sinica 491
Ephedrannin 492
Ephedrine 491
Epiberberine 385
Epifriedelanol 225
Epigoitrin 658
Epimedium brevicornum 496
Epimedium koreanum 496
Epimedium pubescens 496
Epimedium sagittatum 496
Epimedium wushanense 496
Epivogeloside 717
Epoxyganoderiol 567
Eremophilene 829
Ergolide 652
Erigeron breviscapus 504
Eriobotrya japonica 508
Eritadenine 682
Erycibe obtusifolia 465
Erycibe schmidtii 465
Erythrocentaurin 589
Esculentagenic acid 878
Esculentagenin 878
Esculentic acid 877
Esculetin 555
Esculin 555
Ethoxychelerythrine 1243
Eucalyptol 106
Eucommia ulmoides 511
Eucommin 512
Eucommiol 511
Eudesmol 241, 742
Eugenia caryophyllata 519
Eugenol 519
Eukovoside 1225
Eupafurtunin 527
Eupatolitin 207
Eupatorium fortunei 526
Euphol 529
Euphorbia kansui 528
Euphorbia lathyris 531
Euphorbia pekinensis 528
Euscaphic acid 508, 1031
Evocarpine 535
Evodia rutaecarpa 534
Evodia rutaecarpa var. *bodinieri* 534
Evodia rutaecarpa var. *officinalis* 534
Evodiamide 535
Evodiamine 535
Evodol 535
Exocarpium Citri Grandis 342
Exocarpium Citri Rubrum 361
Extractum Acanthopanacis Senticosi 9

f

Fagarine 475
Fagaronine 1243
Falcarindiol 599
Fallacinol 927
Fangchinoline 1176
Fargesin 738
Farrerol 1025
Fenchol 545
Fenchone 545
Fenfangjine 1176
Fennel 545
Feretoside 802
Fernenone 202
Ferruginol 1043
Fisetin 454
Flaxseed 705
Florilenalin 305
Flos Albiziae 53
Flos Buddlejae 267
Flos Carthami 283
Flos Caryophylli 519
Flos Celosiae Cristatae 296
Flos Chrysanthemi 314
Flos Chrysanthemi Indici 311
Flos Daturae 464
Flos Farfarae 1206
Flos Genkwa 460
Flos Inulae 651
Flos Lonicerae 715
Flos Lonicerae Japonicae 715
Flos Magnoliae 738
Flos Magnoliae Officinalis 741
Flos Mume 962
Flos Rhododendri Molli 1024
Flos Rosae Chinensis 1031
Flos Rosae Rugosae 1031
Flos Sophorae 1154
Foeniculum vulgare 545
Foetidinol 319
Foetidissimoside 225
Folium Aconiti Kusnezoffii 23

Folium Apocyni Veneti 153
Folium Artemisiae Argyi 202
Folium Eriobotryae 508
Folium Ginkgo 591
Folium Ginseng 827
Folium Ilicis Cornutae 639
Folium Isatidis 657
Folium Mori 777
Folium Nelumbinis 791
Folium Perillae 850
Folium Polygoni Tinctorii 658
Folium Pyrrosiae 999
Folium Rhododendri Dauruci 1024
Folium Sennae 287
Folium Viticis Negundo 1234
Forsythenside 551
Forsythia suspensa 550
Forsythin 551
Forsythoside 551
Fragransin 782
Fragransol 782
Frangufoline 1262
Fraxin 555
Fraxinellone 475
Fraxinus chinensis 555
Fraxinus rhynchophylla 555
Fraxinus stylosa 555
Fraxinus szaboana 555
Friedel-3-ene 225
Friedelin 9
Fritillaria cirrhosa 561
Fritillaria delavayi 561
Fritillaria hupehensis 561
Fritillaria pallidiflora 561
Fritillaria przewalskii 561
Fritillaria thunbergii 561
Fritillaria unibracteata 561
Fritillaria ussuriensis 561
Fritillaria walujewii 561
Fructus Akebiae 50
Fructus Alpiniae Oxyphyllae 102
Fructus Amomi Rotundus 106
Fructus Anisi Stellati 645
Fructus Arctii 158
Fructus Aristolochiae 176
Fructus Aurantii 342
Fructus Aurantii Immaturus 342
Fructus Bruceae 258
Fructus Cnidii 371
Fructus Corni 404
Fructus Crotonis 428
Fructus Evidoae 534
Fructus Foeniculi 545
Fructus Forsythiae 550
Fructus Galangae 88
Fructus Gardeniae 577
Fructus Gardeniae Preparatus 577
Fructus Hippophae 624
Fructus Jujubae 1259
Fructus Kochiae 671
Fructus Leonuri 688
Fructus Ligustri Lucidi 701
Fructus Lycii 727
Fructus Momordicae 771
Fructus Mori 777
Fructus Mume 962
Fructus Oryzae Germinatus 808
Fructus Perillae 850
Fructus Phyllanthi 872
Fructus Piperis Longi 898
Fructus Piperis Nigri 898
Fructus Psoraleae 974
Fructus Rosae Laevigatae 1031
Fructus Schisandrae Chinensis 1071
Fructus Schisandrae Sphenantherae 1071
Fructus Sophorae 1154
Fructus Toosendan 752
Fructus tribuli 1185
Fructus Trichosanthis 1192
Fructus Tsaoko 106
Fructus Viticis 1234
Furomollugin 1034

g
Gaillardin 651
Galanal 88
Galangin 96
Galanolactone 88
Galbacin 1068
Ganoderal 567
Ganoderic acid 566
Ganoderma 566
Ganoderma lucidum 566
Ganoderma sinense 566
Ganodermanontriol 567
Ganschisandrin 1073
Gansongone 787
Gardenia jasminoides 577
Gardenoside 208, 577
Garlic 63
Gastrodia elata 582
Gastrodin 583
Gastrodioside 583
Gedunin 753
Geissoschizine 1214
Geniposide 577
Genistein 982

Genkwadaphnin 460
Genkwanin 207, 460
Gentiana crassicaulis 588
Gentiana dahurica 588
Gentiana macrophylla 588
Gentiana manshurica 587
Gentiana rigescens 587
Gentiana scabra 587
Gentiana straminea 588
Gentiana triflora 587
Gentiopicral 589
Gentiopicroside 588
Gentrogenin 923
Germacrene D 437
Germacrone-diepoxide 437
Gingerdiol 1250
Gingerdione 1250
Gingerglycolipid 1250
Gingerol 1249
Ginkgetin 592
Ginkgo biloba 591
Ginkgolic acid 593
Ginkgolide 592
Ginsenol 829
Ginsenoside 827, 828
Gitogenin 1186
Gitonin 1185
Glaberide 55
Glabralactone 139
Glabric acid 602
Glabrolide 602
Glabrone 604
Glaucarubinone 45
Glaucine 417
Glehnia littoralis 599
Glycycoumarin 604
Glycyrdione 605
Glycyrol 604
Glycyrrhetic acid 602
Glycyrrhetol 602
Glycyrrhiza glabra 601
Glycyrrhiza inflata 601
Glycyrrhiza uralensis 601
Glycyrrhizin 602
Glyinflanin 605
Glyuranolide 602
Gomisin 1072
Goshuyuamide 535
Gracillin 480
Grandifloric acid 1130
Guaiene 97
Guvacine 169
Guvacoline 169
Gynostemma pentaphyllum 614
Gypenoside 615
Gypsogenic acid 1223

h

Haplopine 1239
Harman 920
Harpagide acetate 511
Hayatine 330
Hecogenin 1186
Hederagenin 50
Hedysarum polybotrys 621
Helenalin 305
Hematoxylin 278
Heracleifolinol 319
Herba Abri 1
Herba Agrimoniae 36
Herba Andrographitis 110
Herba Artemisiae Annuae 192
Herba Artemisiae Scopariae 207
Herba Belladonnae 465
Herba Centellae 299
Herba Centipedae 305
Herba Cissampelotis 330
Herba Cistanches 334
Herba Desmodii Styracifolii 471
Herba Dianthi 473
Herba Ecliptae 488
Herba Ephedrae 491
Herba Epimedii 496
Herba Erigerontis 504
Herba Eupatorii 526
Herba Houttuyniae 629
Herba Inulae 651
Herba Lamiophlomis 680
Herba Leonuri 688
Herba Lycopi 735
Herba Menthae 763
Herba Plantaginis 905
Herba Pogostemonis 917
Herba Portulacae 954
Herba Pyrolae 995
Herba Sarcandrae 1065
Herba Saururi 1067
Herba Schizonepetae 1081
Herba Scutellariae Barbatae 1100
Herba Sedi 1106
Herba Selaginellae 1108
Herba Siegesbeckiae 1129
Herba Verbenae 1224
Herba Visci 1229
Hesperidin 362
Hinesol 241
Hinokiresinol 122
Hippophae rhamnoides 624

Hirsuteine 1214
Hirsutine 1214
Homoarbutin 995
Homoplantaginin 906
Honokiol 741
Houttuynia cordata 629
Hupehenine 563
Huperzia serrata 633
Huperzine 633
Huperzinine 633
Hydnocarpin 716
Hydrastine 413
Hydroxypinusolidic acid 908
Hydroxy-sanshool 1238
Hyoscyamine 464
Hyoscyamus niger 464
Hyperin 153

i

Icariin 496
Icariside 55
Ilex chinensis 639
Ilex cornuta 639
Ilexol 164
Ilexoside 639
Illicium verum 645
Imperatorin 127
Indigo 657
Indigo Naturalis 658
Indirubin 657
Integerrimine 1208
Integriamide 1244
Inula britannica 651
Inula helenium 651
Inula japonica 651
Inula linariifolia 651
Irigenin 247
Irisflorentin 247
Iristectorigenin 247
Isaindigotone 658
Isatis indigotica 657
Iso-aloesin 80
Isoangelol 139
Isobauerenol 1207
Isochaihulactone 272
Isochlorogenic acid 715
Isochondrodendrine 330
Isocryptomerin 1109
Isocryptotanshinone 1042
Isodihydrocadambine 1215
Isofraxidin 10
Isoglabrolide 602
Isoimperatorin 127, 139
Isoiridogermanal 247
Isokarounidiol 1194
Isolicoflavonol 604
Isomangiferin 999
Isomatrine 1145
Isomucronulatol 231
Isonuatigenin 847
Isopaeonisuffral 821
Isopimpinellin 139
Isoprotostemonine 1168
Isopulegol 1238
Isoquercitrin 153
Isorhamnetin 3-O-neohesperidoside 1211
Isorhamnetin 96, 624
Isorhynchophylline 1214
Isotanshinone 1042
Isotetrandrine 1170
Isotoosendanin 752
Isotussilagine 1208
Istanbulin 1065

j

Jalapinolic acid 448
Jaligonic acid 877
Jangomolide 535
Jatrorrhizine 384
Javanicin 261
Jiofuran 1010
Jioglutin 1010
Jujuboside 1260
Julibrine 55
Julibrogenin 53
Julibroside 54
Juncoside 666
Juncunol 666
Juncus effusus 666
Juncusol 666

k

Kadsurenin 896
Kaempferide 96
Kaempferia galanga 668
Kaempferol 96, 97, 153, 227
Kalopanaxsaponin 51
Kanshone 787
Kansuinine 529
Kansuiphorin 529
Karounidiol 1194
Kaurenoic acid 5
Khellactone 861
Khellin 1062
Kirenol 1130
Kochia scoparia 671
Kochianoside 672

Kochioside 671
Kokusaginine 1239
Koparin 454
Kudzubutenolide 983
Kudzusapogenol 983
Kulinone 753
Kulactone 753
Kulolactone 753
Kumatakenin 231
Kumujancine 884
Kumujantine 885
Kuraridinol 1146
Kushenol 1146

l

Lactiflorin 811
Laminaria japonica 675
Lamiophlomiol 680
Lamiophlomis rotata 680
Lappaol 158
Laurifoline 1134
Lentinus edodes 682
Leonticine 417
Leonurine 689
Leonurus japonicus 688
Licochalcone 603
Licocoumarone 604
Licoflavone 604
Liconeolignan 604
Licorice 601
Licorice-saponin 603
Liensinine 792
Lignum Dalbergiae Odoriferae 454
Lignum Santali Albi 1059
Lignum Sappan 278
Ligusticum chuanxiong 692
Ligustilidiol 693
Ligustrazine 693
Ligustroside 702
Ligustrum lucidum 701
Ligustilide 693
Limocitrin 362
Limonene 343
Limonin 362
Linalool 1238
Linamarin 705
Linolenic acid 705
Linum usitatissimum 705
Linusitamarin 705
Linustatin 705
Liqcoumarin 604
Liquiridiolic acid 602
Liquiritic acid 602
Liquiritigenin 454, 603
Liriodendrin 9
Liriodenine 791
Liriope muscari 713
Liriope spicata var. *prolifera* 713
Lithospermic acid 1044
Loganin 404, 717
Lonicera confusa 715
Lonicera hypoglauca 715
Lonicera japonica 715
Lonicera macranthoides 715
Lonicerin 716
Loniceroside 717
Lucidenic acid 566
Lucidin 1033
Lucidin primveroside 1034
Lucyoside 723
Luffa cylindrica 723
Lupenone 202
Luteolin 268
Lycium barbarum 727
Lycium chinensis 727
Lycopus lucidus var. *hirtus* 735
Lyoniresinol 55
Lysicamine 1262

m

Machaerinic acid 53
Macrostemonosides 64
Madecassoside 299
Maesanin 163
Magnoflorine 176
Magnolia biondii 738
Magnolia denudata 738
Magnolia officinalis 741
Magnolia officinalis var. *biloba* 741
Magnolia sprengeri 738
Magnolin 738
Magnolol 741
Mahuannin 492
Manassantin 1068
Mangicrocin 422
Mangiferin 123
Manool 469
Marmesinin 139
Maslinic acid 1261
Matairesinol 158, 550
Matairesinoside 550
Matrine 1145
Mayurone 829
Medicarpin 454
Medioresinol 9
Medulla Junci 666
Melia azedarach 752
Melia toosendan 752

Meliacarpinin 753
Melittoside 1009
Menisdaurilide 1134
Menisperine 758
Menispermum dauricum 757
Mentha haplocalyx 763
Menthol 763
Mentholum 763
Menthone 763
Meranzin 139
Methoxycanthin-6-one 47
Methoxypsoralen 127
Methylcardol 163
Methylcoclaurine 1177
Methylcytisine 1145
Methylene tanshinquinone 1041
Methyleugenol 214
Methylindirubin 661
Methylnonylketone 629
Methylophiopogonanone 804
Methylprotodioscin 222
Methyltyramine 343
Methylvisamminol 1062
Miltionone 1043
Miltirone 1042
Mogroside 772
Mollugin 1034
Momordica cochinchinensis 768
Momordica grosvenori 771
Momordica saponin 769
Momordin 671
Mongholicoside 229
Monotropein 774
Morinda officinalis 774
Morindolide 774
Morofficinaloside 774
Moroidin 297
Morroniside 404
Morus alba 777
Moscatilin 469
Moscatin 469
Moxartenolide 202
Mudanpioside 820
Mulberrin 777
Mulberrochromene 778
Mussaenosidic acid 1235
Myrcene 1238
Myristica fragrans 782
Myristicin 214, 783
Myrsinionoside 1106

n

Nardosinonediol 787
Nardostachin 788
Nardostachys chinensis 787
Nardostachys jatamansi 787
Naringin 343
Neferine 792
Nelumbo nucifera 791
Neoarctin 158
Neobavaisoflavone 975
Neobuddoffiside 268
Neocapillene 207
Neocnidilide 693
Neodarutoside 1130
Neolinustatin 705
Neomangiferin 123
Neotussilagolactone 1207
Nepetin 1225
Neridienone 858
Nerol acetate 526
Nicotinoyl-oxylathyrol-diacetate benzoate 532
Nigakilactone 884
Nigakinone 884
Nigranoic acid 1073
Nilic acid 448
Ningposide 1084
Nishidaside 1235
Nitidine 1243
Nobiletin 343
Nobiline 468
Nodakenetin 139
Nodakenin 139, 798
Nodus Nelumbinis Rhizomatis 791
Nomilin 363
Nomilinic acid 363
Nootkatol 102
Nootkatone 102
Norarjunolic acid 50
Norcapillene 207
Noririsflorentin 247
Nortanshinone 1041
Norvisnagin 320
Notoginsenoside 830
Notopterol 798
Notopterygium forbesii 798
Notopterygium incisum 798
Nuatigenin 847
Nuciferine 791
Nuezhenide 702
Nyasol 122

o

Obacunone 362
Obtusifolin 292
Ochnaflavone 716
Ocimene 1238

Oldenlandia diffusa 801
Oldenlandoside 802
Oleanolic acid 16, 50, 164
Oleum Cinnamomi 323
Oleum Menthae Dementholatum 763
Oleum Rhododendri Dauruci 1024
Oleum Ricini 1027
Oleum Viticis Negundo 1234
Olivil 512
Onjixanthone 920
Ophiopogon japonicus 804
Oridonin 1001
Orientalol 58
Oroselone 371
Oroxylin 111
Oryza sativa 808
Osthole 139
Ostruthin 799
Oxododecanal 629
Oxofangchirine 1176
Oxotuberostemonine 1167
Oxymatrine 1145
Oxypeucedanin 127
Oxyphyllenodiol 102
Oxyphyllenone 102
Oxyphyllol 102
Oxytrogenin 1146

p
Paeonia lactiflora 811
Paeonia suffruticosa 820
Paeonia veitchii 811
Paeoniflorigenone 811
Paeonilactone 811
Paeonimetaboline 812
Paeonisothujone 820
Paeonisuffral 820
Paeonisuffrone 820
Paeonol 820
Paeonolide 820
Paeonoside 820
Palmatine 384
Panasinsanol 829
Panax ginseng 827
Panax Japonicus 831
Panax Japonicus var *bipinnatifidus* 832
Panax Japonicus var *major* 832
Panax notoginseng 830
Panax quinquefolium 831
Panaxynol 829
Panaxytriol 829
Pareirubrine 330
Paris polyphylla var. *chinensis* 847
Paris polyphylla var. *yunnanensis* 847
Patchouli alcohol 918
Pedalitin 1225
Pedunculagin 39
Pedunculoside 639
Peimine 562
Peiminine 562
Pelletierine 992
Pennogenin 847
Penstemoside 680
Pericarpium Arecae 169
Pericarpium Citri Reticulatae 361
Pericarpium Citri Reticulatae Viride 361
Pericarpium Granati 992
Pericarpium Trichosanthis 1192
Pericarpium Zanthoxyli 1237
Perilla alcohol 850
Perilla frutescens 850
Perilla ketone 850
Perillaldehyde 343
Periploca sepium 858
Periplocin 858
Periplocogenin 858
Periplocoside 858
Peroxycalamenene 451
Peucedanol 139
Peucedanum praeruptorum 861
Phellandrene 202, 1238
Phellodendrine 868
Phellodendron amurense 868
Phellodendron chinense 868
Phenylalkynes 207
Pheophorbide 1101
Phillygenin 550
Phlorofucofuroeckol 675
Phyllanthus emblica 872
Physalin 874
Physalis alkekengi var. *franchetii* 874
Physcion 2
Physochlaina infundibularis 465
Phytolacca acinosa 877
Phytolacca americana 877
Phytolaccagenic acid 877
Piceid 928
Picfeltarraegenin 888
Picrasane 45
Picrasidine 884
Picrasinol 884
Picrasinoside 884
Picrasma quassioides 884
Picria fel-terrae 888
Picrocrocin 422
Picrorhiza scrophulariiflora 890
Picroside 890
Pinellia ternata 893

Pinene 97, 1238
Pingpeimine 563
Pinoresinol 9
Pinoresinol diglucopyranoside 512
Piper kadsura 896
Piper longum 898
Piper nigrum 898
Piperine 898
Piperitenone 343
Plantago asiatica 905
Plantago depressa 905
Plantagoside 906
Plantainoside 906
Plantamajoside 905
Platycladus orientalis 908
Platycodigenin 911
Platycodin 911
Platycodon grandiflorum 911
Platycogenic acid 911
Plenolin 305
Plumula Nelumbinis 791
POD-II 924
Pogostemon cablin 917
Pollen Typhae 1210
Polygala sibirica 919
Polygala tenuifolia 919
Polygalacic acid 911, 919
Polygonatum cyrtonema 923
Polygonatum kingianum 923
Polygonatum odoratum 923
Polygonatum sibiricum 923
Polygonum cuspidatum 927
Polygonum multiflorum 938
Polygonum tinctorium 658
Polyphyllin 847
Polyporus 946
Polyporus umbellatus 946
Polyporusterone 947
Ponicidin 1001
Poria 949
Poria cocos 949
Poricoic acid A 949
Portulaca oleracea 954
Praeroside 861, 862
Praeruptorin 861
Pregomisin 1073
Prehispanolone 689
Procyanidin 324
Protoanemonin 989
Protocatechuic acid 11
Protogracillin 480
Protopanaxadiol 827
Protopine 411
Protosappanin 279
Protostemonine 1168
Protostemotinine 1167
Prunella vulgaris 957
Prunus spp. 962
Przewaquinone 1041
Pseudoephedrine 491
Pseudoginsenoside 830
Pseudolaric acid 967
Pseudolarix kaempferi 967
Pseudopelletierine 992
Pseudoprotodioscin 222
Pseudostellaria heterophylla 971
Psoralea corylifolia 974
Psoralen 127
Psoralidin 976
Pterocarpus santalinus 979
Pueraria lobata 981
Pueraria thomsonii 981
Puerarin 982
Puerarol 983
Pulchinenoside 990
Pulsatilla chinenesis 989
Pulsatillic acid 990
Punica granatum 992
Punicalin 992
Pyrola calliantha 995
Pyrola decorata 995
Pyrrosia lingua 999
Pyrrosia petiolosa 999
Pyrrosia sheareri 999

q
Qianhucoumarin 862
Qingdainone 657
Quassin 884
Quercetin 96, 97, 153, 207, 227
Quillaic acid 1222

r
Rabdosia rubescens 1001
Radix Achyranthis Bidentatae 16
Radix Aconiti 23
Radix Aconiti Kusnezoffii 23
Radix Aconiti Kusnezoffii Preparata 23
Radix Aconiti Lateralis Preparata 23
Radix Aconiti Preparata 23
Radix Angelicae Dahuricae 127
Radix Angelicae Pubescentis 139
Radix Angelicae Sinensis 143
Radix Arnebiae 186
Radix Asparagi 222
Radix Astragali 229
Radix Bupleuri 271
Radix Codonopsis 376

Radix Curcumae 435
Radix Dipsaci 485
Radix Echinopsis 1015
Radix et Caulis Acanthopanacis Senticosi 9
Radix et Rhizoma Asari 214
Radix et Rhizoma Asteris 225
Radix et Rhizoma Clematidis 368
Radix et Rhizoma Ephedrae 491
Radix et Rhizoma Gentianae 587
Radix et Rhizoma Ginseng 827
Radix et Rhizoma Ginseng Rubra 827
Radix et Rhizoma Nardostachyos 787
Radix et Rhizoma Notoginseng 830
Radix et Rhizoma Rhei 1017
Radix et Rhizoma Rubiae 1033
Radix et Rhizoma Salviae Miltiorrhizae 1040
Radix et Rhozoma Glycyrrhizae 601
Radix et Rhozoma Glycyrrhizae Preparata Cum Melle 601
Radix Euphorbiae Pekinensis 528
Radix Gentianae Macrophyllae 588
Radix Glehniae 599
Radix Hedysari 621
Radix Hedysari Preparata Cum Melle 621
Radix Inulae 651
Radix Isatidis 657
Radix Kansui 528
Radix Liriopes 713
Radix Morindae Officinalis 774
Radix Ophiopogonis 804
Radix Paeoniae Alba 811
Radix Paeoniae Rubra 811
Radix Panacis Quinquefolii 830
Radix Peucedani 861
Radix Physochlainae 465
Radix Phytolaccae 877
Radix Platycodonis 911
Radix Polygalae 919
Radix Polygoni Multiflori 938
Radix Polygoni Multiflori Preparata cum Succo 938
Radix Pseudostellariae 971
Radix Puerariae Lobatae 981
Radix Puerariae Thomsonii 981
Radix Pulsatillae 989
Radix Rehmanniae 1008
Radix Rehmanniae Preparata 1008
Radix Rhapontici 1015
Radix Sanguisorbae 1056
Radix Saposhnikoviae 1062
Radix Scrophulariae 1084
Radix Scutellariae 1086
Radix Sophorae Flavescentis 1144
Radix Stemonae 1167
Radix Stephaniae Tetrandrae 1176
Radix Trichosanthis 1192
Radix Zanthoxyli 1242
Ramulus Cinnamomi 323
Ramulus et Folium Picrasmae 884
Ramulus Mori 777
Ramulus Uncariae cum Uncis 1213
Ranunculin 989
Raphanus sativus 1006
Receptaculum Nelumbinis 791
Rehmaglutin 1010
Rehmaionoside 1010
Rehmannia glutinosa 1008
Rehmannioside 1009
Remerine 791
Renifolin 995
Reptoside 511
Resveratrol 928
Reticuline 413, 1134
Retinervus Luffae Fructus 723
Rhaponticin 1019
Rhaponticum uniflorum 1015
Rhapontisterone 1015
Rhein 1018
Rhein-anthrone 288
Rheinoside 1018
Rhetsinine 535
Rheum officinale 1017
Rheum palmatum 1017
Rheum tanguticum 1017
Rhizoma Acori Calami 31
Rhizoma Acori Tatarinowii 31
Rhizoma Alismatis 57
Rhizoma Alpiniae Officinarum 96
Rhizoma Anemarrhenae 121
Rhizoma Anemonae Raddeanae 17
Rhizoma Atractylodis 240
Rhizoma Atractylodis Macrocephalae 240
Rhizoma Belamcandae 246
Rhizoma Bolbostemmatis 250
Rhizoma Chuanxiong 692
Rhizoma Cimicifugae 318
Rhizoma Coptidis 384
Rhizoma Corydalis 416
Rhizoma Corydalis Decumbentis 411
Rhizoma Curculiginis 432
Rhizoma Curcumae 435
Rhizoma Curcumae Longae 435
Rhizoma Cyperi 451
Rhizoma Dioscoreae 479
Rhizoma Dioscoreae Hypoglaucae 479
Rhizoma Dioscoreae Nipponicae 479
Rhizoma Dioscoreae Septemblobae 479
Rhizoma Drynariae 342

Rhizoma et Radix Baphicacanthis Cusiae 658
Rhizoma et Radix Notopterygii 798
Rhizoma et Radix Polygoni Cuspidati 927
Rhizoma Gastrodiae 582
Rhizoma Kaempferiae 668
Rhizoma Menispermi 757
Rhizoma Panacis Japonici 832
Rhizoma Panacis Majoris 832
Rhizoma Paridis 847
Rhizoma Picrorhizae 890
Rhizoma Pinelliae 893
Rhizoma Polygonati 923
Rhizoma Polygonati Odorati 923
Rhizoma Smilacis Glabrae 1140
Rhizoma Wenyujin Concisum 435
Rhizoma Zingiberis 1249
Rhizoma Zingiberis Preparatum 1249
Rhizoma Zingiberis Recens 1249
Rhododendron dauricum 1024
Rhododendron molle 1024
Rhodojaponin 1025
Rhodomollein 1025
Rhynchophine 1213
Rhynchophylline 1214
Ricinine 1027
Ricinus communis 1027
Riligustilide 693
Rinderine 526
Rosa chinensis 1031
Rosa laevigata 1031
Rosa rugosa 1031
Rosmarinic acid 1044
Rotundic acid 508
Ruberythric acid 1034
Rubia cordifolia 1033
Rubiadin 774, 1033
Rubidatum 1034
Rubilactone 1034
Ruscogenin 1186
Rutaecarpine 535
Rutaevine 535
Rutin 1154

s
Sabinene 1238
Safflor yellow 283
Saffron 421
Safrole 214
Saikosaponin 275
Salidroside 551, 701
Salvia miltiorrhiza 1040
Salvianolic acid 1044
Salvilenone 1043
Sanguisorba officinalis 1056
Sanguisorba officinalis var. *longifolia* 1056
Sanshool 1238
Santalal 1059
Santalene 1059
Santalin 979
Santalol 1059
Santalum album 1059
Saposhnikovia divaricata 1062
Sappanchalcone 278
Sappanin 279
Sappanone 278
Sarcandra glabra 1065
Sarisan 1068
Sarmentosterol 1106
Sarsasapogenin 121
Sauchinone 1068
Saururin 1068
Saururus chinensis 1067
Savinin 979
Scandoside 802
Schinifoline 1239
Schisandra chinensis 1071
Schisandra sphenanthera 1071
Schisandrin 1072
Schisandrol 1072
Schisandrone 1073
Schisantherin 1072
Schizonepeta tenuifolia 1081
Schizonepetoside 1082
Schizonodiol 1081
Schizonol 1081
Scoparianoside 672
Scoparone 207
Scopolamine 464
Scopoletin 127, 227
Scopolin 127
Scoulerine 412
Scrophularia ningpoensis 1084
Scutebarbatine 1101
Scutellaria baicalensis 1086
Scutellaria barbata 1100
Scutellarin 1100
Secoisolariciresinol 706
Secoxyloganin 717
Sedum sarmentosum 1106
Sedumoside 1106
Segetalic acid 1222
Selaginella pulvinata 1108
Selaginella tamariscina 1108
Semen Alpiniae katsumadai 93
Semen Ameriacae Amarum 962
Semen Arecae 169
Semen Arecae Preparata 169
Semen Cassiae 292

Semen Citri Reticulatae 361
Semen Coicis 380
Semen Crotonis Pulveratum 428
Semen Cuscutae 448
Semen Euphorbiae 531
Semen Euphorbiae Pulveratum 531
Semen Ginkgo 591
Semen Hyoscyanmi 465
Semen Lini 705
Semen Momordicae 768
Semen Myristicae 782
Semen Nelumbinis 791
Semen Persicae 962
Semen Plantaginis 905
Semen Platycladi 908
Semen Pruni 962
Semen Raphani 10006
Semen Ricini 1027
Semen Sesami Nigrum 1120
Semen Sinapis 253
Semen Trichosanthis 1192
Semen Trichosanthis Tostum 1192
Semen Trigonellae 1202
Semen Vaccariae 1222
Semen Ziziphi Spinosae 1259
Senecio scandens 1113
Senecionine 1113
Seneciphylline 1113
Senecrassidiol 829
Senkirkine 1208
Sennidin 287
Serratenediol 633
Sesamin 5, 9
Sesamum indicum 1120
Sesquiphellandrol 1251
Sesquisabinene hydrate 1251
Sesquithujene 1251
Shanzhiside 577
Shegansu 247
Shengmanol 318
Shihunidine 469
Shikonin 186
Shinflavanone 603
Shinjulactone 45
Shionone 225
Shionoside 225, 226
Sibirioside 1084
Siegesbeckia glabrescens 1129
Siegesbeckia orientalis 1129
Siegesbeckia pubescens 1129
Siegesbeckic acid 1130
Siegesbeckiol 1130
Siegesbeckioside 1130
Simiarenol 202
Sinalbin 254
Sinapaldehyde 9
Sinapine 254
Sinapis alba 253
Sinapyl alcohol 12
Sinensal 344
Sinensetin 343
Sinigrin 254
Sinoacutine 1133
Sinomenine 1133
Sinomenium acutum 1133
Sinomenium acutum var. *cinereum* 1133
Sipeimine 563
Sitosterol 5
Skimmianine 1239
Skimmin 127
Skullcapflavone 1087
Smiglanin 1141
Smiglaside 1141
Smilax glabra 1140
Solanidanetriol 563
Songbaisine 562
Sophocarpine 1145
Sophora flavescens 1144
Sophora japonica 1154
Sophoradiol 1
Sophoraflavoside 1146
Sophoranol 1145
Sophoridine 1145
Sorbifolin 1225
Soyasapogenol 1
Spatholobus suberectus 1164
Spica Prunellae 957
Stachydrine 689
Stamen Nelumbinis 791
Stemoamide 1168
Stemona japonica 1167
Stemona sessilifolia 1167
Stemona tuberosa 1167
Stenine 1168
Stephania cepharantha 1170
Stephania tetrandra 1176
Stephanine 1170
Stepharine 758
Stephenanthrine 1177
Stesakine 1170
Stigma Croci 421
Stigmastanol trans-ferulate 380
Stylopine 413
Suffruticoside 820
Sulfoorientalol 58
Sulforaphene 1006
Supinine 526
Suspensaside 551

Sweroside 588
Swertiamarine 588
Swertisin 1260
Synephrine 343
Syringaresinol 9
Syringin 5

t
Tangeretin 343
Tangshenoside 377
Tanshindiol 1041
Tanshinlactone 1043
Tanshinone 1041
Tanshinonic acid 1041
Tectorigenin 247
Tenuifolin 919
Tenuifoliside 921
Terpineol 97, 202
Terrestriamide 1185
Terrestrosin 1185
Tetrahydrocorysamine 412
Tetrahydroxystilbene glucopyranoside 939
Tetrandrine 1176
Tetrahydropalmatine 416
Thallus Eckloniae 675
Thallus Laminariae 675
Thymol methyl ether 526
Tigogenin 1185
Tinctormine 283
Toosendanin 752
Torachrysone 292
Toralactone 292
Tormentic acid 508
Tribulus terrestris 1185
Tribulusamide 1185
Trichilin 753
Trichosanthes kirilowii 1192
Trichosanthes rosthornii 1192
Trigonella foenum-graecum 1202
Trigoneoside 1203
Trimethoxypterocarpan 231
Tryptanthrin 657
Tubeimoside 250
Tuberoside 64
Tuberostemonine 1167
Tuberostemoninol 1168
Tuberostemonone 1168
Tumulosic acid 949
Turkesterone 1015
Turmeric 435
Turmerone 436
Tussilagine 1208
Tussilago farfara 1206
Tussilagolactone 1207
Tussilagone 1206
Tussilagonone 1206
Typha angustifolia 1210
Typha orientalis 1210
Typhaneoside 1211

u
Ulopterol 139
Umtatin 372
Uncaria hirsuta 1213
Uncaria macrophylla 1213
Uncaria rhynchophylla 1213
Uncaria sessilifructus 1213
Uncaria sinensis 1213
Uralenolide 602
Ursolic acid 1261

v
Vaccaria segetalis 1222
Vaccaric acid 1222
Valencene 344
Veranisatin 646
Verbascoside 1225
Verbena officinalis 1224
Verbenalin 1224
Visamminol 320
Viscum coloratum 1229
Viscumneoside 1229
Visnagin 320
Vitex negundo L. var. *cannabifolia* 1234
Vitex trifolia 1234
Vitex trifolia var. *simplicifolia* 1234
Vitexicarpin 1235
Vogeloside 717
Vomifoliol 55

w
Wedelolactone 488
Wilsonirine 413
Wogonin 111, 1087
Worenine 384
Wushanicariin 497

x
Xanthotoxin 127, 139
Xanthoxylin 1239
Xysmalogenin 858

y
Yadanziolide 261
Yadanzioside 259
Yakuchinone 102
Yamogenin 480
Yejuhua lactone 311

Yinyanghuo 497
Yuanhuacin 461
Yuanhuafin 460
Yuzhizioside 50

z
Zanthoxylum bungeanum 1237
Zanthoxylum nitidum 1242
Zanthoxylum schinifolium 1237
Zeaxanthin 727
Zingiber officinale 1249
Zingiberene 1251
Zingiberenol 1251
Ziziphus jujuba 1259
Ziziphus jujuba var. *spinosa* 1259
Zizyphusine 1262